HORMONES IN DEVELOPMENT

HORMONES IN DEVELOPMENT

Edited by:

MAX HAMBURGH

Department of Biology, The City College of the City University of New York; Department of Anatomy, Albert Einstein College of Medicine of Yeshiva University, Bronx, New York

E. J. W. BARRINGTON

Department of Zoology, The University of Nottingham, Nottingham, England

APPLETON-CENTURY-CROFTS
Educational Division
MEREDITH CORPORATION
New York

721-1

Library of Congress Catalog Card Number: 72-116424

PRINTED IN THE UNITED STATES OF AMERICA
390-40417-9

Preface

The National Foundation (New York) sponsored a four-day conference on *Hormones in Development* at Nottingham University, England from September 9 through 12, 1968. The conference centered on problems of the mechanisms of hormone action as it affects processes of growth and differentiation of selected hormone-sensitive target tissues. About 150 participants from 14 different countries attended the conference.

The publication of this volume was stimulated by the proceedings of this conference. All of the manuscripts presented at that meeting are included in this volume, but many of them have since been considerably revised and updated. Other contributions of relevance to the field of fetal endocrinology, besides those presented at the Nottingham meeting, have been included.

The analysis of hormone action in development has, up to now, been carried out on three levels: The "fundamentalist approach" aims to demonstrate that hormones indeed influence and participate in the control of differentiation of various target structures. The demonstration of hormonal control of differentiation relies essentially on procedures that remove the source of hormone secretion or that administer excess hormone during fetal life and on observation of the effects of such treatment on the embryo. On a second level, the "analytical approach" focuses on critical periods during embryogenesis in an attempt to identify the specific developmental processes that are subject to hormonal influence. The third level of study attempts to interpret hormone action in terms of ultimate cellular mechanisms and to fit the effect of hormones on target structures with the "central doctrine of experimental biology": that differentiation reflects the emergence of selective protein synthesis, an event which is presumably preceded by the differential repression or derepression of genes that code for the different proteins.

All of these approaches will be found in this volume. For didactic purposes, however, the contents are organized along more orthodox lines in this book. Each chapter is devoted to a survey of the effects of a single source of hormone on a number of developing target structures, as revealed by a variety of experimental procedures.

Emphasis is placed on analysis of the mechanisms of action of hormones on maturation and differentiation of multiple target structures, such as the effect of corticosteroids on differentiating intestinal epithelium, retina, medulla, and hepatic enzyme patterns. Multiple hormone interactions in growth and differentiation are discussed on a model system such as the developing mammary gland.

Experiments testing the effects of thyroid hormone on a variety of processes promoting neurogenesis in both mammalian and amphibian species are extensively reviewed, as are the mechanisms of action by which parathyroid hormone influences osteogenesis. The maturing central nervous system as a primary target of fetal sex hormone is discussed with particular emphasis on the role sex hormones play on the emergence and patterning of subsequent sexual behavior. Considerable attention is devoted to investigation of the multiple effects of pituitary hormones on fetal metabolism and growth and on the mechanisms by which pituitary trophic substances stimulate ontogeny of other endocrine glands. A chapter on the nerve growth promoting factor (NGF) and epithelial growth promoting factor (EPGF) is also included even though the source of these substances has not yet clearly been identified and they, therefore, represent "hormones" without a known gland.

Consideration is also given to the clinical implication of some of the basic findings obtained from some of the experimental endocrine research presented. The ultimate aim and objective of the editors was to assemble in one single volume a broad range of investigations in the field of fetal endocrinology, one of the least studied areas of developmental biology, so that investigators in the field could review the accomplishments in their respective areas. Between its covers is probably now assembled most of the more recent information on the topic of "Hormones in Development."

It is the hope of the editors that this book may contribute for its readers the intellectual adventure that was started for most of its contributors at Nottingham in 1968, where many of us met as strangers to one another, and all of us left as friends in fact and in spirit.

Contents

Contributors

Henning J. Anderson
Department of Pediatrics, Fuglebakken and Endocrine Clinic, Dronning Louises Children's Hospital, Copenhagen, Denmark

Pietro U. Angeletti
Istituto Superiore di Sanita, Rome, Italy

C. M. Ann Baker
Department of Zoology, University of Adelaide, South Australia

R. Balázs
Neuropsychiatry Unit, Medical Research Council Laboratories, Woodmansterne Road, Carshalton, Surrey, England

E. J. W. Barrington
Department of Zoology, The University of Nottingham, Nottingham, England

A. K. Basu
Department of Biochemistry, All India Institute of Medical Sciences, New Delhi 16, India

Joseph G. Bearn
Department of Anatomy, The Middlesex Hospital Medical School, London, England

Denis Bellamy
Department of Zoology, University College, Cardiff, Wales

Rolf von Berswordt-Wallrabe
Department of Experimental Endocrinology, Main Laboratory of SCHERING AG, 1 Berlin 65, West Germany

T. W. Betz
Department of Biology, Carleton University, Ottawa, Ontario, Canada

John S. Brand
Department of Radiation Biology and Biophysics, University of Rochester School of Medicine and Dentistry, Rochester, New York

Elmer D. Bueker
Department of Anatomy, New York University College of Dentistry, New York, New York

John F. Burkart
Department of Biology, City College of New York, New York, New York

Hela Bursztyn
Departments of Medical Microbiology and Immunology, University of California at Los Angeles School of Medicine, Los Angeles, California

Lise Cedard
Laboratoire de Chimiehormonale de la Maternité de Port-Royal, Paris, France

Wendy A. Cocks
Neuropsychiatry Unit, Medical Research Council Laboratories, Woodmansterne Road, Carshalton, Surrey, England

Stanley Cohen
Department of Biochemistry, Vanderbilt University, Nashville, Tennessee

Gaston Côté
Centre de Recherces, Institut de Cardiologie, Quebec City, Canada

D. A. Crossley
Department of Anatomy, University of Birmingham, Birmingham 15, England

Albert Derby
Department of Biology, Yale University, New Haven, Connecticut

John E. Eaton
Department of Chemistry, Florida State University, Tallahassee, Florida

J. T. Eayrs
Department of Anatomy, University of Birmingham, Birmingham 15, England

Carl W. Ehmann
Alaska Native Medical Center, United States Public Health Service, Anchorage, Alaska

Walter Elger
Department of Experimental Endocrinology, Main Laboratory of SCHERING AG, 1 Berlin 65, West Germany

Sally B. Fand
Department of Pathology, Wayne State University School of Medicine and Veterans Administration Hospital, Detroit, Michigan

Jacob J. Frankel
Pediatric Metabolic and Endocrine Service, Beilinson Hospital, Petah Tikva, Israel

Earl Frieden
Department of Chemistry, Florida State University, Tallahassee, Florida

Tomoko Fujii
Department of Pharmacology, Tokyo Women's Medical School, Tokyo, Japan

Giulio Gabbiani
Institut de Pathologie, Université de Genève, Geneva, Switzerland

P. J. Gaillard
Laboratory for Cell Biology and Histology, University of Leiden, Leiden, The Netherlands

Stanley E. Geel
Departments of Physiology and Anatomy, University of California, Berkeley, California

Jean-Pierre Geloso
Faculté des Sciences, Paris, France

Carlos Jorge Gómez
Química Biológica, Facultad Farmacia y Bioquímica, Universidad de Buenos Aires, Buenos Aires, Argentina

Aubrey Gorbman
Department of Zoology, University of Washington, Seattle, Washington

Michael Granich
Institute of Human Development, University of California, Berkeley, California

S. L. Gupta
Department of Biochemistry, All India Institute of Medical Sciences, New Delhi, India

Katy Haffen
Institut d'Embryologie et de Teratologie Expèrimentale du CNRS, Nogent/Marne 94, France

William Hamilton
Department of Child Health, The University of Glasgow, Glasgow, Scotland

Max Hamburgh
Department of Biology, The City College of the City University of New York, New York, New York; Department of Anatomy, Albert Einstein College of Medicine of Yeshiva University, Bronx, New York

J. Harel
Laboratoire de Physiopathologie, Institut Gustave Roussy, Villejuif, 94, France

J. W. Hekkelman
Laboratory for Cell Biology and Histology, Leiden, The Netherlands

Philippe Hemon
Department of Comparative Physiology, Faculty of Sciences, Paris, France

M. P. M. Herrmann-Erlee
Laboratory for Cell Biology and Histology, University of Leiden, Leiden, The Netherlands

J. Imbenotte
Laboratoire de Physiopathologie, Institut Gustave Roussy, Villejuif, 94, France

R. L. Jacquot
Department of Physiologie Animale, Faculté des Sciences, Rheims, France

B. L. Jailkhani
Department of Biochemistry, All India Institute of Medical Sciences, New Delhi 16, India

P. Jha
Department of Biochemistry, All India Institute of Medical Sciences, New Delhi 16, India

Alfred D. Jost
Laboratory of Comparative Physiology, Faculty of Sciences, 75 Paris 5e, France

Jane C. Kaltenbach
Department of Biological Sciences, Mount Holyoke College, South Hadley, Massachusetts

Junzo Kato
Department of Obstetrics and Gynecology, Faculty of Medicine, University of Tokyo, Tokyo, Japan

Francis T. Kenney
Biology Division, Oak Ridge National Laboratory, Oak Ridge, Tennessee

Ki-Han Kim
Department of Biochemistry, Purdue University, West Lafayette, Indiana

David G. Klein
Laboratory of Biomedical Sciences, National Institute of Child Health and Human Development, National Institutes of Health, Bethesda, Maryland

Takashi Kobayashi
The Red Cross Lying-in Hospital, Tokyo, Japan

S. Kovacs
Institute of Physiology, University Medical School, Pécs, Hungary

Zvi Laron
Pediatric Metabolic and Endocrine Service, Beilinson Hospital and Rogoff Wellcome Medical Research Institute, Petah Tikva, Israel

Jacques Legrand
Laboratoire de Physiologie comparée, Faculté des Sciences, Montpellier 34, France

Rita Levi-Montalcini
Department of Biology, Washington University, St. Louis, Missouri

S. E. Levina
Institute of Developmental Biology, Academy of Sciences of the U.S.S.R., Moscow, U.S.S.R.

Milton Levy
Department of Biochemistry, New York University College of Dentistry, New York, New York

Ladislav Macho
Department of Developmental Endocrinology, Institute of Experimental Endocrinology, Slovak Academy of Sciences, Bratislava, Czechoslovakia

Clyde Manwell
Department of Zoology, University of Adelaide, South Australia

Lorenzo A. Mendoza
Department of Endocrinology, Hospital Vargas, Caracas, Venezuela

M. S. Mitskevich
Laboratory of Hormonal Regulations, Institute of Developmental Biology, Academy of Sciences of the U.S.S.R., Moscow, U.S.S.R.

Pat G. Model
Department of Anatomy, Albert Einstein College of Medicine of Yeshiva University, Bronx, New York

Florence Moog
Department of Biology, Washington University, St. Louis, Missouri

A. A. Moscona
Department of Biology, Univerity of Chicago, Chicago, Illinois

Kathleen M. Mosher
Department of Physiology, Dartmouth Medical School, Hanover, New Hampshire

G. N. Moskovkin
Laboratory of Hormonal Regulations, Institute of Developmental Biology, Academy of Sciences of the U.S.S.R., Moscow, U.S.S.R.

Allan Munck
Department of Physiology, Dartmouth Medical School, Hanover, New Hampshire

N. B. Myant
Medical Research Council Lipid Metabolism Unit, Hammersmith Hospital, London, W.12, England

Evelina Ortiz
Department of Biology, University of Puerto Rico, Rio Piedras, Puerto Rico

Ida S. Owens
Section on Intermediary Metabolism, National Institute of Arthritis and Metabolic Diseases, National Institutes of Health, Bethesda, Maryland

B. M. Nataf
Laboratoire des Isotopes, Institut Gustave Roussy, Villejuif, 94, France

Friedmund Neumann
Department of Experimental Endocrinology, Main Laboratory of SCHERING AG, 1 Berlin 65, West Germany

M. R. Pandian
Department of Biochemistry, All India Institute of Medical Sciences, New Delhi 16, India

E. M. Pantelouris
Department of Biology, University of Strathclyde, Glasgow, Scotland

Athalia Pertzelan
Pediatric Metabolic and Endocrine Service, Beilinson Hospital, Petah Tikva, Israel

Irwin Pesetsky
Department of Anatomy, Albert Einstein College of Medicine of Yeshiva University, Bronx, New York

Luc Picon
Department of Animal Physiology, Université de Paris VII, Paris, France

Régine Picon
Department of Physiologie comparée, Faculté des Sciences, Paris, France

Dorothy Price
Department of Biology, University of Chicago, Chicago, Illinois

Lawrence G. Raisz
Department of Pharmacology, University of Rochester School of Medicine and Dentistry, Rochester, New York

K. N. Rao
Department of Biochemistry, All India Institute of Medical Sciences, New Delhi 16, India

B. B. Rawdon
Department of Animal Biology, University of Natal, Durban, Natal, S.A.

Jerry R. Reel
Department of Experimental Therapeutics, Parke, Davis and Company Research Laboratories, Ann Arbor, Michigan

W. Ann Reynolds
Department of Anatomy, College of Medicine, University of Illinois, Chicago, Illinois

Jacques Roffi
Laboratoire de Physiologie comparée, Faculté des Sciences, Paris, France

K. K. Sen
Department of Biochemistry, All India Institute of Medical Sciences, New Delhi 16, India

Isaac Schenkein
Department of Biochemistry, New York University College of Dentistry, New York, New York

Denise Scheib
Institut d'Embryologie et de Teratologie expèrimentale du CNRS, Nogent/Marne 94, France

Vratislav Schreiber
Laboratory for Endocrinology and Metabolism, Faculty of General Medicine, Charles University, Prague, Czechoslovakia

S. K. Sharma
Department of Biochemistry, All India Institute of Medical Sciences, New Delhi 16, India

Thomas H. Shepard
Central Laboratory for Human Embryology, Department of Pediatrics, School of Medicine, University of Washington, Seattle, Washington

Kenneth A. Slickers
Department of Biochemistry, Purdue University, West Lafayette, Indiana

Michael Solursh
Department of Zoology, University of Iowa, Iowa City, Iowa

M. L. Sopori
Department of Biochemistry, All India Institute of Medical Sciences, New Delhi 16, India

Anna Strázovcová
Department of Histology, Institute of Experimental Endocrinology, Slovak Academy of Sciences, Bratislava, Czechoslovakia

Vladimir Strbak
Department of Developmental Endocrinology, Institute of Experimental Endocrinology, Slovak Academy of Sciences, Bratislava, Czechoslovakia

Noriko Sugimura
Department of Obstetrics and Gynecology, Faculty of Medicine, University of Tokyo, Tokyo, Japan

Heidi H. Swanson
Department of Anatomy, University of Birmingham, Birmingham 15, England

G. P. Talwar
Department of Biochemistry, All India Institute of Medical Sciences, New Delhi 16, India

Jamshed R. Tata
Department of Biochemistry, National Institute for Medical Research, Mill Hill, London N.W. 7, England

Paola S. Timiras
Departments of Physiology and Anatomy, University of California, Berkeley, California

Edward Tokarsky
Department of Biochemistry, New York University College of Dentistry, New York, New York

Yale J. Topper
Section on Intermediary Metabolism, National Institute of Arthritis and Metabolic Diseases, National Institutes of Health, Bethesda, Maryland

Clarence L. Trummel
Department of Pharmacology, University of Rochester School of Medicine and Dentistry, Rochester, New York

Beatriz Tuchweber
Institut de Médicine et de Chirurgie expérimentales, Universite de Montréal, Montréal, Canada

W. E. Tuinebreyer
Department of Endocrinology, Growth and Reproduction, Rotterdam Medical Facility, Rotterdam, The Netherlands

Roger W. Turkington
Departments of Medicine and Biochemistry, Duke University Medical Center, Durham, North Carolina

J. Th. M. Ureeburh
Department of Endocrinology, Growth and Reproduction, Rotterdam Medical Facility, Rotterdam, The Netherlands

Theony Valcana
Departments of Physiology and Anatomy, University of California, Berkeley, California

Edith Van Marthens
Departments of Medical Microbiology and Immunology, University of California at Los Angeles School of Medicine, Los Angeles, California

Antonia Vernadakis
Departments of Psychiatry and Pharmacology, University of Colorado Medical Center, Denver, Colorado

Claude A. Villee
Department of Biological Chemistry, Harvard University, Cambridge, Massachusetts

Anthony E. Voytovich
Cleveland Metropolitan Hospital, Cleveland, Ohio

J. J. van der Werff ten Bosch
Department of Endocrinology, Growth and Reproduction, Rotterdam Medical Facility, Rotterdam, The Netherlands

R. K. A. van Wermeskerken
Laboratory for Cell Biology and Histology, University of Leiden, Leiden, The Netherlands

F. Weil
Albert Einstein College of Medicine of Yeshiva University, Bronx, New York

Charles R. Wira
Department of Physiology, Dartmouth Medical School, Hanover, New Hampshire

Emil Witschi
Biomedical Division, Rockefeller University, New York, New York

Donald A. Young
Department of Physiology, Dartmouth Medical School, Hanover, New Hampshire

J. J. P. Zaaijer
Laboratory for Cell Biology and Histology, University of Leiden, Leiden, The Netherlands

Stephen Zamenhof
Departments of Medical Microbiology and Immunology, and Biological Chemistry, University of California at Los Angeles School of Medicine, Los Angeles, California

Antonia Zanini
Istituto Superiore di Sanita, Rome, Italy

HORMONES IN DEVELOPMENT

1

Hormones in Development: Past and Prospects

Alfred Jost

Laboratory of Comparative Physiology,
Faculty of Sciences, Paris, France

The development of a normal full-grown living organism from a single egg cell is an extraordinary achievement. The young individual may start life hidden in the maternal womb or in the eggshell before receiving parental care and help, or it may develop alone, free and exposed. In spite of these differences, for the biologist its development raises questions of morphological and physiological specialization. The physiologist attempts to uncover the mechanisms involved in the chain of events that links genetic information received by the egg cell to its final expression in development. Hormones have been recognized as taking part in these events. They participate in processes of growth, morphogenesis, and differentiation in both plants and invertebrate and vertebrate animals. The most spectacular instance of hormonal intervention is provided by metamorphosis of insects and of anuran larvae. Hormonal control of development can also be demonstrated in higher vertebrates and in man. The first clue for hormonal control of development was derived from clinical observations, and it is therefore rewarding to review the historical groundwork of developmental endocrinology.

HISTORICAL BACKGROUND

First Clues and Concepts

Studies of the thyroid opened the path. As early as 1850, Curling(1) reported "two cases of absence of thyroid and symmetrical swellings of fat tissue at the side of the neck, connected with defective cerebral development." Fagge(2), in 1871, confirmed these data and connected cretinism with thyroid deficiency, but the final proof that dwarfism and idiocy depended on thyroid secretion was obtained only during the following twenty years, after the first thyroidectomies in humans and in animals were performed. Schiff(3), in 1884, avoided the effects of thyroidectomy in dogs with abdominal thyroid grafts. In the early 1890's, myxoedematous patients were treated with injected thyroid extracts(4) or thyroid given orally(5,6). At the end of the nineteenth century the need for thyroid hormone to assure normal postnatal growth, stature, and mental maturation was recognized.

The role of the pituitary gland in growth, also first suggested by clinicians, slowly emerged from studies of hypophysectomies performed on dogs by experimenters, such as Caselli(7) (1900) and Aschner(8) (1912), culminating in the preparation of growth-promoting extracts in 1921 by Evans and Long(9).

Progress in developmental endocrinology during these early times was also stimulated by embryological, morphological, and biochemical studies of endocrine glands in developing animals that paralleled endocrine research on adult organisms.

The discovery that the mammalian foetus has active endocrines was not immediately accompanied by any special hypothesis about their role in development. It was supposed that foetal endocrine glands are merely held in readiness for their future postnatal function. Occasionally it was implied that they might exert the same regulatory function in developing animals as in adults. As early as 1899, Langlois and Rehns(10) extracted a pressor substance from adrenal glands of sheep, rabbit, and guinea pig foetuses. Camus(11), in 1902, detected secretin in the duodenum of rabbit and guinea pig foetuses. Insulin was extracted from bovine foetal pancreas by Banting and Best(12) in 1922, and growth hormone was obtained from pituitaries of 9- to 11-cm pig foetuses in 1929 by Smith and Dortzbach(13).

Other experimenters concentrated more on the significance of hormones in development. The concept that hormones acting at a "definite period" of development bring about permanent effects was first presented by Bouin and Ancel(14) (1903), who were studying the testicular interstitial cells as the source of male hormone. They observed the presence of active-looking interstitial cells in sections of the testes of 3-cm pig foetuses—i.e., before sexual differentiation of the genital tract—and suggested that the internal secretion of the fetal testis "imprints on the organism" the essential characters of the male sex, so that castration, even if performed at birth, will no longer be able to suppress sexual differentiation. The correctness of this assumption was later verified by Jost(15) in castration experiments on rabbit foetuses. More recent research extended this concept to the influence of androgens upon the function of the hypothalamus in the rat(16) and upon neural structures involved in sexual behaviour in the guinea pig(17). Similarly, early treatment of young rats with thyroid hormones results in permanent alterations of thyroid function(18,19). The hormonal theory of sex differentiation was developed and popularized by Lillie's beautiful papers(20,21) (1916,1917) dealing with the bovine freemartin. This work stimulated experimental research on all classes of vertebrates for half a century.

The discovery by Gudernatsch(22-23) (1912–1914) that thyroid could provoke metamorphosis in anuran tadpoles is to be regarded as another milestone in developmental endocrinology. Gudernatsch's experiments—published in two papers under the modest general title of *Feeding Experiments in Tadpoles* and supplemented by other studies in which the effects of thyroidectomies and hypophysectomies on tadpole metamorphosis were investigated—demonstrated unequivocally the requirement of hormones for amphibian development, as well as stimulating interest in pituitary-thyroid relationships. The dramatic morphological and physiological changes produced by thyroxine in tadpoles has no exact equivalent in higher vertebrates. Nevertheless, ideas concerning the role of hormones in development were largely influenced by the demonstration of their role in amphibian metamorphosis.

Maternofoetal Relations in Humans and Other Mammals

The human and, more generally, the mammalian foetus and newborn closely depend upon the mother during intrauterine development and during lactation. A maternal endocrine adjustment to pregnancy and to lactation is a prerequisite for normal foetal development. This extensive and important chapter of the endocrinology of reproduction cannot be considered here. But it should be recalled that uterine accommodation under ovarian hormones is so important for the young that foetal malformations may result from hormonal imbalances in the mother(24-26).

Since the foetus produces and needs hormones for its development, and since the mother also produces her own hormones, the possibility of exchanges or of reciprocal help in case of hormonal shortage in either organism was a tempting subject for study. Kocher(27), in 1892, was probably the first to suggest that the secretion of the maternal thyroid could act vicariously in the hypothyroid foetus. Such a concept presupposed transfer of thyroid hormone from mother to foetus. It was later contradicted by Dorff's observation(28) in 1934 of prenatal retardation in bone maturation in congenital cretins.

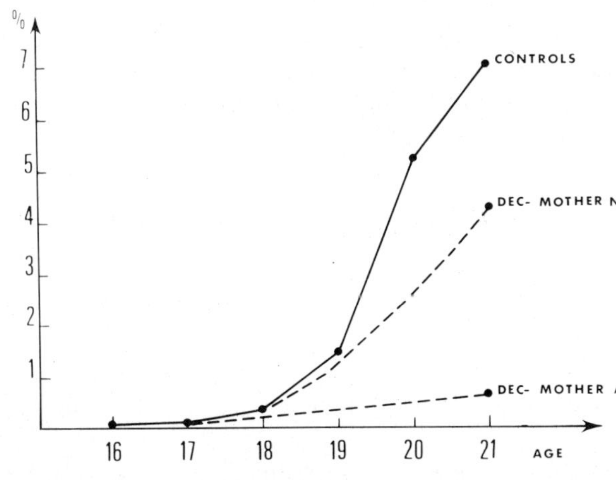

Fig. 1. Glycogen contents in the liver of rat foetuses (percent of fresh weight). The heavy line shows the increase of glycogen contents in controls according to age. In foetuses decapitated on day 17 in mother animals which were adrenalectomized (adrex.) on day 14, the glycogen is low on day 21. In decapitated foetuses of entire mother animals, the amount of glycogen is two-thirds normal. The interrupted lines arbitrarily join the initial (day of decapitation) and final glycogen values. From Favard and Jost. *Arch. Anat. Micr. Morph. Exp.*, 55:603-632, 1966.

The concept of vicarious action of some maternal hormone in the foetal organism supposes at least that the hormone in question crosses the placenta and that it is necessary for the foetus. One example may be cited: In the rat foetus, glycogen storage in the liver(29,30) or in the heart(31) is hormonally controlled. If the foetus is deprived of its own adrenocortical hormones by decapitation, heart glycogen remains normal while liver glycogen is reduced by approximately one third (Fig. 1). If, in addition to foetal decapitation, the pregnant mother is adrenalectomized, both heart and liver glycogens stores are lowered. Corticosteroids injected directly into the decapitated foetus may replace the maternal hormones.

A recent extension of the problem of maternal influences upon the foetus is linked with the finding that some proteins, especially gamma globulins, are transferred from mother to foetus. A very important contribution was the discovery that congenital hyperthyroidism may result from transfer of maternal long-acting thyroid stimulator (LATS) to the foetus(32,33).

Transfer of hormones from foetus to mother and their vicarious effect in the mother were largely accepted after Carlson and Drennan(34), in a 1911 paper entitled *The Control of Pancreatic Diabetes in Pregnancy by the Passage of the Internal Secretion of the Pancreas of the Foetus to the Blood of the Mother*, reported that in pancreatectomized pregnant bitches glycosuria was reduced toward the end of pregnancy. Experiments performed by others during the following years yielded conflicting results. The hypertrophy of the pancreatic islands in foetuses of diabetic women, noted by Dubreuil and Anderodias(35) in 1920, could lend support to the view of Carlson and Drennan, but it could also result from maternally induced foetal hyperglycemia. Schlossmann(36) insisted that there is a high glucose uptake in large dog foetuses during late pregnancy, and Cuthbert et al.(37) noticed that the need for insulin by pancreatectomized bitches is diminished not only during late pregnancy but during lactation as well.

The problem of placental transfer of insulin is obviously difficult to solve, and the theory that foetal hormone could act vicariously for the mother is still controversial. The contribution of the foetoplacental unit to steroid production during pregnancy(38) renewed one aspect of this question.

HORMONES AND BODY GROWTH

There is considerable evidence that indicates that three major hormones are concerned in growth—thyroid hormone, growth hormone, and insulin. Only a few pertinent data will be recalled.

Thyroid. In frog larvae, thyroidectomy prevents metamorphosis but does not stop growth; after a prolonged period of development giant tadpoles may be obtained(39,40).

In mammals, thyroidectomy of young animals stops growth, but it is not certain that prenatal growth depends on thyroid hormones. Athyreotic human babies are of normal size and cease growing more or less completely a few months after birth. At birth they already show signs of hypothyroidism(28,41); only limited amounts of maternal thyroid hormones reach the foetus toward the end of pregnancy(42).

Similarly, rabbit foetuses thyroidectomized in utero continue growing(43) (Fig. 2) despite the reduced level of thyroxine present in foetal blood(44) and other signs of hypothyroidism(45,46). It appears likely that either thyroid hormone is not necessary for prenatal growth at all, or if it is needed, minimal amounts suffice to meet the requirements.

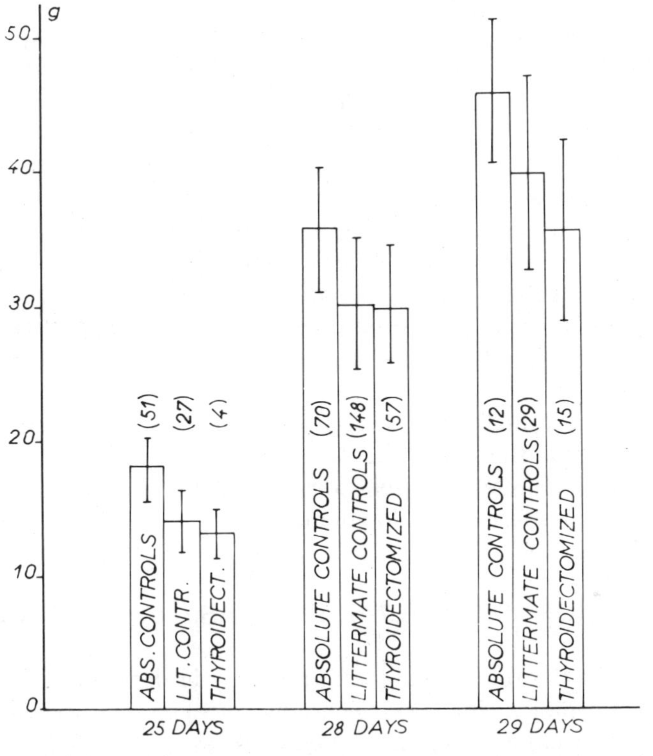

Fig. 2. Body weight of various groups of rabbit foetuses sacrificed on day 25, 28, or 29. The absolute controls refer to foetuses of unoperated females. In females which are submitted to surgery during pregnancy, the weight of intact foetuses (littermate controls) is somewhat reduced; the thyroidectomized foetuses of the same litters show no significant reduction in weight. Thyroidectomy was performed on day 22 (group sacrificed on day 25) or on 23 (the other groups).

Numbers of foetuses are in brackets, and standard deviation is indicated. From Jost. *In* Coloquio sobre a Tireoide, Chagas and Lobo, eds., 1961. Courtesy of Universidade do Brasil.

Figs. 3 to 8. In all figures, filled circles represent control foetuses, and open circles are decapitates. The interrupted lines join arbitrarily the stage at which decapitation was done and the stage at which the foetuses were sacrificed; when foetuses were sacrificed after varying periods of time the interrupted line joins the corresponding values. Figures 3 to 8 from Jost. *Perspect. Biol.,* 11:371-375, 1968.

Hypophysis. In frog larvae, hypophysectomy impairs metamorphosis but permits continuation of growth at a slow pace. According to Smith(47), in *Rana boylei* retardation of growth becomes very pronounced mainly after the midlarval period, and stage of 30 to 32 mm appears as a "critical point" in growth. However, this has not been observed in other species. In hypophysectomized larvae of *Alytes obstetricans*, beef growth hormone or ovine prolactin induces gigantism(48).

In rats and in rabbits, the pituitary gland is not indispensable for growth at early stages. In rats, subnormal growth continues after early hypophysectomy until approximately the end of the fourth week(49, 50). In young rabbits hypophysectomized on day 52 or 74, subnormal growth continues until approximately 100 days of age. In animals hypophysectomized on day 105, growth stops immediately(51). Prenatal growth, on the other hand, continues practically normally in decapitated rabbit foetuses(52-54).

In human infants suffering from hypopituitarism, growth continues for some time after birth(55). In the guinea pig, the situation is different in that hypophysectomy does not stop growth, although the thyroid and the adrenals are depressed (56,57).

Insulin. Babies born to diabetic or prediabetic women are frequently overweight and somewhat oversized. This has been attributed to an overproduction of insulin by the foetal pancreas in response to hyperglycemia(58). These infants secrete an excess of insulin(59). Picon(60) succeeded in inducing an excess of growth in rat foetuses given repeated injections of insulin, but this effect appeared only after the stage of 21 days in prolonged pregnancies. Similarly, excess of growth in foetuses of diabetic women becomes conspicuous only after the twenty-eighth week (61).

A complete interpretation of the hormonal control of body growth requires more detailed studies of the various hormonal and metabolic interrelationships. But from the foregoing examples, it would appear that the hormonal requirement for growth varies during different developmental periods.

THE ROLE OF THE FOETAL PITUITARY IN DEVELOPMENT

The analysis of the role of the foetal pituitary in development affords an example of a rather complex situation concerning correlations between foetal glands, maternofoetal interrelations, and maturational processes. It is well established that the larval or foetal pituitary gland is necessary for the normal development and function of other endocrine glands, such as the thyroid or adrenal cortex(47,54). The modalities of action of the foetal pituitary are still currently under investigation. Only a few selected topics will be considered here.

Significance of the Pituitary Stimulus

The differentiation into specific endocrine glands (thyroid or adrenal cortex) of the early embryonic primordia is independent of the hypophysis, but subsequent growth and functional maturation is controlled by pituitary hormones(54,62). The modalities and significance of the pituitary stimulus may be illustrated by examples pertaining to the rat foetus.

The effect of pituitary removal by decapitation on the growth of the adrenal glands of rat foetuses varies according to the stage at which decapitation is performed(63,64) (Fig. 3). If the pituitary gland is removed on day 17 or before, the adrenal glands actually continue to grow at a very low rate until birth; their ap-

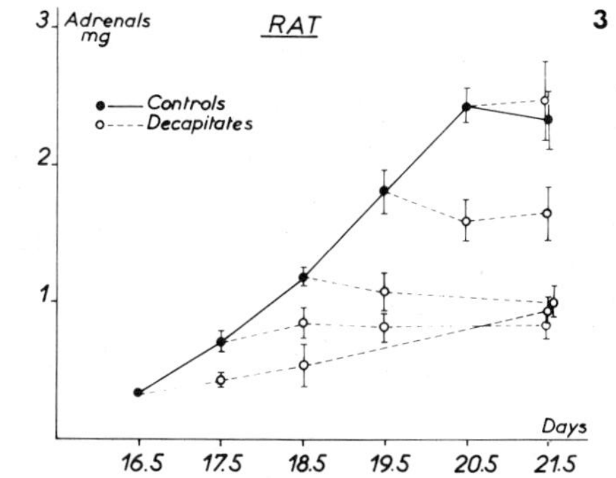

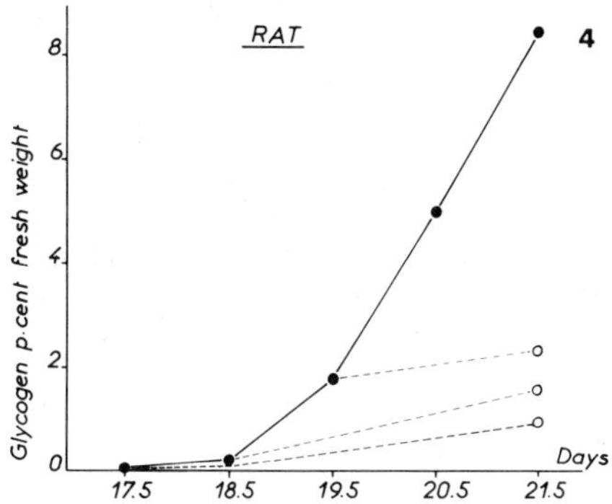

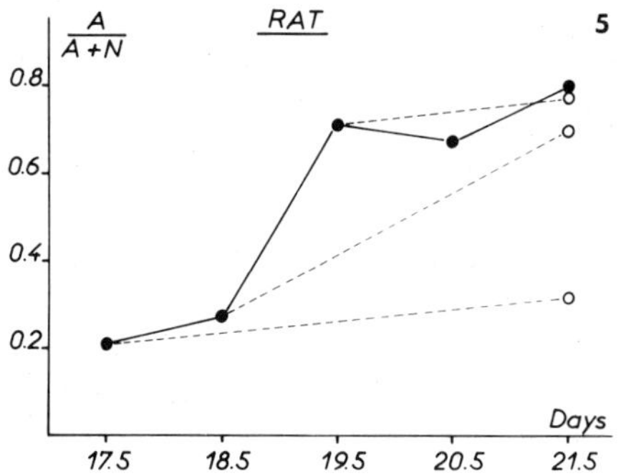

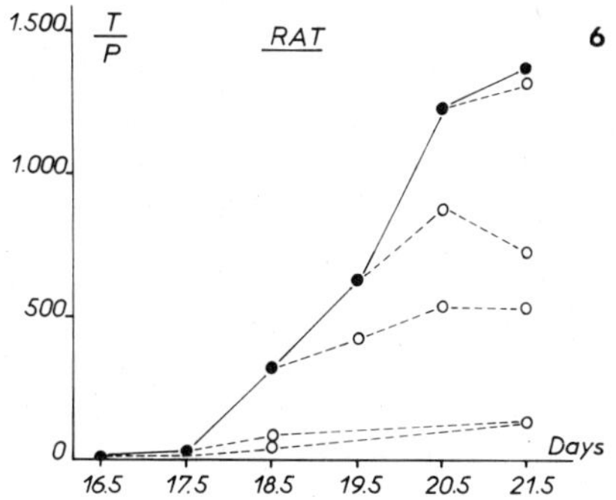

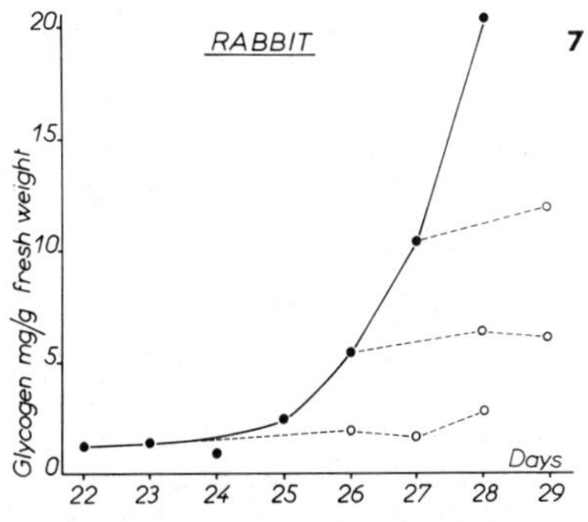

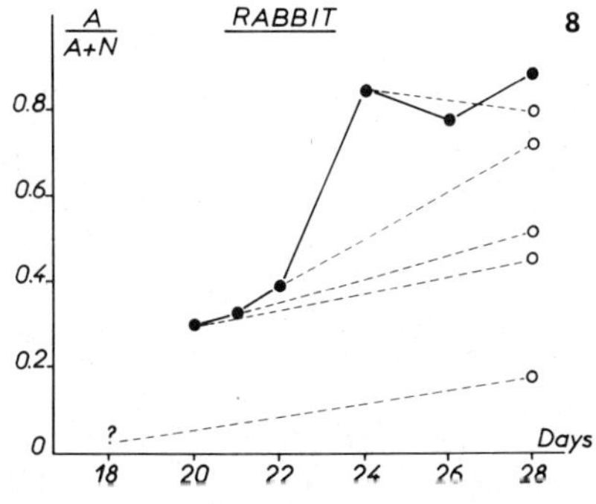

Fig. 3. Weight of one pair of adrenal glands from rat foetuses decapitated at various stages, in comparison with controls. Confidence interval indicated.

Fig. 4. Glycogen contents (percentage of fresh weight) in the liver of rat foetuses from mother animals adrenalectomized on day 14.

Fig. 5. Ratio of epinephrine to epinephrine plus norepinephrine (ratio A/A+N) as measured fluorometrically in one pair of adrenal glands of rat foetuses.

Fig. 6. Collection of ^{131}I by thyroid tissue of rat foetuses (T/P=ratio of the radioactivities per milligram of thyroid tissue/blood plasma). Tracer doses of ^{131}I were injected 24 hours previously to the mother animal. (Data from Geloso [68].)

Fig. 7. Glycogen contents (milligrams per gram fresh weight) in the liver of rabbit foetuses.

Fig. 8. Same as Fig. 5, in rabbit foetuses. Epinephrine was undetectable with the technique used in the 18-day-old rabbit foetuses, but measurable amounts were present on day 28 in foetuses decapitated on day 18.

parent "atrophy" at birth results from impaired growth and not from decrease in size. The difference between the growth curves of the adrenals in control and in decapitated foetuses graphically illustrates the role of the pituitary stimulus. If the pituitary is removed on day 18 or 19, after it has already begun to stimulate the adrenals, the adrenal gland maintains the size attained at that stage (see Fig. 3).

The effects of early or delayed decapitation on some other pituitary-controlled systems provide similar curves. Thus, rate of glycogen deposition in the growing liver, expressed as percentage of glycogen in fresh tissue, remains low after early decapitation; it continues at intermediary values if the foetus is decapitated slightly later (Figs. 4 and 8). Similar results are obtained in measurements of the rate of methylation of norepinephrine to epinephrine in the adrenal medulla, which in rats and rabbits (Figs. 5 and 7) has been shown to be controlled by the adrenal cortex(65-67). A similar curve (Fig. 6) was obtained by Geloso(68) in a study of the uptake of radioiodine by the thyroid gland of rat foetuses.

It appears that the length of time the pituitary gland is allowed to stimulate the target systems determines whether full or partial physiological maturation is attained(62).

Hypothalamic Control of Pituitary Function

In the rat foetus, the encephalon can be removed easily along with the hypothalamus, while the pituitary body remains in situ in the sella turcica (Fig. 9). In foetuses submitted to such surgery on days 16 to 18, the pituitary gland continues

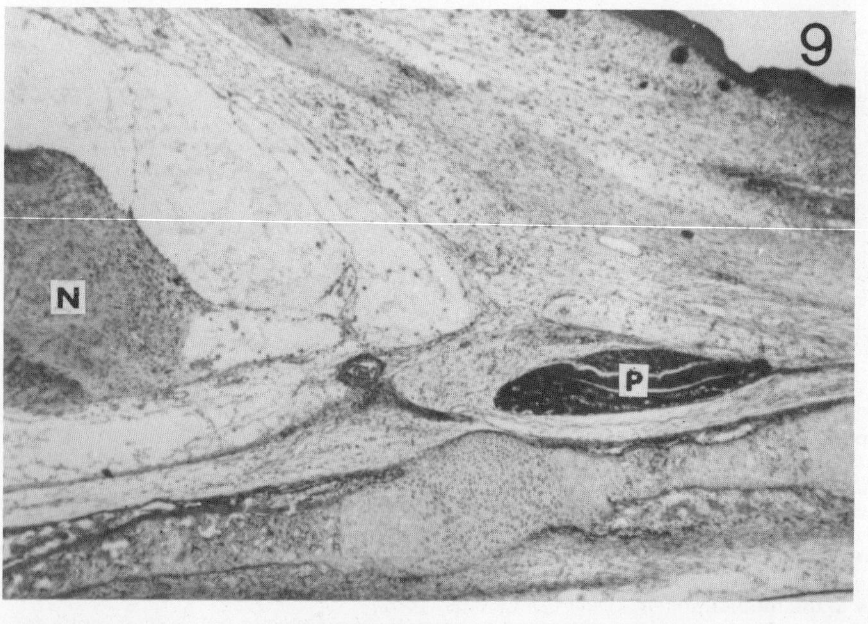

Fig. 9. Longitudinal section through the head of an encephalectomized rat foetus. The nervous structures normally covering the pituitary gland (P) were removed. The interrupted nervous system (N) is seen on the left, in the posterior part of the head. From Jost and Geloso. *C.R. Acad. Sci. (Paris),* 265:625-627, 1967.

to grow and differentiate, but at birth the adrenals are reduced almost to the same extent as in foetuses deprived of their hypophysis by decapitation. Crude hypothalamic extracts or purified corticotrophin releasing factor (both almost inactive in decapitates) correct the size of the adrenals in encephalectomized foetuses(54, 63,69).

By contrast, the thyroid glands of encephalectomized rat foetuses look normal, and respond to maternally administered propylthiouracil with characteristic hyperplasia(70) (Fig. 10). In the absence of the pituitary gland, the antithyroid drug has no effect(44,71). In unpublished experiments, Annette Geloso observed that blood thyroxine is somewhat reduced in encephalectomized foetuses studied under conditions of isotopic equilibration.

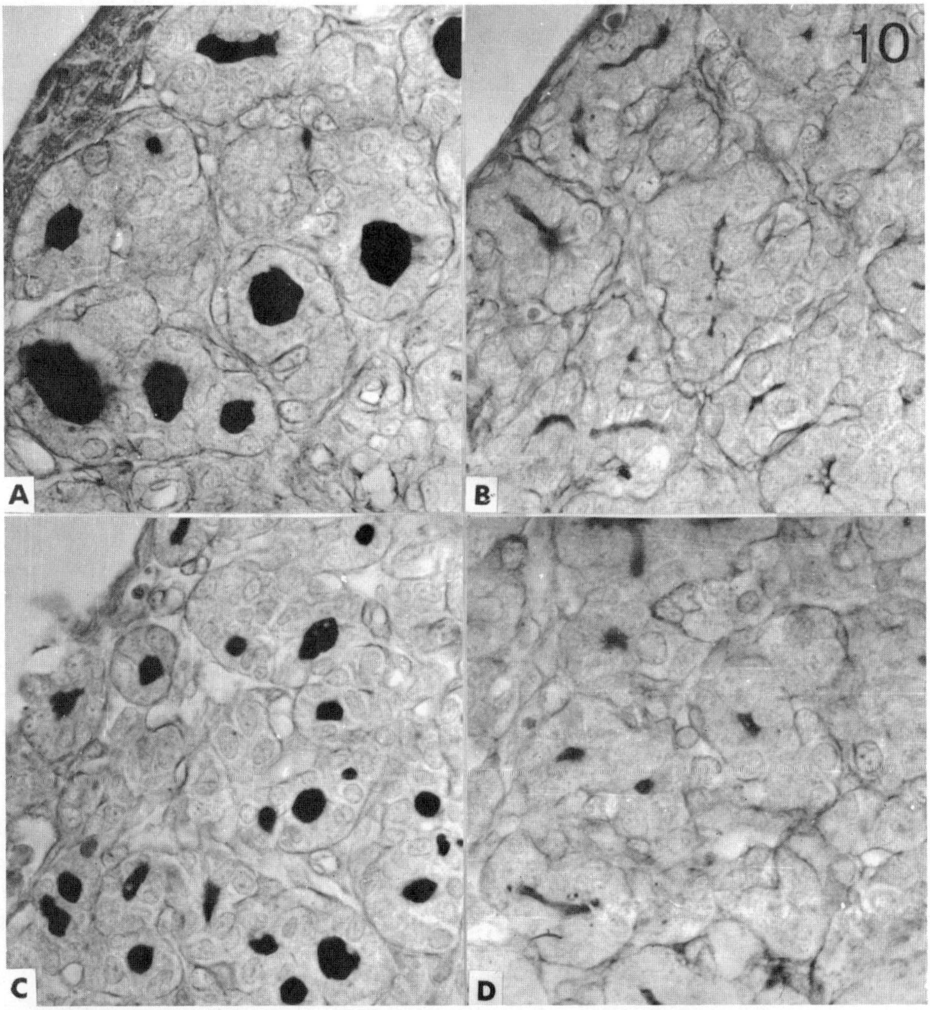

Fig. 10. Section of thyroid glands from 21-day-old rat foetuses (PAS strain). A=control. B, C, and D=foetuses of propylthiouracil (PTU) treated mother animals (50 mg/day for three days). B=entire foetus; C=foetus decapitated on day 18; D=foetus encephalectomized on day 18 (pituitary present). From Jost and Geloso. *C.R. Acad. Sci. (Paris),* 265:625-627, 1967.

The thyrotropic function of the foetal pituitary thus seems to be much less strictly controlled by the hypothalamus than the adrenocorticotropic function. These results give a clue as to the causes of the condition found in human anencephalia in which (1) the pituitary gland is usually present but has no or abnormal connections with the disturbed hypothalamus; (2) the adrenal cortex is poorly developed; and (3) the thyroid gland is histologically close to normal.

THE HUMAN FOETAL ADRENAL CORTEX

The adrenal cortex of the human foetus is exceedingly large, but it shrinks rapidly after birth. The involution concerns mainly the most internal zone (foetal zone) and can be prevented if ACTH is given to the newborn infant(72). In anencephalic monsters, the foetal zone is absent at birth. It has long been known that the size and histological structure of the adrenal cortex are normal in anencephalic foetuses under 20 to 22 weeks of prenatal age(73,74). After this stage, the relative amount and width of the foetal cortex decrease by a process of involution while the permanent cortex becomes somewhat thicker.*

Data from literature on the weight of the adrenals in control(76,77) and in anencephalic infants(78-81) (Fig. 11) may be helpful in analyzing the process of involution in adrenals of anencephalics. It appears that in anencephalia, growth of the adrenals is stopped after approximately the twentieth week, although after that time some histological reorganization of the cortex still takes place. The curve resembles that of early decapitated rat foetuses (see Fig. 3), but in anencephalic monsters the primary cause is probably a hypothalamic defect.

The question of the identity of the hormone that stimulates the adrenal cortex until birth and which is absent in anencephalia has been often debated(72). It should first be recalled that in rat and rabbit foetuses, feedback relations between adrenocorticotropic function of the hypophysis and corticosteroids of maternal or foetal origin are well documented(82). Similar feedback effects exist in the human foetus. Depression of the adrenal glands may result either from an excess of corticosteroids given to the mother(83,84) or from maternal Cushing syndrome(85). Conversely, impaired production of corticosteroids in adrenogenital syndrome permits an increase in ACTH secretion(86) and adrenal hypertrophy.

The intense stimulation of the adrenal cortex in the normal foetus probably results from a feeble feedback inhibition of the foetal pituitary, due to a low level of unbound or free corticosteroids in foetal blood. During human pregnancy, the corticosteroid-binding capacity of the maternal plasma, especially the transcortin globulin, increases steadily under the influence of ovarian hormones(87). It is tempting to consider that the high level of maternal corticosteroid-binding proteins is involved in the foetal adrenocorticotropic function. They might cause an outflux of free corticosteroids from foetal to maternal blood, or if appreciable amounts of

*According to Tuchmann-Duplessis(75), in anencephalia "atrophy of the foetal cortex is all the more important as term approaches. The average weight of the adrenals at term is 30 times less in anencephalia than in controls; it is diminished by 11 times in 8-month foetuses and by 5 times in 7-month foetuses." These data are approximated in Figure 11.

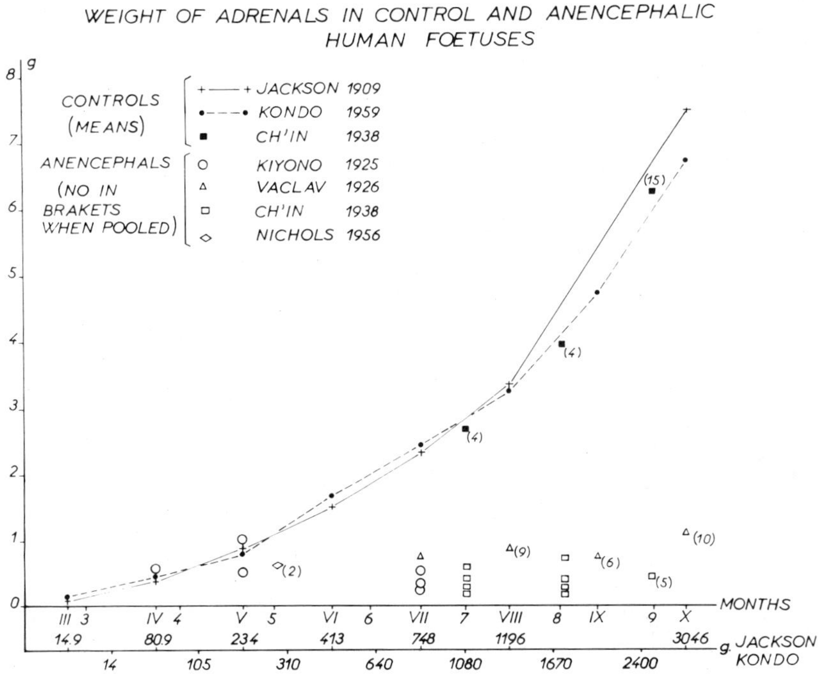

Fig. 11. Weight of the adrenals in control or in anencephalic human foetuses. *Controls:* weights recalculated from Jackson's data(76) for American foetuses; weights for Japanese foetuses from Kondo(77). Isolated values given by Ch'in for Chinese foetuses. *Anencephalic foetuses:* data from Kiyono(78), Vaclav(79), Ch'in(80), and Nichols(81). Lunar months are in roman numbers; regular months in arabic numbers. Body-weight values given by Jackson and Kondo.

binding proteins were transferred from mother to foetus, they could directly reduce the concentration of free corticosteroids in the foetal blood. In either case, at birth an increased level of free corticosteroids would depress the hypophyseal activity.

This last hypothesis is in accord with the data recently observed in newborn rats by Koch et al.(88). In this species, some decrease of the size of the adrenal glands also occurs after birth. The transcortin capacity in the blood of the newborn decreases during the first 10 postnatal days, while free corticosterone increases. After that time the young rat produces its own transcortin, and the percentage of free corticosterone diminishes again.

CORTICOSTEROIDS AND FOETAL HEPATOCYTES

Glycogen deposition in the foetal liver is also under hormonal control(63,89). In the rat foetus, it can be prevented if corticosteroids are suppressed by maternal adrenalectomy and foetal decapitation(30) and reinitiated if corticosteroids are given. In the livers of these decapitated foetuses, the activity of several enzymes involved in glycogen synthesis remains low(90).

CONCLUDING REMARKS

The purpose of this review was to present some general problems in developmental endocrinology and some of the major unsolved questions. The material has been arbitrarily selected, and some data were purposely grouped in order to deal successively with the growth of the whole body, including analytical studies of endocrine correlations and hormonal effects at the cellular level.

It is clear that normal development demands adequate hormonal actions for some morphogenetic events, such as sex differentiation, and for ontogenesis or for full maturation of physiological functions. In addition, hormones have a multifarious role in a developing organism. They help in the immediate maintenance and regulation of the *milieu intérieur*: for instance, parathyroidectomy of the rat foetus in utero determines foetal hypocalcemia(94).

Hormones also participate actively in the changes characteristic of a developing system. These changes involve either growth by simple accretion or much more complex processes, such as morphogenesis or renewed physiological potentialities or correlations. The latter result both from cellular specialization and cellular replacement. Usually these changes occur at definite stages of development as a consequence of earlier events. In several instances, they cannot be anticipated or postponed with present means. The capacity of the effector cells to respond in a certain way during the critical stage depends on their previous history, probably on the degree of structural and biochemical differentiation already attained. At such stages, hormones may impose new permanent and irreversible orientations on physiologically crucial structures.

Since hormonal effects can be either deliberately prevented by experimental procedures or permitted to proceed, the developing system affords an excellent tool for physiological and biochemical studies. Moreover, the detailed analysis of the onset of physiological specialization of organs should give an insight into the process of cellular differentiation. Thus, studies of the biochemical and molecular processes of hormonal actions during development will open up new dimensions of developmental endocrinology. It might be sobering to recall that even at the level of the whole organism, the various hormonal correlations operative during development and their significance for adult life are far from being completely recognized.

REFERENCES

1. Curling, T. B. Two cases of absence of thyroid body and symmetrical swellings of fat tissue at the side of the neck, connected with defective cerebral development. Trans. Med. Chir. London, 33:303–306, 1850.
2. Fagge, C. H. On sporadic cretinism occurring in England. Med. Chir. Trans., 54:155, 1871.
3. Schiff, M. Résumé d'une nouvelle série d'expériences sur les effets de l'ablation des corps thyroïdes. Rev. Med. Suisse Rom., 4:425–445, 1884.
4. Murray, G. R. Note on the treatment of myxoedema by hypodermic injections of an extract of the thyroid gland of a sheep. Brit. Med. J., 2:796–797, 1891.

5. MacKenzie, H. W. G. A case of myxoedema treated with great benefit by feeding with fresh thyroid glands. Brit. Med. J., 2:940, 1892.
6. Fox, E. L. A case of myxoedema treated by taking extract of thyroid by mouth. Brit. Med. J., 2:941, 1892.
7. Caselli, A. Studi Anatomici e Sperimentali sulla Fisiopatologia della Glandola Pituitaria. S. Calderini, Reggio Emilia, 228 pp., 1900.
8. Aschner, B. Über die Funktion der Hypophyse. Pflueger Arch. Ges. Physiol., 146:1-146, 1912.
9. Evans, H. M., and Long, J. A. The effect of the anterior lobe of hypophysis administered intraperitoneally upon growth and the maturity and oestrous cycles of the rat. Anat. Rec., 21:61–63, 1921.
10. Langlois, J. P., and Rehns, J. Les capsules surrénales pendant la période foetale. C.R. Soc. Biol. (Paris), 51:146–147, 1899.
11. Camus, L. Recherches expérimentales sur la "sécrétine." J. Physiol. Path. Gen. (Paris), 4:998–1013, 1902.
12. Banting, F. G., and Best, C. H. Pancreatic extracts. J. Lab. Clin. Med., 7:464–472, 1922.
13. Smith, P. E., and Dortzbach, C. The first appearance in the anterior pituitary of the developing pig foetus of detectable amounts of the hormone stimulating ovarian maturity and general body growth. Anat. Rec., 43:227–297, 1929.
14. Bouin, P., and Ancel, P. Sur la signification de la glande interstitielle du testicule embryonnaire. C.R. Soc. Biol. (Paris), 55:1682, 1903.
15. Jost, A. Recherches sur la différenciation sexuelle de l'embryon de lapin. III. Rôle des gonades foetales dans la différenciation sexuelle somatique. Arch Anat. Micr. Morph. Exp., 36:271–315, 1947.
16. Barraclough, C. A., and Gorski, R. A. Evidence that the hypothalamus is responsible for the androgen-induced sterility in the female rat. Endocrinology, 68:68, 1961.
17. Phoenix, C. H., Goy, R. W., Gerall, A. A., and Young, W. C. Organizing action of prenatally administered testosterone propionate on the tissues mediating mating behaviour in the female guinea-pig. Endocrinology, 65:369–382, 1959.
18. Eayrs, J. T., and Holmes, R. L. Effect of neonatal hyperthyroidism on pituitary structure and function in the rat. J. Endocr., 29:71, 1964.
19. Bakke, J. L., and Lawrence, N. Persistent thyrotropin insufficiency following neonatal thyroxine administration. J. Lab. Clin. Invest., 67:477–482, 1966.
20. Lillie, F. The theory of the freemartin. Science, 43:611, 1916.
21. ——— The freemartin, a study of the action of sex hormones in foetal life of cattle. J. Exp. Zool., 23:371, 1917.
22. Gudernatsch, J. F. Feeding experiments in tadpoles. I. The influence of specific organs given as food on growth and differentiation. A contribution to the knowledge of organs with internal secretion. Arch. Entw. Mech. Org., 35:457, 1912.
23. ——— Feeding experiments in tadpoles. II. A further contribution to the knowledge of organs of internal secretion. Amer. J. Anat., 15:431, 1914.
24. Courrier, R., and Jost, A. Sur l'analyse quantitative de l'endocrinologie de la gestation chez la lapine. C.R. Soc. Biol. (Paris), 130:726, 1939.
25. Carpent, G. Le déséquilibre hormonal gravidique et ses répercussions sur la morphologie du foetus chez le rat. Arch. Anat. Micr. Morph. Exp.,54:459–539, 1962.
26. Jost, A. Hormones et tératogenèse. *In* Proceedings of the European Society of Drug Toxicity, Amsterdam, Excerpta Medica Foundation, International Congress Series No. 64, pp. 35–47, 1963.
27. Kocher, T. Zur Verhütung des Cretinismus und cretinoider Zustände nach neuen Forschungen. Deutsche Z. Chir., 34:556–626, 1892.
28. Dorff, G. B. Sporadic cretinism in one of twins. Report of cases with roentgen demonstration of osseous changes that occured in utero. Amer. J. Dis. Child., 48:1316, 1934.
29. Jost, A., and Jacquot, R. Recherches sur le contrôle hormonal de la charge en glycogène du foie foetal du lapin et du rat. C.R. Acad. Sci. (Paris), 239:98–100, 1954.
30. Jacquot, R. Recherches sur le contrôle endocrinien de l'accumulation de glycogène dans le foie chez le foetus de rat. J. Physiol. (Paris), 51:655–721, 1959.

31. Picon, L., and Bouhnik, J. Action de la cortico-surrénale sur la teneur du coeur en glycogène chez le foetus de rat et le rat nouveau-né. C.R. Soc. Biol. (Paris), 160:288, 1966.
32. Rosenberg, D., Grand, M. J. H., and Silbert, D. Neonatal hyperthyroidism. New Eng. J. Med., 268:292–296, 1963.
33. MacKenzie, J. M. Neonatal Graves' disease. J. Clin. Endocr., 24:660–668, 1964.
34. Carlson, A. J., and Drennan, F. M. The control of pancreatic diabetes in pregnancy by the passage of the internal secretion of the pancreas of the foetus to the blood of the mother. Amer. J. Physiol., 28:391–395, 1911.
35. Dubreuil, G., and Anderodias, S. Ilots de Langerhans géants chez un nouveau-néde mère glycosurique. C.R. Soc. Biol. (Paris), 83:1490, 1920.
36. Schlossmann, H. Beiträge zur Biologie der Plazenta. II. Ist die Plazenta für Insulin durchlässig? Arch. Exp. Pathol. Pharm., 159:213, 1931.
37. Cuthbert, F. P., Ivry, A. C., Isaacs, B. L., and Gray, J. The relation of pregnancy and lactation to extirpation diabetes in the dog. Amer. J. Physiol., 115:480–496, 1936.
38. Diczfalusy, E., Pion, R., and Schwers, J. Steroid biogenesis and metabolism in the human foeto-placental unit at midpregnancy. Arch. Anat. Micr. Morph. Exp., 54:67–84, 1965.
39. Hoskins, E. R., and Hoskins, M. M. Growth and development of Amphibia as affected by thyroidectomy. J. Exp. Zool., 29:1, 1919.
40. Allen, B. M. The influence of thyroid gland and hypophysis upon growth and development of amphibian larvae. Quart. Rev. Biol., 4:325–352, 1929.
41. Wilkins, L. The Diagnosis and Treatment of Endocrine Disorders in Childhood and Adolescence, 2nd Ed. Oxford, Blackwell, 1957.
42. Fisher, D. A., Lehman, H., and Lackey, C. Placental transport of thyroxine. J. Clin. Endocr., 24:393–400. 1964.
43. Jost, A. Physiologie des hormones thyroïdiennes chez le foetus. *In* Coloquio sobre a Tireoide. Chagas, C., and Lobo, L. C., eds. Rio de Janeiro, Universidade do Brasil, pp. 86–112, 1961.
44. ——— Action du propylthiouracile sur la thyroïde du foetus de lapin intact, décapité ou injecté de thyroxine. C.R. Soc. Biol. (Paris), 153:1900–1902, 1959.
45. ——— and Picon, L. Influence de la thyroïdectomie du foetus de lapin sur la teneur de son corps en lipides et en azote. C.R. Acad. Sci. (Paris), 246:1281–1283, 1958.
46. Bearn, J. G. Hormonal control of the metabolism of cholesterol in the rabbit foetus. Nature, 198:1005–1006, 1963.
47. Smith, P. E. The pigmentary, growth and endocrine disturbances induced in the anuran tadpole by the early ablation of the pars buccalis of the hypophysis. Amer. Anat. Memoirs, No. 11, 1920.
48. Rémy, Ch., and Bounhiol, J. J. Gigantisme expérimental obtenu par intervention chirurgicale et traitements hormonaux chez le têtard du crapaud accoucheur. Ann. Endocr. (Paris), 27(Suppl.):377–382, 1966.
49. Collip, J. B., Selye, H., and Thomson, D. L. Beiträge zur Kenntnis der Physiologie des Gehirnanhanges. Arch. Path. Anat. Physiol., 290:23–46, 1933.
50. Walker, D. G., Simpson, M. E., Asling, C. W., and Evans, H. M. Growth and differentiation in the rat following hypophysectomy at 6 days of age. Anat. Rec., 106:539–554, 1950.
51. Vezinhet, A. Effet de l'hypophysectomie sur la croissance pondérale du lapin. C.R. Acad. Sci. (Paris), 266:2348-2351, 1968.
52. Jost, A. Expériences de décapitation de l'embryon de lapin. C.R. Acad. Sci. (Paris), 225: 322–324, 1947.
53. ——— Hormonal factors in the development of the foetus. Cold Spring Harbor. Symp. Quant. Biol., 19:167–180, 1954.
54. ——— Anterior pituitary function in foetal life. *In* The Pituitary Gland. Harris, G. W., and Donovan, B. T., eds. London, Butterworths, 1966.
55. Brasel, J. A., Wright, J. C., and Wilkins, L. An evaluation of seventy-five patients with hypopituitarism beginning in childhood. Amer. J. Med., 38:484–498, 1965.
56. Knobil, E., and Greep, R. O. The physiology of growth hormone with particular refer-

ence to its action in the rhesus monkey and the "species specificity" problem. Rec. Progr. Hormone Res., 15:1–58, 1959.
57. Clayton, B., and Worden, J. M. Growth in the young hypophysectomized guinea-pig. J. Endocr., 20:36–47, 1960.
58. Pedersen, J., and Osler, M. Hyperglycemia as the cause of characteristic features of the foetus and newborn of diabetic mothers. Danish Med. Bull., 8:78–83, 1961.
59. ——— Pedersen, L. M., and Jørgensen, K. R. Insulin and glucose in the plasma from umbilical vein and heel blood of newborn infants of diabetic women. Acta Endocr. (Kobenhavn), 53:310–314, 1966.
60. Picon, L. Effect of insulin on growth and biochemical composition of the rat foetus. Endocrinology, 81:1419–1421, 1967.
61. Cardell, B. S. The infants of diabetic mothers. A morphological study. J. Obstet. Gynaec. Brit. Comm., 60:834, 1953.
62. Jost, A. Full or partial maturation of foetal endocrine systems under pituitary control. Perspect. Biol., 11:371–375, 1968.
63. ——— Problems of fetal endocrinology: The adrenal glands. Recent Progr. Hormone Res., 22:541–569, 1966.
64. ——— and Cohen, A. Signification de l'"atrophie" des surrénales foetales du rat provoqué par l'hypophysectomie (décapitation). Develop. Biol., 14:154–168, 1966.
65. ——— and Roffi, J. Action de la cortisone et de l'hydrocortisone sur la teneur de la surrénale du rat et de la souris en substances hypertensives. C.R. Acad. Sci. (Paris), 246:163–165, 1958.
66. Roffi, J. Influence de la cortico-surrénale sur le contenu en adrénaline et noradrénaline de la surrénale chez le foetus de rat. C.R. Acad. Sci. (Paris), 260:1267–1270, 1965.
67. Margolis, F. L., Roffi, J., and Jost, A. Norepinephrine methylation in fetal rat adrenals. Science, 154:275–276. 1966.
68. Geloso, J. P. Fonctionnement de la thyroïde et corrélations thyréo-hypophysaires chez le foetus de rat. Ann. Endocr. (Paris), 28(Suppl. 1):1–80, 1967.
69. Jost, A., Dupouy, J. P., and Monchamp, A. Fonction corticotrope de l'hypophyse et hypothalamus chez le foetus de rat. C.R. Acad. Sci. (Paris), 262:147–150, 1966.
70. ——— and Geloso, A. Réponse de la thyroïde foetale du rat au propylthiouracile en l'absence d'hypothalamus. Remarques sur les glandes endocrines du foetus anencéphale humain. C.R. Acad. Sci. (Paris), 265:625–627, 1967.
71. ——— Action du propylthiouracile sur la thyroïde de foetus de rat intacts ou décapités. C.R. Soc. Biol. (Paris), 151:1295–1298, 1957.
72. Lanman, J. T. An interpretation of human foetal adrenal structure and function. *In* The Human Adrenal Cortex. Currie, A. R., Symington, T., and Grant, J. K., eds. Edinburgh and London, Livingstone, 1962.
73. Meyer, R. Nebennieren bei Anencephalie. Virchow. Arch. Path. Anat., 210:158, 1912.
74. Benirschke, K. Adrenals in anencephaly and hydrocephaly. Obstet. Gynec., 8:412–425, 1956.
75. Tuchmann-Duplessis, H. Etude des glandes endocrines des anencéphales. Biol. Neonat., 1:8–32, 1959.
76. Jackson, C. M. On the prenatal growth of the human body and the relative growth of the various organs and parts. Amer. J. Anat., 9:119–165, 1909.
77. Kondo, S. Developmental studies on the japanese human adrenals. I. Ponderal growth. Bull. Exp. Biol. (Japan), 9:51, 1959.
78. Kiyono, H. Die pathologische Anatomie der endokrinen Organe bei Anencephalie. Virchow. Arch. Path. Anat., 257:441–476, 1925.
79. Vaclav, J. Aparat endokrinni u anencefalu. (L'appareil des glandes à sécrétion interne chez les anencéphales). Sbornik Lékărský (Praha), 28:399–513, 1927.
80. Ch'in, K. Y. The endocrine glands of anencephalic foetuses. Chin. Med. J., (Suppl. 2): 63–90, 1938.
81. Nichols, J. Observations on the adrenal of the premature anencephalic fetus. Arch. Path., 62:312–317, 1956.
82. Jost, A., Jacquot, R., and Cohen, A. The pituitary control of the foetal adrenal cortex.

In The Human Adrenal Cortex. Currie, A. R., Symington, T., and Grant, J. K., eds. Edinburgh and London, Livingstone, pp. 569–579, 1962.

83. Hottinger, A. Vorübergehende passive Nebennieren. Insuffizienz beim Neugeborenen. Schweiz. Med. Wschr., 89:419–422, 1959.
84. Oppenheimer, E. H. Lesions in the adrenals of an infant following maternal corticosteroid therapy. Bull. Johns Hopkins Hosp., 114:146–151, 1964.
85. Kreines, K., Perin, E., and Salzer, R. Pregnancy in Cushing's syndrome. J. Clin. Endocr., 24:75–79, 1964.
86. Sydnor, K. L., Kelley, V. C., Raile, R. B., Ely, R. S., and Sayers, G. Blood adrenocorticotrophin in children with congenital adrenal hyperplasia. Proc. Soc. Exp. Biol. Med., 82:695, 1953.
87. Slaunwhite, W. R., Jr., and Sandberg, A. A. Transcortin, a corticosteroid-binding protein of plasma. J. Clin. Invest., 38:384–391, 1959.
88. Koch. X., Mialhe-Voloss, C., and Stutinsky, F. Fractions liées et libre de la corticostérone plasmatique au cours du développement postnatal chez le rat. C.R. Acad. Sci. (Paris), 264:2497–2500, 1967.
89. Jost, A. The role of fetal hormones in prenatal development. The Harvey Lectures, 55: 201–226, 1961.
90. Jacquot, R., and Kretchmer, N. Effect of fetal decapitation on enzymes of glycogen metabolism. J. Biol. Chem., 239:1301–1304, 1964.
91. Favard, P., and Jost, A. Différenciation et charge en glycogène de l'hépatocyte du foetus de rat normal ou décapité (étude au microscope électronique). Arch. Anat. Micr. Morph. Exp., 55:603–632, 1966.
92. Girard, J., Jost, A., and Dupouy, J. P. Modifications précoces des hépatocytes de foetus de rat décapités, sous l'influence du cortisol. C.R. Acad. Sci. (Paris), 267:857–859, 1968.
93. Plas, C., and Jacquot, R. Recherches sur la différenciation fonctionnelle du foie chez le foetus de rat. Activité UDPG-transglucosylasique et teneur en glycogène aprés administration d'hydrocortisone. C.R. Acad. Sci., (Paris), 264:374, 1967.
94. Pic, P., and Jost, A. 1969. Unpublished data.

2

Cell Structure and Biosynthesis During Hormone-Mediated Growth and Development*

J. R. Tata

National Institute for Medical Research,
Mill Hill, London, England

INTRODUCTION

Cell structures in higher organisms are in a dynamic state of turnover and not static entities. As cells differentiate and mature, the changing pattern of biosynthetic activity is accompanied by an increasing complexity of intracellular structures. Biochemists are now becoming increasingly aware that the processes of protein synthesis and transport themselves are related to structural components of the cell, especially membranes of the endoplasmic reticulum(1,2).

Most animal cells, at one stage or another, are dependent on hormones for the initiation or maintenance of their normal growth and development. Growth and developmental hormones are now well known to exert a strong influence on the protein synthetic activity of their target tissues(3-5). It is worthwhile emphasizing at the very beginning the danger of overgeneralization concerning the possible mechanisms involved in hormonal regulation of protein synthesis. In this context, one should bear in mind the evolutionary history of hormones, which in many instances reveals that the availability of the hormone may have anticipated some of the more sophisticated regulatory mechanisms of higher animal cells(6). It is, therefore, not unlikely that different target cells for the same hormone may elicit their responses to hormonal stimuli by a combination of different mechanisms. Nevertheless, the earliest studies on growth hormone, androgens, oestrogen, and insect steroid hormones indicated that to a great degree the hormonal effect was medi-

*The work herein described was made possible by much help from many of my colleagues. In particular, I am indebted to Doctors T. M. Andrews, J. A. Armstrong, L. Ernster, P. R. Kerkof, C. C. Widnell, T. S. Work, and G. R. Wyatt.

ated via the nucleus and involved homone-induced synthesis of additional ribonucleic acid (RNA)(3-5). However, relatively little is known about the relationship between hormonal influence on the formation of RNA and protein, on the one hand, and the structural requirements for this influence to be fully realized, on the other.

In this review, I shall attempt to integrate two aspects of cellular structural requirements in the regulation by growth and developmental hormones of the biosynthetic activity of the target cell: (1) the nature and type of RNA synthesized by nuclei, and (2) the coordinated formation of microsomal membranes with hormone-induced synthesis of new enzymes or acceleration of growth.

SEQUENTIAL HORMONAL EFFECTS ON RNA AND PROTEIN SYNTHESIS

A few years ago, when studying the regulation of basal metabolic rate by thyroid hormones under growth-promoting conditions, I was impressed by the rather dramatic effects produced by thyroidectomy and hormone replacement on the amount of mitochondrial respiratory constituents(7). These hormones seemed to regulate both the number and the proportion of cristae and cristae-associated enzymes(7,8). It focused our attention on protein synthesis, and soon work on this problem showed that the increase in mitochondrial respiratory and phosphorylative activity provoked by triiodothyronine was accompanied by independent but concomitant stimulations of amino acid incorporation by microsomes(7) and mitochondria(9). Later work showed that the effects of thyroid hormones on cytoplasmic protein synthesis were preceded by a hormonal stimulation of RNA synthesis in the nucleus(11). The main biochemical responses of the target cells, summarized in Table 1, involve those processes or enzymes whose activities require the presence of membranes or some rigid structural organization (RNA polymerase, protein syn-

TABLE 1

*Acute cellular effects of thyroid hormones on some biochemical responses of the liver and muscle of thyroidectomized rat**

Marked, early stimulation	Late or no change
Synthesis of nuclear and cytoplasmic RNA	G-6-P, lactic and isocitric dehydrogenases
DNA-dependent RNA polymerase	Increase in mitochondrial ubiquinone
Incorporation of amino acids into protein by mitochondria and microsomes	—
Mitochondrial respiration and phosphorylation	NAD phosphorylase in nuclei
Synthesis of membrane phospholipids	Mitochondrial stability, coupling of phosphorylation and respiration
Submitochondrial cytochrome oxidase and succinoxidase	Creatine phosphokinase
Microsomal G-6-Pase, NADPH-cytochrome c reductase	Amino acid activation, α-glucan phosphorylase
Increase in cytochromes a, c, and b_5	UDPG-glycogen glucosyl transferase

*See Tata (10).

thesis, cytochrome oxidase, glucose-6-phosphatase, and NADPH-cytochrome c reductase). On the other hand, most of the enzymes that are active in the soluble state were relatively unaffected, at least during the early action of the hormone. (The preferential effects on membrane-associated functions are not in themselves a specific property of the hormone but merely reflect the increasing structural complexity that is a well-recognized feature of growing or developing cells(23).

A relatively rapid stimulation of nuclear RNA synthesis in the target cells is now a well-known feature of the administration of growth and developmental hormones to immature animals or in animals in which the endogenous hormone was

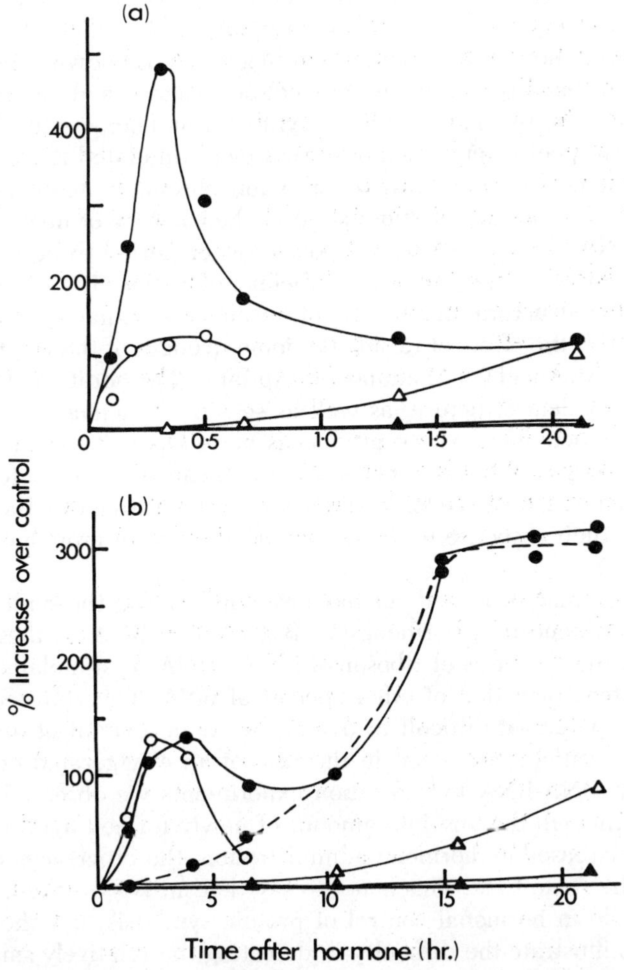

Fig. 1. Time course of stimulations of specific activity of acid-soluble fraction, labelling of nuclear RNA, and RNA synthesis by isolated nuclei in (a) the uterus of ovariectomized rats after a single injection of estradiol-17β and (b) the liver of thyroidectomized rats following a single injection of 3,5,3′-triiodothyronine. Labelling in vivo was achieved by 2- to 10-minute pulses of (a) ^{3}H-uridine and (b) ^{14}C-orotic acid. Open circles—acid-soluble fraction labelled in vivo; solid circles—rapidly labelled nuclear RNA (dashed line represents values obtained after correcting for the increase in acid-soluble fraction); open triangles—RNA polymerase assayed at high ionic strength (product mainly DNA-like RNA). Note the differences in the time scales for the two hormones. (See Refs. 5 and 16.)

experimentally eliminated. The relatively rapid effect on RNA synthesis, preceding that on protein synthesis, has been observed in a variety of hormone-induced growth systems(3-5,10). This feature is illustrated in Figure 1 for the fast-acting oestrogen, as well as for the slower effects of thyroid hormone. A stimulation of synthesis of rapidly labelled nuclear RNA of the uterus can be observed within minutes after the administration of oestrogen, and the relatively delayed stimulation by thyroid hormone is still one of its earliest actions(5,11). It is important, however, to distinguish between the possible changes in the precursor pool sizes or tissue uptake and a true net synthesis of RNA in vivo in the initial phase of hormone action. The first phase of what appears to be a very early increase in the rate of RNA synthesis may therefore be a consequence of variations in uptake or labelling of precursor molecules. Almost every hormone, growth-promoting or not, is known to have some rapid influence on permeability barriers for nucleotides, sugars, and amino acids.

Evidence for the regulation of RNA synthesis, independently of hormonal effects on precursor pools, can be demonstrated by a stimulated RNA polymerase activity in isolated nuclear preparations or when chromatin template efficiency is measured(4,10). The hormonal stimulation of the capacity of nuclear preparations to synthesize RNA in vitro, however, takes a longer period to be manifested after hormone administration than the rate of labelling of nuclear RNA in vivo (see Table 2). An interesting structural implication of hormonal stimulation of RNA synthesis by nuclei in vitro is the effect of raising the ionic strength of the incubation medium (usually by the addition of 0.4 M ammonium sulfate). The addition of the salt causes a disruption of nuclear structure, as well as selectively activating one of the two RNA polymerase activities, whose product is more DNA-like than that formed in the absence of the salt. What is of particular significance in the context of hormone action is that activation of nuclei by high ionic strength causes them virtually to lose or to delay their response to the hormonal stimulus of growth or development (Table 2).

A number of analytical methods have established that the synthesis of all species of RNA is influenced by hormones(4). But whether RNA synthesis is studied in vitro or in vivo, the synthesis of ribosomal RNA (rRNA) is stimulated by hormones to a greater extent than that of other species of RNA. This differential transcriptional sensitivity makes it difficult to detect the small amount of mRNA (presumably hormone-specific) synthesized in the face of an exaggerated enhancement of rRNA. Thus, in DNA-RNA hybridization experiments we observed a paradoxical effect in that although the absolute amount of newly formed hybridizable RNA in the nucleus is increased by hormone administration, the efficiency of hybridization is lowered at the same time. This is not to say that mRNA synthesis does not play an important role in hormonal control of protein synthesis, but the foregoing example serves to illustrate the difficulty in detecting the relatively small amounts of specific mRNA molecules that may be produced in response to a given hormone in the midst of a vast increase in rRNA formation. The situation may, however, vary from one hormone to another—as noted by Drews and Brawerman(13) for growth hormone and hydrocortisone—but, in general, growth and developmental hormones provoke a preferential synthesis of rRNA. It is not possible to say why rRNA synthesis is so sensitive to hormones, but it seems to be a common feature of nucleated

TABLE 2

*Lag period and maximal stimulation of nuclear RNA synthetic capacity measured in different ways after hormone administration in vivo**

		Lag period (time) and maximal stimulation (percent)			
Hormone†	Target	**RNA labelled in vivo‡**	**RNA polymerase: low ionic strength**	**RNA polymerase: high ionic strength**	**Chromatin activation**
Oestrogen	Rat uterus	2-10 min, 500 percent	1 hr, 150 percent	24 hr, 50 percent	2 hr, 70 percent
Growth hormone	Rat liver	1 hr, 50-100 percent	2 hr, 80 percent	No effect	
	Rat muscle		18 hr, 50 percent	18 hr, 50 percent	Inhibition
Hydrocortisone	Rat liver	1 hr, 300 percent	2-4 hr, 75 percent	No effect	4 hr, 30 percent
Testosterone	Rat prostate	90 min, 300-500 percent	2 hr, 120 percent	20 percent	No effect
	Seminal vesicles				
Thyroid hormone	Rat liver	3 hr, 350 percent	12 hr, 200 percent	24 hr, 50 percent	
	Tadpole liver	8 hr, 600 percent	—	—	4 days, 50 percent

*Lag period is the time elapsed between hormone administration and a 10 percent stimulation of RNA synthesis.

†Hormones administered to immature animals or animals made hormone-deficient.

‡For oestrogen, hydrocortisone, and triiodothyronine, data are available on the extent to which the incorporation of the labelled precursor represents a net synthesis of RNA or may be due to specific activity changes in the acid-soluble components (see Tata [10]).

TABLE 3

Hybridization with homologous DNA of RNA synthesized during hormone action

Species	Source of RNA	Hormonal treatment	Specific activity (cpm/μg RNA)	RNA hybridized at plateau (cpm)	Hybridization efficiency (%)
Rat	Rapidly labelled nuclear RNA	Hx, control	2004	198	6.5
		Hx, 2 days after HGH + T_3	3630	242	4.3
Rat	Long-term labelled microsomal RNA	Tx, control	220	21	0.6
		Tx, 3 days after T_3	395	44	0.7
Bullfrog	Rapidly labelled nuclear RNA	Premetamorphosis	30	13	6.9
		Metamorphosis induced	155	33	3.2
Xenopus	Total body long-term labelled RNA	Premetamorphosis	87	23	4.8
		Metamorphosis induced	629	86	2.5

Hx = hypophysectomized; Tx = thyroidectomized; HGH = human growth hormone; T_3 = triiodothyronine.

^{3}H-labelled RNA was used on cellulose nitrate membranes on which denatured homologous DNA was trapped. The hybridization plateau level was determined by hybridizing a constant amount of RNA with different amounts of DNA bound to membranes. Hybridization efficiency is the percent of radioactive RNA input recovered as the hybrid at saturation levels of DNA/RNA.

Data adapted from Wyatt and Tata (35).

cells in which growth or an adaptational process is rapidly taking place. It may reflect the possible mechanism whereby mRNA transfer from the nucleus to cytoplasm is facilitated by a combination with one of the two ribosomal subunits. In any case, the transfer of messenger RNA from the nucleus to the cytoplasm is now thought to involve the generation of structural elements, either "informosomes" or precoded subribosomal units. It has been suggested that mRNA in nucleated cells may be first packaged with a special kind of protein to give 30–40S particles(14) or combined with the smaller of the two ribosomal subunits(15). Both these particles are thought to be the direct precursors of the cytoplasmic polysomes, and consequently their rate of turnover would be of great importance to the regulation of protein synthesis. Whether these transfer mechanisms are crucial or not has to be experimentally verified, but a number of studies with inhibitors of RNA synthesis have shown that the newly synthesized RNA in the nucleus is essential for hormonal stimulation of protein synthesis as well as for the growth and developmental action of hormones.

For many hormones, including androgens, oestrogens, thyroid hormones, and growth hormone, the rate of protein synthesis increases rather abruptly following a relatively long lag period after hormone replacement(4,16,17). The onset of increased protein synthetic activity coincides with the accumulation in the cytoplasm of the hormone-induced newly formed ribosomes. This feature is illustrated in Figure 2 for the stimulation of growth of the uterus by oestrogen and that of liver by triiodothyronine. Similar results were observed with triiodothyronine-induced metamorphosis of bullfrog tadpoles(17). It appears, therefore, that a large part of the hormonal effect on protein synthesis is incorporated into the new ribosomes formed after hormone treatment, but it is not clear whether these "new" ribosomes play a passive role as mere transporters of additional mRNA or whether their intrinsic amino acid incorporation activity is also under hormonal influence. In any case, it is interesting to note that growth hormone and hydrocortisone have been shown to increase the rate of appearance in rat liver cytoplasm of the complex of the 40S ribosomal subunit and mRNA (or perhaps even the mRNP particle) at a time just preceding the stimulation of cytoplasmic protein synthesis or the induction of specific enzymes(10,16). It is, however, unlikely that hormonal regulation of protein synthesis is uniquely a question of the regulation of the number of ribosomes in the cytoplasm.

At the same time as finding additional ribosomes accompanying the stimulation of protein synthesis, we have observed that the polysomes were more firmly bound to microsomal membranes in hormonally stimulated tissues, as judged by the ease with which the particles were released from membrane complexes(10,16). Both these effects are illustrated in Table 4 with hepatic ribosomes from bullfrog tadpoles in which metamorphosis was precociously induced with triiodothyronine(17). A firmer attachment of this type also caused a shift from the smooth to the rough endoplasmic reticulum membranes as the predominant type. Qualitatively similar, but less pronounced, changes were also observed in tissues of young, thyroidectomized and hypophysectomized rats whose rate of growth was stimulated by growth hormone and thyroid hormone(18). The question then arose as to whether the new ribosomes formed upon hormonal stimulus were merely redistributed on existing membranes or whether there occurred alterations in the rate or type of membranes formed at that time. Although free ribosomes detached from all membranes can

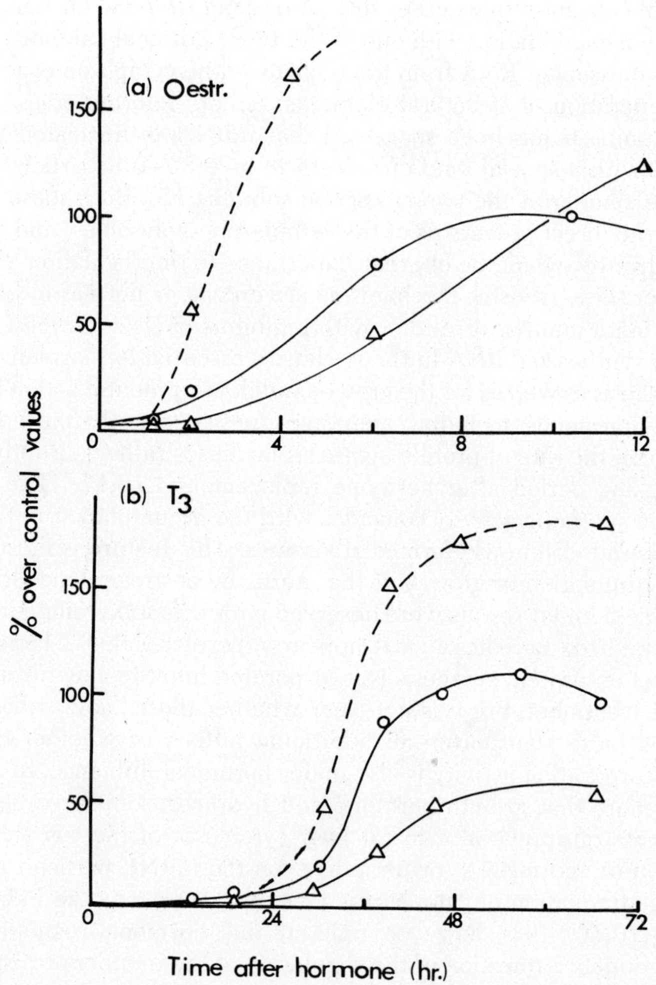

Fig. 2. Increase in protein synthetic capacity of cytoplasmic ribosomes coinciding with the appearance of additional polysomes formed as a result of hormone administration. (a) Effect of a single injection of oestradiol on ovariectomized rat uterus; (b) effect of a single injection of triiodothyronine on thyroidectomized rat liver. Triangles—increase in amount of cytoplasmic ribosomal population (dashed line indicates the increase in radioactivity of the newly labelled polysomes); circles—amino acid incorporation in vitro expressed per unit ribosomal RNA. (See Refs. 5 and 16.)

incorporate amino acids into proteins in vitro, it is believed that in vivo the structural disposition of polysomes is very important, so that virtually all protein synthesis may take place on these ribosomes that are attached to membranes of the endoplasmic reticulum or even other structures(19). This feature enhances the interest in studying the turnover of membranes to which ribosomes are bound at the time ribosome formation was accelerated by growth-promoting hormones.

TABLE 4

Effect of inducing metamorphosis on the relative distribution of newly synthesized RNA in membrane-bound and free ribosomal fractions from liver mitochondria-free supernatants treated with different amounts of Na-deoxycholate

Microsomal RNA was labelled by injecting 11.5 μc of [^3H] uridine per tadpole 22.5 hr before killing them. Each group comprised 10 tadpoles, and one group was treated with 0.7 μg of triiodothyronine 6 days before the experiment. Equal amounts of mitochondria-free supernatants were treated with different quantities of Na-deoxycholate, indicated by their final concentration before the sucrose density gradient centrifugation was carried out. One-ml fractions were then pooled, after determination of ξ_{260}, into the four fractions. The "membrane-bound" ribosomal fraction is that which did not sediment through the 2.0 M sucrose interface (from Tata [17]).

Tadpoles	Concentration of Na-deoxycholate (%)	Specific activity (counts/min/ξ_{260}) and distribution (%) of RNA in fractions*			
		Membrane-bound ribosomes	**Polysomes**	**Dimers + monomers†**	**< 50S**
Control	0.08%	295 (4.7)	227 (21.5)	107 (56.6)	58 (12.0)
	0.16%	418 (2.0)	254 (22.8)	92 (53.0)	41 (8.8)
	0.30%	— (0)	238 (18.7)	122 (62.5)	73 (14.5)
6 days after T_3	0.08%	875 (20.5)	430 (33.5)	492 (26.0)	196 (10.5)
	0.16%	990 (12.8)	518 (35.1)	631 (32.9)	281 (15.1)
	0.30%	1185 (1.5)	550 (25.8)	675 (29.0)	255 (11.9)

*Figures in parentheses indicate the percent of total radioactivity delivered on the gradient that was recovered in each of the fractions.

†Also includes ribosomal precursor particles smaller than 78S monomers.

COORDINATED PROLIFERATION OF MICROSOMAL MEMBRANES AND RIBOSOMES

A convenient way to follow the proliferation of membranes of the endoplasmic reticulum is to measure the incorporation of ^{32}P-phosphate and ^{14}C-labelled choline and glycerol into membrane phospholipids. In order to correlate the effects of hormones on this process with those on biosynthetic activity of the tissue, we have studied the rate of incorporation of radioactive amino acids into nascent microsomal proteins simultaneously with the accumulation of newly synthesized microsomal RNA and phospholipid in the following hormonal systems: (1) growth of liver induced by a single injection of growth hormone, triiodothyronine, and testosterone in hypophysectomized, thyroidectomized, and castrated rats, respectively; (2) growth and maturation of the seminal vesicle of the castrated rat when stimulated by testosterone; (3) precocious initiation of enzyme and protein synthesis in bullfrog tadpole liver during thyroid hormone-induced metamorphosis; (4) accelerated synthesis of thyroglobulin or iodoprotein by thyroid tissue under the influence of thyrotropic hormone (TSH).

It was initially found in these studies that administration of the aforementioned hormones accelerated the synthesis of phospholipids, after different lag periods and to different extents, in the microsomal as well as the mitochondrial and nuclear fractions (Fig. 3). It was, however, in the microsomes that the effect was most prominent.

Figure 4 compares the rates of formation of liver microsomal phospholipids with those of microsomal RNA and of nascent protein on microsomes following the combined administration of growth hormone and triiodothyronine to hypophysec-

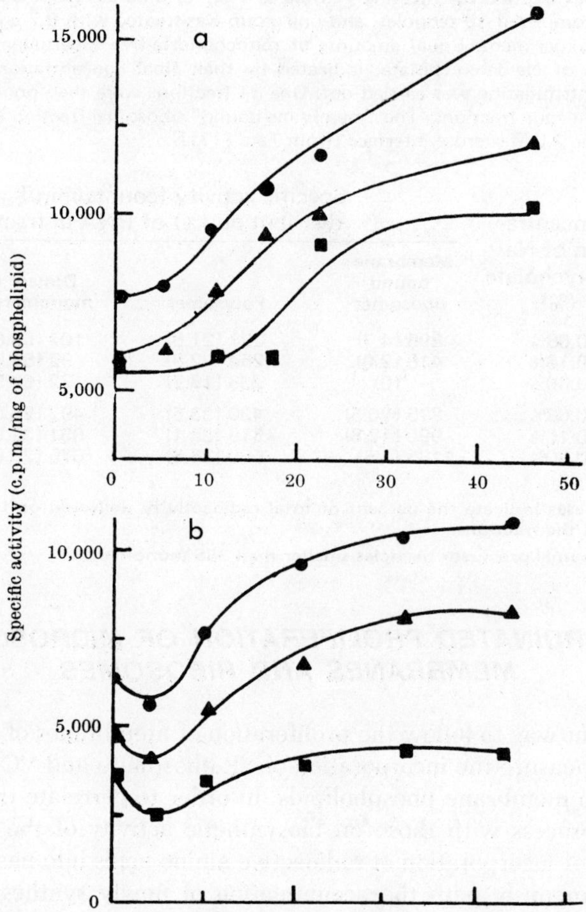

Fig. 3. Incorporation of labelled choline into phospholipids of hepatic nuclei (squares), mitochondria (triangles), and microsomes (circles) as a function of time after the injection of (a) 3,5,3′-triiodothyronine (T_3) to thyroidectomized rats and (b) human growth hormone (HGH) to hypophysectomized rats. Thyroidectomized rats weighing 140–160 g were injected intraperitoneally with 6 μc of labelled choline chloride (methyl ^{14}C, 37.6 mc/mmole) 60 minutes before they were killed, and hypophysectomized rats weighing 110–130 g received 5.2 μc of the radioisotope 40 minutes before death. The two hormones were administered subcutaneously. Each point represents the average of four determinations on material pooled from three rat livers. From Tata. *Nature,* 213:566-569, 1967.

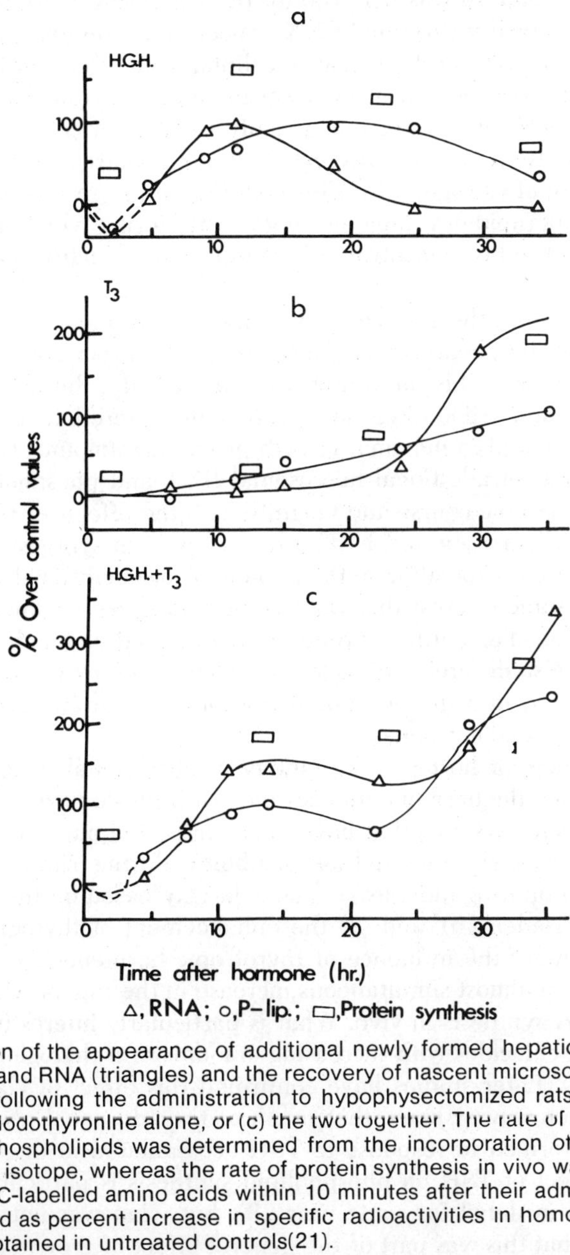

Fig. 4. Coordination of the appearance of additional newly formed hepatic microsomal phospholipids (circles) and RNA (triangles) and the recovery of nascent microsomal protein formed in vivo (squares) following the administration to hypophysectomized rats of (a) growth hormone alone, (b) triiodothyronine alone, or (c) the two together. The rate of formation of microsomal RNA and phospholipids was determined from the incorporation of ^{32}P after about 10 hr exposure to the isotope, whereas the rate of protein synthesis in vivo was derived from the incorporation of ^{14}C-labelled amino acids within 10 minutes after their administration. The results are expressed as percent increase in specific radioactivities in homone-treated animals over the values obtained in untreated controls(21).

tomized rats. Since the two hormones have quite different lag periods, it was quite simple to detect a temporally coordinated enhancement by the hormones of microsomal phospholipid formation with that of the rates of microsomal RNA and protein synthesis. Thus, growth hormone (Fig. 4a) stimulated the protein synthetic capacity of hepatic ribosomes several hours earlier than did triiodothyronine (Fig.

4b), and the same pattern was followed for the accelerated formation of polysomes (much of the additional microsomal RNA appeared as polysomes).

Such coordination of proliferation of cellular structures and biosynthetic activity is not a unique property of growth hormone and thyroid hormones. Although ribosomes and membranes were not actually isolated, a coordinated synthesis of phospholipid and RNA in their respective target tissues may underlie the same process in the case of estrogen(20). This coordination is, in fact, a common feature of cells undergoing rapid development in general, during which different biochemical functions tend to be concentrated in cytoplasmic structures of increasing complexities(23,24).

Where, then, does the specificity of individual hormones lie in terms of such seemingly nonspecific phenomena? There are two indirect approaches which provide a partial answer to this important question. Firstly, the experiment in which growth hormone and triiodothyronine were administered to hypophysectomized rats simultaneously and in maximal growth-promoting amounts (Fig. 4c) produced two stepwise bursts of additional microsomal RNA and phospholipids, each burst corresponding in its time-course and magnitude to the effects of the individual hormones shown in Figures 4a and b. The rate of protein synthesis in vivo was elevated in a manner corresponding to the two bursts of microsomal proliferation, and it is therefore of some interest that the two hormones have additive effects on the growth of the tissue. The nature of proteins synthesized under the influence of the two hormones is also different. It is possible that some degree of hormonal specificity is incorporated into the additional ribosome-membrane units formed in response to the individual hormones.

Other evidence for hormonal specificity in microsomal structures is provided by systems in which the hormone induces the synthesis de novo of specific enzymes or proteins. Two such systems that have been studied in our laboratory are (1) the induction of urea cycle enzymes and serum albumin during triiodothyronine-induced metamorphosis in bullfrog tadpoles, which is largely based on the classical work of Cohen(25) and Frieden(26), and (2) the enhancement of thyroglobulin formation by thyroid cells under the influence of thyrotropic hormone(27). In the latter system, TSH causes an almost simultaneous increase in the rate of whole thyroid phospholipid and RNA synthesis in vivo. What is particularly interesting is that the onset of this change coincides with increasing amounts of a thyroglobulin-like protein being formed(22). Later studies have confirmed this effect in vitro using pig thyroid slices in which it was shown that much of the additional phospholipid, RNA, and iodoprotein formed in response to TSH accumulated in the microsomal fraction(27). The effect of TSH on phospholipid synthesis is particularly striking and has been studied in detail for some time(28), but what was not known until our studies(22) was that this was part of the general effect of growth and developmental hormones in enhancing membrane proliferation.

The artificial induction of metamorphosis in amphibians has also received much attention in our laboratory over the past five years. Figure 5 summarizes some of our findings on major biosynthetic events that occur during the lag period preceding the appearance of newly formed proteins as a result of administration of triiodothyronine to immature *Rana catesbeiana* tadpoles. It can be seen that there oc-

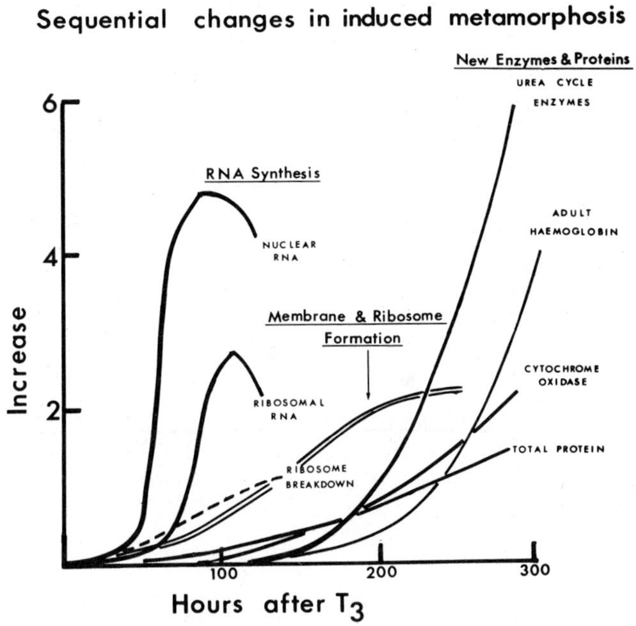

Fig. 5. Schematic representation of sequential stimulation of rates of RNA and phospholipid synthesis in relation to the increases in enzymes or protein synthesized upon the precocious induction of metamorphosis in *Rana catesbeiana* tadpoles with triiodothyronine. Curves 1: rate of rapidly labelled nuclear RNA synthesis; 2: specific activity of labelled RNA in cytoplasmic ribosomes; 3: rate of microsomal phospholipid synthesis; 4: carbamyl phosphate synthetase; 5: cytochrome oxidase per mg mitochondrial protein; 6: appearance of serum albumin in the blood; 7: total liver protein per mg wet weight. The values are expressed as percentage increases over those in the noninduced control tadpoles. The decreasing values in curves 1, 2, and 3 reflect the dilution of specific radioactivity in precursor molecules due to the onset of regression of tissues such as the tail and intestine.

curred, well within the latent period (five to six days) for new proteins to be detected, an acceleration of the rate of nuclear and cytoplasmic RNA synthesis. Of particular interest are the events that occur between the initial burst of RNA synthesis following triiodothyronine administration and the appearance of newly synthesized proteins. Unlike the mammalian systems that we noted earlier, there occur during this period a more complex process of breakdown of "old" RNA, alterations in polysome profiles, and a redistribution of ribosomes attached to membranes of the endoplasmic reticulum. The last process is accompanied by a simultaneous increase in the synthesis of ribosomes and membranes of the endoplasmic reticulum as judged by membrane phospholipid synthesis (Fig. 6a). It is interesting to note that this simultaneous increase in the formation of the two rough endoplasmic reticulum components is accompanied by an enhanced rate of protein synthesis (Fig. 6b). These biochemical changes are so marked that it is easy to observe structural changes by electron microscopy.

Figure 7 shows schematically that during metamorphosis there is a shift in the distribution of ribosomes from around the simple vesicular membrane structures of the immature larvae to the more complex double lamellar structures more common-

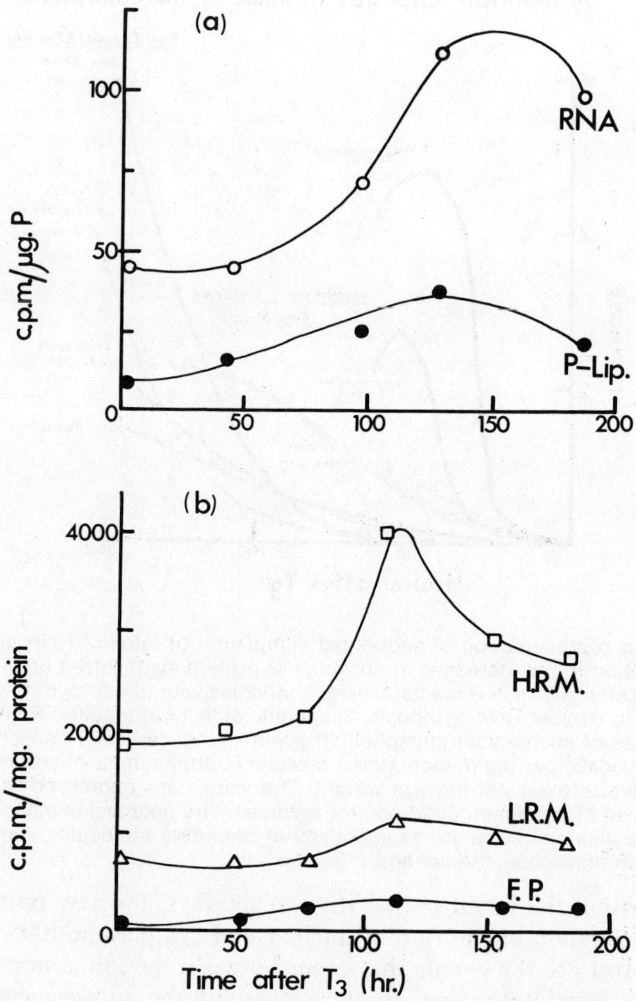

Fig. 6. Coordinated proliferation of endoplasmic reticulum and the increase in protein synthesis in vivo in the heavy rough membranes (densely packed ribosomes attached to membranes)during triiodothyronine-induced metamorphosis of young bullfrog tadpoles. (a) The appearance of newly synthesized (^{32}P-labelled) heavy rough membrane phospholipids (open circles) and RNA (solid circles); (b) the recovery of labelled nascent proteins formed in vivo after a short pulse of radioactive amino acids in heavy rough membranes (squares), light rough membranes (triangles), and free polysomes (solid circles). (See Tata[17] for details.)

ly seen in mature tissues. This shift in structural organization of the endoplasmic reticulum coincided temporally with the changes in the formation of microsomal RNA and phospholipids and the synthesis of proteins by polyribosomes. Similar changes have been observed in the ecdysone-induced maturation of insect wing epidermal cells(29) as well as in the androgen-dependent growth of male accessory sexual tissues (prostate, seminal vesicles) in rodents(30). The significance of such a

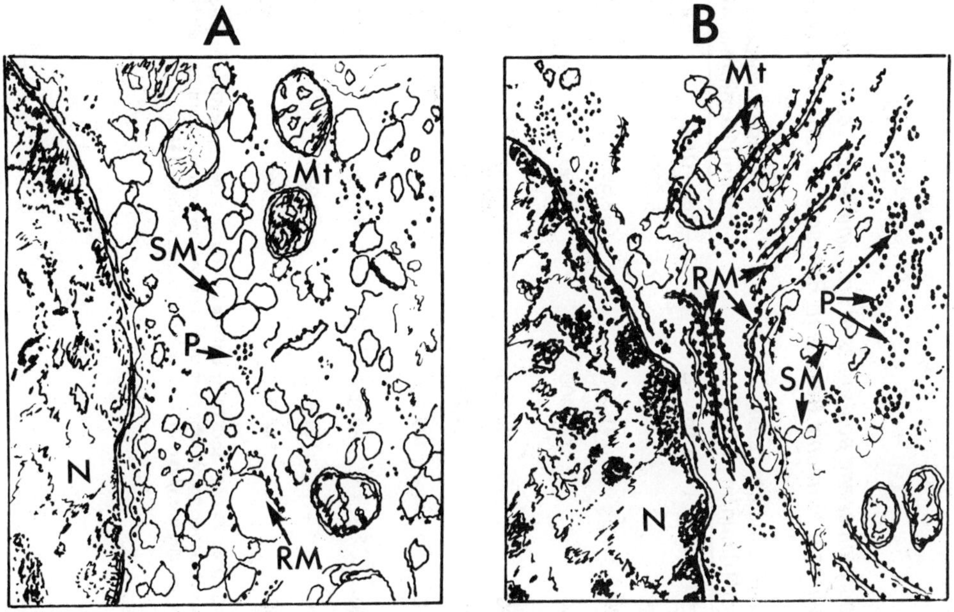

Fig. 7. Schematic representation of electron micrographs showing the reorganization of the cytoplasmic membranes and ribosomes during hormone-induced metamorphosis in bullfrog tadpole liver cells. (a) Hepatocyte in a premetamorphic tadpole showing small groups of ribosomes around simple vesicular structures. (b) Similar cells in tadpoles six days after induction of metamorphosis with triiodothyronine, showing dense ribosomal accumulation often attached to a more differentiated double-lamellar type of reticulum. N—nucleus; Mt—mitochondrion; SM—smooth membrane; RM—rough membranes; P—typical polysomal chains, rosettes, and spirals. (See Tata[16,17] for details.)

coupling of structure and biosynthetic activity may lie in the possibilities that (1) a response to growth and developmental stimulus is contained in newly formed ribosomes, and (2) there is a topographical segregation on membranes of the endoplasmic reticulum of differentially precoded polyribosomes carrying out the synthesis of different groups of proteins.

AMINO ACID INCORPORATION IN VITRO BY FREE AND MEMBRANE-BOUND RIBOSOMES

The significance of the preceding considerations in terms of biosynthetic regulation is largely based on the idea that protein synthesis in vivo requires a suitable structural attachment of the ribosome. However, it is well known that membrane-free ribosomes or polysomes are quite active in incorporating amino acids into protein in vitro. Furthermore, the effects of deprivation or replacement of many growth and developmental hormones are reflected in the amino acid incorporation activity in vitro of free ribosomes (or polysomes) obtained from the target tissue (3-5,16,20). For this reason, it has been important to determine whether the way in which hormonal treatment affects amino acid incorporation by free and mem-

TABLE 5

Incorporation of ^{14}C-phenylalanine and ^{14}C-isoleucine into protein by "free" polysomes and polysomes in light and heavy rough membrane fractions of liver microsomes from normal and hypophysectomized rats with or without hormonal treatment. Response to poly=U and poly=UA

Rats	Polysomes	mg RNA/g tissue	^{14}C-Phe incorporated (cpm/10 min/mg RNA)		^{14}C-Ilue incorporated (cpm/10 min/mg RNA)	
			– Poly=U	**+ Poly=U**	**– Poly=UA**	**+ Poly=UA**
Normal	Free	0.57	13,590	38,090	5,320	6,383
	Light membranes	0.51	18,620	41,840	3,590	5,748
	Heavy membranes	1.62	58,800	128,400	23,800	29,980
Hx	Free	0.44	6,700	30,520	3,120	6,150
	Light membranes	0.46	14,550	44,250	3,560	5,036
	Heavy membranes	0.59	40,100	102,900	18,550	21,190
Hx + HGH + T_3	Free	0.66	26,700	86,300	10,900	20,720
	Light membranes	0.47	22,630	101,130	6,980	11,590
	Heavy membranes	1.40	66,000	144,000	32,100	45,700

Hormonal treatment consisted of three injections of 75 μg of human growth hormone (HGH) at 21, 47, and 74 hr before killing and two injections of triiodothyronine (T_3) at 47 and 74 hr. Light and heavy rough membrane fractions consist of ribosomes attached to membranes of the endoplasmic reticulum and differ from one another by the density of ribosome packing. Abbreviations are as in Table 2, and further details are given in Tata and Williams-Ashman (18).

brane-bound ribosomes in vitro is compatible with the protein synthetic changes occurring in vivo as those that have been previously described.

An example of such a comparison is given in Table 5, which summarizes the results of some experiments on the effect of hypophysectomy and combined treatment with growth hormone and triiodothyronine on the capacity of microsomal fractions to incorporate amino acids into protein, both with endogenous and synthetic mRNA(18). In these experiments, microsomes were separated into their light and heavy rough membrane fractions (distinguishable according to the relative density of ribosomes attached to membranes of the endoplasmic reticulum) and free polysomes, *without the use of detergents.* When the ability to incorporate ^{14}C-phenylalanine and ^{14}C-isoleucine was measured in the presence and absence of synthetic messengers (polyuridylic acid and copolymer of uridylic and adenylic acids), it was found that hormone treatment corrected the lowered incorporation capacity caused by hypophysectomy in all three types of submicrosomal fractions. The effect of hormone deprivation or replacement was of greater magnitude in the heavy rough membranes than in the light rough membranes. To a large extent, the response to synthetic messengers of both the membrane-bound and free ribosomes paralleled that to endogenous messenger, whether growth had been retarded or accelerated. This observation suggests that the newly formed ribosomes, upon hormone administration, were inherently more efficient in handling synthetic or natural messengers. At the same time, the relative stimulation by synthetic polynucleotides was slightly higher in the less active preparations, thus indicating that a small part of hormonal effect could be due to an increase in the polysomal mRNA content. An important problem that now awaits resolution is to determine whether the anabolic and developmental actions of hormones may be restricted to a small population of newly generated ribosomes, or modifying the activity of all the ribosomes in the cytoplasm.

HORMONAL EFFECTS AND SITES OF ACTION

From the foregoing discussion of what little is known about the actions of growth and developmental hormones, it is clear that the biosynthetic functions of the target cells are affected at multiple steps of genetic transcription and translation. Although a selective control of messenger RNA species, as predicted by Karlson(31), still remains one of the simplest ways of explaining hormonal specificity, the extreme sensitivity of ribosomal RNA synthesis makes it difficult to detect unequivocally the small changes in mRNA populations. As summarized schematically in Figure 8, an exaggerated increase in rRNA synthesis and ribonucleoprotein particles during all types of growth processes may be a common feature of nucleated cells reflecting a need to facilitate the transport of mRNA from the nucleus to the cytoplasm. In the case of hormone-induced growth and development, this would lead to an accumulation of polysomes precoded with hormone-specific mRNA. The ribosome-membrane association could conceivably restrict the exchange between ribosomes and messengers made before and after exposure of cells to hormones. The easily separable and additive effects of different hormones on the biosynthetic activity are compatible with this view (see Fig. 4). So far, almost all the attention

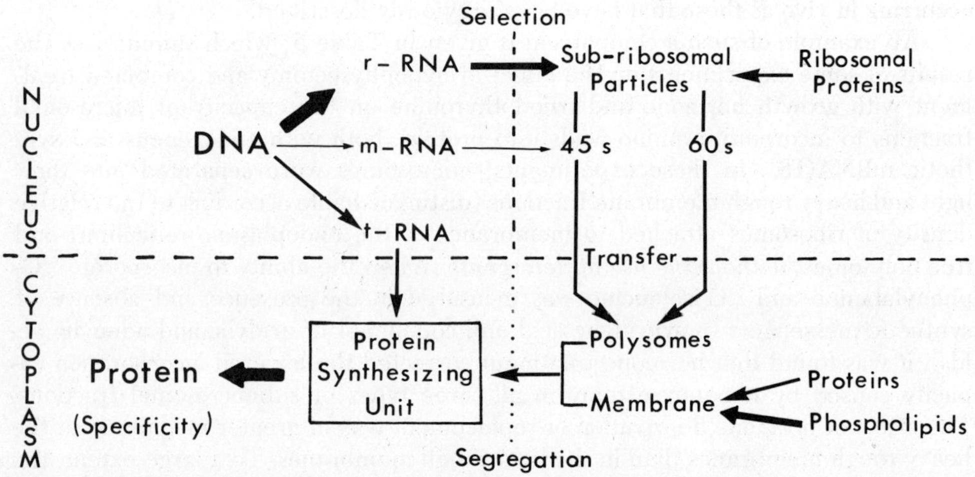

Fig. 8. Scheme attempting to integrate some of the processes regulating protein synthesis in nucleated cells which respond to the growth-promoting stimulus of hormones. It is proposed that the type of growth or development would be determined genetically via mRNA coding for specific proteins but that its expression requires the generation of ribosomes and mRNA protein particles and that their topographical segregation may be achieved by a firm attachment to membranes. The width of the arrows is roughly related to the magnitude of response of the target cell to growth and developmental hormones with respect to the process studied.

on ribosome-membrane attachment has been directed toward the problem of export of proteins in secretory cells, and it would be most worthwhile to investigate the possible function of polysome segregation for this phenomenon. The main advantage to a cell of a topographical segregation of protein-synthesizing units would be to isolate responses to hormonal as well as nonhormonal stimuli, and thus limit to a minimum any perturbations in the orderly synthesis of proteins that are not involved in any one response.

In differentiated cells a large number of species of RNA are restricted to the nucleus and therefore are not utilized for cytoplasmic protein synthesis(32). Virtually nothing is known about the mechanism of selection, and it is therefore impossible to say at the present how, if at all, hormones may affect the processes of discrimination between those species of RNA that do not leave the nucleus and the relatively few that do. But the high degree of restriction of many RNA species to the nucleus makes it imperative to characterize more carefully both the nuclear and cytoplasmic RNA formed under the influence of hormone and not to focus attention only on nuclear RNA, as has been done in the past.

There are other possible regulatory steps susceptible to control by growth and developmental hormones that have not been considered in this review. To name only two, it has been suggested that hormones may directly control the process of translation and that DNA synthesis may play an as yet unsuspected role in the response of the target cells. In the first case, a large part of hormonal activity may involve the modulation of polypeptide chains assembled on relatively stable messengers(33). The best evidence for such a regulation has been obtained for cortisol, insulin, and ACTH. However, with most growth and developmental hormones, direct translational control does not seem to be a key function. One has to distinguish between a direct translational control in which the hormone would act on preexisting ribosomes and an alteration in the amino acid incorporating activity seen after the hormone has caused additional ribosomes to appear, as we have seen in Figures 2, 4, and 6 and Table 5. In the other instance, the work on prolactin-induced casein synthesis in cultured mammary glands has indicated that a limited DNA synthesis, which could be controlled by other hormones, may be a prerequisite for the control of specific protein synthesis during development(34). One cannot overemphasize the importance of extending this promising observation to other growth and developmental systems.

Ultimately, our understanding of how hormones elicit their multiple actions will depend upon how soon we shall unravel the elusive primary site of their action. Despite some recent attempts to correlate the sites of localization and actions of hormones(36), it has not been shown that all the multiple actions of growth and developmental hormones originate from a single primary site of interaction, whether it be the gene(31,37) or some permeability regulator in the cell membrane(12,38). An action at a single site—say, situated at the level of cyclic 3′,5′-adenylic acid(38)—may explain the rapid metabolic effects of some hormones, but it does not explain the relatively slow and complex phenomena associated with growth and developmental hormones. In the phenomena that I have described here, a number of biosynthetic activities and structural modifications of growing cells are coupled and are likely to be remote from the initial sites of action of hormones. Until we can know more about such sites, it is realistic to accept multiple sites of action of growth and developmental hormones. Such an acceptance eliminates many difficulties at present arising from the interpretations based on mutually exclusive single sites of hormone action.

REFERENCES

1. Siekevitz, P., Palade, G. E., Dallner, G., Ohad, I., and Omura, T. The biogenesis of intracellular membranes. *In* Organizational Biosynthesis. Vogel, H. J., Lampen, J. O., and Bryson, V., eds. New York, Academic Press, Inc., pp. 331–362, 1968.
2. Campbell, P. N. The biosynthesis of proteins. Progr. Biophys., 15:3–38, 1965.
3. Korner, A. Growth hormone control of biosynthesis of protein and ribonucleic acid. Recent Progr. Hormone Res., 21:205 236, 1965.
4. Tata, J. R. Hormones and the synthesis and utilization of ribonucleic acids. Progr. Nucl. Acid Res. Molec. Biol., 5:191–250, 1966.
5. Hamilton, T. H. Control by estrogen of genetic transcription and translation. Science, 161:649–661, 1968.
6. Barrington, E. J. W Hormones and Evolution. London, The English Universities Press, 1964.
7. Tata, J. R., Ernster, L., Lindberg, O., Arrhenius, E., Pedersen, S., and Hedman, R. The action of thyroid hormones at the cell level. Biochem. J., 86:408–428, 1963.
8. Gustafsson, R., Tata, J. R., Lindberg, O., and Ernster, L. The relationship between the structure and activity of rat skeletal muscle mitochondria after thyroidectomy and thyroid hormone treatment. J. Cell. Biol., 26:555–578, 1965.
9. Roodyn D. B., Freeman, K. B., and Tata, J. R. The stimulation by treatment in vivo

with tri-iodothyronine of amino acid incorporation into protein by isolated rat-liver mitochondria. Biochem. J., 94:628–641, 1965.
10. Tata J. R. Hormonal regulation of growth and protein synthesis. Nature, 219:331–337, 1968.
11. ——— and Widnell, C. C. Ribonucleic acid synthesis during the early action of thyroid hormones. Biochem. J., 98:604–620, 1966.
12. Hechter, O., and Halkerston, I. D. K. On the action of mammalian hormones. Hormones, 5:697–825, 1964.
13. Drews, J., and Brawerman, G. Messenger RNA patterns in rat liver nuclei before and after treatment with growth hormone. Science, 156:1385–1386, 1967.
14. Georgiev, G. P. The nature and biosynthesis of nuclear ribonucleic acids. Progr. Nucl. Acid Res. Molec. Biol., 6:259–351, 1966.
15. McConkey, E. H., and Hopkins, J. W. Subribosomal particles and the transport of messenger RNA in HeLa cells. J. Molec. Biol., 4:257–270, 1965.
16. Tata, J. R. The formation and distribution of ribosomes during hormone-induced growth and development. Biochem. J., 104:1–16, 1967.
17. ——— The formation, distribution and function of ribosomes and microsomal membranes during induced amphibian metamorphosis. Biochem. J., 105:783–801, 1967.
18. ——— and Williams-Ashman, H. G. Effects of growth hormone and tri-iodothyronine on amino acid incorporation by microsomal subfractions from rat liver. European J. Biochem., 2:366–374, 1967.
19. Henshaw, E. C., Bojarski, T. B. and Hiatt, H. H. Protein synthesis by free and bound rat liver ribosomes in vivo and in vitro. J. Molec. Biol., 7:122–129, 1963.
20. Mueller, G. C. The role of RNA and protein synthesis in estrogen action. *In* Mechanisms of Hormone Action. Karlson, P., ed. Stuttgart, Georg Thieme Verlag, 1965.
21. Tata, J. R. Membrane phospholipid synthesis and the action of hormones. Nature, 213: 566–569, 1967.
22. Kerkof, P. R., and Tata, J. R. Simultaneous acceleration in vivo of the formation of thyroid ribonucleic acid, phospholipid and iodoprotein by thyroid-stimulating hormone. Biochem. Biophys. Res. Commun., 28:111–116, 1967.
23. Wischnitzer, S. The ultrastructure of the cytoplasm of the developing amphibian egg. Advances Morphogenesis, 5:131–179, 1966.
24. Pollak, J. K., and Ward, D. B. Changes in chemical composition and the enzymic activities of hepatic microsomes of the chick embryo during development. Biochem. J., 103: 730–738, 1967.
25. Cohen, P. P. Biochemical aspects of metamorphosis, transition from ammonotelism to ureotelism. The Harvey Lectures, 60:119–154, 1966.
26. Frieden, E. Hormones and the biochemistry of amphibian metamorphosis. Recent Progr. Hormone Res., 23:139–186, 1967.
27. Kerkof, P. R., and Tata, J. R. The subcellular distribution of ^{32}P-labelled phospholipids and RNA and ^{125}I-labelled iodoprotein in pig thyroid slices: Effect in vitro of thyrotrophic hormone and dibutyryl cyclic 3'5'-adenylic acid. Biochem. J., 112:729–773, 1967.
28. Freinkel, N. The intermediary metabolism of thyroid tissue. *In* The Thyroid Gland. Pitt-Rivers, R., and Trotter, W. R., eds. London, Butterworths, 1:131–162, 1964.
29. Ebstein, M., Greenstein, M., and Wyatt, G. R. Personal communication, 1968.
30. Szirmai, J. A., and Van der Linde, P. C. Effect of castration on the endoplasmic reticulum of the seminal vesicle and other target epithelia in the rat. J. Ultrastruct. Res., 12:380–395, 1965.
31. Karlson, P. New concepts on the mode of action of hormones. Perspect. Biol. Med., 6: 203–214, 1963.
32. Shearer, R. W., and McCarthy, B. J. Evidence for ribonucleic acid molecules restricted to the cell nucleus. Biochemistry, 6:283–289, 1967.
33. Tomkins, G. M., and Thompson, E. B. Hormonal control of protein synthesis at the translational level. Achtzehntes Colloquium der Gesellschaft für Physiologische Chemie. Berlin, Springer-Verlag, pp. 107–120, 1967.

34. Lockwood, D. H., Stockdale, F. E., and Topper, Y. J. Hormone-dependent differentiation of mammary gland: Sequence of action of hormones in relation to cell cycle. Science, 156:945–946, 1967.
35. Wyatt, G. R., and Tata, J. R. The hybridization capacity of ribonucleic acid produced during hormone action. Biochem. J., 109:253–258, 1968.
36. Jensen, E. V., Suzuki, T., Kawashima, T., Stumpf, W. E., Jungblut, P. W., and De Sombre, E. R. A two-step mechanism for the interaction of estradiol with rat uterus. Proc. Nat. Acad. Sci. U.S.A., 59:632–638, 1968.
37. Sekeris, C. E. Wirkung der Hormone auf den Zellkern. Achtzehntes Colloquium der Gesellschaft für Physiologische Chemie. Berlin, Springer-Verlag, pp. 126–151, 1967.
38. Sutherland, E. W., Butcher, R. W., Robison, G. A., and Hardman, J. G. The role of adenosine 3′,5′-monophosphate in hormone action. Achtzehntes Colloquium der Gesellschaft für Physiologische Chemie. Berlin, Springer-Verlag, pp.1–29, 1967.

3

Some Hormonal Effects on the Ontogenesis of Enzyme and Protein Patterns

E. M. Pantelouris

University of Strathclyde
Glasgow, Scotland

Regulation of gene action in the ontogenesis of higher organisms should be reflected in alterations of the enzyme and protein patterns. The switchover from foetal to adult hemoglobin and the appearance of lactic dehydrogenase isozyme patterns characteristic of particular stages and tissues are classical examples of such changes (1,2).

There is a tendency to attribute sequential gene action to "effectors" and "repressors" by analogy with the Jacob and Monod model of microbial genetics. Furthermore, as development is closely linked to hormonal changes, the concept has evolved that the "repressors" and "effectors" in higher organisms may in fact be the hormones themselves. In insects, ecdysone, which induces molting, is thought to be responsible also for the appearance of specific chromosomal puffs; the latter are believed to contain newly synthesised messenger RNA and are therefore evidence of activity of the genes themselves(3).

It follows that when research workers come across hormonal modulation of protein and isozyme patterns, they tend to think about mechanisms in the foregoing terms. Examples abound in the literature, and it is not my purpose to simply add to the list but rather to show how some of our own observations have led us to seek further clarification of the aforementioned concepts.

Morphogenetic changes in insect metamorphosis are so dramatic that one would expect equally drastic changes in protein patterns. Some see metamorphosis as progressive development, with an interposed phase of prolonged restraint, the larval stage. The restraint may be interpreted as the "repressor" effect of the juvenile hormone, overruled later by the "effector" action of the ecdysone. Others discern a duality of determination in metamorphosis. The larva and imago are interpreted as polymorphic stages which are, undoubtedly, subject to entirely different selective pressures. Therefore, two separate sets of genes are postulated, as if, in the words of Thomas Browne, there were "two souls operating in these little bodies." A complete switchover from one gene set to the other is envisaged at metamorphosis according to this theory(4).

We investigated the esterase pattern of the third instar larva, pupa, and imago of certain *Drosophila* species, using esterases simply as proteins, neither necessarily representative of the whole spectrum of proteins nor, on the other hand, too uniform a group(5). Starch electrophoresis resolves twelve or so fractions in adult *D. immigrans*. Only two or three of these are undetectable in third instar larva provided the gel is "overstained," and even these two or three may, of course, exist in traces. The most obvious change accompanying metamorphosis is the increased concentration of certain components found mainly in the anterior portion of the larval body. These may therefore be associated with the imaginal discs. Fractions strong in the larva tend to remain strong in the adult also. Some fractions become electrophoretically faster in the course of metamorphosis (Fig. 1).

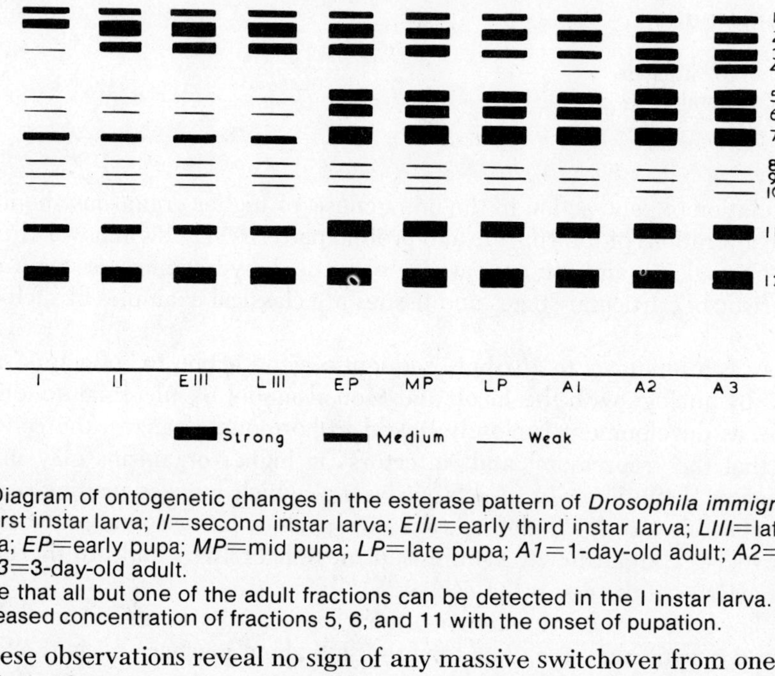

Fig. 1. Diagram of ontogenetic changes in the esterase pattern of *Drosophila immigrans*.
I=first instar larva; *II*=second instar larva; *EIII*=early third instar larva; *LIII*=late third instar larva; *EP*=early pupa; *MP*=mid pupa; *LP*=late pupa; *A1*=1-day-old adult; *A2*=2-day-old adult; *A3*=3-day-old adult.
Note that all but one of the adult fractions can be detected in the I instar larva. Note also the increased concentration of fractions 5, 6, and 11 with the onset of pupation.

These observations reveal no sign of any massive switchover from one genome to another. They are rather compatible with the stepwise release of esterase synthesis from inhibition. It is interesting that in amphibian metamorphosis, Manwell (6) has similarly found six esterase fractions common to tadpole and adult (*Rana catesbeiana*), compared to only two fractions specific to each stage. Other enzyme groups gave comparable results.

In the mammal, overall levels of esterases and acid phosphatases in mouse serum follow a characteristic pattern. The shape of these curves, however, is definitely aberrant in the thymusless mutant, *nude*(7), as shown in Figure 2. We find that not only is the thymus absent in homozygous *Nu Nu* individuals but the adrenal cortex shows increased thickness of the reticular zone. We are now attempting to analyse in this interesting mutant the interaction of thymus, adrenals, and pituitary.

In our experiments, electrophoresis of mouse serum has resolved 19 esterase bands in the adult and has pinpointed a number of ontogenetic changes and sex differences that can be assumed to have hormonal causation(8-10). We have demon-

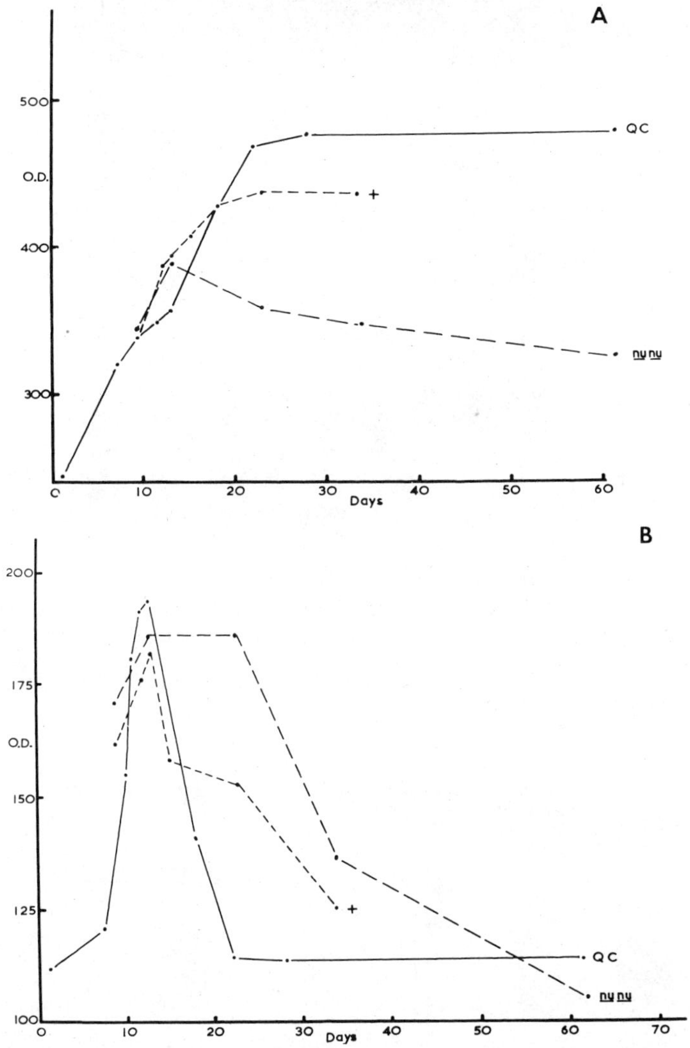

Fig. 2. Changes in the serum esterase levels during development of the mouse. Optical density plotted against age in days.

Left: Nonspecific esterases, substrate 1-naphthyl acetate. *Right:* acid phosphatases, substrate 1-naphthyl phosphate. (Unpublished, E. M. Pantelouris and Helen Martin.)

1: An inbred mouse strain, QC. *2:* Phenotypically normal animals of the stock carrying the mutation, *nude*. *3:* Animals homozygous for the mutation; these are the phenotypically *nude* animals that lack a thymus.

strated a fraction restricted to the gravid female, as well as a zone that is weak in the female (and resolvable into two fractions) though strong (and comprising three fractions) in the male. The 18-day foetus already possesses the main adult fractions but generally in small amounts. There are exceptions in that bands 16–19 are stronger in the foetus than in the adult. There are also six "foetal esterases" that decrease and disappear after birth (Fig. 3).

Similar is the fate of a specific serum protein which is very strong in the foetus but which diminishes in amount with time in an inverse way to the increasing

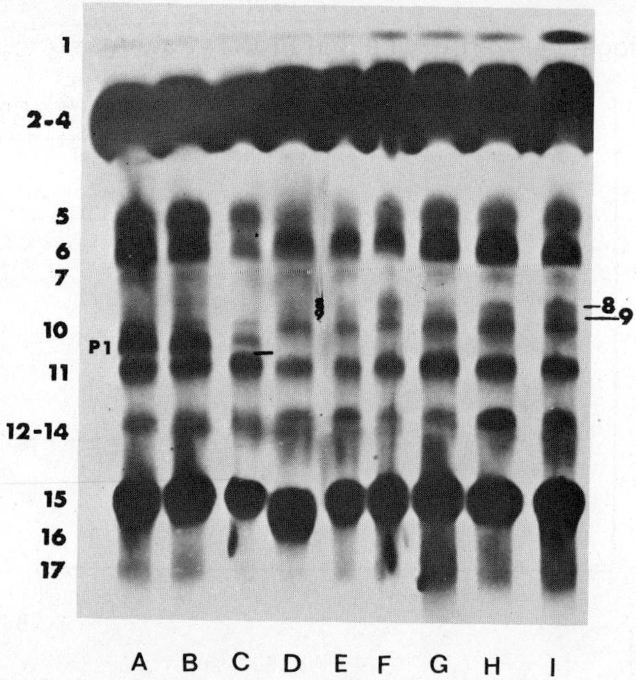

Fig. 3. Esterase zymogram of mouse serum. The thick front zone can be resolved by the use of appropriate buffers; hence it is numbered 1 and 2-4. The samples shown demonstrate the special "pregnancy esterase," *P1*.

A, B, C: females in the 10th, 11th, and 16th day of pregnancy; *D, E:* nongravid females; *F, G, H, I:* lactation days 15, 9, 11, and 11.

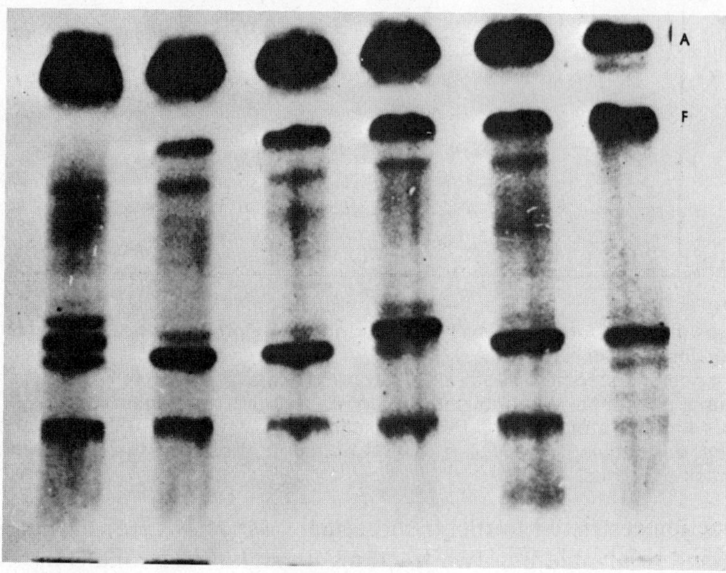

Fig. 4. Serum protein patterns by starch electrophoresis stained with nigrosin. The samples are from mice of different ages—*A:* 19-day; *B:* 8-day; *C* and *D:* 4-day; *E:* 1-day. Note the decrease of the "fetuin," F, which is strong in the foetal stage *(foe)* also when the albumin, A, is relatively weak.

amounts of albumin. Eventually, it disappears altogether by the second week of life. We were able to ascertain that this "fetuin" is a glycoprotein and has esterolytic properties or, alternatively, it binds esterases(10,11) (Fig. 4).

One cannot but conclude that the greater hormonal complexity of the mammal is associated with relatively drastic ontogenetic changes in the esterase pattern. By contrast, and despite the more impressive morphogenetic transformation of metamorphosis, changes in insect esterases are less pronounced.

Out of these and other similar observations indicative of hormonal control of esterase patterns, we selected a particular "fast esterase *b*" to examine more intimately its susceptibility to steroids(12). This is a fraction found in the rat liver microsomes. The normal adult rat has no fast esterase *b*, but the young possess it for a period after birth, from the second to the fifth week of age (Fig. 5).

A series of injections of certain steroids, in particular of 17-acetoxy-6α-trifluoromethylpregn-4-ene-3,20-dione (flumedroxone acetate), caused the reappearance in the adult of this juvenile esterase *b*. In addition, the drug caused the appearance of yet another fraction, the "fast esterase *a*." Here is then, apparently, typical enzyme induction. To us this case was of special interest in that a juvenile esterase could be experimentally "recalled" in the adult. We surmised that a normal hormone acts as a repressor to this fraction in development, and that what the drug does is to reverse the process by acting antagonistically to that hormone.

Though we started using one substrate only (1-naphthyl acetate) for the detection of esterases, we later proceeded to test a battery of alternative substrates. It turned out that esterase *b* was virtually undetectable in the adult liver extract with a whole series of substrates, just as is the case with 1-naphthyl acetate. Surprisingly, however, it showed up in the untreated adult with other substrates, such as 1-naphthyl laurate and 1-naphthyl butyrate. Furthermore, after treatment with flumedroxone acetate, the fast esterase *b* is detectable with *both* series of substrates. Treatment does not seem to induce the formation of this fraction—it is already there but reacts with some substrates only. Rather, treatment appears to increase the affinities of the fraction to cover other substrates as well(Fig. 6).

It is possible to suggest other explanations—for example, that the drug induces de novo a new esterase which just happens to overlap electrophoretically the naturally occurring esterase *b*. We have been unable to split this fraction, however, by the use of alternative buffer systems. Also, dialysis of liver homogenate did not affect the result, so a dialysable cofactor is unlikely. Addition of the steroid to the sample before electrophoresis again had no effect. While we are continuing with tests of this nature, we feel we must consider seriously the hypothesis that treatment does modify substrate affinity.

Concerning the implications of this work, the fact that only small changes in esterase patterns are observed during insect metamorphosis does not necessarily mean that the same should apply to proteins generally. It may be argued that there is a difference of great importance between enzymes, on the one hand, and structural proteins, on the other. As Epstein(13) pointed out, selection will favour mutations that maintain the original conformation of proteins. An even stricter limitation to change may apply to enzymes because of the need to preserve catalytic function. While changes of esterase patterns do not exhibit the massive switchover implied

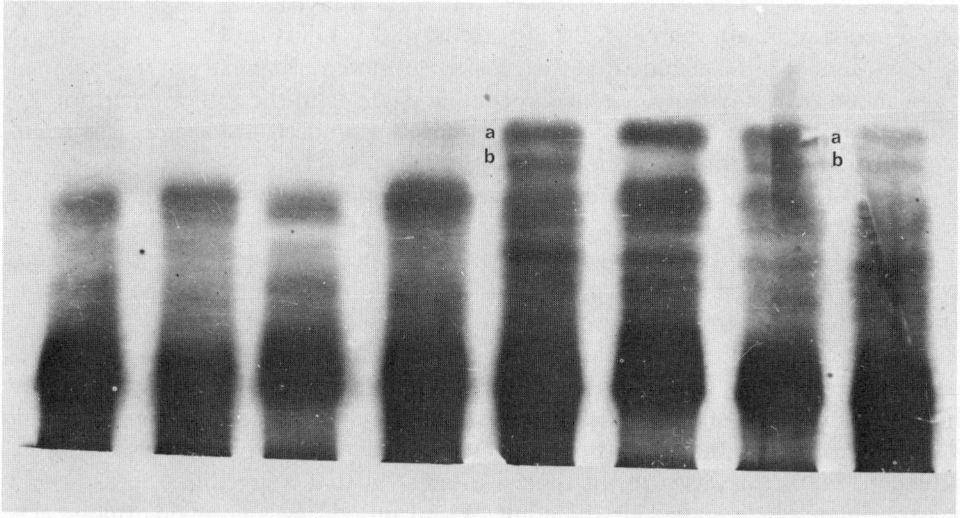

Fig. 5. Starch electrophoresis of esterases in rat liver microsomal fractions. The first four samples are from untreated adult animals. The other four represent siblings of the controls treated with flumedroxone acetate and show the appearance of the "fast esterases" *a* and *b*.

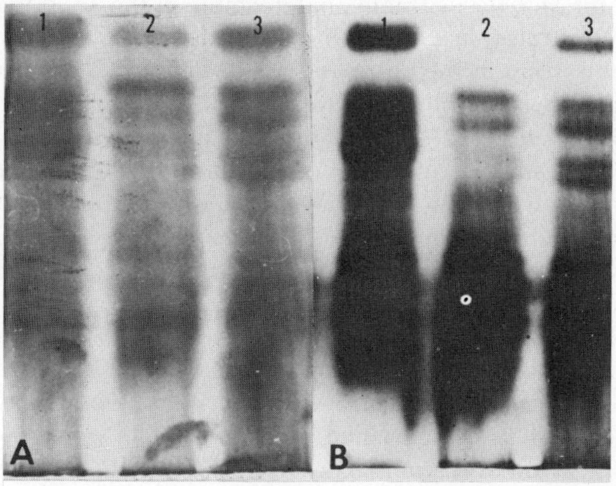

Fig. 6. Electrophorograms of esterases in liver homogenates of treated and untreated rats.
A: Substrate=2-naphthyl stearate. *1:* Treated. *2:* Untreated adult. *3:* Untreated 8-day-old.
B: Substrate=1-naphthyl acetate. *1:* Treated. *2:* Untreated. *3:* Untreated 8-day-old.

by the two-genomes theory of metamorphosis, investigations of other groups of proteins may prove more compatible with that theory.

There is, of course, evidence that steroids influence the production of messenger RNA necessary for the synthesis of specific proteins. This is the typical process of hormonal-type induction as discussed, for example, by Kidson(14). There is, however, good evidence also for the direct action of hormones on enzyme molecules. Tomkins and Yielding(15) described an in vitro effect that bears some similarities to the phenomenon observed in our material. They found that certain steroids alter the relative affinity of a dehydrogenase to glutamic and alanine substrates, in this case by splitting in vitro the crystalline enzyme into quarter units. In our case, the electrophoretic mobilities of the wide- and narrow-specificity versions of the fast esterase *b* were found to be the same, and we do not think that splitting of the molecule is involved.

The particular interest of these rat liver "fast esterases" lies in the fact that the experimentally produced changes occur in vivo and imitate the natural situation prevailing at an earlier stage of development. Hence, the phenomenon observed may be relevant to hormonal action in development. It also gives rise to the more general question of whether other instances of apparently typical hormonal induction may not in fact be direct allosteric effects.

REFERENCES

1. Ingram, V. M. Hemoglobin and its Abnormalities. Springfield, Illinois, Charles C. Thomas, 1961.
2. Markert, C. L., and Upsprung, H. The ontogeny of isozyme patterns of lactate dehydrogenase in the mouse. Develop. Biol., 5:363–381, 1959.
3. Karlson, P. The effects of ecdysone on giant chromosome, RNA metabolism and enzyme induction. *In* Endocrine Genetics. Spickett, S. G., ed. Cambridge University Press, pp. 67–76, 1967.
4. Wigglesworth, V. B. Metamorphosis and body form. *In* Molecular and Cellular Aspects of Development. Bell, E., ed. New York, Harper and Row, pp. 447–478, 1965.
5. Pantelouris, E. M., and Downer, R. G. H. Phenotypic changes of the esterase pattern in insect metamorphosis. J. Insect Physiol., 15:2357–2362, 1969.
6. Manwell, C. Metamorphosis and gene action. I. Electrophoresis of dehydrogenases, esterases, hemoglobins and other soluble proteins of tadpole and adult bullfrogs. Comp. Biochem. Physiol., 16:222–231, 1966.
7. Pantelouris, E. M. Absence of thymus in a mouse mutant. Nature (London), 217:370–371, 1968.
8. ——— and Arnason, A Ontogenesis of serum esterases in *Mus musculus*. J. Embryol. Exp. Morph., 16:55–64, 1966.
9. ——— and Arnason, A. Further observations on mouse serum esterases. Comp. Biochem. Physiol., 20:647–652, 1967.
10. ——— and Arnason, A. Serum proteins of *Apodemus sylvaticus* and *Mus musculus*. Comp. Biochem. Physiol., 21:533–539, 1967.
11. ——— and Hale, P. A. Developmental changes in the plasma protein of the mouse Nature (London), 195:79, 1962.
12. ——— and Hines, W. V. W. Steroid effects on the substrate profile of a liver esterase. Comp. Biochem. Physiol., 26:129–136, 1968.
13. Epstein, C. J. Non-randomness of aminoacid changes in the evolution of homologous proteins. Nature (London), 215:355–359, 1967.
14. Kidson, C. Cortisol in the regulation of RNA and protein synthesis. Nature (London), 213:779–782, 1967.
15. Tomkins, G. M., and Yielding, K. L. Regulation of biological function mediated by changes in protein structure. *In* The Biochemical Aspects of Hormones. Eisenstein, A. B., ed. London, J. A. Churchill, Ltd., pp. 189–199, 1964.

4

Hormonal Control of Lactose Synthetase in Developing Mammary Gland*

Roger W. Turkington
Department of Medicine and Biochemistry,
Duke University, Durham,
North Carolina

Development of the mammary gland during pregnancy involves the growth of a branching ductal network and leads to the formation of alveolar cells, which are uniquely capable of synthesizing the components of milk. The formation of the alveolar cells is characterized by the expression of new genetic information, and represents a definitive transition to a highly differentiated cell type. The alveolar cells formed during the developmental period have a low level of secretory activity, but assume a dramatic increase in secretory function following parturition. Cell proliferation and differentiation, as well as the induction of secretory milk products, appear to represent sequential responses of mammary epithelial cells to specific alterations in the hormonal environment. It is the purpose of this review to demonstrate how multiple hormones may interact with cells in the regulation of a specific enzyme system, lactose synthetase, and how the regulation of this enzyme may determine a pattern of growth during development or of specialized function in the mature organ.

The synthesis and secretion of large amounts of lactose are uniquely characteristic of the lactating mammary gland. The biosynthesis of lactose is catalyzed by three enzymes in the following reactions:

$$\text{UTP}+\text{glucose-1-P} \xrightleftharpoons{\text{UDPG-pyrophosphorylase}} \text{UDP-glucose}+\text{PP} \quad \text{(I)}$$

$$\text{UDP-glucose} \xrightleftharpoons{\text{UDP-galactose-4-epimerase}} \text{UDP-galactose} \quad \text{(II)}$$

$$\text{UDP-galactose}+\text{glucose} \xrightarrow{\text{Lactose synthetase}} \text{lactose}+\text{UDP} \quad \text{(III)}$$

*I gratefully acknowledge the contributions of my collaborators in this work, Dr. Keith Brew, currently at the University of Leeds; Dr. Thomas C. Vanaman; and Professor Robert L. Hill, in whose laboratory at Duke University much of this work was carried out.
This work was supported by USPHS Grant CA-10268 from the National Cancer Institute.

The enzyme system that catalyzes the final biosynthetic step, lactose synthetase (reaction III), consists of two proteins, the A and B proteins, both of which are required for lactose synthetase activity(1). The B protein has been identified as the milk whey protein α-lactalbumin(2). Although both proteins are found in soluble form in milk(3), cell fractionation procedures demonstrate that the A protein is particulate and sediments with the microsomal fraction, while the B protein is easily dissociated from the lactose synthetase particles and appears predominantly in the soluble fraction(4). The separated A protein has been shown to be a galactosyl transferase(5) which catalyzes reaction IV:

$$\begin{array}{c}\text{UDP}-\text{galactose}\\ N\text{-acetyl lactosamine}+\text{UDP}\\ +N\!=\!\text{acetyl glucosamine}\end{array} \qquad \text{(IV)}$$

A similar enzyme is found in a number of mammalian tissues(6,7), and its transfer of galactose to *N*-acetyl glucosamine covalently bound to protein may be important in the fabrication of the oligosaccharide side chains of glycoproteins. In the presence of α-lactalbumin, the substrate-acceptor specificity of the A protein is changed to include glucose, thus permitting the synthesis of lactose. Alpha-lactalbumin has been designated the "specifier protein" of the lactose synthetase system by Brew et al.(5), and represents a new type of regulatory role for a mammalian protein. Kinetic data provide evidence for the formation of a complex between the A and B proteins in the presence of substrates and manganese ion.

In our laboratory the regulation of lactose synthetase during development of the mouse mammary gland has been studied in vivo and in organ cultures of mid-pregnancy mammary tissue. The A protein activity in mammary homogenates was assayed by reaction IV, while B protein activity was assayed by reaction III utilizing the addition of excess purified bovine A protein(5). In each reaction, radioactive galactose was transferred to a neutral sugar product which was separated by ion-exchange chromatography. It was not possible to reconstruct the endogenous lactose synthetase activity, since the dilution of homogenization markedly interferes with the interaction of the endogenous proteins.

Previous studies have demonstrated that the in vitro differentiation of mammary epithelial cells into secretory alveolar cells requires the presence of insulin, hydrocortisone, and prolactin(8-11). The results in Figure 1 show that these hormones can also regulate the levels of the lactose synthetase A and B proteins in vitro(12). Although light increases occur in the presence of insulin or insulin and prolactin, maximum stimulation required the presence of all three hormones. Since the increase in cell number is the same in all the insulin-containing systems(13-15), the increases in enzymatic activity effected by insulin, hydrocortisone, and prolactin represent increased activity per cell.

Insulin serves an important function in this system by inducing mammary epithelial cells to initiate DNA synthesis and subsequently to divide(16). Hormonal effects observed during the period of insulin-induced cell proliferation reflect the activities of a population of cells that are somewhat asynchronously dividing and differentiating. However, since cell division virtually ceases after 72 hours of culture(17), it was possible to study the kinetics of enzyme induction in postmitotic cells in response to a specific hormonal stimulus given after 72 hours of preincu-

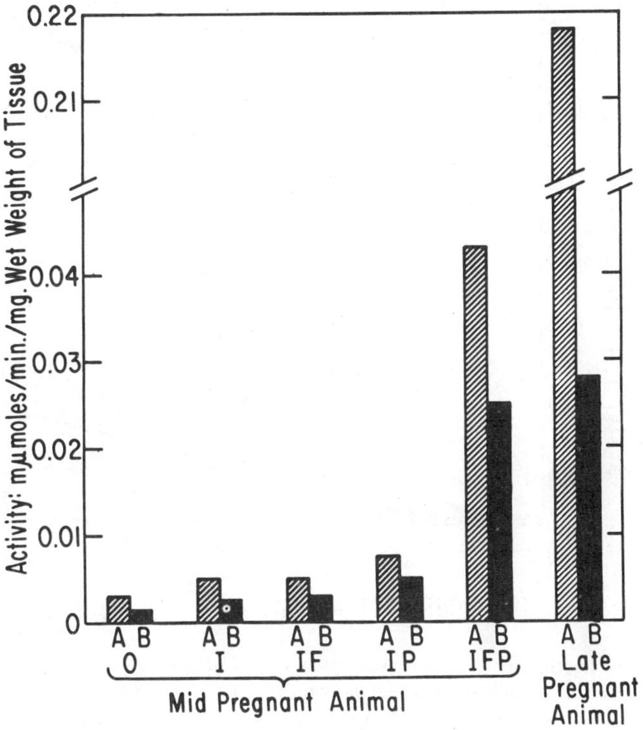

Fig. 1. The effect of different combinations of insulin (I), hydrocortisone (F), and prolactin (P) on the levels of the A and B proteins of lactose synthetase in midpregnancy mouse mammary glands. Tissue was cultured for 48 hours in each of the hormone combinations. Each hormone was present at a concentration of 5 μg/ml. From Turkington et al. *J. Biol. Chem.,* 243, 3382-3387, 1968.

bation. As shown in Figure 2, the addition of prolactin to the culture medium of explants preincubated in a medium containing insulin and hydrocortisone leads to a stimulation of the A and B enzyme activities. The addition of puromycin or cycloheximide at any time after the addition of prolactin prevented further increases in enzymatic activity, indicating that the increased activity of each protein reflects the formation of new protein molecules. Actinomycin D or mitomycin C completely prevented the induction of both proteins, suggesting that RNA synthesis is specifically required(12). The relative activities of the A and B proteins appear to be precisely controlled during the induction process in vitro. After an initial lag period, the A protein is formed at a much greater rate than that for the B protein. Such biphasic kinetics have been observed in the induction of several specific proteins by steroid hormones(18-20).

Induction of the A and B proteins in response to prolactin stimulation occurs only in the cells formed in vitro. If cell division is prevented by colchicine, no increase in enzyme activity is observed (Table 1). Colchicine does not, however, inhibit enzyme induction after cell formation has been completed. The role of hydro-

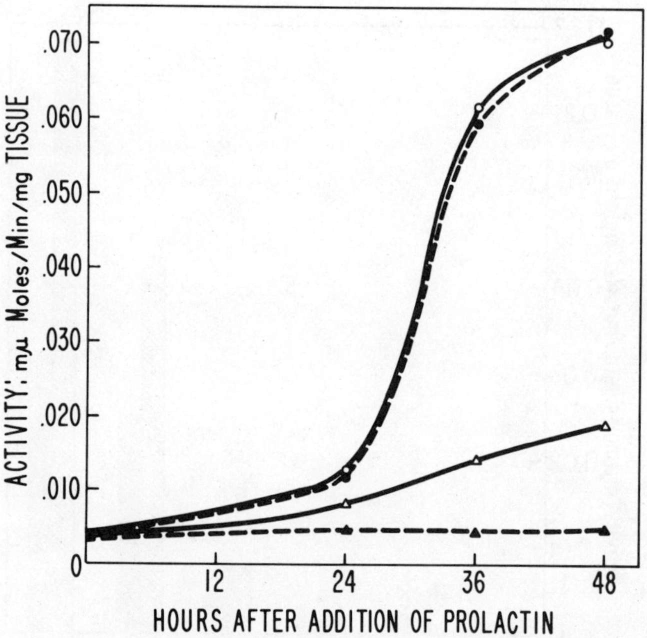

Fig. 2. Kinetics of induction of the A and B proteins of lactose synthetase in mammary explants preincubated in insulin-hydrocortisone medium and then exposed to insulin (I), hydrocortisone (F), and prolactin (P): open circles, A protein, IFP system; open triangles, B protein, IFP system. Explants from the same tissue and similarly preincubated in IF medium were exposed to IFP medium containing progesterone (1×10^{-6} M): solid circles, A protein, IFP-progesterone system; solid triangles, B protein, IFP-progesterone system.

TABLE 1

Influence of preincubation conditions on induction of A and B proteins by prolactin

Incubation system		Activity (mμmoles/min/mg tissue)	
0-72 hr	**72-108 hr**	**A protein**	**B protein**
IF	IF	0.010	0.005
IF	IFP	0.042	0.021
I	IFP	0.010	0.003
IF + colchicine	IFP	0.012	0.003
IF	IFP + colchicine	0.041	0.021

Insulin (I), hydrocortisone (F), and prolactin (P) were each present at a concentration of 5 μg/ml. The concentration of colchicine was 0.05 μg/ml. (Data reproduced from Journal of Biological Chemistry by permission of the publishers.)

cortisone is to elicit "covert" differentiation prior to cell division(21). As shown in Table 1, enzyme induction in response to prolactin fails to occur in cells formed in the absence of hydrocortisone. Although the action of hydrocortisone occurs here in cells that proliferate, it cannot act solely on postmitotic cells to permit a response

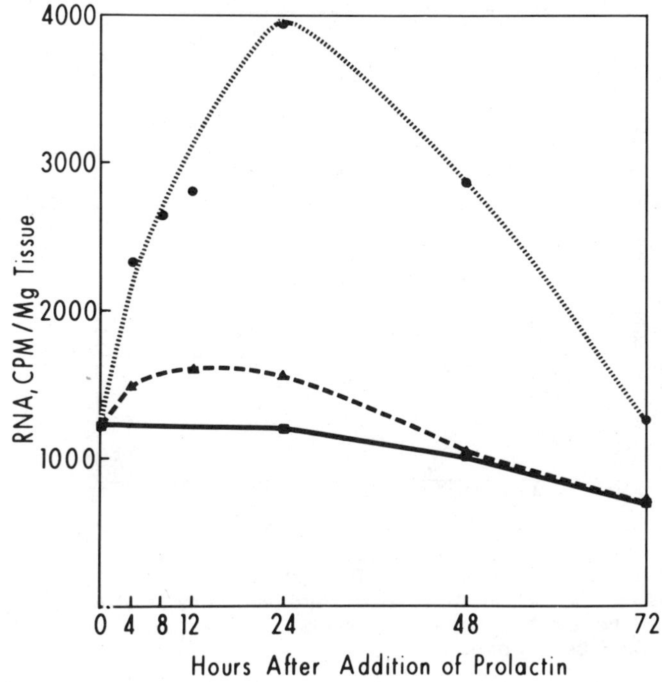

Fig. 3. RNA synthesis in midpregnancy mouse mammary gland explants preincubated for 72 hours in medium containing insulin and hydrocortisone and then exposed to IFP, circles, or FP, triangles, at zero time. Squares, IF control. Each point represents the incorporation of ^{3}H-uridine into trichloroacetic acid-precipitable material during the preceding 4-hour labelling period. From Turkington and Ward. *Biochimica Biophysica Acta,* 174:298, 1969.

to prolactin. The induction of casein and β-lactoglobulin by prolactin also occurs only in daughter cells formed in the presence of insulin and hydrocortisone(21,22).

As shown in Figure 3, prolactin stimulation of explants preincubated for 72 hours on a medium containing insulin and hydrocortisone is followed by a rapid rise in the rate of RNA synthesis, reaching a 200 percent increase after 24 hours. This stimulation fails to occur within 24 hours in explants preincubated for 72 hours on a medium containing only insulin, although it does occur after hydrocortisone treatment during the 72 to 120 hour period. No stimulation occurs in explants in which cell division is prevented by colchicine. The induction of lactose synthetase and casein by prolactin also requires the continuing presence of insulin in the medium. This may be explained by the observation that only a slight increase in RNA synthesis occurs in response to prolactin when insulin is removed from the medium. Since probably not all insulin could be removed from the explants, these results may reflect an absolute requirement for insulin during stimulation of RNA synthesis.

Similar conclusions have been derived from assays of RNA polymerase in nuclei isolated from cultured explants. This enzyme activity was DNA dependent, required all four nucleoside triphosphates and a divalent metal ion for maximal acti-

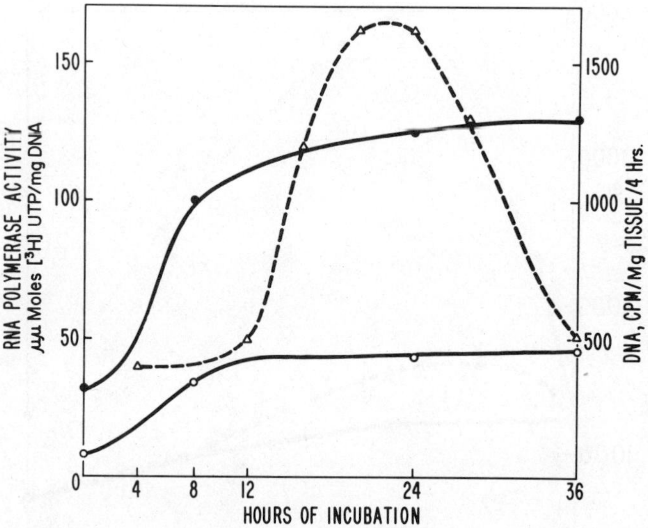

Fig. 4. Time course of the effect of insulin on RNA polymerase activity (open circles, $-(NH_4)_2SO_4$; solid circles, $+(NH_4)_2SO_4$ in the assay mixture) and rates of DNA synthesis (triangles). Each point on the DNA synthetic curve represents the incorporation of ^{3}H-thymidine during the preceding 4-hour labelling period.

vity, and was inhibited by actinomycin D. The reaction product was insoluble in trichloroacetic acid, and was solubilized by ribonuclease(23). As shown in Figure 4, RNA polymerase activity in midpregnancy mammary explants is doubled after 8 hours of incubation with insulin, rising to a level approximately 300 percent above the initial value. Similar increases are observed with or without 0.2 M ammonium sulfate in the reaction mixture. This action of insulin precedes its stimulatory effect on DNA synthesis and represents an early action of insulin on nuclear transcription. Figure 5 shows that a further increase in RNA polymerase activity occurs upon prolactin stimulation of explants preincubated in a medium containing insulin and hydrocortisone. This figure also summarizes a sequence of changes induced by prolactin. Prior to measurable increases in protein synthesis, prolactin, acting synergistically with insulin, stimulates RNA synthesis and nuclear RNA polymerase activity. Although hydrocortisone itself does not alter rates of RNA synthesis or RNA polymerase activity, prolactin stimulates RNA and milk protein synthesis only in epithelial cells which were "covertly" differentiated after proliferation in the presence of hydrocortisone(23). Since increases in RNA synthesis and RNA polymerase activity occur prior to detectable increases in the formation of specific milk proteins, an early effect of prolactin appears to be on nuclear transcription.

Although prolactin induces the formation of both the A and B proteins of lactose synthetase in vitro, the synthesis of these two proteins is asynchronous during pregnancy and lactation(12,24). As shown in Figure 6, the levels of A and B proteins remain quite low during the first half of pregnancy. During the second half of pregnancy the A protein activity rises rapidly to a nearly maximal level, while the B protein activity increases only slightly. With the initiation of lactation, the B

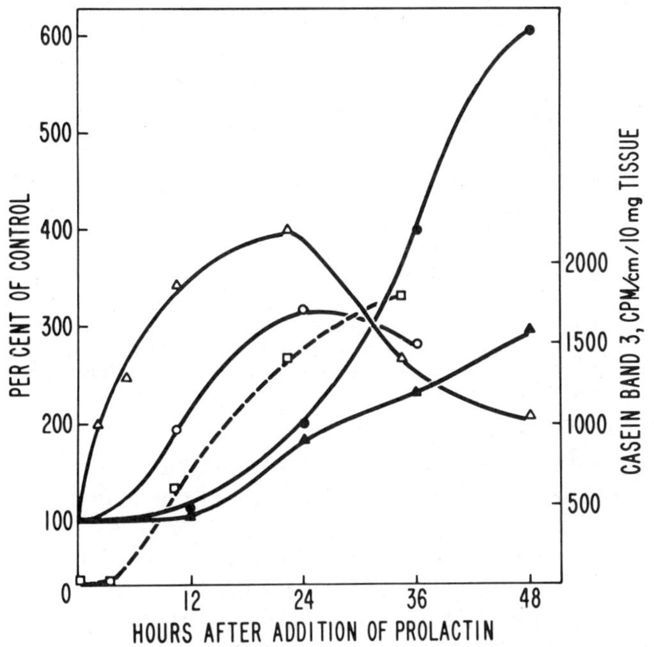

Fig. 5. Sequence of effects elicited by prolactin and insulin in mouse mammary epithelial cells of midpregnancy explants preincubated for 72 hours on medium containing insulin and hydrocortisone. Open triangles, rate of RNA synthesis; open circles, RNA polymerase activity; solid circles, A protein activity; solid triangles, B protein activity. The rate of synthesis of casein (electrophoretic band 3) was determined as previously described (squares)(21,22).

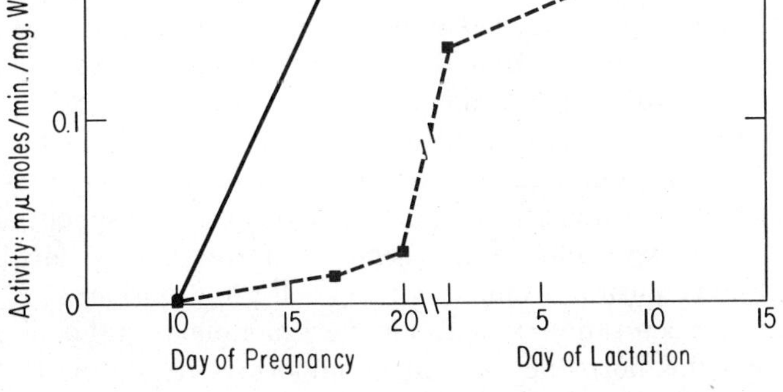

Fig. 6. The levels of A and B proteins in mammary glands from mice at different stages of pregnancy and lactation. Circles, A protein; squares, B protein. From Turkington et al. *J. Biol. Chem.*, 243, 3382-3387, 1968.

TABLE 2

The levels of A and B proteins in mice preparturition and postparturition

Time	mμmoles/min/μg DNA	
	A	B
1 day preparturition	0.380	0.016
2 days postparturition	0.392	0.264

The activities of the A and B proteins were determined as described by Brew et al. (5). DNA was determined by the diphenylamine method. The results are representative of three such experiments.

protein rises sharply, and both proteins are found in high amounts during lactation. The observed rise in B protein at parturition appears to coincide temporally with increases in the activities of several other enzymes and in total protein observed in the mammary glands of parturient rats(26,27). As shown in Table 2, the rapid rise in B protein activity at parturition cannot be explained by an increase in cell number alone. The physiological basis for the asynchronous syntheses of the A and B proteins has not been directly demonstrated. However, it may be noted that an important physiological change at parturition is the rapid fall in plasma concentration of progesterone, occurring presumably as a result of withdrawal of placental synthesis of progesterone and luteotropic hormone(28). As shown in Figure 2, progesterone can inhibit the in vitro induction of B protein by prolactin. At the nearly "physiological" concentration used in these experiments, the induction of the A protein is unaffected(25). These observations are consistent with the concept that induction of the B protein may be inhibited by progesterone, resulting in low levels during pregnancy, whereas induction of A protein is unaffected, allowing its accumulation during pregnancy.

During pregnancy the placenta releases a lactogenic hormone which represents a major portion of the prolactin-like activity in plasma(13,29,30). Table 3 shows that human placental lactogen resembles prolactin in that each hormone can induce the A and B proteins to a similar degree and each stimulates RNA polymerase activity. The nearly equivalent effects of these hormones over a wide concentration range in vitro suggest that the replacement of placental lactogen by pituitary prolactin at parturition does not appear to provide an explanation for the asynchronous development of the two components of lactose synthetase.

Hormonal regulation of lactose synthetase involves the interaction of at least five and possibly six hormones during the growth and development of the mammary gland. Substantial expression of the genetic information for lactose synthesis in vitro has as a prerequisite, insulin-mediated cell proliferation. Growth hormone at "physiological" concentrations can also induce mammary epithelial cells to proliferate in vitro, and this hormone may be an important regulating factor in vivo(16). Hydrocortisone acts to cause the formation of daughter cells in which the synthesis of RNA and lactose synthetase can be stimulated by placental lactogen or prolactin.

TABLE 3

Comparison of the activities of ovine prolactin and human placental lactogen in stimulating lactose synthetase and RNA polymerase in vitro

Explants of midpregnancy mouse mammary gland were preincubated for 72 hours on medium containing insulin and hydrocortisone prior to transfer to the incubation systems below. RNA polymerase activity was assayed 24 hours later, and lactose synthetase was assayed after 48 hours.

	RNA polymerase ($\mu\mu$moles UTP/mg DNA)		Lactose synthetase (mμmoles/min/mg tissue)	
System	$-(NH_4)_2SO_4$	$+(NH_4)_2SO_4$	A	B
Insulin and hydrocortisone	31	71	0.006	0.003
Insulin, hydrocortisone, and prolactin	67	211	0.058	0.015
Insulin, hydrocortisone, and placental lactogen	64	204	0.061	0.022

The relative rates of formation of the A and B proteins may be further modified by progesterone.

The relative levels of the A and B proteins during development appear to reflect specific functional requirements of the alveolar cell. The rapid rise in A protein galactosyl transferase activity during pregnancy may be important for the synthesis of cellular glycoproteins. Electron microscope studies have shown that the addition of prolactin to an incubation medium containing insulin and aldosterone* leads to the formation in mammary epithelial cells of rough endoplasmic reticulum and secretory membranes(32). The formation of such intracellular glycoprotein membranes in response to prolactin stimulation may be partially mediated through the induction of A protein. The A protein may also be essential for the formation of the oligosaccharide side-chains of the casein glycoproteins. A comparison of the relative activities of the enzymes on the pathway of lactose biosynthesis(24) indicates that from the beginning of lactation the formation of lactose must be regulated through the activity of the terminal enzyme, lactose synthetase. Lactose synthesis is promoted at this time by the sharp rise in the intracellular level of the B protein, α-lactalbumin. Lactose synthetase thus represents a mechanism in which differentiated function is hormonally regulated by controlling the synthesis of each of two interacting proteins.

REFERENCES

1. Brodbeck, U., and Ebner, K. E. Resolution of a soluble lactose synthetase into two protein components and solubilization of microsomal lactose synthetase. J. Biol. Chem., 241:762–764, 1966.
2. ——— Denton, W. L., Tanahashi, N., and Ebner, K. E. The isolation and identification of the B protein of lactose synthetase as α-lactalbumin. J. Biol. Chem., 242:1391–1397, 1967.

*The activity of aldosterone is equivalent to that of hydrocortisone in eliciting the changes of alveolar cell differentiation in vitro(31).

3. Babad, H., and Hassid, W. Z. Soluble uridine diphosphate D-galactose:D-glucose β-4-D-galactosyltransferase from bovine milk. J. Biol. Chem., 241:2672–2678, 1966.
4. Brodbeck, U., and Ebner, K. E. The subcellular distribution of the A and B proteins of lactose synthetase in bovine and rat mammary tissue. J. Biol. Chem., 241:5526–5532, 1966.
5. Brew, K., Vanaman, T, C., and Hill, R. L. The role of α-lactalbumin and the A protein in lactose synthetase. A unique mechanism for the control of a biological reaction. Proc. Nat. Acad. Sci. U.S.A., 50:491–497, 1968.
6. McGuire, E. J., Jourdian, G. W., Carlson, D. M., and Roseman, S. Incorporation of D-galactose into glycoproteins. J. Biol. Chem., 240:PC4112–4115, 1965.
7. Turkington, R. W., and Hill, R. L. Unpublished observations, 1968.
8. Rivera, E. M., and Bern, H. A. Influence of insulin on maintenance and secretory stimulation of mouse mammary tissues by hormones in organ-culture. Endocrinology, 69: 340–353, 1961.
9. Juergens, W. G., Stockdale, F. E., Topper, Y. J., and Elias, J. J. Hormone-dependent differentiation of mammary gland in vitro. Proc. Nat. Acad. Sci. U.S.A., 54:629–634, 1965.
10. Turkington, R. W., Juergens, W. G., and Topper, Y. J. Hormone-dependent synthesis of casein in vitro. Biochim. Biophys. Acta, 111:573–576, 1965.
11. Lockwood, D. H., Turkington, R. W., and Topper, Y. J. Hormone-dependent development of milk protein synthesis in mammary gland in vitro. Biochim. Biophys. Acta, 130:493–501, 1966.
12. Turkington, R. W., Brew, K., Vanaman, T. C., and Hill, R. L. The hormonal control of lactose synthetase in the developing mouse mammary gland. J. Biol. Chem., 243:3382–3387, 1968.
13. ——— and Topper, Y. J. Stimulation of casein synthesis and histological development of mammary gland by human placental lactogen. Endocrinology, 79:175–181, 1966.
14. Stockdale, F. E., and Topper, Y. J. The role of DNA synthesis and mitosis in hormone-dependent differentiation. Proc. Nat. Acad. Sci. U.S.A., 56:1283–1289, 1966.
15. Turkington, R. W., and Topper, Y. J. Androgen inhibition of mammary gland differentiation in vitro. Endocrinology, 80:329–336, 1967.
16. ——— Hormone-induced synthesis of DNA by mammary gland in vitro. Endocrinology, 82:540–546, 1968.
17. Lockwood, D. H., Stockdale, F. E., and Topper, Y. J. Hormone-dependent differentiation of mammary gland: sequence of action of hormones in relation to cell cycle. Science, 156:945–946, 1967.
18. O'Malley, B. W. In vitro hormonal induction of a specific protein (avidin) in chick oviduct. Biochemistry, 6:2546–2551, 1967.
19. Varner, J. E., and Chandra, G. R. Hormonal control of enzyme synthesis in barley endosperm. Proc. Nat. Acad. Sci. U.S.A., 52:100–106, 1964.
20. Griffin, M. J., and Cox, R. P. Studies on the mechanism of hormonal induction of alkaline phosphatase in human cell cultures. J. Cell Biol., 29:1–9, 1966.
21. Turkington, R. W., Lockwood, D. H., and Topper, Y. J. The induction of milk protein synthesis in post-mitotic mammary epithelial cells exposed to prolactin. Biochim. Biophys. Acta, 148:475–480, 1967.
22. ——— Induction of milk protein synthesis by placental lactogen and prolactin in vitro. Endocrinology, 82:575–583, 1968.
23. ——— and Ward, O. T. Hormonal stimulation of ribonucleic acid polymerase in mammary gland in vitro. Biochim. Biophys. Acta, 174:291–301, 1969.
24. Kuhn, N. J. Lactogenesis in the rat. Metabolism of uridine diphosphate galactose by mammary gland. Biochem. J., 106:743–748, 1968.
25. Turkington, R. W., and Hill, R. L. Science, 163:1458–1460, 1969.
26. Baldwin, R. L., and Milligan, L. P. Enzymatic changes associated with the initiation and maintenance of lactation in the rat. J. Biol. Chem., 241:2058–2066, 1966.
27. Kuhn, N. J., and Lowenstein, J. M. Lactogenesis in the rat. Changes in metabolic parameters at parturition. Biochem. J., 105:995–1002, 1968.

28. Van der Molen, H. J., and Aakvaag, A. Progesterone. *In* Hormones in Blood. Gray, C. H., and Bacharach, A. L., eds. New York, Academic Press, Inc., pp. 221–305, 1967.
29. Josimovich, J. B., and MacLaren, J. A. Presence in the human placenta and term serum of a highly lactogenic substance immunologically related to pituitary growth hormone. Endocrinology, 71:209–220, 1962.
30. Grumbach, M. M., and Kaplan, S. L. On the placental origin and purification of chorionic "growth hormone-prolactin" and its immunoassay in pregnancy. Trans. N.Y. Acad. Sci., 27:167–188, 1964.
31. Turkington, R. W., Juergens, W. G., and Topper, Y. J. Steroid structural requirements for mammary gland differentiation in vitro. Endocrinology, 80:1139–1142, 1967.
32. Wellings, S. R., Cooper, R. A., and Rivera, E. M. Electron microscopy of induced secretion in organ cultures of mammary tissue of the C3H/Crgl mouse. J. Nat. Cancer Inst., 36:657–672, 1966.

5

Postmitotic Actions of Prolactin and Insulin on Mammary Gland Extracts

Yale J. Topper,[1] Anthony E. Voytovich,[2] and Ida S. Owens[1]

[1]National Institute of Arthritis and Metabolic Diseases, National Institutes of Health, Public Health Service, United States Department of Health, Education and Welfare, Bethesda, Maryland
[2]Cleveland Metropolitan Hospital, Cleveland, Ohio

Casein comprises a family of phosphoproteins whose formation is under endocrine control. The marked increase in the rate of casein synthesis observed in mouse mammary gland explants that have been cultured in the presence of indulin, hydrocortisone, and prolactin reflects the formation of new casein-producing cells from precursor epithelial cells which have no detectable ability to make casein. Actually, it is not known whether the precursor cells must undergo DNA synthesis, mitosis, or both in this environment in order to become producers of casein. Thus, whereas mitosis normally follows the S and G_2 phases of the cell cycle, it is conceivable that tetraploid G_2 cells derived from the precursor cells might be able to make casein. However, we have not arrested the cycle in G_2, and the cells do proceed through mitosis. Operationally, then, it is the daughter cells that are observed to make casein. Some of the postmitotic events in this system will be discussed.

Two types of experiments will be described (all hormones were used at a final concentration of 5 μg/ml). Both involved three incubations. In the first incubation, mammary gland explants, derived from 10- to 12-day-pregnant C3H/HeN mice, were cultured on "Medium 199" (in all cases the medium contained D-fructose instead of D-glucose) in the presence of insulin (I) and hydrocortisone (F) until epithelial cell proliferation ceased (72 hours). The postmitotic explants so formed were then cultured during a second (48 hours) and third (4 hours) period as follows. In the first type of experiment, the postmitotic explants were divided into four groups. Group 1 was cultured on a fresh IF medium (second incubation) and then on an IFP medium (insulin-hydrocortisone-prolactin). Group 2 was cultured on IFP medium during both the second and third incubations. Group 3 was cultured on an FP medium during the second and third incubations. Group 4 was cultured on an FP medium and then on an IFP medium. In all cases ^{32}Pi (100 $\mu c/ml$) was present during the last culture period. A summary of the hormones present in "Medium 199" during the three incubations of the various groups follows: group 1, IF-IF-IFP; group 2, IF-IFP-IFP; group 3, IF-FP-FP; group 4, IF-FP-IFP. At the end of the third incubations, phosphoprotein was isolated from the explants in the presence of rennin, Ca^{++}, and mouse carrier casein. Rennin-Ca^{++}-induced precipitates from each group, normalized with respect to tissue weight, were electrophoresed on starch-urea gel and autoradiographed.

In the second type of experiment, the postmitotic explants were also divided into four groups. Group A was cultured on an F medium during both the second and third culture periods. Group B was incubated on an F medium and then on an IF medium. Group C was cultured on FP medium during both the second and third periods. Group D was incubated on FP medium and then on IFP medium. In all cases $^{32}P_i$ (20 $\mu c/ml$) and sufficient puromycin 5 x 10^{-4}M) to effect 99 percent inhibition of total protein synthesis were present during the final 4-hour incubation period. A summary of the hormones present during the three incubations of the various groups of explants follows: Group A, IF-F-F; group B, IF-F-IF; group C, IF-FP-FP; group D, IF-FP-IFP. At the end of the final incubations, phosphoprotein was isolated from the explants in the presence of rennin, Ca^{++}, and bovine carrier casein, and was counted in a scintillation spectrometer.

It was previously reported that mammary gland explants derived from 10- to 12-day-pregnant C3H/HeN mice lose their ability to synthesize three of the four major casein components after several days of culture on IF medium. It was also observed that casein synthesis reemerged in these postmitotic explants after exposure to both insulin and prolactin. The resurgence of casein production, however, represents a new activity within daughter cells formed in vitro rather than a resumption of synthesis by cells formed in vivo.

Figure 1 is an autoradiograph obtained from the first type of experiment described earlier. It can be seen (system 1) that when prolactin is present *only* during the last 4 hours of the culture period, casein bands 2, 3, and 4 are not synthesized, but casein band 1 and phosphoprotein at the origin and front are synthesized. Thus, it takes more than 4 hours of exposure to prolactin for the explants to produce casein components 2, 3, and 4. System 2 made all the mouse casein components; this is a confirmation of earlier results. System 3, which was not exposed to insulin during the entire postmitotic phase of the culture period, gave results similar to those obtained in system 1; i.e., casein components 2, 3, and 4 were not synthesized, but casein component 1 and phosphoprotein at the origin and front were synthesized. These results confirm the report that postmitotic formation of casein requires insulin as well as prolactin. Four-hour contact with prolactin is insufficient in terms of casein synthesis (system 1), but 4-hour exposure to insulin is enough (system 4).

From these results we can deduce the probable sequence of prolactin and insulin operation in the postmitotic induction of casein formation. More than 4 hours of prolactin treatment is required, but 4 hours of insulin treatment is sufficient.

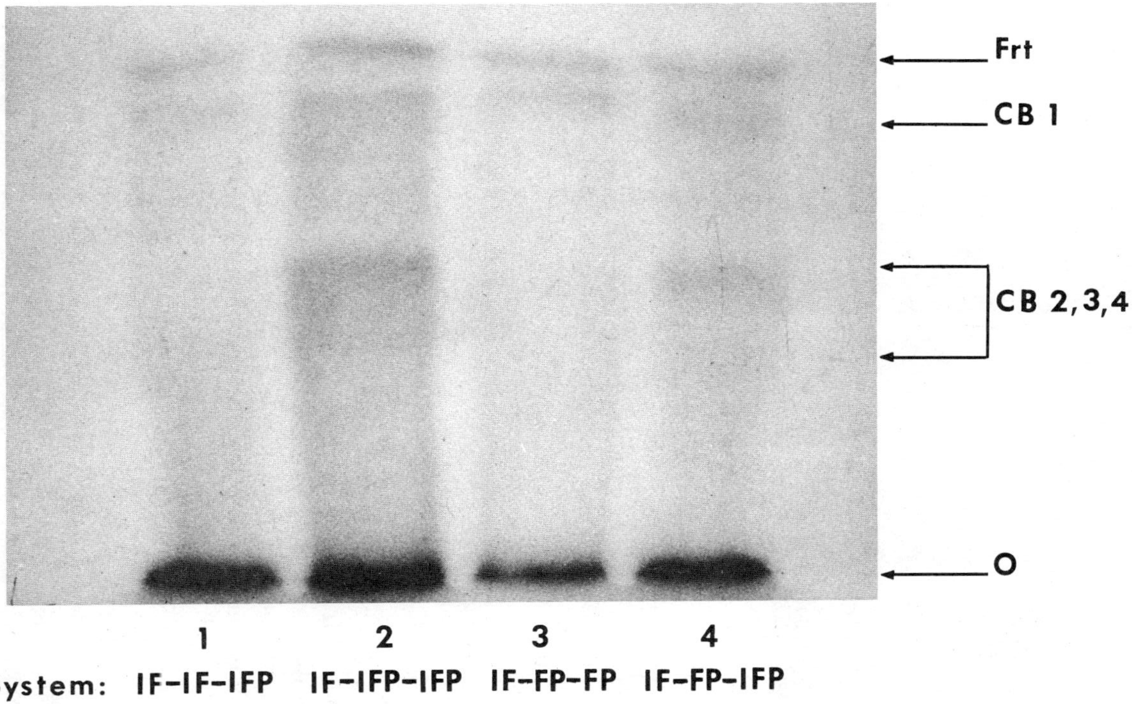

Fig. 1. Autoradiograph of ^{32}P-labeled phosphoproteins made by mammary gland explants derived from 10- to 12-day-pregnant C3H/HeN mice. Three culture periods were employed. In the first, all explants were incubated for 72 hr on "Medium 199" (in all cases the medium contained D-fructose instead of D-glucose) containing I and F (IF medium). The explants were then cultured during a second (48-hr) and third (4-hr) period as follows. Group 1 was cultured on a fresh IF medium (second incubation) and then on IFP medium. Group 2 was cultured on IFP medium during both the second and third incubations. Group 3 was cultured on FP medium during the second and third incubations. Group 4 was cultured on FP medium and then on IFP medium. In all cases ^{32}P (100 μc/ml: final specific radioactivity, 120 μc/μM) was present during the last culture period. At the end of the third incubations, phosphoprotein[5] was isolated from the explants in the presence of rennin, Ca^{++}, and mouse carrier casein.[6] Rennin-Ca^{++}-induced precipitates from each group, normalized to equal amounts of tissue, were electrophoresed on starch-urea gel and autoradiographed.[6] *I*=insulin; *F*=hydrocortisone; *P*=prolactin; *Frt*=front; *CB 1*=casein band 1; *CB 2, 3, and 4*=casein bands 2, 3, and 4; *O*=origin.

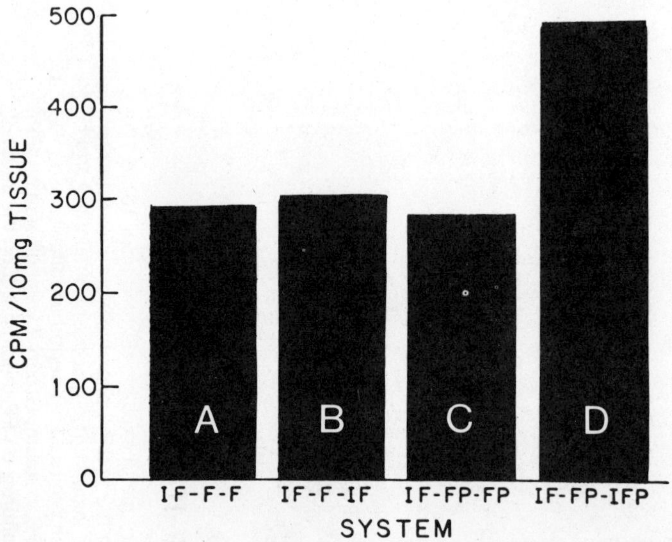

Fig. 2. Introduction of phosphate into preformed polypeptides. Three incubations were used. All explants were first (72 hr) cultured on IF medium. Group A was then incubated on F medium during both the second (48-hr) and third (4-hr) culture periods. Group B was incubated on F medium and then on IF medium. Group C was cultured on FP medium during both the second and third periods. Group D was incubated on FP medium and then on IFP medium. In all cases $^{32}P_i$ (20 μc/ml; final specific radioactivity, 24 μc/μM) and sufficient puromycin (5×10^{-4} M) to effect 99 percent inhibition of incorporation of ^{14}C-algal hydrolysate into TCA-precipitable protein were present during the final 4-hour incubation period. At the end of the third incubations, phosphoprotein[5] found in the soluble fraction of homogenized explants was isolated after 15 minutes of treatment with rennin, Ca++, and bovine casein, and was counted in a scintillation spectrometer. *I*=insulin; *F*=hydrocortisone; *P*=prolactin.

This suggests that prolactin acts before insulin. Since casein band 1 and phosphoprotein found at the origin and front were synthesized in all four systems, the relationships previously described apply only to casein bands 2, 3, and 4.

Figure 2 presents results obtained from the second type of experiment described earlier. Comparison of bars A and B reveals that when prolactin is absent during the second culture period, insulin does not produce an increment in phosphoprotein synthesis during the final culture period. It is clear from a comparison of bars C and D, however, that when prolactin is present during the second incubation, insulin subsequently elicits an increment in phosphoprotein formation. The increment was 32, 60, and 70 percent in three experiments, respectively. The results of these experiments indicate (1) that no phosphoprotein-phosphate exchange occurs in system D, and (2) that the phosphoproteins synthesized in systems A and C undergo neither peptide-bond nor phosphate-ester hydrolysis during a 4-hour period after their formation. The increment between systems C and D must, therefore, be due to increased incorporation of phosphate in system D. The possibility was considered that the increment is due to an increased transport of phosphate into the cell, or to an increased rate of conversion of inorganic phosphate into ATP.

This is extremely unlikely, however, because insulin did not increase phosphate incorporation in system B; i.e., the insulin effect is completely dependent upon prior exposure of the tissues to prolactin.

Since polypeptide synthesis has been blocked by puromycin in all four systems during the third culture period, phosphate groups are introduced into preformed pools of polypeptides. The phosphorylation of the pools formed in the absence of prolactin is not influenced by insulin (bars A and B). Phosphorylation of the pools is also not affected by prolactin in the absence of insulin (bar C). However, phosphorylation of the polypeptide pools formed in the presence of prolactin is enhanced by insulin (bar D). The sequence of action of prolactin and insulin previously described is consistent with these observations. A direct quantitative comparison of the results shown in Figures 1 and 2 is not meaningful, however, because in the first instance polypeptide synthesis occurred during the pulse with ^{32}Pi, whereas in the second instance such synthesis was blocked by puromycin.

It appears, then, that prolactin is responsible either for the sythesis of some polypeptides which are destined to become phosphoproteins, or for the synthesis of a unique protein kinase, or for both. In any case, insulin is required for participation of the prolactin-induced factor(s) in a protein-kinase reaction. This action of insulin is very specific, since the hormone had no effect on phosphorylation of protein by tissue not previously treated with prolactin.

6

Hormonal Dependence of Embryonic Neural Tissue in Culture*

Antonia Vernadakis
Departments of Psychiatry and Pharmacology,
University of Colorado Medical Center,
Denver, Colorado

INTRODUCTION

Studies by Vernadakis and Timiras have indicated that the acetylcholinesterase (AChE) activity of cerebellar explants removed from 16-day-old chick embryos and maintained in culture may depend on the presence of hormonal steroids in the culture medium(1). The present study explores further the sensitivity to hormonal steroids of neural tissues removed from chick embryos at different incubation periods and maintained in organ culture. The specific action of certain hormonal steroids on neural tissues in culture is also reported.

The activity of acetylcholinesterase (AChE) and protein content were used as indices of biochemical sensitivity of neural explants in culture to hormonal steroids. Because enzyme induction by some hormonal steroids in several tissues has been related to their action on RNA synthesis(2-5), RNA concentration was determined in explants cultured in the presence and absence of hormonal steroids.

Although hormonal dependence for biochemical growth of neural explants cannot directly explain hormonal dependence of in vivo developing systems, in vitro studies may provide important insights into the role of hormonal steroids in the growth of the central nervous system during embryonic development.

MATERIALS AND METHODS

Organ-Culture Procedure. Explants of cervical spinal cord removed from 9-day-old chick embryos and cerebellum removed from 16-day-old chick embryos

*This work was supported by the University of Colorado General Research Support Fund. The able technical assistance of Mrs. Judith Shearer is gratefully acknowledged.

were used. The cervical spinal cord was removed from the vertebral column by means of the "toothpaste tube" method of Pomerat et al.(6). The cerebellum was cut horizontally through the dorsal surface and one explant per cerebellum was removed. The explant contained two or three folia and extended from the molecular to the subcortical layer.

Explants were washed in Earle's balanced salt solution(7) and oriented on a triangular stainless-steel organ-culture grid(8). The ventral side of the spinal cord and the molecular surface of the cerebellum rested against the grid. Platforms with explant (one explant per grid) were placed in organ-culture dishes with a center well and an absorbent ring (Falcon Organ-Culture Dish). The medium (0.5 ml) was added to the center well of the organ-culture dish and did not reach the top of the platform. Humidity was maintained by saturating the absorbent ring with distilled water. The culture medium was Eagle's basal medium(9) with Earle's balanced salt solution, with or without hormones.

Explants were cultured at 35° to 36°C for 24 hours; the gas phase was 95 percent O_2, 5 percent CO_2 Freshly prepared culture media were equilibrated with the gas phase prior to transfer of the explants. Organ-culture dishes were stacked in the bottom portion of unsealed containers (plastic bread box, volume 4,260 ml.), and the gas was supplied at a constant flow to an inlet in the lid according to the method of Lostroh(10). A total of 10 experiments were performed and each experiment consisted of 24 to 32 explants: noncultured explants,* explants cultured on basal medium only, and explants cultured in the presence of each hormone.

Hormones. Cortisol, corticosterone, estradiol dipropionate,† and progesterone were prepared in aqueous suspension, 1 mg per ml of Eagle's basic medium solution, to which Tween 80 had been added (2.25×10^{-3} ml/100 ml Eagle's). The same concentration of Tween 80 was present in control cultures containing basal medium only.

Enzyme Analysis. Acetylcholinesterase (AChE)* activity was determined colorimetrically by means of a Beckman DU spectrophotometer, using the rate of hydrolysis of the substrate acetylthiocholine (AcTCh) according to the method of Ellman et al.(11). The enzyme determination was carried out at 37°C. Homogenates consisted of 1 mg of tissue per ml of 0.07 M phosphate buffer, pH 8.0, prepared with 0.07 M Na_2HPO_4, and 0.07 M KH_2PO_4. The final reaction mixture for determining AChE activity consisted of 2.0 ml pH 8.0 buffer, 1 ml homogenate, 100 μl (0.01 M) dithiobisnitrobenzoic acid (DTNB), and 20 μl acetylthiocholine iodide (0.075 M). Enzyme activity was expressed as micromoles of substrate hydrolyzed per minute per gram of wet tissue.

Protein Analysis. An aliquot of the tissue homogenate was utilized for pro-

*Noncultured explants represent fresh tissue removed and analyzed for the various biochemical parameters and served as controls to test the effects of culture condition.

†Cortisol and estradiol dipropionate were made available through the generosity of Ciba Pharmaceutical Company.

*The Commission of Enzymes of the International Union of Biochemistry has recommended that acetylcholine-acetylhydrolase (System No. 3.1.1.7) be the formal name, and that acetylcholinesterase be the trivial name for the enzyme having the higher affinity for acetylcholine than for any other ester. Acylcholine-acylhydrolase (System No. 3.1.1.8), or cholinesterase, is the term to be used for the enzyme(s) that hydrolyzes certain other esters of choline—i.e., butyrylcholine or propionylcholine—at a higher rate than acetylcholine.

tein analysis by the Lowry method(12) with bovine serum albumin used as standard. Calculations of total protein based on this determination were expressed as mg per gram of wet tissue.

Nucleic Acid Extraction. Nucleic acids were extracted according to the method of Schneider(13) as modified by Geel and Timiras(14).

RNA Analysis. Determinations of RNA from portions of the nucleic acid extract were made according to the orcinol procedure of Ceriotti(15) as modified by Geel and Timiras(14).

Statistical Methods. To determine whether the parameters measured in control and test explants differed significantly in their mean, the t test for nonpaired data was applied(16).

RESULTS

Cerebellar Explants. Both AChE activity and protein content were significantly lower in cerebellar explants cultured on basal medium than in noncultured explants (Tables 1 and 2, columns 2 and 4). AChE activity was higher in the explants cultured in the presence of either cortisol (Table 1), corticosterone (Table 2), or estradiol (Tables 1 and 2) (but not in the presence of progesterone) than in explants cultured on basal medium only. Protein content remained at the basal level in explants cultured in the presence of hormonal steroids in the medium.

RNA concentration remained at the noncultured level in the cerebellar ex-

TABLE 1

Acetylcholinesterase activity and protein content of cerebellar explants removed from 16-day-old chick embryos and maintained as organ cultures[a]

Culture medium[b]	Acetylcholinesterase activity (μmoles AcTCh hydrolyzed/min/g wet tissue)	Protein content (mg/g wet tissue)
Noncultured	5.33 ± 0.024[c]	54 ± 1
Basal	4.27 ± 0.21 (< 0.01)[d]	48 ± 1 (< 0.01)[d]
Cortisol	5.04 ± 0.23 (< 0.05)[e]	50 ± 2
Estradiol	5.13 ± 0.26 (< 0.02)[e]	47 ± 2

[a]Values from Vernadakis and Timiras. Experentia, 23:467, 1967.

[b]The medium was Eagle's basal medium with Earle's salts. Hormones, when added, were in the following concentrations: Cortisol 10 μg/ml, estradiol dipropionate 50 μg/ml.

[c]Each value represents the mean ± SE of 12-18 explants.

[d]Numbers in parentheses are P values for comparison to noncultured control group.

[e]Numbers in parentheses are P values for comparison to basal medium group.

TABLE 2

Acetylcholinesterase activity, RNA concentration, and protein content of cerebellar explants removed from 16-day-old chick embryos and maintained as organ cultures

Culture medium[a]	Acetylcholinesterase activity (μmoles AcTCh hydrolyzed/min/g wet tissue)	RNA concentration (μg/mg wet tissue)	Protein content (mg/g wet tissue)
Noncultured	6.50 ± 0.41[b]	5.08 ± 0.25	55 ± 3
Basal	5.41 ± 0.36 (< 0.05)[c]	4.82 ± 0.16	46 ± 2 (< 0.05)[c]
Cortisol	—	5.50 ± 0.18 (< 0.02)[d]	—
Corticosterone	7.90 ± 0.59 (< 0.02)[d]	4.94 ± 0.21	49 ± 2
Estradiol	7.40 ± 0.59 (< 0.01)[d]	5.64 ± 0.26 (< 0.02)[d]	44 ± 4
Progesterone	5.17 ± 0.58	—	46 ± 2

[a]The medium was Eagle's basal medium with Earle's salts. Hormones, when added, were in the following concentrations: Cortisol, 10 μg/ml; corticosterone, 10 μg/ml; estradiol dipropionate, 20 μg/ml; progesterone, 20 μg/ml.

[b]Each value represents the mean of 8-12 explants.

[c]Numbers in parentheses are P values for comparison to noncultured control.

[d]Numbers in parentheses are P values for comparison to basal medium group.

plants cultured on basal medium only (Table 2, column 3). In the explants cultured in the presence of either cortisol or estradiol but not corticosterone, RNA concentration was significantly higher than in explants cultured on basal medium only. The effect of progesterone on RNA concentration was not tested.

Spinal Cord Explants (9-Day-Old Chick Embryos). Both AChE activity and protein content were lower in spinal cord explants cultured on basal medium than in noncultured explants (Table 3). In spinal cord explants cultured in the presence of either cortisol, corticosterone, or estradiol, but not progesterone, AChE was higher than in explants cultured on basal medium only. Protein remained at the basal level in the explants cultured on medium containing hormones.

RNA concentration remained at the noncultured level in the spinal cord explants cultured on basal medium only. RNA was significantly lower in explants cultured in the presence of either cortisol, corticosterone, or estradiol (progesterone was not tested) than in explants cultured on basal medium only

Spinal Cord Explants (16-Day-Old Chick Embryos). Both AChE activity and protein content were lower in spinal cord explants cultured on basal medium than in noncultured explants (Table 4). In spinal cord explants cultured in the presence of either cortisol or estradiol, both AChE activity and protein content remained at the basal level.

TABLE 3

Acetylcholinesterase activity, RNA concentration, and protein content of spinal cord explants removed from 9-day-old chick embryos and maintained as organ cultures

Culture medium[a]	Acetylcholinesterase activity (μmoles AcTCh hydrolyzed/min/g wet tissue)	RNA concentration (μg/mg wet tissue)	Protein content (mg/g wet tissue)
Noncultured	8.83 ± 0.36[b]	4.46 ± 0.27	49 ± 2
Basal	5.97 ± 0.28 (< 0.001)[c]	5.13 ± 0.27	42 ± 2 (< 0.05)[c]
Cortisol	7.52 ± 0.70 (< 0.05)[d]	4.07 ± 0.24 (< 0.05)[c]	37 ± 2
Corticosterone	7.21 ± 0.39 (< 0.02)[d]	3.80 ± 0.54 (< 0.05)[c]	37 ± 2
Estradiol	7.98 ± 0.52 (< 0.01)[d]	4.35 ± 0.15 (< 0.05)[c]	45 ± 1
Progesterone	5.99 ± 0.56	—	40 ± 2

[a]The medium was Eagle's basal medium with Earle's salts. Hormones, when added, were in the following concentrations: Cortisol, 10 μg/ml; corticosterone, 10 μg/ml; estradiol dipropionate, 20 μg/ml; progesterone, 20 μg/ml.

[b]Each value represents the mean of 8-12 explants.

[c]Numbers in parentheses are P values for comparison to noncultured control.

[d]Numbers in parentheses are P values for comparison to basal medium group.

TABLE 4

Acetylcholinesterase activity and protein content of spinal cord explants removed from 16-day-old chick embryos and maintained as organ cultures[a]

Culture medium[b]	Acetylcholinesterase activity (μmoles AcTCh hydrolyzed/min/g wet tissue)	Protein content (mg/g wet tissue)
Noncultured	22.77 ± 1.31[c]	51 ± 3
Basal	19.74 ± 1.22 (< 0.05)[d]	35 ± 4 (< 0.01)[d]
Cortisol	18.58 ± 2.01	39 ± 4
Estradiol	16.49 ± 2.01	36 ± 4

[a]Values from Vernadakis and Timiras. Experentia, 23:467, 1967.

[b]The medium was Eagle's basal medium with Earle's salts. Hormones, when added, were in the following concentrations: Cortisol, 10 μg/ml; estradiol dipropionate, 50 μg/ml.

[c]Each value represents the mean ± SE of 12-18 explants.

[d]Numbers in parentheses are P values for comparison with noncultured control group.

DISCUSSION

Acetylcholinesterase activity, protein content, and RNA concentration were the indices used to test the sensitivity of neural explants to organ-culture conditions and the influence of hormonal steroids thereon. AChE activity was lower in all explants cultured on basal medium than in noncultured explants. This decrease may be attributed to a general decrease in protein synthesis as indicated by the lower protein content found in these explants. AChE activity was maintained at the noncultured level in cerebellar explants removed from 16-day-old chick embryos and in spinal cord explants removed from 9-day-old chick embryos when these explants were cultured in the presence of either cortisol, corticosterone, or estradiol (see Tables 1 and 2). Previous studies have shown that AChE activity remains at the basal level in spinal cord explants removed from 16-day-old chick embryos and cultured in the presence of either cortisol or estradiol (Table 1). It appears, therefore, that hormonal sensitivity may be dependent on the stage of maturation of the neural tissue.

Assuming that the so-called blood-CNS barrier is absent under in vitro conditions, uptake of hormones would be expected to be similar in explants removed from chicks of all ages. The possibility, however, that these hormonal steroids may be altered extraneuronally in the mature neural structures and thus not reach their locus of action cannot be excluded from these findings. Studies in vivo have shown that the metabolism of cortisol varies in different CNS structures (cerebral cortex, cerebellum, brainstem, hypothalamus) of newborn and adult primates (17,18). Glial cells play an important role in the regulation of extraneuronal environment and neuronal function(19). Inasmuch as morphological form, functional activities, and metabolic requirements of glial cells may differ among CNS areas during development and aging(20,21), the differential hormonal sensitivity of neural tissue in culture may be a function of the glial cells.

AChE activity has been used as evidence for the presence of acetylcholine(22), a proposed neurotransmitter(23). Changes, therefore, in AChE activity have been interpreted as representing changes in synaptic transmission. The present findings suggest that synaptic elements in the process of maturation may be dependent on specific hormones for maintenance of their activity in organ culture. Hormonal influence on cholinergic systems has also been demonstrated in developing rats made hypothyroid neonatally. Lack of physiological thyroid hormone reduces AChE activity in the cerebral cortex of developing rats(24).

Maintenance of AChE activity by these hormones is interpreted as a specific effect of these steroids on AChE synthesis rather than as a general effect on protein or RNA synthesis. This interpretation is based on the fact that protein content was not altered in the presence of any of the hormones studied. Furthermore, RNA concentration was lower in the spinal cord from 9-day-old chick embryos cultured in the presence of these hormonal steroids, and corticosterone did not alter RNA concentration but increased AChE activity in cerebellar explants. On the other hand, the differential action of these steroids on RNA concentration of the two CNS structures is consistent with other findings that steroids activate RNA synthesis in some tissues and depress RNA synthesis in others(25).

Progesterone did not influence AChE activity or RNA concentration in any of the neural explants; corticosterone had no effect on RNA concentration of the cerebellar explants but reduced RNA concentration in spinal cord explants from 9-day-old chicks. These findings indicate a hormonal tissue specificity which is substantiated by in vivo studies. For example, deoxycorticosterone exerts no effect on the functional development of the CNS as assessed by responses to brain electroshock stimulation, whereas cortisol and estradiol increase sensitivity of the developing CNS to electrical stimulation(26). Estradiol increases the Na and Cl content only in the developing cerebellum, whereas cortisol increases the Na and Cl content only in the spinal cord(27).

On the basis of the present in vitro findings and previous in vivo observations (27), it is postulated that hormonal steroids may be control factors for the growth of specific neural systems during early CNS development. Whether steroid hormones regulate neuronal-glial mechanisms, influence the extracellular environment, or both, remains to be investigated.

REFERENCES

1. Vernadakis, A., and Timiras, P. S. Effects of estradiol and cortisol on neural tissue in culture. Experentia, 23:467, 1967.
2. Feigelson, P., and Feigelson, M. Metabolic effects of glucocorticoids in various target organs as related to enzyme induction. *In* International Congress of Endocrinology, Amsterdam, Excerpta Medica Foundation International Congress Series No. 83, pp. 57–59, 1965.
3. Kidson, C. Kinetics of cortisol action on RNA synthesis. Biochem. Biophys. Res. Commun., 21:283–289, 1965.
4. Ryan, J. J., Meigs, R. A., Petro, Z., and Morrison, G. Estrogen-induced 16-hydroxysteroid dehydrogenase activity in rat kidney. Science, 142:243–244, 1963.
5. Kirk, D. L., and Moscona, A. A. Synthesis of experimentally induced glutamine synthetase (glutamotransferase activity) in embryonic chick retina in vitro. Develop. Biol., 8:341–357, 1963.
6. Pomerat, C. M., Droger, G. A., and Painter, J. Effect of some barbiturates on tissues in vitro. Proc. Soc. Exp. Biol. Med., 63:322–325, 1946.
7. Earle, W. R. Production of malignancy in vitro. IV. The mouse fibroblast cultures and changes seen in living cells. J. Nat. Cancer Inst., 4:165–212, 1943.
8. Fell, H. B. Recent advances in organ culture. Sci. Progr., 162:212–231, 1953.
9. Eagle, H. Nutrition needs of mammalian cells in tissue culture. Science, 122:501–504, 1955.
10. Lostroh, A. J. Synthetic activity in mouse organs maintained on chemically defined media. Acta. Endocr., 47:331–337, 1964.
11. Ellman, G. L., Courtney, K. D., Andres, V., and Featherstone, R. M. A new standard colorimetric determination of acetylcholinesterase activity. Biochem. Pharm., 7:88–95, 1961.
12. Lowry, O. H., Rosenbrough, N. J., Farr, A. L., and Randall, R. J. Protein measurement with the Folin phenol reagent. J. Biol. Chem., 193:265–275, 1951.
13. Schneider, W. C. Phosphorus compounds in animal tissues. I. Extraction and estimation of desoxypentose nucleic acid and pentose nucleic acid. J. Biol. Chem., 161:293–303, 1945.
14. Geel, S. E., and Timiras, P. S. The influence of neonatal hypothyroidism and of thy-

roxine on the ribonucleic acid and deoxyribonucleic acid concentrations of rat cerebral cortex. Brain Res., 4:135–142, 1967.
15. Ceriotti, G. Determinations of nucleic acids in animal tissues. J. Biol. Chem., 214:59–70, 1955.
16. Fisher, R. A. Statistical Methods for Research Workers. New York, Hafner, 1950.
17. Grosser, B. I., and Axelrod, L. R. Conversion of cortisol to cortisol acetate, cortisone acetate and cortisone by the developing primate brain. Steroids, 11:827–836, 1968.
18. Purdy, R. H., Grosser, B. I., and Axelrod, L. R. The distribution of corticosteroid-21-O-acetyltransferase in the baboon brain. Steroids, 11:837–850, 1968.
19. Galambos, R. Glial Cells. *In* Neurosciences Research Symposium Summaries, Vol. 1. Schmitt, F. O., and Melnechuk, T., eds. Cambridge, Massachusetts, The M.I.T. Press, pp. 375–436, 1966.
20. Vernadakis, A., and Woodbury, D. M. Cellular and extracellular spaces in developing rat brain: Radioactive uptake studies with chloride and insulin. Arch. Neurol., 12:284–293, 1965.
21. Timiras, P. S., Vernadakis, A., and Sherwood, S. Development of plasticity of the nervous system. *In* Biology of Gestation, Vol. 2. Assali, N. S., ed. New York, Academic Press, Inc., 1968.
22. Koelle, G. G. A new general concept of the neurohumoral functions of acetylcholine and acetylcholinesterase. J. Pharm. Pharmacol., 14:65–90, 1962.
23. Eccles, J. C. The Physiology of Synapses. New York, Academic Press, Inc., 1964.
24. Geel, S. E., and Timiras, P. S. Influence of neonatal hypothyroidism and thyroxine on the acetylcholinesterase and cholinesterase activities in the developing central nervous system of the rat. Endocrinology, 80:1069–1074, 1967.
25. Feigelson, M., and Feigelson, P. Metabolic effects of glucocorticoids as related to enzyme induction. *In* Advances in Enzyme Regulation, Vol. 3. Weber, G., ed. pp. 11-27, 1965.
26. Vernadakis. A., and Timiras, P. S. Regulation of brain and spinal cord excitability by cortisol and estradiol in developing rats. *In* Proceedings of the 2nd International Congress on Hormonal Steroids. Amsterdam, Excerpta Medica Foundation, International Congress Series No. 132, pp. 908–912, 1966.
27. Valcana, T., Vernadakis, A., and Timiras, P. S. Influence of estradiol and cortisol on electrolytes in the central nervous system of developing rats. Neuroendocrinology, 2: 326–329, 1967.

7

The Pars Distalis and Avian Development*

T. W. Betz
Department of Biology, Carleton University,
Ottawa, Canada

INTRODUCTION

For the study of developmental endocrinology, avian embryos offer several advantages: The pars distalis and pars nervosa do not fuse during development(1), some of the cells of the pars distalis are distributed in separate regions of the gland (2-8), and the techniques are simple. The avian egg is a fairly clean experimental system because only thyroxine(9-11) and insulin(12) have been found in the yolk in small amounts and estrogens may or may not be present(13-16). More importantly, no one has reported that pars distalis hormones occur in eggs.

The major disadvantage of using these embryos is that purified avian pars distalis hormones are not readily available. Thus, one must use natural sources of trophic hormones for replacement therapy, such as pars distalis grafts which produce several different kinds of hormones, or preparations from other vertebrates in which the problem of apparent species specificity may be encountered. This has been true of mammalian somatotrophins.

The most practical way to "hypophysectomize" avian embryos(17) is to decapitate them partially at stage 10 or 11(18) at about 33 hours of incubation. This involves resection of the head from the midmesencephalon forward(19). These parts are not regenerated, so the embryos lack several structures of ordinary endocrinological importance—the pars nervosa, epiphysis, hypothalamus, and the pars distalis, as well as the upper beak and eyes(19).

*The triocephalic embryos produced by x-irradiation of the anterior half of the head(21,23) are supposed to lack the pars distalis and do not, except for the x-ray lesion, develop any abnormalities during incubation(22,24,25). Stoll and Ancel(21,22,25) looked at serial sections to verify the absence of the pars distalis but did not state how many of the embryos were checked or if the embryos in which the absence of the pars distalis had been verified were the same ones they examined for end organ defects. The only rational explanation is that the partes distales had not been completely suppressed. Maraud et al.(33) have described defects in the development of the thyroid glands of embryos which were "hypophysectomized" by x-rays. Thus, this method can be effective if it is properly controlled. This fact and the high resistance of pars distalis tissue to x-rays lend credence to this argument.

Even though partial decapitation has no apparent effect on the early development of avian embryos(20-28), there are many somatic and endocrine defects(19, 26-32) which are characteristic of later development.* Generally, the rate of development is decreased but duodenal development is arrested(30,34) whereas development of the spleen may be accelerated(35). Also, the rate of maturation of the extraembryonic membranes is subnormal(19,30) and the mortality of these embryos is greatly increased. The few that survive do not pip, hatch, or retract the yolk sac, but they remain in the eggs days beyond the normal time of hatching(19,30,32). In rare instances, the operated embryos may survive for as many as 11 days beyond hatching(36). These defects become more numerous and severe during development, and they eventually result in the death of the embryo. Thus, partial decapitation prolongs the fetal period indefinitely.

It has not been clear whether the defects which occur after partial decapitation are due solely to adenohypophyseoprivia and not in part to the absence of the secretions of the pars nervosa, epiphysis, or hypothalamus(17,19,30,32). However, embryonic pars distalis homografts alleviate or prevent the defects in these embryos. This indicates that other parts of the head do not play important roles in development. Since the effects of partial decapitation would be the same as the effects of hypophysectomy, the terms may be used interchangeably.

There is a reduction in the amount of albumin ingested by partially decapitated embryos(37). The lack of an upper beak interferes with swallowing(37), and the dietary restriction and hormonal deprivation adversely affect the body and liver (35) growth in these embryos. These are the only known defects induced by partial decapitation which are not consistently prevented by replacement therapy with pars distalis grafts.

One of the sources of disparity in reports on the development of avian embryos is their perverse variability, and many investigators have not used clear notations of the length of incubation. Hamburger and Hamilton's series(18) of normal stages, based on morphological criteria rather than the length of incubation, is a great help in reducing the variability of chick embryos at a particular point in development. But many workers still do not use this series and they should consider its merits. Another method for describing the developmental stages of chick embryos is the weight-length index of normality developed by Walker(38). This is a refinement similar to Hamburger and Hamilton's series, and though it is easier to use, for some reason it has not been adopted. There is a need for a study of the Walker body weight-length index of chick embryos at the various stages of Hamburger and Hamilton's series. Davis and Garrison(39) reported the mean body weights of chick embryos at the stages of Hamburger and Hamilton's series. They weighed the embryos after they had been fixed in Bouin's fluid and stored in 70 percent ethanol. If the time in

*I am grateful to the National Research Council of Canada for supporting my research program and the travel grant which enabled me to attend the conference on Hormones in Development. Thanks are due our excellent hosts at the University of Nottingham for their many kindnesses. Parts of the manuscript were written during my tenure as a NIH postdoctoral fellow with Dr. A. E. Wilhelmi of Emory University, Atlanta, Georgia. Dr. David R. Gardner and my wife Britt rendered good advice about the final format. My thanks to Mr. Rak and Mr. Brasch for translating the papers by Sedláček and Nevoral (Ref. 31) and by Mess et al. (Ref. 162).

storage was varied, the body weights probably changed considerably since the weights of 20-day-old chick embryos (stage 46–) can change as much as 3 g during one year(36). Also, the percentage of body weight lost during storage may change with the age of the embryo.

The following is a review of the relationship between the pars distalis hormones and the fetal end organs and some of the effects of hormones on avian embryos.

THE PARS DISTALIS

The first signs of cytological differentiation in the cells which line Rathke's pouch appear on the 3rd day of incubation(40) rather than on the 6th day(1,3,41, 42). Perhaps as early as the 8th day(3,43), but at least by the 10th day(2,3,41,44, 45), of incubation the regional distribution of the pars distalis cells has become apparent. The cells occur in two regions-the cephalic and caudal. The evidence for this is based on normal cytology, on the cytological changes which occur if the end organs are removed, and on bioassay of the regions. Apparently, the cells which produce ACTH, FSH, and TSH(6,7,46) only occur in the cephalic region(4-6,46), and those which produce LH and STH are confined to the caudal region(6,44,46-48). MSH occurs in both regions but is more concentrated in the cephalic region(49-51). Prolactin cells may occur only in the cephalic region as indicated by cytological evidence (1,6). However, another cytological study(8) and regional bioassay(52) indicate their presence in both regions. These results do not permit a final decision regarding the distribution of prolactin cells in the avian pars distalis. The basis for the maintenance of this distribution of cells is not known. But it does not depend on the hypothalamus, and the cells of one region do not influence the cells of the other(46).

There are some indications of the time when hormone synthesis begins in the pars distalis, but we do not know exactly when the trophic hormones are first released or when the titers of these hormones first become high enough to stimulate the end organs. The MSH activity first appears in the pars distalis on the 5th day of incubation(50), and Tixier-Vidal(53) has demonstrated that a TSH activity is present by day 6. If 3-day-old pars distalis rudiments are isolated in vitro, they are able to differentiate this TSH activity even in the absence of the hypothalamus(54). The secretion of ACTH may begin on the 8th day of incubation(55), and the first signs of hyperglycemia induced by hypophysectomy(56) are evident. Chorioallantoic grafts of 10-day-old partes distales show a TSH activity in intact hosts(28), and at the same time the titer of this TSH activity attains a level which stimulates the thyroid glands (53). On day 11, hypophysectomy experiments(27) indicate that a TSH is systemically active because the thyroid glands show the first signs of nonsupport and chorioallantoic transplants of thyroid glands(57) show signs of stimulation indicating initial production of a TSH.

On the 13th day of incubation, the gonads of hypophysectomized embryos first undergo a reduction in weight, and the ovarian cortex and interstitial cells of the testis show an apparent decrease in width and in number(27). As evidenced by changes in steroid dehydrogenase activity in the gonads, recent results(58) indicate that the pituitary-gonadal axis matures by day 13.5 of development. Apparently, a gonadotrophic hormone is released from the pars distalis at least by this time. The adenohypophysis attains the definitive vascular pattern on day 13 of incubation, which is supposed to be a morphological prerequisite for function(42). Case(59) reported reductions in weight and ascorbic acid content of adrenal glands of hypophysectomized chick embryos on day 13. There is also an increase in the rate of accumulation of ^{131}I by the thyroid glands(60), a degranulation of the AF-positive material in the pars distalis cells, and a concomitant decrease in basophilia(2,40,61). A growth-promoting activity, which may be due to STH, appears on the 15th day of incubation in the caudal region of the pars distalis(47). Beginning on day 16, the injection of partes distales stimulates the adrenal cortex and gonads of prepubertal guinea pigs (62). The improved resolution of new techniques has enabled investigators to place progressively the time of inception of the functional differentiation of the pars distalis earlier in development. However, the functional significance of these structural and biochemical changes which occur during development is still to be elucidated.

In partially decapitated embryos with pars distalis grafts (which are histologically normal), the end organs are not over-stimulated even if there are two or three grafts present(19,36,37). The way in which this apparent regulation is accomplished is not known, but feedback inhibition of the secretion of trophic hormones by the hormones of the end organs may be responsible, or perhaps the grafts mutually inhibit their own hormone production. Nevertheless, this phenomenon occurs in the absence of the hypothalamus, which indicates that the hypothalamus is not involved in a significant way. In addition, the ability of the grafted embryonic pars distalis to secrete a TSH is not reduced in the absence of the hypothalamus(63). However, if 16-day-old partes distales are cultivated for three days with hypothalamic tissue of the same age, the beta (FSH) and gamma (TSH) cells are richer in PAS-positive material. Thus, the hypothalamus may stimulate these cells, but it apparently does not induce the release of gonadotrophins because testes which are cultivated with partes distales or with hypothalamic tissue and partes distales show an equal degree of stimulation(64). Other interpretations are also possible.

In post-hatching birds, the synthesis of gonadotrophins apparently depends on the hypothalamus(65-67). In chickens this dependency develops at least as early as 7 weeks of age(66), but in the embryo it has not developed by the 10th day of incubation(19) and we now know that it does not begin until sometime after hatching (36). The electrophoretic species of soluble proteins in the pars distalis cells at the time of hatching are almost the same as those of adult chickens(68), which indicates that the pars distalis can synthesize hormones sometime before hatching.

Apparently, the pars distalis cells which produce ACTH, FSH, LH, MSH, and TSH differentiate during the first half of development, and the cells which produce STH differentiate during the latter half of development. The hypothalamus is not necessary to maintain the differentiation of these cells. We do not know when the prolactin cells differentiate because we do not know of any structure or process in chick embryos which is sensitive to prolactin.

To date, no one has biochemically characterized the factors in the avian embryonic pars distalis. This kind of information would do much to exclude the possibility that the embryonic pars distalis only secretes a single developmental hormone which is necessary for normal avian development. At this point, one should only speak of trophic activities rather than of hormones.

GROWTH

Thyroxine or adrenocorticoids may stimulate(69,70) or retard(19,71,72) the rate of growth and development of chick embryos, depending on the size of the dose and the age of the embryo when treated. There is a concomitant decrease or increase in the length of the fetal period. Presumably, ACTH has the same effect by stimulating the production of adrenocorticoids(71), and one expects that TSH should also affect the growth of chick embryos by stimulating the synthesis of endogenous thyroxine. However, Fugo(27) reported that TSH had no effect on the growth of hypophysectomized embryos. Perhaps these embryos will not increase their growth without the addition of STH, prolactin, ACTH, and albumen. Goitergens(19) and radio necrosis(11) of the thyroid glands retard growth and prolong the fetal period. Strangely enough, if chick embryos are exposed to light during incubation, the rate of growth and development is stimulated and they hatch as much as 24 hours early (73). The weight of these light-treated birds is the same as the controls. Thus, they do not grow more; they just grow faster. The injection of obstetrical pituitrin into embryonated eggs can increase the body weight of chick embryos(74). The increase in body weight is probably caused by the retention of water due to the ADH activity of the preparation, rather than by growth.

Bovine STH, contrary to "positive" reports(75,76), has not been shown to have a significant effect on the growth of chick embryos(77,78), which may be due to a species specificity problem, or perhaps the rate of growth of the embryos was already so high that exogenous STH could not stimulate it further. Other interpretations are possible. Enemar(47) demonstrated by bioassay that a growth promoting activity occurs in the caudal region of the pars distalis by the 15th day of incubation. This activity may be due to STH, since the only other hormones presumably in this region are LH and MSH. Sobel(79) demonstrated that bovine STH alleviates the dwarfing effect of cortisone in chick embryos. One could explain the alleviation by the growth promotion of STH or perhaps by a nonspecific binding of some of the cortisone to the STH. The latter may be the case if the growth rate of the embryos was maximal at the time of cortisone treatment. If so, the STH could not increase the rate. On the other hand, perhaps it was possible to see the effect of STH only after the cortisone had decreased the rate. Hazelwood and Hazelwood (80) demonstrated that crude extracts of the partes distales of chickens have a weak STH activity in rats. Hirsch obtained similar results in rats(81) and chickens (82) which were hypophysectomized. According to Cavallero(83), glucagon is able to cause an increase in the weight of chick embryos. However, Elrick et al.(84), using a more highly purified preparation, could not confirm this result. Hirsch(82) was unable to duplicate the results of Schooley et al.(85), Hublé(86), and Bates et al.(87) which indicated that prolactin can promote growth in birds. Additional evidence that prolactin may not be the normal growth hormone of birds is provided by the results of Enemar(47). He demonstrated that the cephalic region of the pars distalis inhibits the growth of the *Rana temporaria* tadpoles in his bioassay(88), and the only hormones presumably in this region are ACTH, FSH, MSH, TSH, and perhaps prolactin. Recently, Meier et al.(89) have pointed out some difficulties in the copr-sac bioassay for prolactin. Definitive experiments in this area have not been done, probably because there are so many factors which may affect embryonic and fetal growth.

Recent results(90) demonstrate that the topical application of less than 1.6 nanogram of chromatographically pure porcine growth hormone with no other pars distalis activity to intact chick embryos from 33 to 72 hours of incubation results in significant increases in the dry weight and protein content of the body by about 84 hours of incubation (stage 21). The dose is so low that one can exclude a nutritional effect. The increase in dry body weight probably reflects an increase in the amount of protein in the body, since embryos of this age do not have depot fat stores. Thus, embryonic chick growth is not refractory to growth hormone stimulation. More recently, bovine STH was found to behave the same way in chick embryos(36). In both cases, the somites of the treated embryos were increased in size but not in number, which indicates that to some extent the mechanism of somite formation is independent of growth. Thus, there is not a species specificity problem, at least concerning the activity of bovine and porcine growth hormones in chick embryos.

Hinni and Watterson(30) did not report an increase in the growth of hypophysectomized chick embryos which were injected with thyroxine or cortisone, either alone or together. This indicates that one or more additional hormones may be necessary for normal growth and that these remain to be determined. Partial decapitation severely reduces the growth of chick embryos beginning on the 12th day of incubation, and on the 20th day (stage 46–) the embryos have only gained 45 percent of the normal body weight(19,27,37,59,78,91). Normal embryos ingest the albumen in the egg during incubation(92), but partial decapitates which lack an upper beak only swallow one-fifth of the normal amount of albumen(37). Thus, the dwarfing of these embryos may be caused by either hormonal deprivation or dietary restriction, or both may contribute to the reduction in growth. Witschi(93) pointed out that the ingested albumen contributes to normal body weight, and in partial decapitates the reduced ingestion contributes to the lower body weight.

If pars distalis grafts are implanted in partial decapitates, the growth is considerably improved. However, they still gain only 68 percent of the normal body weight(17,19,37), and if the guts are removed before the embryos are weighed, it does not significantly alter the differences between the partial decapitates and the normal embryos.

If the upper beak rudiment(17,37) is resected from otherwise intact embryos, or if the albumen(37) is removed from the eggs, the embryos only grow as much as the partial decapitates with grafts. Thus, one can duplicate the degree to which growth is reduced in partial decapitates with pars distalis grafts by restricting the albumen intake of normal embryos. The reduced growth of the embryos with grafts is probably not due to hormonal deprivation, since the endocrine end organs (adrenal and thyroid glands) and the grafts appear normal(19) and the upper beakless and albumenless embryos have intact partes distales. The increase in size of partial decapitates may indicate the secretion STH or prolactin by the grafts. Nalbandov(94) and Simpson et al.(95) argued that growth in vertebrates depends on STH and that other hormones and dietary intake only play permissive roles in the process. But recently King(96) has shown that thyroxine can produce limited growth in hypophysectomized post-hatching chickens.

However, Johnson(32) offered the alternative hypothesis that the improved growth of partial decapitates with grafts is not due to STH but to the improved albumen ingestion caused by the increased muscular strength and activity of these embryos. Perhaps the increase in strength and activity occurs because of increased levels of thyroxine and adrenocorticoids stimulated by TSH and ACTH from the grafts. A recent study demonstrates that partial decapitates with grafts in which the esophageal opening had been sealed grew significantly more than untreated partial decapitates, even though they did not ingest any albumen(37). It is clear that the improved growth of these embryos was caused by something produced by the grafts and not by albumen ingestion alone, perhaps STH or prolactin, since thyroxine and adrenocorticoids either alone or together may not be sufficient to promote the growth of hypophysectomized chick embryos(30). It has been demonstrated that chorioallantoic grafts of nonendocrine tissues can cause minor but significant dwarfing of otherwise intact chick embryos(97). This effect may be partially responsible for the incomplete repair of the growth of hypophysectomized chick embryos which receive chorioallantoic pars distalis grafts(19,37). The extent to which this and other factors, such as surgical trauma, affect growth in chick embryos is not known. The details of the ingestion of albumen and the utilization of the proteins of the albumen and yolk by chick embryos has been reported and reviewed by Carinci and Manzoli-Guidotti(98).

Albumen is not essential for the survival of chick embryos, as reported by Fritz(99), but it is necessary for normal growth. I am currently studying the effect of injections of albumen and albumen hydrolysates on the growth of partial decapitates with grafts. If the above interpretation is correct, we expect to restore the normal growth of these embryos but, if not, then the grafts for some reason do not produce normal amounts of a STH, perhaps due to a lack of hypothalamic GRF (see McCann et al.(100) for a review). If this is the case, it will represent the first known instance in which a hypothalamic releasing factor is important for the production of a pars distalis hormone in chick embryos.

THE ADRENAL CORTEX

The first signs of histochemical differentiation (dehydroepiandrosterone or pregnenolone) of the adrenal glands of chick embryos occur on the 5th day of incubation(101); on the 6th day aliesterase activity appears(102), and the cortical cells attain the ultrastructure of steroidogenic cells(103). The ultrastructural manifestations of the effects of ACTH and metopirone on the cortical cells has been reported (104). The cortical cells first become cytologically differentiated (osmiophilia, sudanophilia-positive Schiff, Schultz and carbonyl response of Ashbel-Seligman, and birefringence) on the 7th day of incubation(105). Ascorbic acid and cholinesterase activity appear on the 10th day(102,106), as well as cholesterol and neutral lipids (107). The ultrastructure of the cortical cells has been described on the 12th day of incubation(108). Delta^5-3-hydroxysteroid dehydrogenase and delta^5-3-ketoisomerase activity first appear on day 14(108). Adrenal glands from chick embryos begin to synthesize corticosterone, 11-deoxycorticosterone, and aldosterone on day 14 (109), rather than on day 16(110) when they also synthesize progesterone(108).

However, Straznicky et al.(108) reported that corticosterone synthesis does not occur until after hatching. It is hard to construct a clear developmental sequence out of the above information, because many of the papers reporting this information are too abbreviated. Also, the functional significance of these events is not clear.

Most evidence indicates that the adrenocortical-pars distalis axis may mature between the 12th and 14th day of incubation. In 12- to 14-day-old chick embryos, an insulin-induced hypoglycemia causes a reduction in the weight and cholesterol content of the adrenal glands(111). Also on day 13, intravenously administered insulin results in a depletion of adrenal gland ascorbic acid(112). Moog(113), Moog and Ford(71), and Ford(114) showed that daily intravenous injections of ACTH into intact embryos produce an increase in adrenal gland weight by the 12th day, and by the 14th day the lipid content of the glands is increased over the controls. Mazina (115) confirmed and extended these observations by reporting that ACTH causes a slight increase in cholesterol on day 11, followed by a reduction during the period from 12 to 14 days. There was only a slight increase in the weight of the adrenal glands at this time. Freeman(116) has demonstrated that histamine, as well as ACTH, causes ascorbic acid depletion on the 19th day of incubation.

Case(59) reported that hypophysectomy causes a decrease in the lipid content of the adrenal glands of chick embryos and that the ascorbic acid level does not exceed the normal 12-day-old value. In hypophysectomized embryos, the ultrastructural and enzymatic differentiation does not proceed beyond the normal level of 6 days(108). ACTH injections on day 13 restore these glands to normal(108), and pars distalis grafts have the same effect(19,46).

Recent evidence may indicate that ACTH secretion begins on the 8th day of incubation, and that there is a sudden increase in the synthesis or release of this hormone on the 12th day of incubation(55). As yet, there has not been an unequivocal demonstration of an ACTH in avian embryos.

THE GONADS

Weniger(117) and Weniger and Zeis(118) demonstrated that sexually indifferent gonads of 5-day-old chick embryos synthesize both estrone and estradiol after 24 hours in vitro. Feminized testes of chick embryos and normal duck ovaries also secrete these steroids(119).

The gonads begin to form the agranular reticulum which is characteristic of the ultrastructure of steroidogenic cells on the 6th day of incubation(120), and by 6.5 days a formazan reaction appears(121,122).

As early as the 7th day, rather than the 9th day(120,123), the left ovary secretes estrone and estradiol in organ culture(124). At this time the cells of the ovary have large vesicles and large amounts of lipid(120). If pregnenolone is used as a substrate, steroid dehydrogenase activity appears in the cells of the medullary sex cords. Weniger(125) reported that the left ovaries of 12-day-old chick embryos were able to secrete epiestriol when cultured in vitro. On day 13, testosterone appears in the ovary(101), followed by estriol on day 14(126).

In the testes on day 8, the vesicles in the seminiferous cord cells increase in size

(120), and on the 9th day steroid dehydrogenase activity appears(101). There is an increase in lipoids(120) and in the size of the vesicles in the interstitial cells for the first time(120). But the cholesterol(120) and pregnenolone-sensitive steroid dehydrogenase activity(101) do not appear until day 10. The Leydig cells appear on day 12 and have attained their characteristic ultrastructure at least by the 18th day(127). Testosterone first occurs in the interstitial cells on day 19, but as yet there is no estradiol-17-β(101). Haffen and Cèdard(128) have studied the ontogeny of gonadal steroid synthesis in correlation with steroidogenic enzyme histochemistry. However the functional significance of these changes is not apparent.

The evidence indicates that sexual differentiation in birds does not depend on pars distalis hormones. But the time at which the gonads switch from trophic-hormone independence to dependence is not known(129), although it is probably on day 13(127) or 13.5(58) of development. At least the gonads of hypophysectomized chick embryos show the first indications of subnormal development at these times. So far, there has not been an unequivocal demonstration of a gonadotrophin in avian embryos. In post-hatching birds, the interstitial tissue of the testis is stimulated by LH and the seminiferous cords by FSH, and there is no apparent synergism between the two hormones(130,131). There has been considerable progress in the purification of avian gonadotrophins(132). The important observation has been made(133) that both the seminiferous tubule cells and the interstitial cells have steroidogenic enzymes.

There is some evidence for a pars distalis gonadotrophin. Extracts of 18-day-old partes distales of chick embryos stimulated follicular maturation in prepubertal guinea pigs(62). If several glands from 16- and 17-day-old embryos are cultured in vitro in contact with testes of the same age, they cause enlargement and lumenogenesis of the seminiferous cords and an increase in the number of gonial mitoses (134). Unfortunately, these studies do not tell us when the hormone is released from the pars distalis in amounts which affect the gonads. But recent evidence indicates that the pituitary-gonadal axis matures at 13.5 days of development(58).

It is possible to increase the weight of the gonads by injecting ovine pars distalis extracts(135,136) or gonadotrophin preparations into intact(137-139) or hypophysectomized chick(140) and duck embryos(32). This increase is due to an enlargement of the seminiferous cords and an apparent increase in the number of interstitial cells. Cèdard et al.(141) have demonstrated that human chorionic gonadotrophin stimulates estradiol and estrone synthesis by the ovaries of 10- and 18-day-old chick embryos and by the testes of 18-day-old embryos when cultivated in vitro. Cholesterol synthesis is also stimulated by HCG in the 10 day gonads as well(142). These studies show us that the gonads are sensitive to exogenous gonadotrophins, but tell us little of the normal role of endogenous gonadotrophins.

By contrast, van Deth et al.(143) reported that chorioallantoic pars distalis grafts or injections of FSH and LH preparations had no apparent effect on the gonads of intact chick and duck embryos. The failure of the grafts to stimulate the gonads is perhaps predictable, because if several grafts are placed in partial decapitates, they do not over-stimulate the end organs(19,35). But the reason for the failure of the FSH and LH preparations to stimulate the gonads is not clear.

Hypophysectomy of chick(19,27,140) and duck embryos(32,143) reduces the weight and the amount of the various tissue components of the gonads. In the testes of the chick, there is an apparent reduction in the amount of interstitial tissue(19, 27,46,140) and in the size of the seminiferous cords(19,46), and in the duck there is a decrease in the degree of convolution of the seminiferous cords(32). In the left ovary, the thickness of the cortex and the amount of medullary tissue are reduced (19,27,46,140); the follicles are fewer and smaller(140), but the cholesterol content is not subnormal(144). Recent evidence indicates that basal levels of steroid synthesis by the gonads are independent of the pars distalis(145). Hypophysectomy has no apparent effect on the rudimentary right ovary(17,27,32).

van Deth et al.(143) reported that hypophysectomy had no effect on the gonads of either chick or duck embryos, and Johnson(32) could not decide if the amount of interstitial tissue was actually reduced in the testes of duck embryos. In regard to the controversy of whether or not the amount of interstitial tissue is decreased by hypophysectomy, van Deth et al.(143) argued that workers have failed to compare sections from comparable levels of the gonads. However, this was done in more recent work(17,19), and Johnson(32) claimed that the histological differences between the testes of normal and partially decapitated embryos are not particularly striking, and that this is a subjective judgement. Admittedly, this is a subjective judgement, as seen by the discussion of Domm's paper(136), but at least in the testes of hypophysectomized chick embryos, the interstitial cells appear to be reduced in number(17,19,27,46,140) and the seminiferous cords are significantly smaller(46). Johnson(32) pointed out that although the gonads of partial decapitates are reduced in size compared with normal, the degree of reduction is only slightly greater than that of the body. Perhaps, one would expect proportionately smaller components in a smaller gonad. Thus, it is difficult to determine the significance of a reduction in size of gonadal components in partial decapitates. Recent results(46) settle this argument by indicating that cephalic and caudal pars distalis region grafts have different gonadotrophic effects on the testes and ovaries of hypophysectomized chick embryos. Perhaps the cephalic region secretes FSH and the caudal region secretes LH. Nevertheless, the sizes of the bodies of the embryos with cephalic grafts and of those with caudal grafts were not significantly different, yet the differences in testicular and ovarian morphology in the two groups were striking.

A quantitative analysis of the reduction of the components of the testes and steroid titers of partial decapitates must be made before a decision can be reached concerning the role of gonadotrophins in avian embryos.

Perhaps the best evidence so far that the embryonic pars distalis secretes a gonadotrophin which is necessary for normal ovarian function is that injections of PMS prevent the masculinization of the genital tubercle and syrinx of partially decapitated genetic female duck embryos(32,146,147).

THE THYROID GLANDS

The excellent reviews by Dieterlen-Lièvre(148) and Thommes(149) should be consulted. The development of the thyroid glands has been well-studied(40,150-156). As early as day 5 in the chick(157), the thyroids begin to accumulate selectively ^{131}I, and by day 6 of incubation the glands consist of epithelial cords surrounded by blood sinusoids. The alpha cytomembranes first appear in the ultrastruture as flat vesicles in the cytoplasm, but there are some signs of secretory activity (150). On day 7 the amount of ^{131}I which is taken up is greater than that by other tissues(158,159). By day 8 the first signs of lumenogenesis(160) and of PAS-positive colloid droplets appear(40), and the sinusoidal network has reached its maximal development(149). At 8.5 days, the glands begin to synthesize monoiodotyrosine(157), followed by diiodotyrosine on the 9th day. The follicles and thyroxine appear on the 10th day(157), and the capillaries of the gland begin to form(149). If 10-day-old thyroid glands are grafted into amphibian larvae, they cause precocious metamorphosis(161).

By the 11th day, the alpha cytomembranes are very numerous, and they have formed large vesicles which become extracellular and form thyroid follicles(150). The uptake of ^{131}I reaches high levels(157,162), and the curve plateaus on day 12 (158). The reports on ^{131}I metabolism after this stage of development are so variable that they are confusing(163-166). By day 14, the ultrastructure of the thyroid is the same as that found in adults(160).

The equivocal evidence for a TSH in the pars distalis is based on the results of several different experiments. Partial decapitation of chick(19,27,46,57,162,167) and duck embryos(32,146) retards the growth of the thyroid glands and interferes with the formation of colloid. There is also a permanent reduction in the uptake of ^{131}I to the basal 8-day(162,167,168) or 10-day(33) level, and the ultrastructure does not exceed the normal day 8 morphology. The period of rapid mitosis is prolonged and there is a severe reduction in colloid synthesis(162). The definitive vascular pattern does not develop(149). Most of these defects are prevented either by pars distalis grafts(19,46,168) or injections of TSH(27,162,168).

The thyroid glands of normal and partially decapitated embryos, both before and after stage 37 (day 11) when TSH secretion apparently starts(168), are sensitive to exogenous TSH(169). If thyroid glands from partial decapitates are grafted onto normal or hypophysectomized hosts, they are stimulated to the same degree as the thyroid glands of the host(57).

The results of in vitro studies are highly variable. If 8-day-old normal thyroid glands are explanted in culture, they differentiate normally(170,171), and after 8 days they take up ^{131}I. But the medium used in these experiments was not ahormonal because of the chicken-plasma content. Thyroid glands do not do well in ahormonal culture, since they only synthesize diiodotyrosine and thyroxine for about 2 days(170-173). Eight-day-old thyroid glands differentiate to an 11-day-old stage, but not beyond(174,175), unless the medium is enriched. One sees completely normal differentiation only if pars distalis tissue is present(53). The importance of trophic factors in the differentiation of the thyroid gland in vitro has been studied(176), and there is a direct correlation between the richness of the medium and the degree of differentiation. If various organs (pancreas, liver, parathyroid, hypophysis) are added to poor media, then the thyroid glands differentiate very well, apparently because of a nonhormonal trophic effect. Thus, thyroid glands can differentiate without the pars distalis.

Some of the effects of endogenous thyroxine on avian embryos are indicated by experiments with thiourea and thiouracil(177-182), which apparently block the production of thyroxine by the thyroid glands. These drugs cause a decrease in the rate of development and delay the onset of hatching by as much as 10 days(178). The earlier the drugs are given, the greater the retardation(178). These agents also inhibit the rate of growth(179). The proportions of the body and the shape of the growth curve are normal. Apparently, a decrease in metabolic rate is responsible for the reduced growth, rather than a particular sensitivity of certain organs(182). The cells of the thyroid follicles show typical signs of hypertrophy and hyperplasia (180). This also occurs in embryos from eggs laid by hens which had been treated with a goitergen(177), and also in cases where the hens were fed thyroprotein(183, 184). In the latter case, the reasons for this seemingly paradoxical effect are unknown. If thyroxine is given at the same time as thiourea, the effects of the drug are neutralized(178).

The effects of exogenous TSH and thyroxine have been studied. If thyroid glands from adult chickens are grafted onto the chorioallantois of normal embryos, typical signs of hyperthyroidism develop(185). The same is true if TSH is injected (186).

A TSH activity is present in the pars distalis at least by the 6th day of incubation, as demonstrated by the stimulation of juxtahypophyseal implants of thyroid glands from 11-day-old partial decapitates placed into 3.5- to 4-day-old intact hosts (187). Also, 6-day-old partes distales stimulate thyroid glands if they are explanted together in vitro(53).

If thyroxine is injected before eggs are incubated, the metabolic rate of the embryos is reduced(69). However, if it is given on day 4 of incubation, the growth rate is improved and there is a concomitant increase in oxygen consumption. These embryos hatch precociously by as much as 36 hours. On the basis of RQ (respiratory quotient) measurements, the embryos normally use carbohydrates from the beginning of incubation to day 6 as a source of energy. During the period from 6 to 10 days, they switch to lipids. Thyroxine causes an increase in oxygen uptake which starts on the 4th day, and by the 7th day the embryos use as much oxygen as 8-day-old normal embryos. On the 13th day, the rate of respiration is 25 percent greater than normal. The body weight is increased by the same amount. The drop in the RQ is more rapid than normal, indicating that the embryos begin to use lipids on day 9 rather than waiting until day 10(188). Thus, the effect of thyroxine during the first 5 days lasts until the end of incubation. There is some indication that thyroxine of maternal origin in the egg assures the normal growth of chick embryos prior to the secretory maturation of the embryonic thyroid glands(189). If this is the case, then it is the only known instance in which a maternal hormone directly participates in normal avian development.

In summary, functional differentiation of the endocrine end organs apparently occurs before the trophic hormones of the pars distalis begin to be released, which indicates that the trophic hormones are not necessary for the differentiation of the end organs. However, the maintenance of the normal rate of growth and development and the termination of the fetal period apparently depend on normal titers of end organ hormones which are produced only if the end organs are adequately stimulated by the trophic hormones of the pars distalis. The trophic hormones may also directly affect the growth and development of avian embryos.

ADDENDUM

Since this manuscript was completed, it has been found that the subnormal development of the leg muscles of partially decapitated chick embryos is significantly alleviated by chorioallantoic pars distalis grafts(190). Also, the ultrastructural changes in the development of the left ovary of the chick embryo from the 4th to the 21st day of incubation has been described(191). There is also direct evidence for separate hypothalamic releasing factors for FSH and LH in the hen(192,193).

REFERENCES

1. Wingstrand, K. G. The Structure and Development of the Avian Pituitary, Lund, C. W. K. Gleerup. 1951.
2. Rahn, H. The development of the chick pituitary with special reference to the cellular differentiation of the pars buccalis. J. Morph., 64:483–517, 1939.
3. Wilson, M. E. The embryological and cytological basis of regional patterns in the definitive epithelial hypophysis of the chick. Amer. J. Anat., 91:1–49, 1952.
4. Mikami, S. The cytological significance of regional patterns in the adenohypophysis of the fowl. J. Fac. Agric. Iwate Univ., 3:473–545, 1958.
5. ——— The structure of the hypothalamo-hypophysial neurosecretory system in the fowl and its morphological changes following adrenalectomy, thyroidectomy and castration. J. Fac. Agric. Iwate Univ., 4:359-379, 1960.
6. Tixier-Vidal, A. Histophysiologie de l'adénohypophyse des oiseaux. *In* Cytologie de l'adénohypophyse, pp. 255–274. Eds. J. Benoit and C. Da Lage, Paris, Édition du C.N. R.S., 1963.
7. Wingstrand, K. G. Comparative histology of the adenohypophysis of birds. *In* Cytologie de l'adénohypophyse, pp. 243–256. Eds. J. Benoit and C. Da Lage, Paris, Édition du C.N.R.S., 1963.
8. Yasuda, M. Cytological studies of the anterior pituitary in the broody fowl. Jap. Acad. Proc., 29:586–594, 1953.
9. Roche, J., Michel, O., Michel, R., and Marois, M. Sur les modalités de la concentration du radioiode (I^{131}) dans le vitellus de l'oeuf de poule et sur la formation ovarienne de combinaisons iodées. C.R. Soc. Biol., 145:1833–1836, 1951.
10. Blanquet, P., Stoll, R., Maraud, R., Mounier, J., and Meyniel, G. Sur la concentration du radioiode dans l'ovaire de la poule. C.R. Soc. Biol., 151:104–107, 1957.
11. Mraz, F. R. Embryonic response to iodine-131. Poult. Sci., 45:96–99. 1966.
12. Shikinami, Y. Material from different animal substances, especially egg yolk, tending to produce hypoglucemia. Tohoku J. Exp. Med., 10:1–25, 1928.
13. Riboulleau, J. Teneur en substances oestrogènes de l'oeuf vierge et de l'oeuf fécondé des oiseaux. C.R. Soc. Biol., 129:914–916, 1938.
14. Marlow, H. W., and Richert, D. Estrogens of the fowl. Endocrinology, 26:531–534, 1940.
15. Hertelendy, F., and Common, R. H. A chromatographic investigation of egg yolk for the presence of steroid estrogens. Poult. Sci., 44:1205–1209, 1965.
16. Mitskevich, M. S. Hormonal interrelations between mother and fetus in mammals. Gen. Comp. Endocr., Suppl. 1:300–308, 1962.
17. Betz, T. W. Correction of defects induced by "hypophysectomy" in chicken embryos by grafts of embryonic anterior pituitary glands. Ph.D. dissertation, University of Illinois, 1965.
18. Hamburger, V., and Hamilton, H. L. A series of normal stages in the development of the chick embryo. J. Morph., 88:49–92, 1951.

19. Betz, T. W. The effects of embryonic pars distalis grafts on the development of hypophysectomized chick embryos. Gen. Comp. Endocr., 9:172–186, 1967.
20. Thomas, E. Innersekretorische Drüsen bei Feten und Kindern. *In* Handbuch der inneren Sekretion, 2:1291–1410, Ed. M. Hirsch. Leipzig, Curt Kabitzsch, 1933.
21. Stoll, R. Sur les conditions nécessaires pour obtenir des monstres triocéphales expérimentaux avec agénésie de l'hypophyse et de la thyroïde. C.R. Soc. Biol., 126:704–706, 1937.
22. ——— L'agénésie de l'hypophyse et de la thyroïde est sans influence sur le développement de l'embryon de poulet. C.R. Soc. Biol., 130:926–928, 1939.
23. Wolff, E., and Kantor, S. Conditions dans lesquells on réalise l'agénésie de l'hypophyse chez les cyclocéphales obtenus expérimentalement. C.R. Soc. Biol., 126:707–709,1937.
24. ——— and Stoll, R. Le rôle de l'hypophyse dans le développement embryonnaire du poulet, d'apres l'étude des cyclocéphales expérimentaux. C.R. Soc. Biol., 126:1215–1217, 1937.
25. Ancel, P. Sur le rôle de l'hypophyse et de la thyroïde dans le développement embryonnaire de vertébrés amniotes. Actualités Sci. Ind., 603:185–200, 1938.
26. Fugo, N. W., and Witschi, E. Removal of the hypophyseal primordium in the chick during the second day of incubation. Acta Biol. Latvica., 8:73–80, 1938.
27. ——— Effects of hypophysectomy in the chick embryo. J. Exp. Zool., 85:271–297, 1940.
28. Studitskii, A. N. Endocrine correlation in the embryonal development of the vertebrates. Nature, 157:427–430, 1946.
29. Benoit, J. A. A. Le role de l'hypophyse dans de développement des embryons de vertébrés amniotes à sang chaud. Ann. Biol., 1:297–325, 1962.
30. Hinni, J. B., and Watterson, R. L. Modified development of the duodenum of chick embryos hypophysectomized by partial decapitation. J. Morph., 113:381–426, 1963
31. Sedláček, J., and Nevoral, J. Vývoj kuřecích plodů po dekapitaci. Sborník Lekařsky, 67:313–320, 1965.
32. Johnson, N. J. Some effects of "hypophysectomy" of white pekin duck embryos and replacement therapy with pregnant mares' serum. Ph.D. dissertation, University of Illinois, 1966.
33. Maraud, R., Stoll, R., and Blanquet, P. Sur le rôle de l'hypophyse dans la concentration de radioiode par la thyroïde de l'embryon de poulet. C.R. Soc. Biol., 151:572–574, 1957.
34. Bellware, F. T., and Betz, T. W. The dependence of duodenal differentiation in chick embryos on pars distalis hormones. J. Embryol. Exp. Morph. (In press).
35. Betz, T. W. The developmental endocrinology of the spleen in chick embryos. I. The pars distalis. J. Embryol. Exp. Morph. (In press).
36. ——— Unpublished data.
37. ——— The effects of embryonic pars distalis grafts and albumen on the growth of chick embryos. J. Embryol. Exp. Morph., 20:431–436, 1968.
38. Walker, P. A. Identification of normal stages in chick embryos. Growth, 2:145–147, 1938.
39. Davis, J. E. Jr., and Garrison, N. E. Mean weights of chick embryos correlated with the stages of Hamburger and Hamilton. J. Morph., 124:79–82, 1968.
40. Phillips, J. Evidence of early pituitary function in the white leghorn chick. Anat. Rec., 143:69–75, 1962.
41. Aronsson, J. Studies on the cell differentiation in the anterior pituitary of the chick embryo by means of the PAS reaction. Acta Univ. Lund., 48: No.11, 1–20, 1952.
42. Thommes, R. C., and Russo, R. P. Vasculogenesis in the adenohypophysis of the developing chick embryo. Growth, 23:205–219, 1959.
43. Mosier, H. D. The development of the hypothalamo-neurohypophysial secretory system in the chick embryo. Endocrinology, 57:661–669, 1955.
44. Rahn, H., and Painter, B. T. A comparative histology of the bird pituitary. Anat. Rec., 79:297–311, 1941.
45. Payne, F. The cytology of the anterior pituitary of the fowl. Biol. Bull., 82:79–111, 1942.

46. Brasch, M., and Betz, T. W. The hormonal activities associated with the cephalic and caudal regions of the avian pars distalis. (In press).
47. Enemar, A. Ontogeny of the hypophysial growth-promoting activity in the chick. J. Endocr., 37:9–15, 1967.
48. Grignon, G. Développement du complexe hypothalamo-hypophysaire chez l'embryon de poulet. Nancy, Société d'Impressions Typographiques, 1956.
49. Kleinholz, L. H., and Rahn, H. The distribution of intermedin: A new biological method of assay and results of tests under normal and experimental conditions. Anat. Rec., 76: 157–172, 1940.
50. Chen, G., Oldham, F. K., and Geiling, E. M. K. Appearance of the melanophore-expanding hormone of the pituitary gland in developing chick embryo. Proc. Soc. Exp. Biol. Med., 45:810–813, 1940.
51. Groenendijk-Huijbers, M. M. Development of black down feathering in hybrid chick embryos (cross new hampshire ♂ X light sussex ♀) after pituitary implantation. Experientia, 24:501–503, 1968.
52. Nakajo, S., and Tanaka, K. Prolactin potency of the cephalic and caudal lobe of the anterior pituitary in relation to broodiness in the domestic fowl. Poult. Sci, 35:990-994, 1956.
53. Tixier-Vidal, A. Etude chronologique *in vivo* et *in vitro* des corrélations hypophyse-thyroïde chez l'embryon de poulet. Arch. Anat. Micr. Morph. Exp., 45:236–253, 1956.
54. Le Douarin, G., and Ferrand, R. Différentiation fonctionnelle de l'ébauche épithéliale de l'adénohypophyse isolée du plancher encéphalique: activité thyréotrope. C.R. Acad. Sci. (Paris), 266:697–699, 1968.
55. Szedély, By., Endröczi, E., and Szentágothai, J. Erscheinen der argentaffinen Substanz und Beginn der Sekretion adrenocorticotropen Hormons in der Hypophyse von Hühnerembryonen. Acta Biol. Hung., 8:283–287, 1958.
56. Konigsberg, I. R. The effects of early pituitary removal by "decapitation" on carbohydrate metabolism in the chick embryo. J. Exp. Zool. 125:151–169, 1954.
57. Martindale, F. M. Initiation and early development of thyrotrophic function in the incubating chick. Anat. Rec., 79:373–393, 1941.
58. Woods, J. E., and Weeks, R. L. Ontogenesis of the pituitary-gonadal axis in the chick embryo. Gen. Comp. Endocr., 13:242–254, 1969.
59. Case, J. F. Adrenal cortical-anterior pituitary relationships during embryonic life. Ann. Ann. N.Y. Acad. Sci., 55:147–158, 1952.
60. Maraud, R., Stoll, R., and Blanquet, P. Sur le rôle de l'hypophyse dans la concentration du radioiode par la thyroïde de l'embryon de poulet. C.R. Soc. Biol., 151:572–574, 1957.
61. Moscona, H., and Moscona, A. The development in vitro of the anterior lobe of the embryonic chick pituitary. J. Anat., 86:278–288, 1952.
62, Moszkowska, A. Pouvoir corticotrope et gonadotrope de l'hypophyse de l'embryon de poulet. C.R. Soc. Biol., 143:1332–1333, 1949.
63. Mess, B., and Straznicky, K. Reactivity of the subcutaneous pituitary graft in chicken embryos treated with propylthiouracil. Acta Biol. Acad. Sci. Hung., 20:185–192, 1969.
64. Moszkowska, A. Étude *in vitro* de l'hypothalamus et de l'activité préhypophysaire gonadotrope chez l'embryon de poulet. C.R. Soc. Biol., 152:725–728, 1958.
65. Benoit, J. Hypothalamo-hypophyseal control of the sexual activity in birds. Gen. Comp. Endocr., Suppl. 1:254–274, 1962.
66. Ma, R. C. S., and Nalbandov, A. V. Hypophyseal autotransplants and the function of testes, thyroids, and adrenals in birds. *In* Advances in Neuroendocrinology, pp. 306–311. Ed. A. V. Nalbandov. Urbana, Univ. Ill. Press, 1963.
67. Assenmacher, I., and Tixier-Vidal, A. Hypothalamic-pituitary relations. Excerpta Medica Inter. Congr. Ser., 83:131–145, 1964.
68. Manwell, C., and Betz, T. W. The effect of embryonic partial decapitation on the developmental sequences of some proteins in the chicken. J. Embryol. Exp. Morph., 16: 83–89, 1966.
69. Beyer, R. E. The effect of thyroxine upon the general metabolism of the intact chick embryo. Endocrinology, 50:497–503, 1952.

70. Moog, F., and Richardson, D. The functional differentiation of the small intestine. IV. The influence of adrenocortical hormones on differentiation and phosphatase synthesis in the duodenum of the chick embryo. J. Exp. Zool., 130:29–55, 1955.
71. ——— and Ford, E. Influence of exogenous ACTH on body weight, adrenal growth, duodenal phosphatase, and liver glycogen in the chick embryo. Anat. Rec., 128:592, 1957.
72. Badran, A., and Provenza, D. V. Studies of the growth-inhibitory action of cortisone on chick embryos. Teratology, 2:221–234, 1969.
73. Lauber, J. K., and Shutze, J. V. Accelerated growth of embryo chicks under the influence of light. Growth, 28:179–190, 1964.
74. Cunningham, B., and Stanfield, W. Some effects of the injection of pituitrin into hen's eggs. J. Elisha Mitchell Sci. Soc., 44:63–65, 1928.
75. Hsieh, K. M., Wang, T. Y., and Blumenthal, H. T. The diabetogenic and growth-promoting activities of growth hormone (somatotropin) in the developing chick embryo. Endocrinology, 51:298–301, 1952.
76. Wang, T. Y., Hsieh, K. M., and Blumenthal, H. T. The effect of growth hormone on the nucleic acid content of the developing chick embryo. Endocrinology, 53:520–526, 1953.
77. Libby, D. A., Meites, J., and Schaible, P. J. Growth hormone effects in chickens. Poult. Sci., 34:1329–1331, 1955.
78. Vogel, N. W. Growth in chick embryos hypophysectomized by "decapitation." Proc. Penn. Acad. Sci., 39:57–60, 1965.
79. Sobel, H. Antagonistic effects of cortisone and growth-hormone on the developing chick embryo. Proc. Soc. Exp. Biol. Med., 97:495–498, 1958.
80. Hazelwood, R. L., and Hazelwood, B. S. Effects of avian and rat pituitary extracts on tibial growth and blood composition. Proc. Soc. Exp. Biol. Med., 108:10–12, 1961.
81. Hirsch, L. J. Unpublished results, 1955.
82. ——— A study on the growth promoting principle in the domestic fowl. Ph.D. dissertation. University of Illinois, 1961.
83. Cavallero, C. Internal secretions of pancreas. *In* Ciba Foundation Colloquia on Endocrinology. Vol 9. p. 266, Boston, Little, Brown and Co. 1956.
84. Elrick, H., Konigsberg, I. R., and Arai, Y. Effect of glucagon on growth of chick embryo. Proc. Soc. Exp. Biol. Med., 97:542–544, 1958.
85. Schooley, J. P., Riddle, O., and Bates, R. W. Analysis of pituitary support of growth of body and viscera in pigeons. Anat. Rec. 72(Suppl.):90–91, 1938.
86. Hublé, J. Effects of cortisone-acetate on chondrogenesis and ossification in cockerels. Acta Endocr., 25:59–63, 1957.
87. Bates, R. W., Miller, R. A., and Garrison, M. M. Evidence in the hypophysectomized pigeon of a synergism among prolactin, growth hormone, thyroxine and prednisone upon weight of the body, digestive tract, kidney and fat stores. Endocrinology, 71:345–360, 1962.
88. Enemar, A., Essvik, B., and Klang, R. Growth-promoting effects of ovine somatotropin and prolactin in tadpoles of *Rana temporaria*. Gen. Comp. Endocr., 11:328–331, 1968.
89. Meier, A. H., Burns, J. T., and Dusseau, J. W. Seasonal variations in the diurnal rhythm of pituitary prolactin content in the white-throated sparrow, *Zonotrichia albicollis.* Gen. Comp. Endocr., 12:282–289, 1969.
90. Betz, T. W., and McKnight, R. O. The growth promoting activity of porcine growth hormone in chick embryos. Amer. Zool., 9:1090, 1969
91. Love, D. S., and Konigsberg, I. R. Enhanced desoxyribonucleic acid accumulation and retarded protein accumulation in skeletal muscle of "hypophysectomized" chick embryos. Endocrinology, 62:378–384, 1958.
92. Witschi, E. Utilization of the egg albumen by the avian fetus. Contrib. Ornithol Biol. Wissensch. (Heidelberg), Winter, 111–112, 1949.
93. ——— Comments in discussion of A. Raynaud. *In* Comparative Endocrinology, p. 476, Ed. A. Gorbman, New York, John Wiley and Sons, 1959.
94. Nalbandov, A. V. Symposium on growth: Endocrine causes of growth and growth stasis. J. Anim. Sci., 22:558–560, 1963.

95. Simpson, M. N., Asling, C. W., and Evans, H. M. Some endocrine influences on skeletal growth and differentiation. Yale J. Biol. Med., 23:1–27, 1950.
96. King, D. B. Effect of hypophysectomy of young cockerels, with particular reference to body growth, liver weight, and liver glycogjn level. Gen. Comp. Endocr., 12:242-255, 1969.
97. van Alten, P. J., and Fennell, R. A. The effects of chorioallantoic grafts on the developing chick embryo. I. Studies on weight and histology of homologous and heterologous tissues. J. Embryol. Exp. Morph., 7:459–475, 1959.
98. Carinci, P., and Manzoli-Guidotti, L. Albumen absorption during chick embryogenesis. J. Embryol. Exp.Morph., 20:107–118, 1968.
99. Fritz, H. I. Studies on the role of the white in the development of the chick embryo. Ph. D. dissertation. University of California at Davis, 1964.
100. McCann, S. M. Dhariwal, A. P. S., and Porter, J. C. Regulation of the adenohypophysis. Ann. Rev. Physiol., 30:589–640, 1968.
101. Boucek, R. J., Györi, E., and Alvarez, R. Steroid dehydrogenase reactions in developing chick adrenal and gonadal tissues. Gen. Comp. Endocr., 7:292–303, 1966.
102. Sivaram, S. Histochemical studies on the developing adrenal gland of *Gallus domesticus*. Histochemie., 12:316–325, 1968.
103. Grignon, G., Hatier, R., Guedenet, J. C., and Dollander, A. Aspects ultrastructuraux de la glande surrénale au cours du développement chez l'embryon de poulet. C.R. Soc. Biol., 160:1654–1657, 1966.
104. ——— Hatier, R., Guedenet, J. C., and Dollander, A. Étude ultrastructurale de la réponse des cellules de la cortico-surrénale à un traitement par l'ACTH ou la métopirone chez l'embryon de poulet et chez le poussin. J. Microscopie, 6:58a–59a, 1967.
105. Dawson, A. B. Histochemical evidence of early differentiation of the suprarenal gland of the chick. J. Morph., 92:579–595, 1953.
106. Sivaram, S. Histochemistry of the developing adrenal gland of the chick. Anat. Rec., 148:336, 1964
107. ——— Nature and distribution of lipids in the developing adrenal cortex of *Gallus domesticus*. Acta Histochem., 32:253–261, 1969.
108. Straznicky, K., Hajòs, F., and Bohus, B. Relationship between the ultrastructure and cortical activity of the embryonic adrenal gland in the chicken. Acta Biol. Hung., 16: 261–274, 1966.
109. Bonhommet, M., and Weniger, J. P. Sécrétion de corticostéroides par la surrénale embryonnaire de poulet cultivée *in vitro*. C.R. Soc. Biol., 161:2052–2055, 1967.
110. ——— and Weniger, J. P. Sécrétion de cojticostérone par les surrénales de l'embryon de poulet cultivées *in vitro*. C.R. Acad. Sci. (Paris), 264:3034–3036, 1967.
111. Mazina, T. I. Adrenocortical reaction to insulin administration in the chick embryo. Physiol. Zh. SSSR, 49:873–878, 1963.
112. ——— Ascorbic acid content in adrenal glands of chick embryos in ACTH and insulin administration. Biull. Eksp. Biol. Med., 55:54–58, 1964.
113. Moog, F. The development of function in the adrenal cortex. *In* Comparative Endocrinology, Ed. A. Gorbman, New York, John Wiley and Sons, 1959.
114. Ford, E. A study of adrenal weights of the chick embryo under normal conditions and after injections of ACTH. M.A. thesis, Washington University, 1957.
115. Mazina, T. I. The influence of ACTH administration on the weight of adrenals and their cholesterol content in the developing chick embryo. Physiol. Zh. SSSR, 49:589–595, 1963.
116. Freeman, B. M. Depletion of ascorbic acid from the adrenal of the intact embryo of *Gallus domesticus* by adrenocorticotrophic hormone or histamine. Comp. Biochem. Physiol., 24:905–914, 1968.
117. Weniger, J-P. Sur la précocité de la sécrétion d'oestrogènes par les gonades embryonnaires de poulet cultivées *in vitro*. C.R. Acad. Sci. (Ser. D), 266:2277–2279, 1968.
118. ——— and Zeis, A. Formation d'oestrone et d'oestradiol radioactifs, à partir d'acétate de Na-1-^{14}C, par les ébauches gonadiques d'embryon de poulet de 5 à 6 jours. C.R. Acad. Sci. (Ser. D), 268:1306-1309, 1969.

119. Akram, H., and Weniger, J-P. Sécrétion d'oestrone et d'oestradiol par les gonades embryonnaires d'oiseaux. Gen. Comp. Endocr., 12:568–573, 1969.
120. Narbaitz, R., and Adler, R. Submicroscopical observations on the differentiation of chick gonads. J. Embryol. Exp. Morph., 16:41–47, 1966.
121. Scheib, D., and Haffen, K. Étude histochimique de la 3-β-hydroxystéroïdedéshydrogénase des jeunes gonades embryonnaires de poulet. C.R. Acad. Sci. (Paris), 264:161–164, 1967.
122. ——— and Haffen, K. Apparition et localisation des hydroxystéroide déshydrogénases (Δ_5-3β et 17β) dans les gonades de l'embryon et du poussin de la caille (*Coturnix coturnix japonica*). Etude histoenzymologique et comparaisons avec le poulet (*Gallus gallus domesticus*). Gen. Comp. Endocr., 12:586–597, 1969.
123. Ozon, R. Mise en évidence d'hormones stéroides oesgrogènes dans le sang de la poule adulte et chez l'embryon de poulet. C.R. Acad. Sci. (Paris), 261:5664–5666, 1965.
124. Weniger, J-P., Ehrhardt, J-D., and Fritig, B. Sur la formation d'oestrone et d'oestradiol par les gonades de l'embryon de poulet femelle cultivées *in vitro.* C.R. Acad. Sci. (Paris), 264:838–841, 1967.
125. ——— Sécrétion d'épi-oestriol par l'ovaire embryonnaire de poulet cultivé *in vitro*. C.R. Acad. Sci. (Ser. D), 266:2010–2012, 1968.
126. ——— Sécrétion d'oestriol par l'ovaire embryonnaire de poulet cultivé *in vitro*. C.R. Acad. Sci. (Paris), 266:159–160, 1968.
127. Porte, A., and Weniger, J-P. Ultrastructure des cellules interstitielles du testicule d'embryon de poulet de 18 jours. C.R. Soc. Biol., 155:2181–2184, 1961.
128. Haffen, K., and Cèdard, L. Étude, en culture organotypique *in vitro*, du métabolisme de la déhydroépiandrosterone et de la testostérone radioactives, par les gonades et intersexuées de l'embryon de poulet. Gen. Comp. Endocr., 11:220–234, 1968.
129. Willier, B. H. Ontogeny of endocrine correlation. *In* Analysis of Development. Eds. B. H. Willier, P. A. Weiss, and V. Hamburger. Philadelphia, W. B. Saunders Co., 1955.
130. Breneman, W. R. The effect on the chick of some gonadotrophic hormones. Anat. Rec., 64:211–220, 1936.
131. Nalbandov, A. V., Meyer, R. K., and McShan, W. H. The role of a third gonadotrophic hormone in the mechanism of androgen secretion in chicken testes. Anat. Rec., 110: 475–493, 1951.
132. Hartree, A. S., and Cunningham, F. J. Purification of chicken pituitary follicle-stimulating hormone and luteinizing hormone. J. Endocr., 43:609–616, 1969.
133. Woods, J. E., and Domm, L. V. A histochemical identification of the androgen-producing cells in the gonads of the domestic fowl and albino rat. Gen. Comp. Endocr., 7: 559–570, 1966.
134. Moszkowska, A. Activité gonadotrope de l'antèhypophyse d'embryon du poulet. Étude *in vitro*. Arch. Anat. Micr. Morph. Exp., 45:65–76, 1956.
135. Domm, L. V., and Dennis, E. A. The effect of hebin upon the reproductive system of the chick embryo. Anat. Rec., 64:53, 1935.
136. ——— Observations concerning anterior pituitary-gonadal interrelations in the fowl. Cold Spring Harbor Symp. Quant. Biol., 5:241–257, 1937.
137. Breneman, W. R. Some effects on the chick embryo of the injection of certain endocrine extracts. Indiana Univ. Publ. Sci. Ser., 2:5–26, 1935.
138. Venzke, W. G. The response of the gonads of chick embryos to gonadogen. Amer. J. Vet. Res., 2:281–284, 1941.
139. Snedecor, J. G. A study of some effects of sex hormones on the embryonic reproductive system and comb of the white leghorn chick. J. Exp. Zool., 110:205–246, 1949.
140. Vogel, N. W. Pituitary-gonad relationship in the chick embryo. Ph.D. dissertation, Indiana University, 1956.
141. Cèdard, L., Haffen, K., and Guichard, A. Influence de l'hormone gonadotrope chorionique sur la production d'oestrogènes à partir d'acétate de NA et de déhydroepiandrostérone radio-actifs par les gonades embryonnaires de poulet, cultivées *in vitro*. C. R. Acad. Sci. (Ser. D), 267:118–120, 1968.
142. Haffen, K., Cèdard, L. and Guichard, A. Formation de cholestérol radio-actif à partir d'acétate de Na-1-^{14}C par les gonades embryonnaires de poulet, cultivées *in vitro.* C.

R. Acad. Sci. (Ser. D), 268:2914–2917, 1969.
143. van Deth, J. H. M. G., van Limborgh, J., and van Faassen, F. Le rôle de l'hypophyse dans la détermination du sexe de l'oiseau. Acta Morph. Neerland-Scandin, 1:70–80, 1956.
144. Vogel, N. W. Free tissue cholesterol and growth in chick embryos hypophysectomized by "decapitation." Anat. Rec., 127:382, 1957.
145. Akram, H., and Weniger, J-P. Activité hormonale des gonades d'embryons de poulet hypophysectomisés. J. Embryol. Exp. Morph., 21:509–515, 1969.
146. Kinyon, N., and Watterson, R. L. Reduced endocrine activity of ovaries of hypophysectomized duck embryos as indicated by modified development of genital tubercle and syrinx. Physiol. Zool., 31:60–72, 1958.
147. Watterson, R. L., Veneziano. P., and Stenseth, J. H. Further experiments to analyze the apparent reduced endocrine activity of ovaries of hypophysectomized duck embryos. Anat. Rec., 139:284–285, 1961.
148. Dieterlen-Lièvre, F. Le rôle de la thyroide dans le développement embryonnaire des oiseaux et des mammifères. Ann. Biol., 2:17–33, 1963.
149. Thommes, R. C. Vasculogenesis in selected endocrine glands of normal and hypophysectomized chick embryos. I. The thyroid. Growth, 22:243–264, 1958.
150. Stoll, R., Maraud, R., and Lacourt, P. L'ultrastructure de la thyroïde (revue d'ensemble). Ann. Endocr., 22:281–304, 1961.
151. Bradway, W. The morphogenesis of the thyroid follicles of the chick. Anat. Rec., 42: 157–167, 1929.
152. Yoshikawa, T. Supplementary study of the morphogenesis of the thyroid gland in the chick (*Gallus domesticus*). J. Coll. Agric. Imp. Univ. Tokyo, 11:57–74, 1930.
153. Sun, T. P. Histo-physiogenesis of the glands of internal secretion—thyroid, adrenal, parathyroid, and thymus—of the chicken embryo. Physiol. Zool., 5:384–396, 1932.
154. Hopkins, M. L. Development of the thyroid gland in the chick embryo. J. Morph., 58: 585–613, 1935.
155. Venzke, W. G. Morphogenesis of the thyroid glands of chicken embryos. Amer. J. Vet. Res., 10:272–281, 1949.
156. Phillips, J., and Semmelman, A. A. The synthesis and storage of thyrotrophin in the embryo of the chick and rat. Anat. Rec., 134:624, 1959.
157. Trunnell, J. B., and Wade, P. Factors governing the development of the chick embryo thyroid. II. Chronology of the synthesis of iodinated compounds studied by chromatographic analysis. J. Clin. Endocr., 15:107–117, 1955.
158. Stoll, R., and Blanquet, P. Sur l'activité des thyroides de l'embryon de poulet, étudiées à l'aide du radioiode ^{131}I. Ann. Endocr., 13:569–577, 1952.
159. Wollman, S. H., and Zwilling, E. Radioiodine metabolism in the chick embryo. Endocrinology, 52:526–535, 1953.
160. Hilfer, S. R. Follicle formation in the embryonic chick thyroid. I. Early morphogenesis. J. Morph., 115:135–152, 1964.
161. Uhlenhuth, E., and Ebeling, A. Comparative hormonic efficiency of embryonic thyroid. Anat. Rec., 29:398, 1925.
162. Mess, B., Straznicky, K., and Hajos, F. La differenziazione della ghiandola tirode e la sua dipendenza dall'ormone tireotropo (TSH) in embrioni di pollo normali e decapitati. Rass. Neurol. Veg. 20:359–374, 1966.
163. Maraud, R., Stoll, R., Macario, C., and Blanquet, P. Sur la concentration du radioiode ^{131}I par la thyroïde de l'embryon de poulet. Ann. Endocr., 15:607–611, 1954.
164. Stoll, R., Maraud, R., Blanquet, P., Mounier, J., and Meyniel, G. Sur le métabolisme du radioiode chez l'embryon de poulet. Ann. Endocr., 17:15–32, 1956.
165. Blanquet, P., Stoll, R., Maraud, R., and Capot, L. Sur l'élaboration thyroïdienne de thyroxine chez l'embryon de poulet. C.R. Soc. Biol., 147:676–678, 1953.
166. Hansborough, L. A., and Khan, M. The initial function of the chick thyroid gland with the use of radioiodine (I^{131}). J. Exp. Zool., 116:447–453, 1951.
167. Hámori, J., Mess, B., and Székely, G. Onset of thyroidal I^{131} accumulation in normal and decapitated chick embryos. Acta Biol. Hung., 10:207–214, 1959.
168. Mess, B., and Straznicky, K. Functional differentiation of the thyroid gland investigated

by I^{131}-autoradiography in decapitated chick embryos bearing hypophyseal homografts. Acta Biol. Hung., 15:77–85, 1964.
169. ——— and Hámori, J. Bioassay of thyrotrophic hormone in blood by ^{131}I-autography in embryonic chick thyroid. Acta Physiol. Hung., 20:299–303, 1961.
170. Carpenter, E. Differentiation of chick embryo thyroids in tissue culture. J. Exp. Zool. 89:407–431, 1942.
171. ——— Beattie, J., and Chambers, R. D. The uptake of I^{131} by embryonic chick thyroid glands *in vivo* and *in vitro*. J. Exp. Zool., 127:249–269, 1954.
172. Gaillard, P. J. Differentiation phenomena in explants of the embryonic chick thyroid gland. Acta Physiol. Pharmacol. Neerl., 3:439–447, 1954.
173. Gonzales, F. Evidence of the functional differentiation of the embryonic chick thyroid in tissue culture. Texas Rep. Biol. Med., 12:828–832, 1954.
174. Tixier-Vidal, A. Développement *in vitro* de la thyroïde embryonnaire du poulet. Étude comparée de la différenciation suivant l'âge de l'explantation. C.R. Soc. Biol., 150: 1371–1375, 1956.
175. ——— Étude histophysiologique des relations hypophyse et thyroïde chez l'embryon de poulet. Arch. Anat. Micr. Morph. Exp., 47:235–340, 1958.
176. Wolff, E., Haffen, K., and Dieterlen, F. Influence des états nutritionnels sur la différenciation et le fonctionnement des glands endocrines embryonnaires. Ann. Nutr., 14:11–29, 1960.
177. Andrews, F. N., and Schnetzler, E. E. The effect of feeding thiouracil to hens upon the thyroid gland of chicks. Endocrinology, 37:382-384, 1945.
178. Grossowicz, N. Influence of thiourea on development of the chick embryo. Proc. Foc. Exp. Biol. Med., 63:151–152, 1946.
179. Adams, A. E., and Bull, A. L. The effects of antithyroid drugs on chick embryos. Anat. Rec., 104:421–443, 1949.
180. ——— and Buss, J. M. The effect of a single injection of an antithyroid drug on hyperplasia in the thyroid of the chick embryo. Endocrinology, 50:234–253, 1952.
181. ——— Bull, J. M., and Buss, J. M. A comparison of cellular hypertrophy and hyperplasia in thyroids of thiourea-injected chick embryos. Anat. Rec., 120:778, 1954.
182. Vidal, A. Influence de la thiourée sur le développement de l'embryon de poulet. Ann. Endocr., 13:982–990, 1952.
183. Wheeler, R. S., and Hoffmann, E. Goiterous chicks from thyroprotein-fed hens. Endocrinology, 42:326–328, 1948.
184. ——— and Hoffmann, E. Influence of quantitative thyroprotein treatment of hens on length of incubation period and thyroid size of chicks. Endocrinology, 43:430–439, 1948.
185. Willier, B. H. The endocrine glands and the development of the chick. I. The effect of thyroid grafts. Amer. J. Anat., 36:67–103, 1924.
186. Woodside, G. L. Experimentally induced hyperthyroidism in the chick embryo. Anat. Rec., 67:423–430, 1937.
187. Dossel, W. M. Early thyroid-stimulating activity in the chick embryo. Anat. Rec., 133: 269, 1959.
188. Portet, R. Effet de la thyroxine sur les échanges respiratoires de l'embryon de poulet. J. Physiol. (Paris), 52:200–201, 1960.
189. Rumpler, Y. Role des hormones thyroidiennes stockées dans l'oeuf sur le développement pondéral et squelettique de l'embryon de poulet. Bull. Assoc. Anat. (51st réunion), pp. 941–946, 1966.
190. Love, D. S., Stoddard, F. J., and Grasso, J. A. Endocrine regulation of embryonic muscle development; hormonal control of DNA accumulation, pentose cycle activity and myoblast proliferation. Develop. Biol., 20:563–582, 1969.
191. De Simone-Santoro, I. Morphologie ultrastructurale de l'ovaire de l'embryon de poulet au cours de sa différenciation et de son organogenèse. J. Microscop., 8:739–752, 1969.
192. Kamiyoshi, M., Tagami, M., and Tanaka, K. *In vitro* demonstration of follicle-stimulating hormone releasing activity in the chicken. Poult. Sci. 48:1977–1978, 1969.
193. Tanaka, K., Kamiyoski, M., and Tagami, M., *In vitro* demonstration of luteinizing hormone releasing activity in the hypothalamus of the hen. Poult. Sci., 48:1985–1987, 1969.

8

Similarities and Differences in the Mode of Action of Growth Hormone and Estradiol*

G. P. Talwar, S. L. Gupta, M. R. Pandian, S. K. Sharma, K. N. Rao, B. L. Jaikhani, K. K. Sen, M. L. Sopori, A. K. Basu, and P. Jha

Department of Biochemistry, All India Institute of Medical Sciences, New Delhi, India

Growth hormone and estradiol-17β are both growth-promoting hormones, though the action of growth hormone (GH) is of a more generalized nature in that it affects a number of tissues, whereas estradiol influences mainly a limited number of organs. The selective action of estradiol on its target organs is probably related to the presence of specific receptor proteins for this hormone in these tissues, while similar receptors are absent, or present in subminimal amounts, in tissues not affected by estradiol(1-6). If the "receptor" hypothesis is valid as a requirement for the interaction of a hormone with the tissue influenced by it (as it should theoretically be), the implication is that the receptors for estradiol are present in only a few organs, while those for growth hormone are more ubiquitous in their distribution.

The concept of "receptors" has existed in pharmacology for several decades and has evolved from studies on neurohumours and various types of drugs. The notion has, however, remained mostly a kinetic postulate. None of the receptors has been isolated or identified and described in clear molecular terms. The lack of information in this area stems mainly from the fact that receptors for most "effector agents" are membrane-linked structures and are destroyed or disarranged on disruption of the constituted aggregate structure. Evidence has grown in recent years to suggest that membrane-linked adenyl cyclase may well be a component of

*Various aspects of this work have been supported by research grants of The Population Council, Inc., New York; Department of Atomic Energy, Government of India; Council of Scientific and Industrial Research, India; and the World Health Organization, Geneva.

the receptor for catecholamines and perhaps for many other hormones in some organs(7,40).

Estradiol is taken up rapidly from circulation by the uterus(1,2). It was possible to show that the steroid was not present in the uterus in free form but is bound to macromolecular aggregates present in the cytoplasm and nuclei of the uterus(2-5,8,9). These are proteinic components, the protein moiety being essential for binding properties(10-12). The cytoplasmic receptor has a sedimentation constant of 8 to 9.5S(13,14) and is probably a precursor of the nuclear 5S receptor(14, 15). The binding of the steroid is of a noncovalent type, hydrophobic interactions playing an important part. The binding is highly stereospecific and differs in many respects from the binding of the steroid to serum albumin(10). Several workers have recently reported the purification of estradiol-binding proteins from different sources(16-18,41). There is no discrete information so far on the possible types of receptors for growth hormone.

GROWTH HORMONE, ESTRADIOL, AND RNA SYNTHESIS

Both estradiol and growth hormone stimulate the biosynthesis of RNA in their target tissues. There are, however, some discernible differences in the two cases.

Lag Period

In the case of growth hormone, there is a minimum lag period of 30 to 60 minutes before the enhancement of RNA synthesis is detectable(19). In fact, an optimum stimulation of RNA synthesis is observed only 3 to 5 hours after the administration of the hormone(20,21). As it takes only a few minutes for the hormone to be equilibrated and available to the tissue, the long latent period in manifestation of its effect on RNA synthesis would suggest that the hormone action on RNA synthesis may not be a primary effect of this hormone, but a consequence of other physicochemical events triggered by the hormone in the initial phase of its action.

The effect of estradiol on uterine RNA synthesis is, on the other hand, a fairly early manifestation. Activation of nuclear RNA synthesis has been reported at 2 to 10 minutes after the administration of the hormone(22,23). A 500 percent increase in the uterine nuclear synthesis is observed 20 minutes after the administration of estradiol in vivo(24). These observations would suggest that the activation of RNA synthesis is among the early events, if not the primary one, of the action of estradiol.

Action at the Gene Locus

Template Activity of Chromatin. Steroid hormones appear to have an action at the generic locus, though the exact mechanisms are poorly understood. The "puffing" of polytene chromosomal segments caused by ecdysone is a familiar example (25,26). Cortisol induces the synthesis of amino acid transaminases, which is an action sensitive to actinomycin D block, and hence is suggestive of an induced synthesis of a new mRNA(27,28). Estradiol induces the synthesis of two new phosphoproteins in rooster liver(29-31). Progesterone induces the synthesis of avidin in the chick oviduct(32). These and many other observations are indicative of the fact that steroids either directly or indirectly influence the genetic expression in the target tissues.

An increase in the template activity of the chromatin of rat uteri has been reported after estradiol administration(33). The nearest neighbour frequency of nucleotides in the RNA synthesized in the uterus after estradiol treatment is also different from the RNA synthesized in the ovariectomized rat uteri(39). There is, however, no change in the template activity of the liver chromatin after administration of the growth hormone(34,35,37). There is also no new species of RNA detectable in the target tissue after treatment with growth hormone(36,37). The latter conclusion is subject, of course, to the limits of sensitivity of current methods of hybridization of RNA to homologous DNA on membrane filters, which may fail to detect changes in a single or limited number of molecular species of RNA. The ensemble of these results, together with the fact that the action of growth hormone on RNA synthesis is most marked in experiments where the period of incorporation of radioactive precursors is limited to one or two hours, would suggest that this hormone influences essentially the rate of RNA synthesis in the tissue(37). Furthermore, the effect of this hormone on RNA synthesis may not be a direct action of the hormone on the transcription process, but a consequence of other influences of the hormone on the tissue.

EARLY ACTIONS OF GROWTH HORMONE

Growth hormone affects a number of metabolic activities of the body. It has an "insulin-like" and "insulin-antagonistic" action on carbohydrate and lipid metabolism. It promotes a positive nitrogen balance, presumably by its action on RNA and protein metabolism. The first phase of the action of this hormone in vivo or in vitro is its insulin-like action, in which it promotes the uptake and utilization of glucose and also promotes lipogenesis. This phase is followed by the insulin-antagonistic type of action, in which there is a mobilization of fats (instead of lipogenesis) and a reduced uptake and utilization of glucose. The development of the second phase can be blocked by actinomycin D and/or cycloheximide, which implies that synthesis of RNA and protein(s) is involved in the expression of the metabolic actions characteristic of the second phase.

To investigate the initial events in the action of growth hormone, an in vitro system has been employed by us in experiments carried out in collaboration with S. L. Gupta and M. R. Pandian. Growth hormone, when added in vitro to adipose tissue of hypophysectomised animals, enhances (1) the uptake of glucose in the adipose tissue, (2) its utilization, and (3) its conversion into fatty acids(38,42).

It has been observed that

1. Hypophysectomy leads to a diminished capacity of the intact tissue to take up ^{14}C-glucose from the medium and convert it into fatty acids.
2. The adipose tissue from normal animals has an equivalent capacity to convert ^{14}C-glucose into fatty acids irrespective of the level of tissue organization at which studies are made, i.e., the intact tissue from normal rats meta-

bolizes glucose at the same rate as the cell suspensions and homogenates.

3. Hypophysectomy causes a deprivation in the ability of the intact tissue to metabolize glucose into fatty acids at the same rate as cell suspensions and homogenates. This deficiency is compensated by preincubation of the intact tissue with growth hormone in vitro.
4. When growth hormone is added directly to the cell suspensions of the adipose tissue, no increase in conversion of ^{14}C-glucose into fatty acids is observed.
5. When growth hormone is added in vitro to homogenates of adipose tissue from hypophysectomized animals, a marginal stimulation of only 20 to 30 percent is observed in the conversion of glucose into fatty acids, while the addition of growth hormone to intact adipose tissue of the hypophysectomized rats causes a stimulation of severalfold (up to 200 percent) in the conversion of glucose into fatty acids.
6. Localization studies with fluorescein-labelled antigrowth hormone sera, as well as autoradiography with ^{125}I-labelled growth hormone, show the presence of the hormone (or derivatives) along the cell wall structures(42,43).

These observations suggest an action of growth hormone on membrane function (uptake of glucose) of the adipose tissue.

Growth hormone may, however, act at more than one site in the cell metabolism. The fact that the hormone is effective (though to a modest extent of 20 to 30 percent) in the homogenates is indicative of a possible direct effect of this hormone (or a product of its action) on enzyme reaction(s). Furthermore, preincubation of the intact tissue with growth hormone before homogenization improves the response of the tissue to growth hormone, suggesting the synergistic role of some factors elaborated in the initial phase of the interaction of the hormone with the intact adipose tissue.

REFERENCES

1. Jensen, E. V., and Jacobson, H. I. Basic guides to the mechanism of estrogen action. Recent Progr. Hormone Res., 18:387–414, 1962.
2. ——— Metabolic fate of sex hormones in target tissues with regard to tissue specificity. Proc. 2nd Int. Congr. Endocrinol. Excerpta Medica, International Congress Series No. 83, 420–433, 1965.
3. Talwar, G. P., Segal, S. J., Evans, A., and Davidson, O. W. The binding of estradiol in the uterus: A mechanism for derepression of RNA synthesis. Proc. Nat. Acad. Sci. (U.S.A.), 52:1059–1066, 1064.
4. Eisenfeld, A. J., and Axelrod, J. Selectivity of estrogen distribution in tissues. J. Pharmacol. Exp. Ther., 150:469–475, 1965.
5. King, R. J. B., and Gordon, J. The localization of (6,7,^3H)-estradiol-17-β in rat uterus. J. Endocr., 34:431-437, 1966.
6. Talwar, G. P. Mechanism of action of steroid hormones. Proceedings Fifth International Symposium on Comparative Endocrinology. Gen. Comp. Endocr., Suppl. 2:123–134, 1969.
7. Sutherland, E. W., Butcher, R. W., Robison, G. A., and Hardman, J. G. The role of adenosine-3′,5′-monophosphate in hormone action. *In* Wirkungs mechanismen der Hormone. Springer-Verlag Berlin, 1–32, 1967.

8. Gorski, J., Noteboom, W. D., and Nicolette, J. A. Estrogen control of the synthesis of RNA and protein in the uterus. J. Cell Comp. Physiol. 66, Suppl., 1:91–109, 1965.
9. Talwar, G. P., Segal, S. J., Gupta, S. L., Querido, A., Kakar, S. N., Nandi, J. K., Srinivasan, C. N., and Sopori, M. L. Growth hormone, estrogens and RNA synthesis. Proc. 2nd Int. Congr. Endocrinol. Excerpta Medica, International Congress Series No. 83, 14–18, 1965.
10. ——— Sopori, M. L., Biswas, D. K., and Segal, S. J. Nature and characteristics of the binding of estradiol-17β to a uterine macromolecular fraction. Biochem. J., 107:765–774, 1968.
11. Noteboom, W. D., and Gorski, J. Stereospecific binding of estrogens in the rat uterus. Arch. Biochem. Biophys., 111:559–568, 1965.
12. Jensen, E. V., et al. Proc. 5th Int. Symp. Comp. Endocrinol. Gen. Comp. Endocrinol., Suppl. 2, 1969.
13. Toft, D., Shyamala, G., and Gorski, J. A receptor molecule for estrogens: studies using a cell-free system. Proc. Nat. Acad. Sci. (U.S.A.), 57:1740–1743, 1967.
14. Jensen, E. V., Suzuki, T., Kawashima, T., Stumpf, W. E., Jungblut, P. W., and Desombre, E. R. A two-step mechanism for the interaction of estradiol with rat uterus. Proc. Nat. Acad. Sci. (U.S.A.), 59:632–638, 1968.
15. Puca, G. A., and Bresciani, F. Receptor molecule for estrogens from rat uterus. Nature, 218:967–969, 1968.
16. Takikawa, H., and Yoshinaga, S. Isolation of estrogen-binding proteins from follicular fluids of cow ovary. Proc. 3rd. Int. Congr. Endocrinol. Excerpta Medica, International Congress Series No. 157,9, 1968.
17. King, R. J. B. Properties of an acidic estradiol-binding protein from chromatin. Excerpta Medica, International Congress Series No. 157,9, 1968.
18. Soloff, M. S., and Szego, C. M. Isolation of soluble estradiol-binding macromolecules from rat uterus. Excerpta Medica, International Congress Series No. 157,9, 1968.
19. Jefferson, L. S., and Korner, A. A direct effect of growth hormone on the incorporation of precursors into proteins and nucleic acids of perfused rat liver. Biochem. J., 104:826–832, 1967.
20. Talwar, G. P., Panda, N. C., Sarin, G. S., and Tolani, A. J. Effect of growth hormone on RNA metabolism. Incorporation of radio-active phosphate into RNA fractions of rat liver. Biochem. J., 82:173–175, 1962.
21. Widnell, C. C., and Tata, J. R. Additive effects of thyroid hormone, growth hormone, and testosterone on deoxyribonucleic acid-dependent ribonucleic acid polymerase in rat liver nuclei. Biochem. J., 98:621–629, 1966.
22. Means, A. R., and Hamilton, T. H. Evidence for depression of nuclear protein synthesis and concomitant stimulation of nuclear RNA synthesis during early estrogen action. Proc. Nat. Acad. Sci. (U.S.A.), 56:686–693, 1966.
23. ——— and Hamilton, T. H. Early estrogen action—concomitant stimulations within two minutes of nuclear RNA synthesis and uptake of RNA precursors by the uterus. Proc. Nat. Acad. Sci. (U.S.A.), 56:1594–1598, 1966.
24. Hamilton, T. H., Widnell, C. C., and Tata, J. R. Sequential stimulations by estrogen of nuclear RNA synthesis and DNA-dependent RNA polymerase activities in Rat uterus. Biochem. Biophys. Acta, 108:168–172, 1965.
25. Clever, U. Gene activities and gene activations in the hormonal control of molting in insects. Proc. 16th Int. Congr. Zool., Washington, 4:256–263, 1963.
26. Beerman, W., and Clever, U. Chromosome puffs. Sci. Amer. 210:50–58, 1964.
27. Thompson, E. B., Tomkins, G. M., and Curran, J. F. Induction of tyrosine alpha-ketoglutarate transaminase by steroid hormone in a newly established tissue culture cell line. Proc. Nat. Acad. Sci. (U.S.A.), 56:296–303, 1966.
28. Reel, J. R., and Kenney, F. T. Regulation of enzyme synthesis by hydrocortisone in hepatoma cell cultures. Excerpta Medica, International Congress Series No. 157:148, 1968.
29. Schjeide, O. A., and Urist, M. R. Proteins and calcium in serum of estrogenised roosters. Science, 124:1242–1244, 1956.

30. Heald, P. J., and McLachlan, P. M. The synthesis of phosvitin in vitro by slices of liver from laying hen. Biochem. J., 94:32–39, 1965.
31. Greengard, O., Sentenac, A., and Acs, G. Induced formation of phosphoprotein in tissues of the cockerels in vivo and in vitro. J. Biol. Chem., 240:1687–1691, 1965.
32. McGuire, W. L., and O'Malley, B. W. Steroid hormone regulation of chromatin transcription during specific protein induction, Excerpta Medica, International Congress Series No. 157:11, 1968.
33. Barker, K. L., and Warren, J. C. Template capacity of uterine chromatin: control by oestradiol. Proc. Nat. Acad. Sci. (U.S.A.), 56:1298-1302, 1966.
34. Breuer, C. B., and Florini, J. R. Effects of ammonium sulphate, growth hormone and testosterone propionate on ribonucleic acid polymerase and chromatin activities in rat skeletal muscle. Biochemistry, 5:3857–3865, 1966.
35. Talwar, G. P., Gupta, S. L., Sopori, M. L., Biswas, D. K., Sachar, R., Sen, K. K., Rao, K. N., and Segal, S. J. Hormones and RNA synthesis in animal tissues. 7th Intern. Congr. Biochem., Tokyo, Symp. II, 3, 2, 1967.
36. Drews, J., and Brawerman, G. Messenger RNA patterns in rat liver nuclei before and after treatment with growth hormone. Science, 156:1385–1386, 1967.
37. Gupta, S. L., and Talwar, G. P. Effect of growth hormone on RNA metabolism. The template activity of the chromatin and molecular species of RNA synthesized after treatment with the hormone. Biochem. J., 110:401–406, 1968.
38. Goodman, H. M. Growth hormone and the metabolism of carbohydrate and lipid in adipose tissue. Ann. N.Y. Acad. Sci. 148:419–440, 1968.
39. Trachewsky, D., and Segal, S. J. Differential synthesis of RNA in uterine nuclei: evidence for selective gene transcription induced by estrogens. European J. Biochem., 4: 279–285, 1968.
40. Bar, H. P., and Hechter, O. Adenyl cyclase and hormone action. I. Effects of adrenocorticotropic hormone, glucagon, and epinephrine on the plasma membrane of fat rat cells. Proc. Nat. Acad. Sci. (U.S.A.), 63:350–356, 1969.
41. De Sombre, E. R., Puca, G. A., and Jensen, E. V. Purification of an estrophilic protein from calf uterus. Proc. Nat. Acad. Sci. (U.S.A.), 64:148–154, 1969.
42. Talwar, G. P., Jailkhani, B. L., Sharma, S. K., Sopori, M. L., Pandian, M. R., Sundharadass, G., and Rao, K. N. Mechanism of action of growth hormone and estrogens. *In* Control Processes in Multicellular Organisms. Wolstenholme, G. E. W., and Knight, J., eds. Ciba Foundation Symposium, J. & A. Churchill, Ltd., in press.
43. Pandian, M. R., Gupta, S. L., and Talwar, G. P. Studies on the early interactions of growth hormone. Localization and effect in vitro on lipogenesis in adipose tissue. Submitted for publication.

9

The Effect of Hormones on DNA Synthesis and Cell Number in the Developing Chick and Rat Brain*

Stephen Zamenhof, Edith van Marthens, and Hela Bursztyn

Department of Medical Microbiology and Immunology, Department of Biological Chemistry, Mental Retardation Program, and Brain Research Institute, U.C.L.A. School of Medicine, Los Angeles, California

INTRODUCTION

This following discussion deals with the prenatal development of one organ only, the brain (cerebral hemispheres).† The reason for choosing this organ is the prospect of correlations existing between alterations in brain development and behavior; the reason for studying such early development is the evidence that the number of neurons becomes final at or before birth (see below). From this point of view, such studies also may have bearing on other organs in which cell division (DNA synthesis) is virtually completed very early, such as the lung(1).

With reference to the brain, several reports indicate that in the rat the number of neurons does not increase after birth(2-5), with the possible exception of short-axoned neurons(5). In man this termination of neuron proliferation may occur as early as the fifth to seventh month of pregnancy(4), but little work has

*This investigation was supported by USPHS grants No. HD-01909 and NS-08723, and American Cancer Society grant No. P-503A.

The authors are grateful to the Endocrinology Study Section, National Institutes of Health, and to A. E. Wilhelmi, Emory University, for the generous supply of bovine pituitary growth hormone.

†The term "brain," used hereafter, refers to cerebral hemispheres without cerebellum and olfactory lobes in the rat and, in addition, without optic lobes in the chick.

been done on this important subject. In the chick the only report indicates that this termination occurs on or before the 12th day of incubation(6). In all of these species the increase in *total* number of cells after birth is much slower and reflects the proliferation of glia cells and, later, possibly also the cells of the circulatory system.

Normal neuron and glia cells at birth are essentially diploid, and the amount of DNA per diploid cell of a given species is constant(7,8). Thus, determination of brain DNA is a convenient and objective quantitative method for determination of total brain cell numbers in developing brain. Such determinations(8,9) will be used throughout this work.*

The total amount of brain DNA in the developing chick and rat reaches a transient plateau at hatching or birth (see Results, below). At the plateau this amount (i.e., also total brain cell number) is one of the most constant parameters of the developing organism: Later in this report it will be shown that this amount cannot be changed by any of the numerous factors that can change neonatal body weight or neonatal brain weight. Brain weight, the parameter most often reported in the literature, represents a composite of many factors (such as the amount of water, lipid, or protein) and, for individual animals, is not correlated with the amount of DNA or cell number (see Results).

The constancy of the amounts of DNA in neonatal brains suggests a stringent regulatory mechanism(s) that is not easily disturbed. It was therefore unexpected that these amounts in the chick and the rat undergo seasonal variations: The high values coincide with the period of rapid lengthening of daylight (see Results). These results suggest the possibility that the regulatory mechanism(s), which eventually slows down and stops brain DNA synthesis, is under hormonal influences. It is more likely that such influences act *after* organogenesis has been completed; also, the interference with organogenesis (first few days of embryo's development) is likely to produce lethal or teratogenic effects. Thus, in this study we have limited ourselves to the experimental approach starting *after* the first trimester of embryonic life.

To date we have investigated only three types of hormones: (1)thyroxine, whose levels are known to undergo seasonal variations; (2) pituitary growth hormone, which may influence fetal growth; and (3) sex hormones, in view of the known differences in brain weight of mature males and females. Of course, this list is very incomplete; also, the study of the influences of single hormones, instead of their special combinations, may be rather naive. Thus, this presentation is to be considered merely as a beginning progress report.

MATERIALS AND METHODS

Chick embryos. Fertile eggs, White Leghorn strain K137, were supplied biweekly by the Kimber Farms, Pomona, Calif. For the study of genetic differen-

*From the DNA values per brain, the actual total number of brain cells could be calculated by dividing by a (constant) DNA content per brain (6×10^{-6} μg for the rat; $2 \times 10^{-}$ μg for the chick[7,10]. For cerebellum and for mature brains, such constants may have to be corrected in view of the reports of polyploidy in Purkinje cells and in large pyramidal neurons of the hippocampus(11,12).

ces only, strain K155 was also used. The hens' ages and egg weights varied, and for each group as a whole (though not for single individuals) a correlation was found between egg weight and neonatal brain DNA content (see Results). To obviate this difficulty, care was taken to use as controls only eggs of the same weight, laid in the same week by hens of the same age as the experimentals. The eggs were incubated in the Jamesway incubator, type 252.

Rats. The strain used was derived from Sprague-Dawley. The rats have been bred in our laboratory for at least fifteen generations; the females were virgin, 3 months old and weighed 200 to 260 g. The animals were maintained on a pelleted diet ad libitum (Wayne Mouse-breeder Block, Allied Mills, Chicago, Ill.).

Treatments

Chick embryos. On the indicated day of incubation (see Results), the eggs were candled, and sterile solutions in physiological saline were injected with a 25-gauge needle as follows: thyroxine or thiourea (0.1 ml into the albumen); pituitary growth hormone (0.05 ml into the yolk). The control animals received sterile physiological saline injected in the same manner. After the injections, the holes in the eggshell were sealed with paraffin.

Rats. Pregnant females were injected subcutaneously (0.1 ml) or intravenously (tail vein, 1 ml) with aliquots of the solutions. The control animals received sterile physiological saline injected in the same manner.

Dissections

Chicks. The embryos were removed on the days indicated (see Results) or allowed to hatch, and then weighed. Cerebral hemispheres (i.e., without cerebellum and olfactory and optic lobes) were removed, weighed, and then frozen and stored at −15° C for subsequent DNA determination.

Rats. The newborn rats (natural delivery, or caesarian if so indicated) were weighed and then killed by decapitation within 6 hours of delivery. The cerebral hemispheres (without cerebellum and olfactory lobes) were removed, weighed, and frozen as above.

DNA Determination

The brains were individually homogenized and the total DNA in each brain was determined by a modification(8,9) of the diphenylamine colorimetric method.

RESULTS AND DISCUSSION

Normal Brain DNA Contents

The brain DNA contents as a function of prenatal and postnatal age are represented in Figure 1 (chick; see also Margolis[9]) and Figure 2 (rat). It can be seen that in both species these amounts reach a transient plateau at hatching or birth.

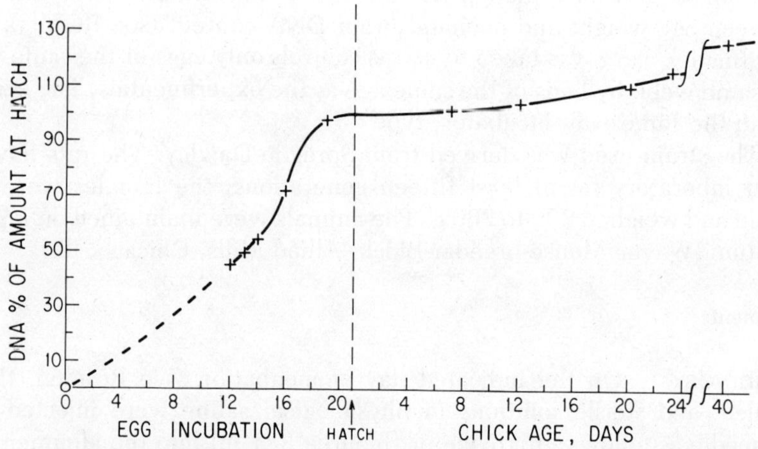

Fig. 1. DNA synthesis in developing chick brain (cerebral hemispheres). Abscissa—incubation (prehatch) and age (posthatch), in days. Ordinate—DNA, in percentage of the amount at hatch. Broken line at the beginning is a mere extrapolation.

This time, then, is a convenient one for estimating DNA, because a time difference of even half a day either way will not cause any significant error.

As explained earlier, the postnatal increase in brain DNA is not due to an increase in the number of brain neurons.

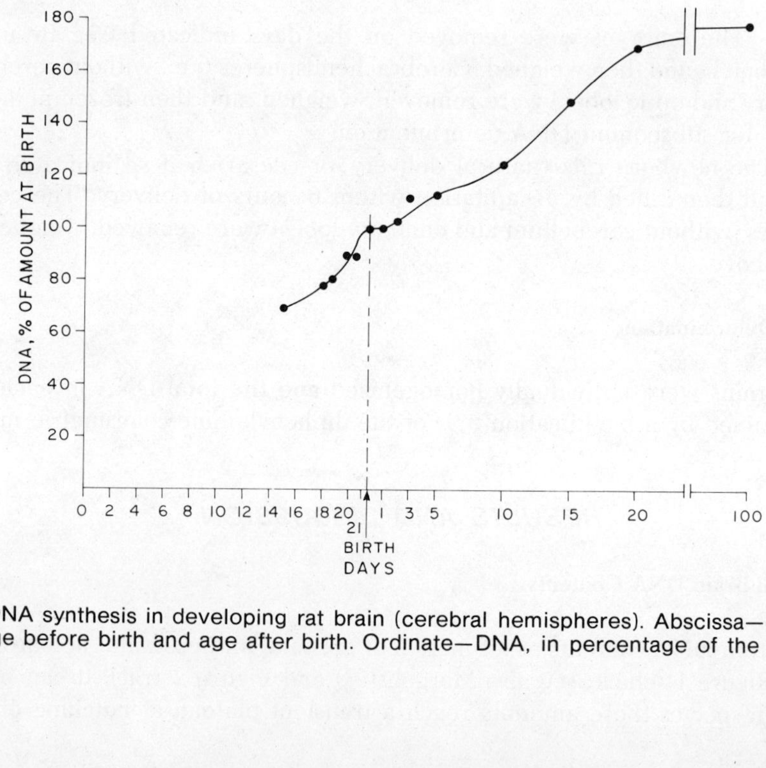

Fig. 2. DNA synthesis in developing rat brain (cerebral hemispheres). Abscissa—embryonal (fetal) age before birth and age after birth. Ordinate—DNA, in percentage of the amount at birth.

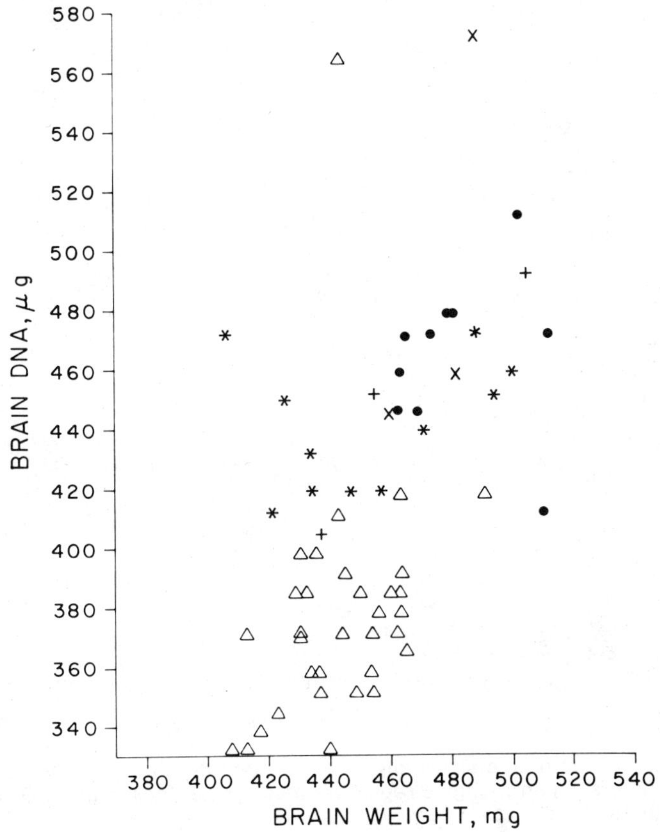

Fig. 3. Individual newly hatched chick brain weights and newly hatched chick brain DNA. Various symbols represent chicks from various batches of eggs.

Figure 3 (chick) and Figure 4 (rat) show that, *for individual animals,* even in the same batch of eggs or in the same litter, respectively, the correlation between brain weight and brain DNA cannot be demonstrated. Thus, as explained earlier, and as will be seen further on, individual brain weights cannot be used as a meaningful parameter of brain development.

The operational procedures (injections) themselves introduced very little injury; physiological saline (injected controls) resulted in a decrease of brain weight or DNA content by no more than 2 percent in newly hatched chicks and produced no detectable change in newborn rats.

The following drastic treatments, which produced changes in neonatal body weights and brain weights, did not produce any significant changes in neonatal brain DNA:

Chick Embryo. ATP (10 mg); cyclic AMP (500 μg); temperature of incubation lower by 2°C or higher by 1.5° than the optimal 37.6°C.

Pregnant Rat. Litter size; location of this particular fetus on uterine horn; 5 per-

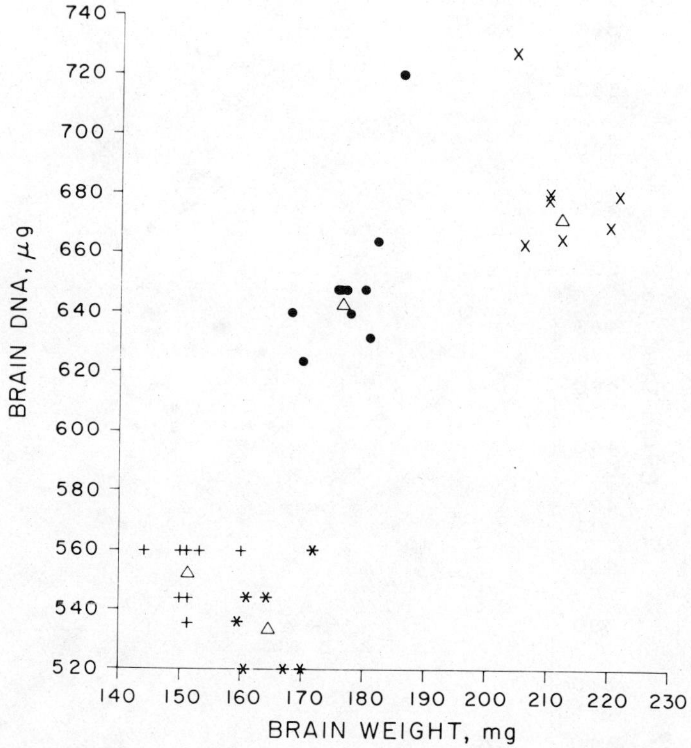

Fig. 4. Individual newborn rat brain weights and newborn rat brain DNA. Various symbols represent various litters. Open triangles represent mean value for the litter. Group (x) is a substrain of the others.

cent glucose in the drinking water; 2-deoxy-D-glucose (0.1–0.8 percent in drinking water or 15 intravenous injections, total of 750 mg; experimental hypoglycemia); serotonin (intravenously, total 30 mg).

It appears that neonatal brain DNA (neonatal brain cell number) is a very constant parameter and that the mechanisms regulating the speed and the termination of its synthesis are not easily influenced. The results to be discussed refer to natural and experimentally produced cases in which this parameter *can* be changed.

Spontaneously Occurring Cases of High Brain DNA

As can be seen from Figures 3 and 4, sporadically animals can be found which have brain DNA well above the range of others from the same batch of eggs or from the same litter. Such occurrences are rare—in 0.25 percent of the cases in the chick and 0.4 percent of the cases in the rat. The causes of such high DNA are completely unknown, but their occurrence indicates that the precise mechanisms of regulation of DNA synthesis are *not* completely inviolable.

Seasonal Variations

As mentioned, for *individual embryos* there is no demonstrable correlation between brain weight and brain DNA (Fig. 3); there is also no demonstrable

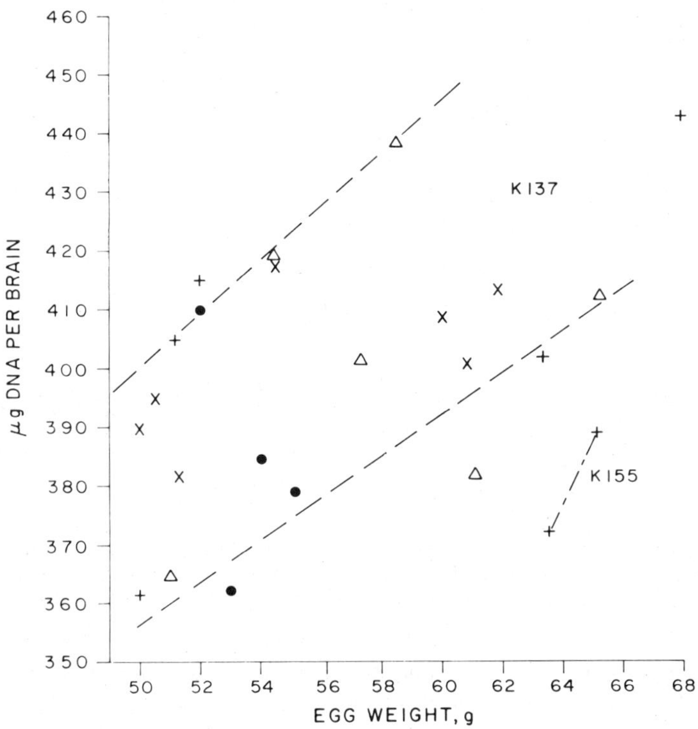

Fig. 5. Correlation between brain DNA and egg weight. Each symbol represents a batch of 8 to 12 eggs. K155 represents another strain. Broken lines enclose main range.

correlation between individual egg weight and brain weight; consequently, there is no demonstrable correlation between individual egg weight and brain DNA. However, if batches of 8 to 12 eggs are considered, a definite correlation between egg weight and brain DNA becomes noticeable (Fig. 5). This result is not unexpected, since the amount of nutrients available may conceivably influence the extent of cell proliferation of an organ. In rats, in which there was also no demonstrable correlation between brain weight and brain DNA in individual neonatals, a correlation appears if the averages of the litter are considered (Fig. 4).

In both chicks and rats the average DNA values are spread over a wide range (e.g., between broken lines on Fig. 5). This wide range appeared puzzling, until a survey was made of the time of the year the eggs were laid or the rats were born. The results are shown in Figure 6 (chick) and Figure 7 (rat).

For the chick, the results are represented as brain DNA per gram of egg weight to differentiate between seasonal variations and the differences in egg weight. It appears that the brain DNA curve (broken line, Fig. 6) has a relationship to the curve representing the number of daylight hours (full line). The relationship appears to be that the maximum speed of change toward longer days corresponds to maximum brain DNA, and the maximum speed of change toward shorter days to minimum brain DNA. In this respect, the correlation resembles the long-known relationship between egg-laying capacity and change of daylight(13). The activity and

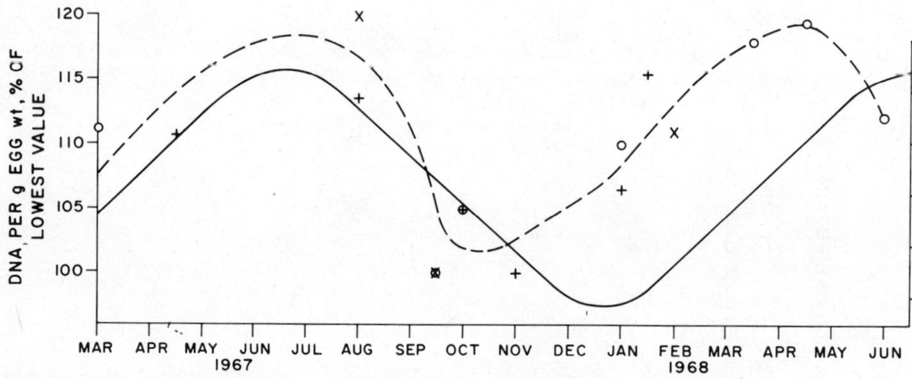

Fig. 6. Seasonal variations in amount of newly hatched chick brain DNA (broken line, individual points) and in hours of Los Angeles daylight (full line) for the months in which eggs were laid. Each point (DNA) represents an average of at least 12 brains from the same batch.

the size of gonads in birds are well known to be dependent on the change of the daylight length, and these phenomena are also known to be hormonally controlled (14). It thus appears likely that brain DNA synthesis is also at least partially under hormonal control.

This information about the final amount of brain DNA to be synthesized per gram of egg weight must exist in the egg at the time it is laid (in the particular season); at present, however, we do not know what form this information is in. Conceivably, it could be just in the form of nutritional value per gram of egg weight,

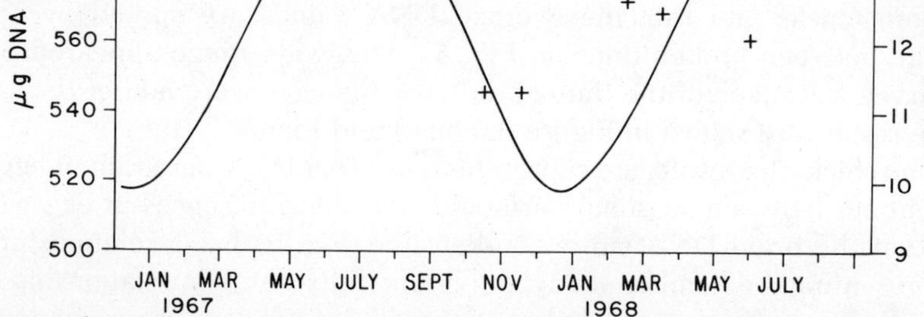

Fig. 7. Seasonal variations in amount of newborn rat brain DNA (individual points, no curve drawn) and in hours of Los Angeles daylight (full line) for the months in which the litters were delivered. Each point (DNA) represents as average of at least 20 brains.

or of enzymes (or their messenger RNAs) in the egg. However, the information could be also in the form of hormones present in the egg, or in the form of instruction about the future development of fetal glands producing these hormones.

Seasonal variations can also be demonstrated in the rat (Fig. 7). With the rat, however, the results are less distinct; in particular, for unknown reasons, the DNA values also fluctuated from one year to another. Nevertheless, here, too, it appears that there is a tendency for the maximum brain DNA to correspond to maximum speed of change toward longer days, and the minimum brain DNA to maximum speed of change toward shorter days. However, at present one cannot exclude the possibility that unaccountable seasonal factors other than daylight influence brain DNA as well. Genetic differences (strain differences) can also be noticed for both the chick (Fig. 5) and the rat (Fig. 4). In general, however, since seasonal changes are known to influence the level of several hormones in the rat, the brain DNA synthesis in this species is again likely to be, at least partially, hormonally controlled.

Thyroxine and Antithyroid Substances

Since thyroxine levels are known to undergo variations with the strains used (15) (genetic variability), the change of temperature and of season (presumably through hypothalamus, TRF, and TSH[16]), this hormone was investigated first.

Prenatal Rats. The influence of thyroxine and antithyroid substances on perinatal development has been a subject of extensive investigations by Hamburgh (17-22), Eayrs(23-25), and others(26-29). In general, it is felt that although maternal thyroxine passes the placental barrier during the third trimester of the gestational period(18), the fetus is unlikely to respond to it before the 18th day of fetal age (rat) because the fetal thyroid itself does not functionally mature before that time (21,22).

In the present study, females were injected subcutaneously with 100 μg of thyroxine in 0.1 ml of 0.1 N NaOH solution adjusted to pH 9. The injections were made daily on the 8th through 13th day, (two animals), or the 15th through 18th day of pregnancy (two animals). The animals littered with a 6- to 12-hour delay. The brain DNA in all animals was the same as in the saline-treated controls.

The antithyroid substance thiourea (see below), in concentrations of 0.002 to 0.001 percent in the drinking water, was given to one female from the 1st to 21st day of pregnancy. Higher concentrations were lethal. The litter was 2 days overdue, which accounted for a 5 percent increase in body weight over term controls. However, the brain weight and brain DNA were each 5 percent lower than in the controls.

In conclusion, it appears that neither thyroxine nor thiourea produces profound effect on brain DNA synthesis of prenatal rats. This result may be due to lateness of the development of fetal thyroid and lateness of the response to fetal thyroxine(21,22,26).

Chick Embryos. In contrast to the rat and altricial birds, the early function of the thyroid in the chick (10th to 11th day) made this species a favored subject of study of the effect of thyroxine and antithyroid substances(26,30-36). It was

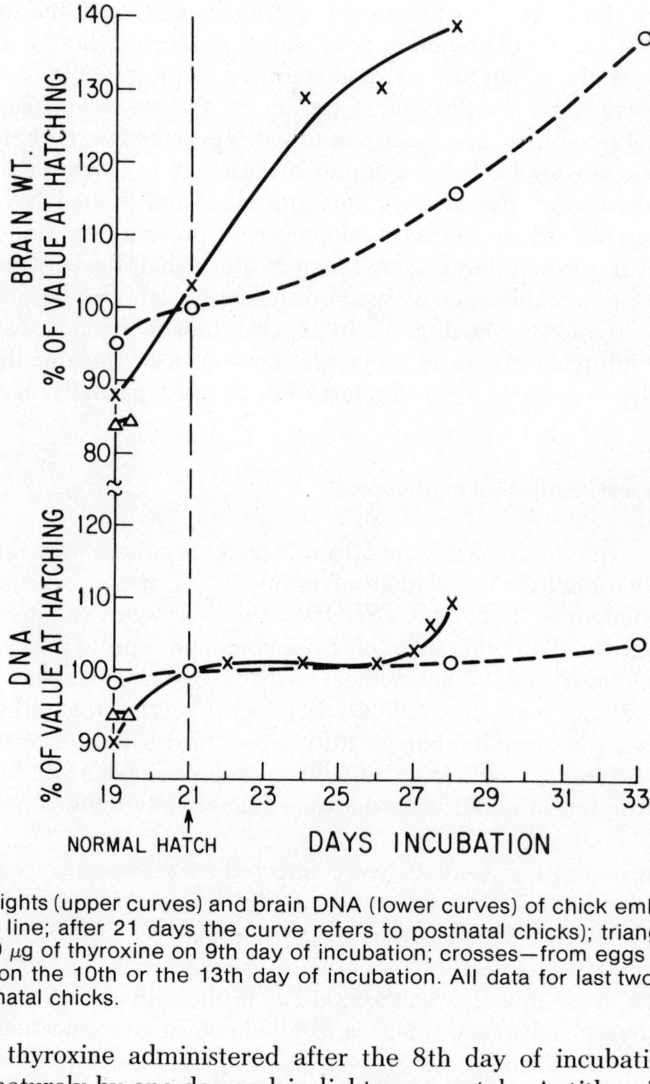

Fig. 8. Brain weights (upper curves) and brain DNA (lower curves) of chick embryos: circles—control (broken line; after 21 days the curve refers to postnatal chicks); triangles—from egg injected with 20 μg of thyroxine on 9th day of incubation; crosses—from eggs injected with 2 mg of thiourea on the 10th or the 13th day of incubation. All data for last two groups are for prenatal or neonatal chicks.

reported that thyroxine administered after the 8th day of incubation results in hatching prematurely by one day and in lighter neonatals. Antithyroid substances delayed hatching by up to 8 days, but the hatched animals were also lighter than controls. The brain studies were not reported in the literature.

In the present work, 0.1-ml aliquots of thyroxine solution as above were introduced once into eggs on the 9th day of incubation as described in the Methods section. The results are represented in Figure 8. Total doses of 20 μg resulted in hatching prematurely by 1 to 2 days; at hatching, brain weight was 10 percent lower than the control of the same age, and 15 percent lower than the control at normal hatching. The changes in brain DNA were more conservative: 5 percent below the control of the same age and 6 percent below the control at normal hatching. This effect was noticeable even with thyroxine doses as low as 2 μg. Thus,

Fig. 9. Left: control chick 10 days old. Right: experimental chick 20 days old after hatching with 8 days delay from egg injected with 2 mg of thiourea on the 15th day of incubation.

thyroxine, while exerting the known acceleration of differentiation, actually depressed DNA synthesis and presumably also brain cell divisions.

Thiourea was chosen as the antithyroid substance because of its previous extensive use in chick embryos(30,31,34,36) and the reports of its lower toxicity and complete reversibility by thyroxine(31,37). The main function of thiourea appears to be the prevention of iodination of tyrosine(37). In the present work, 0.1-ml aliquots of an aqueous solution containing 2 mg of thiourea were introduced once into eggs on the 10th, 13th, or 15th day of incubation. This treatment (Fig. 8) delayed hatching by 7 to 8 days. Brain weights and brain DNA at the 19th day of incubation were considerably lower than controls; they approximately equalled the controls on the 21st day but then continued to increase during the delay in hatching. At hatching, the experimental animals had 16 to 26 percent lower body weights, but they displayed large brains that were up to 39 percent heavier and had up to 10 percent more DNA. The brain weights and brain DNA were, in fact, higher than those of 7-day-old control chicks—i.e., those that hatched 7 days earlier and had for 7 days full access to food. In postnatal life such experimental chicks were grossly retarded (Fig. 9) and did not gain body weight for at least two weeks (Fig. 10).

In discussing the effects of thyroxine and antithyroid substances it is of interest to mention the situation in amphibians(38). In these animals, thyroxine is known to induce differentiation; consequently, tadpoles treated with thyroxine before normal metamorphosis will produce premature dwarf frogs. Conversely, thyroidectomy produces giant tadpoles that do not undergo metamorphosis. In many organs of birds and mammals, mitoses and differentiation may proceed at the same

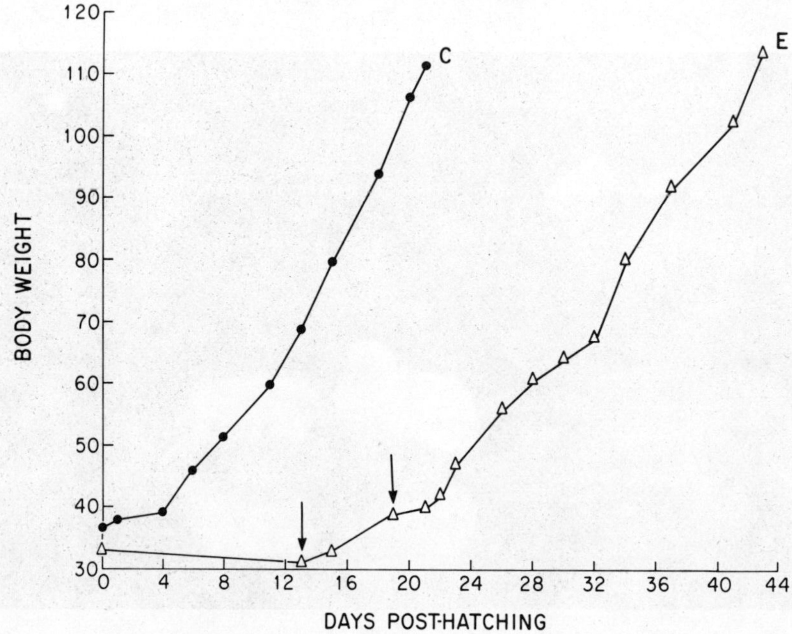

Fig. 10. Gains in body weight of control chick (C), and of experimental chick (E) shown on Figure 8. The experimental chick received two injections (arrows) of bovine growth hormone preparation B-11 (each injection: 3 mg in 0.1 ml saline into each of the two wing veins).

time(28). Nevertheless, with respect to brain neurons, the ability to undergo mitosis is clearly separated in time from differentiation: The latter may make the former impossible. It is, therefore, of interest that in chick embryo brain, thyroxine, by speeding up the differentiation, inhibits the DNA synthesis, whereas antithyroid substance, by slowing down the differentiation, allows for more DNA. In the rat this phenomenon cannot be observed, if only because of the short time (3 days) between onset of fetal thyroid activity and birth.

Thyroxine or Thiourea and Pituitary Growth Hormone

The synergistic effect of thyroxine and pituitary growth hormone after birth has often been reported(25,39–44, reviews in 37,45). Nonpurified and purified growth hormone improved growth of thyroidectomized animals(22,39,40,42,43); the combination of thyroxine and growth hormone was more effective than either hormone alone. The two hormones together may control both the amount and amino acid incorporation of polyribosomes(44,46).

In our experiments (chick), the introduction of bovine pituitary growth hormone into eggs (to be described) *together* with thyroxine or thiourea (as previously described) did not significantly change the results obtained for thyroxine or thiourea alone (see above).

Pituitary Growth Hormone

Mammals. Active pituitary growth hormone has been detected as early as in the 9- to 11-cm-long fetal pig(47), two to three days before birth in mouse(48) and rat(49), and as early as the 15th or even 7th week of pregnancy in human (50,51). The fetal level of this hormone eventually becomes higher than the maternal level(50,52). This fetal level is the highest at birth(53); it would be, indeed, in acromegalic range(52,53).

The role of fetal pituitary growth hormone in such high concentration is unknown. It has been suggested that this hormone is responsible for about 20 percent of the fetal growth(54).

The effects of pituitary growth hormone administered in pregnancy on neonatal body weight(54-61), postnatal brain development(60), neonatal brain weight, and brain DNA(61) have been reported. The latter results were now found to be subject to considerable variability (see also Sontag and Munson[55]), possibly because of the differences in individual effects of growth hormone on prenatal nutritional status. If, as it has been reported, the maternal pituitary growth hormone does not pass through placenta(62,63), then the influence of this hormone might be, indeed, through the change of maternal nutrient supply to the fetus.

We have recently found that the prenatal nutritional status influences brain weight and brain DNA of the neonatal animals: Maternal protein deprivation before and during pregnancy results in significant decreases in neonatal brain weights and brain DNA(64); on the other hand, operational decrease in number of fetuses results in significant increases in neonatal brain weight and brain DNA(65). These nutritional correlations in rats are now under study in our laboratory.

Chick Embryos. In these precocious birds the pituitary can be identified anatomically as early as the 5th day of incubation(66); the onset of endocrine activity (acidophils) was estimated to be the 10th(67) or the 15th day(68, review 26, 66). Crude hypophysectomy at 33 to 38 hours resulted in smaller embryos(69). Bovine growth hormone introduced into eggs after the 13th day was reported to increase neonatal body weight(70); the total DNA, however, was not affected(71). Bovine growth hormone was also reported to increase the weight of chick tibia in tissue culture(72).

Chick pituitary growth hormone appears to be active in the rat(73), although it seems to be less potent(74). However, it is still debatable whether bovine growth hormone is active in adult chicks(75,76).

We have injected bovine growth hormone into chicks (see Fig. 9) stunted by prenatal administration of thiourea (see above). The results (see Fig. 10) strongly suggest that bovine growth hormone is indeed active in dwarf chicks (see also Hulbé[77]).

When we introduced bovine growth hormone (300 μg) into eggs (8th to 13th day of incubation), we found that neonatal brain DNA was 4 to 7 percent higher than the controls, but we consider this a borderline of significance ($p=0.02$). What was significant ($p<0.01$) was the variability in individual responses: The differences between the highest individual values and the mean values (in groups of 8 to 12) were two to three times higher in animals treated with growth hormone than in controls. In one case, the highest value was 49 percent higher than the mean value for this treated group.*

*In the rat, in two cases, the highest values were 11 percent higher than the mean values for their treated groups.

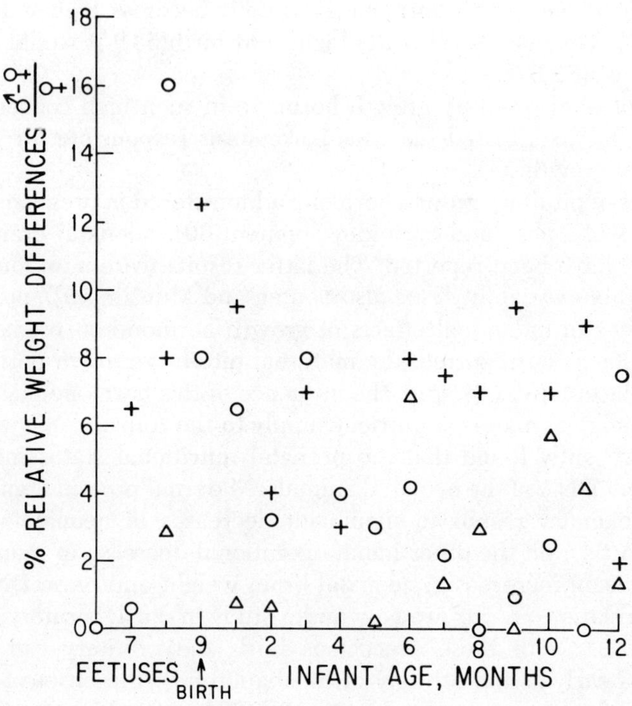

Fig. 11. Relative differences between weights of human male and female organs: brain (circle), heart (cross), and fetuses and infants (triangle). (Calculated from Schultz et al.[81])

Sex Hormones

The production of sex hormones by embryos has been detected as early as on the 10th day of incubation in the chick(78, reviews in 66,79) or on the 15th to 16th day of pregnancy in the rat(80, review in 45). In view of the known effects

TABLE 1

Body weights, brain weights, and brain DNA of male and female rats

Age		Number examined	Body weight (g)*	Brain† Weight (g)	Brain† DNA (μg)
Newborn	Male	8	6.1 ± 0.16	0.1667 ± 0.006	568 ± 26
	Female	10	6.3 ± 0.4	0.1626 ± 0.003	554 ± 24
	Difference,† %		−3	+2.5	+2.5
	Probability		0.02	0.1	0.3
100 days	Male	11	379 ± 39	1.218 ± 0.02	998 ± 51
	Female	10	263 ± 26	1.139 ± 0.05	954 ± 54
	Difference,† %		44.1	6.9	4.5
	Probability		< 0.001	0.01	0.1

*±Standard deviation.

†Male minus female, in percent of the value for female.

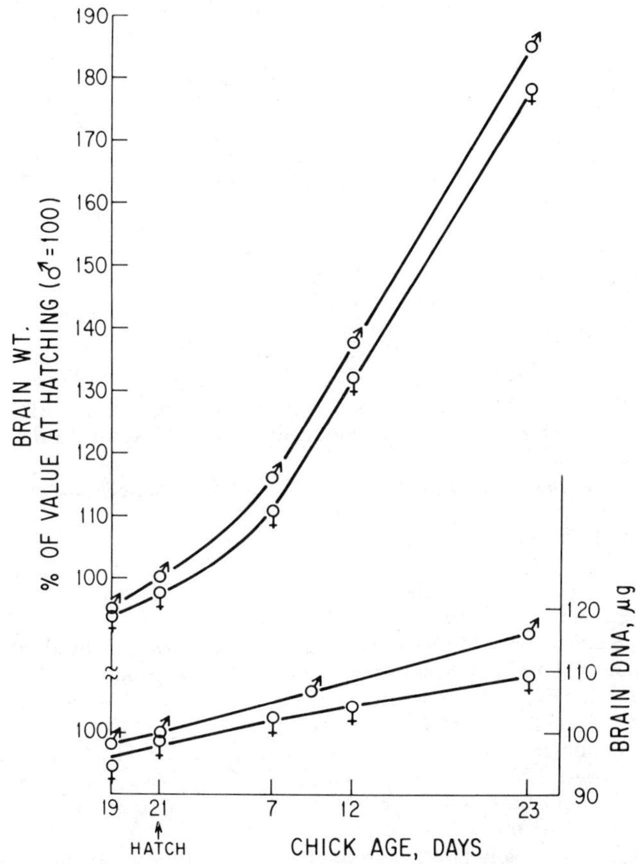

Fig. 12. Relative brain weights (upper curves) and brain DNA (lower curves) of male and female chick embryos and chicks (100 percent assumed for males). Male:female body-weight differences (not shown on drawing): at hatching +2 percent; at 100 days +30 percent.

of sex on differences in organ weights (Fig. 11: human; calculated from 81), it appeared of interest to investigate the possible effect of sex on the number of brain cells.

Our results are represented in Figure 12 (chick) and in Table 1 (rat). It can be seen that in both species there are detectable differences in brain weight due to sex. In the chick, these differences are small around hatching (Fig. 12); in the rat, they become significant only at maturity, when body weight differences are highly significant (Table 1). However, in both species there was no significant sex difference in neonatal brain DNA; in the rat there was also no significant difference in mature animals (see also 82).

For the determination of the number of neurons it is sufficient to consider neonatal animals, for by this time (or even much earlier in the chick, see introductory discussion) the number of neurons is final. Since sex differences made no significant change in the total neonatal amount of DNA in either the chick or the rat (total number of cells), it is not unlikely that the number of neurons also did not differ. If this is indeed true, then both sexes have the same number of brain neurons; and fetal sex hormones, already active at this time, do not exert any differential influence on this number.

REFERENCES

1. Winick, M. and Noble, A. Quantitative changes in DNA, RNA, and protein during prenatal and postnatal growth in the rat. Develop. Biol., 12:451--466, 1965.
2. Brizee, K. R., Vogt, J., and Kharetchko, X. Postnatal changes in glia/neuron index with a comparison of methods of cell enumeration in the white rat. Progr. Brain Res., 4:136–149, 1963.
3. Angevine, J. B., and Sidman, R. L. Autoradiographic study of cell migration during histogenesis of cerebral cortex in the mouse. Nature, 192:766–768, 1961.
4. McIlwain, H. Biochemistry and the Central Nervous System. Little, Brown and Co., Boston, 1966, p. 274.
5. Altman, J., and Das, G. D. Autoradiographic and histological studies on postnatal neurogenesis. J. Comp. Neurol., 126:337–389, 1966.
6. Nurnberger, J. I. Direct enumeration of cells of the brain. *In* Biology of Neuroglia, Windle, W. F., ed., Charles C. Thomas, Springfield, 1958, pp. 193–202.
7. Vendrely, C. L'acide désoxyribonucléique du noyau des cellules animales. Bull. Biol. France Belg., 86:1–87, 1952.
8. Zamenhof, S., Bursztyn, H., Rich, K., And Zamenhof, P. J. The determination of deoxyribonucleic acid and of cell number in brain. J. Neurochem., 11:505–509, 1964.
9. Margolis, F. L. DNA and DNA-polymerase activity in chicken brain regions during ontogeny. J. Neurochem., 16:447–456, 1969.
10. Leslie, I. The nucleic acid content of tissues and cells. *In* The Nucleic Acids, Chargaff, E., and Davidson, J. N., eds. Vol. 2. Academic Press, Inc., New York, 1955, pp. 8, 16,
11. Herman, C. J., and Lapham, L. W. DNA content of neurons in the cat hippocampus. Science, 160:537, 1968.
12. Lapham, L. W. Tetraploid DNA content of Purkinje neurons of human cerebellar cortex. Science, 159:310–312, 1968.
13. Lewis, J. H., Hannas, R. R., and Wene, E. H. New Jersey Agr. Exptl. Sta. Bull., 338:5–96, 1919.
14. Romanoff, A. L. and Romanoff, A. J. The Avian Egg. J. Wiley, New York, 1949.
15. Mendoza, L. A., Hamburgh, M., and Fuld, H. Differences in thyroid activity in several inbred strains of mice. Anat. Rec., 158:275–280, 1967.
16. Hahn, D. W., Ishibashi, T., and Turner, C. W. Alteration of thyroid hormone secretion rate in fowls changed from a cold to a warm environment. Poultry Sci., 45:31–33, 1966.
17. Hamburgh, M., and Flexner, L. B. Biochemical and physiological differentiation during morphogenesis. XXI. Effect of hypothyroidism and hormone therapy on enzyme activities of developing cerebral cortex of the rat. J. Neurochem., 1:279–288, 1957.
18. ——— Sobel, E., Koblin, R., and Rinestone, A. The passage of thyroid hormone across placenta in intact and hypophysectomized rats. Anat. Rec., 144:219–228, 1962.
19. ——— and Bunge, R. P. Evidence for a direct effect of thyroid hormone on maturation of nervous tissue grown in vitro. Life Sci., 3:1423–1430, 1964.
20. ——— and Lynn, E. The influence of temperature on skeletal maturation of hypothyroid rat. Anat. Rec., 150:163–172, 1964.
21. ——— Lynn, E., and Weiss, E. P. Analysis of the influence of thyroid hormone on prenatal and postnatal maturation of the rat. Anat. Rec., 150:147–161, 1964.
22. ——— An analysis of the action of thyroid hormone on development based on *in vivo* and *in vitro* studies. Gen. Comp. Endoc. 10:198–213, 1968.

23. Eayrs, J. T., and Taylor, S. H. The effect of thyroid deficiency induced by methyl thiouracil on the maturation of the central nervous system. J. Anat. (London, 85:350–358, 1951.
24. ——— Age as a factor determining the severity and reversibility of the effects of thyroid deprivation in the rat. J. Endocrinol., 22:409–419, 1961.
25. ——— Protein anabolism as a factor ameliorating the effects of early thyroid deficiency. Growth, 25:175–189, 1961.
26. Mitskevich, M. S. Glands of Internal Secretion in the Embryonic Development of Birds and Mammals. Natl. Sci. Found., Washington, 1959.
27. Kalter, H., and Warkany, J. Experimental production of congenital malformations in mammals by metabolic procedure, Physiol. Rev., 39:69–115, 1959.
28. Myant, N. B. On the possible role of the thyroid in the control of the development of the mammalian brain. Biol. Neonat., 9:148–175, 1965-66.
29. Campbell, H. J., and Eayrs, J. T. Influence of hormones on the central nervous system. Brit. Med. Bull., 21:81–86, 1965.
30. Grossowicz, N. Influence of thiourea on development of the chick embryo. Proc. Soc. Exp. Biol. Med., 63:151–152, 1946.
31. McMenamin, J. W. The role of the thyroid gland in the development of the embryonic chick. UCLA Dissertation, 1949.
32. Adams, A. E., and Bull, A. L. The effect of antithyroid drugs on chick embryos. Anat. Rec., 104:421–443, 1949.
33. Beyer, R. E. The effect of thyroxin upon the general metabolism of the intact chick embryo. Endocrinology, 50:497–503, 1952
34. Vidal, A. Influence de la thiourée sur le developpement de l'embryon de poulet. Annales d'Endocrinol., 13:982–990, 1952.
35. ——— Influence de la thyroxine sur le developpement de l'embryon de poulet. Annales d'Endocrinol., 14:437–443, 1953.
36. Romanoff, A. L., and Laufer, H. The effect of injected thiourea on the development of some organs of the chick embryo. Endocrinology, 59:611–619, 1956.
37. Comsa, J. Les Antithyroidiens Biologiques. G. Doin & Cie, Paris, 1953, p. 83.
38. Weber, R. Biochemistry of amphibian development. *In* The Biochemistry of Animal Development, R. Weber, ed., Vol. 2. Academic Press, Inc., 1967, p. 227.
39. Flower, C. F., and Evans, H. M. The repair of dwarfism following thyroidectomy by the administration of anterior hypophyseal fluid. Anat. Rec., 29:383, 1925.
40. Smith, P. E. Increased skeletal effects in A.P. growth-hormone injections by administration of thyroid in hypophysectomized, thyro-parathyroidectomized rats. Proc. Soc. Exp. Biol. Med., 30:1252–1254, 1933.
41. Evans, H. M., Simpson, M. E., and Penharz, R. I. Relation betweeen the growth promoting effects of the pituitary and the thyroid hormone. Endocrinology, 25:175–182, 1939.
42. Ray, R. D., Simpson, M. E., Li, Ch. H., Asling, C. W., and Evans, H. M. Effects of the pituitary growth hormone and of thyroxin on growth and differentiation of the skeleton of the rat thyroidectomized at birth. Am. J. Anat., 80:479–516, 1950.
43. Gomez, C. J., Ghittoni, N. E., and Dellacha, J. M. Effect of L-thyroxine or somatotrophin on body growth and cerebral development in neo-natally thyroidectomized rats. Life Sci., 5:243–246, 1966.
44. Tata, J. R., and Williams-Ashman, H. G. Effects of growth hormone and tri-iodothyronine on amino acid incorporation by microsomal subfractions from rat liver. European J. Biochem., 2:366–374, 1967.
45. Willier, B. H. Ontogeny of endocrine correlation. *In* Analysis of Development, Willier, B. H., Weiss, P. A., and Hamburger, V., eds. W. B. Saunders Co., Philadelphia, 1955, p. 574.
46. Tata, J. R. Hormonal regulation of growth and protein synthesis. Nature, 219:331–337 1968.
47. Smith, P. A., and Dortzbach, C. The first appearance in the anterior pituitary of the

developing pig foetus of detectable amount of the hormones stimulating ovarian maturity and general body growth. Anat. Rec., 43:277–297, 1929.
48. Enemar, A. The ontogeny of the hypophysial growth-promoting activity of the laboratory mouse, studied by means of tadpole tests. Acta. Univ. Lund., 56:1–34, 1961.
49. Contopoulos, A. N., and Simpson, M. E. Presence of tropic hormones in fetal rat pituitary. Fed. Proc., 16:24, 1957.
50. Kaplan, S. L., and Grumbach, M. M. Serum chorionic "growth hormone-prolactin" and serum pituitary growth hormone in mother and fetus at term. J. Clin. Endocr., 25:1370–1374, 1965.
51. Matsuzaki, F., and Shizume, K. Growth hormone in human fetuses. 3rd Int. Congr. Endocrinol., Excerpta Medica Found., No. 157, Amsterdam 1968, p. 19.
52. Glick, S. M., Roth, J., Yalow, R. S., and Benson, S. A. The regulation of growth hormone secretion. *In* Recent Progress in Hormone Research, G. Pincus, ed., Vol. 21, Academic Press, Inc., New York, 1965, pp. 241–283.
53. Hunter, W. M., and Greenwood, F. C. A radio-immunoelectrophoretic assay for human growth hormone. Biochem. J., 91:43–56 (1964).
54. Heggestad, C. B., and Wells, L. J. Experiments on the contribution of somatotropin to prenatal growth in the rat. Acta. Anat., 60:348–361, 1965.
55. Sontag, L. W., and Munson, P. L. The effect on the weight of the offspring of administration of Antuitrin G to the pregnant rat. Amer. J. Physiol., 108:593–598, 1934.
56. Watts, R. M. The effect of administration of preparations of growth hormone of the anterior lobe of the pituitary upon gestation and the weight of the newborn (albino rats). Amer. J. Obstet. Gynec., 30:174–185, 1935.
57. Barns, H. H. F., and Swyer, G. I. M. Effects on foetal weight of growth-hormone-containing anterior pituitary extracts given to pregnant rats. Brit. Med. J., pp. 914–915, 1952.
58. Hulquist, G. T., and Engfeldt, B. Giant growth of rat fetuses produced experimentally by means of administration of hormones to the mother during pregnancy. Acta Endocrinol., 3:365–376, 1949.
59. Engfeldt, B., and Hulquist, G. T. Administration of crystalline growth hormone to pregnant rats. Acta Endocrinol., 14:181–188, 1953.
60. Clendinnen, B. G., and Eayrs, J. T. The anatomical and physiological effects of prenatally administered somatotrophin on cerebral development in rats. J. Endocrinol., 22: 183–193, 1961.
61. Zamenhof, S., Mosley, J., and Schuller, E. Stimulation of the proliferation of cortical neurons by prenatal treatment with growth hormone. Science, 152:1396–1397, 1966.
62. Gitlin, D., Kumate, J., and Morales, C. Metabolism and maternofoetal transfer of human growth hormone in the pregnant woman at term. J. Clin. Endocrinol., 25: 1599–1608, 1965.
63. Laron, Z., Pertzelan, A., Mannheimer, S., Goldman, J. and Guttman, S. Lack of placental transfer of human growth hormone. Acta Endocrinol., 53:687–692, 1966.
64. Zamenhof, S., van Marthens, E., and Margolis, F. L. DNA (cell number) and protein in neonatal brain: Alteration by maternal dietary protein restriction. Science, 160: 322–323, 1968.
65. van Marthens, E., and Zamenhof, S. Deoxyribonucleic acid of neonatal rat cerebrum increased by operative restriction of litter size. Exp. Neurol., 23:214–219, 1969.
66. Romanoff, A. L. The Avian Embryo. The Macmillan Co., New York, 1960.
67. Rahn, H. The development of the chick pituitary with special reference to the cellular differentiation of the pars bucalis. J. Morph., 64:483–517, 1939.
68. Enemar, A. Ontogeny of the hypophysial growth-promoting activity in the chick. J. Endocrinol., 37:9–15, 1967.
69. Fugo, N. W. Effects of hypophysectomy in the chick embryo. J. Exp. Zool., 85:271–296, 1940.
70. Hsieh, K. M., Wang, T. Y., and Blumenthal, H.T. The diabetogenic and growth-pro-

moting activities of growth hormone (somatotropin) in the developing chick embryo. Endocrinology, 51:298–301, 1952.
71. Wang, T. Y., Hsieh, K. M., and Blumenthal, H. T. The effect of growth hormone on the nucleic acid content of the developing chick embryo. Endocrinology., 53:520–526, 1953.
72. Hay, M. F. The effect of growth hormone and insulin on limb-bone rudiments of the embryonic chick cultivated in vitro. J. Physiol., 144:490–504, 1958.
73. Moudgal, N. R., and Li, C. H. Immunochemical studies of bovine and ovine pituitary growth hormone. Arch. Biochem. Biophys., 93:122–127, 1961.
74. Hazelwood, R. L., and Hazelwood, B. S. Effects of avian and rat pituitary extracts on tibial growth and blood composition. Proc. Soc. Exp. Biol. Med., 108:10–12, 1961.
75. Carter, R., Risner, R., and Yacowitz, H. Some effects of growth hormone preparations in pullets and mature hens. Poul. Sci., 34:1407–1414, 1955.
76. Hazelwood, R. L., and Lorenz, F. W. Effect of fasting and insulin on carbohydrate metabolism of the domestic fowl. Amer. J. Physiol., 197:47–51, 1959.
77. Hulbé, J. Langdurige behandeling van dwerghoenders met een groeibevorderend extract uit der Voorkwat der runderhypophyse. Natuurwetenschap. Tijdschr., 31:201–205, 1949.
78. Gallien, L., and Le Foulgoc, M. T. Détection par fluorométrie et colorimétrie de stérodes sexuels dans les gonades embryonnaires de poulet. Comp. Rend. Soc. Biol., 151: 1088–1089, 1957.
79. Willier, B. H. Development of sex-hormone activity of the avian gonad. Ann. N.Y. Acad. Sci., 55:159–171, 1952.
80. Jost, A. Sur le contrôle de la différenciation sexuelle du lapin. Arch. Anat. Micr. Morph. Exp., 39:577–607, 1950.
81. Schulz, D. M., Giordano, D. A., and Schulz, D. H. Weights of organs of fetuses and infants. Arch. Pathol., 74:80–86, 1962.
82. Nurnberger, J. I., and Gordon, M. W. The cell density of neural tissues: direct counting method and possible applications as a biologic referent. *In* Ultrastructure and Cellular Chemistry of Neural Tissue, Waelsch, H., ed. Harper and Row, New York, 1957. pp. 100–138.

10

The Role of the Foetal Pituitary in Organogenesis

Joseph G. Bearn

Department of Anatomy, The Middlesex
Hospital Medical School, London, England

INTRODUCTION

Great advances have been made in the understanding of the role of foetal hormones in developmental biology over the past two decades. In particular, the development of techniques for the surgical removal of foetal endocrine organs in the mammal by pioneer workers such as Jost in Paris annd Wells in Minnesota over twenty years ago has led to a great increase in experimental surgery in utero in mammals—for example, the rabbit, the rat, the mouse, and the hamster—by workers in many countries. This work, while helping to solve many problems in foetal endocrinology, has inevitably raised many more questions for study, such as the importance of the foetal hypothalamus in the functional development of the foetal pituitary. Furthermore, this experimental work has greatly increased our understanding of clinical abnormalities in human foetuses.

In this review three problems will be discussed: (1) The foetal pituitary and foetal growth; (2) The role of the foetal pituitary-adrenal axis in foetal development; (3) The development of the genital tract in human anencephalic foetuses.

THE FOETAL PITUITARY AND FOETAL GROWTH

As long ago as 1929, Smith and Dortzbach demonstrated that the anterior pituitary of the foetal pig contains a hormone which stimulates growth as judged by weight gain in a test animal. However, whether this hormone exerts a physiological role in the growth of the mammal foetus has still to be established. In any work in this field the role of growth-promoting substances from the placenta and

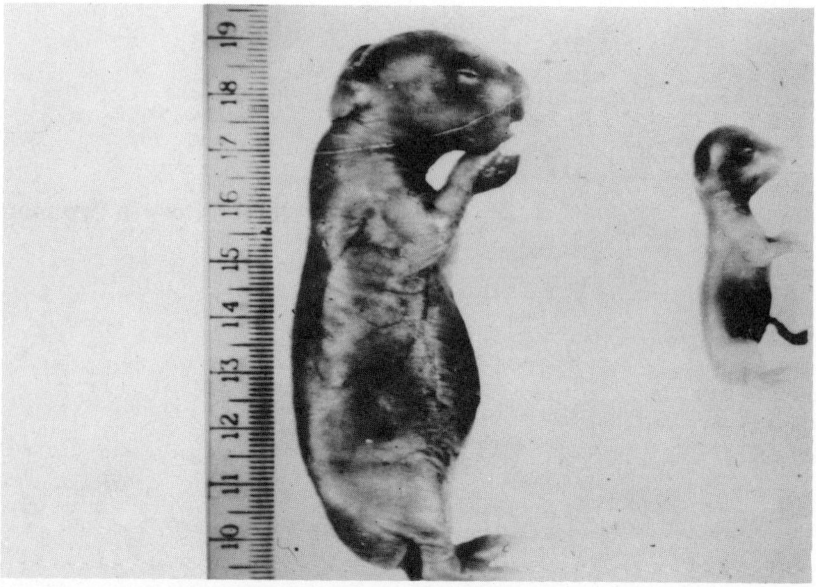

Fig. 1. A 29-day rabbit foetus with a 21-day foetus for comparison, to show the great increase in foetal growth during the last 8 days of intrauterine life.

Fig. 2. "Intraamniotic decapitation" of the rabbit foetus at 21 days. The foetal head is seen delivered from the uterus with the amniotic membrane intact. When the head is removed with a ligature, no loss of amniotic fluid occurs.

the mother must also be considered as well as growth hormone from the anterior pituitary of the foetus.

Figure 1 shows a 29-day rabbit foetus with a 21-day foetus demonstrating the great increase in growth which occurs during these 8 days in utero, just prior to term. In a personal series of observations on normal rabbit foetuses in which the mother was not subjected to any surgical interference, it was found that foetal weight increased from just under 5 g at 21 days to just under 30 g at 29 days.

The observations of Jost (1954) have shown convincingly that the foetal pituitary of the rabbit does not appear to influence foetal growth as judged by weight. However, Heggestad and Wells (1965), in a series of experiments on the rat foetus, concluded that foetal growth hormone is essential for normal body growth during the last three days of intrauterine life. They further demonstrated that the retardation in growth which follows foetal decapitation could be prevented by the injection of growth hormone into the foetus immediately following the operation of "hypophyseoprivus" by foetal decapitation.

This work in the rabbit and the rat has been restudied. A new technique for foetal decapitation has been developed, "intraamniotic decapitation," whereby the foetal head may be removed without opening the amniotic sac (Fig. 2) (Bearn and Pilkington, 1963). Using this method, a large series of rabbit foetuses have been decapitated, and it has been found that rabbit foetuses decapitated at day 21 and recovered by caesarean section for study at day 29 show no significant difference when compared with the headless littermate controls (Table 1). Furthermore, there is no significant difference between these littermate controls and 29-day rabbit foetuses from unoperated rabbits. Figure 3 shows a 29-day foetus which had undergone "intraamniotic decapitation" at day 21 and which was recovered within an intact amniotic sac at 29 days. Figure 4 shows two foetuses decapitated at day 21 and recovered at day 29 with two littermate control foetuses for comparison. The small foetus is a 21-day foetus showing the size at the time of operation.

Alizarin preparations have been made of these foetuses to study the skeletal ossification following decapitation, and there is no difference in ossification between the experimental and control foetuses (Fig. 5). At day 21 the skeletal ossification is very rudimentary.

It is of interest to note in passing that the decapitated foetuses (Fig. 5) clear much less well than control foetuses. This is due to the increase in the subcutaneous

TABLE 1

Headless body weights of rabbit foetuses subjected to intrauterine decapitation (means ± S.E.)

Foetuses	Number	Age at operation (days)	Age at study (days)	Body weight headless (g)
Decapitated	59	21	29	$28.34^1 \pm 0.56$
Littermate controls	95	21	29	$29.19^2 \pm 0.62$

The difference between 1 and 2 is not significant ($p = 0.2$).

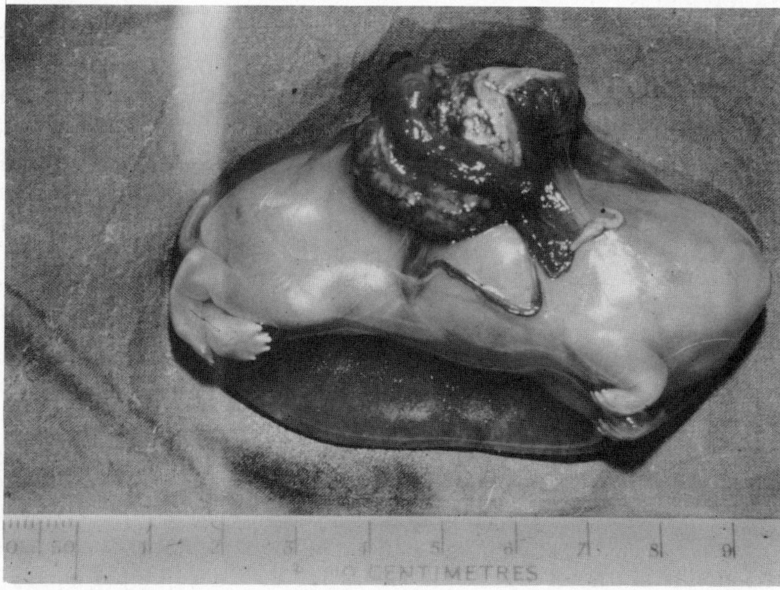

Fig. 3. A 29-day rabbit removed from the uterus within an intact amniotic sac. "Intraamniotic decapitation" was performed at 21 days.

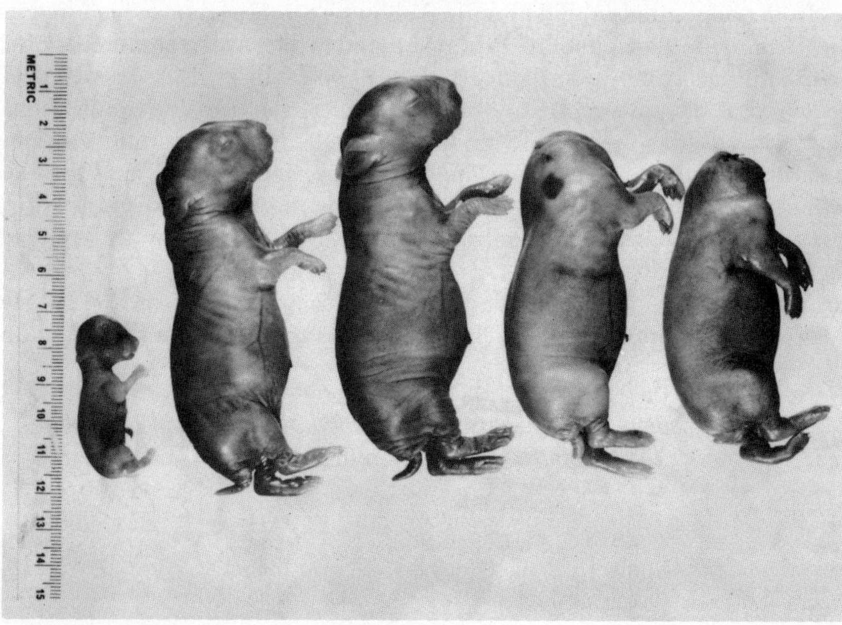

Fig. 4. Two 29-day foetuses decapitated by the "intraamniotic technique" at 21 days. The small foetus is a 21-day rabbit foetus to show the size at the time of the decapitation.

Fig. 5. A 29-day rabbit foetus decapitated at 21 days with a litter-mate control, stained with alizarin to show normal skeletal ossification. Note the decapitated foetus has cleared less well due to the thick pannus of subcutaneous fat.

pannus of fat in the decapitated foetuses. It has often been reported that human anencephalic foetuses show a great increase in this layer of subcutaneous fat, and it may well be that this increase is under a hormonal control acting through the foetal pituitary. Earlier observations (Table 2) show that livers from decapitated foetuses contain significantly more fat than control foetuses (Bearn, 1960).

More detailed studies (Bearn and Pilkington, 1963) have shown that blood cholesterol increases significantly following foetal decapitation, and this effect may be prevented by injecting thyroxine into the foetus at the time of decapitation. Further studies (Bearn et al., 1961; Bearn and Pilkington, 1963) have shown that foetal thyroidectomy similarly results in an increase in blood cholesterol. It is,

TABLE 2

Total fat of liver of decapitated and control rabbit foetuses (means ± S.E.)

Foetuses	Number	Age at investigation	Total liver fat (percent of dry weight)
Decapitated (at 20-21 days)	5	28-29 days	$49.4^{1} \pm 2.3$
Littermate controls	15	28-29 days	$41.3^{2} \pm 1.5$

The difference between 1 and 2 is significant ($p < 0.01$).

therefore, possible that foetal lipids may well be under a hormonal control acting through the pituitary and possibly also the thyroid. Further studies (Bearn et al., 1967) have also indicated that blood triglycerides are likewise considerably increased following foetal decapitation.

Some studies on the growth of the rat foetus have also been made (Bearn, 1968). As reported by Heggestad and Wells (1965), some decapitated rat foetuses may be smaller than headless control foetuses. However, some decapitated foetuses grow to the same extent as littermate controls (Fig. 6). It is therefore suggested that rat foetuses which show retardation of body growth following decapitation (Fig. 6) are not necessarily suffering from the effects of the loss of foetal pituitary growth hormone, since many foetuses clearly grow as well as controls. This retardation cannot, as yet, be explained but may be due to other factors such as interference in the placental circulation.

In support of these observations, de Beer and Gruneberg (1940) reported that in congenital pituitary dwarfism in the mouse, there is little difference in the birth weight of normal or dwarf mice, and growth in the dwarf mice continues at nearly the same rate as in the normals for the first three weeks after birth, when it stops completely.

Furthermore, this work correlates well with observations in human foetuses. Nãnagas (1925) concluded that skeletal growth in human anencephalics is not retarded. Lessof (1964) pointed out that human hypopituitary dwarfs grow normally up to two or three years of age, and Vest et al. (1963) found that infants under

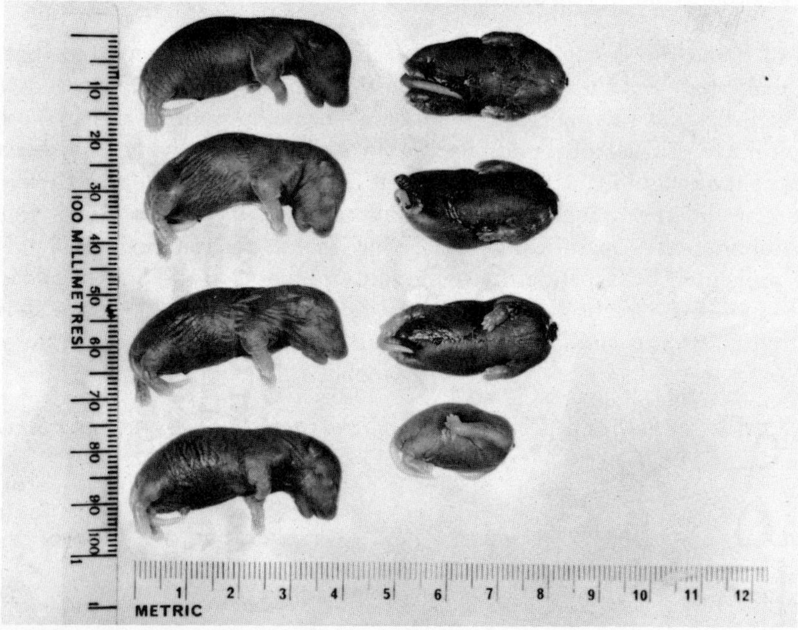

Fig. 6. Four rat foetuses decapitated at day 17½ and recovered at day 21½ with four littermate controls. Three decapitated foetuses have grown as well as the controls, but one is reduced in size.

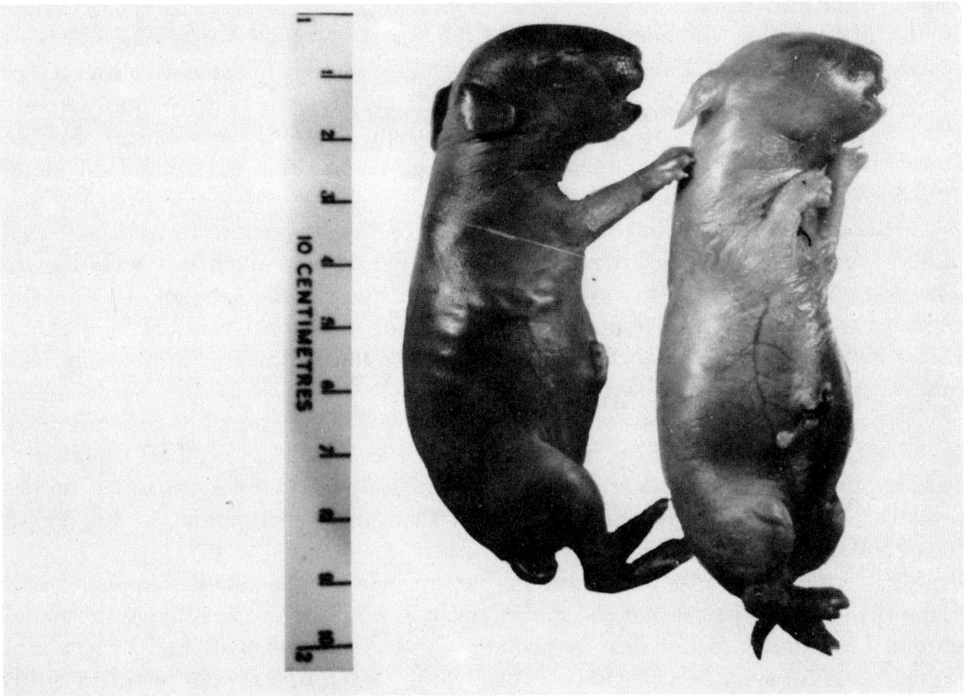

Fig. 7. A 29-day rabbit foetus thymectomized on day 22 (right) with a litter-mate control foetus (left). No reduction in growth follows thymectomy.

three months are less responsive to human growth hormone than older children. Grumbach found that premature human foetuses did not respond to human pituitary growth hormone (Grumbach, personal communication).

It is possible that a growth factor from the placenta may be essential for the great increase in growth during the latter part of intrauterine life (Grumbach, 1969).

The foetal thymus has in the past been suggested as playing an important role in foetal growth. However, in a study of foetuses thymectomised at 22 days and recovered for investigation at 29 days, body growth appeared to be normal (Fig. 7) (Bearn, unpublished observations). This work is part of an investigation to determine the effect of foetal thymectomy on the maturation of the foetal lymphatic system.

THE PITUITARY-ADRENAL AXIS AND FOETAL DEVELOPMENT

The importance of the foetal pituitary of mammals in the development of the foetal adrenals was demonstrated by Jost (1948, 1953) in the rat foetus and in the rabbit foetus. Following hypophysectomy by foetal decapitation, it was found that the subsequent development of the foetal adrenals was markedly retarded. Jost found that this effect could be abolished by the injection of ACTH into the foetus at the time of the decapitation. Similar findings were reported by Wells (1947,

1948) in the rat, using a similar experimental technique. Raynaud (1950) made similar observations in the mouse when the foetal pituitary was destroyed by a narrow beam of x-rays. These original findings have been confirmed on a number of occasions—for example, by Bearn (1960, 1968) and Eguchi (1961)—and the subject has been extensively reviewed by Jost (1966). These observations all demonstrate that the normal development of the foetal adrenal is dependent on foetal ACTH produced by the foetal pituitary.

Although the functional interrelationship of the foetal pituitary-adrenal axis is now well established, its role in foetal development is much less well-known. The first indication of its importance was shown by Jost and Jacquot (1958), who reported that this axis is essential for normal glycogen deposition in the liver of the rabbit foetus from the twenty-fourth day of development onward. Similar findings have been reported for the rat (Jacquot, 1959).

In view of the well-known observations in the adult animal on the influence of the adrenal gland on the lymphatic tissues of the body (reviewed by Dougherty, 1952), a series of investigations have been made in the rabbit and rat foetus on the influence of the foetal pituitary-adrenal axis on the development of the foetal lymphatic organs.

Foetal thymus. Initial observations were made on the foetal thymus; it was found that in both the rat and the rabbit, decapitation in utero to remove the foetal pituitary resulted in a decrease in growth of the foetal adrenals and hyperplasia of the foetal thymus (Bearn, 1960, 1961, 1966, 1967), and it was found that this hyperplasia could be prevented by injecting depot ACTH into the foetus at the time of the operation (Bearn, 1961, 1966, 1967).

Foetal spleen. The next organ studied was the foetal spleen of the rabbit (Bearn, 1967). Similar, but less well-marked, hyperplasia followed foetal decapitation, and this could be prevented by the injection of depot ACTH into the foetus at the time of the operation. It was concluded that the foetal pituitary, acting through the foetal adrenal, provides an inhibitory regulatory control on the growth of the foetal thymus and spleen.

Foetal liver haematopoietic tissue. Similar observations have been made on the foetal rat liver haematopoietic tissue (Bearn, 1967). This work demonstrated that when rat foetuses were hypophysectomised by decapitation in utero at day 17½ of development and the livers and adrenals were studied at day 21½, after

TABLE 3

Weight of rat foetal adrenals following hypophysectomy by decapitation (means ± S.E.)

Foetuses	Number	Age at operation (days)	Age at study (days)	Adrenal weights (mg/100 g headless body weight)
Decapitated	7	17½	21½	31.85[1] ± 1.90
Littermate controls	9		21½	73.10[2] ± 3.71

The difference between 1 and 2 is significant ($p < 0.0005$).

TABLE 4

DNA content of rat foetal liver following hypophysectomy by decapitation (means ± S.E.)

Foetuses	Number	Age at operation (days)	Age at study (days)	Liver DNA (mg/100 g wet weight)
Decapitated	29	17½	21½	738.9[1] ± 54
Littermate controls	49		21½	414.8[2] ± 13.6

The difference between 1 and 2 is significant ($p < 0.001$).

a further 4 days of intrauterine development, the adrenals from the experimental animals were less than half the size of the littermate controls (Table 3). The DNA content of the foetal livers was found to be greatly increased in the experimental foetuses (Table 4), but no change in liver weight was found in experimental foetuses compared with the littermate controls (Table 5).

A histological study was made of the haematopoietic tissue of the experimental and control livers. At the time of operation (day 17½), the foetal liver was found to contain large amounts of haematopoietic tissue scattered throughout the liver lobule. In the control livers, this tissue had declined rapidly by day 21½. However, livers from decapitated foetuses showed a much greater quantity of this tissue than the controls. It is concluded that in the rat foetus the foetal pituitary-adrenal axis, as in the case of the thymus and spleen, is exerting a similar inhibitory regulatory control over the normal decline in the liver haematopoietic tissue during the last 4 days of intrauterine life.

Circulating blood lymphocytes. Observations have now been completed on the circulating lymphocytes of the peripheral blood of the rabbit foetus (Bearn and Hawkey, Bearn and Marks, unpublished observations). In these experiments, a detailed haematological examination was made of the peripheral blood of 29-day rabbit foetuses which had been subjected to hypophysectomy by decapitation on day 22, and the results were compared with littermate controls. In both series of investigations, the decapitated foetuses showed a marked increase in the number of circulating lymphocytes in the region of over 100 percent. These observations

TABLE 5

Weight of rat foetal liver following hypophysectomy by decapitation (means ± S.E.)

Foetuses	Number	Age at operation (days)	Age at study (days)	Liver weight (grams per 100 g of headless body weight)
Decapitated	10	17½	21½	8.96[1] ± 0.316
Littermate controls	14		21½	9.48[2] ± 0.27

The difference between 1 and 2 is not significant ($p = 0.15$).

correlate well with the earlier studies, suggesting that the foetal pituitary-adrenal axis is exerting an inhibitory control on the circulating lymphocytes.

Alkaline phosphatase in the foetal duodenum. One further target organ for the foetal adrenal has been studied. This is the accumulation of alkaline phosphatase in the duodenal epithelium of the foetal rabbit (Bearn, 1966). This work was stimulated by the observations of Moog on the effect of the neonatal adrenal of the mouse on the evolution of alkaline phosphatase in the brush border of the duodenal villi. Observations were first made at daily intervals from day 21 to day 29 in the rabbit foetus on the development of the duodenal villi and the appearance of alkaline phosphatase. It was found that at day 21 the foetal duodenal epithelium is still extremely simple and consists of low columnar cells devoid of villi. By day 22 some folding of the epithelium has occurred, and by day 23 villi first begin to appear; they elongate up to the 29th day. Alkaline phosphatase activity first occurs at day 23, and increases rapidly with each day of development.

Adrenal activity in the foetus was inhibited experimentally by foetal decapitation in utero at day 22 to remove the foetal pituitary. The operated foetuses and littermates for controls were recovered 7 days later. The development of the duodenal epithelium had proceeded normally in the experimental foetuses, but alkaline phosphatase activity, although detectable, was very markedly reduced compared with control foetuses.

Sham operation on the day-22 foetus was undertaken to see whether the result was due to the surgical trauma rather than to the removal of the pituitary. The operation chosen was foetal thymectomy. This is a major procedure on the foetus and therefore comparable to foetal decapitation in terms of severity.

The development of the alkaline phosphatase in the duodenum of the thymectomised foetuses was found to be normal when compared with the control foetuses.

To restore the adrenal activity following foetal decapitation, depot ACTH was injected subcutaneously into the foetus at the time of the decapitation. Foetuses subjected to decapitation and injection of ACTH showed near normal accumulation of alkaline phosphatase.

These experiments suggest that the foetal pituitary-adrenal axis is essential for the normal development of alkaline phosphatase activity in duodenal epithelium.

THE DEVELOPMENT OF THE MALE GENITAL TRACT IN HUMAN ANENCEPHALY

Human anencephalic foetuses have abnormalities of their endocrine organs, the best known being the small adrenal and the small or absent anterior pituitary (Ballantyne, 1904; Covell, 1927; Angevine, 1938). It has been suggested that the cranial and adrenal defects were both due to a common cause, a placental abnormality (Jones, 1957). However, the observations on the pituitary-adrenal axis and the thymus in experimental animals reported earlier strongly support the view that the failure of growth of the adrenals in human anencephaly is secondary to the lesion of the hypothalamic-pituitary system. This is further supported by the occasional observation that the thymus of human anencephalics is found to be hypertrophied (Potter, 1961). It has also been noted by Blizzard and Alberts (1956) that foetuses with congenital absence of the pituitary in an otherwise normal head also show small adrenals and a large thymus. It may, therefore, be concluded that the small foetal pituitary when present in human anencephalics is nonfunctional and causes the small adrenals and large thymus reported in this abnormality.

This therefore raises a further problem for investigation. The classic work of Jost, reviewed in 1953, showing that the foetal testes of mammals are essential for normal male differentiation and that the foetal testes of the rabbit are under foetal pituitary control, poses the problem that male human anencephalics may show impaired differentiation and growth of the genital tract.

To study this problem, 13 male human anencephalic foetuses have been collected. These 13 male foetuses were part of a series of 60 anencephalics, the remaining 47 being phenotypic females. This disturbed sex ratio has been reported on many occasions. It is not due to complete sex reversal of some genetic males because of complete testicular insufficiency with subsequent feminisation of the genital tract, since the nuclear chromatin corresponds to the phenotypic sex in all cases (Bearn, 1959, 1968). However, in spite of this, it is possible that the male anencephalics may show impairment of either differentiation or growth of the genital tract, in view of the observations reported by Jost in 1953 that in the rabbit foetus, the foetal pituitary plays an essential and important role in the regulation of the foetal testes.

A detailed study has been made of these 13 male anencephalic foetuses, with particular reference to the differentiation and growth of the male genital tract (Bearn, 1968). The findings show that differentiation of the male genital tract is normal in all cases. However, at least 9 of the 13 male foetuses show marked hypoplasia of the normally differentiated external genitalia (Figs. 8 and 9). No

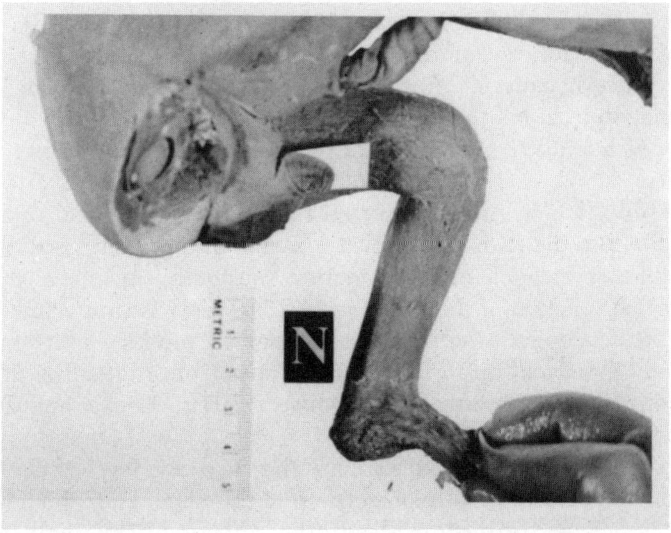

Fig. 8. A normal male human foetus showing normal external genitalia.

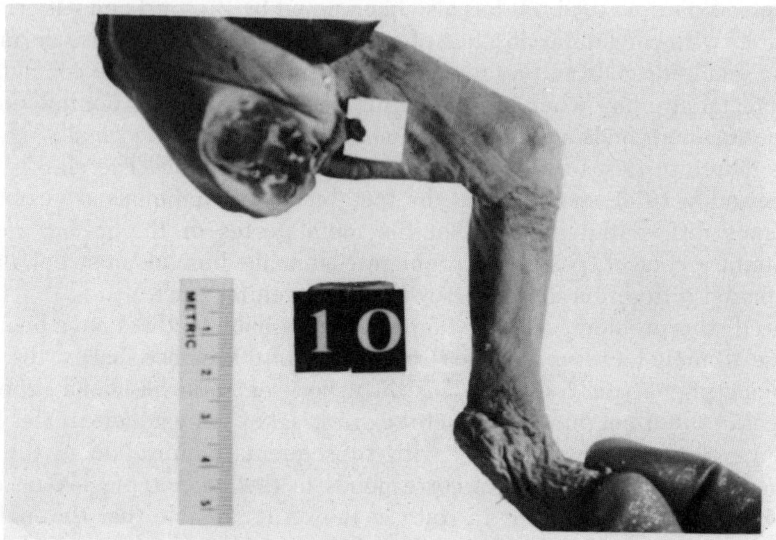

Fig. 9. A human male anencephalic foetus showing marked hypoplasia of normally differentiated external genitalia.

evidence of partial sex reversal was found. It was concluded that the normal differentiation of the male genital tract in these foetuses indicates normal foetal testicular function up to the fourth month, by which time differentiation is completed in the human foetus.

However, the marked hypoplasia of the genital tract found in the majority of the foetuses suggests that a partial testicular deficiency occurs after the fourth month in most anencephalic foetuses. It is suggested that the partial testicular deficiency after the fourth month is due to the hypothalmic-pituitary lesion, resulting in a decrease in anterior pituitary gonadotropic hormone. The normal foetal testicular function in human anencephaly, which is clearly present up to the fourth month and which leads to the normal differentiation, cannot, as yet, be explained with certainty.

It is possible that in the human, the foetal testis functions at an early stage without a gonadotropic stimulus from the foetal pituitary. It is also possible that the foetal pituitary in anencephaly functions normally up to this early stage, resulting in normal function of the foetal testis. This view is supported by the studies made on early anencephalic foetuses prior to the fifth month which show that adrenal development is normal up to this stage. It is not until after the 20th week that the foetal zone undergoes involution (Meyer, 1912; Keene and Hewer, 1927; Hunter, 1934; Benirschke, 1956).

However, the most likely explanation is the very high level of chorionic gonadotropic hormone which is normally present up to this stage and which could compensate for the deficiency in the foetal pituitary function. This high level falls to a very low level after the fourth month and may explain the foetal testicular insufficiency during the latter half of pregnancy, leading to the marked hypoplasia of the normally differentiated external genitalia in most of the male anencephalic foetuses.

REFERENCES

Angevine, D. M. Anatomy of hypophysis and adrenals in anencephaly. Arch. Path., 26:507–518, 1938.

Ballantyne, J. W. Manual of antenatal pathology and hygiene. *In* The Embryo, Edinburgh, Wm. Greene, pp. 332–351, 1904.

Bearn, J. G. Nuclear chromatin of anencephalic foetuses. Lancet, 2:24–26, 1959.

——— Effect of foetal hypophysectomy on the foetal liver fat in the rabbit. Nature (London), 185:849–850, 1960.

——— Le rôle de la glande surrénale foetale dans le developpement du thymus foetal du lapin. C.R. Acad. Sci. (Paris), 250:198–199, 1960.

——— Hormonal control of the development of the thymus of the foetal rabbit. Nature (London), 192:875–876, 1961.

——— Influence of the pituitary-adrenal axis on development of the rat foetal thymus. Proc. Soc. Exp. Biol. Med., 122:273–275, 1966.

——— The influence of the foetal pituitary-adrenal axis on the accumulation of alkaline phosphotase in the duodenal epithelium of the foetal rabbit. J. Anat., 100:449–450, 1966.

——— Role of the foetal pituitary and adrenal glands in the development of the foetal thymus of the rabbit. Endocrinology, 80:979–982, 1967.

——— The influence of the foetal adrenals on the development of the spleen of the rabbit foetus. Acta Anat., 68:97–101, 1967.

——— The influence of the rat foetal pituitary on foetal liver deoxyribonucleic acid and foetal liver haematopoietic tissue. Acta Anat., 68:239–250, 1967.

——— Anencephaly and the development of the male genital tract. Acta. Paediat. Hung., 9:159–180, 1968.

——— The thymus and the pituitary-adrenal axis in anencephaly. A correlation between experimental foetal endocrinology and human pathological observations. Brit. J. Exp. Path., 49:136–144, 1968.

——— and Pilkington, T. R. E. Effect of total thyroidectomy in the rabbit foetus on foetal cholesterol metabolism. Nature (London), 197:1302–1303, 1963.

——— and Pilkington, T. R. E. Hormonal control of the metabolism of cholesterol in the rabbit foetus. Nature (London), 198:1005–1006, 1963.

——— Antonis, A., and Pilkington, T. R. E. Foetal and maternal plasma triglyceride and cholesterol levels in the rabbit after foetal hypophysectomy by decapitation in utero. J. Endocro., 37:479–480, 1967.

——— Koerselman, H. B., and Pilkington, T. R. E. Function of the foetal thyroid of the rabbit with regard to cholesterol metabolism studied by foetal thyroidectomy. Nature (London), 190:822–823, 1961.

Benirschke, K. Adrenals in anencephaly and hydrocephaly. Obstet. Gynec., 8:412–425, 1956.

Blizzard, R. M., and Alberts, M. Hypopituitarism, hypoadrenalism, and hypogonadism in the newborn infant. J. Pediat., 48:782–792, 1956.

Covell, W. P. A quantitative study of the hypophysis of the human anencephalic fetus. Amer. J. Path., 3:17–28, 1927.

de Beer, G. R., and Gruneberg, H. J. Genet., 39:297, 1940.

Dougherty, T. F. Phys. Rev., 32:397, 1952.

Eguchi, Y. Atrophy of the foetal mouse adrenal following decapitation in-utero. Endocrinology, 68:716–719, 1961.

Grumbach, M. The effects of human chorionic growth hormone prolactin on energy metabolism in pregnancy. Unpublished.

Heggestad, C. B., and Wells, L. J. Experiments on the contribution of somatotrophin to prenatal growth in the rat. Acta. Anat., 60:348–361, 1965.
Hunter, R. H. An anencephalic embryo of 25 mm. C. R. length. Ulster Med. J. 3:105–108, 1934.
Jacquot, R. Recherches sur le contrôle endocrinien de l'accumulation de glycogène dans le foie chez le foetus de rat. J. Physiol. (Paris), 51:655–721, 1959.
Jones, I. C. The Adrenal Cortex. London, Cambridge University Press, 1957.
Jost, A. Influence de la décapitation au le développement du tractus génital et de surrénales de l'embryon de lapin. C. R. Soc. Biol., 142:273–275, 1948.
——— Problems of fetal endocrinology: The gonadal and hypophyseal hormones. Recent Progress in Hormone Research, 8:379–418, 1953.
——— Hormonal factors in the development of the foetus. Cold Spring Harbor Symp. Quant. Biol., 19:167–181, 1954.
——— Problems of fetal endocrinology: The adrenal glands. Recent Progress in Hormone Research. G. Pincus, ed. New York, Academic Press, Inc., 22:541–574, 1966.
——— and Jacquot, R. Sur la rôle de l'hypophyse des surrénales et du placenta dans la synthése de glycogène par le foie foetal du lapin et du rat. C. R. Acad. Sci. (Paris), 247:2459–2462, 1958.
Keene, M. F. L., and Hewer E. E. Observations on the development of the human suprarenal gland. J. Anat., 61:302–324, 1927.
Lessof, M. Develop.Med. Child. Neurol., 6:57, 1964.
Meyer, R. Nebennieren bei anenzephalic. Virchows Arch., 210:158–164, 1912.
Nanagas, J. G. A comparison of the growth of the body dimensions of anencephalic human foetuses with normal foetal growth as determined by graphic analysis and empirical formulae. Amer. J. Anat., 35:455–494, 1925.
Potter, E. L. Pathology of the Fetus and Infant, 2nd Ed. Chicago, Year Book Medical Publishers, 1961.
Raynaud, A. Recherches expérimentales sur le développement de l'appareil génital et le fonctionnement des glandes endocrines des foetus de souris (et le mulot). Arch. Anat. Micr. Morph. Exp., 39:518–576, 1950.
Smith, P. E., and Dortzbach, C. The first appearance in the anterior pituitary of the developing pig foetus of detectable amounts of hormones stimulating ovarian maturity and general body growth. Anat. Rec., 43:277–298, 1929.
Vest, M., Girard, J., and Bühler, J. Acta. Endocr. (Copenhagen), 44:613, 1963.
Wells, L. J. Progress of studies designed to determine whether the fetal hypophysis produces hormones that influence development. Anat. Rec., 97:409, 1947.
——— Some experimental evidence of production of adrenotrophin by the fetal hypophysis. Proc. Soc. Exp. Biol. Med., 68:487–488, 1948.

11

Insulin and Fetal Growth in the Rat

Luc Picon

Department of Animal Physiology,
University of Paris,
Paris, France

The possible control of fetal growth by hormones of fetal, placental, or maternal origin not only is of basic biological interest but it has clinical implications due to the frequency of fetal overgrowth in diabetic pregnant women. Two main hypotheses have been put forward to explain this abnormality:

1. Somatotropic (STH) hypersecretion by diabetic women and passage of an excess of hormone into the fetal blood may overstimulate fetal growth.

However, newborns of acromegalic women are of normal size(1). In the rat, STH injection into the pregnant female does not enhance fetal growth(2), and hypophysectomy of the dam does not result in diminished fetal size when litters of operated females and pair-fed control females are compared. Further, conditions consistent with this first hypothesis are lacking:

(a) The amount of plasma STH, as measured by radioimmunoassay, is essentially normal in diabetic women.(5)
(b) Growth hormone does not cross the placental barrier in humans(6,7).
(c) Fetal growth (and also the growth of the newborn during the first weeks of postnatal life) seems largely independent of somatotropic hormone. The weight increase is normal or subnormal in the fetal hypophysectomised rat(8) even if the dam is also hypophysectomised(9). In the decapitated rabbit fetus, weight increase is also normal(8,10), although no maternal or placental somatotropic factor seems to be present(11). Moreover, direct injection of STH into the intact rat fetus does not enhance its growth(12,13).

2. The overgrowth of the fetus may be produced by fetal hyperinsulinism in response to maternal hyperglycemia.

Hypertrophy of the pancreatic islets was, indeed, described as early as 1920 by Dubreuil and Anderodias(14) in a baby born to a diabetic woman. Hyperinsulinemia(15) and enhanced glucose tolerance(16) were also found in such babies. Moreover, the increased lipid content of the body in these newborns agrees well with their hyperinsulinism. Nevertheless, attempts to produce fetal gigantism experimentally, particularly in the pregnant rat made diabetic by pancreatectomy or alloxan, have been generally unsuccessful(17).

Although the fetal pancreatic islets are hypertrophied in such hyperglycemic pregnant rats, an increase of the fetal secretion of insulin, as is the case in the newborns of diabetic woman, may not have been noted. This has led us to test whether, in the fetal rat, insulin exerts a direct action on the growth and biochemical compositon of the body and on the glycemic levels in the blood.

ACTION OF INSULIN ON THE GROWTH AND BIOCHEMICAL COMPOSITION OF THE FETUS

Insulin was injected daily for two or three days into fetuses in utero; controls of the same litters were injected with saline. Injections started with fetuses aged 17½, 18½, 19½, and 21½ days post coitum (pc). The fetuses were removed one day after the last injection at 20½, 21½, 22½, and 23½ days pc, respectively. Fresh weight, dry weight, total protein and lipid content were measured (methods have been fully described by Picon[17]).

The injected insulin is without significant effect in fetuses removed on day 20½ pc (Table 1). In fetuses sacrificed on day 21½ pc, the insulin produces a clear-cut stimulation of growth. The fresh weight, the dry weight, the total content of proteins and, particularly, of lipids in the insulin-treated fetuses are all above the control values. The relative amount of lipids, as evidenced by the lipid:dry weight ratio, is also increased. Insulin action on growth and lipid content of the body is even more marked in fetuses sacrificed on day 22½ and 23½ pc.

Thus the injection of insulin in the rat fetus can produce the overgrowth and obesity observed in newborns from diabetic mothers. This effect of insulin can only be obtained at the end of gestation. In the same way, the overgrowth of fetuses from diabetic mothers is initiated only in the last trimester of the pregnancy(18).

ACTION OF INSULIN ON FETAL GLYCEMIA

Maternal and fetal glycemia were measured using the glucose oxidase method, one hour after injection of insulin (40 milliunits) to the treated fetuses and of saline to the controls of the same litter. These experiments were performed under Nembutal anesthesia on litters aged 19½ to 22½ days pc.

During this period, the maternal glycemia remains approximately stable; the glycemia of the control fetuses is very low at first, but rises and gradually approaches the maternal glycemia so that the maternofetal gradient for glucose steadily diminishes (Table 2 and Fig. 1). In the insulin-treated fetuses, by contrast, the glycemia remains low and significantly different from that of the mother until the end of pregnancy. Comparison between controls and insulin-treated fetuses

TABLE 1

Effect of insulin on the growth and biochemical composition of the rat fetus

Age in days post coïtum		No. of fetuses	Fresh weight (g)		Dry weight (mg)		Lipids (mg)		Proteins (mg)		$\frac{\text{Lipids}}{\text{Proteins}} \times 1000$	
			mean	**S.D.**	**mean**	**S.D.**	**mean**	**S.D.**	**mean**	**S.D.**	**mean**	**S.D.**
20½	controls	7	3.42	±0.38	405	±49	77	±9	291	±36	264	±10
	insulin	7	3.57	±0.27	387	±33	71.5	±4	270	±20	266	±18
21½	controls	18	4.46	±0.33	558	±48	105	±10	410	±39	259	±19
	insulin	18	4.92**	±0.55	619**	±74	120***	±15	445*	±70	275**	±14
22½	controls	11	5.83	±0.48	783	±66	144	±13	571	±55	253	±14
	insulin	11	6.67***	±0.30	920***	±44	196***	±21	635**	±41	309***	±22
23½	controls	15	6.71	±0.53	897	±84	179	±19	645	±64	278	±18
	insulin	11	8.06***	±0.79	1055***	±89	249***	±24	716**	±58	348***	±21

Student's t-test was used to determine statistical significance of the differences between controls and insulin-treated fetuses: * = $p < 0.05$; ** = $p < 0.01$; *** = $p < 0.001$.

TABLE 2

Effect of insulin on the glycemia of the rat fetus

Age in days post coïtum	No. of litters	Glycemia (grams/liter) 1 Dam		2 Control fetuses		3 Insulin-treated fetuses		Statistical significance of the differences between the means		
		mean	**S.D.**	**mean**	**S.D.**	**mean**	**S.D.**	**1-2**	**1-3**	**2-3**
19½	8	0.83	± 0.11	0.29	± 0.08	0.25	± 0.03	***	***	n.s.
20½	6	1.01	± 0.29	0.53	± 0.12	0.38	± 0.07	**	***	*
21½	6	1.16	± 0.22	0.78	± 0.15	0.52	± 0.07	**	***	**
22½	6	1.10		1.00	± 0.12	0.49	± 0.10	n.s.	***	***

Student's t-test was used to determine statistical significance of the differences between the means: * = $p < 0.05$; ** = $p < 0.01$; *** = $p < 0.001$; n.s. = not significant.

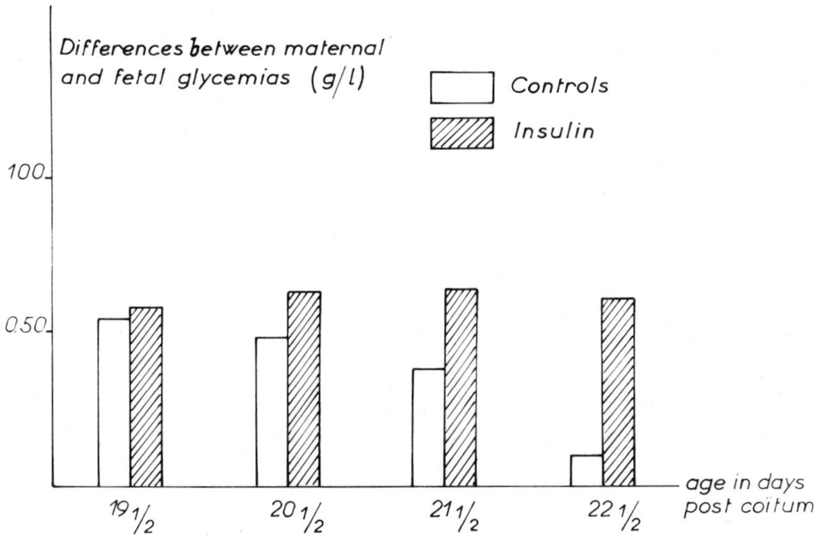

Fig. 1. Differences among the glycemias of controls, of insulin-treated fetuses, and of the dam.

shows that the effect of the hormone on the fetal glycemia becomes significant at 20½ days pc and then gradually increases(19).

Hypoglycemia in insulin-treated fetuses compared to controls of the same litters has already been described in several species: e.g., the dog(20), goat(21), and rabbit(22). However, in the rat fetus of 20½ days, Seller (23) was not able to obtain this result. It must be stressed that 20½ days is the very start of the period during which, according to our results, insulin hypoglycemia can be obtained in the fetal rat. Moreover, ether anesthesia, used by Seller, brought about a general hyperglycemia of the mother and fetuses which perhaps concealed the differences between controls and insulin-treated fetuses.

CONCLUSIONS

1. In the fetal rat, the effect of insulin on body weight can be observed only one day after its effect on glycemia. However, the effect on growth necessarily appears slowly in a cumulative way, whereas the hypoglycemic action of insulin is rapid. This is a satisfactory concurrence between these two effects of insulin and, probably, a correlation between them may be assumed.

2. In view of the increased maternofetal gradient for blood glucose and the lipid retention observed in the insulin-treated fetuses, it seems probable that this hormone increases the transfer of glucose from mother to fetus. Indeed, preliminary results obtained in collaboration with F. Bailly, using radioglucose injected into pregnant rats at 21½ days pc, support this view.

3. On the basis of observations in the rat, it may be assumed that growth and glycemia are partly regulated by the fetal pancreatic islets, at least in species, such as humans, in which fetal secretion of insulin is noticeable.

4. Finally, one wonders if the rise in glycemia observed at the end of gestation in the fetal rat is not important for the integrity of the nervous system in the developing fetus. It would be interesting, from that point of view to study the postnatal development of rats submitted to protacted hypoglycemia during late intrauterine life.

REFERENCES

1. Jackson, W. P. U. The prediabetic syndrome: Large babies and the (pre)diabetic father. J. Clin. Endocr., 14:177–183, 1954.
2. Frazer, J. F. D., Huggett, A. S. G., and Wohlzogen, F. X. Pituitary growth hormone and fetal growth. J. Physiol. (London), 108:44, 1949.
3. Campbell, R. M., Innes, I. R., and Kosterlitz, H. W. Some dietary and hormonal effects on maternal, foetal and placental weights in the rat. J. Endocr., 9:68–75, 1953.
4. Picon, L. Influence de l'hypophysectomie et de la surrénalectomie de la ratte gestante sur le poids des foetus. C. R. Soc. Biol. (Paris), 151:1314–1317, 1957.
5. Glick, M. S., Roth, J., Yalow, R. S., and Berson, S. A. The regulation of growth hormone secretion. Recent Progr. Hormone Res., 21:241–283, 1965.
6. Gitlin, D., Kumate, J., and Morales, C. Metabolism and maternofetal transfer of human growth hormone in the pregnant woman at term. J. Clin. Endocr., 25:1599–1608, 1965.
7. Laron, Z., Pertzelan, A., Mannheimer, S., Goldman, J., and Guttman, S. Lack of placental transfer of human growth hormone. Acta Endocr., 53:687–692, 1966.
8. Jost, A. Anterior pituitary function in fetal life. The Pituitary Gland. Harris, G., and Donovan, B. T., eds. London, Butterworths, 1966.
9. ——— and Picon, L. Hormonal factors in the growth of the foetus. *In* The Biological Action of Growth Substances. XIth Symp. Soc. Exp. Biol. Cambridge, Cambridge University Press, pp. 228–234, 1957.
10. ——— Experiénces de décapitation de l'embryon de Lapin. C. R. Acad. Sci. (Paris), 225:322–324, 1947.
11. ——— and Jacquot, R. Sur le rôle de l'hypophyse, des surrénales et du placenta dans la synthèse du glycogène par le foie foetal du lapin et du rat. C. R. Acad. Sci. (Paris), 247:2459–2462, 1958.
12. Heggestad, C. B., and Wells, L. J. Experiments on the contribution of somatotrophin to prenatal growth in the rat. Acta Anat., 60:348–361, 1965.
13. Picon, L., and Bouhnik, J. Action de l'hormone somatotrope et de la thyroxine sur la teneur en lipides du foetus de rat et du rat nouveau-né. C. R. Soc. Biol. (Paris), 162: 315–318, 1968.
14. Dubreuil, G., and Anderodias, J. Ilots de Langerhans géants chez un nouveau-né issu de mère glycosurique. C. R. Soc. Biol. (Paris), 83:1490, 1920.
15. Stimmler, L., Brazie, J. V., and O'Brien, D. Plasma insulin levels in the newborn infants of normal and diabetic mothers. Lancet, 1:137–138, 1964.
16. Isles, T. E., Dickson, M., and Farquar, J. W. Glucose tolerance and plasma insulin in newborn infants of normal and diabetic mothers. Pediat. Res., 2:198–208, 1968.
17. Picon, L. Effect of insulin on growth and biochemical composition of the rat fetus. Endocrinology, 81:1419–1421, 1967.
18. Cardell, B. S. The infants of diabetic mothers. A morphological study. J. Obstet. Gynaec. Brit. Comm., 60:834–853, 1953.
20. Schlossman, H. Beiträge zur Biologie der Plazenta. II Mitteilung: Ist die Plazenta für

Insulin durchlässig? Naunyn Schmiedeberg's Arch. Exp. Path. Pharm., 159:213–222, 1931.

21. Passmore, R., and Schlossmann, H. The effect of large doses of insulin on the foetal sheep and goat. J. Physiol. (London), 92:459–466, 1938.
22. Davies, J., and Lacy, P. E. Observations on the failure of insulin to pass from the fetus to the mother in the rabbit. Amer. J. Obstet. Gynec., 74:514–517, 1957.
23. Seller, M. J. The effect of glucose and insulin on the pregnant rat and foetus. J. Physiol. (London), 172:353–357, 1964.

[illegible] distribution [illegible]

[illegible] and [illegible]

[illegible] Transactions [illegible] the [illegible]

[illegible] (1943) [illegible]

[illegible]

12

Corticoids and the Enzymic Maturation of the Intestinal Surface: Alkaline Phosphataze, Leucyl Naphthylamidase, and Sucrase*

Florence Moog

Department of Biology, Washington University, Saint Louis, Missouri

In recent years the hypothesis that steroidal hormones, including corticosteroids, act by influencing the configuration of macromolecules has gained increasing support(1,2). An especially attractive aspect of this hypothesis is the possibility that corticoids influence the structure of membrane components, with consequent effects on the catalytic properties of membrane-bound enzymes, on permeability, and on the total economy of the cell. Although this hypothesis does not necessarily exclude the view that corticoids influence protein synthesis directly, many important actions of these hormones appear to be interpretable in terms of effects on membranes. The intestinal epithelium, which is dependent on cortical hormones for both its normal maturation and its maintenance, presents a number of situations that offer insights into the nature of some of the effects that corticoids are able to exert.

The epithelium of the small intestine is a continuous sheet of cuboidal or columnar cells that cover the villi and line the crypts of the intestinal wall. On their exposed surface the membrane of these cells is folded to provide the outer covering of a vast array of microvilli. This epithelial population is a dynamic one, in which new cells are continuously being produced in the crypts, glide up the sides of the

*Original work reported in this paper was supported by grant HD 03940 from the National Institutes of Health (U.S.).

I am grateful to my recent students, Dr. Marilynn Etzler and Dr. Robert D. Grey, for much stimulating research, discussion, and correspondence that have contributed significantly to this essay. I am also indebted to Edward Birkenmeier for the histochemical studies reported in this paper, to Chin-hai Chang for sectioning the material summarized in Figure 9, to Alex Denes for the sucrase determinations, and to Metin Leblebici for varied and skillful technical assistance.

villi, and are extruded at the villus tips; in a variety of mammals from mouse(3) to man(4) the life-span of an epithelial cell is close to two days. The period of transit is also a period of differentiation. Comparison of the proximal with the distal portions of the villi has revealed that the cells of the distal region have longer and more densely packed microvilli(5); are richer in disaccharidases(6), dipeptidases(7), and alkaline phosphatase(7,8); and have a higher capacity for absorbing sugars and amino acids(9). That these may be real differentiative changes, rather than enlargements of properties present in cells emerging from the crypts, is indicated by the discovery that the isozymic form of phosphatase at the tips of the villi is different from that at their bases(8).

To what extent the normal structural, biochemical, and dynamic properties of the adult intestinal epithelium are dependent on the secretions of the adrenal cortex has not been fully explored. In the rat the high mitotic rate of the crypt epithelium does not depend on corticosteroids(10). The capacity to absorb glucose does, however, fall about 40 percent in the rat duodenum after adrenalectomy(11,12), the defect being partly reparable by deoxycorticosterone but not by epinephrine(11). More recently, it has been shown that adrenalectomy also results in malabsorption of triglycerides and fatty acids(13). The alkaline phosphatase activity of the rat duodenum is reduced to a little more than half the normal level after removal of the adrenals, but is fully restored with cortisone or deoxycorticosterone(14). Similar results were obtained in the duodenum of the mouse(15). It was previously found that duodenal phosphatase is unresponsive to exogenous corticoids in intact mice between 15 and 30 days of age(16), but recent studies in my laboratory have shown that the adult duodenal epithelium in the intact mouse regains the ability to respond to injected cortisone by altering the activity and character of its phosphatase (Table 1). Jejunal phosphatase, however, is not affected either by adrenalectomy or by exogenous corticosteroids.

If the part that adrenocorticoids play in maintaining the differentiated state of the adult epithelium remains to be fully delineated, the importance of these secretions in the attainment of the adult state is beyond question. The relation of corticosteroids to intestinal differentiation has been examined most completely in terms of

TABLE 1

Influence of cortisone acetate on intestinal alkaline phosphatase of 8-month-old male mice

Region	Number*	Cortisone acetate	Phosphatase activity† (μg P/mg protein)	PhP/bGP ratio
Duodenum	10	−	4047	2.43
	10	+	6620	3.00
Jejunum	6	−	150	0.68
	6	+	152	0.65

*Duodenal and jejunal activities were determined for six animals in each group, only duodenal activities were determined for four others in each group.

†Phenylphosphatase was used as substrate in these determinations.

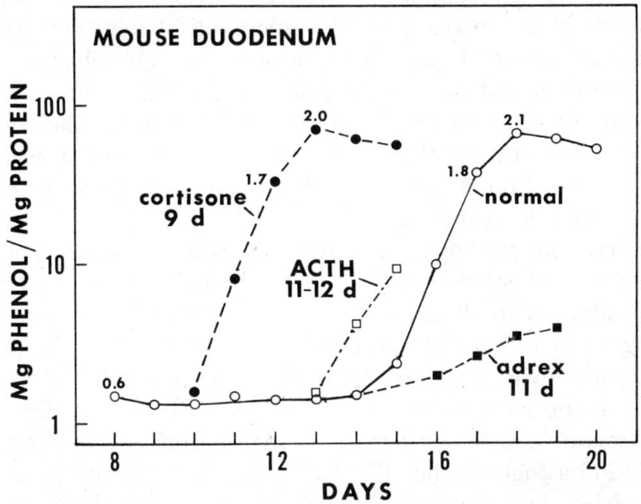

Fig. 1. A summary of the evidence showing that the increase of alkaline phosphotase activity in the duodenum of the mouse is under pituitary-adrenal control. Values given are activities with phenylphosphate as substrate; small figures appended to several points are PhP/bGP ratios (see text). Cortisone acetate, 30 μg/gm body weight injected at 9 days; ACTH, 70 μg/gm total given in 3 doses at 11 and 12 days. (Adapted from Moog[16,33].)

alkaline phosphatase (Fig. 1). This enzyme, which is a component of the microvilli, normally has relatively low activity in the duodenum during the first two weeks after birth, but then undergoes an approximately fiftyfold rise to a maximum at about 20 days, after which activity falls somewhat(17). An increase proceeding at the normal rate to the normal maximum may be elicited as much as a week early by a single subcutaneous dose of cortisone; administration of ACTH also causes a precocious increase, but the normal rise is suppressed by adrenalectomy at 11 or 12 days(16). The effect is readily evoked by other 11-oxycorticosteroids, but not by deoxycorticosterone or sex steroids(18).

Following this lead, a number of workers have added important results to our growing comprehension of the part that corticoids play in intestinal maturation. Clark demonstrated that the ability of the jejunal epithelium of the infant mouse to absorb intact protein molecules is halted prematurely by injection of cortisone(19). Halliday, in repeating this work on the rat, also found that cortisone has the same effect on the duodenal phosphatase of the rat as on that of the mouse(20). Doell and Kretchmer showed that invertase activity, which is scarcely detectable in the intestine of the infant rat, rises to its normal maximum precociously in response to exogenous cortisone(21). All these effects have one important characteristic in common: they proceed slowly, being negligible after 24 hours, but requiring 72 hours or even more to reach their maximum extent.

Perhaps the most significant of these studies on the effects of exogenous cortisone are those of Overton on microvilli. In the duodenum of the chick embryo, in which corticosteroids accelerate a wide range of maturational changes(22), Overton and Shoup found that the microvilli become longer and narrower in parallel with the normal increase of phosphatase activity(23). Subsequently, Overton repeated these observations in the mouse(24). The microvilli of the newborn mouse intestine are uniformly short and broad. In the period when phosphatase normally rises, the microvilli lengthen and become narrower. Since the volume of the microvilli remains constant, Overton suggested that the "maturation of these structures may involve reorganization rather than synthesis." In this case it was also demonstrated that the maturation of the microvilli can be elicited precociously by the administration of cortisone acetate(24).

These observations lead directly to two pertinent questions. The first question is inherent in the suggestion that the morphological alteration of the microvilli, which is concomitant with changes in the catalytic properties of phosphatase, entails a rearrangement rather than synthesis of new elements. The problem thus is to obtain additional evidence that might bear on these alternatives. The second question concerns the locus of the effect. Are only the partly differentiated cells in the crypts susceptible to the differentiation-promoting effects of adrenocorticoids, or are the cells in passage up the villi also responsive? Physiologically, these questions probably relate to a single mechanism. Operationally, however, they need to be examined separately.

CORTICOIDS AND THE ENZYMES OF THE MICROVILLAR MEMBRANE

The nature of the cortocoid effect on the maturation of the intestinal epithelium has been most extensively examined in terms of the influence on alkaline phosphatase. Since the early days of histochemistry, evidence has been accumulating to show that this enzyme is an integral part of the continuous phase of the outer membranous coat of the microvilli. Among the most important recent studies on this matter are those demonstrating that the enzyme is concentrated in the microvilli when these are stripped from the rest of the cell by treatment with EDTA(25); that the activity remains associated with the membrane of the microvilli when this component is separated from the fibrous cores(26); and that phosphatase also remains with the intact membrane after more loosely associated enzymes are removed with papain(27,28). In addition, electron microscopy has now elegantly confirmed the presence of phosphatase in the membrane(29).

These facts make it reasonable to assume that alterations in the properties of intestinal phosphatase are at least indicative of corresponding alterations in the outer membrane. In the intestine of the mouse, especially in the duodenum, nonspecific phosphomonoesterase comprises a series of isozymes differing in chromatographic, electrophoretic, and immunochemical characteristics(30-32). These isozymes also differ in the relative rates at which they hydrolyze phenylphosphate (PhP) and beta-glycerophosphate (bGP), and consequently the ratio of rate of release of phosphorus from the two substrates (PhP/bGP ratio) provides a convenient means of recognizing changes in phosphatase(33). By this means we have been able to establish that the isozymes of the duodenum are separated in space and time; as the cells travel up the villi their rate of cleavage of phenylphosphate

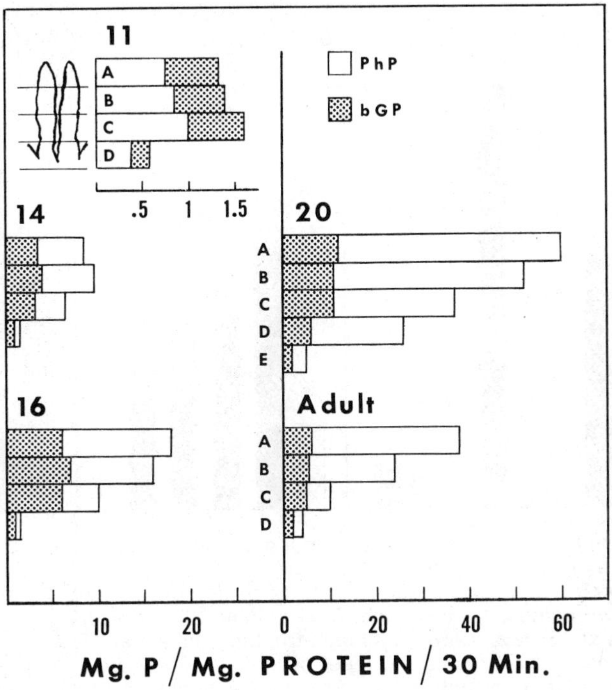

Fig. 2. The shifting distribution of phosphatase activity on the villi of the developing mouse duodenum. These data were obtained by sectioning frozen villi at right angles to their length (see sketch). Clear segments—activity with phenylphosphate as substrate; stippled segments—with beta-glycerophosphate. The 11-day histograms represent the infant condition with bGP activity exceeding that on PhP and uniform activity along the length of the villi. At 14 days and later, PhP activity exceeds that on bGP, with the difference becoming continually greater toward the villi tips (A level). These changes are corticoid-dependent. (Adapted from Moog and Grey[35].)

rises sharply and that of beta-glycerophosphate very slightly, so that the PhP/bGP ratio increases from about 2 in cells at the bases of the villi to about 6 in those near the tips(8) (Fig. 2).

The isozymes of the adult duodenum represents a class of phosphatases that seem to be present in inactive form in the duodena of infant mice(32,34). In normal development these isozymes appear gradually during the end of the second and the beginning of the third postnatal weeks(32) and, as one can deduce from the PhP/bGP ratios of groups of sections cut in parallel with the intestinal wall, little by little assume a graded distribution along the length of the villi(35) (Fig. 2). This differentiation is, of course, dependent on the adrenal cortex(16), and exogenous corticoid, administered in the second postnatal week, elicits a normal pattern of increase of overall PhP/bGP ratio as well as of activity. For the reasons stated before, these differentiative events may be regarded as indicating correlated events occurring within the microvillar membrane.

Do these effects, then, represent synthesis or reorganization? The attempt to

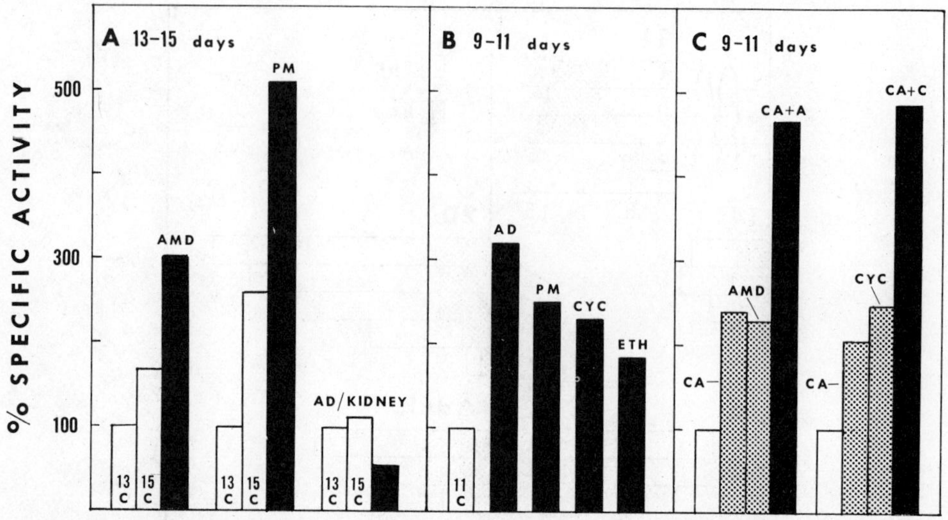

Fig. 3. The influence of antibiotics on the normal and cortisone-induced increase of phosphatase activity in the duodenum of the young mouse. *A*. Effect of actinomycin D (AMD) and puromycin (PM) injected at 13 days. Each group of three columns represents average values from two litters, with nine young in each. First column—13-day control, set at 100 percent; second—controls at 15 days; third—antibiotic-treated animals at 15 days. The last group of columns shows that AMD reduces the level of kidney phosphatase in the same animals in which it enhances duodenal activity. *B*. Effect of actinomycin, puromycin, cycloheximide (CYC), and ethionine (ETH) administered at 9 days. Each drug was given to half the young in two or more litters, the others receiving an equal volume of saline; 11-day control activities were set at 100 percent in each case. *C*. Effect of actinomycin, cycloheximide, and cortisone acetate (CA) given alone or in combination at 9 days. In these experiments two closely matched litters with eight or more young in each were combined, with equal numbers being given saline or one or both agents. Cortisone and antibiotic were injected within 1 hour. Animals were sacrificed at 11 days, control values at that time being set at 100 percent. (Adapted from Moog[36-38].)

resolve this question by the use of actinomycin D and puromycin led to the surprising discovery that these antibotics, far from inhibiting the increase of alkaline phosphatase that is normally going on between 13 and 15 days in the duodenum of the mouse, actually accelerate it(36) (Fig. 3A). These drugs also elicit a rise of alkaline phosphatase activity and an increase of its PhP/bGP ratio between 9 and 11 days, when activity is normally constant and the pituitary-adrenal system relatively quiescent(37,38) (Fig. 3B). That the effect is not due to a stress reaction is indicated by two findings: 1. Actinomycin is effective in enhancing duodenal phosphatase activity in 10-day-old mice that were adrenalectomized 24 hours earlier(38) (Fig. 4); and 2. When 9-day-old mice are given the largest dose of actinomycin, puromycin, or cycloheximide on which they can survive, plus a maximally effective dose of cortisone, the effects on phosphatase activity are additive(38) (Fig. 3C).

Similar effects on duodenal phosphatase activity are also produced by cycloheximide and ethionine(38) (Fig. 3B). Thus, precocious enhancement of phosphatase activity is brought about by four diverse agents that share in common the ability to inhibit protein synthesis, either by reducing synthesis of mRNA or by interfering

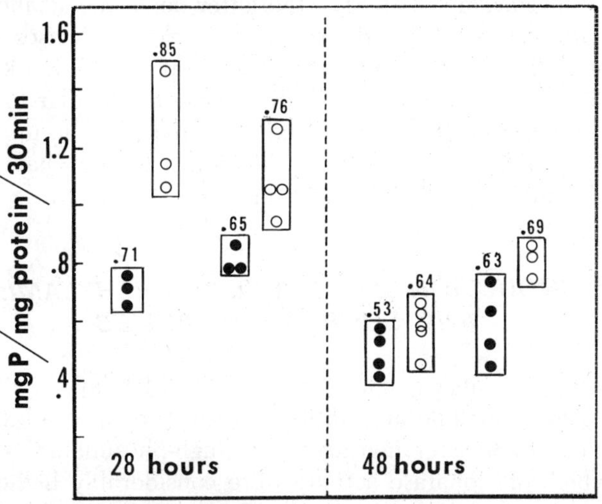

Fig. 4. The influence of actinomycin D on duodenal phosphatase activity in mice adrenalectomized at 9 days. Each pair of blocks represents one litter; small figures above the blocks are averages of Ph/bGP ratios. Left—0.3 μg actinomycin/g body weight at 10 days, with sacrifice after 28 hours. Average increase is 52 percent, $p<0.001$. Right–0.2 μg/g at 10 days, with sacrifice after 48 hours. Average increase is 22 percent, $p<0.04$.

with translation at different points. By contrast, actinomycin D (Fig. 3A), puromycin, and cycloheximide lower alkaline phosphatase activity in the kidney(38), a result consistent with their known inhibitory effects on protein synthesis. In the light of these findings, it seems reasonable to conclude that the normal developmental increase of duodenal phosphatase, and the alteration of its isozymic composition, are not primarily due to de novo synthesis of new phosphatase molecules. Since the developmental changes are dependent on adrenocorticoid stimulation, it follows that in the duodenal epithelium of the mouse, cortical hormones do not exert their maturational effects, on phosphatase at least, by augmenting or eliciting specific syntheses. Many of these same considerations apply to the duodenum of the chick embryo, in which actinomycin and cycloheximide evoke the appearance of active phosphatase on the epithelial surface at a time when histochemical techniques fail to demonstrate any surface activity at all(39,40).

Is the antibiotic-induced enhancement of the corticoid-dependent increase of phosphatase activity a peculiar situation, or is it representative of the kind of reorganizational processes that Overton(24) postulated as a result of her study of the development of microvilli? One way to approach this question is to examine the effects of corticoids and inhibitors of protein synthesis on other enzymes known to be associated with the microvillar membrane. Two enzymes that we have chosen for this purpose are sucrase (invertase) and leucyl naphthylamidase (LNA), which is frequently referred to as leucine aminopeptidase. These enzymes are associated with knob-like protrusions from the microvillar membrane, from which the protrusions can be released by treatment with papain(27,28). Eichholz has recently shown that the LNA-containing particles are separable on Sephadex G-200 from those bearing disaccharidase activities, the latter being the smaller(41). In the rat, sucrase is virtually absent during the first two weeks, then rises swiftly(21); LNA is present in the infant but rises late in the second postnatal week(42). Our studies indicate that the patterns of increase in the mouse are similar to those in the rat. In the adult both enzymes have their maximal activity in the midregion of the intestine (jejunum), and are in this respect in contrast to phosphatase, which is most active in the proximal sixth of the intestine.

THE INDUCTION OF LEUCYL NAPHTHYLAMIDASE AND SUCRASE ACTIVITIES

A series of experiments was carried out in which 10-day-old mice were injected subcutaneously with cortisone and antibiotics singly or in combination. The mice used were from brother-sister mating in our high-phosphatase strain, an SWR/J derivative in which phosphatase activities are considerably higher than those reported from my laboratory before 1967(16,17,33). Litters of nine were used; three were given water, three cortisone acetate or an antibiotic, and three cortisone plus an antibiotic, the two injections being made one hour apart. A second dose of cycloheximide was administered after 24 hours, but the other agents were given only once. After 72 hours the young were killed by decapitation and the entire small intestine, which is about 17 cm long at 13 days, was removed onto a plastic plate. The first 2.5 cm (duodenum) and the middle 4 cm (jejunum) were cut out, weighed

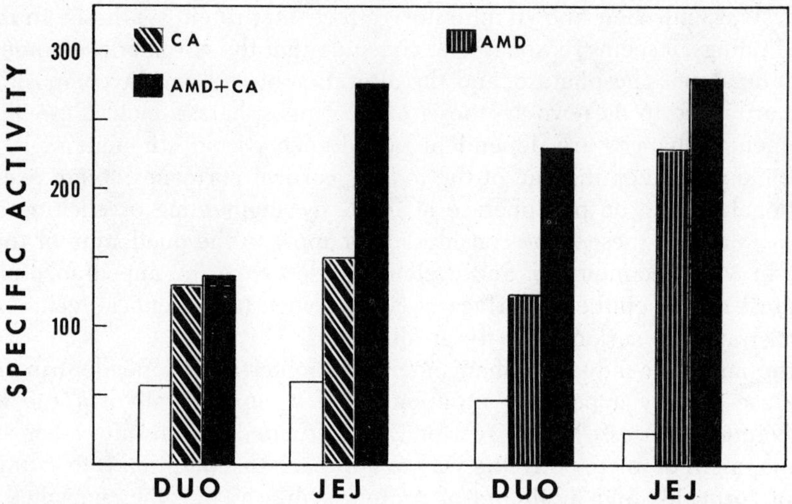

Fig. 5. The influence of cortisone acetate and actinomycin D singly and in combination on sucrase activity in the duodenum and jejunum of the mouse (see text). Each group of three columns represents two litters with nine young in each. Injections were made at 10 days with sacrifice at 13 days. Clear columns are values in water-injected controls. AMD—0.3 μg/g body weight; CA—30 μg/g. In combination experiments AMD was given first, cortisone one hour later. Specific activity is μg glucose/mg protein/hour.

on a torsion balance, and placed in ice-cold distilled water. In some cases a small piece just behind the jejunal segment was quickly frozen and promptly sectioned on a cryostat. Of 91 mice used in the experiments reported here, only 2 died during the experimental period. At the time of sacrifice the treated young were lively and alert; almost all had milk in their stomachs, and the intestines appeared undamaged.

The excised pieces of intestine were homogenized in water, and LNA(43) and sucrase(44) determinations were begun within one hour. Subsequently the protein content was measured(45), and phosphatase activity was assayed with phenylphosphate and beta-glycerophosphate by methods routinely in use in my laboratory(33, 38). The results may be conveniently dealt with for each enzyme separately.

Alkaline Phosphatase. Previous findings that duodenal phosphatase activity is elevated by cortisone(16,33), with the effect being enhanced by actinomycin or cycloheximide(36,38), were confirmed. The elevations were accompanied by the expected increases in PhP/bGP ratio. The only differences from previous findings were in two litters in which phosphatase activity at 13 days was already at the level beyond which responsiveness to cortisone disappears(16); in these cortisone did not produce any elevation above the control level alone, but did so in concert with actinomycin D. Preliminary studies also confirmed previous work showing that phosphatase in the jejunum, where the level of activity does not normally rise in juvenile stages (33), is not affected by cortisone or inhibitors of protein synthesis.

Sucrase. In the rat, sucrase activity is virtually absent until the 17th day, when a rapid increase begins(21). In our white Swiss mice the activity at 13 days is low

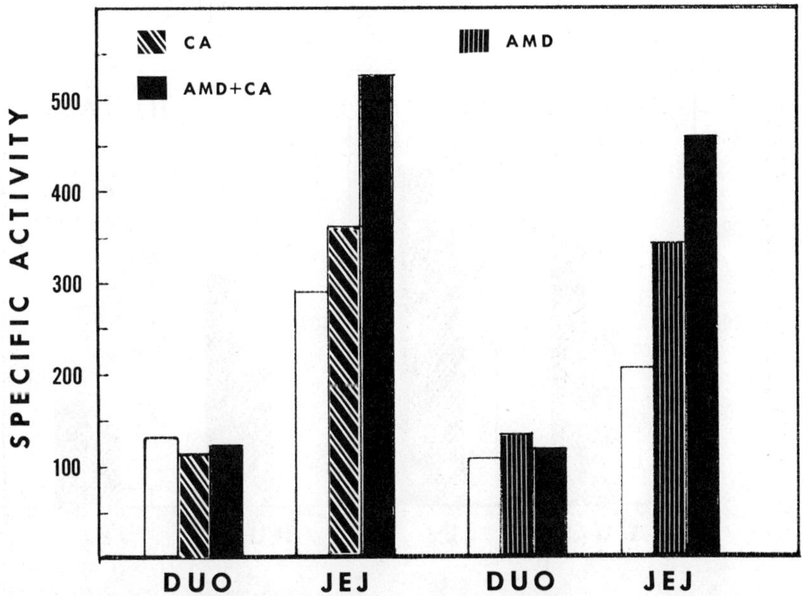

Fig. 6. The influence of cortisone acetate and actinomycin D singly and in combination on leucyl naphthylamidase activity. These data are from the same animals shown in Figure 5. Specific activity is μg beta-naphthylamine/mg protein/30 min.

but variable. In all cases the administration of cortisone at 10 days elevates sucrase activity in both duodenum and jejunum (Fig. 5). In the former area actinomycin either carries the rise further or fails to prevent it; in the latter, it has a marked enhancing effect. Actinomycin alone increases sucrase slightly. Cycloheximide augments the cortisone-induced rise strongly in the duodenum, less so in the jejunum (Fig. 7).

Leucyl Naphthylamidase, LNA. Work in progress in our laboratory shows that in the mouse, as in the rat(42), LNA activity maintains a uniformly low level of activity in the proximal region of the small intestine from infancy to adulthood. Cortisone has very little influence on duodenal LNA activity (Fig. 6). Actinomycin does not enhance the level, but does not lower it either. Cycloheximide does, however, have a positive effect in combination with cortisone (Fig. 7). In the jejunum, in which LNA activity normally increases approximately threefold between 16 and 20 days, the activity rises moderately in response to cortisone or actinomycin alone (Fig. 6). The effect is much stronger when both agents are administered, and cycloheximide and cortisone also have a marked enhancing effect when acting jointly.

In several cases a small piece of intestine from the posterior end of the jejunal segment was frozen, promptly sectioned at 10 μ on a cryostat, and stained for LNA activity according to the technique of Nachlas et al.(46), in which 1-leucyl-4-methoxy-2-naphthylamide is used as substrate(46). The sections were incubated with the substrate for periods of 5 to 40 minutes. These sections revealed that at 13 days LNA activity is largely confined to the crypts and villi bases (Fig. 8A). Under the influence of cortisone, activity increases over most of the surface of the villi (Fig. 8B). Actinomycin, however, causes a marked augmentation of activity in the basal

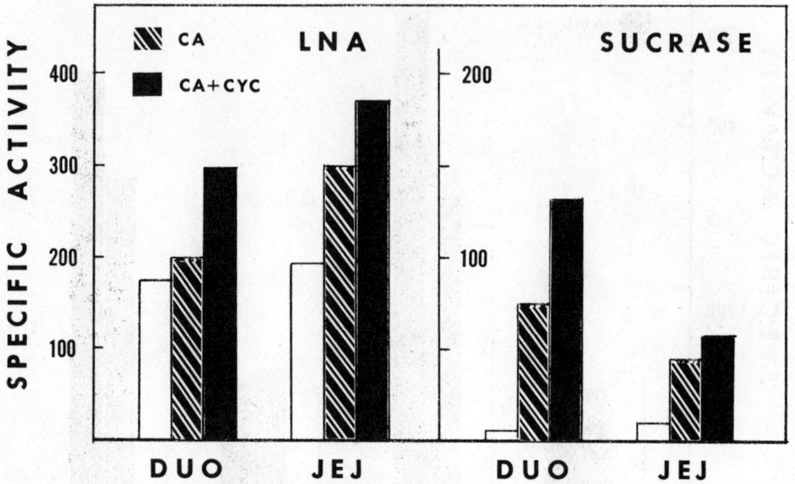

Fig. 7. The influence of cortisone acetate and cycloheximide on LNA and sucrase activity in the duodenum and jejunum of the mouse. Procedure was the same as for Figure 5, except that a second dose of cycloheximide was given 24 hours after the first. CA—30 μg/g; CYC—9 μg/g total. Specific activities as in Figures 5 and 6.

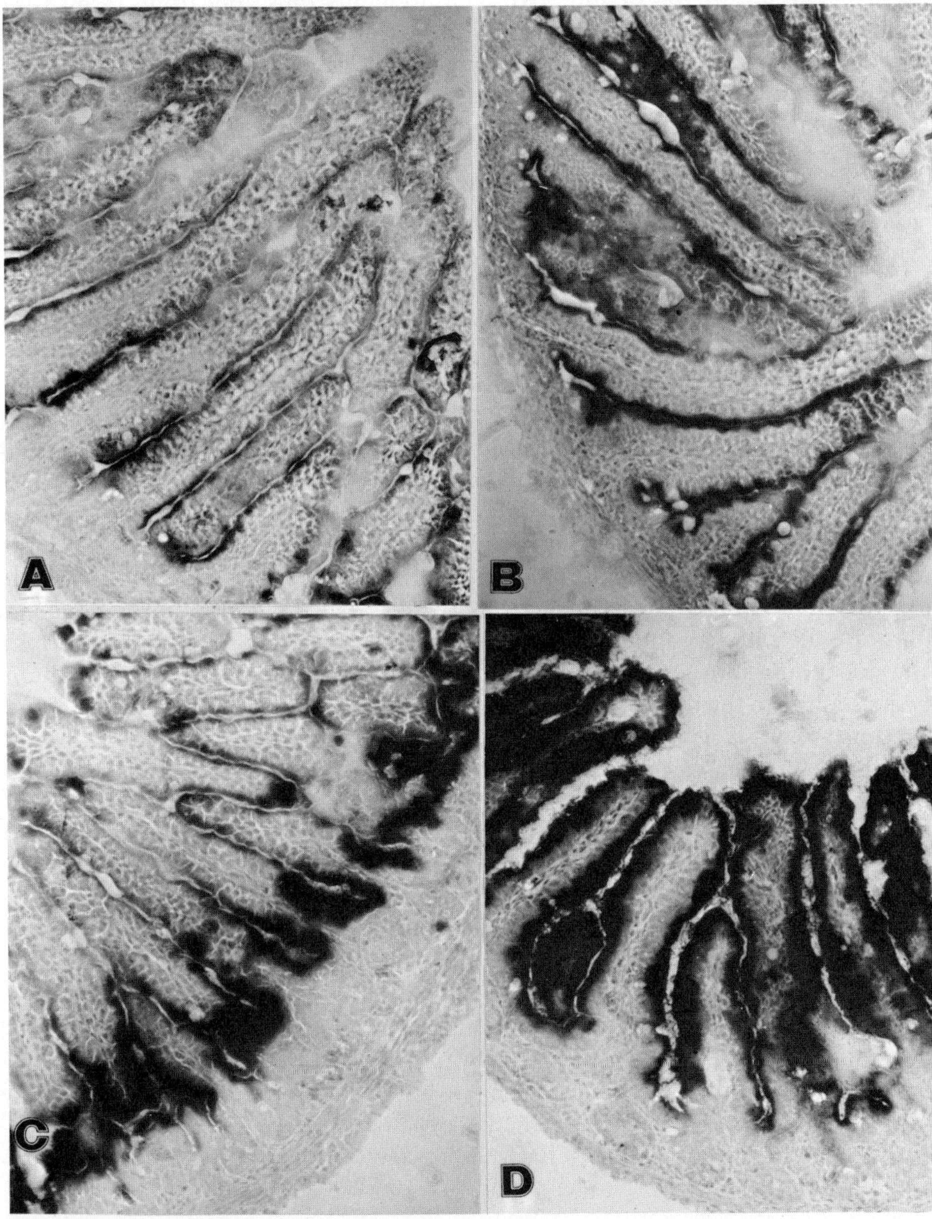

Fig. 8. LNA activity in the jejunum of 13-day normal and experimentally treated mice. The tissues are from some of the mice represented in Figure 6. *A*. Control, incubated 20 min. *B*. Cortisone-injected, incubated 20 min. *C*. Actinomycin-injected, incubated 20 min. *D*. Cortisone plus actinomycin, incubated 10 min. Specific activities of neighboring pieces of jejunum were A, 290; B, 360; C, 333; D, 445.

area only (Fig. 8C). When both agents are administered, activity is strongly enhanced over almost the entire villus (Fig. 8D). The synergism of cortisone and actinomycin suggested by the quantitative measurements thus appears to result from their ability to exert regionally discrete effects.

These findings reveal that three enzymes that are integral components of the microvillar membrane have certain developmental characteristics in common. In mice and rats all three increase in activity in some part of the intestine a little before the animals are normally weaned. The rise of all three enzymes may be brought about precociously by exogenous corticoids. Diverse antibiotics that share the ability to interfere with protein synthesis also elicit increases in the activities of these enzymes and enhance the corticoid-induced increases. That these are not simply nonspecific effects reflecting a stabilization of surface enzymes in the face of accelerated degradation of other proteins is indicated by the facts that activity per unit fresh weight is also increased, that the protein concentration of the fresh tissue is not significantly altered by anyof the experimental treatments reported here, and that the intestinal surface retains its integrity in both gross and microscopic observation (Fig. 8; see also reference 35). That the agents specifically act on normal developmental processes is, on the other hand, revealed by the already established fact that phosphatase is not only increased in activity but also altered in character according to the normal developmental trend. Now we may add the significant finding that corticoids and antibiotics do not enhance activities throughout the intestine, but rather act selectively in accordance with the normal regional pattern. By the same treatment, phosphatase activity is enhanced and transformed in the duodenum but not the jejunum; LNA is strongly affected in the jejunum, but very slightly in the duodenum (though the cycloheximide response is exceptional at present); sucrase activity is elevated in both regions, but more strongly so in the jejunum. In each case the result anticipates a normal differentiative change, for phosphatase rises to a high level in the third postnatal week in the duodenum but remains unchanged in the jejunum(33), whereas the reverse is true for LNA (unpublished results). Sucrase, starting from a trace in both areas, rises to a level three to four times higher in the jejunum than in the duodenum.

For the three enzymes that have been studied, the cortical hormone appears to make possible the realization of a series of differentiations that exist as potentialities in the 10-day epithelium. These differentiative changes are apparently not primarily a result of de novo synthesis of enzyme protein, for they are all promoted by several agents that have the common property of inhibiting synthesis. Whatever their nature, these differentiations seem to be localized in the microvillar membrane, for they are concomitant with the expansion of the outer membrane(24) and its acquisition of the ability to bar the passage of intact protein molecules(19).

WHAT CELLS ARE RESPONSIVE TO CORTICOSTEROID STIMULATION?

The second question that was raised earlier in this review concerns the locus of the adrenocorticoid effects at the tissue level. The principal alternatives are two: cells are responsive to corticoids only when they arise in the crypts and begin to acquire their essential characteristics; or they retain responsiveness throughout their period of transit up the villi. The only piece of evidence bearing specifically on this point is the demonstration by Doell et al., by the immunofluorescence technique, that sucrase appears first in the crypts, whether in normal development or as a result of hydrocortisone administration, and spreads up the villi as new cells move upward(47).

There are, however, no a priori grounds for ruling out the possibility that corticoids, and other agents, act directly on cells above the villi bases. In 18-day-old mice the cytoplasm of cells near the villi tips becomes heavily labeled within one hour after injection of tritiated uridine, and the nuclei become strongly labeled with tritiated thymidine within the same time(48). In adults epithelial cells all over the villi rapidly take up tritiated galactose, which appears in the microvillous border within 30 minutes(49), and Holmes and Crane(50) have recently shown that ^{14}C-leucine is incorporated into brush border membranes of cells from both the villus tips and bases within two hours. Evidently, both the protein and carbohydrate constituents of the membrane are being worked over throughout the life of the cell.

We have attempted to approach this problem in terms of the influence of cortisone on phosphatase activity in two ways. One way is to section the villi at right angles to their length 24 or 36 hours after the administration of cortisone acetate. On the basis of Grey's demonstration(51), on the mice used in our laboratory, that cells marked with tritiated thymidine in nurslings have moved only about a quarter of the distance from villus base to tip in 24 hours and have not yet arrived at the tip in 48, it may be estimated that cells found on the distal third of the villi in the cortisone-treated mice were already on the sides of the villi 36 hours earlier. Nine-day-old mice were used; four or six animals were taken from a single litter, half being given hormone, the others an equal volume of water.

No differences were evident at any level after 24 hours, but 12 hours later all levels were significantly elevated. Since the differences between litters were unexpectedly too great to permit averaging of the specific activities, I have instead set the lowest control value at each level in each litter at 100, and expressed other values in the same group in terms of this standard. Figure 9 sums up the results obtained with 14 mice from four litters. Although the percentage increases are greatest at the D and E (villus base and crypt) levels, the A and B levels, which consist largely of cells that were probably on the villi at the time of cortisone administration, are also strongly affected.

Another approach to the same problem has involved a comparison, by the same technique of cross-sectioning frozen villi, of the influence of cortisone and actinomycin(35). Again 9-day-old mice were used. The time between injection and sacrifice in this series was 60 hours. In this interval the entire intestinal epithelium has probably been replaced, but a distinct difference in phosphatase distribution nevertheless appears between the two treatments (Fig. 10). Although both agents exert their strongest effect at the bases of the villi (C level), only cortisone produces marked increases, relative to normal 11-day values, at more distal levels. The result clearly resembles the effects of these agents on LNA distribution, as revealed by histochemical staining (Fig. 8). It is worth noting that the effect on phosphatase distribution occurs in the duodenum, the similar effect on LNA distribution in the jejunum.

From these results it may be inferred that corticosteroid stimulation or sup-

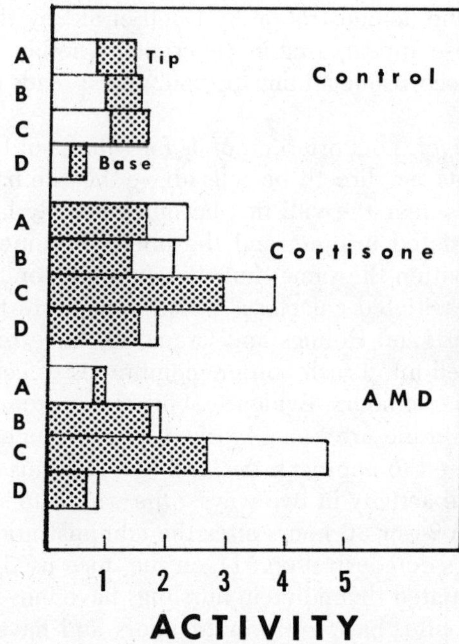

Fig. 9. The influence of cortisone and actinomycin D on the distribution of phosphatase activity along the length of duodenal villi (see Fig. 2). The animals were injected at 9 days and sacrificed 60 hours later. Clear segments—activity with PhP; stippled segments—activity with beta-glycerophosphate. Cortisone causes a strong increase in activity and relative increase on phenylphosphate at all levels; actinomycin has little effect at the tip (A) level, but a strong effect at the villi bases (C). (Adapted from Moog and Grey[35].)

port is essential not only for the normal developmental increase of phosphatase in the duodenum and of LNA in the jejunum, but also for the maintenance of these enzymes over the entire surface of the villus. The implication of the kind of data presented in Figures 8 and 10 is that the increased activity evoked by actinomycin D fails to persist in the absence of an adequate supply of adrenocorticoids; the enzyme molecules may be degraded, or simply lost into the intestinal lumen. Although the way in which the hormone exerts this effect is only a matter of speculation at present, there are reasons for thinking that the effect should involve the integrity of the membrane. The ability of corticoids to stabilize lysosomal surfaces is well known(52). That the plasma membrane itself may be stabilized is indicated by the finding that HeLa cells are protected against deoxycholate lysis by glucocorticoids(53). In this case, screening of six different glucocorticoids revealed that the ability to protect against lysis paralleled the ability to induce alkaline phosphatase activity(53), an event that has been shown not to involve enhanced synthesis of phosphatase molecules in HeLa cells(54). In the maturing intestine the central fact is that the developmental alterations of the surface enzymes are correlated with an apparent reorganization of the microvilli(24), both events being under adrenocortical control. These effects might be brought about in various indirect ways, though

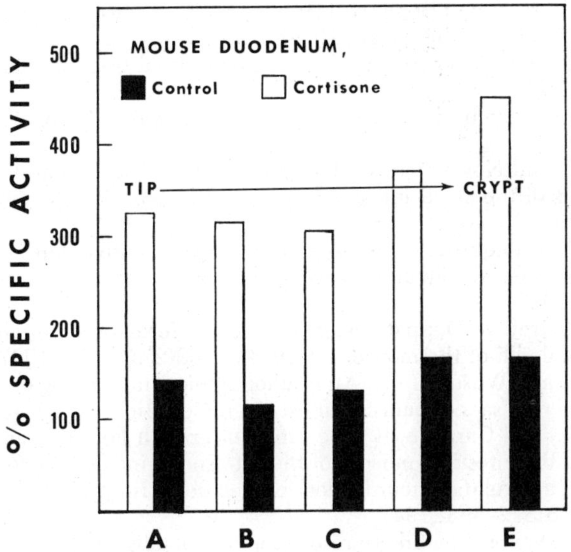

Fig. 10. The influence of cortisone on phosphatase activity along the length of the duodenal villi 36 hours after injection at 9 days (see Fig.2). Seven experimental animals and seven controls from three litters were used. Lowest control value in each group was set at 100 percent (see text). In this series the D level includes sections from villus bases and upper crypts.

they might also reflect a direct interaction of steroid and membrane, such as Willmer(55) envisaged, affecting the packing of the phospholipid constituents and thus the character of the membrane. An intriguing model suggesting what might occur has recently been put forward by Poltorak and Vorobeva, who have found that absorption onto artificial membranes can elevate the activity of purified alkaline phosphatase as much as six times above its level in homogenous solution, the degree of activation depending on the constitution of the membrane(56).

It is, of course, not to be thought that the maturation of the intestinal epithelium can be reduced to two elements: a membrane poised to differentiate and a hormone pouring out to deliver the signal. Different areas of the intestine and different components of its surface clearly respond according to predetermined patterns that bring us back to the vexatious puzzle of differentiation. Amid the welter of diverse events that bring the intestinal epithelium to the structural and functional complexity of its adult status, however, the recognition that corticosteroids elicit a series of correlated changes in the microvilli, their membranous coat, and its component enzymes may provide a unifying thread.

REFERENCES

1. Tomkins, G. M., and Maxwell, E. S. Some aspects of steroid hormone action. Ann. Rev. Biochem., 32:677–703, 1963.

2. Hechter, O., and Halkerson, I. D. K. Effects of steroid hormones on gene regulation and cell metabolism. Ann. Rev. Biochem., 27:133–162, 1965.
3. Lesher, S., Fry, R. J., and Kohn, H. I. Influence of age on transit time of cells of mouse intestinal epithelium. Lab. Invest., 10:291–300, 1961.
4. Bertalanffy, F. D. Cell renewal in the gastro-intestinal tract of man. Gastroenterology, 43:472–475, 1962.
5. Brown, A. L. Microvilli of the human jejunal epithelial cell. J. Cell Biol., 12:623–627, 1962.
6. Dahlquist, A., and Nordstrom, C. The distribution of disaccharidase activities in the villi and crypts of the small intestinal mucosa. Biochim. Biophys. Acta, 113:624–626, 1966.
7. Nordstrom, C., Dahlquist, A., and Joseffson, L. Quantitative determination of enzymes in different parts of the villi and crypts of rat small intestine. J. Histochem. Cytochem., 15:713–721, 1967.
8. Moog, F., and Grey, R. D. Spatial and temporal differentiation of alkaline phosphatase on the intestinal villi of the mouse. J. Cell Biol., 32(2):C1–C6, 1967.
9. Kinter, W. B., and Wilson, T. H. Autoradiographic study of sugar and amino acid absorption by everted sacs of hamster intestine. J. Cell Biol., 25:19–40, 1965.
10. Leblond, C. P., and Carriere, R. The effect of growth hormone and thyroxine on the mitotic rate of the intestinal mucosa of the rat. Endocrinology, 56:261–266, 1955.
11. Soulairac, A. La regulation neuroendocrinienne de l'adsorption intestinale des glucides. Ann. Endocr., 8:372–393, 1947.
12. Verzar, F., and Sailer, E. Glukose-resorption und alkalische Phosphatase des Dünndarmes nach Adrenalektomie bei mit NaCl behandelten Tieren. Helv. Physiol. Pharmacol. Acta, 10:247–258, 1952.
13. Rodgers, J. B., Riley, E. M., Drummey, G. D., and Isselbacher, K. J. Lipid absorption in adrenalectomized rats: the role of altered enzyme activity in the intestinal mucosa. Gastroenterology, 53:547–556, 1967.
14. Verzar, F., Sailer, E., and Richterich, R. Einfluss der Nebennierenrinde auf die alkalische Phosphatase der Dünndarmschleimhaut. Helv. Physiol. Pharmacol. Acta,10:231–246, 1952.
15. Moog, F., Bennet, C. J., and Thomas, E. R. Differential effect of adrenalectomy and adrenocortical hormones on the alkaline phosphatase of the duodenum, jejunum and ileum of the adult mouse. Anat. Rec., 120:777–778, 1954.
16. ——— The influence of the pituitary-adrenal system on the differentiation of phosphatase in the duodenum of the suckling mouse. J. Exp. Zool., 124:329–346, 1953.
17. ——— The differentiation of alkaline phosphomonoesterase in the duodenum of the mouse. J. Exp. Zool., 118:187–208, 1951.
18. ——— and Thomas, E. R. The influence of various adrenal and gonadal steroids on the accumulation of alkaline phosphatase in the duodenum of the suckling mouse. Endocrinology, 56:187–196, 1955.
19. Clark, S. L. The ingestion of proteins and colloidal materials by columnar absorptive cells in suckling rats and mice. J. Biophys. Biochem. Cytol., 5:41–50, 1959.
20. Halliday, R. The effect of steroid hormones on the absorption of antibody by the young rat. J. Endocr., 18:56–66, 1959.
21. Doell, R. G., and Kretchmer, N. Intestinal invertase: precocious development after injection of hydrocortisone. Science, 143:42–44, 1964.
22. Moog, F. Developmental adaptations of alkaline phosphatases in the small intestine. Fed. Proc., 21:51–56, 1962.
23. Overton, J., and Shoup, J. Fine structure of cell surface specializations in the maturing duodenal mucosa of the chick. J. Cell Biol., 21:75–85, 1964.
24. ——— Fine structure of the free surface in developing mouse intestinal mucosa. J. Exp. Zool., 159:195–201, 1965.
25. Holt, J. H., and Miller, D. The localization of phosphomonoesterase and aminopeptidase in brush borders isolated from intestinal epithelial cells. Biochim. Biophys. Acta, 58: 239–243, 1965.

26. Eichholz, A., and Crane, R. K. Studies on the organization of the brush border in intestinal epithelial cells. I. Tris disruption of isolated brush borders and density gradient separation of fractions. J. Cell Biol., 26:687–691, 1965.
27. Oda, T., and Seki, S. Molecular basis of structure and function of the plasma membrane of the microvilli of intestinal epithelial cells. Sixth Int. Congr. Electron Micr., pp. 387–388, 1966.
28. Johnson, C. F. Disaccharidase localization in hamster intestine brush borders. Science, 155:1670–1672, 1967.
29. Hugon, J., and Borgers, M. Ultrastructural localization of alkaline phosphatase activity in the absorbing cells of the duodenum of the mouse. J. Histochem. Cytochem., 14:629–640, 1966.
30. Moog, F., Vire, H. R., and Grey, R. D. The multiple forms of alkaline phosphatase in the small intestine of the young mouse. Biochim. Biophys. Acta, 113:336–349, 1966.
31. Etzler, M. E., and Moog, F. Immunochemical characterization of alkaline phosphatase isozymes of the mouse duodenum. Biochim. Biophys. Acta, 154:151–161, 1968.
32. ——— and Moog, F. Biochemical identification and characterization of the multiple forms of alkaline phosphatase in the developing duodenum of the mouse. Develop. Biol., 18:515–535, 1968.
33. Moog, F. Regional differences in the alkaline phosphatases of the small intestine of the mouse from birth to one year. Develop. Biol. 3:153–174, 1961.
34. Etzler, M. E., and Moog, F. Inactive alkaline phosphatase in the nursling mouse duodenum of the chick embryo and mouse. Biol. Neonat. 9:10–23, 1966.
35. Moog, F., and Grey, R. D. Alkaline phosphatase isozymes in the duodenum of the mouse: attainment of pattern of spatial distribution in normal development and under the influence of cortisone or actinomycin D. Develop. Biol., 18:481–504, 1968.
36. ——— Intestinal phosphatase activity: acceleration of increase by puromycin and actinomycin. Science, 144:414–416, 1964.
37. ——— Acceleration of the normal and corticoid-induced increase of alkaline phosphatase in the duodenum of the nursling mouse by actinomycin D, colchicine, puromycin and ethionine. Adv. Enzyme Regulation, 3:221–236, 1965.
38. ——— The regulation of alkaline phosphatase activity in the duodenum of the mouse from birth to maturity. J. Exp. Zool., 161:353–368, 1966.
39. Grey, R. D., and Moog, F. Elevation of alkaline phosphatase activity in the intestine of the chick embryo by actinomycin D. Nature, 211:418–419, 1966.
40. Moog, F., and Grey, R. D. A system regulating alkaline phosphatase activity in the duodenum of the chick embryo and mouse. Biol. Neonat. 9:10–23, 1966.
41. Eichholz, A. Studies on the organization of the brush border in intestinal epithelial cells. V. Subfractionation of enzymatic activities of the microvillus membrane. Biochim. Biophys. Acta, 163:101–107, 1068.
42. Noack, R., Koldovsky, O., Friedrich, M., Heringova, A., Jirsova, V., and Schenk, G. Proteolytic and peptidase activities of the jejunum and ileum of the rat during postnatal development. Biochem. J., 100:775–778, 1966.
43. Goldbarg, J. A., and Rutenburg, A. M. The colorimetric determination of leucine aminopeptidase in urine and serum of normal subjects and patients with cancer and other diseases. Cancer, 11:283–291, 1958.
44. Dahlquist, A. Assay of intestinal disaccharidases. Anal. Biochem., 22:99–107, 1968.
45. Lowry, O. H., Rosebrough, N., Farr, A., and Randall, R. Protein measurement with the Folin phenol reagent. J. Biol. Chem., 193:265–275, 1951.
46. Nachlas, M. M., Monis, B., Rosenblatt, D., and Seligman, A. M. Improvement in the histochemical localization of leucine aminopeptidase with a new substrate. J. Biophys. Biochem. Cytol., 7:261–264, 1960.
47. Doell, R. G., Rosen, G., and Kretchmer, N. Immunochemical studies of intestinal disaccharidases during normal and precocious development. Proc. Nat. Acad. Sci. U.S.A., 54:1268–1273, 1965.
48. Shorter, R. G., and Creamer, B. Ribonucleic acid and protein metabolism in the gut. Observations in gastro-intestinal cells with rapid turnover. Gut, 3:118–128, 1962.

49. Neutra, M., and Leblond, C. P. Radioautographic comparison of the uptake of galactose-H^3 and glucose-H^3 in the Golgi region of various cells secreting glycoproteins or mucopolysaccharides. J. Cell Biol., 30:137–150, 1966.
50. Holmes, R. J., and Crane, R. K. Protein turnover in the digestive-absorptive surface (brush border membrane) of the rat small intestine. Gut, 8:630, 1967.
51. Grey, R. D. Epithelial cell migration in the intestine of the young mouse. Develop. Biol., 18:505 514, 1968.
52. deDuve, C., and Wattiaux, R. Functions of lysosomes. Ann. Rev. Physiol., 28:435–492, 1966.
53. Melynkovych, G. Glucocorticoid induced resistance to deoxycholate lysis in HeLa cells. Science, 152:1086–1087, 1966.
54. Griffin, M. J., and Cox, R. P. Studies on the mechanism of hormonal induction of alkaline phosphatase in cell cultures. Proc. Nat. Acad. Sci. U.S.A., 56:946–953, 1966.
55. Willmer, E. N. Steroids and cell suraces. Biol. Rev., 36:368–398, 1961.
56. Poltorak, O. M., and Vorobeva, E. S. Adsorption models of biological membranes and alkaline phosphatase activation in different types of adsorption layers. Zhur. Fiz. Khim., 40:1665–1667, 1966.

13

Hormonal Regulation of Enzyme Synthesis: Transcriptional or Translational Control?*

Francis T. Kenney[1] and Jerry R. Reel[2]

[1]Biology Division, Oak Ridge National Laboratory,
Oak Ridge, Tennessee
[2]Department of Experimental Therapeutics,
Parke, Davis, and Co. Research Laboratories,
Ann Arbor, Michigan

It is well established that the synthesis of many enzyme proteins, particularly in the liver, is regulated by one or another of the hormones. This effect is specific, because large changes in rates of synthesis of specific proteins are detected under conditions where total protein synthesis is not affected to an appreciable extent. The complexity of the protein-synthesizing mechanism in animal cells affords many potential sites at which hormonal interaction may alter synthetic rates. However, many of these possibilities can be ignored in view of the requirement for specificity; this restriction means that attention must be directed toward mechanisms capable of changing the rates of synthesis of some proteins while leaving others unaltered. Our considerations on mechanism must, therefore, be elaborated in terms of specific templates for the proteins whose synthesis is altered. The option is immediately apparent: if synthesis of an enzyme is increased ("induction"), either more templates for that enzyme have been made (transcriptional control) or an unchanged number of templates have been made to function more effectively (translational control). Which of these is operative in hormonal regulation?

Our attempts to answer this question have centered on an experimental model system—the inducible enzyme tyrosine-α-ketoglutarate transaminase of rat liver. This enzyme has proven to be a most fortunate choice, particularly for reasons that could not have been predicted when the choice was first made. These include (1) its rapid turnover, which means that any change in the rate of synthesis is reflected

*This research was jointly sponsored by the National Cancer Institute, U.S. Public Health Service, and the U.S. Atomic Energy Commission under contract with Union Carbide Corporation.

almost immediately as a change in the amount of the enzyme and is thus readily detected; (2) its retention and unchanged susceptibility to hormonal regulators in hepatoma cells which can be cultured in vitro; (3) the sensitivity of its synthesis to both of what can now be recognized as the chief modes of hormonal regulation of enzyme synthesis in the liver. These are, respectively, regulation by steroid hormones (in this case, glucocorticoids), and regulation by glucagon or other hormonal effectors of the formation of hepatic cyclic AMP. In this review we shall discuss the first of these two types of hormonal regulators, the steroid hormones, in terms of the experimental findings that allow us to reach some tentative conclusions as to whether steroids affect transcriptional or translational processes in the liver. Details of the experiments described and relevant background can be found in earlier publications from this laboratory(1-6).

INDUCTION BY GLUCOCORTICOIDS

When moderate amounts of hydrocortisone or other glucocorticoids are administered to rats, synthesis of hepatic tyrosine transaminase is increased after a lag period of one or two hours (Fig. 1), leading to a marked elevation of the enzyme level, as originally described by Lin and Knox(7). The steroid is metabolized rapidly in vivo, and this induction will cease after four to six hours unless the hormone treatment is repeated, in which case a plateau is reached and maintained until treatment is discontinued. In cultured cells of the "minimal deviation" hepatoma discovered by Reuber(8) (H35 cells), supplementation of the medium with 10^{-6}M hydrocortisone effects induction of the transaminase that is, in most respects, iden-

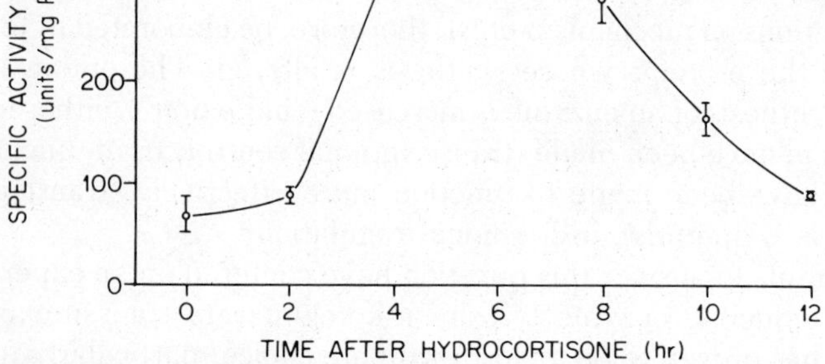

Fig. 1. Induction of tyrosine transaminase by hydrocortisone in hypophysectomized rats. The hormone (2.5 mg per 100 g) was given at zero time; data are means ± S.E. for three animals at each point. From Kenney et al. *In* Regulatory Mechanisms for Protein Synthesis in Mammalian Cells. San Pietro et al., eds. Courtesy of Academic Press, Inc.

tical to that seen in animals (Fig. 2). However, since the cultured hepatoma cells are apparently unable to degrade the steroid, elevated synthesis will continue in this system until the medium is replaced by steroid-free medium. Similar results have been described by Wicks(9) for the response of organ cultures of fetal liver. When the steroid is removed, the enzyme level remains at the plateau level for a period of about two hours. It then is rapidly lowered as synthesis returns to the basal rate and the process of enzyme degradation gains the upper hand (Fig. 2).

In terms of the alternative mechanisms proposed here, these results can be interpreted to mean either that the steroid has promoted synthesis of more transaminase templates or that the functioning of existing templates has been altered. When administered in vivo(2-4), glucocorticoids promote a large burst in the synthesis of all types of liver RNA, but have no detectable effect on total RNA synthesis in cultured hepatoma cells(10,11). Similar results have been reported by Wicks(9) in experiments with fetal liver in organ culture; i.e., glucocorticoids promote synthesis of tyrosine transaminase without altering total RNA synthesis. Thus it is clear that the massive synthesis of ribosomal RNA, for example, which follows hormone treatment in vivo, is not a necessary part of the hormonal induction of enzyme synthesis. However, if synthesis of RNA is prevented by moderate amounts of actinomycin, transaminase induction by hydrocortisone is completely blocked in tissue

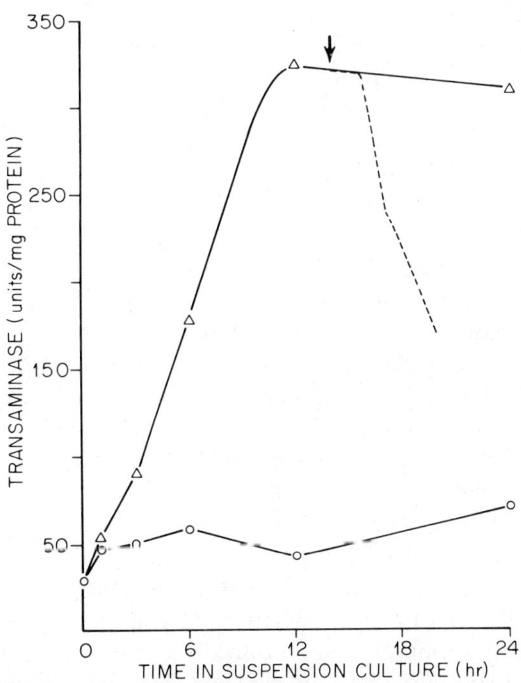

Fig. 2. Induction of tyrosine transaminase by hydrocortisone in suspension cultures of H35 hepatoma cells. Cells grown as monolayers were shaken in 10-ml suspension cultures with and without added hydrocortisone (10^{-6} M). The dotted line shows the drop in enzyme level which occurs after steroid-counting medium is replaced with fresh medium at the time indicated by the arrow. From Kenney et al. *In* Regulatory Mechanisms for Protein Synthesis in Mammalian Cells. San Pietro et al., eds., 1968. Courtesy of Academic Press, Inc.

cultures(10,12) as well as in animal experiments(13). This requirement for RNA synthesis lends support to the conclusion that the steroid acts at the level of transcription, promoting synthesis of new transaminase templates which support elevated synthesis of the enzyme. Such a specific alteration inRNA synthesis would not be expected to be detectable in analyses of total RNA synthesis, but might be found if the appropriate techniques were employed. In fact, a shift in the spectrum of hybridizable RNA s in liver nuclei after hydrocortisone treatment has been reported by Drews and Brawerman(14). In hepatoma cultures, Gelehrter and Tomkins(10) were able to detect a small increase in labeling of the 45S particle, which appears to be the form in which mRNA is transported from nucleus to cytoplasm. These important findings, which we can expect to be elaborated, are consistent with the conclusion that hydrocortisone promotes a specific transcription.

The kinetics of the induction response both in vivo (Fig. 1) and in vitro (Fig. 2) are also consistent with a transcriptional mechanism, *provided that the transaminase is synthesized on short-lived templates.* This requirement is necessary to explain the rapid cessation of induction following removal of the steroid, for if stable templates were formed as the result of a steroid effect on transcription, elevated enzyme synthesis should continue for an extended time after the steroid is withdrawn. Instead, synthesis of the enzyme returns quickly to the basal rate, a change that is detectable as a drop in the enzyme level only about two hours after removal of hormone from cultured cells (Fig. 2). This result, together with some paradoxical effects of actinomycin which could be interpreted as indicating that the transaminase template is stable, led Tomkins et al.(12) to suggest that hydrocortisone is required for effective translation of the induced templates as well as for their formation. It can be seen that resolution of this problem requires an understanding of (1) the stability of the template and (2) the effects of actinomycin on the enzyme level.

We have sought answers to these questions by using hepatoma cells in tissue culture as the experimental system best suited to yield the required precision; the results of one such experiment are presented in Figure 3. In these experiments hydrocortisone was added to the culture medium 24 hours before any further manipulations were made; thus, the transaminase and its rate of synthesis were both at high levels (i.e., induced) when the experiments began. Addition of actinomycin in amounts that block RNA synthesis nearly completely (5 μg/ml) causes the enzyme to increase to a new level that is nearly twice the induced level (Fig. 3, top curve). This paradoxical effect has been termed "superinduction"; it is this effect that has led to the postulation that translational control mechanisms are operative in this system, for it has been assumed that the increased amount of enzyme reflects increased enzyme synthesis. However, when these cells were pulse-labeled with ^{3}H-leucine and the incorporation of isotope into the enzyme was determined in order to measure its rate of synthesis, it could be seen that actinomycin actually causes a progressive inhibition of enzyme synthesis (Fig. 3, middle curve). Such a result, an increase in the enzyme level at a time when synthesis of the enzyme is decreasing, can be understood only if the degradation of the enzyme is also blocked. To measure this parameter we prelabeled the enzyme with ^{14}C-leucine, exposing the cells to this isotope for one hour before actinomycin was added, at which time the ^{14}C-leucine was washed from the cells. Thus, this portion of the experiment constitutes a "chase" analysis of the rate of degradation of the enzyme; the slope of the regres-

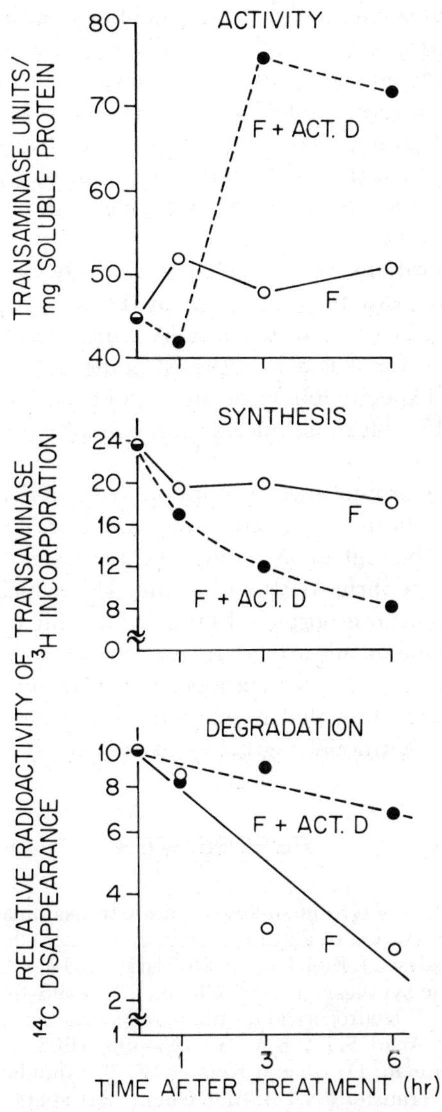

Fig. 3. Effect of actinomycin D on transaminase synthesis and degradation in preinduced HTC cells. Hydrocortisone (10^{-6} M) was added to monolayer cultures 24 hours before; ^{14}C-leucine (0.5 μc per ml), 1 hour before these measurements were begun. Immediately preceding the zero time the ^{14}C medium was decanted and the monolayers were washed with balanced salt solution. At zero time, fresh medium was added containing hydrocortisone (10^{-6} M) alone (open circles) or together with 5 μg per ml of actinomycin (shaded circles). For rates of synthesis measurements, 7.5 μc of ^{3}H-leucine per ml was added 15 minutes before cells were collected for analysis. Radioactivity (^{14}C and ^{3}H) of the enzyme was determined by differential counting after its precipitation with specific antiserum. From Reel and Kenney. *Proc. Nat. Acad. Sci. U.S.A.*, 61:200-208, 1968.

sion lines obtained from sequential measurements of ^{14}C in the enzyme (Fig. 3, lower curve) yields the rate constant for enzyme degradation. It is apparent that actinomycin blocks turnover of the enzyme and that this effect explains the elevation in the enzyme level that occurs under these conditions. Although synthesis is being progressively lowered, the rate is still unusually high for a few hours because of prior exposure to the inducing hormone. This being so, when degradation is blocked, the enzyme level rises despite inhibition of its synthesis. Hence it is clear that this effect of actinomycin does not indicate stability of the transaminase template and cannot be interpreted in terms of translational control mechanisms. The transaminase template is, in fact, quite labile, decaying with a half-life of about three hours (Fig. 3, middle curve).

From these considerations we conclude that the only mechanism consistent with all the experimental observations is one whereby the steroid promotes specific transcription, resulting in an increase in the mRNA which codes for the enzyme(s) induced. This hypothesis—for it is a hypothesis, being not yet established unequivocally—points toward a specific effect of the steroid, perhaps in combination with a receptor protein(15-17), on some nuclear component capable of limiting specific transcriptions.

If steroids induce by promoting transcription, what of the other hormones which increase the rate of synthesis of tyrosine transaminase? These include both those hormones that can be thought of as acting via increased cellular levels of cyclic AMP (glucagon[18], epinephrine[19]) and insulin(18), which lowers hepatic cyclic AMP levels. The data on these hormonal inducers are far from complete and their review is beyond the scope of this article. It can be said, however, that the evidence is clear that these hormones all act by a mechanism distinct from that initiated by hydrocortisone, thus suggesting that regulation of translation may be involved in the action of nonsteroid hormonal regulators of enzyme synthesis.

REFERENCES

1. Kenney, F. T. Induction of tyrosine-α-ketoglutarate transaminase in rat liver. II. Enzyme purification and preparation of antitransaminase. J. Biol. Chem., 237:1605–1609; III. Immunochemical analysis. J. Biol. Chem., 237:1610–1614; IV. Evidence for an increase in the rate of enzyme synthesis. J. Biol. Chem., 237:3495–3498, 1962.
2. ——— and Kull, F. J. Hydrocortisone-stimulated synthesis of nuclear RNA in enzyme induction. Proc. Nat. Acad. Sci. U.S.A., 50:493–499, 1963.
3. Wicks, W. D., Greenman, D. L., and Kenney, F. T. Stimulation of RNA synthesis by steroid hormones. I. Transfer RNA. J. Biol. Chem., 240:4414–4419, 1965.
4. Greenman, D. L., Wicks, W. D., and Kenney, F. T. Stimulation of RNA synthesis by steroid hormones. II. High molecular weight components. J. Biol. Chem., 240:4420–4426, 1965.
5. Hager, C. B., and Kenney, F. T. Induction of tyrosine-α-ketoglutarate transaminase in rat liver. VII. Hormonal effects on synthesis in the isolated, perfused liver. J. Biol. Chem., 243:3296–3300, 1968.
6. Reel, J. R., and Kenney, F. T. "Superinduction" of tyrosine transaminase in hepatoma cell cultures: differential inhibition of synthesis and turnover by actinomycin D. Proc. Nat. Acad. Sci. U.S.A., 61:200–208, 1968.
7. Lin, E. C. C., and Knox, W. E. Adaptation of the rat liver tyrosine-α-ketoglutarate transaminase. Biochim. Biophys. Acta, 26:85–88, 1957.

8. Reuber, M. D. A transplantable, bile-secreting hepatocellular carcinoma. J. Nat. Cancer Inst., 26:891–895, 1961.
9. Wicks, W. D. Induction of tyrosine-α-ketoglutarate transaminase in fetal liver. J. Biol. Chem., 243:900–906, 1968.
10. Gelehrter, T., and Tomkins, G. M. The role of RNA in the hormonal induction of tyrosine aminotransferase in mammalian cells in tissue culture. J. Molec. Biol., 29:59–65, 1967.
11. Reel, J. R., and Kenney, F. T. Manuscript in preparation.
12. Tomkins, G. M., Thompson, E. B., Hayashi, S., Gelehrter, T., Granner, D., and Peterkofsky, B. Tyrosine-transaminase induction in mammalian cells in tissue culture. Cold Spring Harbor Symp. Quant. Biol., 31:349–360, 1966.
13. Greengard, O., Smith, M. A., and Acs, G. Relation of cortisone and synthesis of ribonucleic acid to induced and developmental enzyme formation. J. Biol. Chem., 238: 1548–1551, 1963.
14. Drews, J., and Brawerman, G. Alterations in the nature of ribonucleic acid synthesized during regeneration and after cortisol administration. J. Biol. Chem., 242:801–808, 1967.
15. Toft, D., and Gorski, J. A receptor molecule for estrogens: isolation from the rat uterus and preliminary characterization. Proc. Nat. Acad. Sci. U.S.A., 55:1574–1581, 1966.
16. Sluyser, M. Binding of testosterone and hydrocortisone to rat tissue histones. J. Molec. Biol., 22:411–414, 1966.
17. Fiala, E. S., and Litwack, G. Binding of a ^{14}C-hydrocortisone metabolite in liver supernatant. Biochim. Biophys. Acta, 124:260–266, 1966.
18. Holten, D., and Kenney, F. T. Regulation of tyrosine-α-ketoglutarate transaminase in rat liver. VI. Induction by pancreatic hormones. J. Biol. Chem., 242:4372–4377, 1967.
19. Wicks, W. D. Tyrosine-α-ketoglutarate transaminase: induction by epinephrine and adenosine-3′, 5′-cyclic phosphate. Science, 160:997–998, 1968.
20. Kenney, F. T., Reel, J. R., Hager, C. B., and Wittliff, J. L. Hormonal induction and repression. *In* Regulatory Mechanisms for Protein Synthesis in Mammalian Cells. San Pietro, A., Lamborg, M. R., and Kenney, F. T., eds. New York, Academic Press, Inc., pp. 119–142, 1968.

14

Control Mechanisms in Hormonal Induction of Glutamine Synthetase in the Embryonic Neural Retina*

A. A. Moscona

Department of Biology, University of Chicago,
Chicago, Illinois

INTRODUCTION

The mechanisms which control the synthesis and accumulation of specific products in metazoan cells are of obvious interest in studies on embryonic differentiation. In embryonic cells, these mechanisms can be studied most effectively in circumstances in which changes characteristic of differentiation can be experimentally elicited. Embryonic inductions provide such conditions in that they result in the appearance of specific biosynthetic and morphogenetic changes in competent embryonic cells in response to the effects of appropriate inducers.

Past work on embryonic induction was directed predominantly towards analysis of morphological changes in inductive tissue interactions. The important and extensive information derived from this work (Grobstein, 1959; Wolff, 1968) stimulated much interest in the physiology and molecular biology of inductive processes in embryos (Rutter and Weber, 1965; Tiedemann, 1966). However, attempts to define and elucidate in detail the specific regulatory and biosynthetic events that take place in tissue interactions encountered various difficulties. The immediate and specific biochemical changes that occur in the initial phases of tissue interactions have, thus far, eluded identification; the consequent morphological changes usually appear considerably later and are probably several steps removed from the primary events involved in the mechanism of induction. Furthermore, attempts to isolate

*This work was supported by research grants G-23852 from the National Science Foundation and HD-01253 from the U.S. Public Health Service, Institute for Child Health and Human Development.

from embryonic tissues the actual inducer substances responsible for inductive tissue interactions and to identify them chemically have made slow progress because of the objective difficulties inherent in these systems and the limited amounts of cells accessible for such studies. It has long been felt that in order to study fundamental aspects of induction in embryonic cells, it is essential to work with experimental systems in which a chemically known inducer could be readily presented to a defined and easily accessible population of embryonic cells and would elicit in them a prompt, measurable change characteristic of their differentiation. The hormonal induction of glutamine synthetase (GS) in the embryonic neural retina meets these requirements. The purpose of this article is to review some of the work done in my laboratory on this system up to 1968.

The properties and functions of glutamine synthetase have been studied and described by Meister (1962) and Khedouri and Meister (1965). In our work, the activity of this enzyme is assayed spectrophotometrically in tissue sonicates by the glutamyl transferase reaction (Kirk and Moscona, 1963; Moscona and Hubby, 1963) modified from Rudnick et al. (1954) and Rudnick and Waelsch (1955). GS specific activity is expressed as micromoles of glutamyl-hydroxamate formed per hour per mg tissue protein.

The eye of the chick embryo is a relatively large structure; at embryonic ages relevant to this work the eye is approximately 0.6 cm in diameter. The neural retina can be easily and cleanly removed from it by a simple dissection (Piddington and Moscona, 1965). It is a cellular sheet consisting approximately of 10^8 cells (3 mg protein); it is avascular and is therefore uncomplicated by the presence of blood, vascular cells, or connective tissue elements. Other features add to its convenience for experimental studies and will be mentioned later.

Hormones play an important role in the control of embryonic development by coordinating synthetic and degradative processes through the regulation of enzyme activities in embryonic cells (Tata, 1965). Clarification of the mechanisms of enzyme induction in embryonic cells is of obvious significance to the problem of differentiation (Spiegelman, 1948). Analysis of these mechanisms and of the role of hormones in enzyme induction should advance the overall understanding of inductive processes in embryonic systems and of hormone action.

THE DEVELOPMENTAL PATTERN OF GS ACTIVITY IN THE EMBRYONIC RETINA

In the neural retina of the chick embryo the activity of glutamine synthetase (GS) follows a characteristic developmental pattern (Fig. 1) found only in this tissue (Rudnick and Waelsch, 1955; Moscona and Hubby, 1963). During early embryonic life, GS activity in the retina is low and increases at a slow rate until the 16th day of development when it starts to rise very sharply. It increases approximately a hundredfold in the next five days and then remains at this high level (Rudnick and Waelsch, 1955; Moscona and Hubby, 1963; Piddington and Moscona, 1965). The time of the rapid increase of GS activity in the retina coincides with the period of its final morphological differentiation and functional maturation. Detailed studies have determined that there is a remarkably precise temporal and spatial correspon-

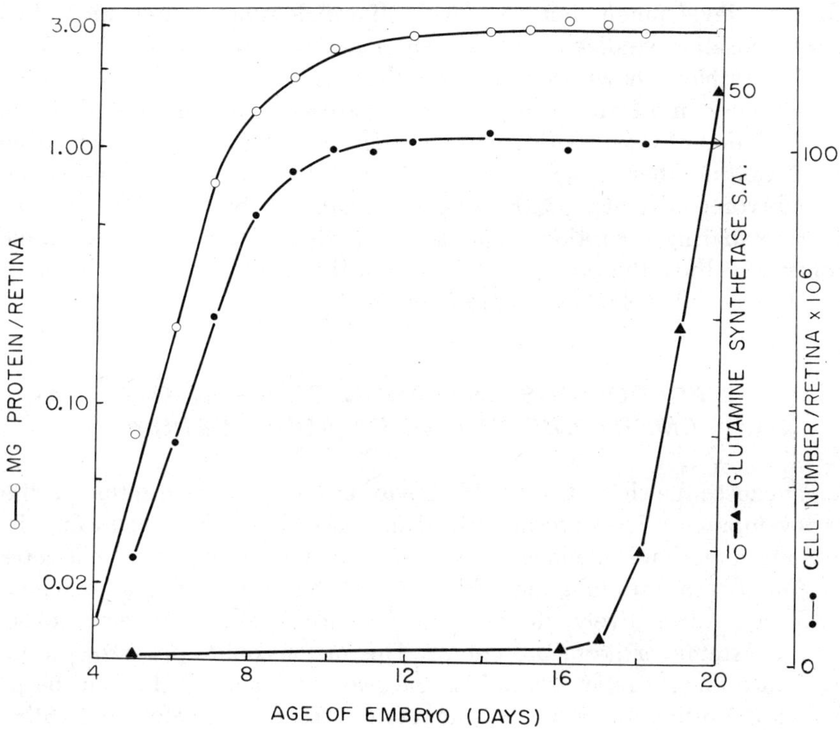

Fig. 1. The pattern of glutamine synthetase (GS) activity in the neural retina of the chick embryo showing the sharp rise of the enzyme after the 16th day of development. Also shown are changes in cell number and protein content of the retina.

dence between the level of GS activity in the embryonic neural retina and the degree of its morphological differentiation. This was established by taking advantage of the fact that morphological differentiation in the retina does not progress uniformly throughout the circumference of the eye; certain areas develop faster than others and in these areas GS activity is consistently highest throughout embryonic development (Piddington and Moscona, 1954; Piddington, 1967). These topographical and temporal correspondences between the level of GS and morphological differentiation in the retina, together with other information (see below), show convincingly that the sharp rise of GS activity is characteristic of the developmental program of this tissue and therefore provides a biochemical marker of its differentiation.

PROTEIN CONTENT AND CELL NUMBER IN THE RETINA

The above conclusions are further supported by the fact that the rapid rise of retinal GS activity which begins in the 16-day embryo is not due to cell proliferation in the retina or to an increase in the total amount of protein in this tissue. As seen in Figure 1, the amount of protein in the neural retina increases until the 12th day of embryonic development, but then levels off and does not change markedly thereafter. Incorporation studies of labeled amino acids into retinas from 12-day and older embryos have shown that although there is no growth, there is continuous protein synthesis and turnover in the retina associated, presumably, with differential accumulation of various proteins as the cells differentiate and attain specialized functions. Cell number in the retina reaches a steady level also at about the 12th day of embryonic development (Fig. 1). Therefore, on the 16th day, i.e., when GS activity rises sharply, the retina is a nongrowing cell population engaged primarily in differentiation; thus, the processes involved in the rise of GS in this tissue are not complicated by cell replication and growth.

PRECOCIOUS INDUCTION OF GS IN THE EMBRYONIC RETINA BY ADULT SERUM

What causes the sharp rise of GS activity in the embryonic retina on the 16th day of development? Theoretically, two major possibilities are conceivable. It could be caused by a mechanism endogenous to the retina cells which does not depend on any specific external stimuli, and which sets off the enzymic rise autonomously on the 16th day. Alternatively, the rise could be triggered by a specific exogenous factor of a systemic or hormonal nature. This factor could appear in the embryo shortly before the 16th day and induce the rise of GS activity. It could be present earlier, but the retina might not be competent to react to it before the 16th day of development. A further possibility is that the retina is potentially competent to respond to the factor long before the latter appears in the embryo; in this case, presentation of the factor to retinas from embryos younger than 16 days should elicit a precocious sharp rise of GS.

The analysis of these questions was markedly advanced by the finding that various adult sera elicited a rapid precocious rise of GS activity in retinas from young embryos, while fetal or early embryonic sera were considerably less effective or ineffective (Moscona and Hubby, 1963; Moscona and Kirk, 1965; Reif and Amos, 1966; Piddington, 1967). This was demonstrated by in vitro experiments, using organ culture procedures, in which retina tissue isolated from early chick embryos was maintained in culture media containing adult or fetal sera. These experiments tested two assumptions: (1) if the rise of GS activity in the 16-day embryonic retina was caused by a systemic factor which appears in later embryonic life, this factor would more likely be present in adult serum than in fetal serum, and therefore adult serum might cause a rise of GS; (2) if retinas from pre-16-day embryos were potentially capable of responding to this factor but normally unexposed to it till the 16th day, then exposure of such early retinas to adult serum might cause a precocious rise of GS activity.

Figure 2 shows the results of experiments in which retinas isolated from 7- through 16-day embryos were cultured at 37°C for 24 or 48 hours in Erlenmeyer flasks on a gyratory shaker (70 rpm) in a medium with 20 percent adult horse serum in Eagle's MEM. In all cases GS activity rose sharply during this culture period

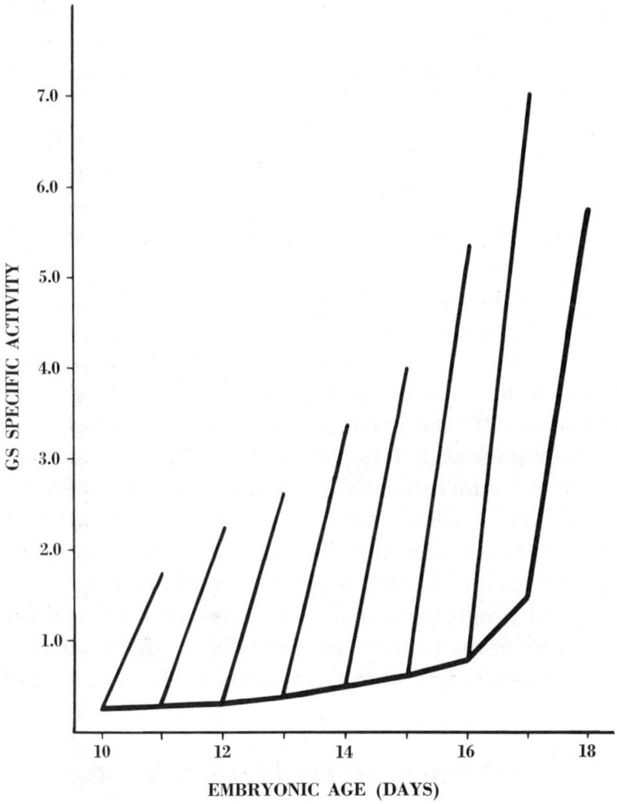

Fig. 2. Precocious induction of GS in neural retina from 10- to 16-day chick embryos cultured in the presence of adult horse serum. Base line—GS activities in the embryo. Vertical lines—the rise in GS activities induced by culturing retina of different embryonic ages in adult serum for 24 hours.

and started to go up soon after the retina was placed in the culture medium with the adult serum. On the other hand, in parallel cultures with fetal bovine serum or with serum from early chick embryos, there was either a slight increase or no significant increase in GS activity during the same culture period. In fact, the younger the chick embryo from which the serum was obtained, the less effective was the serum in inducing the premature rise of the enzyme (Piddington, 1967).

The precocious induction of GS activity in the cultured retina was obtained with adult sera from various animals, including the chicken, horse, rat, and man. Although these adult sera differed somewhat in their inductive efficiency, there was no evidence that the causal factor was species specific. On the other hand, the inductive effect of the serum was remarkably tissue specific in that it induced GS only in the embryonic retina and in no other embryonic tissue tested so far under the same experimental conditions, with the interesting exception of the embryonic optic tectum (Shimada et al., 1967). It should be pointed out that if the retina is cultured for longer than 24 hours in late fetal serum (for example, calf fetal serum), GS activity eventually goes up, presumably because late fetal serum contains a low concentration of the inducing factor which gradually accumulates in the cells. It was also found that addition of fetal bovine serum to medium with adult serum results in higher levels of GS activity induced by the latter, presumably because overall protein synthesis in the retina in vitro is better in culture medium with fetal bovine serum (for these and other effects of fetal serum in this system, see Moscona and Kirk, 1965).

From these findings it was concluded that (1) the embryonic chick retina is competent for the sharp rise of GS several days before this rise begins in normal ontogeny; (2) a precocious rise of GS in the retina can be induced in vitro by a serum factor present in various adult sera. We assumed that this factor accumulates in the serum in later embryonic life and that by the 16th day it reaches a concentration sufficient to induce GS in the retina.

The precocious in vitro induction of GS activity in the embryonic retina by the adult serum is accompanied or followed by the premature appearance of several morphological changes which normally appear in later stages of retinal differentiation (Piddington, 1967). Thus, the increase in the level of GS is not the only aspect of the developmental program of the retina that is accelerated when GS is precociously induced. These additional aspects may not include all of the changes required for the typical progression and completion of normal differentiation in the retina and may not be in the proper synchrony or sequence. It remains to be determined if they are somehow linked to GS induction or are independently controlled.

HORMONAL INDUCTION OF GS

Further analysis of the adult serum effect indicated that the GS-inducing factor was associated with the steroid-containing serum fraction. Various steroids were tested in the organ culture system for their ability to induce a precocious rise of GS activity in retinas from 12-day embryos. The steroids were added to culture media without serum (Tyrode's physiological solution or Eagle's MEM) or with fetal bovine serum (FBS). The duration of culture was 24 hours. It was found that 11β-hydroxycorticosteroids in physiological concentrations were excellent inducers of retinal GS (Moscona and Piddington, 1966, 1967). Their effect was specific for the retina in that they did not cause significant increases of GS activity in other tissues, with the exception of the optic tectum (Shimada et al., 1967). Other hormones—insulin, glucogon, adrenalin, sex hormones and somatotrophin—did not measurably affect the level of GS in the retina, neither alone nor in combinations with the active steroids. Thus, the steroid-mediated induction of GS in the neural retina does not require a complex sequence of hormonal effects, unlike certain other cases of hormonally controlled protein synthesis, for example, casein synthesis by the mammary gland in vitro (Turkington, 1968).

Figure 3 shows dose-response curves of GS induction by various corticosteroids. The most effective group were those with hydroxyls in 11β, 17, and 21 positions, including hydrocortisone, aldosterone, corticosterone, prednisolone, triamcinolone, and dexamethasone. In the absence of the 11β-hydroxyl, or if it was present in

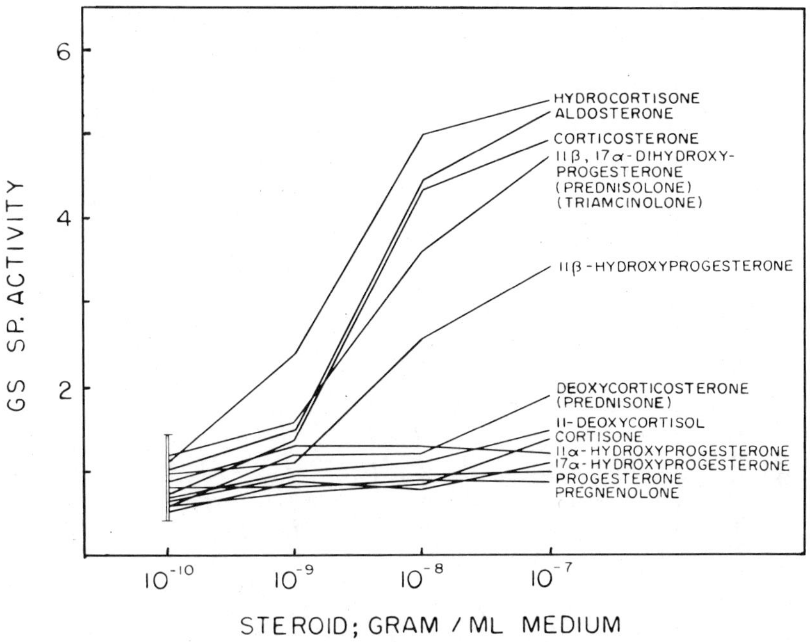

Fig. 3. Induction of GS in the 12-day embryonic retina cultured for 24 hours in the presence of various steroids. Dose-response curves. The culture medium consisted of Eagle's basal medium with 20 percent fetal calf serum (unpublished data).

the *a* configuration or replaced by oxygen, the steroid was ineffective or only partially active as GS inducer. The hydroxyls in the 17 and 21 positions were not essential for the induction but their absence reduced somewhat the inductive effectiveness of the molecule (Moscona and Piddington, 1966). These findings were recently confirmed and extended by Reif-Lehrer and Chader(1969). Further information about the detailed molecular properties of these steroid inducers and their effectiveness should prove useful in subsequent clarification of their mode of function in this system.

It should be stressed that GS induction by the active steroids also takes place readily in retina maintained only in physiological salt solution (Tyrode's) without any serum, which means that the steroid itself is responsible for this induction. Thus, GS induction in the embryonic chick neural retina represents a newly defined biological activity of 11β-hydroxycorticosteroids. Attempts to induce GS in the retina by adding cyclic AMP, dibutyryl cyclic AMP, or epinephrine to the culture medium or by injecting them into the embryo have produced negative results. Addition of these agents to cultures with the steroid inducer did not result in significantly higher GS activities than those elicited by the inducer alone. While these tests do not completely exclude a possible role of endogenous cyclic AMP in the control of GS levels in the retina, this enzyme is not induced by the addition of cyclic AMP to the cells.

Figure 4 shows that the precocious induction of GS by hydrocortisone in cultured retinas from embryos of different ages is very similar to that obtained with

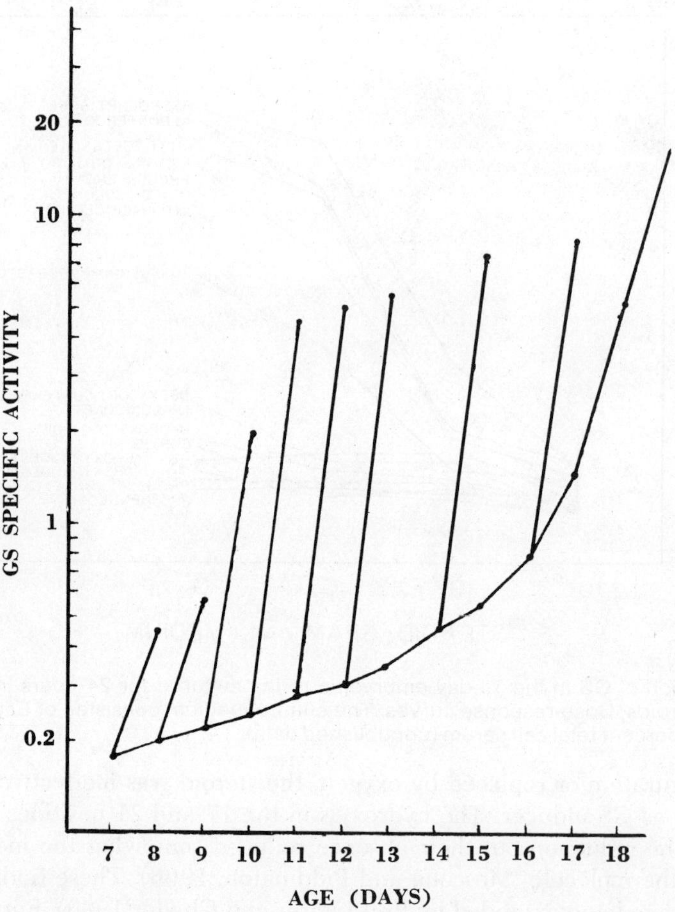

Fig. 4. Induction in vitro of GS by hydrocortisone. Retinas obtained from chick embryos of various ages were cultured in medium with 3×10^{-7}g/ml of the steroid. Broken lines represent the 24-hour increases in GS activity; solid line represents GS activity in the retina in normal embryonic development.

adult serum. As mentioned earlier, the levels of GS induced in the explanted retina by adult serum are higher if fetal serum is also present in the culture medium (Moscona and Kirk, 1965). Similarly, the levels of GS induced by the steroid inducer are higher in the presence of fetal serum in the medium. On the other hand, adult serum does not significantly augment the effect of the steroid inducer if the latter is present at optimal concentration. The effects of fetal serum may be due to its stimulation of protein synthesis in the cultures; possibly, the fetal serum may contain an as yet unidentified factor which in some way amplifies the effect of the steroid inducers and of the adult serum factor (Moscona and Kirk, 1965).

It is particularly important that the active steroids induce GS in the retina not only in culture, but also in the embryo (Fig. 5). A single injection of hydrocortisone (or of the other effective steroids) into early embryos induces a precocious

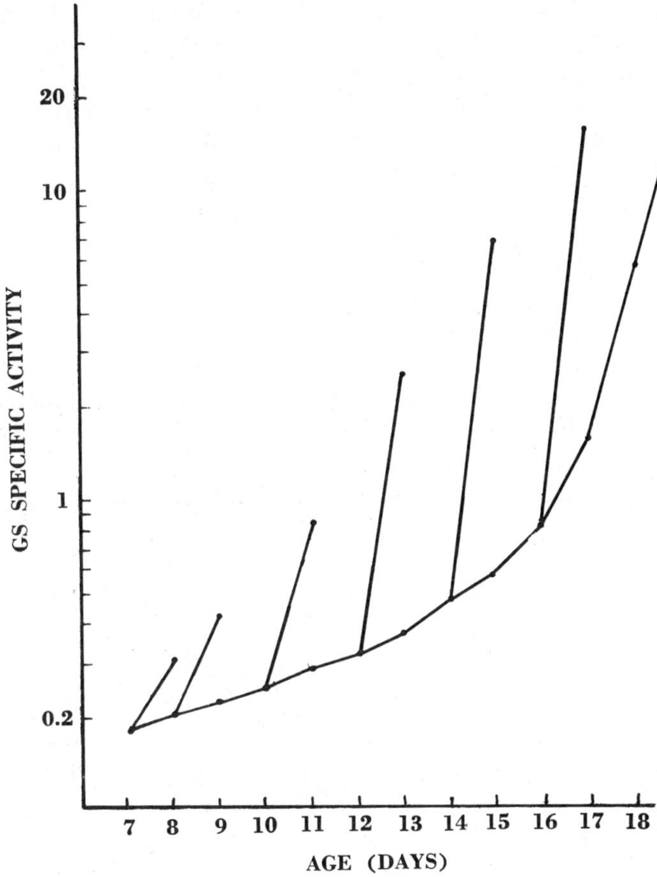

Fig. 5. Induction of retinal GS by hydrocortisone in the chick embryo. Broken lines represent the increases in GS activity 24 hours after one injection of 0.5–1.0 mg of hydrocortisone into the embryos; solid line represents GS activity in the retina in normal embryonic development.

rise of GS activity in the retina, but in no other tissue in the embryo (Moscona and Piddington, 1966; Piddington and Moscona, 1967; Piddington, 1967). Various other hormones tested had no such effect. Furthermore, a precocious rise of GS is obtained if embryos are injected with ACTH to stimulate a premature onset of adrenal cortical function; ACTH does not have this effect on the isolated retina in culture. In this connection it is of interest that the secretory function of the adrenal cortex in the chick embryo is said to begin between the 14th and 15th day of development and thus to precede by a day or two the time of the normal rise of GS activity in the retina.

While there is as yet no *direct* proof that an 11β-hydroxycorticosteroid is actually responsible for the inductive effect of the adult serum or for the normally timed rise of GS in the retina in the embryo, the above information clearly agrees with this assumption. Thus, pending evidence to the contrary, our working assumption is that corticosteroids induce GS in the retina not only under experimental conditions, but are also responsible for the normal rise of this enzyme in the embryonic development of the neural retina. A direct test of this assumption is technically difficult. However, the immediate usefulness of these steroids is clear: they provide chemically defined agents for inducing in the embryonic retina cells an enzymic change characteristic of their differentiation and for studying the processes involved in this cellular response.

THE KINETICS OF GS INDUCTION

The kinetics of GS induction were determined in retinas from 12-day chick embryos cultured for 24 hours in Erlenmeyer flasks in the presence of 10^{-7} g/ml of hydrocortisone in Tyrode's solution or Eagle's MEM, with or without 20 percent fetal bovine serum (Moscona et al., 1968). A measurable increase in GS specific

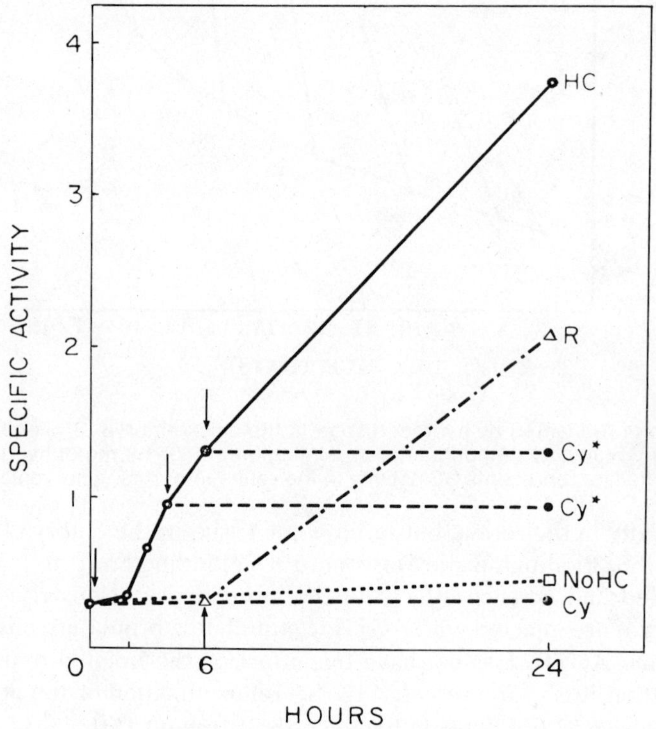

Fig. 6. Kinetics of induction of GS activity in cultures of embryonic retina tissue by HC (hydrocortisone; 8.2×10^{-7} M) and its inhibition by cycloheximide. All cultures were with inducer except the one marked "No HC." Arrows pointing down mark the addition times of cycloheximide (Cy)(2 μg/ml) to cultures with inducer. Arrow pointing up marks transfer of tissue from medium with inducer and cycloheximide to inducing medium without the inhibitor (R). Similar effects were obtained with puromycin. These data represent averages from one set of experiments; the absolute values varied somewhat in different experiments. Star indicates that the same values were obtained when cultures were transferred from medium with inducer to inducer-free medium with cycloheximide.

activity was detected already after 1.5 to 2 hours of incubation at 37°C in the medium with the inducer. Enzyme activity continued to rise sharply for the duration of these cultures. In control cultures without the steroid, GS specific activity remained level or rose only very slightly (Fig. 6). Since there is no detectable change in cell number in these cultures and since the total amount of tissue protein remains unchanged, the rise in GS specific activity represents a true increase of enzyme activity. If the steroid is removed from the culture medium after a short period of incubation, GS specific activity continues to increase for some time, depending on the residual amount of steroid in the tissue and the duration of precultivation. If the tissue is thoroughly washed to remove the inducer, GS activity levels off rapidly, but the enzyme does not return to its preinduction level. In this respect the behavior of GS in the retina differs significantly from that of induced adaptive enzymes in microorganisms and in mammalian adult liver, which return to the preinduction base line following removal of the inducer (Thompson et al., 1966; Kenney, 1967).

The initial kinetics of GS induction in vitro are not grossly modified by the absence of fetal serum from the culture medium, but the subsequent increase is less, so that after 24 hours in culture the level of GS is approximately 20 percent lower if the medium contains no fetal serum. Additions of glutamine or glutamate in physiological concentrations to the culture medium do not exert an inductive effect on GS, nor do they measurably modify enzyme levels during the initial phases of GS induction. In the presence of high concentrations of these metabolites, and particularly of γ-aminobutyric acid, the 24-hour levels of the enzyme were found to be lower (Kirk and Moscona, 1963).

PROTEIN SYNTHESIS AND GS INDUCTION IN THE RETINA

The hormonally induced increase of GS in the embryonic retina requires contінous protein synthesis and is completely prevented by translational inhibitors (Fig. 6) (Kirk and Moscona, 1963; Moscona and Kirk, 1965; Kirk, 1965; Reif-Lehrer and Amos, 1968; Moscona et al., 1968). Inclusion of 2 μg/ml of cycloheximide or puromycin in the culture medium with the inducer inhibits 80 to 90 percent of ^{14}C-amino acids incorporation into proteins and also completely prevents the increase of GS activity. The effect of cycloheximide (and to a lesser extent, of puromycin) is partially reversible by washing and transfer of the tissue to medium with inducer. If protein synthesis in the retina is halted at any time after the beginning of induction, this stops further increase of GS activity, but the enzyme remains at the level attained at the time of the inhibition and does not return to preinduction level.

EVIDENCE FOR ENZYME SYNTHESIS IN GS INDUCTION

The above information shows that protein synthesis is essential for the steroid-induced increase of GS activity; the following findings provide evidence that this increase represents, at least partly, synthesis and accumulation of newly made enzyme protein. This question was tested by a radioimmunochemical procedure in which anti-GS antiserum was used to precipitate radioactive GS from retina sonicates and to estimate the rate of GS synthesis and accumulation. Some of these tests were done with antiserum prepared in rabbits against chick retina GS. The enzyme was partially purified by adsorption (Kirk, 1967) according to a procedure modified from Kogut et al. (1956). In other tests the antiserum was against sheep brain GS (purified according to the method of Pamiljans et al., 1962). Both of these antisera gave essentially similar results with respect to the questions discussed here (Alescio and Moscona, 1969).

The retina was cultured for up to 24 hours in media with and without the inducer in the presence of ^{14}C-amino acids. The tissue was then sonicated in buffer, the enzyme-rich fraction was separated, and it was reacted with an appropriate amount of the anti-GS serum. The precipitate was collected, washed with cold TCA, and its ^{14}C content was determined by scintillation counting. It was assumed that if GS induction was due to accumulation of newly made enzyme, the anti-GS serum would precipitate more ^{14}C-labeled enzyme from induced than from uninduced

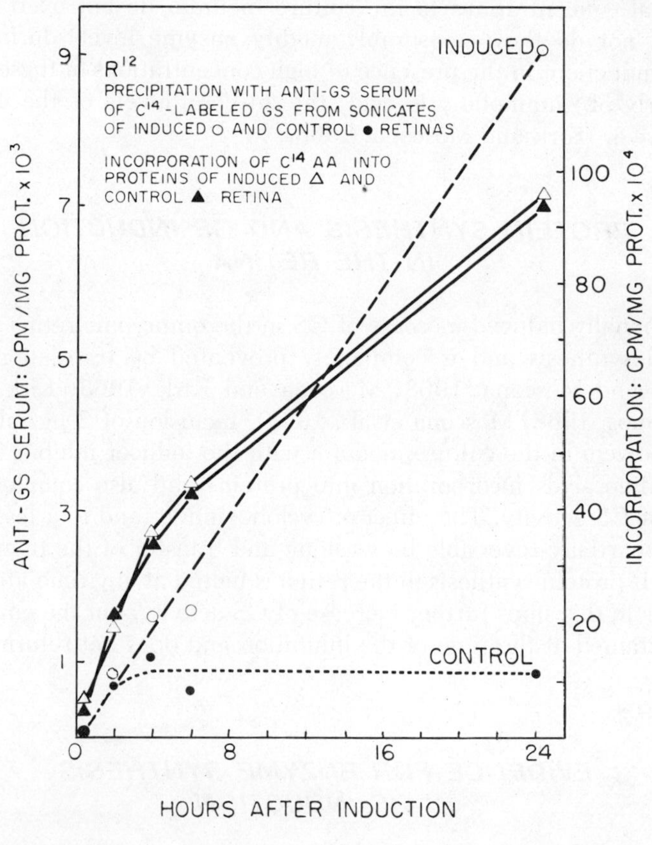

Fig. 7. For explanation see text. (Adapted from Alescio and Moscona, 1969.)

retina. Figure 7 shows that the antiserum precipitated approximately ten times more ^{14}C counts from induced than from uninduced tissue, even though ^{14}C incorporation into total protein was similar in both cases. This tenfold greater number of ^{14}C counts in the induced tissue coincided with a tenfold increase in GS specific activity

These results provide acceptable evidence that the steroid-induced increase in GS activity in the embryonic retina involves synthesis of enzyme protein and is not due exclusively to activation or assembly of preexisting precursors of the enzyme. They do not tell us whether all of the subunits which make up the active GS molecule are newly made upon induction.

RNA SYNTHESIS IN GS INDUCTION

RNA synthesis is essential for GS induction; the enzyme does not go up if transcription in the freshly explanted retina is inhibited at zero hour (i.e., at the time of the addition of the inducer) with 10 μg/ml of actinomycin D (Act D), which blocks more than 95 percent of incorporation of ^{3}H-uridine into retina RNA (Kirk and Moscona, 1963; Kirk, 1965; Moscona et al., 1968). Tests with radioactive hydrocortisone have shown that Act D at this concentration does not prevent the uptake of the steroid by the retina cells. This practically complete inhibition of RNA synthesis does not immediately stop all protein synthesis in the retina (Moscona et al., 1968), yet it completely prevents the induction of GS, as determined by measurements of enzyme activity and by precipitation tests of GS with anti-serum. However, if Act D at this concentration is added to cultures in inducing medium at various times after the beginning of induction, GS continues to go up even though RNA synthesis is stopped. Thus, after 4 hours of induction, complete inhibition of RNA synthesis does not prevent GS from reaching activity levels similar to those in normal induced controls by 24 hours. This effect is progressive in that the later the Act D is added between zero and 4 hours, the higher is the subsequent increase of GS activity and the closer it comes to that in normal induced controls (Fig. 8). This transcription-independent increase of GS activity in the presence of Act D is inhibited by cycloheximide and was shown by immunochemical tests to represent synthesis and accumulation of GS.

The following information further describes this situation:

1. Act D does not cause an increase in GS activity in the noninduced retina.
2. The levels of GS attained in the presence of Act D are not significantly higher than in the appropriate induced controls (in contrast to the "superinduction" by Act D of tyrosine transaminase in hepatoma cells; Tomkins et al., 1966).
3. Between 12 to 24 hours of culture, GS plateaus in the Act D-treated induced retina but does not return to preinduction levels. In the induced controls not treated with Act D, the enzyme continues to go up.
4. It is especially important that complete inhibition of RNA synthesis after 4 hours of induction allows the continued accumulation of GS even if the inducer is removed at that time. However, removal of the inducer at this time without the simultaneous complete inhibition of transcription causes

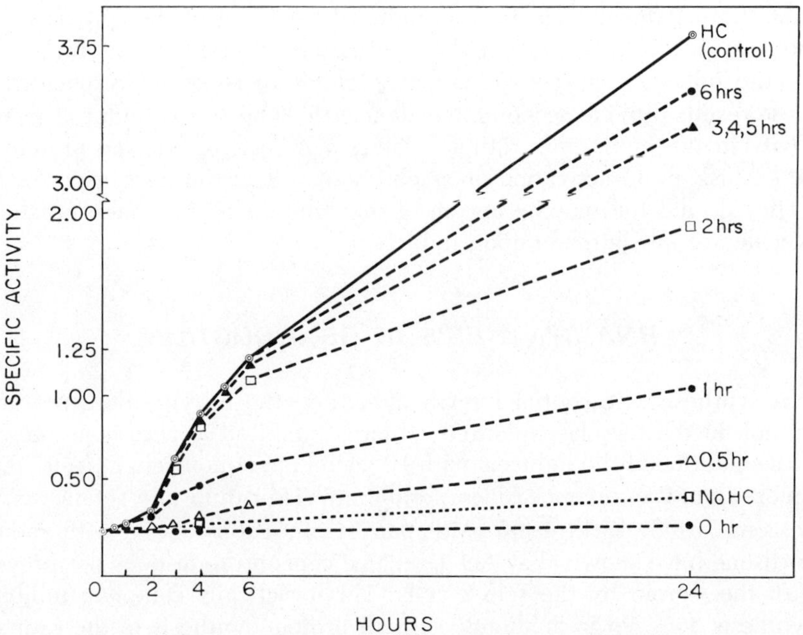

Fig. 8. Effects of Act D (10 μg/ml) added at various times on the induced increase in GS specific activity. All cultures (except control—No HC) were with the inducer (HC). Act D was added at the times listed on the right, next to the specific activity values at 24 hr. In this series of experiments, the results for 3, 4, 5, and 6 hr were below the control (HC) average but fell within the general range of control values. The absolute values varied somewhat in different experiments.

a leveling of the enzyme. Therefore, even in the induced tissue the inducer is persistently required for the increase of GS unless transcription is totally halted. This suggests that in the induced retina the inducer does not directly affect the translation of GS but acts through transcription to produce an effect which can be simulated by total inhibition of RNA synthesis. This important point will be discussed again later.

Taken as a whole, the above results suggest that in the first 4 hours of GS induction (and possibly thereafter), there is accumulation in the retina cells of relatively long-lived active RNA templates for making GS; this requires RNA synthesis. However, it remains to be determined whether the requirement for RNA synthesis during this period is exclusively for these templates or also for some other RNA or RNA-mediated product's involved in their stability or function.

Is protein synthesis required during the first 4 hours of induction for the accumulation of active, stable templates for GS? This was answered by the following experiment. Retina tissue was cultured for 4 hours in medium with the inducer in the presence of cycloheximide (at a concentration which inhibits approximately 90 percent of protein synthesis). Cycloheximide was then washed out and the tissue was transferred into medium without the inducer and with a high dose of Act D, whereupon the enzyme went up (Fig. 9). This tells us (1) that the uninduced cells have the proteins necessary for transcribing and accumulating the RNA required

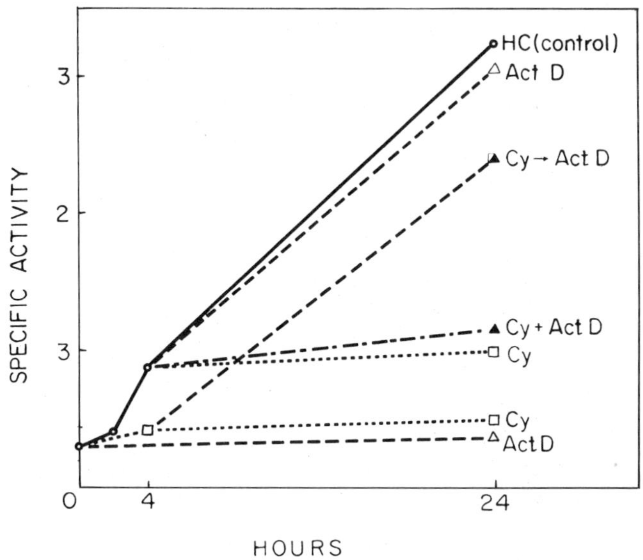

Fig. 9. Summary of results showing that (1) after 4 hr of induction GS activity continues to increase although RNA synthesis is halted (10 μg/ml Act D); (2) inhibition of protein synthesis by cycloheximide (2 μg/ml) during the first 4 hr of induction does not prevent the subsequent transcription-independent rise in GS activity; (3) protein synthesis is essential for the transcription-independent increase of GS activity after 4 hr. All cultures were in medium with inducer. HC (control), cycloheximide (Cy) was added at 0 or 4 hr, as indicated. Cy → Act D, GS activity in cultures transferred after 4 hr in cycloheximide into medium with Act D.

for GS synthesis, and (2) that the conditions for attaining the state of transcription-independent enzyme synthesis are provided by newly made RNA and not by newly made protein (see also Reif-Lehrer and Amos, 1968).

THE EFFECT OF PARTIAL INHIBITION OF TRANSCRIPTION

Since after 4 hours of induction GS in the retina is evidently made on relatively long-lived RNA templates, one could assume that after this stage had been attained further accumulation of the enzyme is not subject to constant genomic control. Such is the case in hemoglobin synthesis in mammalian erythroid cells and presumably also in some other cells that undergo terminal differentiation accompanied by degeneration or inactivation of the nucleus. However, the following results raise doubts whether the same applies to GS in the retina. They show that even though in the induced retina GS accumulation can continue when all transcription is experimentally halted, under normal conditions the accumulation of the enzyme in the induced retina is continuously controlled through gene action.

This is suggested by the finding that partial inhibition of RNA synthesis in the induced retina stops the accumulation of GS and the enzyme levels off, even though at this time complete inhibition of transcription allows further increase of GS (Fig.

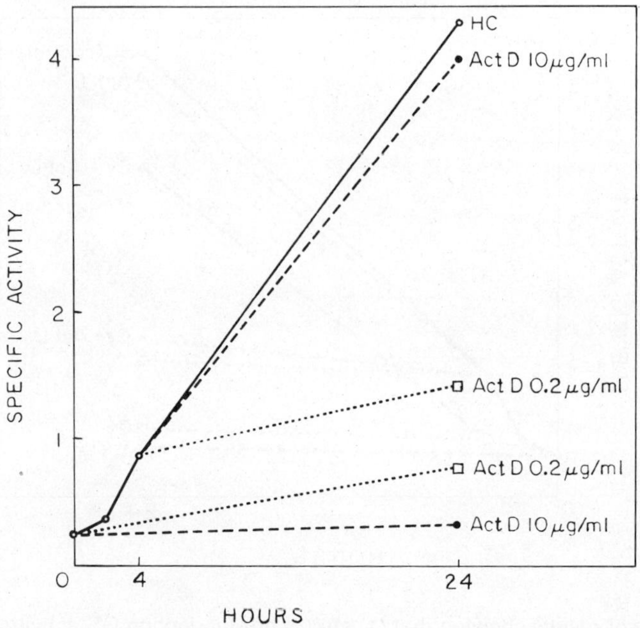

Fig. 10. Reverse effects of high and low doses of Act D on the increase of GS activity when the inhibitors were added to cultures at 0 or 4 hr. All cultures were in medium with the steroid inducer. Added to the cultures at 0 hr, both doses of Act D inhibited GS induction; the lower dose sometimes inhibited incompletely, as in the experiments represented here. Added at 4 hr, the high dose was not inhibitory, whereas the low dose suppressed approximately 70 to 90 percent of GS increase, compared with control cultures.

10). Thus, addition of 0.05 to 0.2 μg/ml of Act D to cultures after 4 hours of induction inhibits only 55 to 60 percent of RNA synthesis, yet it halts the increase of GS although the inducer is present and although by this time the cells contain long-lived templates for making GS (Moscona et al., 1968). The simplest explanation of this is that when RNA synthesis in the induced retina is partially inhibited, some of the transcripts which still come through cause a suppression of GS accumulation. These transcripts are, of course, not made when all RNA synthesis is halted. Their effect is also not evident when gene action in the induced tissue is not partially inhibited with the low dose of Act D.

MULTIPLE CONTROLS FOR GS ACCUMULATION IN INDUCED RETINA

The following hypothesis provides a basis for further exploration of this system. As indicated above, we interpret the contrary effects of the high and low doses of Act D as suggesting that GS accumulation in the retina is subject to constant genomic control exerted through antagonistically acting regulators which function at a post-transcriptional level; their formation requires constant gene action and they are relatively short-lived RNAs or proteins. One regulator (or regulators) suppresses GS accumulation. This suppressor action is not inhibited by the low dose of Act D. The other regulator functions as a desuppressor and it is blocked by the low dose of Act D. The terms "suppressor" and "desuppressor" as used here relate merely to the final effects on the enzyme level of GS and are not meant to imply a specific site of action of these postulated regulators. The proposed hypothesis is concerned with the principle of such a dual or multiple control mechanism of GS induction. It applies regardless of whether these regulators control the rate of GS accumulation at the level of enzyme synthesis, enzyme degradation, or both.

Starting with the evidence that the action of the inducer requires RNA synthesis, the hypothesis suggests that in addition to accumulation of stable templates for GS, the inducer elicits, through gene action, the formation of a short-lived desuppressor; this desuppressor must be continuously formed in order to counteract a suppressor and this requires the persistent availability of the inducer. If the inducer is removed, transcription for the desuppressor stops and GS accumulation is halted, even though stable templates for making GS are present. This suggests that the suppressor functions at a post-transcriptional level. The desuppressor is also stopped by the low dose of Act D even though the inducer is present. However, this low dose of Act D does not stop the suppressor, and, therefore, it halts GS accumulation. Thus, the low dose of Act D simulates in outcome the effect of removal of the inducer because it prevents the inducer from exerting desuppression. The high dose of Act D, which inhibits all RNA synthesis, also blocks the suppressor. However, since the induced retina contains stable templates for making GS, the enzyme continues to be made in the absence of further RNA synthesis for a period dependent on the life-span of the pool of templates, on the availability of precursors, energy, and so forth. The inducer is no longer necessary for this accumulation of GS in the absence of transcription since the suppressor is not made. If the inducer is left in, this does not significantly affect the levels of enzyme attained in this situation, which agrees with the postulate that the inducer functions in this induction through pre-translational processes, even though its ultimate effect results in translational or post-translational regulation of GS accumulation.

Thus, the effect of the high dose of Act D in the induced retina—i.e., the fact that stable templates for GS can continue to function after all RNA synthesis is blocked—conceals the apparently persistent genomic control over GS accumulation exerted through the postulated regulators. This is a heuristically important conclusion derived from the above hypothesis; its validity remains to be tested. It is of interest that evidence in other systems also points to the possible existence of similar regulatory mechanisms (McAuslan, 1963; Garren et al., 1964). Alternative interpretations of the experiments reported above are, of course, conceivable; thus, the notions advanced here serve as a provisional framework to aid in the further analysis of this system.

Concerning the nature of the postulated suppressor, the available experimental findings suggest that (1) it is a relatively short-lived product, RNA or protein; and (2) it functions in the induced retina at a post-transcriptional level. These considerations raised the question whether in the uninduced retina, suppression also plays an important role in controlling GS and maintaining it at a low level. To test this point, we assumed that if it were possible to deplete the postulated suppressor in the uninduced cells while allowing the accumulation of templates for GS, the enzyme should increase in the complete absence of the inducer. We therefore attempted to deplete the suppressor in uninduced retina by blocking its synthesis.

In these experiments, retinas from 12-day embryos were incubated for 5 hours in medium without the inducer but with a high dose of cycloheximide which inhibited more than 90 percent of protein synthesis and approximately 40 percent of RNA synthesis. After 5 hours of incubation, cycloheximide was washed out to allow resumption of protein synthesis, and the retina was transferred to medium with the high dose of Act D to stop all further RNA synthesis and prevent renewal of the suppressor. If stable templates for GS had accumulated during treatment with cycloheximide and the suppressor had been depleted, GS would be expected to go up. The results showed that after the transfer to Act D, GS increased rapidly in these cultures, even though they were never exposed to the steroid inducer. In retinas transferred to medium without Act D, the enzyme did not go up, presumably because of renewed production of the suppressor which prevented the expression of the preformed templates; addition of the steroid inducer allowed their expression. Thus, the accumulation of stable, potentially active templates for GS was made possible by temporarily blocking certain aspects of biosynthesis in the uninduced retina without exposing the tissue to the steroid inducer.

While these preliminary results agree with the predictions from our working hypothesis concerning the role of desuppression in GS induction, their interpretation requires considerable additional study. For example, it is by no means clear whether suppression of the activity of stable templates in the induced retina involves the same mechanism which prevents template accumulation in the uninduced tissue. Further experiments with inhibitors of macromolecular synthesis might provide additional clues towards analysis of this problem.

COMMENTS

Taken as a whole, the available information points towards a tentative model of GS induction in the retina. We assume that in the uninduced retina there is synthesis of the RNA required for making GS, but this RNA does not accumulate. Exposure to the steroid inducer or inhibition with the high dose of cycloheximide allow the accumulation (stabilization) of this RNA, resulting in the availability of a pool of stable templates for GS. This makes it unlikely that the steroid inducer functions by directly initiating the transcription of the structural genes for GS and suggests that it acts through regulatory genes to desuppress the system. The available facts agree with the notion that the levels of GS are controlled both in the uninduced and in the induced retina by regulatory factors, the production of which requires transcription. They also agree with the suggestion that the postulated factors function as suppressors and that the steroid inducer counteracts this suppression. The possible nature of these hypothetical factors, their mode and site of action, and the manner in which they are counteracted are at present unknown.

The above hypothesis reflects our information about the GS system at the time of this writing. It provides guidelines for further analysis, which are to be modified as further information accrues. For example, there is as yet no satisfactory information concerning the types of RNA involved in GS induction; nor do we have detailed data about changes in the rates of GS synthesis and degradation in the retina under the various experimental conditions described here. It will be recalled that in eukaryote cells, increased accumulation of protein may be achieved in different ways: it may be due primarily to an increase in the translational activity of preformed mRNA (Monroy et al., 1965; Slater and Spiegelman, 1966; Cline and Bock, 1966; Gross, 1967; Brown, 1967; Wool and Kurihara, 1967), or to availability of newly formed mRNA. The evidence that GS induction involves the action of regulatory as well as structural genes raises special problems in connection with RNA analysis in this system. Future attention should also be directed toward post-translational regulatory processes. The importance of changes in the rate of turnover in enzyme induction has been amply demonstrated (Schimke et al., 1965; Kenney, 1967), and such changes may play a role in the regulation of GS. While these possibilities fit within the proposed regulatory scheme for GS induction, their examination should aid toward a more detailed understanding of enzyme control in embryonic differentiation.

REFERENCES

Alescio, T., and Moscona, A. A. Immunochemical evidence for enzyme synthesis in the hormonal induction of glutamine synthetase in embryonic retina in culture. Biochem. Biophys. Res. Commun., 34(2):176–182, 1969.

Brown, D. D. The genes for ribosomal RNA and their transcription during amphibian development. *In* Current Topics in Developmental Biology. Moscana, A. A., and Monroy, A., eds. Vol. 2, pp. 47–73. New York, Academic Press, Inc., 1967.

Cline, A. L., and Bock, R. M. Translational control of gene expression. Cold Spring Harbor Symp. Quant. Biol., 31:321–334, 1966.

Garren, L. D., Howell, R. R., Tomkins, G. M., and Crocco, R. M. A paradoxical effect of actinomycin D: the mechanism of regulation of enzyme synthesis by hydrocortisone. Proc. Nat. Acad. Sci. U.S.A., 52:1121–1129, 1964.

Grobstein, C. Differentiation in vertebrate cells. *In* The Cell. Brachet, J., and Mirsky, A. E., eds. Vol. 1, pp. 437–496, New York, Academic Press, Inc., 1959.

Gross, P. R. The control of protein synthesis in embryonic development and differentiation. *In* Current Topics in Developmental Biology. Moscona, A. A., and Monroy, A., eds. Vol. 2, pp. 1–46. New York, Academic Press, Inc., 1967.

Kenney, F. T. Turnover of rat liver tyrosine transaminase: stabilization after inhibition of protein synthesis. Science, 156:525–528, 1967.

Khedouri, E., and Meister, A. Synthesis of D-β-glutamine from β-glutamic acid by glutamine synthetase. J. Biol. Chem., 240:3357–3360, 1965.

Kirk, D. L. The role of RNA synthesis in the production of glutamine synthetase by developing chick neural retina. Proc. Nat. Acad. Sci. U.S.A., 54:1345–1353, 1965.

——— An automated glutamotransferase assay: recovery from serendipity. *In* Automation in Analytical Chemistry, Vol. 3, Technicon Symposium, Chauncey, New York, pp. 1–4, 1967.

——— and Moscona, A. A. Synthesis of experimentally induced glutamine synthetase

(glutamotransferase activity) in embryonic chick retina *in vitro*. Develop. Biol., 8:341–357, 1963.
Kogut, M., Pollock, M. R., and Tridgell, E. J. Purification of penicillin-induced penicillinase of bacillus cereus NRRL 569: A comparison of its properties with those of a similarly purified penicillinase produced spontaneously by a constitutive mutant strain. Biochem. J., 62:391–401, 1956.
McAuslan, B. R. The induction and repression of thymidine kinase in the poxvirus-infected hela cell. Virology, 21:383–389, 1963.
Meister, A. Glutamine synthesis. The Enzymes. Boyer, P. D., Lardy, H., and Myrbak, K. eds. Vol. 6, pp. 443–468,. New York, Academic Press, Inc., 1962.
Monroy, A., Maggio, R., and Rinaldi, A. M. Experimentally induced activation of the ribosomes of the unfertilized sea urchin egg. Proc. Nat. Acad. Sci. U.S.A., 54:107–111, 1965.
Moscona, A. A., and Hubby, J. L. Experimentally induced changes in glutamotransferase activity in embryonic tissue. Develop. Biol., 7:192–206, 1963.
——— and Kirk, D. L. Control of glutamine synthetase in the embryonic retina *in vitro*. Science, 148:519–521, 1965.
——— and Piddington, R. Stimulation by hydrocortisone of premature changes in the developmental pattern of glutamine synthetase in embryonic retina. Biochim. Biophys. Acta, 121:409–411, 1966.
——— and Piddington, R. Enzyme induction by corticosteroids in embryonic cells: steroid structure and inductive effect. Science, 158:496–497, 1967.
——— Moscona, M. H. and Saenz, N. Enzyme induction in embryonic retina: the role of transcription and translation. Proc. Nat. Acad. Sci. U.S.A., 61:160–167, 1968.
Pamiljans, V., Krishnaswamy, P. R., Dumville, G., and Meister, A. Studies on the mechanism of glutamine synthesis; isolation and properties of the enzyme from sheep brain. Biochemistry, 1:153–158, 1962.
Piddington, R. Hormonal effects on the development of glutamine synthetase in the embryonic chick retina. Develop. Biol., 16:168–188, 1967.
——— and Moscona, A. A. Correspondence between glutamine synthetase activity and differentiation in the embryonic retina *in situ* and in culture. J. Cell Biol., 27:247–252, 1965.
——— and Moscona, A. A. Precocious induction of retinal glutamine synthetase by hydrocortisone in the embryo and in culture: age-dependent differences in tissue response. Biochim. Biophys. Acta, 141:429–432, 1967.
Reif, L., and Amos, H. A dialyzable inducer for the glutamotransferase of chick embryo retina. Biochem. Biophys. Res. Commun., 23:39–48, 1966.
Reif-Lehrer, L., and Amos, H. Hydrocortisone requirement for the induction of glutamine synthetase in chick embryo retinas. Biochem. J., 106:425–430, 1968.
——— and Chader, G. J. The steroid induction of glutamine synthetase in chick embryo retinas in culture. A possible role of corticosteroid-binding globulin. Biochim. Biophys. Acta, 192:310–317, 1969.
Rudnick, D., and Waelsch, H. Development of glutamotransferase and glutamine synthetase in the nervous system of the chick. J. Exp. Zool., 129:309–326, 1955.
——— Mela, P., and Waelsch, H. Enzymes of glutamine metabolism in the developing chick embryo. A study of glutamotransferase and glutamine synthetase. J. Exp. Zool., 126:297–321, 1954.
Rutter, W. J., and Weber, C. S. Specific proteins in cytodifferentiation. *In* Developmental and Metabolic Control Mechanisms and Neoplasia, pp. 195–218, Baltimore, Williams and Wilkins, 1965.
Schimke, R. T., Sweeney, E. W., and Berlin, C. M. The roles of synthesis and degradation in the control of rat liver tryptophan pyrrolase. J. Biol. Chem., 240:322–331, 1965.
Shimada, Y., Piddington, R., and Moscona, A. A. Experimentally induced increases in glutamine synthetase in the optic tectum in the embryo and in culture. Exp. Cell Res., 48: 240–243, 1967.

Slater, D. W., and Spiegelman, S. An estimation of genetic messenger in the unfertilized echinoid egg. Proc. Nat. Acad. Sci. U.S.A., 56:164–170, 1966.

Spiegelman, S. Differentiation as the controlled production of unique enzymatic patterns. Sympos. Soc. Exp. Biol., 2:286–325, Cambridge, England, University Press, 1948.

Tata, J. R. Growth and developmental hormones as tools for the study of biosynthetic control mechanisms. *In* Developmental and Metabolic Control Mechanisms and Neoplasia, pp. 335–360, Baltimore, Williams and Wilkins, 1965.

Thompson, E. B., Tomkins, G. M., and Curran, J. F. Induction of tyrosine-*a*-ketoglutarate transaminase by steroid hormones in a newly established tissue culture cell line. Proc. Nat. Acad. Sci. U.S.A., 56:296–303, 1966.

Tiedemann, H. The molecular basis of differentiation in early development of amphibian embryos. *In* Current Topics in Developmental Biology. Moscona, A. A., and Monroy, A. eds. Vol. 1, pp. 85–112, New York, Academic Press, Inc., 1966.

Tomkins, G. M., Thompson, E. B., Hayashi, S., Gelehrter, T., Granner, D., and Peterkofsky, B. Tyrosine transaminase induction in mammalian cells in tissue culture. Cold Spring Harbor Symp. Quant. Biol., 31:349–360, 1966.

Turkington, R. Hormone-dependent differentiation of mammary gland *in vitro*. *In* Current Topics in Developmental Biology. Moscona, A. A., and Monroy, A., eds. Vol. 3, pp. 199–218, New York, Academic Press, Inc., 1968.

Wolff, E. Specific interactions between tissues during organogenesis. *In* Current Topics in Developmental Biology. Moscona, A. A., and Monroy, A., eds. Vol. 3, pp. 65–94, New York, Academic Press, Inc., 1968.

Wool, I. G., and Kurihara, K. Determination of the number of active muscle ribosomes: effect of diabetes and insulin. Proc. Nat. Acad. Sci. U.S.A., 58:2401–2407, 1967.

15

Specific Metabolic and Physicochemical Interactions of Glucocorticoids with Rat Thymus Cells*

A. Munck, D. A. Young, K. M. Mosher, and C. R. Wira

Department of Physiology, Dartmouth Medical School, Hanover, New Hampshire

INTRODUCTION

Over the past few years we have been engaged in an effort to understand the mechanisms of action of glucocorticoids on thymus tissue, with the ultimate goal of understanding the physiological functions of these hormones in the whole organism. Out of the multitude of rapid glucocorticoid effects that are seen in vivo (cf. Munch[1]), we have chosen to study those on the thymus because with incubated thymus cell suspensions we can reproduce in vitro some of the earliest hormone effects observed in vivo, notably those on glucose metabolism(2-8) and on protein and RNA synthesis(3,9-12). These actions may well be prototypes for the general catabolic actions of glucocorticoids, since similar effects, particularly on glucose, are found with skin(13) and adipose tissue(14,15).

Figure 1 summarizes observations we and others have made with thymus cells, and outlines our working hypothesis on the relations between these observations. As soon as cortisol—which is used as a prototype of glucocorticoids—is incubated with thymus cells, two types of reversible physicochemical binding reactions are initiated(16): a virtually instantaneous *nonspecific interaction,* which under steady-state conditions accounts for most of the bound molecules; and a slower *specific interaction.* The terms used refer to the close relation that exists, on the one hand, between specific binding and the *specific effects* produced by glucocorticoids in vitro at physiological concentrations (10^{-8} to 3×10^{-6} M) and, on the other, be-

*This work was supported in part by Research Grants AM-03535 and AM-10998 from the National Institute of Arthritis and Metabolic Diseases. Dr. Munck holds a Research Career Development Award (AM-K3-16) from the same institute.

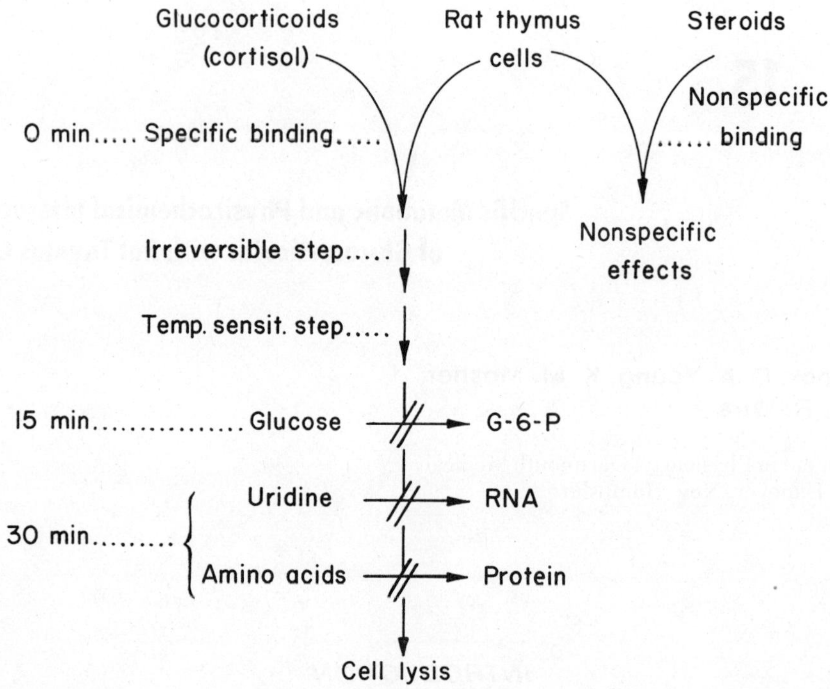

Fig. 1. Summary of the principal physicochemical and metabolic interactions of glucocorticoids with thymus cells in vitro and of the hypothetical relationships between them. The times give a rough indication of the time each process is observed to begin.

tween nonspecific binding and the *nonspecific effects* that all steroids produce in vitro at high concentrations (10^{-5} M and up). These latter effects, which have plagued the field for many years, predominate with nonpolar steroids such as deoxycorticosterone, testosterone, and progesterone(1,4).

After a short lag period, the earliest specific metabolic effect that has been observed so far—a block in the conversion of glucose to glucose-6-phosphate—appears quite abruptly(7). Indirect evidence indicates that this effect is preceded by at least two steps, an irreversible step and a temperature-sensitive step(8). These are distinct from specific binding and may or may not involve the cortisol molecule itself. Somewhat later the first signs of inhibition of protein and RNA synthesis (3,9-12) begin to appear, and eventually visible signs of cell deterioration become detectable(17).

Our working hypothesis(1-3) for the relations between these events is that by binding specifically, cortisol initiates a series of steps that lead to the observed changes in glucose metabolism, and that these changes in turn—as postulated originally by Ingle(18) and Drury(19)—are in large part responsible for the later changes in RNA and protein synthesis, and eventually for cell breakdown. General support for this hypothesis comes from the sequence of events just outlined. Further evidence is detailed in subsequent sections.

SPECIFIC AND NONSPECIFIC BINDING OF GLUCOCORTICOIDS TO THYMUS CELLS IN VITRO (6,16)

Binding of steroids to thymus cells has been measured, generally with tritiated steroids of high specific activity, by determining the relative amounts of steroid in the supernatant and cell phase of a cell suspension. A finding with cortisol that has greatly simplified these experiments is that even after cortisol-^{3}H has been incubated for an hour with thymus cells, the radioactivity bound to the cells in the specific and nonspecific fractions is still in the form of unaltered cortisol-^{3}H. It thus appears that the physiologically active form of cortisol is cortisol itself, a conclusion that is in accord with a great deal of circumstantial evidence(1).

Typical binding experiments with two contrasting steroids, cortisol and cortisone, are shown schematically in Figure 2. It should be kept in mind that cortisone in this system is not converted to cortisol and, therefore, is inactive. When freshly isolated rat thymus cells are added to Krebs-Ringer bicarbonate buffer containing steroid, the steroid rapidly becomes attached to the cells, reaching an equilibrium value of (Sc)/$(S)_e$ that remains more or less unchanged for several hours. This value, which gives the extent to which the steroid bound to the cells appears to be concentrated above the concentration in the medium, turns out to be nearly independent of concentration, $(S)_e$, from 10^{-10} to 10^{-4} M, and quite unrelated to glucocorticoid activity. Its magnitude correlates well, however, with nonspecific

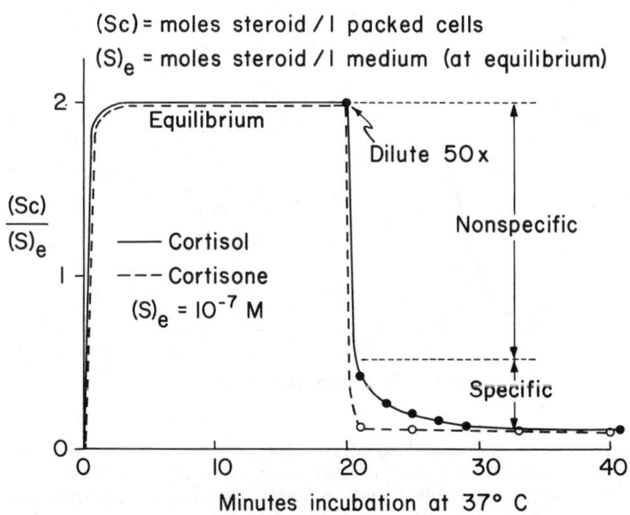

Fig. 2. Outline of a typical pair of binding experiments, one with cortisol and one with cortisone. A thymus cell suspension was incubated at 37°C in Krebs-Ringer bicarbonate buffer, with 5 mM glucose and the tritium-labeled steroid at a concentration $(S)_e$ in the medium of about 10^{-7} M. After an initial equilibration period of 20 minutes, for which the binding curves are indicated schematically, the suspension was diluted fiftyfold into buffer without steroid. The values of (Sc) at various time points were measured by centrifuging aliquots of the suspensions and determining the tritium bound to the cells. (Data from 16.)

activity, predominating with such steroids as progesterone, deoxycorticosterone, and testosterone.

If the equilibrium is suddenly displaced by diluting the suspension fiftyfold, most of the bound steroid dissociates within one minute. This rapidly dissociating or *nonspecific fraction* is found with all steroids and is closely related to nonspecific activity, accounting for the correlation noted of equilibrium binding with nonspecific activity. It probably consists of molecules that are distributed throughout the cell volume, rather than just on the cell surface as we originally supposed.

By one minute after such a dilution, a clear difference can be noted between cortisol and cortisone. Whereas cortisone has practically reached a new equilibrium, cortisol can be seen to possess a distinct fraction that dissociates over about 15 minutes. This slowly dissociating or *specific fraction* is also a prominent feature of the dissociation curves of the only other glucocorticoids we have tested in this way, corticosterone and dexamethasone. For this and other reasons to be given, it is likely that the specific fraction consists of those molecules that initiate specific hormonal effects.

The magnitude of the specific fraction is determined by extrapolating the slow part of the dissociation curve back to the time of dilution. The results of this type

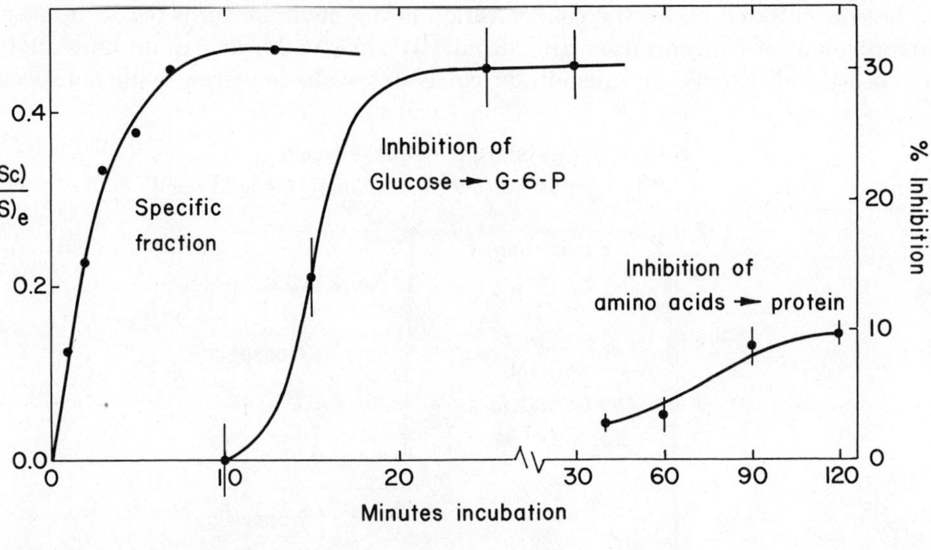

Fig. 3. Time course of specific binding of cortisol at 10^{-7} M and of the effects of 10^{-6} M cortisol on glucose metabolism and protein synthesis. The specific fraction was determined by measuring the amount of slowly dissociating fraction in the dissociation curves from each of a succession of dilutions, carried out as described for Figure 2 at the indicated time points. Effects on glucose metabolism were measured by incubating cells with 10^{-6} M cortisol and no substrate, adding glucose (5.5 μmoles/ml cell suspension) at the times indicated, and 5 minutes later measuring the levels of glucose-6-phosphate in the cells. The effect of cortisol is given as the percent inhibition of the increase in glucose-6-phosphate produced by the 5-minute glucose pulse. Effects on protein synthesis were determined by incubating cells with 10^{-6} M cortisol and glucose (5.5 mM) present from the beginning of the incubation, and adding leucine-^{14}C (0.05 mg/ml, s.a. 10 mc/g) after 30 minutes. The effect of cortisol is given as the percent inhibition of incorporation of ^{14}C into protein at the times indicated.

of measurement, carried out under a variety of conditions, have led us to the following conclusions:

Rate of Association. The specific fraction is bound rapidly enough to account for the earliest metabolic effects that have been observed so far. The time course of 37°C is shown schematically in Figure 3, where it can be compared with that of the metabolic effects.

Saturation. In contrast to the nonspecific fraction, the specific fraction becomes saturated over the physiological range of glucocorticoid concentrations. Since this is approximately the same range over which a graded response is obtained in vivo and in vitro, it may be concluded that the magnitude of the physiological response is limited by the number of "sites" available for specific binding. This number can be roughly estimated at 5,000 per cell.

Competition for Specific Binding—Antiglucocorticoid Activity of Cortexolone. The specific fraction is competed for by all the glucocorticoids we have tested. Relative binding constants measured by competition are given in Table 1, where they can be compared with relative metabolic activities with respect to inhibition of glucose uptake in vitro. It can be seen that for the glucocorticoids, activity is roughly proportional to binding. Among the metabolically inactive steroids, some do not compete for binding at all; others do compete, notably cortexolone.

As shown in Figure 4, cortexolone, which has little known hormonal activity of any kind in vivo or in vitro, is effective in reversing the inhibitory action of cortisol in vitro on glucose uptake(16,20). Such "antiglucocorticoid" activity would be anticipated from the ability of cortexolone to displace selectively the specific

TABLE 1

Competition of steroids for specific cortisol binding to thymus cells: Comparison with specific metabolic activities

Cortisol-^3H at a concentration $(S)_e$ = 6-9nM, together with various unlabeled steroids, was incubated with thymus cells for 20-30 min, and then dissociated by fifty-fold dilution as described for Figure 2. Relative binding activities for the unlabeled steroids were calculated from their ability, relative to unlabeled cortisol, to displace cortisol-^3H from the specific fraction (16). Relative metabolic activities with respect to inhibition of glucose uptake by thymus cells in vitro were obtained from measurements of glucose uptake in 1-hour incubations (5, 16).

Competing steroid	Specific binding activity (Cortisol = 1.0)	Specific metabolic activity (Cortisol = 1.0)
Dexamethasone	4.8	10
9α-Fluoroprednisolone	3.6	18
Prednisolone	1.3	1.2
Corticosterone	0.5	0.4
Deoxycorticosterone	0.5	< 0.02
Progesterone	0.4	< 0.01
Cortexolone	0.3	< 0.01
17α-Hydroxyprogesterone	0.1	< 0.01
Testosterone	< 0.05	< 0.01
Cortisone	< 0.05	< 0.01
11-Epicortisol	< 0.05	< 0.01
Tetrahydrocortisol	< 0.05	< 0.02

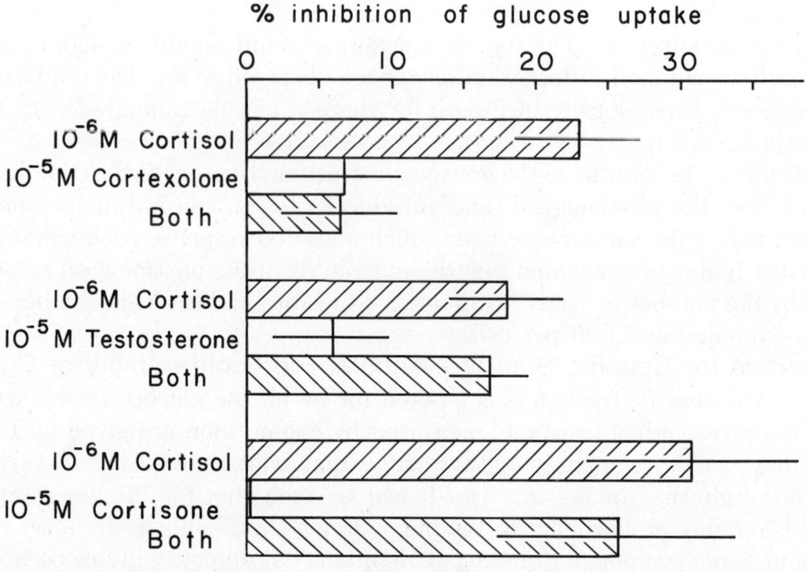

Fig. 4. Counteraction by cortexolone and lack of counteraction by testosterone and cortisone to the inhibitory effect of cortisol on glucose uptake by thymus cells. Glucose uptake was determined in one-hour incubations in which the various steroids were present from the start. (Data from 16,20.)

fraction of cortisol. No such antiglucocorticoid activity is observed with cortisone or testosterone, neither of which displaces the specific fraction.

Cortexolone also counteracts the inhibitory effects of cortisol on levels of glucose-6-phosphate(8) and on protein and RNA synthesis(21) in thymus cells, as well as the stimulatory effects of glucocorticoids on alkaline phosphatase in HeLa cells(22). This latter observation suggests strongly that the characteristics of specific cortisol binding found for thymus cells may apply to other glucocorticoid target tissues.

Relation to ATP Levels. In contrast to nonspecific binding, specific binding depends on the condition of the cells. Practically no specific fraction is present in cells incubated anaerobically without substrate. Under these conditions ATP levels are very low. Oxygen introduced into such a system returns the levels of both specific binding and ATP to normal. Glucose by itself can also maintain binding and ATP levels. These observations suggest a number of possibilities, such as that metabolic energy is necessary to transport hormone molecules to specific intracellular receptors, or that ATP or related substances may be necessary as allosteric effectors to maintain the structure of hormone receptors.

Location of the Specific Fraction. By diluting briefly a cell suspension at 37°C, it is possible to dissociate most of the nonspecific fraction. If the cells are then cooled rapidly to about 3°C, they retain the specific fraction for more than an hour. From preliminary experiments in which such cells have been broken by exposure to a hypotonic medium and then centrifuged in a sucrose gradient, it appears that a large part of the specific fraction may be attached to the cell nucleus.

EFFECTS ON GLUCOSE METABOLISM

As shown in the last column of Table 1, steroids in vitro inhibit glucose uptake by thymus cells with relative activities that on the whole agree well with their activities in vivo(5,16). Discrepancies, such as the lack of activity of cortisone, can be explained by the lack of metabolic transformation of steroids by thymus cells. The decrease in glucose uptake is apparently due to noncompetitive inhibition of the conversion of glucose to glucose-6-phosphate, since the inhibition occurs at all glucose concentrations and is accompanied by a decrease in levels of glucose-6-phosphate and in lactic acid production. Under anaerobic conditions cortisol has no effect(7,16).

By measuring the accumulation of glucose-6-phosphate during 5-minute glucose pulses(7), this inhibition is found to appear with the time course shown in Figure 3. No effect is discernible until cells have been exposed to cortisol for 15 minutes, at about which time the effect begins quite abruptly. If the incubation is continued for 2 hours the inhibition increases to 50 percent or more.

The limiting step in glucose uptake is located somewhere near the periphery of the cell, since under steady-state condition, in the presence or absence of cortisol, the glucose space is no greater than 10 percent of the cell volume. It is unclear whether the step affected is that of glucose transport or phosphorylation, but on the whole our evidence favors transport.

EFFECTS ON PROTEIN AND RNA SYNTHESIS

The time course of inhibition of incorporation of leucine (added in substrate amounts) into protein is shown in Figure 3. This time course, which is similar to that of inhibition of incorporation of uridine into RNA, is strikingly different from that on glucose metabolism(10).*

These effects of glucocorticoids on protein and RNA synthesis practically disappear in the absence of substrates. Of a number of substrates tested, including glucose, pyruvate, and lactate(10,23), glucose is the most effective in maintaining synthesis and in demonstrating glucocorticoid inhibition of synthesis(10). This observation supports the hypothesis that the later inhibitory effects of glucocorticoids on protein and RNA synthesis depend to a large extent on the earlier effects on glucose metabolism.

How the effect on glucose metabolism is transmitted to protein and RNA synthesis is a question to which we are actively seeking an answer. The most likely connecting link at present appears to be ATP (or related nucleotides). Levels of

*Makman et al.(11) claim that exposure of cells to 10^{-6}M cortisol for 10 minutes produces significant inhibition of incorporation of uridine into total cellular RNA. From close reading of their procedure, however, it is clear that in their experiments cells have been incubated with cortisol for 70 minutes by the time incorporation is measured. The 9 percent inhibition they deserve at the time agrees well with our results.

this substance are slightly but significantly lowered by cortisol, with a time course comparable to that for inhibition of protein synthesis(10). It does not seem likely that transport of amino acids into cells plays a significant role in the cortisol effect since the effect is also seen on the incorporation of radioactivity from ^{14}C-pyruvate into protein(10), and since the rate of exchange of amino acids across the cell membrane is much greater than the rate of amino acid incorporation into protein(24).

EVIDENCE FOR A TEMPERATURE-DEPENDENT AND AN IRREVERSIBLE STEP

Burton et al.(17) have reported that early cytolytic effects of glucocorticoids in vitro on nuclear structure, which take several hours to develop, are not manifested at room temperature. We have now found that the effects of cortisol we detect on glucose metabolism within 15–20 minutes at 37°C take more than 2 hours to develop at 20°C(8), but can be demonstrated in a much shorter time at 20°C if cells are preincubated with cortisol for a few minutes at 37°C. Thus, as indicated in Figure 1, a temperature-sensitive step appears to precede the rate-limiting step of glucose metabolism on which the effect of cortisol is eventually exerted, since this latter step can exhibit a normal cortisol effect at 20°C. Furthermore, since the specific binding step is not much slower at 22°C than at 37°C(16), the temperature-dependent step cannot be identical with, but must follow, specific binding.

Removal of cortisol by washing after one or two hours incubation has been shown not to remove the effects on cytolysis(17) or on RNA synthesis(11). Using cortexolone to displace cortisol in the specific fraction at precisely determined times, we find that even after 5 minutes incubation at 37°C, removal of cortisol does not remove the cortisol effects on glucose metabolism that first appear after 15–20 minutes(8). If the initial exposure to cortisol and subsequent removal are carried out at 20°C and the cells are then tested for an effect on glucose metabolism at 37°C, virtually the same results are obtained. It may be concluded that cortisol leads rapidly to an irreversible step that is distinct from and precedes the temperature-sensitive step described here.

NOTE ADDED IN PROOF

Work done since preparation of this manuscript has led to a number of advances.

We have shown that the specifically bound fraction of cortisol is located largely, if not entirely, in the nucleus(25). The nuclear "receptor" responsible for specific binding, and, apparently, for transmitting the cortisol effects, has been extracted from nuclei with 0.6 M KCl and separated Sephadex(26). It is at least partly a protein. At 37°C the rates of dissociation of cortisol from isolated nuclei and from isolated receptors are very similar to the slow rate characteristic of the specific fraction of whole cells. Thus, slow dissociation is an intrinsic property of the cortisol-nuclear receptor complex.

In addition to the irreversible and temperature-sensitive steps through which the action of cortisol passes after formation of the cortisol-receptor complex and before it reaches glucose metabolism, there is an actinomycin D sensitive step. This step, which may be identical with the irreversible step, is demonstrated by the fact that if actinomycin D (at concentrations which reduce RNA synthesis by 90 percent) is added simultaneously with cortisol, it prevents development of the subsequent cortisol effect on glucose metabolism; whereas if it is added 5 minutes after cortisol, it does not prevent the cortisol effect.

We have obtained further evidence that glucose transport, rather than phosphorylation, is the principal step blocked by cortisol. A number of experiments comparing the kinetics of glucose metabolism by normal cells with the kinetics of glucose phosphorylation by damaged or anaerobic cells and by isolated thymus cell hexokinases, show that the capacity of cells to phosphorylate glucose greatly exceeds the maximal rate of glucose uptake by normal cells, strongly indicating that in normal cells the rate-limiting step is transport. The view that cortisol acts on this rate-limiting step is supported by our earlier observations that conditions which markedly increase the rate of glucose uptake by thymus cells—anaerobiosis(6,7,16) and cell damage(4)—also abolish the cortisol effect. In other experiments we have found that 10^{-6}M cortisol inhibits the uptake of 3-0-methyl glucose, which is not metabolized by thymus cells.

The relationship between the early action of cortisol on glucose metabolism and later actions on incorporation of radioactive precursors into RNA and protein has been further clarified(27). Both the stimulatory effects of glucose and the inhibitory effects of cortisol on the incorporation of amino acids into protein develop following glucose addition at 120 minutes, even when RNA synthesis has been previously blocked by the addition of actinomycin D at 100 minutes. Similar studies with puromycin indicate that changes in rates of RNA labeling are independent of effects on protein synthesis. Both the stimulatory effects of glucose and the inhibitory effects of cortisol on protein and RNA labeling have been attributed to effects on glucose-dependent ATP production. While the effects on protein labeling do appear to represent changes in protein synthesis, those on RNA labeling have been attributed to changes in the rate of incorporation of extracellular uridine into the intracellular uridine nucleotides(28).

The results of Rosen et al.(29), relating cortisol inhibition of DNA synthesis in lymphosarcoma cells to earlier effects on the metabolism of glucose to glucose 6-P, also support the hypothesis that a reduced availability of high-energy compounds for biosynthetic processes play a central role in mediating the physiological action of glucocorticoids on lymphoid tissues.

REFERENCES

1. Munck, A. The effects of hormones at the cellular level. *In* Recent Advances in Endocrinology, James, V. T., ed. London, J. and A. Churchill, Ltd., pp. 139–180, 1968.
2. Bartlett, D., Morita, Y., and Munck, A. Rapid inhibition by cortisol of incorporation of glucose *in vivo* into the thymus of the rat. Nature (London), 196:897–898, 1962.

3. Morita, Y., and Munck, A. Effect of glucocorticoids *in vivo* and *in vitro* on net glucose uptake and amino acid incorporation by rat thymus cells. Biochim. Biophys. Acta, 93: 150–157, 1964.
4. Munck, A. Steroid concentration and tissue integrity as factors determining the physiological significance of effects of adrenal steroids *in vitro*. Endocrinology, 77:356–360, 1965.
5. Kattwinkel, J., and Munck, A. Activities *in vitro* of glucocorticoids and related steroids on glucose uptake by rat thymus cell suspensions. Endocrinology, 79:387–390, 1966.
6. Munck, A., and Brinck-Johnsen, T. Specific metabolic and physicochemical interactions of glucocorticoids *in vivo* and *in vitro* with rat adipose tissue and thymus cells. Proceedings of the 2nd International Congress on Hormonal Steroids, Excerpta Medica International Congress Series, 132:472–481, 1966.
7. ——— Metabolic site and time course of cortisol action on glucose uptake, lactic acid output, and glucose-6-phosphate levels of rat thymus cells *in vitro*. J. Biol. Chem., 243: 1039–1042, 1968.
8. Mosher, K. M., and Munck, A. Unpublished results.
9. Makman, M. H., Dvorkin, B., and White, A. Alterations in protein and nucleic acid metabolism of thymocytes produced by adrenal steroids *in vitro*. J. Biol. Chem., 241:1646–1648, 1966.
10. Young, D. A. Glucocorticoid action on rat thymus cells: Interrelationships between carbohydrate, protein and adenine nucleotide metabolism and cortisol effects on these functions *in vitro*. J. Biol. Chem., 244:2210, 1969.
11. Makman, M. H., Dvorkin, B., and White, A. Influence of cortisol on the utilization of precursors of nucleic acids and protein by lymphoid cells *in vitro*. J. Biol. Chem., 243: 1485–1497, 1968.
12. Young, D. A., and Klurfeld, G. Regulation of protein synthesis in rat thymus cells: *in vitro* effects of cortisol and substrates on protein and RNA synthesis, and on α-aminoisobutyric acid transport. Fed. Proc., 27:2310, 1968.
13. Overell, B. G., Condon, S. E., and Petrow, V. The effect of hormones and their analogues upon the uptake of glucose by mouse skin *in vitro*. J. Pharm. Pharmacol., 12: 150–153, 1960.
14. Munck, A. Studies on the mode of action of glucocorticoids in rats. II. The effects *in vivo* and *in vitro* on net glucose uptake by isolated adipose tissue. Biochim. Biophys. Acta, 57:318–326, 1962.
15. Yorke, R. E. The influence of dexamethasone on adipose tissue metabolism *in vitro*. J. Endocr., 39:329–343, 1967.
16. Munck, A., and Brinck-Johnsen, T. Specific and nonspecific physicochemical interactions of glucocorticoids and related steroids with rat thymus cells *in vitro*. J. Biol. Chem., 243:5556–5565, 1968.
17. Burton, A. F., Storr, J. M., and Dunn, W. L. Cytolytic action of corticosteroids on thymus and lymphoma cells *in vitro*. Can. J. Biochem., 45:289–297, 1967.
18. Ingle, D. J. Problems relating to the adrenal cortex. Endocrinology, 31:419–438, 1942.
19. Drury, D. R. Control of blood sugar. J. Clin. Endocr., 2:421–430, 1942.
20. Munck, A. Glucocorticoid activity, specific binding, and competition for cortisol binding sites on thymus cells: cortexolone as potential anti-glucocorticoid. Fed. Proc., 27: 691, 1968.
21. Giddings, S., and Young, D. A. Unpublished results, 1968.
22. Melnykovych, G., and Bishop, C. F. Alkaline phosphatase induction in HeLa cells by prednisolone:reversal by cortexolone. Biochem. Biophys. Res. Commun., 32:233–239, 1968.
23. Makman, M. H., Nakagawa, S., and Dvorkin, B. Studies concerning locus of cortisol action on thymocytes *in vitro*. Fed. Proc., 26:483, 1967.
24. Young, D. A. Unpublished results, 1968.
25. Wira, C., and Munck, A. In vitro binding of ^{3}H-cortisol to rat thymus cell nuclei. Fed. Proc., 28:702, 1969.

26. ——— and Munck, A. Extraction and preliminary identification of physiological glucocorticoid receptors in rat thymus cell nuclei. Fed. Proc., 29:833, 1970.
27. Young, D. A. Glucocorticoid action on rat thymus cells. II. Interrelationships between RNA and protein metabolism, and between cortisol and substrate effect on these metabolic parameters in vitro. J. Biol. Chem., 245:2742, 1970.
28. Swonger, A. K., and Young, D. A. In vitro effects of substrates and cortisol on uridine incorporation into uridine nucleotides and RNA in rat thymus cells. Fed. Proc., 28: 3615, 1969.
29. Rosen, J. M., Fina, J. J., Milholland, R. J., and Rosen, F. Inhibition of glucose uptake in lymphosarcoma P1798 by cortisol and its relationship to the biosynthesis of dexoyribonucleic acid. J. Biol. Chem. (In press).

16

Mechanisms of Action of Corticosteroids on Physiological Maturation of Foetal Adrenal Medulla

Jacques Roffi

Laboratory of Comparative Physiology,
Faculty of Sciences, Paris, France

Much work has been devoted to investigations of the foetal adrenals. Most of this has concerned the foetal adrenal cortex, and it is surprising to note that only a few studies have been made on the foetal adrenal medulla. The early binding of the adrenal medulla and cortex into a single gland during foetal development has remained largely a matter for speculation, and several authors have discussed the possible role of the cortex in the differentiation of medullary cells(1). Ten years ago we approached this problem from a physiological point of view and obtained an increase in the adrenomedullary hormone content of rat and rabbit foetuses treated with cortisone(2).

The presence of catecholamines in foetal adrenals has been demonstrated by various investigators(3). The progressive accumulation of norepinephrine and epinephrine has been studied in the rat(4,5) and in the rabbit(6-8). During foetal development, norepinephrine accumulates first in the adrenals. Then, at a well-defined stage of gestation, the proportion of epinephrine increases rapidly in the two species. This sharp increase occurs just when the foetal cortex begins to manifest obvious signs of physiological activity, both in the rat and in the rabbit(9). The striking chronological coincidence of these two phenomena, suggesting that the adrenal cortex could exert an influence on epinephrine accumulation in foetal adrenals, prompted our investigation of the action of corticoids on the epinephrine-synthesizing system in rat and rabbit foetal adrenals.

In this review there will be presented some results obtained in that field in the past two years(10,11), together with previously unpublished data.

MATERIALS AND METHODS

Our experiments were performed on rat (Sherman strain), rabbit foetuses, and newborn rats. Pregnant females were anaesthetized with ether, and foetuses were deprived of their hypophyses by decapitation in utero, according to the technique of Jost(12). This operation depresses the physiological activity of the foetal adrenal cortices(13).

Rat foetuses were decapitated after 17 ½ days of gestation and, at 19½ days, the mothers were reoperated for intraperitoneal administration of corticotropin (1 IU ACTH in oil) or cortisol acetate (0.5 mg in saline) to some of the decapitated foetuses. At 21 ½ days, after killing the mothers by cervical fracture, the control, decapitated, and decapitated-injected foetuses were removed from the uteri. After incision of the posterior region of the abdominal cavity, both adrenals were dissected out.

Rabbit foetuses were decapitated at 20 days; some of them received at 24 days, in the same way, corticotropin (6 IU) or cortisol acetate (2.5 mg). Foetuses were removed at 28 days under anaesthesia of the mother, and adrenals were removed as in rat foetuses.

Epinephrine and norepinephrine were extracted from the adrenals in 4 percent trichloroacetic acid and were estimated using the fluorometric method of von Euler and Lishajko(14). The results obtained with this chemical technique were confirmed by two biological tests: rat uterus in vitro(15) (epinephrine), and rat blood pressure (epinephrine plus norepinephrine).

In collaboration with F. Margolis, other adrenal extracts were prepared for the determination of the enzymatic activity of phenylethanolamine-*N*-methyl transferase (PNMT), which is responsible for the methylation of norepinephrine into epinephrine in the adrenals of the adult rat(16). These determinations were made according to the method of Wurtman and Axelrod(17).

Finally, norepinephrine, epinephrine, and PNMT activity were assayed in the extra-adrenal chromaffin tissue of newborn rats treated with corticotropin or cortisol acetate. This tissue was removed as described by Lempinen(1): the retroperitoneal region, following prior removal of kidneys and adrenals, was excised as a single block of tissue. The presence therein of the main para-aortic chromaffin body has been verified histochemically.

RESULTS

The Role of the Adrenopituitary Axis in the Accumulation of Epinephrine in Rat and Rabbit Foetal Adrenals

In the rat, fetal decapitation reduces at term the adrenal epinephrine content to about one fourth of that of littermate controls. By contrast, the norepinephrine content increases, apparently doubling the control values. Administration of corti-

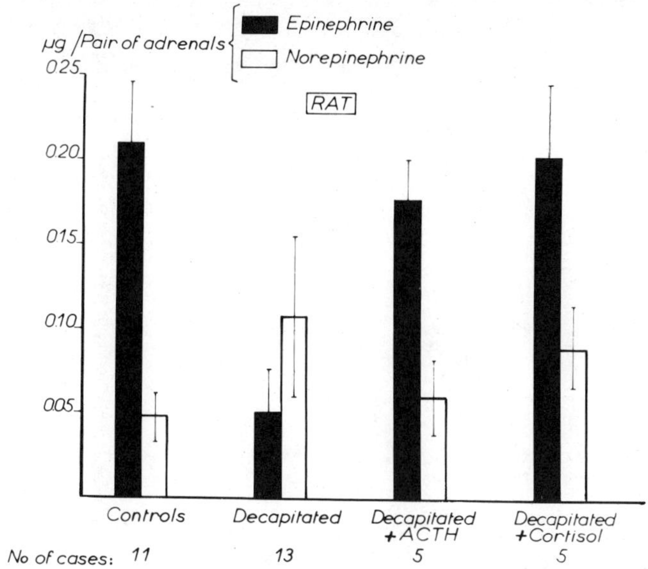

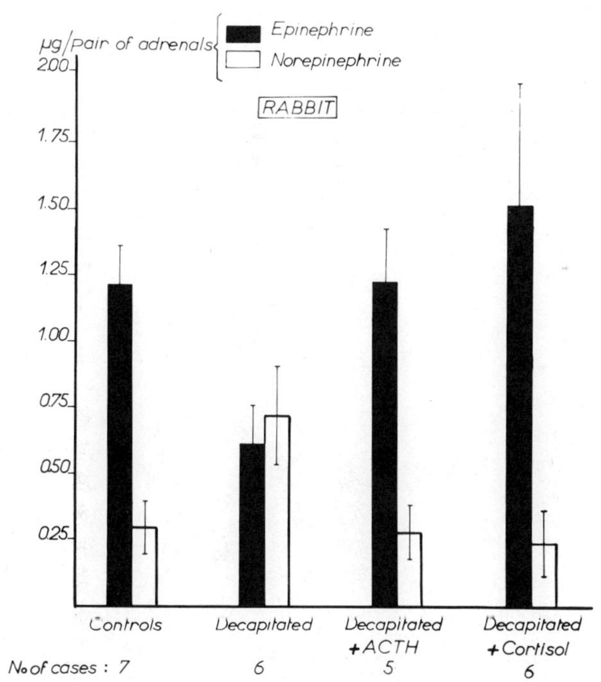

Figs. 1A and B. Effects of foetal decapitation and hormonal administration to decapitates on the epinephrine and norepinephrine content of foetal adrenals.

cotropin, which strongly stimulates the foetal adrenal cortex(18), is sufficient to prevent these differences (Fig. 1).

In the rabbit, decapitation results in a similar pattern: the epinephrine level remains lower than in the controls, while norepinephrine accumulates to above-normal values. The effects of decapitation can be completely obviated, as in the rat, if one injects ACTH or cortisol into the decapitated foetuses (Fig. 1).

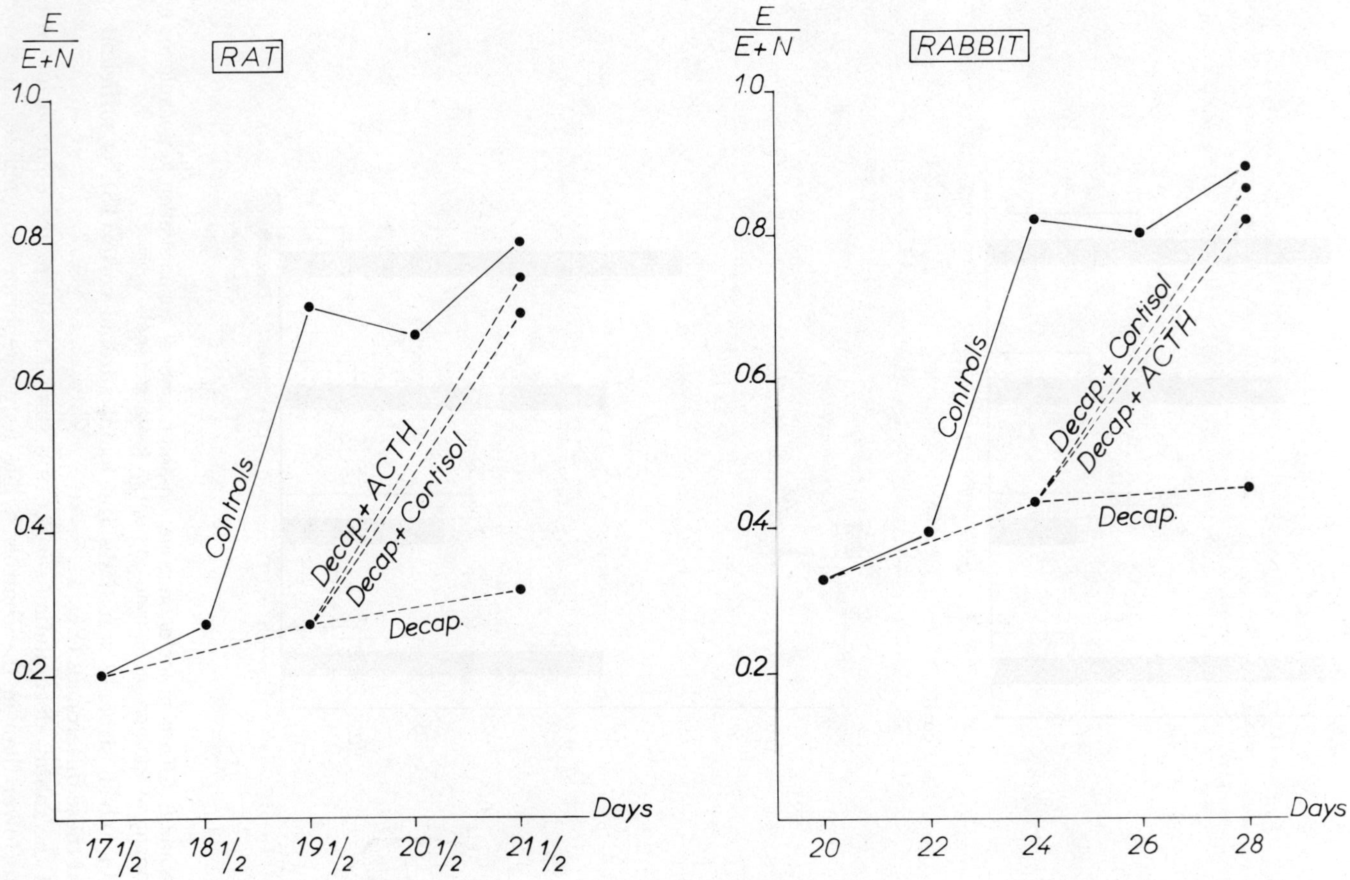

Figs. 2A and B. Development of the epinephrine/epinephrine + norepinephrine ratio in foetal adrenals. Effects of foetal decapitation and hormonal administration to decapitates.

In Figure 2 the ratios of epinephrine to the total epinephrine plus norepinephrine (E+N) in rat and rabbit foetal adrenals are plotted as a function of age. It may be seen that in the control foetuses, a sudden rise of this ratio occurs in both species. After decapitation, the proportion of epinephrine increases very slowly both in the rat and in the rabbit, and remains at a very low level at term compared to the controls. Therefore, injection of ACTH or cortisol 2 days (rat) or 4 days (rabbit) after decapitation allows the proportion of epinephrine to reach near-control values.

Effects of Corticoids on the Activity of the Methylating Enzyme of Norepinephrine in Rat Foetal Adrenals

The PNMT activity has been reported as micromicromoles of the substrate (metanephrine) in vitro synthesized per hour and per pair of adrenals.

Decapitation results in a drastic reduction of this activity, which is depressed at term to about 10 percent of the normal level (Fig. 3). This reduction in PNMT activity is obviated if ACTH is injected into the decapitated foetuses. Cortisol administration raises the level of the enzymatic activity to more than three times the untreated decapitated level. However, despite the small number of determinations, it is pertinent to note that the control values are not reached with cortisol administration as well as they are with ACTH administration.

These results indicate that foetal corticoids regulate the level of PNMT activity in foetal rat adrenals.

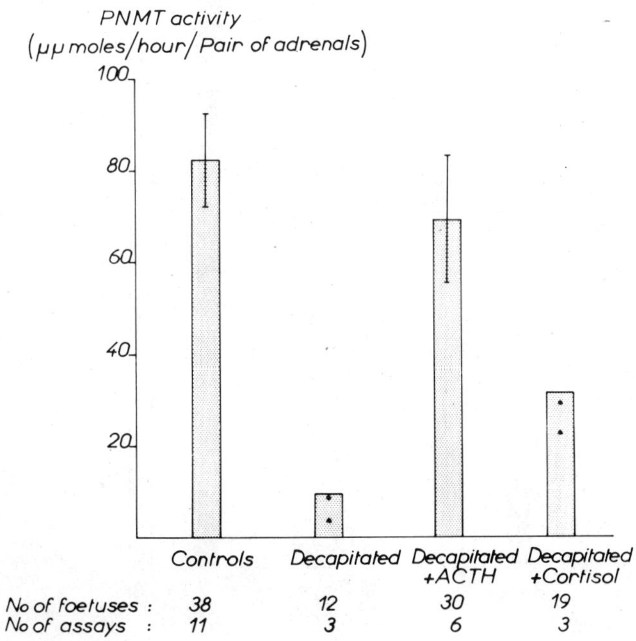

Fig. 3. Activity of the norephinephrine methylating enzyme in the adrenals of the control, decapitated, and decapitated-injected rat foetuses.

Action of Corticoids on the Epinephrine Synthesis in Extra-adrenal Chromaffin Tissue of the Neonate Rat

In an attempt to see if an elevation of circulating glucocorticoids could produce the synthesis of epinephrine in extra-adrenal chromaffin tissue as well as in the adrenal medulla, we have studied the effects of cortisol acetate or ACTH injections in the newborn rat on the epinephrine-norepinephrine contents and PNMT activity of this tissue.

Newborn rats were injected subcutaneously on the first, fourth, and seventh day after birth with 250 μg of cortisol acetate and were killed on the ninth day after birth. Untreated rats served as controls. The results are expressed in Figure 4.

The content of epinephrine in the extra-adrenal chromaffin tissue of the controls falls from 0.08 μg at birth to technically undetectable levels at 9 days. The

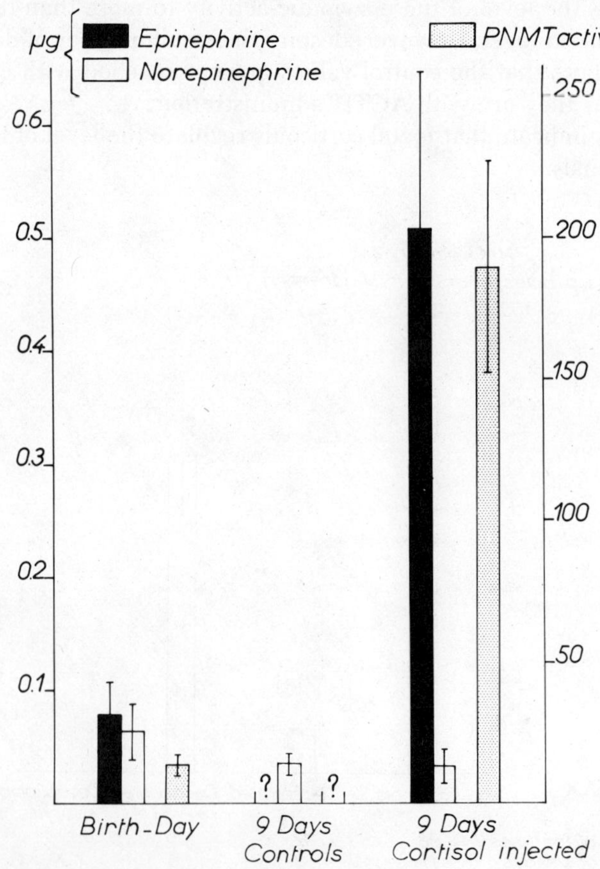

Fig. 4. Effects of cortisol on the epinephrine-norepinephrine content and PNMT activity of the extra-adrenal chromaffin tissue in the newborn rat.

activity of PNMT in this tissue follows a similar pattern and decreases during postnatal development.

In cortisol-treated animals, the epinephrine content is at least fiftyfold higher than in controls of the same age, while PNMT activity is more than thirtyfold higher.

Moreover, in order to evaluate the effect of various doses of cortisol in provoking this increase in epinephrine accumulation, a series of 30 newborn rats were given four injections of 1 to 50 μg of cortisone acetate every second day and then sacrificed at 9 days after birth. A total dose of 4 μg, administered in four injections of 1μg, exerts a clear action on epinephrine increase, and a progressive dose-related response is obtained.

On the other hand, to see whether an elevation of circulating endogenous glucocorticoids produces the same effect as administrated exogenous cortisol, a series of newborn rats was injected with 3 units of ACTH in olive oil daily from birth, with the controls receiving oil alone. This treatment resulted in a marked elevation in epinephrine content of extra-adrenal chromaffin tissue, which attained about threefold the epinephrine content of 4 μg cortisol-treated rats.

DISCUSSION

Various authors have emphasized the importance of the anatomical relationships between the adrenal cortex and medulla in regulating the synthesis of epinephrine. Coupland(19) has suggested that during foetal development, ganglionic chromaffin cells become competent to synthesize epinephrine only when they have penetrated the adrenal cortex. Histochemical findings in favour of this hypothesis were obtained in this laboratory several years ago(20). Shepherd and West(21) had also noted the probable importance of the adrenal cortex in regulating the conversion of norepinephrine to epinephrine in the adrenal medulla. More recently, the role of cortical hormones in this process has been demonstrated in adult rats to be mediated via the methylating enzyme PNMT(17,22).

The present results confirm our previous experimental data(2,23), indicating that the initiation of epinephrine synthesis in rat and rabbit foetal adrenals is controlled by the foetal hypophyseo-adrenocortical system. Decapitation involves a large ablation of the central nervous system of the foetus; nervous damage caused by this operation might therefore be suspected to be responsible for the troubles in catecholamine accumulation in adrenals. If this were so, however, a pure hormonal treatment would not be expected to restore normal conditions.

One should note that the total amount of epinephrine contained in foetal adrenals depends upon the rate of synthesis and also the rate of destruction or secretion of this hormone, since Comline and Silver(3) have demonstrated that the adrenals of calf and lamb foetuses are able to liberate their medullary hormones into the foetal blood. Whatever the differences in catecholamine secretion that may be caused by decapitation, in vitro assays of PNMT show that the normal methylation of norepinephrine in foetal adrenals requires glucocorticoids.

Corticotropin is more active than cortisol in elevating the PNMT activity in the adrenals of decapitated foetuses. This difference could be explained by the fact that

ACTH administration probably results in a higher local concentration of glucocorticoids in the adrenal medulla than does cortisol injections. In adult animals, the adrenal medulla is mainly irrigated by the blood derived from the adrenal cortex (24). If this type of irrigation is realized in foetal adrenals, such a situation would permit delivery of a relatively high concentration of glucocorticoids to the medulla, compared to other tissues, and consequently a high PNMT activity.

Thus, foetal adrenocortical hormones, which are known to influence the storage of glycogen in the foetal liver and heart and to depress the size of the thymus (25), are also responsible for the physiological maturation of the adrenal medulla, which may be considered as another target organ for foetal glucocorticoids.

Our experiments on the extra-adrenal chromaffin tissue are in complete agreement with the previous observations of Lempinen(1), who noted that repeated injections of hydrocortisone prevent the normal degeneration of this tissue in the newborn rat. Ganglionic chromaffin cells can respond as well as medullary cells to elevated levels of circulating glucocorticoids by manifesting elevation in PNMT activity and epinephrine content.

How do glucocorticoids exert their action on PNMT activity? In adult animals cortisone stimulates the synthesis of various enzymatic proteins(26). To learn something about this problem, extracts of extra-adrenal chromaffin tissue from treated and untreated 9-day rats were pooled together. These mixtures showed only the summation of the two activities. This result does not seem compatible with the induction by corticoids of some activating substance, but strongly suggests that glucocorticoids produce an increase in enzymatic protein synthesis in foetal chromaffin cells.

CONCLUSION

The present data indicate that in rat and rabbit foetuses, adrenocortical hormones allow the adrenals to achieve an active synthesis of epinephrine during the last days of gestation. Glucocorticoids manifest their action by controlling the activity of the norepinephrine methylating enzyme. In these conditions, one can conceive the functional usefulness of the precocious juxtaposition of cortex and medulla during development, which seems to provide a particularly clear example of anatomic-physiological adaption in the endocrinology of mammals.

REFERENCES

1. Lempinen, M. Extra-adrenal chromaffin tissue of the rat and the effect of cortical hormones on it. Acta Physiol. Scand., 62 (Suppl. 231):1–91, 1964.
2. Jost, A., and Roffi, J. Action de la cortisone et de l'hydrocortisone sur la teneur de la surrénale du rat et de la souris en substances hypertensives. C.R. Acad. Sci. (Paris), 246:163–165, 1958.
3. Comline, R. S., and Silver, M. Development of activity in the adrenal medulla of the foetus and new-born animal. Brit. Med. Bull., 22:16–20, 1966.
4. Kamoun, A., Miahle-Voloss, C., and Stutinsky, F. Evolution de la teneur en adrénaline et noradrénaline de la surrénale du rat. C.R. Acad. Sci. (Paris), 259:3079–3081, 1964.

5. Chéoux, G., and Roffi, J. Dosage de l'adrénaline et de la noradrénaline dans les surrénales du foetus et du nouveau-né de rat. C.R. Acad. Sci. (Paris), 260:6702–6705, 1965.
6. Roffi, J. Dosage de l'adrénaline et de la noradrénaline dans les surrénales du foetus de lapin au cours de la gestation. J. Physiol. (Paris), 56:434–435, 1964.
7. Brundin, T. Catecholamines in adrenals from fetal rabbits. Acta Physiol. Scand., 63: 509–510, 1965.
8. Coupland, R. E., and MacDougall, J. D. B. Adrenaline formation in noradrenaline-storing chromaffin cells in vitro induced by corticosterone. J. Endocr., 36:317–324, 1966.
9. Roffi, J., Chéoux, G., Cohen, A., and Jost, A. Evolution du taux d'adrénaline dans les surrénales chez les foetus de lapin et de rat. Rapports avec la corticosurrénale. J. Physiol. (Paris), 57:690–691, 1965.
10. Margolis, F., Roffi, J., and Jost, A. Norepinephrine methylation in fetal rat adrenals. Science, 154:275–276, 1966.
11. Roffi, J., and Margolis, F. Synthése d'adrénaline dans le tissu chromaffine extra-surrénalien, chez le rat nouveau-né, sous l'effet de l'hydrocortisone ou de la corticostimuline. C.R. Acad. Sci. (Paris), 263:1496–1499, 1966.
12. Jost, A. Expériences de décapitation de l'embryon de lapin. C.R. Soc. Biol. (Paris), 225:322–324, 1947.
13. ——— Jacquot, R., and Cohen, A. *In* The Human Adrenal Cortex, Currie, A., et al., eds., Edinburgh, Livingstone, pp. 569–579, 1962.
14. Euler, U. S. von, and Lishajko, F. The estimation of catecholamines in urine. Acta Physiol. Scand., 45:122–132, 1959.
15. Gaddum, J. H., and Lembeck, F. The assay of substances from the adrenal medulla. Brit. J. Pharmacol., 4:401–408, 1949.
16. Axelrod, J. Purification and properties of phenylethanolamine-*N*-methyl transferase. J. Biol. Chem., 237:1657–1660, 1962.
17. Wurtman, R. J., and Axelrod, J. Adrenaline synthesis: control by the pituitary gland and adrenal glucocorticoids. Science, 150:1464–1465, 1965.
18. Jost, A. Problems of fetal endocrinology: The adrenal glands. Recent Progr. Hormone Res., 22:541–574, 1966.
19. Coupland, R. E. On the morphology and adrenaline-noradrenaline content of chromaffin tissue. J. Endocr., 9:194–203, 1953.
20. Picon, R. Recherches histochimiques sur la médullo-surrénale foetale du lapin. Arch. Anat. Micr., 53:343–356, 1964.
21. Shepherd, D. M., and West, G. B. Noradrenaline and the suprarenal medulla. Brit. J. Pharmacol., 6:665–674, 1951.
22. Wurtman, R. J., and Axelrod, J. Control of enzymatic synthesis of adrenaline in the adrenal medulla by adrenal cortical steroids. J. Biol. Chem., 241:2301–2305, 1966.
23. Roffi, J. Influence de la corticosurrénale sur le contenu en adrénaline et noradrénaline de la surrénale, chez le foetus de rat. C.R. Acad. Sci. (Paris), 260:1267–1270, 1965.
24. Coupland, R. E. The Natural History of the Chromaffin Cell, London, Longmans, Green and Co., 1965.
25. Jost, A. The function of the foetal adrenal cortex. *In* The Human Adrenal Cortex: Its Function Throughout Life. Ciba Foundation, Wolstenholme, G. E. W. and O'Connor, M. eds., London, Churchill, J. and A. pp. 11–28, 1967.
26. Feigelson, P., and Feigelson, M. Studies on the role of hormonal enzyme induction in glucocorticoids action. *In* Mechanisms of Hormone Action, Karlson, P., ed., New York, Academic Press, Inc., pp. 264–267, 1965.

[illegible]

[illegible]

[illegible]

[illegible]

[illegible]

[illegible]

[illegible]

[illegible]

[illegible]

[illegible]

[illegible]

[illegible]

[illegible]

[illegible]

[illegible]

[illegible]

[illegible]

[illegible]

[illegible]

[illegible]

17

Mechanism of Action of Cortisol in Maturation of Brain Lipid Patterns in Embryonal and Young Chicks*

Michael Granich[1] and Paola S. Timiras[2]

[1]Institute of Human Development, University of California, Berkeley, California
[2]Department of Physiology-Anatomy, University of California, Berkeley, California

Controversy exists concerning the possible role of cortisol—and of other glucocorticoid hormones with similar actions—in the maturation of the central nervous system. It has been reported that when glucocorticoids are administered neonatally to rats, myelination is "considerably" delayed(1). On the other hand, glucocorticoids have been shown to accelerate the development of convulsive responses in rats(2) and to increase the level of cerebrosides in spinal cord lipid(3), both of which imply accelerated myelination.

Normal brain maturation is characterized by an increase in the percentage of cerebroside, total sphingolipid (principally cerebroside and sphingomyelin), and cholesterol, the predominant lipids of myelin(4). We have determined the percentages of these lipids and of total phospholipids in brain extracts of cortisol-treated embryos and those of chicks treated with cortisol during incubation. The lipid patterns thus determined suggest that cortisol may, indeed, delay or accelerate myelination, depending upon dosage as well as on age at treatment and at sacrifice.

A preliminary report of this work has been published(5).

EXPERIMENTAL PROCEDURES

Cortisol (as the free alcohol, Ciba) suspended in 50 μl of water was injected into the air cell(6,7) of embryonated White Leghorn eggs on the 5th and 15th day

*This investigation was supported by USPHS Grants No. GM-09267 and HD-101.

of incubation. Two dose levels were used at each age: 25 μg and 2.5 μg on day 5, and either 50 or 100 μg and 10 or 12.5 μg on day 15. The larger doses specified in each case have been shown to produce significant increases in glycogen content per unit weight of liver, whereas the lower doses produce a small but definite elevation of liver glycogen(8).

The incubation periods of 5 and 15 days were selected for the following reasons: at the 5th day of incubation there appears to be little or no production of hormones by the adrenals or by any other endocrine organ of the embryo(9). Consequently, the effects of hormone treatment on lipid pattern development should be independent of interaction with other hormones.

The 15th day of incubation shortly precedes both the period of most active myelination(10) and the onset of demonstrated function of the pituitary-adrenal axis(11). Treatment on this day thus will permit us to observe the direct effects of cortisol on lipid patterns in the brain at a time when these patterns are rapidly maturing. Caution must be exercised in interpreting such effects, however, inasmuch as they may, at this time, be due to interactions between cortisol and other hormones present in the embryo. Furthermore, even though cortisol may have no direct effect on the lipid pattern, it may induce changes in the hormonal or other environments within the embryo that, in turn, are responsible for any changes in lipid patterns.

Embryos were sacrificed by decapitation on the 17th and 19th day of incubation, and chicks were sacrificed by the same method on the 8th day after hatching. These ages are all within the period of most active myelination. Weights of embryo or chick, left gonad(12), adrenals, and brain were determined. Brain lipids were extracted by homogenization in 19 volumes of chloroform:methanol (2:1) within two minutes following decapitation. Extraneous materials were removed by filtration through sintered glass and by washing with 0.88 percent potassium chloride(13). Total lipid was determined after evaporation of the washed extract at 60°C under nitrogen, followed by further drying to constant weight over potassium hydroxide in vacuo. Total cholesterol, phospholipid, cerebroside, and total sphingolipid were determined(14-17). Sphingomyelin was determined directly on pooled extracts by thin-layer chromatography.(18).

RESULTS AND DISCUSSION

The effects of cortisol on body, brain, and endocrine organs varied according to the day of treatment, the dose, and the age of sacrifice.

In 17-day embryos, weights of body, brain, and left testis were significantly decreased in those that had been treated with the larger (25 μg) dose of cortisol on the 5th day of incubation, whereas the smaller (2.5 μg) dose of cortisol had no apparent effect on body or organ weights (Table 1). In contrast, the weight of the adrenals in embryos of this age seemed to be virtually unaffected by cortisol. Indeed, in cortisol-treated embryos whose body weight was half that of their controls, the essentially normal-sized adrenals were the most prominent organs observed. Neither the large (25 μg) nor the small (2.5 μg) dose effected any significant changes

TABLE 1

Body, brain and endocrine weights in embryos and chicks treated with cortisol at two incubation periods

Treatment day	Dose (μg)	Body (g)	Brain (mg)	Ovary* (mg)	Testis* (mg)	Adrenals (mg)
		(Age at sacrifice: 17th day† of incubation)				
5th	0	21.3 ±0.7	744 ±20	5.48 ±0.52	4.03 ±0.39	3.99 ±0.47
	2.5	20.4 ±0.5	759 ±8	7.34 ±1.39	4.40 ±0.54	4.61 ±0.36
	25	15.8 ±0.8	672 ±21	4.93 ±0.47	2.75 ±0.35	4.13 ±0.58
		p‡ < 0.001	$p < 0.05$		$p < 0.05$	
		(Age at sacrifice: 19th day† of incubation)				
15th	0	25.2 ±0.6	807 ±16	5.36 ±0.20	4.48 ±0.59	3.98 ±0.12
	12.5	24.4 ±1.1	779 ±14	5.97 ±0.92	2.90 ±0.00	5.95 ±0.64
						$p < 0.01$
	50 or 100	21.7 ±1.6	712 ±17	3.78 ±0.57	3.63 ±0.68	5.36 ±0.71
		$p < 0.1$	$p < 0.01$	$p < 0.05$		$p < 0.1$
		(Age at sacrifice: 8th day† after hatching)				
15th	0	44.1 ±1.4	974 ±8	10.47 ±0.95	4.50 ±0.30	12.16 ±1.00
	10	40.2 ±1.3	958 ±13			
				13.23 ±1.63	7.65 ±1.00	9.95 ±2.40
	100	44.8 ±1.7	964 ±12			

*Left gonad.
†±4 hours.
‡Calculated with the t-test for nonpaired data.

in total lipid per unit wet weight of brain or in the percentage distribution of several major classes of lipids (Table 2).

Certain effects of cortisol were similar between embryos treated at 5 days and those treated at 15 days: the larger doses in each case resulted in decreased body, brain, and gonad size. The smaller doses did not affect embryo or gonad weight. Differential effects, however, were observed with respect to lipids and adrenals (Tables 1 and 2). In embryos treated on the 15th day of incubation and sacrificed at day 19, both dose levels produced a statistically significant increase over control values in total lipid content relative to brain weight. In addition, adrenals in the embryos that had received the smaller cortisol dose were significantly larger than those in controls. Similarly, the percentages of total sphingolipid and of cerebroside in brain lipid extracts from embryos treated with the smaller dose were greater than in controls or in embryos treated with the larger doses. When repli-

TABLE 2

Total lipid: brain weight, and major lipids as percent of total brain lipid of embryos and chicks treated with cortisol at two incubation periods

Treatment day	Dose (μg)	Total lipid / Brain weight (mg/g)	Sphingolipid / Total lipid (%)	Cerebroside / Total lipid (%)	Cholesterol / Total lipid (%)	Phospholipid / Total lipid (%)
(Age at sacrifice: 17th day* of incubation)						
5th	0	34.8 ±0.4	3.51 ±0.06	1.80 ±0.19	19.3 ±0.3	74.0 ±0.5
	2.5	34.4 ±0.6	3.66 ±0.11	1.69 ±0.14	19.2 ±0.0	73.8 ±0.4
	25	35.0 ±0.7	3.52 ±0.16	2.07 ±0.15	19.0 ±0.1	73.8 ±1.4
(Age at sacrifice: 19th day* of incubation)						
15th	0	37.2 ±1.0	8.60 ±0.26	2.21 ±0.12	20.0 ±0.2	71.5 ±0.3
	12.5	40.6 ±0.5 p† < 0.02	9.24 ±0.21 $p < 0.1$	2.64 ±0.10 $p < 0.02$	20.1 ±0.3	70.0 ±0.4 $p < 0.02$
	50 or 100	40.2 ±0.4 $p < 0.02$	8.81 ±0.20	2.47 ±0.14	19.4 ±0.2 $p < 0.05$	70.5 ±0.4
(Age at sacrifice: 8th day* after hatching)						
15th	0	59.5 ±0.6	17.71 ±0.33	9.86 ±0.23	22.8 ±0.5	63.5 ±1.1
	10	57.2 ±0.6	18.20 ±0.34	9.64 ±0.25	23.7 ±0.1	64.5 ±0.8
	100	57.3 ±0.8	17.31 ±0.26	9.20 ±0.16 $p < 0.05$	22.9 ±0.1	63.0 ±0.5

* ±4 hours.

† Calculated with the t-test for nonpaired data.

cate thin-layer chromatographic determinations were made on pooled extracts, a significantly greater ($p<0.001$) percentage of sphingomyelin was found in the lower-dose group than in control or higher-dose groups. Therefore, despite the lack of statistical significance in the mean total increase of sphingolipid, we believe that a real increase in the percentage of total sphingolipids did, indeed, occur. Both cerebroside and sphingomyelin shared in this increase. The larger dose of cortisol effected a small but significant decrease in percentage of cholesterol.

In 8-day chicks, only one significant difference was observed between cortisol-treated and control groups: the percentage of cerebroside in total lipid was decreased in chicks that had been treated with 100 μg of cortisol as 15-day embryos.

Regardless of isolated effects of cortisol on body and endocrine weights, the lipid pattern remained essentially characteristic of the age at which the embryo was sacrificed. Glucocorticoid treatment, known to inhibit mitosis in the brain(19), might have been expected to decrease markedly the replication of the rapidly proliferating oligodendroglia, the cells that produce myelin, and to have little effect on the number of neurons, which are essentially nonreplicating. The age-specific constancy of the lipid pattern suggests that the timed appearance of inducers of the enzymes responsible for lipogenesis, rather than the number of oligodendroglia present, may govern the brain lipid pattern during the period of rapid myelination. Hormones are known to be inducers of some enzymes associated with differentiation and development(20), but the nature of the inducers of lipogenic enzymes remains unknown.

The results observed in these experiments seem to indicate little effect on brain lipids when the hormone is administered at 5 days of incubation and the animal is sacrificed at the 17th day. Yet it should not be excluded that some transient effects may have occurred during early development. In contrast, when cortisol is administered on the 15th day of incubation, effects are more pronounced and show a clear relation to the dose-level administered. The decrease in cholesterol and cerebroside concentration, noted in both embryos and chicks that had been treated with large doses of cortisol, may be interpreted as representing delayed myelination. Indeed, it has been reported that high levels of glucocorticoids, due either to exogenous administration of the hormones or to adrenal hypersecretion, as occurs in Cushing's syndrome, result in altered brain lipid patterns(21). On the other hand, embryos treated with the smaller dose of cortisol demonstrated a more mature lipid pattern than the corresponding controls. The cortisol-induced increases in total sphingolipid and cerebroside observed in the chick are in agreement with similar studies conducted in the developing rat brain(3).

The complexity of the hormonal environment at this stage of embryonal development makes it difficult to determine the precise mechanisms underlying the actions of cortisol: it is possible that the observed acceleration in myelination is due to a direct action of the hormone on brain lipids, for it has been shown that cortisol is capable of regulating enzymatic activity in explants of nervous tissue in vitro(22). Furthermore, administration of cortisol at a critical period in the development of the pituitary-adrenocortical axis may stimulate the secretion of adrenocorticotropic hormone and thereby raise the "set point" for cortisol(23). The hypertrophy of the adrenals seen in these experiments is compatible with this hypothesis. However, cortisol may also exert its effects indirectly by stimulating other endocrine glands. It has been shown that thyroxine and triiodothyronine accelerate myelination both in vivo(24) and in vitro(25), whereas hypothyroidism decreases cerebroside and sphingomyelin levels in the brain of infant rats(26). Although the present data demonstrate that cortisol influences brain lipid patterns during development, the mechanisms underlying its effects require further investigation.

REFERENCES

1. Field, E. J. Observations on the development of microglia together with a note on the influence of cortisone. J. Anat., 89:201–208, 1955.
2. Vernadakis, A., and Timiras, P. S. Regulation of brain and spinal cord excitability by cortisol and estradiol in developing rats. Excerpta Medica International Congress of Hormonal Steroids, Series No. 111, pp. 908–912, 1966.

3. Casper, R., Vernadakis, A., and Timiras, P. S. Influence of estradiol and cortisol on lipids and cerebrosides in the developing brain and spinal cord of the rat. Brain Research, 5:524–526, 1967.
4. McIlwain, H. Biochemistry and the Central Nervous System, 3rd ed. Boston, Little, Brown and Co., pp. 197–242, 1966.
5. Granich, M., and Timiras, P. S. Influence of estradiol and cortisol on lipid patterns in the brain of chick embryo. Fed. Proc., 26:709, 1967.
6. Burnet, F. M., and Warner, N. L. Assay of corticosteroids in the chick embryo. Nature, 187:938–939, 1960.
7. Freericks, R. Über die Vitalfärbbarkeit von Hühnerembryonen. Biol. Zbl., 73:155–169, 1954.
8. Granich, M., and Timiras, P. S. Unpublished data, 1967.
9. Hazelwood, P. L. Carbohydrate metabolism. *In* Avian Physiology, Sturkie, P. D., ed. Ithaca, Cornell University Press, 1965.
10. Garrigan, O. W., and Chargaff, E. Studies on the mucolipids and the cerebrosides of chicken brain during embryonic development. Biochim. Biophys. Acta, 70:452–464, 1963.
11. Case, J. F. Adrenal cortical-anterior pituitary relationships during embryonic life. Ann. N.Y. Acad. Sci., 55:147–158, 1951.
12. Romanoff, A. L. The Avian Embryo: Structural and Functional development. New York, The Macmillan Co., pp. 781–862, 1960.
13. Folch, J., Lees, M., and Sloane Stanley, G. H. A simple method for the isolation and purification of total lipids from animal tissues. J. Biol. Chem., 226:497–509, 1957.
14. Zlatkis, A., Zak, B., and Boyle, A. J. A new method for the direct determination of serum cholesterol. J. Lab. Clin. Med., 41:486–492, 1953.
15. Rouser, G., Siakotos, A. N., and Fleischer, S. Quantitative analysis of phospholipids by thin-layer chromatography and phosphorus analysis of spots. Lipids, 1:85–86, 1966.
16. Svennerholm, L. The distribution of lipids in the human nervous system—I. Analytical procedure. Lipids of foetal and newborn brain. J. Neurochem., 11:839–853, 1964.
17. Lauter, C. J., and Trams, E. G. A spectrophotometric determination of sphingosine. J. Lipid. Res., 3:136–138, 1962.
18. Rouser, G., Kritchevsky, G., Galli, C., and Heller, D. Determination of polar lipids: quantitative column and thin-layer chromatography. J. Amer. Oil Chem. Soc., 42:215–227, 1965.
19. Howard, E. Effects of corticosterone and food restrictions on growth and on DNA, RNA and cholesterol contents of the brain and liver in infant mice. J. Neurochem., 12:181–191, 1965.
20. Moog, F. The development of function in the adrenal cortex. *In* Comparative Endocrinology, Gorbman, A., ed. New York, John Wiley and Sons, pp. 624–638, 1959.
21. Kabara, J. J. Brain cholesterol: the effect of chemical and physical agents. Advances Lipid Res., 5:279–327, 1967.
22. Vernadakis, A., and Timiras, P. S. Effects of estradiol and cortisol on neural tissue in culture. Experientia, 23:467–468, 1967.
23. Yates, F. E., Leeman, S. E., Glenister, D. W., and Dallman, M. F. Interaction between plasma corticosterone concentration and adrenocorticotropin-releasing stimuli in the rat: evidence for the reset of an endocrine feedback control. Endocrinology, 69:67–80, 1961.
24. Myant, N. B., and Cole, L. A. Effect of thyroxine on the deposition of phospholipids in the brain *in vivo* and on the synthesis of phospholipids by brain slices. J. Neurochem., 13:1299–1307, 1966.
25. Hamburgh, M. Evidence of a direct effect of temperature and thyroid hormone on myelinogenesis *in vitro*. Develop. Biol., 13:15–30, 1966.
26. Bizzi, A., Garattini, S., Jori, A., and Tacconi, M. T. Effect of drugs on brain lipids and behavior during development. Istituto di Richerche Farmacologiche, Mario Negri, Milan, Italy. Internal report No. 3727, 1968.

18

The Role of Transcriptional Events in Tissue-Specific Responses to Thyroxine in Induced Anuran Metamorphosis

Michael Solursh[1] and Aubrey Gorbman[2]

[1]Department of Zoology, University of Iowa, Iowa City, Iowa
[2]Department of Zoology, University of Washington, Seattle, Washington

INTRODUCTION

There are numerous active studies of the processes of embryonic development and differentiation in terms of successive activation of portions of the genome and sequential appearance of RNAs presumably coded for particular characteristic proteins. In whole embryos, several workers have found that progressing stages of developing embryos contain correspondingly different populations of RNAs(1-4). During the differentiation of a single cell type, there is also evidence for occurrence of RNA molecules characteristic of the differentiated state(5-7). A valuable tool in such studies has been the technique of annealing RNA with membrane-fixed DNA(8,9), since it depends upon base sequences. While this technique is still imperfect, as pointed out by Birnboim et al. and by Church and McCarthy(9,10), it is at this time the technique of choice for characterizing qualitative differences between RNA preparations.

An intriguing application of the DNA-RNA competitive hybridization technique is in the analysis of the roles of hormones in activating parts of the genomes of embryos to evoke their phenotypic contribution. Particularly, the dramatic action of thyroxine in initiating simultaneous changes in many tissues of the metamorphosing anuran embryo requires clarification and definition in terms of nucleic acid metabolism and specific RNA synthesis. Restricted tissue responses to locally applied thyroid hormones(11,12), as well as in vitro responses to these hormones(13) have made it clear that metamorphosing amphibian tissues are responding directly to the hormones. Furthermore, the numerous experiments of O. M. Helff and his students in the 1920's demonstrated specificity of transplanted tadpole tissues responding to thyroid hormone(14). For example, tail skin is resorbed regardless of its location, while back skin is not resorbed when transplanted to the tail. The diversity and specificity of the thyroid hormone responses are further emphasized in more recent studies of particular proteins whose synthesis is evoked in particular cell types, like urea cycle enzymes in the liver or hydrolases in the tail(15). In both of these synthetic systems, not only is the new formation of RNA and protein evoked by thyroxine, but also it is obligatory for further metamorphic changes to proceed.

If it is accepted that thyroxine-evoked amphibian metamorphosis depends on an ordered sequence of protein synthetic events, then some important derivative questions can be asked whose answers may be obtained through use of the DNA-RNA competitive hybridization technique. Where a single stimulant, thyroxine, can evoke so many different cellular syntheses, it is important to know whether the hormone is directly affecting the same DNA transcriptional event in various kinds of responding tissues and whether, as a consequence, diversification of the responses follows a common type of cellular response to thyroxine. Another possibility is that thyroxine affects different parts of the genomes of different tissues more or less directly, so that a variety of transcriptions and RNA syntheses follow hormone treatment. Combinations of these two alternatives also could occur.

Some pertinent research already has been done in this area. According to Tata(16,17), not only is there a general acceleration of RNA synthesis in the thyroxinized amphibian tadpole liver, but also there is a reorganization of the protein synthetic apparatus. This includes an increased rate of ribosome formation, faster turnover of preexisting ribosomes, and firmer association of new ribosomes with newly synthesized intracellular membranes. The latent period after thyroxinization for appearance of new hepatic enzymes is relatively long, but new RNA synthesis occurs early during this interval. Hormone-induced regression of the tail, also, is accompanied by an accelerated synthesis of both RNA and protein. Regression and hydrolase synthesis are blocked by actinomycin D(18,19). This and the increased template efficiency of thyroxine-primed hepatic chromatin(20) emphasize the special importance of prior RNA synthesis in the initiation and progress of metamorphosis.

To explain the diversity of responses evoked by the single agent, thyroxine, in the tadpole, it might be expected that comparison of the qualitative features of RNAs in responding tissues would be highly informative. Accordingly, we undertook such a study using a DNA-RNA hybridization technique. Data derived from such tests could permit more precise assessment of the role of differentiated or specialized transcriptional events in the diversification of sequelae to the common hormonal signal.

MATERIALS AND METHODS

Locally caught *Rana aurora* tadpoles, about 10 cm in length, were injected intraperitoneally with 2.5×10^{-8} moles/g animal of L-thyroxine to induce precocious metamorphosis. At various times after hormone treatment, tissues to be studied were dissected free and frozen in liquid nitrogen. To prepare labeled RNA, animals were killed 8 hours after intraperitoneal injection of 3–5 mc/animal of $^{32}PO_4$ (New England Nuclear Corporation). A concentration of 1.6 g/liter of streptomycin and 1 million units/liter of penicillin was maintained in the water during the labeling period.

Whole RNA was extracted by a modification of the method of Scherrer and Darnell(21). Tissues were triturated from the frozen state in two volumes of a solution containing 0.15 M LiCl, 0.005 M $MgCl_2$, 0.005 M sodium acetate (pH 5.5), 1 percent sodium lauryl sulfate, 500 μg/ml bentonite, and 10 μg/ml sodium dextran sulfate. The extract was shaken with an equal volume of 0.01 M sodium acetate (pH 5.5) saturated phenol containing 0.1 percent 8-hydroxyquinoline at 60°C for 5 minutes. After centrifugation and removal of the aqueous phase, the protein pad was sonicated briefly in one volume of homogenizing solution, and the pooled extracts were reextracted with phenol. RNA was precipitated by addition to the aqueous phase of 0.1 volume of 10 percent NaCl and 2.5 volumes of 95 percent ethanol. After a second ethanol precipitation from 0.01 M sodium acetate (pH 5.5), the RNA solution was digested with 20 μg/ml electrophoretically purified DNAase (Worthington) for one hour at 37°C in 0.01 M $MgCl_2$-0.01 M tris (pH 7.5) containing 10 μg/ml of sodium dextran sulfate. Self-digested pronase (Calbiochem) was added to give a 20 μg/ml concentration, and the solution was incubated for 30 minutes at 37°C. After adjusting the salt concentration to 1× SSC (0.15 M NaCl, 0.015 M Na_3 citrate), the RNA was shaken with phenol at 60°C and at room temperature for 10 minutes each, and was finally precipitated with ethanol. The RNA was further purified by precipitation with cetyl-trimethyl ammonium bromide, according to the method of Ralph and Bellamy(22), and passed through G-25 Sephadex. Concentration was estimated by absorbancy at 260 mμ. All RNA samples used had 260:230 and 260:280 ratios greater than 2.0.

DNA was prepared from tadpole tails and skin by homogenization in 0.5 volumes of 2× SSC. The homogenate was made 2 percent in sodium lauryl sulfate and 0.75 M in sodium perchlorate, followed by extraction, for 30 minutes and 10 minutes respectively, at 0°C with equal volumes of chloroform-isoamyl alcohol (24:1). The DNA was precipitated from the aqueous phase with three volumes of ethanol(23). The DNA was dissolved in SSC and treated with 20 μg/ml of heated RNAase (Sigma) at 37°C for 1 hour. Then the solution was made 1 percent in sodium lauryl sulfate and incubated with 50 μg/ml of self-digested pronase for 2 hours at 37°C. The DNA was extracted twice for 10 minutes at 60°C and at room temperature with water-saturated phenol (pH 7.8), and was then precipitated with isopropanol. The precipitate was dissolved in 0.1× SSC, and the salt concentration was adjusted to 1× SSC before dialysis against 1× SSC. The final preparation was stored at 4°C over chloroform. The T_m 87.7°C with 37.6 percent hyperchromaticity.

The annealing procedure used was that of Gillespie and Spiegelman(8), as modified by Church and McCarthy(24). Heat-denatured DNA was immobilized on 100 sq cm nitrocellulose filters (Scheicher and Schuell type B6, coarse), and discs containing approximately 7 μg of DNA were punched out. Retention of DNA by the filters was estimated by OD_{260} of the filtrate, by the Dische test for DNA, and by measuring DNAase-released 260-mμ absorbing material. Hybirds were formed by incubation of 7 μg of DNA, 5–15 μg of ^{32}P-RNA, and various amounts of unlabeled competitor RNA in 0.2 ml of 2× SSC at 67°C for 12 hours. After annealing, each filter was washed in 5 ml of 2× SSC at 67°C three times for 5 minutes each time, dried, and assayed for radioactivity in a Beckman Liquid Scintillation System. Nonspecific binding was determined by incubation of a similar filter bearing *E. coli* DNA, or no DNA, in each vial. Background counts were less than 0.2 percent of input counts.

RESULTS

In the RNA competition tests, each reaction tube contains a constant, small amount of ^{32}P-labeled RNA, a given amount of DNA on a nitrocellulose filter, and increasing amounts of the various unlabeled RNA's being compared. If the distribution of RNA species in the labeled and cold RNAs is the same, then the fraction of labeled RNA annealed to the DNA should be inversely proportional to the cold RNA concentration. At sufficiently high competitor concentration, all the labeled RNA can be diluted out of the DNA hybrid. If some of the labeled species are not present in the cold RNA preparation, a plateau is reached where greater concentrations of the cold RNA are unable to reduce the annealed counts any further.

In DNA-RNA competitive hybridization analysis of unlabeled RNAs from thyroxine-stimulated liver and back skin, there was no evidence that formation of

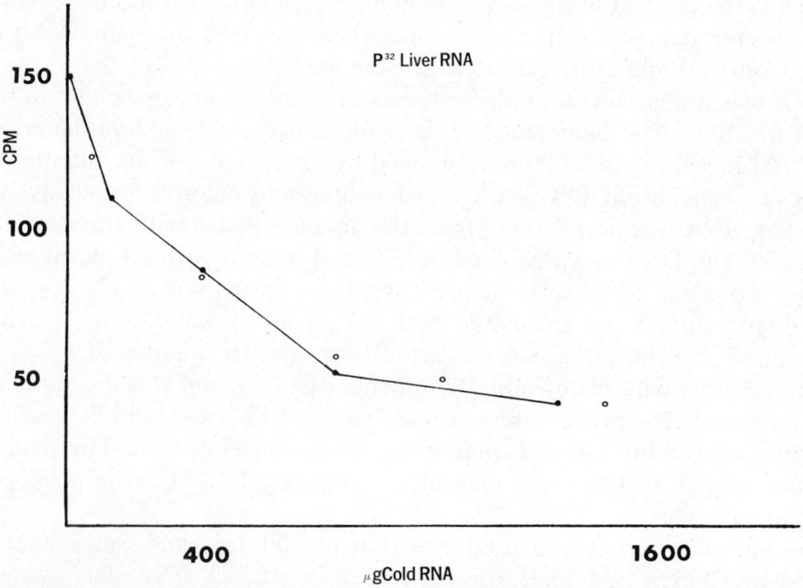

Fig. 1. Comparison of the abilities of unlabeled premetamorphic tadpole liver RNA and liver RNA 5 days after thyroxinization to compete with the binding of 10.9 μg (849 cpm/μg) of ^{32}P-labeled premetamorphic liver RNA to 12.5 μg of tadpole DNA on a 5.8-mm-diameter filter during incubation in 0.2 ml of 2×SSC for 12 hours at 67°C.

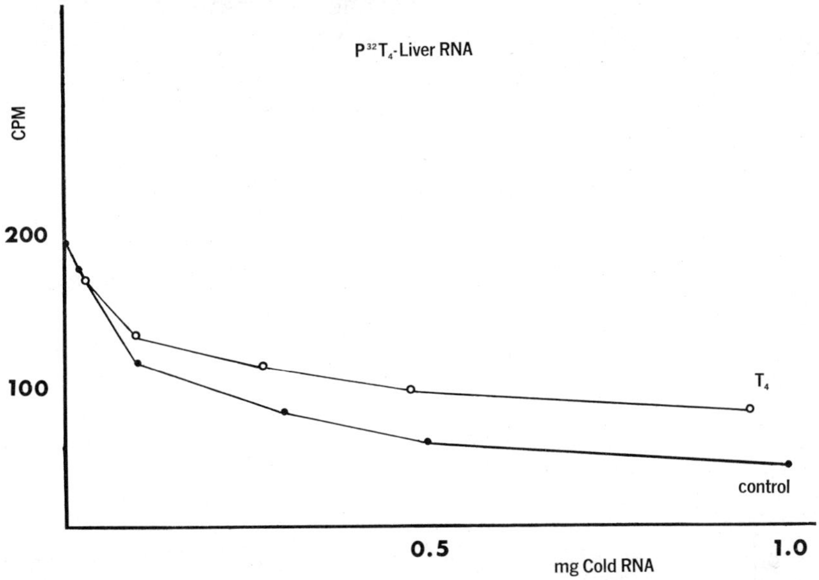

Fig. 2. Comparison of the abilities of unlabeled premetamorphic tadpole liver RNA and liver RNA 5 days after thyroxinization to compete with the binding of 10.9 μg (849 cpm/μ) of ^{32}P-labeled liver RNA 5 days after thyroxinization to 7.7 μg of tadpole DNA, bound to a 5.8-mm-diameter filter, during incubation in 0.2 ml of 2$\times$SSC for 12 hours at 67°C. The homologous RNA does not compete as well as the premetamorphic liver RNA.

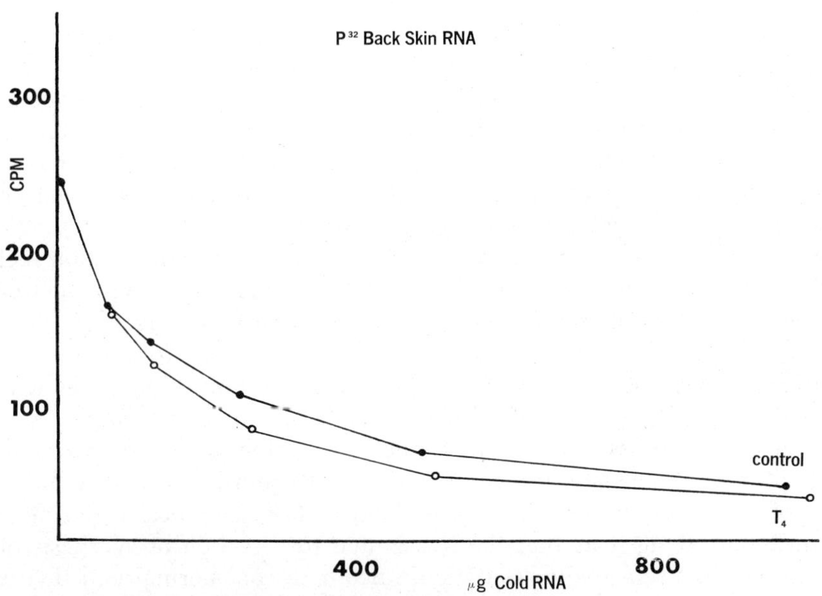

Fig. 3. Comparison of the abilities of unlabeled premetamorphic tadpole back skin RNA and back skin 5 days after thyroxinization to compete with the binding of 6.1 μg of ^{32}P-labeled premetamorphic back skin RNA to 12.5 μg (1,054 cpm/μg) of tadpole DNA during incubation in 0.2 ml of 2$\times$SSC for 12 hours at 67°C.

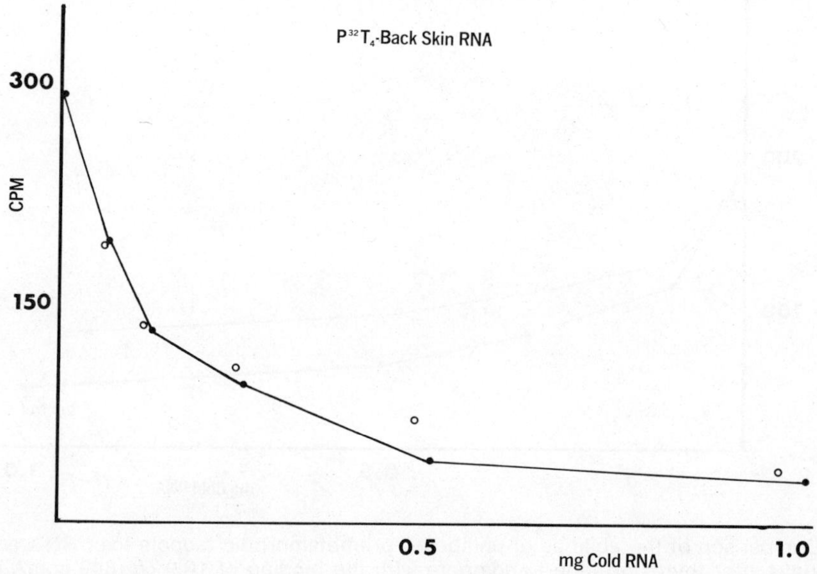

Fig. 4. Comparison of the abilities of unlabeled premetamorphic tadpole back skin RNA and back skin RNA 5 days after thyroxinization to compete with the binding of 5.3 μg (1,107 cpm/ μg of ^{32}P-labeled back skin RNA 5 days after thyroxinization fo 12.5 μg of tadpole DNA during incubation in 0.2 ml 2×SSC for 12 hours at 67°C.

new species of RNA had been evoked (Figs. 1–4). In fact, in one type of experiment (Fig. 2), competition between labeled thyroxinized tadpole liver RNA and the homologous unlabeled RNA was less effective than the competition with premetamorphic liver RNA. This phenomenon has been observed before(1), and possible explanations are offered in the discussion. Here the point can be made that there is a reduction of hybridization efficiency in the thyroxinized liver RNA, and that the technique does not reveal any new species of RNA in these differentiating tissues.

In the tail, a resorbing tissue, on the other hand, the hybridization technique revealed a qualitative change in newly synthesized RNAs after thyroxine treatment of tadpoles. In the tail there are some RNA types that are missing as early as 2 days after hormone treatment, before any morphological changes are apparent (Fig. 5). The difference seen between control and thyroxinized tail RNAs varies with incubation conditions, but it is about 10 percent of the total counts bound. As a measure of the degree of tissue specificity of the competition, liver RNA was competed against ^{32}P-premetamorphic tail RNA, and would compete against as little as 70 percent of the total ^{32}P-tail RNA counts bound. The reciprocal experiment illustrated in Figure 6 suggests that the loss is due to cessation of synthesis of some RNAs, rather than selective degradation, since both cold normal and thyroxinized tail RNAs compete about equally well against labeled thyroxinized tail RNA. To reduce any possible contribution of changing tissue composition in the whole tail, the experiment was repeated using dissected tail skin alone, and similar results were obtained (Fig. 7).

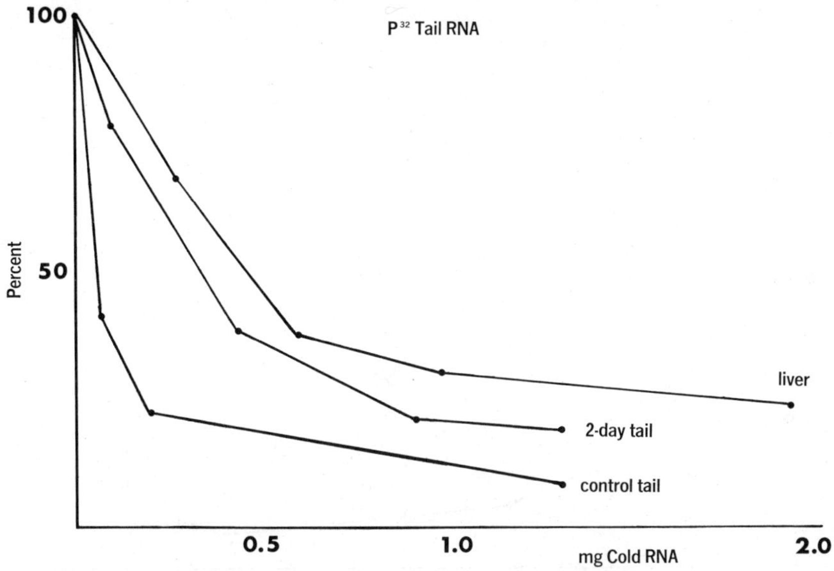

Fig. 5. Comparison of the abilities of premetamorphic liver and tail RNA and tail RNA 2 days after thyroxine treatment to compete with the binding of 14.8 μg of ^{32}P-labeled premetamorphic tail RNA (2,181 cpm/μg) to 7.7 μg of tadpole DNA on a 5.8-mm diameter filter during incubation in 0.2 ml of 2×SSC for 12 hours at 67°C.

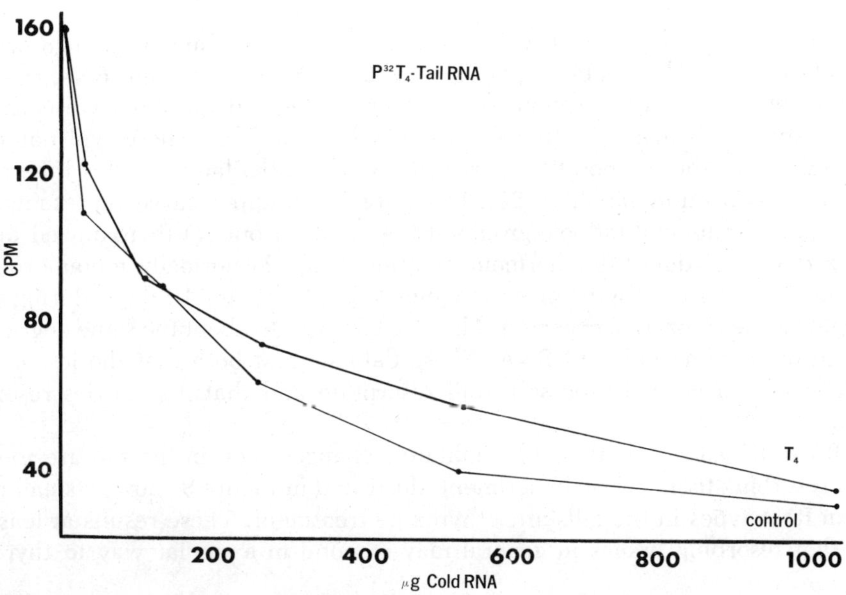

Fig. 6. Comparison of the abilities of unlabeled premetamorphic tail RNA and tail RNA 5 days after thyroxine treatment to compete with the binding of 9.3 μg of ^{32}P-labeled tail RNA 5 days after thyroxine treatment (652 cpm/μg) to 7.7 μg of tadpole DNA during incubation in 0.2 ml of 2×SSC for 12 hours at 67°C.

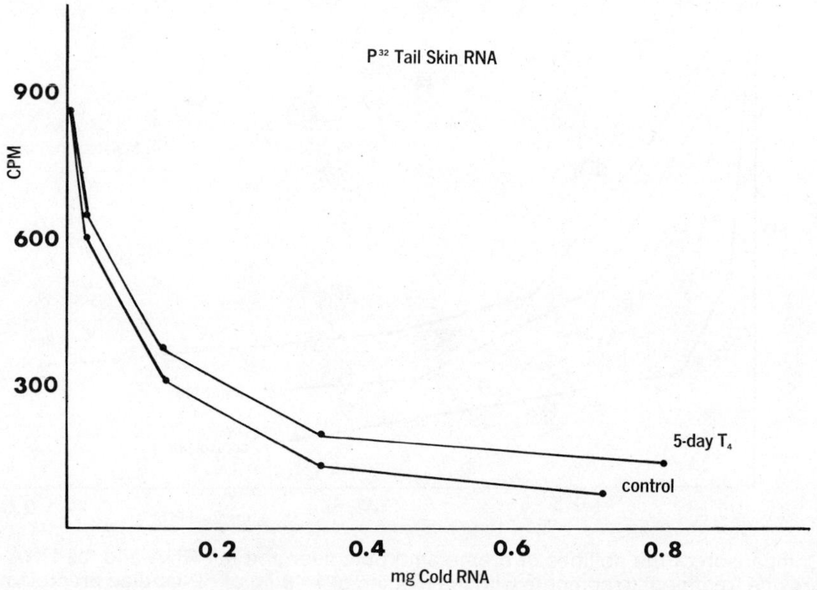

Fig. 7. Comparison of the abilities of unlabeled premetamorphic tail skin RNA and tail skin RNA 5 days after thyroxine treatment to compete with the binding of 6.2 μg of ^{32}P-labeled premetamorphic tail skin RNA (2,170 cpm/μg) to 12.5 μg of tadpole DNA on a 5.8-mm diameter filter during incubation in 0.2 ml of 2×SSC for 12 hours at 67°C.

Two other questions that were asked are: (1) does a similar change occur in the tail during normal metamorphosis, and (2) is the loss of some RNA species a cause or effect of tail resorption? To answer the second question, we conducted experiments with thyroxine-treated tadpoles kept at 5°C. These will not resorb their tails until the temperature is raised, even months later. At this time all the hormone has been metabolized(25). In Figure 8 the upper curve represents three different experimental tadpole groups: RNAs taken from (1) thyroxinized tadpole tails 2 days or 5 days after hormone treatment, or (2) normally metamorphosing tails, or (3) tails from thyroxinized tadpoles kept at 5°C for 10 days, during which time no visible resorption occurred. All three compete to about the same degree with labeled premetamorphic tail RNA. These data suggest both that the loss of some RNA types is a normal process in tail resorption and that it precedes resorption itself.

It can be asked whether the qualitative changes seen in the tail are found in other resorbing tissues. The experiment illustrated in Figure 9 shows a small reduction in RNA types in the gills after thyroxine treatment. These results at least suggest that resorbing tissues in general may respond in a similar way to thyroxine.

DISCUSSION

Provided that essential precautions are taken to ensure high RNA purity, the low RNA-to-DNA ratio experiments described here indicate tissue specificity and

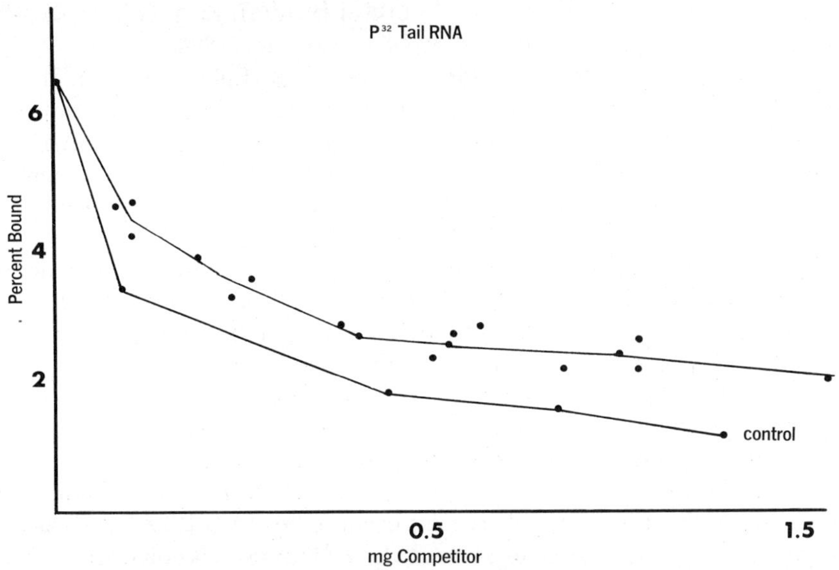

Fig. 8. Various unlabeled RNA preparations are compared in their abilities to compete with the binding of 7.3 μg of ^{32}P-labeled premetamorphic tail RNA (1,231 cpm/μg) to 5.25 μg of tadpole DNA during incubation in 0.2 ml of 2×SSC buffered at pH7 with 0.01 M tris [*N*-tris (hydroxy-methly) methyl-2-amino ethane sulfonic acid, Calbiochem] for 16 hours at 67°C. In the lower curve, unlabeled premetamorphic tail RNA was the competitor. The upper curve consists of points using the following RNAs as the competitor: tail RNA 2 days and 5 days after thyroxine treatment, tail RNA from stage XXI (Taylor-Kollros stages) tadpoles, which are beginning to resorb their tails, and tail RNA from tadpoles kept at 5°C for 10 days after thyroxine injection. In these animals there is no tail resorption unless the ambient temperature is raised.

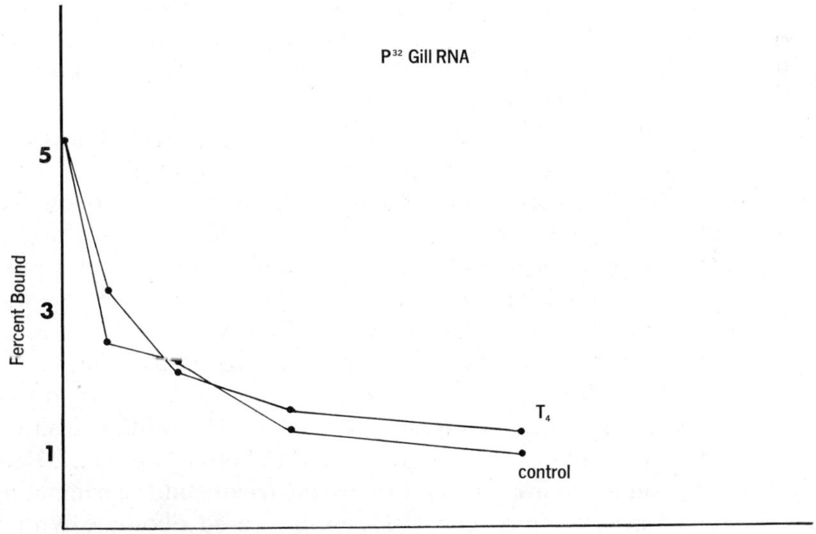

Fig. 9. Comparison of the abilities of unlabeled gill RNA from premetamorphic animals and animals 2 days after thyroxine treatment to compete with the binding of 4.2 μg of ^{32}P-labeled premetamorphic gill RNA (2,560 cpm/μg) to 5.85 μg of tadpole DNA during incubation in 2×SSC-0.01 M-tes (pH 7.0) for 16 hours at 67°C.

can consistently demonstrate a qualitative loss of some RNA species in the tail or tail skin within 2 days after thyroxine treatment. The usual interpretation of such differences in competition curves would be that synthesis of some messenger RNAs is arrested after thyroxine treatment. Several facts resist this interpretation but do not necessarily negate it. It has been shown that most rapidly labeled RNA never leaves the nucleus(26-29). In addition, the binding procedure itself has several important limitations. As shown by thermal denaturation profiles, the degree of locus specificity of the DNA-RNA binding reactions is highly dependent on incubation temperature and ionic strength(9). By manipulating these conditions and the DNA-to-RNA ratios, the competition method can be made insensitive to differences between RNAs present even in two different organs. Furthermore, there are repeated sequences in the DNA of higher organisms(30) so that, during short binding periods, one is measuring RNAs complementary to redundant DNA sequences. Nevertheless, consistent differences, such as we have obtained, can be interpreted to represent a minimal estimate of true differences between RNA preparations(9).

The function or cytoplasmic location of these RNAs is unknown. However, it is of interest at this point to formulate a hypothesis, based on the facts available from actual experiments, to indicate the role of RNAs in the tissue specificity of thyroxine-induced responses. It is, of course, reasonable to believe that various tadpole tissues are preprogrammed to respond in a specific fashion to thyroxine, as are other differentiating tissues to particular metabolic signals. RNA is at least a likely agent for the molecular basis of such a program, and can serve as an element in a working model with which to approach the problem of embryonic competence. There are several examples of tissue-specific transcriptional responses to hormones (31-33). The data reported here appear to provide the first instance of specific inactivation of genes. However, this observation is puzzling in light of the evidence indicating a requirement for the thyroxine-induced acceleration of RNA synthesis in the resorbing tadpole tail in organ culture(18,19).

Although the qualitative change observed by us in the tadpole tail is an early response relative to the long lag period preceding visible thyroxine action, it remains to be proved whether this change is a cause or effect of resorption. The experiment conducted with RNA from tadpole tails prevented from resorbing by low temperature at least suggests that transcriptional changes came before resorption. Perhaps a more direct proof of the RNA-resorption relationship might be a substitution for thyroxine by a particular RNA preparation. However, it will still remain to be shown at what level the primary hormonal effect occurs in the tail.

It appears that at the level of RNA metabolism, various tissues are diversified in their response to thyroxine also in a quantitative sense. One of the several possible explanations for the results seen in Figure 2 is that both competing RNA preparations contain the same species, but after thyroxine treatment the competing species become reduced in relative concentration by increased ribosome synthesis in the liver, and possibly by degradation during cytoplasmic reorganization(16,17). Dilution could occur theoretically to the point that one sees an apparent plateau 34. Results that are not too different are seen in Figure 6 for the tail, both experiments together suggesting dilution of the competing RNA species after thyroxinization. This result differs from that seen with the back skin (Figs. 3 and 4). Experiments in which constant amounts of various ^{32}P-labeled RNAs are compared in their abilities to bind to increasing amounts of DNA indicate that, with liver and tail RNAs, a lower percentage of the added radioactivity is bound after thyroxinization. With the back skin, on the other hand, an increased binding efficiency is observed after thyroxinization. These results are in agreement with the alterations in competitor effectiveness previously described and support the suggestion of diversity between the tail and liver, on one hand, and the back skin, on the other. It has been suggested(15) that the thyroxine-enhanced synthesis of RNA seen in all tadpole tissues studied may be a common denominator in all phases of thyroxine action(16,18,35). Perhaps, as more detailed information accumulates, responses to thyroxine at the level of RNA metabolism will appear as diverse as they now appear at the morphological level.

Since RNA changes have been seen in a variety of embryonic and hormonal systems with the technique used in the present study, one is left to explain the unexpected lack of qualitative differences in RNA synthesis in the liver and back skin. During the short incubation periods used, only RNA complementary to repetitive DNA is examined. Furthermore, the method is sufficiently insensitive so that the possibility of an undetectably small alteration in the synthesis of RNA species in these tissues cannot be excluded. The stimulation of ribosomal RNA synthesis may be so great that detection of any small changes in hybridizable RNA is difficult. However, the same method in the authors' hands has revealed extensive qualitative changes in differentiating tissue-cultured chick cartilage cells(5-7). The fact that many proteins merely increase in amount during metamorphosis from initially low levels may mean that translational regulation may be as important as, or even more important than, qualitative changes in RNA synthesis. The most clear-cut qualitative enzyme changes(36) involve the proliferation of a new cell population rather than transcriptional processes of differentiated cells (e.g., hemoglobin [37]).

As J. R. Tata has suggested(17), the long lag period required for thyroxine action on RNA synthesis makes a direct effect of thyroxine on RNA synthesis appear unlikely, although earlier effects are as yet unknown. Perhaps, RNA transport into the cytoplasm and modulation at the ribosome level may be of relatively greater importance in tissue-specific responses to thyroxine.

REFERENCES

1. Denis, H. Genetic expression in amphibian development. II. Release of the genetic information in growing embryos. J. Molec. Biol., 22:285–304, 1966.
2. Glisin, V. R., Glisin, M. V., and Doty, P. The nature of messenger RNA in the early stages of sea urchin development. Proc. Nat Acad. Sci. U.S.A., 56:285–289, 1966.
3. Whiteley, A. H., McCarthy, B. J., and Whiteley, H. R. Changing populations of messenger RNA during sea urchin development. Proc. Nat. Acad. Sci. U.S.A., 55:519–525, 1966.
4. Flickinger, R. A., Green, R., Kohl, D. M., and Miyagi, M. Patterns of synthesis of DNA-like RNA in parts of developing frog embryos. Proc. Nat. Acad. Sci. U.S.A., 56:1712–1718, 1966.
5. Cahn, R. D., Cahn, M. B., and Solursh, M. Heritability of differentiation *in vitro*: con-

trol of expression at the level of synthesis of DNA-like RNA in cartilage and pigmented retina cell clones. American Zoologist, 7:750, 1967.
6. ——— Factors affecting inheritance and expression of differentiation—some methods of analysis. *In* The Stability of the Differentiated State, Vol. 1, Beerman, W., Reinert, J., and Ursprung, H., eds. Results and Problems in Cell Differentiation Series, New York, Springer-Verlag, pp. 58–84, 1968.
7. Solursh, M., and Cahn, R. D. Unpublished data.
8. Gillespie, D., and Spiegelman, S. A quantitative assay for DNA/RNA hybrids with DNA immobilized on a membrane. J. Molec. Biol., 12:829–842, 1965.
9. Church, R. B., and McCarthy, B. J. The arrangement of base sequences in the DNA of simple and complex organisms. II. The interpretation of DNA/RNA hybridization studies with mammalian nucleic acids. Biochem. Genetics, 2:55–74, 1968.
10. Birnboim, H. C., Pène, J. J., and Darnell, J. E. Studies on HeLa cell nuclear DNA-like RNA by RNA/DNA hybridization. Proc. Nat. Acad. Sci. U.S.A., 58:320–327, 1967.
11. Kollros, J., and Kaltenbach, J. Local metamorphosis of larval skin of *Rana pipiens*. Physiol. Zool., 25:163–170, 1952.
12. Kaltenbach, J. Nature of hormone action in amphibian metamorphosis. *In* Metamorphosis: A Problem in Developmental Biology. Etkin, W. and Gilbert, L. I., eds. New York, Appleton-Century-Crofts, 1968.
13. Weber, R. Induced metamorphosis in isolated tails of *Xenopus* larvae. Experientia, 18: 84–85, 1962.
14. Needham, J. Biochemistry and Morphogenesis. Cambridge, The University Press, 1942.
15. Frieden, E. Thyroid hormones and the biochemistry of amphibian metamorphosis. Rec. Prog. Horm. Res., 23:139–186, 1967.
16. Tata, J. R. Turnover of nuclear and cytoplasmic ribonucleic acid at the onset of induced amphibian metamorphosis. Nature, 207:378–381, 1965.
17. ——— The formation, distribution and function of ribosomes and microsomal membranes during induced amphibian metamorphosis. Biochem. J., 105:783–801, 1967.
18. ——— Requirement for RNA and protein synthesis for induced regression of the tadpole tail in organ culture. Develop. Biol., 13:77–94, 1966.
19. Weber, R. Inhibitory effect of actinomycin D on tail atrophy in *Xenopus* larvae at metamorphosis. Experientia, 21:665–666, 1965.
20. Kim, K. H., and Cohen, P. P. Modification of tadpole liver chromatin by thyroxine treatment. Proc. Nat. Acad. Sci. U.S.A., 55:1251–1255, 1966.
21. Scherrer, K., and Darnell, J. E. Sedimentation characteristics of rapidly labeled RNA from HeLa cells. Biochem. Biophys. Res. Commun., 7:486–490, 1962.
22. Ralph, R. K., and Bellamy, A. R. Isolation and purification of undergraded ribonucleic acids. Biochim. Biophys. Acta, 87:9–16, 1964.
23. Marmur, J. A procedure for the isolation of DNA from microorganisms. J. Molec. Biol., 3:208–218, 1961.
24. Church, R. B., and McCarthy, B. J. Ribonucleic acid synthesis in regenerating and embryonic liver. I. The synthesis of new species of RNA during regeneration of mouse liver after partial hepatectomy. J. Molec. Biol., 23:459–475, 1967.
25. Frieden, E. Thyroid hormones and anuran differentiation: RNA and protein biosynthesis; temperature effects. *In* Proceedings of the Second International Congress of Endocrinology. Intl. Med. Congr., London, England, pp. 34–45, 1964.
26. Harris, H. Nuclear ribonucleic acid. Progr. Nucl. Acid Res., 2:19–59, 1963.
27. Penman, S. RNA metabolism in the HeLa cell nucleus. J. Molec. Biol., 17:117–130, 1966.
28. Shearer, R. W., and McCarthy, B. J. Evidence for ribonucleic acid molecules restricted to the cell nucleus. Biochemistry, 6:283–289, 1967.
29. Attardi, G., Parnas, H., Hwang, M-I., H., and Attardi, B. Giant-size rapidly labeled ribonucleic acid in immature duck erythrocytes. J. Molec. Biol., 20:145–182, 1966.
30. Britten, R. J., and Kohne, D. E. Repeated sequences in DNA. Science, 161:529–540, 1968.

31. Drews, J., and Brawerman, G. Alterations in the nature of ribonucleic acid synthesized in rat liver during regeneration and after cortisol administration. J. Biol. Chem., 242: 801–808, 1967.
32. ——— and Brawerman, G. Messenger RNA patterns in rat liver nuclei before and after treatment with growth hormone. Science, 156:1385–1386, 1967.
33. Hahn, W. E., Church, R. B., Gorbman, A., and Wilmot, L. Estrone and progesterone-induced synthesis of new RNA species in the chick oviduct. Gen. Comp. Endocr., 10: 438–442, 1968.
34. Bolle, A., Epstein, R. H., Salser, W., and Gerduscheck, E. P. Transcription during bacteriophage T_4 development: synthesis and relative stability of early and late RNA. J. Molec. Biol., 31:325–348, 1968.
35. Finamore, F. J., and Frieden, E. Nucleic acids and induced amphibian metamorphosis. J. Biol. Chem., 235:1751–1755, 1960.
36. Manwell, C. Metamorphosis and gene action. I. Electrophoresis of dehydrogenases, esterases, phosphatases, hemoglobins and other soluble proteins of tadpole and adult bullfrogs. Comp. Biochem. Physiol., 17:805–823, 1966.
37. Hollyfield, H. G. Erythrocyte replacement at metamorphosis in the frog, Rana pipiens. J. Morph., 119:1–5, 1966.

19

Metamorphosis and Gene Action: II. Amphienzymes in Amphioxus and Amphibians*

Clyde Manwell and C. M. Ann Baker

Department of Zoology, University of Adelaide, South Australia

INTRODUCTION

The improved resolution of starch gel electrophoresis(1,2) and the use of enzyme histochemical techniques(3,4) have greatly stimulated research on individual and developmental variation of proteins. Not surprisingly, this improved method of enzyme electrophoresis has been applied to amphibian metamorphosis(5). It was observed that approximately one-third of the enzymes are specific to either the tadpole or the frog phase of the life cycle, whereas two-thirds of the major soluble proteins are specific to the different phases. It was thus concluded that, although enzyme changes are a significant part of metamorphosis, changes in the major soluble proteins, some of which may well be enzymes with unknown activities, are quantitatively more important. Thus, many enzymes appear to be present in both phases of life cycle—e.g., are **amphienzymes.** In addition, a few proteins change at times other than metamorphosis, and several proteins are present only during the metamorphic period.

Two different approaches can be used to decide if certain structural genes are active in both phases of the life cycle, as well as to decide if others are expressed only in specific phases(5,6).

1. Isolation and characterization of enzymes with identical substrate specificities from tadpole and adult phases. In particular, if such enzymes are the products of different genetic loci, then they should differ in at least part of their amino acid sequence, such a difference being most readily demon-

*The authors thank Dr. Peter Wickstead of the Marine Biological Laboratory, Plymouth for the larvae of Amphioxus *(Branchiostoma lanceolatum).*

strated by peptide mapping ("fingerprinting"), although preferably being carried to the ultimate point of complete primary structure determination.

2. If two proteins are coded from different genetic loci, then genetic variation of one should not be paralleled by genetic variation of the other.

Each approach has its problems. The first is laborious. For trace enzymes, large amounts of tissues are needed in order to obtain sufficient pure protein. This means that individuals must be pooled.

The second approach is biochemically less sophisticated, if considerably simpler. As small amounts of tissue can be extracted and submitted to electrophoresis, many studies can be done without combining organs or individuals. It has recently been realized that the pattern of genetic variation of many enzymes and other proteins is very easily determined by starch gel electrophoresis(7,8). For a number of proteins, heterozygotes for a polymorphism have the two zones characteristic of the corresponding homozygotes—e.g., serum albumin. For other proteins there may be multiple banding in homozygotes but heterozygotes have simply the sum of the zones present in both homozygotes—e.g., transferrin. For still other proteins, heterozygotes are easily detected by the possession not only of the two zones corresponding to those of appropriate homozygotes but also of one or more *hybrid zones* of intermediate electrophoretic mobility. For example, polymorphism of 6-phosphogluconate dehydrogenase in man, the quail (*Coturnix coturnix*), snails (*Cepaea nemoralis* and *C. hortensis*), and a sea anemone (*Actinia equina*) involves a single hybrid zone in heterozygotes(9,10). For a number of vertebrate lactate dehydrogenases, the pattern of individual variation involves basically three hybrid zones in heterozygotes(7).

There are several complications to the approach either through biochemical characterization or through genetic variation in studying the proteins of different phases in a life cycle:

1. Appearance of a new protein does not necessarily mean gene activity in the sense of messenger RNA formation followed immediately by protein synthesis. There can be a long delay between gene transcription and activation of the messenger RNA-ribosome complex.
2. Some proteins are the product of more than one genetic locus, e.g., the distinct α and β chains of most vertebrate hemoglobins. For a β chain mutation in man the polymorphism is confined to the major adult hemoglobin, Hb A. For an α chain mutant there will be parallel variation of Hb A, Hb A_2 and Hb F.
3. Proteins can have different properties and still have been the product of the same genetic locus if there has been a subsequent difference in chemical modification. For example, transferrin and conalbumin in galliform birds differ electrophoretically and yet always show parallel genetic variation(11). It has recently been shown that these proteins differ not in amino acid sequence but in the carbohydrate moiety attached to sequence(12).

The present paper includes additional data on individual variation of certain proteins in amphioxus and amphibians. Amphioxus is in some reports the most primitive known living chordate. It metamorphoses from a planktonic larva to a benthic adult phase which, though still motile, spends much of its time partly or completely buried in coarse sand. Although the matter is by no means settled, it has been suggested that the metamorphosis of amphioxus is controlled by thyroxine or some related iodine compound(13).

MATERIALS AND METHODS

Amphioxus, *Branchiostoma lanceolatum,* were obtained from the English Channel in the vicinity of Plymouth, England. Larvae were collected in plankton tows during the summer and adults were collected by bottom trawling with appropriate fine-meshed netting(13). The amphibians sampled were edible frogs, *Rana esculenta,* common frogs, *Rana temporaria,* and newts, *Triturus vulgaris.*

Treatment of samples, electrophoresis in vertical starch gel, and histochemical localization of enzymes are detailed in other papers(5,6,8,14). Amphioxus larvae are very small, as well as thin and elongate; thus, it was necessary to pool 30 or 60 larvae at a time. Adult amphioxus were pooled for direct comparison with larvae. But, in addition, variation in a number of enzymes was studied in 40 individual adult amphioxus. Separation of organs in adult amphioxus is difficult; the best that can be done with reasonable rapidity (so as to avoid autolytic modification) and with minimal contamination is the division of adults into body musculature (including notocord and nervous system), viscera (including the digestive system), and, depending on sex, ovary or testis.

RESULTS

Enzymes of Amphioxus, Branchiostoma lanceolatum

Esterases. Amphioxus esterases are resolved best in the pH Ferguson-Wallace discontinuous buffer, although good resolution of some esterases is obtained in pH 6.0 phosphate. The major amphioxus esterases are relatively nonspecific in that almost identical patterns, differing only in relative activity in a few zones, are obtained whether the enzymes are localized by hydrolysis of α-naphthyl acetate, α-naphthyl butyrate, or 5-bromoindoxylacetate. None of these enzymes has activity on α-naphthyl phosphate, there being at least two distinct and active acid phosphatases in adult amphioxus, although no detectable phosphatase activity in extracts of larvae.

The esterase pattern for small larvae (5 mm long), large larvae (8 mm long) and adults is given in Figure 1. There is some individual variation in electrophoretic position of two of the more anodal esterases of adult amphioxus, though most of the esterase zones were identical in a sample of 32 adults screened singly. It is clear that larval amphioxus lack many of the esterases present in adults. Only three zones of esterase activity are detectable in small larvae; the number increases to eight in large larvae. Seven of these eight zones are identical in position to zones found in extracts of adults. There is one conspicuous esterase of larval amphioxus which does not appear to be present, at least in anywhere near the same level of activity, in

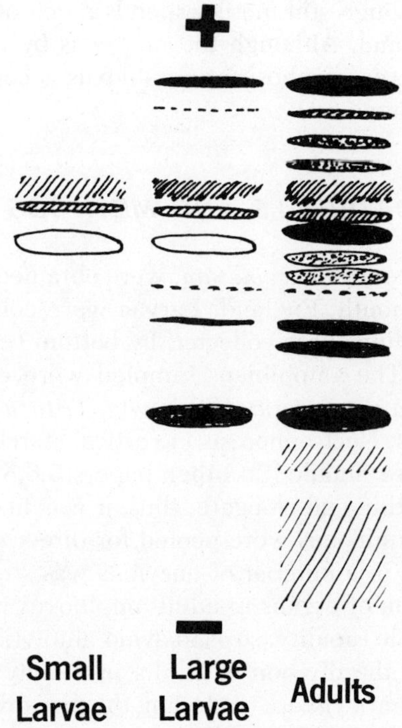

Fig. 1. Esterases of larval and adult amphioxus, *Branchiostoma lanceolatum.* Diagrammatic representation of the anodal half of a starch gel electrophoresed with the Ferguson-Wallace discontinuous buffer, pH 8.0. There are no major esterases in the more cathodal half separating these zones from the sample insertion slots.

adults. In addition, adult amphioxus have at least seven esterases not present in larvae.

Dehydrogenases. No activity could be found in extracts of either larval or adult amphioxus for dehydrogenases of the following substrates added to sliced starch gels at the completion of electrophoresis: lactate, isocitrate (neither NAD- nor NADP-dependent IDH's), glutamate, 6-phosphogluconate, ethanol, glyceraldehyde, α-glycerol-phosphate, or α-glyceraldehye-phosphate. Very strong activity was observed for NAD-dependent malate dehydrogenase (abbreviated NAD-MDH) and glucose-6-phosphate dehydrogenase (abbreviated G-6-P DH), each of which yields a series of multiple zones (isozymes).

NAD-Dependent Malate Dehydrogenases (NAD-MDH's). Extracts of larval amphioxus yield the same NAD-MDH pattern of two major and one trace isozymes as found in extracts of adult viscera (Fig. 2). Adult muscle has these isozymes plus two more major NAD-MDH's which migrate somewhat more slowly. A curious observation is that these NAD-MDH's of amphioxus resolve so well in alkaline starch gels containing borate. In many extracts of other species the supernatant and mitochondrial NAD-MDH's resolve best in pH 6 to 7.5 phosphate gels, which give fairly poor resolution of the amphioxus NAD-MDH's. In addition, resolution of the amphioxus NAD-MDH's is only slightly improved by addition of coenzyme to the

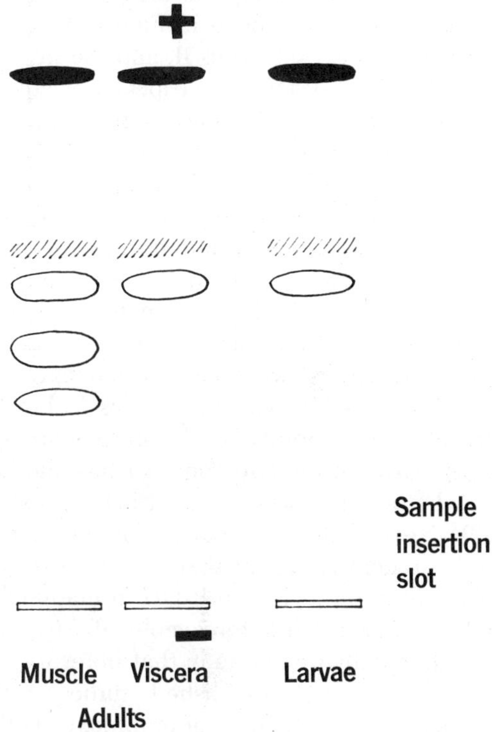

Fig. 2. NAD-dependent malate dehydrogenase isozymes of larvel and adult amphioxus, *Branchiostoma lanceolatum.* The pattern is essentially the same in Smithies' borate buffer, pH 8.4 to 8.5 and Ferguson-Wallace discontinuous, pH 8.0.

gel (50 or 100 mg of NAD per liter), whereas this is mandatory to avoid spurious variation and excessively bad resolution of NAD-MDH's in extracts of a number of species of invertebrates and vertebrates(8,15). No individual variation of NAD-MDH's was observed in 40 adult amphioxus surveyed singly. The distinct NADP-dependent malate dehydrogenases ("malic enzymes"), which require Mg++ or MN++ for full activity, and which yield such nice sharp zones in electrophoresis, could not be localized after electrophoresis of amphioxus extracts, even with the proper precaution of the addition of 20 mg of NADP per liter to both gel and the upper bridge buffers.

Glucose-6-Phosphate Dehydrogenases (G-6-P DH's). These enzymes are extremely active after electrophoresis of amphioxus extracts, multiple zones appearing within 10 minutes after the staining solution is added to the sliced starch gel. The enzymes are specific for glucose-6-phosphate and no zones appear when gels are incubated for as long as 24 hours with glucose, glucose-1-phosphate, or 6-phosphogluconate. Magnesium ion added to incubation mixtures increases activity slightly. The multiple G-6-DH's have an obligatory requirement for NADP and, in addition, best resolution demands the presence of NADP in the gel and bridge buffers (20 mg NADP per liter). There are basically three sets of G-6-P DH isozymes. As variation at each set is distinct from variation at the others, the three sets are the product of distinct genetic loci. This is especially clear for the two rapidly anodal sets of G-6-P DH's and these are shown in Figure 3. The most anodal set of zones is designated Locus I, the next set Locus II, and the slowly moving and poorly resolving set, Locus III. Locus I G-6-P DH's are especially active in extracts of testes or ovary and Locus II G-6-P DH's are especially active in extracts of viscera; however, the specificity is relative and not absolute. Muscle contains primarily the Locus III G-6-P DH. Larvae have active Locus II and Locus III G-6-P DH's but no Locus I. Thus, metamorphosis leads ultimately to the activation of one of the three G-6-P DH loci. The nature of individual variation in larvae was not studied; however, in adults many different electrophoretic patterns were observed (Fig. 3). With one exception, the pattern of individual variation for both Locus I and Locus II G-6-P DH's can be explained by postulating five alleles at Locus I and four alleles at Locus II, heterozygotes having any two zones and homozygotes having one heavy zone. The exception is one individual with four zones at Locus I, although only a single zone at Locus II. A further problem is that there is often considerable difference in staining intensity between the two zones in postulated heterozygotes. Unfortunately, G-6-P DH, whether in sea anemones or in quail extracts, has a tendency to form satellite zones(9). Nevertheless, the pattern of variation of amphioxus G-6-P DH is unusual for a dehydrogenase in that there are no hybrid zones in heterozygotes. Except for sex-linked erythrocyte G-6-P DH in mammals, other G-6-P DH's in vertebrates have one hybrid zone in heterozygotes (9,16)—and in sea anemones the pattern resembles lactate dehydrogenase in that heterozygotes have five zones, three of which are hybrid isozymes (unpublished studies by the senior author on *Actinia equina*). No evidence for sex linkage of amphioxus G-6-P DH has been obtained; as for mammalian liver G-6-P DH and avian erythrocyte G-6-P DH, the amphioxus G-6-P DH loci are autosomal. The absence of hybrid zones in heterozygotes in the variation of amphioxus Locus I and Locus II G 6-P DH's, together with the extremely rapid electrophoretic mobility in starch gel, are suggestive of a smaller and simpler structure than other dehydrogenases. It is possible that, in

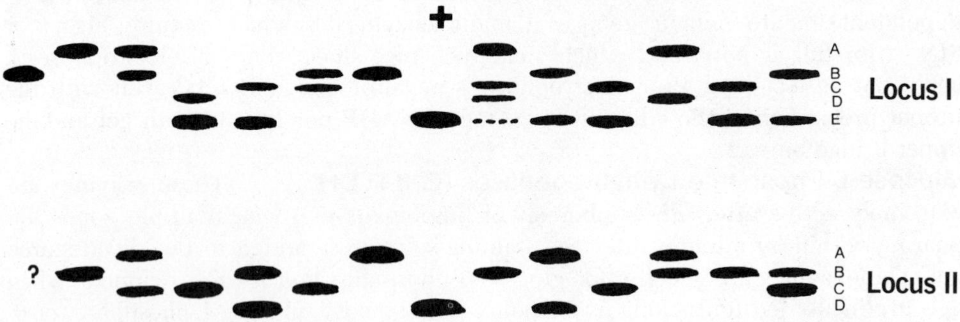

Fig. 3. Individual variation of Locus I and Locus II glucose-6-phosphate dehydrogenases of adult amphioxus, *Branchiostoma lanceolatum*. This diagram represents the most anodal region, immediately cathodal to the "brown line" in a Ferguson-Wallace discontinuous electrophoresis, pH 8.0. The "?" is for an individual who apparently lacks the Locus II G-6-P DH.

contrast to other G-6-P DH's, those of amphioxus lack a stable quaternary structure, a point which is compatible with the observed data but which can only be proven by careful biochemical characterization of the isolated enzymes.

Phosphoglucomutase (PGM). Extracts of adult amphioxus also have a very high activity of phosphoglucomutase. Initially it was not realized that amphioxus PGM is completely inactivated by even brief freezing and thawing. Thus, PGM was studied in only seven individuals. However, the range of individual variation of PGM among those seven adults is considerable. In addition, there is a surprising similarity in the patterns observed in amphioxus PGM and those observed in studies on variation of human PGM (e.g., compare Fig. 4 of this paper with Chart 4 in Harris's review of enzyme polymorphisms in man, 8). Amphioxus and man both have two or three minor, rapidly moving zones of PGM, whose variation is independent of the variation of the more active and more slowly moving zones of PGM. The minor zones are determined by a locus, designated PGM_2 in man, which is distinct from that determining the major zones, designated PGM_1 in man(8). For man, the existence of two distinct PGM loci has been confirmed by the observation of independent segregation in appropriate pedigrees; whether or not the two PGM loci in amphioxus are linked can only be determined by appropriate mating experiments, which have not been done. As with variants of human PGM, the heterozygote is a simple sum of the zones of the corresponding homozygotes. Whereas in man two alleles at PGM_1 are relatively common and the other variants are rare: in only seven adult amphioxus, five different positions for the major zone are observed, suggesting five different reasonably common PGM_1 alleles. The PGM_1 locus appears to be active in most tissues, whereas PGM_2 is most readily observed in gonad and, irregularly, in some viscera extracts.

Aldolase. A single zone of fructose-1, 6-diphosphate aldolase is observed in extracts of adults. It is probable that this necessary enzyme occurs in larvae but, if so, its presence could not be detected after starch gel electrophoresis of extracts.

Peroxidase. Adult amphioxus but not larvae have from one to four zones of peroxidase, migrating fairly rapidly anodally in alkaline buffers (pH 8.0 discon-

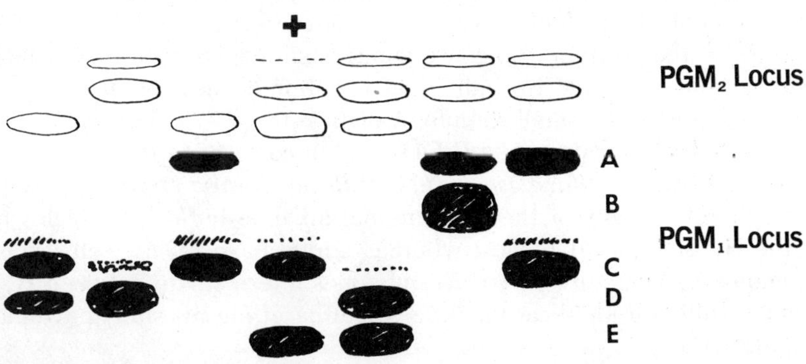

Fig. 4. Individual variation of seven adult amphioxus, *Branchiostoma lanceolatum PGM*$_1$ and PGM_2 phosphoglucomutases. This represents the middle third of a starch gel made with the Ferguson-Wallace discontinuous buffer, pH 8.0.

tinuous and pH 8.5 borate). There is extreme variation in the number and relative activity of the peroxidase isozymes. There is no obvious correlation in the number and activity of peroxidases with sex or nutritional state. The variation defies rational genetic explanation, unless one postulates two different peroxidase loci, each with null alleles in significant frequency. The absence of these isozymes in pooled larval samples suggests that peroxidases become conspicuous only in the adult phase. Though some amphioxus adults have a pink color, which has been casually called myoglobin, no evidence for muscle hemoglobin could be found in these individuals. The peroxidase activity represents a true peroxidase and not hemoglobin.

Dehydrogenases of **Triturus vulgaris, Rana esculenta,** and **R. temporaria**

Newts, *Triturus vulgaris*. In an initial survey of eight individual newts, four larvae and four adults, it was observed that both adults and larvae had only a single LDH isozyme in all tissues. One individual adult was observed with the typical five LDH isozymes expected of a heterozygote for a polymorphism(7,17). However, these studies were done utilizing only the pH 8 Ferguson-Wallace discontinuous buffer, which yields very sharp zones of LDH. When these samples were run in a pH 6.7 phosphate buffer ($\mu = 0.04$), which also yields good resolution of LDH's, it was observed that all newts have distinct heart- and muscle-type lactate dehydrogenases. These are H_4 and M_4—and the intermediate isozymes, corresponding to H_3M, H_2M_2, and HM_3, typical of mammalian, avian, and some "lower" vertebrate LDH's, do not occur in certain amphibians(18,19), including the common newt. The individual newt with five LDH zones in alkaline buffer turned out to have an M chain LDH polymorphism. Thus, in the newt hybrid isozymes do not form readily between H and M chains, although the heterozygote for the M chain polymorphism possessed the typical five zones as found in heterozygotes among birds and mammals. No significant differences were observed between adults and larvae and, in the newt, it is clear that metamorphosis involves no changes in LDH gene expression.

Similarly, larval and adult newts have similar glutamate dehydrogenases and the two zones of NAD-dependent malate dehydrogenase. Larvae have less G-6-P DH than adults but the position of the two zones is similar, even if larvae lack nearly all of the more anodal zone. No individual variation of these other dehydrogenases was observed among the small sample of eight efts.

Frog Lactate Dehydrogenases (LDH). In contrast to the newt, the muscle and heart-type LDH's of *Rana esculenta* and *R. temporaria* are easily differentiated by electrophoresis in any of the conventional alkaline buffers. Tadpoles have primarily the H_4 LDH in all tissues, whereas adults have M_4 as well, especially in muscle and liver. There are only traces of zones intermediate between H_4 and M_4. Thus, our results are identical to those of Salthe et al.(19) studying *Rana pipens* and *R. palustris.*

Bullfrog *R. Catesbeiana* H chain has more of a tendency to combine with the M chain, although there is still only the H_2M_2 molecule formed and little, if any, of the H_3M and HM_3 tetramers(5,19). We have observed, though inconsistently, that refrozen and thawed samples containing H_4 and M_4 frog LDH's tend to increase in the amount of the trace intermediate isozymes, although these never are as prominent as in extracts of appropriate avian or mammalian tissues. Electrophoresis of unfrozen frog tissues often fails to yield any intermediate LDH isozymes between H_4 and M_4, except for some H_2M_2 in bullfrogs.

The bullfrog is also different from the other ranids in the more rapid electrophoretic mobility of the H_4 isozyme and that the M_4 isozyme occurs in appreciable amounts in tadpoles, especially in muscle and liver. Bullfrog adults also have a heterogeneity of the H_2M_2 and M_4 LDH's that is not paralleled in *Rana pipiens*, *R. esculenta*, and *R. temporaria*.

Genetic Polymorphism at the H Chain LDH Locus in *Rana esculenta*. The hypothesis that the H chain is the same in both tadpole and adult phases is proven by finding a particularly clear polymorphism at the H chain locus in *Rana esculenta* (Fig. 5). The pattern of variation is exactly the same as observed in LDH polymorphisms of several vertebrates, including pigeons, mice, and men, for which the genetic interpretation has been confirmed by pedigree studies or appropriate matings(7,8). In both tadpoles and adults of *R. esculenta*, there are three different H chain alleles. Heterozygotes have five zones, including three hybrid isozymes not found in homozygotes. Five of the six predicted phenotypes for a triallelic variation have been observed in the small sample of frogs so far investigated. It is of interest that the sample of eight adults yielded 1 *AA*, 1 *AB* and 6 *AC*, whereas the sample of eight tadpoles yielded 2 *AB*, 2 *AC*, 3 *BB*, and 1 *BC*. Thus, the estimated gene frequencies are in adults $p_A = 0.56$, $p_B = 0.06$, and $p_C = 0.38$; and, in tadpoles $p_A = 0.25$, $p_B = 0.56$, and $p_C = 0.19$.

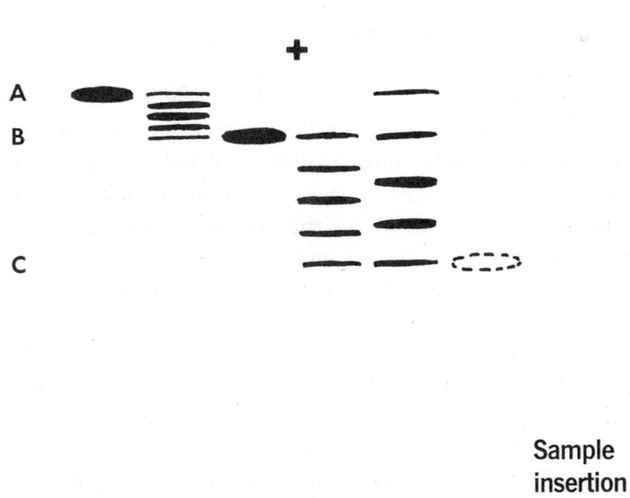

Fig. 5. Polymorphism of the H chain lactate dehydrogenase of tadpoles or adult edible frogs, *Rana esculenta*. The pattern, which has not been exaggerated in terms of the sharpness of the zones, is essentially the same in Aronson-Gronwal modified "tris"-EDTA-borate buffer (pH 8.8), in Smithies' borate (pH 8.4) and in Ferguson-Wallace discontinuous (pH 8.0). The dashed line on the zone for the CC homozygote indicate that this is the expected pattern but it has not yet been observed.

Because of the small sample size, the differences in p_A in the two phases, as well as in p , are not statistically significant. However, the difference in gene frequency of the *B* allele, 0.56 in tadpoles versus 0.06 in frogs, in the two phases is statistically significant. The probability of such an unequal distribution occurring under the null hypothesis of no significant difference can be easily calculated—see Mather(20) for discussion of the use of binomial and multinomial distributions for analysis of significance of differences in small samples. The total sample of 16 *R. esculenta* represents 32 alleles at the H chain locus. The probability of the 10 *B* alleles being distributed 9 in one phase and 1 in the other, or 10 in one phase and 0 in the other, is, from the appropriate binomial expansion:

$$p^{10} + 10\ p^9q + 10\ pq^9 + q^{10},$$

where, because of the equal number of tadpoles and frogs in the sample, $p = q = ½$. The exact value for the above sum of four terms is $22 \times (½)^{10} = 0.0215$. The probability of observing such a difference in the frequency of the *B* allele in tadpoles and frogs due to chance is only about 2 in 100. Thus the data, in spite of the small sample size, are statistically significant at the $\alpha = 0.05$ level.

There must be an extensive differential mortality of tadpoles and frogs to account for such a large gene frequency difference between individuals in different phases of the life cycle. However, frogs lay thousands of eggs. Many tadpoles fail to survive metamorphosis, and froglets are the food of many predators. There is plenty of numerical room for establishment of such different "genetic élites" in any species with a large number of zygotes(7,9).

Survey of 6 adults and 44 tadpoles of *R. temporaria* failed to reveal any LDH variants at either the H chain or the M chain locus.

Frog Malate Dehydrogenases. There are two major NAD-dependent malate dehydrogenases in both tadpole and adult *R. esculenta* and *R. temporaria*. The tadpole enzymes migrate slightly more rapidly anodally than the frog enzymes in pH 8.0 discontinuous buffer. However, the resolution of these enzymes is poor and the electrophoretic mobility depends on the concentration of the extract and the organ extracted. Thus, the tadpole and frog differences are probably not significant. No clear individual variation was observed among 16 *R. esculenta* and 30 *R. temporaria*, although some type of individual variation exists in the position of the more slowly moving NAD-MDH isozyme in *R. esculenta*.

Frog Glutamate Dehydrogenases. In discontinuous buffer, pH 8.0, glutamate dehydrogenase of *R. temporaria* adults has two major zones. Tadpoles consistently have only the faster of these zones. However, in pH 6.7 phosphate buffer the tadpole and frog glutamate dehydrogenases present a different pattern of heterogeneity; thus it is possible that the tadpole and frog GDH's are completely different. No individual variation was observed among 30 *R. temporaria*. Thus, we cannot yet say whether the electrophoretic difference between the tadpole and frog GDH's is the result of different GDH genes being active in the tadpole and frog phases of the life cycle, or the result of some difference in chemical modification. It is well known that vertebrate liver glutamate dehydrogenase is an allosteric enzyme.

DISCUSSION

Enzyme Differentiation Versus Adaptation

This comparative approach indicates that in both amphioxus and amphibians, where metamorphosis is triggered (at least in amphibians) by thyroxine or similar iodinated compounds, some enzymes are unique only to one phase or the other of the life cycle. In general, as differentiation proceeds there is an increase in the number of different enzymes detectable after electrophoresis of extracts, e.g., larval amphioxus have only one possibly unique larval esterase, whereas adult amphioxus have at least seven different unique adult esterases. Similarly, in *Rana temporaria* and *R. esculenta*, though not in *R. catesbeiana*, only traces of the M LDH chain are synthesized prior to metamorphosis. In the newt, however, we have observed no differences between larvae and adults, whether the comparison involves H chain LDH, M chain LDH, malate dehydrogenases, or glutamate dehydrogenase. Additional studies on the same eight individuals have revealed neither any significant hemoglobin differences, nor differences in phosphatases; adults have two sets of esterases present in significant levels of activity which are not present, or barely detectable, in larvae; however, at least six other esterases are identical in adults and larvae. Thus, most of the newt enzymes are amphienzymes and are unaffected by metamorphosis.

The patterns of enzyme differentiation show some degree of variation between species of the same genus. The best example in the present studies is the much greater activity of the M chain locus in bullfrog tadpoles as compared with tadpoles of other related ranids. Similar differences suggesting species differences in "control genes" have been observed in hemoglobin differentiation—e.g., the persistence of larval hemoglobin in adult lampreys of the semineotenous species *Entosphenus lamottei*(6), the difference in the time of appearance of the major and minor adult hemoglobins in the embryos of galliform birds(21), the relative differentiation and absolute differentiation of coelomic and water vascular hemoglobins in sea cucumbers(22,23), and the differences in the presence or absence of hemoglobin components, as well as their number, in embryonic, fetal, and adult mammals(24).

Although enzyme changes in different phases of a life cycle are often studied as examples of hormonal influence on the activation of genes—or at least the activation of messenger RNA-ribosome complexes—it must not be forgotten that some of these differences in protein phenotype reflect differences in the environment. To what extent are such examples of enzyme differentiation altered by differences in the environment? It is known that the relative amounts of the H and M chains of lactate dehydrogenase can be altered by oxygen level independently of other embryological events(25). Accordingly, it is just possible that the tadpole-frog difference in the amount of the M_4 isozyme reflects differences in the availability of oxygen and would be reduced or eliminated if tadpoles and frogs were kept at identical oxygen levels.

A further paradox is posed by the fact that tadpoles of some species have primarily the H LDH chain type, which is considered to be more suited in its kinetic properties to metabolism in obligatorily aerobic tissues(18). However, tadpoles live in environments of variable, and often low, levels of oxygen—for which the absence of a Bohr effect in tadpole hemoglobin is ideally suited(5). Thus, at our present state of understanding—or misunderstanding—tadpoles are better adapted anaerobically than frogs in respect to hemoglobin, but better adapted aerobically than frogs in respect to lactate dehydrogenase!

In general, the greater the morphological difference between phases of a life cycle, the greater the difference in the habitats of larvae and adults—although the correlation is not perfect, especially among some insects and amphibians. It is of interest that the highest frequency of amphienzymes is found in the newt, where adults spend considerable time in the same aquatic environment as the larvae. It may well be that some of the specific enzymes being studied as examples of differentiation are, in fact, really of more significance in terms of adaptation to the difference in larval and adult environment. As suggested earlier(5), from the standpoint of changes in morphology, the major soluble and insoluble proteins are of particular importance, even though de novo synthesis, or activation, of specific modifying enzymes is also necessary for these morphological changes—e.g., the collagenase of tadpole tails.

It is worth bearing in mind that metamorphosis in some marine invertebrates, e.g., tunicates, barnacles, and polychaetes, can be initiated and finished within a few hours, in some cases as little as 15 minutes. It is unlikely that all the necessary protein changes are accomplished by de novo protein synthesis. Rather, reliance must be placed more on synthesis or activation of a few key modifying enzymes, which in turn can alter certain major proteins, which might be additional enzymes or structural proteins. Very likely there is a series of such activation steps involving successive amplification, as in the activation of the "factors" in blood clotting. It is proposed to call such changes in development that involve the chemical modification of already synthesized proteins *transubstantiation*. Transubstantiative changes may involve proteolytic modification, as in the case of collagenase, addition or removal of nonprotein moieties, such as the carbohydrate changes in transferrin, or modification of configuration by combination with a small molecule, e.g., an allosteric effect. The very unusual difference between adult and fetal shark *Squalus* hemoglobin may reflect a transubstantiative change which results in the loss of two tryptic or seven chymotryptic peptides of what are otherwise identical proteins(26).

Consequences of Individual Variation

In biochemical and endocrinological studies it has often been the practice to combine a given organ or gland from many individuals in order to obtain sufficient protein. In relation to comparison of tadpoles and adult frogs, the electrophoretic studies warn that such a procedure may give rise to differences in enzyme activity or other properties even when the enzymes are fundamentally identical in both phases of the life cycle. This can result in three ways:

1. Pooling of a small number of individuals may result, purely by chance, in combining different numbers of various variants for a protein polymorphism common to both phases. This can be minimized by increasing the size of the pooled sample.

2. When there is a change in the gene frequency of alleles for a polymorphic locus active in both phases of the life cycle, this difference in variants will manifest itself in a pooled sample *no matter how large that sample is*. The data on *R. esculenta B* allele of the H chain LDH locus indicates that considerable differences in frequency exist in a sample of tadpoles and frogs from the same locality. This could result from seasonal differences in selection pressure, for adults will have been exposed longer to any progressive changes than will tadpoles. Or, this could result from differential selection arising from the difference in the tadpole and frog environments. The large number of eggs laid by frogs makes it perfectly possible to have very different "genetic élites" in the two phases, the only restrictions being those imposed by the fact that any unusual distribution in frogs must arise from what is already present in tadpoles and vice versa. Such a system could stabilize polymorphisms which would otherwise be unstable. If the different H chain variants of *R. esculenta* differ in their enzyme kinetics, a point which has not been investigated, then pooled samples of tadpoles and frogs would show differences in kinetics because of the differences in the frequency of the variants, although fundamentally the same LDH genes were active in both phases of the life cycle. Similarly, a pooled sample of tadpoles or frogs would yield somewhat different electrophoretic patterns or chromatographic elution profiles.

3. Another observation is that protein polymorphisms are by no means rare. In some species from 20 to 73 percent of various proteins show genetically typable variation(7,8,9). Furthermore, not all such polymorphisms involve only two or three alleles. Several protein polymorphisms are known where from 6 to 13 different alleles all occur in relatively similar gene frequencies, probably representing natural selection for diversity. A small sample from such a population will inevitably miss many of the possible phenotypes. What is worse is that two different samples will tend to contain quite different phenotypes, simply by probability theory and without any necessity for differential selection. Consider the amphioxus Locus I and Locus II G-6-P DH's. With a total of five different alleles at one locus and four alleles at another, there are $15 \times 10 = 150$ different possible phenotypes of this one enzyme.

In view of the number of studies which have compared tadpole and adult frog hemoglobins, especially where tadpole hemoglobins have been pooled and then used for determination of the number of polypeptide chains present, it is instructive to see the diversity of variation that occurs in one population of the bullfrog *Rana catesbeiana*. A diagram of the 11 different hemoglobin patterns observed among 13 tadpoles by Elzinga(27) is provided in Figure 6. In addition, many tadpoles have appreciable amounts of nonhemoglobin protein in hemolysates. By way of contrast, neither Elzinga(27) nor the authors found any individual variation of hemoglobin in surveying over 80 adult bullfrogs. Another type of variation has been observed in studies on *Rana temporaria*. Tadpoles of this species from two different locations in Devon are without detectable hemoglobin, even using the

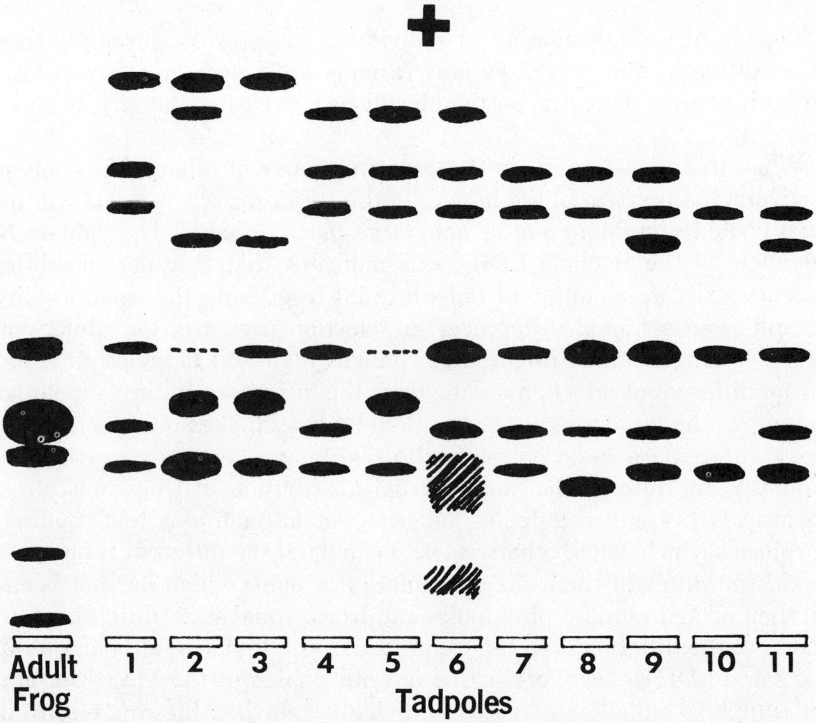

Fig. 6. Individual variation of the hemoglobins of 11 bullfrog *Rana catesbeiana* tadpoles out of a sample of 13. After Elzinga(27). Electrophoresis in a modified Aronson-Gronwal "tris"-EDTA-borate buffer, pH 8.8.

sensitive o-dianisidine test after electrophoresis. Newly metamorphosed individuals have a hemoglobin component which resembles the more anodal component of the multiple hemoglobin of adult *R. temporaria*. The absence of hemoglobin in tadpoles is, however, not a general characteristic of *R. temporaria*, for tadpoles from Yorkshire have hemoglobin.

In summary, individual variation can be either a serious problem or a great benefit in studies on the endocrine control of gene action. It depends entirely on what attempt is made to detect, and when present, to utilize, protein polymorphism.

REFERENCES

1. Smithies, O. Zone electrophoresis in starch gels: group variations in the serum proteins of normal human adults. Biochem. J., 61:629–641, 1955.
2. ——— An improved procedure for starch-gel electrophoresis: further variations in the serum proteins of normal individuals. Biochem. J., 71:585–587, 1959.
3. Hunter, R. L., and Markert, C. L. Histochemical demonstration of enzymes separated by zone electrophoresis in starch gels. Science, 125:1294–1295, 1957.
4. Markert, C. L., and Möller, F. Multiple forms of enzymes: tissue, ontogenetic, and species specific patterns. Proc. Nat. Acad. Sci. U.S.A., 45:753–763, 1959.
5. Manwell, C. Metamorphosis and gene action. I. Electrophoresis of dehydrogenases, esterases, phosphatases, hemoglobins and other soluble proteins of tadpole and adult bullfrogs. Comp. Biochem. Physiol., 17:805–823, 1966.

6. ——— The blood proteins of cyclostomes: a study in phylogenetic and ontogenetic biochemistry. *In* The Biology of Myxine, Brodal, A. and Fänge, R., eds., pp. 372–455. Oslo, Universitetsforlaget, 1963.
7. ——— and Baker, C. M. A. Molecular Biology and the Origin of Species: Heterosis, Protein Polymorphism and Animal Breeding. London, Sidgwick and Jackson (Seattle, University of Washington Press), 1970.
8. Harris, H. Genes and enzymes in man. Cancer Res., 26:2054–2063, 1966.
9. Manwell, C., and Baker, C. M. A. Genetic variation of *iso*citrate, malate and 6-phosphogluconate dehydrogenases in snails of the genus *Cepaea*—introgressive hybridization, polymorphism and pollution? Comp. Biochem. Physiol., 26:195–210, 1968.
10. ———and Baker, C. M. A. Hybrid proteins, heterosis, and the origin of species. I. Unusual variation of polychaete *Hyalinoecia* "nothing dehydrogenases" and of quail *Coturnix* erythrocyte enzymes. Comp. Biochem. Physiol., 10:1007–1028, 1969.
11. Baker, C. M. A. Molecular genetics of avian proteins. VII. Chemical and genetic polymorphism of conalbumin and transferrin in a number of species. Comp. Biochem. Physiol., 20:949–973, 1967.
12. Williams, J. A comparison of glycopeptides from the ovotransferrin and serum transferrin of the hen. Biochem. J., 108:57–67, 1968.
13. Wickstead, J. H. *Branchiostoma lanceolatum* larvae: some experiments on the effect of thiouracil on metamorphosis. J. Mar. Biol. Ass. U.K., 47:49–59, 1967.
14. Manwell, C., Baker, C. M. A., and Betz, I. W. Ontogeny of haemoglobin in the chicken. J. Embryol. Exp. Morph., 16:65–82, 1966.
15. ——— Baker, C. M. A., Ashton, P. A., and Corner, E. D. S. Biochemical differences between *Calanus finmarchicus* and *C. helgolandicus*: esterases, malate and triose-phosphate dehydrogenases, aldolase, 'peptidases,' and other enzymes. J. Mar. Biol. Ass. U.K., 47:145–169, 1967.
16. Ruddle, F. H., Shows, T. B., and Roderick, T. H. Autosomal control of an electrophoretic variant of glucose-6-phosphate dehydrogenase in the mouse (*Mus musculus*). Genetics, 58:599–606, 1968.
17. Manwell, C., and Kerst, K. V. Possibilities of biochemical taxonomy of bats using hemoglobin, lactate dehydrogenase, esterases and other proteins. Comp. Biochem. Physiol., 17:741–754, 1966.
18. Kaplan, N. O. Lactate dehydrogenase—structure and function. Brookhaven Symposia in Biology No. 17, Subunit Structure of Proteins: Biochemical and Genetic Aspects, pp. 131–149, 1964.
19. Salthe, S. N., Chilson, O. P., and Kaplan, N. O. Hybridization of lactic dehydrogenase *in vivo* and *in vitro*. Nature (London), 207:723–726, 1965.
20. Mather, K. Statistical Analysis in Biology, 4th Ed. London, Methuen, 1951.
21. Manwell, C., Baker, C. M. A., Roslansky, J. D., and Foght, M. Molecular genetics of avian proteins. II. Control genes and structural genes for embryonic and adult hemoglobins. Proc. Nat. Acad. Sci. U.S.A., 49:496–503, 1963.
22. ——— and Baker, C. M. A. A sibling species of sea cucumber discovered by starch-gel electrophoresis. Comp. Biochem. Physiol., 10:39–53.
23. ——— Sea cucumber sibling species: polypeptide chain types and oxygen equilibrium of hemoglobin. Science, 152:1393–1396, 1966.
24. ——— and Baker, C. M. A. Evolution of mechanisms for the management of genetic information. *In* Phylogeny of Immunity, Proceedings of the First Workshop in Developmental Immunology. Smith, R. T., Miescher, P. A., and Good, R. A., eds., pp. 3–15. Gainesville, University of Florida Press, 1966.
25. Lindy, S., and Rajasalmi, M. Lactate dehydrogenase isozymes of chick embryo: response to variations of ambient oxygen tension. Science, 153:1401–1403, 1966.
26. Manwell, C. Fetal and adult hemoglobins of the spiny dogfish *Squalus suckleyi*. Arch. Biochem. Biophys., 101:504–511, 1963.
27. Elzinga, M. Hemoglobin Differentiation in the Bullfrog, *Rana catesbeiana*. Ph. D. Dissertation in Physiology, Urbana, University of Illinois, 1964.

20

Triiodothyronine-Induced RNA Synthesis and Turnover in Bullfrog Tadpole Liver

John E. Eaton, Jr., and Earl Frieden

Department of Chemistry, Florida State University, Tallahassee, Florida

INTRODUCTION AND HISTORY

Amphibian metamorphosis is an important postembryonic developmental process, in which nonreproductive structures of the larval form of the amphibian change drastically to an adult form during a discrete period. This is a process quite distinct from postembryonic growth and maturation. Amphibian metamorphosis has become a model system for the study of comparative biochemistry, cellular differentiation, and development, especially since it comprises a series of remarkable biochemical and structural adaptations.

The basic hormonal controls of this process were established by Gudernatsch, Allen, and others in the period from 1912-1930. The extensive anatomical transformation during metamorphosis, reflecting eons of successful evolution were described by biologists even before 1900. In a recent review, *Metamorphosis: A Problem in Developmental Biology*(1). Etkin has described a significant extension of the hormonal control of amphibian metamorphosis. This text contains relevant studies characterizing the biological and cytological aspects of this developmental phenomena. This work has shown a great need for the advancement of our knowledge of the chemical and molecular changes associated with the aforementioned biological transformations which have been associated with amphibian metamorphosis.

DE NOVO NUCLEOTIDE AND RNA SYNTHESIS

During spontaneous and induced metamorphosis, the liver of the tadpole undergoes major metabolic reorganization. Until recently, most of the changes in liver

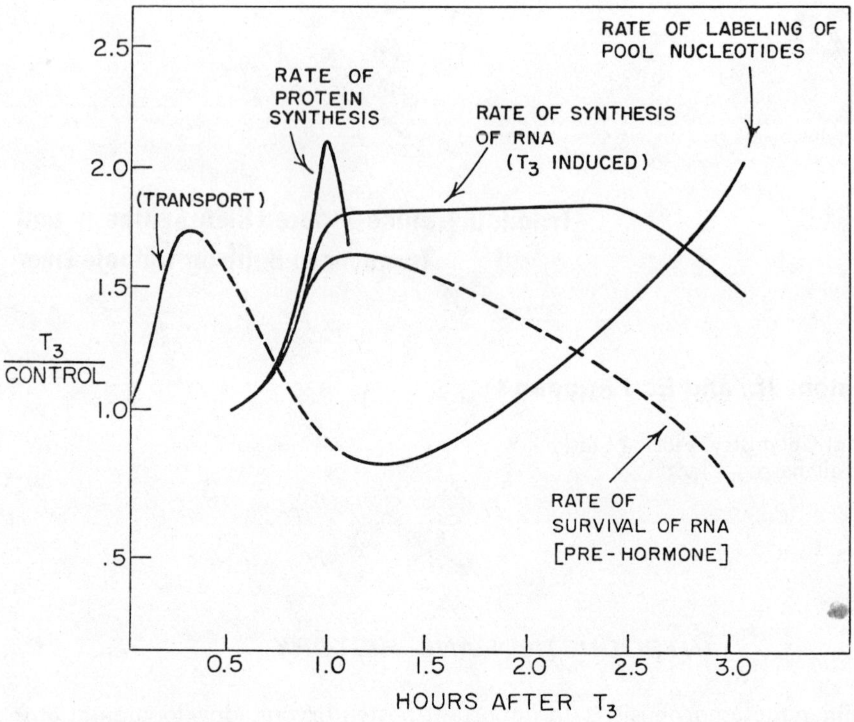

Fig. 1. A proposed sequence of molecular events during the first 3 hours after a single injection of 0.3 nmoles of triiodothyronine. From Eaton and Frieden, *Gen. Comp. Endocr.*, Suppl. 2:398-407, 1969.

metabolism have been studied in terms of the changes in certain key proteins and enzymes. Dramatic increases in numerous serum proteins including ceruloplasmin (2,3) have been shown to take place. These occur as a result of their increased biosynthesis by the liver. Increases in urea cycle enzymes have been documented by Cohen and his collaborators(4). An additional group of proteins including cytochrome oxidase was also reported to increase by Tata and co-workers(5).

Our approach to this problem has been to study the effects of the thyroid hormone prior to the appearance of new protein systems; i.e., during the well-recognized lag period preceding the major thyroid hormone action. The results of recent studies(6,7) on nucleotide, RNA, and protein metabolism in tadpole liver at very early times after T_3 are summarized briefly:

1. There is a rapid increase in the specific activity of the purine-pyrimidine pool in the liver when labeled glycine and phosphate are used as precursors.
2. Using large groups of animals, a study of the amounts of nucleotides and RNA in the liver revealed a rapid increase in both acid-soluble nucleotides and RNA within the first 3 hours.
3. This is accompanied by an apparent rapid and significant increase in the

specific activity of both liver RNA and proteins from appropriate precursors such as adenosine and phenylalanine.

4. Using nonmetabolites, as cytosine arabinoside and cycloleucine, transport was measured and found to be affected by T_3 within 30 minutes.
5. Pulse-chase experiments have shown a fourfold increase in the rate of degradation of RNA labeled prior to hormone administration.
6. Using phosphate labeling, the base composition of the RNA isolated from T_3-treated animals after 2 to 4 hours differed significantly from that of the controls.

In Figure 1, we show a proposed sequence of molecular events during the first 2 to 3 hours after T_3 is injected into tadpoles taking into account the data which have been outlined above. During the first hour, T_3 promotes transport of nucleotides and amino acids into the liver. At the same time, there is a steadily increasing acceleration of protein and RNA biosynthesis reaching its maximum at 60 minutes. Further interpretation of the data becomes more complicated because increased prehormonal RNA degradation and de novo nucleotide biosynthesis now impinges on the continuing macromolecular biosynthesis initiated earlier. The net result is the induction of new protein and RNA biosynthesis in the very early period after T_3 injection, eventually leading to major synthetic events involving most of the intracellular structures of the liver.

INHIBITION STUDIES ON RNA SYNTHESIS

After the gross examination of the major events of the first 3 hours of the lag period, we concluded that closer examination of the pool and RNA constituents was necessary.

The base composition of the RNA synthesized after T_3 injection was studied, using the method of Volkin and Astrachan(8) in which the $^{32}PO_3$ in each nucleotide of labeled RNA is measured. It was found that the RNA after 3 and 12 hours of hormone treatment showed a GC content of 51 percent compared to the control GC content of 62 percent. These values agree well with values of tadpole DNA (approximately 50 percent GC) and ribosomal RNA (61 percent GC), reported by Nakagawa and Cohen(9). The isolation of an "operational messenger" from 36-hour T_3-treated tadpoles with a GC content of 49 percent was reported by Nakagawa and Cohen(10).

The G content of RNA from T_3-treated animals at 3 and 12 hours are 25 and 24 percent, respectively, whereas the control values are 36 and 33 percent. The A content goes from 22 to 27 and 33 percent at 3 and 12 hours for T_3-treated animals.

Recently, Maitra and Hurwitz(11,12) have shown a preferential use of purine nucleotides at the initiation points of RNA. AMP was found to be highly favored in several organisms. GMP is known to be the initiating termini of the transfer RNAs. The 3′-termini of transfer RNAs is adenosine and the turnover of the AMP in several test systems is known to be rapid(13,14,15). This could explain the very rapid effects using labeled adenosine cited earlier.

In order to test the effects of several inhibitors on nucleotide synthesis, the A to G conversion rates, and the rate of RNA synthesis, we used the following experimental design. A dual-labeling technique was used to minimize experimental variability (Fig. 2). Adenine-8-^{14}C was injected 6 hours prior to T (0.3 nmoles/g) administration. The inhibitor was then injected 4 hours prior to T treatment. Adenosine-^{3}H (G) was pulsed for 20 minutes prior to sacrifice. This plan provides a measure of synthesis and conversion both prior to and after inhibitor and a T_3 administration.

The labeling pattern of the RNA was carefully examined (Fig. 3A). RNA, when base hydrolysed, yields mainly 2′, [3′]-monophosphates. However, of necessity, the 3′-terminus, having no phosphate, will be released as the nucleoside. Since the mechanism of base hydrolysis requires a cyclic-nucleotide intermediate, the 5′-termini is released as a 5′, 2′ [3′]-diphosphate. When AR—^{3}H + A^{14}C are used, labeled AR, GR, AMP, GMP, PAP, and PGP are obtained. Using a modified Dowex-1-formate system(16,17), these compounds were separated, analyzed, and examined for their isotopic distribution. For the nucleotides and nucleosides from RNA, the duration of the chromatography is 12 to 14 hours. In order to analyze the dual-labeling data most efficiently (over 10,000 samples were counted) an automated flow system for spectral and scintillation analysis was constructed(18). The output is a punched paper tape which is then put on a computer and the results graphed, integrated, and tabulated (Eaton, unpublished computer program).

In order to examine the pool constituents to estimate the A to G conversion, the following experiments were performed (Fig. 3B). Because the adenosine-^{3}H is both ring and sugar labeled, the usual acid-hydrolytic conversion to adenine and guanine could not be used. Samples of the acid-soluble pools were treated with potato apyrase(19) (ATP-diphosphohydrolase) and a nonspecific acid phosphatase. The apyrase nonspecifically cleaves all di- and triphosphates to monophosphates, which are then dephosphorylated to nucleosides. The nucleosides are then chromatographed in a modified Dowex-1-formate system(20).

Table 1 presents evidence for the enzymic conversion of the pool purine nucleotides to the corresponding nucleosides. The purpose of the charcoal was to ascer-

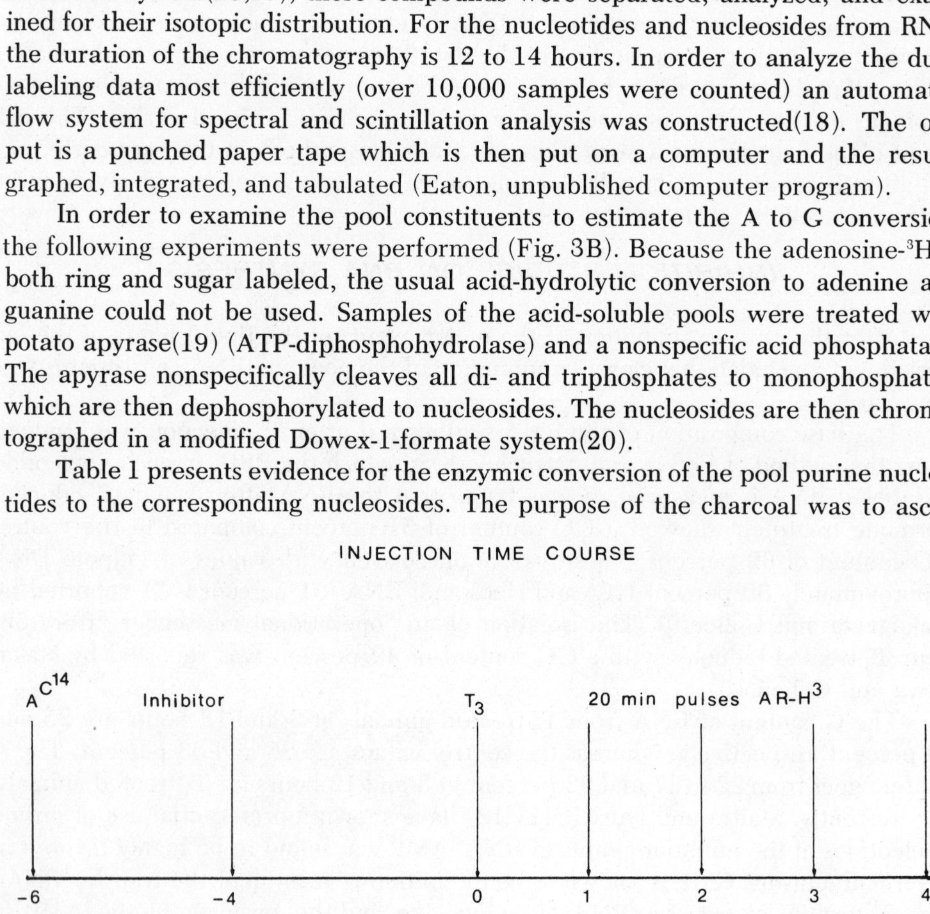

Fig. 2. Schematic representation of the injection protocol for dual labels, inhibitor, and hormone administration. The pulse label-AR-H3 was injected 20 minutes prior to sacrifice. Sacrifice times are given in the text.

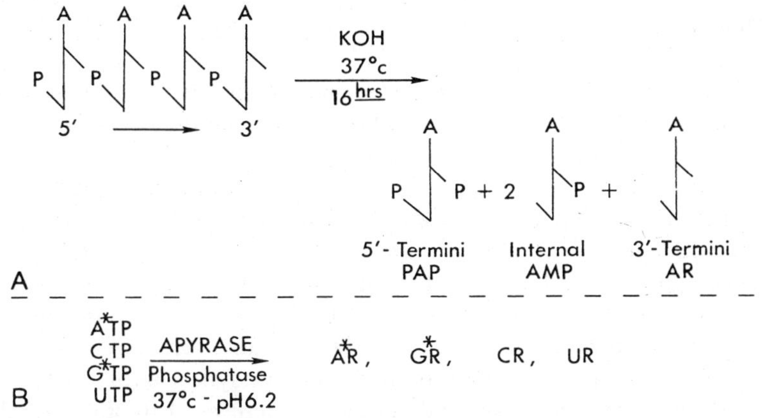

Fig. 3.*A.* Products of base hydrolysis of RNA. *B.* Procedure for dephosphorylation of the nucleotide components of the acid-soluble pool. The products of either A or B were then separated using a Dowex-1-formate ion-exchange chromatography elution system.

tain whether the tritiated nucleoside moiety was still intact or if any hydrolysis had occurred. More than 98 percent of both isotopes were adsorbed indicating no labeled ribose or base had been degraded. Dowex-1 at pH 4.0 was used for the selective adsorption of phosphorylated compounds, the bases and nucleosides remaining in solution. Again, more than 98 percent was recovered on the resin. Enzyme treatment followed by Dowex-1 at pH 4.0 showed that more than 93 percent of the nucleotide pool was susceptible to the dephosphorylation reactions. The only exception to this treatment was the ethionine treated groups where all labels could be adsorbed by charcoal but only 70 percent to Dowex-1. Only 50 percent was susceptible to enzyme treatment suggesting the presence of ^{14}C-S-adenosylethionine.

TABLE 1

Conversion of pool nucleotides to nucleosides

Results of the Enzymic Dephosphorylation of Pool Nucleotides to Nucleosides.

Inhibitor	Charcoal[a] adsorbed	Dowex-1[b] pH 4.0 adsorbed	Apyrase + phosphatase
None	> 98%	> 98%	93%
Actinomycin D	> 98%	> 98%	95%
Thioguanine	> 98%	> 98%	95%
Ethionine	> 98%	70%	50%

[a]Charcoal treatment was used to measure the stability of the nucleoside moiety.

[b]Dowex-1 measured amount phosphorylated compounds in pool.

RNA Inhibitors

The following inhibitors were chosen because of their different sites and modes of action: actinomycin D, thioguanine, and ethionine. Actinomycin D has its major effect on the nucleus and the transcription of mRNA on DNA templates(21). Recent studies have shown actinomycin D to have a stimulatory effect on the cytoplasm, but only after long periods of time(22). Thioguanine has been used for many years as an anticancer drug and, similar to other mercapto- and azapurines, it inhibits the conversion of adenine-containing compounds to the corresponding quanine analogs. This does not interfere with the basic energy source of the cell, ATP, but decreases dramatically the amount of GTP available for RNA synthesis. When phosphorylated, thioguanine inhibits competitively the synthesis of RNA, thus creating RNAs which cannot form normal tertiary structures. Ethionine inhibits in a different manner. Villa-Trevino et al. and Shull et al.(23,24) have shown that the inhibitition of RNA synthesis in rat liver is caused by the formation of a stable S-adenosylethionine molecule. After 4 hours, 80 percent of the available ATP exists in the form of S-adenosylethionine. Accordingly, inhibition of RNA would then be affected in two ways: First, the conversion to guanine-containing compounds would be inhibited, thus neither ATP or GTP would be available for RNA biosynthesis; and, second, the major energy source of the organism is trapped in a useless form, thus inhibiting overall metabolism.

RESULTS

Although all experiments employed a dual-label technique, the results will be discussed as "long-term" or ^{14}C experiments, and "pulse" or ^{3}H experiments. In all experiments cited, maximum total standard error is about 20 percent, with less variation in many experiments.

Figure 4 shows the effects of T_3 and the aforementioned inhibitors on the incorporation of AR-^{3}H into RNA. The inhibitor dose levels were as follows: actinomycin D, 2 μg/g; thioguanine, 200 μg/g; ethionine, 200 μg/g. When no inhibitor is present, T_3 doubles the incorporation during the first 90 minutes. Actinomycin D and ethionine inhibit both the control and T_3 groups, the T_3 stimulation being completely eliminated. As expected, thioguanine shows a small but statistically significant increase at 60 minutes.

In the "long-term" experiments, the effects of T_3 are first noted at 2 hours (Fig. 5). Since the label is present for 6 to 8 hours before sacrifice, a longer period of time is required before a statistically significant T_3 effect can be detected. As with the short-term effects, actinomycin D and ethionine completely inhibit RNA synthesis. Thioguanine, having a longer time period to inhibit A to G conversion and RNA synthesis, shows no ^{14}C increase due to T_3 as compared to the positive AR-^{3}H results.

It is possible that the effects on RNA induced by T_3 could be due to rapid changes in pool size and distribution. Therefore, we studied the conversion rates of

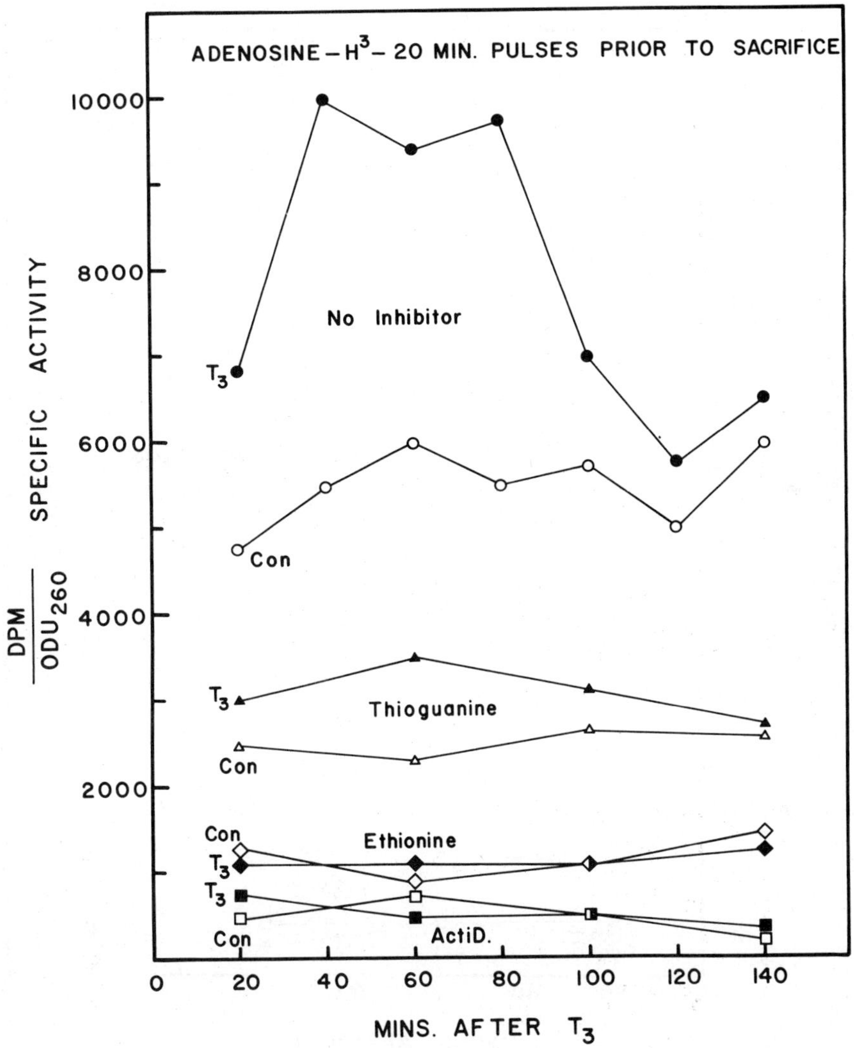

Fig. 4. The specific activity of RNA pulse labeled with ³H-adenosine for 20 minutes versus time in minutes after injection of hormone. Inhibitor doses: thioguanine, 200 μg/g; ethionine, 200 μ g/g; actinomycin D, 2 μg/g.

the A-^{14}C and AR-^{3}H in the acid-soluble pool. Table 2 shows the effects of T_3 on the distribution of pool radioactivity in the no-inhibitor experiment. As the total dpm of both ^{3}H and ^{14}C appeared not to be affected by hormone, the results are expressed as the ratio of T_3 dpm to control dpm for either AR or GR at each time period. Both ^{3}H and ^{14}C results have been tabulated in Table 2. There appears to be a rapid conversion of AMP to GMP, returning to normal after 140 minutes. There is little difference due to the length of label, implying that the steady-state condition for pool constituents is reached rapidly.

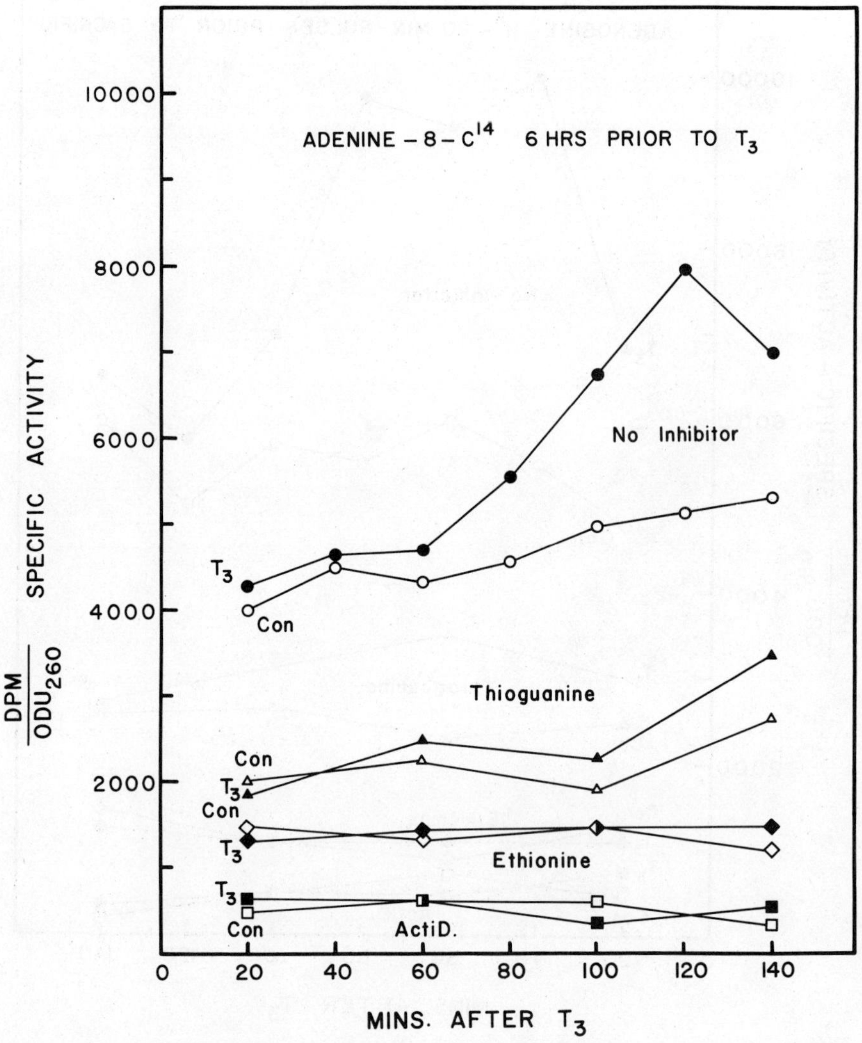

Fig. 5. The specific activity of RNA labeled for 6 hours prior to hormone treatment versus time in minutes after injection of hormone. Inhibitor doses: thioguanine, 200 μg/g; ethionine, 200 μg/g; actinomycin D, 2 μg/g.

Let us now examine the labeling patterns of the RNA labeled for both "pulse" and "long-terms." Table 3 shows the results obtained from the base hydrolysed RNA, which was chromatographed on Dowex-1-formate automated columns. The results are expressed as the ratio of T_3 dpm to control dpm for each of the six compounds at the different time periods. There is a AR 3′-termini effect at 20 minutes as well as a PAP 5′-termini effect. The major effect is seen in the initiating positions. Taking into account the rapid conversion of AR to GR, whereby T_3/control is 0.6 for AR-^{3}H, there is an eightfold increase due to T_3 at the 5′-termini PAP after

TABLE 2

Labeling of Pool $\frac{T_3DPM}{Con\ DPM}$ [a]

Distribution of Label in the Acid-Soluble Pool.

		H^3		C^{14}	
T_3 time		AR	GR	AR	GR
20′	T_3/Con	0.6	1.6	0.6	2.0
60′	T_3/Con	0.6	1.5	0.5	2.0
100′	T_3/Con	0.6	1.5	0.5	2.3
140′	T_3/Con	0.9	1.1	0.8	1.5

[a]Adenosine-^{3}H—20-minute pulses, adenine-^{14}C—6 hours prior to T_3. Nucleosides resulting from enzymic dephosphorylation were separated on Dowex-1-formate and counted.

60 minutes. There is a sixfold internal AMP increase and a fivefold increase at the 3′-termini. The T_3 effect on PGP is twofold after 60 minutes and the internal GMP effect is 0.7-fold, correcting for the pool label distribution. This is in good agreement with the base composition data, which showed a large decrease in GMP, a large increase in AMP, and an overall increase in RNA content of the liver.

An examination of the long-term experiments shows results similar to the short-term experiments (Table 4). After using the necessary factor for the pool conversion (see Table 2), the major effects are a 2- to 3-fold increase in label at the AMP and PAP positions. The large conversion factor for G compounds (Table 2) again shows that there is a decrease in the label found as G compound. The only major effect for G compounds is found in PGP at 140 minutes.

Rather than show the same series of figures for each of the three inhibitors, the effects of each inhibitor are outlined below:

1. Actinomycin D. The major effect is on the nuclear synthesis as less than 2 percent of the AR-^{3}H recovered in the RNA was found in the 5′-termini

TABLE 3

Position labeling of RNA $\frac{T_3DPM}{Con\ DPM}$ [a]

Distribution of Tritium in RNA After 20-minute Pulse.

	3′-Termini		Internal		5′-Termini	
T_3	AR	GR	AMP	GMP	PAP	PGP
H^3–20′	1.7	0.0	0.5	1.0	1.4	0.9
H^3–60′	3.0	0.0	3.5	1.3	5.1	4.1
H^3–100′	3.9	0.0	1.1	1.0	3.5	2.5
H^3–140′	1.8	0.0	1.4	1.0	4.0	3.8

[a]Adenosine-^{3}H—20-minute pulses prior to sacrifice. The RNA was base hydrolyzed, the radioactive products separated on Dowex-1-formate and analyzed by liquid scintillation counting.

TABLE 4

Position labeling of RNA $\frac{T_3DPM}{Con\ DPM}$ [a]

Distribution of ^{14}C in RNA, Labeled for 6 Hours.

	3′-Termini		Internal		5′-Termini	
T_3	AR	GR	AMP	GMP	PAP	PGP
C^{14}–20′	0.0	1.0	1.0	1.1	1.1	1.1
C^{14}–60′	0.0	1.2	1.0	0.7	0.8	1.2
C^{14}–100′	0.0	1.6	1.5	1.3	1.6	1.5
C^{14}–140′	0.0	2.0	1.1	1.1	1.6	2.3

[a]Adenine-^{14}C—6 hours prior to T_3. The RNA was base hydrolyzed, the radioactive products were separated on Dowex-1-formate and analyzed by liquid scintillation counting.

positions compared with 50 percent for the noninhibitor experiment. In both ^3H and ^{14}C experiments, there was greater than 90 percent inhibition after 4 hours of inhibitor treatment.

2. Thioguanine. The crossover of adenosine to guanosine is inhibited by 90 percent. The internal AMP and 5′-termini PAP, though reduced in activity by 50 percent, show a 2- to 3-fold T_3 effect. There is no effect seen at PGP.
3. Ethionine. Effects of ethionine are similar to actinomycin D. The conversion of adenosine to guanosine is inhibited to a large extent. Pool studies show the production of an adenosine containing species which is not susceptible to dephosphorylation as predicted for S-adenosylethionine.

SUMMARY AND CONCLUSIONS

Recent work from this laboratory has shown a multiplicity of effects on tadpole liver within 3 hours after the administration of T_3 (25,26). Our present rationale for effects on cellular development and differentiation supports the idea of the transfer of information from DNA via RNA to protein. Thus, having found a large RNA stimulation in these experiments, we have studied base composition and position labeling.

The acid-soluble pool distribution shows a rapid conversion of injected adenine and adenosine to guanine-containing compounds. This is a 1- to 2-fold effect, but an examination of the RNA at the 3′, 5′ and internal positions reveal that as much as three to five times the isotope is found in the A-containing positions of T_3-treated animals. The same large effect on the G-containing compounds was not observed. This is in accord with the base composition data of a newly synthesized RNA which shows a large increase in AMP and a corresponding decrease in GMP.

This fact, along with the large increases at the initiating positions of the RNA chains leads us to the conclusion that within the first few hours after T_3 administration, a considerable number of new RNA chains are synthesized, with a base composition differing significantly from that of the control. Typical inhibitors, e.g., actinomycin D, thioguanine, and ethionine which act at different sites as points of inhibition, provide further evidence that T_3 is causing the early production of new RNA, rather than just affecting cytoplasmic pool levels. This new RNA, in conjunction with the other early effects cited previously(27,28), appear then to be preparing the liver at the transcriptional level for the manifold biochemical and morphological changes of the metamorphic climax.

REFERENCES

1. Etkin, W., and Gilbert, L. Metamorphosis: A Problem in Developmental Biology, New York, Appleton-Century-Crofts, 1968.
2. Herner, A. E., and Frieden, E. VII. Changes in serum proteins during spontaneous and induced metamorphosis. J. Biol. Chem., 235:2845, 1960.
3. Inaba, I., and Frieden, E. Changes in ceruloplasmin during anuran metamorphosis. J. Biol. Chem., 242:218, 1967.
4. Kim, K. H., and Cohen, P. P. Biochemical aspects of metamorphosis: changes in RNA associated with the transition from ammonotelism to ureotilism. Amer. Zool., 8:243, 1968.
5. Tata, J. R. The formation and distribution of ribosomes during hormone induced growth and development. Biochem. J., 104:1, 1967.
6. Eaton, J. E., and Frieden, E. Molecular changes during anuran metamorphosis; early effects of triiodothyronine on nucleotide and RNA metabolism in bullfrog tadpole liver. Proc. 7th Gunma Symp. Comp. Endocr., 5:43, 1968.
7. ——— and Frieden, E. Primary mechanisms of thyroid hormone control of amphibian metamorphosis. Int. Symp. Comp. Endocr., New Delhi, India. Gen. Comp. Endocr., Suppl. 2:398, 1969.
8. Volkin, E., and Astrachan, L. Phosphorous incorporation in *E. Coli* ribonucleic acid after infection with bacteria phage T_2. Virology, 2:149, 1956.
9. Nakagawa, H., and Cohen, P. P. Studies on ribonucleic acid synthesis in tadpole liver during metamorphosis induced by thyroxine. II. Analysis of ribonucleic acid fractions. J. Biol. Chem., 242:642, 1967.
10. *Ibid.*
11. Maitra, U., and Hurwitz, J. The role of DNA in RNA synthesis. IX. Nucleoside triphosphate termini in RNA polymerase products. Proc. Nat. Acad. Sci. U.S.A., 54:815, 1965.
12. ——— Nakata, Y., and Hurwitz, J. The role of deoxyribonucleic acid in ribonucleic acid synthesis. J. Biol. Chem., 242:4908, 1967.
13. Holt, C. E., Joel, P. B., and Herbert, E. Turnover of terminal nucleotides of soluble ribonucleic acid intact reticulocytes. J. Biol. Chem., 241:1819, 1966.
14. Cannon, M. Instability of the terminal adenosine of soluble ribonucleic acid of *Escherichia Coli* in vivo. Biochem. Biophys. Acta, 87:154, 1964.
15. Rosset, R., and Monier, R. On the instability of transfer—RNA terminal nucleotide sequence in yeast. Biochem. Biophys. Res. Commun. 10:195, 1963.
16. Hurlbert, R. B., Schmitz, H., Brumm, A. F., and Potter, V. R. Nucleotide metabolism: II. chromatographic separation of acid soluble nucleotides. J. Biol. Chem., 209:23, 1954.
17. Rosset, R., Monier, R., and Julien, J. Les ribosomes d'un RNA ribosomique de faible poids moleculaire. Bull. Soc. Chim. Biol., 46:87, 1964.
18. Eaton, J. E. Automated specific activity analysis. International Symposium on Current Topics in Liquid Scintillation Counting. (To be published.)
19. Krishnan, P. B. Methods of Enzymology, Vol. 2, 59, New York, Academic Press, Inc., 1955.

20. Eaton, J. E. Automated specific activity analysis. International Symposium on Current Topics in Liquid Scintillation Counting. (To be published.)
21. Stewart, G. A., and Farber, E. The rapid acceleration of hepatic nuclear ribonucleic acid breakdown by actinomycin, but not by ethionine. J. Biol. Chem., 243:4479, 1966.
22. Garren, L. D., Howell. R. R., Tomkins, G. M., and Crocco, M. A paradoxical effect of actinomycin D: the mechanism of regulation of enzyme synthesis by hydrocortisone. Proc. Nat. Acad. Sci. U.S.A., 52:1121, 1964.
23. Villa-Trevino. S., Schull, K. H., and Farber, E. The inhibition of liver ribonucleic acid synthesis by ethionine. J. Biol. Chem., 241:4670, 1966a.
24. Schull, K. H., McConomy, J., Vogt, M., Castrillo, A., and Farber, E. On the mechanism of induction of hepatic adenosine triphosphate deficiency by ethionine. J. Biol. Chem., 241:5060, 1966.
25. Eaton, J. E., and Frieden, E. Molecular changes during anuran metamorphosis; early effects of triiodothyronine on nucleotide and RNA metabolism in bullfrog tadpole liver. Proc. 7th Gunma Symp. Comp. Endocr., 5:43, 1968.
26. ——— and Frieden, E. Primary mechanisms of thyroid hormone control of amphibian metamorphosis. Int. Symp. Comp. Endocr., New Delhi, India. Gen. Comp. Endocr., Suppl. 2:398, 1969.
27 ——— and Frieden, E. Molecular changes during anauran metamorphosis; early effects of triiodothyronine on nucleotide and RNA metabolism in bullfrog tadpole liver. Proc. 7th Gunma Sump. Comp. Endocr., 5:43, 1968.
28. ——— and Frieden, E. Primary mechanisms of thyroid hormone control of amphibian metamorphosis. Int. Symp. Comp. Endocr., New Delhi, India. Gen. Comp. Endocr., Suppl. 2:398, 1969.

21

An In Vitro Quantitative Analysis of the Response of Tadpole Tissue to Hormone Treatment*†‡

A. Derby

Department of Biology, Yale University,
New Haven, Connecticut

INTRODUCTION

Metamorphosis in the typical anuran has been divided into three stages: premetamorphosis, a period in which growth occurs with little change in form of the animal; prometamorphosis, a period of differential hind leg growth and continued body growth but at a reduced rate; climax, a period that is initiated by the emergence of the forelegs (this event is designated as the day of *E*) and terminates with the complete resorption of the tadpole's tail(8).

Amphibian metamorphosis is under endocrine control (for reviews see 9,10,22) and has been shown to be dependent upon the synthesis and release of hormone by the thyroid gland. The capacity of this gland to synthesize and release an active product changes during metamorphosis.

At the present time, the views regarding the mode of thyroid hormone action in regulating the pattern of metamorphic events are controversial. Kollros(20) has interpreted the action of thyroxine in metamorphosis to be based on a threshold concept. He suggests that the tissues involved in metamorphosis possess different

*Submitted in partial fulfillment of the requirements for the degree of Doctor of Philosophy in the Graduate Division of the City College of the City University of New York.

†This study was supported by Training Grant 5T1 Hd-0016-01-03 of the National Institute of Child Health and Human Development, National Institutes of Health, and by National Science Foundation Grant GB-5913.

‡I am indebted to Dr. William Etkin for his encouragement and guidance throughout this work and to the advice and criticism of Dr. Max Hamburgh. Part of this research was conducted in Dr. Etkin's laboratory in the Department of Anatomy, The Albert Einstein College of Medicine, New York. I want to thank Mr. R. Stern for his assistance in Dr. Etkin's laboratory.

thresholds and that the stimulation of metamorphic events occurs to the extent that the increasing concentration of circulating thyroid hormone matches the threshold levels. He also suggests that tissues may change their threshold sensitivity to thyroxine at different stages of metamorphosis. Etkin's interpretation of the role of thyroxine in metamorphosis is a quantitative concept in which thyroxine is thought to act in a fashion common to many chemical reactions. That is, in a stoichiometric manner, i.e., the higher the concentration the faster the reaction proceeds. Each tissue is characterized by a different specific total requirement of thyroid hormone to complete a metamorphic event. The rate of response of tissues varies in proportion to the concentration of hormone available to it(9,26).

Recently, evidence has been presented indicating that another hormone, prolactin, may play an important role in metamorphic changes. Etkin and Lehrer(13) demonstrated that pituitaries implanted into hypophysectomized tadpoles produced excessive growth of these animals. The authors suggested that these grafts might be releasing prolactin and that this substance acts as a growth factor in tadpoles. It has been demonstrated(14,15) that, in the rat, prolactin secretion is normally under hypothalamic inhibition and that grafts released from inhibition produce excesses of this hormone. In a similar manner, a transplanted tadpole pituitary would be freed from hypothalamic inhibition and as a result release high levels of prolactin. Several groups of investigators(3,4,23,24,12) have injected mammalian prolactin preparations into normal and hypophysectomized tadpoles and observed that prolactin produced increases in growth. In addition, these authors have reported that administration of growth hormone also increased growth in tadpoles but it is not as effective as prolactin. This growth effect is thought to be due in part to contamination of growth hormone preparations by prolactin. It has also been demonstrated (4,12,18) that prolactin can inhibit metamorphosis. It has been observed(11) that pituitary grafts placed into the tails of tadpoles near climax interfere with their normal climax pattern so that tail resorption and other climax events are partially delayed. The area around the graft stands out as a lobe of unresorbed tissue. The local action of pituitary grafts leading to inhibition of tail resorption could be explained by assuming that prolactin released by these grafts interferes with the action of thyroxine peripherally, at the level of target tissues. Support for this concept comes from the work of Bern et al.(4). They injected tadpoles with prolactin and then placed these tadpoles into solutions of DL-thyroxine. At the concentration of thyroxine used, tail length and depth of fins was reduced in tadpoles treated only with thyroxine. However, simultaneous treatment with thyroxine counteracted the tail-resorbing influence of thyroxine. Therefore, these authors have suggested that prolactin functions peripherally to antagonize the action of thyroxine. The experiments performed in Etkin's laboratory(12,18) have suggested that prolactin inhibits metamorphosis in yet another manner, by acting as a goitrogenic agent that interferes with thyroid action. As a result of the action of prolactin, the levels of functional hormone in circulation are reduced and these tadpoles enter metamorphic stasis. The more recent work of this laboratory(11,19) suggests that prolactin may function in both capacities.

It was the aim of this investigation to study the role of two hormones (thyroxine and prolactin) in metamorphic change. Specifically, this study was designed to answer the following major guestions. Is the response of target tissues to thyroxine quantitative in nature? Do target tissues of metamorphosing animals change in their sensitivity to thyroid hormone? Answers to these first two questions would be of significance in understanding the mode of action of thyroid hormone in metamorphosis. The last major question concerns the role of prolactin in metamorphosis. There is now evidence that prolactin may be a growth-promoting factor in amphibians, and prolactin has been reported to inhibit metamorphic change. The question that arises is: Does prolactin inhibit metamorphic change by interfering with the action of thyroxine peripherally, at the level of target tissues?

MATERIALS AND METHODS

A modification of Weber's(25) method for culturing tail tips of *Xenopus laevis* was devised following several experiments in which tail tips of *Rana pipiens* tadpoles were cultured. Stated briefly, this method consists of cutting out rectangular pieces, called discs, from the fins of *R. pipiens* tadpoles (see 6, for greater details). These discs are composed of a simple integument that encloses a loose connective tissue and they are completely devoid of the striated muscle present in whole tail tip preparations. Tail discs heal over readily and rapidly by epithelial growth and provide a uniform population of test objects that maintain themselves well in vitro for several weeks. The medium employed was Hanks balanced salt solution (GIBCO) diluted with sterile distilled water. Penicillin-streptomycin was added to the medium. Discs were cultured en masse (15 per dish) and then only completely healed discs were chosen and assigned to experimental categories. Experimental discs were maintained in small petri dishes (60 x 15 mm) containing 10 ml of the appropriate medium.

Tail resorption was measured quantitatively by placing the petri dish containing a disc onto the stage of a Bausch and Lomb Tri-Simplex Microprojector. The projector was set so that it projected the magnified image of the disc downwards onto a recording sheet where the outline of the disc was copied. The area of this drawing was then measured with a planimeter and the data were expressed in terms of percent shrinkage per day. Figure 1 shows discs as visualized by this method.

EXPERIMENTAL SERIES

The Response of Tail Discs to Thyroxine

Procedure. *Experiment 1. The Response of Tail Discs Taken From Tadpoles in Early Prometamorphosis to Thyroxine.* Tail discs were cut from tadpoles in early prometamorphosis and were treated with six concentrations of L-thyroxine (T_4), 3, 9, 27, 81, 250, and 750 parts per billion (ppb) of thyroxine. (A solution of 1 ppb of T_4 contains one microgram of T_4 per liter.) Discs placed in Hanks balanced salt solution without hormone supplement served as controls.

Experiment 2. The Response of Tail Discs Taken from Animals in Late Prometamorphosis to Thyroxine. In order to determine whether the responsiveness of

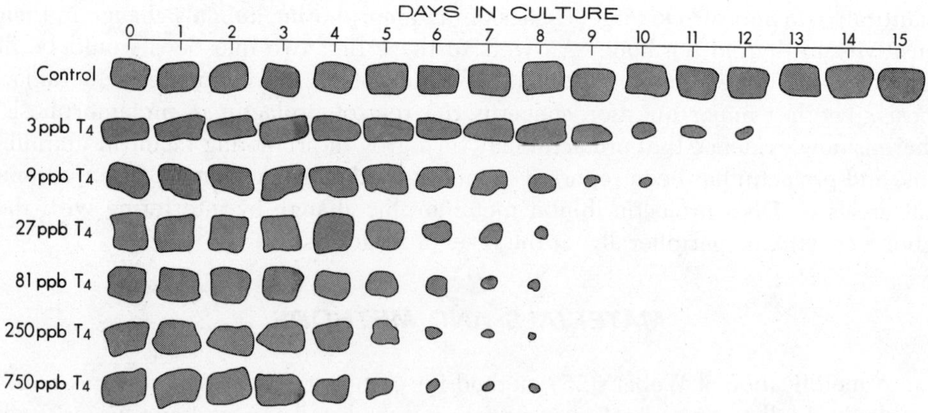

Fig. 1. Early prometamorphic discs treated with different concentrations of thyroxine as recorded by projection. From Derby. *J. Exp. Zool.*, 168:147-156; 1968.

discs to T_4 changes with advancing stages of metamorphosis, tail discs were cut from animals in late prometamorphosis (animals were chosen that were close to the day of emergence of the forelegs, $E-3$ to $E+0$). These discs were treated with the same six concentrations of thyroxine used in Experiment 1.

Experiment 3. Tail Resorption During Climax. This experiment was designed to study tail resorption during climax. In *R. pipiens*, tail resorption begins after the emergence of the forelegs. The fins show rapid resorption on the second and third days after the emergence of the forelegs. It has been proposed that the levels of T_4 rise abruptly above the late prometamorphic level about the time of forelimb emergence although the precise day of this rise has not as yet been clarified(9). In an attempt to demonstrate on what day during this period such a rise occurs, discs were cut from animals at $E-1$, $E+0$, and $E+1$. These discs were not placed en masse for healing; instead each was placed singly into a petri dish containing Hanks medium. As will be seen in the results, the preliminary healing period of these discs had to be eliminated since $E+1$ discs resorb rapidly.

Experiment 4. The Response of Discs to Thyroxine Treatment for Varying Periods of Time. This experiment was designed to study the dependence of tail resorption on the continuous presence of thyroid hormone. Discs were cut from animals in early prometamorphosis and placed into solutions of 3 ppb T_4 for 1, 3, 5, and 7 days. At the end of these periods discs were rinsed in and transferred to Hanks medium for the remainder of the experiment. Two controls were employed; discs in Hanks only and discs in 3 ppb T_4 continuously.

Results. *Experiment 1. The Response of Tail Discs Taken from Tadpoles in Early Prometamorphosis to Thyroxine.* Early prometamorphic discs placed into control medium maintained themselves well for at least 15 days in culture (Fig. 2). Control discs may maintain themselves for as long as 30 days.

Early prometamorphic discs treated with thyroxine demonstrate that a quantitative relationship exists between hormone concentration and the rate of tissue resorption. Discs treated with 3 ppb of T_4 begin to show resorption after about 8 to 9

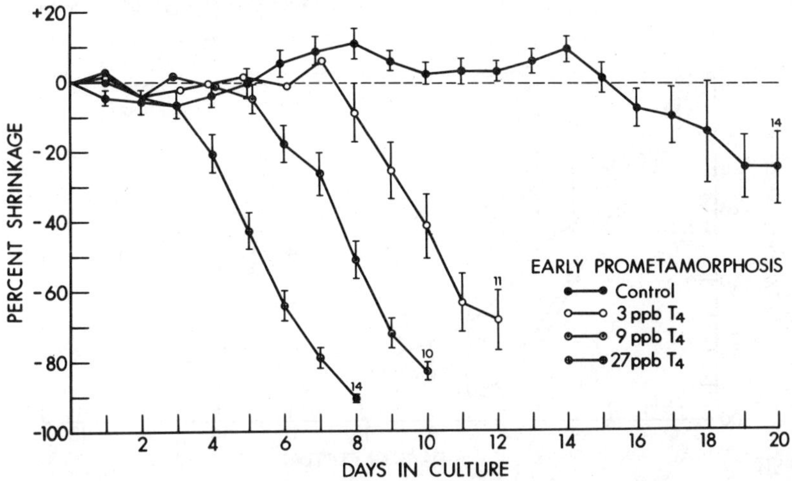

Fig. 2. Experiment 1, early prometamorphic discs treated with thyroxine. The vertical lines indicate the standard error of the mean and the numbers above these lines indicate the number of discs still present out of the original 14 discs in the group. From Derby. *J. Exp. Zool.*, 168: 147-156, 1968.

days in culture and resorption is completed about 3 to 4 days later. At this concentration, some variability exists as to the day when shrinkage begins. The majority of discs start this process on days 8 or 9; however, a few may begin shrinkage on day 7 and in one instance a disc only began to shrink on day 12. Discs treated with 9 ppb T_4 initiate resorption about days 5 to 6 and also complete this process in about 3 to 4 days. At 9 ppb, the variability with reference to what day shrinkage begins is reduced. Discs treated with 27 ppb of T_4 begin resorption on days 3 to 4 and this process is also completed in 3 to 4 days. Higher concentrations do not differ significantly from 27 ppb. Thus discs treated with 81, 250, and 750 ppb T_4 show rates of resorption similar to that observed for 27 ppb T_4. At these concentrations (27 to 750 ppb T_4) there is little variability and resorption is initiated and completed uniformly by almost all discs.

Experiment 2. The Response of Tail Discs Taken from Animals in Late Prometamorphosis to Thyroxine. Control discs from animals in late prometamorphosis show resorption from day 1 in culture; this is in contrast to the control discs taken from animals in early prometamorphosis (Experiment 1) that do not show resorption for at least 15 days. The resorption of these discs from late prometamorphic donors tends to level off after several days and the discs persist at this reduced size (Fig. 3).

Addition of exogenous thyroid hormone to discs of this experiment also demonstrated that a quantitative relationship exists between hormone concentration and the rate of resorption of tail discs. Treatment with 3 ppb T_4 results in complete resorption by day 9. This rate of resorption is similar to the rate obtained by giving 9 ppb T_4 to discs from animals in early prometamorphosis (Fig. 2). In 9 ppb T_4, discs show a resorption curve similar in form to that observed with 27 ppb T_4 in early prometamorphosis; resorption is completed in 6 to 7 days. This dosage (9 ppb) ap-

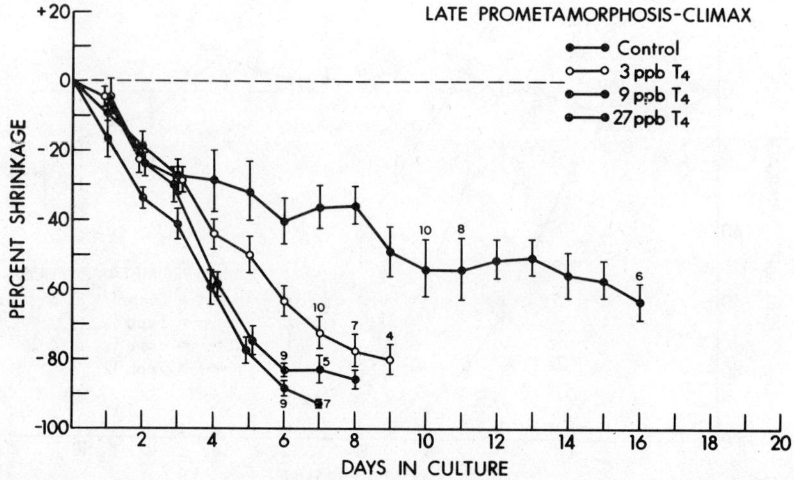

Fig. 3. Experiment 2, discs taken from animals at late prometamorphosis and treated with thyroxine. From Derby. *J. Exp. Zool.*, 168:147-156, 1968.

pears to be the lowest concentration giving the maximum response at this stage of metamorphosis. The curves for 27, 81, 250, and 750 ppb all show rates of resorption similar to that of 9 ppb T_4.

Experiment 3. Tail Resorption During Climax. If the level of hormone rises around the day of *E* it is expected that tail discs cut on specified days near this time might show different rates of resorption, in control medium, based on their higher endogenous level of hormone content. Discs that were cut from animals at

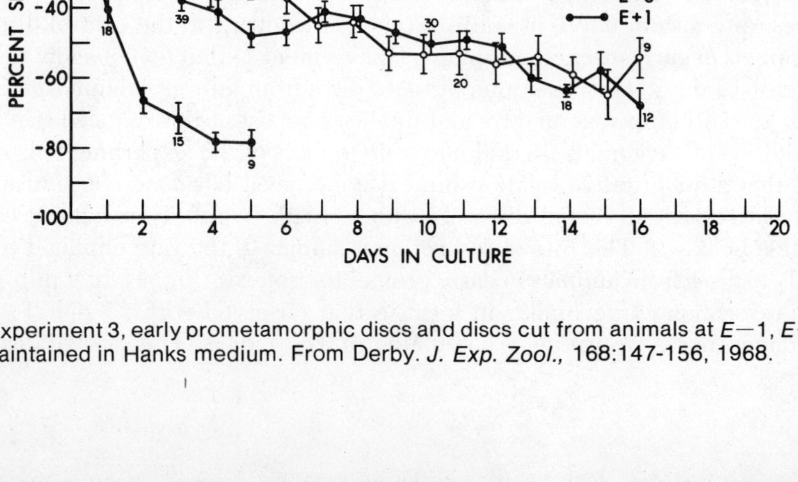

Fig. 4. Experiment 3, early prometamorphic discs and discs cut from animals at $E-1$, $E+0$, and $E+1$; maintained in Hanks medium. From Derby. *J. Exp. Zool.*, 168:147-156, 1968.

$E+1$ show a dramatic increase in their rate of resorption as compared to discs from animals at $E-1$ and $E+0$ that behaved as did the controls (that were essentially the same, i.e., $E-3$ to $E+0$) in Experiment 2 (Fig. 4). The rate of resorption of discs at $E+1$ is about the same as that observed for the fins of intact animals during climax.

Experiment 4. The Response of Discs to Thyroxine Treatment for Varying Periods of Time. In this experiment, early prometamorphic discs maintained in 3 ppb T_4 continuously show a curve for resorption similar to that obtained in Experiment 1 (Fig. 5). Discs placed in 3 ppb T_4 for just one day do not differ from the controls maintained in Hanks continuously. Discs placed into 3 ppb T_4 for 3 days show initiation of resorption on day 5 but this rate is low and soon levels off and the discs maintain themselves at their reduced size. Discs placed into 3 ppb T_4 for 5 days give a similar response but the degree of resorption is slightly greater. Discs placed into 3 ppb T_4 for 7 days show further resorption than the previous groups, these discs give an initial rate of resorption that parallels that of discs maintained in thyroxine continuously. However, this group then shows a sudden leveling off in disc size whereas discs maintained in thyroxine continuously are completing their resorption.

Conclusions. *The Response of Tail Discs to Thyroxine—Experiments 1 through 4.* One of the major intents of these experiments was to study, quantitatively, the response of tail tissue, taken from animals at different stages of metamorphosis, to thyroxine and to investigate whether a quantitative relationship exists between thyroid hormone and its mode of action. The demonstration of such a relationship would be of significance in distinguishing between the two concepts

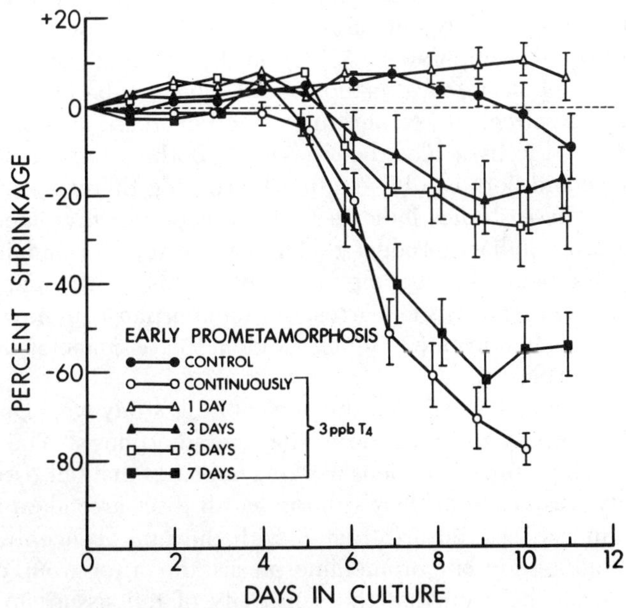

Fig. 5. Experiment 4, early prometamorphic discs placed into 3 ppb of T_4 for different periods of time and then maintained in Hanks medium. From Derby. *J. Exp. Zool.*, 168:147-156, 1968.

that have been proposed for the mode of thyroid hormone action in metamorphosis (threshold or stoichiometric).

The results of Experiments 1 and 2 demonstrate that a quantitative relationship does exist between the level of thyroid hormone and the rate of resorption of tail discs. The range of this relationship is related to the stage of metamorphosis being tested. In early prometamorphosis, this range is observed over 3 to 27 ppb of T_4 while with discs taken from animals in late prometamorphosis the range is reduced so that 9 ppb T_4 appears to be the lowest dosage tested that gives the maximum response. The results of Experiment 4 provide further evidence that such a quantitative relationship exists, for when early prometamorphic discs are placed into 3 ppb of T_4 for varying lengths of time (1, 3, 5, and 7 days) and then removed from this medium and placed into Hanks medium these discs initiate but fail to complete resorption. A quantitative relationship is observed between the amount of time the tissue had been in contact with hormone and the degree of resorption achieved.

In addition, the results of Experiment 4 indicate that the action of this hormone is dependent upon the continuous presence of the hormone. It can be concluded therefore that the hormone does not merely act as a triggering mechanism.

In these experiments it is important to note that changing the concentration of thyroxine does not affect the length of the period in which resorption occurs (generally 3 to 4 days). Changing the concentration of thyroxine changes the length of the period that precedes resorption (latent period). The quantitative nature of these results indicates that we may be dealing with a stoichiometric phenomenon involving molecular interactions of hormone and some other molecule.

Additional support for the concept that thyroxine exerts its effects of metamorphosis by taking part in the molecular changes of these tissues can be provided by the studies of the effects of low temperature on metamorphosis. Frieden et al.(17) studied the effect of low temperatures (5°C) on thyroxine (T_3) induced tail resorption. No T_3 effect was observed after up to 100 days incubation at 5°C. In more recent work(16,1) the effects of temperature on tail decrease and shifts in ammonia to urea excretion by the liver have been studied. Both of these responses are induced by thyroxine and both can be arrested by shifting thyroxine treated animals from 25° to 5°C. These studies indicate that the action of thyroxine involves the formation of an intermediary product that is responsible for inducing metamorphic change and it appears that it is the formation of this intermediary product that is dependent upon temperature. The arrest of metamorphosis by low temperatures should prove to be a valuable clue to understanding the basic molecular changes accompanying metamorphosis.

The next major question asked at the outset of this study was: Do target tissues change in their sensitivity to thyroxine during metamorphosis? The results of Experiments 1 and 2 and those of previous workers indicates that tail tissues do change in their sensitivity (response) to thyroid hormone. If discs are taken from tadpoles in early prometamorphosis and are treated with the same concentration of T_4 as discs taken from tadpoles in late prometamorphosis, the latter group exhibit a more rapid response. It can be seen that the sensitivity of tail tissues to thyroxine increases during the prometamorphic period.

An important question that arises is: Why does the same tissue change in its sensitivity to thyroid hormone during metamorphosis? Perhaps an answer to this question can be found in a further examination of the results of these experiments. In Experiment 1, early prometamorphic discs maintained in control medium do not demonstrate resorption of any kind until day 15 in culture. It appears that these tissues were not previously exposed to a concentration of thyroid hormone that was equivalent to or greater than 3 ppb of T_4, because the results of Experiments 1 and 4 indicate that if discs had been exposed to 3 ppb T_4 for only 3 days, some resorption of these discs would have occurred prior to day 15. The response of these control discs is therefore in accord with the generally accepted concept that the level of circulating thyroid hormone is extremely low during early prometamorphosis. This level is less than the equivalent of 3 ppb T_4 in the external medium.

Late prometamorphic discs, Experiment 2, placed into control medium demonstrate partial resorption. This is interpreted to mean that these tissues had been exposed to or possessed a sufficient level of thyroid hormone at the time of isolation to program them for some resorption. The leveling off suggests that removal of these tissues from a further source of hormone (cutting off the tail) was responsible for the cessation of further resorption. The resorption demonstrated by these control discs would be in accord with the previously mentioned concept that the level of thyroid hormone in circulation increases during the prometamorphic period.

The results of Experiment 3 demonstrate that discs taken from animals at $E+1$ undergo rapid and complete resorption in control medium while discs taken from animals at $E-1$ and $E+0$ do not. This indicates that at about the time of E (beginning of climax) there is a rise in the level of thyroid hormone in circulation so that by $E+1$ these tissues have been exposed to or possess sufficient hormone to program them for complete resorption. These results are also in agreement with the general concept of thyroid activity during metamorphosis.

In accord with a quantitative interpretation for the mode of thyroid hormone action in metamorphosis, it can be seen that discs taken from animals in early prometamorphosis do not show resorption in vitro because they have not been exposed to a sufficient level of thyroid hormone to induce resorption. The resorption of discs taken from animals in late prometamorphosis and during climax is interpreted to mean that these tissues had been exposed to sufficient thyroid hormone during the prometamorphic and climax periods to program them, at the time of isolation, for resorption.

It is therefore suggested that the observed increase in sensitivity of tail discs to T_4 at later stages of metamorphosis results from the fact that such tissues possessed or had been previously exposed to more thyroid hormone than identical tissues isolated at earlier stages of metamorphosis. Therefore, treating tail discs from tadpoles in late prometamorphosis with thyroxine is actually equivalent to adding these amounts of hormone to the previously existing level so that resorption would result from the sum of these two factors.

The results of Experiment 4 provide evidence to support this interpretation. Early prometamorphic discs placed into T_4 for varying periods of time and then maintained in control medium produced a series of curves showing increased degrees of resorption dependent upon the exposure of these discs to T_4. The curve for discs placed into T_4 for 5 days simulates the curve produced by the controls in Experiment 2 (late prometamorphic discs in control medium) in which resorption occurred initially but then leveled off.

An interesting observation is that tail discs in vitro appear to be more sensitive to thyroxine than the same tissues in vivo. Tail discs give a pronounced response to 3, 9, 27, and 81 ppb T_4, whereas intact tadpoles generally require concentrations of hormone greater than 200 ppb of T_4. In addition, responses of tail tissue, in vivo, to even 81 ppb T_4 are very slow and do not reach maximal levels until higher concentrations of hormone are used(9). This increased sensitivity of tail discs to T_4 will be analyzed in greater detail later in this paper.

The Regional Response of Tail Fins to Thyroxine

Clausen(5) demonstrated that different regions of the tadpole's tail undergo histolysis at different rates. He found that there was an anterior to posterior gradient of response of these tissues with respect to their rate of histolysis (anterior regions responded most rapidly). It was felt that this problem could be approached by use of the tail disc preparation.

Procedure. *Experiment 5. The Regional Response of Tail Fins Taken from Animals at Early Prometamorphosis.* This experiment was performed to study the regional sensitivity of tail tissues to thyroxine and to find out whether different geographical areas of the tail respond identically to the same concentration of T_4. In this experiment, early prometamorphic tadpoles were used. Initially, five regions of the tail were tested: the anterior dorsal fin, the middorsal fin, tail tips, the midventral fin, and the anterior ventral fin. In subsequent runs of this experiment a sixth region was added, the portion of the ventral fin that includes the anal canal piece. After healing, these discs were placed into either 9 ppb T_4 or in Hanks medium. This experiment consisted of three runs, in the first run only the first five areas were tested but in the last two runs all six areas were examined.

Results. The results of the three runs of this experiment were similar and therefore these data have been grouped together. The controls from all six regions maintained themselves equally well and they are represented by a single line (Fig. 6). However, regional differences in sensitivity to thyroxine were observed for two areas, the anterior dorsal fin and the anal canal discs. These two regions exhibited a more rapid response to thyroxine than did the other regions tested. The anal canal discs were the most sensitive region to T_4 and by day 5 these discs had undergone 58 percent shrinkage; anterior dorsal fin discs had undergone 47 percent shrinkage, and the other four regions tested varied from 24 to 28 percent shrinkage. All of these differences between the anal canal discs and the less sensitive regions are statistically significant at the <0.001 level, while the differences between the anterior dorsal fin and these regions varies from <0.05 to <0.01.

Conclusions. The results of this experiment are in general agreement with the studies performed in the past in which it was shown that the anterior regions of the tail are more sensitive to T_4 than the more posterior regions. One of the reasons that this experiment was actually performed was to test whether dorsal and ventral fins could be used as equivalent test objects so that it would be possible to have each donor animal supply more discs for use in experiments. Based on the results of this experiment, it was determined that discs taken from either the dorsal

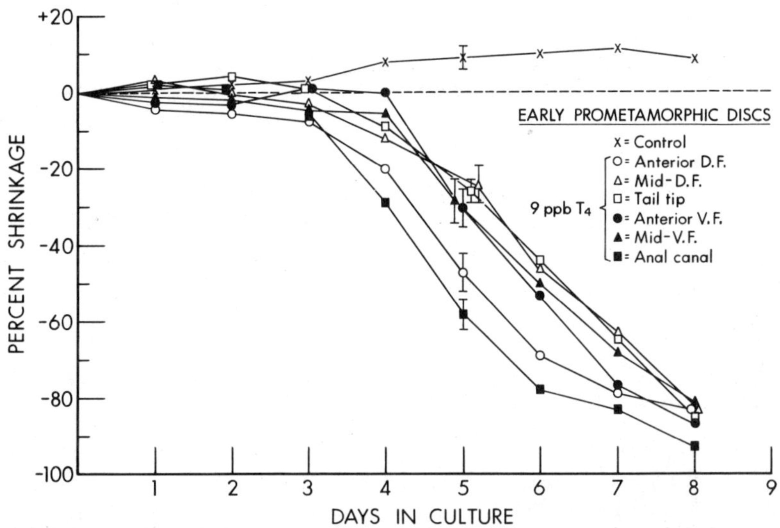

Fig. 6. Experiment 5, response of six regions of the tadpole's tail to thyroxine treatment.

or ventral fin could be used as equivalent test objects as long as the more anterior portions of both fins were not used.

This experiment also makes the point that the tail, like the entire animal, represents a mosaic with respect to its sensitivity to thyroid hormone. Here, however, we are not faced with explaining how thyroid hormone induces a different type of metamorphic response but rather, with determining what mechanism is responsible for inducing the same response, resorption, at different rates in the same tissue.

The Activity of Discs After a Long Period In Vitro

Discs cut from animals in early prometamorphosis and maintained in Hanks medium do well for long periods of time; based upon their general appearance and maintenance of size. The question was raised: How would discs that had been in culture for long periods of time respond to treatment with thyroid hormone?

Procedure. *Experiment 6. The Response of Discs to Thyroxine after Long Periods in Vitro.* The 50 control discs from the first run of Experiment 5 were divided into two groups (25 discs per group). These discs had already been in culture for 10 days and Experiment 5 was terminated by this time. One group of discs was placed into 9 ppb T_4 and the other group continued to be maintained in Hanks medium.

Results. Discs maintained in vitro for long periods are still capable of responding to thyroxine by undergoing resorption (Fig. 7). These discs responded in a predictable fashion based on the concentration of thyroxine used. The rate of resorption of these discs was similar to that of the original experimental discs (Experiment 6) right after healing. In Experiments 6 and 7, thyroxine-induced resorption of tail discs was completed after approximately 8 days.

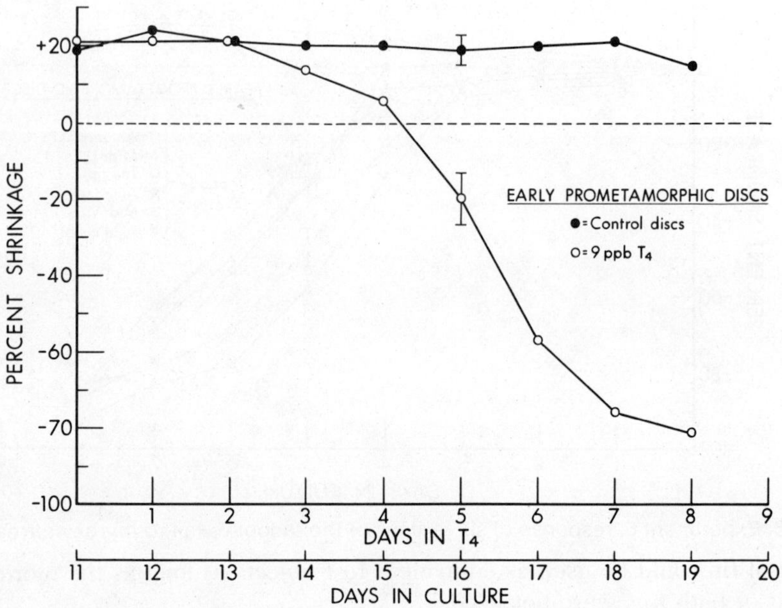

Fig. 7. Experiment 6, early prometamorphic discs maintained in control medium for 10 days prior to treatment with thyroxine.

Conclusion. It has been demonstrated that discs survive in Hanks medium for extended periods of time. At the present time nothing is known about the biochemical status of discs during this period. The results of this experiment indicate that discs maintained in culture for 11 days are still capable of responding to the addition of T_4 to their medium by undergoing resorption. This resorption follows the same pattern observed for discs treated with the same concentration of hormone right after healing. Therefore, it may be suggested that discs maintained in vitro, under the conditions of these experiments, are still capable of performing the biochemical processes involved in resorption. It would be of interest to know more about the metabolic status of these tissues in vitro. Another point of interest is to consider how these discs are capable of maintaining themselves in vitro, in view of the simplicity of the culture medium. It may be that since most of the disc is composed of loose connective tissue this material possesses a pool of nutrients sufficient to maintain the relatively few cells that are present in a disc. Also, the thinness of discs could allow for adequate gas exchange.

Thyroxine Interactions With Pituitary Grafts and Anterior Pituitary Hormone Preparations

This study was undertaken to examine the nature of the inhibitory action of prolactin and pituitary grafts on metamorphosis; that is, do prolactin and pituitary grafts inhibit the action of thyroxine at the level of target tissues? Specifically, can prolactin and pituitary grafts inhibit thyroxine-induced tail resorption in vitro?

Procedure. *Experiment 7. Inhibition of T_4-Induced Tail Resorption by Tadpole Pituitaries.* Tadpoles in early prometamorphosis were decapitated and their pituitary glands were rapidly dissected out and were implanted into discs(7). Cerebrum-implanted and sham-operated discs were used as controls. In addition, pituitaries were split into two parts, the anterior lobe and the posterior lobe, and these pieces were implanted into separate discs as a means of evaluating the contribution of each of these lobes to the observed effect. The concentrations of thyroxine tested were 3 and 6 ppb.

Experiment 8. Inhibition of Spontaneous Resorption of Climax Discs by Pituitary Grafts. This experiment was performed to examine the capacity of an implanted pituitary to inhibit the spontaneous resorption of discs cut from animals at $E+1$ of climax. A set of three discs (one pituitary, one cerebrum, and one sham-operated) was prepared from each dorsal fin and these discs were maintained in Hanks medium. The pituitaries and tails used in this experiment were both taken from the same $E+1$ donors.

Experiment 9. Inhibition of Spontaneous Tail Resorption by Pituitary Hormone Preparations. This experiment was designed to test the effect of three anterior pituitary hormone preparations in inhibiting the spontaneous tail resorption demonstrated by discs cut from tadpoles at $E+1$. Three discs were cut from the dorsal fin of a tadpole at $E+1$ and each disc was placed, singly, into a solution containing either ovine, prolactin, growth hormone (GH), follicle-stimulating hormone (FSH), or Hanks medium. The sequence of assignment of discs was changed from

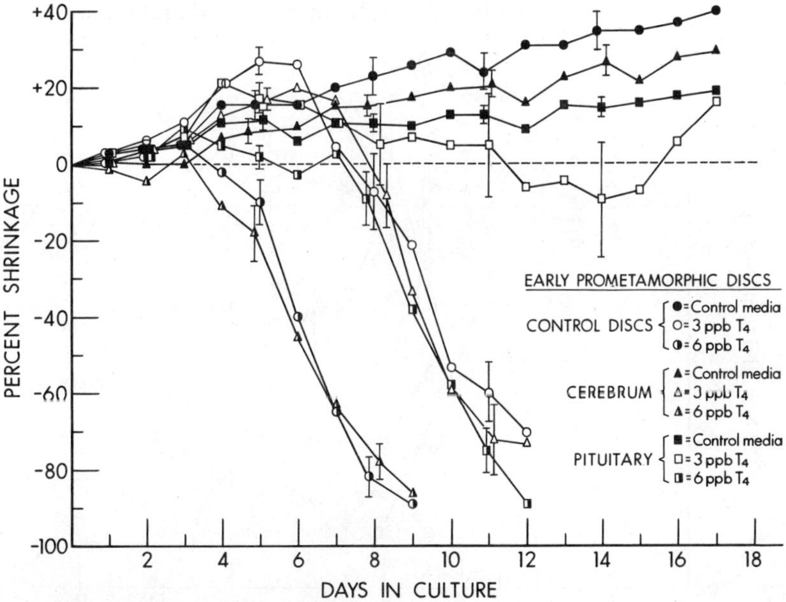

Fig. 8. Experiment 7, discs taken from animals in early prometamorphosis. Discs contain pituitary or cerebrum implants, the last group are sham-operated controls. Discs are maintained in one of two concentrations of thyroxine (3 ppb or 6 ppb) or in Hanks medium. From Derby and Etkin. *J. Exp. Zool.*, 169(1):1-8, 1968.

animal to animal. This experiment was conducted in two groups with nine animals in each group. In group one, growth hormone was tested and compared to prolactin, while in group two, FSH was tested. FSH was used in this study as a means of testing the specificity of the responses observed with group one.

The hormones used were obtained as a gift from the Endocrine Section of NIH. Ovine prolactin (NIH=p.s.=8), ovine GH (NIH=GH=s8), and ovine (NIH=FSH-1).

Results. *Experiment 7. Inhibition of T_4-Induced Resorption by Tadpole Pituitaries.* The results of this experiment indicate that all of the controls maintained in Hanks survived equally well in culture (Fig. 8). Sham-operated and cerebrum-implanted discs placed into 3 ppb T_4 completed resorption by day 12 in culture. Pituitary implanted discs placed into 3 ppb T_4 maintained their size during this period. Sham-operated and cerebrum-implanted discs placed into 6 ppb T_4 completed resorption by day 9 in culture while pituitary discs in 6ppb T_4 demonstrated a three-day delay in the initiation of and completion of resorption so that these discs were resorbed by day 12. From these results it would appear that the implanted pituitary had the capacity to block the action of approximately 3 ppb of T_4. In addition, the implanted anterior lobe of the pituitary alone is also capable of inhibiting thyroxine-induced resorption (Fig. 9).

Experiment 8. Inhibition of Spontaneous Resorption of Climax Discs by Pituitary Grafts. The results of this experiment show that an implanted pituitary can also inhibit the spontaneous resorption of tail discs cut from animals at $E+1$ (Fig. 10). All discs undergo some initial resorption, part of this occurs regularly and is associated with the process of healing but only unimplanted and cerebrum-implant-

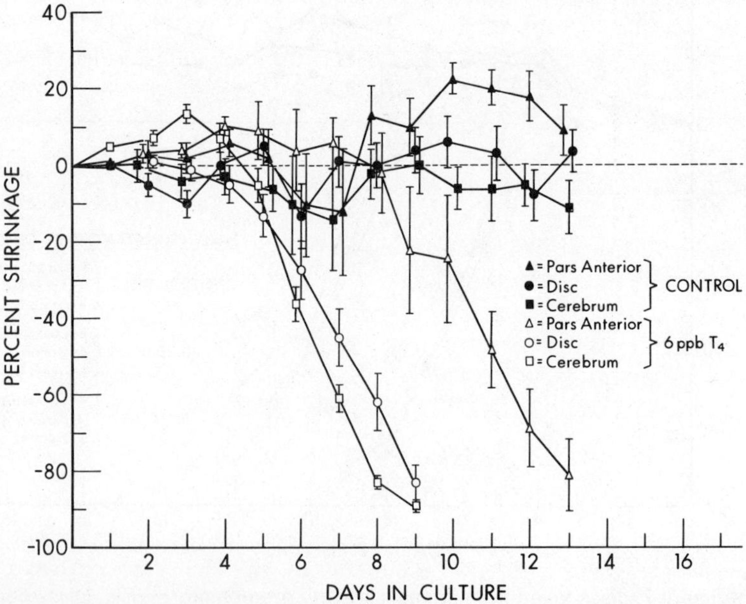

Fig. 9. Experiment 7, early prometamorphic discs containing pars anterior or cerebrum implants and sham-operated controls maintained in 6 ppb T_4 or in control medium. From Derby and Etkin. *J. Exp. Zool.*, 169(1):1-8, 1968.

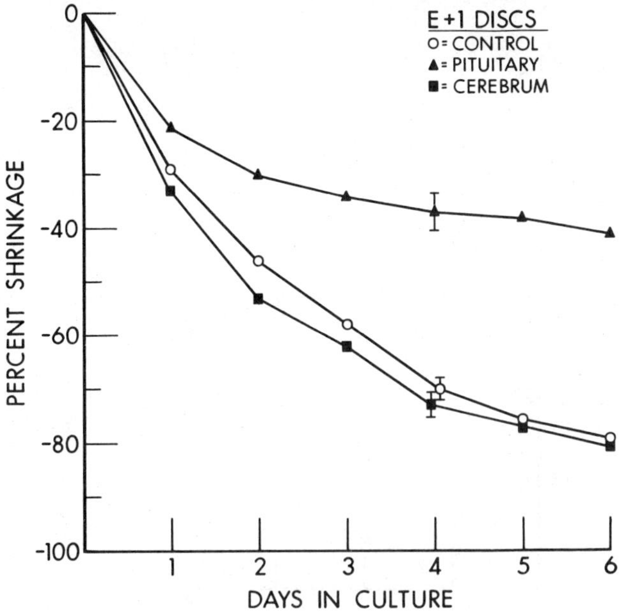

Fig. 10. Experiment 8, discs taken from tadpoles at $E+1$. Each tail supplied three discs; a pituitary implant, a cerebrum implant, and a sham-operated disc. Discs were maintained in control medium. From Derby and Etkin. J. Exp. Zool., 169(1):1-8, 1968.

ed discs completed resorption. Discs containing pituitary implants show initial resorption but this resorption levels off. At day 6 in culture, pituitary implanted discs have only undergone 41 percent reduction in size while the other two groups have undergone 80 percent reduction.

As an interesting sidelight to these experiments it should be mentioned that these discs possess pigment cells and discs containing pituitary glands demonstrate expansion of these pigment cells. The rapidity of this effect indicates that substances released by these implants (presumably MSH, possibly ACTH) are capable of rapid diffusion through these tissues. Further experimentation has indicated that the implanted pituitary is releasing hormone continuously and that these hormones are also destroyed rapidly in these tissues. This effect of the implanted pituitary gland on the pigment cells present in a disc provides a visible indication of the functional integrity of the graft.

Experiment 9. Inhibition of Spontaneous Tail Resorption by Pituitary Hormone Preparations. Treatment of discs cut from animals at $E+1$ with prolactin and GH demonstrated that both of these hormones reduce the degree of spontaneous resorption normally exhibited by such discs (Fig. 11). However, the effect of prolactin is more uniform than that of GH for the variance in resorption exhibited by discs treated with GH is greater than that for discs treated with prolactin. With prolactin, the difference between experimental and control discs gives a p value of <0.01 while the GH effect has a p value of of <0.05. Discs treated with FSH by contrast underwent resorption at the same rate as discs treated with Hanks medium.

It should be mentioned that treatment of discs with these hormone preparations also induced expansion of the pigment cells present in discs. This effect may be

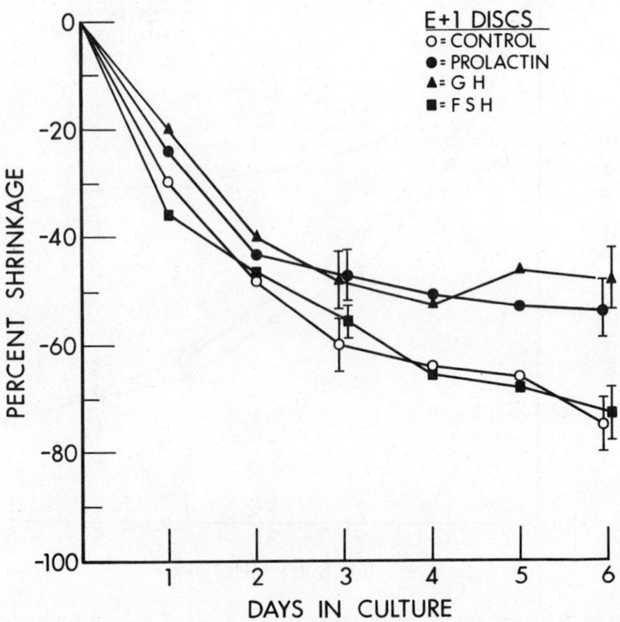

Fig. 11. Experiment 9, discs taken from tadpoles at $E+1$ maintained in solutions containing prolactin, GH, FSH, or control medium. From Derby and Etkin. *J. Exp. Zool.*, 169(1):1-8, 1968.

due to contamination of these hormones by MSH or ACTH. Expansion often persists overnight, and in some cases varying degrees of expansion may continue for as long as 24 to 36 hours. This effect does not continue for days as in the case of implanted pituitary glands. Also, these hormones do not all appear to produce the same degree of expansion of the pigment cells; GH gives the greatest effect followed by prolactin and FSH.

Conclusion. These experiments were conducted to test the concept that a pituitary graft releases a substance (possibly prolactin) that acts peripherally to inhibit the action of thyroxine. Such a concept was suggested by previous studies(11) in which pituitary glands were implanted into the tails of tadpoles near climax and retardation of metamorphosis occurred. Similarly(4,12,19), it was shown that injections of mammaliam prolactin could also inhibit metamorphic changes. It was anticipated that if the pituitary graft acted in this fashion in vivo, perhaps thyroxine-induced resorption of tail discs in vitro would be retarded.

The results of these experiments show that a pituitary implanted into a tail disc inhibits thyroxine-induced tail resorption. The inhibitory effect of this implant is quantitatively equivalent to reducing the level of thyroxine in the medium by about 3 ppb. These findings suggest that an antagonism exists between the actions of a pituitary factor, produced by the graft, and thyroid hormone on tail tissues. The status of these tissues at any stage of metamorphosis may be dependent upon the relative concentrations of these two hormones in the tadpole's circulation. Other lines of evidence also suggest that amphibian metamorphosis may be dependent upon the reciprocal action of both of these factors(12).

Pituitary implants were also capable of inhibiting the spontaneous resorption of discs cut from animals at $E+1$. It was proposed earlier that such discs were already programmed for rapid resorption based upon their previous exposure to thyroid hormone and to the level of thyroid hormone present in these tissues at this stage of metamorphosis. The question that arises is: How does the pituitary factor accomplish this inhibition? Perhaps it acts by binding or inactivating the thyroxine present in the tissues. It has been demonstrated that tail disc resorption is dependent upon the continuous presence of thyroid hormone; for if discs that are in the process of resorption are transferred from a solution of thyroxine to Hanks medium, resorption continues but levels off (Experiment 4). Another possible mechanism for the mode of action of the pituitary factor might be to suppress one or more of the changes induced by thyroxine at the biochemical level. For example; it has been demonstrated(21) that thyroxine leads to an increased rate of collagen destruction in tail tissues as compared to the rate of collagen synthesis that continues to occur. The pituitary factor might act by suppressing the increased rate of collagen destruction induced by thyroxine. It might also be that the pituitary factor does not interfere directly with the action of thyroid hormone but it may operate at other sites. In the example just cited, the pituitary factor might influence tail resorption by inducing an increase in the rate of collagen synthesis and as a result cancel out the effect of thyroxine induced collagen breakdown. Similar mechanisms could be proposed for other changes that have been reported to take place during tail resorption; increased levels of acid hydrolases, cathepsins, hyaluronidase, and changes in RNA synthesis(26).

The results of the experiment in which purified mammalian hormone preparations were used have not been helpful in clarifying the question as to the identity of the pituitary growth factor; specifically whether it is similar to mammalian prolactin or GH. In this study, both hormones were capable of inhibiting thyroxine-induced tail resorption. The fact that FSH does not exert this action indicates that this inhibition is probably specific and not simply the result of adding protein to these tissues.

It was mentioned earlier (but not discussed at the time) that early prometamorphic tail discs maintained in culture are more sensitive to T_4 than these tissues in vivo. Several explanations are possible. It may be that the intact animal requires higher levels of thyroid hormone because of the capacity of these animals to bind, metabolize, or excrete this substance. This capacity would probably be greatly reduced in tail discs in vitro. However, another possibility exists based on the role of the pituitary factor in metamorphosis. In the intact animal, this factor is probably released by the pituitary continuously. It may be that by isolating pieces of tail from the tadpole, the levels of pituitary factor in the tissues is now reduced, and as a result, an agent that interferes with the action of T_4 peripherally is no longer present. Therefore these tissues become more sensitive to thyroxine. These two explanations are not offered as alternatives, for the increased sensitivity of tail discs to T_4 in vitro could result from a combination of both of these factors.

A similar interpretation could be advanced to account for the difference in sensitivity of tail fins to T_4 at different stages of metamorphosis. Discs taken from animals in early prometamorphosis might be less sensitive to treatment with thyroxine than similar pieces taken from animals in late prometamorphosis because tissues from animals in early prometamorphosis were exposed to or possessed a higher level of prolactin. Therefore, higher levels of T_4 would be required to induce resorption of early prometamorphic discs as compared to discs taken from animals in late prometamorphosis. Finally, it appears reasonable to expect that the differences in sensitivity of tail discs to thyroxine at these two stages of metamorphosis result from the reciprocal changes that are believed to be occurring in the levels of both of these hormones, that is, the decreasing concentration of prolactin and the increasing levels of thyroxine.

Finally, the results of this study demonstrated that pituitary implants were more effective in inhibiting tail resorption than are mammalian hormones. However, such results are difficult to interpret.

SUMMARY AND CONCLUSIONS

An in vitro system, consisting of tadpole tail fins maintained in Hanks balanced salt solution, has been employed to study the effects of hormones (prolactin and thyroxine) on metamorphic change. The results of these experiments demonstrate that a quantitative relationship exists between thyroxine and its action in inducing tail resorption. The response of tail fins to thyroxine varies, over a given range, directly with the concentration of thyroxine tested. The action of thyroid hormone in this process is dependent upon the continuous presence of this hormone.

Tail discs change in their sensitivity to thyroxine at progressive stages of metamorphosis; sensitivity to hormone increases during the prometamorphic period. The results support the concept that the increased sensitivity of tail discs to thyroxine is due to the previous in vivo exposure of these tissues to a low but rising level of thyroid hormone during prometamorphosis and a sudden increase in hormone one to two days before the emergence of the forelimbs. In addition, changes in sensitivity of tail discs to thyroxine might reflect changes in the level of a pituitary factor (prolactin) present in these tissues at the time of isolation.

Pars anterior of the tadpole's pituitary implanted into a tail disc release a substance that interferes with the action of thyroxine in vitro. A quantitative, reciprocal relationship exists between the pituitary factor and thyroid hormone with respect to their effects on tail resorption.

Simultaneous treatment of tail discs with prolactin or growth hormone and thyroxine also results in inhibition of thyroxine-induced tail resorption. These results make it difficult to determine what the substance is that the tadpole's pituitary is releasing. Is it a prolactin-like or GH-like molecule? The results of the present investigation and of other works would favor the concept that the pituitary factor is a prolactin-like hormone. Finally, it is proposed that the interaction of these two hormones is important in controlling amphibian metamorphosis.

REFERENCES

1. Ashley, H., Katti, P., and Evans, E. Urea excretion in the Bullfrog tadpole: Effect of temperature, metamorphosis, and thyroid hormone. Develop. Biol., 17:293–307, 1968.
2. Berman, R. H., Strohman, R. C., Nicoll, C. S., and Bern, H. Growth-promoting effects of mammalian prolactin and growth hormone in tadpoles of *Rana catesbeiana*. Amer. Zool., 4:324 (Abstr.), 1964.
3. ——— Bern, H. A., Nicoll, C. S., and Strohman, R. C. Growth-promoting effects of mammalian prolactin and growth hormone in tadpoles of *Rana catesbeiana*. J. Exp. Zool., 156:353–360, 1964.
4. Bern, H. A., Nicoll, C. S., and Strohman, R. C. Prolactin and tadpole growth. Proc. Soc. Exp. Biol. Med., 126:518–520, 1967.
5. Clausen, H. J. Rate of histolysis of anuran tail skin and muscle during metamorphosis. Biol. Bull., 59:199–210, 1930.
6. Derby, A. An *in vitro* quantitative analysis of the response of tadpole tissue to thyroxine. J. Exp. Zool., 168:147–156, 1968.
7. ——— and Etkin, W. Thyroxine induced tail resorption *in vitro* as affected by anterior pituitary hormones. J. Exp. Zool., 169(1):1–8, 1968.
8. Etkin, W. Growth and resorption phenomena in anuran metamorphosis. Physiol. Zool., 5:275–300, 1932.
9. ——— Metamorphosis. *In* Physiology of the Amphibian, Moore, J. A., ed., New York, Academic Press, Inc., pp. 427–468, 1964.
10. ——— Hormonal control of metamorphosis. *In* Metamorphosis, Etkin, W., and Gilbert, L. I., eds., New York, Appleton-Century-Crofts, pp. 313–348, 1968.
11. ——— Derby, A., and Gona, A. Prolactin-like antithyroid action of pituitary grafts in tadpoles. 5th Int. Symp. Comp. Endocr. Gen. Comp. Endocr., Suppl. 2:253–259, 1969.
12. ——— and Gona, A. Antagonism between prolactin and thyroid hormone in amphibian development. J. Exp. Zool., 165:249–258, 1967.
13. ——— and Lehrer, R. Excess growth in tadpoles after transplantation of the adenohypophysis. Endocrinology, 67:457–466, 1960.
14. Everett, J. W. Functional corpora lutea maintained for months by autografts of rat hypophysis. Endocrinology, 58:786–796, 1956.
15. ——— The control of the secretion of prolactin. *In* The Pituitary Gland, Harris, G. W., and Donovan, B. T., eds., Berkeley, Univ. Calif. Press, Vol. II, pp. 166–194, 1966.
16. Frieden, E. Biochemistry of amphibian metamorphosis. *In* Metamorphosis, Etkin, W., and Gilbert, L. I., eds., New York, Appleton-Century-Crofts, pp. 349–398, 1968.
17. ——— Wahlber, A., and Howard, E. Temperature control of the response of tadpoles to triiodothyronine. Nature (London), 205:1173–1176, 1965.
18. Gona, A. Prolactin as a goitrogenic agent in amphibia. Endocrinology, 81:748–754, 1967.
19. ——— Radioiodine studies on prolactin action in tadpoles. Gen. Comp. Endocr., 11: 278–283, 1968.
20. Kollros, J. Mechanisms of amphibian metamorphosis: Hormones. Amer. Zool., 1:107–114, 1961.
21. Lapiere, C. M., and Gross, J. Animal collagenase and collagen metabolism. *In* Mechanisms of Hard Tissue Destruction. Publ. No. 75 AAAS, Washington, D.C., pp. 663–694, 1963.
22. Lynn, W. G., and Wachowski, H. E. The thyroid gland and its functions in coldblooded vertebrates. Quart. Rev. Biol., 26:123–168, 1951.
23. Rémy, C., and Bounhiol, J. J. Croissance exagérée des têtards de crapaud accoucher entiers ou privés de leurs hypophyse et subissant traitment par la prolactine. C.R. Soc. Biol. (Paris), 159:1532, 1965.

24. ——— and Bounhiol, J. J. Gigantisme expérimental obtenu par intervention chirugicale et traitements hormonaux chez le têtard du crapaud accoucher. Ann. Endocr. (Paris), 27: Suppl., 377–382, 1966.
25. Weber, R. Induced metamorphosis in isolated tails of *Xenopus laevis*. Experientia, 18: 84–85, 1962.
26. ——— Biochemistry of amphibian metamorphosis. *In* Biochemistry of Animal Development, Weber, R., ed., New York, Academic Press, Inc., Ch. 5, pp. 227–295, 1967.

22

Histochemical Patterns in the Tail of the Metamorphosing Tadpole*†

Jane C. Kaltenbach

Department of Biological Sciences,
Mount Holyoke College,
South Hadley, Massachusetts

Resorption of the tail of the metamorphosing tadpole has long been studied with respect to morphological changes, but only recently has attention been focused on biochemical aspects of tail regression (e.g., see reviews 1,2). Studies have emphasized the increase in lysosomal enzyme activity which accompanies tail shortening. However, only a few reports have been concerned with the distribution of such enzymes within the tissues and cells of the tail(3,4). The purpose of the present histochemical investigation was twofold: (1) to identify localizations of various enzymes and carbohydrates within specific tissues and regions of the tail of *Rana pipiens* tadpoles; and (2) to determine changes in distribution and in activities or amounts of such substances during normal and thyroxine-induced metamorphosis.

Proximal and distal portions of tails were removed from *R. pipiens* larvae in 10 to 15 different developmental stages, stages according to Taylor and Kollros(5), and from young tadpoles immersed in 1 x 10^{-7} M solution of DL thyroxine for 6 to 10 days, i.e., until considerable tail shortening had occurred. Tails were frozen in OCT compound (Lab Tek Co.) or liquid nitrogen; cross sections, occasionally longitudinal ones, were cut in a cryostat. Prior to freezing tails were often embedded in rat liver which was used as a supporting substance and as a control for various enzyme reactions; some tails were fixed in cold, 10 percent neutral formalin containing sucrose (10 g/100 ml formalin) for better localization of acid and alkaline phosphatase and esterase.

*This investigation was supported in part by a faculty grant from Mount Holyoke College.

†Sincere thanks are extended to the following students for their important contributions to this investigation: Miss Anne Fry, esterase and alkaline phosphatase; Miss Beatrice Bacher, acid phosphatase; Miss Viiu Leius, acid phosphatase and magnesium-activated ATPase(32); Miss Elizabeth Teggiachi; polysaccharides(33). The able technical assistance of Miss Viiu Leius and Miss Fatemeh Kaveh is also greatly appreciated.

Standard histochemical methods were used to identify the following enzymes and carbohydrates: acid phosphatase, pararosanilin technique of Barka(6); nonspecific esterase, pararosanilin technique of Barka and Anderson(7); aminopeptidase, L-leucyl-beta-napthylamide hydrochloride method of Nachlas et al.(8); alkaline phosphatase, azo dye method of McKay et al.(9); magnesium-activated ATPase[1], thiamine pyrophosphatase, and 5′-nucleotidase, lead method of Wachstein and Meisel(10); calcium-activated ATPase, Gomori method modified by Padykula and Herman(11); succinic dehydrogenase, nitro-BT method of Nachlas et al.(12); polysaccharides, PAS technique of McManus(13); and acid mucopolysaccharides, Steedman's alcian blue method modified by Mowry(14) and the toluidine blue procedure of Lillie(15). Times of incubation varied from 5 to 30 minutes. For controls, sections were incubated in media lacking substrate. In addition, control sections for nonspecific esterase were exposed to 10^{-5} M eserine before and during incubation in the substrate mixture in order to inhibit cholinesterases(7). Polysaccharides and acid mucopolysaccharides were also identified on Carnoy-fixed, paraffin-embedded tissues.

Histochemical reactions in the tails of tadpoles in various larval stages were studied; in particular, comparisons were made between patterns of localization of enzymes and carbohydrates in growing tails of young larvae and in resorbing tails of metamorphosing larvae. Tail resorption is a relatively rapid process, beginning at metamorphic stage XXI and lasting less than two weeks, a small portion of the entire larval period (Figs.1 and 2).

The staining patterns for acid phosphatase, nonspecific esterase, and aminopeptidase were similar (Figs. 3, 4, and 5). During early larval stages enzyme activity was detected only in an occasional epidermal and connective tissue cell (Fig. 4a); (a positive acid phosphatase reaction was also found in nerve cell bodies of spinal ganglia). However, as the tail shortened during metamorphic climax, activity of these enzymes increased dramatically. Reactions became intense in the epidermis and connective tissue, especially in phagocytes which developed in large numbers near resorbing muscle cells (Figs. 3a, b; 4c, d, e) and along the dorsal and ventral surfaces of the fins (Figs. 3c, d; 5); no enzyme activity was found in intact striated muscle fibers. A long-standing question regarding the origin of phagocytes

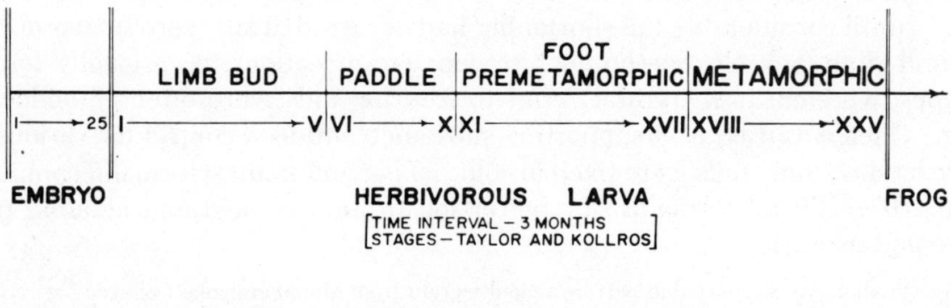

Fig. 1. Development of *Rana pipiens.*

[1]The following abbreviations are used in this paper: ATPase, adenosine triphosphatase; PAS, periodic acid Schiff.

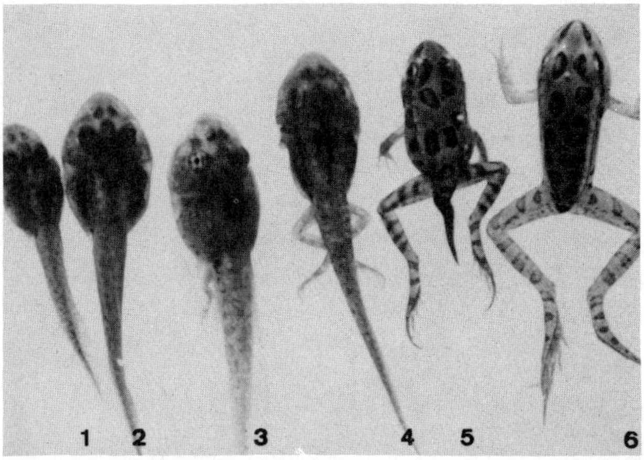

Fig. 2. Photograph of *R. pipiens* larvae in different stages of development. 1, limb bud stage; 2, paddle stage; 3, foot stage; 4, 5, 6, metamorphic stages XVIII, XXII, XXV. Stages according to Taylor and Kollros(5).

near muscle fibers (Fig. 6), e.g., whether from muscle cells themselves (16), blood cells (see discussion in 16) or from mesenchymal cells(4), was not investigated in the present study.

In addition to the above locations, (Fig. 6) esterase activity was also found in the region of motor end plates, i.e., at the ends of muscle fibers near the connective tissue septa (Fig. 4b). The reaction in this location did not appear after exposure of the sections to eserine, an inhibitor of cholinesterases. In contrast epidermal and connective tissue cells were still reactive after treatment with eserine, a characteristic of nonspecific esterase.

The present results are in accord with histochemical reports for acid phosphatase and cathepsin-like esterase in *Xenopus* tail rudiments(3,4) and for aminopeptidase in rats, i.e., in areas of active phagocytosis associated with damaged muscle cells(17). Moreover, the present study substantiates biochemical reports of increases in acid phosphatase(18 to 22) and cathepsin(21 to 24) in resorbing tails during normal or thyroxine-induced metamorphosis.

With the alkaline phosphatase technique, mesenchymal and epidermal cells, especially in the fins, were reactive both in early larval and metamorphic stages (Fig. 7). Striated muscle was negative, although its microvasculature displayed weak activity. During normal and induced metamorphosis the epidermis over the entire tail became positive and the phosphatase reaction in general became stronger, as indicated by the effectiveness of shorter incubation times. This was in accord with biochemical reports which showed a rise in alkaline phosphatase activity in regressing tails(18,19). However, the increase in activity seemed to be less dramatic than that of acid phosphatase, esterase, or aminopeptidase.

Magnesium-activated ATPase was localized in the epidermis and connective tissue but not in striated muscle. Yet the distribution pattern was somewhat different than that of the aforementioned enzymes. In early larval stages, activity was restricted to the subepidermal connective tissue, the outer sheath of the notochord, the septa (myocommata) between myomeres, and a few epidermal cells (Fig. 8a).

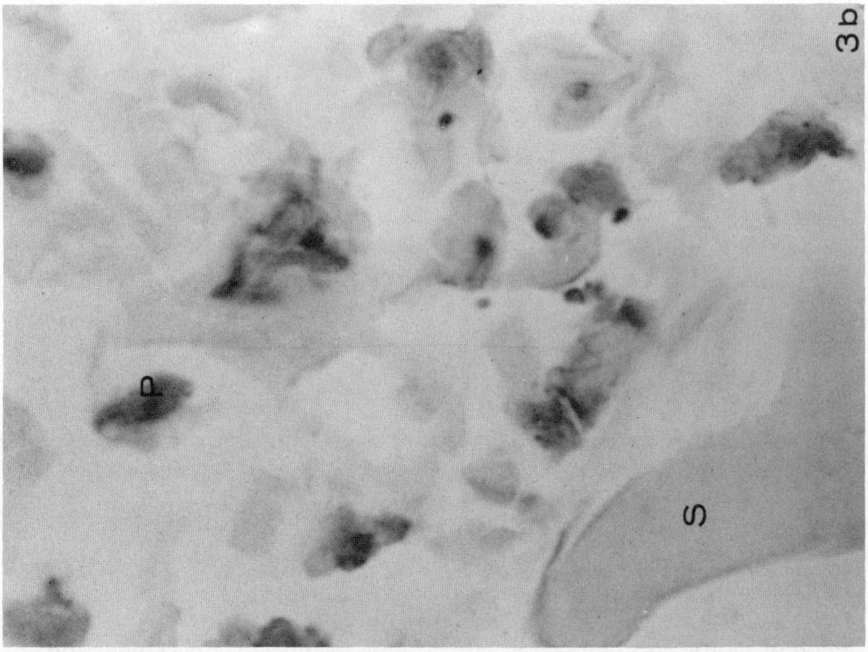
3b
P
S

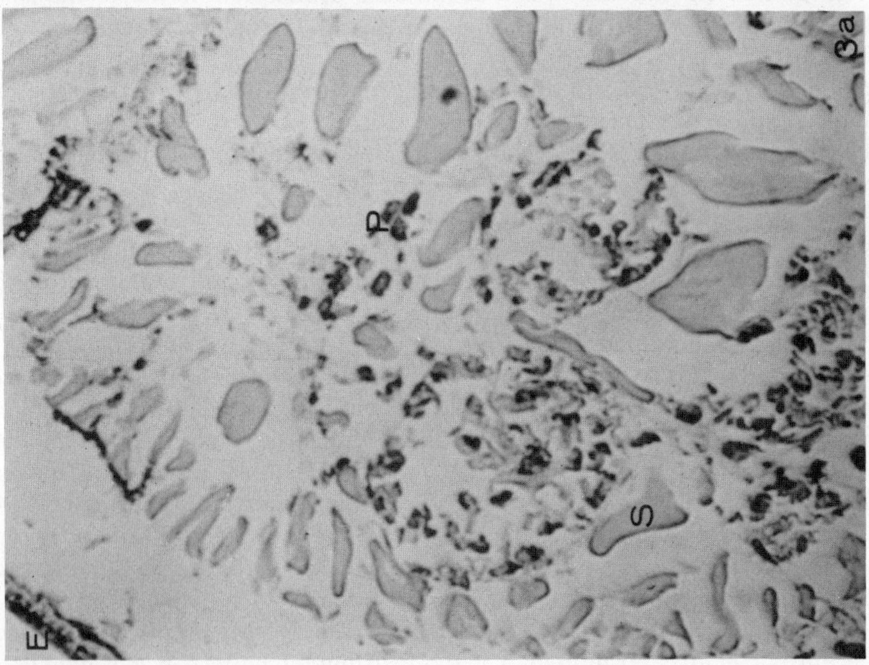
3a
P
S
E

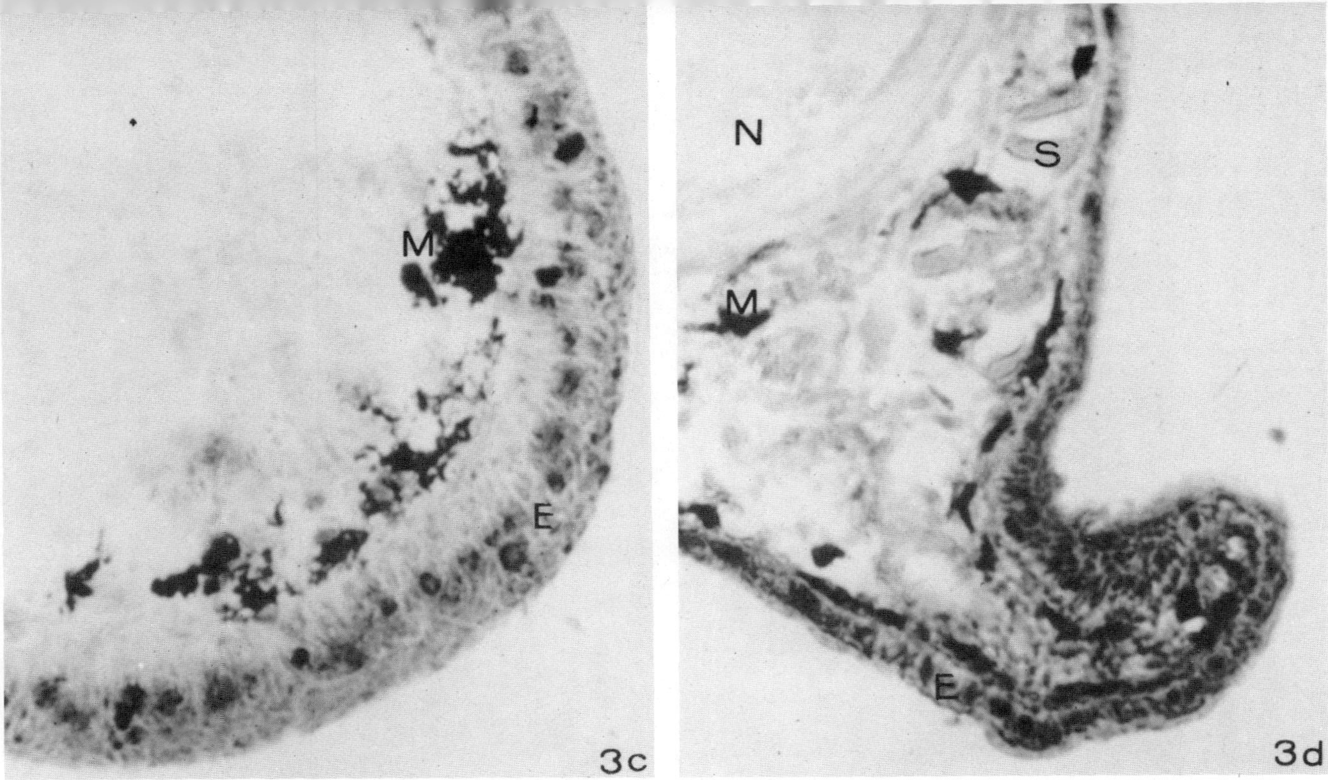

Fig. 3. Cross sections of tails showing reactions for acid phosphatase. *a, b,* from slide of B. Bacher; *c, d,* from slides of V. Leius. *E*, epidermis; *M,* melanin; *N,* notochord; *P*, phagocyte; *S*, striated muscle. *a.* Stage XXII. Epidermis and phagocytes show enzyme activity. Striated muscle cells are negative. *b.* Stage XXII. Higher magnification of a portion of Figure 3*a*. Phagocytes give a positive reaction. Striated muscle is negative. *c.* Stage XXII. Epidermal cells of the ventral fin show acid phosphatase activity. *d.* Thyroxine-treated.Epidermal cells and some cells in the connective tissue of the resorbing ventral fin show a positive acid phosphatase reaction. Striated muscle cells and notochord are negative.

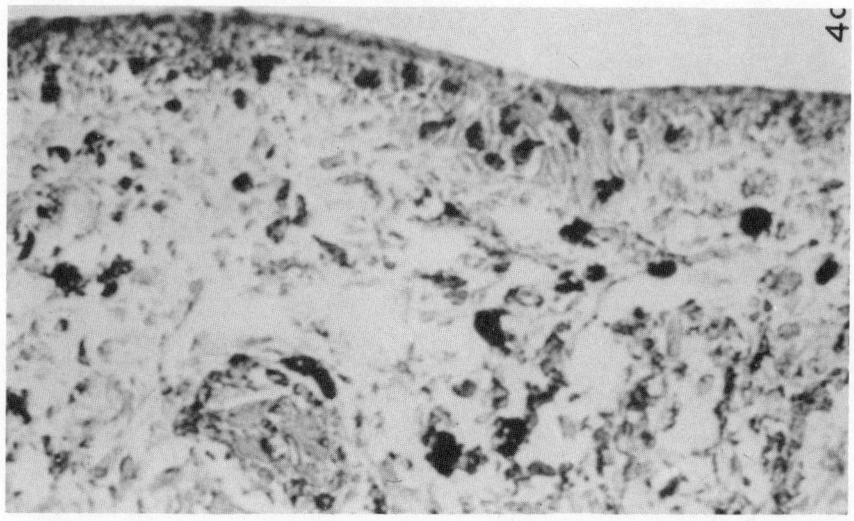
4c
4a
4b

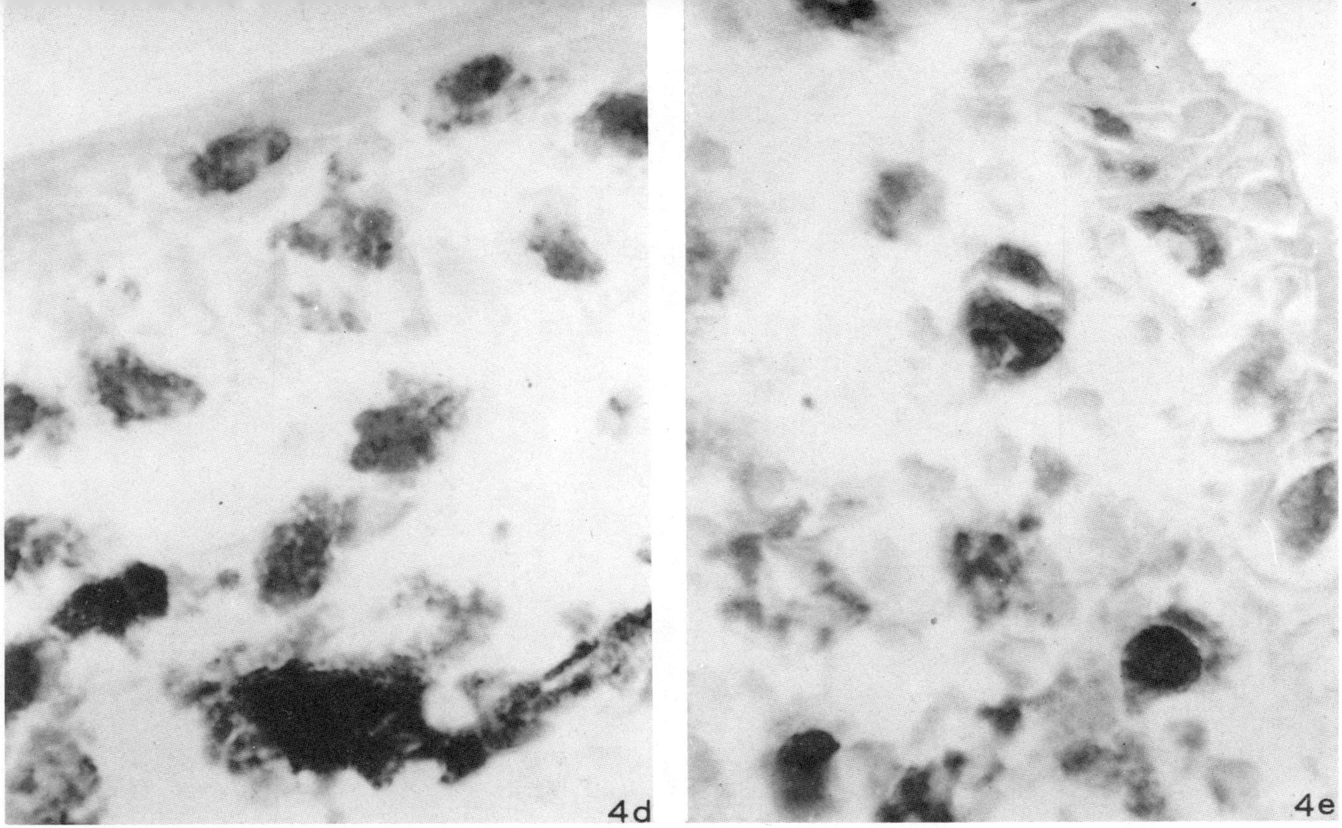

Fig. 4. Cross sections of tails showing reactions for esterases. Unpublished photographs of A. Fry. *a.* Stage IV. Only an occasional epidermal cell has nonspecific esterase activity. *b.* Stage IV. Enzyme reaction at ends of striated muscle fibers in region of motor end plates was attributed to cholinesterases; after treatment with eserine, no reaction appeared in this location. *c.* Stage XXIII. Many epidermal cells as well as phagocytic cells near muscle fibers react positively for nonspecific esterase. *d.* Stage XXIV. Positive reaction for nonspecific esterase is located in epidermal cells and underlying phagocytes. *e.* Thyroxine-treated. Similar to Figure 4*d.*, i.e., a positive reaction for nonspecific esterase is located in epidermal cells and underlying phagocytes.

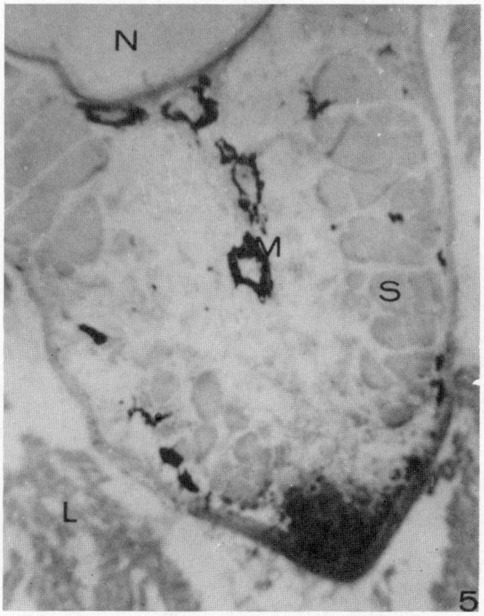

Fig. 5. Cross section of ventral portion of tail, embedded in rat liver for support. Thyroxine-treated. A positive aminopeptidase reaction is found in the epidermis and connective tissue of the resorbing fin. Striated muscle (*S*) and notochord (*N*) are negative. *L*, rat liver; *M*, melanin.

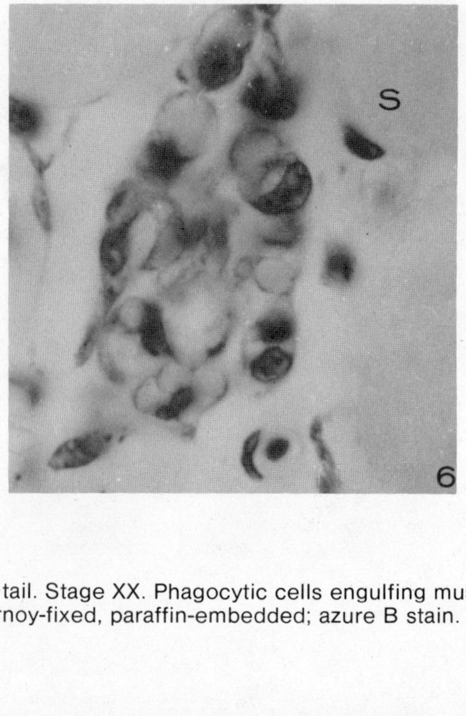

Fig. 6. Cross section of tail. Stage XX. Phagocytic cells engulfing muscle fragments. *S*, intact striated muscle cell. Carnoy-fixed, paraffin-embedded; azure B stain. From unpublished study of Kaltenbach.

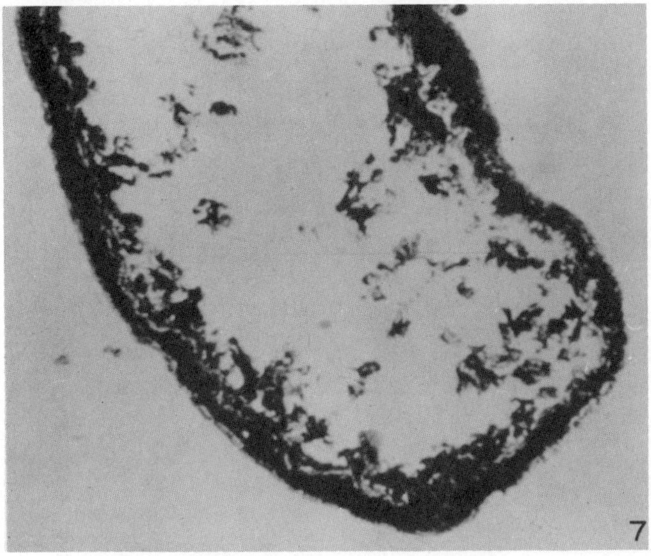

Fig. 7. Cross section of ventral fin of tail. Stage XVIII. Alkaline phosphatase activity is located in the epidermis and in scattered mesenchymal cells. From slide of A. Fry.

However, as tail resorption proceeded during both normal and induced metamorphosis, a positive reaction also appeared in the connective tissue between individual muscle fibers, in regions of phagocytic cells near degenerating muscle, and in many epidermal cells (Figs. 8b, c). Earlier reports indicated that activity increased slightly in homogenates of resorbing tails from triiodothyronine-treated tadpoles(25) but decreased in similar preparations of tails from normally metamorphosing tadpoles(25,26). The staining pattern for thiamine pyrophosphatase was very similar to that of Mg^{++}-activated ATPase.

Very little 5′-nucleotidase activity (pH 7.2) was demonstrable even in tails of metamorphosing tadpoles (Fig. 9). Activity was only observed in the epidermis covering the dorsal and ventral surfaces of the resorbing fins (Figs. 9a, b) and in the walls of some blood vessels (Fig. 9c). Yet the supporting pieces of rat liver in which the tails were embedded gave a strong reaction for this enzyme. Two 5′-nucleotidases probably exist; a neutral one involved in transport processes, and an acid one associated with lysosomes(27). Localization of the latter one remains to be determined in the tail; it should be interesting to compare its distribution with that of the neutral 5′-nucleotidase and with acid phosphatase.

In contrast to the localization of the enzymes discussed so far, calcium-activated ATPase and succinic dehydrogenase were present in striated muscle fibers. Throughout development, i.e., during early larval as well as metamorphic stages, all muscle cells stained more or less uniformly for the ATPase (Figs. 10a, b). Connective tissue at the edges of resorbing fins and in short tail rudiments of metamorphosing tadpoles also gave a positive ATPase reaction (Figs. 10a, b). On the other hand, succinic dehydrogenase was found exclusively in striated muscle, the

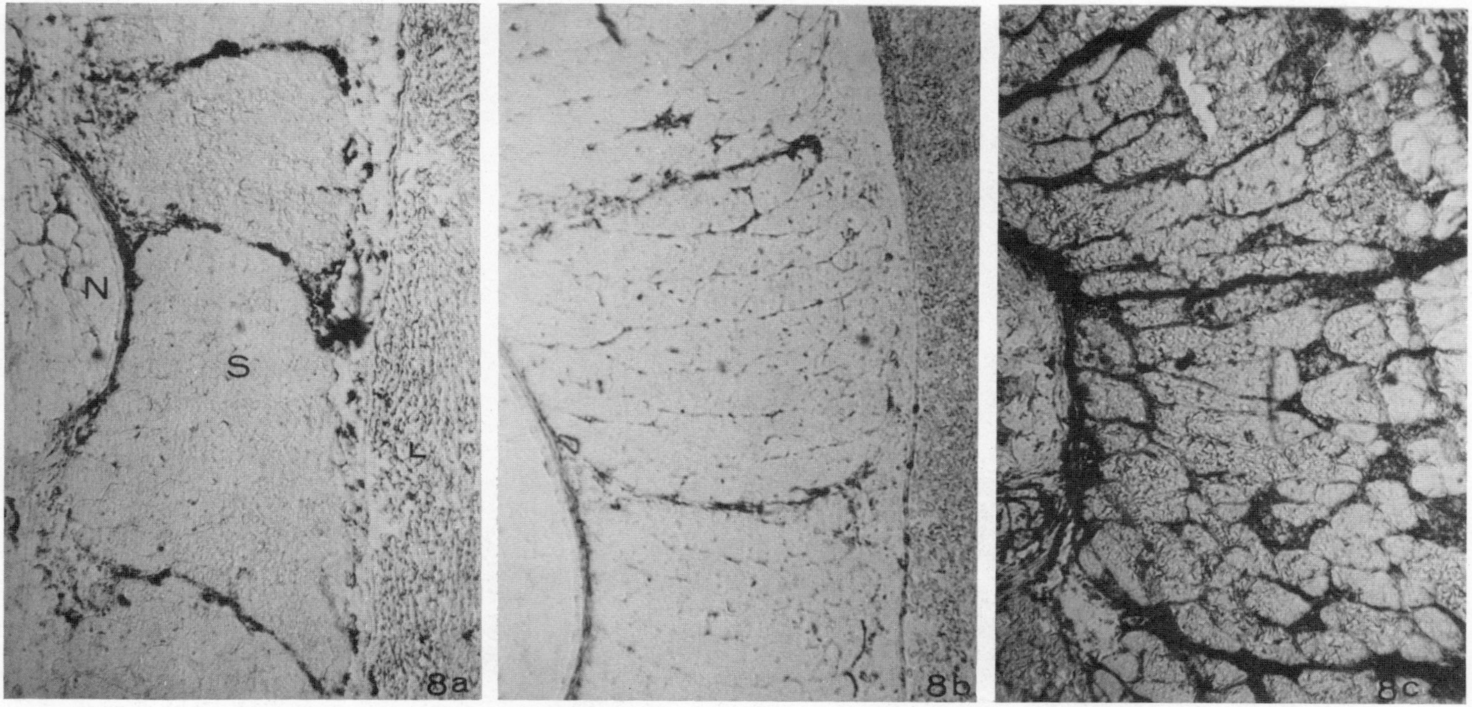

Fig. 8. Cross sections of tails showing reaction for magnesium-activated ATPase. From slides of V. Leius. *a.* Stage IX. Enzyme activity is located mainly in the connective tissue sheath around the notochord and in the septa between bundles of muscle cells. The rat liver (*L*) in which the tail is embedded also gives a positive reaction. Striated muscle cells (*S*) and notochord (*N*) are negative. *b.* Thyroxine-treated. Enzyme activity is localized as in stage IX; in addition activity occurs between individual muscle cells. *c.* Stage XXIII. Enzyme localization is similar to that in Figure 3*b;* activity is also found in regions in which connective tissue has replaced degenerating muscle cells.

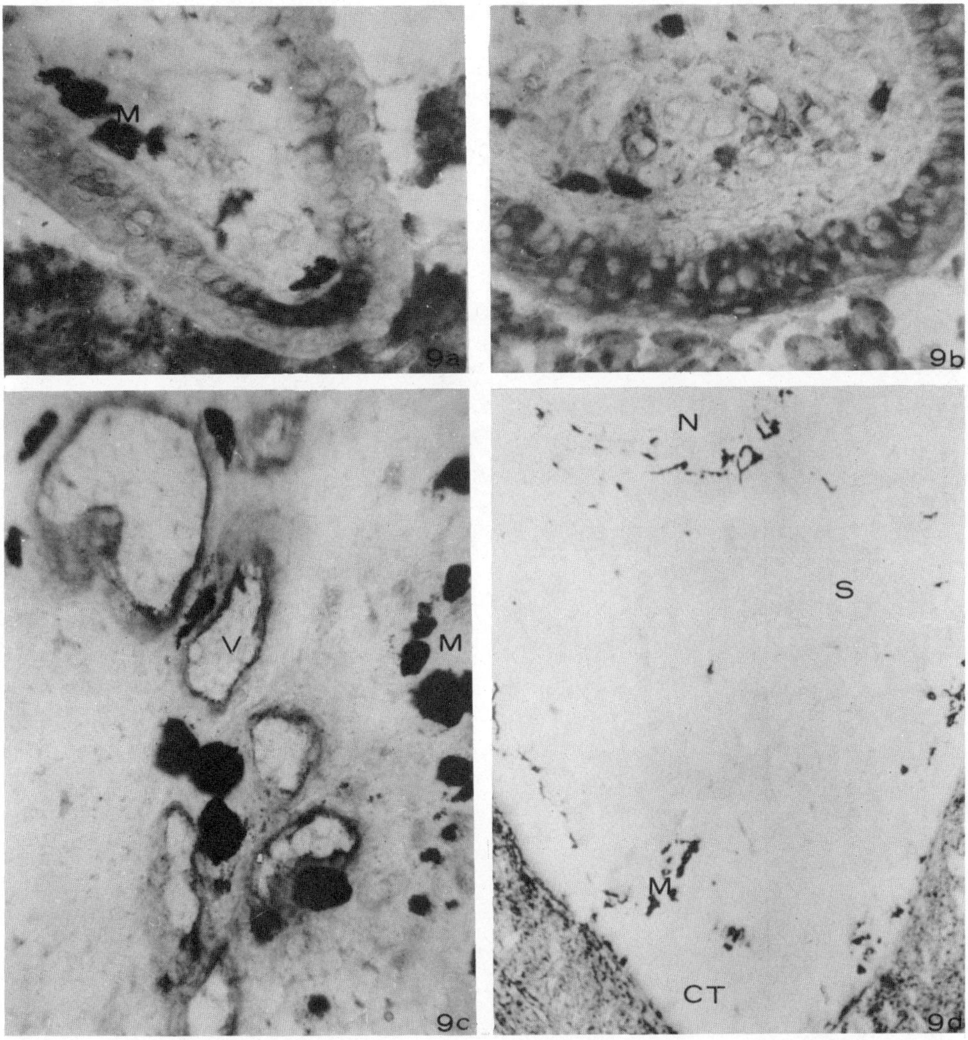

Fig. 9. Cross sections of ventral portions of tails, embedded in rat liver for support. Reaction for 5′ nucleotidase activity (pH 7.2). *a.* Stage XVIII. Basal epidermal cells of the ventral fin, as well as cells in the surrounding rat liver, show enzyme activity. *M,* melanin. *b.* Stage XX. Epidermis of ventral fin and rat liver give positive reaction for this enzyme. *c.* Stage XXII. Walls of some blood vessels (*V*) display enzyme activity. *M,* melanin. *d.* Stage XXII. With the above exceptions tail tissues show little or no activity; e.g., notochord (*N*), striated muscle (*S*), connective tissue (*CT*) are negative. Rat liver is positive. *M,* melanin.

reaction for this enzyme being especially strong in peripheral fibers (Fig. 11). As the tail shortened, activity appeared to decrease, in accord with earlier biochemical studies on succinic dehydrogenase in tails of metamorphosing tadpoles(28,29.).

Each of the enzymes surveyed was present in the rat liver used for support of the tails during sectioning, but no enzyme activity was detected either in the tadpole tail or the rat liver following incubation in substrate-free media. Localizations

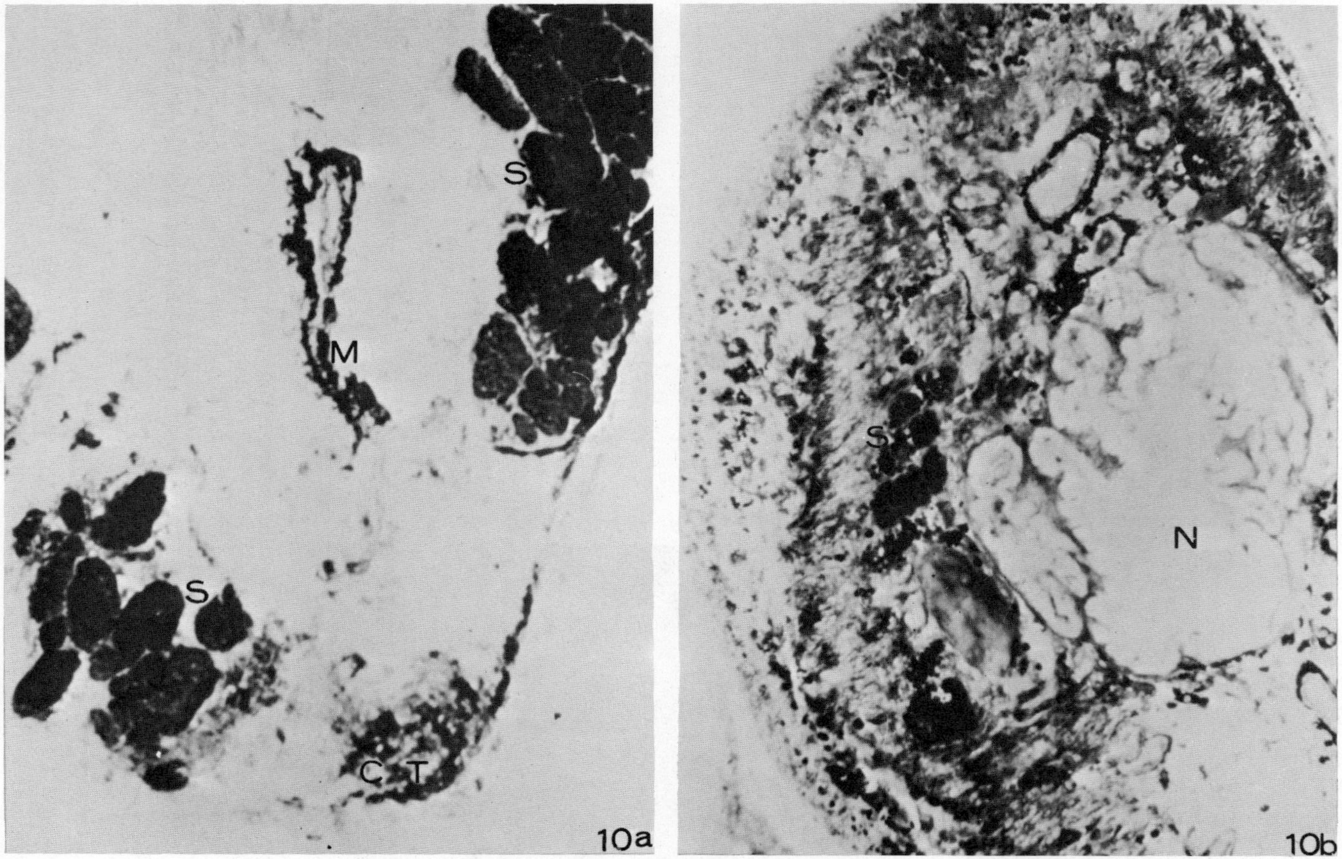

Fig. 10. Cross sections of tails. Reaction for calcium-activated ATPase. *a.* Thyroxine-treated. Enzyme activity is localized mainly in striated muscle cells (*S*) and in the connective tissue (*CT*) of the resorbing fin. *M*, melanin. *b.* Stage XXIII. A positive enzyme reaction is found in the few remaining cells (*S*) and throughout the surrounding connective tissue. *N*, notochord.

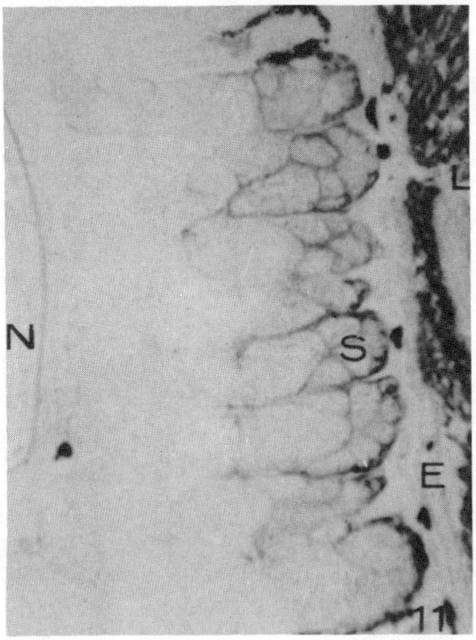

Fig. 11. Cross section of tail. Stage XXII. A positive reaction for succinic dehydrogenase is found in peripheral striated muscle cells (*S*) and in rat liver (*L*) used for support of the tail. Epidermis (*E*) and notochord (*N*) are negative.

of acid and alkaline phosphatases and esterase were more precise in tails fixed before freezing than in fresh-frozen tails.

As demonstrated by diastase digestion and the PAS technique glycogen was found in tail muscle of larvae in premetamorphic stages; it disappeared during normal and induced metamorphosis (Figs. 12a,b,c). The notochordal wall, connective tissue fibers, as well as the mucous coat covering the epidermis and the basement membrane underlying it contained polysaccharides other than glycogen; this was true during early larval as well as metamorphic stages. In paraffin sections, glycogen was located in the median parts of individual muscle cells, but in frozen sections it was more or less uniformly distributed within the cells. Differences were previously observed in the localization of glycogen in paraffin and frozen sections of rat muscle(30).

On the other hand, acid mucopolysaccharides were not found in tail muscle but were found in the intercellular substance of connective tissue (especially pro minent in the fins) and in the mucous coat covering the epidermis (Fig. 13a). This was shown by a positive alcian blue stain and a metachromatic reaction with toluidine blue. Although the mucous coat retained its staining qualities during metamorphosis, the reactions in the connective tissue decreased until almost none remained in the small tail rudiment (Fig. 13b). Treatment of sections with hyaluronidase prior to staining indicated that hyaluronic acid was responsible for the reaction in the connective tissue but not in the mucous coat. Biochemical studies previously showed that 96 percent of the acid mucopolysaccharaides in *R. catesbeiana* tail fins was hyaluronic acid(31).

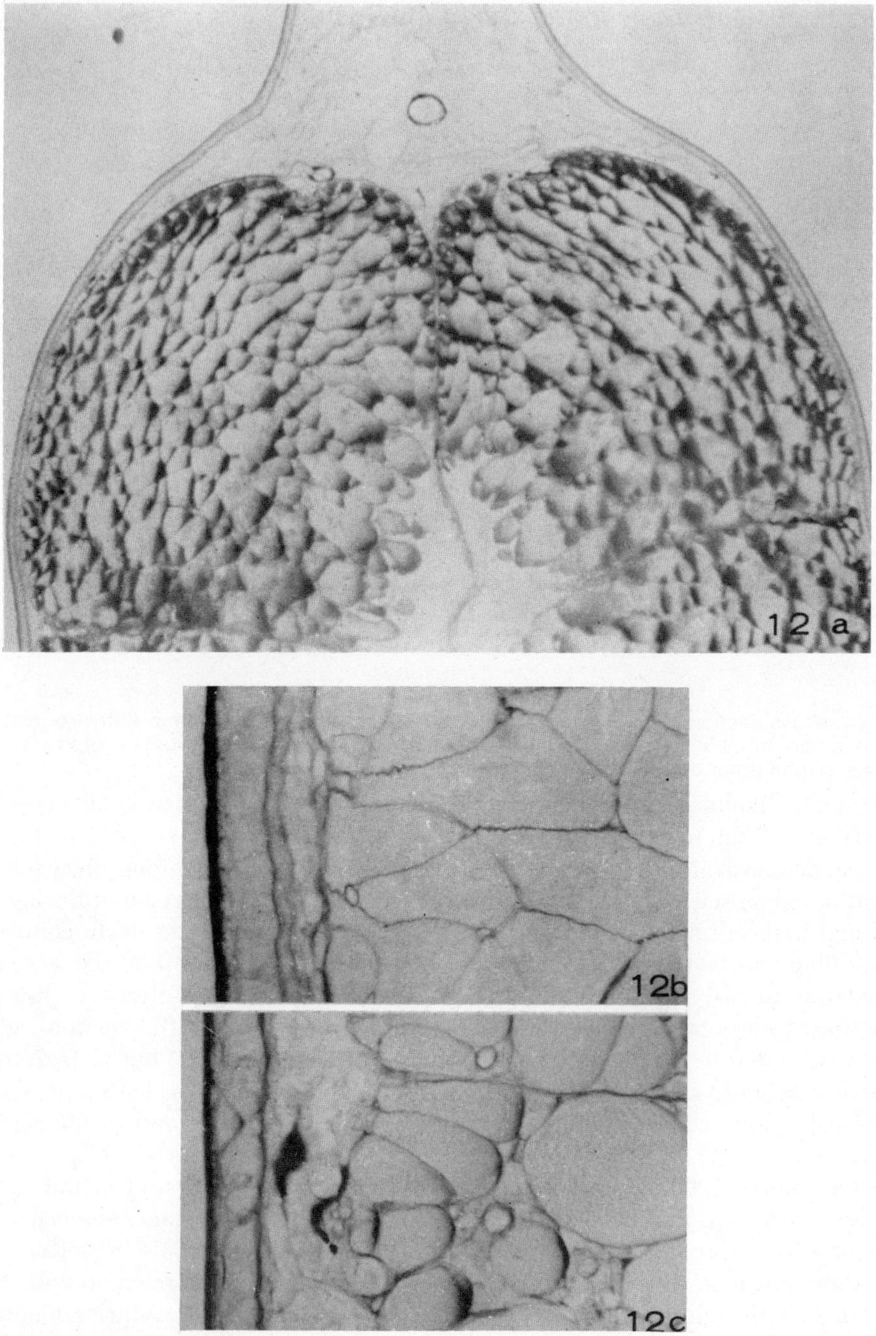

Fig. 12. Cross sections of tails, fixed in Carnoy's solution and embedded in paraffin. PAS reaction for polysaccharides. *a.* Stage VIII. A positive PAS reaction is apparent in striated muscle cells. *b.* Stage VIII. Section of tail treated with saliva prior to application of the PAS technique. No reaction occurred in the muscle cells indicating the polysaccharide shown in the muscle in Figure 12*a.* is glycogen. A positive reaction shows in the mucous coat of the epidermis (visible at this magnification), indicating that it contains polysaccharides other than glycogen. *c.* Stage XXII. Almost no glycogen is left in the muscle cells. The mucous coat of the epidermis gives a positive PAS reaction.

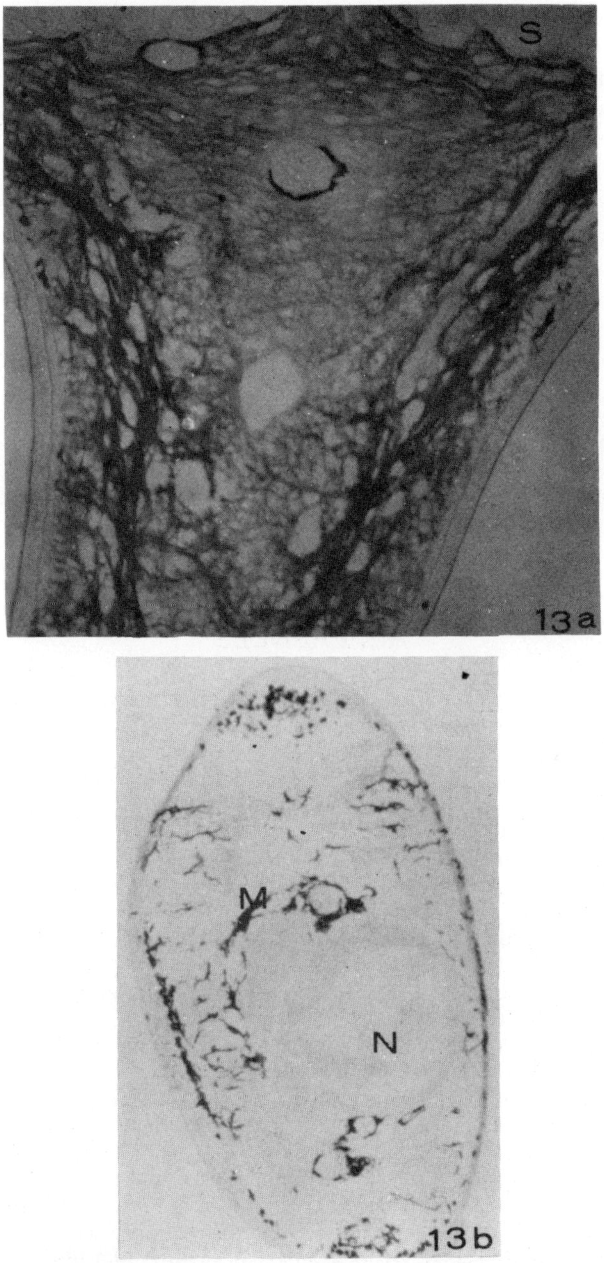

Fig. 13. Cross section of tails stained with alcian blue for acid mucopolysaccharides. *a.* Stage VIII. Acid mucopolysaccharides are located in the intercellular regions of the ventral fin. Striated muscle (S) gives no alcian blue reaction. *b.* Stage XXIII. Almost no alcian blue reaction is found. *M,* melanin; *N,* notochord.

In conclusion the present survey of enzymes indicates that acid phosphatase, nonspecific esterase, aminopeptidase, alkaline phosphatase, magnesium-activated ATPase, thiamine pyrophosphatase, and 5′-nucleotidase (low activity) are localized in epithelial and connective tissues and that they increase as tail resorption proceeds. They were not detected in striated muscle cells. On the other hand, muscle cells do contain calcium-activated ATPase and succinic dehydrogenase, as well as glycogen (but not acid mucopolysaccharides). The mature frog has no problems with respect to tail enzymes and consequently has a happy smile.

REFERENCES

1. Weber, R. Biochemistry of amphibian metamorphosis. *In* Weber, R., ed., The biochemistry of Animal Development. New York, Academic Press, Inc., Vol. 2, pp. 227–301, 1967.
2. Frieden, E. Biochemistry of amphibian metamorphosis. *In* Etkin, W., and Gilbert, L. I., eds., Metamorphosis: A Problem in Developmental Biology. New York, Appleton-Century-Crofts, pp. 349–398, 1968.
3. Salzmann, R., and Weber, R. Histochemical localization of acid phosphatase and cathepsin-like activities in regressing tails of *Xenopus* larvae at metamorphosis. Experientia, 19:352–354, 1963.
4. Weber, R. Ultrastructural changes in regressing tail muscles of *Xenopus* larvae at metamorphosis. J. Cell Biol., 22:481–487, 1964.
5. Taylor, A. C., and Kollros, J. J. Stages in the normal development of *Rana pipiens* larvae. Anat. Rec., 94:7–23, 1946.
6. Barka, T. A simple azo-dye method for histochemical demonstration of acid phosphatase. Nature, 187:248–249, 1960.
7. ——— and Anderson, P. J. Histochemistry: Theory, Practice, and Bibliography. New York, Hoeber Medical Division, Harper and Row, pp. 261–264 and Bibliography, 1963.
8. Nachlas, M. M., Crawford, D. T., and Seligman, A. M. The histochemical demonstration of leucine aminopeptidase. J. Histochem. Cytochem., 5:264–278, 1957.
9. McKay, D. G., Hertig, A. T., Adams, E. C., and Danziger, S. Histochemical observations on the germ cells of human embryos. Anat. Rec., 117:201–209, 1953.
10. Wachstein, M., and Meisel, E. Histochemistry of hepatic phosphatases at a physiologic pH with special reference to the demonstration of bile canaliculi. Amer. J. Clin. Path., 27:13–23, 1957.
11. Padydula, H. A., and Herman, E. Factors affecting the activity of adenosine triphosphatase and other phosphatases as measured by histochemical techniques. J. Histochem. Cytochem., 3:161–169, 1955.
12. Nachlas, M. M., Tsou, K. C., DeSouza, E., Cheng, C. S., and Seligman, A. M. Cyto-

chemical demonstration of succinic dehydrogenase by the use of a new p-nitrophenyl substituted ditetrazole. J. Histochem. Cytochem., 5:420–436, 1957.
13. McManus, J. F. A. Histological and histochemical uses of periodic acid. Stain Techn., 23:99–108, 1948.
14. Mowry, R. W. Alcian blue technics for the histochemical study of acidic carbohydrates. J. Histochem. Cytochem., 4:407, 1956.
15. Lillie, R. D. A brief method for the demonstration of mucin. J. Tech. Methods Bull. Int. Ass. Med. Museums, 12:120–121, 1929.
16. Metchnikoff, E. Atrophie des muscles pendant la tranformation des batraciens. Ann. Inst. Pasteur (Paris), 6:1–12, 1892.
17. Jasmin, G., and Gareau, R. Histopathological study of muscle lesion produced by paraphenylenediamine in rats. Brit. J. Exp. Pathol., 42:592–596, 1961.
18. Yanagisawa, T. Alkaline and acid glycerophosphatase activities in the tail during the metamorphosis of two anuran tadpoles, *Bufo vulgaris formosus* and *Rhacophorus schlegerii arborea*. Rep. Liberal Arts Sci. Faculty Shizuoka Univ. Nat. Sci., 4:20–26, 1953.
19. ——— Effects of thyroid extract on phosphatase activities of liver, intestine and tail of amphibian tadpole. Rep. Liberal Arts Sci. Faculty Shizuoka Univ. Nat. Sci., 5:33–41, 1954.
20. Weber, R., and Niehus, B. Zur Aktivität der sauran Phosphatase im Schwanz der Xenopuslarven während Wachstum und Metamorphose. Helv. Physiol. Pharmacol. Acta, 19:103–117, 1961.
21. Eeckhout, Y. Le comportement des hydrolases acides de la queue du têtard de *Xenopus laevis* pendant la métamorphose. Arch. Int. Physiol. Biochim., 72:316–318, 1964.
22. ——— Aspects biochimiques de la métamorphose. Rev. Questions Sci., 377–393 (3), 1966.
23. Weber, R. On the biological function of cathepsin in tail tissue of *Xenopus* larvae. Experientia, 13:153–155, 1957a.
24. ——— Die Kathepsinaktivität im Schwanz von Xenopuslarven während Wachstum und Metamorphose. Rev. Suisse Zool., 64:326–336, 1957b.
25. Frieden, E., and Mathews, H. Biochemistry of amphibian metamorphosis. III. Liver and tail phosphatases. Arch. Biochem. Biophys., 73:107–119, 1958.
26. Niehus, B., and Weber, R. Kennzeichnung und Verhalten der Mg^{++}-abhängigen Adenosintriphosphatase im Schwanz der Xenopuslarve während Wachstum und Metamorphose. Helv. Physiol. Pharmacol. Acta, 19:344–359, 1961.
27. Hardonk, M. J. 5′-nucleotidase. I. Distribution of 5′-nucleotidase in tissues of rat and mouse. Histochemi, 12:1–17, 1968.
28. Yamamoto, K., and Suzuki, M. Changes in activity of succinic and glutamic dehydrogenases and proteolytic enzyme in amphibian tadpole tissues during normal metamorphosis. Endocr. Jap., 4:262–267, 1957.
29. ——— Changes in activities of succinic and glutamic dehydrogenases, aspartic-a-ketoglutaric transaminase and proteolytic enzyme in tadpole tail tissue of *Rana japonica* during induced metamorphosis. Endocr. Jap., 7:8–12, 1960.
30. Nachmias, V. T., and Padykula, H. A. A histochemical study of normal and denervated red and white muscles of the rat. J. Biophys. Biochem. Cytol., 4:47–54, 1958.
31. Lipson, M. J., and Silbert, J. E. Acid mucopolysaccharides of tadpole tail fin and back skin. Biochim. Biophys. Acta, 101:279–284, 1965.
32. Leius, V. K. The distribution of acid phosphatase and adenosine triphosphatase in the tail of the metamorphosing *Rana pipiens* tadpole. Masters thesis, Mount Holyoke College, 1968.
33. Tegiacchi, E. A. Glycogen in liver and striated muscle of the *Rana pipiens* tadpole during normal and thyroxine-induced metamorphosis. Masters thesis, Mount Holyoke College, 1961.

23

Localization of Radioactive Thyroxine in Metamorphosing Tissues*

W. Ann Reynolds

Department of Anatomy, College of Medicine,
University of Illinois, Chicago, Illinois

INTRODUCTION

The anuran amphibian is totally dependent upon thyroid hormone for maturation from a larval into an adult state. Experiments with thyroxine pellet implants (recently reviewed by Kaltenbach[1]) have established that during metamorphosis, cells respond directly and specifically to thyroid hormone. In all instances, the cells adjacent to the pellet respond most rapidly and profoundly to the hormone with more distant cells responding proportionately to the amount of hormone reaching them. The nature of the response is always that typical of the cell. For example, a pellet adjacent to the lumbosacral spinal cord elicits a precocious size increase in the motor neurons(2). Another indication of the direct response of individual cells to thyroid hormone is the asynchrony in development that results when high levels of hormone are administered to the entire tadpole. For example, the motor neurons of the spinal cord will differentiate rapidly to a level typical of a juvenile frog whereas the limb they innervate has attained only the foot paddle stage(3). Thus, at the tissue or cellular level, thyroxine appears to act by directly eliciting certain responses. The spectrum of observable cell responses is exceedingly wide, including growth(2,4) or decrease in size(5), division(5,6), cytodifferentiation(3), and even death(7). Equally diverse are the numerous biochemical changes that have been demonstrated in direct response to thyroid hormone. These include a localized increase in oxygen consumption(9) and photopigment conversion(10) in response to thyroxine pellet implants. The culture of isolated tissues from tadpoles in media containing thyroxine has provided additional evidence of the direct action of thyroid hormone. Using cultured tail tips, significant changes

*Supported by grant GMHD-13979 from the United States Public Health Service.

in enzyme activity as well as protein and RNA synthesis in response to thyroxine have been documented(11,12).

The utilization of radioactive tracers provides an obvious means for examining the interaction between thyroid hormone and metamorphosing tissues. One early and ingenious experiment(12) involved implanting fragments of thyroid glands containing ^{131}I under the skin of tadpoles. Pronounced localized responses to the endogenously-labeled hormone occurred when diffusion away from the implant site was minimal as indicated by autoradiography. Except for the recent work of Frieden and colleagues(13,14), previous investigations dealing with radioactive tracer uptake in tadpoles have employed radioiodine-131(15,16,17,18,19). There is good agreement among these investigators that ^{131}I is concentrated primarily by the thyroid gland. In addition, localization as indicated by autoradiography, occurred in pigmented tissues, perhaps as a result of binding with available tyrosine.(14). The total body uptake and excretion of thyroxine, triiodothyronine and iodide, using ^{131}I as a label, have been examined in anuran tadpoles(13,14). These studies encompassed such parameters as the effects of temperature and the mode of administration upon hormone uptake and retention, extent of deiodination in vivo and distribution of radioactivity between carcass, viscera, and tail.

Relatively recently, thyroxine has become available which is labeled with carbon-14 or iodine-125, isotopes with long half-lives. The beta emission of ^{14}C and, in effect, of ^{125}I, permit good autoradiographic resolution as well as quantitative studies on hormone uptake. Thyroxine is the compound of choice since it is the circulating form of thyroid hormone in the tadpole(20), rather than other analogs. The studies to be reported are the first in a series aimed at determining the extent of association and uptake of thyroxine by different amphibian tissues and cells during metamorphosis. The first part of the investigation involves quantitative determinations of the rate of uptake of labeled hormone in intact anuran tadpoles as well as within specific tissues. The second objective was to search for tissue and cellular localization of labeled hormone as demonstrated by autoradiography.

MATERIALS AND METHODS

Rana pipiens tadpoles were reared from eggs obtained from adult frogs (Lemberger Co., Oshkosh, Wisconsin) by artificially inducing ovulation. The larvae were maintained in fiberglass tubs in aerated tap water and fed a diet of boiled lettuce. All animals were staged according to the Taylor-Kollros series(21) and weighed before and after experimentation. Tadpoles ranging between stages IV and XV were employed in the experiments. Each experiment involved clutchmates who were matched with respect to weight and stage whenever control series, various dosage levels, or different isotopes were employed.

In most instances, triplicate experiments were performed, using thyroxine-^{125}I in the first, thyroxine-^{14}C in the second, and tyrosine-^{14}C in the third. Tyrosine cannot induce metamorphosis but is a part of the biosynthetic pathway of thyroxine. Comparisons of the extent of uptake of labeled thyroxine with that of tyrosine-^{14}C should indicate if selective, specific localization of thyroxine is occurring.

It is well-known that many tissues are actively able to deiodinate thyroid hormones and that this process may be closely related to the mode of action of the hormone. Thus, radioiodine-125 is advantageous in that it can be followed after its removal from the thyroxine molecule. The advantage of thyroxine-^{14}C is essentially the opposite in that the molecule can be followed after deiodination has occurred. Thyroxine-^{125}I (labeled in the 3,5,3′,5′ position) was purchased from Mallinckrodt Nuclear, St. Louis, Missouri and arrived dissolved in 50 percent ethylene glycol with a specific activity of from 10.0 to 46.5 mc/mg. Immersion hormone levels were prepared at concentration levels ranging from 1.0 to 11.6 μg/l with radioactivity levels of from 0.01 to 0.33 μc/ml. DL-thyroxine-2-^{14}C and dL-tyrosine-3-^{14}C, purchased from New England Nuclear Corp., Boston, arrived as powders and were dissolved in distilled water shortly before use. The specific activity of thyroxine-^{14}C ranged from 0.004 to 0.014 mc/mg, and immersion levels of the hormone were prepared in concentrations of from 250 to 1825 μg/l (0.003 to 0.012 μc/ml). The tyrosine-14C was supplied with a specific activity of 0.038 mc/mg and the immersion concentration was 0.8125 μg/ml (0.0125 μc/ml). Carrier was not used in these experiments because the specific activity of thyroxine-^{14}C was so low. In addition, even in experiments involving ^{125}I, certain tissues could be obtained only in minute amounts (less than 2 μg) making the addition of unlabeled hormone undesirable because of potentially low counting rates.

The basic experimental procedure involved immersing small tadpoles (0.3 g or less) in 50 ml and large tadpoles (more than 0.3 g) in 100 ml of aerated, distilled water containing the labeled material in glass dishes. After a maximum exposure period of 24 or 48 hours, during which the immersion solution was renewed at least once, feeding was resumed and the tadpoles were thereafter maintained in aerated tap water which was changed every 48 hours. Thyroxine retains its effectiveness in bringing about metamorphosis while in solution for at least 24 hours(22). Samples of the immersion solution were collected for counting at the time of animal counts, dissection, and fixation. The water temperature during all experiments was maintained at 20°C.

Total Body Uptake and Loss of Hormone

These experiments were performed only with ^{125}I-thyroxine, and only tadpoles at stages III to IX (0.2 to 1.0 g) could routinely be used for total body measurements of radioactivity because of limitations in the dimensions of the counting chamber. At selected intervals, each tadpole was rinsed in several changes of aerated tap water and then counted in a well-type scintillation detector with a 3-inch crystal and an efficiency of 3.2 x 10^5 cpm (Packard Instrument Co., Downers Grove, Ill.). All counts were corrected for background and decay. In order to observe any differences between live and dead animals with respect to thyroxine uptake and loss, tadpoles were sacrificed by immersion in 5 percent ethyl urethane solution for 5 minutes just prior to being placed in the hormone solutions. The dead animals were then handled and counted just as were the lives ones.

Tissue Uptake and Loss of Hormone

After the appropriate immersion sequence and interval, each tadpole was rinsed in aerated tap water and then anesthetized with ethyl urethane. An anterio-posterior incision through the skin and peritoneum was made in the ventral body wall and peritoneal fluid was collected in a calibrated capillary tube. Next, the ventricle of the heart was punctured permitting collection of up to 70 μl of blood in a calibrated capillary tube. Finally, individual tissues were dissected free, immediately wrapped in preweighed aluminum foil to prevent dessication, weighed and prepared for counting. Tissues containing ^{125}I were simply placed directly in tubes for gamma counting. Tissues containing ^{14}C were solubilized in hydroxide of Hyamine or NCS Solubilizer (Nuclear-Chicago) accompanied by shaking or sonification (Biosonik, Bronwill Scientific, Rochester, N. Y.) or by heating at 50°C for 2 to 3 hours when necessary. After the addition of the appropriate scintillators, each tissue preparation was counted in a Packard 3375 Liquid Scintillation Counter and each count was then corrected for efficiency and background. Results of the individual tissue counts are expressed as:

$$\frac{\text{cpm}/\mu\text{g wet tissue weight}}{\mu\text{c labeled hormone in immersion solution}}$$

Each of these values was derived from the averages of tissues from four or more animals.

Autoradiography

Larvae were anesthetized in ethyl urethane prior to dissection. Immediately upon removal, tissues were frozen in isopentane at the temperature of liquid nitrogen and then kept at −30°C until sectioning. Frozen sections were cut at 10 μ on a cryostat in the dark and placed on slides precoated with Kodak NTB-2 nuclear track emulsion following the techniques of Rogers(23). Care was taken to insure that no thawing occurred in the frozen sections until they were developed. For paraffin embedding, tissues were fixed in buffered formalin or Carnoy's fixative followed by embedding and sectioning at 8 μ. After mounting, the sections were deparaffinized and coated by dipping in liquid emulsion (Kodak NTB or NTB-2) in the dark(25). Both paraffin and frozen sections were exposed at −30°C to minimize diffusion and chemography for periods of from 2 days to 6 weeks. After test slides indicated sufficient silver halide crystals had formed to permit accurate observations, all of the slides prepared in conjunction with an experiment were developed simultaneously in Kodak Dektol or D-11 developer. Thus, some notion of the extent of localization of label after different periods of exposure to the hormone could be obtained.

Control measures included testing for chemography and fading of the latent image. Frozen and paraffin sections were prepared from nontreated animals and subjected to autoradiographic procedures identical to those used for treated specimens. No positive chemography was observed after placement of emulsion. Frozen and paraffin sections were exposed to daylight for a second in order to fog the slides completely. After various exposure times, the slides were developed in the usual manner and examined for negative chemography and fading of the latent image. Fading of the latent image within emulsion as a function of time was not observed to any appreciable extent. However, loss of silver grains directly over the sections did occur occasionally. This phenomenon was more pronounced and occurred more often in frozen sections than in paraffin sections. After the paraffin sections had been mounted, every second slide was coated with 1 percent formvar in ethylene dichloride and every fourth slide with 2 percent parlodion in amyl acetate. If the control measures indicated the absence of observable negative chemography, then the uncoated slides were studied because the membranes tended to impede staining. Since the purpose of the frozen sections were primarily to substantiate the observations made with paraffin sections, the difficult and rather unsatisfactory process of interposing a membrane between section and emulsion was not performed routinely.

After developing, both frozen and paraffin sections were stained either with toluidine blue or nuclear fast red.

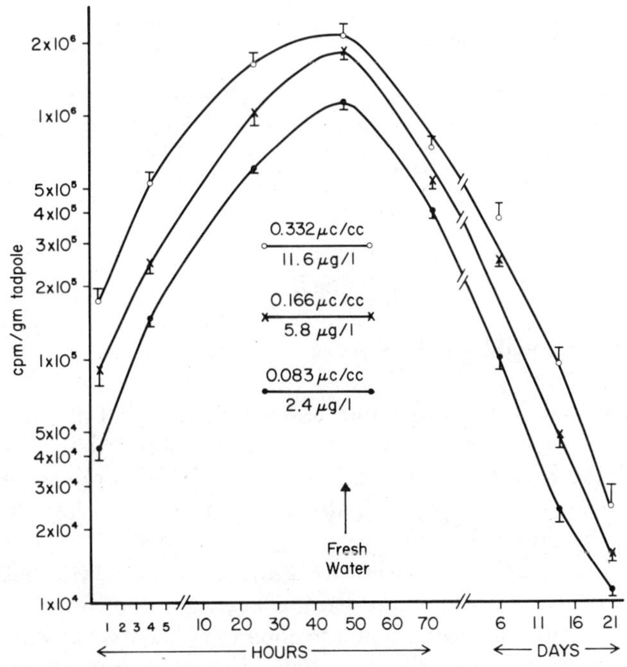

Fig. 1. The uptake and loss of thyroxine-^{125}I is compared for three different dosages. Animals were immersed in the hormone solution for 48 hours before being returned to fresh water. Although initially proportional, by 24 hours the extent of uptake is no longer directly related to the immersion level of thyroxine. Each point represents the average value and standard error derived from measurements upon at least 10 animals. Radioactivity levels (cpm/ml) of the different dosages are indicated by horizontal lines.

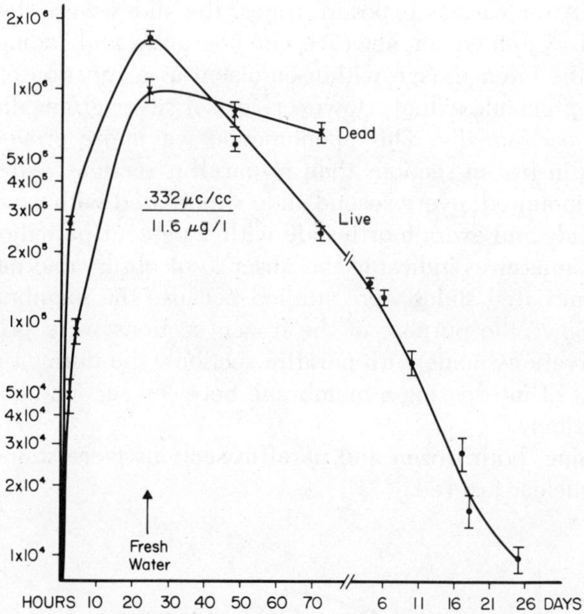

Fig. 2. The uptake and loss of thyroxine-^{125}I by live and dead tadpoles after immersion in a hormone concentration of 11.6 μg/1 for 24 hours. Note similarities in the rate of uptake between the live and dead animals. Each point represents an average derived from measurements on at least 8 animals. The level of radioactivity at this dose level (cpm/ml) is indicated by a horizontal line.

RESULTS

Total Body Uptake and Loss of Hormone

The rate and extent of uptake and loss of thyroxine-^{125}I at three different immersion levels are depicted in Figure 1. It can be noted that a doubling in the amount of available hormone is not reflected by a doubled rate of uptake past 30 minutes of immersion. However, the patterns of uptake and loss do closely parallel each other at the different dose levels.

It is interesting that the rate of hormone uptake differs little between live and dead tadpoles (Fig. 2). This phenomenon has been alluded to previously(14) without including data. Although uptake of hormone to concentrations above that of the surrounding medium appears to be essentially a passive process, only a small loss of label occurred in dead tadpoles upon return to fresh water. Live tadpoles lost label rapidly after being returned to fresh water. Over one-half of the label disappeared within 24 hours and less than 0.5 percent of the maximum level of radioactivity was present 24 days after the return to fresh water. However, the rate of loss after the first 48 hours of being in fresh water became slower, requiring days for a loss of one-half the existing label rather than hours.

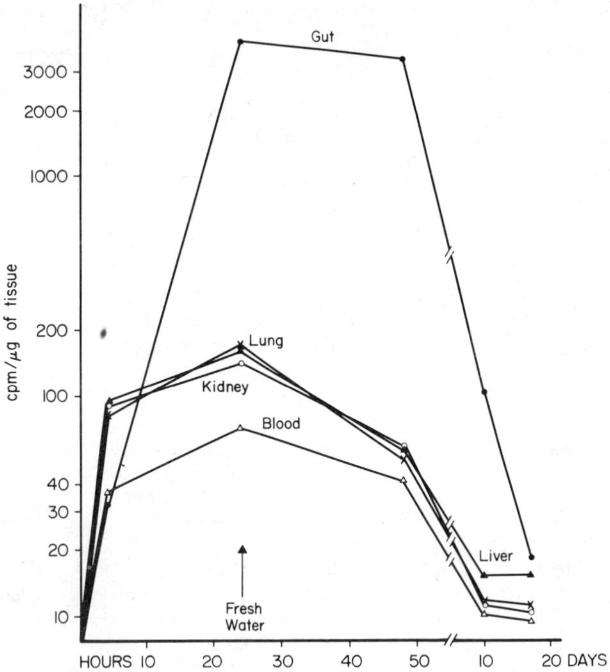

Fig. 3. The accumulation and loss of thyroxine-^{125}I by specific tissues upon immersion in thyroxine-^{125}I solutions for 24 hours after which the larvae were returned to fresh water. The tissues included here all exhibited accumulations above circulating levels.

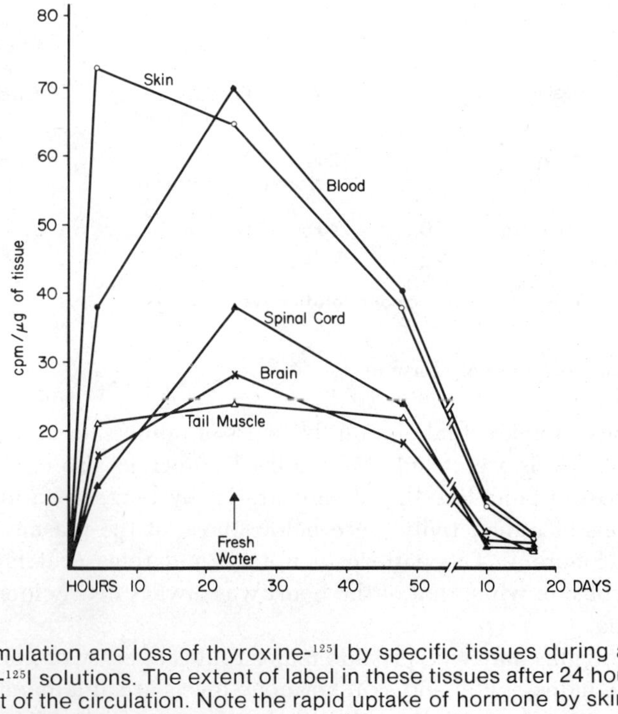

Fig. 4. The accumulation and loss of thyroxine-^{125}I by specific tissues during and after immersion in thyroxine-^{125}I solutions. The extent of label in these tissues after 24 hours of immersion was less than that of the circulation. Note the rapid uptake of hormone by skin which also was true of gill tissue suggesting these tissues are a major route of hormone entry.

TABLE 1

Localization of ^{14}C- and ^{125}I-thyroxine in tadpole tissues as indicated by autoradiography

Tissue	Extent of label	Label first observed (hours after immersion)	Maximum label[a] (hours after immersion)	Cellular localization	Label distribution
Gills	Moderate	0.5	6	No	Accumulated on outer and inner surfaces of filaments
Liver	Heavy	0.5	24	Possibly within hepatocyte cytoplasm	Throughout sinusoids
Skin	Light	0.5	4	Diffuse	In epidermis, rare in dermis
Blood	Light	1	24	Rare	Diffuse
Cartilage	Moderate	1	6	Chondrocyte nuclei	Highly specific
Limb bud	Light	1	—	Diffuse	No specific associations
Muscle	Light	1	6	No	Associated with muscle sheaths
Thymus	Heavy	1	6	No	Evenly dispersed label throughout gland
Brain	Moderate	2	24	Probable	Associated with ependymal and other cellular areas
Spinal cord and ganglia	Light	2	—	Diffuse	Modest ependymal association
Stomach	Heavy	2	24	Possibly in epithelium	Throughout epithelium, mucosa and muscularis to serosa
Intestine	Heavy	6	24	Possibly in epithelium	Throughout epithelium, mucosa and muscularis to serosa
Kidney	Moderate	6	48	Diffuse	Evenly dispersed through tubules

[a]Maximum exposure time to labeled hormone solution was 48 hours.

Tissue Uptake and Loss of Hormone

Tissues whose uptake of label after 24 hours of immersion in thyroxine-^{125}I had surpassed plasma levels are included in Figure 3. Pancreas also exhibited a pattern of uptake and loss of label like that demonstrated by lung, liver, and kidney. Tissues whose levels of radioactivity were below those of the plasma after 24 hours are graphed in Figure 4. The pattern of uptake and release of label in the gills resembled that of skin while that of the heart was always nearly identical to that of the blood plasma.

The skin and gills may well provide major routes of entry of hormone from the medium into the tadpole. Gill and skin hormone levels are at a maximum within 30 minutes after immersion as indicated both by tissue counts and autoradiography

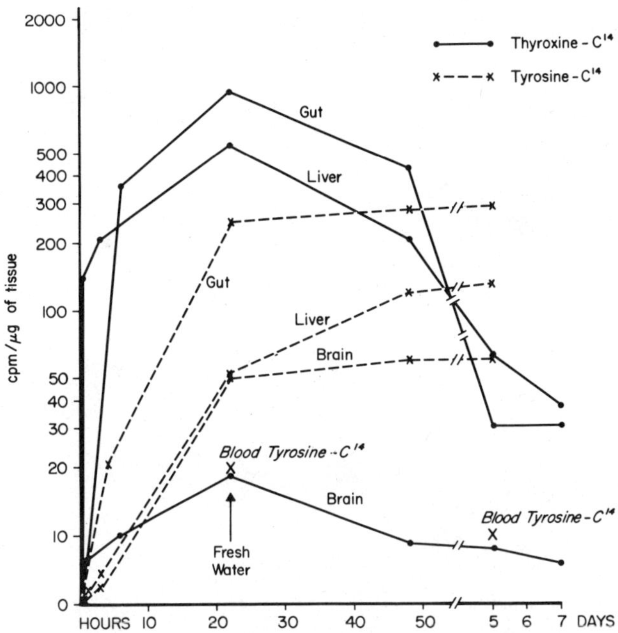

Fig. 5. A comparison of the patterns of accumulation and loss of thyroxine-^{14}C and thyrosine-^{14}C. Whereas the loss of label is precipitate in tissues from animals once immersed in thyroxine-^{14}C, thyrosine-^{14}C persists at maximum concentration for days after the larvae have been placed in fresh water.

(Fig. 4, Table 1). Gut levels of radioactivity lag behind those of blood and liver in the first few hours of immersion, making it unlikely that the gastrointestinal tract is a major means of hormone entry into the body. Lipner and Hazen(25) found that the gut was not an essential route for ^{131}I uptake in the tadpole by placing a ligature around the esophagus.

It is apparent that the gastrointestinal tract is most active in sequestering labeled hormone (Fig. 3). The liver and kidney also exhibit relatively high levels of radioactivity shortly after immersion, followed by a rapid loss of hormone, as befits their probable roles in thyroxine deiodination(26,27) and iodide excretion(15).

The specific activity of thyroxine-^{125}I was much greater than that of thyroxine-^{14}C. Thus, the actual thyroxine levels in immersion solutions containing thyroxine-^{14}C were, of necessity, exceedingly high. Increasing concentrations of thyroxine in the immersion solution are not reflected in proportionately increasing levels of total body accumulation (Fig. 1). Therefore, the uptake of thyroxine-^{14}C and of thyroxine-^{125}I are not directly comparable and no attempt was made to calculate them on a comparative basis. On the other hand, the general patterns of uptake in specific tissues demonstrated with the two labels are relevant to each other. The basic patterns of incorporation of label into a specific tissue followed by rapid loss of radioactivity upon return to fresh water were similar in experiments involving either thyroxine-^{125}I or thyroxine-^{14}C (Figs. 3, 4, and 5). For this reason, only

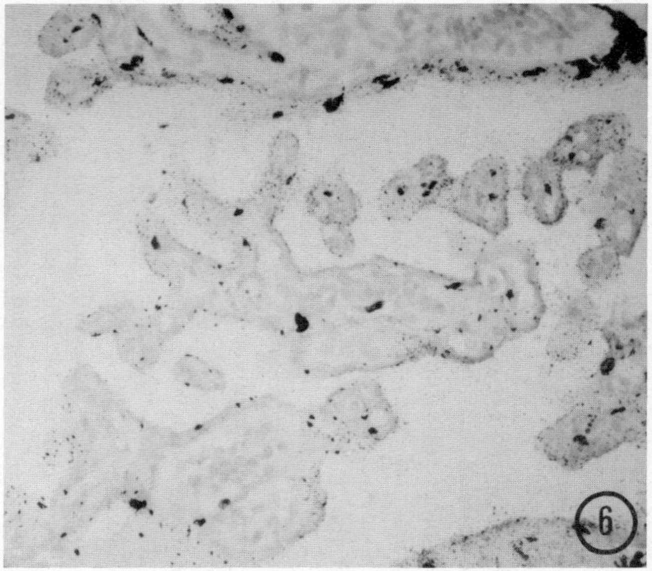

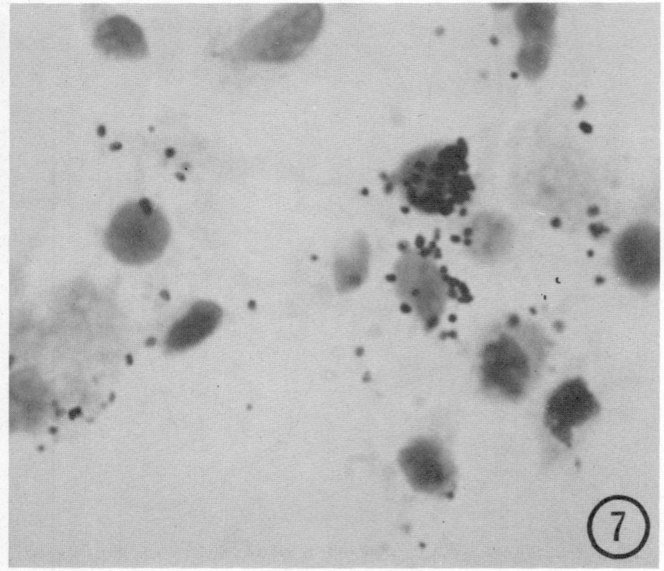

Fig. 6. Gill filaments of a stage IV larva after immersion for 30 minutes in thyroxine-^{14}C (0.016 μc/ml; 1,250 μg/1). Exposure 2 days. Note extensive amount of label lining the surfaces of the gill filaments. ×159.

Fig. 7. Hepatocytes of liver from a stage VI tadpole immersed for 30 minutes in thyroxine-^{14}C (0.012 μc/ml; 1,825 μg/1). Exposure 7 days. Label is concentrated in the region of the hepatocyte nuclei. ×999.

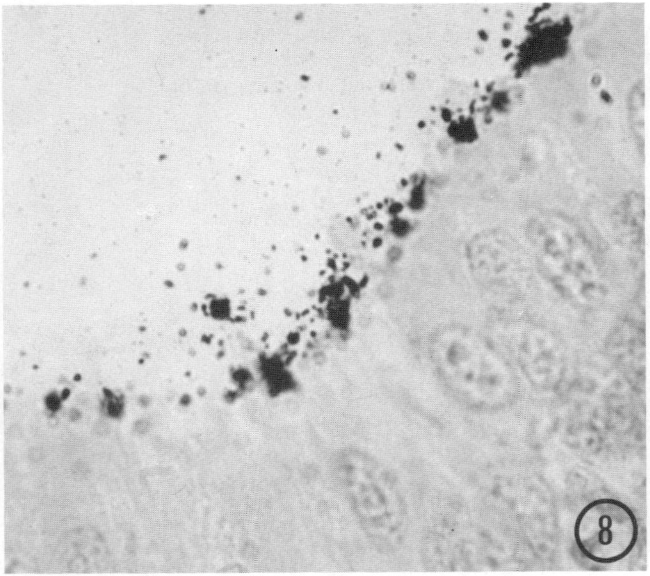

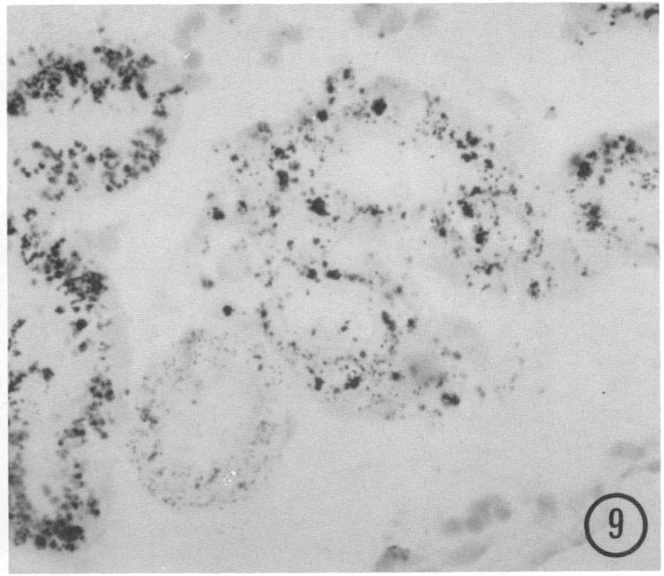

Fig. 8. Section from the distal end of the esophagus, where it meets the stomach. The stage IV animal was immersed for 1 hour in thyroxine-^{14}C (0.016 μc/ml; 1,250 μg/1. Exposure 2 days. The accumulation of grains at the surface of the columnar epithelium after such a short exposure time to hormone suggests some entry into the gastrointestinal tract via this route. $\times$ 999.

Fig. 9. Tubules of the mesonephric kidney of a stage VII larva immersed in thyroxine-^{14}C (0.012 μc/ml; 1,825 μg/1) for 48 hours. Frozen section, exposure 7 days. Grains are extensive in number throughout the tubules. $\times$399.

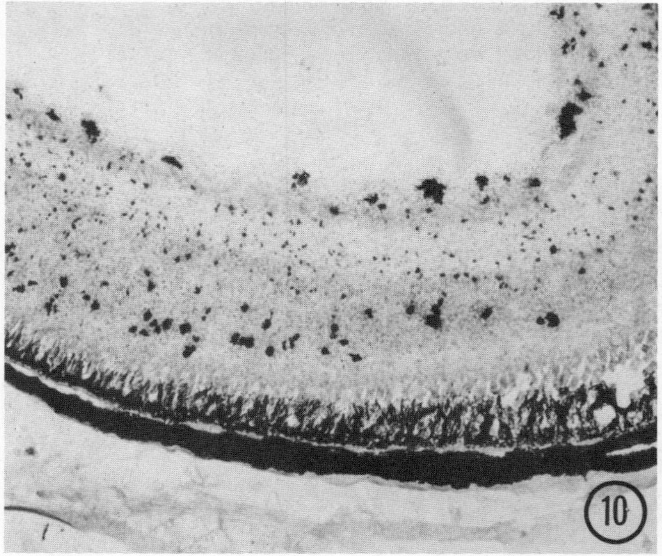

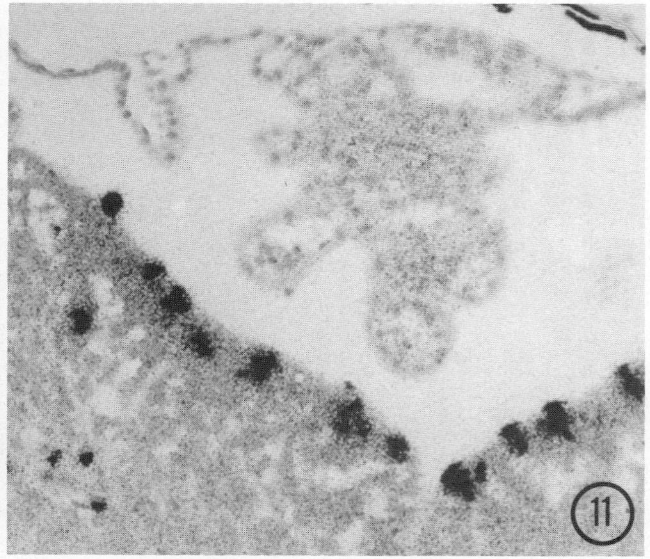

Fig. 10. Portion of the retina from a stage IV larva immersed in thyroxine-^{14}C (0.016 μc/ml; 1,250 μg/1) for 6 hours. Exposure 2 days. In addition to accumulations adjacent to the vitreous humor, label is associated with the neuronal layers. Only individual grains are seen in the plexiform layers. ×399.

Fig. 11. Section through the medulla from a stage V larva immersed in thyroxine-^{125}I (0.25 μc/ml; 6.0 μg/1) for 1 hour. Exposure 2 days. Extensive label is associated with the ependyma, suggesting labeled hormone may reach the central nervous system by way of the cerebrospinal fluid. Note the numerous although dispersed grains within the choroid plexus. ×159.

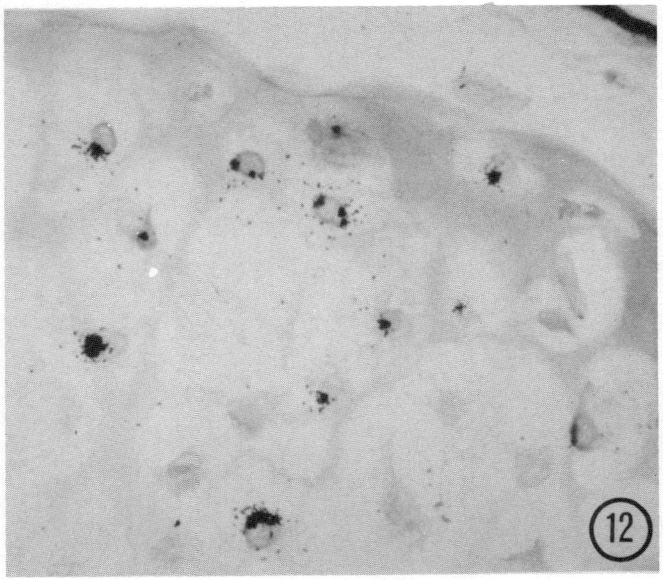

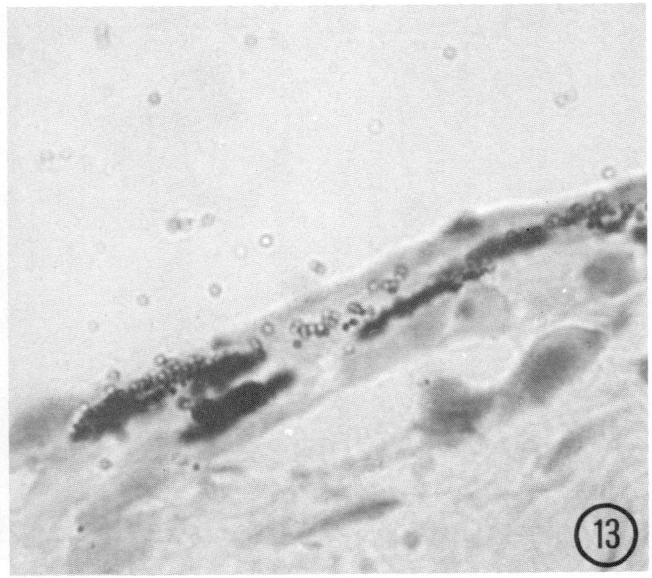

Fig. 12. Cartilage located immediately ventral to the section of medulla seen in Figure 11. Label is concentrated in the region of the chonodrocyte muclei. ×399.

Fig. 13. Skin from a stage VII animal immersed for 48 hours in tyrosine-^{14}C (812.5 μg/ml; 0.0125 μc/ml). Exposure 7 days. Concentrations of label can be seen in association with pigment cells. ×999.

three representatives: gut, liver, and brain of the experiments involving thyroxine-^{14}C, have been included in Figure 5. The exceptions are provided by tissues involved in the enterohepatic circulation, including the gut and liver. Maximum gut accumulations of radioactivity were proportionately lower and liver levels slightly higher in the animals immersed in thyroxine-^{14}C (Fig. 5) than were comparable values for experiments involving thyroxine-^{125}I (Fig. 3). Deiodination of thyroxine by tadpole liver has been well established(26,27). Deiodination would be likely to lower the amount of radioactivity present when the label is ^{125}I but should have little effect on thyroxine-^{14}C. On the other hand, the gastrointestinal tract of the tadpole apparently acts as an iodide pump during metamorphosis(25). Thus, any iodide-125 released by iodination would likely be retained in the gut, adding to the total amount of label present.

Tyrosine-^{14}C is taken up by tissues at a slower rate than is thyroxine-^{14}C. Upon return to fresh water, tyrosine-^{14}C is not lost from tissues (Fig. 5), although plasma levels do decrease (from a maximum of 20.1 cpm/μl to 7.9 cpm/μl some 72 hours later). The patterns of incorporation exhibited by only three tissues are included in Figure 5; similar patterns were found for the other tissues involved in the study including skin, gills, muscle, spinal cord, liver, heart, and kidney.

Autoradiography

The observations derived from the autoradiographic preparations are summarized in Table 1. In general, they correlate well with the amount of radioactivity measured in the specific tissues.

One-half hour after immersion in labeled thyroxine, label may be seen apparently in the process of entering the gill filaments (Fig. 6). Label is readily visible within the liver (Fig. 7) a short time after immersion although discriminating between pigment and label is a problem in the first hour after exposure to the isotope. After maximum accumulation, the quantity of label present obviates the problem. In addition, pigment granules possess a brownish hue and lie in a different plane of focus than do the black silver grains. The extensive accumulation of labeled hormone by the gastrointestinal tract is reflected in autoradiographs. In fact, if adequate exposure time was allowed to obtain sufficient label for evaluation in tissues such as brain and cartilage, the walls of the stomach and intestine became quite blackened. In specimens subjected to less exposure (Fig. 8), label is usually concentrated at either or both luminal and serosal surfaces. These regions of localization suggest that transport through the epithelial, mucosal and muscular layers is occurring although the direction of the transport is not apparent from the autoradiographic preparations. Label is almost never observed within the tubules of the kidneys in autoradiographs made from paraffin sections. It is readily apparent, however, in frozen preparations (Fig. 9) made from tadpoles exposed for 24 hours or more to the hormone. The patterns of label within the nervous system (Figs. 10 and 11) vary from localized with heavy accumulations to exceedingly diffuse distributions. There was little apparent association of either thyroxine ^{14}C or ^{125}I with specific groups of cells in the spinal cord. Distinct patterns of localization were more obvious within the brain and eyes. For example, within the retina, localization could always be observed in association with the nuclear layers (Fig. 10).

Within the first few hours of hormone immersion, extensive accumulations of label were found in the ependymal cell layers throughout the brain (Fig. 11). Later, label diminished in the ependymal layers and was instead liberally distributed among various cell layers. The association of labeled hormone with specific cells was noted in several instances including liver (Fig. 7) cartilage (Fig. 12) and at certain times, within the intestinal epithelium and certain nuclear regions of the brain.

Tyrosine-^{14}C initially exhibited patterns like those of labeled thyroxine with respect to gills and skin, again suggesting that these tissues are primary routes of entry for the labeled substances. However, after 24 hours of exposure to tyrosine-^{14}C, extensive association of label with pigment cells could be observed (Fig. 13). Since *Rana pipiens* larvae are highly pigmented throughout the various tissues, numerous sites for concentration of tyrosine-^{14}C exist. As has been pointed out (Fig. 5), the incorporation of tyrosine-^{14} persists after the return of larvae to fresh water.

No relevant differences in uptake or loss of labeled hormone from specific tissues were found in association with the different metamorphic stages (IV through XV) involved in these experiments. Hence, tissue counts and autoradiographic observations from animals in all of these stages were pooled. Until larval stage X, endogenous thyroid hormone production as indicated by ^{131}I incorporation into the thyroid gland(28), by protein bound iodine(30) and by thyroidal thyroxine(31) is minimal. After stage X, it is therefore reasonable to expect some dilution of the incorporation of labeled hormone by the increasing levels of endogenous thyroid hormone. However, the thyroxine concentrations in the immersing fluids, especially those for thyroxine-^{14}C, are in excess of those minimally needed (0.003 to 0.10 $\mu g/l$) to bring about metamorphic changes characteristic of the tadpole(31) at this period. Thus, any dilution by endogenous hormone after stage X is probably being obscured by the presence of such large amounts of exogenous hormone.

DISCUSSION

Of perhaps greater interest than the rapid and extensive uptake of thyroxine is the exceedingly rapid rate at which thyroxine is initially excreted from live tadpoles. This loss of hormone appears to be an active process in contrast to the uptake of hormone which occurs at approximately the same rate in live and dead animals. Over one-half of the maximum accumulation of radioactivity disappeared within 24 hours after the return to fresh water. Thereafter, label disappeared at a much diminished rate. The half-survival time of thyroxine-^{131}I injected into *Rana grylio* tadpoles was found to be 3 days at 25°C(14). The dose was approximately 0.8 $\mu g/g$ body weight, which would instantly supply more hormone than was available to a tadpole in all of the immersion fluid at any of the concentrations used for determining total body uptake in the present study. The sudden presence of a relatively high hormone concentration from ingestion as contrasted with the slow accumulation of hormone during immersion may account for the observed differences in the rate of hormone loss.

This investigation is concerned with the ability of specific tissues to localize thyroxine, and hence the percentage distributions or radioactivity between different tissues were not computed. However, the accumulation of thyroxine by both the liver and the gastrointestinal tract confirm the existence of an enterohepatic circulation of thyroxine for the amphibian. The long-established enterohepatic pathway of thyroxine in mammals, intimately involved with the metabolism and excretion of the hormone, has been reviewed by Tata(32).

Combined autoradiographic and chromatographic studies(33,34) in the mammal after the administration of thyroxine-^{131}I have demonstrated a localization that was principally iodide in the salivary gland, skin, lymph node, and urinary tract; triiodothyronine predominated in liver, kidney, pancreas, small intestine, adrenal, and spleen. Labeled thyroxine persisted in that form in parts of the gray matter of the brain; both thyroxine and triiodothyronine were found in the hypophysis. Thus, with a few exceptions, the patterns of localization as revealed by autoradiography and direct tissue counting are not dissimilar between the mammal and the amphibian even though the roles played by the same hormone in the two different vertebrate classes outwardly appear to be diverse.

These investigations were partly stimulated by an earlier observation(35) that tadpoles immersed in thyroxine solutions at 4°C for months failed to undergo metamorphosis. However, upon being returned to fresh water at 20 to 25°C, precocious metamorphosis, typical of that induced by exogenous thyroxine, occurred. Similar results were reported by Frieden and associates(13), where premature tail shortening was observed upon rewarming when less than 1 percent of an original dose of triiodothyronine was present. These investigators interpreted this to mean that the hormone had imprinted the responding cells. An alternate hypothesis, based on the low levels of thyroxine required to exceed the threshold for a given metamorphic event(36), is that most of the injected hormone is in surplus and is excreted, a process requiring more time in a poikilothermic animal at 5°C, and the small amount of hormone actually taken up by the cells and remaining in them induces metamorphosis upon rewarming. It seemed likely that information on the quantities of hormone taken up by and persisting in specific amphibian tissues after immersion in hormone solutions, related to hormone levels in the blood, could support or negate the hypothesis. A corollary to the hypothesis was the notion that cells acquiring a readiness to respond to thyroxine as they do in sequence during metamorphosis(37), might also develop the capacity to concentrate the hormone sequentially.

Labeled hormones are apparently actively taken up by target organs in numerous systems that have been investigated. Cytoplasmic, nuclear and chromosomal concentrations of labeled auxins occur in plant root tips with the intranuclear localizations persisting long after the cytoplasmic ones(38). Labeled progesterone is concentrated to different degrees by the various layers of the rat uterus(39) while estrogen is localized primarily by nuclei of epithelial cells, cells of the lamina propria and muscle cells(40). In fact, nuclear binding sites numbering between 2,000 and 2,500 have been estimated for 17-β-estradiol(41). Human kidney cells in tissue culture localize in the nucleus about three-fourths of the thyroxine-

^{125}I taken up from the medium(42). These authors suggest that the genetic matter and the nuclear envelope may be the site of concentration and thus, action.

At first glance, the results of the present study would appear to negate the hypothesis that thyroxine is concentrated by cells as a prelude to metamorphosis. Individual tissue determinations of radioactivity as well as the autoradiographic observations suggest that, at most, concentrations in but a few tissues exceed circulating levels by any magnitude. Further, once the supply of exogenous hormone is diminished, tissue levels fall precipitately, along with total body levels. This first study was of necessity a survey of tissue uptake and loss over relatively long periods of time. A closely timed investigation, involving both autoradiography and the measurement of the net uptake of thyroxine within specific nuclear regions of the brain, should shed some light on the matter. The brain does exhibit a varied pattern of localization and its alterations during metamorphosis have been investigated extensively. Another candidate for further study would be axial cartilage (Fig. 7), which demonstrates a highly specific localization of label and which ossifies in an anterior to posterior sequence during metamorphosis(43). The rapidity with which labeled hormone appears in amphibian tissues and is in turn released is somewhat startling for a poikilotherm. Label can first be seen in autoradiographs in less than an hour but, with the exception of gut, muscle, brain, and spinal cord, one-half or less of the maximum level of radioactivity has disappeared within 24 hours after the return to fresh water. Some 8 hours are required for maximum binding of 17-β-estradiol to chromatin within nuclei of the uterus of the rat(44), a rapidly metabolizing homiotherm. In summation, until further studies are completed, the possibility cannot be ruled out that thyroxine is actively concentrated by specific cells as a first step during metamorphosis. If this is the case, however, the present investigation suggests that thyroxine is incorporated relatively rapidly and in small amounts.

An alternative hypothesis is suggested by these results as well as the work of others. Perhaps in the metamorphosing amphibian, thyroxine acts directly by quickly triggering or imprinting cells so that they may begin metamorphosis. If the temperature were above 5°C, observable metamorphosis would then occur in each tissue after the latent period had been exceeded. There is evidence to support and to negate this view. Single injections of triiodothyronine of thyroxine into the perivitelline space of embryonic axolotls were followed by the onset of metamorphosis some 9 days after the embryos had been placed in fresh water(45). Metamorphosis was complete at 25 to 35 days. Following the injection of thyroxine-^{125}I or thyroxine-^{14}C into *Ambystoma tigrinum* larvae, an initial peak hormone level is reached at 3 hours followed by a second peak in the liver and skin at 24 hours(46). Thereafter, hormone levels drop off rapidly and metamorphosis was able to proceed to completion without continuous hormone treatment.

Actinomycin-D suppresses most of the stimulatory effects of triiodothyronine upon metamorphosis in the axolotl if administered 8 hours or less before hormone injection(47). If administered 72 hours after hormone injection, the antibiotic has little or no effect, suggesting that changes initiated by the hormone have taken place within this interval. Similarly, the action of thyroxine upon tadpole liver can be inhibited by thiouracil administered during the latent period(48). Even though thyroxine seems to be incorporated and released by cells in a matter of hours, a latent period of as much as 24 hours to several days ensues before indications of metamorphosis can be found at the ultrastructural and biochemical levels(49,50).

If thyroid hormone does indeed act by triggering responses within susceptible tissues, hormone concentration is an additional complexity that must also be taken into account. Hormone concentration is relevant both with respect to threshold and to rate of response for a given tissue if temperature is held constant(36,51). Different tissues have notably different threshold levels at which they will first begin to undergo observable metamorphic changes. In general, the thresholds are highest for events normally encountered late in metamorphosis(36,37). Appropriately, thyroid gland release of hormone increases as metamorphosis progresses(29,30). The rate of response to thyroid hormone also increases with increasing hormone levels. Thyroxine dosages as low as 0.002 μg/l may bring about a given event in a hypophysectomised tadpole(51), the rate of response increasing with the dose level up to a maximum rate at about 100 μg/l(43,51). Therefore, triggering of cells in the induction of metamorphosis would require a concentration of hormone equal to the threshold level. Secondly, the "hormonal imprint" would somehow have to be related to the dose level and reflected in the ensuing rate of metamorphosis.

The mode of thyroid hormone action upon cells is not completely understood (reviewed by Wolff and Wolff, 52, and by Barker, 53). It has been suggested that some form of iodine itself, resulting from deiodination, is the effector. The tadpole liver is able to deiodinate thyroxine(26), an ability which increases as metamorphosis proceeds(27). On the other hand, deiodination and oxidative deamination of thyroxine by tadpole skin are not apparent until late in metamorphosis, making the importance of these pathways in metamorphosis questionable(54). No attempt has been made to correlate the extent of hormone uptake with subsequent deiodination during metamorphosis. Even though patterns of label distribution were similar in tissues whether the label was ^{14}C or ^{125}I, undoubtedly deiodination and perhaps other structural changes in the thyroxine molecule are occurring in vivo. Deiodination of about one-half of the accumulated thyroxine after 12 days of constant immersion in thyroxine solutions has been reported(14) but no attempt was made to search for the formation of analogs as well. Attempts are being made in this laboratory to analyze the metabolism of labeled thyroxine in conjunction with continuing studies on tissue uptake and autoradiographic localization.

The alterations undergone by thyroxine in becoming "active" are but one facet of its mode of action. Because of the variety of responses engendered by thyroxine in different tissues, including changes in oxidative metabolism, mitochondrial permeability, and protein synthesis, the hormone is either enabling preexisting systems to begin functioning or acting upon the ultimate control of all cell processes. Enhanced RNA turnover and increased DNA-dependent RNA polymerase activity, resulting in increased enzyme production, could account for the multiplicity of events effected by thyroxine at the cellular level (reviewed by Tata, 55). Indeed, in the amphibian, an accelerated synthesis of RNA and protein accompany induced metamorphosis in liver(56) and in isolated tail tips(11). Tata(50) recently demonstrated that an extensive proliferation of cytoplasmic membranes occurs simultaneously with ribosome formation and protein synthesis during the induction of metamorphosis by thyroid hormones. However, a latent period of 40 hours precedes RNA synthesis leading Tata to believe that the actual effect of the hormone is to initiate an unknown chain of reactions which in turn head to the generalized responses of protein synthesis, membrane and ribosome formation. The determination of the intracellular site of action of thyroid hormone may point to the nature of this initial chain of reactions.

Labeled hormone localization in specific cells or tissues can be misleading because it is predicated on the assumption that a hormone persists at or primarily moves to its site of action or metabolism. In fact, the present study suggests that the means by which specific tissues rid themselves of thyroid hormone may be as informative as the uptake of hormone which seemingly occurs passively even in dead tadpoles.

REFERENCES

1. Kaltenbach, J. C. Nature of hormone action in amphibian metamorphosis. *In* Etkin, W., and Gilbert, L. I., eds., Metamorphosis: A Problem in Developmental Biology, New York, Appleton-Century-Crofts, pp. 399–441, 1968.
2. Beaudoin, A. R. The development of lateral motor column cells in the lumbo-sacral cord in *Rana pipiens*. II. Development under the influence of thyroxin. Anat. Rec., 125:247–259, 1956.
3. Reynolds, W. A. The effects of thyroxine upon the initial formation of the lateral motor column and differentiation of motor neurons in *Rana pipiens*. J. Exp. Zool., 153: 237–250, 1963.
4. Kollros, J. J., and McMurray, V. M. The mesencephalic V nucleus in anurans. II. The influence of thyroid hormone on cell size and cell number. J. Exp. Zool., 131:1–26, 1956.
5. Weiss, P., and Rossetti, F. Growth response of opposite sign among different neuron types exposed to thyroid hormone. Proc. Nat. Acad. Sci. U.S.A., 37:540–556, 1951.
6. Reynolds, W. A. Mitotic activity in the lumbo-sacral spinal cord of *Rana pipiens* larvae after thyroxine or thiourea treatment. Gen. Comp. Endocr., 6:453–465, 1966.
7. Race, J. Thyroid hormone control of development of lateral motor column cells in the lumbo-sacral cord in hypophysectomized *Rana pipiens*. Gen. Comp. Endocr., 1:322–331, 1961.
8. Barch, S. H. Oxygen consumption of normal and thyroxin-stimulated *Rana pipiens* skin. Physiol. Zool., 26:223–231, 1953.
9. Wilt, F. H. The organ specific action of thyroxin in visual pigment differentiation. J. Embryol. Exp. Morph., 7:556–563, 1959.
10. Weber, R. Zur Aktivierung der Kathepsine in Schwanzgewebe von Xenopuslarven bei spontaner und *in vitro* induzierter Rückbildung. Helv. Physiol. Pharmacol. Acta, 21: 277–291, 1963.
11. Tata, J. R. Requirement for RNA and protein synthesis for induced regression of the tadpole tail in organ culture. Develop. Biol., 13:77–94, 1966.
12. Kaltenbach, J. C. Localized metamorphic changes in the skin of *Rana pipiens* larvae by subcutaneous implants of mouse and frog thyroid glands containing radioactive iodine. Anat. Rec., 108:526–527, 1950.
13. Frieden, E., Wahlborg, A., and Howard, E. Temperature control of the response of tadpoles to triiodothyronine. Nature, 205:1173–1176, 1965.
14. Yamamoto, K., Kanski, D., and Frieden, E. The uptake and excretion of thyroxine, triiodothyronine and iodide in bullfrog tadpoles after immersion or injection at 25° and 6°C. Gen. Comp. Endocr., 6:312–324, 1966.

15. Dent, J. N., and Hunt. E. L. An autoradiographic study of iodine distribution in larvae and metamorphosing specimens of anura. J. Exp. Zool., 121:79–97, 1952.
16. Money, W. L., Lucas, V., and Rawson, R. W. The turnover of radioiodine (I^{131}) by the *Rana pipiens* tadpole. J. Exp. Zool., 128:411–421, 1955.
17. Hunt, E. L., and Dent, J. N. Iodine uptake and turnover in the frog tadpole. Physiol. Zool., 30:87–91, 1957.
18. Kaye, N. W., and LeBourhis, E. E. Uptake and turnover of a single injected dose of I^{131} in tadpoles of *Rana clamitans*. Zoologica, 43:73–76, 1958.
19. Lynn, W. G., and Dent, J. N. Phenylthiouria treatment and binding of radioactive iodine in the tadpole. Biol. Bull., 113:160–169, 1957.
20. Race, J., and Cameron, J. A method for the detection of the thyroid hormones in the plasma of amphibian larvae by thin layer chromatography. Amer. Zool., 6:302, 1966.
21. Taylor, C. A., and Kollros, J. J. Stages in the normal development of *Rana pipiens* larvae. Anat. Rec., 94:7–24, 1946.
22. Kollros, J. J. Immersion as a method of thyroxine administratation in amphibian metamorphosis studies. Develop. Biol., 7:1–10, 1963.
23. Rogers, A. W. Techniques of Autoradiography. Amsterdam, Elsevier Publishing Co., 335 pp., 1967.
24. Prescott, D. M. Autoradiography with liquid emulsion. *In* Prescott, D. M., ed., Methods in Cell Physiology, New York, Academic Press, Inc., Vol. 1., pp. 365–370, 1964.
25. Lipner, H., and Hazen, S. Extrathyroidal iodine pump in tadpoles *(Rana grylio)*. Science, 138:898–899, 1962.
26. Galton, V. A., and Ingbar, S. H. Observations on the relation between the action and the degradation of thyroid hormones as indicated by studies in the tadpole and the frog. Endocrinology, 70:622–632, 1962.
27. Yamamoto, K. Changes in thyroxine deiodinase of the frog, *Xenopus laevis* Daudin, during metamorphosis and growth. Gen. Comp. Endocr., 4:360–369, 1964.
28. Kaye, N. W. Interrelationships of the thyroid and pituitary in embryonic stages of the frog, *Rana pipiens*. Gen. Comp. Endocr., 1:1–20, 1961.
29. Saxen, L., Saxen, E., Toivonen, S., and Salimaki, K. Quantitative investigation on the anterior pituitary-thyroid mechanism during frog metamorphosis. Endocrinology, 61: 35–44, 1957.
30. Shellabarger, C. J., and Brown, J. R. The biosynthesis of thyroxine and 3:5:3:-triiodothyronine in larval and adult toads. J. Endocr., 18:98–101, 1959.
31. Etkin, W. The mechanisms of anuran metamorphosis. I. Thyroxine concentration and the metamorphic pattern. J. Exp. Zool., 71:317–340. 1935.
32. Tata, J. R. Distribution and metabolism of thyroid hormones. *In* Pitt-Rivers, R., and Trotter, W. R., eds., The Thyroid Gland. Washington, D.C., Butterworth, Inc., Vol. 1, pp. 163–186, 1964.
33. Ford, D. H., Corey, K. R., and Gross, J. The localization of thyroid hormones in the organs and tissues of the guinea pig: An autoradiographic and chromatographic study. Endocrinology, 61:426–447, 1957.
34. ——— and Gross, J. The metabolism of I^{131}-labelled thyroid hormones in the hypophysis and brain of the rabbit. Endocrinology, 62:416–436, 1958.
35. Reynolds, W. A. Metamorphic stasis in the larval amphibian. Proc. Ind. Acad. Sci., 74: 351–352 (Abstr.), 1965.
36. Kollros, J. J. Mechanisms of amphibian metamorphosis: Hormones. Amer. Zool., 1: 107–114, 1961.
37. Moser, H. Ein Beitrag zur Analyse der Thyroxinwirkung im Kaulquappenversuch und zur Frage nach dem Zustandekommen der Frukbereitschaft des Metamorphose-Reaktionssystems. Rev. Suisse Zool., 57:(Suppl. 2) 1–144, 1950.
38. Liao, S. H., and Hamilton, R. H. Intracellular localization of growth hormones in plants. Science, 151:822–824, 1966.
39. Rogers, A. W., Thomas, G. H., and Gates, K. M. Autoradiographic studies on the dis-

tribution of labelled progesterone in the uterus of the rat. Exp. Cell Res., 40:668–670, 1966.
40. Stumpf, W. E., and Roth, L. J. High-resolution autoradiography of H^3-estradiol with unfixed, unembedded 1.0 μ freeze-dried frozen sections. *In* Advances in Tracer Methodology. New York, Plenum Press, Ch. 4, pp. 113–125, 1968.
41. Maurer, H. R., and Chalkley, G. R. Some properties of a small nuclear binding site of estradiol. J. Molec. Biol., 27:431–441, 1967.
42. Siegal, E., and Tobias, C. A. End-organ effects of thyroid hormone; subcellular interactions in cultured cells. Science, 153:763–765, 1966.
43. Reynolds, W. A. Unpublished observation.
44. Hamilton, T. H. Control by estrogen of genetic transcription and translation. Science, 161:649–661, 1968.
45. Prahlad, K. V., and Delanney, L. E. A study of induced metamorphosis in the axolotl. J. Exp. Zool., 160:137–146, 1965.
46. ——— Personal communication. 1968.
47. ——— and Delanney, L. E. Actinomycin D: Effect on induced metamorphosis in the Mexican axolotl. J. Exp. Zool., 165:425–432, 1967.
48. Paik, W. K., and Cohen, P. P. Biochemical studies on amphibian metamorphosis. II. The effect of thiouracil on thyroxine-stimulated protein synthesis in tadpole liver. J. Biol. Chem., 236:531–535, 1961.
49. Kemp, N. E. Metamorphic changes of dermis in skin of frog larvae exposed to thyroxine. Develop. Biol., 7:244–254, 1963.
50. Tata, J. R. The formation, distribution and function of ribosomes and microsomal membranes during induced amphibian metamorphosis. Biochem. J., 105:783–800, 1967.
51. Kollros, J. J. Thyroxine and temperature thresholds in anuran metamorphosis. Anat. Rec., 125:624, 1956.
52. Wolff, E. C., and Wolff, J. Mechanism of action of thyroid hormones. *In* Pitt-Rivers, R., and Trotter, W. R., eds. The Thyroid Gland. Washington, D.C., Butterworth, Inc., Vol. 1, pp. 237–282, 1964.
53. Barker, S. B. Possible interrelationships between biological actions of thyroxine and its metabolism. Gunma Symp. Endocr., 3:181–201, 1966.
54. Dowling, J. T., and Rezevska, D. Thyroxine metabolism by amphibian skin during metamorphosis and molting, Gen. Comp. Endocr., 6:162–169, 1966.
55. Tata, J. R. Biological action of thyroid hormones at the cellular and molecular levels. *In* Litwack, G., and Kritchevsky, D., eds., Actions of Hormones on Molecular Processes. New York, Academic Press, Inc., pp. 58–131, 1964.
56. ——— Turnover of nuclear and cytoplasmic ribonucleic acid at the onset of induced amphibian metamorphosis. Nature, 204:378–381, 1965.

... [illegible] labeled [illegible] in [illegible] ... J. [illegible] Biol. [illegible]

40. [illegible], R. E. and [illegible], [illegible] high resolution [illegible] ... and [illegible] embedded [illegible] ... sections ... In [illegible] ... Methods ... Plenum Press ... 1968.

41. [illegible], [illegible] and [illegible], [illegible] ... [illegible] ... 1968.

42. [illegible] and [illegible] ... [illegible] ... [illegible] ... 1968.

43. [illegible], [illegible] ... [illegible]

44. [illegible], [illegible] ... [illegible] ... 1968.

45. [illegible] and [illegible] ... [illegible] ... 1968.

46. —— Personal communication, 1968.

47. [illegible] ... [illegible] ... [illegible] ... 1966.

48. [illegible] ... [illegible] ... [illegible] ... [illegible] ... 1968.

49. [illegible] ... [illegible] ... 1966.

50. [illegible] ... [illegible] ... [illegible] ... 1968.

51. [illegible] ... [illegible] ... 1968.

52. [illegible] and [illegible] ... [illegible] ... 1968.

53. [illegible] ... [illegible] ... 1968.

54. [illegible] and [illegible] ... [illegible] ... 1968.

55. [illegible] ... [illegible] ... [illegible] ... 1968.

56. —— ... [illegible] ... 1968.

24

Biochemistry of Anuran Pancreas Development During Thyroxine-Induced Metamorphosis*

Ki-Han Kim and Kenneth A. Slickers†

Department of Biochemistry, Purdue University, Lafayette, Indiana

INTRODUCTION

Cellular death in embryonic development is a part of the early growth in many animals; it also plays a role in the morphogenesis and the differentiation of organs and tissues. Thyroxine (T_4) induced anuran metamorphosis is one of the many examples of death-mediated, ontogenic changes caused by recognized hormones(1,2).

Thus far the main attention on this problem of necrosis has been focused on the analysis of cellular death per se. Tail resorption as a metamorphic event has been extensively studied in an effort to understand the mechanism of thyroxine action in this histolysis. However, resorption of the tail represents merely an effort by the organism to rid itself of a superfluous appendage, since modification or regeneration of the appendage dependent upon biosynthesis does not occur. Only in so far as the breakdown products may be reused by the organism does tail resorption serve other utilitarian purposes. On the other hand, pancreas histolysis appears to require certain predetermined cells to die, in order for those cells remaining or subsequently regenerating to proceed to differentiate in their gene expressions.

The tadpole pancreas regresses extensively before being entirely remodeled during either natural or thyroxine-induced metamorphosis; thus processes of degeneration and differentiation are all involved in the same organ. This cytodifferentiation of the pancreas involves several developmental events occurring in strict chronological order and to finite extents. It thus provides a system for the analysis of the biochemical processes underlying cellular necrosis. Likewise, it is possible to examine the functional significance of this necrosis in the subsequent genetic expressions of the differentiated organ.

*Journal paper no. 3515 of the Purdue Agriculture Experiment Station. This investigation was supported in part by grant from the Purdue Cancer Committee.
†Recipient of NSF graduate fellowship.

This article reports some of our recent observations relevant to the biochemistry of anuran pancreatic metamorphosis and to the significance of its regression with respect to the phenotypic expressions in the pancreas. It is our purpose to ascertain that necrosis occurs, that this necrosis is either caused by or at least accompanied by an increase in the acid hydrolase activities, and that some functional purposes are served by this death in view of observed changes in an exocrine activity.

CAUSE OF PANCREATIC REGRESSION-NECROSIS VERSUS DEHYDRATION

Using *Rana pipiens*, Kuntz(3) reported that the pancreas undergoes a reduction in size during metamorphosis. Kaywin(4) and Jane(5) also observed the reduction in size, and their cytological studies suggested that this phenomenon was the result of acinar cell degeneration. A more recent paper by Race et al.(6) suggests, however, that this reduction was mainly the result of dehydration that occurs during metamorphosis. In order to definitely establish the principal contribution to pancreatic regression—necrosis or dehydration—the relationship of the degenerating organ to its macromolecular constituents was studied. The kinetics of the regression, as well as quantitative and qualitative aspects of DNA, RNA, and protein turnover were observed throughout the regression in both control and thyroxine-treated *Rana catesbeiana* tadpoles.

All animals were first matched with regard to weight and premetamorphic status (it being sufficient for our purposes if no hind legs were visible). Treatment involved immersing the animals in either water or 2.6×10^{-8}M thyroxine solution.

Figure 1a shows the regression expressed as the decrease in pancreas weight relative to total body weight. Twelve days of thyroxine treatment (stage XX to XXI, by which time the *Rana catesbeiana* animals show developmental character-

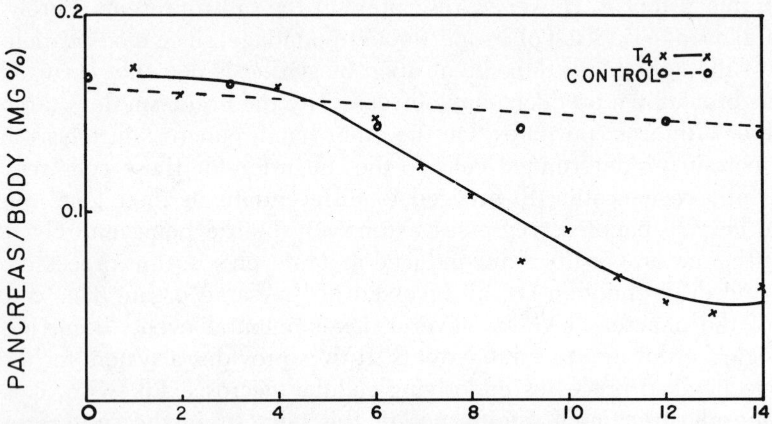

Fig. 1a. Relative weight of pancreas versus thyroxine treatment. Solid line: mg pancreas per 100 mg body weight of thyroxine=treated (T_4) animals. Dashed line: relative pancreas weight of control animals. Each point represents the average value of 5 to 7 animals.

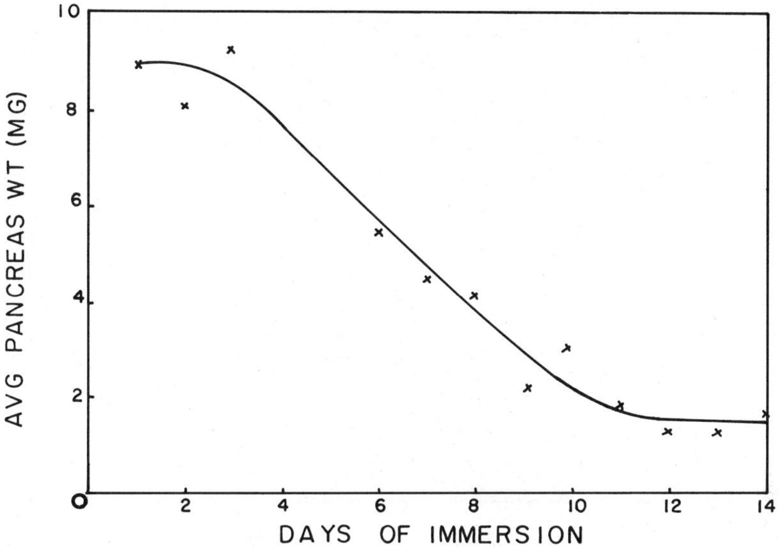

Fig. 1b. Average pancreas weight versus thyroxine treatment. The average weight decrease of the pancreases taken from animals previously matched in their total body weight is plotted against days of thyroxine treatment. Each point represents an average of 5 to 7 animals.

istics quite analogous to those of *Rana pipiens* to which the staging ordinarily applies)(7) results in a marked decrease in the ratio of pancreas weight to body weight (70 to 80 percent), whereas the ratio of these weights in control animals is nearly stationary. The average pancreas weight decreases from 9 mg to about 2 mg during 12 days of thyroxine treatment (Fig. 1b).

If animals are injected with 10 μg T_4 per gram body weight (GBW), the total length of time required to effect the same decrease in the relative size of the pancreas and to achieve the same enzyme changes described below is halved (reduced to 6 days). It is also observed that a dose-response curve on increasing thyroxine concentrations thus administered changes only the magnitude and not the temporal sequence of these parameters. Figure 2 records these data; curves 5 through 8 represent the time course of increase in acid phosphatase activity and will be discussed later.

Since the ratio of pancreas weight to body weight in untreated animals is relatively constant, we have examined the DNA content per pancreas using tadpoles of the same weight and at various stages of thyroxine treatment. As shown in Figure 3, the percentage weight decrease in the amount of DNA per pancreas is also about 75 percent at the peak of regression. The kinetics of DNA degradation and of pancreas weight decrease thus follow the same pattern and proceed to the same extent.

Similar studies on total cellular RNA and protein show 80 and 60 percent degradation of these macromolecules respectively as illustrated in Figures 4 and 5.

In contrast to tail regression, the regression of the pancreas is only partial and is followed by regeneration and differentiation of the organ. Although the majority

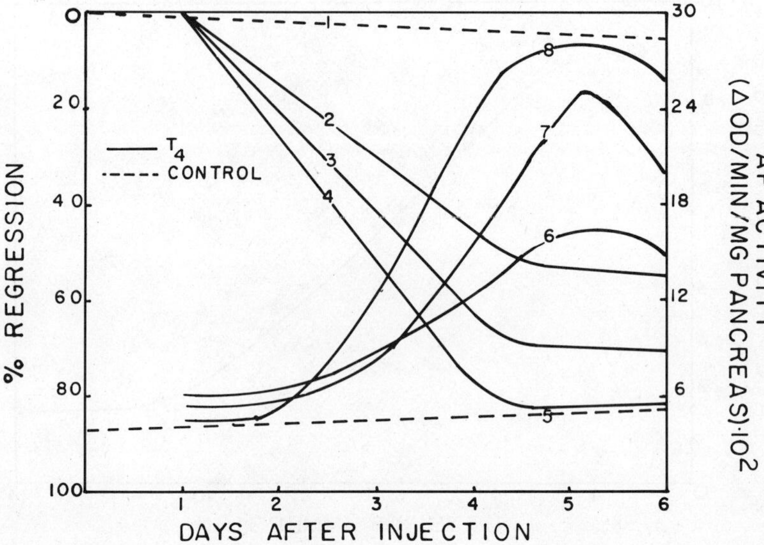

Fig. 2. Relative pancreas weight and acid phosphatase activity versus various levels of thyroxine treatment (injected hormone). Solid lines: relative pancreatic weight decreases following injections of different thyroxine concentrations; 2, 0.1 μg T_4/gram body weight (GBW); 3, 1.0 μg T_4 GBW; and 4, 10 μT_4/GBW. Lines 6, 7, and 8 respectively represent the kinetics of acid phosphatase activation of the 0.1,1.0, and 10 μg T_4/GBW groups. Dashed lines: relative pancreas weight decrease(1) and acid phosphatase activation(5) of control animals.

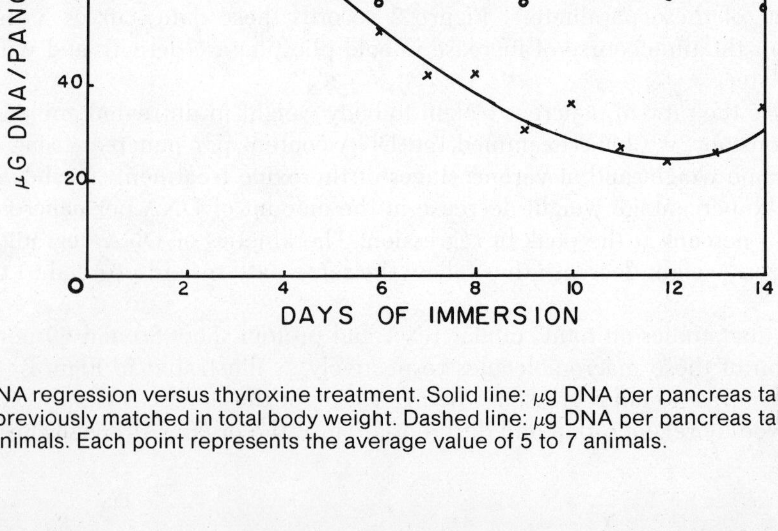

Fig. 3. DNA regression versus thyroxine treatment. Solid line: μg DNA per pancreas taken from animals previously matched in total body weight. Dashed line: μg DNA per pancreas taken from control animals. Each point represents the average value of 5 to 7 animals.

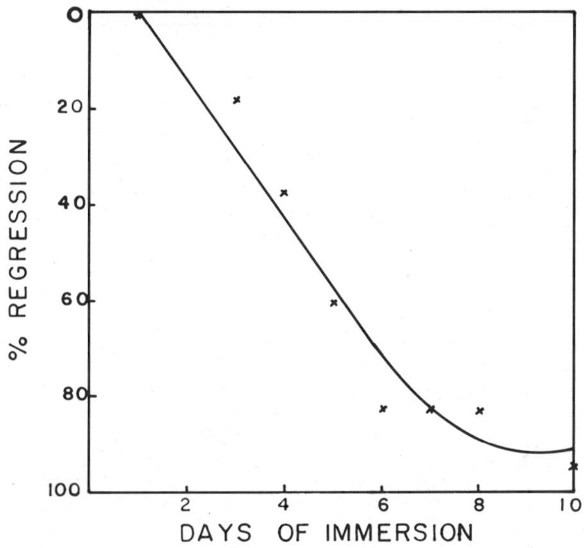

Fig. 4. Relative decrease in RNA during thyroxine treatment. Percent regression was calculated on the basis of the amount of RNA in the pancreas of control animals. Each point represents the average value obtained from samples of 5 to 7 animals previously matched in body weight. RNA determination was carried out as described(15).

of the pancreas cells are destined to die under the influence of thyroxine, it was of interest to study the capacity of the organ for macromolecular synthesis even during the regression period.

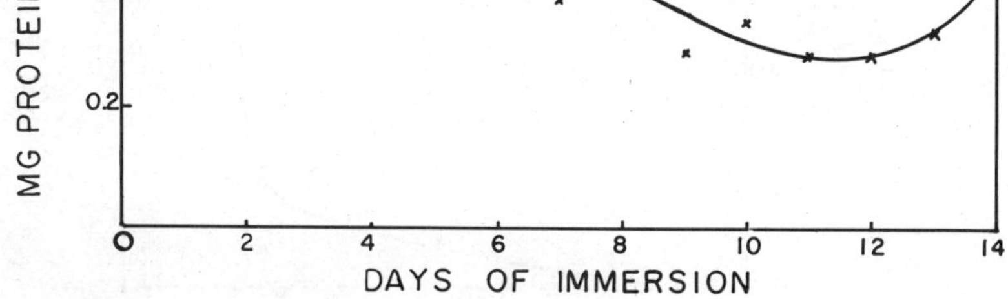

Fig. 5. Decrease in soluble protein during thyroxine treatment. The amount of soluble protein obtained from extraction in 0.2 M acetate buffer(pH 4.0, containing 0.2 percent cetyl trimethylammonium bromide) is plotted against days of hormone treatment. Animals (5 to 7 per sample) were matched in their body weights prior to hormone treatment.

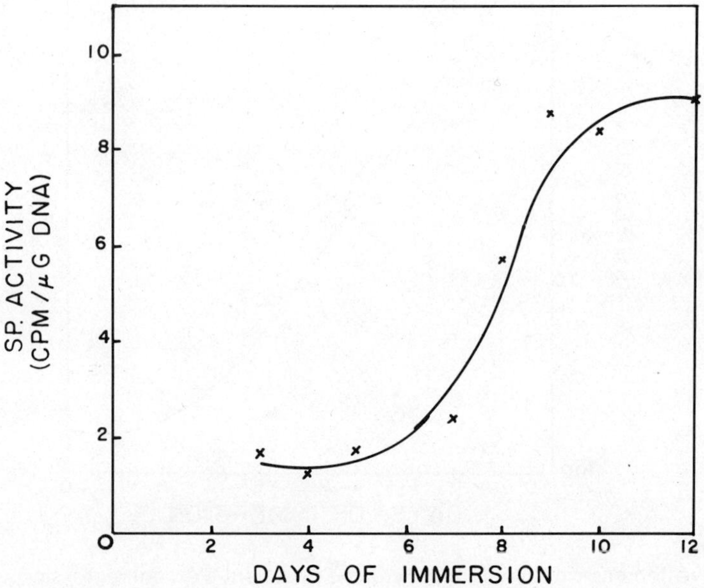

Fig. 6. RNA turnover versus thyroxine treatment. Five to seven animals were pulse labeled with tritiated uridine 3 hours prior to sacrifice. RNA was prepared as described(14). Solid line: RNA turnover of thyroxine-treated animals. Dashed line: RNA turnover of control animals.

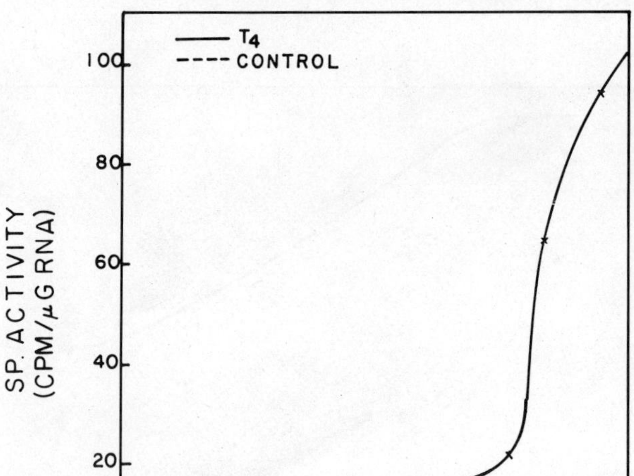

Fig. 7. DNA turnover versus thyroxine treatment. Tritiated thymidine was used to pulse label the animals. See Figure 6 for details.

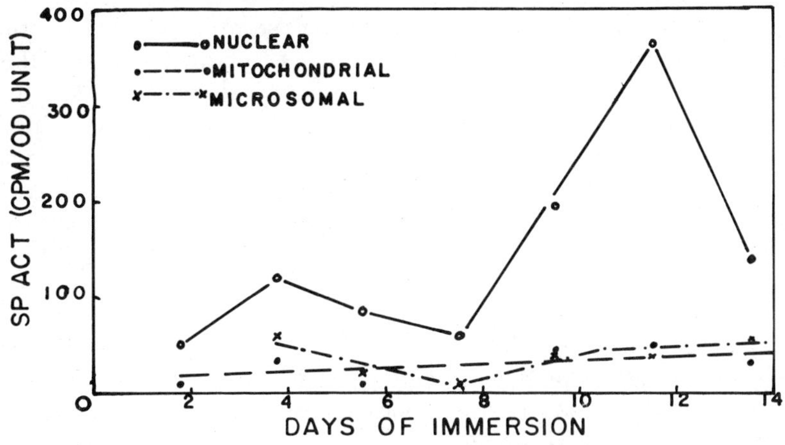

Fig. 8. RNA turnover versus thyroxine treatment. Various subcellular fractions and RNA were prepared as described(14). Turnover rate of sRNA is not indicated here since the amount of sRNA was extremely small and no detectable turnover was recorded. Solid line: nuclear RNA. Dashed line: microsomal RNA. Alternating dashes and dots: mitochondrial RNA. Data taken from Bollin(13).

Pulse labeling with tritiated thymidine or tritiated uridine for a period of 3 hours prior to sacrifice was followed by isolation of labeled DNA and RNA. Turnover rates, expressed as specific activity, against days of thyroxine treatment are shown on Figures 6 and 7. Until the sixth day of treatment, no significant turnover

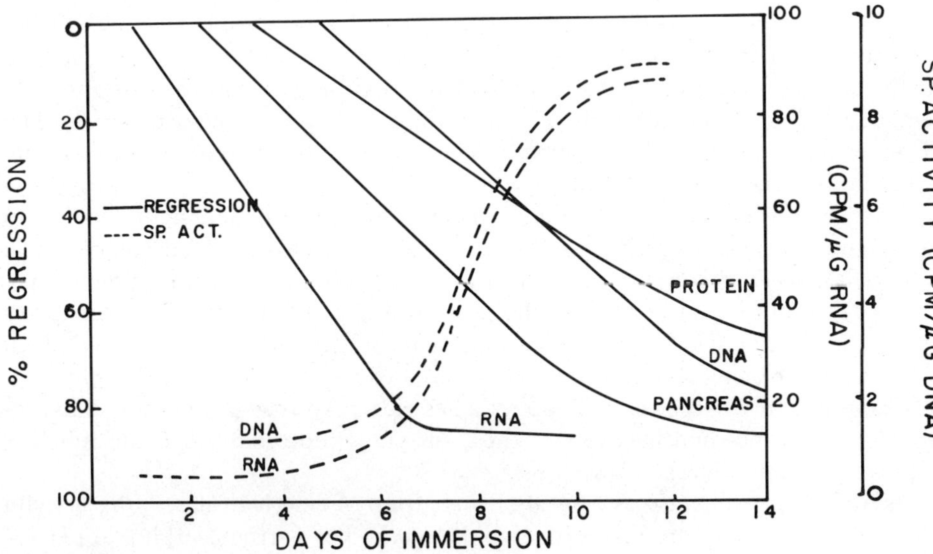

Fig. 9. Composite picture of developmental processes versus thyroxine treatment.

was observed. Following that, however, active turnover of both species of nucleic acids occurs. The turnover of RNA is mainly restricted to the nuclear fraction (Fig. 8). The potential significance of these turnover rates, and the function of the newly synthesized nucleic acids with regard to the subsequent differentiation of the organ remain to be examined. The inhibition of this RNA turnover by the injection of 1 μg actinomycin D per gram body weight at various stages of thyroxine treatment, however, had no effect on the subsequent regression.

Figure 9 gives an overall view of macromolecular degradation and turnover at the various stages of hormone treatment. The earliest notable event after initiation of hormone treatment is the gradual loss of cellular RNA. Both sRNA and rRNA are subject to this degradative process. Simultaneous turnover of nucleic acids occurs on about the eighth day, after the regression of the organ is well under way, in marked contrast to the situation in tail regression where a continuous turnover of RNA and protein is necessary for thyroxine-induced regression(8,9). These studies establish that pancreatic regression is the result of actual cell death and not of dehydration. This conclusion is further supported by histological studies in our laboratory as well as in those of others(4,5).

ACID PHOSPHATASE AND PANCREATIC REGRESSION

After establishing that pancreatic regression is the result of cellular necrosis, we have examined both qualitatively and quantitatively the behavior of acid hydrolases during this period of regression.

It is often supposed that acid hydrolases, localized in lysosomal particles, are causative agents in cellular necrosis. A possible example of this thesis might be found in the breakdown of the intersegmental muscles of saturniid silk moths(10). In this case of programmed histolysis, cathepsin and acid phosphatase are considered directly responsible for the necrosis. Even here, one must have more evidence as to the origins of these enzymes before judging that they are really causes and not effects of cell death. Another example which suggests that the increase in lysosomal hydrolases is the cause of degradation is found in the case of muscle degeneration of vitamin E-deficient dystropic rabbits(11).

On the other hand, there is evidence against the release of acid hydrolases from lysosome-like particles as being the primary cause of necrosis, in tail resorption in the tadpole(8,9). Furthermore, there seems to be definite need for de novo synthesis of these enzymes in order for the increased activity to occur. In most cases, however, increased acid hydrolase activities can be explained equally as well on the basis of their being a manifestation and not a cause of cellular death.

Regardless of whether or not one attaches causative roles to these acid hydrolases, the occurrence and behavior of these enzymes represent a specific function in the breakdown of larval tissue.

Using acid phosphatase as typical of this group of acid hydrolases, the activity of the enzyme was measured (according to the method of Torriani[16] but at pH 4.6 instead of pH 4.0) throughout pancreatic regression. Total enzyme activity (including that solubilized from the particulate fraction by the use of 0.2 M acetate buffer, pH 4.0, containing 0.2 percent cetyl trimethylammonium bromide) increased three-

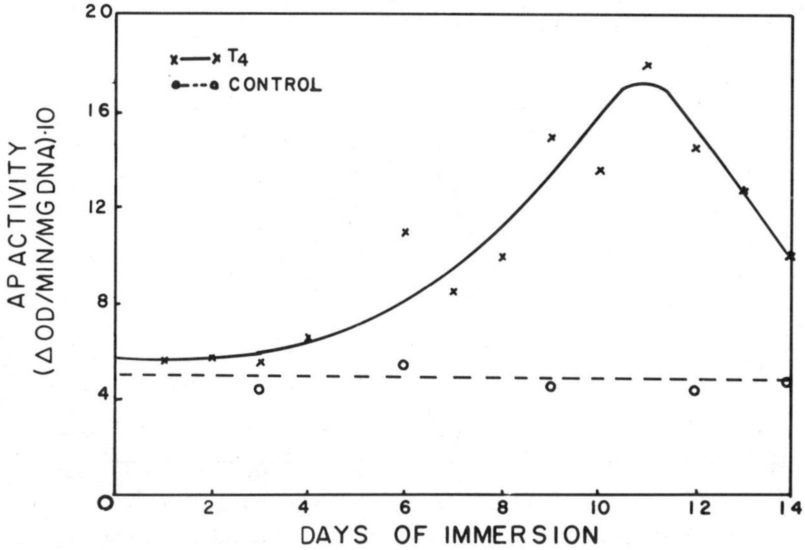

Fig. 10. Acid phosphatase versus pancreatic regression. Solid line: relative pancreatic weight decrease. Dashed line: kinetics of acid phosphatase activation.

to fourfold on a per mg DNA basis (Fig. 10), and up to fivefold on a per mg pancreas basis (Fig. 11), after 12 days of thyroxine treatment. Figure 11 reveals the coordinate manner in which enzyme activity increases and pancreas weight decreases, as well as the thyroxine dependence of these responses (see also Fig.2).

Puromycin (20 μg/GBW), cycloheximide (20 μg/GBW), and actinomycin D (5 μg/GBW) all fail to inhibit this increase in the activity when injected at concentrations sufficient to inhibit 90 percent of protein synthesis in the former two cases (12). It is therefore reasonable to conclude that this activity arises not from de novo protein synthesis, but from activation of preexisting enzyme. The results of the following experiment further substantiate that enzyme activation accounts for the increased specific activity.

Disc gel electrophoresis of the pancreas extracts from different points in the regression curve reveal qualitatively four distinct isozymes (migrating toward the anode at pH 7.5, 10 percent acrylamide, stained with a solution containing 1 mg/ml each of α-napthyl acid phosphate and fast garnet red). Figure 12 shows the results of such an electrophoretic determination, wherein *equal activities* of the extracts were layered on each gel. It is first of all clear that the intensity of the various bands modulates at different stages of regression. However, the most significant generalization with reference to the enzyme activation theory is that even though the same amount of enzyme activity is layered on the acrylamide gels, the activity as assayed on the gel is significantly higher in those samples taken from the more regressed pancreas. Furthermore, increasing amounts of the same enzyme preparation applied to the gels gave proportionately increased stains; and combined samples taken from control and 12-day treated animals gave additive assays both in aqueous and gel assays.

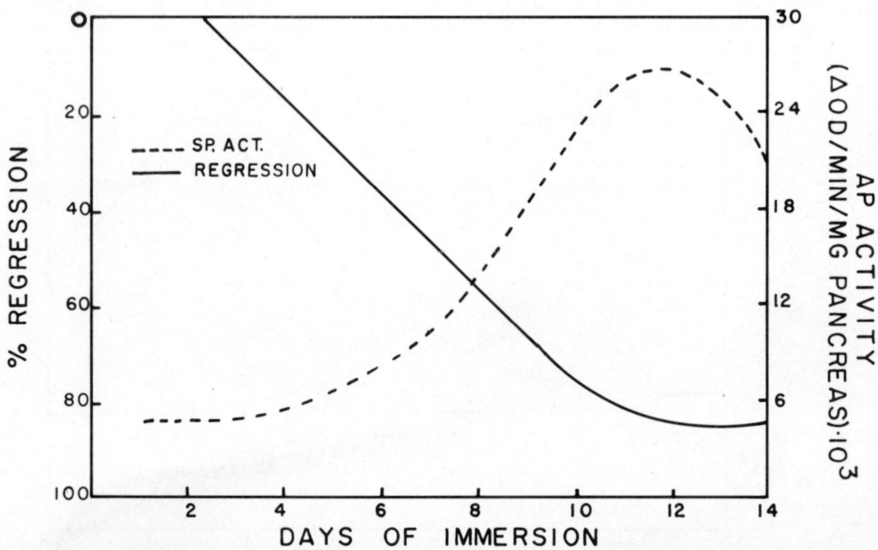

Fig. 11. Acid phosphatase versus thyroxine treatment. In this figure, activity of acid phosphatase is expressed on a per mg DNA basis. Solid line: acid phosphatase activity in the thyroxine-treated animals. Dashed line: acid phosphatase activity in control animals.

These experiments, therefore, suggest that electrophoresis may dissociate an enzyme-inhibitor complex present in tissue extracts from thyroxine-treated animals. Furthermore, the ability of the electrophoresis to separate enzyme from inhibitor is a direct consequence of the effect of thyroxine on the sample pancreas from which the extract was taken. In a similar way, the increased activity, measured in aqueous solutions after thyroxine treatment has progressed several days, may reflect the fact that the inhibitor-enzyme complex is now more labile. Finally, since there

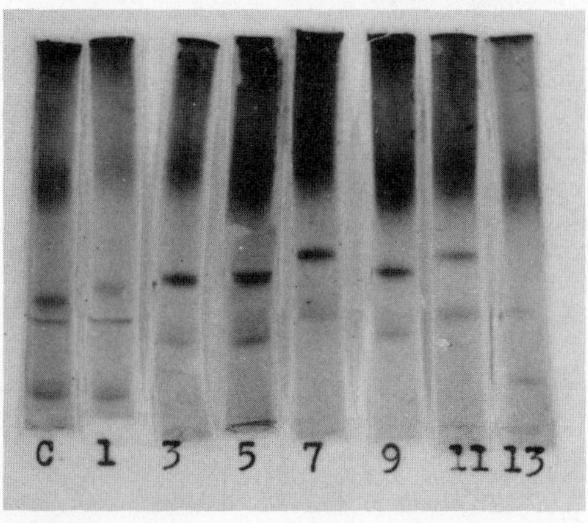

Fig. 12. Electrophoretic pattern of equal amounts of acid phosphatase activity from extracts as different stages of thyroxine treatment. *C* is an extract from control animals, and the numbers 1 to 13 represent the number of days spent under thyroxine treatment.

is no need for protein synthesis at any point in this process, inhibitors of protein synthesis would not be expected to prevent the increased activity. Indeed, the inhibitors of protein synthesis and RNA synthesis had no effect either on the regression or on the increase in acid phosphatase activity.

EXOCRINE ACTIVITY: α-AMYLASE

The pancreas is particularly suited to the analysis of gene action at the protein level, because both its endocrine principles and its exocrine secretions are well characterized. In order to correlate the significance of pancreas development with its functions, we have first examined the enzymatic make-up of the zymogen granules from premetamorphic and from metamorphosed pancreases. All major enzyme activities generally associated with these granules have been observed in both types of pancreas, thus posing a question of why the energy and information content of the larval pancreas should be mostly destroyed during metamorphosis.

However, careful examination of the α-amylases from premetamorphic and from metamorphosed tadpoles indicates that a clear difference exists between the enzymes with respect to their electrophoretic mobility.

Electrophoresis (pH 7.3, on 1 percent agar plates, stained with 0.1 percent soluble starch and iodine-iodide solutions) of α-amylase extracted from the tadpole pancreas shows three cathodal isozymes. Frog pancreas extracts show but one band of α-amylase activity migrating at a position somewhere between those occupied by tadpole isozymes II and III after 1-hour electrophoresis, 1 mA/cm as seen in Figure 13a, b, and c. In the first of these figures, α-amylase prepared from premetamorphic tadpoles gives three distinct isozymes, whereas the preparation from tadpoles treated 13 days in thyroxine gives but faint bands of isozymes II and III, and no detectable isozyme I. After 15 days of thyroxine treatment (stage XX), α-amylase is no longer detectable, while extracts from froglets show a new α-amylase. A similar phenomenon is illustrated in Figure 13b; here the tadpoles are undergoing natural metamorphosis, and again the new α-amylase appears only at the froglet stage. α-Amylase from adult frogs shows the same properties as that from froglets. The relative mobility of the various α-amylase isozymes is shown in Figure 13c. The new α-amylase is shown to migrate between tadpole isozymes II and III. The level of α-amylase in the pancreas during thyroxine-induced metamorphosis is recorded

Fig. 13a. Electrophoretic pattern of α-amylase extracted from thyroxine-treated metamorphosing tadpoles. *C*: control animals (stage VI); frog: froglet (stage XXV); 13 and 15: number of days of thyroxine treatment (stages XX and XXI respectively).

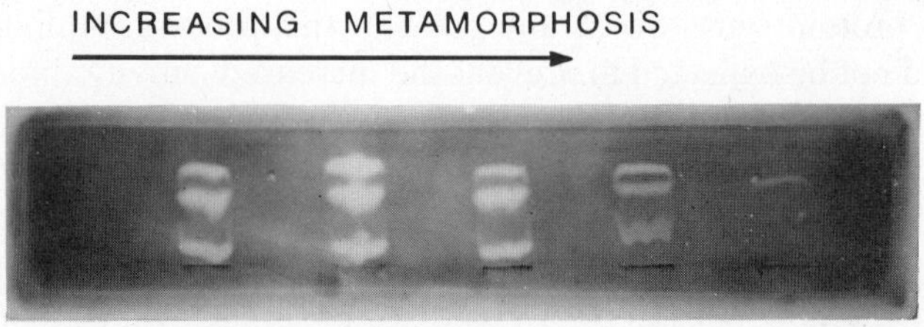

Fig. 13b. Electrophoretic pattern of α-amylases from naturally metamorphosing tadpoles. (1) stage VI; (2) stage XIX; (3) stage XX; (4) stage XXI; (5) completely metamorphosed froglet, stage XXV. Origin is at bottom.

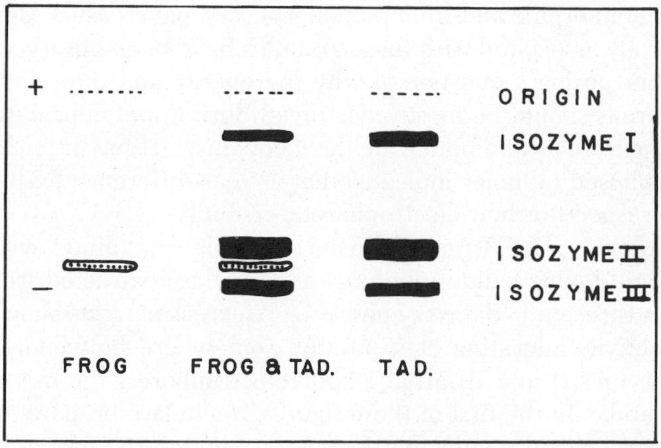

Fig. 13c. Electrophoretic patterns of α-amylases. (1) Extracted from froglet; (2) combined extracts from froglet and tadpole (stage VI); (3) extracted from stage VI tadpole.

in Figure 14; the activity decreases to an undetectable level after the twelfth day of hormone treatment.

With the exception of electrophoretic mobility, no differences in the properties of the two types of enzymes could be found with respect to pH optimum, inherent stability, substrate affinity, or sensitivity to various inhibitors.

Thus these studies indicate molecular as well as morphological differentiation of the pancreas. Among the other characteristics of pancreatic differentiation which have been reported are an increase in the amount of cytoplasm, an increase in cytoplasmic basophilia, and the accumulation of histochemically demonstrable insulin(14).

CONCLUDING REMARKS

The existence of an overall program regulating both the temporal sequence and the extent of various processes is characteristic of cytodifferentiation and of developmental events in biological systems. With our present knowledge we are

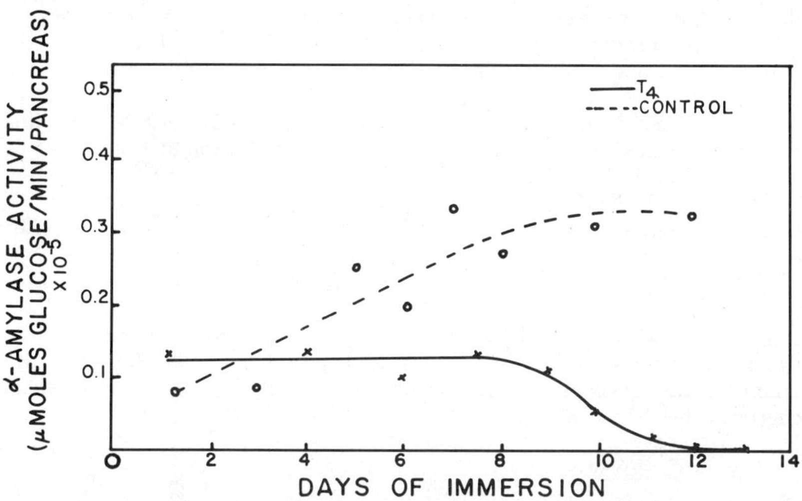

Fig. 14. α-amylase versus thyroxine treatment. Solid line: level of α-amylase during thyroxine treatment. Dashed lines: Level of α-amylase in control animals. Animals were previously matched in total body weight; activity is expressed on a per pancreas basis. Data taken from Bollin(13).

far from understanding the mechanism of the execution of these processes. There are still many unanswered questions: What is the triggering mechanism for the cell necrosis? What is the specificity of the cells towards the hormone action? Of what significance are the products of necrosis during the development; are the degradative products used at all for the regeneration?

Despite the existence of these questions, it is still clear both from these present studies and those of others that cellular death preceding development is not a casual and accidental happening, but rather is as carefully programmed as are other morphogenic events. The death of the pancreatic cells seems to serve various utilitarian goals in the overall process of the formation of the new adult organ.

REFERENCES

1. Saunders, J. W., Jr. Death in embryonic systems. Science, 154:604–612, 1966.
2. ——— and Fallon, J. F. Cell death in morphogenesis. *In* Major Problems in Developmental Biology, Locke, ed., Academic Press, Inc., New York, 1966.
3. Kuntz, A. Anatomical and physiological changes in the digestive system during metamorphosis in *R. pipiens* and *A. tigirnum.* J. Morph., 38:581–598, 1924.
4. Kaywin, L. A cytological study of the digestive system of anuran larvae during accelerated metamorphosis. Anal. Rec., 64:413–440, 1936.
5. Jane, R. G. Studies in the amphibian digestive system. J. Morph., 61:581–611, 1937.
6. Race, J., Jr., Robinson, C., and Terry, K. J. The influence of thyroxine on the normal development of the pancreas in *R. pipiens* larvae. J. Exp. Zool., 162:181–192, 1966.
7. Taylor, A. C., and Kollros, J. J. Stages in the normal development of *Rana pipiens* larvae. Anat. Rec., 94:7–23, 1946.
8. Tata, J. R. Requirement for RNA and protein synthesis for induced regression of the tadpole tail in organ culture. Develop. Biol., 13:77–94, 1966.

9. Weber, R. Inhibitory effect of Actinomycin D on tail atrophy in Xenopus larvae at metamorphosis. Experientia, 21:1–5, 1965.
10. Lockshin, R. A., and Williams, C. M. Programmed cell death. J. Invest. Physiol., 11: 831–844, 1965.
11. Zalkin, H., Tappel, A. L., Caldwell, K. A., Shibko, S., Desai, I, D., and Holliday, T. A. Increased lysosomal enzymes in muscular dystrophy of vitamin E-deficient rabbits. J. Biol. Chem., 237:2678–2682, 1962.
12. Slickers, K. A., and Kim, K. Unpublished results.
13. Bollin, E., Jr., and Kim, K. Pancreatic RNA metabolism in metamorphosing *Rana catesbeiana*. MS thesis, Purdue University, January 1970.
14. Frye, B. E. Metamorphic changes in the blood sugar and the pancreatic islets of the frog. J. Exp. Zool., 155:215–224, 1964.
15. Nakagawa, H., Kim, K., and Cohen, P.P. Studies on RNA synthesis in tadpole liver during metamorphosis induced by thyroxine, J. Biol. Chem., 242:635–641, 1967.
16. Torriani, A. Influence of inorganic phosphate on the formation of phosphatases by *Escherichia coli*. Biochim. Biophys. Acta, 38:460–468, 1960.

25

Thyroxine-Dependent Glial Activation in the Developing Central Nervous System of Amphibians*

Irwin Pesetsky and Pat G. Model

Department of Anatomy, Albert Einstein College of Medicine, Bronx, New York

INTRODUCTION

Normal differentiation and growth of the vertebrate central nervous system (CNS) is impossible in the absence of thyroid hormone(1,2). The profound mental retardation associated with this deficiency has long been recognized in man. However, relatively little is known about the cellular mechanisms which respond to the presence or absence of the hormone. Our studies have dealt with developing amphibians whose thyroidal state has been altered. Like mammals, these animals display neurological deficits when deprived of thyroid hormone(2,3). In contrast to mammals, however, their overall well-being is not diminished. Tadpoles deprived of their thyroid glands early in life can never become frogs, but they continue to feed and grow, becoming much larger than normal larvae(3,4). Such giant, well-proportioned tadpoles have normal or supranormal cranial dimensions, but their brains are consistently smaller in size than those of controls(5). Under these conditions, the CNS retains characteristics normally observed only in very young larvae: cells are closely packed, ventricles are relatively large in size, and walls are thin (6). In the absence of thyroid hormone, the growth of many neurons is markedly retarded, i.e., cells of the mesencephalic V nucleus(7), neurons of the lateral motor column(8), and the pair of normally "giant" Mauthner neurons(3). The administration of exogenous hormone, however, restores growth in these cells (3,7,8). By placing thyroxine-bearing pellets into the brain, Kollros and his associates (see review, 2) have furnished convincing evidence that thyroid hormone may act *directly* on tissues of the CNS. However, it is not clear whether this action is directly upon neurons themselves or whether neuroglial elements are also involved.

*Supported by the following grants from N.I.H., U.S.P.H.S.: NB04555, NB07512, MH06418.

In the brain, neurons and glial fibers are intimately interwoven, and there is evidence suggesting that these cell types are metabolically linked(9). Biochemical investigations indicate that thyroid hormones affect brain maturation by their influence on protein(10) and lipid synthesis in the brain(11). In interpreting these data, it must be recognized that in brain tissue samples used in such studies, the glia occupy far more volume (up to 90 percent) than do neurons. Therefore, thyroxine-mediated changes in enzyme synthesis(12) or incorporation of amino acids into protein(10) might well be attributed to glia rather than exclusively to neurons. There are also data in the literature which may be interpreted as suggesting that in the larval amphibian brain, glial responses accompany or precede thyroxine-elicited developmental alterations in nerve cells. In frogs and tadpoles, a major portion of the glial population is represented by the ependymal cells lining the ventricles of the brain(13). In *Rana* larvae, thyroxine-dependent morphological changes in nerve cells are generally not detected in preparations treated with the hormone for less than 5 days(2,14,15). However, within 36 hours after thyroxine administration, ependymal glia are thrown into a series of rapid divisions. The mitotic index, according to Weiss and Rossetti(16), reaches a peak at 3 days of treatment and then falls toward normal levels. It is also relevant to note that a similar pattern of ependymal mitotic activity has been observed in *Bufo* tadpoles during metamorphic climax(17), a period of several days during which the thyroid gland is most active. Thus, whether metamorphosis occurs normally in fully grown larvae or whether it is precociously induced by treating immature tadpoles with exogenous thyroxine, ependymal glial cells in the brain display proliferative activity.

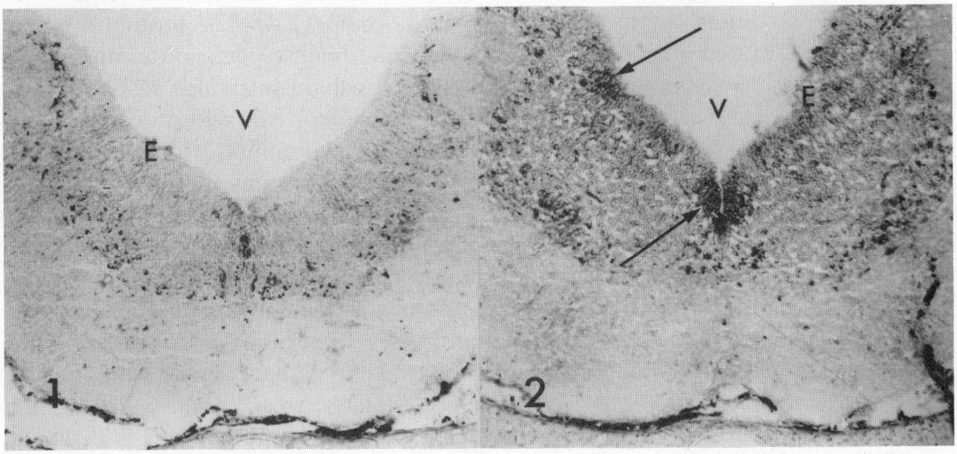

Fig. 1. NADPH (TPNH) diaphorase activity in a cross section through the hindbrain of a normal, untreated stage VIII *Rana pipiens* larva. Incubation, 55 minutes. Note intensely black pigment granules. *V*, fourth ventricle; *E*, ependyma. ×100. From Pesetsky. Gen. Comp. Endocr., 5:411-417, 1965. Courtesy of Academic Press, Inc., New York.

Fig. 2. NADPH (TPNH) diaphorase activity in the hindbrain of a similar larva treated with L-thyroxine, 222 μg/liter, for 6 days. This section was incubated "back-to-back" with the tissue shown in Figure 1. Enzyme activity is increased throughout the section, but is especially intense in the ependyma. Staining is most conspicuous in cells to either side of the median sulcus. *V*, fourth ventricle; *E*, ependyma. ×100. From Pesetsky. *Gen. Comp. Endocr.*, 5:411–417, 1965. Courtesy of Academic Press, Inc., New York.

OXIDATIVE ENZYME ACTIVITY OF EPENDYMOGLIA

Earlier studies from this laboratory demonstrated that thyroxine-stimulated neuron growth and maturation in larval *Rana pipiens* were accompanied by changes in oxidative enzyme activity in the glia(18). In young tadpoles or in larvae surgically thyroidectomized in embryonic stages prior to the onset of thyroxine-synthesis, ependymoglial cells displayed low levels of glucose 6-phosphate dehydrogenase (G6-PD) and NADPH diaphorase activity (Fig. 1). In matched animals immersed in solutions of L-thyroxine, 222 μg/liter for 6 days, the activity of both these enzymes was sharply increased. Although thyroxine-treatment resulted in increased enzymic activity in small mantle neurons and choroid plexus epithelium, the most prominent effect was on ependymal glial cells (Fig. 2), particularly those bordering the median ventral raphe.

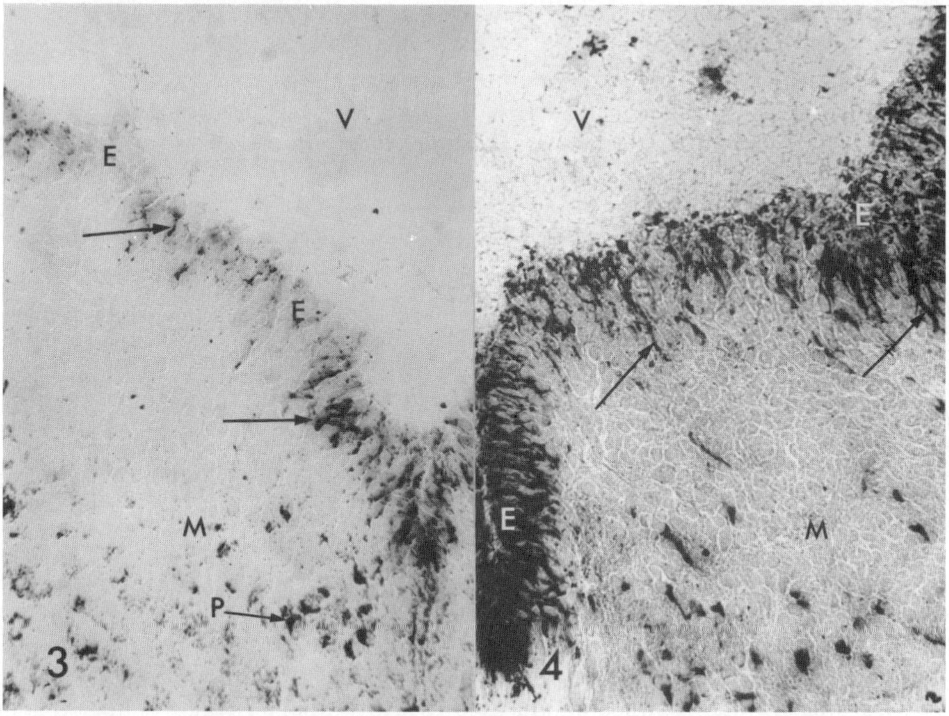

Fig. 3. TPPase activity in ependymal glia in hindbrain of a thyroidectomized *R. pipiens* larva, stage VI, 6 months old. Incubation, 60 minutes. *P* indicates pigment granules in neurons below the median sulcus. *V*, fourth ventricle; *E*, ependyma; *M*, mantle. No counterstain. ×300. From Pesetsky. Gen. Comp. Endocr., Suppl. 2:238-244, 1969. Courtesy of Academic Press, Inc., New York.

Fig. 4. TPPase activity in a normal larva treated with L-thyroxine, 222 μg/liter for 7 days. Incubation, 60 minutes "back-to-back" with preparation shown in Figure 1. No counterstain. ×300. From Pesetsky. *Gen. Comp. Endocr.*, Suppl. 2:238-244, 1969. Courtesy of Academic Press, Inc., New York.

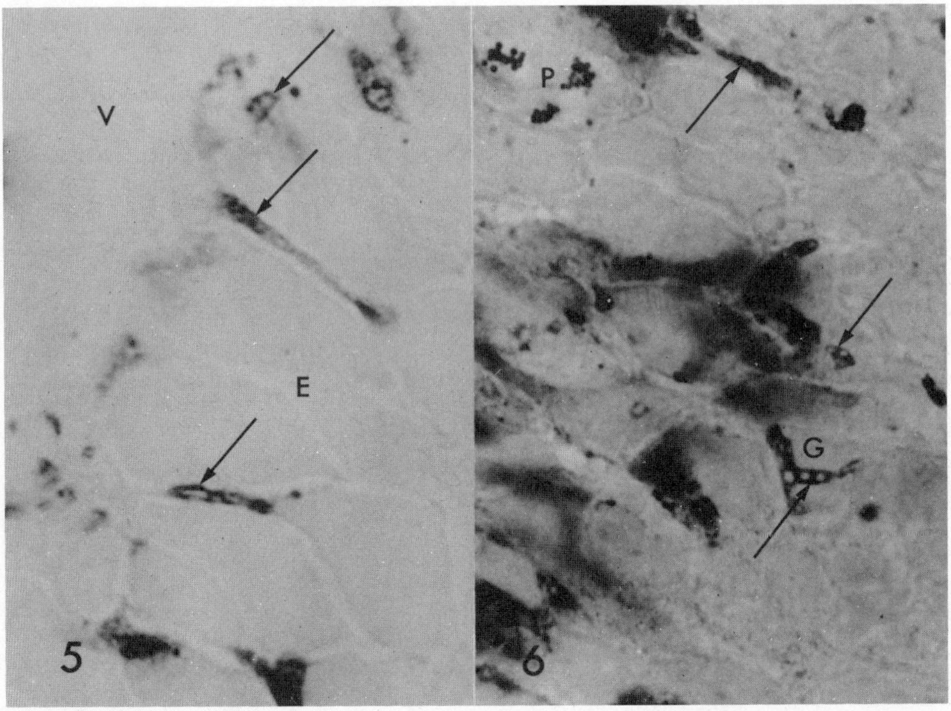

Fig. 5. TPPase activity in centripetal processes and apices of ependymal astroglial cells in hindbrain. *R. pipiens,* normal untreated control larva, stage VII. Incubation, 60 minutes. No counterstain. ×1,800. From Pesetsky. *Gen. Comp. Endocr.,* Suppl. 2:238–244, 1969, Courtesy of Academic Press, Inc., New York.

Fig. 6. TPPase activity in segments of centrifugally directed glial processes of a thyroxine-treated thyroidectomized larva. Treatment, 222 μg/liter 7 days. *P* is one cell away from the ventricular lumen and indicates three nearby clusters of pigment granules. Note enlarged Golgi lamellae and vesicles at *G.* No counterstain. ×1,200. From Pesetsky. *Gen. Comp. Endocr.,* Suppl. 2:238-244, 1969. Courtesy of Academic Press, Inc., New York.

Nucleoside Phosphatase Activity

In histochemical studies on the brains of young or thyroidectomized tadpoles (19), the activity of thiamine pyrophosphatase (TPPase) or inosine diphosphatase (IDPase) was low and confined almost exclusively to capillary walls, epithelial cells of the choroid plexus, and ependymal cells on the ventricular surface of the brain (Fig. 3). These enzymes generally reside in the Golgi apparatus(20). In ependymal cells, tortuous, filiform organelles were delineated by the black or brown products of the histochemical reaction (Fig. 5). Little or no activity was seen in neurons.

In the brains of young or thyroidectomized larvae treated with thyroxine for 5 to 8 days, nucleoside diphosphatase activity in ependymal cells was markedly increased. In addition, perhaps in part as a consequence of hormone-stimulated ependymal proliferation, a far greater number of ependymal cells displayed ac-

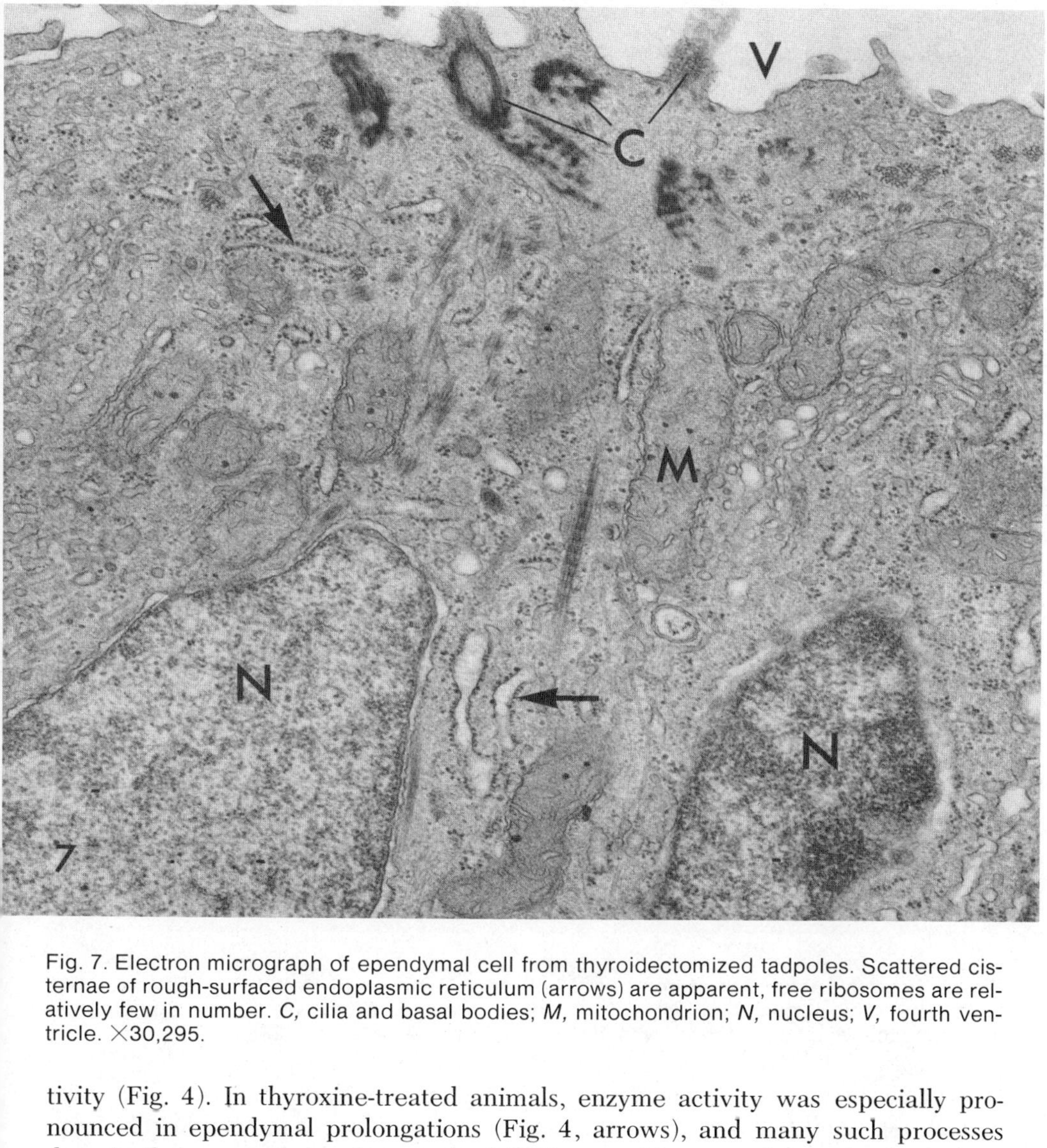

Fig. 7. Electron micrograph of ependymal cell from thyroidectomized tadpoles. Scattered cisternae of rough-surfaced endoplasmic reticulum (arrows) are apparent, free ribosomes are relatively few in number. *C*, cilia and basal bodies; *M*, mitochondrion; *N*, nucleus; *V*, fourth ventricle. ×30,295.

tivity (Fig. 4). In thyroxine-treated animals, enzyme activity was especially pronounced in ependymal prolongations (Fig. 4, arrows), and many such processes deep in the neuronal mantle contained TPPase-positive vacuolar networks (Fig. 6, arrows). No additional activity was seen in neurons, but capillaries and choroid plexus staining was more intense than that found in controls. In all these preparations, blood vessel activity was intense, and possible endothelial staining could not be differentiated from activity attributable to glial foot plates.

Ultrastructure of Ependymoglial Cells

Our studies with the electron microscope were performed primarily on *Rana pipiens* larvae which had been surgically thyroidectomized 1 to 1½ years earlier in

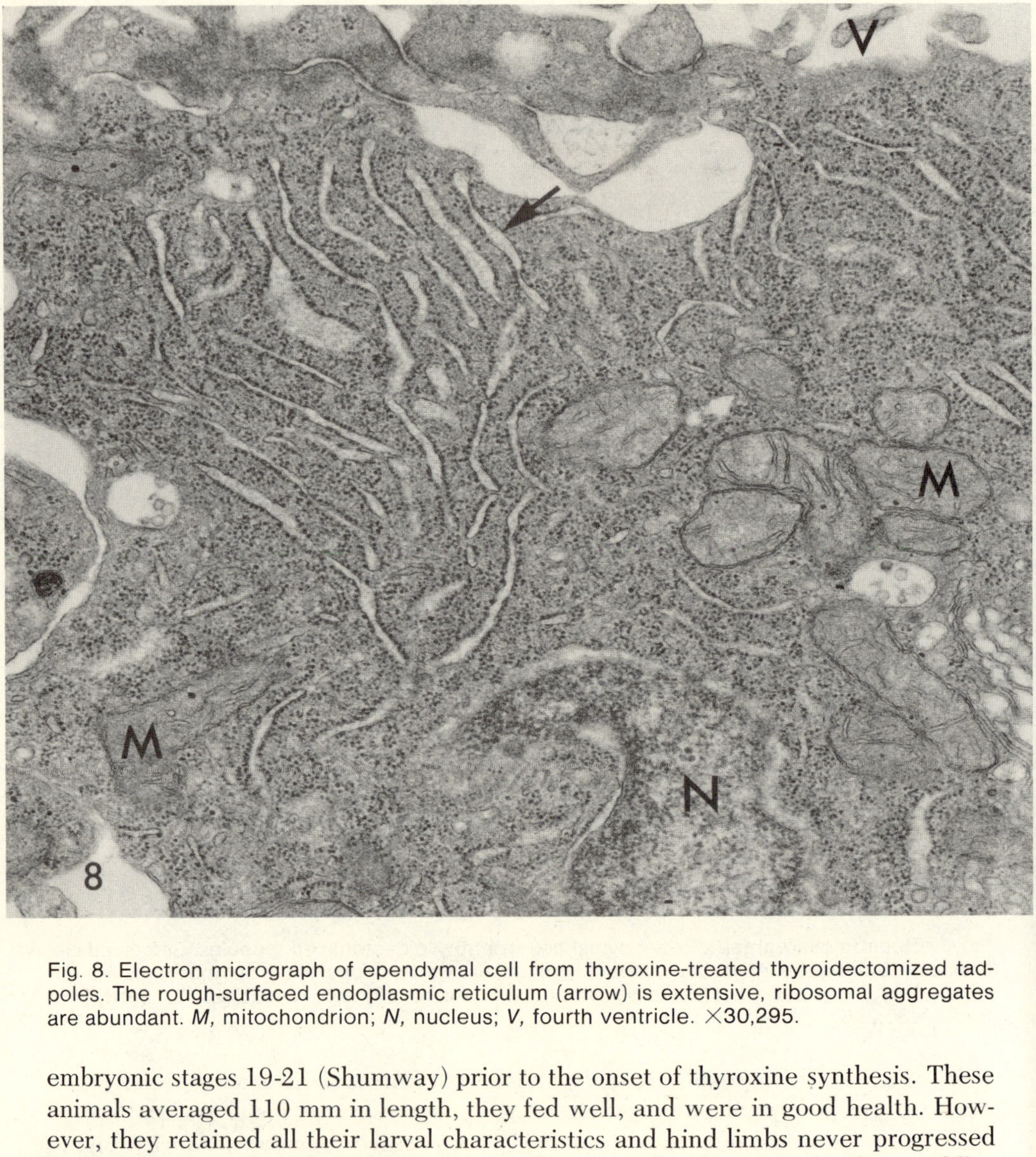

Fig. 8. Electron micrograph of ependymal cell from thyroxine-treated thyroidectomized tadpoles. The rough-surfaced endoplasmic reticulum (arrow) is extensive, ribosomal aggregates are abundant. *M,* mitochondrion; *N,* nucleus; *V,* fourth ventricle. ×30,295.

embryonic stages 19-21 (Shumway) prior to the onset of thyroxine synthesis. These animals averaged 110 mm in length, they fed well, and were in good health. However, they retained all their larval characteristics and hind limbs never progressed beyond the stage characterized by the development of a flattened digitless paddle-like foot. At the ultrastructural level, we found(21) that ependymal cells lining the ventricular surfaces of the brains of thyroidless tadpoles were characterized from the normal primarily by a paucity of ribosomes and by a pronounced scarcity of granular endoplasmic reticulum (rER) profiles (Fig. 7). Many cells contained basal clusters of dilated smooth-surfaced vesicles, some apparently associated with the Golgi apparatus. The Golgi complex was extensive, and lipid droplets and lyosomes were frequently encountered. These cells contained round to elongate, moderately dense mitochondria with transverse cristae and occasional dense bodies.

L-thyroxine, 3 μg per day, I.M. was administered to matched thyroidectomized larvae for 7 to 9 days. Ependymal cells in these animals elaborated extensive arrays of rER lamellae; the cytoplasm became filled with polysomes (Fig. 8). The frequency of lipid droplets and lysosomes was reduced, but little or no change was seen in mitochondria or Golgi.

DISCUSSION

Ependymal glia in lower vertebrates are astroglia. This glial class includes such cell types as astrocytes, Bergman glia of the cerebellum, and retinal Müller cells. Features which are common to the various types of astroglia include processes and terminal footplates which connect the vascular and meningeal surfaces of the brain with the juxtaneuronal interior(13,22). In addition, the astroglia are rich in glycogen, and at the electron microscope level they contain typical gliofibrils(13).

In lower vertebrates, basal processes from ependymal cells span the entire thickness of the brain wall, establishing or approximating contacts with blood vessels and neurons(13). By virtue of their exposure to the cerebrospinal fluid-filled spaces of the brain and their proximity to capillaries and neurons, these ependymal cells are anatomically well situated to carry on the variety of nutritive and regulatory functions postulated by Oksche(13), Friede(22), and Tschirgi(23). These investigators have suggested that astroglial cells are instrumental in transporting nutrient substrates from blood to neurons(13) or in transporting and regulating the passage of water and electrolytes between these mutually isolated compartments of the brain(22,23).

There is evidence suggesting that glial transport to neurons is unnecessary and that electrolytes and small molecules may reach neurons by diffusion through the intercellular interstices of the brain(24). However, the special metabolic demands of developing neurons or of neurons recovering from injury may require glial activation.

The changes described above which occur in ependymal glia of thyroxine-stimulated tadpole brains are similar, in many respects, to the glial reactions which have been observed in mammalian brain during the recovery of neurons from chromatolysis. When, for example, peripheral efferent nerve fibers in mammals are cut, glial cells in the vicinity of the affected neurons in the brain proliferate. These glia subsequently increase their complements of oxidative enzymes(25) and develop high levels of nucleoside phosphatase activity(26,27). TPPase and IDPase activity have been implicated in cellular transport phenomena(27). Increased activity of these enzymes in glia and proliferation of the glia have been interpreted as indicative of increased capacity for transport and as glial responses to changes in functional and synthetic activity of regenerating neurons(26,27). High respiratory enzyme activity under these conditions may be related to the energetics of transport. Extensive increases in quantity of granular reticulum have been reported in satellite glial cells surrounding neurons recovering from chromatolysis(28).

The presence of extensive granular reticulum is typical of cells actively engaged in the production of protein for extracellular transport. Export proteins secreted by perineuronal glia in mammals might, according to Dixon(28), be utilized by these regenerating neurons. The proliferation of rough ER in thyroxine-stimulated ependymal glia of amphibians could be similarly interpreted.

The mechanisms of nerve regrowth resemble, in many respects, those of normal development. Normal CNS development requires thyroid hormone. It is therefore not surprising to find in the brain of thyroxine-treated thyroprivic tadpoles, patterns of glial response which are similar to those accompanying mammalian axon regeneration. The observations reported here concerning glial reactions in larvae of the frog *Rana pipiens* are consonant with the view that thyroxine-dependent neural maturation in developing amphibians may be accompanied or preceded by alterations in capacity for neuroglial synthesis and transport.

REFERENCES

1. Eayrs, J. T. Effects of thyroid hormones on brain differentiation. *In* Brain-Thyroid Relationships, Cameron, M. P., and O'Connor, M. O., eds. Boston, Little, Brown and Co., pp. 60–71, 1964.
2. Kollros, J. J. Endocrine influences in neural development. *In* Growth of the Nervous System, Wolstenholme, G. E. W., and O'Connor, M. O., eds. London, J. A. Churchill, Ltd., pp. 179–192, 1968.
3. Pesetsky, I. The role of the thyroid in the development of Mauthner's neuron. A karyometric study in thyroidectomized anuran larvae. Z. Zellforsch., 75:138–145, 1966.
4. Just, J. J., and Kollros, J. J. Effects of hypophysectomy, thyroxin, and pituitary hormones on growth of *Rana pipiens* larvae. Amer. Zool., 8:762, 1968.
5. Pesetsky, I. Developmental arrest in the central nervous system of thyroidectomized *Rana pipiens* larvae. Amer. Zool., 6:253, 1966.
6. Allen, B. M. Brain development in anuran larvae after thyroid or pituitary gland removal. Endocrinology, 8:639–651, 1924.
7. Kollros, J. J. and McMurray, V. M. The mesencephalic Vth nucleus in anurans. II. The influence of thyroid hormone on cell size and cell numer. J. Exp. Zool., 131:1–26, 1956.
8. ——— Order and control in neurogenesis (as exemplified by the lateral motor column). Develop. Biol., Suppl. 2:274–305, 1968.
9. Hydén, H. Dynamic aspects on the neuron-glia relationship. A study with micro-chemical methods. *In* The Neuron, Hydén, H., ed. Amsterdam, Elsevier Pub. Co., pp. 179–217, 1967.
10. Gelber, S., Campbell, P. L., Deibler, G. E., and Sokoloff, L. Effects of L-thyroxine on amino acid incorporation into protein in mature and immature rat brain. J. Neurochem., 11:221–229, 1964.
11. Cuarón, A., Gamble, J., Myant, N. B., and Osorio, C. The effect of thyroid deficiency on the growth of the brain and on the deposition of brain phospholipids in foetal and newborn rabbits. J. Physiol., 168:613–630, 1963.
12. Hamburgh, M., and Flexner, L. B. Biochemical and physiological differentiation during morphogenesis. XXI. Effect of hypothyroidism and hormone therapy on enzyme activities of the developing cerebral cortex of the rat. J. Neurochem., 1:279–288, 1957.
13. Oksche, A. Die pränatale und vergleichende Entwicklungsgeschichte der neuroglia. Acta Neuropath., Suppl. IV, 4–19, 1968.
14. Pesetsky, I. The thyroxin-stimulated enlargement of Mauthner's neuron in anurans. Gen. Comp. Endocr., 2:229–235, 1962.
15. Reynolds, W. A. The effects of thyroxine upon the initial formation of the lateral motor column and differentiation of motor neurons in *Rana pipiens*. J. Exp. Zool., 153:237–249, 1963.
16. Weiss, P., and Rossetti, F. Growth responses of opposite sign among different neuron types exposed to thyroid hormone. Proc. Nat. Acad. Sci. U.S.A., 37:540–556, 1951.

17. Baffoni, G. M., and Elia, E. L'attività mitotica durante la morfogenesi cerebellar in un Anfibio anuro. Rendic. Accad. Naz. Lincei (ser. VIII), 22:109–114, 1957.
18. Pesetsky, I. Thyroxin-stimulated oxidative enzyme activity associated with precocious brain maturation in anurans. A histochemical study. Ben. Comp. Endocr., 5:411–417, 1965.
19. ——— Altered thyroid state and changes in nucleoside phosphatase activity in ependymoglial cells of anuran amphibians. Gen. Comp. Endocr., Suppl. 2:238–244, 1969.
20. Novikoff, A. B., and Goldfischer, S. Nucleoside diphosphatase activity in the Golgi apparatus and its usefulness for cytological studies. Proc. Nat. Acad. Sci. U.S.A., 47:802–810, 1961.
21. Pesetsky, I., and Model, P. V. Thyroxin-stimulated ultrastructural changes in ependymoglia of thyroprivic amphibian larvae. Exp. Neurol., 25:238–245, 1969.
22. Friede, L. Enzyme histochemistry of neuroglia. *In* Biology of Neuroglia, Progress in Brain Research *15*, DeRobertis, E. D. P., and Carrea, R., eds. Amsterdam, Elsevier Pub. Co., pp. 35–47, 1965.
23. Tschirgi, R. D. The blood-brain barrier. *In* Biology of Neuroglia, Windle, W. F., ed. Springfield, Charles C. Thomas, Pub., pp. 130–138, 1958.
24. Kuffler, S. W., Nicholls, J. G., and Orkand, R. K. Physiological properties of glial cells in the central nervous system of amphibia. J. Neurophysiol., 29:768–787, 1966.
25. Hamberger, A., and Sjöstrand, J. Respiratory enzyme activities in neurons and glial cells of the hypoglossal nucleus during nerve regeneration. Acta Physiol. Scand., 67:76–88, 1966.
26. Sjöstrand, J. Changes of nucleoside phosphatase activity in the hypoglossal nucleus during nerve regeneration. Acta Physiol. Scand., 67:219–228, 1966.
27. Barron, K. D., and Tuncbay, T. O. Phosphatase histochemistry of feline cervical spinal cord after brachial plexectomy. Hydrolysis of β-glycerophosphate, thiamine pyrophosphate and nucleoside diphosphate. J. Neuropath. Exp. Neurol., 23:368–386, 1964.
28. Dixon, J. S. Changes in the fine structure of satellite cells surrounding chromatolytic neurons. Anat. Rec., 163:101–110, 1968.

26

Thyroid and Developing Brain: Anatomical and Behavioral Effects

J. T. Eayrs

Department of Anatomy, University of Birmingham, Birmingham, England

INTRODUCTION

One has to look further than the phenomenon of cretinism in the human species to find evidence for the profound influence which is exerted by the thyroid during the development of the brain. The untreated disorder is characterized by a mental retardation which may range from imbecility to moderate subnormality and which cannot be attributed solely to the reduced rate of metabolism manifest in the slow reactions found in myxoedema. It is reasonable, therefore, to look, by way of explanation of the underlying cause, for some developmental abnormality whose influence is superimposed upon the metabolism dysfunction. The possibility that thyroid deficiency could give rise to such abnormalities is enhanced by the demonstration in submammalian species of an apparently direct action of the hormone upon the maturation of neural tissues(1,2,3) as described by earlier contributions to this symposium.

It has been widely, though not undisputedly, claimed that the degree of success with which the mental symptoms of cretinism can be treated by hormone therapy is related to the age at which therapy is instituted, thus implying that the developing nervous system is maximally sensitive to the influence of the thyroid during a critical phase of its maturation(4,5). Contraindications, however, have come from clinicians who have been unable to find any satisfactory correlation between the nature of medication and the recovery achieved(6) and from the suggestion that failure to respond to treatment may be due to the presence of a primary amentia of congenital origin rather than to any influence of the thyroid upon maturational processes(7).

With a view to clarifying these issues, the nature of the influence of thyroid hormone on the mammalian nervous system has been the subject of several investigations based on structural, electrophysiological and behavioral phenomena. Most of the findings in these fields have already been reviewed(8,9,10) and the purpose of this communication is, accordingly, not so much to repeat in detail what is readily available in the literature, as briefly to sketch in the background which has led to those more recent experiments which will be presented by others and form the main body of this section of the symposium. It will thus seek first, to determine the part played by the thyroid in the etiology of the mental retardation of cretinism; second, to elucidate some of the reasons for the impairment; and finally to take a look at some aspects of the influence of hyperthyroidism during the formative period of life.

The work to be outlined has been based entirely upon studies with the laboratory rat, a species which is readily available, producing litters of a size which permit the application of statistical techniques involving littermate controls. Moreover its central nervous system at birth is so immature as to enable a direct study of the effects of procedures applied at a stage of development when the primate fetus is relatively inaccessible to experimental investigation. Deprivation of thyroid hormone can readily be achieved by surgical removal of the gland, by giving a single dose of radioiodine, or by the chronic administration of a goitrogen. All three methods have been used and, for convenience, animals subjected to any of these treatments will be referred to as "Thyroidectomized."

THYROID DEFICIENCY AND BEHAVIOR

From the behavioral point of view, the development of the rat is marked by the appearance of automatic responses to the presentation of specific stimuli whose timing can be predicted within narrow limits. Thyroidectomy is responsible for a delay in the programming of these responses but does not prevent their eventual emergence(11). It may thus be inferred that, as has been suggested earlier for amphibia(1), the patterning of the neural circuitry responsible for the mediation of the normal stimulus-response relationship underlying such behavior does not depend upon the influence of thyroid hormone. On the other hand, adaptively organized behavior is severely disrupted by the experimental treatment. The results of two experiments, set out side by side for comparison in Table 1, illustrate this finding in so far as the number of errors made by thyroidectomized rats on a simple T-maze (11) or in learning to perform a task involving escape-avoidance to a conditioned auditory stimulus(12) have proved significantly greater than that made by their normal littermates. In both experiments, the rate of performance of the experimental animals was reduced. Such findings confirm the view that thyroid deficiency, per se, can give rise to behavioral impairment without, however, providing conclusive evidence for an irreversible abnormality which could be attributed to the endocrine state. Indeed the effects of treating these animals with thyroid hormone would seem to argue against such irreversibility and to support the view, noted earlier, that an intercurrent congenital abnormality must also be present where therapy fails. Thus medication of neonatally thyroidectomized rats, from 25 days old in the case of the maze-learning trials or from 5 months old for the conditioned avoidance study, has clearly restored behavior to normal levels (Table 1).

TABLE 1

Performance of neonatally thyroidectomized rats on certain adaptive tasks

The data represent the differences in performance between experimental and control subjects (i.e., thyroidectomized minus control).

Elevated T-maze (11)	Thyroidectomized at birth		Escape-avoidance (12)	Thyroidectomized at birth	
	No therapy	**Replacement therapy from 25 days old**		**No therapy**	**Replacement therapy from 5 months old**
Mean number of errors	11.3	–5.1	Mean number of errors	32.6	–0.2
Time to learn (minutes)	10.7	–3.4	Time for 60 trials (seconds)	198.0	6.0
Time for correct run (seconds)	27.1	–0.5	Response latency (seconds)	2.0	–0.3

These results, however, may be contrasted with those of a third experiment(13) using the Hebb-Williams closed-field test which is generally regarded as providing a sensitive index of impairment to cerebral function(14). The findings (Table 2) confirm, in terms of error scores, that thyroidectomy in the newborn impairs the capacity for adaptive behavior. They show, in addition, that the degree of impairment is reduced as thyroidectomy is delayed until, in this species, the experimental procedure is without effect as from an age lying between 10 and 25 days. Conversely, replacement therapy in animals thyroidectomized at birth proved effective only when begun before the tenth day of age. It may thus be inferred that absence of thyroid hormone, operating during a critical period of development, can give rise to irreversible damage to the nervous system which is manifest by behavioral impairment. In the rat, this period extends over the first 12 days or so of postnatal life after which neither thyroidectomy in the normal animal nor medication of the neonatally thyroidectomized individual is effective. While the available data are inadequate for the formulation of firm conclusions, it would seem that this period of development must occur in the human fetus around the fourth to sixth month of gestation and, if it be permissible to extrapolate from rat to man, the inconsistent results of thyroid therapy referred to earlier may be attributable to the unknown thyroidal state of the fetus during this period of intrauterine life. The reasons for the opposing results of these two groups of experiment remain a matter for speculation but it is likely that they are a function of differences in the mode of testing and scoring. The closed-field test relies on the presentation of a number of problems of varying order of difficulty and has the added advantage that, for the investigation of a condition which is characterized by an increased reaction time, rate of performance plays no part in the behavioral assessment. In the T-maze, however, the single problem presented was simple while success in the performance of the easy discrimination posed in the escape-avoidance test was largely dependent on speed of response. In such circumstances it is conceivable that the increased alertness conferred by giving thyroid hormone to the thyroidectomized subjects would mask the presence of a developmental abnormality revealed by more critical appraisal.

THYROID DEFICIENCY AND HISTOGENESIS

It is reasonable to suppose that such structural features as might be responsible for these behavioral manifestations would at least be found in the cerebral cortex and so far it is this part of the brain, and no other, which has been studied from the point of view of quantitative histology. Abnormalities which have so far been revealed have included: (1) a reduction in the weight of the brain but consistent with that of the body(15); (2) a change in the shape of the brain involving an increase in width and height relative to length, which may be related to a failure in endochondral ossification at the base of the skull(15); (3) an altered pattern of vascularity in so far as a reduced number of blood vessels is associated with an increase in their mean size which, however, proves insufficient to restore the surface area available for metabolic exchange(16); (4) an impaired myelination(17,18); (5) a reduction in the size of neuronal perikarya accompanied by an increase in

TABLE 2

Effect of thyroid deficiency related to age of onset and medication, on performance in closed-field test (13)

Age thyroidectomy performed (days)	Mean differences in error scores		Age therapy started (days)
Birth	228	—	—
5	123	0	5
10	24	8	10
24	4	67	24
—	—	65	70

packing density(15); and (6) a hypoplasia of the cortical neuropil(19). The last finding deserves elaboration since it has been proposed that the relationship between axons and dendrites is of special significance for the organization of behavior. Thus, measured throughout the several cortical laminae of the sensory-motor cortex of the 24-day-old rat, the density of axons is reduced in the animal thyroidectomized

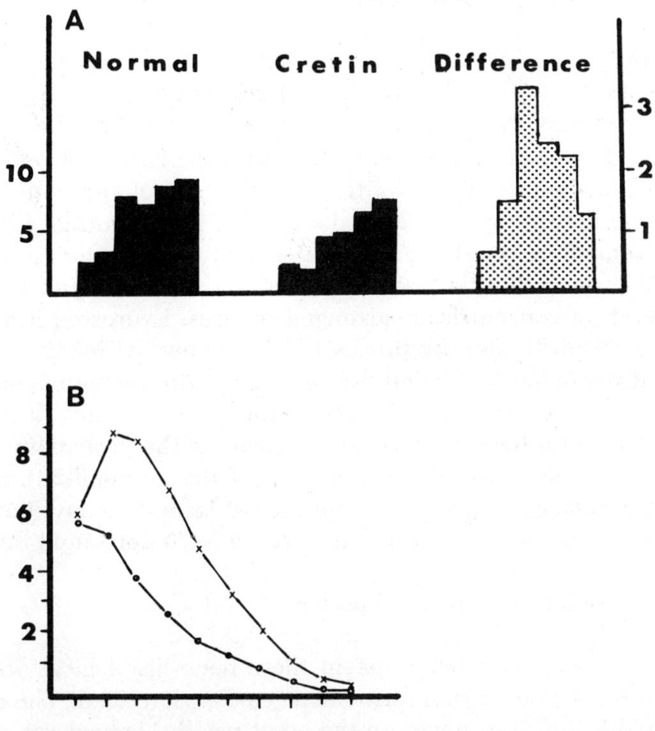

Fig. 1. Effect of neonatal thyroidectomy on growth of cortical neuropil. *A*: Axonal component-density of axons (axons/mm²/10⁴) in successive cortical laminae (solid) and difference between normal and cretinous rats (dotted). *B*: Dendritic component-number of dendrites of successive distances from centre of perikaryon. Normal: x ——— x; Cretinous: o ——— o; Abscissa divided into 18μ intervals.

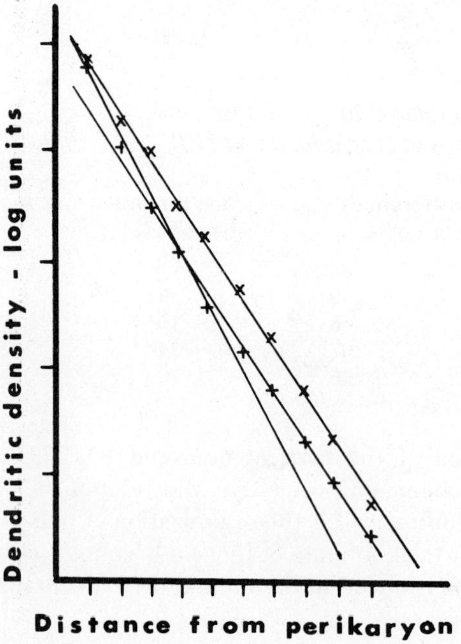

Fig. 2. Decay of dendritic field in normal (x——x) and cretinous (+——+) rats. The computed regression line for the normal shows an excellent fit to the data. Alternative lines are given for the cretin which provide the best fit to the proximal (most reliable) and peripheral (less reliable) plots.

at birth, not proportionally in all laminae but selectively in lamina 4 which receives the specific afferent projections from the thalamus (Fig. 1A). Dendritic density is likewise reduced. Plotted in terms of the mean number of dendrites present at equally spaced intervals from the centre of the perikarya of pyramidal cells in the 24-day-old rat, Figure 1B shows that, in the case of the cretin, this number falls off from the first interval measured whereas, in the normal littermate, there is an initial increment before decay. It is convenient to represent such data in three-dimensional form such that the ordinate measures the density of dendrites which penetrate a series of concentrically arranged spheres. Expressed in this form (Fig. 2) the decay in dendritic density proves to be exponential for the normal rat but curvilinear and displaced to the left for the cretin. An advantage of this mode of presentation is that the regression equations for the best-fitting decay curves permit the application of a formula designed to estimate the probability of interaction between the axonal and dendritic components of the neuropil(20). This increases rapidly during normal growth between the ages of 15 and 36 days but so much less so in the cretin as to account for a decrement of some 70 percent in this measure(9).

Correlation Between Structure and Function

It remains to consider to what extent these neurohistological findings can be held to account for the behavioral phenomena earlier described. The data, summarized in Figure 3A, which compare, in the adult rat, the behavioral effects of thyroidectomy at various ages with several measures of structure in the neuropil, encourage the belief that some relationship exists between the two(21). Thus the pattern of behavioral impairment, as related to the age at which thyroidectomy was performed, appears to be well matched by both the rate of decay of the dendritic

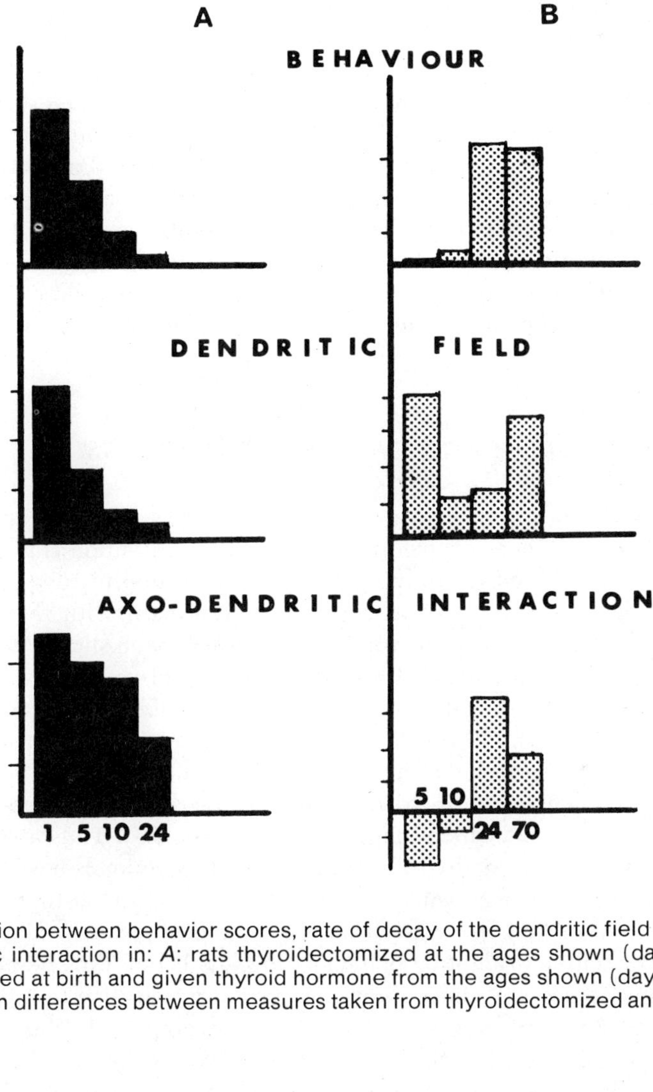

Fig. 3. Correlation between behavior scores, rate of decay of the dendritic field and probability of axodendritic interaction in: *A*: rats thyroidectomized at the ages shown (days) and *B*: rats thyroidectomized at birth and given thyroid hormone from the ages shown (days). The data indicate the mean differences between measures taken from thyroidectomized animals and littermate controls.

field and the factor of connectivity (axons measured in the right hemisphere and dendrites in the left of the same rat). This is not, of course, to imply that only such structural modification as can be measured by quantitative neurohistological techniques are involved. In fact the contrary would seem to apply for it is found (Fig. 3B) that there is no significant correlation between these same behavioral and structural measures in animals thyroidectomized at birth and first receiving thyroid hormone at progressively later ages afterwards. It would thus seem that, in the presence of thyroid hormone, there occurs a recovery of structure as displayed by classical neurohistological techniques without a corresponding amelioration of behavioral capacity. This plainly suggests the persistence of some further developmental abnormality not so far revealed by the methods applied.

It is reasonable to infer from the evidence offered that an error in the synthesis of protein may be invoked as one of the causes underlying the mental retardation of cretinism. Histological evidence for such a deficiency is provided in terms of the hypoplasia of the cortical perikarya and their associated neuropil already noted. This has recently been confirmed by use of biochemical techniques(22,23,24) and it seems likely that an irreversible component related to the maturation of as yet only partially identified enzyme systems may well be involved also. Work directed towards the elucidation both of the enzymological aspects of the problem(25,26, 27,28) and the nature of the disturbance to the metabolism of protein(23,24,29, 30,31) continue but any detailed discussion falls outside the scope of this introductory paper and will be considered in communications which follow in this volume (Balázs et al., Ch. 27, Geel and Timeras, Ch. 29; Gómez, Ch. 31; Valcana and Timras, Ch. 33.)

INFLUENCE OF HYPERTHYROIDISM

The effects of the presence of an excess of thyroid hormone during the early stages of development have received considerably less attention than have the sequelae of deficiency. An early study by Tusques(32) failed to reveal more than a hyperplasia of the cerebellar folia as a result of treatment with thyroxine during the immediate postnatal period but more recently it has been shown that such treatment is associated also with an acceleration in the time of first appearance of those landmarks of innately organized behavior which are retarded by thyroidectomy(33, 34,35). But despite this initial precocity, adult rats made hyperthyroid during the first days of life perform less well than their littermate controls when confronted with the Hebb-Williams closed-field test(33), the degree of impairment being directly related to the age at which the hormone is given(36). Thus a significant difference in error score between experimental and control animals has been observed only when the treatment was administered up to the tenth postnatal day; giving thyroid hormone after that age was either ineffective or gave rise to a modest improvement in capacity for adaptive behavior (Fig. 4). It will thus be noted that, by this behavioral criterion, the central nervous system is sensitive to the influence of thyroid hormone given in excess during the identical period when a deficiency proves deleterious. This critical sensitivity is not, however, confined to the maturation of nervous tissues for, while early postnatal treatment dramatically accelerates such maturational changes as the opening of the eyes and the development of the pinnae, an associated reduction in rate of growth is maintained long after the administration of the hormone has been discontinued in those animals treated before the tenth day of age but not in those made hyperthyroid later in life (Fig. 4). These findings, taken in conjunction with changes in the bodily appearance of the experimental animals(36), suggest the hypothesis that by analogy with the now well-known effects of giving sex steroids during the same critical period of development, the presence of excessive titres of thyroid hormone interferes with the correct maturation of the hypothalamo-pituitary-thyroid axis and so renders the animal hypothyroid later in life. Experiments designed to test this hypothesis, however, have failed to provide more than slight supporting evidence. Measure-

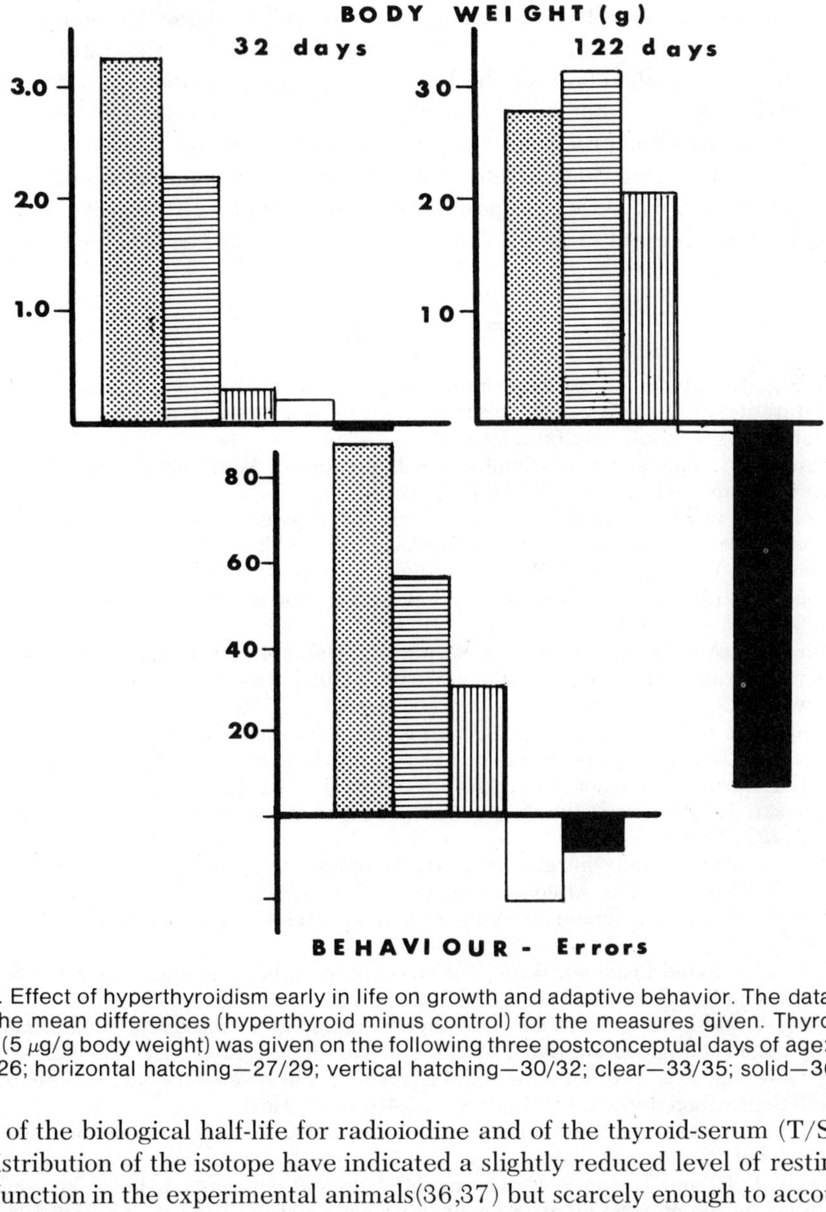

Fig. 4. Effect of hyperthyroidism early in life on growth and adaptive behavior. The data represent the mean differences (hyperthyroid minus control) for the measures given. Thyroid hormone (5 μg/g body weight) was given on the following three postconceptual days of age: dotted —24/26; horizontal hatching—27/29; vertical hatching—30/32; clear—33/35; solid—36/38.

ment of the biological half-life for radioiodine and of the thyroid-serum (T/S) ratio for distribution of the isotope have indicated a slightly reduced level of resting thyroid function in the experimental animals(36,37) but scarcely enough to account for the significant changes observed in growth, bodily configuration, and adaptive behavior. Some support for an impairment to adenohypophysial function comes from the histological findings of Eayrs and Holmes(38) who demonstrated a persistent degranulation of pituitary acidophils following administration of triiodothyronine to infant rats and from experiments indicating a reduced pituitary content of TSH and a failure of goitrogen to give rise to the characteristic hyperplasia of thyroid tissue(39). By contrast, our own work has shown that though the mean weight of the thyroid in the experimental rats was less than that in the controls the hyperplasia and increase in the T/S ratio which followed the administration of goitrogen was of such dimension as to minimize the probability of a defect in the mechanisms controlling the secretion of thyroid-stimulating hormone by the pituitary gland. Some alternative explanation for the long-lasting effects of neonatal hyperthyroidism must clearly be sought, one possibility being that of selective errors in the phasing of maturational processes. As far as the central nervous system is concerned, it would seem probable that current studies of the biochemical and enzymological changes associated with cretinism are likely to indicate at least the site of action of thyroid hormone present in excess during the critical period of development.

REFERENCES

1. Kollros, J. J. Studies of the development of the corneal reflex in amphibia. II. Localised maturation of the reflex mechanism effected by thyroxin-agar implants into the hindbrain. Physiol. Zool., 16:269–279, 1943.
2. Pesetsky, I. The thyroxine-stimulated enlargement of Mauthner's neuron in anurans. Gen. Comp. Endocr., 2:228–235, 1962.
3. Weiss, P., and Rossetti, F. Differential growth response of nerve cells to thyroid hormone. Proc, Nat. Acad. Sci. U.S.A., 37:540–556, 1951.
4. Brown, A. W., Bronstein, I. P., and Kraines, R. Hypothyroidism and cretinism in childhood. VI. Influence of thyroid therapy on mental growth. Amer. J. Dis. Child, 57:517–523, 1939.
5. Smith, D. W., Blizzard, R. M., and Wilkins, L. The mental prognosis in hypothyroidism of infancy and childhood. Pediatrics, 19:1011–1022, 1949.
6. Lewis, A. A study of cretinism in London. Lancet, 1:1505–1509; 2:5–9, 1937.
7. Goodkind, R. P., and Higgins, H. L. Hypothyroidism in infants and children: with reference to the ultimate prognosis concerning intelligence and to the withdrawal of thyroid therapy as a diagnostic measure. New Eng. J. Med. 224:722–726, 1941.
8. Eayrs, J. T. Influence of the thyroid on the central nervous system. Brit. Med. Bull., 16: 122–127, 1960.
9. ——— Thyroid and central nervous development. *In* Scientific Basis of Medicine Annual Reviews, London, Athlone Press, pp. 317–339, 1966.
10. Ford, D. H. Central nervous system—thyroid interrelationships. Brain Res., 7:329–349, 1968.
11. Eayrs, J. T., and Lishman, W. A. The maturation of behaviour in hypothyroidism and starvation. Brit. J. Anim. Behav., 3:17–24, 1955.
12. ——— and Levine, S. Influence of thyroidectomy and subsequent replacement therapy upon conditioned avoidance learning in the rat. J. Endocr., 25:505–513, 1963.
13. ——— Age as a factor determining the severity and reversibility of the effects of thyroid deprivation in the rat. J. Endocr., 22:409–419, 1961.
14. Rabinowitz, M. S., and Rosvold, H. E. A closed-field intelligence test for rats. Canad. J. Psychol., 5:122–128, 1951.
15. Eayrs, J. T., and Taylor, S. H. The effect of thyroid deficiency induced by methyl thiouracil on the maturation of the central nervous system. J. Anat., 85:350–358, 1951.
16. ——— The vascularity of the cerebral cortex in normal and cretinous rats. J. Anat., 88:164–173, 1954.
17. Barrnett, R. J. Quoted in Pincus, V. and Thimann, K. B., eds. The Hormones: Physiology, Chemistry and Applications. New York, Academic Press, Inc., Vol. 2, p. 235, 1948.
18. Balázs, R., Brooksbank, B., Davison, A., and Wilson, D. A. The effect of neonatal thyroidectomy on myelination in the rat brain. Brain Res., 15:219–232, 1969.
19. Eayrs, J. T. The cerebral cortex of normal and hypothyroid rats. Acta Anat., 25:160–183, 1956.
20. Uttley, A. M. The probability of neural connexions. Proc. Roy. Soc., 144B:229–240, 1955.

21. Eayrs, J. T. La influencia de la edad de tiroidectomía sobre el desarrollo da la corteza cerebral. Arch. Mex. Anat., 7:29–30, 1966.
22. Geel, S. E., and Timiras, P. S. The influence of neonatal hypothyroidism and of thyroxine on the ribonucleic acid and deoxyribonucleic acid concentrations of rat cerebral cortex. Brain Res., 4:135–142, 1967.
23. ——— Valcana, T., and Timiras, P. S. Effect of neonatal hypothyroidism and of thyroxine on 1-(^{14}C)leucine incorporation in protein *in vivo* and the relationship to ionic levels in the developing brain of the rat. Brain Res., 4:143–150, 1967.
24. Balázs, R., Kovács, S. Teichgräber, P., Cocks, W. A., and Eayrs, J. T. Biochemical effects of thyroid deficiency on the developing brain. J. Neurochem., 15:1335–1349, 1968.
25. Hamburgh, M., and Flexner, L. B. Biochemical and physiological differentiation during morphogenesis. XXI. Effect of hypothyroidism and hormone therapy on enzyme activities of the developing cerebral cortex of the rat. J. Neurochem., 1:279–288, 1957.
26. Geel, S. E., and Timiras, P. S. Influence of neonatal hypothyroidism and of thyroxine on the acetylcholinesterase and cholinesterase activities in the developing central nervous system of the rat. Endocrinology 80:1069–1074, 1967.
27. García Argíz, C. A., Pasquini, J. M., Kaplan, B., and Gómez, C. J. Hormonal regulation of brain development. II. Effect of neonatal thyroidectomy on succinate dehydrogenase and other enzymes in developing cerebral cortex and cerebellum of the rat. Brain Res., 6:635–646, 1967.
28. Robinson, N., and Eayrs, J. T. Histochemical study of the cerebral cortex in rats thyroidectomised at birth. Brain Res., 9:351–362, 1968.
29. Gelber, S., Campbell, P. L., Deibler, G. E., and Sokoloff, L. Effects of 1-thyroxine on amino acid incorporation into protein in mature and immature brain. J. Neurochem., 11:221–229, 1964.
30. De Guglielmone, A. E. R., and Gómez, C. J. Influence of neonatal hypothyroidism on amino acids in developing rat brain. J. Neurochem., 13:1017–1025, 1966.
31. Nievel, J. G., Robinson, N., and Eayrs, J. T. Protein synthesis in the brain of rats thyroidectomised at birth. Experientia, 24:677–678, 1968.
32. Tusques, J. Recherches expérimentales sur le rôle de la thyroïde dans le developpement du système nerveux. Biol. Med., 45:395–413, 1956.
33. Eayrs, J. T. Effect of neonatal hyperthyroidism on maturation and learning in the rat. Anim. Behav., 12:195–199, 1964.
34. Shapiro, S., and Norman, R. J. Thyroxine: effects of neonatal administration on maturation, development and behavior. Science, 155:1270–1281, 1967.
35. Khamsi, F., and Eayrs, J. T. A study of the effects of thyroid hormone on growth and development. Growth, 30:143–156, 1966.
36. Eayrs, J. T. Developmental relationships between brain and thyroid. *In* Endocrinology and Human Behaviour. Michael, R. P., ed., London, Oxford University Press, Ch. 14, 1968.
37. Schreiber, V., and Kmentova, V. Thyroxine in the early postnatal period: minimal changes in thyroid function. Neuroendocrinology, 1:121–128, 1965/6.
38. Eayrs, J. T., and Holmes, R. L. Effect of neonatal hyperthyroidism on pituitary structure and function in the rat. J. Endocr., 29:71–81, 1964.
39. Bakke, J. L., and Lawrence, N. Persistent thyrotropin insufficiency following neonatal thyroxine administration. J. Lab. Clin. Med., 67:477–482, 1966.

27

Biochemical Effects of Thyroid Hormones on the Developing Brain*

R. Balázs,[1] W. A. Cocks,[1] J. T. Eayrs,[2] and S. Kovacs[3]

[1]Neuropsychiatry Unit, Medical Research Council Laboratories, Carshalton, Surrey, England
[2]Department of Anatomy, University of Birmingham, Birmingham, England
[3]Institute of Physiology, University Medical School, Pécs, Hungary

INTRODUCTION

Thyroid deficiency arising early in life has long been known to give rise to a severe and frequently irreversible mental retardation. Investigations on experimental animals, summarized in the preceding contribution to this conference and elsewhere(1), have shown that an impairment of behavior is associated with structural changes involving a hypoplasia of both the axonal and dendritic components of the cortical neuropil. Hyperthyroidism in infancy results in advanced maturation of innately organized responses, followed by a reduced capacity for adaptive behavior later in life(2). The biochemical correlates of these phenomena have, however, received relatively less attention though, during the past two years or so, a number of papers dealing with this aspect of the subject have appeared(3-10). The present communication sets out the findings of further experiments designed to elucidate the effect of the thyroidal state of the individual on the biochemical maturation of the brain.

*The authors gratefully acknowledge the collaboration of the following: Dr. B. W. L. Brooksbank and Professor A. N. Davison in the studies dealing with lipid metabolism, Mr. J. A. Cocks and Dr. A. J. Patel in those concerned with glucose metabolism and metabolic compartmentation, respectively, and Mr. A. L. Johnson for the statistical analyses.

EFFECTS OF THYROID HORMONES ON POSTNATAL CELL FORMATION

As is well documented, the development of the brain in the rat takes place to a great extent after birth. According to our data (Fig. 1) the weight of the cerebrum increases about eight times, and that of the cerebellum about 25 times, during the first 5 weeks of life. This enlargement is due not only to the growth of cells existing at birth but also to an appreciable increase in numbers which can be estimated, assuming that the cells in the brain are mainly diploid, in terms of the DNA content of the tissue(11). The postnatal increase in DNA content of the cerebrum in the rat has proved to be about 80 percent and the cell number approaches the adult level by 14 days of age. On the other hand, the number of cells in the cerebellum is very small at birth compared with that in the adult and the rapid phase of cell multiplication continues until about 21 days of age (Fig. 2).

Our findings have shown that postnatal cell formation in the brain is selectively influenced by the thyroid state of the animal. Thyroid deficiency in infancy results

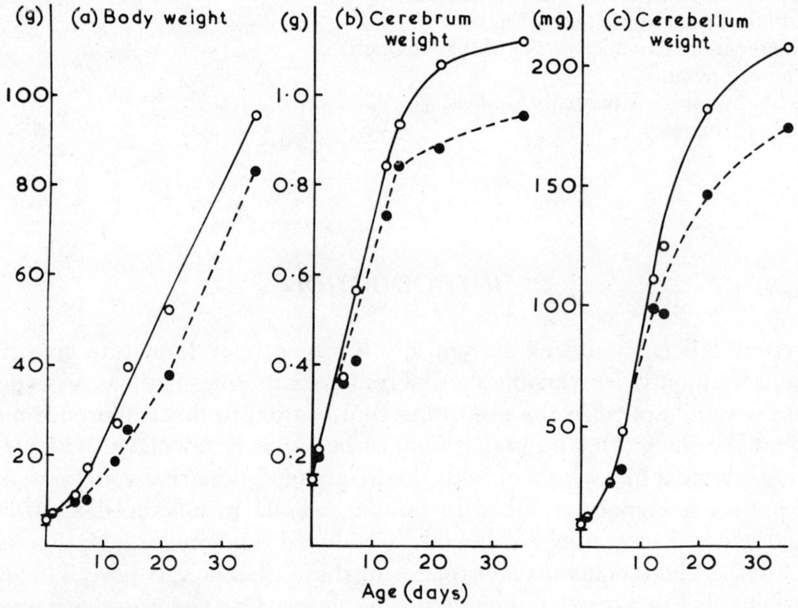

Fig. 1. Effect of treatment with thyroid hormone (T_3, 25 μg at birth, followed by 0.5 μg every second day) on body weight (*a*), growth of the cerebrum (*b*), and growth of the cerebellum (*c*). Littermate controls—circles; T_3-treated rats—dots. The data were statistically analysed by analysis of variance with a two-way classification (experimental or control versus age). The mean values are plotted and the number of observations was 79 in a and b, and 75 in c. The individual data for body weight and the weight of the cerebellum were transformed logarithmically before analysis and the variance of the transformed data was 0.0019 in a and 0.002 in c. The variance in b was 0.0015. The curves representing the experimental group were significantly different from those of the controls ($P<0.001$). The values for newborn (12 animals) refer to untreated rats; these results were not included in the analysis of variance and the mean ±S.E.M. is given.

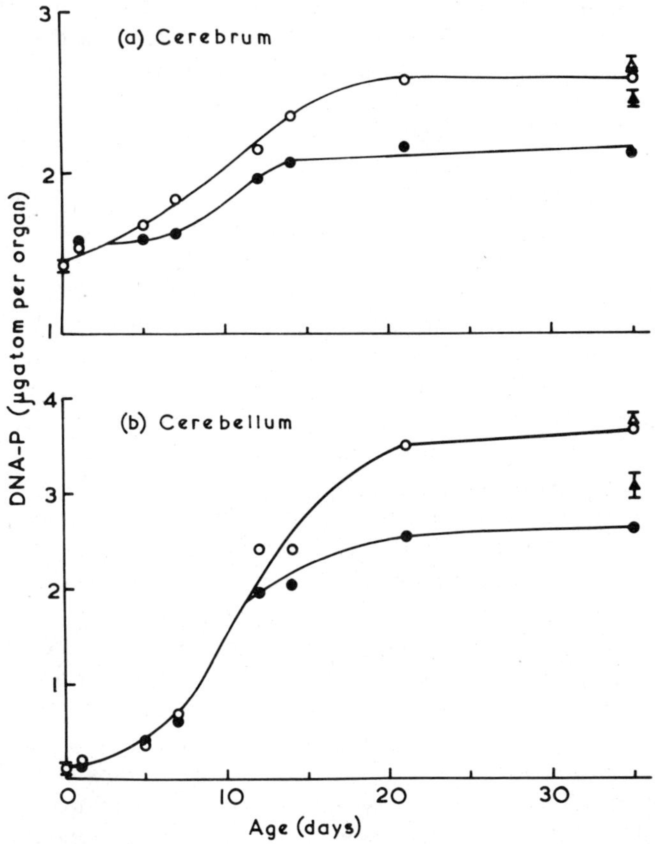

Fig. 2. Effect of thyroid (T_3) treatment on postnatal cell formation in the cerebrum (*a*) and cerebellum (*b*); T_3-treated rats--dots, littermate controls—circles. The number of observations and the statistical analysis were described in the legend to Figure 1. The variance in a was 0.0095; in b the individual results (×10) were transformed logarithmically before analysis and the variance of the transformed data was 0.0029. The curves representing total DNA of the experimental animals both in a and b were significantly different from those of the controls ($P<0.01$). The significance of differences between means was then estimated; significant differences were obtained in the cerebrum after 7 days and in the cerebellum after 12 days. The values for newborn (12 animals) refer to untreated rats, and the triangles to a separate experiment, untreated neonatal thyroidectomy (open triangles) and hypothyroid rats (solid triangles) receiving replacement treatment until 35 days of age (3 rats in each group); these results were not included in the analysis of variance and the mean ± S.E.M. is given. The results for the animals given replacement treatment were significantly different from those of untreated rats (Student's *t* test).

in a retardation of postnatal cell formation in the cerebellum in so far as the DNA content in 14-day-old thyroidectomised rats was about 25 percent less than that of the controls, though reaching the normal level at the age of 35 days (Table 1). On the other hand, the DNA content of the cerebrum was not significantly altered(9). The cell density in both the cerebrum and the cerebellum (DNA per unit weight) was significantly increased compared with that of the controls at the age of 35 days.

These results, together with the observation that the ratios of RNA and protein to DNA are reduced in the brains of the experimental animals(9), indicate that thyroid deficiency in infancy affects the size of the cells rather than their total number. Comparable results have also been obtained in other laboratories(4,6).

Animals treated with thyroid hormones during infancy (25 μg triiodothyroxine [T_3] at birth followed by 0.5 μg every two days) were smaller than their littermate controls. The weight of the brain was also less (Fig. 1), mainly as a result of a reduction in postnatal cell formation, and this effect was most pronounced in the cerebellum (Fig. 2). The sharp rise in the DNA content of the cerebellum continued in the normal animal until 21 days of age, but was greatly reduced in the hyperthyroid animal after the age of 12-14 days. The triangles in Figure 2 represent the results of a further experiment in which rats were thyroidectomized at birth (open triangles) and compared with those receiving replacement treatment during the experimental period (shaded triangles). The growth of the latter was only slightly reduced compared with that of normal animals, but the DNA content of both the cerebrum and the cerebellum was significantly lower than that of either thyroidectomized or normal control rats.

Cell size in the brain was apparently not affected by giving thyroid hormones during infancy: The amount of DNA per unit weight and the ratios of RNA and protein to DNA were not decreased when compared with the values obtained for the control animals.

In summary it may be said that, judged by changes in the cerebellum, an organ in which the final assembly of cells develops to a great extent after birth, thyroid hormone plays a part in regulating the normal rate of cell multiplication though its deficiency does not influence the number of cells ultimately attained. The effect of thyroid deficiency is thus to extend the period in which new cell formation takes place in the cerebellum and these results are consistent with the morphological observations of Legrand(12) which showed that the inward migration of the fetal external granular layer of the cerebellum cortex is delayed in animals made hypothyroid in infancy and is accelerated by giving thyroid hormone.

Hyperthyroidism in infancy results in a significant reduction of the DNA content of the brain, but, in contrast to the effects of thyroid deficiency, the average cell size does not seem to be reduced. Further work is required to answer the following questions: (1) Does the cerebral DNA content of thyroid-treated animals reach the normal level at an age older than 35 days? (2) Is the reduction in the total cerebral cell number affected by treatment with different thyroid hormones, and what is the dose-dependence?(3) Until what age is the thyroid effect reversible, and (4) What are the cell types whose formation is blocked in the brain of hyperthyroid animals?

EFFECTS ON CELL CONSTITUENTS

Morphological(13) and biochemical (Table 1) observations have shown that the average size of cells in the brain is reduced as a result of neonatal thyroidectomy and investigations have accordingly been carried out to determine whether all cell constituents are affected to the same extent. To this end certain enzymic activities

TABLE 1

Effect of neonatal thyroidectomy on the amount of DNA and on the cellular contents of RNA and protein in the cerebrum and cerebellum

Age (days)	Tissue (No. of animals[b])	DNA (μg atom P/g)		DNA (μg atom P/organ)		RNA/DNA		Protein/DNA	
		Normal	**Thyroidectomized**	**Normal**	**Thyroidectomized**	**Normal**	**Thyroidectomized**	**Normal**	**Thyroidectomized**
				A. Cerebrum					
14	Cerebral cortex (12/12)	2.73 ± 0.056	2.59 ± 0.076			3.11 ± 0.106	3.04 ± 0.085	34.3 ± 1.06	32.6 ± 1.44
	Hemispheres (2/2)	2.53	2.57	2.50	2.44	2.92	2.76	39.2	37.1
35	Cerebral cortex (9/8)	2.11 ± 0.031	*2.53* ± 0.081[c]			2.55 ± 0.147[f]	2.20[f]	50.9 ± 1.41	*40.6* ± 1.21[c]
	Hemispheres (7/7)	2.26 ± 0.063	*2.62* ± 0.111[e]	2.60 ± 0.089	2.53 ± 0.147	2.23 ± 0.042	*1.91* ± 0.07[d]	43.5 ± 2.67	*35.6* ± 1.40[e]
				B. Cerebellum					
14	(12/12)	20.5 ± 0.37	*18.5* ± 0.67[e]	2.36 ± 0.087	*1.74* ± 0.084[c]	0.570 ± 0.010	0.574 ± 0.0133	4.85 ± 0.185	4.68 ± 0.211
35	(9/9)	17.1 ± 0.24	*20.3* ± 0.46[c]	3.65 ± 0.065	3.51 ± 0.187	0.395 ± 0.0115	*0.333* ± 0.0152[d]	6.58 ± 0.177	*5.25* ± 0.072[c]

[a] Male albino rats were thyrcidectomized by the administration of 150 μc ^{131}I (i.p. or s.c.) at the age of 1 or 2 days; normal rats were littermates (males). The data are taken from the paper of Balázs et al. (9), which gives the details of the experimental conditions and methods. The amount of DNA and RNA was expressed in μg atom P and that of protein in mg. The values are the mean ± S.E.M.

[b] The number of animals are given in brackets; the first figure refers to the number in the normal and the second to that in the experimental group. The values in italics indicate that the difference between the experimental and control group is significant ($P < 0.05$).

[c] P values were < 0.001.

[d] $P < 0.01$.

[e] $P < 0.02$.

[f] Three determinations in normal and two in experimental group.

TABLE 2

Some enzymic activities in the cerebral cortex of 24-day-old rats thyroidectomized at birth[a]

Glutamate decarboxylase		Glutamate dehydrogenase		Lactate dehydrogenase	
Normal	**Thyroidectomized**	**Normal**	**Thyroidectomized**	**Normal**	**Thyroidectomized**
		(μmoles/g wet wt/min)			
0.657	*0.548*[c]	20.9	21.4	116	110
±0.0295	±0.0135	±0.59	±0.67	±3.1	±4.6
		(μmoles/μg atom DNA-P/min)			
0.291	*0.220*[b]	9.48	*8.59*[d]	51.4	44.4[e]
±0.0142	±0.0065	±0.316	±0.237	±1.86	±2.21

[a]The data are taken from the paper of Balázs et al. (9), which gives the details of the experimental techniques. Number of animals: 5 normal and 9 experimental. The values in italics indicate that the difference between the experimental and the control group is significant.
[b]P values were < 0.001.
[c]P < 0.01.
[d]P < 0.05.
[e]P < 0.1.

have been used as biochemical markers for subcellular fractions: lactate dehydrogenase for the cell sap, glutamate dehydrogenase (GDH) for mitochondrial and glutamate decarboxylase (GAD) for detached nerve endings (synaptosomes) (Table 2). Only in the activity of GAD was a significant reduction observed. The activities of the other enzymes also decreased when expressed on a DNA basis, which reflects the general reduction in cell size. Our extended findings(9), however, taken in conjunction with those of Argíz et al.(5), would seem to indicate that even in the absence of medication the normal level of enzyme activity is approached later in life. This contrasts with a persistence of structural defects as shown by neurohistological techniques(14). Irreversible reduction in the activity of succinate dehydrogenase (SDH) has been observed by Hamburgh and Flexner(15) in the brains of rats thyroidectomized at birth, and these results have recently been confirmed(5). The different effects of neonatal thyroidectomy on two mitochondrial enzymes, GDH and SDH, suggest that synaptosomal and "free" mitochondria are not similarly affected by thyroid deficiency(9).

There is evidence that two populations of mitochondria can be isolated on sucrose gradients from brain homogenates(16-19). Mitochondria contained in the synaptosomes sediment more slowly than the "free" mitochondria which are derived from nerve cells and glia. The enzymic composition of the two types of mitochondria are different. Considering only the two mitochondrial enzymes which are affected differentially by thyroid deficiency it has been claimed that about 75 percent of the mitochondrial GDH activity is contained in the "free" mitochondria, whereas about 75 percent of the total SDH activity is in the synaptosomal mitochondria(16). These observations suggest, therefore, that it is probably the mitochondria in the nerve terminals which are affected in the brain of hypothyroid animals, rather than the mitochondria in the cell bodies. Histochemical determination of SDH in the cerebellum has confirmed these conclusions since the reduction in the activity of this enzyme in the hypothyroid state is confined to the region of dendritic arborization of the Purkinje cells (and hence to that of major synaptic activity) rather than to the cell bodies proper(12).

The following evidence indicates, therefore, that the development of nerve terminals is affected by thyroid deficiency in infancy: (1) a retardation of dendritic arborization(20), (2) a selective reduction in the enzymes GAD and SDH which are relatively concentrated in the synaptoplasm and synaptosomal mitochondria, respectively.

PROTEIN AND RNA SYNTHESIS

It has earlier been reported that the rate of incorporation of amino acid into cerebral protein is decreased in thyroid deficiency during infancy(7). Similar results have been obtained in our laboratory(8) and, in view of evidence that the different subcellular structures in the brain are not equally affected by the lack of thyroid hormones, the incorporation of amino acids into the protein of subcellular fractions of brain homogenates has been investigated.

[1-^{14}C]Leucine (5 to 20 μc/100 g body weight) was selected as the precursor, since C-1 of leucine is lost when this amino acid is metabolized and more than 80 percent of the total ^{14}C in the cerebral acid soluble fraction can be recovered in leucine. It is possible, therefore, to normalize the incorporation of ^{14}C into cerebral protein by expressing the results as relative specific radioactivities, i.e., the specific radioactivity of protein (d.p.m./mg protein) divided by the ^{14}C content of the acid soluble fraction (d.p.m./g wet weight), as a measure of the [^{14}C]leucine pool. Following thyroidectomy at birth, the rate of incorporation, measured at 14 and 35 days of age, was reduced in all the subcellular fractions (Fig. 3) and accordingly it may be inferred that no single fraction can be identified as containing a mechanism specifically affected by thyroid deficiency. The same conclusions can be drawn from the results of the subcellular distribution of protein and RNA (Fig. 4). The only difference between the experimental and the control groups was a significant reduction in the protein content of the myelin-containing subfraction of the crude mitochondria. This question will be further considered in the next section.

The apparent defect in cerebral protein synthesis seems to be confined to thyroid deficiency in infancy. When thyroid deficiency was induced in 30-day-old rats, either by treatment with methylthiouracil or by surgical removal of the thyroid gland with or without subsequent administration of ^{131}I, the incorporation of [1-^{14}C]leucine into protein in the cerebral cortex or cerebellum was not significantly different from the control values (Fig. 5).

Protein and RNA metabolism are intimately connected, and there is evidence that thyroid hormones affect nucleic acid metabolism in organs other than the brain(21). The effect of neonatal thyroidectomy on cerebral RNA metabolism has, therefore, been investigated using [6-^{14}C]orotic acid as the precursor of RNA-pyrimidines. Findings relating to acid-soluble pyrimidine nucleotides have shown that the amounts of both uracil and cytidine nucleotides per gram of brain weight and the conversion of [^{14}C]orotic acid into pyrimidine nucleotides is similar in the

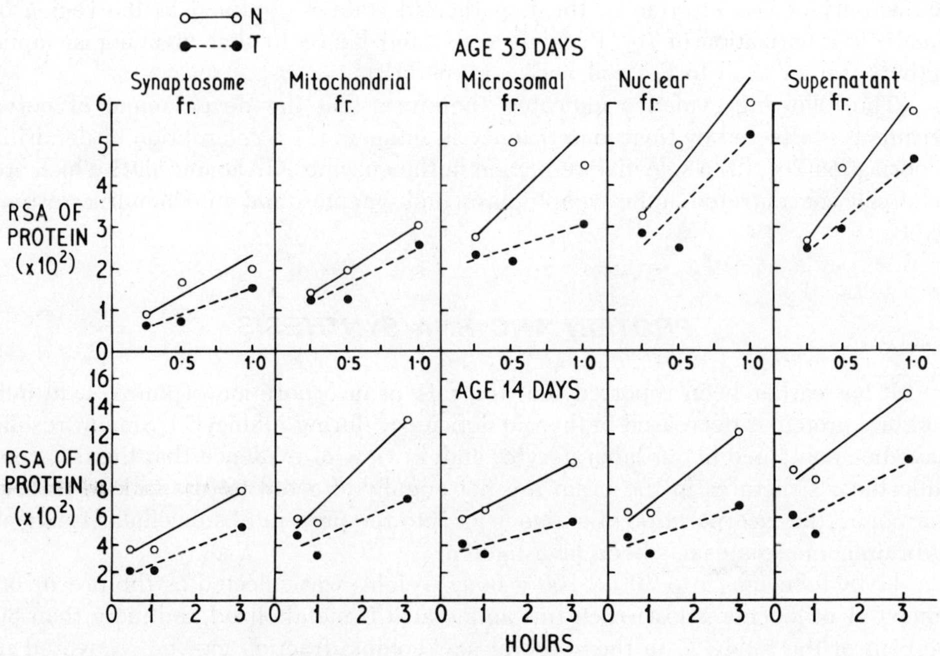

Fig. 3. Effect of neonatal thyroidectomy on the incorporation of [1-^{14}C]leucine into proteins of subcellular fractions of homogenates of cerebral cortex. Rats were given 20 and 10 μc of [1-^{14}C]leucine (specific radioactivity 6.6 μc/mmole) per 100 g of body weight intraperitoneally at the 14th and 35th day of age, and they were killed at the times indicated on the abscissa. Incorporation is expressed as relative specific radioactivity:

$$RSA = \frac{\text{specific radioactivity of protein (d.p.m./mg protein)}}{{}^{14}\text{C content in the acid-soluble pool (d.p.m./g wet weight)}}$$

Normal—open circles; thyroidectomized—solid circles.

experimental and the control animals(9). Furthermore, the incorporation of labeled pyrimidine nucleotides into the cerebral RNA of the purified nuclear and other subcellular fractions is not affected by neonatal thyroidectomy (Fig. 6).

The dissimilar effects of thyroid deficiency on RNA and protein metabolism suggest that, in the developing brain, thyroid hormones are mainly involved at the level of translation of the genetic message in proteins. Evidence is accumulating that, in other tissues also, hormones affect the processes of protein synthesis at this level(21,22).

It has previously been reported(23) that the administration of thyroxine to adult rats increases the incorporation of [^{14}C]leucine into protein in a number of organs but is without significant effect on the brain. In our experience, the rate of cerebral protein synthesis was not increased by chronic treatment with thyroid hormone as early as 21 and 35 days of age (Table 3). These experiments are still in progress with a view to obtaining information on the rate of amino acid incorporation in cerebral protein at even earlier ages.

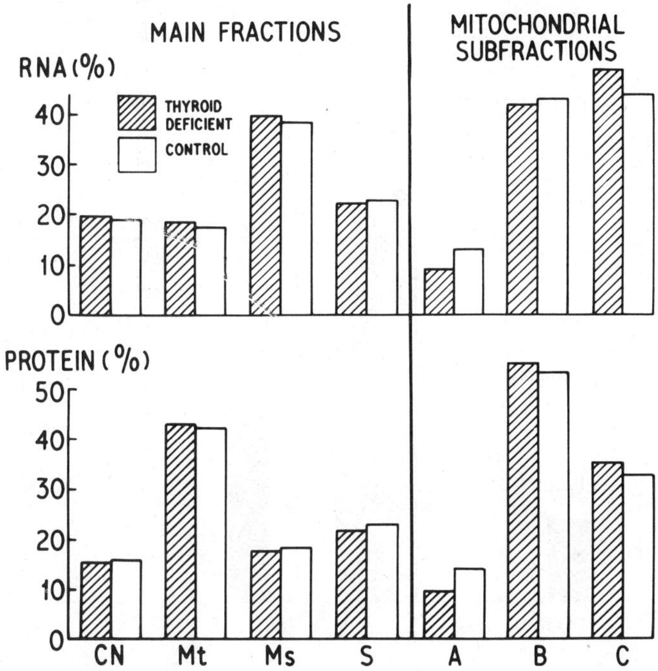

Fig. 4. Effect of neonatal thyroidectomy on the subcellular distribution of protein and RNA in the brain. Thyroid-deficient animals (shaded columns) and littermate controls (open columns) were killed at 35 days of age. The subcellular fractionation was done as described by Balázs and Cocks(45). The fractions were as follows: *CN*—crude nuclear; *Mt*—crude mitochondrial; *Ms*—microsomal; *S*—supernatant fraction. The mitochondrial subfractions were according to the technique and nomenclature of Whittaker(19): *A*—myelin; *B*—synaptosomal; and *C*—"free" mitochondrial fractions. The distribution of RNA and protein is expressed in percentage of the total amounts recovered in the main fractions and in the mitochondrial subfractions respectively.

EFFECT OF THYROID DEPRIVATION DURING DEVELOPMENT ON MYELIN FORMATION

As mentioned earlier, the yield of protein in the myelin fraction of brain homogenates is reduced as a result of neonatal thyroidectomy (Fig. 4). During the fractionation of the crude mitochondrial fractions on sucrose gradients it has been noticed that the myelin fraction derived from the brain of hypothyroid animals is more diffuse and smaller in quantity than that derived from a corresponding fraction obtained from normal animals. In further work these observations have been extended in the form of analyses of the lipid composition of the brain at different times after neonatal thyroidectomy(24,24a).

Cerebrosides are more intimately associated with myelin than are phospholipids and cholesterol(25). This conclusion is borne out by the observations, *inter alia,* that only relatively low proportions of the total phospholipids (about

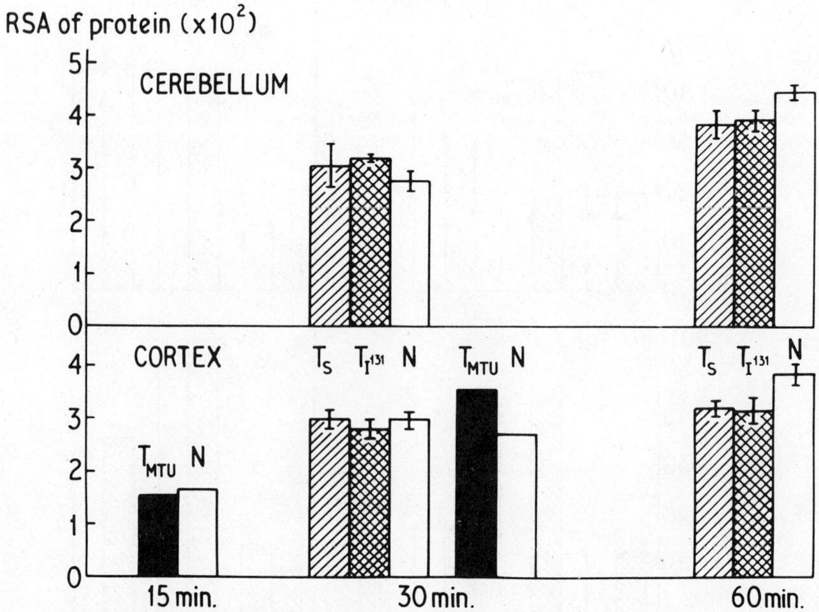

Fig. 5. Effect of thyroid deficiency in adult life on the rate of incorporation of [1-^{14}C]leucine into cerebral protein. Hypothyroidism was affected by the surgical removal of thyroid gland with (T^{131}) or without (T_s) ^{131}I treatment, and by methyl-thiouracil administration (T_{MTU}, 2 ani-only). The experiments were performed at 3 weeks after hypothyroidism had been induced. Littermate controls (*N*). RSA as in legend to Figure 3.

TABLE 3

Effect of thyroid treatment during postnatal life on the incorporation of [1-^{14}C] leucine into cerebral protein[a]

Age (days)	Experimental conditions (no. of expts)	Total ^{14}C (d.p.m./g × 10^{-3})	^{14}C in protein (% of total ^{14}C)	Spec. radioactivity (d.p.m./mg protein)
		a. Cerebrum		
21	Normal (3)	38.1 ± 1.35	67.1 ± 2.43	245 ± 8.2
	Hyperthyroid (3)	39.1 ± 3.62	61.4 ± 2.25	228 ± 17.8
35	Normal (5)	19.5 ± 1.75	62.8 ± 0.94	106 ± 9.7
	Hyperthyroid (6)	23.0 ± 1.21	61.5 ± 0.89	121 ± 10.7
		b. Cerebellum		
21	Normal (3)	42.1 ± 1.25	67.6 ± 2.70	259 ± 14.0
	Hyperthyroid (3)	39.5 ± 3.61	64.2 ± 1.42	235 ± 31.5
35	Normal (5)	21.7 ± 1.89	66.3 ± 1.29	133 ± 13.3
	Hyperthyroid (6)	24.7 ± 1.06	63.1 ± 1.52	124 ± 8.7

[a]Rats were treated with thyroid hormones (T_3) from birth during the experimental period. The 21-day-old animals were injected with 10 μc and the 35-day-old rats with 5μc/100 g body wt. [1-^{14}C] leucine s.c., and they were killed at 15 minutes after the injection.

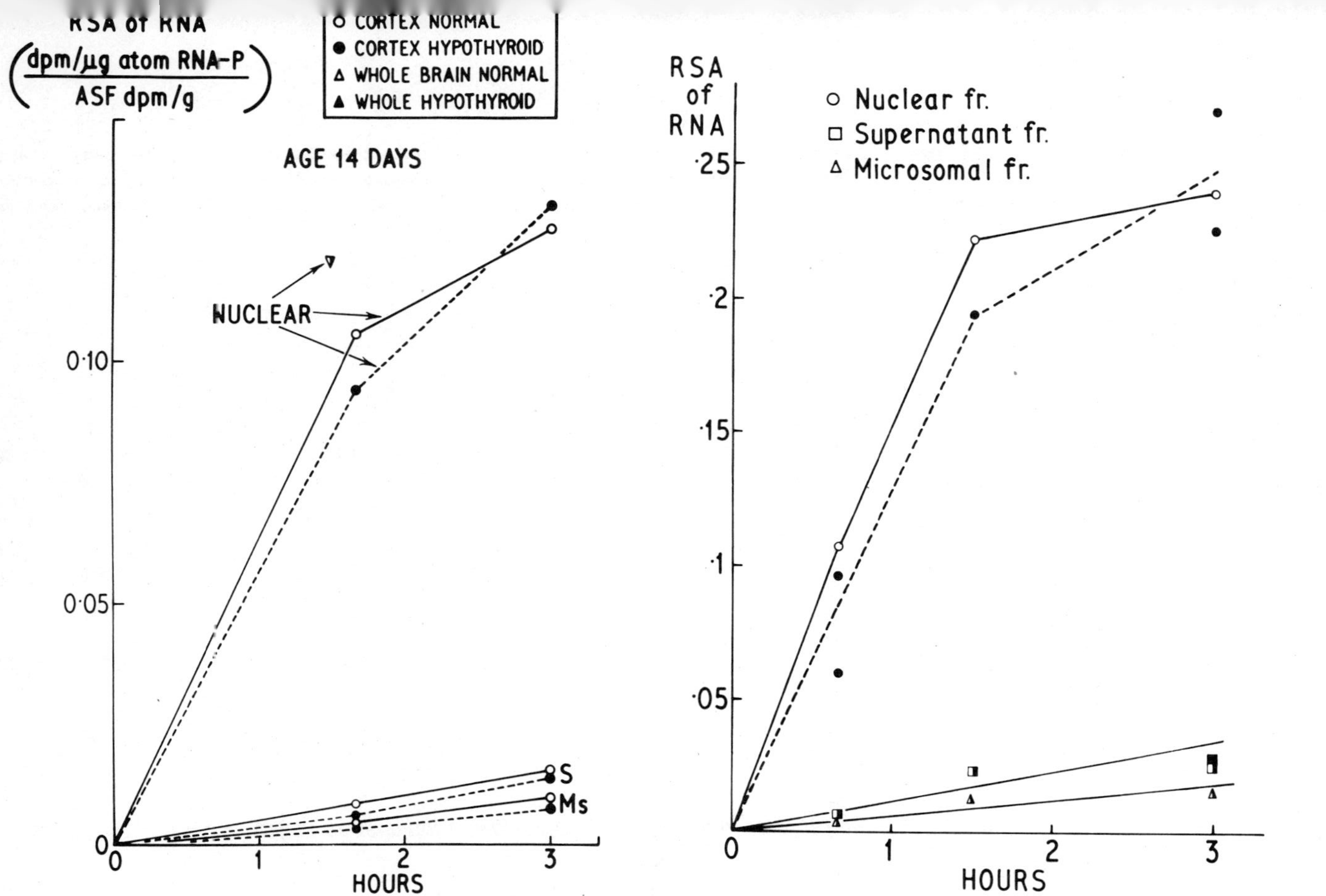

Fig. 6. Effect of neonatal thyroidectomy on the incorporation of labeled pyrimidine nucleotides into RNA of subcellular fractions of brain homogenates. 3 µc of [6-^{14}C]orotic acid (31 mc/mmole) were injected intracerebrally at the age of 14 (Fig. 6a) and 35 (Fig. 6b) days, and the animals were killed at the times indicated after the injection. Open symbols indicate normal, solid symbols hypothyroid animals. Subcellular fractionation was as described by Balázs and Cocks(43); nuclear fraction refers to the purified nuclear fraction as given in(43). *S*—supernatant and *Ms*—microsomal fractions.

TABLE 4

Effect of neonatal thyroidectomy on the deposition of lipids in the brain

	Brain weight (g)				Phospholipids				Cholesterol				Cerebroside			
	Normal	**Hypothyroid**	**(%)[b]**	**P[c]**	**Normal**	**Hypothyroid**	**(%)**	**P**	**Normal**	**Hypothyroid**	**(%)**	**P**	**Normal**	**Hypothyroid**	**(%)**	**P**
								a. mg/g wet weight								
14[a]	0.876 ±0.0111 (9)	0.867 ±0.0170 (9)	99		33.8 ±0.59 (5)	30.8 ±1.77 (5)	91		6.52 ±0.352 (8)	5.95 ±0.165 (8)	91					
23	1.11 ±0.027 (6)	0.988 ±0.0379 (6)	89	< 0.05	50.6 ±0.67 (6)	36.0 ±1.01 (6)	71	< 0.001	9.31 ±0.216 (6)	7.17 ±0.343 (6)	77	< 0.001	4.17 ±0.266 (6)	2.95 ±0.304 (6)	71	< 0.02
35	1.15 ±0.038 (6)	0.977 ±0.0152 (5)	85	< 0.02	45.6 ±1.62 (6)	42.4 ±2.72 (5)	93		10.8 ±0.32 (6)	10.0 ±0.25 (5)	93		8.42 ±0.438 (5)	6.00 ±0.322 (3)	71	< 0.02
								b. mg per cerebrum								
23					56.4 ±1.16	35.4 ±0.78	63	< 0.001	10.4 ±0.22	7.12 ±0.532	69	< 0.001	4.61 ±0.206	2.94 ±0.345	64	< 0.01
35					52.3 ±2.70	41.4 ±2.84	79	< 0.05	12.4 ±0.59	9.77 ±0.356	79	< 0.01	9.53 ±0.517	5.90 ±0.382	62	< 0.01

[a]The lipids were analyzed from the whole cerebrum, except at the 14-day-old age group, in which cerebral cortex was used.

[b]The mean value obtained in the hypothyroid group expressed a percentage of that in the normal animals.

[c]When the difference between the experimental and control groups were significant ($P < 0.05$) the value of P is given.

1/5), and cholesterol (about 1/3) are associated with the myelin fraction isolated from brain homogenates whereas about 60 percent of the cerebrosides are recovered from this fraction in 5-week-old-rats(25). In the brain of hypothyroid rats, the amount of cholesterol and phospholipids per gram wet weight was reduced only for a transient period (at about 23 days of age), whereas the cerebroside content per unit weight was also decreased by 30 percent in the cerebrum of 35-day-old hypothyroid rats, at which age the concentration of the other lipids did not differ significantly from the control values (Table 4,a). When the deposition of lipids in the cerebrum was calculated (Table 4, b) it was evident that the amount was re-

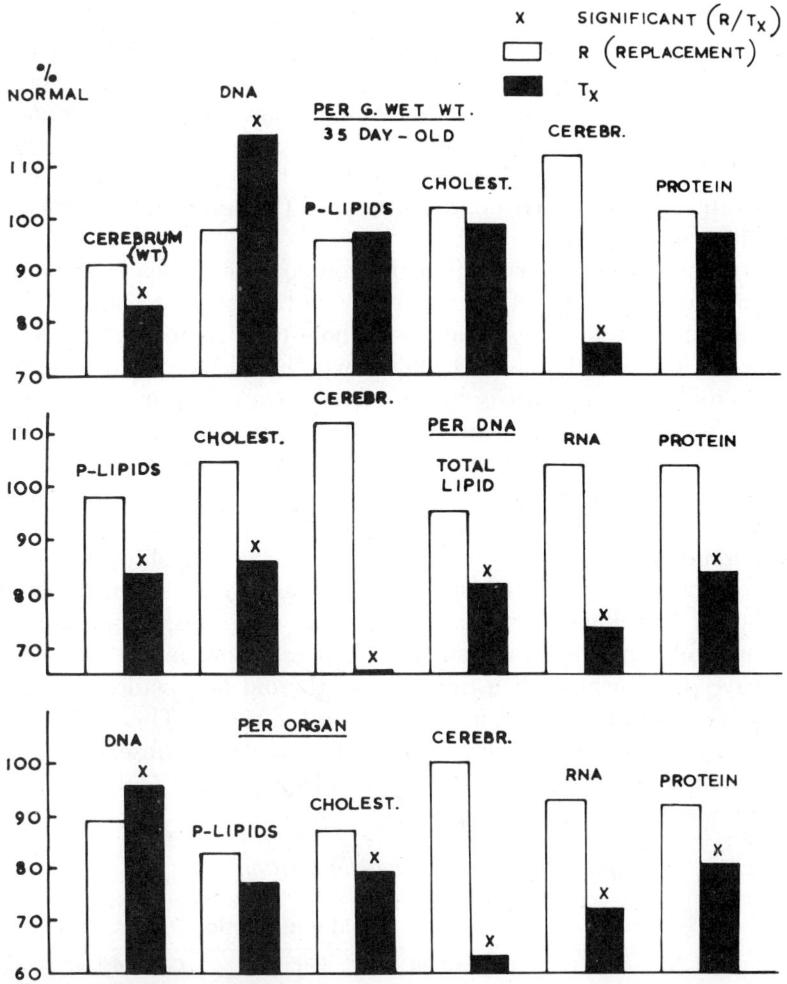

Fig. 7. Effect of replacement therapy on the deposition of lipids and other cell constituents. Animals were thyroidectomized at birth (Tx, solid columns) or thyroidectomized and treated with thyroid hormones during the experimental period (*R*=replacement, open columns). The data are expressed as percentage of the values obtained in a normal group comprising littermate controls. Three animals were in each group. The data in the three rows are expressed on the basis of wet weight, DNA content, and per organ, respectively. Significant differences between replacement and control groups are marked with an *x*. P-lipids—phospholipids; cholest.—cholesterol; cerebr.—cerebrosides.

duced both at 23 and 35 days of age. The decrease in the amount of phospholipids and cholesterol at 35 days was about the same as that in brain weight (about 20 percent), but the deposition of cerebrosides was reduced by 40 percent.

Confirmation that the defect of myelination is related to thyroid deficiency comes from experiments in which the lipid content of the brain was analyzed at 35 days of age in three groups of animals: (1) thyroidectomized at birth; (2) thyroidectomized at birth but receiving replacement treatment during the experimental period; and (3) littermate controls. The values given in Figure 7 are expressed as percentages of the control data and significant differences have emerged as a result of giving thyroid hormone to the thyroidectomized animals. Untreated thyroid deficiency resulted in a reduction in the weight of the cerebrum, an increase in cell density, and a reduction in lipid concentration involving only the cerebrosides. However, when the results were expressed on the basis of DNA the amounts of all the lipids, as well as that of RNA and protein, were significantly decreased. Calculated on the basis of the total cerebral content Figure 7 shows that the constituents most affected by thyroid deprivation were the cerebrosides which were reduced by about 40 percent.

It has recently been observed that the composition of myelin changes during development(25). At 16 days of age the myelin of the rat brain contains more phospholipids and less cerebrosides relative to cholesterol than at 30 days of age or later. It was of interest, therefore, to know whether the reduced myelination observed in thyroid deficiency was solely a quantitative effect or comprised in addition a qualitative change, such as might involve the persistence of "immature" myelin. Analysis of the cerebral myelin-containing fraction of 43-day-old rats has shown that the relative proportions of lipids in the cerebral myelin of thyroidectomized animals were comparable with those in the controls (Table 5).

The marked and specific decrease in the content of cerebrosides in the brain of rats thyroidectomized at birth thus indicates that myelination is impaired under these conditions. The change is a quantitative one, since the composition of myelin in the young adult experimental animals is similar to that in the controls. Cuarón et al.(26) have previously studied the effect of thyroid deprivation during maturation on phospholipid metabolism in the brain of the rabbit. Their failure to detect an impaired myelination may well be related to the present observation that phospholipids are affected by thyroid deficiency only for a transient period. Further-

TABLE 5

Lipid composition of myelin fraction[a]

	Lipid molar ratio		
	Cholesterol	**P-lipid**	**Cerebroside**
Normal	10	8.2	3.0
Thyroidectomized	10	8.0	2.9

[a]43-day-old rats (three animals in each group); the fraction containing myelin was prepared from that of the crude mitochondria of brain homogenates by the method of Cuzner and Davison (25). The lipids were determined as described in (24a), and the relative amounts found in myelin were expressed in μ moles.

more these lipids are not specific markers for myelin. Our current investigations have not so far included the effects of giving thyroid hormone though there is some evidence that this treatment is associated with an acceleration of myelination.* Hamburgh(27), for instance, has observed an increased myelination in cultures of cerebellar tissue in the presence of thyroxine, while both Schapiro(28) and Myant and Cole(29) have respectively reported a slight increase in cerebral cholesterol and phospholipid content in young rats treated with thyroid hormone.

EFFECT OF THYROID HORMONES ON THE DEVELOPMENT OF THE ORGANIZATION OF METABOLIC PATHWAYS RELATED TO CEREBRAL ENERGY METABOLISM AND ON METABOLIC COMPARTMENTATION

The neurochemical maturation of brain has generally been studied by determining changes in individual constituents or enzymic activities during development. Such data are very important and by increasing the scope of the measurements there is reason to hope that the quantitative information thus obtained will lead to the understanding of the qualitative aspects of the transition from the immature to the mature brain. We have tried a somewhat different approach, however, employing indices, previously described by Gaitonde and Richter(30) and by Berl(31), to measure the coordination of a number of metabolic reactions. In this section, experiments will be reported in which these indices have been used to assess the effects of hormones on the maturation of nervous tissues.

Conversion of Glucose-Carbon into Amino Acids

There is convincing evidence that the main substrate for cerebral energy metabolism is glucose(32) and the fate of labeled glucose has recently been thoroughly investigated in our laboratory(32,33). It has been shown that, in the adult brain a few minutes after the administration of [^{14}C]glucose, 50 to 60 percent of the total ^{14}C has already been converted into amino acids associated with the tricarboxylic acid cycle. Gaitonde and Richter(30) have shown that, in the brain of the rat, the capacity for rapid and extensive conversion of [^{14}C]glucose into amino acids develops during a short period of maturation extending between 10 and 21 days. The rapid conversion of glucose-carbon into amino acids is the result of a number of factors characteristic of the metabolism of the mature brain: The enzymic activities of the glycolytic pathway and of the tricarboxylic acid cycle, for instance, are high while those critical for the reversal of glycolysis are negligible. The levels of glucose and glycogen as well as the enzymic activities related to glycogen synthesis are also relatively low. On the other hand, the content of some amino acids, especially that of glutamate, is large when compared with that in other tissues or in immature brain. The oxidation of substrates other than glucose, which are important in the

*In the meantime we have investigated the effects of T_3 treatment on the metabolism of cerebrosides and sulphatides during the period of rapid myelination in rat brain. The results did not show conclusively an advancement of myelination under these conditions. A preliminary account of these studies has has recently been given(24b).

metabolism of other tissues, is of minor significance in the adult brain. These facts indicate, therefore, that the conversion of [^{14}C]glucose into amino acids can be used as an index for the organization of many metabolic reactions characteristic of the mature brain. In the following experiments we have accordingly investigated the effects of thyroid hormones on the conversion of [^{14}C]glucose into amino acids during the critical period of brain maturation(34,34a). Rats were killed at 20 minutes after the intraperitoneal injection of 10 to 20 μc of [^{14}C]glucose per 100 g body weight. At this time more than 90 percent of the ^{14}C in the brain was recovered in the acid-soluble fraction. The amino acids were separated from the neutral compounds and the carboxylic acids (major radioactive constituents were glucose and lactate, respectively) on a cation exchange resin. The ^{14}C content of the amino acid fraction is expressed in Figure 8 as the percentage of the total radioactivity in the acid-soluble fraction. The results confirmed the observations of Gaitonde and Richter(30) concerning the relationship between age and the fate of [^{14}C]glucose in the normal brain. The distribution of the acid-soluble ^{14}C changed with age; at 10 days about 20 percent of the ^{14}C was recovered in the amino acid fraction, whereas this fraction contained almost 60 percent of the ^{14}C at 19 days of age. At 19 days the conversion of glucose-carbon into amino acids approached that reported by Vrba et al.(33) for the adult brain, i.e., at that age this neurochemical index reveals nearly complete maturation.

The extensive conversion of [^{14}C]glucose into amino acids was advanced in the brain of rats treated with thyroid hormones; at 12 days of age about 40 percent of the total label was incorporated into amino acids, compared with about 20 percent in the brain of normal and thyroidectomized rats, respectively. The conversion of glucose-carbon into amino acids was retarded in the brains of thyroid-deficient rats. At 19 days of age only about one third of the ^{14}C was in amino acids, compared with about one half measured in the controls or hyperthyroid animals. These results show, therefore, that treatment with thyroid hormones in infancy accelerates the development of the organization of cerebral energy metabolism characteristic of the adult, whereas this index of brain maturation was reduced during the experimental period in the thyroid-deficient animals. Further work is necessary to investigate whether or not the normal adult level of conversion of glucose-carbon into amino acids will eventually be reached in the brain after neonatal thyroidectomy.

Effects of Hormones on Metabolic Compartmentation of Glutamate

Another characteristic feature of brain metabolism has been discovered by Waelsch's group(35,36), who observed that, at short times after the intracisternal injection of glutamate, the specific radioactivity of glutamine was higher than that of glutamate. Since glutamine is the product of glutamate, these observations indicate that a small and highly active pool of glutamate is the precursor of glutamine; i.e., glutamate is compartmented in the brain. The original observations have been confirmed and extended by different laboratories including our own(37,37a). It is now known that the metabolic compartmentation of glutamate can be demonstrated by use of a number of precursors(36). The results strongly indicate that the tricarboxylic acid cycle and metabolites associated with its intermediates are also

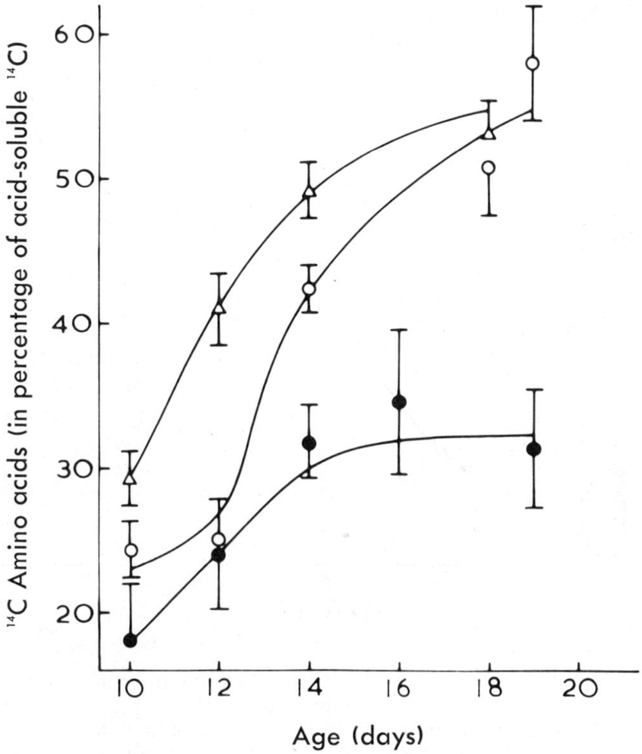

Fig 8. Effect of thyroid hormones on the development of the conversion of glucose-carbon into amino acids characteristic for the adult brain. Rats of different ages (indicated on the abscissa) were injected with 10 to 20 μc of [U-^{14}C]glucose per 100 g body weight intraperitoneally. They were killed 20 minutes after the injection. The neutralized acid-soluble fraction obtained from the cerebrum was passed through a column of Zeocarb-225 H^+-form resin to separate neutral compounds and carboxylic acids (effluent) from the amino acids (retained by the resin). The amino acids were stripped from the resin by elution with 1N NH_4OH. Details of the technique are described by Gaitonde and Richter(30). The ^{14}C Combined in amino acids was expressed as a percentage of the ^{14}C content of the acid-soluble fraction. The data were analyzed by analysis of variance; the mean values ±S.E.M. are indicated. Normal—open circles; thyroid deficient—solid circles; T_3-treated—triangles. The difference compared with normal was significant in the thyroid-deficient group from the age of 14 days and in the T_3-treated group at 12 and 14 days of age. The results were taken from Cocks et al.(34a).

compartmented(36,36a). It has recently been suggested that a model consisting of a minimum of three compartments is compatible with the experimental data(42). Each "compartment" probably results from the summation of a number of different units with more or less similar metabolic patterns. The combination constituting what is defined functionally as a "small" or a "large" compartment(38), in view of the glutamate pool size, will depend on the site of entry and metabolism of the substrate considered. The present series of experiments in our laboratory suggest that much of the neuronal structure is associated with the large compartment, whereas the small compartment includes glial cells(34a,39a,42). It has been shown by Berl(31) that the metabolic compartmentation of glutamate develops during the

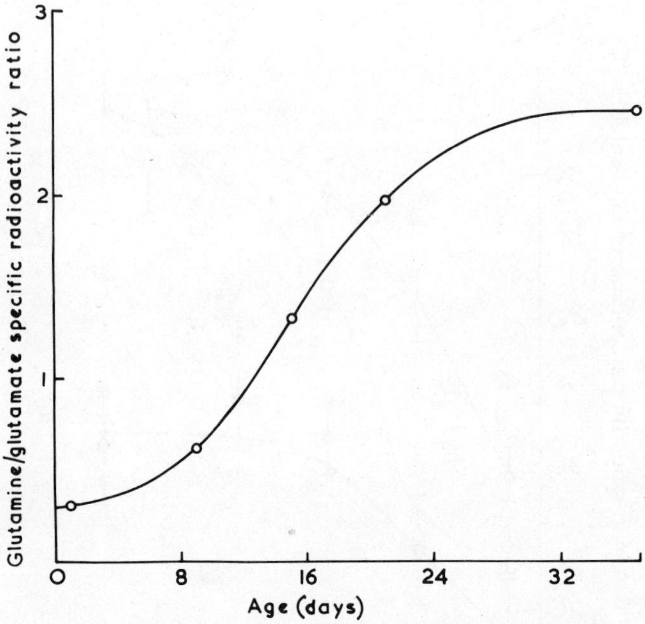

Fig. 9. The development of metabolic compartmentation of glutamate in rat brain. Rats at the age indicated on the abscissa were injected with 10 μc of [U-^{14}C]leucine per 100 g body weight subcutaneously, and were killed 10 minutes later. Glutamate and glutamine were isolated from the acid-soluble fraction by ion exchange chromatography according to the method of Minard and Muschahwar(46). The specific radioactivity (d.p.m./μmoles) of glutamine is expressed relative to that of glutamate (R.S.A.). The results were taken from Patel and Balázs(39a).

maturation of cat brain. The development of the metabolic compartmentation of glutamate has been studied in the rat(39,39a).

[U-^{14}C]Leucine is one of the substrates which shows metabolic compartmentation in brain(40), i.e., the specific radioactivity of glutamine relative to glutamate is greater than one, even before the peak incorporation into glutamine takes place. In the brain, [U-^{14}C]leucine is converted into [^{14}C]acetyl-CoA, which enters the tricarboxylic acid cycle and will subsequently label the amino acids. The specific radioactivity of glutamine relative to that of glutamate (RSA) in the brain at 10 minutes after subcutaneous injection of 10 μc of [U-^{14}C]leucine per 100 g body weight is plotted as a function of the age of the rat in Figure 9. Metabolic compartmentation of glutamate is evident when the RSA of glutamine is greater than one, and Figure 9 shows that the signs of such compartmentation develop during the same period of brain maturation as does the rapid and extensive conversion of [^{14}C]glucose into amino acids (cf. Fig. 8). In fact the two curves obtained from data derived from the brain of identical strains of rats can be satisfactorily superimposed one upon the other. Since metabolic compartmentation of glutamate is characteristic of the adult brain and since its development is probably associated with structural changes in the tissue, we have used the phenomenon as an index of the effects of hormones upon the maturation of the brain.

TABLE 6

Effect of thyroid treatment during development on the maturation of the metabolic compartmentation of glutamate[a]

Specific activity of glutamine relative to that of glutamate	
Hyperthyroid	**Normal**
0.85 ± 0.058	0.65 ± 0.028

$p < 0.05$

[a]Rats were treated with thyroid hormones during postnatal life. The experimental animals and littermate controls were injected with 10 μc/100 g body wt. (U-^{14}C) leucine at the age of 9 days and they were killed at 10 minutes after injection. Glutamate and glutamine were separated from the acid-soluble fraction as described by Patel and Balázs (39).

These studies are still in progress and therefore only the results obtained in the brain of 9-day-old animals treated with T_3 are available(41). The RSA of glutamine in the brains of the experimental animals was higher at that age than in the controls (Table 6). In two 15-day-old thyroid-deficient animals, the RSA of glutamine in their brains was lower than in the controls.*

In summary, two phenomena, which encompass a number of metabolic reactions, have been selected as indices of the biochemical maturation of cerebral tissues. It seems that the age-dependent increase in the conversion of glucose-carbon into cerebral amino acids is the result of both quantitative and qualitative changes in the energy metabolism of the tissue. There is evidence that the maturation of the metabolic pattern characteristic of the adult brain reflects the development of the neuronal processes(34a). As a result of the development of a metabolic compartment associated with neuronal structures (large compartment), the apparently homogeneous metabolic pattern of brain tissue is replaced by the heterogeneous pattern characteristic of the adult(39a). The present observations show that the neurochemical maturation of the brain is influenced by thyroid hormones. The results are consistent with the view that T_3 treatment accelerates, whereas thyroid deficiency retards, the development of the metabolic compartment associated with neuronal structures(41).

CONCLUSIONS

Our results related to the effects of thyroid hormones on the biochemistry of the developing brain have been summarized in Table 7.

1. Thyroid deficiency. The growth of the brain is retarded and this effect is evident after the 14th day of life. The smaller brain weight is related to a

*In the meantime these studies have been completed(41). It was found that thyroidectomy at birth resulted in a marked retardation of the manifestation of metabolic compartmentation. The value of the RSA of glutamine in the brain of 25-day-old thyroid-deficient rats was similar to that obtained in controls at 12 to 15 days after birth. T_3 treatment led to an overall advancement of 2 to 3 days during the period when metabolic compartmentation develops.

TABLE 7

Effects of thyroid hormones on the developing brain

	Thyroid administration	Thyroid deficiency
Brain weight	Reduced (mainly after 14 days)	Reduced (after 14 days)
Postnatal cell formation:		
a. Cerebrum	Reduced	No effect
b. Cerebellum		Reduced at 14 days, but catches up at 35 days
Cell size	No effect	Reduced (at 35 days)
Myelination	Advanced (?)	Reduced
Dendritic formation		Reduced
Enzymes related to synaptosomes		Reduced (GAD and SDH)
Rate of protein synthesis	No effect (in whole brain at 21 and 35 days)	Reduced
Rate of RNA synthesis		No effect
Organization of metabolic reactions characteristic for the adult:		
a. [^{14}C] glucose → [^{14}C] amino acids	Advanced	Retarded
b. Metabolic compartmentation of glutamate	Advanced	Retarded

decrease in average cell size since, except for the cerebellum, the total number of cells is similar to that measured in the controls. In the cerebellum a small but significant retardation in cell formation is detectable at 14 days of age. The decrease in cell size is associated with the following changes. Myelination, as indicated by a specific and lasting reduction in the cerebroside content per unit weight, is impaired. A reduction in the number of nerve terminals is indicated by such evidence as a decrease in dendritic arborization, axonal density, and in the enzymic constituents (GAD and SDH) characteristic of nerve endings. The possible reversibility of this effect has not yet been fully investigated. A concomitant defect in the synthesis of cerebral protein, which is evident only after neonatal thyroidectomy, would seem to involve the translation of the genetic message since the rate of RNA synthesis is not correspondingly affected. Neurochemical maturation, including the coordination of metabolic pathways related to energy metabolism and the development of metabolic compartmentation, is retarded.

2. Hyperthyroidism during infancy. The growth of the brain is reduced, an effect related to a decrease in postnatal cell formation which is most pronounced in the cerebellum after 12 to 14 days of age. Average cell size is not decreased. Neurochemical maturation is advanced.

Thyroid deficiency in early life gives rise in man to a severe mental retardation and in experimental animals to impaired behavioral performance(1). It is tempting to suggest that the impairment of behavior arises as a result of structural changes in brain(20), including the retardation or reduction in the formation of nerve terminals and the impaired myelination observed in the present studies though, as has been suggested in the previous communication (Eayrs, Ch. 26), these seem unlikely to provide the only explanation. Hyperthyroidism during infancy results in an advanced maturation of innately organized responses(2) associated with an acceleration of the biochemical maturation of brain. Impairment of the adaptive, cortically mediated, behavior observed in hyperthyroid rats(2) may well derive from a deficiency in postnatal cell formation. Altman(43) has reported that postnatal cell formation in the brain involves neurons as well as glial cells. According to him, the microneurons formed after birth—whose formation, therefore, may be affected by thyroid treatment during maturation—may modify the structural and functional organization of the developing nervous system to an extent depending on environmental influences. More information, however, must be collected concerning the specificity of hormone treatment and dosage as well as the correlation with behavioral impairment before this suggestion can be tested.

Finally, it has to be emphasized that we do not yet know whether or not the effects observed after the experimental procedures are due to direct or indirect action of thyroid hormone. Additional factors have to be considered: an imbalance in the endocrine system as a whole during a period of development when the CNS is still sensitive to the effects of other hormones, nutritional factors, changes in body temperature,—to mention only a few. Throughout this presentation, reference to the effects of thyroid on the development of the brain has been made in terms of the experimental treatment administered. Further work is necessary to clarify the issue of casual relationships.

REFERENCES

1. Eayrs, J. T. Thyroid and central nervous development. *In* Scientific Basis of Medicine Annual Reviews. London, Athlone Press, pp. 317–399, 1966.
2. ——— Effect of neonatal hyperthyroidism on maturation and learning in the rat. Anim. Behav., 12:195–199, 1964.
3. De Guglielmone, A. E. R., and Gómez, C. J. Influence of neonatal hypothyroidism on amino acids. J. Neurochem., 13:1017–1025, 1966.
4. Pasquini, J. M., Kaplún, X. García Argíz, C. A., and Gómez, C. J. Hormonal regulation of brain development. I. The effect of neonatal thyroidectomy upon nucleic acids, protein and two enzymes in developing cerebral cortex and cerebellum of the rat. Brain Res., 6:621–634, 1967.
5. García Argíz, C. A., Pasquini, J. M., Kaplún, B., and Gómez, C. J. Hormonal regulation of brain development. II. Effect of neonatal thyroidectomy on succinate dehydrogenase and other enzymes in developing cerebral cortex and cerebellum of the rat. Brain Res., 6:635–646, 1967.
6. Geel, S. E., and Timiras, P. S. The influence of neonatal hypothyroidism and of thyroxine on the RNA and DNA concentrations of rat cerebral cortex. Brain Res., 4:135–142, 1967.
7. ——— Valcana, Th., and Timiras, P. S. Effect of neonatal hypothyroidism and of thyroxine on L-(^{14}C)leucine incorporation in protein *in vivo* and the relationship to ionic levels in the developing brain. Brain Res., 4:143–150, 1967.
8. Balázs, R., and Gaitonde, M. K. Factors affecting protein metabolism in the brain. Biochem. J., 106:1–2P, 1968.
9. ——— Kovács, S., Teichgräber, P., Cocks, W. A., and Eayrs, J. T. Biochemical effects of thyroid deficiency on the developing brain. J. Neurochem., 15:1335–1349, 1968.
10. ——— Kovács, S., Teichgräber, P., Cocks, W. A., and Eayrs, J. T. Effects of neonatal

thyroid deficiency on the developing brain. First Int. Meeting Int. Neurochem. Soc., Strasbourg, p. 2, 1967.
11. Enesco, M., and Leblond, C. P. Increase in cell number as a factor on the growth of the organs of the young male rat. J. Embryol. Exp. Morph., 10:530–562, 1962.
12. Legrand, J. La maturation du cervelet chez le rat blanc hypothyroidien. *In* Regional Development of the Brain in Early Life, Minkowski, A., ed. Oxford, Blackwell Scientific Publications, pp. 485–493, 1967.
13. Eayrs, J. T., and Taylor, S. H. The effect of thyroid deficiency induced by methyl thiouracil on the maturation of the central nervous system. J. Anat., 85:356–358, 1951.
14. ——— La influencia de la edad de thyroidectomía sobre el desarrollo de la corteza cerebral. Arch. Mex. Anat., 7:29–30, 1966.
15. Hamburgh, M., and Flexner, L. B. Biochemical and physiological differentiation during morphogenesis. XXI. Effect of hypothyroidism and hormone therapy on enzyme activities of the developing cerebral cortix of the rat. J. Neurochem., 1:279–288, 1957.
16. Salganicoff, L., and Koeppe, R. E. Subcellular distribution of pyruvate carboxylase, diphosphopyridine nucleotide isocitrate dehydrogenases, and malate enzyme in rat brain. J. Biol. Chem., 243:3416–3420, 1968.
17. van Kempen, G. M. J., van den Berg, C. J., van der Helm, H. J., and Veldstra, H. Intracellular localization of glutamate decarboxylase, GABA-transaminase and some other enzymes in brain tissue. J. Neurochem., 12:581–588, 1965.
18. Balázs, R., Dahl, D., and Harwood, J. R. Subcellular distribution of enzymes of glutamate metabolism in rat brain. J. Neurochem., 13:897–905, 1966.
19. Whittaker, V. P. The application of subcellular fractionation techniques to the study of brain function. Progr. Biophys., 15:39–96, 1965.
20. Eayrs, J. T. The cerebral cortex of normal and hypothyroid rats. Acta Anat., 25:160–180, 1955.
21. Tata, J. R. The formation and distribution of ribosomes during hormone-induced growth and development. Biochem. J., 104:1–16, 1967.
22. Korner, A. RNA and hormonal control of protein synthesis. Progr. Biophys. Molec. Biol., 17:61–98, 1967.
23. Michels, R., Cason, J., and Sokoloff, L. Thyroxine: effects on amino acid incorporation into protein *in vivo*. Science, 140:1417–1418, 1963.
24. Balázs, R., Brooksbank. B. W. L., Davison, A. N., Wilson, D. A., and Richter, D. Effect of neonatal thyroidectomy on myelation in rat brain. J. Physiol. (London), 201:28–29, 1969.
24a. ——— Brooksbank, B. W. L., Davison, A. N., Eayrs, J. T., and Wilson, D. A. The effect of neonatal thyroidectomy on myelination in the rat brain. Brain Res., 15:219–232, 1969.
24b. ——— Biochemical effect of thyroid hormones in the developing brain. *In* Cellular Aspects of Growth and Differentiation in Nervous Tissues, Pease, D. C., ed. U.C.L.A. Forum in Medical Sciences, 1970, in press.
25. Cuzner, M. L., and Davison, A. N. The lipid composition of rat brain myelin and subcellular fractions during development. Biochem. J., 106:29–34, 1968.
26. Cuarón, A., Gamble, J., Myant, N. B., and Osorio, C. The effect of thyroid deficiency on the growth of the brain and on deposition of brain phospholipids in foetal and newborn rabbits. J. Physiol. (London), 168:613–630, 1963.
27. Hamburgh, M. An analysis of the action of thyroid hormone on development based on *in vivo* and *in vitro* studies. Gen. Comp. Endocr., 10:198–213, 1968.
28. Schapiro, S. Metabolic and maturational effects of thyroxine on the infant rat. Endocrinology, 78:527–532, 1966.
29. Myant, N. B., and Cole, L. A. Effect of thyroxine on the deposition of phospholipids in the brain *in vivo* and on the synthesis of phospholipids by brain slices. J. Neurochem., 13:1299–1307, 1966.
30. Gaitonde, M. K., and Richter, D. Changes with age in the utilization of glucose carbon in liver and brain. J. Neurochem., 13:1309–1318, 1966.

31. Berl. S. Compartmentation of glutamic acid metabolism in developing cerebral cortex. J. Biol. Chem., 240:2047–2054, 1965.
32. Balázs, R. Carbohydrate metabolism in the central nervous system. *In* Handbook of Neurochemistry, Lajtha, A., ed. Plenum Publishing Co., Ltd., Vol. 3, 1970.
33. Vrba, R., Gaitonde, M. K., and Richter, D. The conversion of glucose carbon into protein in the brain and other organs of the rat. J. Neurochem., 9:465–475, 1962.
34. Cocks, A. J., Balázs, R., and Eayrs, J. T. Effect of thyroid hormones on the biochemical maturation of rat brain. Biochem. J., 111:18, 1969.
34a. ——— Balázs, R., Johnson, A. L., and Eayrs, J. T. Effect of thyroid hormone on the biochemical maturation of rat brain: conversion of glucose-carbon into amino acids. J. Neurochem., 17: in press, 1970.
35. Berl, S., Lajtha, A., and Waelsch, H. Amino acid and protein metabolism. VI. Cerebral compartments of glutamic acid metabolism. J. Neurochem., 7:186–197, 1961.
36. ——— and Clarke, D. D. Compartmentation of amino acid metabolism. *In* Handbook of Neurochemistry, Lajtha, A., ed. Plenum Publishing Co., Ltd., Vol. 2, pp. 447–471, 1969.
36a. van den Berg, C. J., Krzălĭc, L. J., Mela, P., and Waelsch, H. Compartmentation of glutamate metabolism in brain. Evidence for the existence of two different tricarboxylic acid cycles in brain. Biochem. J., 113:281–290, 1969.
37. Balázs, R., Machiyama, Y., and Richter, D. Metabolism of GABA in cerebral cortex. *In* First Int. Meeting Int. Neurochem. Soc., Strasbourg, p. 13, 1967.
38. Garfinkel, D. A stimulation study of the metabolism and compartmentation in brain of glutamate, aspartate, the Krebs cycle, and related metabolites. J. Biol. Chem., 241: 3918–3929, 1966.
39. Patel, A. J., and Balázs, R. Development of metabolic compartmentation in rat brain. Biochem. J., 111:17–18, 1969.
39a. ——— and Balázs, R. Manifestation of metabolic compartmentation during the maturation of the rat brain. J. Neurochem., 17, 1970.
40. Roberts, S., and Morelos, B. S. Regulation of cerebral metabolism of amino acids. IV. Influence of amino acid levels in leucine uptake, utilization and incorporation into protein in vivo. J. Neurochem., 12:373–387, 1965.
41. Patel, A. J., Balázs, R., and Richter, D. Effect of thyroid hormone on the development of metabolic compartmentation of glutamate in rat brain. *In* Second Int. Meeting Int. Soc. Neurochem., Paoletti, R., Fumagalli, R., and Galli, C., eds. Milan, Tamburini Editore, 1969.
42. Balázs, R., Machiyama, Y., Hammond, B. J., Julian, T., and Richter, D. The operation of the γ-aminobutyrate bypath of the tricarboxylic acid cycle in brain tissue *in vitro*. Biochem. J., 116:445–467, 1970.
43. Altman, J. Autoradiographic and histological studies of postnatal neurogenesis. II. A longitudinal investigation of the kinetics, migration and transformation of cells incorporating triated thymidine in infant rats, with special reference to postnatal neurogenesis in some brain regions. J. Comp. Neurol., 128:431–474, 1966.
44. Khamsi, F., and Eayrs, J. T. A study of the effect of thyroid hormones on growth and development. Growth, 30:143–156, 1966.
45. Balázs, R., and Cocks, W. A. RNA metabolism in subcellular fractions of brain tissue. J. Neurochem., 14:1035–1055, 1967.
46. Minard, F. N., and Mushahwar, I. K. The effect of periodic convulsions induced by 1,1-dimethylhydrasine on the synthesis of rat brain metabolites from (2-^{14}C)glucose. J. Neurochem., 13:1–11, 1966.

28

Comparative Effects of Thyroid Deficiency and Undernutrition on Maturation of the Nervous System and Particularly on Myelination in the Young Rat

Jacques Legrand

Laboratory of Comparative Physiology,
Faculty of Sciences, University of
Montpellier, Montpellier, France

INTRODUCTION

A clear-cut effect of thyroid hormone on the histogenesis of the central nervous system of the rat has been shown on cerebral cortex(1) and on cerebellar cortex(2,3); this action can be detected after the first postnatal week and concerns those parts of the neuraxis which mature late (i.e., after birth).

In the cerebellar cortex, the most striking effects of thyroid deficiency initiated during the fetal period or at birth are as follows: (1) a delay and abnormalities in the development of the Purkinje cell arborization (Figs. 5 to 8 compared with Figs. 1 to 4); (2) a retarded differentiation of Purkinje cell bodies which retain their immature appearance after 14 postnatal days (Figs. 16 to 18 compared with Figs. 13 to 15); (3) a delayed migration of the external granule cells towards the inside of the cerebellar cortex (Fig. 21 compared with Fig. 19) and transient persistence of the external granular layer after 21 postnatal days; and (4) a delay in the appearance of cerebellar glomeruli and an abnormal development of these structures. The most persistent abnormalities observed are reduced growth and diminished branching of cell processes, so that the dendritic spread of the Purkinje cells remains hypoplastic even after the migration of granule cells has been completed. As a result, the histological maturation of the whole cerebellar cortex is modified.

Thyroxine-replacement therapy, when initiated before the end of the second postnatal week, corrects these alterations from normal development. In young rats fed propylthiouracil starting at the end of gestation or at birth, thyroxine given during only one of the two first postnatal weeks permits a harmonious development of the cerebellar cortex: the dendritic spread of the Purkinje cells is normal after

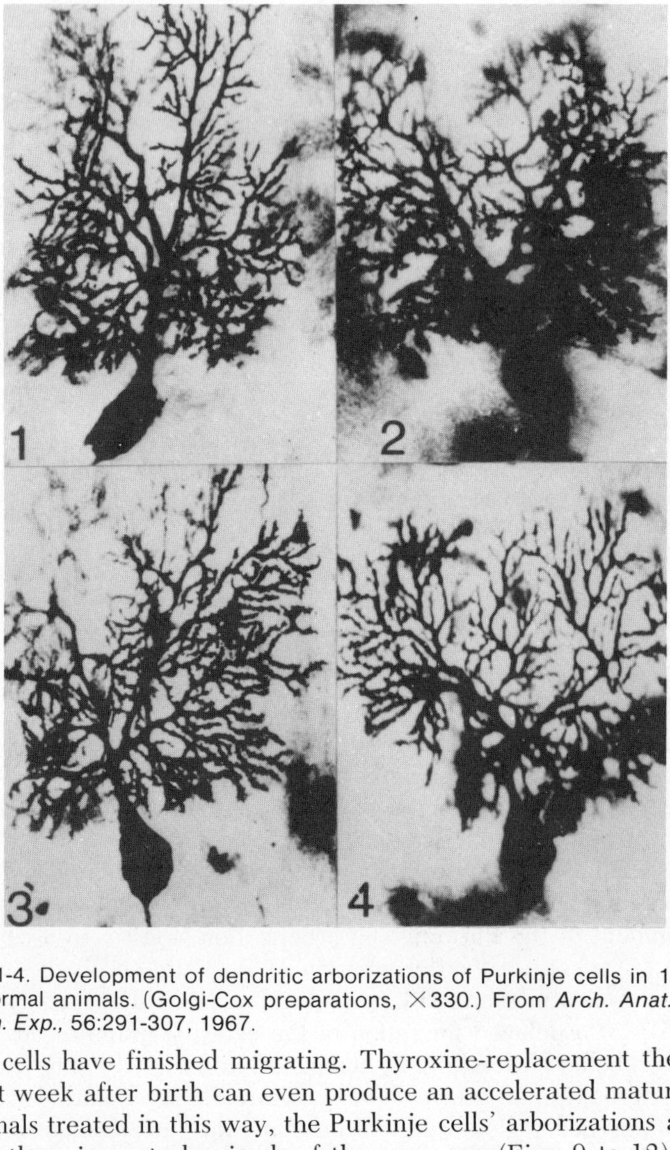

Figs. 1-4. Development of dendritic arborizations of Purkinje cells in 14-day-old normal animals. (Golgi-Cox preparations, ×330.) From *Arch. Anat. Micr. Morph. Exp.*, 56:291-307, 1967.

the granule cells have finished migrating. Thyroxine-replacement therapy during only the first week after birth can even produce an accelerated maturation. In 14-day-old animals treated in this way, the Purkinje cells' arborizations are more advanced than those in control animals of the same age (Figs. 9 to 12). When initiated after the end of the second postnatal week, thyroxine treatment does not permit a normal maturation of the cerebellar cortex: the dendritic arborization of Purkinje cells remains hypoplastic after the granule cells have migrated.

It seems that the second postnatal week is a critical period for the development of the cerebellum as well as for behavioral maturation(4) and for the development of some enzyme activities in the cerebral cortex(5). Thyroid hormones must act before the end of this critical period if a normal histological maturation of the cerebellar cortex is to be achieved.

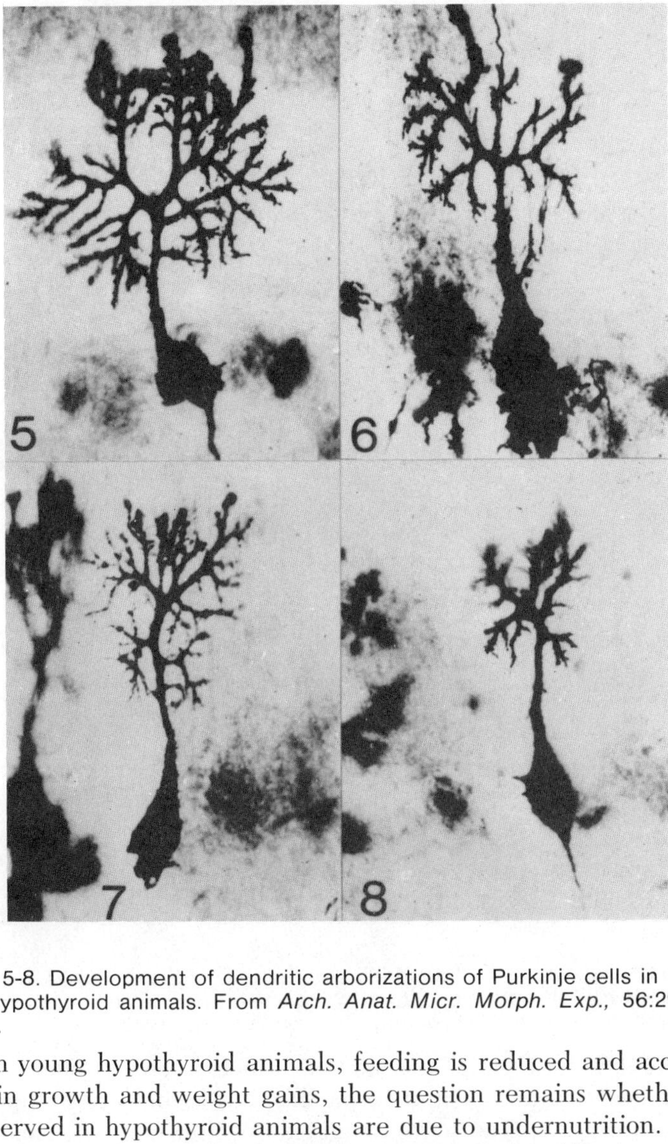

Figs. 5-8. Development of dendritic arborizations of Purkinje cells in 14-day-old hypothyroid animals. From *Arch. Anat. Micr. Morph. Exp.*, 56:291-307, 1967.

Since, in young hypothyroid animals, feeding is reduced and accompanied by a decrease in growth and weight gains, the question remains whether the abnormalities observed in hypothyroid animals are due to undernutrition.

In a previous study we have shown that in underfed suckling rats, the histological maturation of the cerebellar cortex is nearly normal(2) (Fig. 20 compared with Fig. 19 and Fig. 21). Underfeeding is accompanied neither by a delay in the migration of granule cells nor by a retardation in the differentiation of Purkinje cells; the cerebellar glomeruli also appear at the normal age, and we concluded that the abnormalities observed in the cerebellum of hypothyroid rats were specific.

Since thyroid function influences the maturation of the nervous system, it seems reasonable to assume that myelination, which is one of the most distinctive features of nervous system development, may also be under thyroid control. Barrnett(6) has reported a decrease of the myelin content of the brain and spinal cord in 30- and

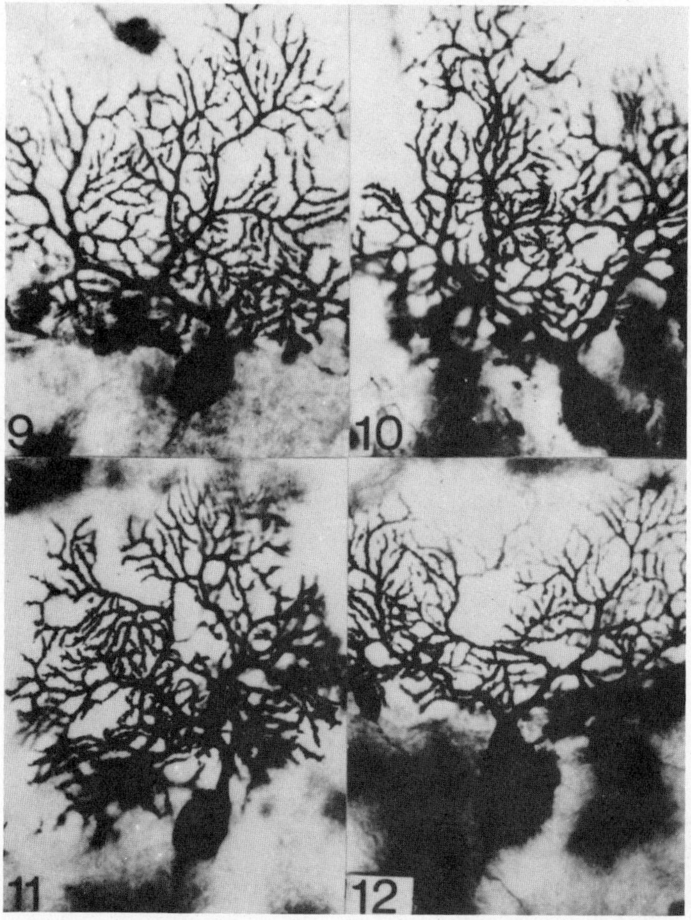

Figs. 9-12. Development of dendritic arborizations of Purkinje cells in 14-day-old hypothyroid animals injected with thyroxine during the first postnatal week only. From *Arch. Anat. Micr. Morph. Exp.*, 56:291-307, 1967.

40-day-old rats that were born from mothers which received thiouracil throughout gestation and lactation and continued on this antithyroid drug after weaning. It also has been shown(7,8) that thyroxine added to cultures of maturing rat cerebella accelerates the emergence of myelin. On the other hand, biochemical studies(9) have led to the conclusion that myelination is not specifically affected by supplying excess thyroxine to normal newborn animals.

Undernutrition during the suckling period affects myelination of the nervous system(10,11,12) also. It was therefore of interest to compare the influences of thyroid deficiency and underfeeding on myelination of the nervous system in young suckling rats. The specific objective of the study was to examine whether hypothyroidism on the one hand and undernutrition on the other hand exert different effects on myelination as has been shown to be the case with respect to maturation of the cerebellum(2).

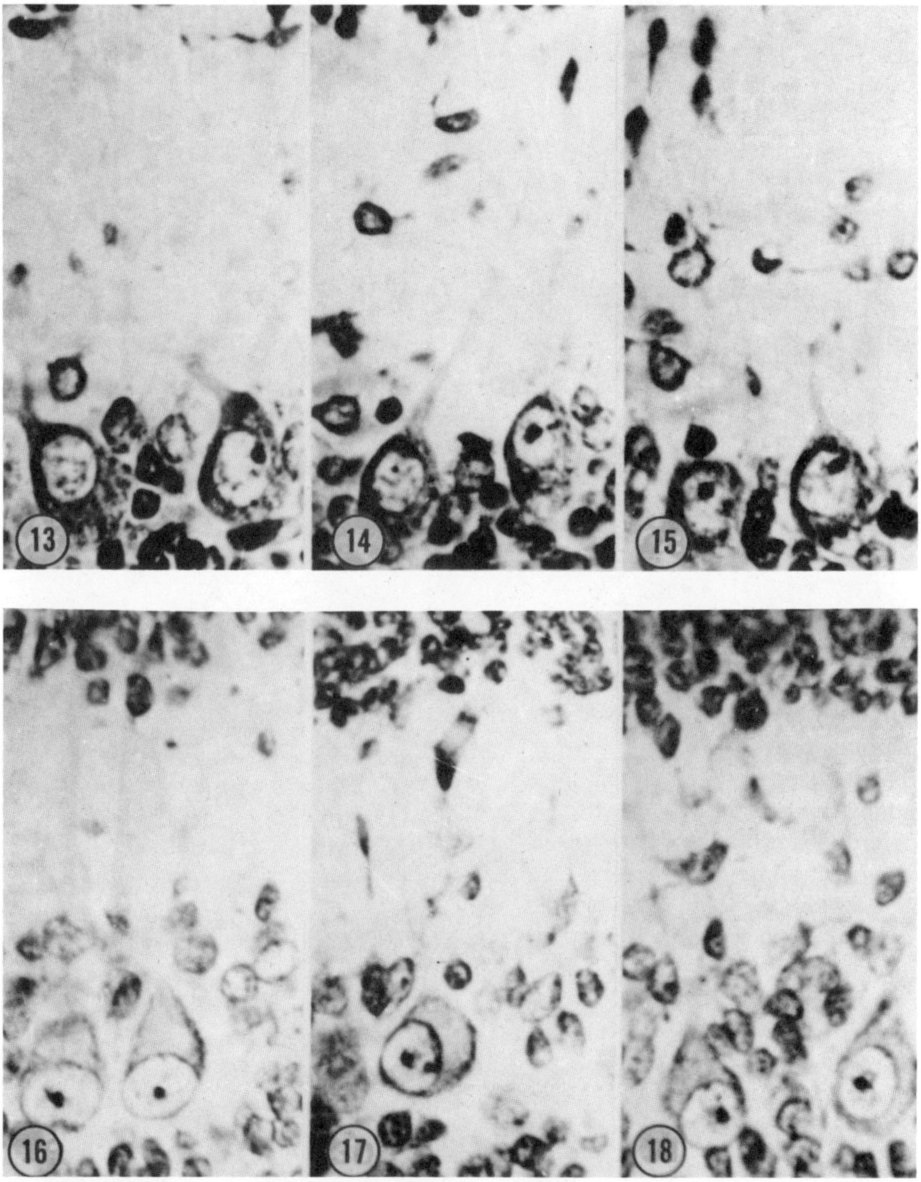

Figs. 13-18. Differentiation of perikarya of Purkinje cells. (Toluidine blue, ×720.) Figures 13 to 15: 14-day-old normal animals; Figures 16 to 18: 14-day-old hypothyroid animals. From *Arch. Anat. Micr. Morph. Exp.*, 56:205-244, 1967.

In the rat, the nervous system myelinates entirely postnatally. Among the nervous structures which myelinate in the first days following birth are the brachial and sciatic nerves and the spinal cord(13). In a preliminary unpublished work, we did not observe any delay in the appearance of the first myelin sheaths in the cer-

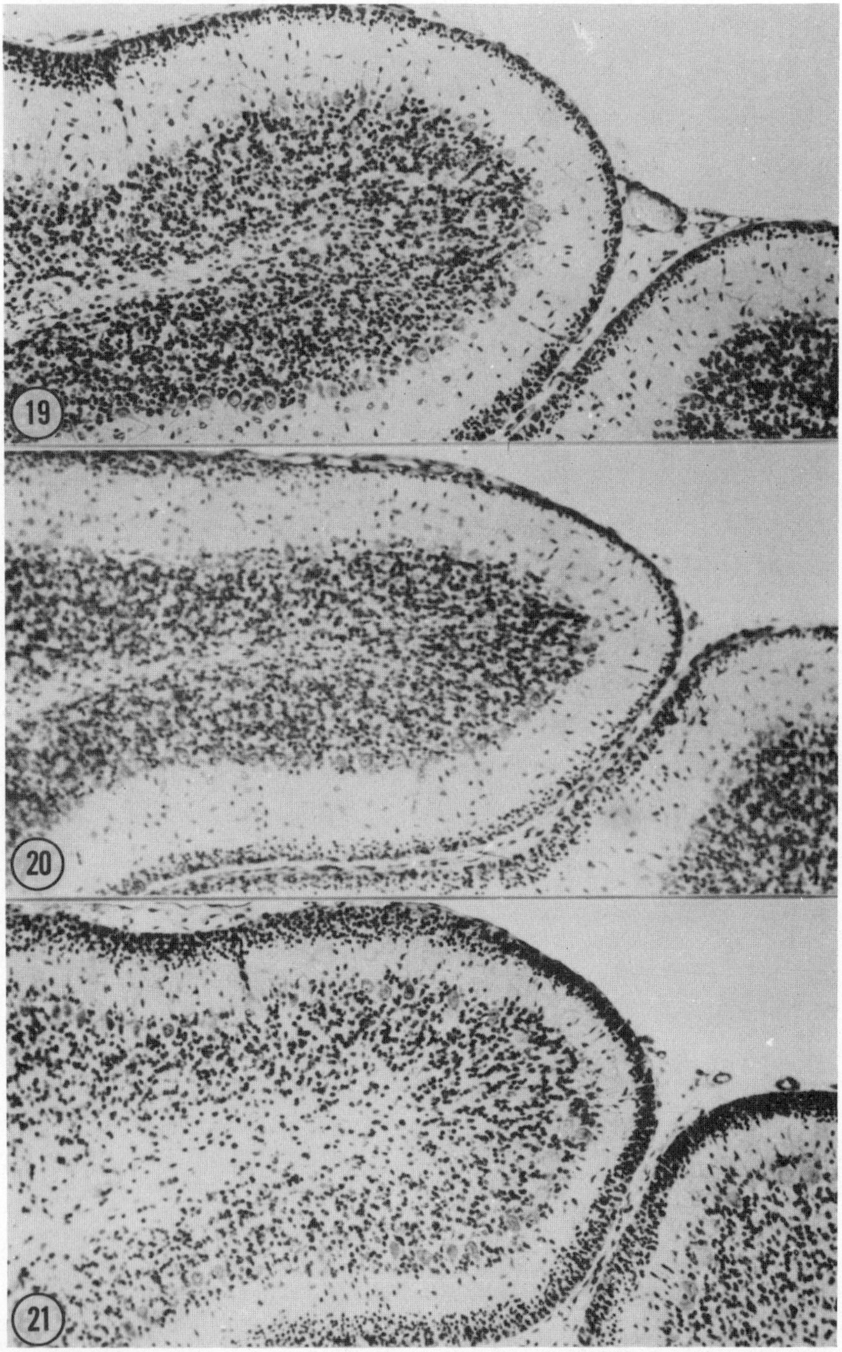

Figs. 19-21. Organization of the cerebellar cortex in 14-day-old animals. (Hemalum-eosine, ×130.) Figure 19: normal animal; Figure 20: underfed animal : Figure 21: hypothyroid animal. From *Arch. Anat. Micr. Morph. Exp.*, 56:205-244, 1967.

vical spinal cord and sciatic nerve of rats of several days of age made hypothyroid at the end of gestation or immediately following birth. Since abnormalities in the development of the cerebellum in such hypothyroid animals are not detectable before the beginning of the second postnatal week(2), we reexamined the influence of hypothyroidism on myelination of the same structures in animals killed during the second postnatal week.

MATERIALS AND METHODS

All animals were killed at 12 days of age. Young rats were made hypothyroid by administering propylthiouracil to the mothers (50 mg per day) directly into the stomach, from the end of gestation (19th day *post coitum*). From the first day following birth, adequate nourishment was assured by housing 5 to 8 littermates with their mother. These were considered as controls. Undernourishment was induced by keeping about 20 suckling rats with a single mother. From time to time, some of them were taken away from the mother to reduce competition for lactation sites. This treatment leads at 12 days to a mean weight loss of about one third of the mean weight obtained by the controls at 12 days of age (Table 1).

For the study of myelination, thin paraffin transversal sections (3 μ) of identical areas of the sciatic nerve and cervical spinal cord were prepared. Sections were stained with Heidenhain's iron hematoxylin and differentiated in a 1 percent solution of ferric alum.

Observations were made in four adjacent areas of the sciatic nerve and of the white matter of the spinal cord near the ventral groove; these areas were drawn with the aid of an ocular micrometer and a camera lucida. About 500 fibers were drawn per animal, both in the nerve and in the spinal cord. In each experimental group (5 animals per group), the distribution of the fibers was studied in relation to their inner diameter. The mean number of myelinated fibers per unit of area, their mean inner and outer diameters, and the mean thickness of their myelin sheath were calculated.

RESULTS

Mean values are reported in Table 1 together with the standard errors.

In the sciatic nerve, the mean inner diameter of myelinated fibers differs significantly between hypothyroid and underfed animals. In hypothyroid animals, fibers are on an average smaller than in underfed animals; nevertheless, they are more extensively myelinated. The fibers are myelinated to the same extent in hypothyroid animals as they are in controls (Figs. 24 and 25 compared with Figs. 22 and 23). In underfed rats, the myelin sheath is thinner than in controls or hypothyroid rats (Figs. 26 and 27).

In the cervical spinal cord, the mean inner diameter of myelinated fibers differs significantly between hypothyroid and underfed animals; the mean number of myelinated fibers per unit area also differs in a significant manner between these two experimental groups. In hypothyroid animals, the mean number of myelinated

TABLE 1

Myelin formation in normal, starved, and hypothyroid young rats[a]

Experimental group	Number of animals	Mean weight (g)	Sciatic nerve: Mean number of myelinated fibers per unit of area	Sciatic nerve: Mean inner radius (μ)	Sciatic nerve: Mean thickness of myelin sheaths (μ)	Cervical spinal cord: Mean number of myelinated fibers per unit of area	Cervical spinal cord: Mean inner radius (μ)	Cervical spinal cord: Mean thickness of myelin sheaths (μ)
Controls	5	22 (21-22.9)	121.0 ± 2.7	1.194 ± 0.020	0.554 ± 0.039	142.0 ± 1.9	1.217 ± 0.020	0.254 ± 0.047
Hypothyroid	5	17.6 (16.4-20.1)	103.2 ± 2.1[b]	1.343 ± 0.040[b]	0.501 ± 0.024	149.0 ± 3.3[b]	1.145 ± 0.035[b]	0.160 ± 0.029[b]
Underfed	5	14.1 (13-15)	107.3 ± 5.5[b]	1.620 ± 0.050[b,c]	0.219 ± 0.029[b,c]	135.1 ± 5.6[b,c]	1.344 ± 0.014[b,c]	0.083 ± 0.008[b,c]

[a]The values are given ± the S.E.

[b]Significant difference between the experimental group and the controls.

[c]Significant difference between the underfed animals and the hypothyroid ones (t test, $P < 0.05$).

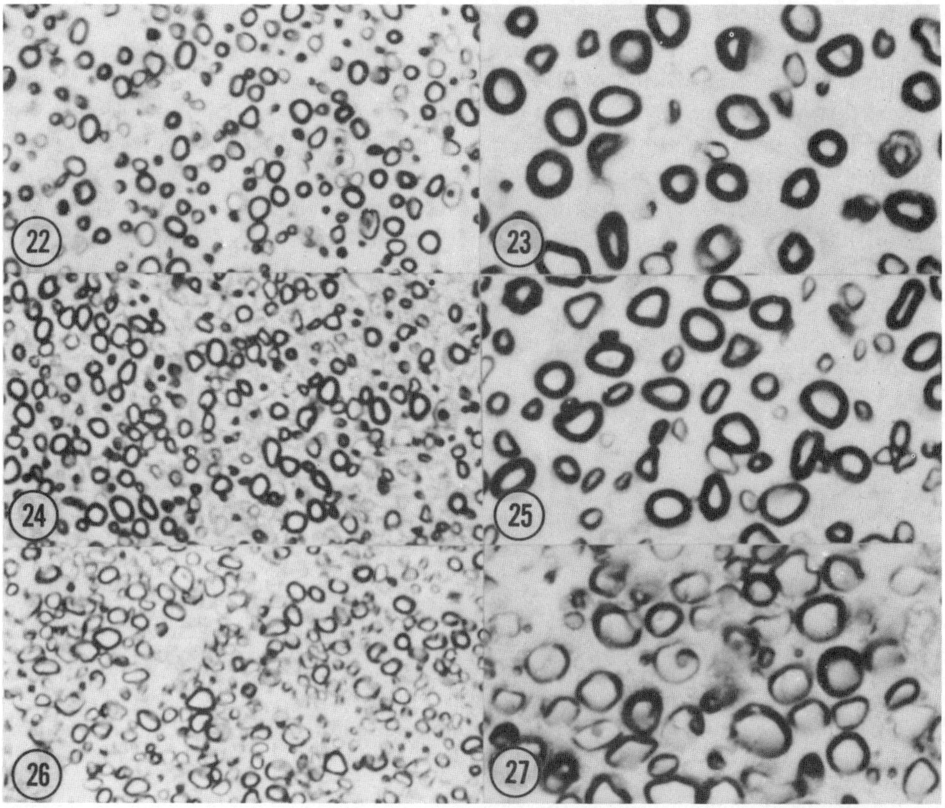

Figs. 22-27. Myelination of the sciatic nerve in normal, hypothyroid, and underfed 12-day-old rats. (Iron-hematoxylin, ×400 and ×1,000.) Figures 22 and 23: normal animals; Figures 24 and 25: hypothyroid animals; Figures 26 and 27: underfed animals.

fibers is greater; small fibers are more numerous, so that the mean inner diameter of fibers is decreased, but the fibers in this group are more heavily myelinated than in underfed animals.

DISCUSSION

Because we have not compared populations of fibers with the same mean inner diameter, we cannot know if hypothyroidism has acted upon myelination or not. However, we can conclude that myelination is more markedly decreased in underfed animals than in hypothyroid ones.

Thyroid deficiency, at least during the period studied, does not markedly affect myelination of the sciatic nerve fibers which have differentiated before birth or immediately afterwards. Undernutrition, on the contrary, acts to a great extent upon myelin deposition even in the fibers which are already well differentiated at birth before underfeeding is initiated. Thyroid deficiency apparently does not lead to sufficient degree of undernutrition because thyroid deprivation does not decrease the myelination of the fibers which are differentiating soon after birth. This observation is cited as a further argument for the view that the observed effects of hypothyroidism on the maturation of the nervous system, and particularly on the histological differentiation of the cerebellar cortex, during the second postnatal week are most likely not mediated through undernutrition and represent direct and specific effects exerted on target structures.

The converse effect, namely that undernutrition can lead to decrease in thyroid function has also been shown(14). Since it was demonstrated(2) that starvation does not appreciably alter the differentiation of the cerebellar cortex, it is unlikely that the underfed rats in our experiments suffered from any degree of hypothyroidism. The two treatments obviously affect neurogenesis at different sites and by different mechanisms.

REFERENCES

1. Eayrs, J. T. Thyroid and central nervous development. Sci. Basis Med. Ann. Rev., 317–339, 1966.
2. Legrand, J. Analyse de l'action morphogénétique des hormones thyroidiennes sur le cervelet du jeune rat. Arch. Anat. Micr. Morph. Exp., 56:205–244, 1967.
3. ——— Variations, en fonction de l'age, de la réponse du cervelet à l'action morphogénétique de la thyroide chez le rat. Arch. Anat. Micr. Morph. Exp., 56:291–307, 1967.
4. Eayrs, J. T. Age as a factor determining the severity and reversibility of the effect of thyroid deprivation in the rat. J. Endocr., 22:409–419, 1961.
5. Hamburgh, M., and Flexner, L. B. Biochemical and physiological differentiation during morphogenesis. XXI. Effect of hypothyroidism and hormone therapy on enzyme activities of the developing cerebral cortex of the rat. J. Neurochem., 1:279–288, 1957.
6. Barrnett, R. J. Some aspects of the physiology of the experimental cretin-like animal. Thesis, Yale Univ. School of Medicine, 1948.
7. Hamburgh, M., and Bunge, R. P. Evidence for a direct effect of thyroid hormone on maturation of nervous tissue grown *in vitro*. Life Sci., 3:1423–1440, 1964.
8. ——— Evidence for a direct effect of temperature and thyroid hormone on myelinogenesis *in vitro*. Develop. Biol., 13:15–30, 1966.
9. Myant, N. B. On the possible role of the thyroid in the control of the development of the mammalian brain. Biol. Neonat., 9:148–165, 1966.
10. Dobbing, J. The influence of early nutrition on the development and myelination of brain. Proc. Roy. Soc., B 159:503, 1964.
11. Culley, W. J., and Mertz, E. T. Effect of restricted food intake on growth and composition of preweanling rat brain. Proc. Soc. Exp. Biol. Med., 118:233–235, 1965.
12. Benton, J. W., Moser, H. W., Dodge, P. R., and Carr, S. Modification of the schedule of myelination in the rat by early nutritional deprivation. Pediatrics, 38:801–807, 1966.
13. Legrand, J. Recherches histologiques sur la date d'apparition de la myéline dans la moelle épinière et dans les nerfs brachial et sciatique du rat blanc. Arch. Anat. Micr. Morph. Exp., 50:189–198, 1961.
14. Mulinos, M. G., and Pomerantz, L. Pseudo-hypophysectomy. A condition resembling hypophysectomy produced by malnutrition. J. Nutr., 19:493–504, 1940.

29

The Role of Thyroid and Growth Hormones on RNA Metabolism in the Immature Brain*†

Stanley E. Geel and Paola S. Timiras

Department of Physiology-Anatomy, University of California, Berkeley, California

INTRODUCTION

The consequences of a thyroid hormone deficiency during infancy on cerebral structure and function have been well documented. The structural changes in cerebral cortical(1) and cerebellar architecture(2,3) are manifested in electrophysiological(4-6) and behavioral alterations(7,8). Recent studies in this(9,10,11) and other laboratories(12,13) suggest that the foregoing changes may be explained in terms of altered cerebral protein metabolism (see Ref. 14).

The present study is a further attempt to characterize the role of thyroid hormones in the development of the central nervous system and to identify the molecular mechanisms underlying the cerebral dysfunction occurring in cretinism. The incorporation of a labeled pyrimidine precursor into brain RNA was investigated in immature hypothyroid rats and in a simialr group receiving thyroxine treatment. Furthermore, since changes in thyroid function are accompanied by alterations in pituitary growth hormone secretion and content(15-19), the possibility that a growth hormone deficit could account for the altered cerebral nucleic acid metabolism in hypothyroidism was investigated.

MATERIALS AND METHODS

Animals. Female Long-Evans rats were selected for uniform weight at birth and randomly allotted to treatment groups. Litters were restricted to five animals

*This investigation was supported by USPHS Grant Number GM-09267.

†The authors are grateful to Miss Arlene Luckock for skilled technical assistance and to Dr. L. Rosenberg for the supply of growth hormone.

and were equally represented in five treatment groups. Water and food were administered ad libitum and animals were housed in a room maintained at constant temperature.

Experimental hypothyroidism. The detailed procedure for inducing neonatal hypothyroidism by radiothyroidectomy has been previously described(9).

Hormone therapy. One group of hypothyroid animals received daily subcutaneous injections of L-thyroxine (5 $\mu g/100$ g body weight) in saline (pH 8.0-8.5) in a volume of 0.05 ml from the 6th day of age, and another group, from the 15th day of age. A third hypothyroid group received daily intraperitoneal injections of bovine growth hormone (50 $\mu g/100$ g body weight) from the 6th day of age in a volume of 0.05 ml. Bovine growth hormone was prepared according to a modification of the method of Papkoff and Li(20). A normal group and a hypothyroid group were maintained as uninjected controls.

Subarachnoidal injection. [6-^{14}C]orotic acid (International Chemical and Nuclear Corporation, California) was diluted with unlabeled orotic acid to a specific activity of 10 mc/mM and dissolved in 0.9 percent NaCl with a minimum addition of 0.1 N NaOH. At 25 days of age, 3 μc in 0.02 ml (0.3 μmole) was injected into the cisterna magna under light ether anesthesia. No leakage of radioactivity occurred due to the close proximity of the atlanto-occipital ligament in this region(21). Injection of Trypan blue by this route showed that significant dispersion of the test substance occurred over the entire brain.

Tissue preparation. At various intervals following injection of the label, animals were killed by decapitation and the brain was rapidly removed. The diencephalon and thin slices of cerebral cortical gray matter were removed, rinsed in cold saline, blotted free of moisture, and weighed. Tissues from two animals were combined for each determination and a 10 percent homogenate prepared by hand in a clear glass homogenizer (Kontes) equipped with a loose-fitting teflon pestle. The medium consisted of 0.32 M sucrose, 0.001 M $MgCl_2$, 0.0004 M $M_2K_4HPO_4$, 0.0004 M KH_2PO_4 (pH 6.7-6.8) (22). An aliquot of the total homogenate was removed for determination of acid-soluble radioactivity. The homogenate was centrifuged at 800 to 900 *g* for 10 minutes at 4°C and the pellet resuspended in the medium and centrifuged. The combined supernatants constituted the cytoplasmic fraction. The crude nuclear pellet was further purified by two additional washings in the medium before resuspension. Examination of the crude nuclear fraction under phase-contrast microscopy revealed reasonable purity and intact nuclei. Using the diphenylamine method of Burton(23), the cytoplasmic fraction was shown to be free of DNA contamination.

Extraction of nucleic acids. RNA was determined according to a modified Schmidt-Thannhauser method essentially as described by Fleck and Munro(24). To the crude nuclear and cytoplasmic fractions, 0.5 volumes of cold 0.6 N perchloric acid (PCA) were added and centrifuged in the cold. The sediments were washed twice in 0.2 N PCA and the supernatants discarded after centrifugation. The PCA supernatants from an aliquot of whole homogenate were retained for the measurement of acid-soluble (pool) radioactivity.

In one set of experiments the effect of lipid removal was tested by consecutive washings with cold 90 percent ethanol containing 2 percent sodium acetate, chloroform-methanol (2:1 v/v), and ether. No effect on recovery of radioactivity was observed, while some loss of RNA occurred.

The acid-insoluble nuclear and cytoplasmic fractions were suspended in 0.3 N KOH and incubated for 1 hour at 37°C. After cooling, the hydrolysates were acidified with 0.7 N PCA (final concentration, 0.2 N PCA) and centrifuged. The DNA and protein precipitate was subjected to two further washings with 0.2 N PCA, and the combined supernatants which constituted the RNA fraction were read at 260 mμ in a Beckman DU spectrophotometer. An extinction coefficient of 34.2 cm^2/mg was adopted for the calculation of RNA concentration(25).

The DNA and protein precipitate was washed with cold 90 percent ethanol containing 2 percent sodium acetate, and DNA was extracted with 5 percent trichloroacetic acid (TCA) at 90°C for 15 minutes. After centrifugation the supernatant was saved and combined with the supernatants from two additional washings of the sediment with 5 percent TCA. DNA was determined according to the diphenylamine method of Burton(23). Calf thymus DNA (Sigma Chemical Co.) was used as a standard.

Measurement of radioactivity. Samples of the RNA extract were neutralized by the addition of saturated KOH in the cold to precipitate maximum amounts of $KClO_4$. A 2-ml aliquot of the clear supernatant was suspended in 15 ml of scintillation fluid consisting of two volumes of toluene containing 0.4 percent 2,5-diphenyloxazole (PPO), 0.01 percent 1,4-bis-2-(5-phenyloxazolyl)-benzene (POPOP), and one volume of Triton X-100(26). Radioactivity was determined in a Packard Tri-Carb spectrometer (Model 314 E) with a counting efficiency of 70 percent. Correction for quenching was performed according to the channels ratio method and results were expressed as disintegrations per minute (dpm).

RESULTS

Body growth. Body weights were comparable between control and hypothyroid animals until 10 to 12 days of age, after which growth rate declined in hypothyroid animals. At 25 days of age, their mean body weight was equivalent to that of 18-day-old normal animals (Fig. 1). Daily injections of physiological levels of L-thyroxine (T_4) from the 6th day of age prevented any reduction in body growth. When thyroxine was administered from day 15, the growth rate, while lower than controls, was significantly increased by day 25 over hypothyroid values. In addition, the physical appearance of these animals was identical to that of normal controls. Growth hormone (GH) administered to hypothyroid animals from day 6 had no effect on body growth. These animals appeared distinctly cretinous at 25 days of age and were indistinguishable from hypothyroid individuals.

Nucleic acid content. Neonatal hypothyroidism resulted in a significant decrease in cerebral cortical nuclear and cytoplasmic RNA content when expressed per unit wet tissue and per unit DNA (Table 1). DNA content, on the other hand, was significantly increased, which suggests an increase in cell density per unit wet

TABLE 1

Effect of neonatal hypothyroidism, thyroxine, and growth hormone therapy on nucleic acid content of cerebral cortical fractions at 25 days of age[a]

Group		RNA (μg/mg wet tissue)		DNA (μg/mg wet tissue)		RNA/DNA	
Normal	Nuclear	0.621 ± 0.015		0.934 ± 0.028		0.672 ± 0.009	
	Cytoplasmic	1.319 ± 0.022				1.420 ± 0.068	
Hypothyroid[b]	Nuclear	0.598 ± 0.014		1.074 ± 0.033	$P < 0.01$	0.569 ± 0.010	$P < 0.001$
	Cytoplasmic	1.210 ± 0.017	$P < 0.01$			1.165 ± .036	$P < 0.01$
Hypothyroid + T_4 early[c]	Nuclear	0.613 ± 0.027		0.969 ± 0.059		0.652 ± 0.095	
	Cytoplasmic	1.337 ± 0.031				1.422 ± 0.080	
Hypothyroid + T_4 late[d]	Nuclear	0.616 ± 0.017		0.957 ± 0.025		0.673 ± 0.013	
	Cytoplasmic	1.271 ± 0.052				1.322 ± 0.067	
Hypothyroid + GH[e]	Nuclear	0.594 ± 0.017		1.058 ± 0.026	$P < 0.01$	0.585 ± 0.018	$P < 0.001$
	Cytoplasmic	1.191 ± 0.017	$P < 0.01$			1.131 ± 0.039	$P < 0.01$

[a]Values are means ± S.E.M. P values refer to significant differences between experimental groups and normal controls.
[b]Pregnant mothers maintained on a low iodine diet and female offspring injected with 100 μc ^{131}I at one day of age.
[c]5 μg L-thyroxine/100 g body weight daily from 6th day of age.
[d]5 μg L-thyroxine/100 g body weight daily from 15th day of age.
[e]50 μg bovine growth hormone/100 g body weight daily from 6th day of age.

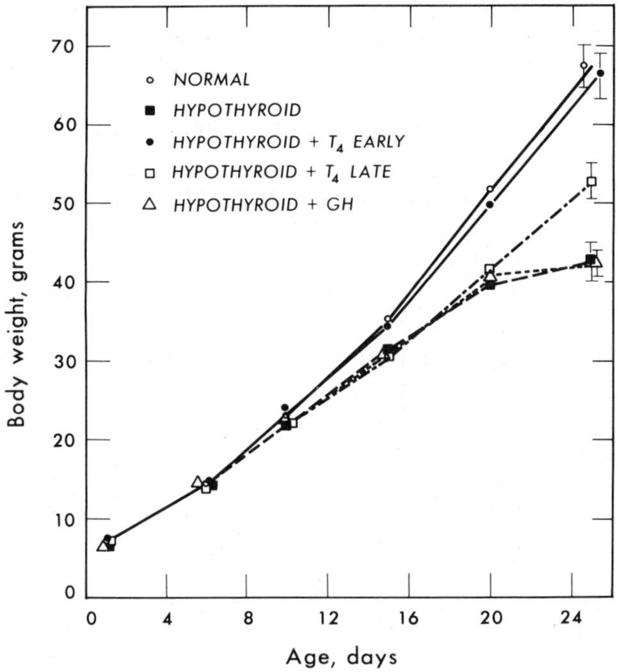

Fig. 1. Influence of thyroid function and growth hormone on body growth. Experimental conditions as in Table 1.

TABLE 2

Effect of neonatal hypothyroidism, thyroxine, and growth hormone therapy on nucleic acid content of diencephalic fractions at 25 days of age[a]

		(μg/mg wet tissue)		
Group		RNA	DNA	RNA/DNA
Normal	Nuclear	0.608 ± 0.012	1.112 ± 0.015	0.546 ± 0.019
	Cytoplasmic	1.235 ± 0.013		1.088 ± 0.026
Hypothyroid	Nuclear	0.595 ± 0.009	1.171 ± 0.013	0.512 ± 0.022
	Cytoplasmic	1.231 ± 0.023		1.043 ± 0.031
Hypothyroid + T_4 early	Nuclear	0.596 ± 0.021	1.137 ± 0.016	0.529 ± 0.023
	Cytoplasmic	1.276 ± 0.028		1.128 ± 0.030
Hypothyroid + T_4 late	Nuclear	0.614 ± 0.018	1.062 ± 0.041	0.596 ± 0.022
	Cytoplasmic	1.196 ± 0.036		1.135 ± 0.057
Hypothyroid + GH	Nuclear	0.588 ± 0.017	1.190 ± 0.035	0.537 ± 0.015
	Cytoplasmic	1.238 ± 0.049		1.037 ± 0.049

[a]Experimental conditions as in Table 1.

weight. These findings are consistent with previous results on nucleic acid content of whole homogenates(9,12,13). Thyroxine therapy, whether initiated from the 6th or the 15th day of age, was effective in restoring nucleic acid levels to normal. Growth hormone administered to hypothyroid animals from the 6th day of age failed to have any effect on nucleic acid content. In contrast to the cerebral cortex, diencephalic nucleic acid levels were unaffected by early thyroid hormone deficiency or hormone therapy (Table 2).

Incorporation of [6-^{14}C]orotic acid into brain RNA. The time course of incorporation of labeled orotic acid into RNA fractions of the cerebral cortex of normal and hypothyroid rats is presented in Fig. 2. While the pattern of incorporation of the label into cerebral cortical fractions is similar between hypothyroid and control rats, the RNA specific activity in the hypothyroid group was approximately two times greater than normal controls at all intervals and was paralleled by a similar increase in acid-soluble (pool) radioactivity. Similar results were obtained for the in-

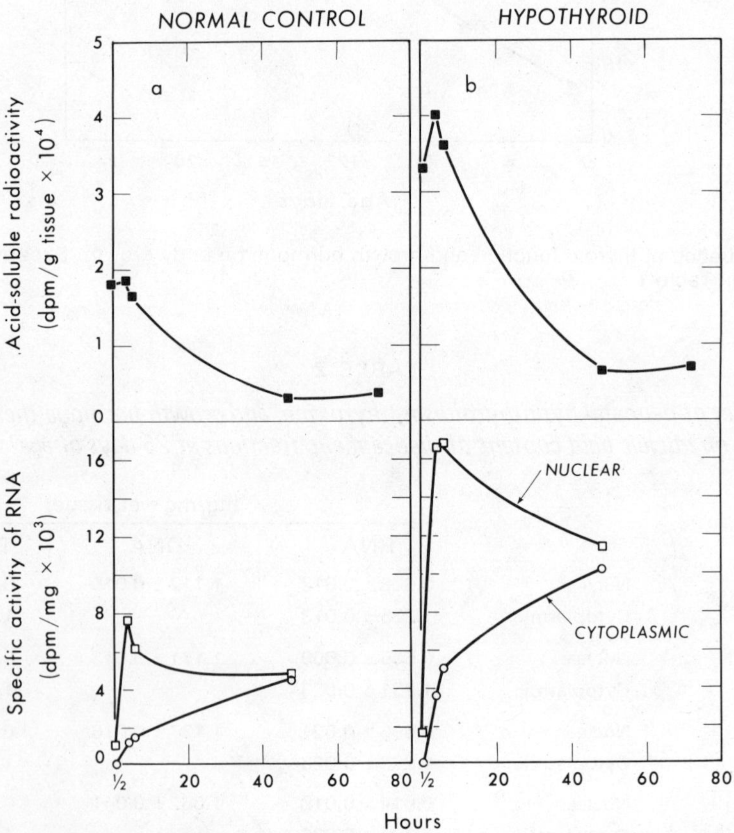

Fig. 2. Labeling of cerebral cortical acid-soluble fraction (x——x), nuclear RNA (□——□), and cytoplasmic RNA (o——o) following the injection of [6-^{14}C]orotic acid into the subarachnoid space of 25-day-old rats. Normal control—*a:* hypothyroid—*b:* Points are means of 2 or 3 experiments employing two animals per determination.

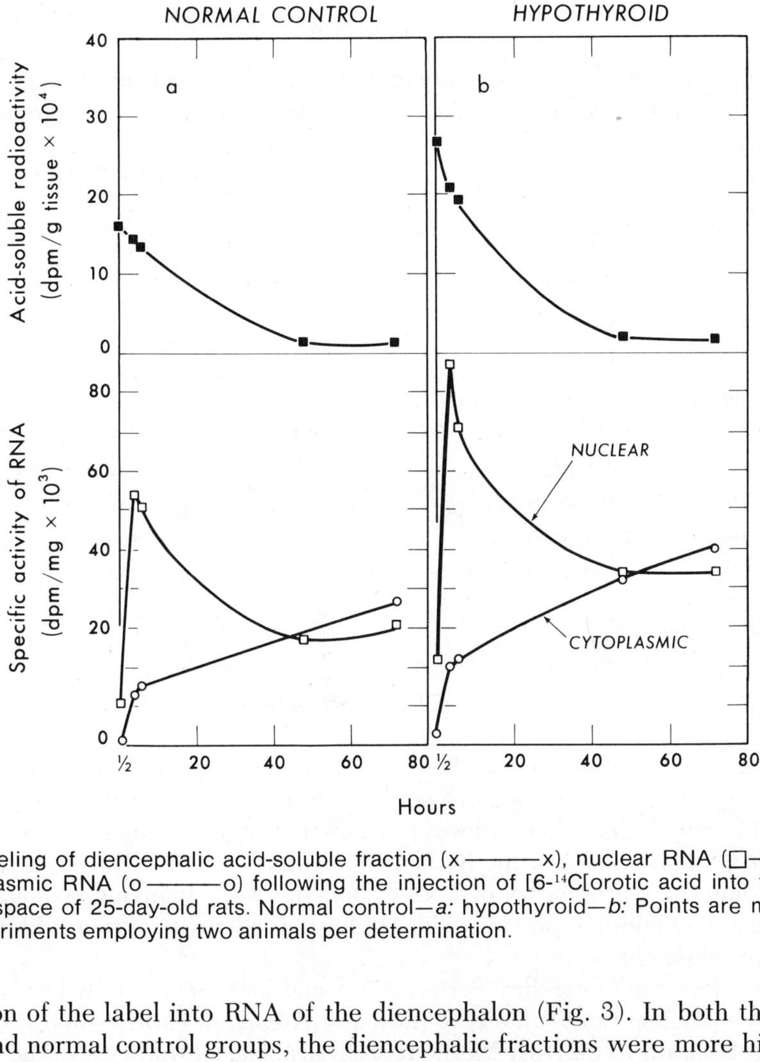

Fig. 3. Labeling of diencephalic acid-soluble fraction (x———x), nuclear RNA (□———□), and cytoplasmic RNA (o———o) following the injection of [6-^{14}C[orotic acid into the subarachnoid space of 25-day-old rats. Normal control—*a:* hypothyroid—*b:* Points are means of 2 or 3 experiments employing two animals per determination.

corporation of the label into RNA of the diencephalon (Fig. 3). In both the hypothyroid and normal control groups, the diencephalic fractions were more highly labeled than cerebral cortical fractions, which may be attributed to the more rapid diffusion of the label to this region. The kinetics of incorporation of labeled orotic acid into RNA and acid-soluble fractions of the growth hormone-treated hypothyroid group (not shown) was similar to that of the hypothyroid animals.

When RNA specific activity is corrected for by the acid-soluble radioactivity or expressed as a function of isolated uridine nucleotide specific radioactivity (unpublished observations), no difference in the rate of RNA synthesis is apparent for all groups. These results are in agreement with those of Balázs et al.(13).

The ratio of incorporation of the label into nuclear RNA to that in cytoplasmic RNA for all intervals is presented in Table 3. The results indicate that the rate of synthesis of nuclear RNS and the transport of newly synthesized RNA into the cytoplasm is similar in all groups.

TABLE 3

Ratio of incorporation of radioactivity into nuclear RNA to that in cytoplasmic RNA at various intervals following subarachnoidal injection of [6-^{14}C] orotic acid[a]

Group		Hours post-injection 0.5	4	6	48	72
Normal	Cerebral Cortex	—	5.6	3.5	1.4	0.9
	Diencephalon	6.4	5.4	3.3	1.1	0.8
Hypothyroid	Cerebral Cortex	—	5.2	3.9	1.3	0.9
	Diencephalon	5.7	5.8	3.7	1.1	0.9
Hypothyroid + T_4 early	Cerebral Cortex	—	5.1	3.6	1.2	1.1
	Diencephalon	7.3	4.5	3.3	1.1	0.9
Hypothyroid + T_4 late	Cerebral Cortex	—	5.5	3.9	1.2	1.1
	Diencephalon	5.7	4.1	3.5	1.1	0.9
Hypothyroid + GH	Cerebral Cortex	—	6.0	3.7	1.2	1.0
	Diencephalon	5.0	5.3	3.8	1.2	0.8

[a]Experimental conditions as in Table 1.

DISCUSSION

The period of ontogeny in the rat, from birth to about 21 days of age, is critical for the functional development of the CNS. It is precisely during this period that thyroid hormones have been shown to exert their most striking effects on the morphological development of the brain(2,27). The results of this study provide some insight into the biochemistry of CNS dysfunction accompanying neonatal hypothyroidism and in addition provide evidence for the specificity of thyroid hormones for brain development.

The decrease in cerebral cortical RNA and increase in DNA levels of thyroid hormone-deficient rats (Table 1) confirm and extend our earlier studies(9) and those of other laboratories(12,13). On the basis of histological studies(28), the increase in cerebral cortical DNA suggests an increase in the packing density of cerebral cortical cells. Daily thyroxine therapy, whether initiated from the 6th or 15th day of age, was effective in restoring the altered cerebral cortical nucleic acid levels to normal. It is significant that neonatal hypothyroidism has no effect on the nucleic acid content of diencephalic fractions (Table 2) for this attests to the regional specificity of thyroid hormones on brain development. Phylogenetically, the diencephalon is an earlier developing area compared to the cerebral cortex, which may account for its unresponsiveness to the hormone during the neonatal period.

The observation that GH was ineffective in restoring growth and nucleic acid content of hypothyroid rats to normal values (Fig. 1 and Table 1) suggests that the lack of GH attendant to thyroid hormone deprivation does not contribute to the altered cerebral development observed in cretins. Such evidence is consistent with reports of the ineffectiveness of hypophysectomy at an early age and of exogenous GH on brain growth(29,30) and cerebral cortical morphology(31,32). On the other hand, the inability of exogenous GH to correct the alterations in nucleic acid levels of hypothyroid rats may be attributed to the small dose administered, particularly in view of the short half-life of plasma GH(33). Furthermore, the results do not discount the fact that any effect of GH in hypothyroid animals may be dependent upon the presence of thyroid hormones which are known to synergize with GH in the promotion of somatic development(34).

The extent of incorporation of [^{14}C]orotic acid into cerebral RNA was considerably greater at all time intervals in the hypothyroid group, and the hypothyroid animals receiving GH (not shown) compared with normal controls (Figs. 2 and 3). However, when RNA specific activity is expressed as a function of the total acid-soluble radioactivity, or more specifically of the acid-soluble uridine nucleotides (unpublished observations), no difference in the rate of RNA synthesis is apparent between all groups. The similar rate of RNA synthesis in hypothyroid rats, which is accompanied by a decreased net synthesis compared to controls, is in agreement with the observation of Balázs et al.(14) and would appear to result from differences in "utilization" or degradation of RNA between the two groups.

The reasons for the enhanced incorporation of labeled orotic acid into all fractions in the hypothyroid and growth hormone-treated groups merit further discussion. The increased labeling of RNA fractions may result from a number of factors, which include, among others, a modification of the endogenous pool size. The pyrimidine nucleotide and orotic acid pool size, however, is reported to be very small (21,35-38), and it can be assumed that the amount injected closely approximates that which is incorporated. Dawson(39) followed intracerebral injection of labeled orotic acid into young rats with "chaser" levels of unlabeled orotic acid by the same route and was unable to alter the rate of turnover of RNA or radioactivity in the soluble phase. Furthermore, experiments (unpublished) reveal no difference in the size of the pyrimidine nucleotide pool between hypothyroid and control animals.

It has been shown that both preformed pyrimidine bases from the blood and those derived from de novo synthesis contribute to the nucleotide pool for RNA synthesis in brain tissue(36,40). Should the availability of RNA precursors from the blood be limiting in hypothyroid individuals, the supply of pyrimidines would originate largely by de novo synthesis from orotic acid, which might account for the greater specific activity of the nucleotide pool compared to controls. This conclusion is supported by the observation that uridine nucleotides in the liver are more rapidly degraded in hypothyroid rats(13), thereby restricting the supply to the blood.

Finally, the possibility exists that differences in the transport or diffusion of the injected labeled material occur between hypothyroid and control rats. Preliminary experiments indicate that the greater labeling of all fractions in hypothyroid animals is essentially normalized by a single acute injection of thyroxine.

In the final analysis it would seem reasonable to conclude that while the rate of cerebral RNA synthesis is unaltered in hypothyroid rats, the translation of newly synthesized RNA species in the protein biosynthetic scheme may be defective. Furthermore, the metabolism of the pyrimidine nucleotides, or at least the pathway by which they reach the cerebral nucleotide pool, is possibly altered as a result of hypothyroidism. Thyroxine appears to be specific for normal RNA metabolism.

REFERENCES

1. Eayrs, J. T. The cerebral cortex of normal and hypothyroid rats. Acta Anat. (Basel), 25: 180–183, 1955.
2. Hamburgh, M., Lynn, E., and Weis, E. P. Analysis of the influence of thyroid hormone on prenatal and postnatal maturation of the rat. Anat. Rec., 150:147–162, 1964.
3. Legrand, J. Influence de l'hypothyroidisme sur la maturation du cortex cérébelleux. C.R. Acad. Sci. (Paris), 261:544–547, 1965.
4. Bradley, P. B., Eayrs, J. T., and Schmalbach, K. The electroencephalogram of normal and hypothyroid rats. Electroenceph. Clin. Neurophysiol., 12:467–477, 1960.
5. ——— Eayrs, J. T., Glass, A., and Heath, R. W. The maturational and metabolic consequences of neonatal thyroidectomy upon the recruiting response in the rat. Electroenceph. Clin. Neurophysiol., 13:577–586, 1961.
6. Harris, R., Rovere, M. D., and Prior, P. F. Electroencephalographic studies in infants and children with hypothyroidism. Arch. Dis. Child., 40:612–617, 1965.
7. Eayrs, J. T., and Lishman, W. A. The maturation of behavior in hypothyroidism and starvation. Brit. J. Anim. Behav., 3:17–24, 1955.
8. ——— and Levine, S. Influence of thyroidectomy and subsequent replacement therapy upon conditioned avoidance learning in the rat. J. Endocr., 25:505–513, 1963.
9. Geel, S. E., and Timiras, P. S. The influence of neonatal hypothyroidism and of thyroxine on the RNA and DNA concentrations of rat cerebral cortex. Brain Res., 4:135–142, 1967.
10. ——— Valcana, T., and Timiras, P. S. Effect of neonatal hypothyroidism and of thyroxine on L-(^{14}C) leucine incorporation in protein *in vivo* and the relationship to ionic levels in the developing brain of the rat. Brain Res., 4:143–150, 1967.
11. ——— and Timiras, P. S. Influence of neonatal hypothyroidism and of thyroxine on the acetylcholinesterase and cholinesterase activities in the developing central nervous system of the rat. Endocrinology, 80:1069–1074, 1967.
12. Pasquini, J. M., Kaplún, B., García Argíz, C. A., and Gómez, C. J. Hormonal regulation of brain development. I. The effect of neonatal thyroidectomy upon nucleic acids, protein and two enzymes in developing cerebral cortex and cerebellum of rat. Brain Res., 6:621–634, 1967.
13. Balázs, R., Kovács, S., Teichgräber, W. A., Cocks, W. A., and Eayrs, J. T. Biochemical effect of thyroid deficiency on the developing brain. J. Neurochem., 15:1335–1349, 1968.
14. Geel, S. E., and Timiras, P. S. The role of hormones in cerebral protein metabolism. *In* Protein Metabolism of the Nervous System. Lajtha, A., ed. New York, Plenum Press, 1970, in press.
15. Contopoulos, A. N., Simpson, M. E., and Koneff, A. A. Pituitary function in the thyroidectomized rat. Endocrinology, 63:642–653, 1958.
16. Solomon, J., and Greep, R. O. The effect of alterations in thyroid function on the pituitary growth hormone content and acidophil cytology. Endocrinology, 65:158–164, 1959.
17. Schooley, R. A., Friedkin, S., and Evans, E. S. Re-examination of the discrepancy between acidophil numbers and growth hormone concentration in the anterior pituitary following thyroidectomy. Endocrinology, 79:1053–1057, 1966.
18. Daughaday, W. H., Peake, G. T., Birge, C. A., and Mariz, I. K. The influence of endocrine factors on the concentration of growth hormone in rat pituitary. *In* Pécile, A., and Müller, E. E., eds. Proc. Int. Symp. Growth Hormone, Milan, Italy, Sept. 11–13, 1967. Amsterdam, Excerpta Medica, p. 238, 1968.
19. Iwatsubo, J., Miyai, K., Abe, H., Kumahara, Y., Omori, K., Okada, Y., and Fukuchi, M. Human growth hormone secretion in primary hypothyroidism before and after treatment. J. Clin. Endocr., 27:1751–1754, 1967.

20. Papkoff, H., and Li, C. H. The isolation and characterization of growth hormone from anterior lobes of whale pituitaries. J. Biol. Chem., 231:367–377, 1958.
21. Adams, D. H. Some observations on the incorporation of precursors into ribonucleic acid of rat brain. J. Neurochem., 12:783–790, 1965.
22. Sporn, M. B., Wanko, T., and Dingman, W. The isolation of cell nuclei from rat brain. J. Cell Biol., 15:109–120, 1962.
23. Burton, K. A study of the conditions and mechanisms of the diphenylamine reaction for the colorimetric estimation of DNA. Biochem. J., 62:315–323, 1956.
24. Fleck, A., and Munro, H. N. The precision of ultraviolet absorption measurements in the Schmidt-Thannhauser procedure for nucleic acid estimation. Biochim. Biophys. Acta, 55:571–583, 1962.
25. Bradford, H. F., Brownlow, E. K., and Gammack, D. B. The distribution of cation-stimulated adenosine triphosphatase in subcellular fractions from bovine cerebral cortex. J. Neurochem., 13:1283–1297, 1966.
26. Patterson, M. S., and Greene, R. C. Measurement of low energy beta-emitters in aqueous solution by liquid scintillation counting of emulsions. Anal. Chem., 37:854–857, 1965.
27. Eayrs, J. T. Influence of the thyroid on the central nervous system.Brit. Med. Bull., 16: 122–127, 1960.
28. ——— Protein anabolism as a factor ameliorating the effects of early thyroid deficiency. Growth, 25:175–189, 1961.
29. Walker, D. G., Asling, C. W., Simpson, M. E., Li, C. H., and Evans, H. M. Structural alterations in rats hypophysectomized at 6 days of age and their correction with growth hormone. Anat. Rec., 114:19–47, 1952.
30. ——— Simpson, M. E., Asling, C. W., and Evans, H. M. Growth and differentiation in the rat following hypophysectomy at 6 days of age. Anat. Rec., 106:539–554, 1950.
31. Diamond, M. C. The effects of early hypophysectomy and hormone therapy on brain development. Brain Res., 7:407–418, 1968.
32. Gregory, K. M., and Diamond, M. C. The effects of early hypophysectomy on brain morphogenesis in the rat. Exp. Neurol., 20:394–414, 1968.
33. Garcia, J. F., and Geschwind, I. I. Investigation of growth hormone secretion in mammalian species. *In* Pécile, A., and Müller, E. E., eds. Proc. Int. Symp. Growth Hormone, Milan, Italy, Sept. 11–13, 1967. Amsterdam, Excerpta Medica, p. 267, 1968.
34. Simpson, M. E., Asling, C. W., and Evans, H. M. Some endocrine influences on skeletal growth and differentiation. Yale J. Biol. Med., 23:1–27, 1950.
35. Lolley, R. N , Balfour, W. M., and Samson, Jr., F. E. The high energy phosphates in developing brain. J. Neurochem., 7:289–297, 1961.
36. Wells, W., Gaines, D., and Koenig, H. Studies of pyrimidine nucleotide metabolism in the central nervous system. I. Metabolic effects and metabolism of 6-azauridine. J. Neurochem., 10:709–723, 1963.
37. Mandel, P., and Edel-Harth, S. Free nucleotides in the rat brain during post-natal development. J. Neurochem., 13:591–595, 1966.
38. Oja, S. S. Postnatal changes in the concentration of nucleic acids, nucleotides and amino acids in the rat brain. Ann. Acad. Sci. Fenn. [Med.], 125:1–369, 1966.
39. Dawson, D. M, Turnover of ribosomal RNA in the rat brain. J. Neurochem., 14:939–946, 1967.
40. Santos, J. N., Hempstead, K. W., Kopp, L. E., and Miech, R. P. Nucleotide metabolism in rat brain. J. Neurochem., 15:367–376, 1968.

30

Thyroid-Dependent Processes in the Developing Nervous System* †

Max Hamburgh,[1] Lorenzo A. Mendoza,[2] John F. Burkart,[3]‡ and Franklin Weil[4] §

[1]Department of Biology, City College of New York, and Department of Anatomy, Albert Einstein College of Medicine, New York, New York, New York
[2]Department of Endocrinology, Hospital Vargas, Caracus, Venezuela
[3]Department of Biology, City College of New York, New York, New York
[4]Albert Einstein College of Medicine, New York, New York

INTRODUCTION

The analysis of hormone action in development up to now has been carried out on three levels: (1) The "fundamentalist approach" aims to demonstrate that hormones are indeed influencing and participating in the control of differentiation of various target structures. The demonstration of hormonal control of differentiation relies on procedures that either remove the source of hormone secretion or that administer excess hormone during fetal life and then observe the effects of such treatment on the miniature adult. (2) On a second level, "the analytical ap-

*This investigation was supported by a research grant from the National Institute of Neurological Diseases and Blindness NIH B 1716 and by training grant 5T1 HD 0016-02-04 of the National Institute of Child Health and Human Development.

†Grateful acknowledgment is made to the following for their assistance in various aspects of this paper: Miss Cassandra Kirk for her technical assistance; Smith, Kline and French for supplying thyroxine used in this investigation; Dr. S. Cohen, Dept. of Biochemistry, Vanderbilt University, School of Medicine for supplying Epithelial Growth Factor; and Mr. M. Kurtz, medical photographer, for his assistance with the photographic work.

‡Submitted in partial fulfillment of the requirement for the M.A. degree in Biology of the Graduate Division of the Department of Biology, City College of New York.

§Medical Student Summer Fellowship, Albert Einstein College of Medicine.

proach" of fetal endocrinology focuses on critical periods during embryogenesis in an attempt to identify the specific developmental processes that are subject to hormonal interference during the maturation of selected embryonic target cells or structures. (3) The third level of study attempts to interpret hormone action in terms of ultimate cellular mechanisms and to fit the effect of hormones on target structures with the "central doctrine of experimental biology" according to which differentiation reflects the synthesis of selective proteins, an event which is presumably preceded by transcription and subsequent translation of the genetic code. Hormones, it has been asserted, may intervene by the differential repression or depression of genes that code for the different proteins(1,4).

Developmental biologists in their quest for the hormonal control mechanisms that presumably operate during the maturational period of a tissue cannot be satisfied by the mere demonstration that a given hormone exerts a stimulatory effect on the synthesis of specific proteins or on RNA. The next question invariably raised is, "Protein synthesis for what?" Are the new proteins, or their RNA codes whose synthesis is initiated by hormone action, mobilized for mitotic proliferation, or do they represent the specific proteins of the cell's biochemical household that lead to cell differentiation, or are they required for some other objective prerequisite to differentiation like the removal of larval cells and tissue mediated by lysosomal enzymes as was demonstrated for the metamorphosing tadpole(5,10)?

One of the most extensively investigated problems of developmental endocrinology is the relationship of the thyroid hormone to the developing nervous system. In the absence of clear-cut evidence about the manner in which this hormone influences differentiation of the central nervous system, there is no shortage of speculation. Previous reports, some of which are reviewed in this symposium, have described transitory increases of DNA in developing whole brain of hypothyroid chicks(11), in developing cerebral cortex in young postnatal rats(12,15), and transitory increases in cell number in the maturing rat cerebellum(16,18). These reports encouraged us to test the proposition that lack of thyroxine may prolong the proliferative phase in parts of the developing central nervous system, and that both the transitory as well as the permanent changes imposed by thyroid deprivation on the maturing CNS may be a consequence of errors in "timing" or "phasing" of the proliferative phase.

The effects of thyroid hormone on cell proliferation and timing of DNA synthesis in the developing nervous system was studied by following the uptake of tritiated ^{3}H thymidine into cells of the cerebellum of normal and hypothyroid rats by means of autoradiography.

The study was extended to test the proposition advanced previously by one of us(19) that possibly the period of "thyroidsensitivity" of the developing central nervous system is terminated by the subsequent emergence of other physiologically active factors or hormones that may act in a similar or identical manner on some parts of the maturing nervous or sensory system as does thyroxine during earlier stages of development. The frequently cited observation(19,20,28,29) that many aspects of behavioral and neurohistological maturation are merely delayed and retarded but not suppressed under conditions of thyroid deprivation might make this hypothesis deserving of further testing.

Gómez et al.(21) were the first to show that the action of thyroid hormone upon some aspect of cerebral development may be replaced by somatotropin. Hamburgh(19) has shown that the maturation of the thermoregulatory mechanism is also thyroid dependent, but that thyroid hormone can be replaced by growth hormone. The efficacy of the epithelial growth factor (EGF) to reverse the delay in maturation imposed upon the newborn rat by continued thyroid deprivation, was chosen in the present study to test the possibility that the physiological action of thyroxine may be shared by biological molecules otherwise unrelated to thyroid hormones.

The choice of EGF was dictated by reports by Cohen(22,24) that injection of this substance into newborn rats or mice accelerates opening of the eyelids, tooth eruption, and hair growth similar to the well-known effects of excess thyroxine administered to maturing animals.

MATERIALS AND METHODS

Pregnant rats of the Charles River Breeding Laboratories Caesarian Derived (CD) strain were used for all experiments. Animals were housed individually in an air-conditioned room and provided with a Rockland mouse diet (ground) and water ad libitum.

Experimental Procedure

Hypothyroidism was induced through administration of propylthiouracil (PTU) beginning during intrauterine life and continued until weaning. Propylthiouracil treatment was initiated during the last week of gestation by offering to pregnant rats a goitrogenic diet containing 0.2 percent propylthiouracil mixed into powdered Rockland mouse diet, starting on day 15 of pregnancy and continued until weaning or beyond. Day 15 was chosen for initiation of the experimental treatment because available evidence suggests that, in the rat, the fetal thyroid does not functionally mature before the 18th day of fetal age, but that during the third trimester of the gestational period, maternal thyroxine passes the placental barrier(25). Controls consisted of pregnant rats maintained on an unsupplemented diet and receiving daily injections of 0.9 percent saline from birth until weaning.

Hyperthyroidism was induced by daily administration of L-thyroxine starting at birth to rats maintained on normal powdered diet or on PTU since the 15th day of gestation. Dosages were as follows: 1 μg thyroxine to newborn rats until age 7 days, 2 μg to rats aged 7 to 14 days, 4 μg to rats between ages 14 to 21 days, and 5 μg to rats above 21 days of age. This concentration was established as the maximum nonlethal dose for young rats. The hormone was injected in 0.05 to 0.25 ml of fluid. The fluid volume was held minimal in order to prevent any interference with osmotic equilibrium in the young animals.

Uptake of ^{3}H Thymidine by Rat Cerebellum

Unanesthesized 15- and 21-day-old rats were injected subcutaneously with 1 μc thymidine ^{3}H (specific activity 27.0 c/mmole, obtained from Schwarz Bioresearch Inc.) per gram of body weight. Six hours after injection of the tritiated thymidine the rats were sacrificed by decapitation, and brains were rapidly removed and fixed in Bouins solution. The tissue was washed in 70 percent ethanol, dehydrated, and embedded in paraffin. Sections were cut at 5 μ, mounted on microscope slides, and coated with Kodak NTB 3 bulk emulsion. The emulsion was exposed to the tritium-labeled tissue section for 2 weeks and stored at 4°C. After development in D19, the emulsion slides were stained with hematoxylin and eosin, cleared in xylol, and coverslips were applied with Permount.

Treatment with Epithelial Growth Factor

Lyophilized epithelial growth factor (EGF) (obtained from the laboratory of Dr. Stanley Cohen,[6] Vanderbilt University School of Medicine) was dissolved in 0.9 percent physiological saline (3,000 μg/10 ml fluid). Injections of EGF were administered with a microsyringe subcutaneously into newborn rats, average weight 5 to 5.5 g starting on day 2 and continuing until day 10 to 11. The following dosages were administered:

Rats 2 to 4 days old: 15 μg in 0.05 ml fluid
Rats 5 to 7 days old: 30 μg in 0.10 ml fluid
Rats 8 to 13 days old: 45 μg in 0.15 ml fluid

Since variations in weight between control, hypothyroid, and hyperthyroid animals did not emerge until after the 10th day of age, dosages of EGF injected were not adjusted to grams/body weight.

RESULTS

The development of the rat cerebellum during the first 3 weeks of age is characterized by extensive migration of cells derived from the transitory external zone past the molecular area to form the layer of granular cells of the cerebellum. In the rat, the thickness of this external zone attains its maximum at about 8 days after birth and subsequently diminishes progressively. It has completely disappeared at day 21, at which time the surface of the cerebellum has acquired the characteristic aspect of the adult.

In the cerebellar cortex of rats, the most immediate and striking effect of thyroid deficiency initiated during the late fetal period or immediately at or shortly after birth includes the persistence of the external granular zone long after this layer of cells has disappeared in normal littermates.

Among hypothyroid rats, the external superficial granular layer of the cerebellar cortex is maintained in the "fetal" condition as late as day 28 of age. After that date, the cerebellar cortex of hypothyroid rats begins to assume the normal architecture even though the thyroid hormone may not be made available to the rat(16,19).

When thymidine ^{3}H was injected on postnatal day 14 and the animals killed 6 hours later, labeled cells were confined mainly to the outer rows of the external

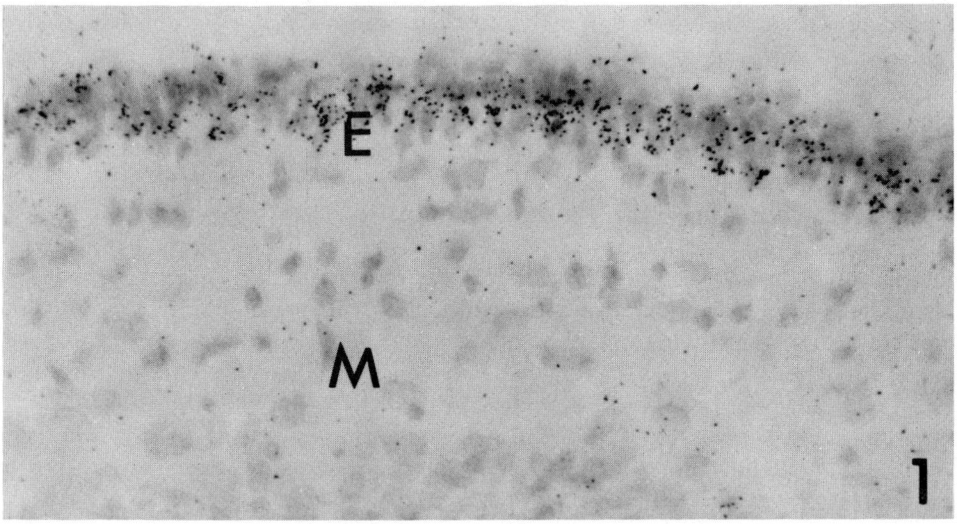

Fig. 1. Radioautograph of cerebellem of 15-day-old rat. Sacrificed 6 hours after injection of 1 μc/gm body weight of tritiated thymidine and exposed for 2 weeks. Note grains in external granular layer. ×440.

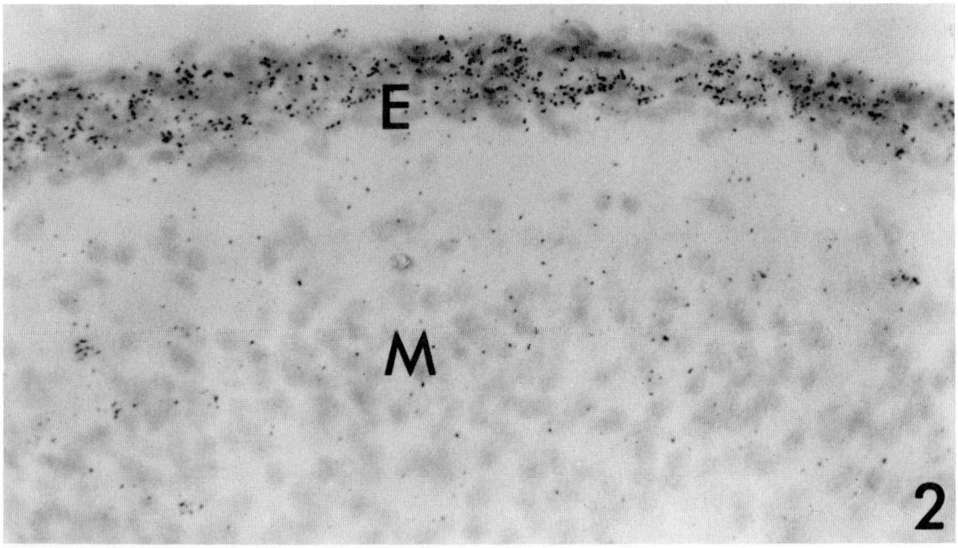

Fig. 2. Radioautograph of cerebellum of 15-day-old hypothyroid rats treated as above. ×440.

granular layer. Moderate numbers of labeled cells were scattered through the white matter and the granular layer proper. There was no difference in intensity of number of labeled cells between normal control rats and rats that were deprived of thyroid hormone since before birth (see Figs. 1,2).

When thymidine ^3H was injected on postnatal day 21 and the control animals killed 6 hours later only few labeled cells were found in the deep white matter and very rare ones were in the granular layer. (See Figs. 3,5). The external granular zone had disappeared at that time and there were practically no labeled cells present in the periphery of the cerebellar cortex.

A radically different picture was presented by cerebella obtained from hypothyroid rats. When thymidine ^3H was injected on postnatal day 21 into rats that had been deprived of thyroxine since before birth and killed 6 hours after administration of the tritiated thymidine, the cells of the external granular layer were still heavily labeled. In addition, many labeled cells were scattered in the granular layer, the molecular layer and the underlying white matter. (Figs. 4,6).

The daily subcutaneous injection of approximately 0.2 μg of thyroxine per gram of body weight into newborn rats resulted in opening of the eyelids in 12 to 13 days instead of the normal 14 to 16 days in control animals (see Table 1).

In hypothyroid (PTU-treated) rats the opening of the eyelids was delayed until day 16 to 18. Simultaneous injection of EGF approximately 1 μg per gram of body weight per day into PTU-treated newborn rats was even more effective than

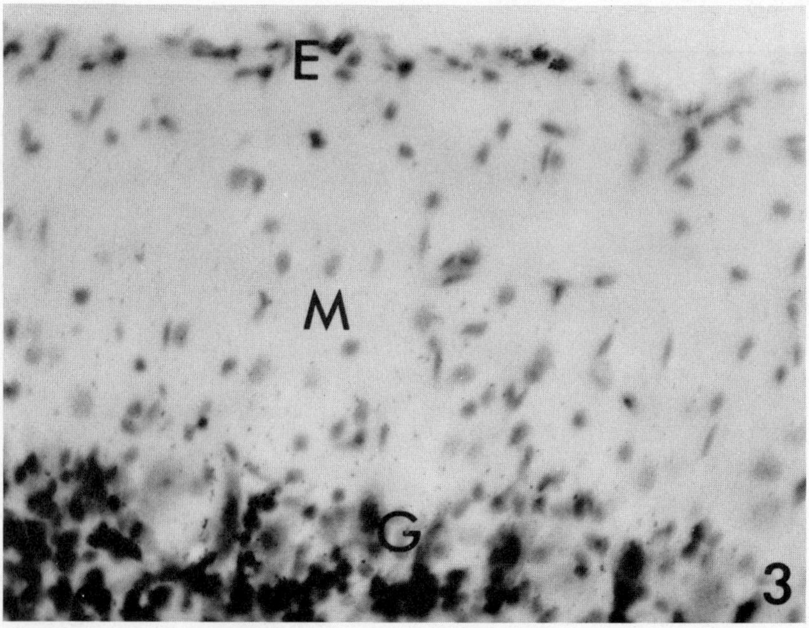

Fig. 3. Radioautograph of cerebellum of 21-day-old rat. Sacrificed 6 hours after injection of 1 μc/gm body weight of tritiated thymidine and exposed for 2 weeks. Note absence of grains ×440. *E*=external granular layer, *M*=molecular layer, *G*=internal granular layer.

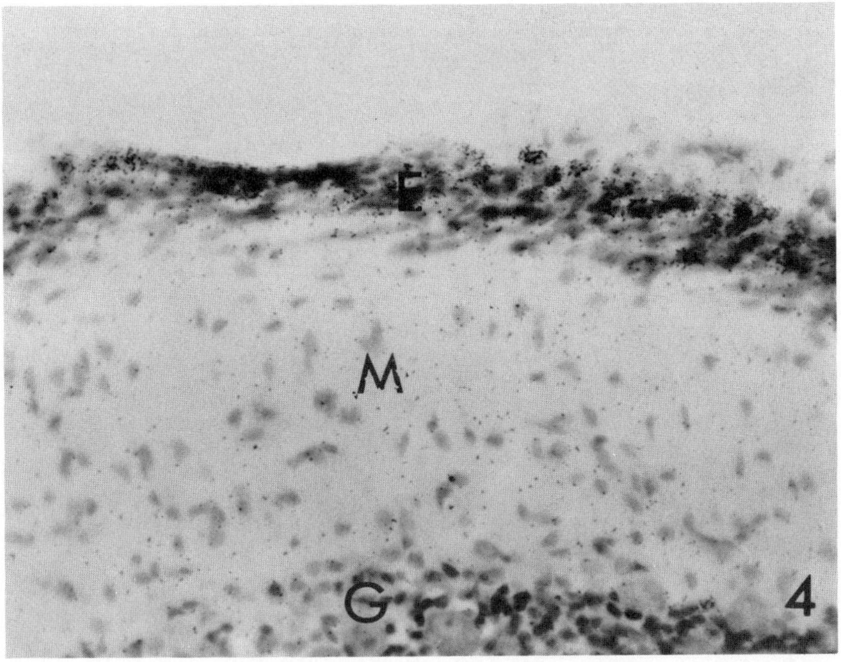

Fig. 4. Radioautograph of cerebellum of 21-day-old hypothyroid rat. Hypothyroidism was induced by mixing propylthiouracil into powdered diet starting on the 15th day of pregnancy and continued until day 21 of postnatal age. Sacrificed 6 hours after injection of 1 μc of tritiated thymidine per g of body weight and exposed for 2 weeks. Note high concentration of grains in the external granular zone. $\times$440.

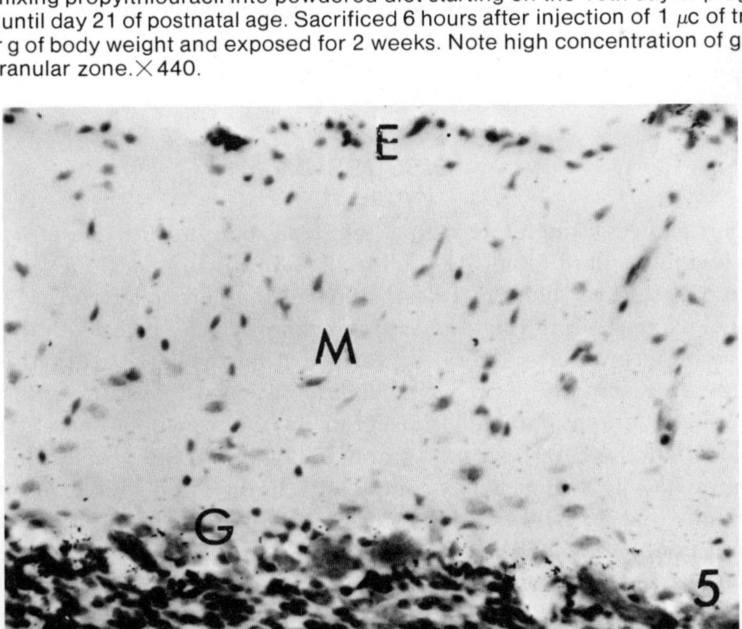

Fig. 5. Radioautograph of cerebellum of animal treated as in Fig. 3. $\times$440.

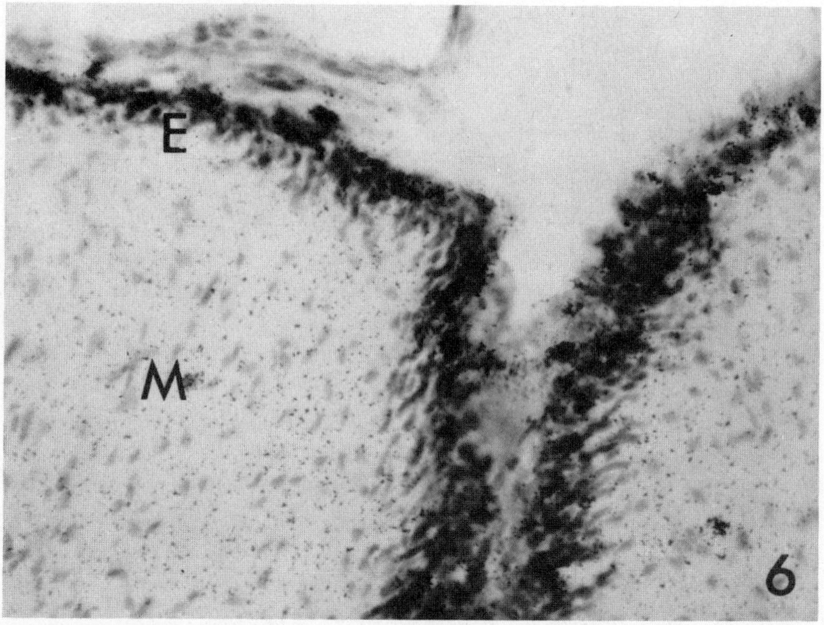

Fig. 6. Radioautograph of cerebellum of animals treated as in Fig. 4. ×440.

thyroxine-replacement therapy (0.2 μg thyroxine per gram of body weight per day) in reversing the propylthiouracil-induced delay and resulted in opening of the eyelids in 11 days instead of the 16 to 18 days that was noted in hypothyroid animals (see Table 1).

DISCUSSION

The persistence of the fetal external granular zone in cerebella of hypothyroid rats could be the result of either: (1) failure of the cells to migrate away from their source of origin toward their final destination in the cerebellar cortex: (2) failure of disposal of excess cells of the fetal external zone through cell death; or (3) delay in the cessation of proliferation of cells derived from the fetal external layer. The results of the present study permit us to choose between these three alternatives.

The radioautographs obtained from cerebella of normal and hypothyroid young rats indicate that the fetal external granular layer that still persists in hypothyroid rats after day 21, when it has completely disappeared in control littermates, consists of cells that are still actively taking up tritiated ^{3}H thymidine, and therefore are presumably engaged in DNA synthesis most likely preparatory to mitosis.

On the basis of these observations, we propose the hypothesis that thyroxine may push cells into the "differentiation" phase by pulling them out of the proliferative or mitotic phase. By acting as a timer, the optimal amount of thyroxine may turn off proliferation, and thus allow differentiation to begin. Excess amounts of

TABLE 1

Effect of injection of lyophilized powder of epithelial growth promoting factor into newborn rats

	Treatment	Total no. of animals	Day of eye opening: Day 11	Day 12	Day 13	Day 14	Day 15	Day 16	Day 17	Day 18	Day 19
Group 1[a]	Controls	75	0 (0%)	0 (0%)	0 (0%)	37 (49.33%)	19 (25.33%)	16 (21.33%)	3 (4.00%)	0 (0%)	0 (0%)
Group 2[b]	PTU	52	0 (0%)	0 (0%)	0 (0%)	1 (1.92%)	5 (9.62%)	11 (21.15%)	18 (34.62%)	17 (32.69%)	0 (0%)
Group 3[c]	Thyroxine	46	0 (0%)	23 (50.00%)	16 (34.78%)	7 (15.22%)	0 (0%)	0 (0%)	0 (0%)	0 (0%)	0 (0%)
Group 4[d]	PTU + THY	37	2 (5.41%)	8 (21.62%)	7 (18.92%)	4 (10.81%)	5 (13.51%)	3 (8.11%)	8 (21.62%)	0 (0%)	0 (0%)
Group 5[e]	PTU + EGF	12	11 (91.67%)	1 (8.33%)	0 (0%)	0 (0%)	0 (0%)	0 (0%)	0 (0%)	0 (0%)	0 (0%)
Group 6[f]	EGF	6	0 (0%)	6 (100%)	0 (0%)	0 (0%)	0 (0%)	0 (0%)	0 (0%)	0 (0%)	0 (0%)

[a]Group 1 represents controls injected with 0.9 percent saline.

[b]Thyroid deprivation was started prior to birth by addition of 0.2 percent propylthiouracil mixed into powdered food.

[c]Group 3 consists of animals given daily subcutaneous injection of 0.2 μg thyroxine per g of body weight since birth.

[d]Group 4 consists of animals treated as Group 2 but given daily thyroxine-replacement therapy starting at birth using the same dosages as in Group 3.

[e]Group 5 consists of animals treated as Group 2 but given daily subcutaneous injections of epithelial growth factor (EGF) (15 to 45 μg) from day 2 to 11 of postnatal age.

[f]Group 6 consists of untreated control animals given daily injections of EGF from day 2 to 11 of postnatal age. The dosage given was half of that used for Group 5.

thyroxine may turn off the "proliferative phase" too soon, while thyroid deficiency prolongs the period of proliferation and thus delays the onset of processes of differentiation. The wrong timing may then lead to both permanent as well as to transitory abnormalities.

A consequence of the prolonged proliferation of "extragranular" cells, and their delayed migration through the molecular layer past the row of perikarya of Purkinje cells, seems to be a resulting hypoplasia of the dendritic spread of the Purkinje network in cerebella of hypothyroid rats(16,18). It is conceivable that the differentiation of the Purkinje network depends on some "inductive" stimulus exerted by the granule cells as they migrate past the row of perikarya of Purkinje neurons prior to taking their proper place in the internal granular zone(26,27).

The possibility that an error in the phasing of maturational processes of the developing nervous system may lead to permanent defects in the CNS was also advanced by Eayrs(28) on the basis of behavioral studies of rats made hyperthyroid during the neonatal period. Eayrs(20) reported that rats made hyperthyroid during the first few days after birth perform less well than their littermate controls when given the Hebb-Williams closed field test 2 months after reaching adulthood. Eayrs(20) also reported that these differences in performance between experimental and control animals were observed only when exogenous thyroid treatment was administered between birth and 10 days of age. When thyroid medication was delayed beyond the 10th day, the treatment was ineffective. It has been proposed that the presence of excess thyroid titer may interfere with the maturation of the hypothalamic-pituitary axis, thus rendering the animal hypothyroid in later life. Experiments to test this hypothesis have failed to produce supporting evidence(20). It is more likely that excess hormone given during the critical period of brain development upsets the biological time clock of neurogenesis thus eliminating some factor necessary for adequate intellectual performance.

Additional support for the idea that the effect exerted by thyroid hormone on the developing brain may be mediated by the hormone's interference in the timing of the proliferative phase came from studies reported by Zamenhof et al.(11). Stimulated by the observation that DNA synthesis in chick eggs is subject to seasonal variations, Zamenhof treated chick eggs with the goitrogen thiourea on day 10 to 13 of incubation. At hatching, the experimental animals displayed larger brain weights (39 percent heavier) with 10 percent increase in total DNA. Conversely, thyroxine introduced into 9-day-old chick eggs depressed total amounts of DNA (5 to 6 percent) and presumably cell division. By day 21 post-hatching, the difference in DNA content between controls and experimentals had disappeared. The magnitude of the difference in total DNA between "control" and "hypothyroid" brains is admittedly small. Possibly, a much greater difference in DNA content could have been revealed if such studies had focused on specific areas of the brain whose maturational period coincides with the critical period of thyroid sensitivity. For the rat, such sensitive periods for brain maturation have been identified by Eayrs and Taylor(30) and Eayrs and Horn(31) and by Hamburgh and Flexner(32). Geel and Timiras(14,15) also reported a significant increase in total DNA content of cerebral cortex of rats thyroidectomized at birth, but failed to find any comparable effect when only the diencephalon of thyroidectomized rats was analyzed. This makes sense and is exactly what one expects, for it attests to the regional specificity of thyroid hormone on brain development.

The hypothesis that thyroxine may push cells into the "differentiative" phase by pulling them out of the proliferative or mitotic phase is a very attractive one, particularly in view of the frequently advanced postulate that growth and differentiation are mutually exclusive events. It is, of course, fascinating to speculate that the proliferative period leading to the increase in DNA content might, in turn, affect the timing of RNA synthesis and thus delay the production of important enzyme proteins. This interpretation finds some support from biochemical work recently reported by Pasquini et al.(33) who followed changes in nucleic acid content taking place during postnatal development of cerebral cortex and cerebellum of both normal and thyroidectomized rats at birth. They noted that neonatal thyroidectomy produced a significant increase in DNA content of cerebral cortex and to a lesser extent in cerebellum. This condition, in turn, led to a substantial decrease of RNA and protein in both tissues.

Geel and Timiras, in a series of publications(14,15), and Geel et al.(34) also reported that neonatal thyroidectomy of rats resulted in a significant increase in DNA content of cerebral cortex, accompanied by a decrease in RNA content.

The demonstration presented in the second part of this study, that the action of thyroid hormone on some aspect of development can be replaced by EGF, is of interest but does not imply that normally thyroxine acts by speeding the release of this factor, nor that the two substances promote the same end effects by influencing identical biochemical processes during development.

We do not as yet know whether EGF has a real function in the biological household, as has been proven for instance for the nerve growth promoting factor by Levi-Montalcini and Booker(35) who have shown that destruction of that substance by antiserum prevents the maturation of sympathetic chain ganglia. The EGF joins the growing list of biological substances that share physiological effects with those of thyroxine during development. Frequently the observation is cited that many of the innately organized responses of an animal are retarded but not suppressed by thyroidectomy, because they do eventually appear in the repertory of the thyroidectomized animal(19,28,29,36) even in the continued absence of thyroid hormones. This might be interpreted as due to the subsequent emergence of such other biologically active molecules as EGF or growth hormone, or still others yet to be identified, that are capable of mimicking the action of thyroxine on developing tissues of the sensory and the central nervous system.

REFERENCES

1. Villee, C. A. Growth of seminal vesicle and prostate: Role of RNA following testosterone stimulation. *In* Hormones in Development. Hamburgh, M., and Barrington, E. J. W., eds. New York, Appleton-Century-Crofts, 1970.
2. Tata, J. R. Turnover of nuclear and cytoplasmic ribonucleic acid at the onset of induced amphibian metamorphosis. Nature, 207:378–381, 1965.
3. ——— Requirement for RNA and protein synthesis for induced regression of the tadpole tail in organ culture. Develop. Biol., 13:77–94, 1966.
4. ——— Cell structure and biosynthesis during hormone-mediated growth and develop-

ment. *In* Hormones in Development. Hamburgh, M., and Barrington, E. J. W., eds. New York, Appleton-Century-Crofts, 1970.
5. Etkin, W. Metamorphosis. *In* Analysis of Development. Willier, B. H., Weiss, P. A., and Hamburger, M., eds. Philadelphia, W. B. Saunders Co., pp. 631–663, 1955.
6. ——— Metamorphosis. *In* Physiology of Amphibian, Moore, J. A., ed. New York, Academic Press, Inc., pp. 427–468, 1964.
7. Weber, R. Behavior and properties of acid hydrolases in regressing tails of tadpoles during spontaneous and induced metamorphosis *in vitro*. Ciba Found. Symp. Lysosomes, De Reuck, A. V. S., and Cameron, M. P., eds. London, J. A. Churchill Ltd., pp. 282–305, 1963.
8. ——— Biochemistry of amphibian metamorphosis. *In* Biochemistry of Animal Development, Weber, R., ed. New York, Academic Press, Inc., pp. 227–295, 1967.
9. Derby, A. An *in vitro* quantitative analysis of the response of tadpole tissue to thyroxine. J. Exp. Zool., 168:147–156, 1968.
10. ——— An *in vitro* quantitative analysis of the response of tadpole tissue to hormone treatment. *In* Hormones in Development. Hamburgh, M., and Barrington, E. J. W., eds. New York, Appleton-Century-Crofts, 1970.
11. Zamenhof, S., Van Marthens, E., and Bursztyn, H. The effect of hormones on DNA synthesis and cell number in the developing chick and rat brain. *In* Hormones in Development. Hamburgh, M., and Barrington, E. J. W., eds. New York, Appleton-Century-Crofts, 1970.
12. Balázs, R., Kovács, S., Teichgräber, P., Cocks, W. A., and Eayrs, J. T. Biochemical effects of thyroid deficiency on the developing brain. J. Neurochem., 15:1335–1349, 1968.
13. ——— Cocks, W. A., Eayrs, J. T., and Kovacs, S. Biochemical effects of thyroid hormones on the developing brain. *In* Hormones in Development. Hamburgh, M., and Barrington, E. J. W., eds. New York, Appleton-Century-Crofts, 1970.
14. Geel, S., and Timiras, P. The influence of neonatal hypothyroidism and of thyroxine on the ribonucleic acid and deoxyribonucleic acid concentration of rat cerebral cortex. Brain Res., 4:135–142, 1967.
15. ——— and Timiras, P. The role of thyroid and growth hormones on RNA metabolism in the immature brain. *In* Hormones in Development. Hamburgh, M., and Barrington, E. J. W., eds. New York, Appleton-Century-Crofts, 1970.
16. Legrand, J. Analyse de l'action morphogenetique des hormones thyroidiennes. Arch. Anat. Micr. Morph. Exp., 56:205–244, 1967.
17. ——— Variations en fonction de l'age de la response du cervelet a l'action morphogenetique de la thyroide chez le rat. Arch. Anat. Micr. Morph. Exp., 56:291–308, 1967.
18. ——— Comparative effects of thyroid deficiency and undernutrition on the maturation of the nervous system and particularly on myelination in the young rat. *In* Hormones in Development. Hamburgh, M., and Barrington, E. J. W., eds. New York, Appleton-Century-Crofts, 1970.
19. Hamburgh, M. An analysis of the action of thyroid hormone on development based on *in vivo* and *in vitro* studies. Gen. Comp. Endocr., 10:198–213, 1968.
20. Eayrs, J. T. Thyroid and developing brain: Anatomical and behavioral effects. *In* Hormones in Development. Hamburgh, M., and Barrington, E. J. W., eds. New York, Appleton-Century-Crofts, 1970.
21. Gómez, C. J., Ghittoni, N. E., and Dellacha, J. M. Effect of L-thyroxine or somatotrophin on body growth and cerebral development in neo-natally thyroidectomized rats. Life Sci., 5:243–246, 1966.
22. Cohen, S. Isolation of mouse submaxillary gland protein accelerating incisor eruption and eyelid opening in the newborn animal. J. Biol. Chem., 237:1555–1562, 1962.
23. ——— The stimulation of epidermal proliferation by a specific protein (EGF). Develop. Biol., 12:394–407, 1965.
24. ——— Studies on the mechanisms of action of epidermal growth promoting factor (EGF). *In* Hormones in Development. Hamburgh, M., and Barrington, E. J. W., eds. New York, Appleton-Century-Crofts, 1970.

25. Hamburgh, M., Sobel, E., Koblin, R., and Rhinestone, A. The passage of thyroid hormone across the placenta in intact and hypophysectomized rats. Anat. Rec., 144:219–228, 1962.
26. ——— Analysis of the postnatal developmental effects of "Reeler," a neurological mutation in mice. A study in developmental genetics. Develop. Biol., 8:165–185, 1963.
27. Miale, I., and Sidman, R. L. An autoradiographic analysis of histogenesis in the mouse cerebellum. Exp. Neurol., 4:277–296, 1961.
28. Eayrs, J. T. Role of thyroid hormone in the differentiation of the nervous system. Excerpta Medica Int. Congr. Series, 83:779–784, 1964.
29. ——— Thyroid and central nervous system development. *In* Scientific Basis of Medicine Annual Review, pp. 317–339, 1966.
30. ——— and Taylor, S. H. The effect of thyroid deficiency induced by methyl thiouracil on the maturation of the central nervous system. J. Anat., 85:350–358, 1951.
31. ——— and Horn, G. The development of the cerebral cortex in hypothyroid and starved rats. Anat. Rec., 121:53–61, 1955.
32. Hamburgh, M., and Flexner, L. B. Biochemical and physiological differentiation during morphogenesis. XXI. Effect of hypothyroidism and hormone therapy on enzyme activities of the developing cerebral cortex of the rat. J. Neurochem., 1:279–288, 1957.
33. Pasquini, J. M., Kaplun, B., García Argíz, C. A., and Gómez, C. J. Hormonal regulation of brain development. I. The effect of neonatal thyroidectomy upon nucleic acids, protein and two enzymes in developing cerebral cortex and cerebellum of the rat. Brain Res., 6:621–634, 1967.
34. Geel, S. E., Valcana, T., and Timiras, P. Effect of neonatal hypothyroidism and of thyroxine on C^{14}-leucine incorporation in protein *in vivo* and the relationship to ionic levels in the developing brain of the rat. Brain Res., 4:143–150, 1967.
35. Levi-Montalcini, R., and Booker, B. Destruction of sympathetic ganglia in mammals by an antiserum to the nerve growth promoting factor. Proc. Nat. Acad. Sci., U.S.A., 42:384–391, 1960.
36. Hamburgh, M., and Vicari, E. Effect of thyroid hormone on nervous system maturation. Anat. Rec., 127:302, 1957.

[illegible], and [illegible], R.: An autoradiographic analysis of [illegible] in the [illegible]. Exp. [illegible] 1963.

[illegible] ——— and [illegible]: The [illegible] of the [illegible] in [illegible]. [illegible]

[illegible], and [illegible]: Biochemical and morphological [illegible] of the [illegible]. [illegible]

[illegible]: [illegible] of the [illegible] and [illegible] development. [illegible] and [illegible] of the [illegible]. [illegible] 1967.

[illegible], and [illegible]: [illegible] of the [illegible]. Brain Res. [illegible]

[illegible]: [illegible] on the nervous system [illegible]. Amer. J. Anat. [illegible] 1957.

31

Hormonal Influences of the Biochemical Differentiation of the Rat Cerebral Cortex*

Carlos J. Gómez

Department of Biological Chemistry,
Faculty of Pharmacy and Biochemistry,
University of Buenos Aires,
Buenos Aires, Argentina

INTRODUCTION

Clinical studies on congenital hypothyroidism have repeatedly demonstrated the association of this disease with mental retardation and neurological symptoms of varying degree. Although several contradictory results published in the literature led to some doubts on the role of thyroid function in brain maturation(1), the well-documented clinical studies of Smith et al.(2) clearly demonstrated that in humans there is a critical period during which the thyroid function is essential for normal mental development.

It has long been established that in the amphibia the nervous system is one of the most sensitive to the maturational effects of the thyroid hormones(3,4), and, at least superficially, the processes of growth and maturation of the brain in mammals resemble in this respect those of metamorphosis in amphibia. In mammals, the growth, differentiation, and maturation of almost all systems are stimulated by thyroid hormones(4,5), but two of them are specially sensitive: the skeletal and nervous system. The problem of hormonal influence on neural development is complicated by the fact that, in some aspects, thyroid hormones act synergistically with other growth-promoting factors (e.g., growth hormone), and also nutritional factors are obviously important(5,6). Nevertheless, it is generally conceded that no tissue suffers more severely from the absence of thyroid hormone during early life than the brain.

*This research was supported by grants 1003 and 1003a from the Consejo Nacional de Investigaciones Cientificas y Tecnicas, Argentina.

In man, the period during which thyroid is essential for brain development, has been estimated to occur in utero or shortly after birth(2). It is well known that in small mammals like cats, rabbits, and rats, the maturation of the cerebral cortex occurs largely during the first 30 postnatal days; histological(7-9) and biochemical studies(10,11) carried out with these species have given a good body of evidence supporting the view that this process is mainly characterized by an active formation of membranes. During this period, the activity of most enzymes increases markedly(12,13), the metabolism of glucose changes to attain the characteristics of adult cerebral cortex(14,15), and the active formation of different membranes might also be associated with the development of the metabolic compartmentation of amino acids(16).

It is worthwhile to note that it is during this period that protein synthesis in the brain attains the highest activity during the life of the animals(17,18). Assuming that protein synthesis might be a peacemaker in the formation of membranes, and in view of the well-established observation that protein synthesis is controlled by hormones having effects on growth and development, it seemed appropriate to ascertain the role of thyroid and growth hormone on several aspects of the biochemical differentiation of the rat cerebral cortex.

The rat is the most suitable species for our purpose for several reasons: (1) It presents at the moment of birth a markedly undifferentiated cerebral cortex; and (2) It is relatively easy to induce an athyroid state in this species by radiothyroidectomy at birth(21). Also, Eayrs and his group(1,22,23) have demonstrated the role of thyroid hormones on the neuroanatomical, neurophysiological, and behavioral development of the rat, which serves as excellent background for trying to find the biochemical counterpart of their findings.

METHODS

The studies were conducted on male and female rats of a highly inbred Wistar strain. In the experiments in which the metabolic and biochemical changes produced by neonatal thyroidectomy were compared with those in normal controls, entire litters, each reduced to eight rats at birth, were used as experimental or control groups. Each rat of the experimental litter was made hypothyroid within 2 hours of birth by a single intraperitoneal injection of 120 μc ^{131}I according to Goldberg and Chaikoff(21). Experimental and control litters were kept with their mothers until sacrificed or until the 20th day of age, when they were weaned and maintained on a standard diet and water ad libitum. The animals were killed by decapitation at the scheduled date, the brain rapidly removed and cerebral cortex removed for metabolic studies or biochemical determinations.

The severity of hypothyroidism was determined at the end of the experiments in one or two rats randomly selected from each experimental litter, which were killed 20 hours after injection of 10 μc of ^{131}I. A section of tissue from the neck region and a similar mass of leg musculature were excised and radioactivity was measured in a well-type scintillation counter; neonatally radiothyroidectomized animals usually showed a thyroid-to-muscle radioactivity ratio of 1:3;when higher values were obtained, the animals from the whole litter were discarded.

Replacement Therapy Experiments

L-triiodothyronine (T_3) and L-thyroxine (T_4) (kindly supplied by Glaxo Laboratories, Argentina), and highly purified bovine growth hormone (BGH) with a biological activity of 1.6 to 1.8 I.U./mg protein were prepared in our department by Professor J. M. Dellacha, using the method of Dellacha and Sonnemberg(24). T_3 and T_4 were dissolved in 0.001 N NaOH whereas BGH was dissolved in 0.05 M carbonate-bicarbonate buffer pH 9.5 immediately injected. Both were subcutaneously administered.

The first set of experiments were designed to compare the effects of the administration of either T_4 or BGH on the metabolic alterations produced in the cerebral cortex by neonatal thyroidectomy. For this purpose, whole litters of neonatally radiothyroidectomized rats were injected on the 5th day, and daily thereafter until the 19th day, with: (1) 0.2 μg T_4/g body weight in 0.05 ml, or (2) 1 μg BGH/g body weight in 0.05 ml. During the experimental period, normal controls and untreated radiothyroidectomized rats received similar handling and equivalent injections of pure solvent. On the 20th day of age, the rats of all groups were killed, and cortical slices prepared as indicated below. A similar experiment was also carried out, starting the hormone administration on the 10th day; the animals were sacrificed on the 25th day after birth and brains prepared for examination.

The second set of experiments consisted of comparing the effects of the hormones on the chemical and enzymatic alterations produced by neonatal thyroidectomy. Two series of experiments were carried out; in the first one, T_3 (0.1 μg/g body weight in 0.05 ml) or BGH (1 μg/g body weight in 0.05 ml) was injected on the 10th day and daily thereafter until the 19th day in half of the litter, whereas the other half as well as normal controls received only pure solvent and similar handling. In the second series, hormone administration was started on the 15th day and maintained until the 24th day; in this case another group was studied, in which therapy consisted in the administration of both T_3 and BGH at the dose indicated above. The animals of all groups were killed one day after the end of treatments, and aliquots of cerebral cortex homogenates were used for chemical and enzymatic determinations.

Metabolic Studies

After killing the animals by decapitation the brain was rapidly excised and cortical slices 0.4 to 0.5 mm in thickness were obtained; the slices were washed in a precooled incubating medium containing no glucose and then placed in chilled Warburg vessels containing in the main compartment 2 ml of incubating medium and 0.2 ml 20 percent (w/v) NaOH in the center well. The composition of the medium in mM was: 115 NaCl, 5 KCl, 1.8 $CaCl_2$, 1.3 $MgSO_4$, 4.0 KH_2PO_4, 33 tris-HCl buffer pH 7.4, and 3.3 glucose. When the effect of potassium was studied, the concentration of KCl was increased by 100 mM. After gassing with pure oxygen and equilibrating at 37°C for 10 minutes, the oxygen consumption was measured for 60 minutes by the standard Warburg procedure.

Biochemical Determinations

Ten percent (w/v) homogenates of cerebral cortex were prepared in 0.32 M sucrose. After overnight freezing, the homogenates were thawed and aliquots used to carry out the following chemical and enzymatic determinations: DNA and RNA were determined by the method of Munro and Fleck(25) slightly modified(12), using yeast RNA (Sigma Co.) and thymus DNA (Mann Research Laboratories) as standards. Succinate dehydrogenase (SDH) (EC 1.3.99.1) was determined by the method of Slater and Bonner(26) and modified by De Robertis et al.(27), the absorbancy was measured at 420 mμ wavelength in a Beckman model DB spectrophotometer, and 1 unit was defined as the amount of enzyme producing a change of absorbancy of 0.100 per minute. Aspartate aminotransferase (AAT) (EC 2.6.1.1) was determined by the method of Karmen et al.(28) with some minor modifications; all suspensions were made 1.42 percent in Na-deoxycholate just prior to assay(29) and determinations were carried out in a Beckman model DB spectrophotometer at 37°C; the activity was expressed as micromoles of oxalacetate formed per minute. GABA Transaminase (GABAT) (EC 2.6.1) was measured by the method of Salvador and Albers(30) using an Aminco spectrofluorometer. Triton-X-100 at a final concentration of 0.5 percent was added just before the determination(31); the activity was expressed as micromoles of succinil semialdehyde (SSA) formed per hour.

Statistical Analysis

When changes in normal controls and neonatally thyroidectomized rats were studied, the results were compared by means of the Student's t test, and differences were considered statistically significant when P values were less than 0.05. In the studies in which the effects of hormone administration to hypothyroid rats were compared, the results were studied by variance analysis according to Snedecor(32) and the probability determined using the maximal limits of significance of F distribution for $P<0.01$ after Van der Waerden(33). The minimal significance difference between means (MSD) was determined according to Tuckey(34), considering statistically significant at the level of $P<0.05$ all differences between means greater than the MSD.

RESULTS AND CONCLUSIONS

Metabolic Changes

Normal cortical slices metabolizing glucose and incubated in an isosmotic medium show a low and continued increase of oxygen uptake up to the 15th day (Fig. 1). A different situation occurs when the concentration of potassium ions in the incubating medium is increased; potassium stimulates the respiration from day

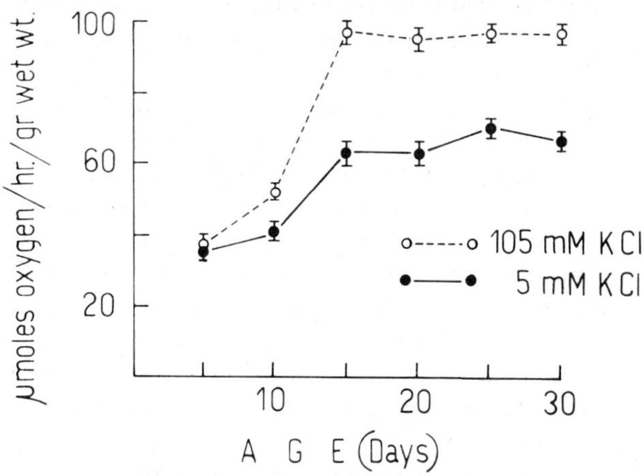

Fig. 1. Respiratory rate of normal cortical slices incubated in a low and a high potassium medium. Experimental conditions as detailed in the text. Results are expressed as micromoles oxygen/hour/g wet weight; points and circles represent the means and vertical bars S.E.M. from eight individual experiments for each age and experimental condition.

10 onward, but the magnitude of such stimulation shows a striking change between 10 and 15 days, such that on the 15th day the oxygen uptake in high potassium nearly doubled the values obtained in the isosmotic medium.

The pattern of changes of the cortical response to the potassium effect is similar to that found by Greengard and McIlwain(35) studying the cortical response to applied electrical pulses. In our laboratory, we have also found that potassium stimulates the conversion of glucose into amino acids of the glutamate group(15); this effect increases with age in a similar way to the respiratory response although the percent stimulation in this case is lower than that of respiration (see Table 1). The metabolic events involved in the cortical response to either high potassium or electrical pulses have been recently elucidated(36-38); these probably start with membrane depolarization and activation of cation-stimulated adenosine triphosphatase, which leads to an increased availability of ADP and phosphate which in turn will increase the rate of glucose utilization. But the full expression of the response also requires functional mitochondria on which ADP will stimulate the turnover of the tricarboxylic acid cycle. These concepts are in agreement with observations that the increase of the cortical response to high potassium levels during development is accompanied by striking increases in the activity of cortical cation-stimulated adenosine triphosphatase and several mitochondrial enzymes(12,13).

It is worthwhile to note that these changes coincide in time with the histophysiological maturation of the cerebral cortex, although it should be kept in mind that a temporal coincidence would not necessarily imply a causal connection. The respiratory rate of cortical slices in a high potassium medium may be considered a useful index of the degree of development attained by the cerebral cortex. This suggestion is supported by the fact that hypo- or hyperthyroidism, induced early in life in the rat, produce only small and transitory changes in the respiratory rate of cortical slices in an isosmotic medium(14,39,40). This contrasts with our observations, demonstrating that neonatal thyroidectomy in the rat produces a striking depression of the respiratory rate from day 10 onward, when cortical slices are incubated in a high potassium medium (Fig. 2).

In this connection it is worth noting that in work undertaken in our laboratory(15) we have found, as is shown in Table 1, that throughout the different stages of development the cerebral cortex from cretinoid rats fails to respond to high potassium with an increased synthesis of amino acids from labeled glucose, as do the normal controls.

From our results it can be concluded that the most striking alteration produced in the brain by thyroid dysfunction starting at birth is the lack of response to high potassium in both respiration and the labeling of amino acids. Taking into consideration the known mechanisms involved in the cortical response to high potassium (36-38), it is suggested that our findings agree with results from histological studies(1,22,23) showing that the most important alteration produced by neonatal thyroidectomy is a marked impairment in the formation of membranes. Our observations demonstrate that hypothyroidism also causes a marked depression in the activity of enzymes tightly bound to membranes, such as cation-stimulated adenosine triphosphatase(13). As will be shown below, studies with SDH, GABAT, and AAT strengthen this hypothesis, since it has been suggested that growth-promoting and developmental hormones like thyroid hormones exert metabolic effects that are independent of their effects on growth processes(19). We propose that the ob-

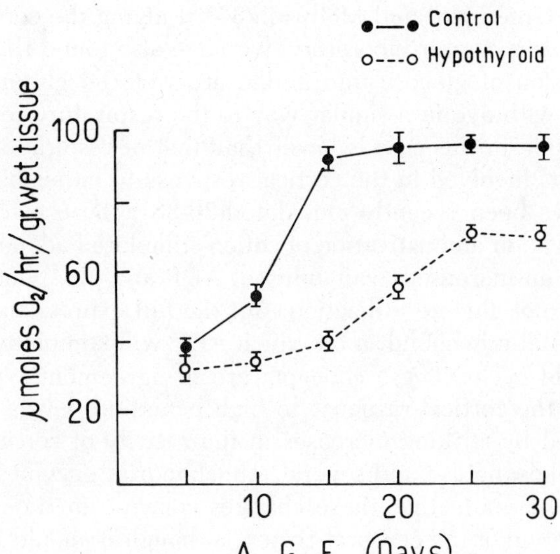

Fig. 2. Respiratory rate in a high potassium medium of cortical slices from normal control and hypothyroid rats. Animals were prepared as indicated in the text, and results are expressed as micromoles oxygen/hour/g wet weight. Points and circles represent the means and vertical bars the S.E.M. from eight individual experiments for each group and age. Results were compared by means of the Student's *t* test and differences were statistically significant from day 10 onward (*P* from <0.01 to 0.001).

TABLE 1

Cortical response to high potassium in the conversion of labeled glucose into radioactive amino acids[a]

Age (days)		Radioactivity incorporated into amino acids: 5 mM KCl	105 mM KCl	Percent stimulation
8	*N*[b]	1467	1711	17
	T[c]	882	659	0
16	*N*	1342	1667	24
	T	1173	1320	11
30	*N*	1827	2330	31
	T	1749	1908	9
60	*N*	1904	2480	30
	T	1927	1987	3

[a]Results from Gómez and Ramirez de Guglielmone (15). Radioactivity is expressed as millimicroatoms of carbon from glucose incorporated into amino acids/hr/100 mg wet weight. Radioactivity represents the sum of millimicroatoms of C from glucose incorporated in glutamate, aspartate, glutamine, alanine, and GABA.
[b]N = normal controls.
[c]T = thyroidectomized rats.

served changes in glucose metabolism by cortical slices reflect the developmental effects of thyroid hormones.

Effect of Neonatal Thyroidectomy on Chemical and Enzymatic Changes

Since in the cerebral cortex the DNA concentration per cell is constant(41), we consider that the expression of results in terms of DNA content is the most consistent frame of reference, and we therefore decided to express our results using this criterion, which has also been adopted by several other investigators(42-44).

Figure 3 shows that during normal development, the DNA content of the cerebral cortex decreases at a high rate from birth to the 20th postnatal day without further significant changes. This agrees with histological studies(7,8), which show a marked decrease of the proportional volume of the perikarya and a concomitant increase of that of dendrites occurring during the same period. The cerebral cortex from cretinoid rats presents from the 15th day onward a DNA content significantly higher than that from normal controls; this is in line with the histological observations showing that one of the most important alterations produced by neonatal thyroidectomy in the rat is a marked impairment of dendrite formation which indicates a defect in cell growth processes(1,22,23).

Figure 4 shows that during normal development of the cerebral cortex the RNA

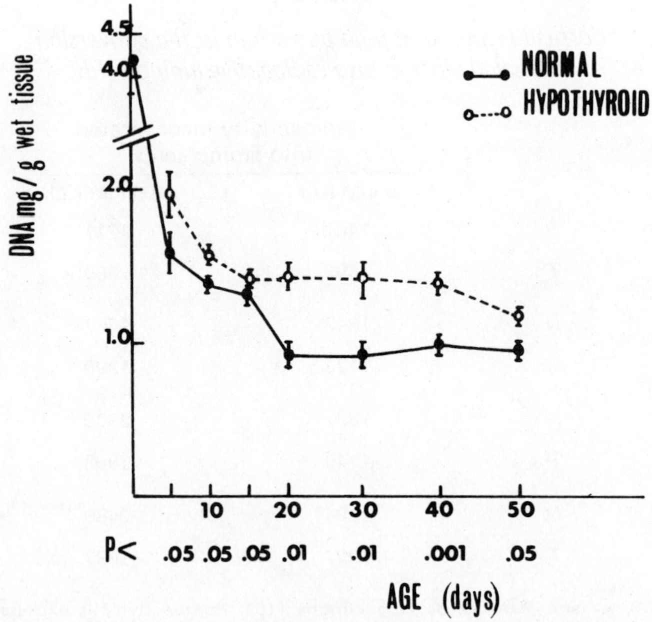

Fig. 3. DNA content in the cerebral cortex from normal control and hypothyroid rats. Animals were prepared as indicated in the text, and results are expressed as mg DNA/g wet weight. Points, circles, and vertical bars as in Figure 2. Results were compared by means of the Student's *t* test, and when statistically significant, *P* values are indicated in the figure.

content per cell increases in a striking way up to day 30 and diminishes thereafter at a high rate; the increase might be correlated with the histological changes occurring during that period(7,8), and presumably reflects the increase in size of the perikarya and the expansion of the dentritic field. On the other hand, since it has been suggested that the concentration of RNA per cell may be considered a useful index of the rate of protein synthesis(46) taking into account the generally accepted role of RNA in protein synthesis(19,20), it is interesting to point out that the period during which RNA concentration per cell is the highest coincides with that at which the protein synthesis in brain occurrs at the highest rate(17,18). Since neonatal thyroidectomy produces a marked depression of the RNA content of the cerebral cortex from the 10th day onward, it is logical to assume that this alteration would indicate the existence of a diminished rate of protein synthesis as a result of the lack of thyroid function from birth.

The latter assumption appears to be supported by our results on the enzymatic changes in developing cerebral cortex from normal control and cretinoid rats. Thus, Figure 5 shows that during normal development there is an overall increase in the activity of SDH, an enzyme tightly bound to mitochondrial membranes(13), with the highest rate between 10 days and 20 days after birth, whereas the cerebral cortex from cretinoid rats presents a significant depression from the 10th day onward, confirming previous findings of Hamburgh and Flexner(45).

Figure 6 shows that the level of GABA transaminase (GABAT), another mitochondrial enzyme, rises during normal development at an almost constant rate up

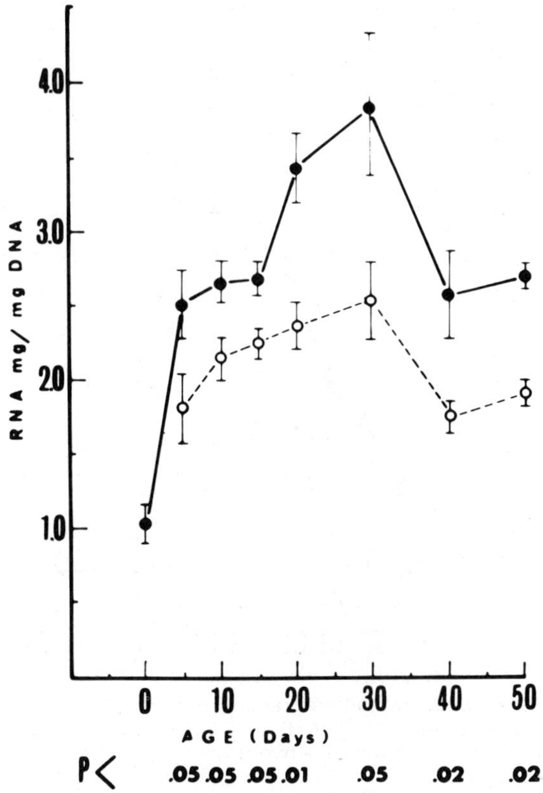

Fig. 4. RNA content in the cerebral cortex from normal control and hypothyroid rats. Results are expressed as mg RNA/mg DNA. Points, circles and vertical bars as in Figure 2. Comparion of result as indicated in Figure 3, and when significant *P* values are indicated in the figure. ——normal controls; ----—hypothyroid.

to the 40th postnatal day, when values are approximately 17 times as high as those found on the 5th day. On the contrary, the lack of thyroid function from birth produces a highly significant depression in the activity of GABAT, which is first apparent on the 10th day and then maintained throughout the studied period.

Finally, we have studied the pattern of changes of AAT, which presents a bimodal subcellular distribution with a particulate and soluble part(12). Measuring the total activity of this enzyme, the results presented in Figure 7 show that this enzyme increases throughout the studied period with the highest rate between 15 and 20 postnatal days, in marked contrast to the changes observed in the cortex from cretinoid rats, where a significant decrease occurs from the 20th postnatal day onward.

The results reported here together with other observations from our laboratory(13) in which we found that also glutamate decarboxylase and cation-stimulated adenosine triphosphatase are also markedly affected by neonatal thyroidectomy, and those from Geel and Timiras(44) showing similar effects on cortical acetylcholinesterase, clearly indicate that the lack of thyroid function from birth produces altered developmental patterns of enzymes which are constituents of dif-

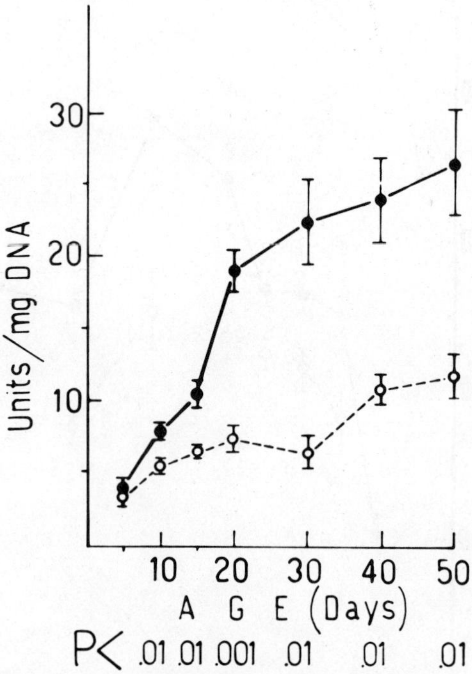

Fig. 5. Succinate dehydrogenase in the cerebral cortex from normal control and hypothyroid rats. Results are expressed as units/mg DNA. Points, circles, vertical bars, and statistical analysis as in Figure 3. ——normal control; -----hypothyroid.

ferent kinds of membranes like mitochondria, nerve endings, and dendrites. These facts, together with the marked depression of the concentration of RNA per cell just presented, strongly suggest that the role of thyroid hormones on brain development would be the reflection of their general effects on protein synthesis in the immature brain(46,47). Two recent papers support that viewpoint, demonstrating by in vivo experiments that neonatal thyroidectomy in the rat produces a significant depression in the incorporation of labeled amino acids into the protein of the cerebral cortex(48), and also into the protein of all subcellular fractions isolated from rat brain(49). Although the way by which thyroid hormones stimulate protein synthesis in the immature brain has not been yet clearly elucidated, it is worth noting that factors which depress protein synthesis, like undernutrition(6) and methionine sulphoximine(50), when acting during the period of early development, elicit histological alterations of the cerebral cortex similar to those produced by thyroid deprivation from birth.

Effects of Administering Thyroid Hormones or BGH to Neonatally Thyroidectomized Rats

Clinical studies(2) and behavioral experiments on rats(51) have given good evidence of the existence of a critical period during which thyroid hormones exert their maturational effects. The studies of Hamburgh and Flexner(45) on cortical

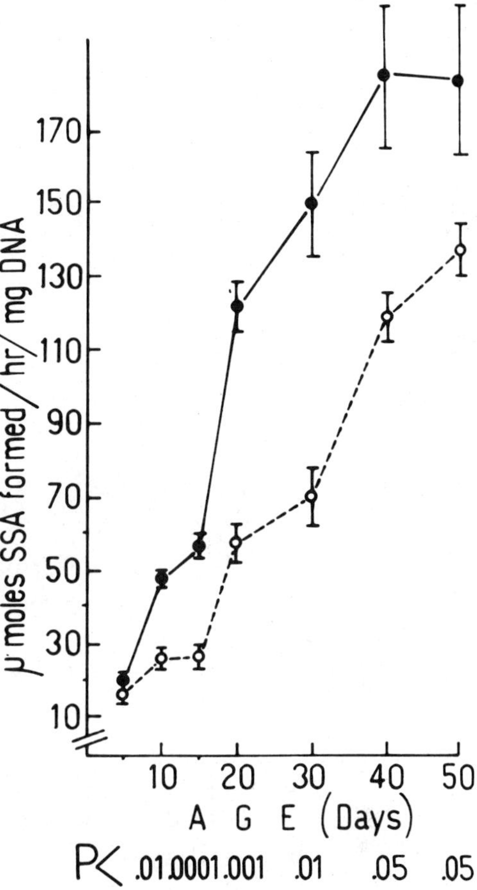

Fig. 6. GABA transaminase in the cerebral cortex from normal control and hypothyroid rats. Results are expressed as micromoles succinyl semialdehyde formed/hour/mg DNA. Points, circles, vertical bars, and statistical analysis as in Figure 3. —— —normal control; ---- —hypothyroid.

SDH, and our observations on the changes of aspartate concentration in brain(52), also suggest the existence of a similar critical period in some aspects connected with the biochemical differentiation of the brain. It seemed therefore appropriate to explore further in order to establish whether or not other biochemical alterations produced by neonatal thyroid deprivation occur within this "critical period."

Neonatal thyroidectomy in the rat produces profound alterations in the pituitary gland; these are mainly characterized by a marked degranulation of eosinophil cells(21), which appear to be those that synthesize or store growth hormone(53, 54). These facts, together with the marked alteration in body growth produced by this condition(12) suggest the possibility of a deficiency of growth hormone in neonatally thyroidectomized rats. We therefore designed experiments to clarify two points: (1) the period during which replacement therapy is effective for correcting the changes produced by the lack of thyroid from birth, and (2) the capacity of thyroid hormones and BGH for rectifying the biochemical changes observed in cretinoid rats.

The results presented in Table 2(14) show that the respiration of cortical slices from cretinoid rats treated either with T_4 or BGH is indistinguishable from

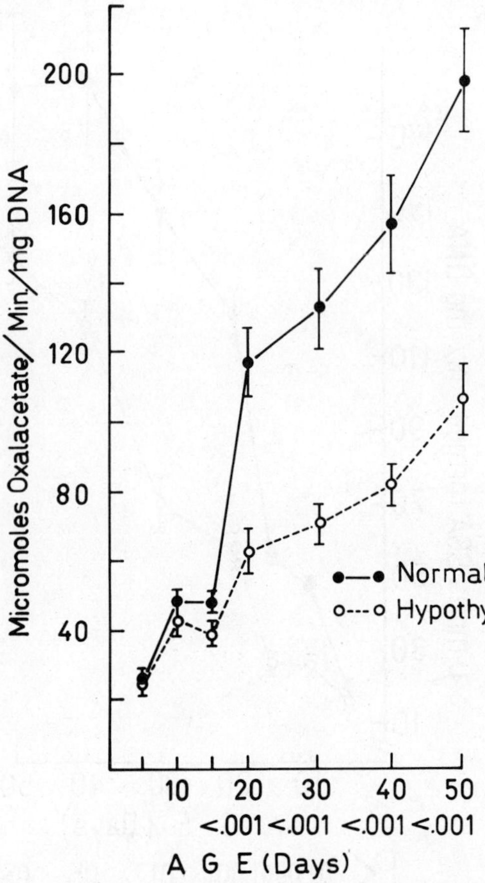

Fig. 7. Aspartate aminotransferase in the cerebral cortex from normal control and hypothyroid rats. Results are expressed as micromoles oxalacetate formed/minute/mg DNA. Points, circles, vertical bars, and statistical analysis as in Figure 3. ——normal control; ---- —hypothyroid.

that of normal controls, the results being similar when the hormone administration was started either on the 5th or the 10th day of age. The ameliorating effect of BGH confirms earlier observations from our laboratory(55).

In the experiments carried out to compare the effects T_3 or BGH in cretinoid rats on the chemical and enzymatic changes produced by thyroid deprivation from birth(56), the results demonstrate that in all cases the correcting effects are directly related to the age at which treatments begin. Thus, the results presented in Table 3 show that both T_3, and BGH administered from the 10th day after birth are effective in correcting the changes of DNA content of the cerebral cortex. When treatment was delayed until the 15th day, T_3, BGH, and also additive treatment failed to restore the increased DNA content to normal levels.

A similar situation occurs with RNA; as is shown in Table 4. The ameliorating effects of T_3 or BGH in cretinoid rats on the chemical and enzymatic changes produced by thyroid deprivation from birth(56), the results demonstrate that in all cases the correcting effects are directly related to the age at which treatments begin. Thus, the results presented in Table 3 show that both T_3 and BGH adminis-

TABLE 2

Effect of T_4 or BGH on the respiratory response to high potassium by cortical slices[a]

Experimental condition	Group A	Group B
Normal control	95.5 ± 3.5	97.3 ± 3.1
Hypothyroid	54.6 ± 1.2	72.1 ± 2.2
Hypothyroid + T_4	105.5 ± 4.4	102.7 ± 3.5
Hypothyroid + BGH	89.7 ± 3.5	91.5 ± 3.1
MSD	12.2	10.8

[a]The results are expressed as micromoles oxygen/hour/g fresh tissue and are the means ± S.E.M. of 6 to 8 individual determinations. Treatment and handling of animals under the different experimental conditions are detailed under Methods. In group A, the treatments begin on the 5th day of age and sustained until the 19th day. In group B, the treatments begin on the 10th day and maintained until the 24th day. In both cases the studies were carried out on the next day after the end of hormone administration. Results were compared by variance analysis as indicated under methods, and differences greater than the minimal significant difference between means (MSD) are statistically significant at the level of $\underline{P} < 0.05$.

tered from the 10th day after birth are effective in correcting the changes of DNA content of the cerebral cortex. When treatment was delayed until the 15th day, T , BGH, and also additive treatment failed to restore the increased DNA content to normal levels.

A similar situation occurs with RNA; as is shown in Table 4. The ameliorating effects of T_3 when treatment begins on the 10th day confirms previous findings of Geel and Timiras(57), who began the replacement therapy with T_4 on the 6th day after birth.

Table 5 shows, that both hormones are also effective in restoring to normal levels the decreased activity of SDH in the cerebral cortex of cretinoid rats if treat-

TABLE 3

Effect of T_3 or BGH on DNA content of the rat cerebral cortex[a]

Experimental condition	Group A	Group B
Normal control	0.97 ± 0.02	1.06 ± 0.02
Hypothyroid	1.23 ± 0.03	1.37 ± 0.02
Hypothyroid + T_3	1.07 ± 0.03	1.32 ± 0.06
Hypothyroid + BGH	0.98 ± 0.04	1.32 ± 0.04
Hypothyroid + T_3 + BGH		1.41 ± 0.04
MSD	0.17	0.14

[a]The results are expressed as mg DNA/g fresh tissue and are the means ± S.E.M. from 13 to 18 individual determinations in group A and from 16 to 22 determinations for group B. In group A, the different treatments begun on the 10th day were sustained until the 19th day, studies being carried out on the 20th. In group B, the treatments were initiated on the 15th day and sustained until the 24th day, the studies being carried out the next day. The results were compared by variance analysis as indicated in the text, and differences greater than the minimal significant difference between means (MSD) are statistically significant at the level of $\underline{P} < 0.05$.

TABLE 4

Effect of T_3 or BGH on RNA content of the rat cerebral cortex[a]

Experimental condition	Group A	Group B
Normal control	3.13 ± 0.11	2.75 ± 0.07
Hypothyroid	2.09 ± 0.13	1.96 ± 0.05
Hypothyroid + T_3	2.81 ± 0.11	2.02 ± 0.05
Hypothyroid + BGH	3.06 ± 0.16	2.11 ± 0.08
Hypothyroid + T_3 + BGH		2.05 ± 0.06
MSD	0.47	0.26

[a]The results are expressed as mg RNA/mg DNA and are the means ± S.E.M. from 13 to 18 individual determinations in group A and from 16 to 22 determinations for group B. Treatment and handling of animals of group A and B, and comparison of results by variance analysis as in Table 3.

TABLE 5

Effect of T_3 or BGH on succinate dehydrogenase activity of the cerebral cortex[a]

Experimental condition	Group A	Group B
Normal control	24.1 ± 1.3	29.9 ± 0.4
Hypothyroid	13.3 ± 0.4	17.6 ± 0.2
Hypothyroid + T_3	22.6 ± 0.8	20.8 ± 0.3
Hypothyroid + BGH	23.0 ± 0.8	18.2 ± 0.6
Hypothyroid + BGH + T_3		19.1 ± 0.2
MSD	3.6	1.9

[a]The results are expressed as IU/mg DNA and are the means ± S.E.M. from 13 to 18 determinations in group A and for 16 to 22 individual determinations in group B. Treatment and handling of animals of group A and B, and comparison of results by variance analysis as in Table 3.

TABLE 6

Effect of T_3 or BGH on the GABA transaminase activity of the rat cerebral cortex[a]

Experimental condition	Group A	Group B
Normal control	110.2 ± 2.5	120.4 ± 0.5
Hypothyroid	59.2 ± 1.8	67.5 ± 1.0
Hypothyroid + T_3	100.0 ± 3.2	81.2 ± 2.3
Hypothyroid + BGH	106.0 ± 4.4	80.3 ± 1.8
Hypothyroid + T_3 + BGH		78.3 ± 2.7
MSD	13.1	7.0

[a]The results are expressed as micromoles succinyl semialdehyde formed/hour/mg DNA and are the means ± S.E.M. of 13 to 18 individual determinations in group A and 16 to 22 in group B. Treatment and handling of animals of group A and B, and comparison of results by variance analysis as in Table 3.

TABLE 7

Effect of T_3 or BGH on the aspartate amino transferase activity in rat cerebral cortex[a]

Experimental condition	Group A	Group B
Normal control	110.8 ± 3.0	115.9 ± 2.2
Hypothyroid	60.2 ± 2.5	70.4 ± 1.1
Hypothyroid + T_3	95.3 ± 3.1	81.9 ± 1.2
Hypothyroid + BGH	95.4 ± 1.6	71.0 ± 2.1
Hypothyroid + T_3 + BGH		75.6 ± 1.8
MSD	9.8	6.6

[a]The results are expressed as micromoles oxalacetate formed/minute/mg DNA and are the means ± S.E.M. from 13 to 18 individual determinations in group A and for 16 to 22 individual determinations in group B. Treatment and handling of animals of group A and B, and comparison results by variance analysis as in Table 3.

ment begins on the 10th day of age, but not when the hormone administration is delayed until the 15th of age.

Table 6 shows that when treatments begin on the 10th day after birth, both hormones are effective in restoring to normal values the decreased activity of GABAT in brains of hypothyroid rats, but when hormone administration is delayed until the 15th day, neither treatment is effective.

Table 7 shows that when treatments are initiated on the 10th postnatal day both hormones produce a substantial and significant increase of AAT activity as compared with untreated cretinoid animals. Although the values attain about 90 percent of those of normal controls, the differences are still significant. On the other hand, hormone treatment is ineffective in restoring the activity of AAT to normal values when it is delayed until the 15th day.

DISCUSSION

As a whole, the results reported here permit two major conclusions. The demonstration that replacement therapy with thyroid hormones must be started at the earlier stages of postnatal development comfirms previous clinical studies in man(2), and experimental observations in the rat obtained from studies on behavioral and learning performance(51) and biochemical changes(45,52). It seems reasonable to estimate that thyroid hormone exerts its maturational effect on the rat brain within the first 10 to 12 postnatal days.

It has recently been established that thyroid hormone stimulates protein synthesis in the developing brain(46-49), but the mechanism involved is still controversial. Although it has been suggested that the stimulatory effect of thyroid hormone on protein synthesis in the brain is secondary to its regulatory action on amino acid transport(48), there exists abundant evidence giving strong support to the view that nucleic acids play an important role in the control of processes of growth and development(19,20). Stemming from studies on other tissues, mainly liver, it has been demonstrated that the stimulation of cytoplasmic protein synthesis is preceded by changes in the synthesis and utilization of a variety of nuclear and cytoplasmic RNAs(19,20). Whether the stimulatory effect of thyroid hormones on protein synthesis in the developing brain is accomplished at the level of transcription(19) or at the level of translation(49) has not been yet clearly elucidated. The synthesis of different RNAs in brain cell nuclei show characteristic changes through postnatal development(18). Such changes appear to be important factors in the control of protein synthesis, which tends to decrease with age(18). Although nothing definite is yet known about the primary site of hormonal action(19), it is possible to speculate that after a certain age the genome would become irresponsive to the action of the hormone, thus preventing a rectification of the biochemical alterations caused by early thyroidectomy.

The second conclusion stems from the demonstration that BGH administered to cretinoid rats has the same ameliorating effects as the administration of thyroid hormones. It is difficult to interpret the meaning of this finding. Although it is well known that growth and development of a tissue may depend on more than one hormone, and that thyroid hormones, growth hormone, and androgens have additive effects on the growth and the ability to synthesize RNA of liver, kidney, and bone (see Tata, [19]), such effects on developing brain have not yet been studied.

Our results would appear to indicate that BGH might replace thyroid hormones in controlling the development of the rat brain. In this connection, it is worth noting that the cortical growth of the offspring is stimulated by administering growth hormone to pregnant rats(58,59), and also that growth hormone administered to neonatally thyroidectomized rats appears to ameliorate partially the histological alteration of the cerebral cortex(60). However, Eayrs(60) considers that the latter effect, which is mainly due to an increase in the mean length of dentritic processes rather than to an increase in the number of branching points, reflects the action of growth hormone on incremental growth. Eayrs(60) denies that growth hormone has a role in the maturational aspects of cerebral development. However, it is difficult to maintain this assumption in view of our findings which demonstrate a marked restoring effect of BGH on the altered levels of brain nucleic acids and on the depressed activity of enzymes bound to several membranes following thyroidectomy. It is worthwhile to note, however, that the growth and development of bone requires both growth and thyroid hormones, and whereas growth may be induced by one in the absence of the other, only thyroid hormone promotes bone differentiation(5,19). Whether or not the same effect occurs in the developing brain has not yet been demonstrated.

Finally, the results presented here raise the question of whether the effects of early thyroid dysfunction on the maturation of the brain are mediated directly or indirectly. It must be said that in mammals the only evidence of a direct action of thyroid hormone on brain development has been provided by tissue culture studies in cerebellum(61,62), whereas there is also some evidence suggesting that thyroid hormone exerts some control on the synthesis or release of growth hormone by the pituitary gland(63). Although it is well known that the neonatal rat has a pituitary as well as a responsive thyroid(64), it is interesting to point out that whereas adult thyroidectomized rats present an enhanced protein synthesis in the pituitary gland(65), this process is markedly depressed in the pituitary from cretinoid rats (49). This presumably reflects a critical role of thyroid on pituitary metabolism during early postnatal life, and at the same time permits us to emphasize the necessity of more work to ascertain clearly the interrelationships between the growth-promoting and maturational effects of thyroid and pituitary hormone.

REFERENCES

1. Eayrs, J. T. Thyroid and central nervous development. Sci. Basis Med. Ann. Rev., 317–339, 1966.
2. Smith, D. W., Blizard, R. M., and Wilkins, L. The mental prognosis of hypothyroidism of infancy and childhood. Pediatrics, 19:1011–1022, 1957.
3. Etkin, W. Metamorphosis. *In* Physiology of Amphibia, Moore, J., ed. New York, Academic Press, Inc., pp. 427–468, 1964.
4. Gorbman, A. Thyroid hormones. *In* Comparative Endocrinology, Von Euler, H. S., and Heller, H., eds. New York, Academic Press, Inc., Vol. 1, pp. 291–324, 1963.
5. Rall, J. E., Robins, J., and Lewallen, C. G. The Thyroid. *In* The Hormones, Pincus, G., Thimann, K. V., and Atwood, T. B., eds. New York, Academic Press, Inc., pp. 159–439, 1964.
6. Eayrs, J. T., and Horn, G. The development of the cerebral cortex in hypothyroid and starved rats. Anat. Rec., 121:53–62, 1955.
7. ——— and Goodhead, G. Postnatal development of the cerebral cortex in the rat. J. Anat. (London), 93:385–402, 1959.
8. Schade, J. P., Van Backer, H., and Colon, E. Quantitative analysis of neuronal parameters in the maturating cerebral cortex. Prog. Brain Res., 4:150–175, 1964.
9. Purpura, D. P., Shofer, R. J., Housepian, E. M., and Noback, C. R. Comparative ontogenesis of the structure-function relations in cerebral and cerebellar cortex. Progr. Brain Res., 4:187–221, 1964.
10. Sperry, W. M. The biochemistry of the brain during early development. *In* Neurochemistry, Elliott, K. A. C., Page, I. H., and Quastel, J. H., eds. Springfield, Charles C. Thomas, Publisher, pp. 385–402, 1962.
11. Himwich, W. A. Biochemical and neurophysiological development of the brain in the neonatal period. Int. Rev. Neurobiol., 4:117–158, 1962.
12. Pasquini, J. M., Kaplun, B., García Argíz, C. A., and Gómez, C. J. Hormonal regulation of brain development. I. Effect of neonatal thyroidectomy upon nucleic acids, protein and two enzymes in development cerebral cortex and cerebellum of the rat. Brain Res., 6:621–634, 1967.
13. García Argíz, C. A.,Pasquini, J. M., Kaplun, B., and Gómez, C. J. Hormonal regulation of brain development. II. Effect of neonatal thyroidectomy on succinate dehydrogenase and other enzymes in developing cerebral cortex and cerebellum of the rat. Brain Res., 6:645–646, 1967.
14. Ghittoni, N. E., and Gómez, C. J. Respiration and aerobic glycolysis in the rat cerebral cortex. Influence of hypothyroidism. Life Sci., 3:979–986, 1964.
15. Gómez, C. J., and Ramirez de Guglielmone, A. E. Influence of neonatal thyroidectomy on glucose-amino acids interrelations in developing rat cerebral cortex. J. Neurochem., 14:1119–1128, 1967.
16. Berl, S. Compartmentation of glutamic acid metabolism in developing cerebral cortex. J. Biol. Chem., 240:2047–2054, 1965.
17. Lajtha, A. Protein metabolism of the nervous system. Int. Rev. Neurobiol., 6:1–98, 1964.
18. Yamagami, S., Fritz, R. R., and Rappoport, D. A. Biochemistry of developing rat brain. VII. Changes in the ribosomal system and nuclear RNA's. Biochim. Biophys. Acta, 139: 532–547, 1966.

19. Tata, J. R. Hormones and the synthesis and utilization of ribonucleic acids. Prog. Nucl. Acid Res., 5:191–250, 1966.
20. Korner, A. Ribonucleic acids and hormonal control of protein synthesis. Progr. Biophys., 17:61–98, 1967.
21. Goldberg, R. C., and Chaikoff, I. L. A simplified procedure for thyroidectomy in the newborn rat without concomitant parathyroidectomy. Endocrinology, 45:64–70, 1949.
22. Eayrs, J. T. The possible significance of neuropil for the mediation of cortical function. *In* Regional Neurochemistry, Kety, S. S., and Elkes, J., eds. New York, Pergamon Press, Inc., pp. 423–436, 1961.
23. ——— Effects of thyroid hormones on brain differentiation. *In* Brain-Thyroid Relationships, Cameron, M. P., and O'Connor, M., eds. London, Churchill, Ltd., pp. 60–74, 1964.
24. Dellacha, J. M., and Sonnemberg, M. Purification of bovine growth hormone. J. Biol. Chem., 239:1515–1520, 1964.
25. Munro, H. N., and Fleck, A. The precision of ultraviolet absorption measurement in the Schmidt-Thannhauser procedure for nucleic acid stimulation. Biochim. Biophys. Acta, 55:571–583, 1962.
26. Slater, E. S., and Bonner, W. C. The effect of fluoride on the succinic oxydase system. Biochem. J., 52:185–196, 1952.
27. De Robertis, E., Pellegrino de Iraldi, A., Rodriguez de Lores Arnaiz, G., and Salganicoff, L. Cholinergic and non-cholinergic nerve endings in rat brain. I. Isolation and subcellular distribution of acetylcholine and acetylcholinesterase. J. Neurochem., 9:23–35, 1962.
28. Karmen, A., Wroblewsky, F., and La Due, J. S. Transaminase activity in human blood. J. Clin. Invest., 34:126–133, 1955.
29. Sellinger, O. Z., Rucker, D. L., and De Balbian Verster, F. J. Cerebral lysosomes. I. A comparative study of lysosomal n-acetyl-βD-glucosaminidase and mitochondrial aspartic transaminase of rat cerebral cortex. J. Neurochem., 11:271–280, 1964.
30. Salvador, R. A., and Albers, R. W. The distribution of glutamic-γ-aminobutyric transaminase in the nervous system of the rhesus monkey. J. Biol. Chem., 234:922–925, 1959.
31. van Kempen, G. M. J., van den Berg, C. J., van der Helm, H. J., and Veldstra, H. Intracellular localization of glutamate decarboxylase γ-aminobutyrate transaminase and some other enzymes in brain tissue. J. Neurochem., 12:581–588, 1965.
32. Snedecor, S. W. Statistical Methods. Ames, Iowa State University Press, 1956.
33. van der Waerden, B. L. Mathematische Statistik, Berlin, Springer-Verlag, 1957.
34. Tuckey, J. W. Some selected and easy methods of statistical analysis. Trans. N.Y. Acad. Sci., Series II, 16:88–97, 1953.
35. Greengard, P., and McIlwain, H. Metabolic response to electrical pulses in mammalian cerebral tissues. *In* Biochemistry of the Developing Nervous System, Waelsch, H., ed. New York, Academic Press, Inc., pp. 251–260, 1955.
36. Minakami, S., Kakinuma, K., and Yoshikawa, H. The control of respiration in brain slices. Biochim. Biophys. Acta, 78:808–811, 1963.
37. Tobin, R. B., and McIlwain, H. The effects of oligomycin on the respiration and glycolisis of electrically-stimulated brain slices. Biochim. Biophys. Acta, 105:191–192, 1965.
38. ——— and Slater, E. C. Effects of oligomycin on the respiration of tissue slices. Biochim. Biophys. Acta, 105:214–220, 1965.
39. Fazekas, J. F., Graves, F. B., and Alman, R. W. The influence of the thyroid on cerebral metabolism. Endocrinology, 48:169–174, 1951.
40. Hamburgh, M., Lynn, E., and Weiss, E. P. Analysis of the influence of thyroid hormone on the prenatal and postnatal maturation of the rat. Anat. Rec., 150:147–159, 1964.
41. Heller, I. H., and Elliott, K. A. C. Deoxyribonucleic acid in brain and human brain tumors. Canad. J. Biochem., 32:584–592, 1954.
42. McCaman, R. E., and Aprison, M. H. The synthetic and catabolic enzyme systems for acetylcholine and serotonin in several discrete areas of developing rabbit brain. Progr. Brain Res., 9:220–233, 1964.

43. Mandel, P., and Edel-Hart, S. Free nucleotide in the rat brain during postnatal development. J. Neurochem., 13:591–595, 1966.
44. Geel, S. E., and Timiras, P. Influence of neonatal hypothyroidism and of thyroxine on the acetylcholinesterase and cholinesterase activities in the developing central nervous system of the rat. Endocrinology, 80:1069–1074, 1967.
45. Hamburgh, M., and Flexner, L. B. Biochemical and physiological differentiation during morphogenesis. XXI. Effect of hypothyroidism and hormone therapy on enzyme activities of the developing cerebral cortex of the rat. J. Neurochem., 1:279–288, 1957.
46. Gelberg, S., Campbell, P. L., Deibler, J. E., and Sokoloff, L. Effect of L-Thyroxine on amino acid incorporation into protein in mature and immature rat brain. J. Neurochem., 11:221–229, 1964.
47. Michels, R., Cason, J., and Sokoloff, L. Thyroxine: effects on amino acids incorporation into protein in vivo. Science, 140:1417–1418, 1963.
48. Geel, S. E., Vulcana, T., and Timiras, P. Effect of neonatal hypothyroidism and of Thyroxine on L-(^{14}C)-leucine incorporation in protein in vivo and the relationship to ionic levels in the developing brain. Brain Res., 4:143–150, 1967.
49. Balász, R., and Gaitonde, M. K. Factors affecting protein metabolism in the brain. Biochem. J., 106:1–2P, 1968.
50. Schade, J. P., and Pascoe, E. G. Maturational changes in cerebral cortex. III. Effects of methionine sulphoximine on some electrical parameters dendritic organization of cortical neurons. Progr. Brain Res., 9:132–154, 1964.
51. Eayrs, J. T. Age as a factor determining the severity and reversibility of the effects of thyroid deprivation in the rat. J. Endocr. 22:409–419, 1961.
52. Ramirez de Guglielmone, A. E., and Gómez, C. J. Influence of neonatal hypothyroidism on amino acids in developing rat brain. J. Neurochem., 13:1017–1025, 1966.
53. Herlant, M. The cells of the adenohypophysis and their functional significance. *In* International Review of Cytology, Bourne, G. H., and Danielli, J. F., eds. New York, Academic Press, Inc., Vol. 17, pp. 299–382, 1964.
54. Schooley, R. A., Friedkin, S., and Evans, E. S. Re-examination of the discrepancy between acidophil numbers and growth hormone concentration in the anterior pituitary gland following thyroidectomy. Endocrinology, 79:1053–1057, 1966.
55. Gómez, C. J., Ghittoni, N. E., and Dellacha, J. M. Effect of L-thyroxine or somatotrophin on body growth and cerebral development in neo-natally thyroidectomized rats. Life Sci., 5:243–246, 1966.
56. Krawiec, L., Pasquini, J. M., García Argíz, C. A., and Gómez, C. J. Unpublished results.
57. Geel, S. E., and Timiras, P. The influence of neonatal hypothyroidism and of thyroxine on the ribonucleic acid and deoxyribonucleic acid concentration of the rat cerebral cortex. Brain Res., 4:135–142, 1967.
58. Clendinnen, B. G., and Eayrs, J. T. The anatomical and physiological effects of prenatally administered somatotrophin on cerebral development in rats. J. Endocr., 22:183–193, 1961.
59. Zamenhof, S., Mosley, J., and Schuller, E. Stimulation of the proliferation of cortical neurons by prenatal treatment with growth hormone. Science, 152:1396–1397, 1966.
60. Eayrs, J. T. Protein anabolism as a factor ameliorating the effects of early thyroid deficiency. Growth, 25:175–189, 1961.
61. Hamburgh, M., and Bunge, R. T. Evidence for a direct effect of thyroid hormone on maturation of nervous tissue grown in vitro. Life Sci., 3:1423–1430, 1964.
62. ——— Evidence of a direct effect of temperature and thyroid hormone on myelinogenesis in vitro. Develop. Biol., 13:15–30, 1966.
63. Eartly, H., and Leblond, C. P. Identification of the effects of thyroxine mediated by the hypophysis. Endocrinology, 54:249–271, 1954.
64. Schapiro, S. Metabolic and maturational effects of thyroxine in the infant rat. Endocrinology, 78:527–532, 1966.
65. Tonoue, T., and Yamamoto, K. Effect of thyroidectomy and thyroxine supplement on amino acid incorporation of rat anterior pituitary. Endocrinology, 81:1029–1032, 1967.

32

Some Effects of Thyroid Hormone on the Development of the Central Nervous System in Early Ontogenesis

M. S. Mitskevich and G. N. Moskovkin

Laboratory of Hormonal Regulations,
Institute of Developmental Biology,
Academy of Sciences of the U.S.S.R.,
Moscow, U.S.S.R.

INTRODUCTION

The close functional relationship between the nervous and endocrine systems, which insures the proper adaptive reactions to provide for the changing needs of the organism, has been extensively investigated in adults. Relatively little has been said, however, concerning this relationship in early ontogenesis. Nevertheless, the data available suggest that each system influences the other during the course of development. In some aspects of development, this effect is functional and transitory; in other aspects of neurogenesis, hormones are absolute prerequisites for normal development.

A number of investigators have established that the functioning of the endocrine glands and the release of hormonal secretions are already under hypothalamic control during early ontogenesis. The reciprocal relationship, namely the dependence of the developing central nervous system (CNS) on hormones produced in the course of fetal development or during the first days or weeks of postnatal life, has been established.

It has been shown that during critical periods in early ontogenesis, sexual, adrenal, and thyroid hormones exert a definite influence on the CNS which, in the absence of normal secretions of these hormones, manifests itself in disturbances of brain function at later developmental stages.

The earlier clinical observations of the changes in brain structure in cretins and myxedematous patients(2-7) have been supplemented by numerous experi-

TABLE 1

Number of neurons in the fifth layer of cortex motor area in normal and hypothyroid rabbits and guinea pigs

Rabbits			Guinea pigs		
	Cell number per unit square of section			Cell number per unit square of section	
Age	**Control**	**Hypothyroid**	Age	**Control**	**Hypothyroid**
Fetuses			Fetuses		
28 days	20.15 ± 0.27	27.97 ± 0.34	44 days	29.08 ± 0.48	24.48 ± 0.46
			52 days	14.37 ± 0.42	12.15 ± 0.58
Newborns	13.45 ± 0.28	17.24 ± 0.40	Newborns	11.90 ± 0.39	9.52 ± 0.29
15 days	8.30 ± 0.15	10.38 ± 0.14	5 days	10.36 ± 0.19	9.10 ± 0.19
30 days	4.78 ± 0.13	7.68 ± 0.15	30 days	9.31 ± 0.21	8.01 ± 0.19
45 days	3.48 ± 0.10	5.38 ± 0.09	45 days	7.95 ± 0.28	5.99 ± 0.23

mental data on the action of thyroid hormone on the development of the CNS(8-29). The most dramatic morphological and enzymatic changes in nervous tissues of hypothyroid rats have been found to occur during the first weeks of postnatal life. The data obtained in our laboratory using rabbits and guinea pigs resemble those obtained with rats. In rabbits and guinea pigs, however, the changes in the brain following early thyroid deprivation were already present during fetal development (30,31). This is not unexpected since thyroid activity in these species reaches a high level during the second half of prenatal life(32).

In hypothyroid rabbit fetuses, changes in the width to length ratio of the cerebrum were already present at the 23rd day of intrauterine life. At the 28th day, the size of cortical neurons was found to be decreased, and their number per unit square, increased. In hypothyroid guinea pig fetuses, on the other hand, the number of cortical neurons decreased without any accompanying change in their size. These data are summarized in Table 1.

As shown in Table 1, in guinea pigs, unlike rabbits and rats, the number of neurons is decreased in hypothyroidism from the 44th day of intrauterine life onwards. These changes coincide in time with the appearance of morphological equivalents of functional maturation of neurons(33). As the size of cerebral hemispheres is the same in control and hypothyroid guinea pigs, it can be assumed that in hypothyroid animals fewer neuroblasts differentiate into cortical neurons.

In the present paper, some additional observations are presented concerning the influence of thyroid hormone on the development of the CNS in rats as revealed by means of enzyme histochemistry, interferometry, and electron microscopy.

MATERIALS AND METHODS

Studies were carried out on newborn and early postnatal rats made hypothyroid by intraperitoneal administration of methylthiouracil (MTU) to the mother

During the first postnatal week, dosages ranged between 70 and 100 mg MTU daily for each mother. A goiterous reaction, as revealed by thyroid weight and histology of the thyroid glands of the young rats, was taken as the criterion for hypothyroidism. In some instances, mothers and litters were surgically thyroidectomized.

Experimental Procedure

Experimental animals were divided into four groups:
Group I. Hypothyroidism was induced by administration of MTU starting on the day of birth. Animals were sacrificed at 9, 14 to 16, and 24 to 25 days of age.
Group II. Hypothyroidism was induced by administration of MTU starting on the day of birth and continuing until 13 to 14 days of age. Animals were sacrificed at 1 month of age.
Group III. Hypothyroidism was induced by surgical thyroidectomy of mothers and litters at 16 or 24 days of age. Litters were sacrificed at 2 months of age.
Group IV. Hyperthyroidism was induced by daily injections of 4 μg of thyroxine starting from the day of birth. Rats were sacrificed at the 24th day of age.

Controls for Groups I and II consisted of litters from mothers receiving saline. Controls for the third group consisted of sham-operated litters.

Histochemical Procedures

The enzymes chosen for study in this investigation were succinic dehydrogenase (SDH), α-glycerophosphate dehydrogenase (α-GPDH), not linked with NAD, NADH-diaphorase, NADPH-diaphorase, and monoamine oxidase (MAO). Histochemical determinations of the enzymes activities were carried out in cerebral cortex, in the nuclei of medulla oblongata and Pons Varolii, and in hemispheres of cerebellum and its nuclei. Incubations were made according to Pearse(34) and Burstone(35). Both frontal and sagittal sections (20 μ thick, temperature = −10°C) of cerebral cortex and cerebellum were taken for histochemistry. This permitted us to follow the changes in enzyme activities in various parts of the cerebrum, which differ in function and in time of maturation.

MAO was determined only in 2-month-old rats since this enzyme cannot be readily revealed in the CNS until this age.

Interferometry of Neurons

In order to follow changes in the dry weight of the cerebral cortex during hypo- and hyperthyroidism, brains were fixed according to Brodsky(37). Serial sections 5 μ thick were stained with cresol violet by the method of Nissl. The squares of cells, nuclei, and cytoplasm on stained preparations were measured by a planimeter. The dry weight was calculated from the results of interferometry and planimetry according to a formula described by Barer(36). One hundred cells from each of three animals in all age groups were measured.

Studies of Ultrastructure of Neurons

Brains from 16 and 24 to 25-day-old rats were prepared for electronmicroscopy. Animals were fixed by perfusion through the aorta for 45 minutes with 4 percent formalin buffered at pH 7.0 to 7.2 with 0.1 M phosphate buffer. Small blocks of brain tissue were immersed subsequently into 1 percent osmic acid buffered by the method of Palade, dehydrated through graded ethanol series, and embedded in Araldite. Ultrathin sections were cut with LK-B ultratome, floated on distilled water, and picked up on uncoated copper grids. Contrast staining was made according to Reynolds(38). Electron micrographs were taken by the Tesla-413 BS and UEMV-100 electron microscopes at the original magnifications of 9,050 and 16,900 times.

RESULTS AND CONCLUSIONS

Histochemical Studies of Enzyme Activities in Brains of Hypothyroid Rats

The enzymes chosen for the present study, i.e., SDH, α-GPDH, NADH-diaphorase, and MAO, were demonstrated in the cerebral cortex, nuclei of medulla oblongata and Pons Varolii, and the cerebellum and its nuclei in Groups I, II, and III.

The demonstration of NADPH-diaphorase enables us to follow the progress in vascularization. In adults, this enzyme can be revealed in the cytoplasm of neuroglial and nerve cells as well as in their processes and capillaries, respectively.

Changes in enzyme activities were found in rats starting at the 9th day postpartum (see Table 2). Later, i.e., on the 14th to 15th days, the intensity of the enzyme reactions that were studied was more pronounced in the control animals.

NADPH-diaphorase was found in basket cells of control animals, but it was absent in hypothyroid ones.

The vascularization of the cerebral cortex was below the control during hypothyroidism. At the 24th to 25th days, the activity of SDH and NADPH-diaphorase

TABLE 2

Changes in intensity of histochemical reaction in CNS of the hypothyroid (HTh) and normal (N) rats on the 9th postnatal day

	Brain areas							
	Cerebral cortex		Nuclei of medulla oblongata and Pons Varolii		Cerebellar cortex		Nucleus dentatus	
Enzymes	N	HTh	N	HTh	N	HTh	N	HTh
SDH	++	+	++++	++++	++	+	++++	++++
α-GPDH	+	+	–	–	++	+	–	–
NADPH-diaphorase	++	+	++++	++++	++	+	++++	++++
NADH-diaphorase	++	++	++++	++++	++	+	++++	++++

in the cerebral cortex in hypothyroid animals was still below the control. In the cerebellum by this time, there were practically no differences in enzyme activities between control and hypothyroid animals, exclusive of NADPH-diaphorase. This reaction was observed in the basket cells not only in the control but also in hypothyroid animals, although it was not so pronounced in the latter.

The data obtained in Groups II and III did not differ from the control.

Thus, it can be concluded that in hypothyroidism the activity of SDH and NADPH-diaphorase in cortical neurons decreases soon after birth as compared with the controls. This decrease is already pronounced at the 9th day postpartum. In the cerebellum of hypothyroid animals, all the enzymes investigated exibited reduced activities shortly after birth, but by the 25th day their activities had recovered in all cerebellar structures, with the exception of basket cells.

It is of interest that α-GPDH disappears from Purkinje cells of control animals during the period between 16 and 24 days; this is one of the indicators of the maturation of cerebellum but it can still be revealed in hypothyroid animals. Savik(43) was the first to observe this phenomenon. A similar situation was reported by Sakharova(44) who reported the disappearance of acetylcholine esterase during maturation of Purkinje cells from cat cerebellum.

Thus, it can be assumed that the retention of α-GPDH in Purkinje cells of hypothyroid rats indicates a retardation in the metabolic differentiation of these cells during hypothyroidism.

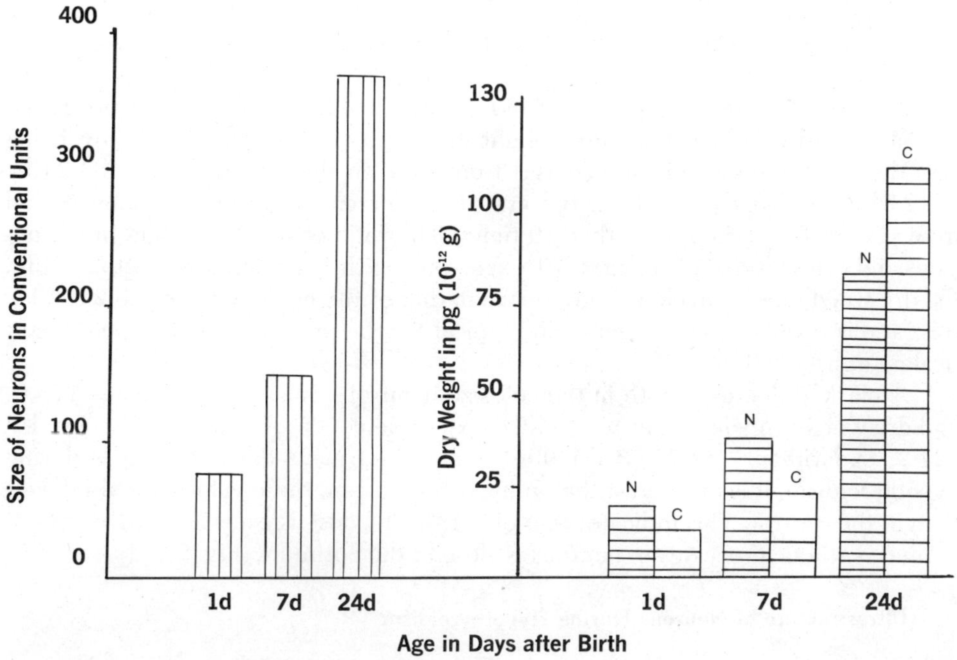

Fig. 1. Histogram of the squares and dry weight of cortical neurons in normal young rats. Note the marked increase in the squares and the dry weight of neurons during the first weeks of postnatal life. *N*—nucleus; *C*—cytoplasm.

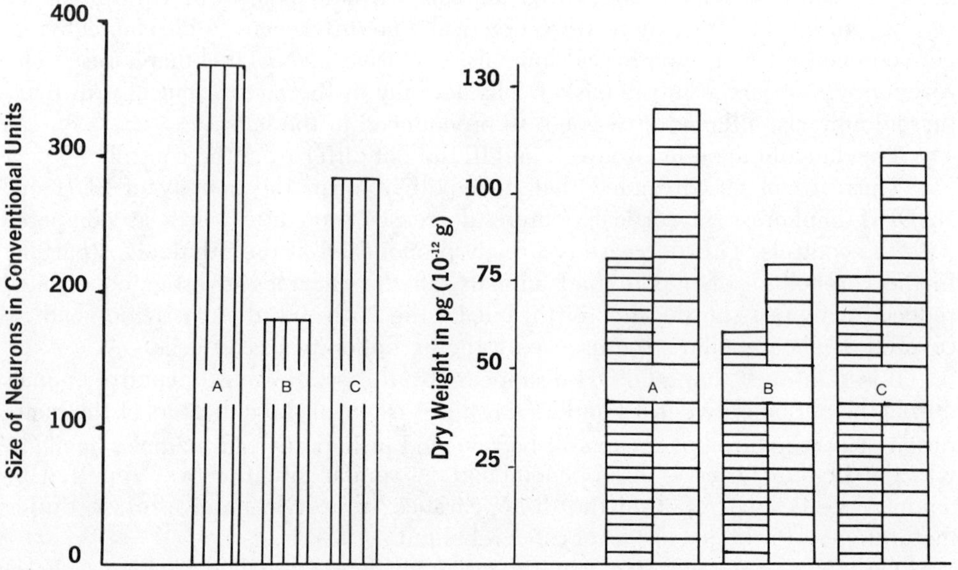

Fig. 2. Histogram of the squares and dry weight of cortical neurons in 24-day-old control, hypo- and hyperthyroid rats. A significant decrease in the squares and dry weight of neurons can be seen in hypo- and hyperthyroid rats, more pronounced in the former. *A*—normal, *B*—hypothyroid, *C*—hyperthyroid rats.

Interferometry of Neurons in the Control, Hypo-, and Hyperthyroid Animals

The data obtained are presented in Figures 1 and 2. As shown in Figure 1, the size of normal cells and their dry weight in pg (10^{-12}g) increases with advancing age: the dry weight of nuclei increases from 17.4 on the day of birth to 84.24 by the 24th day, the dry weight of the cytoplasm increases during the same period from 11.7 to 124.2, i.e., more than 10 times. The ratio of the dry weight of the nucleus and cytoplasm also changes with age. During the first week of postnatal life, the dry weight of the nucleus is greater than that of the cytoplasm; at the 24th day this ratio is reversed. This seems to be typical for the process of maturation of cortical neurons(45) (Fig. 1).

A marked decrease both in the cell size (control=369.5; experiment=173.6) and dry weight (nucleus, control=84.24; experiment=65.37; cytoplasm, control=124.2; experiment=82.77, $P < 0.001$) is observed in 24-day-old animals during hypothyroidism. It is of interest that in hyperthyroidism, these values are also lower than in the control. This indicates that either the T_4 dose is excessive or the rate of normal synthetic processes in neurons is close to the optimum (Fig. 2).

Ultrastructure of Neurons During Hypothyroidism

16-Day-Old-Animals. The attachment of cytoplasmic free ribosomes to the membranes of the endoplasmic reticulum is retarded in hypothyroid animals. The membranes of endoplasmic sacs are almost completely devoid of ribosomes.

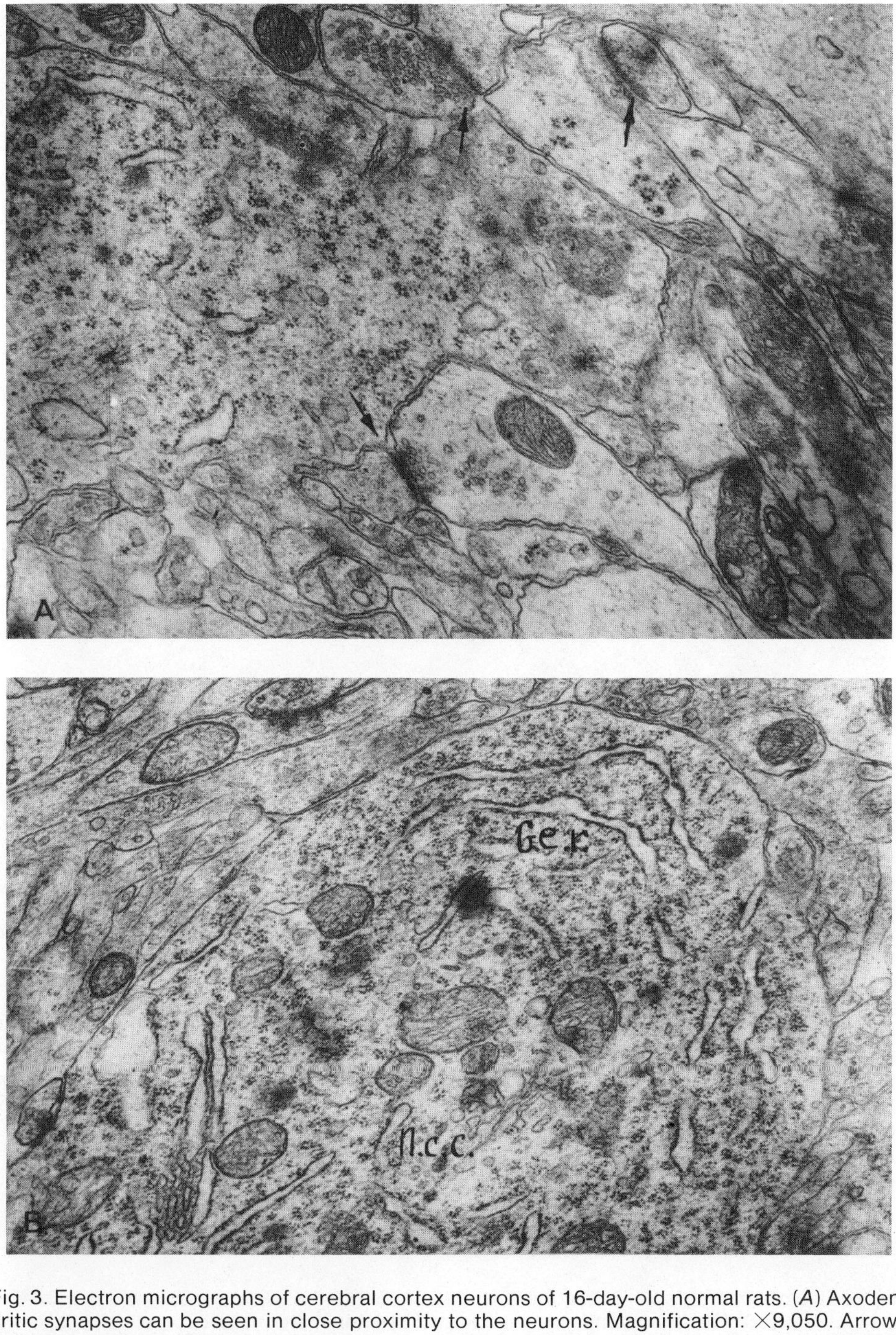

Fig. 3. Electron micrographs of cerebral cortex neurons of 16-day-old normal rats. (*A*) Axodendritic synapses can be seen in close proximity to the neurons. Magnification: ×9,050. Arrows indicate synapses. (*B*) Well-developed granular endoplasmic reticulum can be seen. Magnification: ×10,800. *G.e.r.*—granular endoplasmic reticulum. *N.c.c.*—nervous cell cytoplasm.

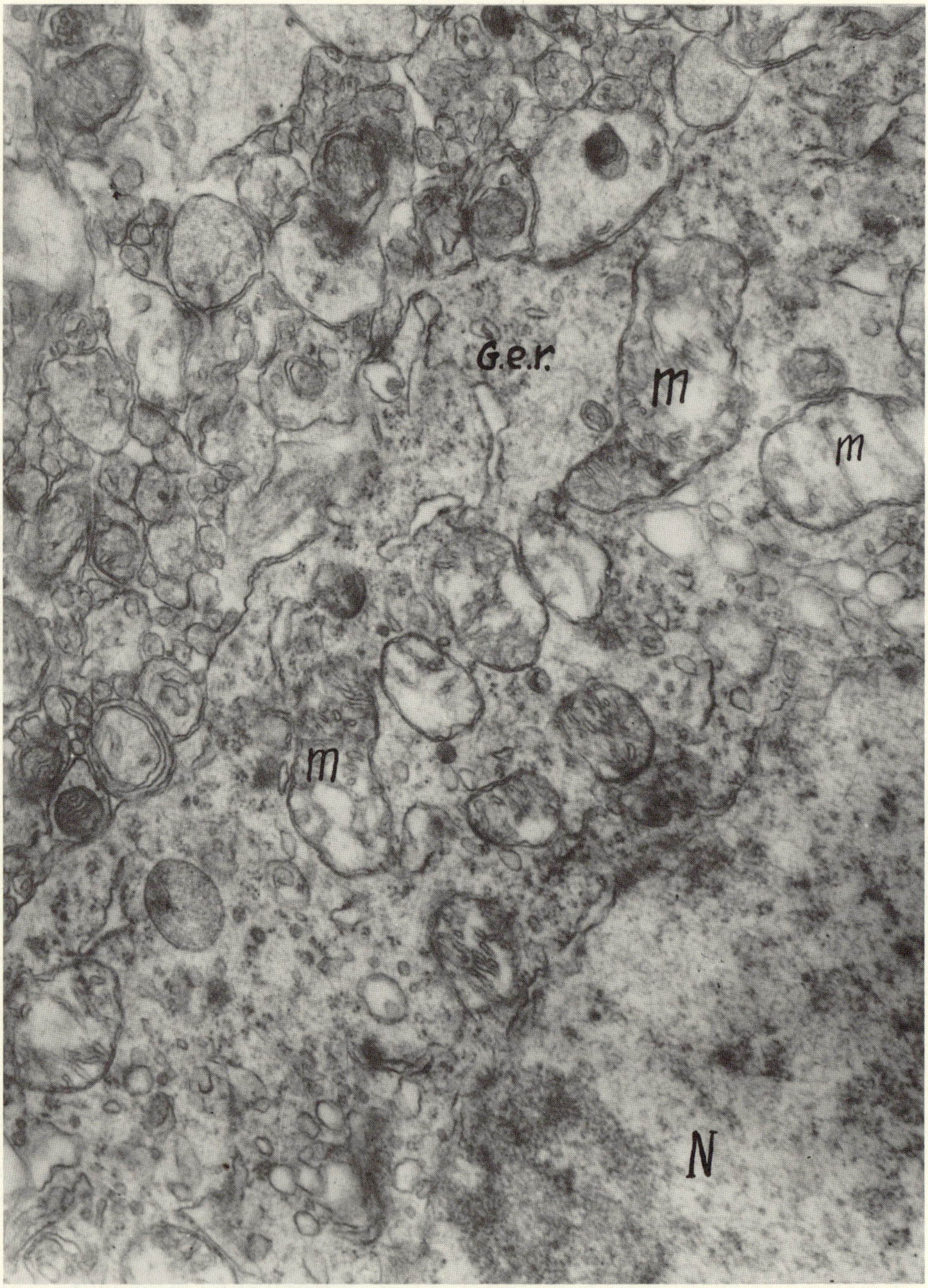

Fig. 4. Electron micrograph of cerebral cortex neurons of 16-day-old hypothyroid rats. Magnification: ×9,050. The membranes of endoplasmic reticulum are almost completely devoid of ribosomes. Endoplasmic reticulum is underdeveloped. *N*—nucleus; *M*—mitochondria; *G.e.r.*—granular endoplasmic reticulum.

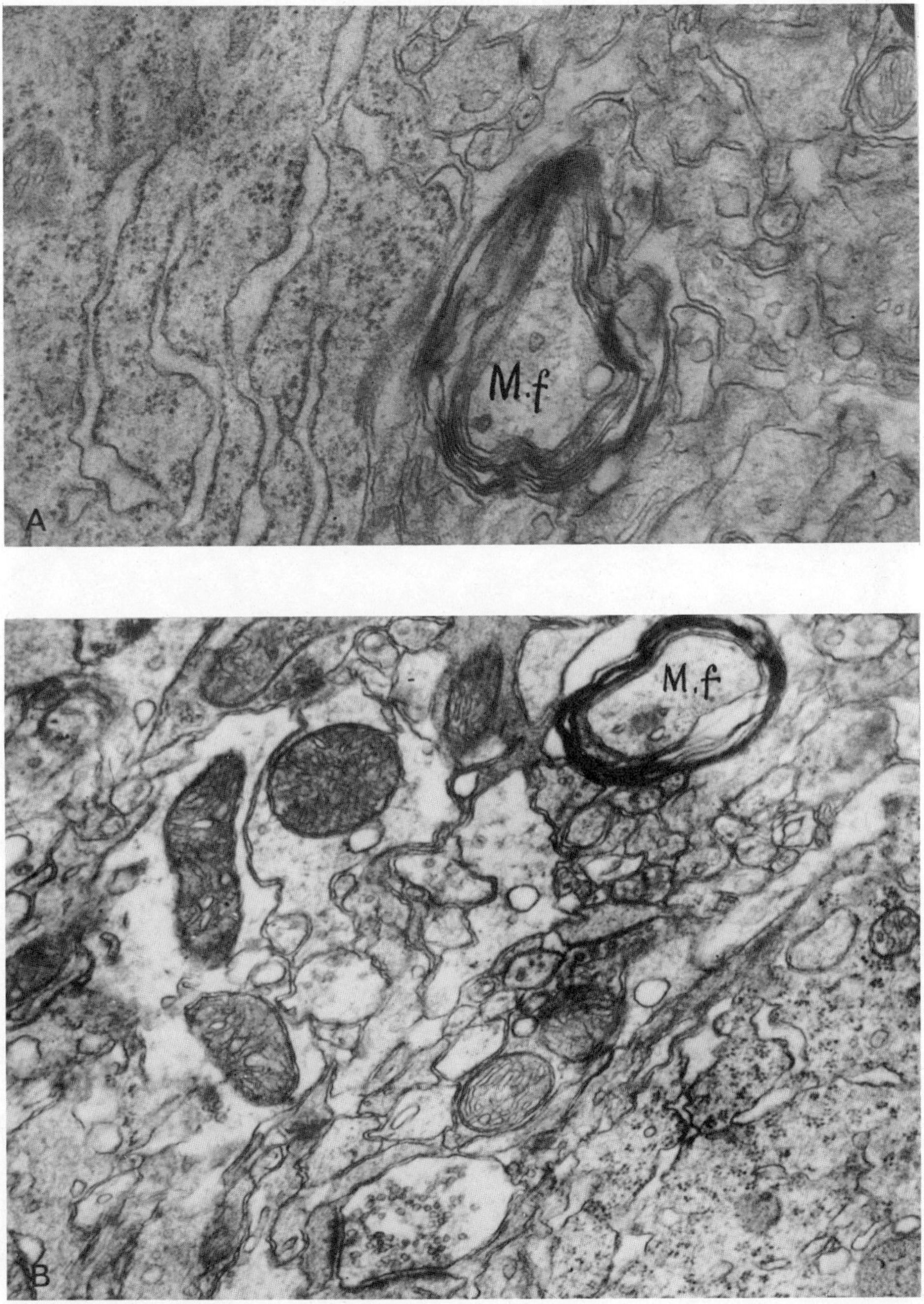

Fig. 5. Electron micrographs of the cerebral cortex of a 24-day-old normal rat. Magnification: ×16,900. Myelinated fibers in close proximity to the body of neurons may be observed. *M.f.*—myelinated fibers.

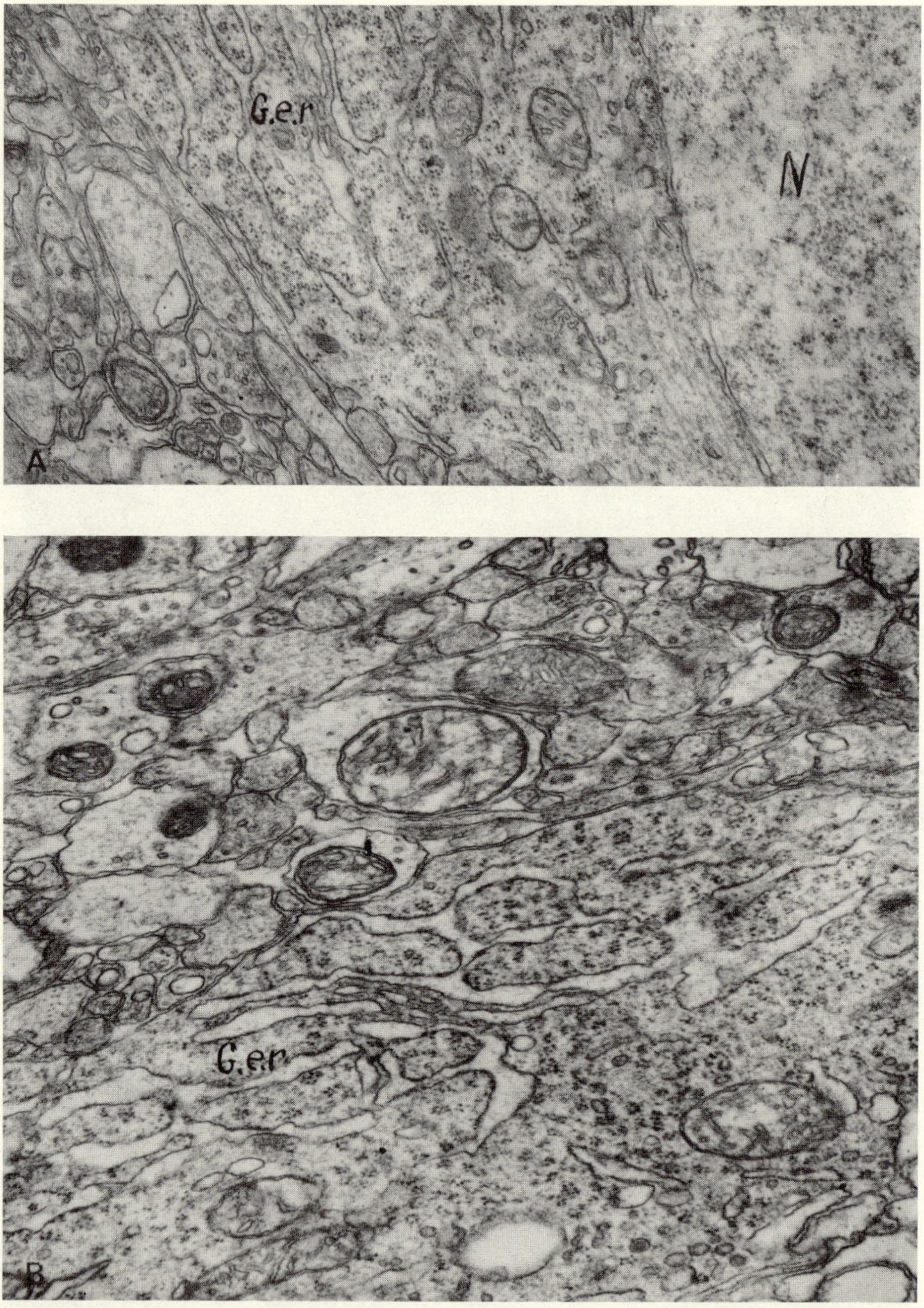

Fig. 6. Electron micrographs of the cerebral cortex of a 24-day-old hypothyroid rat. Magnification: ×9,050. Well-developed granular endoplasmic reticulum can be seen in neurons. Myelinated fibers in close proximity of neurons are not observed. *G.e.r.*—granular endoplasmic reticulum; *N*—nucleus.

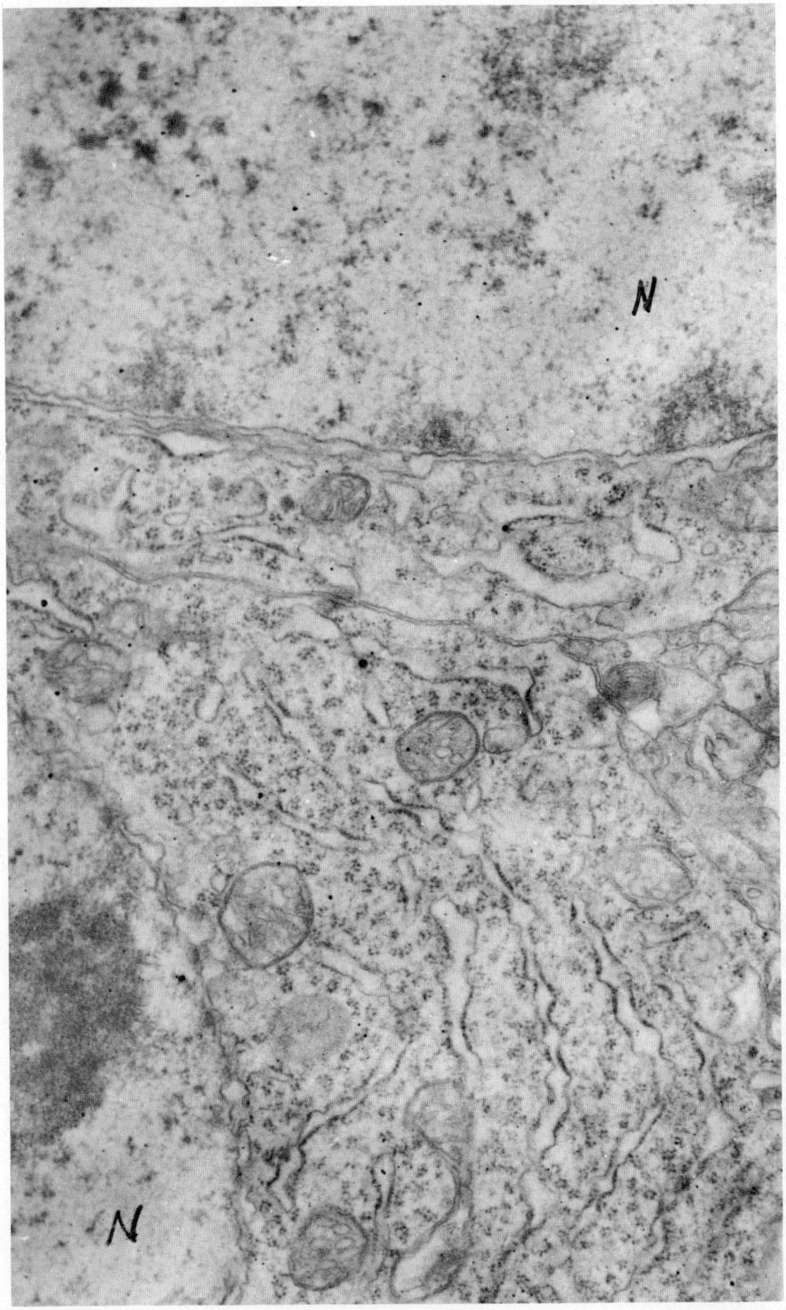

Fig. 7. Electron micrograph showing a cortical neuron of the control rat at 16 days of age. Note irregular distribution of osmiophilic nuclear substance (nuclear substance is aggregated in clumps). *N*—nucleus. Magnification: 10,800 × 3.

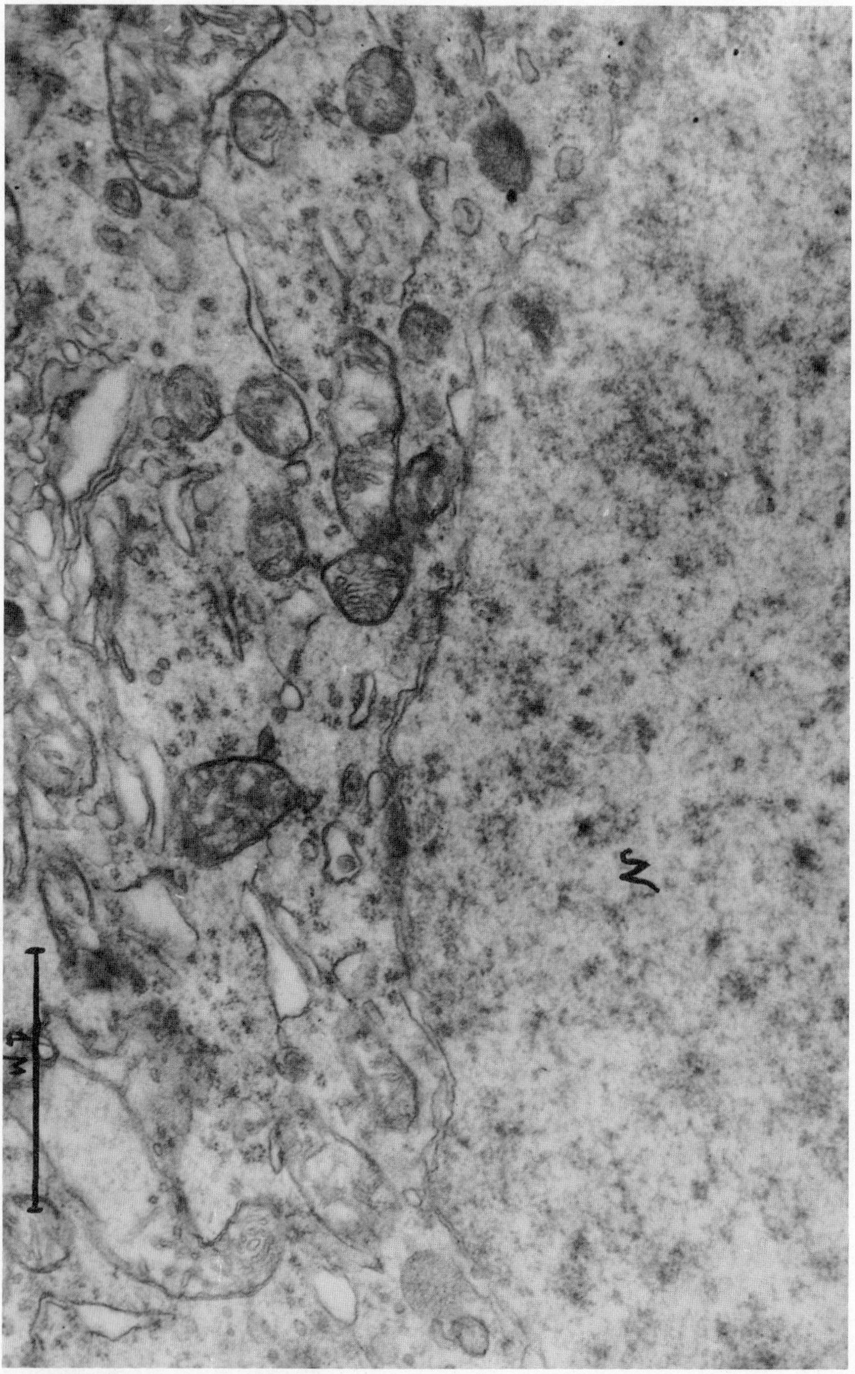

Fig. 8. Electron micrograph showing a cortical neuron of the hypothyroid rat at 16 days of age. Note more homogeneous distribution of osmiophilic nuclear substance than in controls. *n*—nucleus. Magnification: 9,050 × 3.

The cerebral cortex of hypothyroid animals contains fewer axosomatic and axodendritic synapses than does the cortex of normal control animals. Myelinated fibers in the proximity of neurons are not observed in the cortex of hypothyroid rats, whereas they can be found in normal rats (Figs. 3,4).

24-to 25-Day-Old-Animals. In the cerebral cortex of hypothyroid animals, myelinated fibers are scarce and the number of myelin lamellae is less than usual, while in the cerebral cortex of normal animals a great number of myelinated fibers can be found with regular periodical units of myelin present. Axosomatic synapses in hypothyroid animals occur more rarely than in the controls. No differences in the fine structure of the nucleus, nucleolus, and ergastoplasm between neurons of hypothyroid and control rats can be found at this age any more. The Golgi apparatus is well developed both in control and hypothyroid animals. Neuronal mitochondria vary in form, size, and structure to such an extent that it is difficult to reveal any differences that may exist between control and hypothyroid animals (Figs. 5,6).

The nuclei of cortical neurons in hypothyroid 16-day-old animals are characterized by a more homogeneous distribution of osmiophilic nuclear substance as compared with controls, pointing to a delay of neuron differentiation during hypothyroidism. Within the homogeneous nuclear substance, aggregations are present. They are particularly abundant in regions adjoining the inner nuclear membrane. In neurons of control animals of the same age, nuclear substance is aggregated in clumps (Fig. 7 and 8).

DISCUSSION

The following results regarding enzyme activities have been found in the brain during hypothyroidism:

Permanent changes in enzyme activities were observed in cerebral cortex and included a pronounced decrease in SDH and NADPH-diaphorase activities throughout the experimental period (up to 24 to 25 days. Transitory changes in the activities of all the enzymes studied were found in cerebellar cortex. Enzyme activities in the nucleus dentatus and in the nuclei of the medulla oblongata and Pons Varolii were not affected by thyroid deprivation during the maturational period. The different responses to lack of thyroid hormone by different brain areas may be explained by differences in the time of maturation and onset of function of these regions of the CNS. It is well known that thyroidectomy in adult animals does not bring about a change in O_2 consumption by the brain, though it does produce a decrease in the basal metabolic rate(46). Nevertheless, the administration of T_4 or radiothyroidectomy to immature rats leads to an increase or decrease of O_2 consumption, respectively(47,23).

The results of the histochemical demonstration of SDH in hypothyroid rats during the first postnatal weeks are in agreement with biochemical determinations reported by Hamburgh and Flexner(22). Enzyme changes induced by thyroid manipulation have been reported by several investigators(39-42). Our experiments limiting administration of MTU during the first 13 postnatal days demonstrated the reversibility of the goiterous reaction under these experimental conditions (Group II); the thyroid gland recovered quickly and restoration of enzyme activities to normal levels followed within days. These observations are best interpreted by assuming that the first two weeks constitute a critical stage in the development of the CNS of the rat, during which synthetic processes in the cerebral cortex are particularly active.

The use of interferometry enabled us to follow the changes in the dry weight of neurons of hypo- and hyperthyroid rats.

Since dry weight of cells mainly represents protein, the data obtained suggest that hypothyroidism leads to the inhibition of protein synthesis in maturing neurons. These results are in good agreement with those describing inhibition of incorporation of labeled amino acids into proteins(49,50) and decrease in RNA content in nervous tissue(17) of hypothyroid young animals.

The ultrastructural changes in neurons during hypothyroidism revealed by electron microscopy demonstrated a retardation in neuron differentiation in 16-day-old rats that manifested itself in the changes of nuclear and cytoplasmic structure, i.e., more homogeneous distribution of osmiophilic nuclear substance and the retardation of the attachment of cytoplasmic free ribosomes to the membranes of the endoplasmic reticulum. The membranes of endoplasmic sacs in hypothyroid animals were almost completely devoid of ribosomes.

Furthermore, examinations of the cerebral cortex of hypothyroid rats at the electron microscope level revealed a decrease in the number of axodendritic and axosomatic synapses which may be related to the reduction in the functional activity of the CNS, as is implied by electroencephalographic studies in hypothyroid animals(12,51-53).

REFERENCES

1. Levine, S., and Mullins, R. F. Hormonal influences on brain organization in infant rats. Science, 152:1585–1592, 1966.
2. Mott, F. W. The changes in the central nervous system in hypothyroidism. Proc. Roy. Soc. Med., 10:51–59, 1917.
3. Marie, P., Tretiakoff, C., and Stumfer, E. Etude anatomo-pathologique des centres nerveux system (Dans un cas de myxoedeme congenital avec cretinisme). Encephale, 15:601–608, 1920.
4. Lotmar, F. Histopathologische Befunde in Gehirnen von Kongenitalen Myxoedem (Thyreoaplasie). Z. Ges. Neur. Psych., 119:491–513, 1929.
5. ——— Entwicklungsstörungen in der Kleinhirnrinde beim endemischen Kretinismus. Z. Ges. Neur. Psych., 136:412–435, 1931.
6. ——— Histopathologische Befunde in Gehirnen von endemischen Kretinismus, Thyreoaplasie, und Kachexia thyreopriva. Z. Ges. Neur. Psych., 146:1–53, 1933.
7. Benda, C. E. Mongolism and Cretinism: A Study of the Clinical Manifestations and the General Pathology of Pituitary and Thyroid Deficiency. London, Heinemann, 1947.
8. Barrnett, R. J. Some aspects of the experimental cretin-like animal. Thesis Yale Univ. School of Medicine, 1948.
9. ——— Experimental production of cretin-like rats. Yale J. Biol. Med., 22:313–322, 1950.

10. Eayrs, J. T. Hormones and the maturation of the central nervous system with special reference to the rat. Brit. J. Anim. Behav., 1:144–153, 1953.
11. ——— Thyroid hypofunction and the development of the central nervous system. Nature, 172:403–404, 1953a.
12. ——— Influence of the thyroid on the central nervous system. Brit. Med. Bull., 16: 122–127, 1960.
13. ——— Age as a factor determining the severity and reversibility of the effects of thyroid deprivation in the rat. J. Endocr., 22:409–419, 1961.
14. ——— Endocrine influence on cerebral development. Arch. Biol., 75:529–565, 1964.
15. ——— Role of thyroid hormone in the differentiation of the nervous system. *In* Proceedings of the Second International Congress of Endocrinology, London, August 17-22, 1964. Taylor, S., ed. Excerpta Medica Foundation, Part II, p. 779–784, 1965.
16. ——— and Taylor, S. H. The effect of thyroid deficiency induced by methyl thiouracil on the maturation of the central nervous system. J. Anat., 85:350–358, 1951.
17. Geel, S. E. and Timiras, P. S. Influence of neonatal hypo-thyroidism and of thyroxine on the acetylcholinesterase and cholinesterase activities in the developing central nervous system of the rat. Endocrinology, 80:1069–1074, 1967.
18. Hamburgh, M. Enzyme activities in the developing nervous system of normal and hypothyroid rats. Anat. Rec., 121:304, 1955.
19. ——— Evidence for direct effect of thyroid hormone on nerve cell maturation in vitro. Anat. Rec., 151:357, 1965.
20. ——— Evidence for a direct effect of temperature and thyroid hormone on myelinogenesis in vitro. Develop. Biol., 13:15–30, 1966.
21. ——— An analysis of the action of thyroid hormone on development based on in vivo and in vitro studies. Gen. Comp. Endocr., 10:198–213, 1968.
22. ——— and Flexner, L. B. Effect of hypothyroidism and hormone therapy on enzyme activities of the developing cerebral cortex of the rat. J. Neurochem., 1:279–288, 1957.
23. ——— Lynn, E., and Weiss, E. P. Analysis of the influence of thyroid hormone on prenatal and postnatal maturation of the rat. Anat. Rec., 150:147–161, 1964.
24. Horn, G. Thyroid deficiency and inanition: the effects of replacement therapy on the development of the cerebral cortex of young albino rats. Anat. Rec., 121:63–79, 1955.
25. Klee, C. B., and Sokoloff, L. Mitochondrial differences in mature and immature brain: influence on rate of amino acid incorporation into protein and response to thyroxine. J. Neurochem., 11:709–716, 1964.
26. Legrand, J. Maturation du cervelet et deficience thyroidienne dounees chronologiques. Arch. Anat. Micr. Morph. Exp., 52:205–214, 1963.
27. ——— Influence de l'hypothyroidisme sur la maturation du cortex cerebelleux. C.R. Acad. Sci. (Paris), 261:544–547, 1965.
28. Pesetsky, I. The thyroxine-stimulated enlargement of Mauthner's neuron in Anurans. Gen. Comp. Endocr., 2:229–235, 1962.
29. ——— Thyroxine-stimulated oxidative enzyme activity associated with precocious brain maturation in Anurans. A histochemical study. Gen. Comp. Endocr., 5:411–417, 1965.
30. Eljakova, G. V. Development du cortex cerebral du cobaye dans les conditions d'hypothyroidie. Dokl. Akad. Nauk U.R.S.R., 125:229–231, 1959.
31. ——— Modifications du cortex cerebral du Lapin au cours de l'hypothyreos. Dokl. Akad. Nauk U.R.S.R., 129:1154–1157, 1959.
32. Mitskevich, M. S. Glands of internal secretion in the embryonic development of birds and mammals. Published by the Acad. Sci. U.S.S.R. Moscow. English edition published for the Nat. Sci. Found. and Dept. Health, Education and Welfare, U.S.A., by Israel Program for Scientific Translations, 1959.
33. Peters, V. B., and Flexner, L. B. Quantitative morphological studies on the developing cerebral cortex of the fetal guinea-pig. Amer. J. Anat., 86:133–161, 1950.
34. Pearse, A. G. E. Histochemistry. Theoretical and Applied. London, J. A. Churchill, Ltd., 1960.

35. Burstone, E. The Histochemistry of Enzymes. New York, Academic Press, Inc., 1963.
36. Barer, R. Refractometry and interferometry of living cells. J. Opt. Soc. Amer., 47:545–556, 1957.
37. Brodsky, V. J. The cell trophic. Moscow, Publ. House "Nauka" U.S.S.R. Acad. Sci., 1966.
38. Reynolds, E. S. The use of lead citrate of high pH as an electron-opaque stain in electron microscopy. J. Cell Biol., 17:208–212, 1963.
39. Phillips, A. H., and Langdon, R. G. The influence of thyroxine and other hormones on hepatic TPN-cytochrome reductase activity. Biochim. Biophys. Acta, 19:380–382, 1956.
40. Tata, J. R., Ernster, L., Lindberg, O., Arrhenius, E., Pedersen, S., and Hedman, R. The action of thyroid hormones at the cell level. Biochem. J., 86:408–424, 1963.
41. Raymond, M., Trucho R., Kosner, B., and Antissier, N. Influence de divers derives phenoliques iodes sur la monoamineoxydase cerebrale de rat. C.R. Soc. Biol. (Paris), 159:632–637, 1965.
42. Hoffman, W. W., Richert, D. W., Westerfeld, W. W. Effect of thiouracil-type drugs on the α-glycerophosphate dehydrogenase response to thyroxine analogs. Endocrinology, 78:1189–1197, 1966.
43. Savik, Z. F. Activity of some dehydrogenases in the cerebellum of rat during the postnatal ontogenesis. *In* The Structure and Function of Nervous System. Moscow, 1965.
44. Sakharova, A. B. The dynamics of cholinesterase in the cerebellum cortex cells during the ontogenesis of the cat. Cytology (Russia), 8:54–59, 1966.
45. Eayrs, J. T., and Goodhead, G. Postnatal development of the cerebral cortex in the rat. J. Anat., 93:385–402, 1959.
46. Schapiro, S. Metabolic and maturational effects of thyroxine on the infant rat. Endocrinology, 78:527–532, 1966.
47. Reiss, J. M., Reiss, M., and Wyatt, A. Action of thyroid hormones on brain metabolism of newborn rats. Proc. Soc. Exp. Biol. Med., 93:19–22, 1956.
48. Sokoloff, L., and Kaufman, S. Effect of thyroxine on amino acids incorporation into protein. Science, 129:569–570, 1959.
49. ——— and Klee, C. B. The effect of thyroid on protein synthesis in brain and other organs. *In* Endocrines and the Central Nervous System, Levine, R., ed. Assoc. Res. Nerv. Ment. Dis., 43:371–386.
50. Geel, S. E., Valcana, T., and Timiras, P. S. Effect of neonatal hypothyroidism and of thyroxine on L (C^{14}) leucine incorporation in protein in vivo and the relationship to ionic levels in the developing brain of the rat. Brain Res., 4:143–150, 1967.
51. Bertrand, I., Delay, J., and Guillian, J. L'electroencephalogramme dans le myxoedeme. C.R. Soc. Biol. (Paris), 129:395–398, 1938.
52. Clare, M. H., and Bishop, G. H. Dendritic circuits: the properties of cortical paths involving dendrites. Amer. J. Psychiat., 111:818–825, 1955.
53. Bradley, P. B., Eayrs, J. T., and Schmalbach, K. The EEG of normal and hypothyroid rats. Electroenceph. Clin. Neurophysiol., 12:467–477, 1960.

33

Effect of Thyroid Hormones on Ionic Metabolism of the Developing Rat Brain*

Theony Valcana and Paola S. Timiras

Department of Physiology-Anatomy,
University of California,
Berkeley, California

INTRODUCTION

The ionic composition of the rat brain undergoes marked changes during its early postnatal development. Between the 10th and 25th days, a critical period in brain development of the rat, a marked increase in K^+ content in the cerebral cortex is accompanied by a marked decrease in Na^+ and Cl^- contents(1).

Earlier studies in adult rats have shown that thyroid hormones affect brain ionic contents as well as brain excitability(2). More recently, studies in rats thyroidectomized at one day of age have demonstrated an increase in Na^+ content and a decrease in K^+ content in the cerebral cortex of 22-day-old rats—changes which have been interpreted as resulting from an altered Na-K pump consequent to hypothyroidism(3).

Based on the foregoing experiments, studies were conducted in our laboratory to investigate the effects of neonatal hypothyroidism and daily thyroxine treatment on the Na^+ and K^+ contents and their redistribution in the developing brain, and to determine whether these changes are associated with alterations in the Na^+-K^+-activated ATPase, an enzyme system related to the active transport of Na^+ and K^+ (see Skou, 4). In addition, because thyroid hormones are involved in the metabolism of Mg^{++} and Ca^{++} ions(5,6,7) which play an important role in membrane properties(8), excitability(9), ATPase activity in vitro(10) and many other enzyme systems, these ions were also studied. Some of the results of this research have been published elsewhere(11).

*This investigation was supported by USAEC Contract AT(11-1)-34, Project 82.

ELECTROLYTE CHANGES

Ionic contents were determined during early postnatal life in the rat cerebral cortex and cerebellum, two brain tissues representing different maturational states during development. Ionic changes were also followed in plasma. The effects of exogenous thyroxine on hypothyroid and control animals with respect to the above parameters was also studied.

The Na^+ and K^+ changes that occur in the cerebral cortex and cerebellum during development are represented in Figure 1 for control and hypothyroid animals. In both control and hypothyroid animals, K^+ content increases and Na^+ content decreases during development in each tissue. In the hypothyroid animals, however, the Na^+ content was higher and K^+ content lower than controls at all age intervals stu-

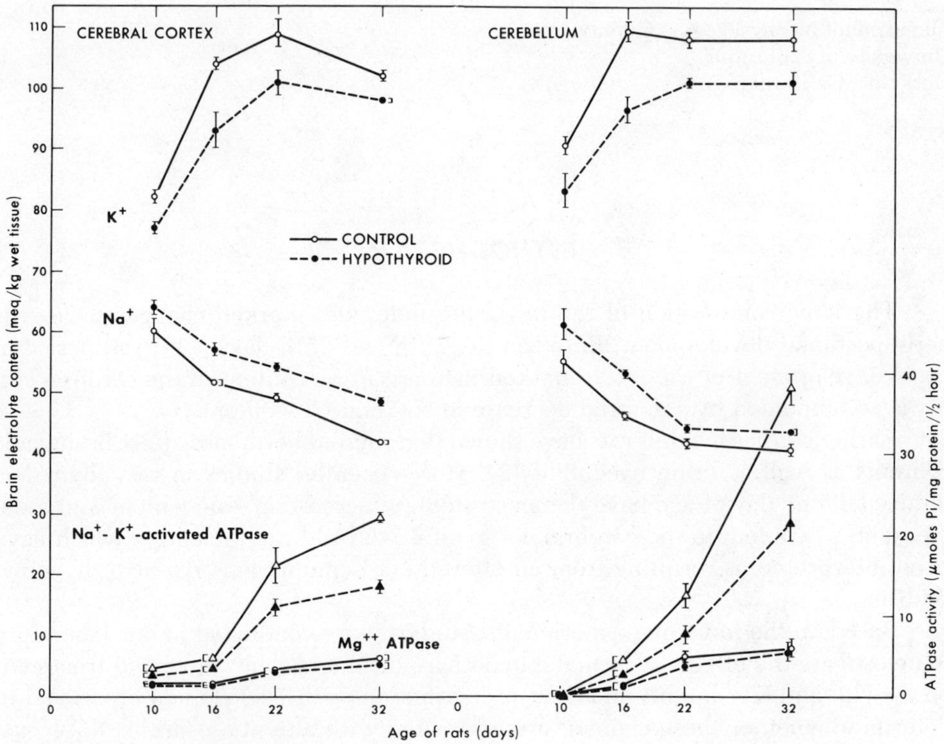

Fig. I. Effects of neonatal hypothyroidism on Na^+ and K^+ contents, Mg^{++} ATPase and Na^+-K^+-activated ATPase activity in the cerebral cortex and cerebellum of the rat during grain development. ATPase was extracted with acid buffered deoxycholate from brain homogenates(44). Mg^{++} ATPase activity was determined in a 2 ml volume of incubation mixture containing 200 μmoles histidine pH 7.2, 6 μmoles tris-ATP, and 6 μmoles Mgn; the incubation mixture for Na - K^+-activated ATPase contained, in addition, 200 μmoles Na^+ and 40 μmoles K^+. Incubation: 30 minutes at 37°C. For ATPase activity, each point represents mean $\pm$ S.E. of duplicate determinations from 16 enzyme preparations and for Na^+ and K^+ contents, each point represents mean $\pm$S.E. of eight or more samples.

died. Together with the increase in Na^+ and decrease in K^+ observed in hypothyroidism, water, and Cl^- contents were also significantly higher and Mg^+ content was lower (Tables 1 and 2). These changes were evident as early as 10 days of age and were sustained throughout the experimental period, although at 32 days of age the increase in H_20 in both cerebral cortex and cerebellum was not statistically significant. Although it has been reported that thyroidectomy in the adult rat decreases brain intracellular Na^+ and increases extracellular Na^+ with no change in K^+ and Cl^- content(12,13), its effects on Na^+, K^+, and Mg^{++} content in muscle are similar to those described herein(14).

The dissimilarities in the ionic changes occurring with hypothyroidism in the adult and developing brain are of interest and could perhaps be related to a differential response of the nervous tissue to this hormone with age as well as to the development of the blood brain barrier to thyroxine.

Ca^{++} levels in the cerebral cortex showed no significant differences between control and hypothyroid animals at any age studied; in the cerebellum, however, a significant decrease was observed in 10-day-old hypothyroid animals.

When hypothroid animals were treated with thyroxine from day 6 until sacrifice at day 22 changes observed in brain electrolytes returned to control values (Tables 1 and 2). Control animals similarly treated with thyroxine showed no significant differences from nontreated controls in ionic contents in either tissue. Yet, inasmuch as the general trend was toward decreased Na^+ content and increased K^+ content, the Na^+, K^+, and H_20 contents were investigated in the cerebellum at 16 days, when this tissue is actively developing. The H_20 and Na^+ contents were found to be significantly lower at this time. K^+ content was increased, but not significantly (Table 2). A relatively high thyroxine dose (40 μg/100 g body weight/day) administered to adult control animals caused an increase in total Na^+ content in the brain (2,13) but was without effect on total K^+ content.

The changes observed in plasma are summarized for all experimental groups in Table 3. Neonatal hypothyroidism induced a decrease in water content and K^+, Mg^{++} and Ca^{++} contents in plasma. These changes were first observed in 22-day-old animals. No significant changes were observed in plasma Na^+ and Cl^- contents at any age interval studied. Again, treatment of hypothyroid rats from day 6 until sacrifice at day 22, returned all observed changes to control values.

Identical thyroxine treatment of control animals produced an increase in plasma water content, a decrease in Na^+, Cl^-, Mg^{++} and Ca^{++} contents, and an increase in K^+ content. The changes observed in H_20, Na^+, and K^+ are in agreement with those reported by Timiras et al.(2), who administered thyroxine and triiodothyronine to adult control rats for a period of 12 days. The role of the adrenal and kidney with respect to ionic changes in plasma must not be overlooked. Adrenal weight, for example, is markedly affected by thyroid activity(15,16) and the plasma and brain changes consequent to thyroxine treatment have been interpreted as resulting partially from adreno-cortical changes(2). In the absence of a plasma Na^+ and Cl^- change with neonatal hypothyroidism (Table 3), suggesting an unaltered extracellular compartment, we may conclude that the plasma changes do not contribute to the alterations observed in brain tissues. Furthermore, ionic changes in the brain precede those in plasma, which are not evident until 22 days of age.

TABLE 1

Effect of neonatal hypothyroidism and thyroxine on water and electrolyte contents in the cerebral cortex of the developing rat[a]

Experimental groups	Age in days	% H_2O	meq/kg wet tissue				
			Cl^-	Na^+	K^+	Mg^{++}	Ca^{++}
Control	10	85.37 ± 0.20	40.40 ± 0.61	59.87 ± 2.32	82.18 ± 0.81	13.05 ± 0.25	3.06 ± 0.19
Hypothyroid[b]	10	86.17 ± 0.18	42.95 ± 0.73	63.95 ± 1.59	77.05 ± 0.56	12.00 ± 0.23	2.96 ± 0.18
		$P < 0.05$	$P < 0.05$	$P = 0.05$	$P < 0.001$	$P < 0.01$	
Control	16	82.24 ± 0.49	38.15 ± 1.96	51.13 ± 0.49	103.92 ± 1.21	13.78 ± 0.11	3.51 ± 0.19
Hypothyroid	16	83.35 ± 0.38	45.39 ± 1.25	56.84 ± 1.08	92.83 ± 2.87	12.67 ± 0.37	3.66 ± 0.29
		$P < 0.05$	$P < 0.01$	$P < 0.001$	$P < 0.02$	$P < 0.01$	
Control	22	81.90 ± 0.06	32.42 ± 0.48	49.33 ± 0.68	108.82 ± 2.39	14.75 ± 0.23	2.87 ± 0.11
Hypothyroid	22	82.80 ± 0.09	34.28 ± 0.99	54.37 ± 0.75	100.71 ± 1.89	13.78 ± 0.22	2.67 ± 0.08
		$P < 0.001$	$P < .05$	$P < 0.001$	$P < 0.01$	$P < 0.02$	
Hypothyroid + Thyroxine[c]	22	81.90 ± 0.09	32.00 ± 0.69	48.11 ± 1.45	109.41 ± 1.25	14.88 ± 0.19	2.85 ± 0.13
Control + Thyroxine	22	81.78 ± 0.25	31.37 ± 1.46	48.83 ± 0.73	110.81 ± 0.84	14.65 ± 0.45	2.56 ± 0.11
Control	32	82.75 ± 0.96	28.02 ± 0.81	42.15 ± 0.35	102.23 ± 0.81	13.31 ± 0.12	3.25 ± 0.45
Hypothyroid	32	82.90 ± 0.60	34.44 ± 0.94	48.39 ± 0.70	97.93 ± 0.26	12.89 ± 0.06	3.11 ± 0.32
			$P < 0.001$	$P < 0.001$	$P < 0.001$	$P < 0.01$	

[a] Na^+, K^+, Mg^{++}, and Ca^{++} were determined spectrophotometrically. Cl^- was determined by the electrometric method of Cotlove et al. (37). Values are mean ± S.E. from 8 or more animals; one sample per animal, 2 determinations per sample. P values refer to significance of differences between hypothyroid and normal controls.

[b] Hypothyroidism was induced at one day of age according to Goldberg and Chaikoff (36).

[c] 10 μg of sodium L-thyroxine/100 g of body weight/day from the 6th day of age until sacrifice.

TABLE 2

Effect of neonatal hypothyroidism and thyroxine on water and electrolyte contents in the cerebellum of the developing rat[a]

Experimental groups	Age in days	% H_2O	meq/kg wet tissue Cl^-	Na^+	K^+	Mg^{++}	Ca^{++}
Control	10	84.78 ± 0.45	37.19 ± 0.71	55.00 ± 1.88	90.53 ± 1.50	15.46 ± 0.36	3.18 ± 0.37
Hypothyroid[b]	10	86.18 ± 0.15	40.32 ± 0.60	61.30 ± 2.36	82.85 ± 3.27	14.01 ± 0.52	2.15 ± 0.18
		$P < 0.05$	$P < 0.01$	$P < 0.01$	$P < 0.01$	$P < 0.05$	$P < 0.05$
Control	16	82.59 ± 0.20	29.53 ± 1.71	45.76 ± 0.62	109.60 ± 1.46	14.87 ± 0.21	3.45 ± 0.18
Hypothyroid	16	83.92 ± 0.25	36.52 ± 2.64	53.37 ± 0.62	95.78 ± 2.31	13.99 ± 0.24	3.18 ± 0.24
		$P < 0.01$	$P < 0.05$	$P < 0.001$	$P < 0.001$	$P < 0.02$	
Control + Thyroxine[c]	16	81.32 ± 0.23	—	42.67 ± 0.42	112.06 ± 0.86	15.14 ± 0.21	3.73 ± 0.33
		$P < 0.01$		$P < 0.001$			
Control	22	81.06 ± 0.15	24.46 ± 1.14	41.43 ± 0.63	107.90 ± 1.21	15.04 ± 0.21	2.85 ± 0.12
Hypothyroid	22	82.07 ± 0.21	26.48 ± 0.63	44.25 ± 0.80	101.06 ± 0.76	14.29 ± 0.19	2.69 ± 0.16
		$P < 0.01$	$P < 0.05$	$P < 0.01$	$P < 0.001$	$P < 0.01$	
Hypothyroid + Thyroxine	22	81.00 ± 0.12	24.12 ± 0.58	41.68 ± 1.04	108.98 ± 1.95	14.51 ± 0.11	3.05 ± 0.21
						$P < 0.05$	
Control + Thyroxine	22	80.81 ± 0.09	24.27 ± 1.23	40.96 ± 0.97	110.54 ± 1.55	15.25 ± 0.14	2.92 ± 0.09
Control	32	82.53 ± 0.42	24.45 ± 0.60	40.44 ± 0.92	108.35 ± 2.11	13.74 ± 0.23	3.18 ± 0.36
Hypothyroid	32	82.68 ± 0.23	26.83 ± 0.80	43.55 ± 0.60	100.93 ± 1.71	13.11 ± 0.20	3.78 ± 0.42
			$P < 0.05$	$P < 0.02$	$P < 0.02$	$P = 0.05$	

[a]Values are means ± S.E. from 8 or more animals; one sample per animal, 2 determinations per sample. P values refer to significance of differences between experimental and control groups.

[b]Hypothyroidism was induced at one day of age according to Goldberg and Chaikoff (36).

[c]10 μg of sodium L-thyroxine/100 g of body weight/day from the 6th day of age until sacrifice.

TABLE 3

Effect of neonatal hypothyroidism and thyroxine on water and electrolyte contents of plasma in the developing rat[a]

Experimental groups	Age in days	% H_2O	meq/kg wet tissue				
			Cl^-	Na^+	K^+	Mg^{++}	Ca^{++}
Control	10	–	96.25 ± 2.48	138.50 ± 2.79	6.39 ± 0.48	1.67 ± 0.05	4.29 ± 0.05
Hypothyroid[b]	10	–	96.06 ± 1.35	134.54 ± 4.88	6.72 ± 0.77	1.80 ± 0.06	4.34 ± 0.10
Control	16	–	102.00 ± 0.39	140.47 ± 2.65	5.49 ± 0.15	1.69 ± 0.02	4.86 ± 0.12
Hypothyroid	16	–	103.51 ± 1.29	139.81 ± 0.75	5.45 ± 0.17	1.61 ± 0.06	4.93 ± 0.13
Control	22	93.66 ± 0.11	107.89 ± 0.78	143.03 ± 0.20	5.85 ± 0.19	1.63 ± 0.05	5.00 ± 0.11
Hypothyroid	22	93.00 ± 0.14	106.95 ± 1.02	142.25 ± 0.67	5.02 ± 0.33	1.38 ± 0.03	4.55 ± 0.01
		$P < 0.001$			$P < 0.05$	$P < 0.001$	$P < 0.01$
Hypothyroid + Thyroxine[c]	22	93.29 ± 0.20	106.75 ± 0.82	142.54 ± 0.62	6.00 ± 0.31	1.71 ± 0.09	5.05 ± 0.18
Control or Thyroxine	22	94.20 ± 0.13	100.23 ± 1.25	136.0 ± 1.91	6.55 ± 0.35	1.41 ± 0.03	4.25 ± 0.13
		$P < 0.01$	$P < 0.05$	$P < 0.01$	$P < 0.01$	$P < 0.01$	$P < 0.01$
Control	32	–	106.17 ± 1.32	145.34 ± 1.08	5.71 ± 0.18	1.76 ± 0.06	5.49 ± 0.08
Hypothyroid	32	–	105.35 ± 1.27	143.19 ± 2.13	4.50 ± 0.20	1.46 ± 0.04	4.93 ± 0.08
					$P < 0.001$	$P < 0.001$	$P < 0.001$

[a] Na^+, K^+, Mg^{++}, and Ca^{++} were determined spectrophotemetrically. Cl^- was determined by the electrometric method of Cotlove et al. (37). Values are mean ± S.E. from 8 or more animals. P values refer to significance of differences between experimental and control groups.

[b] Hypothyroidism was induced at one day of age according to Goldberg and Chaikoff (36).

[c] 10 μg of sodium L-thyroxine/100 g of body weight/day from the 6th day of age until sacrifice.

BRAIN ATPASE CHANGES

The development of Mg^{++}ATPase and Na^+-K^+-activated ATPase extracted with DOC from brain tissues of controls and hypothyroid animals is shown in the lower part of Figure 1. In both cerebral cortex and cerebellum, the specific activity of Mg^{++}ATPase and Na^+-K^+-activated ATPase increases with age, but the latter shows a marked increase during the period of brain development in which the Na^+ and K^+ contents undergo a marked redistribution (Fig. 1). A distinct increase in Na^+-K^+-activated ATPase in the rat brain at this time has also been reported by Samson and Quinn(17). These parallel events are consistent with the role ascribed to this enzyme in brain tissue and in other tissues in which it is found(18,19,20) and attests to the findings of Flexner(21), among others, that enzymes accumulate in anticipation of the ultimate function they serve.

Hypothyroidism resulted in a decrease in the specific activity of Na^+-K^+-activated ATPase in the cerebral cortex and cerebellum but did not appear to significantly affect the specific activity of the Mg^{++} ATPase(Fig. 1). When hypothyroid rats were treated with thyroxine, the enzyme activity reverted to control values in both tissues (Table 4). The enzymatic activity of control animals treated daily with thyroxine (10 μg/100 g body weight) was higher than untreated control animals, but not significantly so (Table 4).

The specific activity of Na^+-K^+-activated ATPase and Mg^{++} ATPase of DOC-treated heavy microsomal fraction isolated from the cerebral cortex of control and hypothyroid rats is presented in Table 5. The Na^+-K^+-activated ATPase decreases with hypothyroidism, whereas Mg^{++} ATPase does not significantly differ between control and hypothyroid animals.

The decrease in Na^+-K^+-activated ATPase from brain homogenate may result both from decreased enzyme synthesis and depressed dendritic branching consequent to hypothyroidism(22). The ATPase prepared by this method represented 30 to 40 percent of the total enzyme present in the brain homogenate. However, the de-

TABLE 4

Effect of neonatal hypothyroidism and thyroxine on ATPase activity in the cerebral cortex of 22-day-old rats[a]

	ATPase activity (μmoles PO_4/mg protein/1/2 hour)	
Experimental groups	Na^+-K^+-Activated MgATPase	MgATPase
Control (9)	27.41 ± 1.54	3.57 ± 0.37
Hypothyroid (6)	18.10 ± 2.74 $P = 0.01$	3.97 ± 1.03
Hypothyroid + Thyroxine (3)	27.56 ± 4.54	4.31 ± 0.55
Control + Thyroxine (9)	30.81 ± 3.81	3.82 ± 0.54

[a]Values are mean ± S.E. from the number of enzyme preparations indicated in parentheses. Enzyme preparation and assay procedures were as described in the legend to Figure 1. *P* value refers to significant difference between hypothyroid and normal control.

TABLE 5

ATPase activity of cerebral microsomes in control and hypothyroid 22-day-old rats[a]

Experimental groups	μmoles phosphate/mg microsomal protein/1/2 hour	
	Na^{+}-K^{+} Mg^{++}ATPase	Mg^{++}ATPase
Control (13)	46.03 ± 1.81	11.70 ± 0.90
Hypothyroid (13)	38.61 ± 1.94	11.25 ± 0.96
	$P < 0.01$	

[a]Values are mean ± S.E. from a number of enzyme preparations shown in parentheses. P value refers to significant difference between hypothyroid and normal control. Tissue was homogenized in 0.1 M histidine, 0.002 M DOC solution, pH 6.5 and was centrifuged at 15,000 g for 10 minutes at 2°C. The supernatant was further centrifuged at 30,000 g for 1 hour and the precipitate was resuspended in 0.1 M histidine, pH 6.5, rehomogenized and centrifuged at 15,000 g for 15 minutes. The supernatant of the final centrifugation constitutes the enzyme preparation. For enzyme assay procedure, see legend to Figure 1.

crease in the specific activity of Na^{+}-K^{+} -activitated ATPase of the heavy microsomal fraction points to a decrease either in the synthesis of this enzyme and/or in its properties of association to the cellular membranes. The decrease observed in Mg^{++} content in the brain of hypothyroid animals perhaps bears a relation to the decreased enzyme activity present in hypothyroidism. Recently, Abrams and Baron (23) have shown that Mg^{++} ions increased the strength of the reattachment of ATPase to *Streptococcus fecalis* membranes from which it had been extracted. It is entirely possible that Mg^{++} alters the integrity of the membrane with respect to this enzyme.

CONCLUSIONS

The $Na^{+}$and K^{+}changes occurring in the cerebral cortex and cerebellum after neonatal hypothyroidism may reflect intracellular changes resulting from a depressed Na-K pump. Such an interpretation is supported by the observation of a decrease occurring concomitantly in the specific activity of Na^{+}-K^{+}-activated ATPase in these tissues. Indirect corroboration is provided by the findings of Matty and Green that thyroxine accelerates active sodium transport across isolated skin and bladder membranes(24,25). On the other hand, the decrease in Mg^{++} in hypothyroid animals may indicate an alteration in membrane permeability to the passive diffusion of Na^{+} and K^{+} down their electrochemical gradients, which would lead to a "leakier" cell. Mg^{++} is implicated in normal monovalent cation permeability, and the stability of the internal molecular arrangement of the cellular membrane(8,26).

Assuming, then, that active Na^{+} and K^{+} transport is depressed and/or that passive permeabilities to $Na^{+}$and $K^{+}$are increased as a consequence of hypothyroidism, the increase observed in Cl^{-}may represent an intracellular change generated to satisfy conditions of electroneutrality. The same phenomena may underlie the parallel increase observed in brain water, inasmuch as both active transport and passive permeabilities regulate total intracellular cation content, and, therefore, water content of cells(27,28).

Many of the biochemical and physiological changes occurring in the developing brain as a result of hypothyroidism could be explained in terms of the effects of thyroxine on inorganic ions and the interaction between ions and enzyme systems. Because a large proportion of energy in the brain is devoted to active transport(29), it would be justifiable to explain the decrease in O_2 consumption(30) in terms of the Na^+, K^+, Mg^{++}, and ATPase changes reported herein. Blond and Whittam(31) have shown that Na^+ and K^+ stimulation of ATPase activity in kidney cortex homogenate leads to a parallel increase in oxygen consumption, which appears to be regulated through the control of the enzyme on ADP concentration.

Another aspect of the study of brain development in hypothyroidism is the role that ionic changes might play on protein synthesis. In vivo studies have shown the incorporation of isotopic leucine into cerebral microsomal protein to be markedly decreased in hypothyroidism(3) and in vivo as well as in vitro studies have shown the protein synthesizing system in the brain to be sensitive to ionic changes(32,33, 34,35). Further, Mg^{++} plays an important role in ribosomal organization and in many enzymatic steps related to protein synthesis (see Petermann, 38). Tata has proposed that the mechanisms governing both the attachment of ribosomes to membranes as well as their turnover are implicated in the growth-promoting effects of thyroid hormones(39).

In view of the fact that Na^+-K^+-activated ATPase has been found to be highly localized in the synaptosomal fraction of the brain(40), its decrease with hypothyroidism, together with the ionic changes, may partially contribute to the alterations in electrophysiological parameters of the developing hypothyroid brain(41,42,43).

REFERENCES

1. Vernadakis, A., and Woodbury, D. M. Electrolyte and amino acid changes in brain during maturation. Amer. J. Physiol., 203:748–752, 1962.
2. Timiras, P. S., Woodbury, D. M., and Agarwall, S. L. Effect of thyroxine and triiodothyronine on brain function and electrolyte distribution in intact and adrenalectomized rats. J. Pharmacol. Exp. Ther., 115:154–171, 1955
3. Geel, S., Valcana, T., and Timiras, P. S. Effect of neonatal hypothyroidism and of thyroxine on L[^{14}C] leucine incorporation in protein *in vivo* and the relationship to ionic levels in the developing brain of the rat. Brain Res. 4:143–150, 1967.
4. Skou, J. C. Enzymatic basis for active transport of Na^+ and K^+ across cell membrane. Physiol. Rev., 45:596–617, 1965.
5. Vitale, J. J., Mark Hegsted, D., Nakamura, M., and Conors, P. The effect of thyroxine on magnesium requirement. J. Biol. Chem., 226:597–601, 1957.
6. Aikawa, J. K. Effect of thyroxine and propylthiouracil on magnesium metabolism in the rabbit. Study with Mg^{28}. Proc. Soc. Exp. Biol. Med., 104:594–497, 1960.
7. Care, A. D. Significance of the thyroid hormones in calcium homeostasis. Fed. Proc., 27:153–155, 1968.
8. Hoffman, J. F. Cation transport and structure in the red-cell plasma membrane. Circulation, 26:1201–1213, 1962.

9. Cerf, J. A. L'action physiologique et pharmacologique des ions alcino-terreux sur les fibres nerveuses. *In* Bacq, Z. M., ed. Ions Alcalino-terreux, Handbuch der experimentellen Pharmakologie. Berlin, Springer, Suppl. 17, Vol. I, pp. 164–285, 1963.
10. Skou, J. C. The influence of some cations on adenosine triphosphatase from peripheral nerves. Biochem. Biophys. Acta, 23:394–401, 1957.
11. Valcana, T., and Timiras, P. S. Effect of hypothyroidism on Na^+-K^+-activated MgATPase and ionic metabolism in the developing brain of the rat. Fed. Proc., 27:470, 1968.
12. Woodbury, D. M. Effect of hormones on brain excitability and electrolytes. Recent Progr. Hormone Res., 10:65–107, 1954.
13. Timiras, P. S., and Woodbury, D. M. Effect of thyroid activity on brain function and brain electrolyte distribution in rats. Endocrinology, 58:181–192, 1956.
14. Pandazi, A. A., Herrington, J. K., Schlueter, D. P. Sodium, potassium and magnesium distribution after thyroidectomy. Fed. Proc., 18:117, 1959.
15. Timiras, P. S., and Woodbury, D. M. Adrenocortical function after administration of thyroxine and triiodothyronine in rats. J. Pharmacol. Exp. Ther., 115:144–153, 1955.
16. Evans, E. S., Rosenberg, L. L., Evans, A. B., and Koneff, A. A. Relative sensitivity of different biological responses to small quantities of thyroxine and triiodothyronine. Endocrinology, 74:770–779, 1964.
17. Sampson, F. E., Jr., and Quinn, D. J. Na^+-K^+-activated ATPase in rat brain development. J. Neurochem., 14:421–427, 1967.
18. Deul, D. H., and McIlwain, H. Activation and inhibition of adenosine triphosphatases of subcellular particles from the brain. J. Neurochem., 8:246–256, 1961.
19. Skou, J. C. Preparation from mammalian brain and kidney of the enzyme system involved in active transport of Na^+ and K^+. Biochem. Biophys. Acta, 58:314–325, 1962.
20. Bonting, S. L., Caravaggio, L. L., and Hawkins, N. M. Studies on sodium-potassium adenosine triphosphatase. IV. Correlation with cation transport sensitive to cardiac glycosides. Arch. Biochem. Biophys., 98:413–419, 1962.
21. Flexner, L. B. Enzymatic and functional patterns of the developing mammalian brain. *In* Waelsch, H., ed. Biochemistry of the Developing Nervous System, New York, Academic Press, Inc., pp. 281–300, 1955.
22. Eayrs, J. T. The cerebral cortex of normal and hypothryroid rats. Acta Anat., 25:160–183, 1955.
23. Abrams, A., and Baron, D. Reversible attachment of adenosine triphosphatase to streptococcal membranes and the effect of magnesium ions. Biochemistry, 7:501–507, 1968.
24. Matty, A. J., and Green, K. Active sodium transport in response to thyroxine. Life Sci., 9:487–489, 1962.
25. Green, K., and Matty, A. J. The effect of thyroid hormones on water permeability of the isolated bladder of the toad *Bufo-Bufo*. J. Endocr., 28:205–211, 1964.
26. Hoffman, J. F., and Tosteson, D. C. Magnesium—an internal cohesive for the plasma membrane of erythrocytes. Abstr. Buophys. Soc., Pittsburgh, R7, 1959.
27. Tosteson, D. C. Regulation of cell volume by sodium and potassium transport. *In* Hoffman, J. F., ed. The Cellular Functions of Membrane Transport, Englewood Cliffs, New Jersey, Prentice-Hall, Inc., pp. 3–22, 1964.
28. Robinson, J. D. Regulating ion pumps to control cell volume. J. Theor. Biol., 19:90–96, 1968.
29. Whittam, R., and Blond, D. M. Respiratory control by an adenosine triphosphatase involved in active transport in brain cortex. Biochem. J., 92:147–158, 1964.
30. Hamburgh, M. An analysis of the action of thyroid hormone on development based on *in vivo* and *in vitro* studies. Gen. Comp. Endocr., 10:198–213, 1968.
31. Blond, D. M., and Whittam, R. Effects of sodium and potassium ions on oxidative phosphorylation in relation to respiratory control by a cell-membrane adenosine triphosphatase. Biochem. J., 97:523–531, 1965.
32. Mase, K., Takahashi, Y., and Ogata, K. The incorporation of [^{14}C] glycine into the protein of guinea pig brain cortex slices. J. Neurochem., 9:281–288, 1962.
33. Merei, F. T., and Gallyas, F. Protein metabolism of the CNS at high plasma sodium

level studied with [^{35}S] methionine. J. Neurochem., 11:265–270, 1964.
34. Lahiri, S., and Lajtha, A. Cerebral amino acid transport *in vitro*. I. Some requirements and properties of uptake. J. Neurochem., 11:77–86, 1964.
35. Roberts, S., and Zomzely, C. E. Regulation of protein synthesis in the brain. *In* Peeters, H., ed. Protides of the Biological Fluids. Amsterdam, Elsevier, pp. 91–102, 1966.
36. Goldberg, R. C., and Chaikoff, I. L. A simplified procedure for thyroidectomy of the newborn rat without concomitant parathyroidectomy. Endocrinology, 45:64–70, 1949.
37. Cotlove, E., Trantham, H. V., and Bowman, K. L. An instrument and method for automatic, rapid, accurate, and sensitive titration of chloride in biologic samples. J. Lab. Clin. Med., 51:461–468, 1958.
38. Petermann, M. L. The Physical and Chemical Properties of Ribosomes. Amsterdam, Elsevier, 1964.
39. Tata, J. R. The formation and distribution of ribosomes during hormone-induced growth and development. Biochem. J., 104:1–16, 1967.
40. Abdel-Latif, A. A., Brody, J., and Ramahi, A. Studies on sodium-potassium adenosine triphosphatase of the nerve endings and appearance of electrical activity in the developing rat brain. J. Neurochem., 14:1133–1141, 1967.
41. Bradley, P. B., Eayrs, J. T., and Schmalbach, K. The Electroencephalogram of normal and hypothyroid rats. Electroenceph. Clin. Neurophysiol., 12:467–477, 1960.
42. ——— Eayrs, J. T., Glass, A., and Heath, R. W. The maturational and metabolic consequences of neonatal thyroidectomy upon the recruiting response in the rat. Electroenceph. Clin. Neurophysiol., 13:577–586, 1961.
43. Hatotani, N., and Timiras, P. S. Influence of thyroid function on the postnatal development of the transcallosal response in the rat. Neuroendocrinology, 2:147–156, 1967.
44. Hayashi, M., and Auditore, J. V. Simple preparation and properties of a Na^+-K^+-activated MgATPase from mouse brain. J. Neurochem., 11:671–677, 1964.

34

The Role of Thyroid Hormone in the Fetal and Postnatal Development of Mammals

N. B. Myant

Medical Research Council, Lipid Metabolism
Unit, Hammersmith Hospital,
London, England

INTRODUCTION

Thyroid hormone is required for normal growth and development of immature mammals and seems to be concerned particularly in the maturation of certain organ systems, notably the central nervous system, the skeleton, and the reproductive system. I propose to describe very briefly what we know about the requirement for thyroid hormone in the fetal and postnatal development of mammals, and then to discuss the possible mechanisms that may underlie the effects of thyroid hormone on mammalian development.

THE THYROID AND FETAL DEVELOPMENT

The problem as to how far thyroid hormone is required for normal fetal development is complicated by the fact that the fetus receives a supply of thyroid hormone from two sources. At some stage before birth (the precise stage depending on the species), the fetal thyroid begins to secrete hormone. However, from an even earlier stage the fetal tissues also receive a supply of maternal hormone by transport across the placenta. For this reason, failure to demonstrate defective development of a fetal tissue by removing the fetal thyroid in utero does not prove that development of the tissue in question is independent of the presence of thyroid hormone. Nevertheless, experiments in which an antithyroid drug is fed to pregnant animals, so that both fetal and maternal thyroid are inhibited, suggest that in most species thyroid hormone plays little part in the early stages of fetal development, but that it may be required towards the end of fetal life in some species.

Thiouracil given to the pregnant rat causes a slight delay in the appearance of centers of ossification in the fetus(1), but this is best regarded as one aspect of a general retardation of growth rather than of differentiation, since the degree of ossification at a given body weight of the fetus is not affected by the treatment(2). Removal of the head and thyroid of rabbit fetuses has no effect on their growth or general development(3), nor does treatment of pregnant rabbits with methylthiouracil have any significant effect on the weight or phospholipid content of the fetal brains(4).

Thyroid hormone seems to be required for some of the later stages of fetal development in man, since congenital aplasia of the thyroid leads to delayed ossification of those bones which begin to ossify shortly before birth(5). This last observation shows not only that a supply of thyroid hormone is required for normal fetal development of the skeleton in man; it also shows that the hormone secreted by the fetal thyroid cannot, in this species, be replaced by hormone normally supplied by the mother.

There is also some inconclusive evidence that the human brain may be damaged by prenatal thyroid deficiency due either to congenital aplasia of the thyroid or to destruction of the fetal thyroid by radioiodine given to the mother. It would not be surprising if the development of the human brain does require thyroid hormone in utero, since myelination of the central nervous system begins several weeks before birth in man. However, cretinism is seldom diagnosed within the first few weeks of life. When, therefore, symptoms suggestive of cerebral damage due to congenital thyroid deficiency are found in infants, it is impossible to tell whether the symptoms arose at birth or in the early weeks of postnatal life. Russell et al.(6) have described an infant whose thyroid was destroyed by radioiodine in utero and who died aged 11 weeks. Autopsy showed "some loss of nerve cells in the superficial layer" of the cerebral cortex. Since this infant was treated with thyroxine from the age of 17 days, the damage to the brain presumably occurred either in utero or during the first few days of postnatal life. But there is no way of distinguishing between these two possibilities. Unfortunately, it is not possible to assess in vivo the physical or functional state of the central nervous system in the newborn infant, in the way that the degree of development of the skeleton can be assessed.

The question whether the human brain can be damaged by thyroid deficiency in utero is of more than academic interest. If irreversible damage can be caused by thyroid deficiency before birth, then treatment begun in postnatal life, however soon after birth, can never be fully effective in infants who have lacked thyroid hormone before birth. The only hope of achieving success would be to give the mother massive doses of thyroid hormone during pregnancy, so as to maintain an adequate level of hormone in the fetal tissues. This, of course, is feasible only if there is presumptive evidence for fetal thyroid deficiency, as in pregnant women who have previously born athyreotic infants or who have been given large doses of radioiodine during their pregnancy(7).

These species differences in the response to deficiency of thyroid hormone in utero would be expected if the hormone is necessary only at a rather late stage in mammalian development. In rats and rabbits, the two species in which it has been difficult to demonstrate any effect of thyroid deficiency in fetal life, the young are born in an immature state and so may leave the uterus before reaching the stage of development for which thyroid hormone is required. In man, on the other hand, the young are born in a more mature state, and the fetus may therefore remain in utero long enough to become susceptible to thyroid hormone deficiency.

Since this is a conference on the role of hormones in development it may be appropriate to point out that by far the greater part of fetal development appears to take place without the need for thyroid hormone. In some species, it is true, there may be a partial requirement for thyroid hormone towards the end of fetal development. But at no stage in the life of the fetus does thyroid hormone play a decisive role comparable, say, with that of sex hormones in determining the development of the fetal sex apparatus (see 8). It is also worth noting that thyroid hormone seems to play no part in controlling the stages of development in mammals that are homologous with the changes that take place when a tadpole turns into a frog. But, as I have suggested elsewhere(9), amphibian metamorphosis may have arisen as a special adaptation in which differentiation is suppressed during a period of rapid growth and is then released under the influence of thyroxine. In its effects on the tissues of amphibian larvae, the thyroid gland may perhaps be regarded as an inductor, in the sense used by embryologists, but one whose evocator has unusually widespread effects and which is secreted at a distance from its site of action.

THE THYROID AND POSTNATAL DEVELOPMENT

General Effects

The marked delay in growth and maturation that takes place in cretinism has been the starting point for a great deal of work on the role of the thyroid in the postnatal development of experimental animals. Removal of the thyroid from newborn animals leads to retardation of growth and of general maturation. In rats thyroidectomized at birth, for example, there is a delay in the change in the shape of the skull from the immature to the adult form, the eyes do not open at the normal age, and the fur fails to acquire the qualities characteristic of the adult. By contrast, thyroid hormone injected into newborn rats accelerates general maturation, as was shown by Hoskins(10) more than 40 years ago.

Central Nervous System

Since defects of mental capacity and behavior are among the most striking abnormalities in cretinism, much of the experimental work in this field has been concerned with the developing central nervous system. Eayrs(11) has shown that if rats are made thyroid-deficient at birth, the central nervous system fails to develop normally. The cell bodies and axons of the cortical neurons are abnormally small and have few dendrites, so that the network of dendritic connections in the intercellular spaces of the grey matter is less dense than normal. There is also a reduction in the number of blood capillaries in the intercellular spaces of the cerebral cortex. Legrand et al.(12) have described histological effects of a rather different kind in the cerebellar cortex of thyroid-deficient newborn rats. Normally, the granular layer on the surface of the rat's cerebellum disappears by the 21st day after birth. If, however, a rat is thyroidectomized before the 14th day, the granular layer persists beyond 21 days.

Both the abnormality in the cerebral cortex described by Eayrs(11) and the abnormal persistence of the granular layer of the cerebellum can be reversed by treating the thyroid-deficient rat with thyroxine, provided that the treatment is begun soon enough. If the treatment is delayed beyond a certain critical period (about 24 days in the case of the cerebral cortical lesions and 14 days in the case of the cerebellar abnormality), the abnormality is not corrected. This partial irreversibility is in keeping with the current view that what underlies differentiation is a continuously changing pattern of gene expression, in the course of which some genes may be active only for a limited period. If, as seems likely, the genes controlling the synthesis of some constituents of the central nervous system belong to this class, having only transient activity, then there would be a critical period beyond which damage sustained during development could not be repaired.

Attempts to find plausible biochemical counterparts to the effects of thyroid deficiency on the histological structure of the brain have not been very successful. Hamburgh and Flexner(13) have shown that destruction of the rat's thyroid at birth prevents the increase in the activity of cerebral succinic dehydrogenase that normally begins on about the 10th day. Myant and Cole(14), on the other hand, found that thyroxine injected into immature rats, during the period of active myelination, had no significant effect on the ratio of myelin to nonmyelin phospholipids in the whole brain, though there was an increase in the total amount of brain phospholipid. Myant and Cole(14) also failed to find any effect of thyroxine on the relative rates of incorporation of ^{32}P into each of four classes of phospholipid in the brain. A study of the influence of thyroid hormone on the time of appearance of the enzymes concerned specifically in the synthesis of myelin lipids in the brain might well be rewarding. I shall return to this point in a moment.

Behavior

The effects of thyroid deficiency on the brains of immature rats are accompanied by defects of behavior and mental ability, similar in general to those characteristic of cretinism. Young rats that have been thyroidectomized at birth give a general impression of sluggishness. They show little of the physical activity shown by their normal littermates and they lack the exploratory behavior that begins after weaning. When tested more objectively, their innate reactions to various stimuli are found to be delayed and they perform poorly in tests of learning ability.

IS THYROID HORMONE A DIFFERENTIATING AGENT IN MAMMALS?

From this partial survey it will be clear that thyroid hormone is in some way necessary for the completion of certain stages in the maturation of mammals. What is not so clear is the extent to which, in promoting or permitting normal development, it is acting specifically as a differentiating agent.

This is a difficult question to answer because in mammals the differentiation that takes place during development is always accompanied by growth, and there is no stage in the life of a mammal comparable with the rapid metamorphosis of amphibians and insects that enables us to observe differentiation without growth. As is well known, undernutrition in early postnatal life may cause irreversible development defects, especially in the central nervous system. Deficiency of thyroid hormone, through its general effects on metabolism, could act in an analogous way by interfering with the nutrition of developing tissues. This could lead either to a delay in maturation as a whole, or it might cause localized developmental defects. Indeed, Eayrs(15) has suggested that the failure of the neurons of the cerebral cortex to develop normally in thyroid-deficient rats may be due in part to an inadequate supply of blood capillaries in the grey matter. No doubt, some of the effects of thyroid deficiency on the differentiation of mammalian tissues are of this kind. The study of such effects is unlikely to give us any very deep insight into the mechanisms underlying differentiation. However, the positive influence of thyroid hormone in bringing about the precocious maturity of immature mammals points much more clearly to a true differentiating action. Equally relevant are the experiments of Hamburgh(16), showing that thyroxine stimulates the formation of myelin in cultures of cerebellum from newborn rats and mice. It seems to me that we may gain more valuable information by studying the biochemical mechanisms underlying these positive effects of thyroid hormone on mammalian development than by studying the effects of thyroid deficiency.

Differentiation and Enzyme Induction

This brings me to the question of enzyme induction (and here I use "induction" in the biochemical sense, and not in the sense in which the word is used by embryologists). The fact that thyroxine leads to induction of various adaptive enzymes in the liver is not necessarily relevant to the question as to how far it is a differentiating hormone. Phenobarbital also induces the formation of microsomal enzymes in the liver, but we should not for that reason regard it as a differentiating agent. The formation of new proteins due to activation of previously quiescent genes is almost certainly a necessary condition for differentiation, but there is more to differentiation than this. In the first place, differentiation of the kind we are considering here is to a large extent irreversible, whereas the formation of adaptive enzymes is completely reversible and is unstable in the sense that it requires the continued presence of the inducing agent. You cannot turn a frog into a tadpole by removing its thyroid, but the gluconeogenic enzymes induced by cortisone become repressed when the cortisone is withdrawn. Differentiation and enzyme induction also differ as to the nature of their specificity. If we are right to regard differentiation as resulting from a changing pattern of gene expression, then there must be a specific pattern appropriate for each stage in the development of every tissue. A natural differentiating agent would then be a substance which is formed in the body and which evokes a more or less stable pattern of gene expression, and hence of enzyme synthesis, characteristic of a particular stage in the normal development of any tissue. An inducing agent, on the other hand, brings about the reversible formation of one or more enzymes related to a particular metabolic requirement.

I realize that I am on controversial ground here, and that reversible enzyme induction, and the irreversible unfolding of a pattern of protein synthesis that we see during differentiation, may turn out to be fundamentally similar. Certainly, one must admit that the irreversibility of differentiation is not absolute, since a given pattern of protein synthesis that arises during differentiation may subsequently disappear. In the meantime, I think the distinction is worth making. If we accept this point of view, then it would be relevant to our discussion to ask the question: "Does thyroxine evoke, in the immature mammal, specific and more or less stable patterns of enzyme synthesis that lie on our path towards greater differentiation?" It would be interesting to know, for example, whether thyroxine can evoke the formation of enzymes in the brain that are required specifically for the synthesis of myelin, at a time before myelination normally begins. The enzymes concerned in the formation of the sphingosine moiety of sphingomyelin would be particularly interesting to study from this point of view. Experiments of this kind would probably give more useful information about the way in which thyroid hormone influences mammalian development than would experiments to test the influence of thyroid hormone on the synthesis of a random selection of enzymes, or on the synthesis of proteins in general.

Approaching the problem along these lines, Greengard and Dewey(17) have shown that thyroxine injected into rat fetuses evokes a pattern of enzyme activity in the liver that does not normally appear until some time after birth. This suggests a differentiating effect of the kind I have tried to define. However, it is not clear from these experiments whether this effect of thyroxine is irreversible, as in the evocation of urea-cycle enzymes in the livers of thyroxine-treated tadpoles, or whether the increased activity of glucose 6-phosphatase and NADPH-cytochrome c reductase, that Greengard and Dewey have described, depends on the continued presence of the injected hormone and is therefore a manifestation of inducibility. Or should we perhaps recognize that differentiation in mammals consists in the appearance not only of new patterns of protein synthesis, but also of new patterns of inducibility?

REFERENCES

1. Weiss, R. M., and Noback, C. R. The effects of thyroxin and thiouracil on the time of appearance of ossification centres of rat fetuses. Endocrinology, 45:389–395, 1949.
2. Jost, A., Moreau, G., and Fournier, C. Date et ordre d'apparition des premiers centres d'ossification chez le foetus de rat normal ou soumis au propylthiouracile. Arch. Anat. Micr. Morph. Exp., 49:431–457, 1960.
3. ——— Sur le développement de la thyroïde chez le foetus de lapin décapité. Arch. Anat. Micr. Morph. Exp., 42:168–183, 1953.
4. Cuarón, A., Gamble, J., Myant, N. B., and Osorio, C. The effect of thyroid deficiency on the growth of the brain and on the deposition of brain phospholipids in foetal and new-born rabbits. J. Physiol. (London), 168:613–630, 1963.
5. Wilkins, L. The Diagnosis and Treatment of Endocrine Disorders in Childhood and Adolescence, 2nd ed., Springfield, Illinois, Charles C Thomas, Pub., 1957.
6. Russell, K. P., Rose, H., and Starr, P. The effects of radioactive iodine on maternal and fetal thyroid function during pregnancy. Surg. Gynec. Obstet., 104:560–564, 1957.

7. Carr, E. A., Jr., Beierwaltes, W. H., Raman, G., Dodson, V. N., Tanton, J., Betts, J. S., and Stambaugh, R. A.Effect of maternal thyroid function on fetal thyroid function and development. J. Clin. Endocr., 19:1–18, 1959.
8. Jost, A. The role of fetal hormones in prenatal development. Harvey Lect., 55:201–226, 1961.
9. Myant, N. On the possible rôle of the thyroid in the control of the development of the mammalian brain. Biol. Neonat., 9:148–165, 1966.
10. Hoskins, M. M. The effect of acetyl thyroxin on the newborn white rat. J. Exp. Zool., 48:373–394, 1927.
11. Eayrs, J. T. The influence of the thyroid on the central nervous system. Brit. Med. Bull., 16:122–127, 1960.
12. Legrand, J., Kriegel, A., and Jost, A. Déficience thyroïdienne et maturation du cervelet chez le rat blanc. Arch. Anat. Micr. Morph. Exp., 50:507–519, 1961.
13. Hamburgh, M., and Flexner, L. B. Biochemical and physiological differentiation during morphogenesis. XXI. Effect of hypothyroidism and hormone therapy on enzyme activities of the developing cerebral cortex of the rat. J. Neurochem., 1:279–288, 1957.
14. Myant, N. B., and Cole, L. A. Effect of thyroxine on the deposition of phospholipids in the brain *in vivo* and on the synthesis of phospholipids by brain slices. J. Neurochem., 13:1299–1307, 1966.
15. Eayrs, J. T. Effects of thyroid hormones on brain differentiation. *In* Brain-Thyroid Relationships with Special Reference to Thyroid Disorders, Ciba Foundation Study Group No. 18, Cameron, M. P., and O'Connor, M., eds. London, J. & A. Churchill, pp. 60–71, 1964.
16. Hamburgh, M. Evidence for a direct effect of temperature and thyroid hormone on myelinogenesis *in vitro*. Develop. Biol., 13:15–30, 1966.
17. Greengard, O., and Dewey, H. K. The developmental formation of liver glucose 6-phosphatase and reduced nicotinamide adenine dinucleotide phosphate dehydrogenase in fetal rats treated with thyroxine. J. Biol. Chem., 243:2745–2749, 1968.

35

Observations on Thyroxine Treatment in Relation to the Thyroid Gland and Skeletal Development in Alevins of the Brown Trout (*Salmo trutta* L.)

E. J. W. Barrington[1] and B. B. Rawdon[2]

[1]Department of Zoology, The University of Nottingham, Nottingham, England
[2]Department of Animal Biology, University of Natal, Durban, Natal

INTRODUCTION

The involvement of the thyroid hormones in the regulation of growth and metamorphosis in amphibian larvae is very well known, but it is far from clear whether these hormones are similarly involved in comparable processes in fish. Metamorphosis is not uncommon in teleosts(1), and there is some presumption that endocrine regulation is involved in this, but suggestions that the thyroid hormones may be specifically concerned are still based on very tenuous evidence. It is certain, however, that exogenous thyroxine has morphogenetic effects upon these animals. Woodhead(2) reared alevins of brown trout in 1×10^{-6} thyroxine, and found that this treatment stimulated the differentiation of the fins, caused thickening of the epidermis and dermis, and produced abnormal growth of the head, with an increase in the interorbital connective tissue. However, the head length, length of the upper jaw, and height of the median fins were reduced in the treated fish, and there was some reduction in body length, mainly at the caudal end. Woodhead concluded that thyroxine had a strong effect in speeding up rates of development, but that the organs showed different degrees of response.

These findings, being based largely upon gross morphometric observations, do not lend themselves to close analysis of the effect of the hormone. Because of this, and because we were finding evidence of an influence of thyroxine upon the development of the skeleton of yearling trout(3), it seemed worthwhile to examine how far effects on skeletal development could be identified in thyroxine-treated alevins.

MATERIALS AND METHODS

Eyed ova of *Salmo trutta* provided by a local trout hatchery were reared on gauze trays. In order to ensure as much uniformity of response as possible, experimental groups were made up of alevins that had hatched from the same batch of eggs, and within a period of 24 hours. They were transferred to glass aquaria immediately after hatching, and were maintained in well-aerated water which was changed every third day. These aquaria were kept under continuous low lighting, at a temperature of $8.0° \pm 1°C$. Experiments were timed to last for not more than 31 days after hatching, because during this period the animals live on the contents of the yolk sacs. They do not require feeding, so that complications resulting from differences in appetite between groups of experimental fish were avoided. Sodium l-thyroxine was added to the water in a concentration of 1×10^{-7}, and thiourea at a concentration of 0.03 percent. Animals were fixed in Bouin's fluid, and sectioned at 8 μ. Slides were usually stained by the Azan procedure or by the periodic acid-Schiff (PAS) procedure. Radiochromatography of thyroidal extracts prepared from the whole lower jaws of alevins that had been immersed for 5 days in tap water containing 200 μc of ^{131}I per liter, was carried out by the resin column procedure described by Barrington and Thorpe(4). For determination of total bound ^{131}I in the samples, the activities of aliquots of the various column eluates were measured by counting with a Geiger-Muller end-window detector (7 mg/cm^2). Procedure for autoradiography was essentially as described by Pearse(5). Statistical significance was determined by Student's *t* test for small samples.

OBSERVATIONS

Lachrymal Bone

Our first observations have been particularly concentrated upon the lachrymal bone, partly because of its clear morphological delineation during early development, and also because it does not begin to develop until after hatching, during the period of hormone treatment.

Stages in the development of this bone are illustrated in Figures 1 to 5. They conform with Devillers'(6,7) account of the development of dermal bones in *Salmo irideus* and *S. trutta* and need only be briefly described. The lachrymal bone develops in *S. trutta* in close association with the anterior part of the infraorbital lateral line, just anterior and ventral to the eye. In newly-hatched alevins, the position that it will occupy can already be defined with reference to certain neuromasts. Using these as a landmark, it can be seen that at hatching there is a layer of undifferentiated mesenchymal cells closely applied to the basement membrane of the epidermis in the region to be occupied later by the bone. By 10 days after hatching, these cells have given rise to a small blastema; this continues to grow (Fig. 2) and soon begins to secrete a material which, following Lorch, we shall call preosseus matrix (Fig. 3). It stains brightly with aniline blue, and also reacts positively to

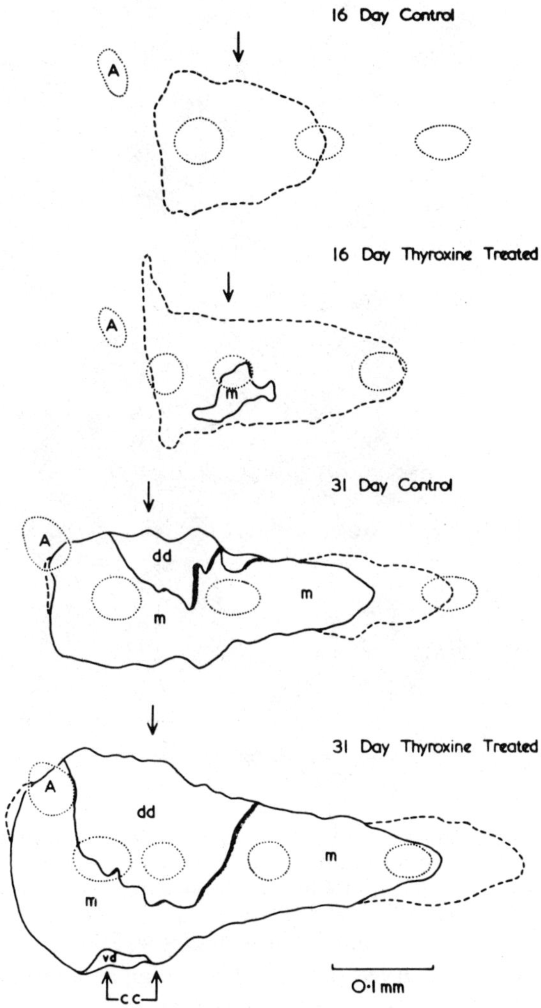

Fig. 1. Surface views of the lachrymal rudiment in typical 16 and 31-day control and thyroxine-treated alevins, prepared by graphic reconstruction from serial transverse sections. Neuromasts are outlined by dotted lines, blastemata (where not masked by "bone") by broken lines, and "bone" (i.e., preosseus matrix and calcified tissue) by continuous lines. *A*, anterior neuromast of the series; *cc*, limits of closed canal; *dd*, dorsal *dermo* element; *m*, *membrano* element; *vd*, ventral *dermo* element.

alcian blue, toluidine blue, and the PAS procedure, so that it is clearly rich in carbohydrate.

With continued development the blastema enlarges, while the preosseus matrix increases and is moulded into a flat lamella lying close beneath the epidermis. This will form the main plate of the bone, or, in Devillers' terminology, the *membrano* component. At this stage it is ossifying, although it is not always possible to distinguish with certainty between the ossified bone and the unossified matrix. With the Azan technique, however, the bone tends to stain red or purplish.

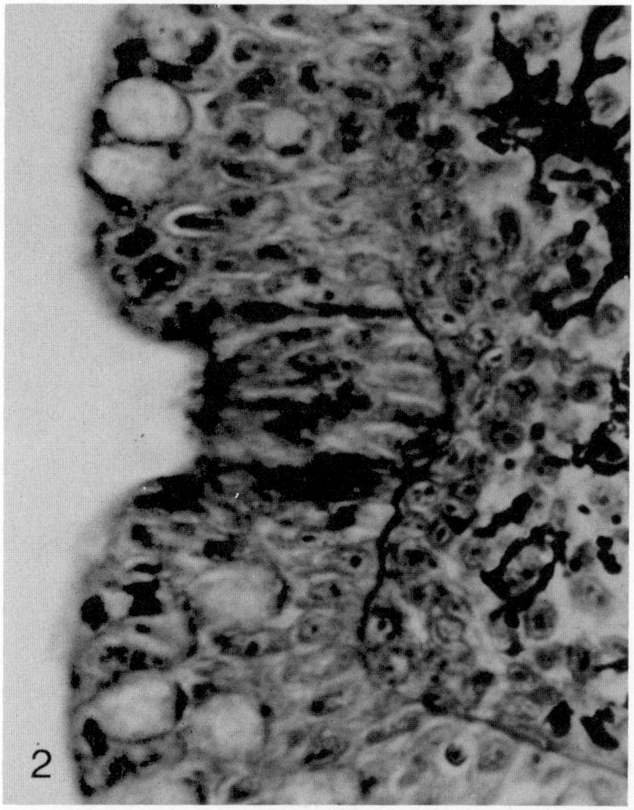

Fig. 2. Transverse section through the epidermis and underlying dermal tissue of a 16-day alevin of *salmo trutta*. The epidermis, with a neuromast lying centrally, is to the left; the cells concentrated immediately beneath it form the lachrymal blastema, in which no secretion is yet visible. The black areas are pigment. Azan. ×820.

Soon the neuromasts invaginate to form the lateral line canal. During this process preosseus matrix appears alongside the developing canal to give rise to the dorsal and ventral *dermo* components(5,6) of the lachrymal bone (Fig. 1). Both *dermo* components fuse with the *membrano* component, and eventually fuse with each other after the canal has closed. In this way the definitive lachrymal bone assumes the form of a plate bearing a bony canal which surrounds the anterior region of the infraorbital lateral line canal. It is now a typical canal bone.

Experiments

Preliminary experiments gave clear indication of an influence of thyroxine upon the development of this bone, but we report here only the results of a definitive experiment in which a total of 27 control, 27 thyroxine-treated, and 10 thiourea-treated alevins were examined. The distribution within age groups are given, with other data, in Tables 2 and 6.

Dealing first with the control and thyroxine-treated alevins, one difference was

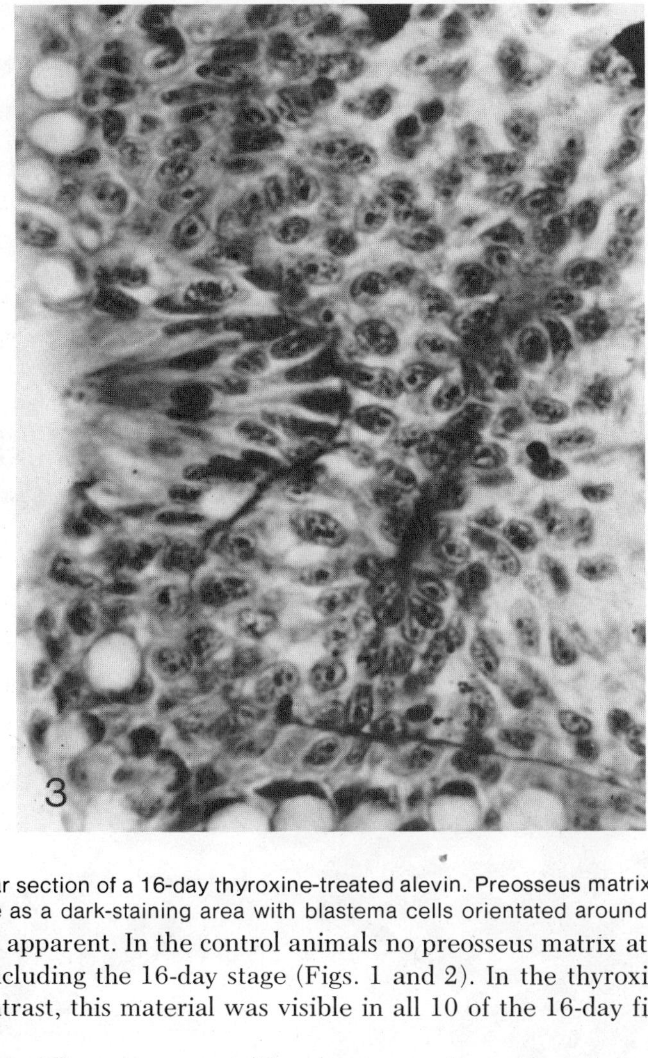

Fig. 3. A similar section of a 16-day thyroxine-treated alevin. Preosseus matrix *membrano* element is visible as a dark-staining area with blastema cells orientated around it. Azan. ×820.

immediately apparent. In the control animals no preosseus matrix at all was visible up to and including the 16-day stage (Figs. 1 and 2). In the thyroxine-treated animals, by contrast, this material was visible in all 10 of the 16-day fish (Figs. 1 and 3).

A further difference was readily apparent on inspection of the 31-day stages. In these, the developing "bone" (the term is used here in quotes to include both preosseus matrix and definitive bone) was obviously more substantial (Fig. 1) than that of the controls. Moreover, in four of the five hormone-treated alevins, the lateral line canal had closed over in the region of the developing bone (Fig. 4), whereas in all of the five control animals it was still an open groove (Fig. 3).

In order to secure more precise and quantitative evidence of the effect of the hormone, however, it was deemed necessary to measure the size of the blastema and of the "bone." These measurements were made by counting, on both sides of the head, the total number of 8 μ sections in which the blastema and "bone" could be seen. This technique takes no account of the volume of the rudiments, but its reliability as an estimate of growth was checked in two alevins in the following way.

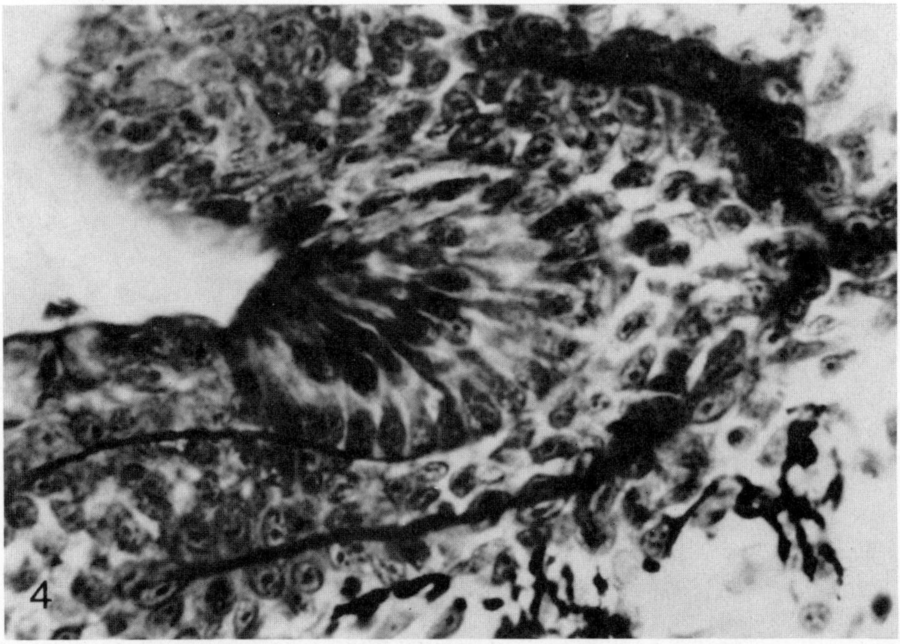

Fig. 4. A similar section of a 31-day control alevin. The neuromast is beginning to invaginate; dorsal and ventral *dermo* elements are visible above and below it. Azan. ×820.

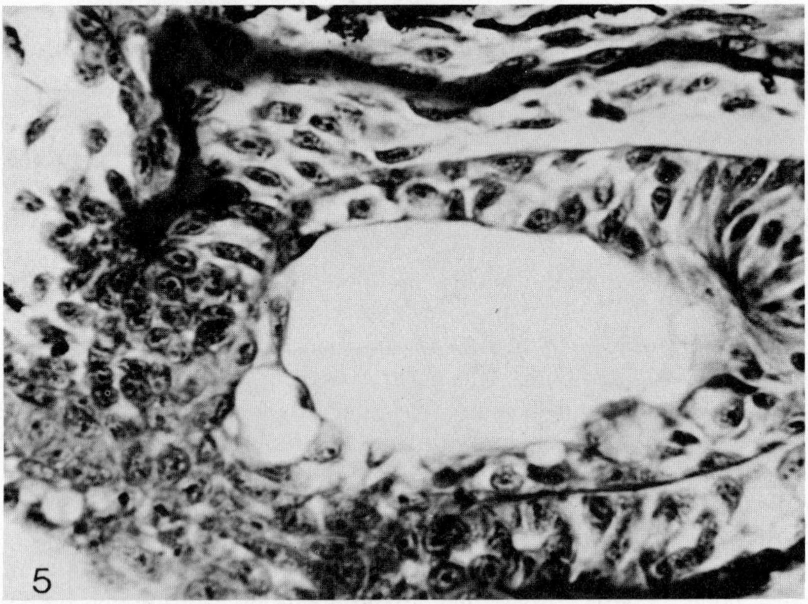

Fig. 5. A similar section of a 31-day thyroxine-treated alevin. The neuromast (at extreme right of the field) is within a canal, which is closed off from the epidermis, but which is not yet separated from it. Note that the development of the dorsal *dermo* element is more advanced than in Figure 4. Azan. ×820.

Using one control and one thyroxine-treated animal, a series of outline drawings of the blastemata of the left side were traced on paper of uniform thickness with the aid of a camera lucida. In each animal, 40 consecutive sections were thus drawn, covering the whole of the lachrymal region, and the blastemata outlines were then cut out and weighed. This weight gave a measure of the volume of each blastema. Lengths were measured by counting sections as already described.

A comparison of the results obtained with the two methods is given in Table 1. This shows that the percentage difference between control and thyroxine-treated alevins, estimated by counting of sections, agreed closely with the difference as determined by the weighing procedure. Probably the latter gives a more accurate measure, but it is so time-consuming as to be impracticable for a lengthy study. We have, therefore, felt justified in using the counting method as a routine procedure.

The results of the section counts are given in Table 2. The blastema is significantly ($P<0.01$) longer in the hormone-treated alevins at 16 and at 31 days, and the only clear departure from this amongst the single pairs is at day 10. Further, the "bone" is significantly ($P<0.01$) longer in the hormone-treated alevins at day 31, and the single pairs of alevins conform with this.

It is thus clear that the rates of development of the lachrymal bone, and of the invagination of its associated neuromasts, are sensitive to thyroxine during these early stages. However, Devillers(9) has shown that the neuromasts act as tertiary organizers in the development of the canal bones of teleosts; it is thus conceivable that the effect of the thyroxine might be exerted primarily on the neuromasts, and that its influence upon the lachrymal bone might be secondary to this. In order to examine this possibility we have therefore examined the rate of development of the parietal bone, which is not associated with the neuromast system and does not form a bony canal.

Parietal Bone

This bone develops slightly in advance of the lachrymal, over the posterior part of the auditory capsule. At 5 days after hatching a small blastema is present closely applied to the underlying perichondrium of the capsule, from which it can be distinguished by its larger oval nuclei. Subsequently (e.g., at 8 and 11 days) the blastema extends anteriorly, being separated in this region from the auditory capsule by a blood space. Secretion of preosseus matrix begins at 14 days or soon after (Fig. 7), and is followed by ossification.

TABLE 1

Comparison of two methods of measuring the lachrymal blastema[a]

	Blastema volume	Blastema length
Control	0.2074	21
Thyroxine treatment	0.3141	31
Percent difference	51.4	47.6

[a]For further explanation, see text.

TABLE 2

State of development of the lachrymal blastema on the left and right sides of control and thyroxine-treated alevins[a]

Age (days after hatching)	Treatment	No. of animals	Left blastema	Right blastema	Left "bone"	Right "bone"
4	C	5	0	0	0	0
	T	5	0	0	0	0
7	C	1	0	0	0	0
	T	1	8	9	0	0
10	C	1	15	15	0	0
	T	1	0	0	0	0
13	C	1	26	21	0	0
	T	1	41	34	0	0
16	C	10	22.3 (5.3)	22.1 (5.3)	0	0
	T	10	33.3 (6.4)	32.3 (6.0)	9.7 (5.9)	9.8 (5.7)
19	C	1	37	43	4	6
	T	1	49	47	29	34
22	C	1	44	42	11	22
	T	1	47	43	37	21
25	C	1	70	63	54	44
	T	1	–[b]	75	55	52
28	C	1	36	40	21	25
	T	1	79	66	73	59
31	C	5	45.6 (6.1)	46.2 (6.4)	36.0 (5.8)	35.2 (7.1)
	T	5	74.8 (11.6)	74.6 (9.3)	60.6 (14.2)	61.4 (10.5)

[a]The figures show the number of 8 μ sections in which blastema and "bone" (preosseus matrix or bone) were present. Standard deviations are given in brackets.
[b]Serial sections incomplete.

The state of development of the parietal bone was determined at 16 days in two distinct groups of alevins, from eggs hatched in two different years, 1964 and 1967. In the 1964 group (Table 3), the control animals showed some variation, the condition of development ranging from a small blastema without matrix, to a blastema with matrix that was already moulded into a definite lamella. The condition

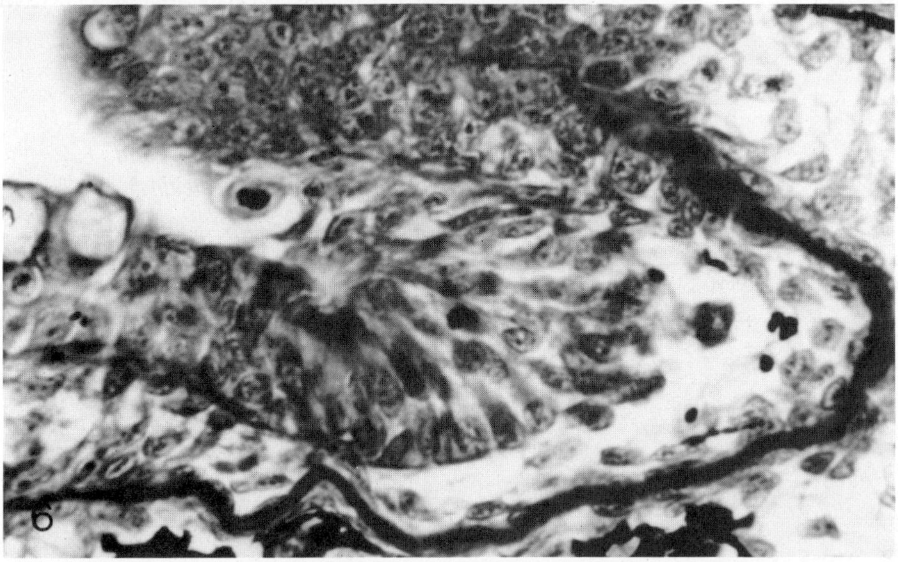

Fig. 6. A transverse section, similar to that shown in Figure 4, of a 31-day thiourea-treated alevin. The appearance of the neuromast and the lachrymal bone is not significantly different from that seen in the control specimen (Fig. 4). Azan. ×820.

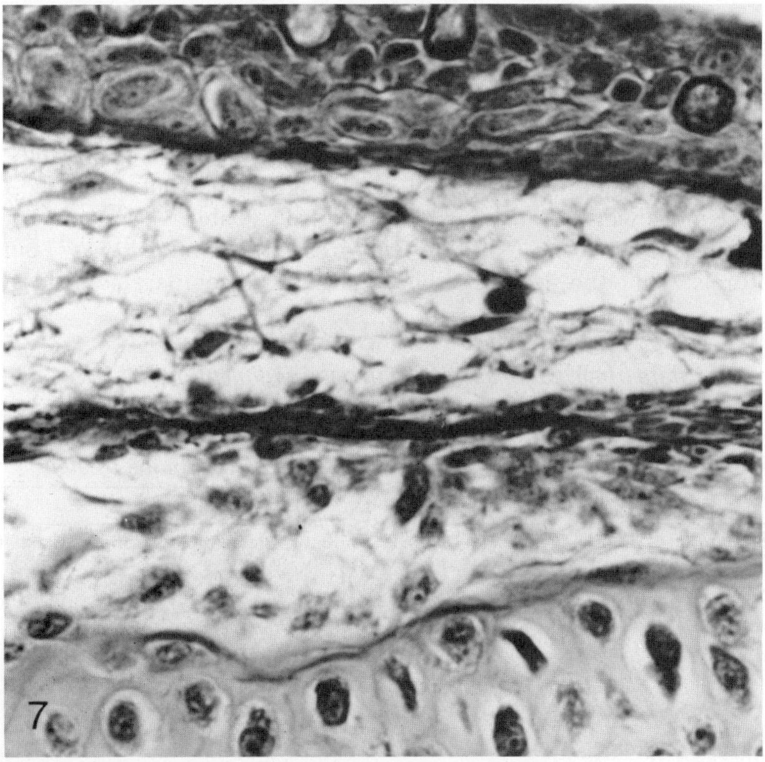

Fig. 7. A transverse section through the auditory region of a 16-day thyroxine-treated alevin. The developing parietal and its blastema extend horizontally between the epidermis (above) and the cartilage of the auditory capsule (below). PAS ×820.

TABLE 3

State of development of the parietal rudiment on the left and right sides of 16-day control and thyroxine-treated alevins[a]

Controls			Thyroxine-treated		
Animal	**Left**	**Right**	**Animal**	**Left**	**Right**
9	++	++++	36	++++	++++
10	+	+	37	++++	++++
11	++++	++++	38	++++	++++
12	+++	+++	39	++++	++++
13	++++	not available	40	++++	+++
14	++	+	41	++++	++++
54	++	++	58	+++	++++
55	++++	++++	59	++++	++++
56	+++	+++	60	++++	++++
57	++	++++	61	++++	++++

[a]Symbols: +, small blastema; ++, small blastema with a little preosseus matrix; +++, matrix increased, but not moulded as a lamella; ++++, lamella of preosseus matrix or bone.

in the thyroxine-treated alevins was in advance of this, considering the group as a whole. In each of the ten alevins a lamella was present on at least one side, and in eight of these it was present on both sides.

Although conditions were arranged in 1967 to be, as far as possible, the same as in 1964, development was not quite so far advanced, but again the thyroxine-treated specimens were in advance of the controls (Table 4). For example, matrix or bone was visible in 8 of the 10 thyroxine-treated specimens (Fig. 7) but in only 3 of the controls, while the size of the blastema showed comparable differences.

TABLE 4

State of development of the parietal rudiment on the left and right sides of 16-day control and thyroxine-treated alevins[a]

Controls					Thyroxine-treated				
	Left		Right			Left		Right	
Animal	**Blastema**	**"Bone"**	**Blastema**	**"Bone"**	**Animal**	**Blastema**	**"Bone"**	**Blastema**	**"Bone"**
1	14	7	22	15	11	trace	0	0	0
2	0	0	0	0	12	18	12	19	12
3	0	0	0	0	13	12	trace	19	9
4	12	0	13	0	14	26	11	33	17
5	15	trace	14	0	15	24	12	19	14
6	0	0	0	0	16	0	0	0	0
7	9	0	17	9	17	15	0	17	11
8	0	0	0	0	18	15	trace	14	8
9	9	0	11	5	19	36	25	21	12
10	trace	0	0	0	20	38	23	17	15

[a]The figures show the number of 8 μ sections in which blastema and "bone" (preosseus matrix or bone) were present.

The variability in these specimens precludes quantitative statistical analysis, while at 31 days the condition of the bone was too variable to permit recognition of any difference between the control and the experimental groups. However, the differences at 16 days are sufficiently clear-cut to justify the conclusion that the time of appearance of the parietal rudiment is advanced in the thyroxine-treated specimens. Thus the effect of the hormone upon the early development of the dermal bone cannot be exerted solely through its effect upon the neuromasts. Does this, however, mean that this early phase of skeletal development is under the control of endogenous thyroxine? This question raises problems which require further study, but certain points can be made at this stage.

Thyroid Gland

Firstly, thyroid follicles are clearly visible in the alevins throughout the period of development on which we are reporting, and they contain stainable colloid (Fig. 8). That they are engaged in iodine binding is made clear by the faint but definite images that overlie them in autoradiograms even at 7 days after hatching (Fig. 11). It does not follow, of course, that this binding is leading to the production of thyroxine, but the effects of thiourea treatment certainly suggest that it is. In alevins that have been reared from hatching in solutions of thiourea, these follicles show hypertrophy (Fig. 9), and are virtually devoid of colloid. Moreover, autoradiograms of these specimens show no trace at all of any image of bound iodine. These are the characteristic responses of active thyroid tissue to goitrogen treatment.

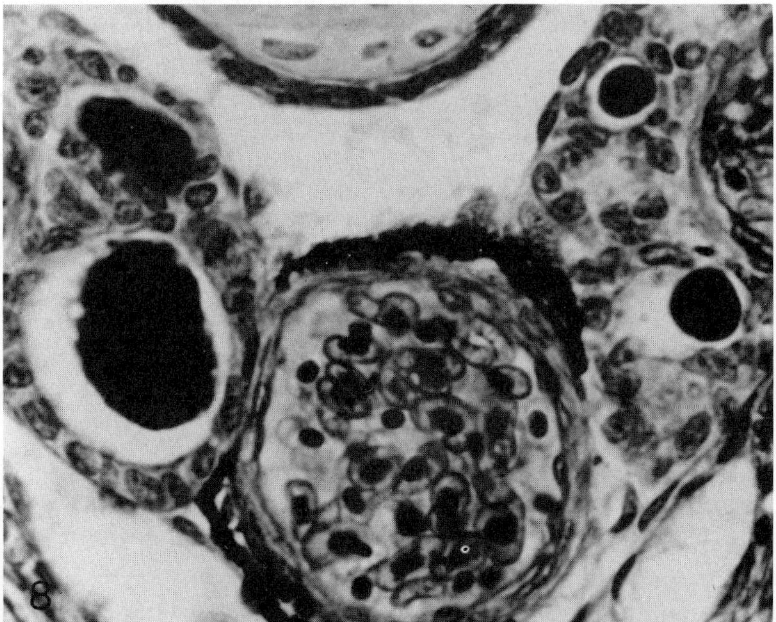

Fig. 8. Transverse section through the thyroid region of a 20-day control alevin. Thyroid follicles, with abundant colloid, are seen on either side of the ventral aorta; the lower edge of the median (copula) cartilage lies above the aorta. Azan. ×820.

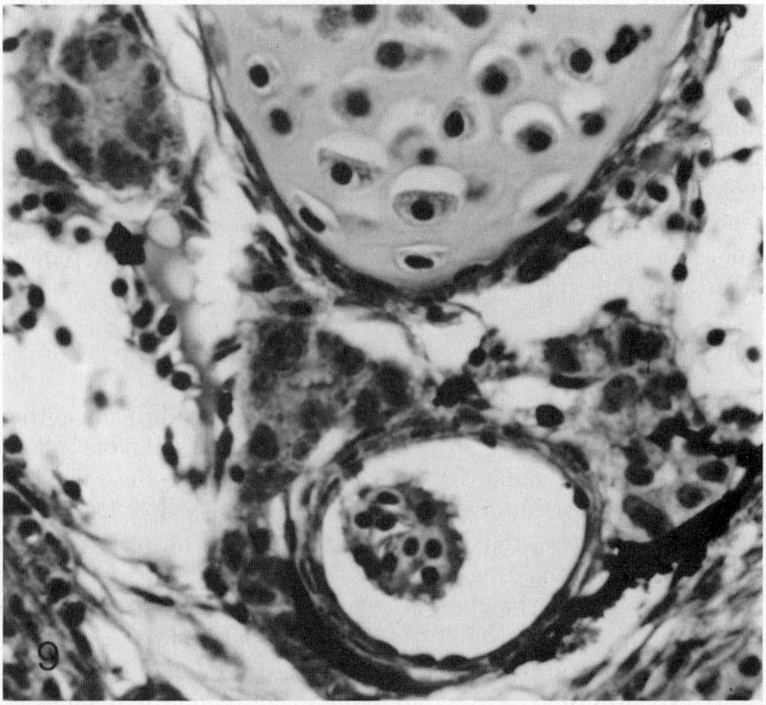

Fig. 9. Similar section of a 21-day alevin, immersed in 0.03 percent thiourea solution from hatching. The thyroid cells show a typical goitrogenic hypertrophic response, and the follicles lack colloid. PAS. ×820.

Further evidence that this response to thiourea is a consequence of the elimination not only of iodine binding, but also of thyroxine formation, is the fact that alevins treated simultaneously with thiourea and thyroxine have thyroid follicles with a normal epithelium, and with abundant colloid (Fig. 10). Moreover, there is a clear image of bound iodine in their autoradiograms. In these latter animals, the absorbed exogenous hormone has presumably been sufficient to abolish the goitrogenic response. There are thus good grounds for supposing that the thyroid of the alevin is an active secretory organ during the first 31 days after hatching, and that it is forming thyroxine.

We hoped that it might be possible to strengthen this conclusion by obtaining positive chromatographic evidence for the presence of thyroid hormones in these early alevins. For this purpose samples of fish were removed at fortnightly intervals from stocks maintained at the hatchery. Each sample was brought to the laboratory and immersed in water containing ^{131}I for 5 days. The results of chromatography and of counts for total bound ^{131}I are shown in Table 5.

It will be seen that monoiodotyrosine (MIT) and diiodotyrosine (DIT) were identifiable from at least the 5-day stage onwards, but that thyroxine did not appear on the chromatograms until 60 days and triiodothyronine until 89 days. This cannot be taken to mean, however, that the two hormones are necessarily absent in the earlier stages. Table 5 shows that there is a marked increase in the total

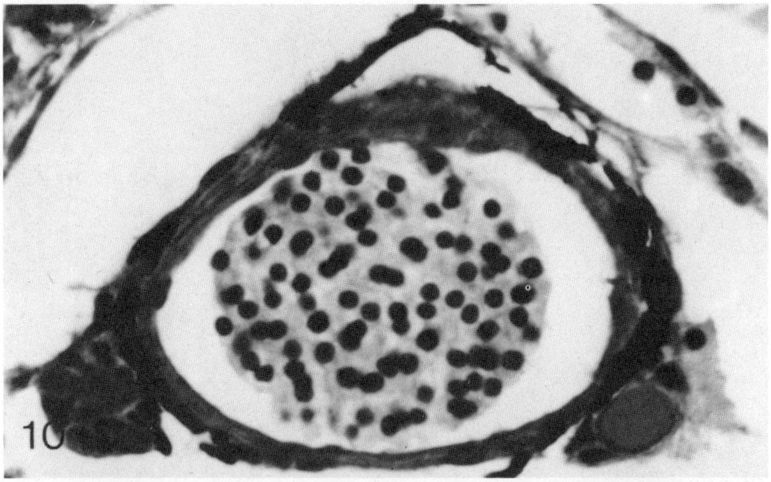

Fig. 10. Similar section of a 35-day alevin, immersed in 0.03 percent thiourea solution together with thyroxine (1×10^{-7}) from hatching. The follicles contain colloid. PAS. $\times 820$.

radioactivity of the samples throughout the period studied, whether the activity is expressed as counts per minute per fish, or as counts per minute per unit dry weight. In part, this increase may well reflect growth of the thyroid gland during this period, although the effect of this is impossible to assess without expressing this growth in relation to the total growth of the animal, and this we have not attempted to do. But what is very clear from the autoradiograms is that the images obtained from the fish after feeding has begun under hatchery conditions are very much more intense that those obtained while they are still subsisting on their yolk. Figure 12 shows the dense black images obtained over the thyroid follicles at 60 days of age. They contrast strikingly with the tenuous images seen at 7 (Fig. 11) and 33 days.

TABLE 5

Results of radiochromatography, total bound iodine counts (including unknowns), and traces of iodide, in counts per minute per fish, of lower jaw hydrolysates of Salmo trutta *aged 5 to 89 days after hatching*[a]

Age (days after hatching)	5	7[b]	18	24	33	47	60	75	89
Number of animals	500	338	404	243	264	253	293	295	201
MIT	+	—	+	+	+	+	+	+	+
DIT	+	—	+	+	+	+	+	+	+
T_4	—	—	—	—	—	—	+	+	+
T_3	—	—	—	—	—	—	—	—	+
Iodide	+	—	—	+	+	+	+	+	+
Unknown	—	—	—	+	+	+	+	+	+
Total bound ^{131}I activity	not recorded	12.5	22.4	89.4	120.4	397.8	571.6	1301.8	1974.0

[a]Animals were immersed for 5 days in tap water containing 200 μc/1 of ^{131}I. MIT, monoiodotyrosine; DIT, diiodotyrosine; T_3, triiodothyronine; T_4, thyroxine.

[b]Radiochromatograms were not prepared for day 7.

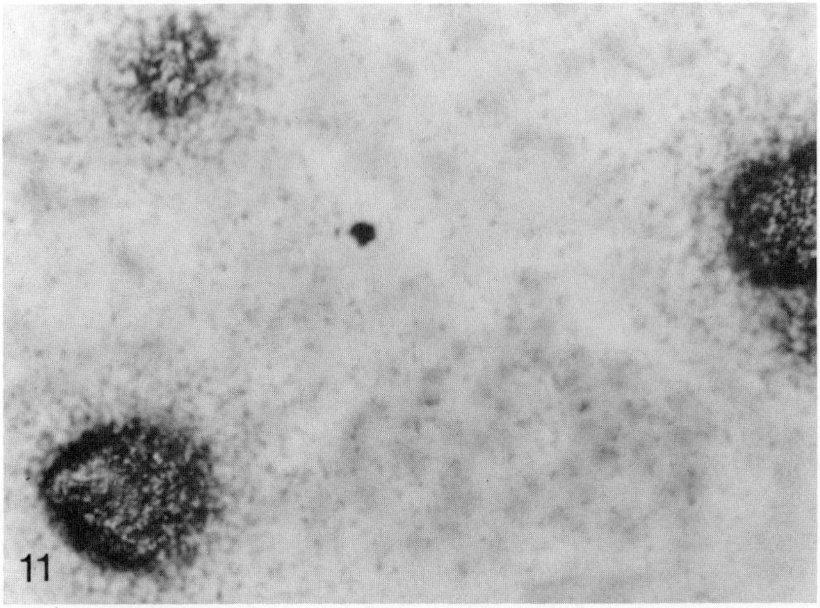

Fig. 11. Autoradiogram of a transverse section through the thyroid region of a 7-day alevin. Images indicative of bound iodine overlie the thyroid colloid (compare Fig. 8). In comparison with Figure 12, the images in the specimen from which Figure 11 was prepared were much less dense than they appear to be in the photograph. ×820.

It seems evident, therefore, that part of the reason for the increased radioactivity in the later samples was an increased rate of iodine binding in the thyroid glands of the older alevins.

The density of the images on the radiochromatograms should be considered in the light of these facts. The image of thyroxine at 75 days (Fig. 16) is very faint in contrast with the dense images of monoiodotyrosine and diiodotyrosine (Fig. 15). The difference is the more striking in that the material applied to the chromatogram shown in Figure 14 represented nearly five times as many animals as that used for the chromatogram in Figure. 13. Reference to the chromatogram obtained from 24-day-old alevins (Figs. 13 and 14) shows that the images of MIT and DIT are very faint indeed at that stage. It seems clear that, even allowing for the large number of animal equivalents used for Figure 14, it would have been impossible to identify thyroxine in the 29-day sample, assuming (if it were present at all) that its titer relative to that of the iodotyrosines would have been similar to the ratio observed at 75 days.

Taking all these observations together, we conclude that there is presumptive evidence that the thyroid gland is synthesizing thyroid hormones during the first 4 weeks following hatching, but that the amount produced is too slight to be detectable on our chromatograms. Whether any thyroxine thus produced is actually released into the blood is another matter; on this we have at present no information at all.

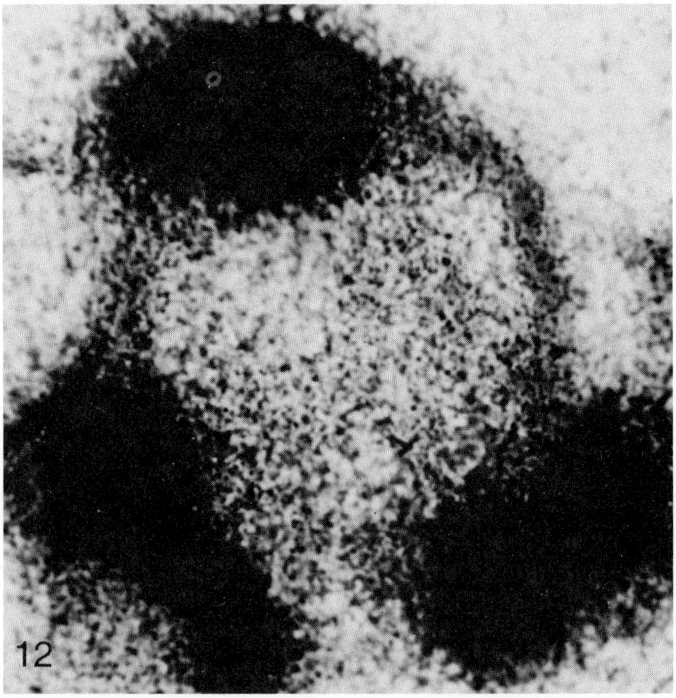

Fig. 12. Autoradiogram of a similar section of a 60-day animal. The follicles are much larger, and the images in the original specimen were much denser than those of Figure 11. ×820.

Thiourea

What can be asked at this stage, however, is whether or not the arrest of thyroidal biosynthesis by thiourea is reflected in the course of development of the lachrymal bone. Some evidence on this is presented in Table 6. It will be seen that the

TABLE 6

State of development of lachrymal rudiment in control and thiourea-treated alevins[a]

Age (days after hatching)	10		31	
Treatment	Control	Thiourea	Control	Thiourea
No. of animals	10	5	5	5
Left blastema	22.3 (5.3)[b]	22.3 (3.5)	45.6 (6.1)	47.0 (7.7)
Right blastema	22.1 (5.3)	20.8 (3.2)	46.2 (6.4)	48.8 (8.0)
Left "bone"	0	0	36.0 (5.8)	39.6 (6.7)
Right "bone"	0	0	35.2 (7.0)	40.4 (6.3)

[a]Compare with Table 2.

[b]Standard deviations are given in brackets.

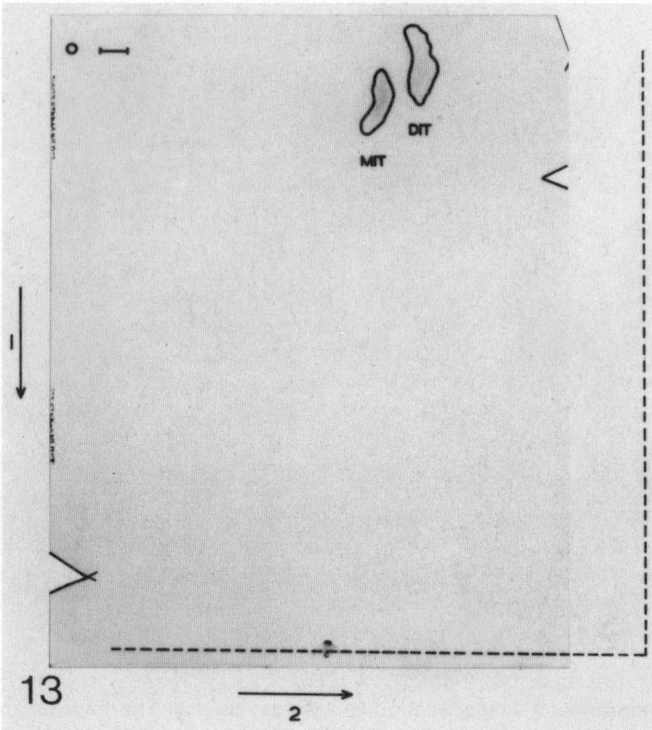

Fig. 13. Radiochromatogram of the 10 percent acetic acid eluate from 24-day alevins. The equivalent of 6.7 animals was applied to the paper. Exposure: 40 days. MIT and DIT are faintly indicated.

goitrogen-treated alevins at day 16 resemble the control specimens; there is no difference between the lengths of the blastemata in the two groups, and no "bone" has appeared in either group. Similarly, at day 31 there is no marked difference between the control animals and the thiourea-treated ones in either the length of the blastema, or in the length of the "bone." The "bone" is, in fact, slightly larger in the goitrogen-treated alevins (Table 6), and there are also indications that the development of the lateral line system of the head is more advanced in these specimens. So far, however, neither difference rates as significant, and this part of our analysis clearly needs much more exploration. We conclude for the present that lack of synthetic activity in the thyroid gland in the early alevin has no pronounced effect on the development of the lachrymal bone.

DISCUSSION

The present study has shown that exogenous thyroxine promotes the synthesis of preosseus matrix and the further development of the lachrymal bone during the first 31 days after hatching. It also advances the condition of the parietal rudiment at 16 days, but variability makes it impossible to establish whether or not this particular effect is still apparent at 31 days.

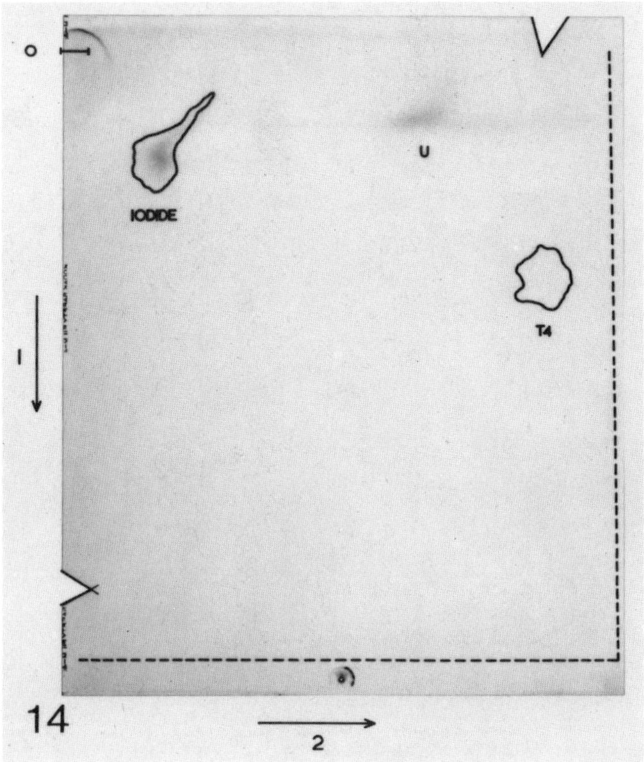

Fig. 14. Radiochromatogram of the 50 percent acetic acid eluate from 29-day alevins. The equivalent of 77.6 animals was applied. Exposure: 20 days. Iodide is faintly indicated; thyroxine is not detectable.

Whether or not thyroxine is actually being synthesized in the alevin during these early stages is less clear. Even in higher vertebrates the course of development of the synthetic capacities of the thyroid gland has been a matter for some disagreement. Trunnell and Wade(10) concluded that the products of biosynthesis appeared successively during the development of the chick: monoiodotyrosine at 8½ days, diiodotyrosine at 9¼ days, and thyroxine at 9¾ days. In contrast to this, however, Blanquet et al.(11) found that thyroxine was present in the chick from the beginning of organic iodine binding, while Waterman and Gorbman(12) drew the same conclusion for the rabbit.

Particularly interesting in the present connection, however, are certain amphibian studies. Shellabarger and Brown(13), studying *Xenopus laevis*, were able to detect thyroxine in the thyroid gland of the tadpole at the "critical" period of rapid metamorphosis, but could not detect it in the "precritical" phase, although monoiodotyrosine and diiodotyrosine were both present. They do not comment on the significance of these results. It is, however, clear from their Table 1 that the apparent absence of thyroxine could quite well have been a consequence of its being present at levels below the sensitivity of their technique, just as we have assumed for our trout alevins. That this is actually so in amphibian tadpoles is certainly implied by current views of the influence of thyroxine on metamorphosis, for these

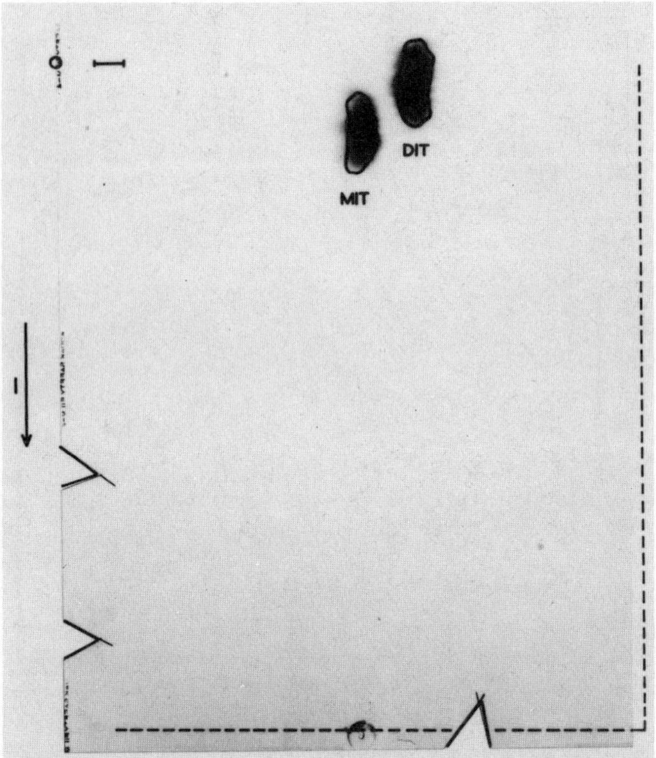

15

Fig. 15. Radiochromatogram of the 10 percent acetic acid eluate from 75-day fry. The equivalent of 5.6 animals was applied. Exposure: 12 days. MIT and DIT are very strongly indicated. Compare with Figure 13, and note the difference in exposure time.

assume that thyroxine is present from an early stage of larval development, and that its titer builds up during later stages(14). Indeed, Flickinger(15) has shown, in confirmation of this, that thyroxine is detectable in the premetamorphic (16mm) frog larva (presumably *Rana pipiens*, although he nowhere states this). He also finds that monoiodotyrosine and diiodotyrosine appear before thyroxine, and in this he is in agreement with Shellabarger and Brown(13), although he finds an earlier onset of thyroidal biosynthesis, with MIT and DIT detectable in the 11-mm tadpole. However, in stating that a similar sequential appearance of these thyroidal products was demonstrated by Waterman and Gorbman(16) in the rabbit fetus, he overlooks the later publication(12) of these authors, in which they find, as already mentioned, that thyroxine is detectable at the first onset of iodine binding in that animal.

The results of this type of analysis must, of course, be limited by the sensitivity of the methods used, and this may account for some of the disagreements in the literature. At this stage we can only say that our own results indicate, on balance, that thyroxine does appear in the alevin during the first 4 weeks after hatching.

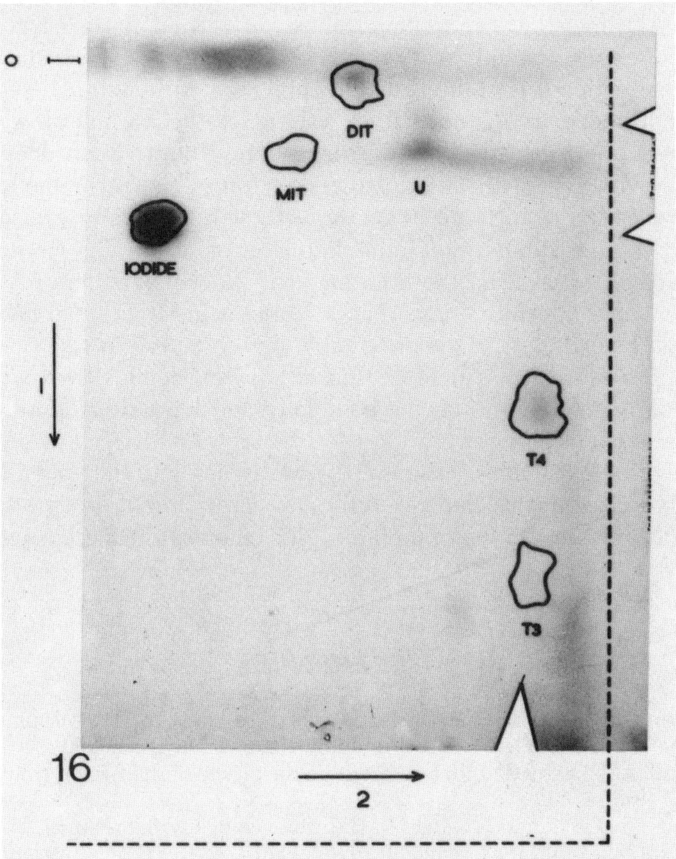

Fig. 16. Radiochromatogram of the 50 percent acetic acid eluate from 75-day fry. The equivalent of 26.6 animals was applied. Exposure: 13 days. Thyroxine is faintly indicated, as well as iodide and DIT. Compare with Figure 14, noting the differences in exposure time and animal equivalents.

Further studies are needed, however, to determine the precise age at which it is first synthesized, and to establish whether or not this coincides with the onset of iodine binding.

In any case, it must be emphasized that our present findings do not necessarily mean that skeletal development in the young alevin is under the control of such endogenous thyroxine as may be present. Indeed, the fact that thiourea has no effect, during the period of experimentation, upon the development of the lachrymal bone speaks against its development being dependent, either directly or indirectly, upon thyroxine. Our observations are in agreement in this respect with those of Woodhead(2), who was unable to find any morphometric effect of goitrogen upon the external characters of alevins.

At the same time, we cannot exclude the possibility that endogenous thyroxine, supposing it to be present, might normally exert a trigger-like action upon skeletal development, and might thus set some developmental process in train before the goitrogen could exert its effect. That a situation of this sort can exist in teleosts is suggested by the fact that the metamorphosis of certain flatfish larvae is a "triggered" reaction, which cannot be stopped or reversed once it has started(17).

In this connection it is of interest that thyroxine, although it is secreted by the rat fetus, is not necessary for normal growth or for the normal appearance of the primary centers of ossification(18). Further, thyroid deprivation does not appreciably interfere with the maturation of the skeleton of the rat if the hormones are withheld during the intrauterine stage(19), although it leads to transitory delay in ossification if there is deprivation immediately after birth. The circumstances are, of course, very different from those in our own experiments, and it would be premature to look for uniformity in the factors influencing teleostean and mammalian development. Nevertheless, it seems possible that in *Salmo trutta*, as in the rat, certain embryonic rudiments, which are sensitive to exogenous thyroxine during the early stages of development, may yet be able to develop independently of that hormone in normal conditions.

Certainly, the existence of a sensitivity to thyroxine in the young alevin creates the possibility that adaptive changes in this sensitivity might provide a basis for the evolution of thyroid regulation of metamorphic changes during early teleostean development.

REFERENCES

1. Barrington, E. J. W. Metamorphosis in lower chordates. In Metamorphosis, Etkin, W., and Gilbert, L. I., eds. New York, Appleton-Century-Crofts, 1968.
2. Woodhead, A. D. Effects of thyroid drugs on the larvae of the Brown trout, *Salmo trutta*. J. Zool., 149:394–413, 1966.
3. Barrington, E. J. W., and Rawdon, B. B. Influence of thyroxine upon the uptake of ^{35}S-labelled sulphate into the branchial arch skeleton of the Rainbow trout (*Salmo gairdnerii*). Gen. Comp. Endocr., 9:116–128, 1967.
4. ——— and Thorpe, A. The identification of monoiodotyrosine, diiodotyrosine, and thyroxine in extracts of the endostyle of the ascidian, *Ciona intestinalis L.* Proc. Roy. Soc., B 163:136–149, 1965.
5. Pearse, A. G. E. Histochemistry, Theoretical and Applied. London, Churchill, 1960.
6. Devillers, C. Recherches sur le crâne dermique des Téléostéens. Ann. Paleont., 33:1–94, 1947.
7. ——— Le crâne des poissons. *In* Traité de Zoologie 13, Fasc. 1, P.-P. Grassé, ed. Paris, Masson, pp. 551–687, 1958.
8. Lorch, I. J. The distribution of alkaline phosphatase in the skull of the developing trout. Quart. J. Micr. Sci., 90:183–207, 1949.
9. Devillers, C. Etude expérimentale du role morphogénétique des fossettes sensorielles (pit-organs) de *Salmo fario*. C.R. Acad. Sci. (Paris), 223:1180–1181, 1946.
10. Trunnell, J. B., and Wade, P. Factors governing the development of the chick embryo. II. Chronology of the synthesis of iodinated compounds studied by chromatographic analysis. J. Clin. Endocr., 15:107–117, 1955.
11. Blanquet, P., Stoll, R., Maraud, R., and Capot, L. Sur l'élaboration de thyroxine par la thyroïde de l'embryon poulet. Bull. Soc. Chim. Biol. (Paris), 35:627–632, 1953.
12. Waterman, A. J., and Gorbman, A. Development of the thyroid gland of the rabbit. J. Exp. Zool., 132:509–538, 1956.
13. Shellabarger, C. J., and Brown, J. R. The biosynthesis of thyroxine and 3:5:3′-triiodothyronine in larval and adult toads. J. Endocr., 18:98–101, 1959.
14. Etkin, W. Hormonal control of amphibian metamorphosis. *In* Metamorphosis, Etkin, W., and Gilbert, L. I., eds. New York, Appleton-Century-Crofts, 1968.

15. Flickinger, R. A. Sequential appearance of monoiodotyrosine, diiodotyrosine, and thyroxine in the developing frog embryo. Gen. Comp. Endocr., 4:285–289, 1964.
16. Waterman, A. J., and Gorbman, A. Onset of function of the foetal rabbit thyroid gland. Anat. Rec., 117:621–613, 1953.
17. Wickstead, J. H. *Branchiostoma lanceolatum* larvae; some experiments on the effect of thiouracil on metamorphosis. J. Mar. Biol. Ass. U.K., 47:49–59, 1967.
18. Jost, A. Physiologie des hormones thyroïdiennes chez le foetus. *In* Colloquio Sobre a Tireoide, Chagas, C., and Lobo, L. C., eds. Rio de Janeiro, Universidade do Brasil, 1961.
19. Hamburgh, M. An analysis of the action of thyroid hormone on development based on *in vivo* and *in vitro* studies. Gen. Comp. Endocr., 10:198–213, 1968.

36

Some Metabolic Effects of Thyroxine During Postnatal Development of the Rat

Philippe Hemon

Laboratory of Comparative Physiology,
Faculty of Sciences, Paris, France

INTRODUCTION

Two enzyme activities known to be under thyroid control in the adult rat liver were studied in the developing rat liver: α-glycerophosphate oxidase[1] (α-GP X), a mitochondrial enzyme which plays a role in the regulation of aerobic glycolysis, and malic enzyme[2] (ME), a cytoplasmic enzyme involved in the biosynthesis of fatty acids from carbohydrates (see 1,2). In addition, we studied the activity of these enzymes in the brown adipose tissue because of its high metabolic activity and its special importance in the young rat. Moreover, both enzyme activities are known to be under thyroid control in the white adipose tissue of adult rats also(3-5).

In the first part of this study, all animals were fed laboratory chow at weaning. Distinct changes of enzyme activities appeared during the weaning period. This observation led us to repeat the same measurements on animals receiving different semi-synthetic diets at weaning. The results obtained with laboratory chow diets on the liver have been previously published(1,2). Other results were published recently(6).

ANIMALS WEANED ON LABORATORY CHOW

Liver Enzyme Activities. The α-GP Ox activity (Fig. 1) was high during the suckling period and later decreased. Thyroidectomy decreased its activity, an effect which was reversed by T_4 injections. These results agree with those of Schapiro and Percin(7).

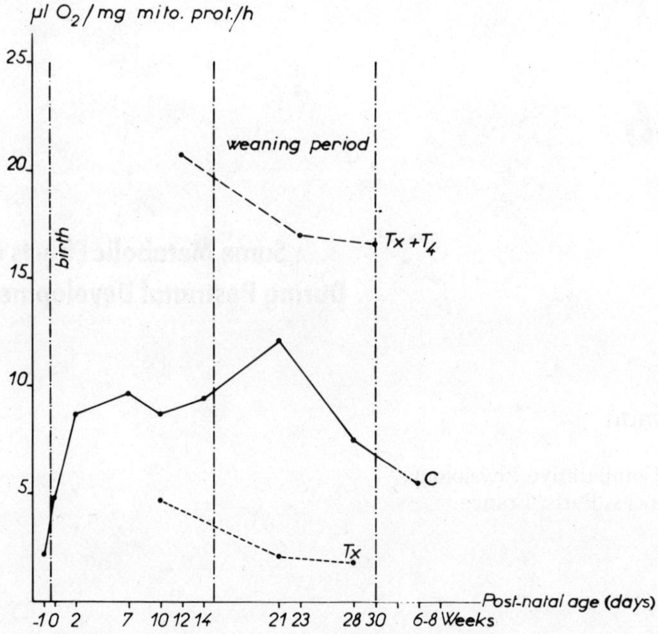

Fig. 1 Liver α-GPOx activity (lab chow). C=control animals; Tx=hypothyroid animals; Tx+T_4= thyroxine-injected hypothyroid animals.

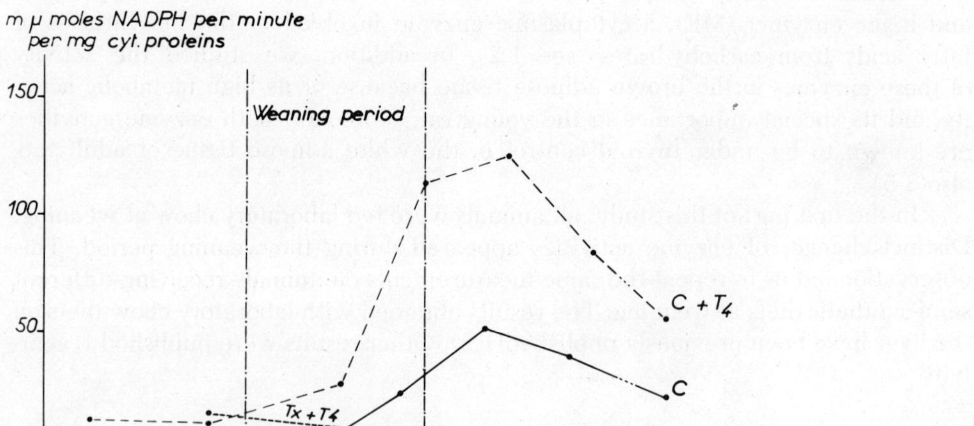

Fig. 2. Liver ME activity (lab chow). C=control animals; Tx=hypothyroid animals; C+T_4=thyroxine-injected control animals; Tx+T_4 =thyroxine-injected hypothyroid animals.

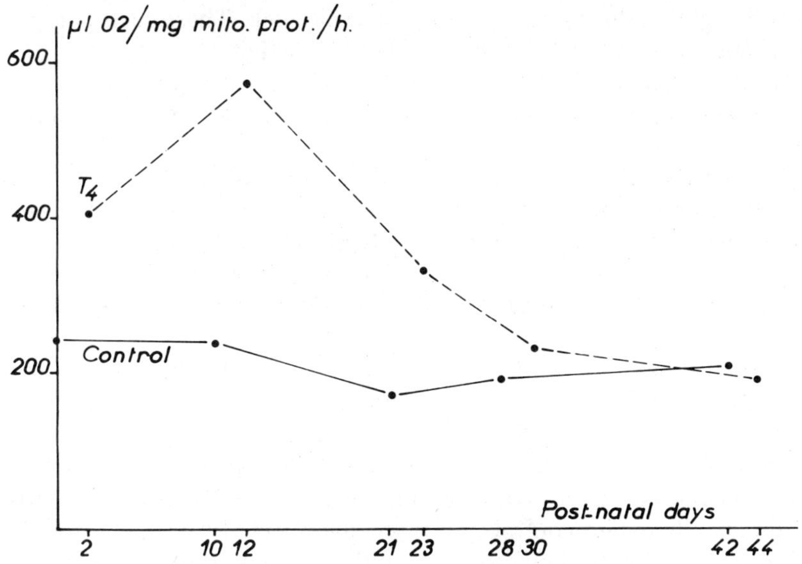

Fig. 3. BAT α-GP Ox activity (lab chow). T_4=thyroxine-injected animals.

The ME activity (Fig. 2) was almost undetectable during the suckling period but it increased sharply during the weaning period (see also 8), as did sensitivity to T . In hypothyroid animals, the liver ME activity failed to increase during the "weaning period" and the sensitivity of the enzyme to T_4 was low: hypothyroid ani-

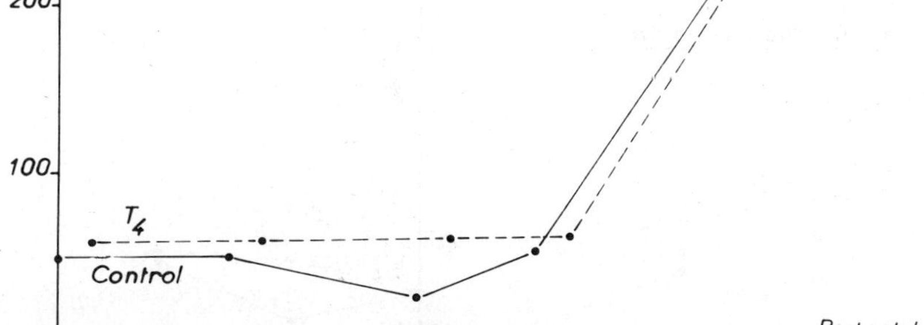

Fig. 4. BAT ME activity (lab chow). T_4=thyroxine-injected animals.

mals are still suckling during the fourth postnatal week. These facts suggest an interaction of hormone and diet in the control of ME activity.

Brown Adipose Tissue Activities. The α-GP Ox activity (Fig. 3) was very high at all stages. Activity was increased by T_4 injection during the suckling period but not later.

The ME activity (Fig. 4) was present at birth and started to increase at the age of one month. T_4 injection accelerated the increase slightly, but differences in enzyme activity were statistically significant only in suckling young.

ANIMALS WEANED ON SEMI-SYNTHETIC DIETS

Two different semi-synthetic diets were used: a high-carbohydrate diet (HC) and a high-lipid diet (HL). The caloric patterns of these diets were respectively similar to the caloric patterns of laboratory chow and milk.

α-GP Ox Activities. No important differences in α-GP Ox activities (Fig. 5) were observed in either liver or brown adipose tissue between the two groups on different diets or between control and thyroxine-injected rats. Liver α-GP Ox activity was always higher in animals kept on either of these semi-synthetic diets than it was in rats fed with laboratory chow.

ME Activities. In the HC fed rats, liver ME activity (Fig. 6) increased sharply starting at weaning (21 days of age), as did sensitivity to T_4. On the HL diet, the liver ME activity increased to a lesser extent and the enzyme was much less sensitive to T_4.

The effect of diet on ME activity was even more striking in brown adipose tis-

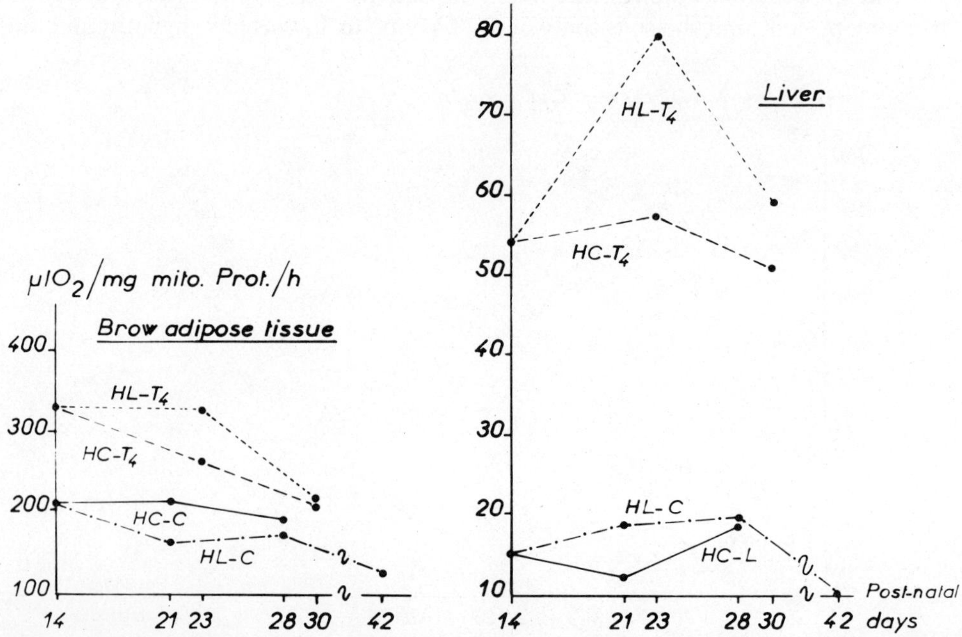

Fig. 5. α-GP Ox activities (semi-synthetic diets). HC-C=high carbohydrate diet, control animals; HC-T_4=high carbohydrate diet, thyroxine-injected animals; HL-C=high lipid diet, control animals; HL-T_4=high lipid diet, thyroxine-injected animals.

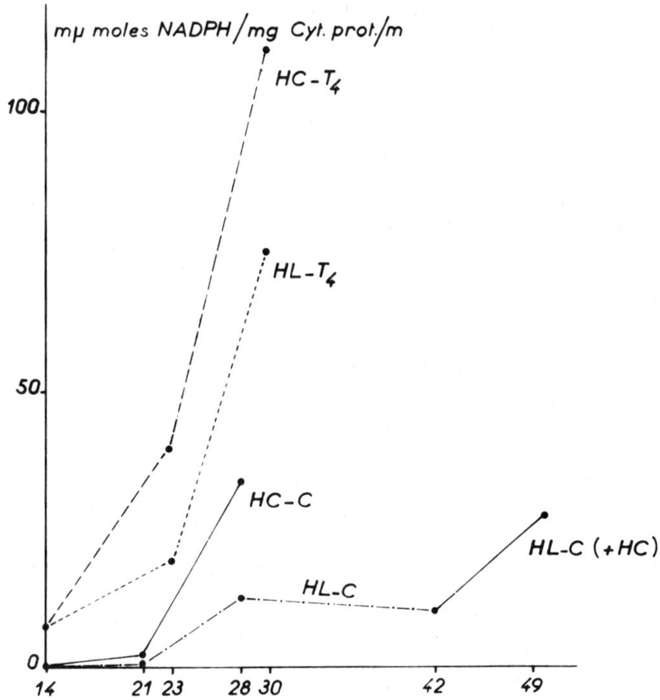

Fig. 6. Liver ME activity (semi-synthetic diets). HC-C=high carbohydrate diet, control animals; HC-T_4=high carbohydrate, thyroxine-injected animals; HL-C=high lipid diet, control animals; HL-T_4=high lipid diet, thyroxine-injected animals; HL-C(+HC)=animals fed with high-lipid diet up to 42 days and with high-carbohydrate diet from 42 days to 49 days.

sue (Fig. 7). As long as the animals were fed with the milk-equivalent diet, the ME activity stayed at the level typical for normal suckling rats. T_4 increased activity slightly, but this increase was significant only in HL fed rats.

DISCUSSION

These results stress the necessity of considering diet changes during development in the interpretation of the metabolic effects of T_4. Florsheim et al.(9) have interpreted the decrease of liver α-GP Ox activity starting at weaning as a reflection of a decrease in thyroid function. But such a decrease was not observed in our experiments with semi-synthetic diets. The decrease observed in animals fed with laboratory chow could result from the presence of an antithyrotoxic factor in this food(10).

The liver ME activity, as well as its sensitivity to T_4, depends among other things on the fat content of the food during the weaning period, the whole phenomenon being an adaptation to an increased lipogenic capacity similar to the situation described in the adult(11).

The effect of T_4 varies with the target organ: liver enzymes respond strongly

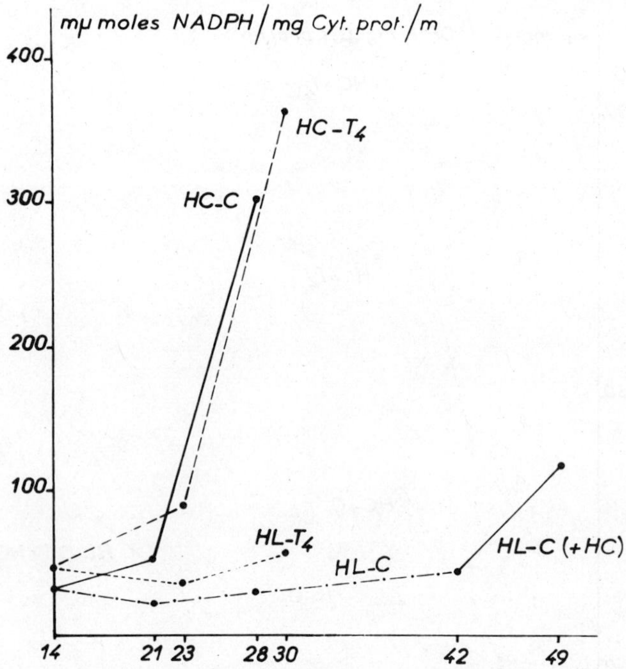

Fig. 7. BAT ME activity (semi-synthetic diets). HC-C, HC-T_4, HL-C, HL-T_4, and HL-C(+HC) as in Figure 6.

and quickly to the presence or absence of thyroid hormone, while the enzymes of the maturing brain are not affected by thyroid deprivation or excess supply of exogenous hormone(2-7).

The role of T_4 in the control of these enzymes in brown adipose tissue has to be defined. In our experiments, an acute treatment with thyroxine increased the activities of the two enzymes studied only slightly. Other experimental conditions such as hypothyroidism or chronic hyperthyroidism might affect them more clearly(12).

We conclude that the physiological significance of the observations presented here is still poorly understood. Therefore it seems impossible, at present, to find any relation between these effects of thyroid hormones and their role in growth and differentiation.

REFERENCES

1. Hemon, P. L'activité α-glucérophosphate oxydase du foie de rat au cours du développement. Biochim. Biophys. Acta, 132:175–178, 1967.
2. ——— Malate dehydrogenase (decarboxylating) (NADP) and α-glycerophosphate oxidase in the developing rat. Biochim. Biophys. Acta, 151:681–683, 1968.
3. Lee, Y. P., and Lardy, H. A. Influence of thyroid hormones on L-α-glycerophosphate dehydrogenases and other dehydrogenases in various organs of the rat. J. Biol. Chem., 240:1427–1436, 1965.

4. Wise, E. M., and Ball, E. G. Malic enzyme and lipogenesis. Proc. Nat. Acad. Sci. U.S.A., 52:1255–1263, 1964.
5. Brown, J., McLean, P., and Greenbaum, A. L. Influence of thyroxine and luteinizing hormone on some enzymes concerned with lipogenesis in adipose tissue, testis and adrenal gland. Biochem. J., 101:197–203, 1966.
6. Hemon, P., and Berbey, B. Changes of enzyme activities with diet and thyroxine during postnatal development of the rat. Biochim. Biophys. Acta, 170:235–243, 1968.
7. Schapiro, S., and Percin, C. J. Thyroid hormone induction of α-glycerophosphate dehydrogenase in rats of different ages. Endocrinology, 79:1075–1078, 1966.
8. Ballard, F. J., and Hanson, R. W. Changes in lipid synthesis in rat liver during development. Biochem. J., 102:952–958, 1967.
9. Florsheim, W. H., Faircloth, M. A., Corcoran, N. L., and Rudko, P. Perinatal thyroid function in the rat. Acta Endocr. (Kobenhavn), 52:375–382, 1966.
10. Ruegamer, W. R., Wagner, B. J., Barston, M., and Keran, E. E. Mechanism by which certain dietary substances interfere with the peripheral activity of thyroxine. Endocrinology, 81:49–53, 1967.
11. Tarentino, A. L., Richert, D. A., and Westerfeld, W. W. The concurrent induction of hepatic α-glycerophosphate and malate dehydrogenase by thyroid hormone. Biochim. Biophys. Acta, 124:295–309, 1966.
12. Young, J. W. Effects of D- and L-thyroxine on enzymes in liver and adipose tissue of rat. Amer. J. Physiol., 214:378–383, 1968.

37

Effect of Thyroxine and Calcitonin on Experimental Hyperparathyroidism*

B. Tuchweber,[1] G. Gabbiani,[2] and G. Côté[3]

[1]Institute of Experimental Medicine and Surgery, University of Montreal, Montreal, Canada
[2]Institute of Pathology, University of Geneva, Geneva, Switzerland
[3]Institute of Cardiology, Quebec, Canada

INTRODUCTION

Extensive investigations have clearly demonstrated that parathyroid hormone increases resorption of bone salts, liberating calcium and phosphorus into the blood. At the same time, it removes phosphorus from the extracellular fluid by decreasing the renal tubular reabsorption of phosphate. In addition to the phosphaturic effect, it decreases the urinary excretion of calcium and magnesium(1-9). Parathyroid hormone exerts several effects on bone: (1) it promotes resorption; (2) it inhibits synthesis(10-15); and (3) when given chronically, it promotes bone apposition(16).

Until recently the maintenance of normal calcium levels was explained by changes in the rate of secretion of parathyroid hormone. This theory(17) is no longer adequate as there is now evidence of the existence of a second hormone: calcitonin. The source and nature of calcitonin as well as its mode of action both in experimental animals and in man have been discussed in recent reviews(18-21). Calcitonin lowers plasma calcium and phosphorus levels in nephrectomized, parathyroidectomized, and eviscerated rats; its primary target is bone.

*This work was supported by the Canadian Heart Foundation, the Medical Research Council of Canada (Block Term Grant MT-1829) and the Ministère de la Santé, Quebec.

The authors wish to thank Professor I. MacIntyre for the supply of calcitonin and the British Drug House for the thyroxine used in these experiments. They also gratefully acknowledge the technical assistance of Mrs. C. Jubert and D. Daboval and Misses F. Dionne and C. Wandowska.

It is known that thyroxine increases the rate of bone formation and resorption, and raises the levels of plasma and urinary calcium(22-24). It appears therefore that the actions of thyroxine and parathyroid hormone on calcium metabolism are qualitatively similar(24). However, the mechanism by which thyroxine effects a more rapid turnover of circulating calcium is not yet clear.

Overdosage with parathyroid extract produces metastatic calcification, especially in the kidneys, cardiovascular system, lungs, and gastrointestinal tract(25). Similar lesions are seen in the rat after bilateral nephrectomy and are probably due to endogenous hyperparathyroidism(25-28).

Hypophysectomy, glucocorticoids, and stressors (e.g., restraint) inhibit the metastatic calcification as well as the osteitis fibrosa normally produced by parathyroid extract intoxication(25,29,30).

The purpose of this paper is to review our work concerning the action of calcitonin and thyroxine on the lesions induced by parathyroid extract overdosage and endogenous (uremic) hyperparathyroidism(31-35).

EFFECT OF THYROXINE ON PARATHYROID EXTRACT OVERDOSAGE

In the rat, parathyroid extract produces severe renal calcinosis and mild lesions in the heart. Histologically, calcification was located in the proximal and distal tubules of the kidney, in the myocardium and in the elastic fibers of the aorta. Severe osteitis fibrosa with osteoclast formation was also observed. Pretreatment with

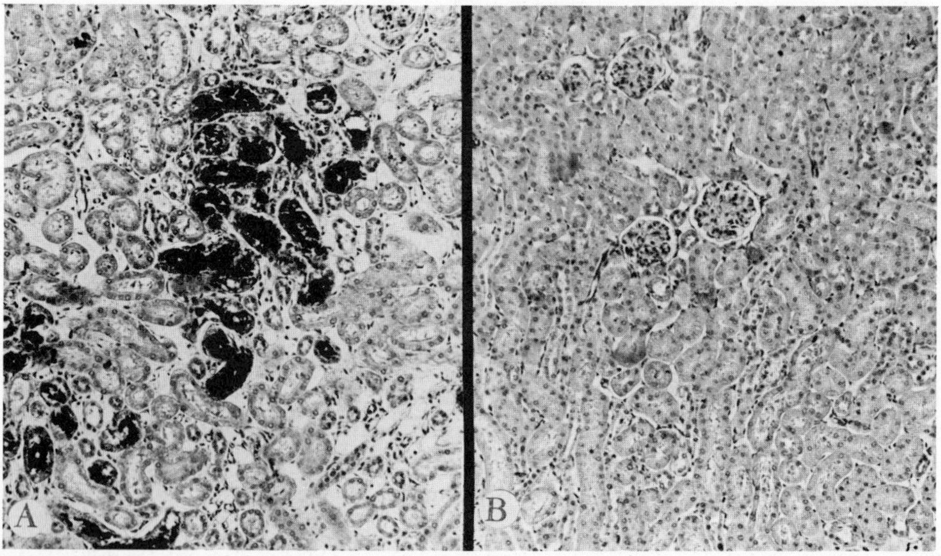

Fig. 1. Inhibition by thyroxine of the renal lesions induced by parathyroid extract overdosage. *A*. Calcification of the renal tubules in a rat treated with parathyroid extract. *B*. Prevention by thyroxine (alcohol-formol, von Kóssa, ×120). From Gabbiani et al. *Endocrinology*, 81:798, 1967.

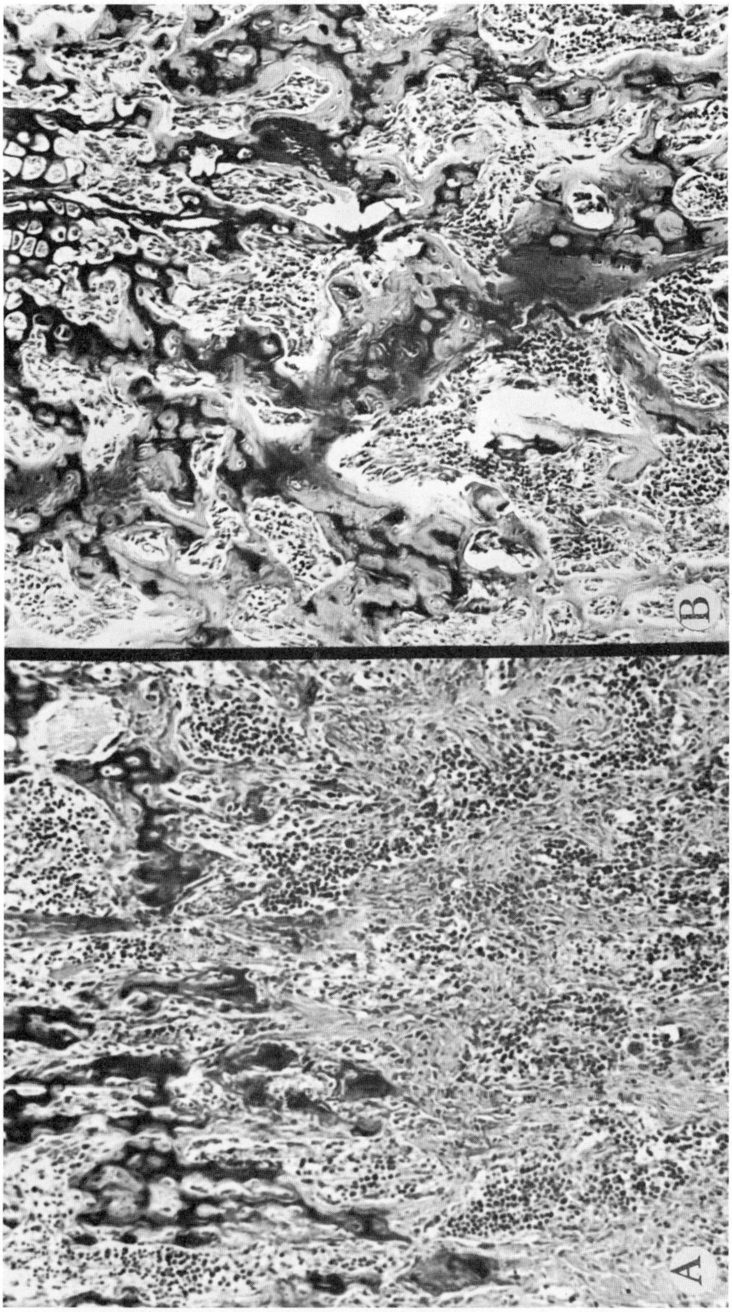

Fig. 2. Inhibition by thyroxine of the bone changes induced by parathyroid extract overdosage. *A*. Osteitis fibrosa in a rat treated with parathyroid extract. *B*. Inhibition by thyroxine (neutral formalin, PAS, ×120). From Gabbiani et al. *Endocrinology*, 81:798, 1967.

TABLE 1

Prevention by thyroxine of parathyroid extract overdosage in the absence of the thyroid and parathyroid glands

Group	Treatment[a]	Lesions (scale 0-3) Heart	Kidney	Bone
1	Thyroparathyroidectomy	1.6	3.0	2.0
2	Thyroparathyroidectomy + thyroxine	0.2	0.4	0
3	Parathyroidectomy	0.9	3.0	2.0
4	Parathyroidectomy + thyroxine	0	0.2	0

[a]In addition, all animals received 60 IU of parathyroid extract subcutaneously twice daily for 2 days.

TABLE 2

Prevention by thyroxine of parathyroid extract overdosage in the absence of the adrenals

Group	Treatment[a]	Lesions (scale 0-3) Heart	Kidney	Bone
1	None	0.7	2.6	2.0
2	Thyroxine	0	0	0
3	Desoxycorticosterone + cortisol	0.4	2.4	2.0
4	Desoxycorticosterone + cortisol + thyroxine	0.1	0.2	0
5	Desoxycorticosterone + cortisol + adrenalectomy	1.8	2.4	2.0
6	Desoxycorticosterone + cortisol + adrenalectomy + thyroxine	0	0.5	0

[a]In addition, all animals received 80 IU of parathyroid extract subcutaneously twice daily for 2 days.

TABLE 3

Action of thyroxine and calcitonin on parathyroid extract overdosage

Group	Treatment[a]	Lesions (scale 0-3) Heart	Kidney	Bone
1	None	2.8	2.6	2.0
2	Thyroxine 100 μg	0	0.2	0.2
3	Calcitonin	0.7	1.4	1.6
4	Thyroxine 25 μg	0.2	1.0	1.0
5	Thyroxine 25 μg + calcitonin	0	0.1	0.2

[a]In addition, all animals were thyroparathyroidectomized on the first day and received parathyroid extract (60 IU subcutaneously) twice daily for 2 days.

thyroxine completely prevented these lesions (Figs. 1 and 2). Thyroxine exerts its protective action in normal, adrenalectomized, or thyroparathyroidectomized rats (Tables 1 and 2). Hence, this action is not mediated through the adrenals, thyroid, or parathyroid glands.

Metabolic studies indicated that in animals receiving parathyroid extract, thyroxine decreases calcemia and increases phosphatemia and phosphaturia without affecting the urinary excretion of calcium and the plasma clearance of calcium and phosphorus(31). Our results agree with previous reports which indicated that thyroxine accelerates the turnover of circulating calcium(3,23,24) and with recent findings on the interactions of these hormones on serum calcium and phosphorus(21, 36).

EFFECT OF CALCITONIN AND THYROXINE ON PARATHYROID EXTRACT OVERDOSAGE

The morphologic manifestations resulting from parathyroid extract overdosage are inhibited by calcitonin and thyroxine and when these hormones are administered concurrently, there is a summation of their actions (Table 3). Under our conditions, calcitonin counteracts the hypercalcemia produced by parathyroid extract but does not affect the renal calcium and phosphorus excretion. These results agree with the findings of other investigators on the metabolic effects of calcitonin(21, 37-43). Preliminary experiments suggested that thyroxine does not modify the hypercalcemic effect of parathyroid hormone in thyroparathyroidectomized animals. However, we have now consistently observed that the hormone decreases hypercalcemia in both intact and thyroparathyroidectomized rats.

EFFECT OF CALCITONIN AND THYROXINE ON ENDOGENOUS (UREMIC) HYPERPARATHYROIDISM

Nephrectomy elicits bone changes which are similar to osteitis fibrosa and are inhibited by thyroparathyroidectomy(26). Furthermore, parathyroid activity may result from the progressive retention of phosphate which in turn could induce reciprocal lowering of calcemia(44). After nephrectomy, changes were observed in the parathyroid cells(26-28) and an increase in the diffusible nonionized calcium fraction of the blood was noted(45). We have compared the effects of calcitonin and thyroxine on the lesions caused by bilateral nephrectomy. Our experiments (Table 4) show that the calcification in the heart and stomach and bone resorption were not

TABLE 4

Action of thyroxine and calcitonin on changes induced by bilateral nephrectomy

Group	Treatment	Lesions (scale 0-3)				
		Heart	Kidney	Bone	Stomach	Mortality[a] (%)
1	None	0	0	0	0	0
2	Thyroxine	0	0	0	0	0
3	Calcitonin	0	0	0	0	0
4	Nephrectomy	0.9	—	1.2	1.3	60
5	Nephrectomy + thyroxine	0.4	—	1.5	1.7	80
6	Nephrectomy + calcitonin	0	—	0	0	30

[a]On the third day after nephrectomy.

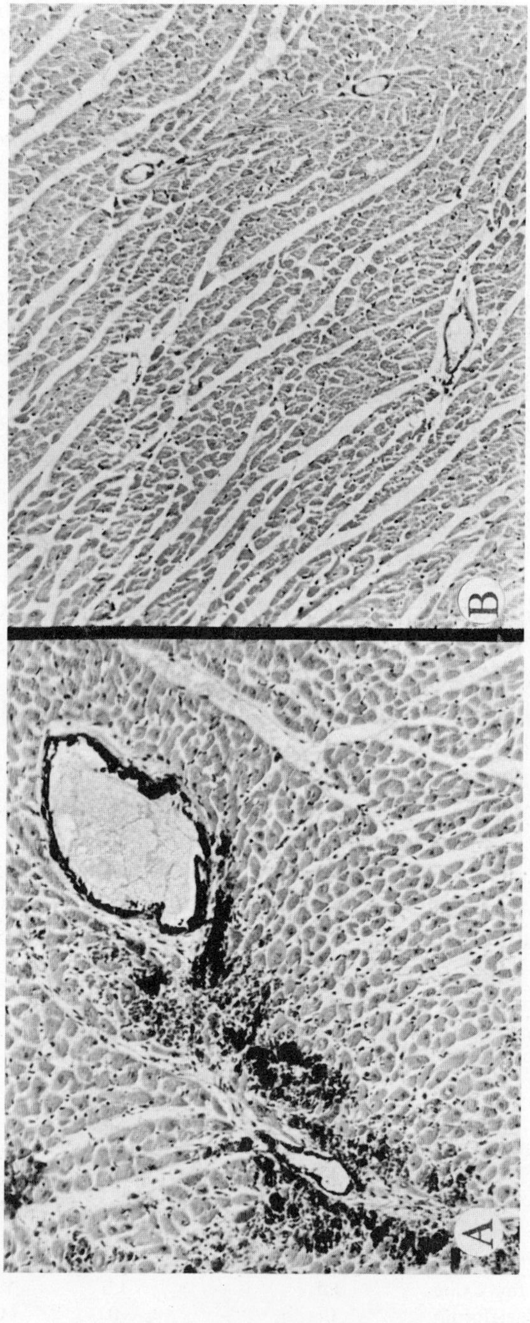

Fig. 3. Inhibition by calcitonin of the cardiac lesions induced by bilateral nephrectomy. *A.* Intense calcification between myocardial fibers and around coronary arteries in a rat after bilateral nephrectomy (von Kóssa, ×120). *B.* Absence of cardiac calcification in an animal receiving additional treatment of calcitonin. From Côté et al. *Acta Endocr.* (Kobenhavn), 59:362-370, 1968.

modified by thyroxine, but were completely inhibited by calcitonin which also prolonged the survival time of the animals.

Histologically, there was marked osteitis fibrosa with giant cell formation and resorption of trabeculae in the subepiphyseal region. Calcification was located in the muscular layer of the stomach and in the elastic fibers of the coronary arteries and the aorta. None of these changes appeared in nephrectomized animals treated with calcitonin (Figs. 3 and 4). Metabolic studies indicate that the changes induced by compensatory overactivity of the parathyroid are also inhibited by calcitonin, thus confirming that removal of the kidney does not alter the plasma calcium lowering effect of the hormone(43,46).

ACTION OF THYROXINE ON PARATHYROID EXTRACT OVERDOSAGE IN THYROPARATHYROIDECTOMIZED AND NEPHRECTOMIZED RATS

As expected, thyroparathyroidectomized and nephrectomized rats showed no signs of endogenous hyperparathyroidism (e.g., soft-tissue calcification and osteitis fibrosa). Mild lesions were observed in animals receiving additional parathyroid hormone treatment. As described above, thyroxine inhibits the intense lesions caused by parathyroid extract in thyroparathyroidectomized rats but not in nephrectomized animals (Table 5). Presumably, thyroxine requires the presence of the kidney to be effective against the changes caused by parathyroid extract overdosage.

EFFECT OF THYROXINE AND DINITROPHENOL ON PARATHYROID EXTRACT OVERDOSAGE

It is known that thyroxine interferes with oxidative phosphorylation(47,48) and causes swelling of mitochondria in vitro(48). Lehninger(49) reviewed these effects and concluded that the hormone acts on an enzymatic system responsible for

TABLE 5

Action of thyroxine on parathyroid extract overdosage in normal thyroparathyroidectomized and nephrectomized rats

Group	Treatment[a]	Lesions (scale 0-3): Heart	Kidney	Bone	Stomach	Mortality[b] (%)
1	None	0	0	0	0	0
2	Nephrectomy	0	—	0	0	30
3	Parathyroid extract	1.7	2.3	1.4	0	20
4	Thyroxine + parathyroid extract	0.1	0.2	0	0	0
5	Nephrectomy + parathyroid extract	0.4	—	0.2	0.4	60
6	Thyroxine + parathyroid extract + nephrectomy	1.3	—	1.5	2.4	60

[a]All animals were thyroparathyroidectomized.
[b]On the third day after parathyroid extract or nephrectomy.

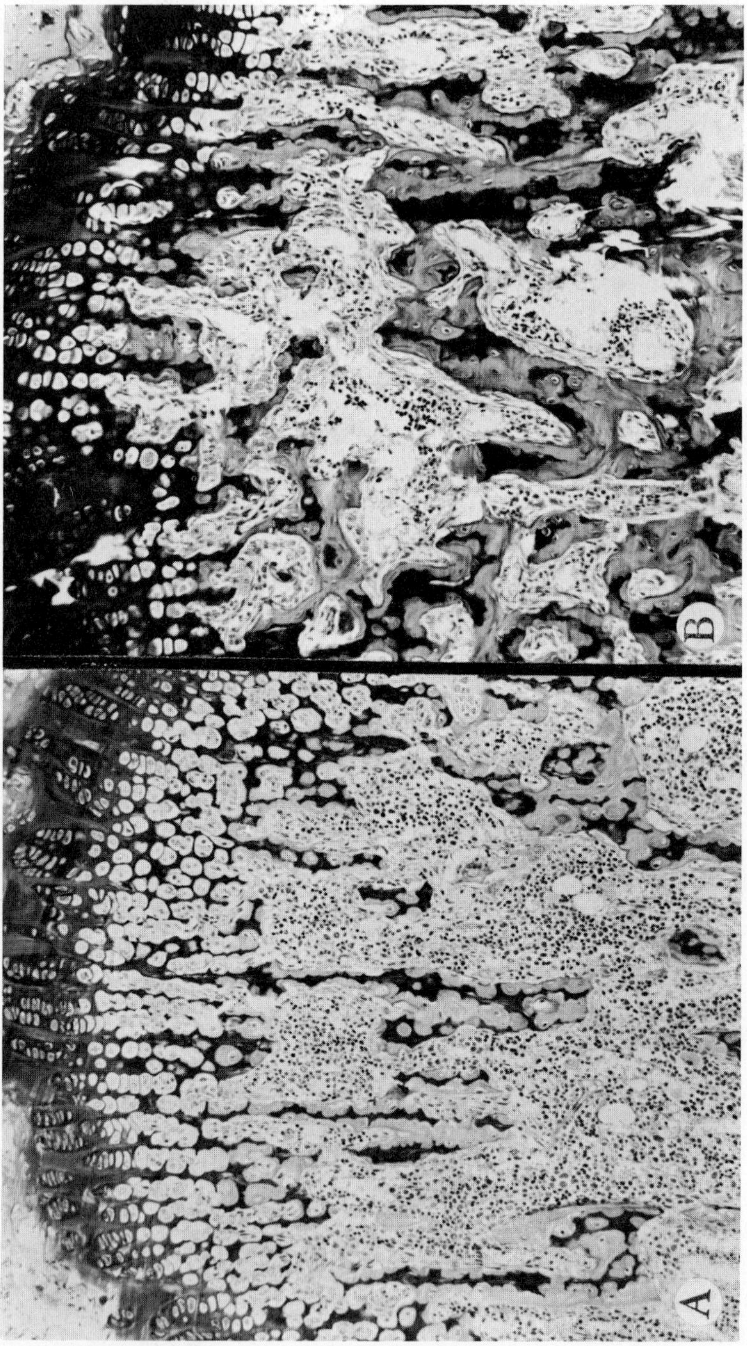

Fig. 4. Inhibition by calcitonin of the bone lesions induced by bilateral nephrectomy. *A*. Osteitis fibrosa in a rat bilaterally nephrectomized (PAS, ×120). *B*. Absence of lesions after treatment with calcitonin. From Côté et al. *Acta Endocr.* (Kobenhavn), 59:362-370, 1968.

the permeability of the mitochondrial membrane. Parathyroid hormone also causes uncoupling of oxidative phosphorylation and swelling of mitochondria(50,51).

The question arose whether the protective effect of thyroxine was due to its specific hypermetabolic effect. We showed that dinitrophenol, an agent uncoupling oxidative phosphorylation, does not duplicate the action of thyroxine. Thus it is clear that thyroxine does not act only through its calorigenic action but an antagonism between parathyroid hormone and thyroxine at the level of mitochondria cannot be excluded.

To summarize, pretreatment with thyroxine or calcitonin inhibits the soft-tissue calcification and osteitis fibrosa induced by parathyroid hormone overdosage. When the hormones are administered concurrently, their actions are summated. Both thyroxine and calcitonin decrease hypercalcemia but only thyroxine counteracts the effect of parathyroid hormone on blood phosphorus levels. Calcitonin inhibits the lesions induced by bilateral nephrectomy while thyroxine does not, although it is very effective against the lesions induced in intact animals by parathyroid extract.

REFERENCES

1. Munson, P. L., Hirsch, P. F., and Tashjian, A. Parathyroid gland. Ann. Rev. Physiol., 25:325–368, 1963.
2. Ellsworth, R., and Nicholson, W. M. Further observations upon the changes in electrolytes of the urine following the injection of parathyroid extract. J. Clin. Invest., 14:823–827, 1935.
3. Nordin, B. E. C. The effect of intravenous parathyroid extract on urinary pH, bicarbonate and electrolyte excretion. Clin. Sci., 19:311–319, 1960.
4. Kleeman, C. R., Bernstein, D., Rockney, R., Dowling, J. T., and Maxwell, M. H. Studies on the renal clearance of diffusible calcium and the role of the parathyroid glands in its regulation. Yale J. Biol. Med., 34:1–30, 1961.
5. Ito, Y., Tsurufuji, S., and Shikita, M. Acute effect of parathyroidectomy on renal tubular reabsorption of phosphorus in rats. Endocr. Jap., 8:68–70, 1961.
6. Beutner, E., and Munson, P. L. Time course of urinary excretion of inorganic phosphate by rats after parathyroidectomy and after injection of parathyroid extract. Endocrinology, 66:610–616, 1960.
7. MacIntyre, I., Boss, S., and Troughton, V. A. Parathyroid hormone and magnesium homoestasis. Nature (London), 198:1058–1060, 1963.
8. Arnaud, C., Rasmussen, H., and Anast, C. Further studies on the interrelationship between parathyroid hormone and vitamin D. J. Clin. Invest., 45:1955–1964, 1966.
9. Talmage, R. V. Studies on the maintenance of serum calcium levels by parathyroid action on bone and kidney. Ann. N.Y. Acad. Sci., 64:326–335, 1956.
10. Gaillard, P. J. Parathyroid gland tissue and bone 'in vitro.' Exp. Cell Res. Suppl., 3: 154–169, 1955.
11. ——— Parathyroid gland and bone in vitro. VI. Develop. Biol., 1: 52–181, 1959.
12. ——— Parathyroid and bone in tissue culture. *In* The Parathyroids. Greep, R. O., and Talmage, R. V., eds. Springfield, Illinois, Charles C. Thomas, Pub., pp. 20–48, 1961.
13. ——— Observations on the effect of parathyroid products on explanted mouse limb-bone rudiments. *In* Gaillard, P. J., Talmage R. V., and Budy, A. M., eds. The Parathyroid Gland. Ultrastructure, Secretion and Function. Chicago, The University of Chicago Press, pp. 145–152, 1965.
14. Raisz, L. G. Inhibition by actinomycin D of bone resorption induced by parathyroid hormone or vitamin A. Proc. Soc. Exp. Biol. Med., 119:614–617, 1965.

15. ——— and Niemann, I. Early effects of parathyroid hormone on bone in tissue culture. Abstr. 19 (58th Ann. Meeting, Amer. Soc. for Clin. Invest., Atlantic City, N.J., May 2) p. 87, 1966.
16. Selye, H., Mortimer, H., Thomson, D. L., and Collip, J. B. Effect of parathyroid extract on the bones of the hypophysectomized rats: a histologic study. Arch. Path. (Chicago), 18:878–880, 1934.
17. McLean, F. C., and Urist, M. R. Bone: An Introduction to the Physiology of Skeletal Tissue. Chicago, University of Chicago Press, p. 182, 1955.
18. MacIntyre, I. Calcitonin. Sci. Basis Med. Ann. Rev., 7:242–252, 1968.
19. Gittes, R. F. Thyrocalcitonin: A new look at urinary calcium homeostasis. Urol. Int., 22:285–300, 1967.
20. Copp, D. H. Hormonal control of hypercalcemia. Amer. J. Med., 43:648–655, 1968.
21. Milhaud, G. La thyrocalcitonine, les connaissances actuelles et les applications cliniques.Rev. Praticien, 18:825–831, 1968.
22. Nordin, B. E. C. Hormones and calcium metabolism. *In* Calcified Tissues. Fleisch, H., Blackwood, H. J. J., and Owen, M., eds. New York, Springer-Verlag New York Inc., pp. 226–241, 1965.
23. Engfeldt, B., and Hjertquist, S. O. The relation between the functions of the thyroid and the parathyroids. Acta Endocr. (Kobenhavn), 9:118–128, 1952.
24. Krane, S. M., Brownell, G. L., Stanbury, J. B., and Corrignan, H. The effect of thyroid disease on calcium metabolism in man. J. Clin. Invest., 35:874–887, 1956.
25. Selye, H., Gabbiani, G., and Jean, P. Inhibition by hypophysectomy of organ lesions normally produced by parathyroid hormone or dihydrotachysterol. Lab. Invest., 11: 1332–1339, 1962.
26. ——— Mechanism of parathyroid hormone action. Arch. Path. (Chicago), 34:625–632, 1942.
27. Baker, B. L. The structural response of the parathyroid glands to ureteral ligation or bilateral nephrectomy. Anat. Rec., 93:125–143, 1945.
28. Davis, R., and Enders, A. C. Light and electron microscope studies on the parathyroid gland. *In* The Parathyroids. Greep, R. O., and Talmage, R. V., eds. Springfield, Illinois, Charles C. Thomas, Pub., pp. 76–92, 1961.
29. Bradford, R. H., Howard, P., Joel, W., and Shetlar, M. R. Antagonistic effects of parathyroid extract and cortisone. Arch. Path. (Chicago), 69:382–389, 1960.
30. Selye, H., Gabbiani, G., and Tuchweber, B. Protection by restraint against parathyroid extract intoxication in the absence of the adrenals. Amer. J. Physiol., 207:573–576, 1964.
31. Gabbiani, G., Tuchweber, B., and Côté, G. Action of thyroxine on experimental parathyroid extract overdosage. Endocrinology, 81:798–802, 1967.
32. Côté, G., Gabbiani, G., Tuchweber, B., and Déziel, C. Effect of thyrocalcitonin and thyroxine on experimental metastatic calcification. Acta Endocr. (Kobenhavn), 59:362–370, 1968.
33. Gabbiani, G., Tuchweber, B., Côté, G., Pahk, U. S., and Selye, H. Influence of thyroparathyroid apparatus on experimental soft-tissue calcification. *In* Parathyroid Hormone and Thyrocalcitonin (Calcitonin). Talmage, R. V., and Bélanger, L. F., eds. Amsterdam, Excerpta Medica Foundation, p. 458, 1967.
34. Côté, G., Gabbiani, G., and Tuchweber, B. Inhibition de l'hyperparathyroidisme expérimental. L'Union Méd. Canad., 96:1489, 1967.
35. Gabbiani, G., Tuchweber, B., Côté, G., and Lefort, P. Action of thyroxine and calcitonin on experimental soft-tissue calcification. Calcified Tissue Res., 2:30, 1968.
36. Milhaud, G., Tsien-Ming, L., and Moukhtar, M. S. Synergie et antagonisme de la thyroxine, de la parathormone et de la thyrocalcitonine sur la calcémie. C.R. Acad. Sci. (Paris), 264:846–849, 1967.
37. Copp, D. H., Cameron, E. C., Cheney, B. A., Davidson, A. G. F., and Henze, K. G. Evidence for calcitonin—a new hormone from the parathyroid that lowers blood calcium. Endocrinology, 70:638–649, 1962.

38. ——— and Henze, K. G. Parathyroid origin of calcitonin. Evidence from perfusion of sheep glands. Endocrinology, 75:49–55, 1964.
39. Hirsch, P. F., Gauthier, G. F., and Munson, P. L. Thyroid hypocalcemic principle and recurrent laryngeal nerve injury as factors affecting the response to parathyroidectomy in rats. Endocrinology, 73:244–252, 1963.
40. Foster, G. V., Baghdiantz, A., Kumar, M. A., Slack, E., Soliman, H. A., and MacIntyre, I. Thyroid origin of calcitonin. Nature (London), 202:1303–1305, 1964.
41. Soliman, H. A., Robinson, C. J., Foster, G. V., and MacIntyre, I. Mode of action of calcitonin. *In* Calcified Tissues. Fleisch, H., Blackwood, H. J. J., and Owen, M., eds. New York, Springer-Verlag New York Inc., pp. 242–243, 1965.
42. Gittes, R. F., and Irwin, G. L. Roles of thyroxine and thyrocalcitonin in the response to hypercalcemia in rats. Endocrinology, 79:1033–1039, 1966.
43. Evanson, J. M., Garner, A., Holmes, A., Lumb, G. A., and Stanbury, S. W. Interrelations between thyrocalcitonin and parathyroid hormone in rats. Clin. Sci., 32:271–278, 1967.
44. Albright, F. Glandular physiology and therapy. The parathyroids—physiology and therapeutics. J.A.M.A., 117:527–533, 1941.
45. Freeman, S., Breen, M., and Meintzer, R. *In* The Parathyroids. Greep, R. O., and Talmage, R. V., eds. Springfield, Illinois, Charles C. Thomas, Pub., pp. 262–274, 1961.
46. Hirsch, P. F., Voelkel, F. F., and Munson, P. L. Thyrocalcitonin: hypocalcemic hypophosphatemic principle of the thyroid gland. Science, 146:412–413, 1964.
47. Maley, G. F., and Lardy, H. A. Efficiency of phosphorylation in selected oxidations by mitochondria from normal and thyrotoxic rat livers. J. Biol. Chem., 215:377–388, 1955.
48. Lehninger, A. L. Physiology of mitochondria. Relationship between mitochondrial structure, oxidative phosphorylation and the action of thyroxine. *In* Enzymes: Units of biological structure and function. Gaebler, O. H., ed. Henry Ford Hosp. Int. Symp. New York, Academic Press, Inc., p. 217, 1956.
49. ——— Thyroxine and the swelling and contraction cycle in mitochondria. Ann. N.Y. Acad. Sci., 86:484–493, 1960.
50. Utzumi, K., Sallis, J. V., and DeLuca, F. D. Parathyroid hormone-induced mitochondrial swelling. J. Biol. Chem., 241:1128–1131, 1966.
51. Vitale, J. J., Nakamura, J., and Hegsted, D. M. The effect of magnesium deficiency on oxidative phosphorylation. J. Biol. Chem., 228:573–576, 1957.

38

The Mode of Action of Parathyroid Hormone and its Effects on Cultivated Embryonic Mouse Radii

J. W. Hekkelman, M. P. M. Herrmann-Erlee, R. K. A. van Wermeskerken, and P. J. Gaillard

Laboratory of Cell Biology and Histology,
University of Leiden, Leiden
The Netherlands

INTRODUCTION

By cultivating fragments of parietal embryonic mouse bone with pieces of parathyroid gland Gaillard(1) showed the occurrence of a typical lacunar resorption of the bone tissue. The same effect was found when the bone fragments were cultivated with parathyroid extract (PTE). Histological examination revealed the presence of many giant multinuclear osteoclasts during the resorption process while the typical osteoblasts had disappeared. These experiments raised our interest in the mechanism of action of parathyroid hormone (PTH) and its relation to the developmental processes in embryonic bones.

To obtain a better insight into the rather complicated resorption process, a biochemical analysis of the mode of action of PTH was undertaken. This study was based on the observations of Neuman and Neuman(2) and several other authors that the effect of PTH on bone tissue causes a solubilization of calcium and an increased production of lactate and citrate. From the biochemical point of view, the latter effect looked especially interesting because it is known that citrate has a strong tendency to form a complex with calcium ions. It was supposed that the stimulated production of citrate could be caused by a decreased rate of oxidation of this metabolite. For this reason we started with a study of the first two enzymes involved in the oxidation of citrate: aconitase and isocitrate dehydrogenase.

INTERMEDIARY METABOLISM

In order to get a sufficient enzyme activity we used the diaphyseal parts of young adult male rabbits. Injection of PTE (100 IU/kg body weight) 16 hours prior

TABLE 1

Effect of PTE (1 IU/ml) and PTH on the NADP and NADPH content of 18-day-old mouse embryo calvaria

Time of cultivation	NADP Percentage[a] decrease	NADP P	NADPH Percentage decrease	NADPH P
1 hr 50 min[b]			(10) 15.4 ± 6.8	< 0.05
3 hr	(9) 29.8 ± 7.0	< 0.01	(6) 16.6 ± 4.9	< 0.05
6 hr	(11) 24.0 ± 4.8	< 0.05	(7) 18.9 ± 4.8	< 0.05
16 hr	(6) 41.3 ± 3.3	< 0.001	(6) 49.0 ± 6.6	< 0.01

[a]The percentages are the means ± S.E. of values obtained from the number of paired experiments given in brackets.
[b]0.5 μg/ml PTH (Aurbach preparation).

to death caused no change in the aconitase activity of the bone extract(3). However, it was found that this enzyme is inhibited by calcium ions, so it is not impossible that the hormone-induced increase in calcium concentration results in an inhibition of aconitase and hence in a stimulated citrate production.

With regard to the NADP-dependent isocitrate dehydrogenase we found a complicated situation. At first a decrease of the enzyme activity was found after administration of the hormone, but it turned out that this effect was due to a decreased extractability, and not to a decrease in the amount of enzyme(4). It appeared that the extractability of isocitrate dehydrogenase depends on the concentration of the coenzyme NADP. These results suggested that the hormone action causes a diminished NADP level in the cell.

Consequently, the effect of PTE on the NADP(5) and NADPH(6) levels was studied. These experiments were performed using cultivated mouse embryo calvaria in which the levels of coenzymes were estimated by ultramicrotechniques. From Table 1 it appears that already after 3 hours a significant decrease in the amount of coenzyme can be found. This finding points to a decreased activity of isocitrate dehydrogenase in situ under the influence of the hormone, which might result in an increased production of citrate.

Recapitulating our results, an action of PTH on the oxidation of citrate is most probable and can be explained by an inhibition of aconitase by calcium ions as well as by a decreased concentration of the coenzyme NADP. The first effect is certainly of a secondary nature so we continued our research by investigating the metabolism of NADP.

ALKALINE PHOSPHATASE

Owing to technical problems we were not able to study the synthesis of NADP in extracts of bone tissue, so we concentrated our efforts on the breakdown reactions. The breakdown of NADP was studied using extracts of diaphyseal rabbit

TABLE 2

NADP and NAD breakdown in bone tissue extracts[a,b]

	pH 7.4	pH 8.8
NAD breakdown	5.4 ± 0.8	13.9 ± 1.5
NADP breakdown	147 ± 27	692 ± 85
Formation of NAD from NADP	139 ± 22	712 ± 74

[a]Mean activities from 6 young adult male rabbits are expressed as mμmoles/minute/g bone ± S.E.

[b]From Hekkelman. Proceedings of the 4th European Symposium on Calcified Tissues, Noordwijk, 1966, Excerpta Medica International Congress Series No. 120. Reproduced by courtesy of the publisher.

bone(7). Table 2 shows that NADP is dephosphorylated to NAD by an enzyme which may be called an alkaline phosphatase. The high activity of the NADP breakdown as compared with the NAD breakdown appeared to be a unique feature of bone tissue when compared with marrow, kidney, liver, and brain tissue. Efforts to separate the enzyme responsible for this reaction from the well-known unspecific alkaline phosphatase have been completely unsuccessful. Experiments with inhibitors have not shown any difference in behavior of these activities and we therefore assume that the unspecific alkaline phosphatase is responsible for the NAD phosphatase activity.

Fractionated extraction and centrifugation experiments showed that about 70 percent of the enzyme is bound to cellular membranes. This in in agreement with current electron microscopical data which show that alkaline phosphatase is localized on the cell membrane. These reasons, together with the very high activity of the enzyme, make it improbable that the NADP breakdown is its only function. It has been postulated(7) that the alkaline phosphatase is part of a transport system for phosphate or phosphate and calcium.

This hypothesis is supported by the finding that in all probability the phosphatase also has an ATPase activity. Efforts to separate the two activities by electrophoresis, extraction, and centrifugation were unsuccessful while the effect of inhibitors on both activities was rather similar. In addition, each activity is inhibited by the substrate of the other as is shown in Figure 1. This is in agreement with the view that both activities belong to one active center.

Let us go back now to the effects of the hormone. If PTH indeed causes an increased breakdown of NADP this could be the result of a change in localization or activity of the phosphatase. The determination of the maximal enzyme activity in tissue homogenates will not be very helpful for getting information on this point because in this way possible changes in localization cannot be found. For this reason we studied the enzyme activity in cell fractions rather than in cell extracts. As a result of the relatively rapid appearance of the decrease in NADP level (Table 1) we have to look for an effect appearing within a few hours at the maximum. In order to avoid diffusion problems and the interference from calcium or phosphate ions from the bone, we studied the effect of PTE on the phosphatase activity of intact

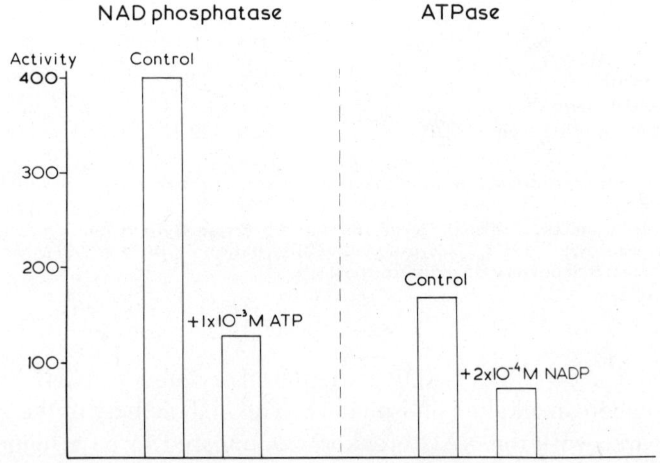

Fig. 1. Mutual inhibition by the substrates of NADPhophatase and ATPase. Activity expressed as mμmoles/minute/g bone. From Hekkelman and Herrmann-Erlee. *In* Parathyroid Hormone and Thyrocalcitonin, Excerpta Medica, Amsterdam, 1968. Reproduced by courtesy of the publisher.

cells isolated from embryonic rat calvaria(8). These cells were incubated in a medium containing 0.8 percent sodium chloride, 0.1 percent glucose, 0.8 mM magnesium chloride, 3 mM potassium chloride, and 0.02 M Tris buffer, pH 8.5. In 1 ml of this medium with 1 IU PTE/ml or the PTE-solvent 200,000 to 300,000 cells were suspended and incubated at 37°C in an Erlenmeyer flask of 5 ml under mild shaking. After 30 minutes, 0.1 ml 2.2 mM NADP dissolved in the same medium was added, and after incubation for another 30 minutes, the suspension was centrifuged for 5 minutes at 300 *g*. In the supernatant, the amount of NAD formed was measured spectrofotometrically as described earlier(7). The cells were homogenized by sonification for 15 seconds in 1 ml of the same medium and centrifuged for 15 minutes at 25,000 *g*. The residue, containing all types of cellular membranes and mitochondria, was resuspended in 1 ml of the same medium with PTE or placebo and after 5 minutes preincubation 0.1 ml NADP was added. After 30 minutes, the suspension was cooled, and centrifuged for 15 minutes at 25,000 *g*. The amount of NAD formed was then measured in the supernatant. In this way the first determination of the NAD phosphatase activity started when the cells were

TABLE 3

Effect of PTE (1 IU/ml) on the alkaline phosphatase activity of isolated bone cells[a]

	Controls	PTE	*P*
Cells in suspension (30 min PTE)	11.4 ± 0.4	11.1 ± 0.6	n.s.
Cell fragments (60 min PTE)	3.3 ± 0.2	2.1 ± 0.3	< 0.01

[a]Mean activities from 9 experiments are expressed as mμmoles/minute/10^6 cells ± S.E.

treated for 30 minutes with PTE. In the second determination, the cells were incubated for 60 minutes with hormone. The results are shown in Table 3.

It appeared that after a 60-minute treatment with PTE a significant decrease in the phosphatase activity occurred, while after 30 minutes, no effect could be found. In the latter case, however, a possible effect may be masked by the presence of solubilized enzyme which in this phase is not removed by centrifugation. The result, together with our knowledge concerning the localization of the enzyme, points towards an early effect of the hormone on the cell membrane.

METABOLISM OF RNA

A different approach to the study of the mode of action of PTH was followed by studying the hormone effects on the ribonucleic acid metabolism(9). Actions on this level will have a direct bearing on protein synthesis, and consequently on the synthesis of enzymes, which will be reflected in metabolic changes. In this study the uptake and incorporation in RNA of tritium-labeled uridine was measured. Calvaria of 20-day-old rat embryos were cultivated in a medium consisting of 15 percent human serum in Hanks salt solution. After 12 hours of precultivation with PTE or a purified PTH (1 IU/ml) a significant increase in the uridine incorporation into the low-molecular-weight-fraction as well as in the RNA fraction was found. To study whether the increased uptake of uridine as reflected by the increased radioactivity of the low-molecular-weight-fraction is a direct result of the hormone action or is caused by the stimulated RNA synthesis we performed a series of experiments in which the RNA synthesis is blocked by the addition of actinomycin D. After 11 hours of cultivation with 1 IU PTE/ml the calvaria were incubated for 1 hour in a similar medium to which 1 μg/ml actinomycin D was added. In addition, a pulse-labeling with tritiated uridine was performed during 1 hour in the presence of actinomycin D. It appears from Table 4 that the incorporation of label in RNA is rather low as compared with the uptake in the low molecular fraction and that nevertheless a considerable stimulating effect of the hormone is found. It follows from this result that the effect on the incorporation of label in the low-molecular-weight-fraction cannot be caused by the effect of RNA synthesis. Consequently it is probable that PTH has a direct effect on the uptake of uridine into the cell. The

TABLE 4

Effect of PTE (1 IU/ml) on the rate of incorporation of uridine-GL-^3H into embryonic rat calvaria in the presence of actinomycin D (1 μg/ml)[a]

Fraction	Unit	Controls	PTE-treated	Percentage increase	*P*
Low-molecular	dpm/μg DNA/h	12,342 ± 541	17,359 ± 766	41.0 ± 4.5	< 0.01
RNA	dpm/μg RNA/h	473 ± 15	533 ± 29	12.9 ± 6.2	< 0.01

[a]Results are given as means ± S.E. of 6 paired observations.

results of these experiments favor the view that PTH has an effect on transport mechanisms in the cellular membrane.

CELL AGGREGATION

Information concerning hormone actions on cellular membranes can also be obtained by studying effects on cell aggregation. Toward this end, suspensions of cartilage cells were prepared from the epiphyses of the humeri of 16-day-old mouse embryos(10). A suspension of 3 ml of 400,000 to 500,000 cells per ml in a medium consisting of 15 percent human serum in Hanks' salt solution with 60 μg DNAase (Sigma) per ml were incubated in a 25-ml Erlenmeyer flask at 30°C on the center of a rotating table (50 rpm) for 24 hours. The experimental flasks contained 2 IU PTE/ml and the control ones a corresponding amount of PTE-solvent. After fixation, the cells were studied by the usual histological methods. It appeared that in both the control and experimental flasks numerous aggregates occurred. However, the aggregates in the experimental flasks were smaller and more variable in size than in the control ones. The latter aggregates showed a strong tendency to become confluent while the experimental aggregates remained easily discernible. The characteristic difference in appearance of the aggregates formed in PTE-containing medium as compared with control ones can be clearly seen in Figures 2 A and B. It is likely that this effect is caused by a change in the structure of the periphery of the cell. These results are in agreement with the opinion that PTH in some way influences the cellular surface.

Taken together, the results of the biochemical and biological experiments de-

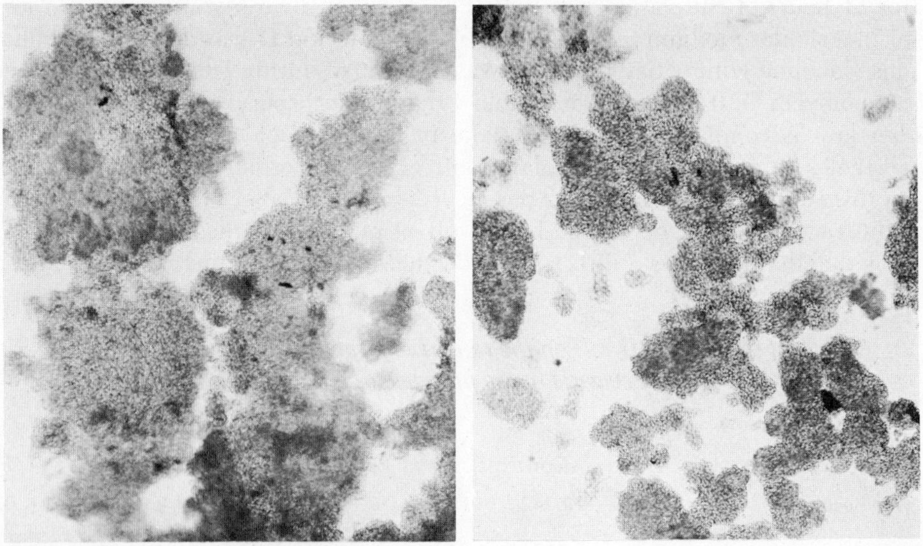

A B

Fig. 2. Aggregates of cartilage cells after 24 hours of incubation in rotating flasks. Unstained living cultures, magnification ×45. *A*. control, *B*. 2 IU PTE/ml added.

scribed in this paper produce evidence supporting the view that PTH has an early membrane effect which causes a number of metabolic changes. However, these experiments are too simple to obtain information on the hormone effect on developmental processes in the sense that they are dealing with only one type of cell or that differences between cells are not taken into account. For this reason we will now consider the morphological effects of the hormone on an embryonic organ.

ORGAN CULTURE EXPERIMENTS

To investigate the influence of PTH on a developing organ, tissue culture experiments were performed using 15-day-old embryonic mouse radii(1). These limb bone rudiments comprise different types of tissue in different developmental stages: viz., the subperiosteal layer of bone in the process of formation; the cartilaginous epiphyses with cells in different stages of maturity, and the connective tissues inside the shaft and surrounding the primordium. Cultivation for 2 days

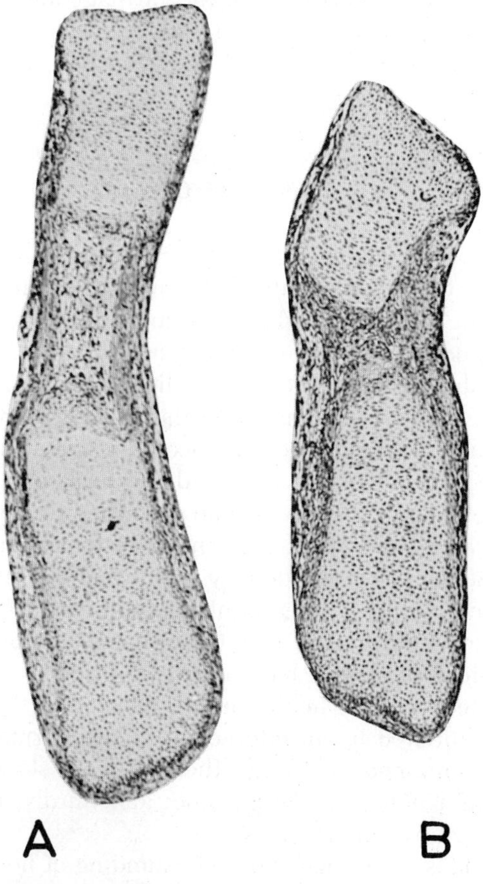

Fig. 3. Azan-stained sections of radii of 15-day-old mouse embryos cultivated for 48 hours. Magnification ×50. *A*. control, *B*. 2 IU PTE/ml added.

with 0.1 to 2.0 IU PTE/ml in a medium consisting of 15 percent human serum in Hanks salt solution showed histological changes in all three types of tissue. It characteristic histological phenomena occurred(1). As regards the bone, three crite-
connective tissue, an increased cell density in the shaft can be observed.
ria became evident: (1) the disappearance of typical osteoblasts combined with the loss of bone formation; (2) the occurrence of bone resorption, a process by which bone matrix as well as bone mineral is dissolved; and (3) an increase in the number of multinucleated osteoclasts. The cartilage also showed three typical phenomena: (1) a striking difference in the shape of the terminal cartilages which are more conically shaped with the apex turning towards the center of the rudiment; (2) in the same region the azurophilia of the cartilage matrix is decreased; and (3) the zone of hypertrophic cells, the most mature part of the cartilage, has undergone a process of transformation in the direction of a "connective tissue." As regards the connective tissue, an increased cell density in the shaft can be observed.

With these seven criteria, a percentage of the observed positive reactions can be calculated providing a means for estimating the hormone concentration in the culture medium. Thus it was found that for the cartilage and the connective tissue, 50 percent positive reactions were obtained with a concentration of approximately 0.03 IU PTH/ml, while the bone tissue already reached this score with approximately 0.01 IU PTH/ml.

CONCLUSIONS

At present it is not yet possible to draw a complete picture of the relations of the morphological effects of PTH on cultivated embryonic limb bone rudiments with the metabolic effects of the hormone described in the first part of this paper. Some of the hormone-induced phenomena are not too difficult to explain, e.g., the resorption of bone has already been discussed earlier in this paper, and the inhibition of bone formation, together with a change in the secretory activity of the chondrocytes (a decreased amount of acid mucopolysaccharides expressing itself as a loss of azurophilia), may be a result of the impaired synthesis of extracellular material indicated by our aggregation experiments and described by several authors. On the other hand, nothing can be said about the conic transformation, and loss of mature cartilage cells, and the increased cell density in the shaft. To bridge this gap in our understanding we have to take into account that the translation of the effects observed in tissue culture into terms of metabolic changes is complicated by the fact that the hormone effects probably have different origins. A rough division into three catagories of effects may be helpful in this respect. In the first place, the rapid metabolic effects, initiated by an interaction of the hormone with the cellular membrane, can be mentioned; secondly, the relatively slowly appearing effects resulting from an interaction with the genome; and thirdly, the complex indirect consequences of the first two interactions.

These considerations imply that the understanding of hormonal effects on developmental processes will not be complete when the primary site (or sites) of action of the hormone are elucidated. After this crucial phase, at which we have not yet arrived, the long and laborious work of unravelling all the complex interrelations of metabolic systems with developmental processes will still be a challenge for further research.

REFERENCES

1. Gaillard, P. J. Parathyroid and bone in tissue culture. *In* R. O. Greep and R. V. Talmage, eds. The Parathyroids, Springfield, Illinois, Charles C. Thomas, Pub., p. 20, 1961.
2. Neuman, W. F., and Neuman, M. W. The chemical dynamics of bone mineral. Chicago, University of Chicago Press, p. 143, 1958.
3. Hekkelman, J. W. Bone metabolism and the action of parathyroid extract. Thesis, University of Leiden, 1963.
4. ——— Enzymological studies on the influence of parathyroid extract on bone metabolism. *In* P. J. Gaillard, R. V. Talmage, and A. M. Budy, eds. The Parathyroid Glands, Chicago, University of Chicago Press, p. 221, 1965.
5. Herrmann-Erlee, M. P. M. The in vitro effect of parathyroid extract on the amount of NADP in embryonic mouse calvaria. Biochem. Biophys. Res. Commun., 23:1–4, 1966.
6. ——— The in vitro effect of parathyroid extract and cyclic AMP on NADP and NADPH concentrations in embryonic mouse calvaria. Acta Endocrinologica, Supp. 138: 189, 1969.
7. Hekkelman, J. W., and Herrmann-Erlee, M. P. M. The possible role of alkaline phosphatase in the chain of actions of parathyroid hormone on bone cell metabolism. *In* R. V. Talmage and L. F. Bélanger, eds., Parathyroid Hormone and Thyrocalcitonin, Amsterdam, Excerpta Medica Congress Series, 159:273–281, 1968.
8. ——— and Moskalewski, S. The effect of parathyroid extract on the alkaline phosphatase activity of the surface of isolated bone cells. Exp. Cell Res., 58:283–288, 1969.
9. Van Wermeskerken, R. K. A. The action of parathyroid extract on the incorporation of ^{3}H-uridine by bone in tissue culture. Proc. Kon. Nederl. Akad. Wet. (C), 71:85–98, 1968.
10. Gaillard, P. J., Moskalewski, S., Verhoog, M. J., and Wassenaar, A. M. The effect of parathyroid extract on the aggregation and the secretion of isolated chondrocytes. Proc. Kon. Nederl. Akad. Wet. (C), 72:521–536, 1969.

39

Hormonal Factors Influencing Bone Resorption in Tissue Culture*

Lawrence G. Raisz,[1] David C. Klein,[2] Clarence L. Trummel,[2] and John S. Brand[3]

[1]Departments of Pharmacology and Medicine, University of Rochester School of Medicine and Dentistry, Rochester, New York
[2]Department of Pharmacology, University of Rochester School of Medicine and Dentistry, Rochester, New York
[3]Department of Radiation Biology and Biophysics, University of Rochester School of Medicine and Dentistry, Rochester, New York

INTRODUCTION

Organ cultures of embryonic bone were first used to study resorption and to demonstrate its stimulation by vitamin A and parathyroid hormone by Fell(1) and Gaillard(2). We adapted their methods to a quantitative bioassay which could measure stimulation of bone resorption by small amounts of parathyroid hormone, although it was not sufficiently sensitive to detect the concentrations in blood(3). The rate of bone resorption is presumed to be the major controlling system for the regulation of extracellular ionized calcium concentration. If this were the only function of bone resorption, a simple parathyroid feedback might suffice for its control. However, resorption of the skeleton for growth and remodeling is quantitatively much more important; this presumably requires a regulation independent in serum ionized calcium concentration. We have used organ cultures to examine factors which influence bone resorption and its hormonal responses hoping to learn how dual regulation of this system might be achieved. Since the major regulator of resorption

*This paper is based on work performed partially under Grants AM 06205 and 2TI-DE 3-2, and partially under contract with the U.S. Atomic Energy Commission at the University of Rochester Atomic Energy Project and has been assigned Report No. UR 49-1017.
†Burroughs Wellcome Scholar in Clinical Pharmacology.

appears to be parathyroid hormone, particular attention has been paid to factors influencing the parathyroid response. In addition we have examined changes which occur in tissue culture in the rate of "control resorption," that is, the slower but qualitatively indistinguishable resorption that occurs in fetal rat bone cultured without parathyroid hormone. This phenomenon presumably is analogous to the slow, but structurally appropriate, rate of bone turnover and remodeling which occurs in parathyroidectomized animals.

PARATHYROID HORMONE (PTH) AND THYROCALCITONIN (TCT)

Parathyroid hormone has a rapid but small initial effect on the release of calcium from bone to the surrounding medium in tissue culture, and a subsequent augmented effect on resorption which presumably involves not only an increased rate of calcium release but cell transformation and the removal of matrix(4). This sequence is not completely defined, but the initial step may be independent of new RNA synthesis while the subsequent transformation appears to depend upon stimulation of RNA synthesis. This stimulation may be specific for osteolytic cells, particularly osteoclasts, and their precursors(5).

Once the stimulation of bone resorption by large doses of parathyroid hormone is initiated in tissue culture, it can proceed at an increasing rate until all of the bone disappears. The relative effects on mineral and matrix can be followed by measuring the release of previously-incorporated radioisotopes, either ^{45}Ca or hydroxyproline ^{3}H from the bone into a chemically-defined culture medium(6) (Fig. 1). The data for ^{45}Ca release are complicated by a large component of exchange between the labeled Ca of the bone and the stable Ca of the medium but, on the basis of stable calcium analysis and bone specific activity, the difference between PTH-treated and control bones can be ascribed largely to net resorption. The difference between PTH-treated and control bones reaches a peak on the third day. By the fourth day PTH-treated bones are nearly completely resorbed and lose calcium or hydroxyproline at a less rapid rate, while the control bones begin to show an increasing rate of resorption. It has been assumed that mineral removal preceeds matrix removal in vivo although the two processes must occur close together. In tissue culture we can measure a small initial lag between the effects of parathyroid hormone on calcium and matrix removal, although eventually proportional amounts of calcium and matrix are lost from the bone (see Fig. 1). One possible reason for this lag is the time required for the synthesis, secretion, and action on matrix of the collagenolytic enzymes released by bone cells.

The discovery of calcitonin has not only added to our understanding of feedback control of serum calcium concentration, but has provided new information about the control of bone resorption. Calcitonin is produced by the ultimobranchial or C cells which are found in the thyroid of mammals, but can constitute a separate organ in other vertebrates. Calcitonin is secreted in response to increased serum calcium concentration and probably plays an important role in preventing the adverse effects of hypercalcemia(8). Its role in the physiologic control of growth and development remains uncertain, although calcitonin is probably present in the serum at

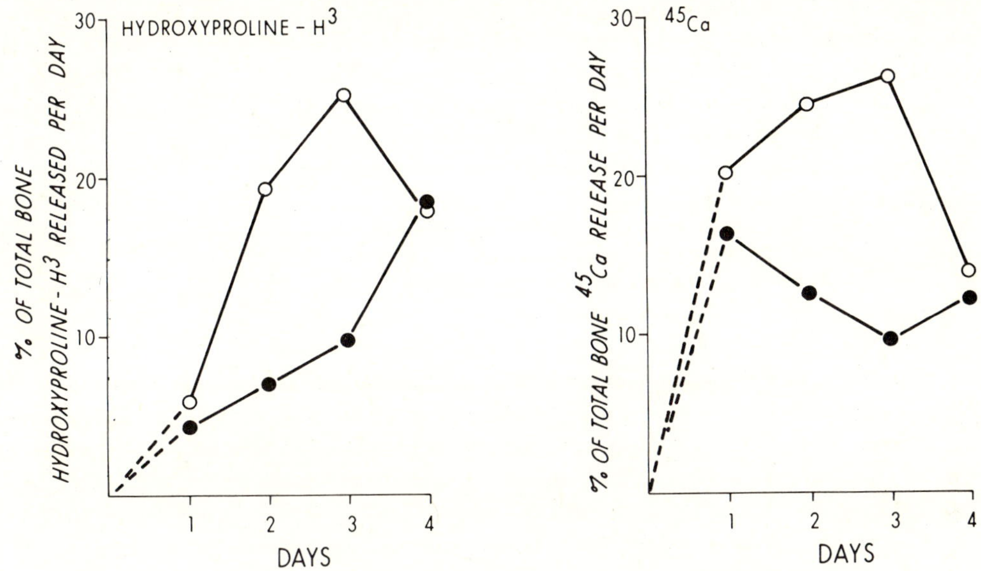

Fig. 1. Time course of the release of previously incorporated hydroxyproline-^{3}H or ^{45}Ca from fetal rat bone in tissue culture. The bone shafts of the radius or ulna were dissected from 19-day fetal rats obtained from mothers which had been injected with either 1 mc of proline-^{3}H or 0.5 mc of ^{45}Ca on the previous day. Each culture consisted of four bones in 0.5 ml of chemically-defined medium which was changed each day. The radioactivity of isolated hydroxyproline or ^{45}Ca was measured in the media and in the bones at the end of 4 days cultivation. Points represent the mean percent of total incorporated hydroxyproline-^{3}H or ^{45}Ca released per day from 4 to 8 cultures. The difference between PTH-treated cultures (1 μg/ml, open circles) and paired control cultures (closed circles) was significant on the second, third, and fourth day for ^{45}Ca release and on the second and third day for hydroxyproline-^{3}H release. The curve for ^{45}Ca represents the composite of loss by exchange which occurs largely in the first 2 days and net resorption which increases with time. On the basis of dead bone experiments and of stable calcium analysis, losses by exchange represent about 20 percent of total bone ^{45}Ca.

normal calcium concentrations and therefore could modify bone metabolism continuously. The major action of calcitonin is to inhibit bone resorption. When calcitonin is added to tissue cultures of embryonic bone which are treated with parathyroid hormone, there is a marked inhibition of the effect of calcium release, but at least initially no effect on the PTH-stimulated removal of labeled hydroxyproline (Fig. 2). Ultimately hydroxyproline release is decreased, presumably because, with decreased calcium release, no more demineralized or partially demineralized matrix is available for resorption. In vivo thyrocalcitonin administration results in an early decrease in urinary hydroxyproline excretion. In some studies there appears to be a lag between the effects of the hormone on calcium and hydroxyproline in vivo. These data and other pieces of evidence suggest that thyrocalcitonin does not act at the same site as parathyroid hormone. Calcitonin is equally effective in inhibiting bone resorption when stimulated by other agents such as vitamin A, vitamin D, or dibutyryl-3′,5′-cyclic adenosine monophosphate. The effects of calcitonin on bone resorption in tissue culture are not persistent. After several days treatment, bones

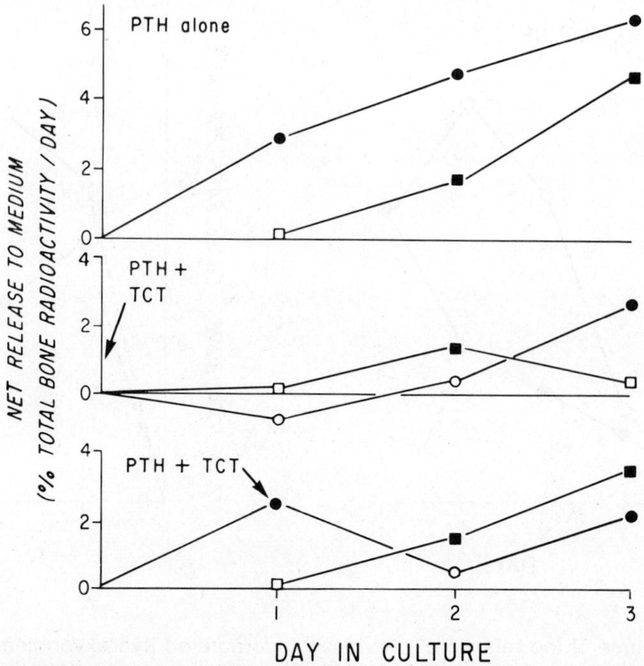

Fig. 2. The effect of calcitonin on PTH-stimulated bone resorption. Data are for an experiment identical to that in Figure 1 except that the medium was 50 percent serum in Basal Medium Eagle. The values are for net release; that is, the difference in percent of total radioactivity released per day between hormone-treated and paired control cultures. Circles indicate ^{45}Ca release and squares hydroxyproline-^{3}H release. Closed points indicate significant difference from untreated controls. In the upper graph, purified parathyroid hormone (PTH, 1 μg/ml) was added alone. There was significant increase in ^{45}Ca release on the first day but hydroxyproline-^{3}H release does not increase above controls until the second day. In the middle graph, PTH was added with partially purified rat thyroid calcitonin (TCT, 0.3 μg/ml) throughout. In the lower graph the PTH was added throughout and TCT was added only after the first day. TCT inhibited ^{45}Ca release for 2 days when added initially but only for 1 day when added on the second day. Significant inhibition of hydroxyproline release was seen only after 3 days' continuous treatment with TCT at a time when the bones were actually beginning to escape from the TCT inhibition of ^{45}Ca release.

show a decreased responsiveness to the inhibitory effects of the calcitonin, although this can be overcome by increasing the dose(7,9). Calcitonin does not affect RNA synthesis(4). It appears to decrease the formation of bone-resorbing cells particularly osteoclasts(10); in addition histologic evidence has been presented that osteoblasts are increased in TCT-treated bones(11).

In vivo, parathyroid hormone has a relatively prolonged effect involving cell transformation, while thyrocalcitonin has a relatively transient inhibitory effect. The half-life of thyrocalcitonin has been estimated as only 4 minutes, even shorter than the 20-minute half-life of parathyroid hormone. However, the biological effects of these hormones may not depend upon the concentration in the blood. Prolonged changes in resorption of embryonic bone can be observed in tissue culture after only brief periods of exposure to large doses of parathyroid hormone (Fig. 3). One possible explanation for this phenomenon is that parathyroid hormone pro-

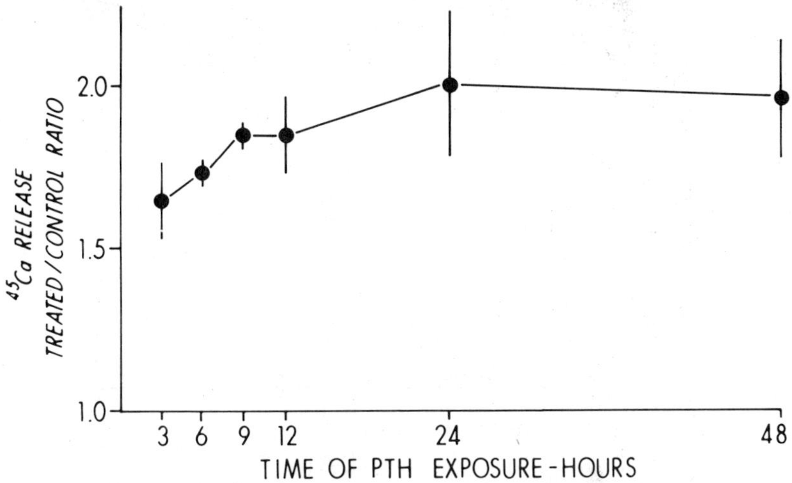

Fig. 3. Effect of varying the duration of initial exposure on the parathyroid hormone response. Groups of fetal rat bones were precultured for 18 to 24 hours in chemically-defined medium without hormone. At time zero paired bones were transferred to cultures with or without parathyroid hormone (PTH, 1 μg/ml). After the indicated times of exposure the PTH-treated bones were washed and transferred to fresh control medium; the control pairs were also washed and transferred to fresh control medium. The bones for the 48-hour continous PTH points were washed and transferred to fresh PTH-containing medium at 24 hours. The cumulative release of ^{45}Ca from 0 to 48 hours was compared in PTH-treated and control bones. Points represent the mean ± S.E. of 4 pairs of cultures. Forty-eight and 24-hour PTH exposures gave the same increase in ^{45}Ca release. The responses after 9 to 12 hours of exposure were not significantly different from 24 hours. The pooled values for 3 and 6 hours are significantly less than for 9 and 12 hours ($P < 0.05$).

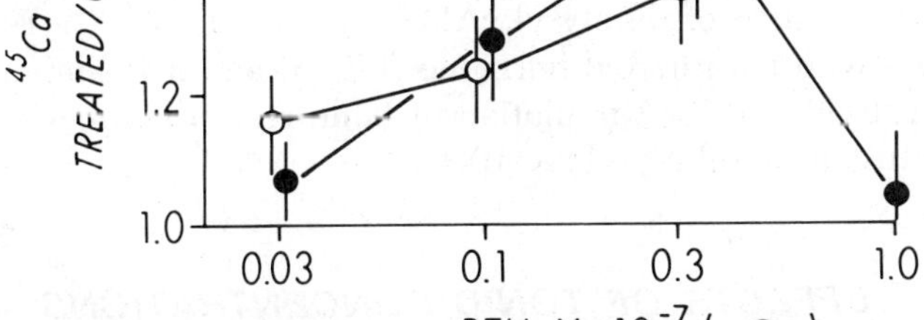

Fig. 4. Effect of dibutyryl-3′,5′-cyclic adenosine monophosphate on bone resorption compared with parathyroid hormone. Paired fetal rat bones were cultured in chemically-defined medium for 48 hours with or without the indicated concentrations of dibutyryl cAMP or PTH. Points represent the means ± S.E. for treated/control ratio of release of previously-incorporated ^{45}Ca from 4 pairs of bone cultures.

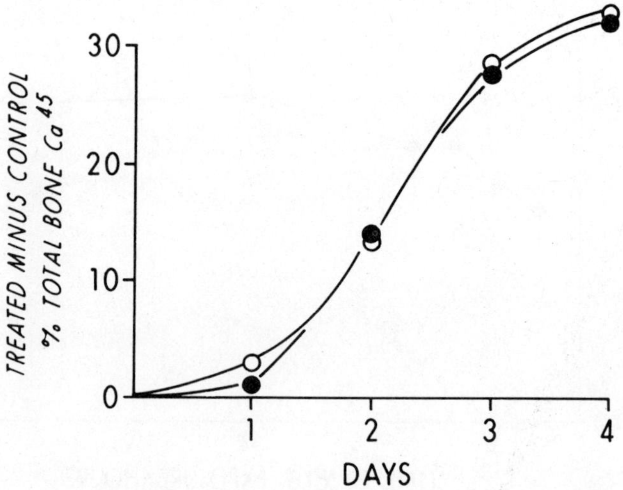

Fig. 5. Comparison of a time course of ^{45}Ca release from bones treated with parathyroid hormone or dibutyryl cAMP. Cumulative net release as percent of total bone radioactivity is shown for 4 groups of cultures treated with dibutyryl cAMP 10^{-4}M (closed circles), or parathyroid hormone 10^{-7}M (open circles).

duces an intermediate step or a "second messenger" in the bone cells which then induces the changes leading to increased resorption. One such "second messenger" is 3′,5′-cyclic adenosine monophosphate (cAMP), and the formation of this material by adenyl cyclase does appear to mediate hormone-induced responses in a number of other systems(12). Parathyroid hormone was shown to stimulate adenyl cyclase activity and to increase the urinary excretion of cAMP in renal cells by Chase and Aurbach(13). Subsequently(14) they showed that the adenyl cyclase activity of homogenized embryonic bone could be stimulated directly by parathyroid hormone. In tissue culture we have found that cAMP does not have any effect on bone resorption but that a dibutyryl derivative does increase the release ^{45}Ca from embryonic bone. The dose-response curve is similar to that for PTH at low doses but at higher doses the effect is lost perhaps because of some toxic or inhibitory effect (Fig. 4). The time course of dibutyryl cAMP stimulation of ^{45}Ca release is similar to that observed with parathyroid hormone (Fig. 5) and it is similarly inhibited by thyrocalcitonin. Evidence for stimulation of bone resorption in vivo has been obtained using infusions of dibutyryl cAMP(15).

EFFECTS OF TONIC CONCENTRATIONS

Little is known of the actual mechanism by which parathyroid hormone increases the movement of calcium from bone to blood. One recent hypothesis is that the hormone increases the permeability of bone cells to calcium(16) and that removal is the result of active extrusion of this calcium by existing calcium transport mechanisms in the cell. A corollary of this hypothesis is that the induction of new

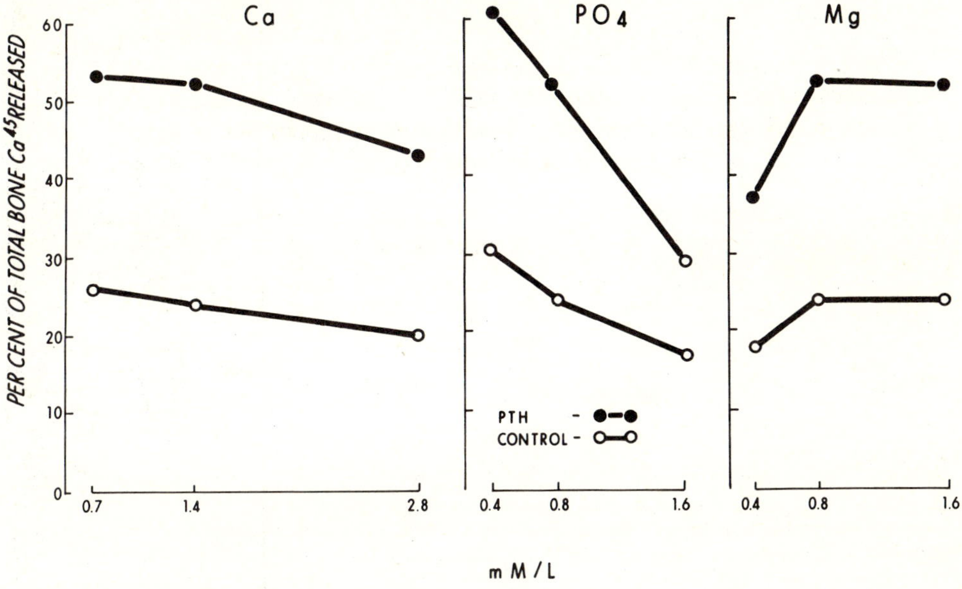

Fig. 6. Comparison of the effects of changing calcium, phosphate, and magnesium concentration on ^{45}Ca release in control and PTH-treated cultures. Values are mean percent of total bone ^{45}Ca released during 48 hours in culture at the indicated ion concentrations. The middle points in each chart are for the same bones in a "standard" medium. Parathyroid hormone (1 μg/ml) caused a significant ($P<0.01$) increase in ^{45}Ca release except for the group cultured at the highest phosphate concentration (open circle).

cell transport machinery need not be due to the hormone itself but to the change in ionic composition of the cell(17). Such mechanisms may be involved in the remarkable sensitivity of bone resorption in tissue culture to variations in the ionic concentration of the medium. We have found that ^{45}Ca release from both control and PTH-stimulated bones can be inhibited by increasing the concentrations of phosphate or calcium or by decreasing the concentrations of magnesium or hydrogen ion(3,18,19). The effects of changing phosphate concentration are striking (Fig. 6) and appear to be due to direct inhibition of the cellular processes involved in bone resorption rather than to physicochemical changes in bone solubility or a primary effect on bone formation. Inhibition of PTH-stimulated bone resorption is observed when phosphate is increased even if calcium-phosphate product is held constant by decreasing medium calcium concentration (Table 1). Moreover, the decrease in ^{45}Ca release observed with increasing phosphate is not accompanied by a decrease in bone specific activity. If resorption of the labeled bone were maintained in high phosphate media, and the ^{45}Ca released were redeposited together with a large amount of stable calcium, the observed decreases in medium ^{45}Ca should be associated with a decreased specific activity in the bone. Phosphate inhibits the active resorption of calcium from living bones more than the dissolution of mineral from heat-killed bones, and increasing phosphate inhibited not only calcium by hydroxyproline release from PTH-stimulated bones(18). The inhibitory effect of phosphate

TABLE 1

Effect of changing PO_4 concentration with constant Ca or constant Ca × PO_4 on release of previously incorporated ^{45}Ca from embryonic bones cultured in a chemically defined medium with parathyroid hormone (PTH, 1 μg/ml) or PTH plus thyrocalcitonin (TCT, 0.3 μg/ml)[a]

Group	Ion concentration (mM/1)		^{45}Ca release (% total radioactivity)			Bone specific activity (cpm/μg Ca)	
	Ca	PO_4	PTH	PTH + TCT	difference	PTH	TCT
I	1.5	0.6	42	24	18	3,000	2,900
Ca constant	1.5	1.8	27	15	12	2,800	3,100
II	3.0	0.4	37	17	20	2,400	2,800
Ca × PO_4 constant	0.8	1.8	28	17	11	3,400	3,200

[a]Values are means for four bones. Partially purified rat TCT was used with a potency of 5 to 10 MRC U/mg.

differs from that of calcitonin in this regard and also is more sustained; there is no tendency for escape from phosphate inhibition. The two agents can produce additive effects in tissue culture, and there is some evidence that this also can occur in vivo. The availability of two agents which appear to lower serum calcium by different mechanisms may have important implications for therapy.

In contrast to the effects of phosphate, the effects of calcium on ^{45}Ca loss appear to be parallel for PTH-treated bones, untreated living controls, and heat-killed bones(19). The effect of magnesium was of some interest. Increasing concentrations did not alter ^{45}Ca release but, at concentrations similar to those encountered in severe magnesium deficiency, resorption was decreased in both control and PTH-treated bones, with no change in dead bone (Fig. 6). This may be part of a general cellular effect of magnesium deficiency; nevertheless it could explain the observation that magnesium-depleted humans showed decreased responsiveness to both the osseous and renal effects of parathyroid hormone(20). The effects of hydrogen ion resemble those of magnesium in that an effect on resorption is seen only at abnormally low concentrations. At pH values of 7.5 or above, bones show no PTH response while increasing hydrogen ion concentration from pH 7.4 to 6.9 does not appear to alter bone resorption(3,19).

The implications of these ion effects on bone resorption in tissue culture for the regulation of bone turnover in vivo are not known. Certainly the situation is more complicated in intact animals. Ions may simultaneously affect either bone formation or resorption and may act either directly or indirectly by altering the secretion of hormones. For example, increased phosphate concentrations could not only inhibit bone resorption in vivo but also increase mineralization of existing bone matrix and soft tissue deposition of calcium phosphate. These effects would all decrease the ionized calcium concentration in serum and hence increase parathyroid hormone secretion (and perhaps decrease calcitonin secretion). Since the major effect of parathyroid hormone on calcium is to increase resorption, it is possible that the end result of phosphate loading in the intact animal would be an increase in bone resorption due to compensatory overactivity of the parathyroid. Just such a response has been observed in animals maintained on high-phosphate, low-calcium diets(21). Further study of the effects of phosphate on bone turnover may help to explain the positive correlation between serum phosphate and growth rate observed in several mammalian species. A major implication of our studies is that the interpretation of hormonal responses in bone requires knowledge of concurrent ionic composition. Hence measurements of concentration or even rates of secretion of parathyroid hormone and calcitonin may not give sufficient information about bone resorption unless the responsiveness of the bone cells is also assessed.

OTHER AGENTS WHICH AFFECT BONE METABOLISM

Thus far we have considered only the effects of parathyroid hormone and calcitonin on bone resorption and their interaction with the ionic environment. There are many other agents which have important effects on bone metabolism including

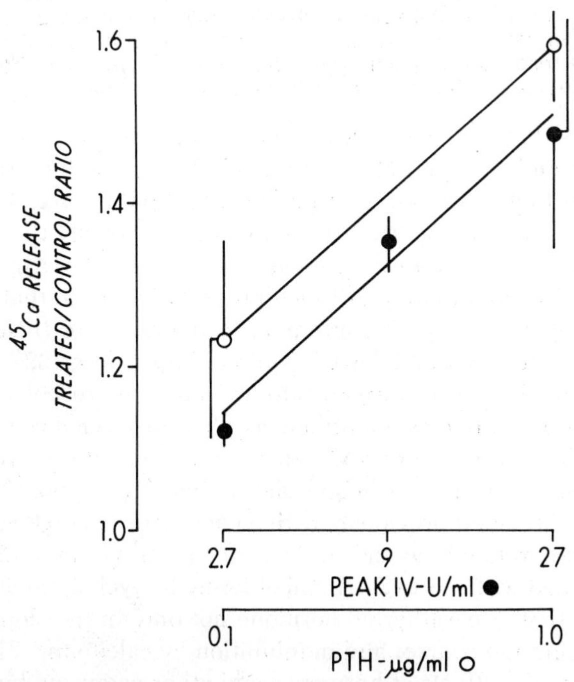

Fig. 7. Comparison of dose-response curves for Peak IV vitamin D_3 metabolite and parathyroid hormone. The metabolite of vitamin D was isolated from Peak IV of silicic acid column chromatograms of extracts of hog serum after the administration of vitamin D_3, and has been identified as 25-hydroxycholecalciferol. Its activity was 60 IU/μg so that the doses indicated here are from 0.045 to 0.45 μg/ml or about 10^{-7} to 10^{-6}M. Doses of PTH used were 10^{-8} to 10^{-7}M. Values are means ± S.E. for treated/control ratio of release of ^{45}Ca for 4 paired cultures in 48 hours.

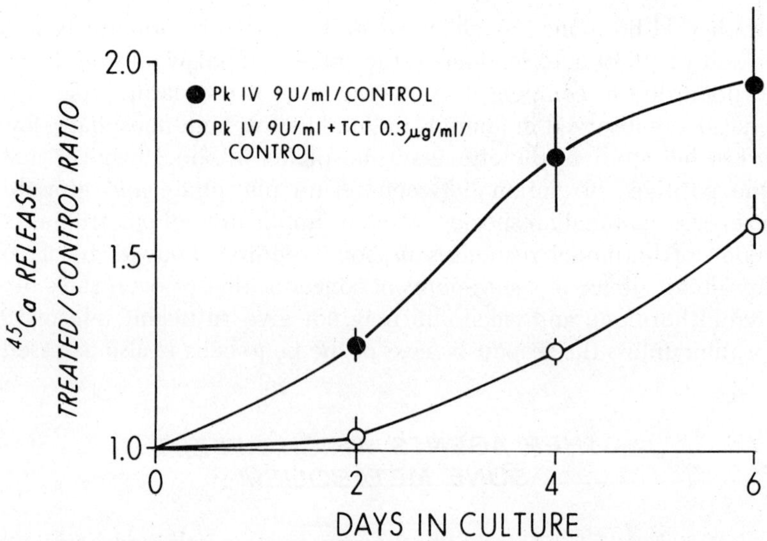

Fig. 8. Time course of response to vitamin D_3 metabolite and its inhibition by calcitonin. The Peak IV metabolite, 25-hydroxycholecalciferol, was added fresh every 2 days and the cumulative release of ^{45}Ca in paired treated and control bones determined. Rat calcitonin (TCT, 0.3 μg/ml) inhibited the response to vitamin D completely at 2 days but subsequently resorption increased. Values are mean ±S.E. for treated/control ratios of 4 paired cultures.

hormones such as thyroxine or glucocorticoids, vitamins A, C, and D, or trace ions such as fluoride and pyrophosphate. Probably the most important influence on bone resorption is that of vitamin D. Vitamin D and parathyroid hormone share the ability to affect calcium transport, although characteristically the effects of vitamin D are most marked on the intestine, while the effects of parathyroid hormone are more striking on bone and kidney. Although there is evidence that vitamin D stimulates bone resorption in vivo(17), attempts to demonstrate this in tissue culture have been inconsistent(3), and have required large doses(22). DeLuca(23) has shown that vitamin D_3 is transformed into an active, more polar metabolite originally called Peak IV and now identified as 25-hydroxycholecalciferol. This compound is somewhat more potent and rapid-acting than the native vitamin in the intestine, but much more potent in stimulating bone resorption(23). In contrast to the erratic results obtained previously with a concentration of 400 IU/ml of vitamin D, a consistent dose-related increase in ^{45}Ca release from embryonic bone in tissue culture was obtained with 0.9 to 27 IU/ml of 25-hydroxycholecalciferol (Fig. 7). The effect resembled that of parathyroid hormone not only in the slope of the dose-response curve, but in time course; and in inhibition by calcitonin (Fig. 8)

This review of the effects of hormones and other agents on bone resorption has necessarily been largely descriptive; only a few biochemical steps are known. Nevertheless the large number of agents available for study should begin an era in which the pharmacology of bone will receive increasing attention using these agents for therapy to explore physiological and pathological mechanisms.

REFERENCES

1. Fell, H. B. The physiology of skeletal tissue in culture. Sci. Basis Med. Ann. Rev., 6: 28–45, 1956–57.
2. Gaillard, P. J. Parathyroid and bone in tissue culture. *In* R. O. Greep and R. V. Talmage, eds., The Parathyroids. Springfield, Illinois, Charles C Thomas, Publ., pp. 20–48, 1961.
3. Raisz, L. G. Bone resorption in tissue culture: Factors influencing the response to parathyroid hormone, J. Clin. Invest., 44:103–116, 1965.
4. ——— and Niemann, I. Early effects of parathyroid hormone and thyrocalcitonin on bone in organ culture. Nature, 214:486–487, 1967.
5. Owen, M., and Bingham, P. J. The effect of parathyroid extract on RNA synthesis in osteogenic cells in vivo. *In* R. V. Talmage and L. F. Belanger, eds. Parathyroid Hormone and Thyrocalcitonin, Proc. 3rd Parathyroid Conference. Amsterdam, Excerpta Medica Foundation, pp. 216–225, 1968.
6. Raisz, L. G., Brand, J. S., Au, W. Y. W., and Niemann, I. Interactions of parathyroid hormone and thyrocalcitonin on bone resorption in tissue culture. *In* R. V. Talmage and L. F. Belanger, eds. Parathyroid Hormone and Thyrocalcitonin, Proc. 3rd Parathyroid Conference. Amsterdam, Excerpta Medica Foundation, pp. 370–380, 1968.
7. ——— Au, W. Y. W., Friedman, J., and Niemann, I. Inhibition of bone resorption in tissue culture by thyrocalcitonin. *In* S. Taylor, ed., Calcitonin, Proc. of the Symposium on Thyrocalcitonin and the C Cells. London, William Heinemann Medical Books, Ltd., pp. 215–222, 1968.
8. ——— Au, W. Y. W., Friedman, J., and Niemann, I. Thyrocalcitonin and bone resorption. Studies employing a tissue culture bioassay. Amer. J. Med., 43:684–690, 1967.
9. Friedman, J., Au, W. Y. W., and Raisz, L. G. Responses of fetal rat bone to thyrocalcitonin in tissue culture. Endocrinology, 82:149–156, 1968.
10. Reynolds, J. J., Dingle, J. T., Gudmundsson, T. V., and MacIntyre, I. Bone resorption in vitro and inhibition by calcitonin. *In* Thyrocalcitonin and the C Cells. London, William Heinemann Medical Books, Ltd., pp. 223–229, 1968.
11. Gaillard, P. J., and Thesingh, C. W. Bone culture studies with thyrocalcitonin. *In* Thyrocalcitonin and the C Cells. London, William Heinemann Medical Books, Ltd., pp. 238–245, 1968.
12. Robinson, G. A., Butcher, R. W., and Sutherland, E. W. Cyclic AMP Ann. Rev. Biochem., 37:149–174, 1968.
13. Chase, L. R., and Aurbach, G. D. Renal adenyl cyclase: Anatomically separate sites for parathyroid hormone and vasopressin. Science, 159:545–547, 1968.
14. ——— Fedak, S. A., and Aurbach, G. D. Activation of skeletal adenyl cyclase by parathyroid hormone in vitro. Endocrinology, 84:761–768, 1969.
15. Rasmussen, H., Pechet, M., and Fast, D. Effect of dibutyryl cyclic adenosine 3′,5′ monophosphate theophylline, and other nucleotides upon calcium and phosphate metabolism. J. Clin. Invest., 47:1843–1850, 1968.
16. Borle, A. B. Calcium metabolism in HeLa cells and the effects of parathyroid hormone. J. Cell Biol., 36:567–582, 1968.
17. Rasmussen, H. The parathyroids. *In* Williams, R. W., ed. Textbook of Endocrinology. Philadelphia, W. B. Saunders Co., pp. 847–965, 1968.
18. Raisz, L. G., Brand, J. S., Klein, D. C., and Au, W. Y. W. Hormonal regulation of bone resorption. *In* Gual, C., and Ebling, F. J. G., eds. Progress in Endocrinology. Excerpta Medica Foundation, Amsterdam, pp. 696–703, 1969.
19. ——— and Nieman, I. Effects of phosphate, calcium and magnesium on bone resorption and hormonal responses in tissue culture. Endocrinology, 85:446, 1969.
20. Estep, H., Shaw, W., Watlington, C. O., Hobe, C., Holland, W., and Tucker, H. St. G. Hypocalcemia due to reversible PTH resistance. Clin. Res., 26:265, 1968.

21. Jowsey, J., and Raisz, L. G. Experimental osteoporosis and parathyroid activity. Endocrinology, 82:384–396, 1968.
22. Goldhaber, P. Some chemical factors influencing bone resorption in tissue culture. *In* R. F. Sognnaes, ed., Mechanisms of Hard Tissue Destruction. Amer. Ass. Advance. Sci. Pub. No. 75, pp. 609–636, 1963.
23. Trummel, C. L., Raisz, L. G., Blunt, J. W., and DeLuca, H. 25-hydroxycholecalciferol: metabolite of vitamin D which stimulates bone resorption in tissue culture. Science, 163:1450, 1969.

40

Early Secretion of Androgenic Hormones by Human Fetal Gonads and Adrenal Glands in Organ Culture and Possible Implications for Sex Differentiation*

J. J. P. Zaaijer[1] and D. Price[2]

[1]Laboratory for Cell Biology and Histology,
University of Leiden, Leiden, The Netherlands
[2]Biology Department, University of Chicago,
Chicago, Illinois

INTRODUCTION

For several mammalian species, convincing evidence has been presented that fetal testicular androgens have an important function in the normal differentiation of the reproductive ducts(1,2). Research on a possible influence of fetal androgenic hormones on the normal differentiation of the reproductive ducts in man has its inherent limitations in experimental approach. However, this basic question can be asked: are fetal androgens available in the human fetus at the time of sex duct differentiation?

With histochemical staining methods and chemical analyses, results have been obtained which indicate that at the end of the first trimester of fetal life testes and adrenal glands are already enzymatically equipped for steroidosynthesis, and capable of in vitro conversion of added precursors into androgenic steroids(3,4,5,6). There is, however, little evidence as yet that the secretion of androgenic steroids normally occurs in vivo at these early fetal ages.

In the experiments to be reported in this paper, fetal gonads and adrenal glands from human fetuses between the estimated ages of 7½ and 20 weeks have been tested for the secretion of androgens by an organ culture bioindicator method. This method(7) is based upon culturing the fetal organs in contact with rat ventral pros-

*This research has been aided by the Nederslandse Organisatie voor Zuiver-Wetetenschappelijk Onderzoek Z.W.O.

tate tissue, which is androgen-dependent for the maintenance of normal structure and secretory function.

MATERIALS AND METHODS

Testicular, ovarian, adrenal, and metanephric pieces taken from 4 male fetuses, 9½, 12, 15, and 20 weeks old, and 4 female fetuses, 7½, 9½, and two of 13 weeks old, were explanted. The age of the fetuses was estimated on the basis of crown-rump (C-R) and crown-heel (C-H) measurements, in combination with clinical data in some cases.

Depending on its size, each organ was cut into three to six fragments, of which one was fixed as an uncultured control. The metanephric explants served as cultured nonendocrine control tissue. The total number of explants of fetal material was 140.

The prostate tissue was taken from the ventral prostate gland of male rats 22 to 25 days after birth. A number of prostate explants were cultured alone, others were used to constitute the combined fetal gland-prostate explants.

TABLE 1

Androgenicity test of human fetal organs

Donor fetus:					Results of tests:	
Sex	Estimated age (weeks)	Length (mm)	Organs explanted with prostate	Total no. of explants	Positive	Negative
Male	9½	[a]C-R: 40	testis	3	3	0
		[b]C-H: 47	adrenal	4	3	1
			metanephros	4	0	4
Male	12	C-R: 65	testis	10	10	0
		C-H: 85	adrenal	9	9	0
Male	15	C-R: 105	testis	10	10	0
		C-H: 160	adrenal	9	9	0
Male	20	C-R: 170	testis	8	8	0
		C-H: 245	adrenal	6	6	0
			metanephros	8	0	8
Female	7½	C-R: 25	ovary	3	0	3
			adrenal	7	5	2
			metanephros	3	0	3
Female	9½	C-R: 40	ovary	4	0	4
		C-H: 49	adrenal	5	5	0
			metanephros	4	0	4
Female	13	C-R: 74	ovary	10	4	6
		C-H: 100	adrenal	12	10	2
Female	13	C-R: 75	ovary	11	7	4
		C-H: 105	adrenal	6	6	0
			metanephros	4	0	4
			prostate cultured alone	18	0	18

[a]Crown-rump.
[b]Crown-heel.

All explants were cultured at 35°C in Falcon Plastic organ culture dishes on a medium composed of cock plasma, chick embryo extract, and Hanks balanced saline solution(7). Pieces of lens paper served to keep the explants from being submerged. After 3 days of culture the explants were transferred to a fresh medium; handling of the tissue was avoided by transporting the explants on their lens paper raft. The total time of culture was 5 to 6 days.

Histotechnical procedures were: fixation in Bouin's fluid, embedding in paraffin, serial sectioning at 8 μ, and staining with Harris hematoxylin.

Stages in Normal Morphogenesis of the Reproductive Tract and the Adrenal Gland (Figure 9)

At 3½ weeks of fetal age the Wolffian (mesonephric) ducts are complete after they have reached the cloaca. The gonad primordia first appear in the genital ridge at 4½ weeks; they are not yet histologically differentiated as a testis or an ovary. The adrenal primordia become distinct at 5½ weeks. At 6 weeks, the Müllerian (paramesonephric) ducts begin their development; from this moment on until sex duct differentiation has been completed, the embryo possesses two sets of reproductive ducts. At 7 weeks the testis is first recognizable as such by its seminiferous cords; Leydig cells appear in the testis interstitium at 8½ weeks. At approximately 8 weeks the adrenal gland has developed an outer and an inner zone; the definitive cortex and the fetal zone. Medullary cells are just beginning to invade the gland at that time. The main differentiation processes of the sex duct take place between the fetal ages of 9½ and 12½ weeks. During this period the Müllerian ducts retrogress in the male fetus and the Wolffian ducts retrogress in the female fetus.

RESULTS (TABLE 1)

Within 24 hours after explantation, the two constituents of the fetal organ-prostate explants were firmly connected and were surrounded by a continuous connective tissue membrane. In histological examination all explants proved to be in good condition at the end of the culture period.

Prostate. Before explantation the rat ventral prostate tissue had the normal histological appearance of the androgen-stimulated gland: large and medium-sized acini lined with columnar epithelium, and sparse interacinous connective tissue.

In all prostate explants cultured alone, the histological structure after 5 to 6 days of culture was typical for prostate tissue which has retrogressed in the absence of androgens: acini mostly small and with low epithelium; interacinous tissue abundant.

Testis. At 9½ weeks the testis showed many, not sharply defined, seminiferous cords and some Leydig cells. At 12 weeks the testis cords were abundant and many Leydig cells were present. At 15 and 20 weeks the numerous testis cords were sharply defined and an extensive interstitium contained a large number of Leydig cells.

In all explants of prostate cultured with testicular tissue of all the ages investi-

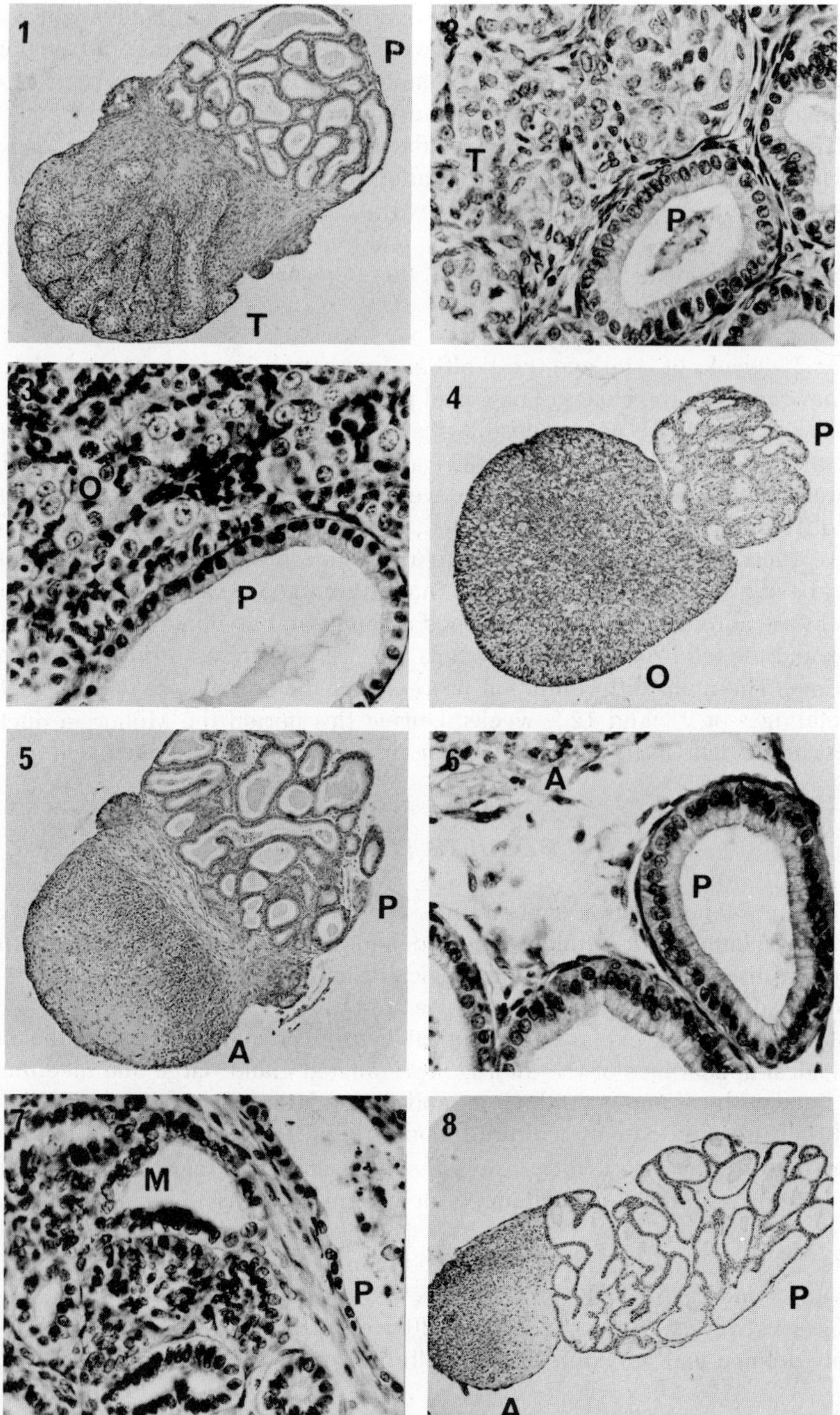
1
P
T
2
T
P
3
O
P
4
P
O
5
P
A
6
A
P
7
M
P
8
P
A

gated, the histological structure of a prostate stimulated by androgens was maintained (Figs. 1 and 2).

Ovary. At 7 ½ weeks the ovary showed little differentiation. At 9 ½ weeks some agglomerations of oocytes were present which, at 13 weeks, had considerably increased in number.

For the ovarian tissue taken from the 7½- and 9½-week-old fetuses, the androgenicity test was negative (Fig. 4). For about half the number of explants from the two 13-week-old fetuses, moderate androgenic activity during the culture period was apparent: some large acini with a moderately high columnar epithelium (Fig. 3). For the other explants in this age-group there was little or no indication of androgenicity.

Adrenal gland. At 7 ½ weeks the definitive cortex and fetal zone areas were not yet distinguishable. At 9 ½ weeks the definitive cortex consisted of a narrow band of small basophilic cells which surrounded a well-developed fetal zone with large nonbasophilic cells. With increasing age of the fetuses, the area of the definitive cortex became wider and the boundary between fetal zone and definitive cortex was more clearly demarcated.

The adrenal explants from the older fetuses (15 and 20 weeks old) all proved to be androgenic (Fig. 5). In the younger age-groups, a few explants showed only slight androgenic activity or none at all; but for all the other explants, from both male and female donors, the result of the androgenicity test was definitely positive (Figs. 6 and 8). In all cases the adrenal explants contained both definitive cortex and fetal zone tissue.

Fig. 1. Testis explant (*T*) from a 15-week-old fetus, cultured with rat ventral prostate (*P*) for 5 days. Testis cords and Leydig cells are present. Normal histological structure of stimulated prostatic tissue is maintained: acini large, with columnar epithelium. For comparison with retrogressed prostate see Figure 4. ×40.

Fig. 2. Testis explant *(T)* from a 9½-week-old fetus, cultured with rat ventral prostate *(P)* for 6 days. Structure of stimulated prostate is maintained: acini with columnar epithelium. For comparison with retrogressed prostate tissue see Figure 7. ×250.

Fig. 3. Explant of ovary (*O*) from a 13-week-old fetus, cultured with rat ventral prostate (*P*) for 6 days. Clusters of oocytes are present. Normal prostatic structure partly maintained: acini with moderately high epithelium. ×250.

Fig. 4. Explant of ovary (*O*) from a 9½-week-old fetus, cultured with rat ventral prostate (*P*) for 6 days. Prostatic tissue retrogressed:small acini with low epithelium, interacinous connective tissue well developed. ×40.

Fig. 5. Adrenal explant (*A*) from a 15-week-old male fetus, cultured with rat ventral prostate (*P*) for 5 days. Definitive cortex and fetal zone tissue are present. Normal structure of stimulated prostate maintained: large acini with columnar epithelium. ×40.

Fig. 6. Adrenal explant (*A*) from a 13-week-old female fetus, cultured with rat ventral prostate (*P*) for 5 days. Maintenance of normal prostatic structure: acini with tall columnar epithelium. ×250.

Fig. 7. Explant of metanephros (*M*) from a 13-week-old female fetus, cultured with rat ventral prostate (*P*) for 6 days. Prostatic tissue retrogressed: acinus lined with very low epithelium. ×250.

Fig. 8. Adrenal explant (*A*) from a 9½-week-old female fetus, cultured with rat ventral prostate (*P*) for 6 days. Definitive cortex and fetal zone tissue are present. Prostatic tissue has maintained its normal histological structure: acini large, interacinous tissue sparse. ×40.

Metanephros. For explants of metanephros no androgenic activity was found (Fig. 7).

DISCUSSION

From histochemical data it seems evident that both fetal testes and adrenal glands are enzymatically equipped for the biosynthesis of steroid hormones from the time just before the beginning of differentiation of the reproductive ducts (Fig. 9). The enzyme 3 β-hydroxysteroid dehydrogenase (3 β-HSD) which is essential in the biosynthesis of nearly all biologically active steroid hormones, has been detected with histochemical staining methods in the testis from approximately 9 weeks on(8,9). In the adrenal, the activity of this enzyme is very weak at 9 weeks but it becomes strong by 12 weeks(10,11). At a stage about 2 ½ weeks prior to the time at which 3 β-HSD activity is first found in the testis and the adrenal the enzyme has been demonstrated in the mesenchyme of the genital ridge of a fetus of unknown sex(12).

The histochemical evidence of potential steroid formation by the adrenal gland at the beginning of the third month of pregnancy is corroborated by electronmicroscopic findings. In the cells of the fetal zone an agranular or smooth endoplasmic reticulum has been demonstrated from approximately 8 weeks on(13,14). This pattern of cytoplasmic structure has been associated in several mammalian species with steroidogenic activity(15).

Early steroid biosynthesis in testis and adrenal gland has been confirmed by chemical analyses demonstrating the capability of the testicular tissue to produce androgenic steroids in vitro and by the detection of androgenic steroids in adrenal tissue. The earliest stage at which in vitro formation of androgenic (C_{19}) steroids from added precursors by testis slices or homogenates could be found is about 9 weeks.(16) At this same age, C_{19} steroids were detected in extracts of adrenal tissue from both male and female fetuses, but hydrocortisone could not be found before 16½ weeks(17). From studies on anencephalic fetuses it has been concluded that the adrenal-pituitary axis is not established before the age of approximately 17 weeks(18).

The ovary is considered to be hormonally quiescent until quite late in development. No potential activity in steroid production has been reported until beyond the stage at which sex duct differentiation is in essence completed. A slight activity of 3β-HSD in ovarian follicle cells as early as the beginning of the fourth month has been mentioned(19). A positive reaction for the enzyme is evident in the interstitium at 19 weeks(6).

Indicative as these findings on enzyme activity, electronmicroscopic structure, and in vitro steroidosynthesis may be, they do not provide decisive evidence for enzyme activity under in vivo conditions or active secretion by intact cells(14,20, 21). Perfusion experiments in the fetal-placental unit, as described by Diczfalusy (22) and Solomon(23), among others, have a much closer relation to the in vivo situation, but these experiments have been mainly concerned with the fetus at midpregnancy. They therefore do not provide data with respect to the young developmental stages which include differentiation processes of the sex ducts.

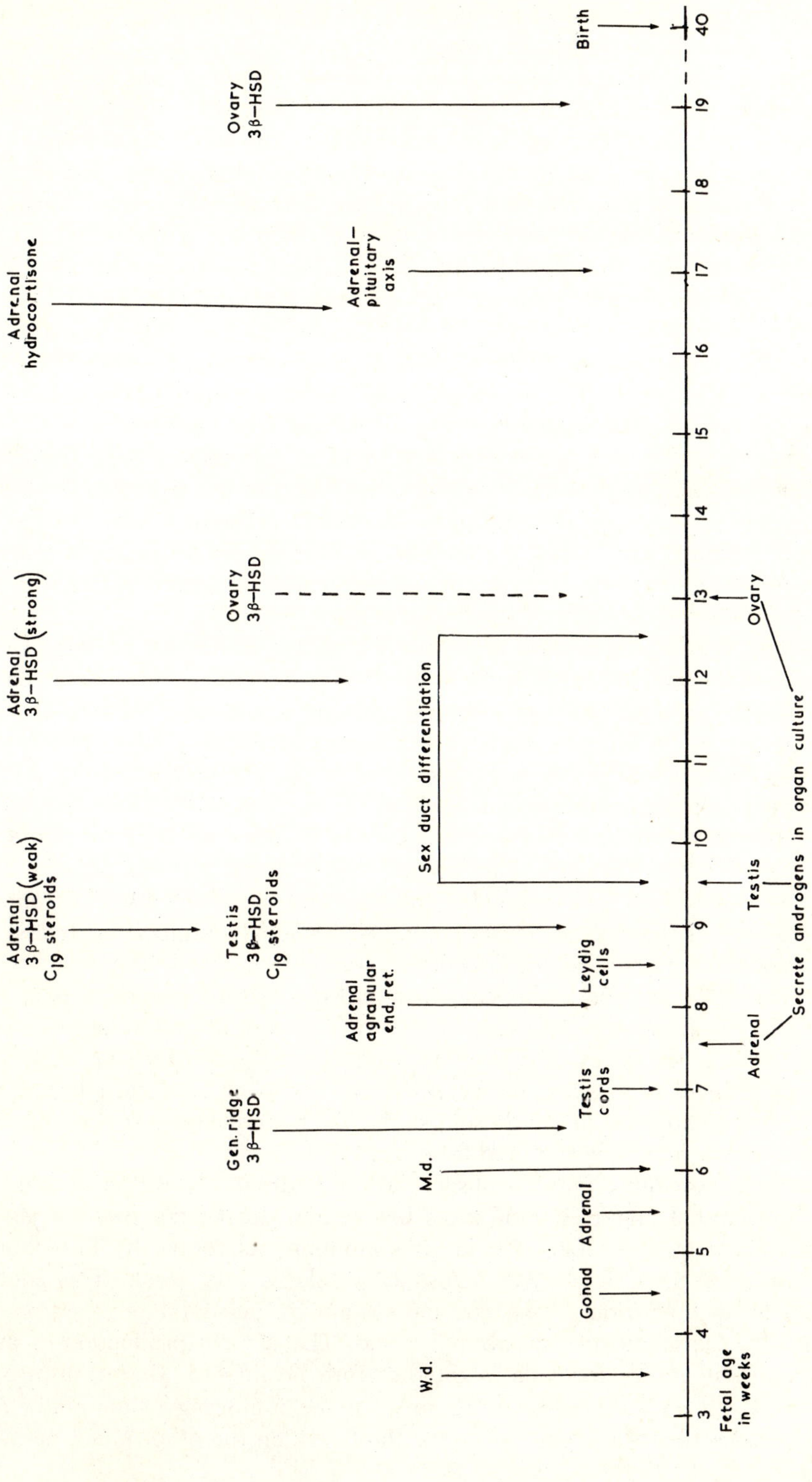

Fig. 9. Stages in the morphological differentiation of the reproductive tract and adrenal gland in relation to evidence for onset of functional activity of gonads and adrenals. Sources for data on functional activity given in text.

In a few cases, organ culture methods have been combined with chemical methods to approximate in certain respects the conditions in vivo. The advantages of this type of experimental approach are apparent. Provided that the culture period is of sufficient duration, the chemical analyses deal with intact cells, and active secretion into the culture medium can be tested. The culture experiments most pertinent here, because they involved young developmental stages, are those described by Bloch et al.(24). In these experiments, adrenal and ovarian tissue from an 11-week-old fetus was cultured for 4 to 8 days. Chemical analyses did not show C_{19} steroid formation by the adrenal tissue from endogenous sources nor from added precursors. The adrenal did, however, convert progesterone to hydrocortisone. The definitive cortex was maintained in good condition during the culture period but the fetal zone tissue degenerated. From the ovary, some conversion products were obtained but no C^{19} steroids. The experiments provided evidence for the capability of intact adrenal and ovarian cells to secrete steroids in organ culture.

In the organ culture experiments described in this paper, the experimental design included the opportunity to examine the effect of the secretion products on a target organ, in this case the androgen-dependent prostate tissue. For the evaluation of the results of this test for androgenicity it should be taken into account that the prostate tissue requires for its maintenance in organ culture the presence of an androgen throughout the entire culture period(7).

The results of applying this bioindicator method to human gonads (Table 1 and Fig. 9) show that the testis secretes androgens in organ culture from the time of onset of sex duct differentiation. According to the bioindicator method, the ovary does not show any androgenic activity at these early stages. The secretion of androgens by the ovary in organ culture soon after the completion of sex duct differentiation confirms histochemical findings at this stage(19). For the adrenal gland, androgen secretion in organ culture could be demonstrated for tissue from both male and female fetuses. The bioindicator test was already positive at the fetal age of 7½ weeks. This coincides with the time at which the electronmicroscopic structure of the cells indicates potential steroidosynthesis. It precedes, however, the chemical indications for C_{19} steroid production. The discrepancy between the outcome of the present bioindicator test and the culture experiments of Bloch et al.(24), in which the adrenal tissue yielded no C_{19} steroids, might be attributable to the fact that, in the latter case, the fetal zone tissue degenerated in culture. The possible differences in function between the two zones of the fetal adrenal were, however, not investigated in the present study; the bioindicator test was performed with combined definitive cortex and fetal zone tissues.

There is a possibility that the observation of androgen secretion in organ culture by the adrenal gland of both sexes before and during the time of sex duct differentiation does not reflect the in vivo situation. As shown by histochemical methods, adrenal 3β-HSD activity before 12 weeks is very weak. This could be attributed to an inhibition of the enzyme system by placental progesterone(25), which might accumulate in the adrenal tissue. The in vivo production of androgenic steroids by young adrenals might therefore be limited. Under culture conditions, however, an inactivation or removal of the inhibiting factor might result in an active secretion of androgens(5). For the testis, on the other hand, no in vivo inhibition of 3β-HSD activity has been postulated. It is thus probable that the androgenic secretion phenomena found in organ culture do represent what is going on in vivo and that the testis secretes androgens at early stages in the human fetus. These androgens would then be available to act as morphogenetic agents in sex duct differentiation.

CONCLUSIONS

Testes from fetuses at the time of sex duct differentiation (9½ to 12 weeks) secrete androgens in organ culture, whereas ovaries do not. These findings complement the chemical data of others and strengthen the evidence that testicular androgens might be available to the male fetus at this time, thereby substantiating the concept of an important role of these androgens in sex duct differentiation in the human.

Androgen secretion by adrenal glands from male and female fetuses is evident in organ culture. Androgen secretion may be limited in vivo through the presence of an enzyme-inhibiting factor.

REFERENCES

1. Price, D., and Ortiz, E. The role of fetal androgen in sex differentiation in mammals. *In* Organogenesis. DeHaan, R. L., and Ursprung, H., eds., New York, Holt, Rinehart & Winston, Inc., pp. 629–652, 1965.
2. Jost, A. Steroids and sex differentiation of the mammalian foetus. Proc. 2nd Int. Congr. Hormonal Steroids; Excerpta Medica Int. Congr. Ser., 132:74–81, 1967.
3. Villee, D. B., and Villee, C. A. Synthesis of corticosteroids in the fetal-placental unit. Proc. 2nd Int. Congr. Endocr. Excerpta Medica Int. Congr. Ser., 83:709–714, 1965.
4. Bloch, E. Hormone production by the foetal adrenocortical gland. Proc. 2nd Int. Congr. Endocr. Excerpta Medica Int. Congr. Ser., 83:785–789, 1965.
5. ——— In vitro steroid synthesis by gonads and adrenals during mammalian fetal development. Proc. 2nd Int. Congr. Hormonal Steroids. Excerpta Medica Int. Congr. Ser., 132:675–679, 1967.
6. Cavallero, C., and Magrini, U. Histochemical studies on 3 β-hydroxysteroid dehydrogenase and other enzymes in the steroid-secreting structures of human foetus. Proc. 2nd Int. Congr. Hormonal Steroids. Excerpta Medica Int. Congr. Ser., 132:667–674, 1967.
7. Zaaijer, J. J. P., Price, D., and Ortiz, E. Organ culture studies of hormone secretion in endocrine glands of fetal guinea pigs. I. Androgenic secretion as demonstrated by a bioindicator method. Proc. Kon. Nederl. Akad. Wet. (Biol. Med.), C 69:389–399, 1966.
8. Baillie, A. H., Niemi, M., and Ikonen, M. 3 β-Hydroxysteroid dehydrogenase activity in the human foetal testis. Acta Endocr. (Kobenhavn), 48:429–438, 1965.
9. Cavallero, C., Magrini, U., Dellepiane, M., and Cizelj, T. Étude histochimique du cortex surrénalien et du testicule chez le foetus humain. Ann. Endocr. (Paris), 26:409–418, 1965.
10. Bloch, E., Tissenbaum, B., Rubin, B. L., and Deane, H. W. Δ^5-3βhydroxysteroid dehydrogenase activity in human fetal adrenals. Endocrinology, 71:629–632, 1962.
11. Niemi, M., and Baillie, A. H. 3β-Hydroxysteroid dehydrogenase activity in the human foetal adrenal cortex. Acta Endocr. (Kobenhavn), 48:423–428, 1965.
12. Baillie, A. H., Ferguson, M. M., and Hart, D. McK. Histochemical evidence of steroid metabolism in the human genital ridge. J. Clin. Endocr., 26:738–741, 1966.

13. Ross, M. H. Electron microscopy of the human foetal adrenal cortex. *In* The Human Adrenal Cortex. Currie, A. R., Symington, T., and Grant, J. K. eds. Edinburgh and London, Livingstone, pp. 558–569, 1962.
14. Johannisson, E. The foetal adrenal cortex in the human. Acta Endocr. (Kobenhavn), 58, (Suppl.):130, 1968.
15. Christensen, A. K. Fine structure of steroid-secreting cells. Proc. 2nd Int. Congr. Hormonal Steroids. Excerpta Medica Int. Congr. Ser., III:179–180, 1966.
16. Bloch, E. Metabolism of 4-^{14}C-progesterone by human fetal testes and ovaries. Endocrinology, 74:833–845, 1964.
17. ——— Benirschke, K., and Rosemberg, E. C_{19} steroids, 17 α-hydroxycorticosterone and a sodium retaining factor in human fetal adrenal glands. Endocrinology, 58:626–633, 1956.
18. Benirschke, K. Adrenals in anencephaly and hydrocephaly. Obstet. Gynec., 8:412–425, 1956.
19. Goldman, A. L., Yakovac, W. C., and Bongiovanni, A. M. Development of activity of 3β-hydroxysteroid dehydrogenase in human fetal tissues and in two anencephalic newborns. J. Clin. Endocr., 26:14–22, 1966.
20. Benirschke, K., Bloch, E., and Hertig, A. T. Concerning the function of the fetal zone of the human adrenal gland. Endocrinology, 58:598–625, 1956.
21. Solomon, S., Bird, C. E., Ling, W., Iwamiya, M., and Young, P. C. M. Formation and metabolism of steroids in the fetus and placenta. Rec. Progr. Hormone Res., 23:297–347, 1967.
22. Diczfalusy, E. Steroid metabolism in pregnancy and in the foeto-placental unit. Proc. 2nd Int. Congr. Hormonal Steroids. Excerpta Medica Int. Congr. Ser., 132:82–95, 1967.
23. Solomon, S. The formation and metabolism of neutral steroids in the human fetus and placenta. Proc. 2nd Int. Congr. Hormonal Steroids. Excerpta Medica Int. Congr. Ser., 132:653–662, 1967.
24. ——— Romney, S. L., Klein, M., Lippiello, L., Cooper, P., and Goldring, I. P. Steroid synthesis by human fetal adrenals and ovaries maintained in organ culture. Proc. Soc. Exp. Biol. Med., 119:449–452, 1965.
25. Villee, D. B. The role of progesterone in the development of adrenal enzymes. Proc. 2nd Int. Congr. Hormonal Steroids. Excerpta Medica Int. Congr. Ser., 132:680–686, 1967.

41

Sexual Differentiation in Epiphysial and Hypothalamic Regulation of the Secretion of Follicle-Stimulating Hormone and Luteinizing Hormone by Fetal Hypophyses Cultured in Vitro

S. E. Levina

Institute of Developmental Biology,
Academy of Sciences of the U.S.S.R.,
Moscow, U.S.S.R.

INTRODUCTION

Previous reports have shown that sex differences in the gonadotropin content of the human fetal hypophysis emerge as early as the fourth month of fetal life for follicle-stimulating hormone (FSH) and the fifth month for luteinizing hormone (LH). Only prolactin has been shown to be secreted in an identical manner in the fetuses of both sexes(1). The present study attempts to identify some of the factors which induce sex differences in the regulation of the secretion of FSH and LH by the fetal hypophysis.

MATERIALS AND METHODS

Human fetal hypophyses were cultivated in medium 199 (37°C, 95 percent O_2, and 5 percent CO_2) for 16 hours. Brain homogenates were added to control dishes. In the experimental cultures, either hypothalamic homogenates (one hypothalamus per one hypophysis) or epiphyses (with a very small piece of surrounding tissue) were added.

The fetuses were stored in the refrigerator for 3 to 12 hours from the time of birth to the time of culturing. The fetal hypophysis was able to survive that long and could be successfully cultured if the fetus had not been dead for some time before delivery. The state of the fetus was checked by histological examination of the adrenals.

The hypophysis was divided into two symmetrical parts: one was placed in the control dish and the other was placed in the experimental culture. The total amount of medium in each dish was 2 ml.

FSH was detected by bioassay on infant mice using a test for increase in ovarian weight(2). After adding 0.4 ml of choriogonin (Richter-Hungary, 40 IU/0.1 ml) to each culture dish, the medium was assayed using four mice and the average ovarian weight was determined. As a standard preparation in the determination of FSH, Humegon (HMG-21.R.P.) was used, which had been provided courtesy of the National Institute of Medical Research of London. FSH data are shown in units of HMG.

LH was detected immunologically by measuring hemagglutination inhibition (3). Choriogonin was used for the preparation of rabbit antiserum and as a standard preparation. LH data are reported in IU of choriogonin (HCG). Forty-eight female and 60 male fetuses between 16 and 40 weeks of development were examined.

RESULTS

None of the control male fetal hypophyses released LH to the medium during the periods of embryogenesis examined, and only some hypophyses obtained from these fetuses (period of between 24 and 28 to 29 weeks of age) released FSH. The hypophyses of about half of the female fetuses (19 to 29 week) secreted either FSH, LH, or both.

The effects of different tissues on the secretion of FSH and LH by fetal hypophyses developing in vitro are shown in Figure 1 and Table 1. It can be seen from Figure 1 that the hypothalami from female fetuses (between 19 and 29 weeks) and from male fetuses (between 24 and 28 to 29 weeks) are able to stimulate secretion of FSH when cultured with the fetal hypophyses. The epiphyses of female fetuses also stimulated FSH secretion (Fig. 1). In contrast, the epiphyses of only three male fetuses (24 to 28 weeks) promoted FSH release slightly (not shown in Fig. 1).

The FSH concentration of the culture media after the addition of male fetal hypothalami (24 to 29 weeks) was about 0.2 IU/ml (HMG units), and after the addition of female fetal hypothalami (19 to 29 weeks), about 0.3 IU/ml. After the addition of female epiphyses, the FSH concentration was at least 0.15 IU/ml.

The ability to promote the secretion of LH was found only in the hypothalami of female fetuses during the 19- to 29-week period and in those of male fetuses during the 24- to 29-week period; the addition of epiphyses from the fetuses of both sexes was ineffective (Table 1). During the second trimester of embryogenesis, most of the female hypothalami (19 to 29 weeks) possessed activity comparable to that exhibited by male specimens (24 to 29 weeks), but some female hypothalami demonstrated considerably greater activity (more than 2 IU/ml). The lat-

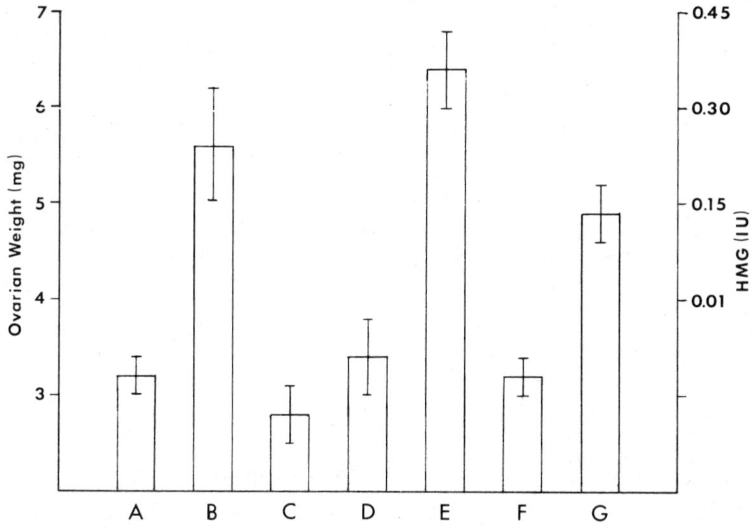

Fig. 1. Bioassay determination of FSH content in medium after culturing fetal hypophyses with the following (number of fetuses examined in parentheses):
A—Fetal brain tissue(44);
B—Male fetal hypothalamus, 24-29 weeks(7);
C—Male fetal hypothalamus prior to and after this period(6);
D—Male fetal epiphysis (8, not including three anomalous 24- to 29-week-old fetuses—see text);
E—Female fetal hypothalamus, 19-29 weeks(9);
F—Female fetal hypothalamus prior to and after this period(6);
G—Female fetal epiphysis(8);

ter were obtained only from some fetuses between the 20th and 23rd weeks of development.

In the absence of stimulation, some female hypophyses secreted FSH and/or LH into the medium spontaneously during the second trimester of gestation, whereas some of the male hypophyses (seven between 24 and 29 weeks) secreted only FSH. The effects on FSH secretion upon adding hypothalamic or epiphysial tissue to the media in which these spontaneously active hypophyses were cultured are shown in Figure 2, and on LH secretion, in Table 1. The addition of these tissues to the culture did not influence secretion in some cases and inhibited FSH and LH secretion in others.

DISCUSSION

It was found that the hypophyses of half of the female fetuses in the second trimester of embryogenesis have the ability to secrete FSH and/or LH into the medium spontaneously. However, this ability is not based on any sex-related differences in the hypophysial cells themselves(1,4,5,6). The spontaneous activity of the fetal hypophysis can be interpreted as the result of a previous high level of activity of the hypothalamus in female fetuses. The ability to secrete LH may be a

TABLE 1

Secretion of LH by the human fetal hypophysis in vitro after the addition of fetal hypothalamus or epiphysis

	Hypophyses from female fetuses		Hypophyses from male fetuses	
Addition to medium	IU/ml of medium[a]	IU/mg of hypophysis[a]	IU/ml of medium[a]	IU/mg of hypophysis[a]
I. In the cases in which the hypophysis does not secrete into the medium spontaneously:				
Fetal brain tissue (81)[b]	—	0.14 ± 0.059	—	0.05 ± 0.007
Male hypothalamus, 24-29 weeks (11)	0.43 ± 0.094	0.10 ± 0.069	0.14 ± 0.049	0.04 ± 0.002
Male hypothalamus prior to and after this period (12)	—	0.28 ± 0.61	—	0.05 ± 0.003
Female hypothalamus, 19-29 weeks (19)	0.41 ± 0.079	0.17 ± 0.089	0.17 ± 0.039	0.03 ± 0.009
Female hypothalamus prior to and after this period (10)	—	0.21 ± 0.056	—	0.04 ± 0.003
Female hypothalamus, 20-23 weeks (4)	3.82 (2.5 to 10.2)		2.06 (1.0 to 3.5)	
Female epiphysis (14)	0.06 ± 0.004	0.18 ± 0.057	—	0.05 ± 0.001
Male epiphysis (15)	—	0.26 ± 0.053	—	0.04 ± 0.006
II. In the cases in which the hypophysis is able to secrete into the medium spontaneously:				
Fetal brain tissue (25)	0.96 ± 0.24	0.39 ± 0.083	—	—
Male and female hypothalamus (11)	0.74 ± 0.09	0.27 ± 0.069	—	—
Male and female epiphysis (21)	0.89 ± 0.09	0.17 ± 0.043	—	—

[a]In IU of choriogonin.
[b]Number of fetuses in parentheses.

sign of sexual differentiation of the hypothalamus that occurs prior to the fourth month of prenatal development. As to the secretion of FSH, Figure 1 shows that the FSH-stimulating ability of the female hypothalamus begins earlier (19 to 29 weeks of development) than that of the male hypothalamus (24 to 29 weeks). This is also to be considered as evidence for sexual differentiation of this function of the hypothalamus occurring prior to this time. If the spontaneous activity of the hypophysis resulted from postmortem changes in the hypophysial cells, it would have been equally distributed throughout all ages and both sexes.

The early activity of the female hypothalamus may also be manifested in the differences in LH concentration in the hypophysis. It can be seen that the concentration of LH in the tissue of female hypophyses is higher than in male ones (Table 1).

The female epiphysis may also have an influence on the sexual differences in the level of hypophysial FSH secretion. This tissue was found to stimulate FSH secretion under the experimental conditions, whereas the male epiphysis, with three possible exceptions (see above), did not. In the three exceptional cases in

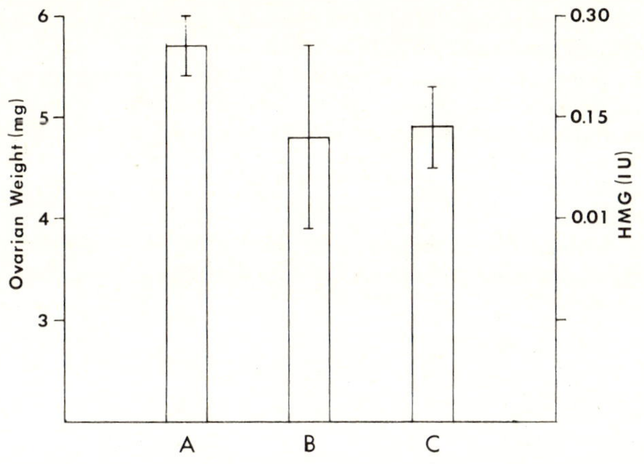

Fig. 2. FSH content in medium after culturing those fetuses which secreted spontaneously with the following (number of fetuses examined in parentheses):
A—Fetal brain tissue(38);
B—Male and/or female hypothalami(12);
C—Male and/or female epiphyses(26).

which the male epiphyses appeared to stimulate the secretion of FSH, the epiphyses were obtained from fetuses of between 24 and 28 weeks of age. During this same period, the hypophyses of some of the male fetuses were able to secrete FSH into medium spontaneously. Despite this apparent correlation, the physiological role of the epiphysis in continuous regulation of FSH secretion seems dubious.

Moszkowska described two epiphysial fractions: F_2, which induced hypophysial FSH secretion, and F_3, which blocked it(7,8). The author also demonstrated inhibition of FSH and LH secretion in the embryonic chicken hypophysis (16 to 17 days of incubation) on addition of epiphyses to the medium(9).

It is of interest to note that the tissues examined stimulated not only the release but also the synthesis of FSH and LH in the hypophysis(10). Sometimes more hormone was found in the medium than the hypophysis from a fetus of the age and sex studied could contain (especially from fetuses in the first half of embryogenesis).

REFERENCES

1. Levina, S. E. Endocrine features in development of human hypothalamus, hypophysis and placenta. Gen. Comp. Endocr., 11:151–159, 1968.
2. Brown, P. S. The assay of gonadotropin from urine of nonpregnant human subjects. J. Endocr., 13:59–64, 1955.
3. Wide, L. An immunological method for the assay of human chorionic gonadotropin. Acta Endocr. (København), 41(Suppl.):70, 1962.
4. Levina, S. E. Secretion of luteinizing hormone and its regulation in early human ontogenesis. Probl. Endokr. Gormonoter., 14:56–60, 1968.
5. Harris, G. W., and Jacobsohn, D. Functional grafts of the anterior pituitary gland. Proc. Roy. Soc. [Biol.], 139:263–276, 1952.

6. Knigge, K. M. Gonadotropic activity of neonatal pituitary gland implanted in the rat brain. Amer. J. Physiol., 202:387–391, 1962.
7. Ebels, J., Moszkowska, A., and Scemama, A. Étude in vitro des extraits épiphysaires fractionnés. C.R. Acad. Sci. (Paris), 260:5126–5129, 1965.
8. Moszkowska, A. Étude des extraits épiphysaires fractionnés. Biol. Med. (Paris), 56:403–412, 1967.
9. ——— Activité de l'antehypophyse en culture organotypique. Biol. Med. (Paris), 51: 293–298, 1962.
10. Kobayashi, I., Kobayashi, T., Kigawa, T., Mizuno, M., Amenomori, Y., and Watanabe, J. Gonadotropin releasing activity in rat hypothalamic extract. Endocr. Jap., 13:430–437, 1966.

42

A Maturing Enzyme System in Foetal Testes*,†

William Hamilton

Department of Child Health, The University of Glasgow, and The Royal Hospital for Sick Children, Yorkhill, Glasgow, Scotland

INTRODUCTION

The paediatrician and the obstetrician alike are concerned with the satisfactory development and growth of the foetus. A foetus, influenced adversely by drugs administered to the mother or by its own endogenous secretions, becomes a paediatric problem. Progestens administered to the expectant mother are known to lead to masculinization of the female foetus, while congenital adrenal hyperplasia (CAH) results in varying degrees of masculinization in the female and macrogenitosomia in the male. In CAH, severe salt loss may be an accompaniment.

While rapidly advancing skeletal maturation is an undesirable feature of untreated CAH, affected foetuses do not have an advanced bone age at birth. Also, thyroid deficiency in the postnatal period results in failure of linear growth and delayed skeletal maturation, but the athyreotic foetus is not dwarfed at birth although ossification is retarded.

Foetal linear growth, skeletal maturation, and genital and gonadal development are clearly complexly interrelated, so that studies for correlation are required to elucidate the interplay of mechanisms.

The present study was designed to correlate the development of testicular steroidogenesis with skeletal maturation in the human foetus. Only the biochemical results will be given here since a true assessment of foetal ossification will require a much larger study of foetal material.

*I am indebted to Dr. C. L. Hewett (Organon Laboratories Ltd., Lanarkshire, Scotland) and Dr. C. J. W. Brooks (Department of Chemistry, The University of Glasgow) for generous gifts of steroids.
†This work has been made possible by a grant from the Scottish Home and Health Department. I am indebted also to the Rankin Fund trustees, of the University of Glasgow. and to the Board of Management of the Royal Hospital for Sick Children, Yorkhill, Glasgow, for their generous financial support.

MATERIALS AND METHODS[1]

Testes from 6 fresh foetuses ranging in gestational age from 8 to 18 foetal weeks were used. In each case, one testis was prepared for histological study and the other, in 1 mm slices, incubated in 4 ml Krebs-Ringer bicarbonate-glucose 0.1 M buffer, pH 7.4, in an atmosphere of 95 percent oxygen and 5 percent carbon dioxide at 37.5°C for 2 hours. Added to three flasks was 1 μc pregnenolone-7α-^{3}H (SA 500 mc/mM), and to the other three flasks, 1 μc cholesterol 7α-^{3}H (SA 500 mc/mM). Both radioactive compounds were previously purified by thin-layer chromatography (TLC), pregnenolone-7α-^{3}H in cyclohexane:ethyl acetate, 80:20, and cholesterol-7α-^{3}H in cyclohexane:ethyl acetate:water, 600-400-1. All the incubations were commenced within 20 minutes of the material being removed from the uterus.

The incubations were terminated by the addition of 3 ml benzene:chloroform, 6:1, and incubation mixture further extracted three times with benzene:chloroform (3 ml). The pooled extracts were reduced to small volume under a stream of nitrogen and applied to Whatman No. 1 filter paper. The chromatogram was developed in the Bush A system (petroleum ether 60 to 80:methanol:water, 5:4:1) for 3 hours and zones corresponding to standards removed, eluted, and the eluate prepared for thin-layer chromatography in the solvent systems chloroform:ethanol, 9:1, chloroform:ethanol, hanol, 99:1, and cyclohexane:ethyl acetate, 1:1.

Radioactive zones on the TLC plates were detected by the Panax TLC Scanner and counting was done in a Nuclear Chicago (Mark 1) scintillation spectrometer, the samples being dissolved in toluene-POP-POPOP.

Chemical methods for oxidation, reduction, acetylation, and saponification were those of Griffiths et al.(1). All solvents were purified according to Bush(2).

The nature of an unknown compound was assumed confirmed if the radioactivity remained with the cold barrier or its chemical derivatives in three chromatographic systems.

RESULTS

The testes from the three youngest foetuses (8, 12, and 12 weeks) were incubated with pregnenolone-7α-^{3}H and those from the older foetuses (16, 16, and 18 weeks) were incubated with cholesterol-7α-^{3}H. Radioactive areas on the TLC plates were compared with Rf values established for likely compounds. Visualization of the Δ4-3oxo- grouping was made under ultraviolet light at 242 mμ and of the Δ5-3β-OH- grouping under ultraviolet light at 365 mμ or by spraying with antimoney trichloride in chloroform.

[1]Abbreviations used: progesterone (4-pregnene-3,20-dione); 16α, 17α-dihydroxyprogesterone (16α, 17α-dihydroxy-4-pregnene-3,20-dione); 17α-hydroxyprogesterone (17α-hydroxy-4-pregnene-3,20-dione);20α-hydroxyprogesterone (20α-hydroxy-4-pregnene-3-one); pregnenolone (3β-hydroxy-5-pregnene-20-one); 17α-hydroxypregnenolone (3β, 17 α-dihydroxy-5-pregnene-20-one); 16-dehydropregnenolone (3β-hydroxy-5,16-pregnadiene-20-one); Δ4-pregnenolone (3β-hydroxy-4-pregnene-20-one); dehydroepiandrosterone (3β-hydroxy-5-androstene-17-one); androstenedione (4-androstene-17-one); Δ5-androstenediol (3β,17β-dihydroxy-5-androstene); testosterone (17β-hydroxy-4-androstene-3-one); Δ4-cholestenone (4-cholestene-3-one).

TABLE 1

Percentage conversion of pregnenolone-7α-^{3}H by human foetal testes in vitro

Compound	Age estimated in foetal weeks		
	8	**12**	**12**
pregnenolone	19.2	19.21	18.9
17α-hydroxypregnenolone	5.4	7.2	7.56
17α-hydroxyprogesterone	1.2	3.1	2.8
20α-hydroxyprogesterone	4.6	5.2	5.8
16α,17α-dihydroxyprogesterone	7.36	4.2	3.6
dehydroepiandrosterone	12.2	10.1	11.2
Δ5-androstenediol	1.0	0.2	0.3
Δ4-androstenedione	10.6	12.4	11.8
testosterone	20.5	21.4	19.96
unidentified	18.2	18.4	18.3

TABLE 2

Percentage conversion of cholesterol-7α-^{3}H by human foetal testes in vitro

Compound	Age estimated in foetal weeks		
	16	**18**	**18**
cholesterol	42.1	40.8	46.4
Δ4-6-ketocholestenone	3.6	4.2	4.0
Δ5,16-pregnadiene-20-one	2.1	2.4	2.16
Δ4-pregnenolone	0.04	0.08	0.1
pregnenolone	0.08	1.0	2.3
testosterone	1.0	0.9	0.89
unidentified (A)	22.6	24.0	26.0
unidentified (B)	26.5	25.7	18.9

Under the conditions of these incubations, the percentage conversions of pregnenolone-7a-^{3}H and cholesterol-7a-^{3}H to the various compounds are shown in Tables 1 and 2. Recoveries from the Bush A chromatogram were greater than 85 percent; those from the TLC plates were greater than 90 percent. The percentages are based on the counting of aliquots from the first TLC plate in each instance, when constant specific actpvity was achieved.

DISCUSSION

It will be noted that no reference is given in this paper for the age of the foetuses. It is appreciated that the least accurate estimate of foetal maturation is according to the last menstrual period. Acevedo et al.(3) give crown-rump and foot lengths with the gestational age of their foetal material while Bloch(4) adds to these the weight of the gonad, but clearly, in these developmental studies it is necessary to distinguish between foetal age and foetal maturity. In current investigations, only 25 foetuses have been examined and this is considered insufficient data on which to state irrefragably the criteria for maturity scored according to skeletal development. This will be the subject of another communication.

In Tables 1 and 2, the results of the enzymatic activity of the six foetal testes are shown. All the incubations were commenced within 20 minutes of the sac being removed from the uterus, the material being kept at 0°C until the incubation was commenced. Maximum enzymatic activity was therefore retained.

From Table 2 it will be seen that when pregnenolone-7α-^{3}H was used as substrate, the following enzymes were active:- Δ5-3β-ol dehydrogenase/isomerase; 17α-hydroxylation for pregnenolone and progesterone; 20α-hydroxylase for progesterone and presumably for pregnenolone since dehydroepiandrosterone was detected; 16α-hydroxylase for progesterone or 17α-hydroxyprogesterone; C21-side-chain splitting enzymes for pregnenolone and progesterone; and 17β-hydroxylase for both dehydroepiandrosterone and androstenedione. It is worthy of note that the youngest foetal testis produced the greatest conversion to Δ5-3β, 17β-androstenediol and it would appear that a pathway from pregnenolone to testosterone exists where the Δ5-3β-ol structure in Rings A and B is retained until the final dehydrogenation and isomerization takes place. It might be argued that the 3β-ol dehydrogenase/isomerase system is a late-maturing enzyme system and that the pathway to testosterone, in foetuses older than 8 foetal weeks, is preferentially via progesterone to Δ4-androstenedione followed by 17β-hydroxylation.

The compound, 16α,17α-dihydroxyprogesterone, was obtained in greater quantities again from the youngest foetal material. The nature of this reaction is still under study, but Oshima et al.(5) have shown that 16α-hydroxylase is specific, at least in adult human testes, for progesterone alone. Against this Acevedo and co-workers(3) have isolated from foetal testicular incubations Δ4-16α-hydroxyandrostenedione and 16α-hydroxydehydroepiandrosterone when using progesterone-7α-^{3}H and pregnenolone-7α-^{3}H respectively.

When the foetal testes were incubated using cholesterol-7α-^{3}H the only readily indentifiable compounds were cholesterol, pregnenolone, and testosterone (Table 2). It will be noted that under the conditions of the incubations, 40 to 46 percent of the cholesterol was unmetabolized, while the yields of pregnenolone (0.08 to 2.3 percent) and testosterone (0.89 to 1.0 percent) were very low. The compounds, Δ4-6-ketocholestenone and Δ5, 16-pregnadiene-20-one were unexpected findings of doubtful significance.

Hydroxylation at C6 is known to occur in the oxidation of cholesterol(6) so that while dehydrogenation/isomerization with 6β-hydroxylation of cholesterol is a possible potential of the foetal testicular enzymes, oxidation of the 6β-OH group may be due to artifact. Alternatively, Δ4-cholestenone blocks adrenal steroidogenesis(7) and it may be that 6β-hydroxylation of Δ4-cholestenone prevents this effect. Several authors(8,9,10,11) have isolated 6β-hydroxysteroids from urine and it may be that 6β-hydroxylation is an early and essential protective step in steroid biosynthesis.

Using a 4-hour incubation period with ATP, NAD^+, and $NADP^+$, Gower and Ahmad(12) isolated three 16-dehydro compounds from incubations of boar testes with pregnenolone-4-^{14}C, although according to other evidence which they quote, progesterone is nearer to Δ^{16}-steroids than is pregnenolone. Labeled progesterone was not isolated in this present work. The possibility exists that these 16-dehydrosteroids are by-products of in vitro incubations and do not occur in vivo.

The two unidentified compounds obtained in high yield are neither 17α-hydroxypregnenolone (A) nor 17α-hydroxyprogesterone (B) being in all systems slightly less polar than these compounds. In oxidation with either chromium trioxide or Kiliani's reagent, the unknown substances were lost in these preliminary experiments. Pierrepont et al.(13) have shown that 7α-tritium is lost with oxidation.

From these combined studies it is clear that foetal testes require a pool of pregnenolone from which to synthesize testosterone. Only after the foetal age of 12 weeks can testosterone be derived from cholesterol. It would appear, therefore, that the foetal testes are unlikely to be autonomous in the production of testosterone.

Further, histologically, there is a regression in size and number of the Leydig cells as the foetus ages until at birth they are almost absent, those remaining disappearing shortly after birth. Thus it is possible that the younger foetal testes are more active enzymatically than those at a later age. Since differentiation of the genital duct system and the external genitalia occur between the ninth and eleventh week it is reasonable that the foetal testes should be enzymatically competent prior to this age.

This work therefore suggests that the younger foetal testes should be further investigated using a variety of precursors under varying incubation conditions.

SUMMARY

Human foetal testes were incubated with pregnenolone-7α-^{3}H and from the gestational age of 8 weeks testosterone was synthesized preferentially via progesterone, 17α-hydroxyprogesterone, androstenedione with subsequent 17β-hydroxylation. An alternative pathway from pregnenolone to Δ5-3β,17β-androstenediol followed by dehydrogenation/isomerization of Rings A and B was noted. 16α-hydroxylation was active presumably for progesterone since 16α,17α-dihydroxyprogesterone was isolated.

With cholesterol-7α-^{3}H as substrate, only small yields of pregnenolone and testosterone were obtained along with 6-ketocholestenone and Δ5,16-pregnadiene-20-one. The significance of these findings is discussed.

REFERENCES

1. Griffiths, K., Grant, J. K., and Whyte, W. G. Steroid biosynthesis in vitro by cryptorchid testes from a case of testicular feminization. J. Clin. Endocr., 23:1044–1051, 1963.
2. Bush, I. E. The Chromatography of Steroids. Oxford, Pergamon Press, 1961.
3. Acevedo, H. F., Axelrod, L. R., Ishikawa, E., and Takaki, F. Studies in fetal metabolism. II. Metabolism of progesterone-4-C^{14} and pregnenolone-7α-H^3 in human fetal testes. J. Clin. Endocr., 23:885–890, 1963.
4. Bloch, E. Metabolism of 4-^{14}C-progesterone by fetal testis and ovaries. Endocrinology, 74:833–845, 1964.
5. Oshima, H., Sarada, T., Ochi-ai, K., and Tamaoki, B-I. Intracellular distribution and properties of steroid 16α-hydroxylase in human testes. J. Clin. Endocr., 27:1249–1254, 1967.

6. Fieser, L. F., and Fieser, M. Steroids. New York, Reinhold Publishing Corp., 1959.
7. Steinberg, D., and Fredrickson, D. S. Inhibitors of cholesterol biosynthesis and the problem of hypercholesterolemia. Ann. N.Y. Acad. Sci., 64:579–589, 1967.
8. Burstein, S., Dorfman, R. I., and Nadel, E. M. 6β-hydroxycortisol—a new steroid in human urine. Arch. Biochem. Biophys., 53:307–308, 1954.
9. Pasqualini, J. R. Identification and isolation of 4-pregnene-6β,17α,21-triol-3,11,20-trione (6β-hydroxycortisone) from human urine. C.R. Acad. Sci. (Paris), 255:3258–3260, 1962.
10. ——— 6β-hydroxycorticosterone in human urine. Arch. Biochem. Biophys., 106:15–19, 1964.
11. Reynolds, J. W., and Ulstrom, R. A. 6β-hydroxy compound S; report of its isolation from a case of the hypertensive form of congenital adrenal hyperplasia. Biochem. Biophys. Acta, 57:606–608, 1962.
12. Gower, D. B., and Ahmad, N. Studies on the biosynthesis of 16-dehydro steroids. The metabolism of (4-^{14}C) pregnenolone by boar adrenal and testis tissue in vitro. Biochem. J., 104:550–556, 1967.
13. Pierrepont, C. G., Galley, J. McI., Griffiths, K., and Grant J. K. Steroid metabolism of a Sertoli cell tumor of the testis in a dog with feminization and alopecia and of the normal canine testis. J. Endocr., 38:61–70, 1967.

43

Prenatal Damage in Hypothyroidism

Henning J. Andersen

Children's Hospital, Fuglebakken and
Endocrine Clinic, Dronning Louises
Children's Hospital,
Copenhagen, Denmark

The clinical picture of congenital hypothyroidism and the literature have recently been reviewed(1-3). Two major forms may be distinguished: *nongoitrous*, mainly due to dysgenesis of the thyroid which may be complete (athyreosis), or hypoplastic, eventually in an ectopic, sublingual position. In *goitrous hypothyroidism*, the gland is enlarged due to increased thyrotropic stimulation, secondary to hormonodysgenesis. This disorder, in contrast to the nongoitrous hypothyroidism, often occurs in several members of a sibship. Both types are found associated with signs of fetal hypothyroidism.

In studying the effects of nature's own experiment on the human fetus, several difficulties are encountered. The disease is rare and mostly unpredictable. Nearly all the hypothyroid babies look normal at birth and the diagnosis is usually not made until the second or third months of life or even later. In these cases the effects of postnatal hypothyroidism are added to eventual prenatal damage. Fortunately, now and then the disorder is diagnosed at or just after birth, and such cases are useful in the extrapolation from postnatal to prenatal conditions.

Very few autopsies of newborn hypothyroid children seem to have been reported. The disorder by itself is not fatal and is rarely associated with other congenital, fatal anomalies.

Lesions due to lack of thyroid hormone must either be reversed specifically by thyroid medication or, if they are irreversible, as for example the dental enamel hypoplasias, they must be independent of the type of thyroid hypofunction that the child is suffering from.

In Pendred's syndrome, for example, congenital high-tone deafness and goitrous hypothyroidism are often found in the same patient, but this type of deafness is not found in other forms of prenatal hypothyroidism and is not influenced by thyroid medication.

The birth length and birth weight of newborn hypothyroid babies are usually normal or even above normal(4). In our experience small-for-date hypothyroid babies are exceptionally rare.

The bone development in the more severe cases of neonatal hypothyroidism is always delayed in spite of a normal birth weight and birth length, as pointed out by Dorff(5) in his elegant observation on bone development in an athyreotic and a normal twin. The distal bone center of the femur is present in all normal newborn with a birth weight of 3,000 g or more(6), but is missing in severe congenital hypothyroidism irrespective of type. To my knowledge the same is true only in babies born of diabetic mothers(7). An x-ray picture of the knee, therefore, is a rapid and reliable diagnostic procedure. Very often the proximal center of the tiia and the center of the cuboid are missing or grossly fragmented. Other parts of the skeleton, for example the skull and the spine, show retarded development(3). The localized deformities found in the corpora of the lower thoracic and upper lumbar vertebrae are probably due to postnatal influences on abnormal soft bone tissue. The retardation and dysgenesis of the skeleton are reversed completely by thyroid medication except in cases grossly neglected for years.

The number, size, and shape of the teeth in patients with congenital hypothyroidism are normal, whereas enamel hypoplasias of prenatal or early postnatal origin are found in thc deciduous teeth inmore than 80 percent of our children with neonatal hypothyroidism(3). An explanation for this may be that the fetal thyroid secretion normally begins after the formation of the dental anlage but earlier than the enamel formation. These enamel hypoplasias do not disappear on thyroid treatment and may therefore, although they are not specific for hypothyroidism, be useful later in life in timing the onset of thyroid insufficiency. Besides dysmineralization, a general retardation of the development of the teeth may be demonstrated by x-ray of the jaw bones of newborn hypothyroid children. The retarded dentition is reversible on thyroid medication. Animal experiments(8) seem to indicate that abnormal amounts of mucopolysaccharides around the tooth buds may hamper normal eruption. Such changes do not seem to have been reported in hypothyroid babies, but increased amounts of acid mucopolysaccharides, increased number of mast cells, and edema have been demonstrated in the skin of young hypothyroid babies (9).

Among other changes in the soft tissues may be mentioned the large protruding tongue nearly always present after a few months in severe congenital hypothyroidism but sometimes also found at birth.

Signs of cardiovascular damage are usually not dominating in the hypothyroid child at birth. In the few newborn whom we have studied, the cardiac shadow and the EKG have been normal. A slight mottling of the skin may be visible in such cases, as the forerunner of the very marked cardiovascular hypofunction found a few months later. The basis for the mottled skin may be circulatory insufficiency together with the short and poorly developed capillary loops demonstrated in the skin of hypothyroid babies. A prolonged neonatal jaundice is found in approximately 10 percent of the cases of both hypothyroidism in the newborn and has in some cases led to the diagnosis(10-12). The reason for the jaundice is thought to be immaturity

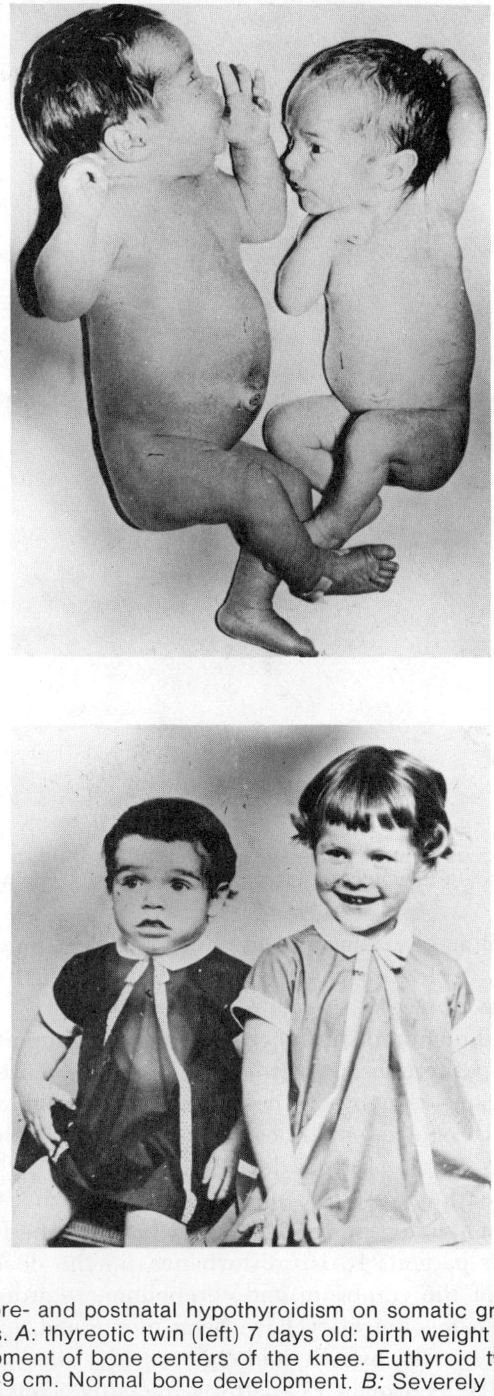

Fig. 1. The effect of pre- and postnatal hypothyroidism on somatic growth illustrated bv two different pairs of twins. *A*: thyreotic twin (left) 7 days old: birth weight 4,500 g, birth length54 cm. Retarded development of bone centers of the knee. Euthyroid twin (right) birth weight 2,550 g, birth length 49 cm. Normal bone development. *B:* Severely hypothyroid, untreated twin (left) 3 years old, height age 21 month, bone age 12 months. Euthyroid twin (right) of normal development. From Andersen. *In* Nordisk Laerebog i Paediatri, Copenhagen, Munksgaard, 1967, p. 287.

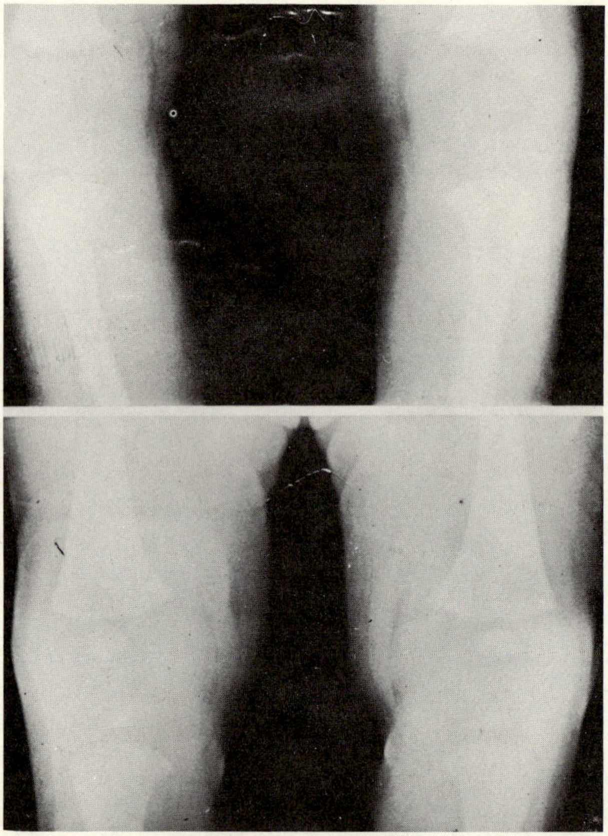

Fig. 2. Retarded bone development (absence of distal femur and proximal tibia centers) in a newborn athyreotic child, birth weight 3,600 g (upper) compared with normal newborn (lower).

of the bilirubin-coupling mechanism in the liver, but this subject apparently needs further examination. In the few cases we have seen, thyroid medication seems to accelerate the disappearance of the jaundice.

Neurological and mental disturbances are the most important lesions found in severe congenital hypothyroidism regardless of what has caused the disorder. This is not surprising because the thyroid insufficiency, impossible to diagnose before birth and difficult to recognize sometimes for months after birth, strikes the nervous system at an important time of rapid development.

As pointed out earlier, very few autopsies are performed on hypothyroid children. I have not been able to find reports on newborn. In the few reports available on autopsies of older patients(13-16) disturbances in the development and structures of the cortex of the cerebrum and cerebellum, retarded capillary development, and incomplete myelinization have been mentioned.

The EEG in very young hypothyroid children may show a slow alpha rhythm but some patients have a normal EEG pattern. Recently a retardation in the development of the normal sleep pattern in EEG's of newborn and older babies has been

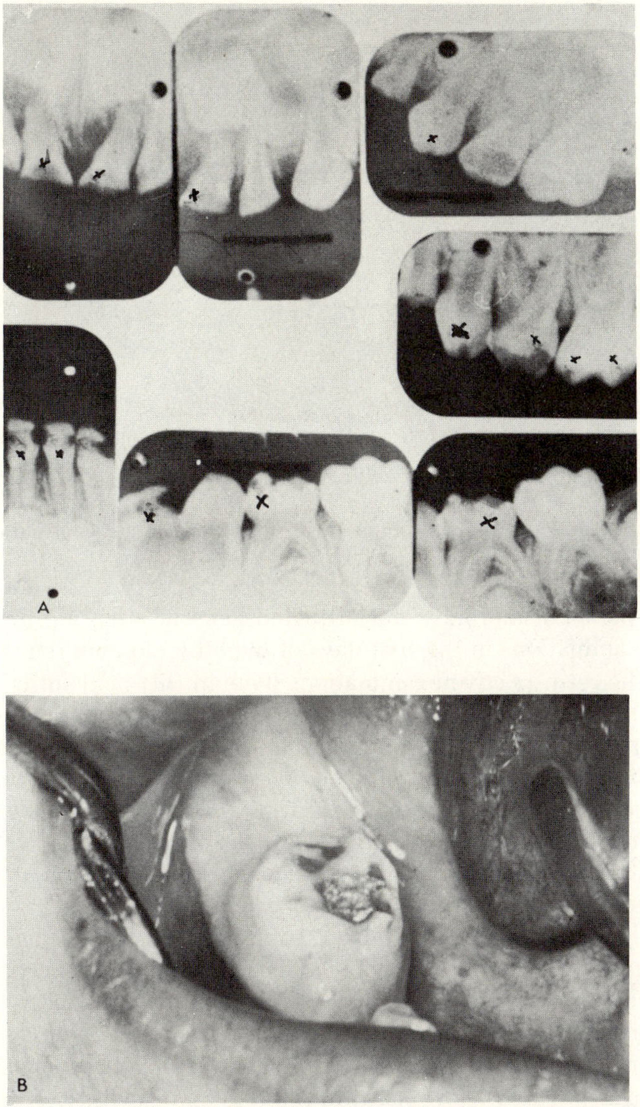

Fig. 3. *A:* Enamel dysplasia (*x*) of fetal origin in a 3-year-old athyretic child. *B:* Enamel dysplasia in a 4-year-old severely hypothyroid child. From Andersen. *In* Endocrine and Genetic Disorders of Childhood, I. Gardner, ed. Courtesy of W. B. Saunders Co.

reported. These findings are interpreted as signs of retarded and incomplete connection between the brain stem and cortex(17).

The protein content of the spinal fluid is elevated in early hypothyroidism.

Prolongation of the ankle and knee jerk during the relaxation period is found even in very early hypothyroidism and is used as a diagnostic criterion(18). I have no personal experience with this examination.

The clinical signs of neuromental disturbances in the newborn hypothyroid

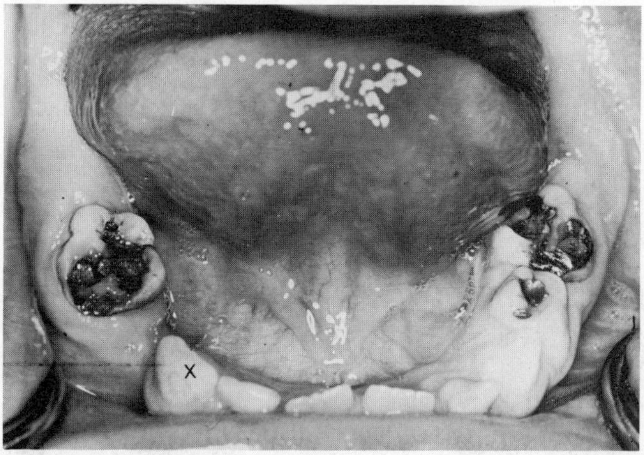

Fig. 4. Enamel dysplasia of perinatal origin (*x*) in right lower in a 7-year-old girl with severe, congenital hypothyroidism.

child seem to be slight and have not attracted general attention. Larger series of neurological examinations in the first days of hypothyroid children do not seem to exist. Most of the retrospective examinations done in this field suffer from the heterogeneity of the patients and also of the treatment. Among 50 patients with hypothyroidism diagnosed and treated within the first 3 months of life we have two with epilepsia and one with a moderate degree of spasticity which may be explained as effects of pre- and postnatal hypothyroidism. On the whole, gross neurological disturbances are not found. However, more than half of the children do have signs of cerebellar dysfunction, an ataxic gait, dysdiadochochinesis, "general clumsiness," etc. for a shorter or longer period of life.

At birth there is nothing peculiar or striking in the mental development of a newborn severely hypothyroid baby, but after a few months it becomes more and more lethargic, ending, if untreated, in the typical stuporous idiocy.

In an attempt to evaluate the role played by prenatal hypothyroidism in mental development, we have tried to differentiate our hypothyroid patients into as uniform groups as possible with regard to the type and severity of the thyroid failure, and the adequacy of the treatment(4). From these and other studies(1,2,19-21) it is obvious that in spite of adequate treatment starting within the first 3 months, prenatal hypothyroidism in a number of cases results in retarded mental development.

At the time of birth or in the early postnatal period, it is not possible to predict the mental outcome of a hypothyroid patient even if the degree of thyroid insufficiency can be established and adequate treatment secured. Individual factors, presumably including the genetic background for the patient's mental development, apparently play a role.

In longitudinal studies of the IQ development(19) in hypothyroid children we have found that the IQ's in most children reach a plateau after about 2 years of

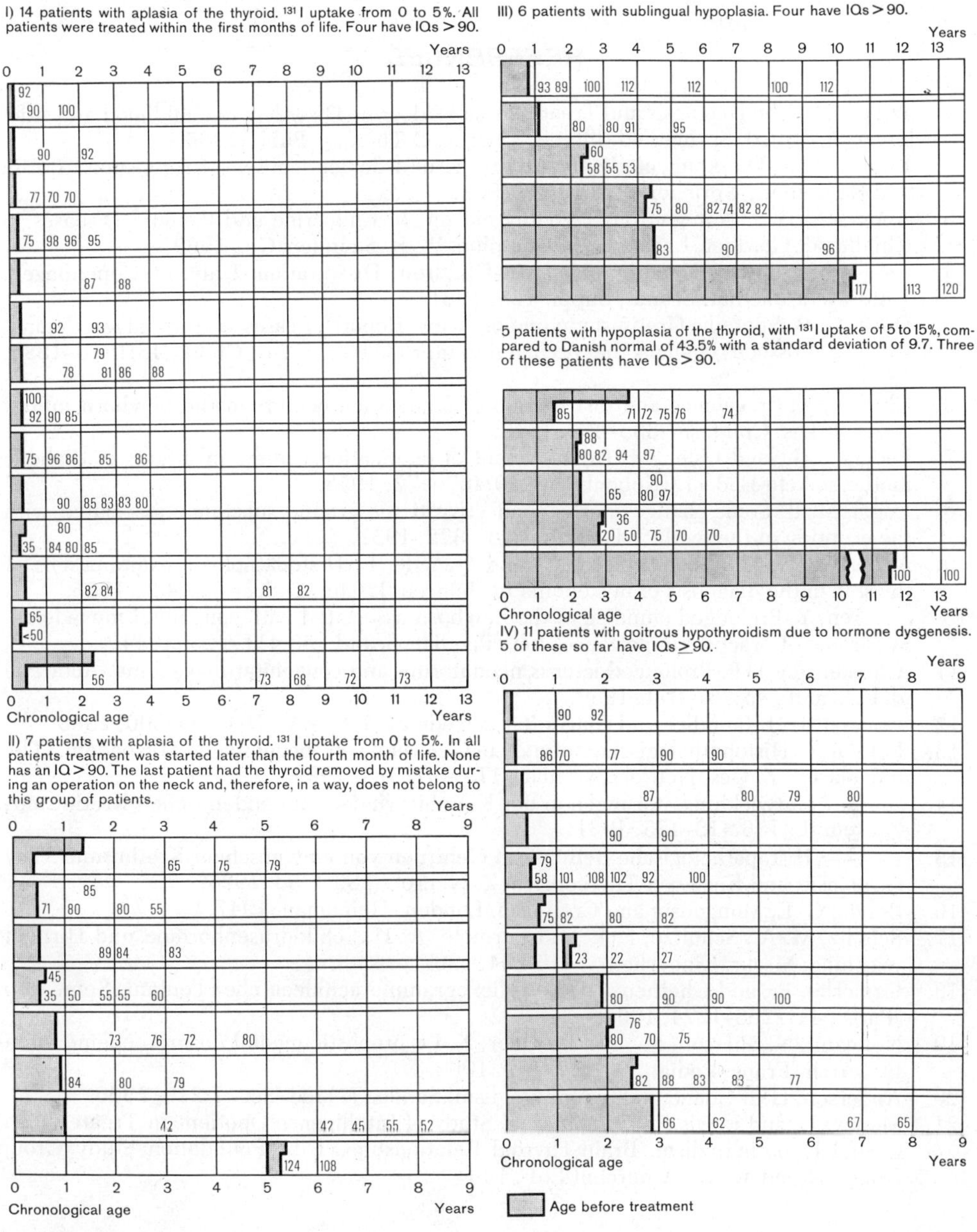

Fig. 5. Mental development in children with various degrees of hypothyroidism. From Andersen. Sandoz J. Med. Sci., 7:210, 1966.

adequate treatment. At that time we feel that the ultimate mental prognosis can be made with some certainty. Other investigators(21) however have found more individual variability in the pattern of sequential IQ's.

REFERENCES

1. Wilkins, L. The Diagnosis and Treatment of Endocrine Disorders in Childhood and Adolescence, 3rd ed. Springfield, Illinois, Charles C Thomas, Publ., 1965.
2. König, M. P. Die kongenitale Hypothyreose und der endemische Kretinismus. Berlin and Heidelberg, Springer-Verlag Inc., 1968.
3. Andersen, H. J. Non-goitrous Hypothyroidism. *In* Endocrine and Genetic Diseases in Childhood. Gardner, L. I., ed. Philadelphia, W. B. Saunders Co., 1969.
4. ——— Studies of Hypothyroidism in Children. Dissertation, Univ. of Copenhagen, 1960. Acta Paediat. Scand., Suppl. 125, 1961.
5. Dorff, G. B. Sporadic Cretinism in one of twins. Report of cases with roentgen demonstration of osseous changes that occurred in utero. Amer. J. Dis. Child., 48:1316–1325, 1934.
6. Christie, A. Prevalence and distribution of ossification centers in the newborn infant. Amer. J. Dis. Child., 77:355–361, 1949.
7. Pedersen, J., and Osler, M. Development of ossification centers in infants of diabetic mothers. Acta Endocr. (Kobenhavn), 29:467–472, 1958.
8. Engel, M. B. Some changes in the connective tissue ground substance associated with the eruption of the teeth. J. Dent. Res., 36:322, 1951.
9. Andersen, H. J., Asboe-Hansen, G. and Quaade, F. Histopathologic examinations of the skin in the diagnosis of myxedema in children. J. Clin. Endocr., 15:459, 1955.
10. Åkerren, Y. Prolonged jaundice in the newborn associated with congenital myxedema. Syndrome of practical importance. Acta Paediat. Scand., 43:411–425, 1954.
11. Christensen, J. F. Prolonged icterus neonatorum and congenital myxedema. Acta Paediat. Scand., 45:367–371, 1956.
12. Doucett, J. H. Jaundice and congenital myxedema. J.A.M.A., 194:299–300, 1965.
13. Lotmar, F. Histopathologische Befunde in Gehirnen von congenitalem Myxoedem. (Thyreoaplasie). Z. Ges. Neurol. Psychiat., 119:491–513, 1929.
14. ——— Entwicklungsstörungen in der Kleinhirnrinde beim endemischen Kretinismus. Z. Neurol., 136:412–435, 1931.
15. ——— Histopathologische Befunde in Gehirnen von endemischem Kretinismus, Thyreoaplasie und Kachexia Thyreopriva. Z. Neurol., 136:1–53, 1933.
16. Benda, C. E. Mongolism and Cretinism. London, Heineman, 1947.
17. Schultz, M. A., Schultze, F. J., and Parmele, A. H. Schilddrüsenhormon und Hirnentwicklung. Mschr. Kinderheilk., 115:284, 1967.
18. Canlorbe, P., and Chefneux, A. Le rèflexogramme achillèen chez l'enfant. Sem. Hop. (Paris), 41:1253–1274, 1965.
19. Neimann, N., Pierson, M., and Berthier, X. Le pronostic mental du myxoedème infantile. Arch. Franc. Pediat., 20:147–159, 1963.
20. Andersen, H. J. Congenital hypothyroidism. Sandoz J. Med. Sci., 7:210, 1966.
21. Money, J., and Lewis, V. Longitudinal Study of Intelligence Quotient in Treated Congenital Hypothyroidism. Brain-Thyroid Relationships. Ciba Foundation Study Group No. 18., London, J. A. Churchill Ltd., 1964.

44

Protein-Bound Iodine Secretion by Human Thyroid in Organ Culture with and without Pituitary Stimuli

Sally B. Fand[1] and Carl W. Ehmann[2]

[1]Department of Pathology, Wayne State University School of Medicine, and Special Histochemistry, Veterans Administration Hospital, Dxetroit, Michigan
[2]Alaska Native Medical Center, U.S. Public Health Service, Anchorage, Alaska

INTRODUCTION

While in man, and indeed in all mammals, it is generally agreed that pituitary stimulation promotes the production and release of thyroid hormones and their precursors, the demonstration of the role of thyroid-stimulating hormone (TSH) in vitro has been difficult(1,2). After adding TSH to rat fetal thyroid glands in organ culture, Nataf et al.(3) found an increase in radioactive ^{131}I uptake by the gland, appparently independent of protein synthesis, leading them to suggest that TSH stimulated ^{131}I incorporation into thyroxine. In measuring the uptake and incorporation of iodine in the cultured tissue, they address themselves only to the production of thyroid hormone and not to its release. This is true of other investigators as well.

We were interested in testing the effect of a pituitary stimulus on the rate of release of thyroxine from thyroids grown in organ culture. We reasoned that withholding iodine from in vitro thyroid fragments would prevent the formation of new thyroxine. Since only 1 percent of total thyroid iodine content is free iodine(4) any new T_4 formed from this would be insignificant. Thus protein-bound iodine (PBI) measurements in the media bathing well-maintained thyroid glands would demonstrate the *release* of these substances from the gland and not their *production*. This approach might permit the demonstration of an in vitro effect of TSH in the human gland.

Other investigators have found that TSH secretion by cultured pituitaries decreases with time, in the absence of hypothalamic stimuli(5). One might hypothesize that the isolated pituitary, with its decreased rate of TSH secretion, would stimulate thyroxine release from cultured thyroid fragments decreasingly over time. We set out to measure PBI release from cultured human thyroid fragments and to assess the influence of TSH as related to time.

MATERIAL AND METHODS

For each of five experiments, anterior pituitary and thyroid glands removed at autopsy were cut into pieces approximately 1 mm in diameter. Equal numbers of pieces were used throughout an experiment to control tissue volume. Fragments were placed on agar strips suspended above the media on a wire mesh grid. Figure 1 illustrates an organ culture preparation of pituitary alone. Each small culture dish was placed in a covered petri dish lined with water-saturated filter paper. The dishes were incubated at 32° C in the presence of 95 percent O and 5 percent CO_2. Sterile technique was practiced in the preparation of these cultures. Culture setups were:

Dish	Symbol
Pituitary alone	P
Thyroid alone	Th
Thyroid alone with thyroid-stimulating hormone in the media	Th-TSH
Pituitary and thyroid in combination (The fragments were arranged in pairs on the agar, each member of the pair touching the other).	P-TH

Control fragments were retained for later histological study(6).

The culture media consisted of synthetic medium 199, 0.4/mg/ml of streptomycin sulfate, and 15 to 150 IU/ml of penicillin. The pituitary and/or thyroid fragments were grown in this medium without hormones and also in the presence of 0.5 mIU of TSH (thyrotropar).

On the day following the preparation and again 3 days after that, randomly selected pieces were removed from the agar strips and fixed for histological examina-

TABLE 1

PBI measurements[a] before and after treatment of culture media with alkaline exchange resin

Conditions of culture	Not treated	Treated
Pituitary alone	0.17	0.17
Thyroid + TSH	1.92	1.41
Thyroid + Pituitary	2.00	1.62

[a]PBI in μg.

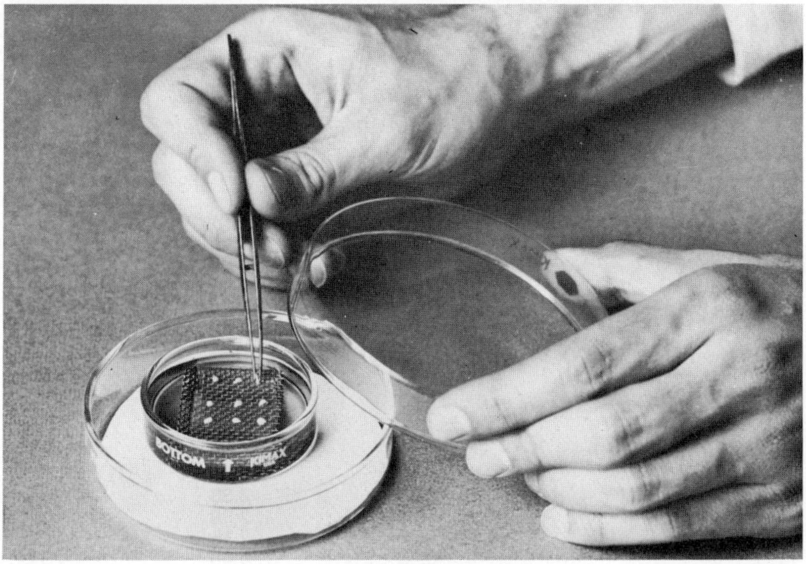

Fig. 1. Photograph illustrating positioning of fragments on agar strips suspended above media on wire mesh grid. Please note that the small culture dish is placed in a petri dish lined with water saturated filter paper and is covered during the incubation.

tion. The medium was removed from each dish, its volume measured, and PBI determinations run.Fresh medium was placed in each dish before its return to the incubator.

To remove inorganic iodine containing substances that might be released by simple necrosis of tissue, alkaline exchange resin was added prior to measuring PBI. Table 1 shows that resin treatment does reduce the measured PBI and hopefully serves to remove this source of variation.

Autopsy records were reviewed to insure that the specimens came from patients that were euthyroid and not receiving hormone therapy or iodine compounds.

RESULTS

Histological evaluation of microscopic sections of pituitary and thyroid fragments showed consistent differences among control, day 1 cultures, and day 2 to 4 cultures. In all cases, the pituitary day 1 cultures showed good maintenance of nuclei and cytoplasmic granules and thyroid fragments of the same age likewise showed maintenance of cellular and follicular elements. By the fourth day of culturing, tissue fragments of both pituitary and thyroid showed some degeneration. A correlation between the amount of protein-bound iodine secreted by the cultured thyroid fragments and their maintenance, amount of colloid, or epithelial cell height was not obvious.

The results presented in Table 2 show that the average daily PBI at the end of the day 2 to 4 period of culturing equals or exceeds that released in the first day of

TABLE 2

Effect of conditions of culture on PBI content (in μg) of day 1 cultures, day 2-4 cultures (average/day) and their ratio

Conditions of culture	Day 1	Day 2-4/3	Ratio
Thyroid alone			
1	0.84	0.78	0.93
2	1.48	1.82	1.23
3	4.91	3.02	0.61
4	1.78	2.21	1.24
5	0.25	0.45	1.80
Thyroid + TSH			
2	0.88	1.45	1.65
3	1.41	2.98	2.11
4	2.88	2.75	0.96
Thyroid + Pituitary			
1	1.58	2.07	1.31
2	2.69	2.38	0.88
3	1.62	2.30	1.42
4	0.76	2.15	2.83
5	0.42	0.49	1.17

culturing. While the average day 2-to-4/3 release of PBI substances ranged from 0.6 to 2.8 that of day 1, this ratio varies from a mean of 1.16 for thyroid cultured alone to 1.57 with added TSH and 1.52 with pituitary present contiguously. Thus the release of PBI substances continues at a good rate after the first day of culturing, and this is enhanced in the presence of TSH.

DISCUSSION

The experimental design used permits separating out thyroxine release from thyroxine production. Production is prevented by withholding added iodine-containing substances, and thus only release is measured.

As many previous investigators have shown, TSH causes increased secretion of thyroglobulin in the intact animal by the thyroid. To date, there has been no convincing demonstration of an in vitro effect of TSH in the human; it is also true that no studies of which the authors are aware explore the parameters of thyroid function concerning release. It seemed to us that the very factors which may make difficult the demonstration of an in vitro effect of TSH on the production side (e.g., requirements for a totally simulated environment, dependence on adequate blood flow to remove secretory products, etc.) might favor the disclosure of the biochemical events involved in the release mechanism and increase the sensitivity and reliability of organ culture measurements of its action.

Under this hypothesis, PBI release by thyroid fragments cultured with pituitary fragments might be expected to be greater than that of thyroid fragments cultured alone. One way to measure this is by the ratio of "late PBI" (day 2-4/3) to "initial PBI" (day 1) and we find the mean of this ratio is increased in both groups with pituitary stimulation. This suggests that an effect of TSH is being demonstrated and that the stimuli continue to act through the fourth day of culturing.

CONCLUSIONS

1. Media PBI is measured as an index of thyroxine secretion (release) by human thyroid fragments in organ culture.
2. Both exogenous TSH and homologous contiguous pituitary fragments are associated with an increased "late PBI" to "initial (day 1) PBI" ratio when compared to thyroid cultured alone.
3. The results suggest that an effect of TSH is being demonstrated in vitro and that the stimuli continue to act through the fourth day of culturing.

REFERENCES

1. Petrovic, A. Recherches sur la prehypophyse en culture organotypique: evolution structurale et action sur les organes-effecteurs associes *in vitro*. Arch. Anat. (Strasb.), 47:121–291, 1963.
2. Roche, J., Pavlovic, M., and Michel, R. Culturing the thyroid glands of young rats *in vitro*, and the biosynthesis of thyroid hormones. Biochim. Biophys. Acta, 24:489–495, 1957.
3. Nataf, B. M., Chaikoff, I. L., and Freeman, W. E. Mechanism of action of TSH studies by organ culture of fetal rat thyroid gland. Proc. Soc. Exp. Biol. Med., 124:7–12, 1967.
4. Turner, C. D. General Endocrinology. Philadelphia, W. B. Saunders Co., 1962.
5. Sinha, D. K., and Meites, J. Stimulation of pituitary thyrotropin synthesis and release by hypothalamic extract. Endocrinology, 71:1002–1006, 1966.
6. Fand, S. B., and Holland, R. H. Punched-out agar petri plates for systematic fixation and embedding of multiple small tissue samples. Stain Technology, 42:151–153, 1967.

45

Growth and Development in the Syndromes of Familial Isolated Absence of HGH or Pituitary Dwarfism with High Serum Concentration of an Immunoreactive but Biologically Inactive HGH

Speculations on the Effects of Pre- and Postnatal HGH Deficiency

Zvi Laron,[1] Athalia Pertzelan,[2] and Jacob Frankel[3]

[1]Pediatric Metabolic and Endocrine Service,
Rogoff Wellcome Medical Research Institute,
Beilinson Hospital, Petah Tikva, and
Department of Pediatrics, Tel Aviv University
Medical School, Tel Aviv, Israel
[2]Pediatric Metabolic and Endocrine Service,
Beilinson Hospital, Petah Tikva, Israel
[3]Pediatric Metabolic and Endocrine Service,
Beilinson Hospital, Petah Tikva, and
Department of Psychology, Bar Ilan
University, Ramat Gan, Israel

INTRODUCTION

The observation that many anencephalic babies were born with body weights within the normal range, and that children who were diagnosed at later age to have hypopituitarism, developed normally during the first year of life, led to the assumption that pituitary growth hormone is not essential for prenatal and immediate postnatal growth(1,2). Information obtained in recent years that birth weight is not a direct index of hormonal activity, but mainly related to intrauterine nutrition, the carbohydrates playing a major role(3,4), and the fact that some children with hypothalamic pituitary insufficiency have impaired growth starting soon after birth(5,6), made many pediatricians question the accuracy of the above postulate.

As an introduction to the discussion of the problem as to whether human growth hormone (HGH) participates in pre- and postnatal growth and development, I should like to review our knowledge of HGH in those periods of life.

HGH IN PRE- AND POSTNATAL PERIODS

The development of quantitative microcomplement fixation tests(7), sensitive and specific radioimmunoassays for the measurement of plasma or tissue growth hormone concentrations(8,9), and routine measurement of body length at birth, added much information on this subject.

Thus, growth hormone was found present even in pituitaries of 12-week-old fetuses(7). Maternal growth hormone is not transferred through the placenta, as was documented by the injection to the mother before delivery of a tracer dose of ^{131}I labeled HGH or large quantities of nonlabeled hormone(10,11) Table 1. The higher concentration of HGH in the arterial cord plasma than in the venous cord plasma (Table 1) is indicative of active secretion of pituitary GH at birth. This is shown by the fact that the serum concentration of HGH in the newborn continues to rise during the first two days of life to acromegalic values(12) Figure 1. In small and premature babies the rise is higher and the decline slower(12,13). It is not as yet certain whether the increased secretion of HGH in the newborn is a physiological reaction(14) or a temporary derangement in the regulatory mechanism of pituitary GH secretion as could be inferred from the paradoxical increase in plasma GH observed after induced hyperglycemia(12,13) Table 2, or by the lack of suppression after intravenous corticosteroid administration (Laron and Nitzan, unpublished).

At an age of 2 to 4 weeks the fasting plasma HGH decreases to very low concentrations and rises upon the usual stimuli, such as hypoglycemia, arginine, stress, and so on.

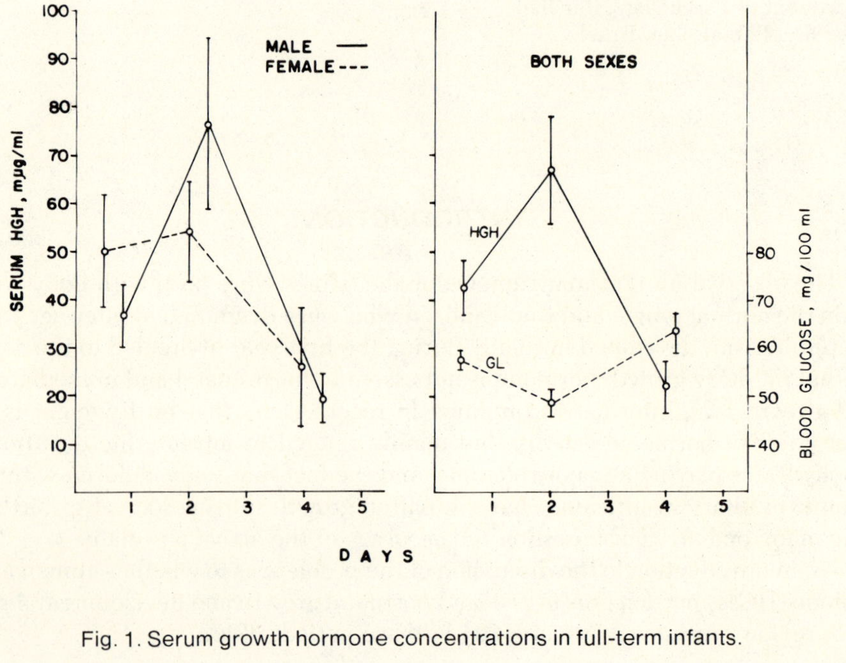

Fig. 1. Serum growth hormone concentrations in full-term infants.

TABLE 1

Growth hormone concentration in the mother's plasma before and after the injection of nonlabeled HGH; cord plasma and newborn infants[a]

Plasma HGH – ng/ml

		Before	After injection (minutes)						Cord		Infant 1-2 hours
No.	Patient	0	15	30	45	60	90	120	Ven.	Art.	age
1	Z.M.	4	860				*320*		1	35	46
2	E.M.		820	550	490	*430*			15	12	38
3	M.S.	39	1150	1150	1145	500	220	*200*	2	17	8
4	B.B.	22	1230	1025	400	320	*466*		15	27	15
5	N.O.		1000	600	*54*				3.3	15	29
6	I.D.	13		*560*[b]					14	14	8
7	A.S.	25		*760*						9	13
8	R.Sh.	25		1780	1300		*1125*		2.7	56	70

[a]From Laron et al. Acta Endocr. (Kobenhavn), 53:687-692, 1966.
[b]The plasma concentration at the time of delivery is in italic.

TABLE 2

Response of serum GH to glucose infusion in two newborns[a]

Sex	Age (hr)	Birth weight (g)		Time after infusion started (min) 0	30	60	90	120
Male	4	2,960	Blood glucose (mg/100 ml)	70	281	466	391	299
			Serum GH (ng/ml)	10	18	12	15	18
Female	8	3,650	Blood glucose (mg/100 ml)	73	380	428	431	390
			Serum GH (ng/ml)	10	31	37	16	14

[a]From Laron et al. Israel J. Med. Sci., 2:770-773, 1966.

Growth hormone effects many metabolic processes (of protein, fat, carbohydrates, and possibly calcium)(15,16,17), enhances cell growth, and induces linear growth even on intermittent administration. This latter finding points towards intermediary mechanisms in GH action, one of which may be the so-called serum "sulfation factor"(18).

PHYSICAL GROWTH AND DEVELOPMENT

As a direct study of how growth hormone affects growth in the first period of life is still difficult, we have attempted to look at this problem by an indirect approach, analyzing the effects of early HGH deficiency in the newborn and infant.

Reviewing the literature we found that several babies with abnormal development of the pituitary or hypothalamus(19,20,21) had a low birth length (Table 3). So had two out of three sibs with familial pituitary dwarfism from Zürich due to HGH deficiency(22 and personal communication by A. Prader) (Table 4). For comparative purposes normal birth length of Israeli children, Jews or Arabs, is given in Table 5(23).

We have studied four children with proven isolated HGH deficiency of the familial type; three of these in whom birth length was known, had been short at birth (Table 6). We also found a birth length of less than 48 cm in 8 out of 11 known birth measurements in children with the syndrome of pituitary dwarfism with high serum concentration of immunoreactive but biologically inactive HGH(24) (Table 7). In both syndromes the only defect known is total lack of growth hormone or lack of

TABLE 3

Birth weight and length in developmental abnormalities of the pituitary and hypothalamus

Ref. No.	Author	Year	Diagnosis	Sex	Gestation (weeks)	Birth weight (g)	Birth length (cm)
19	Reid	1960	Cong. absence of pituitary	M	38	3,900	49
20	Haworth et al.	1961	Cebocephaly	F	38	2,720	44
21	Kind	1962	Anencephaly	F	36 +	1,640	40
21	Kind	1962	Anencephaly	M	36 +	1,980	40

TABLE 4

Birth weight and length in familial pituitary dwarfism due to deficiency

Ref. No.	Author	Year	Relationship	Sex	Gestation (weeks)	Birth weight (g)	Birth length (cm)
22	Dibbern	1962	Fam. sib. 1	F	38 +	3,500	< 48
22	Dibbern	1962	sib. 2	M	36	4,000	49
22	Dibbern	1962	sib. 3	F	38	3,500	45

TABLE 5

Mean length (cm) at birth in various communities in Israel (23)

	Birth					
Origin of parents: geographical area	Males			Females		
	N	Mean	SD	N	Mean	SD
East & West Europe	94	50.7	1.45	84	50.3	1.86
Israeli born	18	50.5	—	21	50.5	—
Irag & Iran	52	50.3	1.74	59	50.2	1.41
Yemen	99	50.0	1.51	63	49.7	1.49
North Africa	27	50.9	1.81	26	49.9	1.22
Israeli Arabs	28	50.4	1.32	21	50.2	1.45

TABLE 6

Birth weight and length in familial isolated deficiency of HGH

Pt.	Relationship	Sex	Gestation (weeks)	Birth weight (g)	Birth length (cm)
R.I.	cousins	M	38	3,500	?
E.I.		M	38	3,400	48
E.S.	sibs	M	38	1,900	46
Es.S.		F	38	1,600	46

TABLE 7

Birth weight and length in patients with the syndrome of pituitary dwarfism and high serum level of growth hormone

Pat. No.	Sex	Gestational age	Delivery	Birth weight (g)	Birth length (cm)
1	F	Full term	Normal	3,400	46[a]
2	F	Full term	Normal	3,200	49
3	M	Full term	Normal	3,740	47
4	M	Full term	Normal	3,050	50
5	F	Full term	Normal	2,900	46[a]
6	M	Full term	Normal	3,000	45[a]
7	M	Full term	Normal	3,000	49
8	F	Full term	Normal	3,100	46[a]
9	M	Full term	Normal	3,250	46[a]
10	F	Full term	Normal	3,200	47
11	M	Full term	Normal	3,200	44[a]

[a]2 SD below the mean birth weight for respective ethnic community.

the active hormone, starting early in utero, as these are inherited diseases. Clinically, the children of both syndromes look alike and present the same signs and symptoms. Among the main features are growth retardation, starting immediately after or very soon after birth (Figs. 2,3). Another typical symptom due to HGH deficiency is hypoglycemia, which starts after birth and continues for many years. Figure 4 illustrates the tendency for spontaneous hypoglycemia in patients with the high inactive serum HGH. Other manifestations typically due to HGH deficiency and starting soon after birth are obesity, acromicria and small gentalia (Fig. 5), and afterwards delayed and faulty dentition(24) (Fig. 6). Motor development is mostly delayed, as shown in Table 8; so is skeletal maturation. The anterior fontanella very often closes between ages 3 and 6 years. The etiological implication of HGH is proven by the fact that early initiation of HGH therapy, in the patients with isolated HGH deficiency, normalizes the above enumerated signs and symptoms. There is a catch-up of linear growth, dental and skeletal maturation, and acral growth, while the hypoglycemia disappears and the patient loses body fat. In short, the whole appearance and feeling of the patient changes.

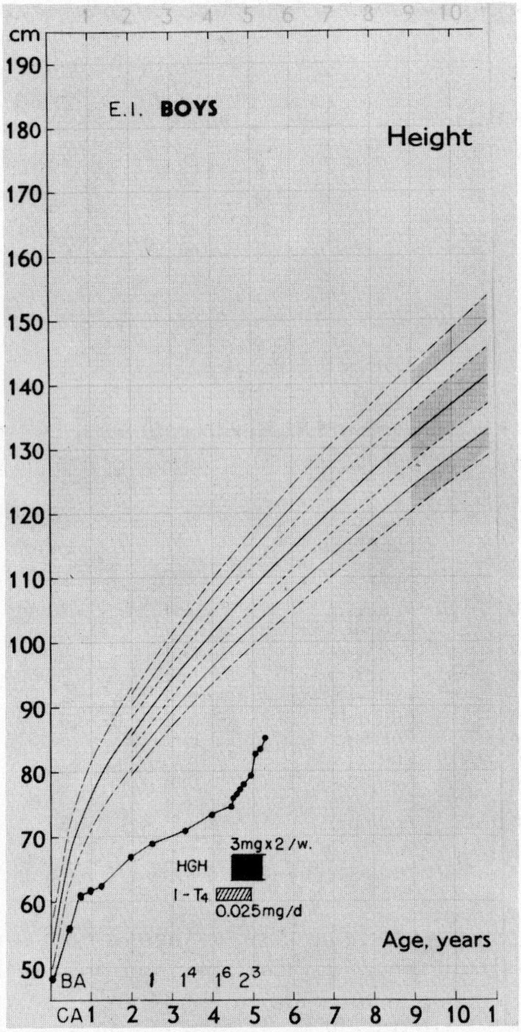

Fig. 2. Growth chart of a child with the familial type of isolated HGH deficiency.

PSYCHOLOGICAL INVESTIGATIONS

In a series of ongoing psychological investigations, a variety of children and adolescents with pituitary dysfunction have been studied(25). Of these, the present discussion focuses on subjects with pituitary insufficiency due to familial lack of HGH or pituitary dwarfism with high serum HGH. These patients were seen in our endocrine unit in the course of routine clinical treatment and follow-up. Each patient

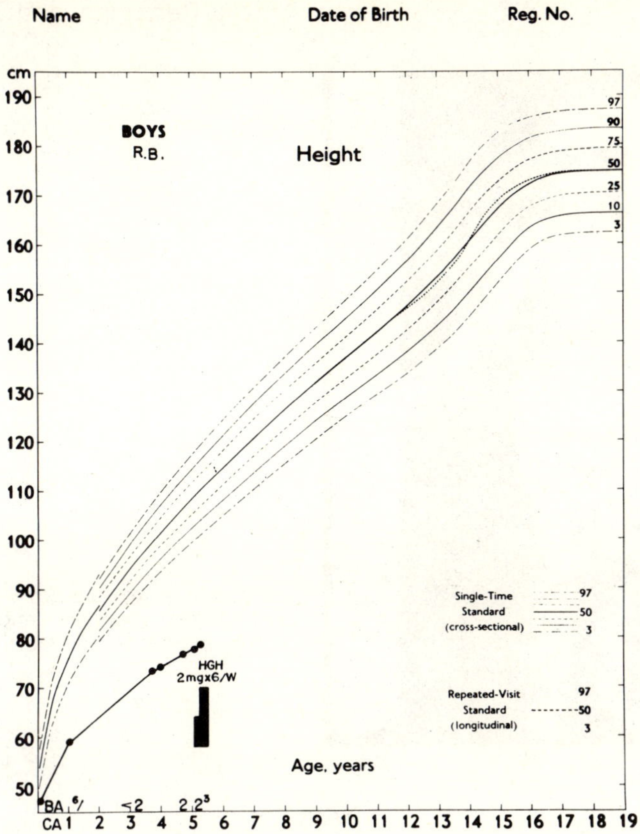

Fig. 3. Growth chart of a child with the familial type of pituitary dwarfism and high serum concentration of an immunologically reactive but biologically inactive HGH.

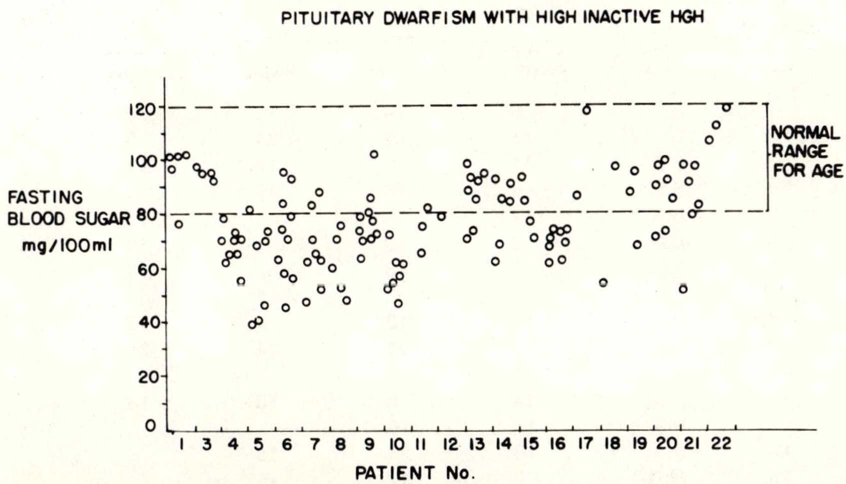

Fig. 4. Fasting blood sugar concentration measured on various occasions in 21 patients with the syndrome of pituitary dwarfism and high serum concentration of an immunologically reactive, biologically inactive HGH. From Laron et al. *Israel J. Med. Sci.,* 4:883-894, 1968. By permission of the editor of the Israel Journal of Medical Sciences.

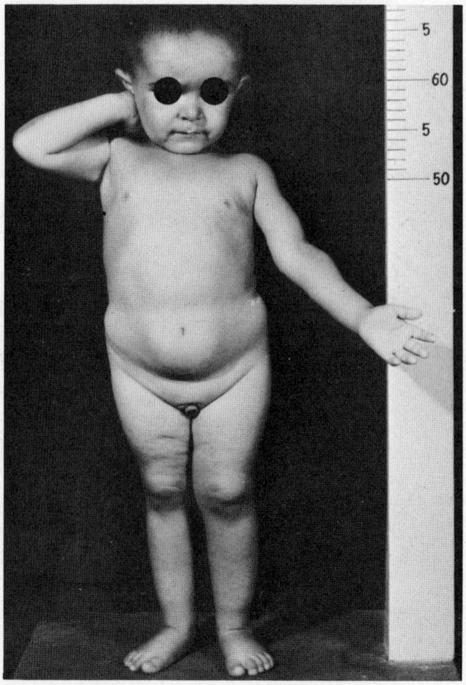

Fig. 5. A 7-year-old boy with the syndrome of pituitary dwarfism denoted by the small stature, obesity, and facial features.

TABLE 8

Developmental milestones in children with the syndromes of "familial HGH deficiency" (Group I) and "pituitary dwarfism with high serum HGH" (Group II)

Patient			Age in months				Fontanels closed—years
Group	No.	Name	Sitting	Walking	Talking	Teething	
I	1	R.I.	6	14	< 12	Early	
	2	E.I.	11	18	< 24	8	$< 2\frac{6}{12}$
	3	E.S.	11	18	18	11	$< 3\frac{6}{12}$
	4	B.S.	8	16	12	9	$> 3\frac{3}{12}$
	5	B.R.	12			11	$< 3\frac{2}{12}$
	6	I.S.	12	18	Early	12	$> 2\frac{1}{12}$
II	7	S.S.		> 18			$> 6\frac{2}{12}$
	8	R.S.		< 18	> 12		$> 4\frac{7}{12}$
	9	Sh.S.	> 12	12	24	> 12	$> 6\frac{3}{12}$
	10	R.J.		24	< 24	24	$> 7\frac{8}{12}$
	11	J.J.	21	> 21			$> 6\frac{2}{12}$
	12	A.M.	14	18	18	13	
	13	H.A.	15	15		7	
	14	E.A.	18	24	36	8	
	15	M.D.	9	12	15		
	16	J.L.	7		9		
	17	M.L.	6	12	12	12	
	18	R.C.		36		36	
	19	Z.S.	8	18			$< 3\frac{1}{12}$

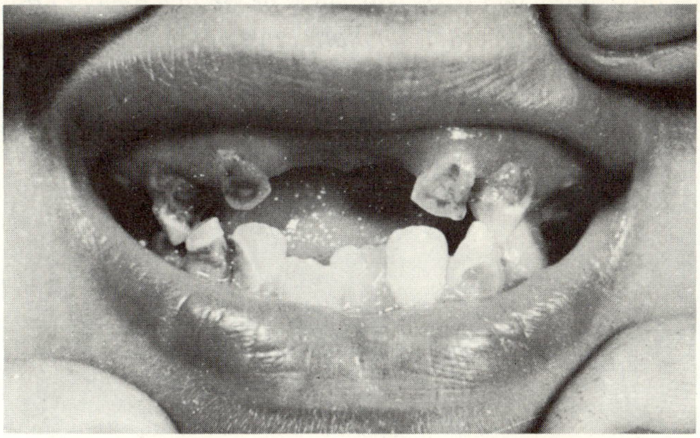

Fig. 6. Discolored and broken teeth and irregular growth of permanent teeth in a 8½-year-old girl with the syndrome of pituitary dwarfism and high serum concentration of immunologically reactive, biologically inactive HGH.

was given, on an individual basis, a series of tests combining into a neuropsychological battery. These consisted of: Wechsler Intelligence Scales for Children (or Adults), Bender Visual Motor Gestalt Test, Human Figure Drawings, House-Tree-Person, and the Money Map Test(26). In addition, a teacher's detailed report was obtained on each subject. The patient population consisted of 10 males and 11 females, ranging in age from 1 7/12 to 16 11/12 years. The findings summarizing the intelligence and visuomotor data appear in Tables 9 and 10. Intelligence data based upon follow-up testing are given in Tables 11 and 12.

Intelligence Data: Initial Testing. Table 9 reveals that our subjects, regardless of age, are well below the expected intelligence for a normal population. Only six patients (or 31 percent) scored in the average range of intellectual functioning defined as an IQ between 90 and 110. The majority scored at or below one standard deviation (S.D.=15). The school performance is in line with the IQ, i.e., skewed toward the fair, poor, and failing side.

TABLE 9

Intelligence data in 19 children with the familial type of isolated HGH deficiency or familial type of pituitary dwarfism and high serum concentration of immunoreactive but biologically inactive HGH

IQ	Classification	% Expected[a]	% Found	No. cases
130 +	Very superior	2.2	0.0	0
120-129	Superior	6.7	0.0	0
110-119	Bright normal	16.1	0.0	0
90-109	Average	50.0	31.6	6
80-89	Dull normal	16.1	26.3	5
70-79	Borderline	6.7	10.5	2
69 and below	Mentally retarded	2.2	31.6	6

[a]According to Wechsler. Wechsler Intelligence Scale for Children, 1949. The Psychological Corporation, New York, p. 16.

TABLE 10

Visuomotor data in terms of performance on Bender test in 14 children with the familial type of isolated HGH deficiency or familial type of pituitary dwarfism and high serum concentration of immunoreactive but biologically inactive HGH

Patient	Age (years)	Mean no. errors[a] expected for age in normal population	No. errors found	No. errors found after correction for IQ
A	$12^2/_{12}$	0.0[b]	3	+1.5
B	$16^{11}/_{12}$	0.0	1	+1.0
C	$9^4/_{12}$	1.7	10	+10.0
D	$6^4/_{12}$	8.4	10	+1.6
E	$12^0/_{12}$	0.0	3	+1.5
F	$10^2/_{12}$	1.6	8	+3.2
G	$11^{11}/_{12}$	0.0	4	+2.3
H	$8^6/_{12}$	2.5	10	–
I	$15^9/_{12}$	0.0	3	+3.0
J	$14^6/_{12}$	0.0	1	–
K	$12^{11}/_{12}$	0.0	6	+4.4
L	$6^8/_{12}$	6.4	12	+3.6
M	$6^{11}/_{12}$	8.4	13	+4.6
N	$13^6/_{12}$	0.0	3	+1.4

[a]Based on Koppitz (35).
[b]Beginning with age 10 to 11 years the Bender test should be without serious error due to maturation of central nervous system and visuomotor coordination.

TABLE 11

Intelligence data – test-retest in 14 children with the familial type of isolated HGH deficiency or familial type of pituitary dwarfism and high serum concentration of an immunoreactive but biologically inactive HGH

IQ on initial examination	IQ on follow-up examination	Change	Interval (mos.)
100	108	+8	12
80	79	–1	18
100	100	0	5
95	98	+3	14
91	79	–12[a]	13
74	75	+1	17
69	73	+4	17
74	88	+14[a]	12
67	76	+9	19
67	84	+17[a]	19
62	69	+7	18
46	44	–2	18
89	90	+1	12
81	80	–1	17
Mean 78.21	81.64	3.42	15.07

[a]No particular explanation is found for these sharp changes, the two positive and the other negative. Subsequent retesting may indicate whether these are artifacts or variations due to some as yet unknown factor.

TABLE 12

IQ comparison of affected subjects with pituitary dwarfism and high serum concentration of an immunoreactive but biologically inactive HGH and their heterozygote nondwarfed siblings

	Affected subjects N = 7	Siblings N = 7
No. males	2	1
No. females	5	6
Mean age	10.9	12.8
Mean Wechsler verbal IQ	77.43	79.57
Mean Wechsler performance IQ	69.43	88.28
Mean Wechsler full-scale IQ	71.28	82.28

Visuomotor Data. Table 10 also shows the subjects' performance on a test recognized as a measure of the patient's ability to combine visual geometric figures with motor function and the extent of successfully coordinating the two. This function is strongly related to central nervous system maturation, and is sensitive to damage or impairment in CNS functioning. It is clear from our data that in this task as well, our subjects rated substantially below age norms.

Intelligence Data: Follow-up Testing. Our subjects were reevaluated at least once, many twice and three times. Table 11 points to the remarkable consistency of the scores, even at intervals of one year or more. This confirmatory finding suggests the validity of our original results.

Sibling Data. Table 12 lists the intelligence scores of seven siblings of the affected patients. These siblings are heterozygotes and not dwarfed. The overall trend is for siblings scores to be above those of their affected brothers and sisters though (1) the differences are not extensive, and (2) in some cases they are almost identical or even below those of the patients.

CONCLUSIONS

Only three studies(27,28,29) in the literature discuss the relationship between intelligence and pituitary deficiency without specific reference to the question of HGH. The three concur in that intelligence is normal as well as normally distributed for hypopituitary dwarfs. In addition, their school performance is judged to be adequate(28). Our findings are obviously at odds with these reports and the degree of the disparity is rather extensive. When our findings based upon a uniform population with isolated active HGH deficiency are taken in toto they suggest a recognizable pattern of organic or central nervous system impairment. This is supported by (1) the lowered IQ, (2) the specific pattern of IQ subtest scores on the Wechsler (to be elaborated in futher publications), and (3) the Bender test results. While there is the possibility of a population bias in our sample, no clear-cut relationship was observed between the known socioeconomic level of our patients and their scores. Further, the sibling data, taken on the whole, are suggestive that the socio-economic variable by itself is insufficient to account for our findings.

Accepting the information that HGH is secreted early in fetal life, we refer to the question put by Blizzard(30): What is the role of pituitary growth hormone in the fetus if it is not to promote growth? However, no data to affirmatively support this supposition is available in the literature.

Our studies seem to suggest that in addition to its role in cellular and organ metabolism, growth hormone really affects pre- and postnatal growth and development. The evidence is the high incidence of low birth length in babies with isolated HGH deficiency (or presence of inactive hormone), early postnatal growth retardation, and retardation in skeletal and dental development. The interesting question is whether there is a relationship between the lack of HGH and the retardation in motor development and future psychological deficits. As prolonged neonatal hypoglycemia has been repeatedly shown in recent years to cause irreversible changes in the central nervous system(31,32,33,34) it is possible that the motor retardation and intellectual deficits are totally or partly due to this metabolic symptom which is itself due to HGH deficiency. However it is also feasible that pituitary HGH, like thyroxine, plays a role in the development of the brain, and that a deficit of this hormone manifests itself at birth in transient or permanent lesions in the brain which can be either augmented by postnatal hypoglycemia or treated by the administration of HGH.

Theoretically it should be possible to answer this speculative assumption in the future by treating one group of those babies with HGH, and another one with frequent feedings and hyperglycemic agents other than HGH.

The discrepancy between the intellectual deficit found by us and the absence of the same found by other investigators(27,28,29) can possibly be explained by the fact that our population is characterized by intrauterine HGH deprivation, whereas there is reason to believe that the others are dealing with natal or postnatal hypopituitarism.

REFERENCES

1. Talbot, N. B., and Sobel, E. H. Endocrine and other factors determining the growth of children. Advances Pediat., 2:238, 1947.
2. Seckel, H. P. G. Concepts relating the pituitary growth hormone to somatic growth of the normal child. Amer. J. Dis. Child., 99:349–379, 1960.
3. Laron, Z., Mannheimer, S., Nitzan, M., and Goldmann, J. Growth hormone, glucose and free fatty acid levels in mother and infant in normal, diabetic and toxaemic pregnancies. Arch. Dis. Child., 42:24–28, 1967.
4. Ziegler, E. Mechanism responsible for overweight infants of diabetic mothers. J. Pediat., 72:908–909, 1968.
5. Francés Antonin, J. M. Hypothalamo-hypophysärer Zwergwuchs mit spontaner Pubertät. Helv. Paediat. Acta, 16:267–276, 1961.
6. Soyka, L. F., Ziskind, A., and Crawford, J. J. Treatment of short stature in children and adolescents with human pituitary growth hormone (Raben). New Eng. J. Med., 271: 754–764, 1964.
7. Makler, M. T. Growth hormone in human development. Nature (London), 217:1149–1150, 1968.
8. Glick, S. M., Roth, J., Yalow, R. S., and Berson, S. A. Immunoassay of human growth hormone in plasma. Nature (London), 199:784–787, 1963.

9. Laron, Z., and Mannheimer, S. Measurement of human growth hormone. Israel J. Med. Sci., 2:115–119, 1966.
10. Gitlin, D., Kumate, J., and Morales, C. Metabolism and maternofetal transfer of human growth hormone in the pregnant woman at term. J. Clin. Endocr., 25:1665–1670, 1965.
11. Laron, Z., Pertzelan, A., Mannheimer, S., Goldman, J., and Guttman, S. Lack of placental transfer of human growth hormone. Acta Endocr. (Kobenhavn), 53:687–692, 1966.
12. ——— Mannheimer, S., Pertzelan, A., and Nitzan, M. Serum growth hormone concentration in full term infants. Israel J. Med. Sci., 2:770–773, 1966.
13. Cornblath, M., Parker, M., Reisner, S. H., Forbes, A. E., and Daughaday, W. H. Secretion and metabolism of growth hormone in premature and full term infant. J. Clin. Endocr., 25:209–218, 1965.
14. Grumbach, M. M. Growth hormone and growth. Pediatrics, 37:245–248, 1966.
15. Matsuzaki, F., and Raben, M. S. Growth hormone. Ann. Rev. Pharmacol., 5:137–150, 1965.
16. Root, A. Growth hormone. Pediatrics, 36:940–950, 1965.
17. Najjar, S., and Blizzard, R. M. Current concepts regarding human growth hormone (somatotropin). Pediat. Clin. N. Amer., 13:437–457, 1966.
18. Daughaday, W. H., and Parker, M. L. Sulfation factor measurement as an aid in the recognition of pituitary dwarfism. J. Clin. Endocr., 23:638–650, 1963.
19. Reid, J. D. Congenital absence of the pituitary gland. J. Pediat., 56:658–664, 1960.
20. Haworth, J. C., Medouy, H., and Lewis, A. J. Cebocephaly with endocrine dysgenesis. J. Pediat., 59:726–733, 1961.
21. Kind, C. Das endokrine System der Anencephalen mit besonderer Berücksichtigung der Schilddrüse. Helv. Paediat. Acta, 17:244–258, 1962.
22. Dibbern, H. H. Familial occurrence of pituitary dwarfism. Inangural dissertation for M. D. degree, Medical. Faculty of Zürich, pp. 1–15, 1962.
23. Zaizon, R., and Laron, Z. Body length and weight at birth and one year of age in different communities in Israel. Acta Paediat. Scand., 55:524–528, 1966.
24. Laron, Z., Pertzelan, A., and Karp, M. Pituitary dwarfism with high serum levels of growth hormone. Israel J. Med. Sci., 4:883–894, 1968.
25. Frankel, J. J. and Laron, Z. Psychological aspects of pituitary insufficiency in children and adolescents with special reference to growth hormone. Israel J. Med. Sci., in press.
26. Money, J. A. Standardized Road-Map Test of Direction Sense. Baltimore, Maryland, The Johns Hopkins Press, 1965.
27. Pollitt, E., and Money, J. Studies in the psychology of dwarfism. I. Intelligence quotient and school achievement. J. Pediat., 64:415–421, 1964.
28. Rosenbloom, A. L., Smith, D. W., and Loeb, D. G. Scholastic performance of short-statured children with hypopituitarism. J. Pediat., 69:1131–1133, 1966.
29. Drash, P. W., Greenberg, N. E., and Money, J. Intelligence and personality in four syndromes of dwarfism. *In* Human Growth. Cheek, D. B., ed. Philadelphia, Lea & Febiger, pp. 568–581, 1968.
30. Blizzard, R. M. Differentiation, morphogenesis and growth with emphasis on the role of pituitary growth hormone. *In* Human Growth, Cheek, D. B., ed. Philadelphia, Lea & Febiger, pp. 41–59, 1968.
31. Haworth, J. C., and McRae, K. N. Neonatal hypoglycemia: A six-year experience. J. Lancet, 87:41–46, 1967.
32. Knobloch, H., Sotos, J. F., Sherard, E. S., Jr., Hodson, W. A., and Wehe, R. A. Prognostic and etiologic factors in hypoglycemia. J. Pediat., 70:876–884, 1967.
33. Anderson, J. M., Milner, R. D. G., and Strich, S. J. Effects of neonatal hypoglycaemia on the nervous system: a pathological study. J. Neurol. Neurosurg. Psych., 30:295–310, 1967.
34. Blattner, R. J. Central nervous system damage and hypoglycemia. J. Pedat., 72:904–906, 1968.
35. Koppitz, E. M. The Bender Gestalt Test for Young Children. New York, Grune & Stratton, Inc., p. 188, 1964.

46

Some Hormonally Controlled Events of Liver Differentiation in the Perinatal Period

Robert Jacquot

Laboratory of Animal Physiology,
Faculty of Sciences, Reims, France

INTRODUCTION

One salient feature of the physiology of the foetus seems to be the existence of what could be called "critical periods" in development.

These periods are well defined in a given species and are characteristic for the structure, organ, or function concerned. During their short length of time, a differentiation and a maturation take place. Very often the process is hormonally controlled. Several review articles have been devoted to this topic(1,2). The purpose of this paper is to present some work done in our laboratory on the physiology of the foetal liver. As many other papers of this symposium deal with the same organ, we will focus exclusively on our data and we will not try to give an extensive bibliography.

One must notice that the expression "foetal liver" covers in fact two major components of this liver, namely the hematopoietic tissue and the hepatic tissue proper.

Each component has its own physiology. Therefore we shall study first the maturation of the hepatocyte, then the evolution of the hematopoietic tissue. In order to clarify the presentation, most of the figures will not show the standard errors, but statistical computations were done and their details can be found in the original papers.

MATURATION OF THE HEPATOCYTE

The changes occurring in glycogen metabolism as gestation progresses represent only one aspect of the complex differentiation of the hepatocyte, but this aspect is relatively easy to follow. In fact, as early as 1857, Bernard had shown that in several species of mammals, the foetal liver starts to store glycogen suddenly at a very precise stage of gestation, characteristic for each species. This has been largely confirmed since (bibliography in 3). We have considered two aspects of this problem in the rat foetus, namely how the hepatocyte acquires the ability to store its glycogen and to utilize it.

Glycogen Storage

In the rat, the foetal liver begins to store glycogen around day 18 p.c. (post coitum): 0.7 percent (wet weight) at day 18, 5 to 6 percent at day 20, and 7 to 8 percent at day 21. As shown by earlier studies(3,4), this rapid increase is hormonally controlled. If a foetus is completely deprived of corticoids before the onset of the storage process, its liver does not store glycogen. Such a deprivation is achieved in practice by maternal adrenalectomy (for the maternal corticosteroids cross the placenta) and by removal, in utero, of either the foetal adrenals or the foetal pituitary (by decapitation). The administration of cortisol (or of ACTH in cases of foetal decapitation) restores normal glycogen levels in livers of foetuses previously deprived of corticosteroids. It has been shown(5) that other hormonal factors from hypophyseal (STH, LTH) or placental origin (STH-like) are also required for normal evolution. But the present paper will only consider the action of the corticosteroids.

In normal foetuses, when glycogen accumulation begins, the activities of several enzymes concerned with glycogen metabolism increase considerably. Phosphoglycomutase (PGM) (EC 2. 7. 5. 1.), phosphorylase (EC 2. 4. 1. 1.) and "glycogen synthetase" (UDPG-Tr) (EC 2. 4. 3. 11) activities, as measured by current methods, follow an evolution very similar to the evolution of the glycogen content(6,7). In foetuses deprived of corticosteroids before day 18, glycogen contents, PGM and UDPG-Tr activities remain low, but phosphorylase activity is not affected and increases normally. Administration of cortisol to these foetuses restores normal levels of glycogen contents and of PGM and UDPG-Tr activities without affecting the phosphorylase activity which is anyway normal (Fig. 1)(8). We have recently reinvestigated this point in greater detail(9). Foetuses were deprived of corticosteroids at day 18 p.c. (in utero decapitation and adrenalectomy of the mother) and all of them studied at the same age, day 20, after having received for several lengths of time 0.6 mg of cortisol per foetus. Some of them were also given 3 μg of actinomycin D together with cortisol. Glycogen contents and UDPG-Tr activity were measured and the results are shown in Figure 2. It seems quite clear that somehow cortisol "induces" UDPG-Tr activity and that glycogen storage ensues. The whole process is suppressed by actinomycin D.

One may therefore propose the following tentative description for the normal development of glycogen storage ability. Around day 17 or 18 p.c., the level of circulating corticosteroids increases sharply in the foetus, mainly because the foetal adrenals increase their production under pituitary control(10). These steroids, in synergy with other hormones from hypophyseal or placental origin(5), induce in a way not yet elucidated a sharp increase in the activities of enzymes necessary for glycogen synthesis and especially of UDPG-Tr. Phosphorylase activity, which is

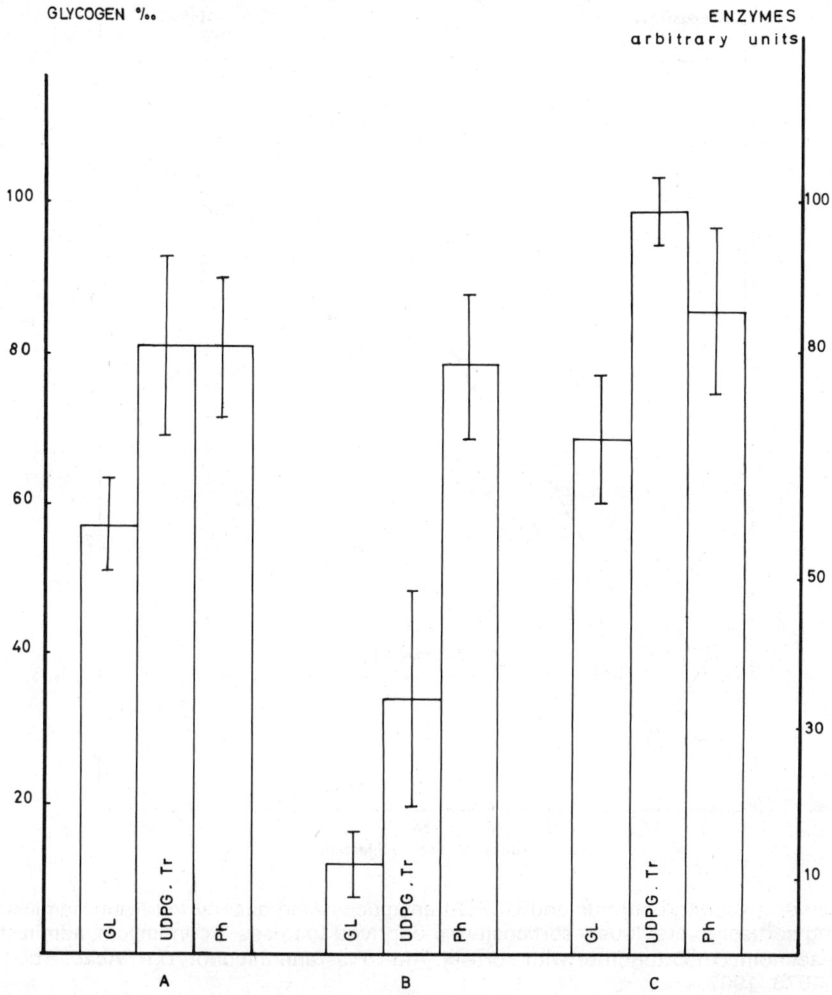

Fig. 1. Liver glycogen contents (GL), phosphorylase (Ph) and UDPG-transglucosylase (UDPG-Tr) activities at day 21 p.c. *A*: control foetuses. *B*: corticosteroid-deprived foetuses. *C*: same as *B*, but cortisol injected. After Plas and Jacquot. *C.R. Acad. Sci. (Paris), 262:1878-1881, 1966.*

probably related to glycogen breakdown, is independent of this control and reaches high levels even in the absence of corticosteroids. As glucose-6-phosphatase activity appears only just before term, glycogen mobilization is probably very small in spite of high phosphorylase activity. Thus, around day 20, the foetal liver has acquired full ability for glycogen synthesis and storage, but a second maturation is needed to allow the utilization of these reserves.

Glycogen Utilization

It is well known that liver glycogen rapidly disappears in the first few hours following birth (see for instance 11). This means that the second maturation just

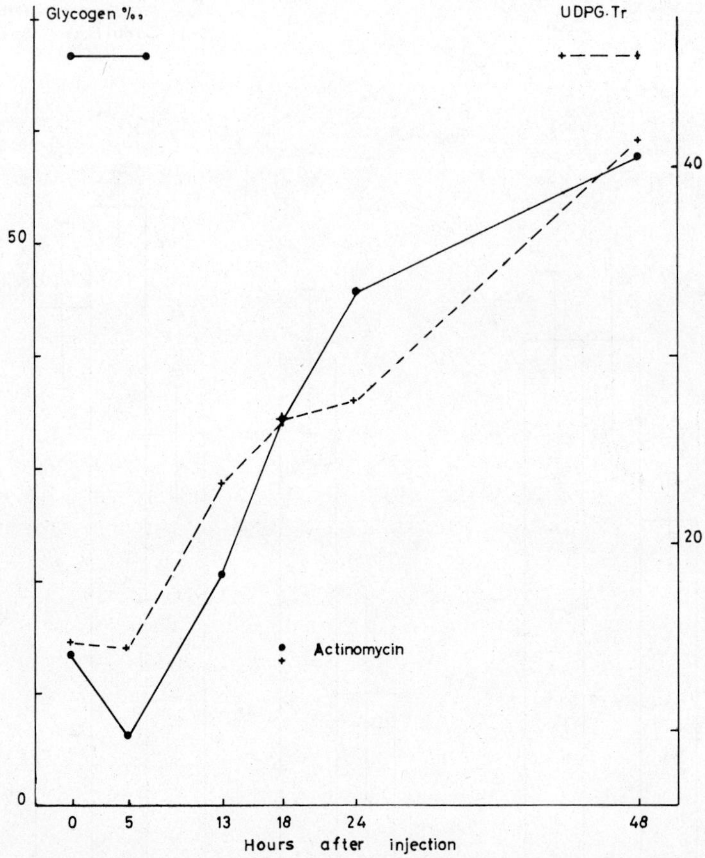

Fig. 2. Liver glycogen contents and UDPG-transglucosylase activity following administration of 0.6 mg cortisol to previously corticosteroid-deprived foetuses. Actinomycin: administration of 3 μg actinomycin D together with cortisol. After Plas and Jacquot. *C.R. Acad. Sci.* (Paris), 264:374-376, 1967.

described has taken place. We have investigated this point in the rat foetus by physiological and enzymological approaches.

The Foetal Glycemia and its Regulation. The experiments reported here were performed by Felix to test the extent to which the foetus is able to regulate its glycemia; preliminary data have been published(12). Only the main results will be summarized here. As the maternal and foetal blood sugar levels can vary quickly during blood collection, it has been necessary to define a standard experimental procedure allowing the estimation of the glycemias under conditions as close as possible to the conditions prevailing in vivo. Then a statistical study of the maternal and foetal glycemias ("true glucose") during the last four days of pregnancy was made in the following groups:

1. normal mothers, fed.
2. normal mothers, starved for 72 hours.
3. adrenalectomized mothers, fed.
4. adrenalectomized mothers, starved for 30 hours.

In each experimental group (for instance adrenalectomized mothers, 18 days of pregnancy, starved for 30 hours) the maternal glycemias form a very homogenous statistical population. In contrast there is a variability in the foetal blood sugar levels even between members of the same litter. We have no explanation for this variability and it is important to notice that the values given hereafter represent mean values.

In the fed mothers, whether normal or adrenalectomized, maternal blood sugar levels do not change much between days 18 and 21 p.c. They remain around 80 mg/100 ml. The mean foetal glycemias for these two experimental groups fluctuate between 40 and 60 mg/100 ml. Figure 3 shows the results for fed normal mothers: the increase in the foetal glycemias between days 20 and 21 is statistically significant. The results for fed adrenalectomized mothers are very similar and are therefore not reported here.

Starvation in both normal and adrenalectomized mothers reduces maternal glycemias below 60 mg/100 ml at all the stages of gestation. In both groups the foetal glycemias at 18, 19, and 20 days p.c. are very low (15 to 20 mg/100 ml), but at day 21 they reach values in the range of 40 to 60 mg/100 ml (individual values are often higher than the corresponding maternal one). Figure 4 presents the re-

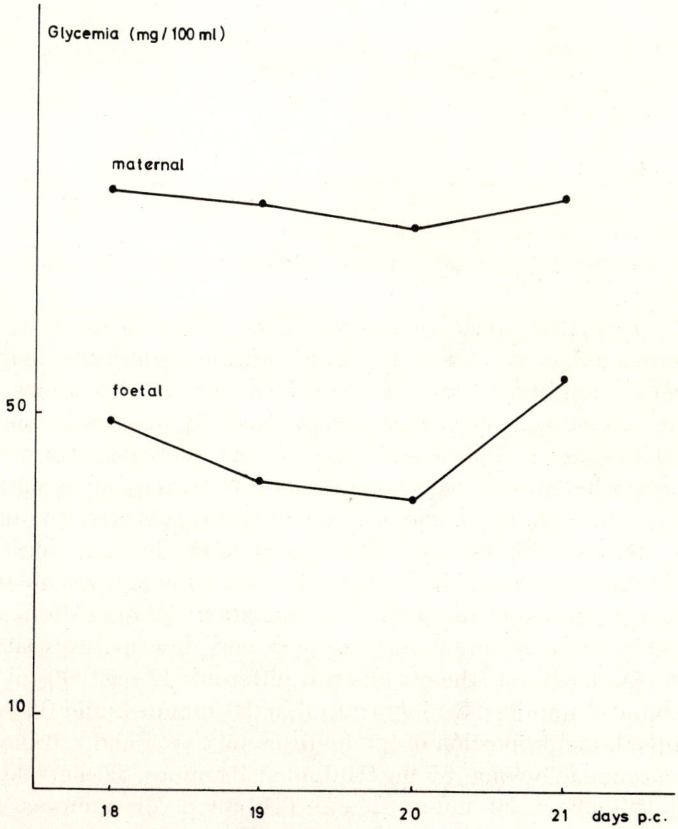

Fig. 3. Foetal and maternal glycemias in fed normal mothers. (Data from 12).

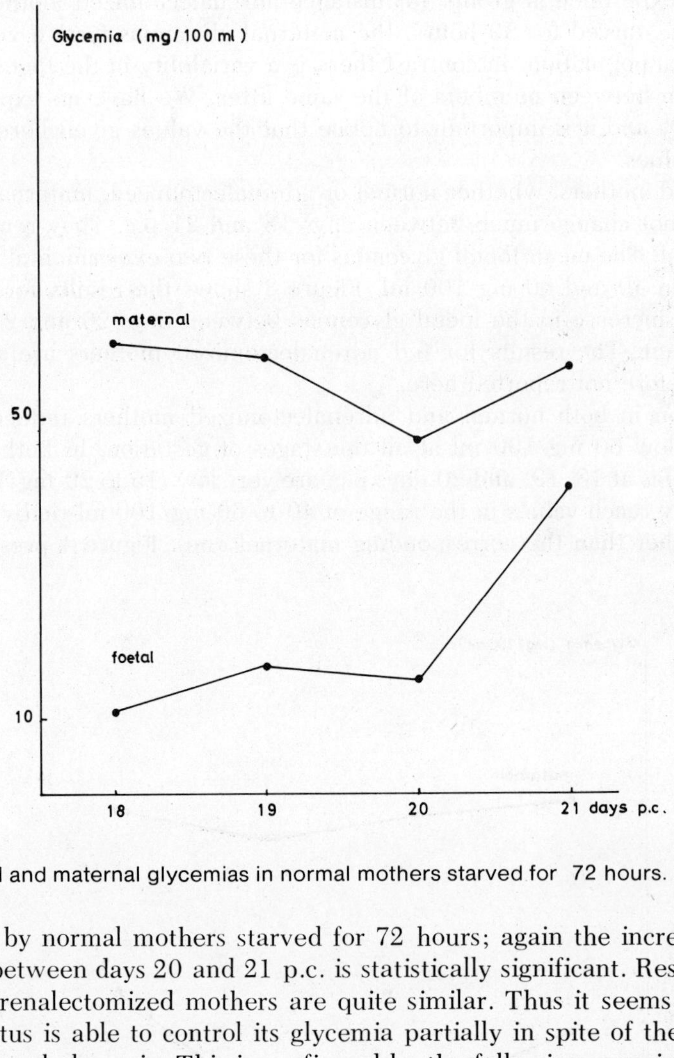

Fig. 4. Foetal and maternal glycemias in normal mothers starved for 72 hours. (Data from 12).

sults given by normal mothers starved for 72 hours; again the increase in foetal glycemias between days 20 and 21 p.c. is statistically significant. Results given by starving adrenalectomized mothers are quite similar. Thus it seems that the 21-day-old foetus is able to control its glycemia partially in spite of the fluctuations of the maternal glycemia. This is confirmed by the following experiment. Normal fed mothers are killed at each stage of gestation by a sharp blow on the neck; exactly one minute later, blood of one or two foetuses is collected, the other foetuses remaining in utero. The procedure is then repeated on the surviving foetuses at 6, 10, and 20 minutes. At days 18 and 19 p.c. the mean foetal glycemia drops quickly following the interruption of the maternal circulation: 50 mg/100 ml at 1 minute, less than 20/100 ml at 20 minutes; moreover very few foetuses survive for the entire period. At day 20 p.c. the evolution is different: 42 mg/100 ml at 1 minute, 28 mg/100 ml at 6 minutes, 30 mg/100 ml at 10 minutes, and 34 mg/100 at 20 minutes—and a large proportion of the foetuses survive. Finally at day 21 p.c., the mean values are the following: 55 mg/100 ml at 1 minute, 52 mg/100 ml at 6 minutes, 65 mg/100 ml at 10 minutes, and 110 ml at 20 minutes—and all the foetuses are alive at the end of the experiment. These results show that after day

20 p.c., the foetus is able to modify its glycemia in response to an interruption in the maternal blood supply. Such a response probably requires a rapid mobilization of liver glycogen. Therefore we have reinvestigated the glucose-6-phosphatase activity of the foetal liver.

Glucose-6-Phosphatase Activity. Glucose-6-phosphatase (G-6-Pase) (Ec 3. 1. 3. 9) is classically considered as a key enzyme for the conversion of glycogen to circulating glucose by the hepatic cell. G-6-Pase activity is absent or very low in foetal liver in most species until term. Figure 5 shows the evolution during the last 4 days of pregnancy of G-6-Pase activity in livers of rat foetuses from normal mothers, normal mothers starved for 72 hours, adrenalectomized mothers, and adrenalectomized mothers starved for 30 hours. It is quite clear that maternal starvation causes a precocious increase in G-6-Pase activity. Activities, in Figure 5

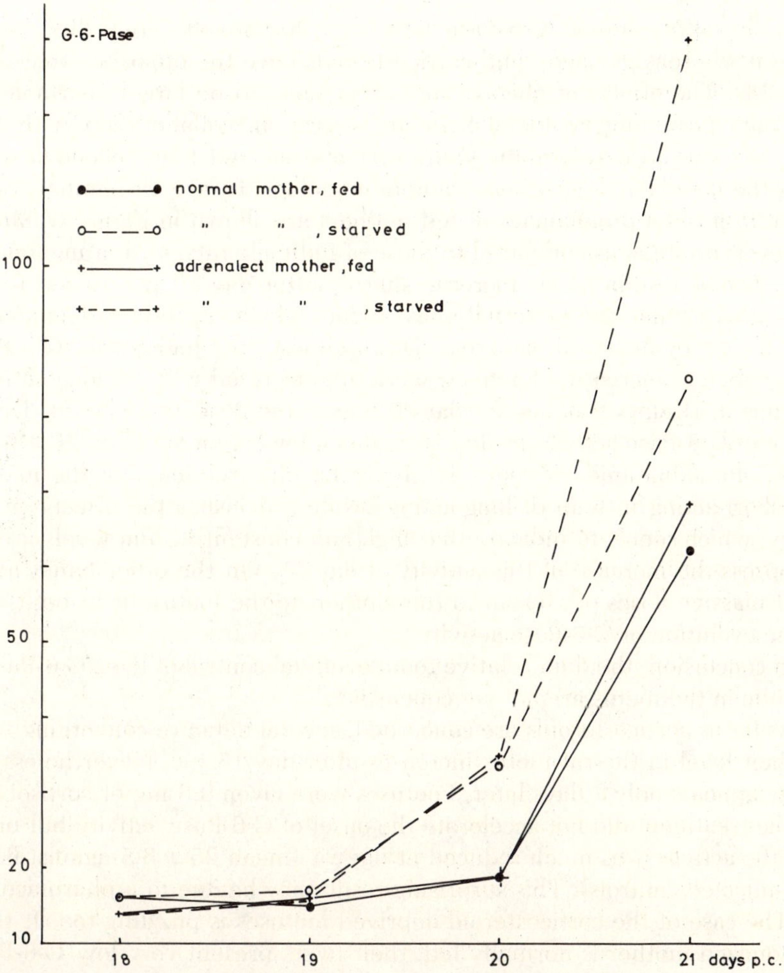

Fig. 5. Glucose-6-phosphatase activity of foetal liver during the last 4 days of gestation under several conditions (cf test).

as well as in the text, are expressed as μmoles inorganic phosphate per gram of fresh liver (30 minutes incubation at 37°C).

The metabolic pathways for gluconeogenesis seem to be lacking in the foetus (cf. for instance 13) so that synthesis of glycogen probably proceeds mainly from circulating glucose. At the end of gestation, the intensity of this synthesis is such that one might conceive that it would produce an increasing tendency to hypoglycemia. With the appearance of G-6-Pase activity this tendency can be corrected. According to this speculation it seems logical that G-6-Pase activity should be enhanced when the mother is starving. Similarly, 21-day-old foetuses respond to the interruption of maternal blood circulation by a hyperglycemia, for their livers show G-6-Pase activity; 20-day-old foetuses, with a lower activity, are just able to maintain their blood sugar level, and younger foetuses, devoid of G-6-Pase activity, cannot. In conclusion, the ability for glycogen mobilization seems to appear after or at day 20 p.c.

In the adult animal G-6-Pase activity is hormonally controlled: insulin decreases it whereas glucagon and corticosteroids have the opposite effect (cf. for instance 14). The effects of glucose starvation seem to be largely hormonally mediated. Data pertaining to foetal liver are scarce. In collaboration with Felix and Sutter, we started a systematic study of maternal and foetal blood insulin levels during the last days of gestation (radioimmunologic assay). Preliminary results(15) relative to normal pregnancies of fed mothers are shown in Figure 6. Maternal insulin levels are high as compared to those of fed male rats, with a maximum at day 19 p.c. Foetal insulin levels increase sharply after day 19 p.c. to reach very high values, greater than the maternal ones—a fact which suggests endogenous production of insulin by the foetal pancreas. Despite these very high levels, G-6-Pase activity appears and increases. Foetuses were administered 2 IU of long-acting insulin each time at 19 days 9 hours, 20 days 9 hours, and 20 days 18 hours. Under these conditions G-6-Pase activity at day 21 is much lower (mean value 26 ± 6.5 against 79 ± 8.5 for saline-injected controls). Reducing the frequency of the injections, or using short-acting instead of long-acting insulin, abolishes the effects on G-6-Pase activity, which seems to indicate that high and constant insulin levels are required to suppress the increase of this activity at day 21. On the other hand, administration of massive doses of alloxan to the mother, to the foetus, or to both did not affect the evolution of G-6-Pase activity.[1]

In conclusion, the data relative to an eventual control of liver G-6-Pase activity by insulin in the foetus are not yet conclusive.

As far as corticosteroids are concerned, several data are converging to indicate that their level in the rat foetus increases after day 18 p.c. Nevertheless G-6-Pase activity appears only 3 days later. Foetuses were given 0.6 mg of cortisol at day 18 p.c.; the treatment did not accelerate the onset of G-6-Pase activity but on the contrary, the activity was much reduced at day 21 (mean 25 ± 8.5 against 61 ± 10 for saline-injected controls). This surprising result may be due to a pharmacological effect. The case of the corticosteroid-deprived foetuses is puzzling too. If their adrenalectomized mother is normally fed, their livers present very low G-6-Pase activ-

[1]The pregnant rat is surprisingly resistant to this drug as far as blood sugar levels are concerned: sublethal doses of 200 to 250 mg/kg (single intraperitoneal injection) are needed to get an effect while 50 to 100 mg/kg are active in nonpregnant females.

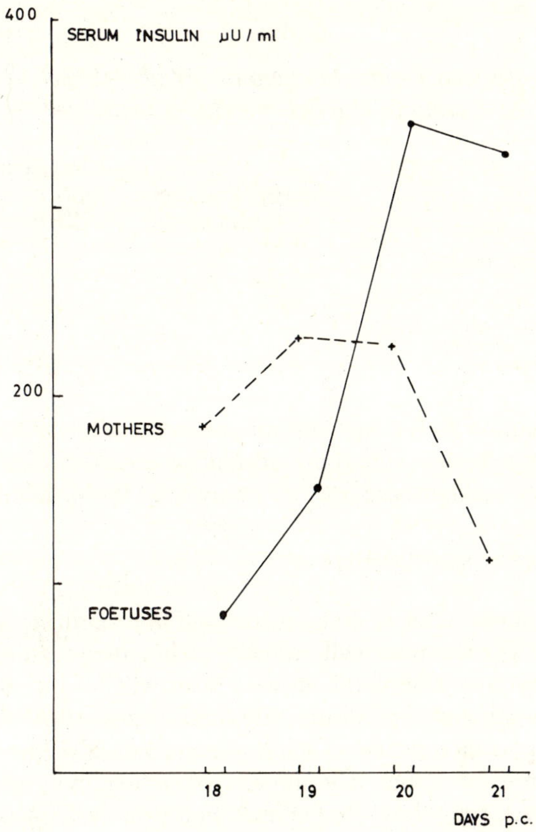

Fig. 6. Foetal and maternal serum insulin levels during the last 4 days of normal pregnancy. After Felix et al. *Horm. Metabol. Res.*, 1:41, 1969.

ity at day 21 (mean 20 ±5 against 52 ±10 for intact littermates). Cortisol does not restore the activity (18 ± 5), although it restores a normal glycogen content, as indicated earlier. On the other hand, if the adrenalectomized mother is starved for 30 hours, the corticosteroid-deprived foetuses can acquire some G-6-Pase activity as shown in Table 1. Therefore the presence of corticosteroids is probably not an absolute requisite for the development of G-6-Pase activity. The question obviously needs further study.

Recently Greengard and Dewey have shown that administration of glucagon to the rat foetus induces a precocious increase in liver G-6-Pase activity; foetuses younger than 20 days p.c. do not react(16). The parallelism between these data and the effects of maternal starvation is striking and suggests that normally the increase in G-6-Pase activity is largely due to glucagon. We have not yet estimated glucagon blood levels in the foetus. An attempt to block glucagon secretion was made by continuous maternal hyperglycemia. One mother received by gastric tubing 14 times (every hour) 1 gram of glucose. She was killed at 21 p.c., 2 hours after the last tub-

TABLE 1

Influence of corticosteroids and of maternal starvation on G-6-Pase activity in foetal liver

	Control foetuses (21 days p.c.)	Corticosteroid-deprived foetuses (21 days p.c.)
Fed adrenalectomized mothers	65 ± 15	20 ± 6
Adrenalectomized mothers, starved for 30 hours	130 ± 22	45 ± 9

ing. Maternal glycemia was 401 mg/100 ml, mean foetal glycemia 350 mg/100 ml. Nevertheless G-6-Pase levels were quite normal (mean 60 ±10).

To conclude: it is evident, as far as we know, that the problem is not yet solved.

In Vitro Differentiation of the Hepatocyte

Recently our collaborator, C. Plas, succeeded in separating hepatic from hematopoietic tissue and getting pure cell cultures of hepatocytes originating from foetuses aged 15 to 18 days. The detailed procedure will be reported elsewhere. Six hours after the beginning of the culture, the hepatocytes stick to the bottom of the petri dish and begin to incorporate labeled glucose into glycogen whatever the age of the foetus they originate from. The amount of radioactivity present in the glycogen pool increases for roughly 48 hours and then remains constant or slightly subsides, even if the milieu is renewed. If cortisol is added to the milieu, this amount continues to increase at a high rate for several days. Such a sustained incorporation reflects a real net synthesis, as is confirmed by direct chemical estimation and by the cytological aspect: in a matter of 3 days, hepatocytes belonging to 15-day-old foetuses present the spect of hepatocytes near term. Moreover, polyploidy begins rapidly (in vivo it appears only postpartum).

The eventual effects of insulin have not yet been tested, but addition of insulin antiserum to the milieu does not alter the process.

The problem obviously deserves further investigation.

EVOLUTION OF THE HEMATOPOIETIC TISSUE

It is well known that the liver is for some time a hematopoietic organ during the development of the foetus; this function disappears more or less completely around term in most species (bibliography in 17). This aspect of liver physiology has been investigated in our laboratory by Nagel. Two techniques were used to estimate the importance of the hematopoietic tissue: numbering hematopoietic cells on histological slices or counting the nuclei (both hepatic and hematopoietic) in 0.25 M sucrose homogenates. The first method, accurate but tedious, shows that the number of hepatocytes per unit volume of liver remains roughly constant between days 15 and 21 p.c. under the different experimental conditions used. The second method, which is much faster, measures the number of nuclei per unit volume of liver tissue and thus allows the evaluation of the changes in the hematopoietic population as the number of true hepatic nuclei remains constant(17). The main results can be summarized as follows (Fig. 7).

1. In normal mothers, the hematopoietic population decreases sharply after days 18 to 19 p.c. but is still present at term.
2. Stressing the mother by laparotomy or blank adrenalectomy results in an

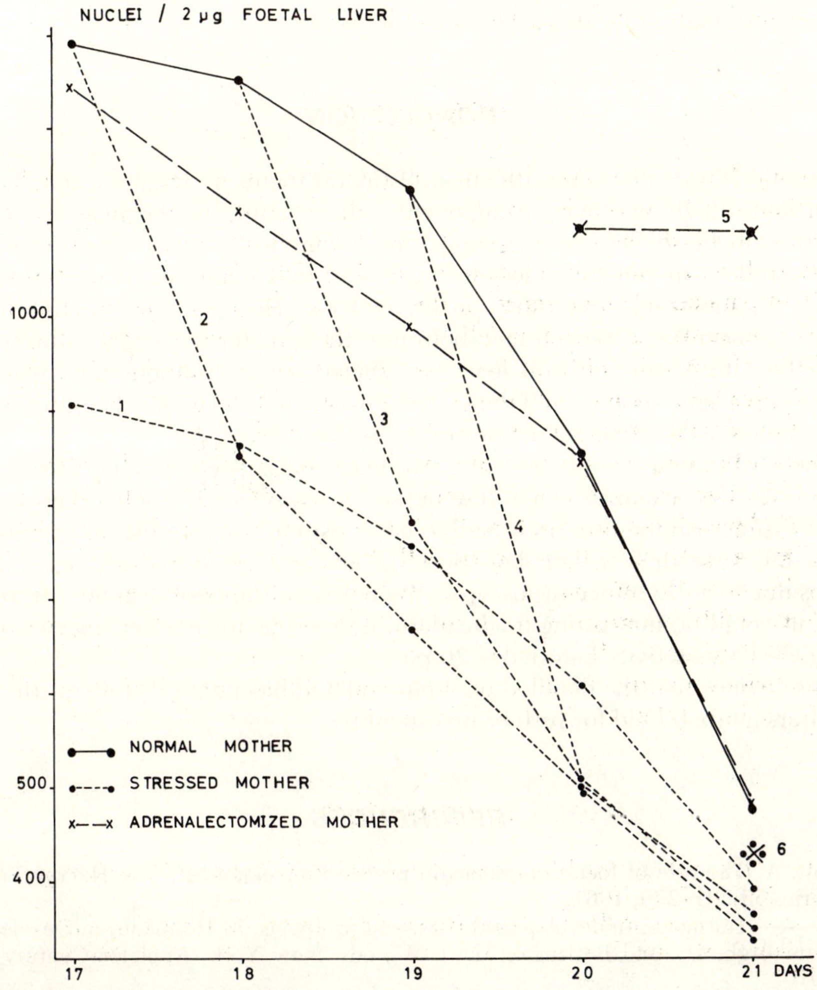

Fig. 7. Evolution of the hematopoietic population in foetal liver during the last 4 days of gestation under several conditions. Laparotomy of the mother at: (1) 14 days, (2) 17 days, (3) 18 days, (4) 19 days. (5) corticosteroid-deprived foetuses. (6) corticosteroid-deprived foetuses receiving cortisol. After Nagel and Nagel and Jacquot. *Arch. Anat. Micr. Morph. Exp.*, 57:91-98, 1968; 57:99-105, 1968; 58:99-104, 1969.

immediate and sharp decrease at whatever stage the operation is performed.

3. Previous adrenalectomy of the mother suppresses the effects of maternal stress upon foetal liver hematopoietic tissue.
4. In foetuses precociously deprived of corticosteroids, liver hematopoietic component does not decrease whereas the intact littermates show a normal evolution. Administration of cortisol to previously corticoid-deprived foetuses causes a rapid disappearance of their liver hematopoietic cells.

These data, which are more completely discussed by Nagel (17,18,19), clearly show that the hematopoietic component of the foetal liver is under an endocrine control in which corticosteroids from foetal and maternal origin play a great part. The exact mechanisms of this action are still unknown.

CONCLUSION

Around 18 to 19 days p.c., the liver of the rat foetus undergoes a complex maturation that affects two major components: the hepatocytes begin to accumulate glycogen, and the hematopoietic population begins to disappear. Such synchrony probably reflects a common mechanism. Foetal and maternal corticosteroids are certainly of paramount importance in this process. However, the situation is very complex. Concerning glycogen metabolism in the hepatocyte (which is but one aspect of the life of this cell), at least two successive maturations take place: the first one is glycogen storage, and the second, glycogen utilization. The latter maturation is not under the same endocrine control as the former.

The synchronous evolution of the two liver components may be disrupted experimentally. For instance, when the mother is stressed at 14 or 15 days p.c., the livers of 17-day-old foetuses are found to have lost most of their hematopoietic population, but nevertheless their hepatic cells have not yet stored glycogen. This is perhaps due to a difference in susceptibility between the two receptors. In relation to this, it would be interesting to elucidate the reasons for the hepatocyte's not acquiring G-6-Pase activity before day 20 p.c.

It is obvious that the detailed physiological and biochemical study of the foetal liver offers a broad field for further investigation.

REFERENCES

1. Jost, A. The role of foetal hormones in prenatal development. The Harvey Lectures, Series 55:201–226, 1961.
2. ——— Hormones in development: Past and prospects. *In* Hormones in Development. Hamburgh, M., and Barrington, E. J. W., eds. New York, Appleton-Century-Crofts, 1970.
3. ——— and Jacquot, R. Recherches sur le contrôle hormonal de la charge en glycogène du foie foetal du lapin et du rat. C.R. Acad. Sci. (Paris), 239:98–100, 1954.
4. Jacquot, R. Recherches sur le contrôle endocrinien de l'accumulation de glycogène dans le foie, chez le foetus de rat. J. Physiol. (Paris), 51:655–721, 1958.
5. Jost, A., and Jacquot, R. Sur le rôle de l'hypophyse, des surrénales et du placenta dans la synthèse du glycogène par le foie foetal du lapin et du rat. C.R. Acad. Sci. (Paris), 247:2459–2462, 1958.

6. Jacquot, R., and Kretchmer, N. Effect of foetal decapitation on enzymes of glycogen metabolism. J. Biol. Chem., 239:1301–1304, 1964.
7. ——— Hormones and liver metabolism in the foetus. Proc. 2nd Int. Congr. Endocr., Excerpta Medica Foundation, Series 83:790–793, 1964.
8. Plas, C., and Jacquot, R. Recherches sur la differenciation fonctionnelle du foie chez le foetus de rat. Activités phosphorylasique et UDPG-transglycosylasique, C.R. Acad. Sci. (Paris), 262:1878–1881, 1966.
9. ——— and Jacquot, R. Recherches sur la differenciation fonctionnelle du foie chez le foetus de rat. Activité UDPF-transglucosulasique et teneur en glycogène après administration d'hydrocortisone. C.R. Acad. Sci. (Paris), 264:374–376, 1967.
10. Cohen, A. Corrélations entre l'hypophyse et le cortex surrénal chez le foetus de rat. Arch. Anat. Micr. Morph. Exp., 52:277–407, 1963.
11. Dawkins, M. J. R. Glycogen synthesis and breakdown in fetal and newborn rat liver. Ann. N.Y. Acad. Sci., 111:203–211, 1963.
12. Felix, J. M. Glycémies foetales et maternelles chez le rat. Ann. Univ. A.R.E.R.S., 6:57–60, 1968.
13. Ballard, F. J., and Hanson, R. W. Phosphoenolpyruvate carboxykinase and pyruvate carboxylase in developing rat liver. Biochem. J., 104:866–871, 1967.
14. Nordlie, R. C., and Snoke, R. E. Regulation of liver microsomal inorganic pyrophosphate-glucose phosphotransferase, glucose-6-phosphatase and inorganic pyrophosphatase. Biochim. Biophys. Acta, 148:222–232, 1967.
15. Felix, J. M., Jacquot, R., and Sutter, B. C. J. Insulinémies maternelles et foetales chez le rat. Horm. Metabol. Res., 1:41, 1969.
16. Greengard, O., and Dewey, H. K. Initiation by glucagon of the premature development of tyrosine aminotransferase, serine dehydratase, and glucose-6-phosphatage in fetal rat liver. J. Biol. Chem., 242:2986–2991, 1967.
17. Nagel, J. Le Tissu hématopoiétique dans le foie foetal du rat en fin de gestation. I. Evolution normale au cours du dernier quart de la gestation. Arch. Anat. Micr. Morph. Exp., 57:91–98, 1968.
18. ——— and Jacquot, R. Le tissu hématopoiétique dans le foie foetal du raten fin de gestation. II. Influence de la laparotomie et de la surrénalectomie maternelles. Arch. Anat. Micr. Morph. Exp., 57:99–105, 1968.
19. ——— and Jacquot, R. Le tissu hématopoiétique dans le foie foetal du rat en fin de gestation. III. Influence de l'hypophysectomie foetale. 58:99–104, 1969.

47

Mechanisms of Sexual Differentiation

Experiments with Xenopus laevis*

Emil Witschi

The Population Council, Biomedical Division,
The Rockefeller University,
New York, New York

INTRODUCTION

When on leave in the fall of 1948 I went to the University of Tübingen, I only intended to continue some earlier work on sex development in the European brown frog *Rana temporaria.* But on passing through Basel, friends at the CIBA laboratories presented me with a teakettle full of males and females of *Xenopus* which, for curiosity's sake, I added to my already burdensome baggage. I soon felt repaid for the trouble. The African toads immediately started breeding while the native frogs *(R. temporaria)*, in defiance of scientific coaxing, tarried until the approach of the natural reproductive season, i.e., past New Year. I was also favorably impressed by the ease with which *Xenopus* larvae pass through the critical metamorphosis period. Twenty years later, all breeders in our aquariums are descendants of four pairs from that teakettle. Being very robust animals, some lines have come down through as many as six generations.

Already in Tübingen it was realized that steroid hormones do not affect sex differentiation in *Xenopus* the same way as in ranids and hylids, but rather as in urodeles and birds. Initial results were communicated in 1950 at a Paris colloquium(1). Under administration of estrogens genic male larvae differentiate as females. However, androgens masculinize only the secondary sex characteristics of genic females, and do not reverse gonadal differentiation.

On the basis of earlier experiments with *Ambystoma* we surmised that gonads

*Essential parts of this work have been supported by Ford Foundation Grant 63-41 to the University of Basel. Thanks for generous gifts of chemical supplies are due to CIBA-Basel (17β-estradiol) and to Sandoz A. G. Basel (anesthetic for cold-blooded vertebrates MS 222).

of genic females might become masculinized by heterosexual parabiosis or by implantation of testes. This expectation received experimental confirmation later by Chang(2).

After the first year at Tübingen, students and associates have participated in continued studies of sexual differentiation in *Xenopus*. Their work is integrated in this summary report on present status. Credit therefore is shared with C. Y. Chang (Peking), S. J. Segal (New York), G. Chieffi (Naples), K. Mikamo (Geneva), D. C. Johnson (Kansas City), E. Dale (Atlanta), and H. Breuer (Bonn; with collaborators K. Dahm, G. S. Rao, and B. P. Lisboa). In Paris, Dr. L. Gallien has performed experiments that parallel and extend some of ours.

GENIC AND CHROMOSOMAL DIFFERENTIATION

Xenopus laevis is a distinctly gonochoristic species. Occasionally in sections of testes one sees aberrant ovocytes within seminal tubules, but so far only one true hermaphrodite has been found with a separate testicular segment in otherwise ovarian gonads (see Michalowski [3]). The sex ratio in laboratory-bred control stock populations of 3,620 was 100 females to 104.4 males. The female being the heterozygous sex, either the *W* chromosome may be more often carried into the polar body than Z, or the chromosomal *ZW* condition may have a lower survival capacity than *ZZ* during early developmental stages.

The practicability of sex conversion makes it possible to keep laboratory breeding stocks of *ZZ*, *ZW*, and *WW* genotypes of either sex. Table 1 illustrates the possible breeding combinations with some observed results.

The conformity of obtained and expected results in all breeding combinations removes every doubt about the reality of a chromosomal *ZW-ZZ* mechanism even though morphologic studies have failed to identify the sex chromosomes(4).

Considerations similar to those now found applicable to mammalian work distinctly suggest a feminizing potency for the *W* chromosome. The *Z* chromosome and/or the autosomal complexes therefore must be expected to have some masculinizing value. However, the sex chromosomes of *Xenopus* exert barely more than a

TABLE 1

Breeding combinations and composition of offspring

Parents (P)	Offspring (F_1): Proportions expected	Offspring (F_1): Observed
1. ♀*ZZ* X ♂*ZZ*	100% *ZZ*♂	2,194♂
2. ♀*WW* X ♂*WW*	100% *WW*♀	340♀
3. ♀*ZZ* X ♂*WW*	100% *ZW*♀	318♀
4. ♀*WW* X ♂*ZZ*	100% *ZW*♀	1,377♀
5. ♀*ZZ* X ♂*ZW*	50% *ZW*♀ + 50% *ZZ*♂	111♀ + 118♂
6. ♀*ZW* X ♂*ZZ*	50% *ZW*♀ + 50% *ZZ*♂	1,771♀ + 1,849♂
7. ♀*WW* X ♂*ZW*	50% *ZW*♀ + 50% *WW*♀	621♀
8. ♀*ZW* X ♂*WW*	Identical expectation	Not bred
9. ♀*ZW* X ♂*ZW*	25% *WW*♀ + 50% *ZW*♀ + 25% *ZZ*♂	418♀ + 144♂

kind of direction-giving influence. They have not a prerogative or any specific importance for the development of primary and secondary sex characteristics. *ZZ* females have been bred with *ZZ* males continuously over five generations. Despite the constant absence of *W* chromosomes they have developed and retained all the physical and behavioral characteristics exhibited by normal "wild type" *ZW* females. So far also three consecutive generations of *WW* males have been bred with *WW* females. Likewise the absence of *Z* chromosomes throughout their lineage has not diminished the masculine instincts nor affected the development of male secondary sex characters and spermatogenesis. Therefore, the assumption is unavoidable that complete sets of sex factors are also located in autosomes and/or in non-sex-determining portions of the *Z* and *W* chromosomes.

THE POSTGENETIC MECHANICS OF SEX DIFFERENTIATION

In a discussion on how in the course of embryonic development sexual differentiation comes about, attention must be focused not only on *the activation of genes but also on the establishment of target organs*. Their relationship is not of a one-two nature, but that of reciprocal interactions. Sometimes the target organ is an essential factor in shaping the environment of the gene and may assume the role of an activator. This indeed is an essential tenet of the corticomedullary induction theory(5). However in this case, induction is not to be considered as a sole prime mover but as a summary expression for a multiplicity of morphologic and biochemical processes. The inductor tissues are already the products of several genetically-controlled evolvements(6).

Development of the Gonads

The major target organ, the *gonad*, encompasses in fact a multitude of reactive tissues with varied responses. Nieuwkoop and Faber(7) have given a condensed outline of gonad morphogenesis in *Xenopus*. At standard stage 24a,[1] about one to two dozen primordial germ cells still heavily laden with yolk platelets lay assembled at the dorsal root of the mesentery. Presently they separate into two strands that move laterally. By the end of the first week (St. 25a) a pair of *primitive gonad*

[1]Vertebrate Standard Stages (abbr. St.) are used in the text, with adscripts for subdivisions. In the last column equivalent Nieuwkoop-Faber (N-F) Stages.

Standard Stages	*Length (mm)*	*Characteristics (external)*	*Age in weeks (at 20°C)*	*N-F(7) Stages*
1-24	1-15	egg to larva	1st	1-47
25a-c	16-20	round limb buds	2nd	48, 49
25c-d	21-40	round limb buds	3rd	50, 51
26	41-50	segmented limb buds	3rd/4th	52
27a-c	50-60	hand and foot	4th	53-55
28a-d	55-65	straight legs	5th	56
29, 30	60-70	bending joints	6th	57
31a-d	70	metamorphosis of body	7th	58-62
32, 33	body 20	regression of tail	8th	63-66

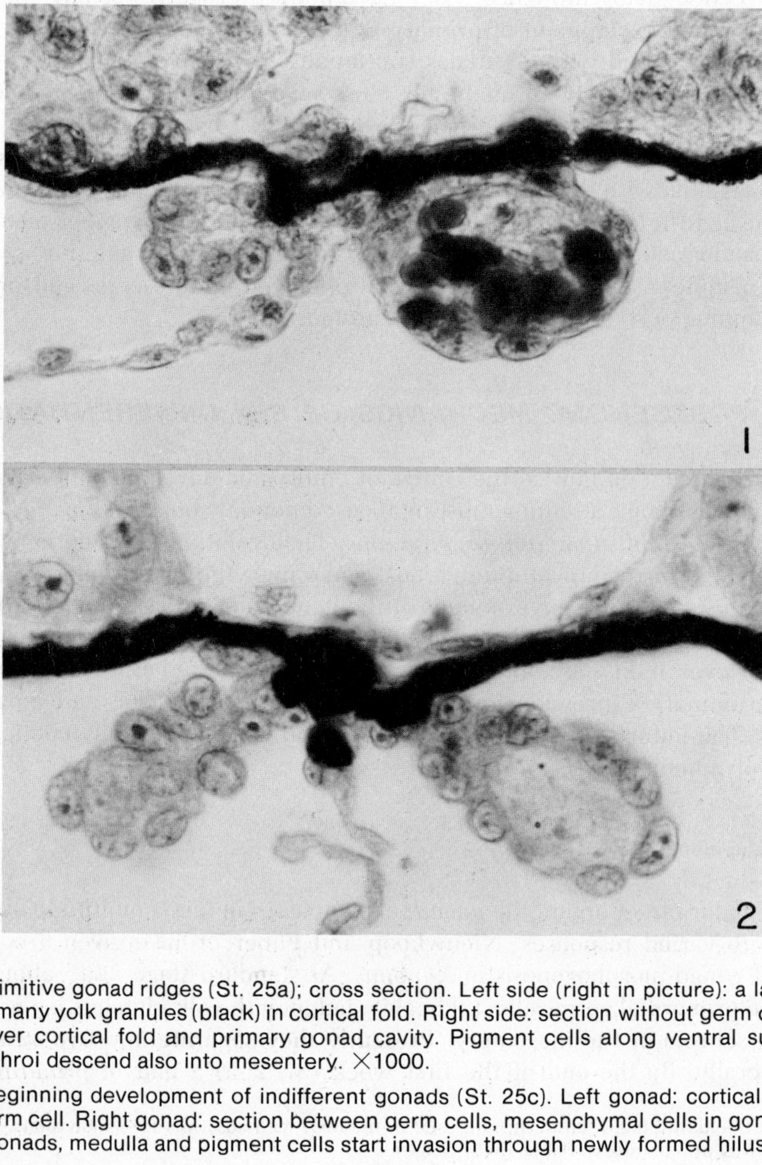

Fig. 1. Primitive gonad ridges (St. 25a); cross section. Left side (right in picture): a large germ cell with many yolk granules (black) in cortical fold. Right side: section without germ cell shows single-layer cortical fold and primary gonad cavity. Pigment cells along ventral surfaces of mesonephroi descend also into mesentery. ×1000.

Fig. 2. Beginning development of indifferent gonads (St. 25c). Left gonad: cortical cells surround germ cell. Right gonad: section between germ cells, mesenchymal cells in gonad cavity. In both gonads, medulla and pigment cells start invasion through newly formed hilus. ×1,000.

ridges have formed which protrude into the body cavity. The germ cells are only thinly covered by squamous epithelial cells (Fig. 1).

During the second week, mesonephric blastema cells migrate through the hilum, accompanied by mesenchyme and pigment cells (Fig. 2). As a result, by the end of the second week (St. 25c), *indifferent gonads* have become established. Each has acquired a cortex and 14 to 18 nodular medulla cords. Fine blood vessels enter through the hilum into the sparse mesenchyme of the primary gonad cavity.

Differentiation of *testes* and *ovaries* proceeds throughout the third week (St.

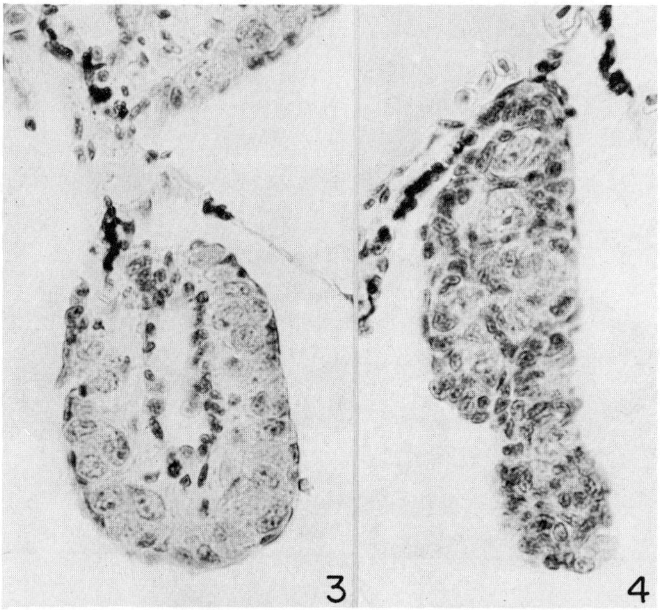

Fig. 3. Ovary (St. 29); cross-section. Cortex contains all germ cells (ovogonia); medulla, an ovarian sac. Pigment cell in hilus, mesenchyme and capillaries in primary gonad cavity. ×450.

Fig. 4. Testis (St. 29); cross-section. In the large hilar part, germ cells are in the medulla, testis tubules are differentiating. Capillaries beneath single-layered cortex. Crest part is not completely differentiated; one gonium still in cortex. ×450.

25d and 26). In the females the medullary cords remain slender and the germ cells, now much smaller after dissolution of the yolk platelets, remain in the cortex, while in the males they take places deeper in the thickening medullary cords. Morphologic differences become more distinct later. At St. 28 the ovarian medulla cords develop central cavities which soon widen to spacious ovarian sacs (Fig. 3, St. 29). In males medullary cells start arranging in epitheloid fashion, forming primitive testicular tubules (Fig. 4). By the mode of their development ovarial sacs and testicular tubules (rete and seminal tubules) prove to be homologous with mesonephric tubules. Meiosis starts in female larvae at St. 30 shortly before the breakthrough of the arms (St. 31). At the end of metamorphosis, the ovaries through accumulation and growth of ovocytes and inflation of the ovarian sacs become much larger than the testes (Figs. 5 and 6). First spermatocytes are usually found only 2 or 3 months after metamorphosis.

Conversion of Genic Males by Estrogenic Hormones

Hormonal feminization is possible during the week or 10 days in which physiologic sex differentiation ordinarily takes place. Earlier and later estrogen administrations produce no conversions. If exposures are of short duration and cover only a part of the sensitive period, hermaphrodite gonads will develop(9).

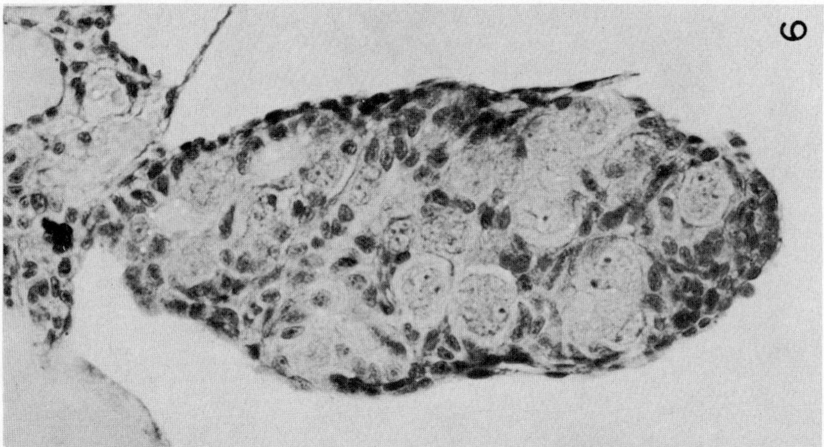

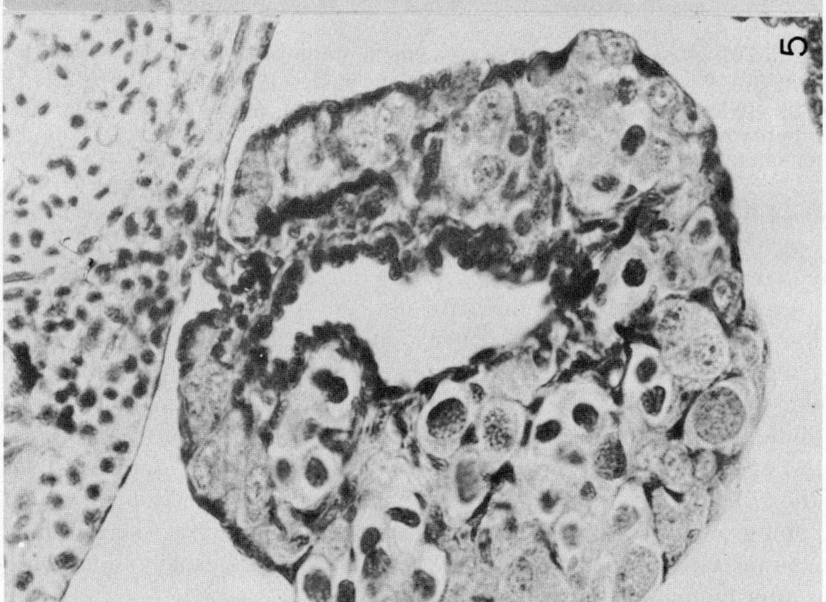

Fig. 5. Ovary (St. 33); cross-section. Ovogonia and ovocytes. ×450.

Fig. 6. Testis (St. 33); cross-section. Tubular organization of medulla progressing; spermatogonia. ×450.

TABLE 2

Feminization experiments with only one or with both parents homozygous

Parents	Control F_1	Estradiol-treated F_1
1. *ZW*♀ × *ZZ*♂	249♀ + 284♂	925♀ (50% *ZZ* + 50% *ZW*)
2. *ZZ*♀ × *ZZ*♂	278♂	69♀ (all *ZZ*)

It is not only of theoretical but also of practical interest that after treatment in the short sensitive period, physiologic sex conversion is complete and permanent. This makes it possible to use *conversion females* (ZZ) as breeders, exactly like true females. Actual records show persistent egg production in some cases over 10 years. If mated with ZZ males, conversion females produce 100 percent male offspring (Table 1, case 1).

Only relatively few estrogens have been tested for capacity to convert male (ZZ) larvae to females. It can be assumed that besides estradiol (25 to 50 μg/l), equilin (1 to 100 μg/l) and equilenin (10 to 100 μg/l), some further phenolic C_{18} steroids would produce the same effect. Equilenin and equilin are slightly more effective than estradiol. Also diethyl stilbestrol is competent, though only within a narrowly restricted dosage (100 μg/l)(10). This is noteworthy since in *Rana sylvatica* stilbestrol is entirely ineffective, while equilenin again is superior to estradiol(1).

Table 2 presents patterns of feminization experiments with bisexual and unisexual larval sibs. The advantages of working throughout with homozygous breeders are obvious. Every offspring is of known (here of ZZ) constitution (case 2). Evaluation of experimental results no longer relies on statistic certification. Any deviation from the male sex would be an experimental effect. To the contrary, in the "all female" estrogenized offspring from a *ZW* parent (case 1), the desired homozygous (ZZ) females must be sorted out by time-consuming breeding tests.

Conversion of Genic Females by Testicular Antagonist

As mentioned in the introduction, the neutral C_{19} steroids—testosterone and androsterone—support only the development of secondary sex characteristic including sex behavior but do not interfere with ovarian differentiation in genic females. Rather to the contrary, they may have a partially feminizing effect on testis development(11).

While the impact of estrogens is restricted to the short period of physiologic sex differentiation affecting indifferent gonad segments at the moment when they enter the differentiation stage, the testicular antagonist depresses cortical development at any larval and postmetamorphic stage. However, actual conversion and sex reversal may possibly also be restricted to early larval stages. Further studies are to resolve this. Two procedures have been used to expose genically female gonads (*ZW* or *WW*) to testicular antagonist: *parabiosis* and *testis implantation*. Para-

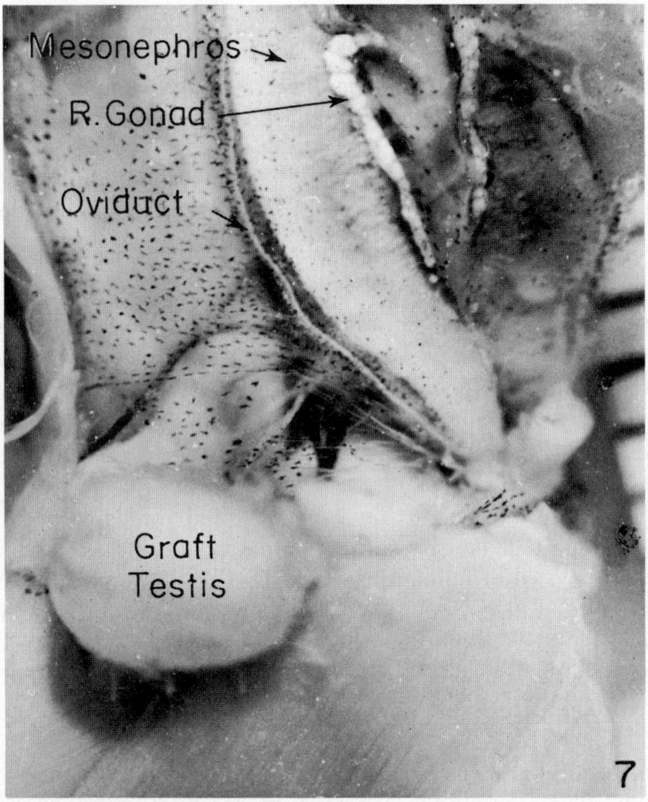

Fig. 7. *Xenopus WW.* Testis implanted on day 21 (St. 27a) attached to muscle of right thigh. Preserved at 228th day. Gonads less than 0.5 percent of normal ovarian size. Oviducts fine threads, regressing. (Specimen T_2)×9. *Testicular antagonism; incomplete sex conversion, female to male.*

biosis may be performed at early embryonic stages(2) or after metamorphosis (unpublished). Implantation is the preferred method for production of converts. A small testis of a newly metamorphosed *ZZ* male is pushed into the body cavity of an anesthetized larva, of female genetic sex (*ZW* or *WW*)(12). So far it has not been attempted to secure the graft in some desired position. At laparotomy, at the age of 5 months or later, it is most often found attached to the body wall or to a fat body. Figure 7 shows a case where after preservation of the animal it was discovered in the right groin. Considerable depression of cortical development follows in every instance of successful attachment, but compensatory growth and differentiation of medullary cords into testicular nodules seems to eventuate only after early grafting (St. 27 or younger larvae). In the case illustrated in Figure 7, the gonads are still ovaries but less than 0.5 percent the normal size. The cortex contains scattered ovogonia; ovocytes degenerate at early synaptic stages. A few auxocytes persist, indicating that some active ovarian development had occurred before establishment of the graft.

The production of *WW* males involves a series of breeding and grafting work, extending over several years(13). To sort out *ZZ*, *WW*, and *ZW*, males from heterogenous implantation groups, it is necessary to testbreed animals that acquire nuptial pads with conversion females. In a first step, *ZW* conversion males must be separated from *ZZ* males by breeding individuals from mixed groups with *ZZ* females (Table 3).

Then *ZW* males are bred with *ZW* or *WW* females and their entire offspring is testis-implanted. Finally, there follows the task of recognizing WW conversion males. Table 4 deals with the simpler second alternative. Breeding of samples of 20 or even as few as 10 offspring is adequate to decide with high probability whether the paternal constitution is *ZW* or *WW*.

It is to be mentioned here that no indication of transfer of germ cells from the *ZZ* implant to the host gonad has ever been found. If it would occur, exceptional males should appear among the offspring of *WW* fathers bred to *ZZ* mothers (case 3 in Table 1), which so far has never been observed.

As mentioned above, male larvae that were treated with estradiol only during part of the sensitive period (3rd week) develop hermaphrodite gonads. At the beginning of these experiments, the possibility had been envisaged that here the antagonism might not be very pronounced, since the testicular and ovarian segments are of the same animal. More recent extensive observations over longer periods show that antagonism gains expression in all hermaphrodites (Fig. 8). In fact, in most animals having been treated over a short two-day period, the ovarian segments become entirely eliminated within the first year. Only the abnormal shape and small size of the testes will then testify to the earlier hermaphrodite phase of development. The case shown in Figure 8 is one of the few in which some eggs have developed

TABLE 3

Testbreeding of possible ZZ *or* ZW *males with* ZZ *females*

Parent ♂	Sex of F_1	Conclusion
1	63♂	P ♂1 is a normal *ZZ*-type
2	98♂	P ♂2 is a normal *ZZ*-type
3	57♂	P ♂3 is a normal *ZZ*-type
4	29♀ + 30♂	P ♂4 is a conversion *ZW*-type
5	44♀ + 40♂	P ♂5 is a conversion *ZW*-type
6	28♀ + 37♂	P ♂6 is a conversion *ZW*-type

TABLE 4

Testbreeding of possible ZW *or* WW *males with* ZZ *females*

Parent ♂	Sex of F_1	Conclusions
1	26♀ + 8♂	P ♂1 is a conversion *ZW*-type
2	17♀ + 14♂	P ♂2 is a conversion *ZW*-type
3	19♀ + 17♂	P ♂3 is a conversion *ZW*-type
4	20♀	P ♂4 is a conversion *WW*-type
5	24♀	P ♂5 is a conversion *WW*-type
6	89♀	P ♂6 is a conversion *WW*-type

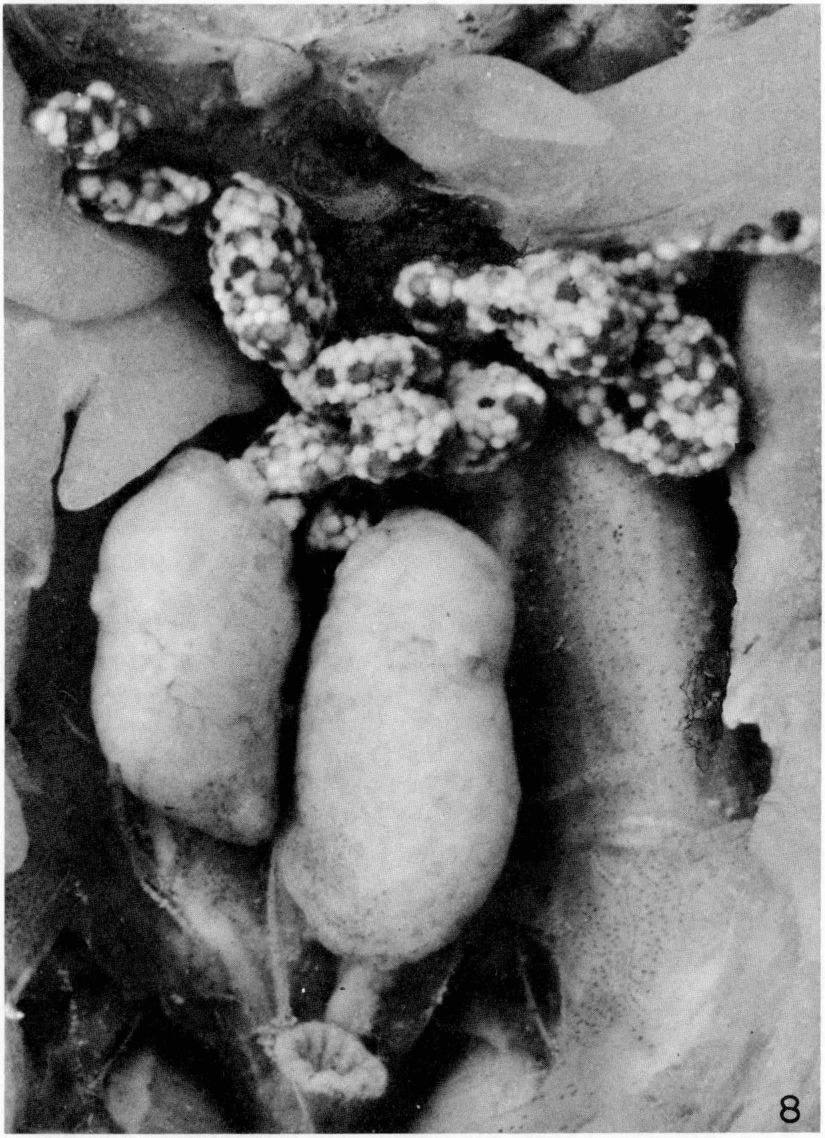

Fig. 8. *Xenopus ZZ*. During days 21-23 (St. 26) immersed in 17β-estradiol 50 μg/l. Upper halves of gonads (8 to 10 segments) converted to ovarian lobules; lower halves testicular, in accord with genic constitution. Preserved at 1 year and 78 days. (Specimen Eh 2)×5. *Sex conversion by estradiol, male to female; testicular antagonism.*

to the pigmented stage. A complete study of a large amount of material of this type is in progress. It should reveal interesting details about the working of the antagonist.

Having gained the possibility of procuring purely male and purely female cultures at almost any desired time, it was tempting to take up investigations on the biogenesis of steroid hormones in amphibian sex development.

Initiation of Endogenous Steroid Production and of Steroid-Responsiveness in Target Tissues

The most characteristic indicators of androgens and of estrogens in circulation are the *nuptial pads* (on forelimbs) and the oviducts. The former make their appearance in males 6 to 10 months old, i.e., about the time when complete spermatogenesis has become established. Funnels and upper portions of the *oviducts* begin to develop in both sexes during climactic metamorphosis (St. 30 to 33). When the tails of the little frogs have become fully resorbed, the tips of the oviducts reach the upper level of the gonads. Within the following 2 months they continue down-growth along the lateral edge of the mesonephric bodies until they reach the cloaca, with ends fused, at the age of 4 months. In males they persist only if castration is performed not later than at 7 months. If this operation is delayed another 3 months, the ducts have regressed up to about midway and only the upper parts are permanently retained. Castration at the end of the first year meets with the condition of complete regression which is characteristic of normal male development at this age.

Less obvious secondary sex characteristics are the large *larynx* of the male and the protruding *cloacal labia* of the female. Both structures start sex-specific enlargement following maturity of the gonads at the age of 6 to 8 months and later.

There is only little evidence of spontaneous production of ovarian hormones before the sixth to the eighth month, which is the time when yolk platelets appear in the largest ovocytes. The oviducts of females, which after the third month have only slightly larger diameters than those of males, then lengthen again, become convoluted and their walls begin to mature histologically. Males and females that were castrated at 7 months or earlier remain practically identical throughout adult life. They have nearly straight filamentous oviducts, barely protruding cloacal lips, and lack nuptial pads. The larynges are not enlarged. In summary, well recognizable effects of gonadal hormones appear in both sexes only at about the sixth or seventh month of age. In close agreement with these observations are the findings by Gallien and Chalumeau-Le Foulgoc(14). They identified traces of estrogenic steroids in alcohol extracts of gonads of 5-month-old female *Xenopi*, and apprecia- larger amounts at 8 and 11 months.

Hormone responsiveness of the primordia of secondary sex characteristics is, however, established long before the normal start of sex-dimorphic development. Injected androgens cause differentiation of the *pad forming areas of arms and fingers* immediately after termination of metamorphosis in both sexes. Size and shape of the responding skin areas are not determined nor altered by hormonal factors. The larynges enlarge precociously. Administered androgens also activate the copulation instincts, the sexes clasping indiscriminately(9).

The formation of primordial funnels and upper ducts at stages 31 to 33 is not hormone determined. However, even the earliest rudiments respond to exogenous estrogen stimulation(10). By such treatment, they become hyperplastic; the cells of walls and budding tips become cytologically adultized and therefore are rendered incapable of proliferating further.

After castration at the juvenile phase (7 months or earlier), the secondary sex characteristics retain their responsiveness to hormonal stimulation. In both sexes the oviducts assume functional condition following estrogen administration. Likewise, nuptial pads develop with androgen injections. The reactions are independent of genic sex constitution and of sex type at the time of castration.

Direct chromatography(15) and indirect chemical methods(16) have provided evidence of adrenal steroidogenesis starting at early larval stages, immediately following histologic organization of cortical islands (St. 25d). The entire larval production is corticoid; sex steroids have not been identified. The latter negative finding agrees with the slow histologic development of the larval gonads and with the apparent unimportance of androgens and estrogens in premetamorphic development.

Catabolism of Sex Steroids

Steroid tests with larval amphibians have been conducted by addition of hormones to the aquarium water. For quantitative work it is essential to obviate the possibility of bacterial interference. To that end the animals are pretreated in baths of 0.1 percent sodium sulfadiazine, and penicillin is added to the incubation fluids.

It was realized early that liver in vitro catabolizes testosterone and estradiol at rates nearly equal to those of intact larvae (St. 26 to 29). Larval liver was found to be quantitatively six to seven times as efficient as adult liver in converting administered hormones to less active compounds(17), reflecting the well known fact that metabolic rates in general are particularly high at larval stages. In experiments in 1955 to 1956, neither qualitative nor quantitative differences were found between the activities of female and male livers of same age(17).

In a recent experiment, male ZZ-larvae were incubated with tritiated estradiol. Treatment was given on the 17th day, i.e., at the stage when the hormone is inducing sex conversion. It was possible to identify estrone very early in the livers but not in the gonad region. Obviously the change of steroid structure occurs only in the liver and is not an essential part of the mechanics of sex conversion by estradiol.

In collaboration with the laboratory of Professor Breuer, the steroid metabolism of Xenopus was reinvestigated, making use of radiocarbon-marked hormones and some of the newer, more sophisticated methods of chromatography(18,19). Table 5 lists the [4-^{14}C] C_{19} and C_{18} steroids that were offered to larvae or incubated with adult liver slices and the subsequently identified metabolites. Some of the latter must have gone undetected since large parts of the residues (sometimes more than half) were not, or were only incompletely, analyzed.

The animal material includes several stages from 8-day larvae (St. 25a) to adult males and females. Even though analyses were not equally complete in all cases, they indicate clearly enough that the metabolites from testosterone and from 17β-estradiol are the same in both sexes and over the full range of larval and adult stages. Adult liver was not incubated with estradiol. An experiment of Lisboa with estrone yielded only estradiol and no estriol (unpublished). The importance of this finding remains unclear. It is difficult to bring it into accord with the reported estradiol metabolism of larvae (Table 5).

TABLE 5

Metabolites after incubation with C_{19} and C_{18} [4-^{14}C] steroids

Recovery of substrates 0 to 70 percent in 34 analyses

Substrates	Biologic materials	Metabolites
Dehydroepi-androsterone	Larvae stage 25c	Δ5-androstenediol, var. (no Δ4-3-ketosteroids)
Testosterone	Larvae Stages 25b to 31 Adult liver slices	Androstenedione Androsterone Androstanediol Androstanedione Hydroxy compounds var. polar fractions
17β-Estradiol	Larvae Stages 25a to 25d	Estrone Estriol 6α-Hydroxyestrone 15α-Hydroxyestrone
Estrone	Adult liver slices	Estradiol

Not yet resolved is the question of a possible *quantitative difference* between male and female steroid catabolisms. On the basis of early results(18) we had opined that more unchanged testosterone is recoverable from male than from female groups, and vice versa that females catabolize estradiol more sparingly than males. But of 16 now available pairs of assays only 10 sustain that rule. Even though it would not be inconceivable that the genic sex differences might have an influence on rate of liver function, independent of gonad physiology, direct evidence is missing or at least insufficient. In an experiment with adult livers, the recovery values of testosterone were 53 percent for *ZZ* males, 47 percent for *ZW* females, and 46 percent for *ZZ* conversion females (Lisboa, unpublished). If the differences are at all meaningful, then physiologic rather than genic sex should seem to be decisive. Repeated experiments with larvae vary considerably by substrate concentrations and by differences of age. Hence, they do not yet give a critical decision about this question. At any rate, a sex difference could be only one of minor degree. This material also is not quite suited for demonstration of *qualitative sex differences* in relative proportions of metabolites. Recovered quantities always are small, and few suggestive observations were confirmed by repetition (for another discussion, see 19).

A discussion of the enzyme systems that are already present in the livers of 1-week old larvae, i.e. before the organization of gonads, will be given elsewhere. Of particular interest is the absence of a hydroxysteroid dehydrogenase in 13-day larvae (St. 25c), as must be concluded from the absence of androstenedione and testosterone among metabolites after incubation with dehydroepiandrosterone (Table 5).

Biogenesis of Sex Hormones

Adopting techniques that have given interesting results in the mammalian field(20,21), the possible steroidogenic capacities of *Xenopus* larvae were tested by offering probable precursor substances marked by radiocarbon. The incubation procedures were those already used in the study of liver catabolism. However, interest was now directed toward the developing gonads and the biogenesis of sex hormones. Beginning with St. 29, testes and ovaries were dissected out—free of adhering somatic tissues—for separate testing. At earlier stages the entire larvae were incubated and extracted(22). The results are presented in Table 6.

In a first set of experiments 17α-hydroxyprogesterone was incubated with 10-day (St. 25b) and 17-day (St. 25d) larvae for various lengths of time, from 3 to 24 hours. These early stages demanded special attention since it is here that initiation of sex differentiation actually occurs. The substrate was continuously metabolized, mainly to corticoids that were not further determined. Since previous work(23) has shown that cortisol is released normally from adrenals at the end of St. 25 and the beginning of St. 26, it is interesting that it appears now (St.25d) as a metabolite of administered hydroxyprogesterone.

In the present series, C_{19}-steroidogenesis starts at St. 29. A small fraction of radioactivity was recovered in the form of androstenedione. This is sufficient evidence of the presence, now, of side-chain splitting desmolase (17α-hydroxyprogesterone C_{17}-C_{20}-lyase). The yield doubles in the later stages of testicular development (St. 31, 33, frogs of 3 months), but remains low in the ovaries. Neither testosterone nor estradiol were obtained at these metamorphic stages. Examining again the histologic state of ovarian and testicular development (Fig. 3 to 6), one misses cellular elements of distinctly endocrine character. However, considering that quantitatively, steroid production is extremely low, it seems not improbable that the walls of the ovarian sacs and the sparse interstitium of the testicular medulla may be competent to exert such endocrine function. Doubts have been expressed by other authors. Pesonen and Rapola(24) applying the Wattenberg steroid-3β-ol-dehydrogenase test on adult females and males of *Xenopus* observed

TABLE 6

Steroidogenesis from [4-^{14}C] labeled precursor substances

Material	Substrate	Metabolites	%S[a]	Enzymes
Larvae 25c	Acetate	No C_{19} Steroids		No C_{17}-C_{20}-desmolase
Larvae 25c	Cholesterin	idem		idem
Larvae 25c	Pregnenolone	idem		idem
Larvae 25b	17α-Hydroxyprogesterone	idem, Corticoids		idem, Hydroxylases
Larvae 25d		idem, Cortisol		idem
Testes of:				
Stage 29/30	17α-hydroxyprogesterone	Androstenedione	0.030	C_{17}-C_{20}-desmolase:
Stage 31	idem	idem	0.072	17α-Hydroxyprogesterone
Stage 33	idem	idem	0.064	C_{17}-C_{20}-lyase
Frog 3 months	idem	idem	0.065	
Frog adult	idem	idem and Testosterone		idem
Ovaries of:				
Stage 29/30	17α-hydroxyprogesterone	Androstenedione	0.029	C_{17}-C_{20}-desmolase
Stage 31	idem	idem	0.026	idem
Stage 33	idem	idem	0.035	idem
Frog adult	idem	idem and 17β-estradiol		idem

[a]Percent of substrate radioactivity per animal.

strongly positive reactions in the adrenals but only trace deposits of formazan granules in testes and none whatsoever in ovaries. They consider it possible that sex-steroid synthesis might be nonexistent in amphibian gonads. Such a hypothesis is untenable now in view of the above-mentioned effects of castration and gonad transplantation in *Xenopus*, and the evidence reported above of desmolase activity in gonads, resulting in the production of androstenedione, testosterone, and estradiol. Independently, Rao and Dale (unpublished) obtained evidence of sex-hormone biogenesis in testicular and ovarian tissues from adult animals (Table 6).

Reference should also be made to the work by Dale and Dorfman(25) on androstenedione and testosterone synthesis by bullfrog testes from progesterone substrate in vitro. Ozon(26), in a very complete study of steroid metabolism in a newt (*Pleurodeles waltlii*), finds similar chains of conversions. In the testis: progesterone → 17α-hydroxyprogesterone → androstenedione → testosterone. In the ovary this biosynthetic pathway is extended by aromatization of testosterone, with resultant 17β-estradiol formation. It seems safe enough to assume that steroid biogenesis in the adult testes and ovaries of *Xenopus* follows essentially identical patterns.

DISCUSSION

Sex Chromosomes

Since males as well as females can develop and become functional with any of the three diploid sex chromosome combinations, one must conclude that *Z* and *W* carry similar arrays of male *(M)* and female *(F)* sex-determining genes. The latter kind obviously prevail in the *W* chromosomes and, under nonexperimental conditions, direct ovarian type differentiation. This is the case in *ZW* diploids and also in *WW* diploids and various polyploids (see the review by Beatty, 27). Haploids described by Gallien(28) were all of *Z* type. Accordingly, they developed testes (sometimes partially feminized by parabiotic union with females). It is of particular interest that the entire functional importance of the seemingly very powerful genes is restricted to the period of less than 10 days during which initial sex differentiation occurs. The histologic appearance of the gonads at this time changes little. The major developments in secondary sex characteristics follow much later, i.e., after having become independent of either male or female genic constitution.

Steroid Organs

Even though steroid hormones may occasionally be produced in trace amounts by cells outside the gonad-adrenal complex, they are essentially secretions of tissue complexes which derive from the primitive intermediate mesoderm. Their morphologic separation is accompanied by functional differentiation. With respect to steroidogenesis, the *adrenal* is first to become productive, synthesizing cortisol as early as St. 26. The *gonadal medulla* gives indications of androstenedione manufacture in mimimal amounts only at St. 29. No evidence of estrogen synthesis or activity was found before the end of metamorphosis. Nevertheless, the two gonad components—cortex and medulla—assume important other functions at early embryonic stages. The *primitive cortex*, first a narrow strip inserted in the celomic epithelium, guides the migration of the primordial germ cells from the crest of the endoderm to the mesial edges of the mesonephric bodies (St. 20 to 24). Apparently the cortical cells are releasing a chemical substance into the surrounding tissues which attracts germ cells until they reach the gonad site and become invested in the cortex itself. Such attraction substances, for which the designation *telopherones* is proposed, have been long recognized, particularly in studies on human(29) and avian embryos(30). Nothing is known about their chemical nature.

Activating the Sex Genes

Every cell of a normal tadpole carries identical genes, including the sex genes. However, the female-determiners (*F*) become activated only in the cortical-cell combine and the male-determiners (*M*) only in the medullary system of differentiating gonads. This activation occurs in the restricted time while larvae pass through St. 25c to 26. At temperatures of 20 to 22°C, this occurs during the third week after fertilization. Like other events in embryogenesis, activation arises "spontaneously" at a hereditarily fixed stage. Beyond doubt it is causally linked with specific physicochemical events, possibly with a definite phase in overall RNA synthesis(31). On the basis of the work by Alfrey and Mirsky(32) we can assume that intranuclear enzymes start the process of acetylation of the specific histones which so far had held the *F* or the *M* genes in the dormant condition.

The possibilities of parallel and of consecutive cortical and medullary development—both leading to hermaphrodism—are realized in some other amphibian species. However, in *Xenopus* immediately following activation, the epistatic sex inhibits the development of the hypostatic gene complex. Thus, in a genic female (*ZW*) the activated *F* genes not only produce RNA sequences, enzymes, and inductors of ovarian differentiation, but also repressors (possibly ribonucleases) which cross over from the cortex to the medulla and prevent the activation and/or RNA production of the *M* genes. The same or similar antagonists prevent copious growth and male-type differentiation of the medullary cords. Instead, transformation into ovarian sacs (Fig. 3) is induced.

In genic males *(ZZ)* activation of the *M* genes follows a similar course. However, antagonism, besides inhibiting activation of the *F* genes, also leads to disaggregation of the cortex. Germ cells and somatic elements become detached from the surface epithelium and are transferred to the medulla (Fig. 4). The complicated relationships were more fully discussed and diagrammatically illustrated in an earlier publication(6).

Mechanisms of Sex Reversal

Recent work on uterine growth stimulation in the rat by estradiol provides some information that also seems applicable to the case of sex conversion by the same hormone in ZZ larvae of *Xenopus*. Literature and pertinent facts are reviewed and their bearing on developmental biology discussed by Segal(33). Of prime importance is the discovery that cells of target organs withdraw the hormone from circulation and accumulate it at a particularly high rate. In the case of the gonad of *Xenopus*, estrogen receptors most likely are cortical cells in which activation of the *d* genes is set to start. Experiments with tritiated estradiol are intended to test this assumption. If the animal is of the *ZZ* constitution, ordinarily F-gene activation would be inhibited by antagonists from the medulla. However, as shown by Libby (34), estradiol increases the rate of histone acetylation in nuclei of target cells. Apparently, in the case of differentiating *Xenopus* larvae such increase is sufficient to reverse the constitutional $M > F$ power ratio and bring about conversion of sex development, i.e., feminization of genic males. The precision with which this estradiol process works is impressive. Withdrawal of the hormone before completion of the differentiation period is immediately followed by return to purely genic control, i.e., testicular development (Fig. 8).

The female to male conversion by testis implants is less understood. Testes suppress ovarian development at any developmental stage, even in the adult. Whether the emitted *antagonist* is always the same agent as the one acting during normal male differentiation at larval stages is not clear. Certainly it is not a steroid hormone. Observations in parabiosis suggest rather a proteinic nature. Testicular development is permitted, but probably not stimulated, by the disappearance of the cortical inhibitor. Usually only one or a few medullary cords start compensatory testicular development. Their growth becomes secondarily stimulated by gonadotropins after removal of the implant testis.

REFERENCES

1. Witschi, E. Génétique et physiologie de la différentiation du sexe. Arch. Anat. Micr. Morph. Exp., 39:215–246, 1950.
2. Chang, C. Y. Parabiosis and gonad transplantation in *Xenopus laevis* Daudin. J. Exp. Zool., 123:1–28, 1953.
3. Michalowsky, J. Über das Vorkommen von jugendlichem Hermaphroditismus bei *Xenopus laevis* Daud. Biol., 78:498–502, 1959.
4. Mikamo, K., and Witschi, E. The mitotic chromosomes in *Xenopus laevis* (Daudin): normal, sex reversed and female WW. Cytogenetics, 5:1–19, 1966.
5. Witschi, E. Hormones and embryonic induction. Arch. Anat. Micr. Morph. Exp., 54: 601–611, 1965.
6. ——— Biochemistry of sex differentiation in vertebrate embryos. *In* The Biochemistry of Animal Development. Weber, R., ed. New York, Academic Press, Inc., Vol. 2, pp. 193–225, 1967.
7. Nieuwkoop, P. D., and Faber, J. Normal Table of *Xenopus laevis* (Daudin). Amsterdam, North-Holland Publishing Co., 1956.
8. Altman, P. L., and Dittmer, D. S. Growth: VII Prenatal Vertebrate development. Biol. Handbooks, Fed. Amer. Soc. Exp. Biol., Washington, 1962.
9. Chang, C. Y., and Witschi, E. Genic control and hormonal reversal of sex differentiation in *Xenopus*. Proc. Soc. Exp. Biol. Med., 93:140–144, 1956.
10. Segal, S. J. The effects of stilbestrol on the differentiation of the gonads and secondary sex characters of *Xenopus laevis*. Thesis, Library Zool. Dept., Univ. Iowa, 1951.
11. Gallien, L. Inversion experimentale du sexe chez un anoure inferieur *Xenopus laevis* Daudin. Bull. Biol. France Belg., 90:1–21, 1956.
12. Mikamo, K., and Witschi, E. Functional sex reversal in genetic females of *Xenopus laevis* induced by implanted testes. Genetics, 48:1411–1421, 1963.
13. ——— and Witschi, E. Zuchtversuche mit geschlechtskonvertierten Krallenfroeschen. Experientia, 19:536–537, 1963.

14. Gallien, L., and Chalumeau-Le Foulgoc, M. T. Mise en evidence de steroides oestrogènes dans l'ovaire juvenile de *Xenopus laevis* Daudin. C.R. Soc. Acad. Sci. (Paris), 251:460–462, 1960.
15. Witschi, E., and Dale, E. Steroid hormones at early developmental stages of vertebrates. Gen. Comp. Endocr., Suppl. 1:356–361, 1962.
16. Rapola, J. The adrenal cortex and metamorphosis of *Xenopus laevis* Daudin. Gen. Comp. Endocr., 3:412–421, 1963.
17. Chieffi, G. Il metabolismo degli ormoni sessuali da parte degli anfibi: Ricerche in vitro e in vivo. Boll. Zool., 23:121–130, 1956.
18. Breuer, H., Dahm, K., Mikamo, K., and Witschi, E. Differences in steroid metabolism of male and female larvae of *Xenopus laevis*. 2nd Int. Congr. Hormone Steroids, Amsterdam, Excerpta Medica Found. Ser., No. 111:215, 1966.
19. Rao, G. S., Breuer, H., and Witschi, E. Metabolism of estradiol-17β by male and female larvae of *Xenopus laevis*. Experientia, 24:1258, 1968.
20. Acevedo, H. F., Axelrod, L. R., Ishikawa, E., and Takaki, F. Steroidogenesis in the human fetal testis. J. Clin. Endocr., 21:1611–1613, 1961.
21. ——— Axelrod, L. R., Ishikawa, E., and Takaki, F. Studies in fetal metabolism. II. Metabolism of progesterone-4-C^{14} and pregnenolone-7α-H^3 in human fetal testes. J. Clin. Endocr., 23:885–890, 1963.
22. Rao, G. S., Breuer, H., and Witschi, E. In vitro conversion of 17α-hydroxyprogesterone to androstenedione by mashed gonads from metamorphic stages of *Xenopus laevis*. Gen. Comp. Endocr., 12:119–123, 1969.
23. Dale, E. Steroid excretion by larval frogs. Gen. Comp. Endocr., 2:171–176, 1962.
24. Pesonen, S., and Rapola, J. Observations on the metabolism of adrenal and gonadal steroids in *Xenopus laevis* and *Bufo bufo*. Gen. Comp. Endocr., 2:425–432, 1962.
25. Dale, E., and Dorfman, R. Conversion of progesterone-4-C^{14} to testosterone by testicular tissue of the American bullfrog. Gen. Comp. Endocr., 9:313–318, 1967.
26. Ozon, R. Synthèse in vitro des hormones stéroides dans le testicule et l'ovaire de l'amphibien urodèle *Pleurodeles waltlii* Michah. Gen. Comp. Endocr., 8:214–227, 1967.
27. Beatty, R. A. Chromosome deviations and sex in vertebrates. *In* Intersexuality in Vertebrates Including Man. Armstrong, C. N., and Marshall, A. J., eds. New York, Academic Press Inc., pp. 17–143, 1964.
28. Gallien, L. Développement d'individus haploides adultes élevés en parabiose chez le triton *Pleurodeles waltlii* Michah. J. Embryol. Exp. Morph., 18:401–426, 1967.
29. Witschi, E. Migration of germ cells of human embryos from the yolk sac to the primitive gonadal folds. Carnegie Inst., Contrib. Embryol., 32:67–80, 1948.
30. Dubois, R. L'attraction exercée par le jeune épithélium germinatif sur les cellules germinales de l'embryon de poulet est un phénomène d'ordre chemiotactique. C.R. Acad. Sci. (Paris), 262:2623–2626, 1966.
31. Brown, D. D. RNA synthesis during early development. *In* Developmental and Metabolic Control Mechanisms and Neoplasia. Baltimore, Williams & Wilkins Co., pp. 219–236, 1965.
32. Alfrey, V. G., and Mirsky, A. E. Mechanisms of synthesis and control of protein and ribonucleic acid synthesis in the cell nucleus. Cold Spring Harbor Symp., Quant. Biol., 28:247–262, 1963.
33. Segal, S. J. Regulatory action of estrogenic hormones. Develop. Biol., Suppl. 1:264–280, 1968.
34. Libby, P. R. Histone acetylation by cell-free preparations from rat uterus: in vitro stimulation by estradiol-17β. Biochem. Biophys. Res. Commun. 31:59–65, 1968.

48

Growth of Seminal Vesicle and Prostate: Role of RNA Following Testosterone Stimulation*

Claude A. Villee[1] and Tomoko Fujii[2]

[1]Department of Biological Chemistry, Harvard Medical School, Boston, Massachusetts
[2]Department of Pharmacology, Tokyo Women's Medical School, Tokyo, Japan

The seminal vesicle of the 3-week-old rat responds to the administration of testosterone with a dramatic increase in weight and in protein content. We had shown previously that testosterone increased the biosynthesis of RNA in the seminal vesicle and prostate of 3-week-old rats as measured by the incorporation of ^{3}H-cytidine into RNA(5). The RNA of the seminal vesicle, when extracted and instilled into the lumen of the seminal vesicle of other 3-week-old rats is effective in stimulating growth and the synthesis of protein (Fig. 1.) Only RNA extracted from the seminal vesicle of testosterone-treated 3-week-old rats was effective in stimulating growth. In contrast, RNA from the liver or prostate of these immature rats and RNA extracted from any of the tissues of control uninjected 3-week-old rats was ineffective in stimulating growth of the seminal vesicle(4,10).

In further experiments (Fig. 2) we were able to demonstrate that RNA extracted from the liver, prostate, seminal vesicle, or kidney of adult male rats also stimulated growth when instilled into the seminal vesicle of 3-week-old rats. The stimulatory effect of the RNA was eliminated if the RNA was treated with ribonuclease or if the RNA was heated for 15 minutes at 100°C before it was injected. Control experiments with testosterone labeled with ^{14}C eliminated the possibility that the RNA preparations injected were contaminated with testosterone. Evidence from several other types of mammalian cell systems is accumulating to indicate that RNA macromolecules can pass across a cell membrane(2,8) and transfer information leading to the induction of enzymes(6), the formation of specific antibodies

*Aided by Grant HD-01232 from the National Institute of Child Health and Human Development.

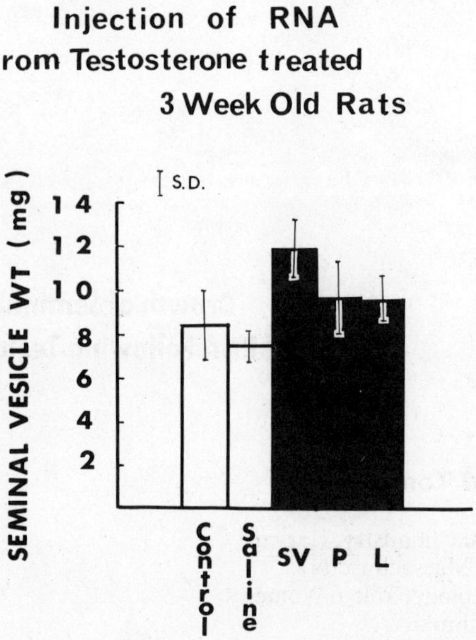

Fig. 1. Weights of seminal vesicles of 21-day-old rats into which RNA was instilled 48 hours previously. RNA was extracted from the seminal vesicle(*SV*), prostate (*P*), or liver (*L*) of testosterone-treated 3-week-old rats.

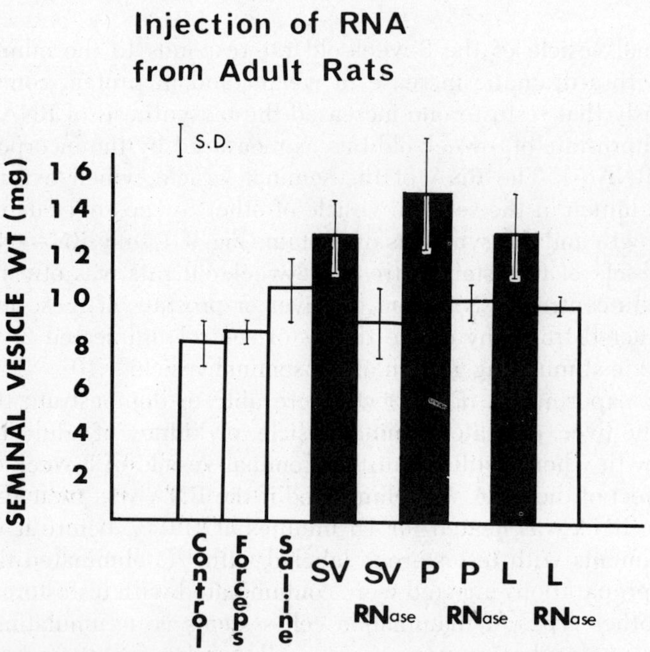

Fig. 2. Weights of seminal vesicles of 21-day-old rats into which RNA was instilled 48 hours previously. RNA was extracted from the seminal vesicle (*SV*), prostate (*P*), or liver (*L*) of adult male rats. Some RNA samples were treated with ribonuclease (RNAase).

(1,3), or the formation of specific cell structures which mimic the effects of a hormone(7).

We had reported previously(10) that histological sections of the seminal vesicle revealed evidence of hypertrophy following instillation of RNA into the lumen of the gland. In collaboration with Professor Jack Davies of Vanderbilt University

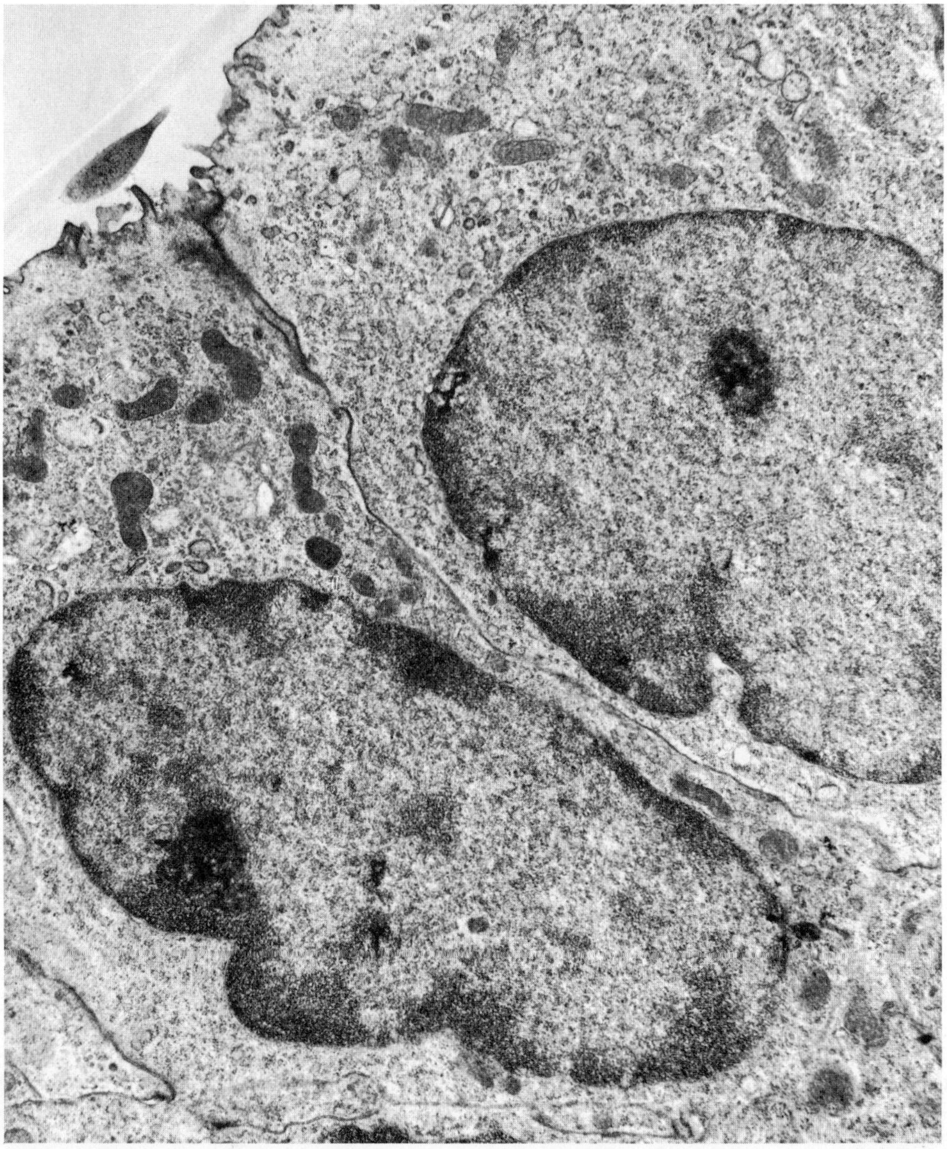

Fig. 3. Electron micrograph of cells of the seminal vesicle of a 3-week-old rat. Saline was instilled into the lumen of the seminal vesicle 24 hours before the rat was killed. The cells are simple with a poorly developed microvillous border. The cytoplasm contains small sparse mitochondria and scattered RNA granules. ×21,000.

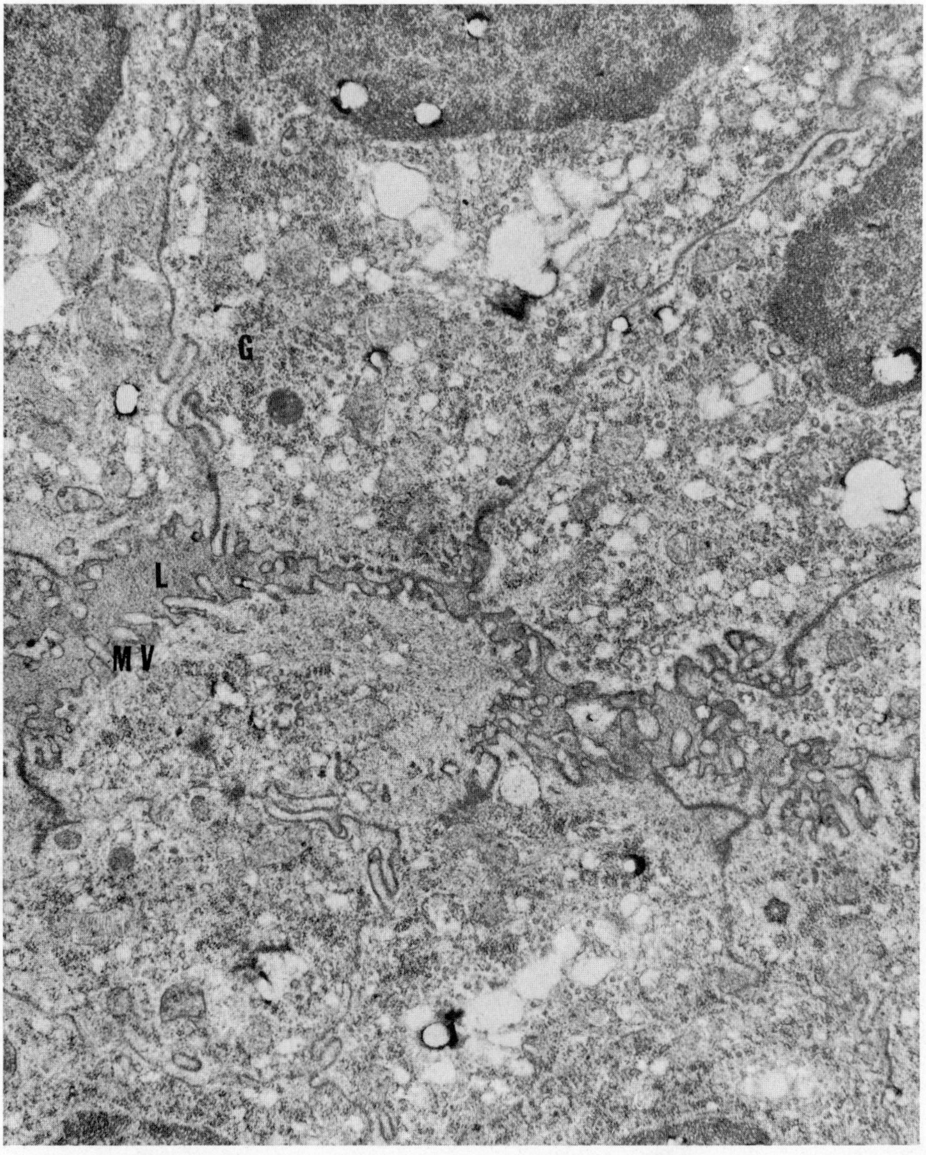

Fig. 4. Electron micrograph of cells of the seminal vesicle of a 3-week-old rat. RNA extracted with phenol and sodium dodecyl sulfate from adult male liver was instilled into the lumen of the seminal vesicle 24 hours before the rat was killed. The microvillous border (*MV*) shows increased complexity, there is secretion in the lumen (*L*), and there is an accumulation of glycogen (*G*). ×21,000.

we are examining the effect of RNA instilled into the seminal vesicles on the fine structure of the cells of this gland. Electron micrographs of the cells of saline-injected control glands of 3-week-old rats (Fig. 3) showed a rather simple structure with a relatively large nucleocytoplasmic ratio and a poorly developed microvillous border. The cytoplasm contained relatively small, sparse mitochondria and scat-

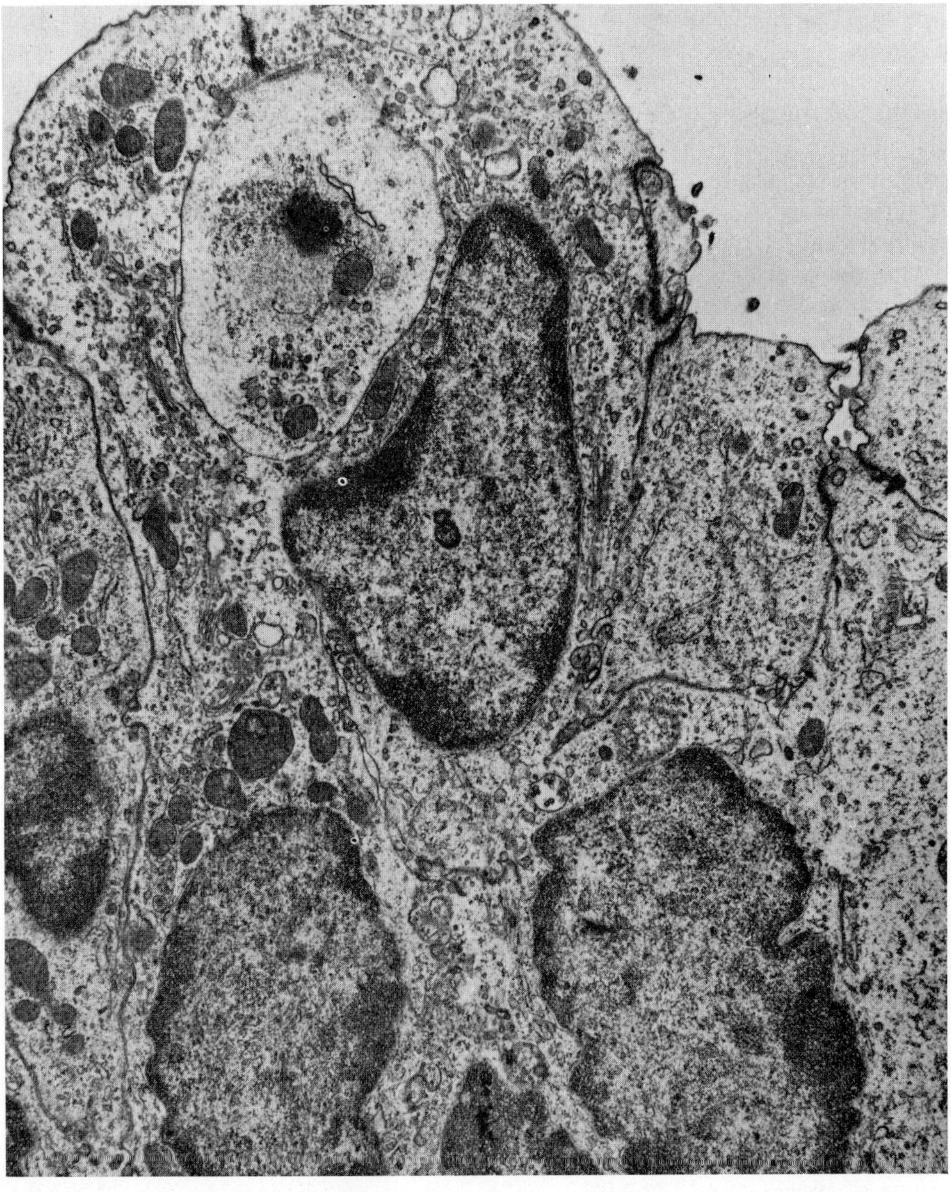

Fig. 5. Electron micrograph of cells of the seminal vesicle of a 3-week-old rat 60 hours after the instillation of RNA in the lumen of the gland. There is a considerable increase in the Golgi and in the membranous elements of both smooth and rough endoplasmic reticulum. ×21,000.

tered RNA granules. The cells of the gland 24 hours after treatment with RNA (Fig. 4) showed increased complexity of the microvillous border, accumulation of glycogen, secretion within the lumen, and suggestions of increased number of RNA granules and greater convolution of the adjacent plasma membrane. The cells 60 hours after RNA treatment (Fig. 5) showed a considerable increase in the Golgi apparatus and in the membranous elements of both smooth and rough endoplasmic reticulum.

In continuing our own studies, we have modified our test system and now use rats that are castrated 35 days after birth. Fourteen days after castration, when the rats are 7 weeks old, RNA is injected into one lobe of the seminal vesicle and saline is injected into the other. Forty-eight hours later the rats are killed and the seminal vesicles are weighed and analyzed. This method of testing decreases the variation from animal to animal since each one is serving as its own control. RNA is extracted from the tissues using phenol:sodium dodecyl sulfate at room temperature(5).

Table 1 summarizes the effects of treating these 7-week-old castrated rats with RNA from intact adult male rats and, in a separate series of experiments, with RNA from testosterone-treated adult castrate males. In both series the instillation of RNA obtained from the seminal vesicle produced 30 percent increase in weight and a 25 percent increase in the amount of protein in the lobe treated with RNA compared to the saline-injected control lobe. The increase in weight and in protein content produced by the instillation of RNA from the liver or prostate were less marked, and averaged about 10 percent. The experiments with testosterone-treated castrate males were reassuring in indicating that it was testosterone and not some other factor in the testis that was producing the RNA effective in the seminal vesicle.

Table 2 depicts the time course of the effect of RNA from the seminal vesicle of testosterone-treated adult castrate rats on the rate of protein synthesis in the seminal vesicles of 7-week-old castrate rats. In these experiments 12 μc of tritium-labeled protein hydrolysate was injected intraperitoneally 6 hours before the animals were killed so that we could compare the specific activity of the protein in the

TABLE 1

Effect of RNA from tissues of intact adult male rats or from testosterone-treated adult castrate male rats on the weight and protein content of 7-week-old rats castrated at 5 weeks[a]

RNA from tissues of intact adult rats	Seminal vesicle weight (mg)		Protein (mg per gland)	
	Saline	RNA	Saline	RNA
Liver	9.2 ± 1.0	10.2 ± 1.0	—	—
Prostate	8.0 ± 1.1	9.6 ± 0.9	—	—
Seminal vesicle	10.2 ± 1.0	[b]13.7 ± 1.0	1.10 ± 0.11	[b]1.55 ± 0.13
RNA from tissues of testosterone-treated adult castrate rats				
Liver	8.7 ± 1.0	10.0 ± 1.2	1.15 ± 0.07	1.35 ± 0.08
Prostate	8.8 ± 0.9	10.1 ± 0.8	1.25 ± 0.04	1.40 ± 0.06
Seminal vesicle	10.3 ± 1.0	[c]14.0 ± 1.1	1.50 ± 0.05	[c]1.95 ± 0.07

[a]Values are expressed as mean ± standard deviation.
[b]$P < 0.025$.
[c]$P < 0.01$.

TABLE 2

Time course of the effect of RNA from the seminal vesicles of testosterone-treated castrate adult rats on the synthesis of protein in the seminal vesicles of 7-week-old rats castrated at 5 weeks of age[a]

Time after treatment (hours)	No. of rats	Seminal vesicle weight (mg)		Protein (mg per gland)		Specific activity of protein (cpm/mg protein)		Increase (%)
		Saline	**RNA**	**Saline**	**RNA**	**Saline**	**RNA**	
24	6	10.4 ± 0.8	11.9 ± 0.8	1.05 ± 0.07	1.29 ± 0.01	1,094 ± 118	992 ± 117	
48	9	11.0 ± 1.0	13.4 ± 1.0	1.19 ± 0.20	1.35 ± 0.12	550 ± 63	734 ± 108	33
72	10	12.2 ± 1.0	13.8 ± 0.9	1.24 ± 0.09	1.47 ± 0.10	439 ± 39	[b]531 ± 58	21

[a]Twelve μc of ^3H-labeled protein hydrolysate was injected intraperitoneally 6 hours before the animals were killed.
[b]$\underline{P} < 0.025$.

seminal vesicles of rats injected with RNA and with saline. The effect of RNA on the weight of the seminal vesicle reached its maximum in 48 to 60 hours, just as does the effect of testosterone itself when injected in vivo. An increased specific activity of the protein of the seminal vesicle was clearly evident in the animals that had been treated with RNA for 72 hours.

The observation that RNA from the liver of adult male rats or from testosterone-treated castrate males was effective in stimulating the growth of the seminal vesicles, prompted us to test the effectiveness of RNA from the livers of female rats. Table 3 shows that RNA from the livers of intact female rats was without effect in stimulating the growth of the seminal vesicles. Neither the weight nor the protein content of the seminal vesicle was increased by treatment with RNA from the livers of control or testosterone-treated adult female rats. Testosterone was administered in four injections, 48 hours apart, with 5 mg of testosterone in each injection. Thus the genetic sex of the animal appears to be of importance in determining whether its liver will respond to testosterone and produce RNA that is biologically active in the seminal vesicle.

In other experiments a synthetic polynucleotide, polyuridylic acid, was instilled into one lobe of the seminal vesicle and saline into the contralateral lobe to determine whether poly U would increase growth. The poly U had no effect at all on the weight of the seminal vesicle (Table 3). However, when 12 μc of tritium-labeled phenylalanine was injected 2 hours before the rats were killed, the incorporation of phenylalanine into proteins was greater in the lobe injected with poly U. This can be taken as evidence that the poly U entered the cells of the seminal vesicle and was effective on their ribosomes. This confirms that these large polynucleotides are transported into the cell.

Several attempts have been made by Dr. D. Z. Ewton to devise a system to assay the activity of RNA on the seminal vesicle in vitro. Either intact or minced

TABLE 3

Effects of RNA from livers of control and testosterone-treated adult female rats and of polyuridylic acid on the growth of the seminal vesicle of 7-week-old rats[a,b]

Substance instilled into seminal vesicle	No. of rats	Body weight (g)	Seminal vesicle weight (mg)		Specific activity of protein (cpm/mg protein)	
			Saline	**RNA**	**Saline**	**Poly U**
RNA from control liver, 100 μg	5	200 ± 3	9.2 ± 0.9	8.9 ± 0.8	—	—
RNA from liver of testosterone-treated females, 100 μg	9	201 ± 12	12.9 ± 1.0	12.3 ± 1.0	—	—
Polyuridylic acid, 100 μg	9	186 ± 6	10.5 ± 0.9	10.6 ± 1.0	573 ± 30 (1.02)[c]	662 ± 52 (1.06)[c]

[a]12 μg of ^{3}H-phenylalanine, specific activity, 1 c/m mole, was injected intravenously 2 hours before the animals were killed.

[b]Values expressed as mean ± standard error.

[c]Figures in parentheses indicate the quantity of protein in each seminal vesicle.

lobes of the seminal vesicles from castrate adult or immature rats were incubated 2 to 6 hours with or without RNA extracted from the seminal vesicle or liver. The addition of RNA to the medium had no effect on the rate of protein synthesis as measured by the incorporation of labeled amino acids. In a second type of assay, seminal vesicle lobes were incubated 1 to 4 days in organ culture with or without polyuridylic acid. The subsequent incorporation of labeled phenylalanine into protein was not increased by poly U. Even when the production of endogenous messenger RNA was depressed by actinomycin D, the addition of poly U did not increase the incorporation of phenylalanine into protein. Perhaps the exogenous RNA did not enter the cell in an amount large enough to be effective; however under similar conditions in organ culture the adrenal and ovary did respond to exogenous RNA(11,12).

We have been attempting to fractionate the RNA to determine the nature of the material that is biologically effective. It is technically difficult to get enough RNA from the seminal vesicle to perform such experiments, hence we have used RNA from the liver, even though liver RNA is at best only one-third as effective as the RNA from the seminal vesicle. As a test system we used the seminal vesicles of 3-week-old rats since they gave a better response to liver RNA than did seminal vesicles of the 7-week-old castrate rats. In these experiments, then, RNA was injected into both lobes of the seminal vesicles of these immature rats and the weights of both lobes were measured. RNA was prepared from the livers of adult male rats and fractionated by sedimentation on a gradient of 5 to 20 percent sucrose in 0.01 M tris-HCl buffer pH 7.4. Approximately 500 μg of RNA was layered on each gradient and centrifuged in a swinging bucket rotor #39 to 36,000 rpm for 5 hours in the cold. After centrifugation one of the tubes was collected dropwise and the optical density of its fractions was measured. The RNA from the other five centrifuge tubes was then fractionated into three parts according to the sedimentation pattern of the first tube. The tubes comprising each peak were pooled and the RNA was precipitated with potassium acetate and ethanol and kept at $-10°C$ for 48 hours. The precipitated RNA was washed with ethanol and ethanol ether, taken to dryness, and dissolved in a minimal amount of saline for instillation (Table 4). The heaviest peak, peak 1, and the lightest peak, peak 3, had only minimal stimulatory effects, whereas peak 2, with a sedimentation constant of 16 to 18S did stimulate growth of the seminal vesicles.

When liver homogenates were treated with deoxycholate at a final concentration of 0.5 percent, much more of the rapidly labeled RNA was released at both 30 and 60 minutes after the injection of ^{3}H-cytidine. When the material released by deoxycholate was instilled into a seminal vesicle (Table 4), peak 2 RNA of the deoxycholate-treated homogenate produced an even more marked stimulation of growth than did peak 2 from the control liver.

Seminal vesicles, ventral prostates, and livers were homogenized, deoxycholate was added to a final concentration of 0.5 percent and each homogenate was then centrifuged at 100,000 *g* for 2 hours. The RNA in the supernatant fraction includes a highly labeled fraction. RNA was extracted from the supernatant fraction with phenol and sodium dodecyl sulfate. The growth-promoting activity of the resulting RNA was then measured in the seminal vesicle of the 7-week-old castrate rat. Table 5 shows that the RNA from the seminal vesicle was biologically active,

TABLE 4

Effect of RNA fractions from the livers of adult male rats on the growth of the seminal vesicle in 3-week-old rats[a]

Source of RNA	Peak	No. of rats	Body weight (g)	Seminal vesicle weight (mg)	Increase (%)
Control liver	Saline	6	53.8 ± 2.4	11.6 ± 0.4	
	I	8	52.5 ± 1.4	12.6 ± 0.6	
	II	7	54.1 ± 1.3	[b]14.0 ± 1.0	22
	III	6	54.0 ± 1.5	11.5 ± 0.6	
Deoxycholate-treated liver	Saline	10	55.5 ± 5.0	11.5 ± 0.6	
	I	11	54.1 ± 2.2	11.7 ± 0.6	
	II	10	56.6 ± 2.0	[c]15.5 ± 0.7	32
	III	7	54.4 ± 1.5	12.9 ± 0.9	

[a]Values are expressed as mean ± standard error.
[b]$P < 0.025$ vs. saline controls.
[c]$P < 0.001$ vs. saline controls.

TABLE 5

Biological activity of the RNA from the 100,000 g *supernatant fraction of control and DOC-treated homogenates of liver, ventral prostate, and seminal vesicle measured by their effects on the growth of the seminal vesicle of 7-week-old castrate rats*

RNA from	Deoxycholate treated	No. of rats	Body weight (g)	Seminal vesicle weight (mg) Saline	Seminal vesicle weight (mg) RNA
Liver	–	6	214	12.1 ± 0.8	11.0 ± 0.8
	+	10	195	10.2 ± 0.6	11.2 ± 1.0
Prostate	–	4	188	10.2 ± 1.9	14.7 ± 2.3
	+	6	177	12.5 ± 2.0	13.4 ± 2.2
Seminal vesicle	–	7	206	11.3 ± 0.8	11.3 ± 0.6
	+	9	195	11.8 ± 0.7	13.8 ± 0.7[a]

[a]$P < 0.05$.

whereas RNA from the control homogenate which was not extracted with deoxycholate and thus is similar to the usual "soluble" RNA, was ineffective. Thus the kind of RNA that is effective in stimulating growth is the kind of RNA that is rapidly labeled following injection of the animal with labeled cytidine. Dextran sulfate is effective in stabilizing the rapidly labeled heavy RNA of nuclear origin(9). The addition of dextran sulfate (Sigma) with a molecular weight of 500,000 to a final concentration of 0.2 percent, 5 minutes after treating the homogenate with deoxycholate, yielded RNA from liver that would stimulate growth of the seminal vesicle.

These experiments thus provide further evidence that RNA is involved in mediating the stimulatory effects of testosterone in the seminal vesicle. They suggest that the most effective RNA is that derived from the seminal vesicle itself, RNA released from whole homogenates with deoxycholate and dextran sulfate, extracted with cold phenol and sodium dodecyl sulfate, and then sedimented on a sucrose density gradient at 16 to 18S. The inference, which of course is not established by these experiments, is that testosterone enters the nucleus of the cells of target organs such as the seminal vesicle and stimulates the production of a particular species of RNA. The experiments do establish that RNA can be extracted from the seminal vesicle, partially purified, and then instilled into the seminal vesicle of control rats and produce the effects on growth and protein synthesis which are ordinarily seen in response to testosterone itself.

The seminal vesicle of the 3-week-old rat responded equally well to RNA from the seminal vesicle, prostate, liver, and kidney of the adult male rat. In contrast, the seminal vesicle of the 7-week-old castrate rat responded significantly better to RNA from the seminal vesicle than to RNA from the liver or prostate. It is of interest to recall that in Niu's experiments, only uterine RNA from adult mice stimulated the induction of alkaline phosphatase in the uterus of castrate adult mice. In Segal's experiments only uterine RNA from estrogen-treated castrate rats initiated estrogen-like morphological changes in the uterus of castrate adult rats. All of these results suggest that well differentiated tissues respond only to organ specific RNA, but less well differentiated tissues in immature animals may respond to nonspecific RNA. In addition, one could postulate that some specific receptor site is established in the cells of the seminal vesicle after sexual maturation or after it has been exposed to a certain level of androgens.

In these experiments the RNA that was most active biologically in stimulating growth was the same RNA fraction that was rapidly labeled following the injection of labeled cytidine. Rapidly labeled RNA is considered by many investigators to be messenger RNA and the precursor of the cytoplasmic ribosomal RNA. Our results, then, may suggest that the biologically active RNA is, at least in part, equivalent to the rapidly labeled RNA. The actual molecular size of these fractions is difficult to estimate because of the difficulty in demonstrating that there is no degradation or aggregation of the RNA during the extraction process.

REFERENCES

1. Bishop, D. C., Pisciotta, A. V., and Abramoff, P. Synthesis of normal and immunogenic RNA in peritoneal macrophage cells. J. Immunol., 99(4):751–759, 1967.
2. Chin, P. H., and Silverman, M. S. Transfer of myeloma protein synthesis with RNA isolated from the C_3H plasma cell tumor (X5563). I. Uptake of isotopically labeled tumor RNA by lymphoid cells. J. Immunol., 99(3):476–488, 1967.
3. Cohen, E. P., Newcomb, R. W., and Crosby, L. K. Conversion of nonimmune spleen cells to antibody-forming cells by RNA; strain specificity of the response. J. Immunol., 95(4):583–590, 1965.
4. Fujii, T., and Villee, C. A. Stimulatory effects of RNA on the growth of the seminal vesicle of immature rats. Proc. Nat. Acad. Sci. U.S.A., 57:1468, 1967.
5. ——— and Villee, C. A. Effect of testosterone on ribonucleic acid metabolism in the

prostate, seminal vesicle, liver and thymus of immature rats. Endocrinology, 82:463–467, 1968.

6. Mansour, A. M., and Niu, M. C. Functional studies with uterine RNA. Proc. Nat. Acad. Sci. U.S.A., 53(4):764–770, 1965.
7. Segal, S. J., Davidson, O. W., and Wada, K. Role of RNA in the regulatory action of estrogen. Proc. Nat. Acad. Sci. U.S.A., 54(3):782–787, 1965.
8. Shanmugam, G., and Bhargava, P. M. The uptake of homologous ribonucleic acid by rat liver parenchymal cells in suspension. Biochem. J., 99:297–307, 1966.
9. Tamaoki, T., and Mueller, G. C. A rapidly labeled ribonucleoprotein in HeLa cell nuclei. Biochim. Biophys. Acta, 108(1):813–94, 1965.
10. Villee, C. A., and Fujii, T. RNA-mediated growth-promoting effects of testosterone. *In* Testosterone. Tamm, J., ed. Stuttgart, Georg Thieme, pp. 96–101, 1968.
11. Villee, D. B. Ribonucleic acid: Control of Steroid synthesis in endocrine tissue. Science, 158:652, 1967.
12. ——— Greenough, L., and Rettig, J. Ribonucleic acid control of steroid synthesis in human adrenals and testes. Science, 159:1365, 1968.

49

In Vitro Studies of the Relation of Fetal Sex Hormones to Sex Differentiation in the Guinea Pig*

D. Price,[1] E. Ortiz,[2] and J. J. P. Zaaijer[3]

The Population Council, Biomedical
Division, The Rockefeller University,
New York, New York
[1]Biology Department, University of Chicago,
Chicago, Illinois
Biology Department,[2]University of Puerto Rico,
Rio Piedras, Puerto Rico
[3]Laboratory for Cell Biology and Histology,
University of Leiden, The Netherlands

INTRODUCTION

Research on the relation of sex hormones to sex differentiation in placental mammals has progressed through a series of major phases in methodology. In approximately chronological order, these have been the administration of male or female sex hormones to pregnant females, gonadectomies and hypophysectomies of fetuses in vivo and the application of organ culture methods. Histochemical staining techniques have been used to study enzyme patterns in fetal gonads, and precise chemical analyses have confirmed and extended the findings. Administration of antiestrogenic and antiandrogenic agents to pregnant females have opened a new parameter. Cytogenetics has made large contributions and electron microscopy has entered the field. A survey of even a few of the reviews(1,6) points up how much has been learned about the hormonal basis of sex differentiation in a number of species and also what important problems still remain unsolved.

It is common knowledge that in all probability sex differentiation in the female is anhormonal and that fetal testicular hormone or hormones exert an organiza-

*Our research on the guinea pig has been aided by research grants HD-03481 and AM 03628 from the National Institutes of Health, United States Public Health Service and by the Nederlandse Organisatie voor Zuiver-Wetenschappelijk Onderzoek.

tional influence on the morphogenesis of the male reproductive tract. Further, the effects, once established are irreversible. Hypotheses as to the relation of sex hormones to sex differentiation of gonads, reproductive ducts, accessory glands, and external genitalia have been proposed. None of these, not even the most plausible, has been completely and satisfactorily corroborated. It has been definitely established that fetal gonads, under certain experimental conditions and at certain ages are capable of steroidogenesis and that fetal testes can synthesize steroidal hormones, some of which are classified as androgens and are produced normally by postnatal testes. But these hormones, when introduced in synthetic form into the fetus do not substitute for fetal testicular hormones. Does the fetal testis secrete a stimulating hormone and a second hormone that is inhibitory? The resolution of the question of the duality (or multiplicity) of fetal testicular hormones and the chemical characterization of these hormones may be major steps in the solution of the problem. However, it seems apparent that the greatest problem of all is how fetal sex hormones exert their effects on the cells of end organs and, indeed, precisely when they begin to do so in the exact timing of differentiation of the reproductive system.

In the meantime, some basic questions on normal sex differentiation remain largely unanswered. From them, a few were selected for study in the guinea pig. Fetal testes can secrete steroidal androgens under experimental conditions, but how early do they normally begin secretion in the fetus, when do these endogenous hormones begin to circulate, and what relationship does their circulation bear to the timing of developmental events in the reproductive system? Above all, precisely when is their presence mandatory for normal sex differentiation of the male? Can a dual action of fetal testicular hormones be demonstrated by the device of subjecting fetal female reproductive tracts to testicular hormones? Is the morphogenesis of the female reproductive tract as hormone-independent as some of the evidence indicates?

These problems have been approached by the use of an organ-culture method in the fetal guinea pig. This method has been successful in the rat(7,8) and in the mouse(9). The special advantages of the guinea pig fetus for such studies have been discussed(2). This paper is a review and synthesis of a series of our published results.(10,14)

ORGAN CULTURE EXPERIMENTS

General Design and Method

The experiments were designed to study two closely interrelated aspects of the problem of fetal hormones and sex differentiation. In the first series, gonads and other organs were removed from fetuses of various ages and were cultured and tested by a sensitive bioindicator method for the presence of androgenic hormones. In the second series, where the emphasis was on the study of sex differentiation of the reproductive ducts, internal reproductive tracts of males and females were excised and cultured under various hormonal conditions.

The culture method was a modified watchglass technique employing Falcon plastic culture dishes with a medium consisting of chick embryo extract, cock plasma, and Hanks saline solution. Lens paper rafts were placed on top of the semisolid medium and the explants were arranged in position on the rafts to which they quickly became affixed. Culturing was done at a temperature of 35°C and the period of culture varied with the type of experiment. Transfers were made every three days by lifting the raft, rinsing it in saline solution, and placing it on fresh medium with the explants still undisturbed. Observations were made during culture, and at the end of the culture period the explants were fixed and prepared for histological study, as previously described.

Bioindicator Test for Androgen

This test is based upon the findings that the ventral prostate gland of the postnatal rat is highly androgen-dependent and markedly androgen-responsive(15). In vivo it can be stimulated by endogenous androgen from testes, adrenals, and ovaries and by a number of exogenous androgenic hormones. In vitro, prostate tissue responds to synthetic testosterone and hydrocortisone(10,16,17).

As a first step in developing the test, explants of the prostate gland of rats (21 to 28 days of age) were cultured alone for periods of 3 to 6 days. The results were compared with the pattern of retrogression in histological structure of the gland in vivo when androgen sources are removed. In this way, the pattern and rate of retrogression in vitro served as a check on androgen in the medium and the responsiveness of the prostate in vitro assured a sensitive test response to the presence of androgen. This responsiveness was tested further by culturing prostate on a medium to which various dosages of testosterone or hydrocortisone had been added.

A standard test period of 5 to 6 days was established and the androgenicity of organs to be examined was judged by definite criteria in the histological structure of the prostate explant. In each experiment, prostate explants were cultured alone as controls; 125 such explants were made.

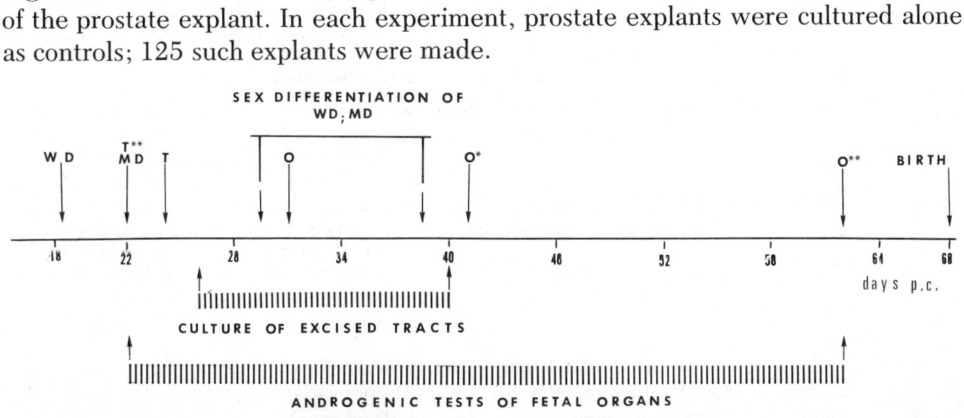

Fig. 1. Experiments in vitro in relation to the timing of some important developmental stages in the reproductive tract. *WD:* Wolffian duct completely formed; *MD:* Müllerian duct beginning to develop; *T***: testis highly androgenic at this age and all subsequent ages tested by the bioindicator test; *T:* testis cords beginning to differentiate; *O:* ovarian cortex beginning to differentiate; *O**: androgenicity first detected in the ovary, organ mildly androgenic; *O***: ovary highly androgenic.

Fetal organs were tested at ages ranging from 22 to 62 days of gestation (Fig. 1). A total of over 400 cultures of organs was studied. These included testes, ovaries, and adrenals, which were considered possible sources of androgen, and also, fetal bladder, mesonephros, metanephros, and anterior pituitary, which were presumed to be nonandrogenic.

Explanation of Fetal Reproductive Tracts

The organ-culture method as applied in this research provided a way to study the differentiation of reproductive tracts that were isolated from the possible influence of placental and maternal hormones. In addition, observations could easily be made of the Wolffian and the Müllerian ducts as they developed in the presence or absence of other fetal organs, as desired (Table 1). Two stages in development of the ducts were selected for explantation: the ages of 26 to 27 days of gestation, when the reproductive ducts of males and females are in an ambisexual stage, and

TABLE 1

Wolffian and Müllerian ducts in explants of male and female reproductive tracts cultured with fetal gonads and adrenal glands[a]

Age in days during culture period: Initial	Final	Wolffian duct	Müllerian duct
Female			
		With or without its ovaries:	
26-27	35-38	Retrogressed	Retained; oviductal segment with increase in diameter and slight development of fimbriae and epithelial folds
29-30	38-39	Retrogressed	Retained; oviductal segment with increase in diameter and length, marked coiling, marked fimbriae, enlarged ampulla with peg cells, deep epithelial folds; isthmus demarcated
		With its ovaries replaced by testis, adrenal, or by ovary of 60-day fetus:	
26-27	37-38	Maintained and stimulated	Retained; oviductal segment with increase in diameter and slight development of fimbriae and epithelial folds
29-30	38-40	Maintained completely or partially or retrogressed	Retained; oviductal segment with increase in diameter and length, marked coiling, marked fimbriae, enlarged ampulla with peg cells; deep epithelial folds; isthmus demarcated
Male			
		With testis, adrenal, or with ovary of 60-day fetus:	
26-27	35-38	Maintained completely	Retrogressed
29-30	38-40	Maintained completely	Retrogressed
		Without testis or with testis replaced by ovary of 29-day fetus:	
26-27	35-38	Retrogressed	Retrogressed
29-30	38-39	Partially retrogressed or retrogressed	Retrogressed

[a]Results compiled from the following previous reports: Price et al., Anat. Rec., 157:27, 1967; Ortiz et al., Proc. Kon. Nedere. Akad. Wet. (Biol. Med.), 70:475, 1967; Price et al., Ch. 1, The Mammalian Oviduct, The Univ. of Chicago Press, 1969.

29 to 30 days, when sexual differentiation is beginning. Reproductive tracts at these two stages were explanted and cultured 9 to 11 days, so that the experiments spanned in vitro the period in vivo from the ambisexual stage to 38 to 40 days, when the early crucial stages in sexual differentiation of the ducts are essentially completed (Fig. 1). In all, 45 reproductive tracts were explanted at these two stages. The developmental history of the Wolffian and Müllerian ducts was followed during culture by tracings of the projected images of the tracts and by photography, and the explants were fixed at the termination of the experiment. Results were compared with normal differentiation of the reproductive tracts studied in 50 fetuses ranging in age from 18 to 38 days.

RESULTS AND DISCUSSION

Prostate explants cultured by themselves retrogressed in the culture period of 5 to 6 days. The rate of retrogression and the pattern of changes in histological structure were essentially similar to the well-known events that occur in the gland in vivo when androgen is withdrawn. There was no evidence that any androgen possibly residual in the medium was affecting the prostate in any way, as judged by definite histological criteria. On the other hand, normal histological structure was maintained in prostate explants when testosterone or hydrocortisone was added to the medium at explantation. Thus, the androgen-responsiveness of the rat prostate in vitro under our experimental conditions has made this tissue as dependable and sensitive a test system for detection of fetal androgenic hormones as it is for endogenous or exogenous androgens in vivo in the postnatal animal. In point of fact, the prostate gland of the rat has been one of the most useful organs in the construction of an operational definition of an androgen.

By utilizing an in vitro system involving organ culture, an important advantage was gained in that the relationship between the organ to be tested and the test tissue was close and direct. Whatever hormone was present in the fetal organs, and diffused into the medium, reached prostatic tissue without being metabolized by any other organ. In this system, the conditions for detecting androgen in the glands were optimal.

In applying the bioindicator test, prostate explants cultured with fetal bladder, mesonephros, metanephros, and anterior pituitary retrogressed as they did when cultured alone, and thus gave negative tests. Fetal testis explants were quite another story. In vivo gonads are cut off from the urogenital ridge between 22 and 23 days and potential testes and ovaries are indistinguishable. Medullary cords begin to form in the testis by 24 to 25 days and differentiation proceeds rather rapidly. In culture, gonad primordia explanted at 22 to 23 days and later identified as testes by their differentiation in vitro, strongly stimulated the associated prostate tissue. The testis explants apparently maintained normal histological structure in the prostate from the time of explantation, indicating the presence of androgen at an extraordinarily early stage in development. Older testes explanted at all subsequent ages (25 to 62 days) were also highly androgenic.

Application of histochemical staining techniques demonstrated the localization of 3β-hydroxysteroid dehydrogenase activity in interstitial tissue in fetal testes at 29, 31, 45, and 50 days of age(2,18). The presence of this enzyme suggested that the testes had the capacity for steroidogenesis. This has been borne out by more recent experiments in which testis homogenates from fetuses of midgestation and late gestation were able to convert progesterone and pregnenolone to testosterone and androstenedione(4). The site of metabolism of these substrates may be presumed to be the interstitial cells. In the testes that were explanted at 22 to 23 days, interstitial cells were not identifiable in routine histological preparations. However, interstitial cells can be detected in fetal testes at 23 to 24 days in electron-microscopical preparations(19). It is significant that these cells already have areas of smooth endoplasmic reticulum similar to the endoplasmic reticulum presumed to be the site of androgen biosynthesis in interstitial cells of adult guinea pig testes(20). It appears that in undifferentiated testes, the cells that are destined to be interstitial cells are already capable of androgen biosynthesis before the testis has begun its morphological differentiation.

Bioindicator tests of the ovary were particularly interesting. Sixty-two tests of fetal ovaries ranging in age from 22 to 36 days gave no evidence that the ovary was active in the biosynthesis of androgenic substance. Some ovaries explanted at 41 and 46 days had weakly androgenic effects on prostatic tissue, but all those explanted at 62 days were strongly androgenic. This indicates that the ovary does not become metabolically active until late in fetal life, which is further suggested by the negative findings of 3β-hydroxysteroid dehydrogenase activity in ovaries of fetuses 31, 45, and 50 days of age(18).

Androgenic tests of adrenal glands from male and female fetuses were all strongly positive, from 22 days, when the primordia of the gland consisted only of clusters of undifferentiated cortical cells, to 62 days, when morphogenesis was far advanced and cortical zonation established. With histochemical staining, 3β-hydroxysteroid dehydrogenase activity was demonstrated in the cortical parenchyma at all ages tested, between 28 and 50 days(18), indicating the capacity for steroidogenesis. Further evidence was the finding of hydrocortisone in the adrenal of 30-day fetuses(21) and the synthesis of cortisol, corticosterone, and 11β-hydroxyandrost-4-ene-3,17-dione by adrenal homogenates (midgestation and late gestation fetuses), when incubated with carbon-14 or tritium labeled progesterone or pregnenolone(4).

It is clear from the results of others and from our own findings that in the guinea pig fetus, the testis and the cortical components of the adrenal gland are capable of synthesizing steroids, some of which are androgens. From our bioindicator tests, we conclude that in all probability, they are normally secreting androgenic hormones even at very early stages. In unpublished observations, we have detected androgenicity in whole urogenital ridges (containing nephric tubules, Wolffian ducts, and early primordia of gonads and adrenals) explanted at 19 to 20 days of age. Apparently, functional differentiation of adrenocortical cells and of cells in the testicular blastema of the urogenital ridge long precedes morphological differentiation.

The studies on fetal testes, ovaries, and adrenals of the guinea pig bring this species essentially into line with findings in the fetal armadillo, mouse, rat, rabbit, calf, and human. These studies, reviewed and discussed recently(3,6,10,11,13) leave no doubt that these fetal organs are capable of steroidogenesis as are the fetal testes of the horse(22). Further evidence has been given for the young calf fetus(23).

With some notable exceptions, many of these studies have dealt with fetuses in relatively late stages of gestation and therefore, beyond the critical periods in sex differentiation. The sequential pattern of the stages in male and female sex differentiation in a number of species including man has been found quite similar, but in species with longer gestation periods sex differentiation is accomplished relatively early(2,14). In the guinea pig, the critical stages occur within the first half of gestation and sex differentiation of the reproductive ducts covers a span of 9 days (Fig. 1). This time span was particularly advantageous for study of duct differentiation in organ culture.

Female reproductive tracts were explanted at the ambisexual stage or three days later at the beginning of duct differentiation (Table 1). In vivo, the Wolffian ducts are equally well developed in the two sexes at 26 to 27 days but by 29 to 30 they become slender in females (Fig. 2) and retrogression may begin, although this is variable. At 34 to 36 days, they have retrogressed cranially (Fig. 3) and only short solid cords fusing with the urogenital sinus remain; by 37 to 38 days they are usually absent. In vitro, Wolffian ducts retrogressed essentially at the same rate as in vivo in tracts with or without ovaries present (Table 1 and Figs. 4 and 5). But in explanted tracts with their ovaries replaced by fetal testes or adrenals, androgen from these organs maintained and stimulated female Wolffian ducts, unless the ducts were too old and were already conditioned to retrogress (Figs. 6, 7, and 8). Pieces of ovary from 60-day-old fetuses were equally effective in stimulating the Wolffian ducts.

The results show clearly that female Wolffian ducts become androgen-dependent immediately following the ambisexual, i.e., they become conditioned to retrogress in the absence of an androgen source, as, indeed, they normally do in vivo.

In the male fetus in vivo the duct development progresses from the ambisexual stage (Fig. 9) to the formation of dilatations, the seminal vesicle primordia, by 30 to 31 days, enlargement of diameter of the duct and coiling of the epididymal region at 34 to 35 days, followed by pronounced epididymal coiling and prominent seminal vesicles by 37 to 38 days. In male tracts explanted at the ambisexual stage (Fig. 9) or 3 days later (Fig. 12), the Wolffian ducts were maintained completely when testes were present (Figs. 10 and 13) or when the testes were replaced by other androgen sources—the adrenal or pieces of ovary from a 60-day fetus (Table 1) When no androgen source was present, the duct retrogressed completely in all tracts explanted at 26 to 27 days (Fig. 11 and Table 1) but only partially in some explants of older tracts (Fig. 14).

Results show that the male Wolffian duct, as is the case with the female duct, becomes androgen-dependent immediately after the ambisexual stage and, as in the female, can be maintained by fetal testicular, adrenocortical and ovarian (60-day) androgens. In the absence of an androgen source, retrogression occurs, but if explantation is delayed until the duct is almost stabilized, retrogression is not so complete.

The history of the Müllerian ducts in males in vivo and in our experiments in vitro is simple. In normal sex differentiation in the fetus, the cranial region is

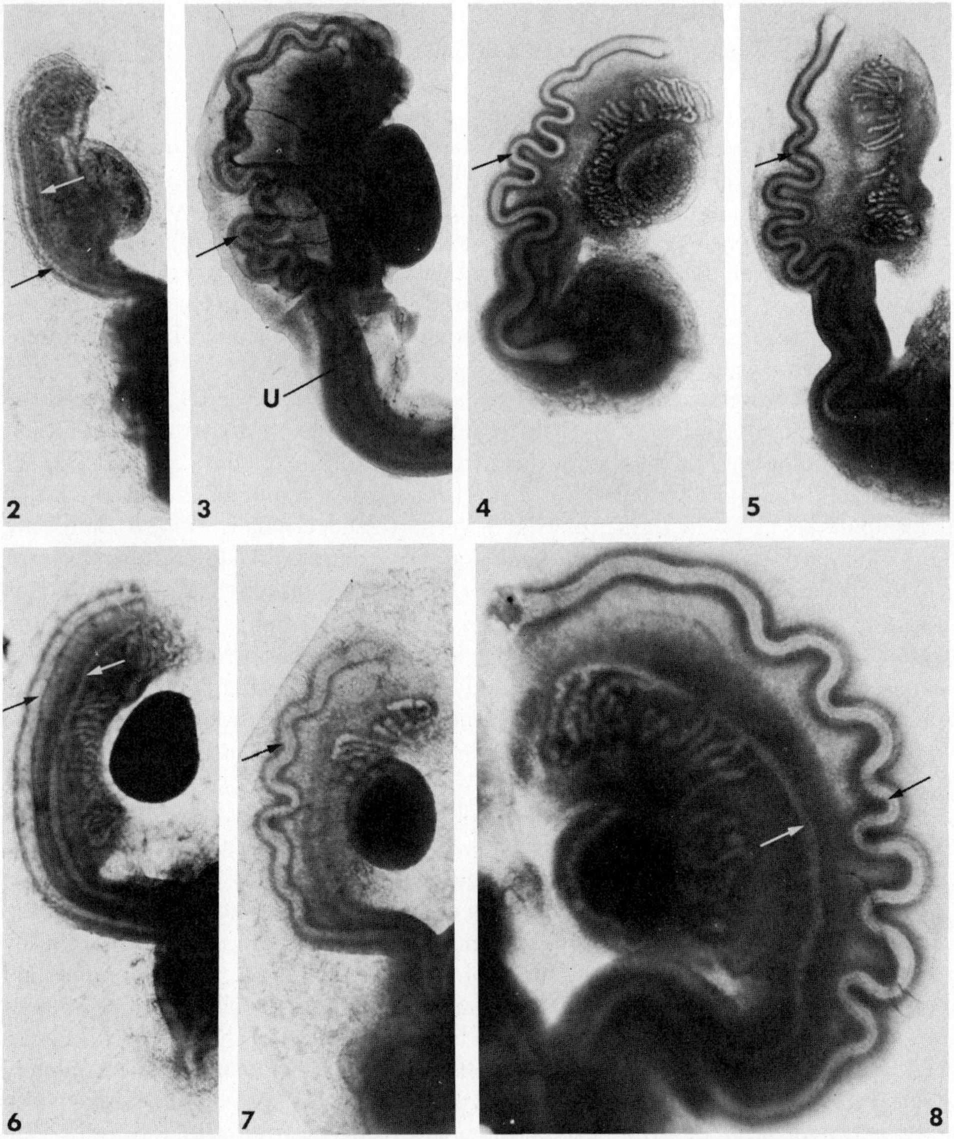

Figs. 2-8: One side of female reproductive tracts photographed on the culture medium to compare development of the ducts in vivo and in vitro; culture period, 9 days. Black arrows indicate Müllerian duct; white arrows, Wolffian duct. Figures 2, 4, 5, and 6 are reprinted from Figures 17, 18, 21, and 22, respectively, in Price et al., Anat. Rec., 157:27–41, 1967. Figures 3 and 7 from Price et al. *The Mammalian Oviduct,* 1969. Courtesy of The University of Chicago Press. Magnifications: Figs. 2 to 7, ×13; Fig. 8, ×26. Figure 2: From 29-day-old fetus with ovary left in situ; not cultured but used as control to show development of the ducts at the time of explantation. Figure 3: From 36-day-old fetus with ovary left in situ; not cultured but used as control to show development in vivo. Compare with Figure 2 and note development of the Müllerian duct with extensive coiling of the oviductal segment, demarcation of the uterine segment *(U),* and complete retrogression of the Wolffian duct. Figure 4: Cultured with ovary left in situ. Initial age, 29 days; final age, 38 days. Note marked coiling of the oviductal segment of the Müllerian duct, similar to the coiling in vivo, Figure 3. The Wolffian duct has retrogressed and appears only as a caudal segment of condensed tissue to the right of the oviduct. Figure 5: Cultured with ovary removed. Initial age, 29 days; final age, 38 days.

slender by 29 to 30 days; by 34 to 35 days anteroposterior retrogression has begun but caudal segments of about one-third the original length remain; at 37 to 38 days only short caudal segments are present but they typically have a wide lumen.

In explanted male tracts the Müllerian ducts retrogressed except for short caudal segments (as they normally do in vitro) in the presence of androgen from testes, adrenals or 60-day-old ovary. Retrogression occurred also when fetal testes were removed and all other androgen sources were lacking. These experiments demonstrated that the absence of testicular androgen, even at the ambisexual stage, did not result in the retention of the male Müllerian ducts nor did the presence of hormone from adrenals or 60-day-old ovaries hasten or delay retrogression (Table 1).

The early developmental stages in the Müllerian ducts of females in vivo and in vitro have been described(12) and the details of the differentiation of the oviductal segment have been reported recently(14). The oviduct is of particular interest for a study of the degree of hormone-independence of the Müllerian duct and its derivatives.

In the 29-day-old female fetus, the Müllerian ducts are simple and the oviductal region is not demarcated (Fig. 2) but by 30 to 31 days slight undulations set off this segment from the uterus. These undulations are more prominent at 34 to 36 days as coils (Fig. 3), fimbriae are developing around the ostium and histogenesis of the epithelium is beginning. Differentiation progresses rapidly so that at 38 to 40 days coils are further developed, infundibular fimbriae are more complex, the epithelium contains peg cells and ciliated cells, crypts foreshadowing the epithelial folds are present, and the infundibulum and ampulla are demarcated. Growth in diameter is marked from 29 days to 40 days.

The results of culturing female tracts not only show the stability and hormone-independence of the Müllerian duct but they also provide evidence of great ability for oviductal differentiation under all experimental conditions (Table 1). Morphogenesis and histogenesis were more limited in tracts explanted at the younger stage and no coiling developed, but in the older stage, morphogenesis and histogenesis were similar in kind if not in degree to normal development of the oviduct in vivo. In tracts explanted at 29 to 30 days (Fig. 2) and cultured 9 days with ovaries of the same age (Fig. 4), without ovaries (Fig. 5), with testes (Fig. 7), or with adrenal (Fig. 8), the growth in length and diameter and the coiling pattern compared quite faborably with the normal oviduct of a 36-day-old fetus (Fig. 3). Other aspects of oviductal differentiation progressed equally well (Table 1). The forma-

Note similar degree of development in the oviduct and of regression in the Wolffian duct as in the explant cultured with ovary, Figure 4. Figure 6: Cultured with ovaries replaced by one testis. Initial age, 28½ days; final age, 37½ days. Both sides of this tract showed marked increase in size and development of the Wolffian and Müllerian ducts during culture (compare with Fig. 2) and the absence of coiling of the oviduct. Figure 7: Cultured with ovaries replaced with one testis on each side of the tract. Initial age, 29 to 30 days; final age, 38 to 39 days. In contrast with the previous case (Fig. 6), both sides of this tract showed coiling of the oviduct and complete retrogression of the Wolffian duct. Figure 8: Cultured with ovaries replaced by one piece of adrenal gland on each side of the tract. Initial age, 29 to 30 days; final age 38 to 39 days. Note extensive coiling of the oviduct, marked development of infundibilum and fimbriae and maintenance of Wolffian duct. The contralateral ducts appeared equally well developed. Figs.

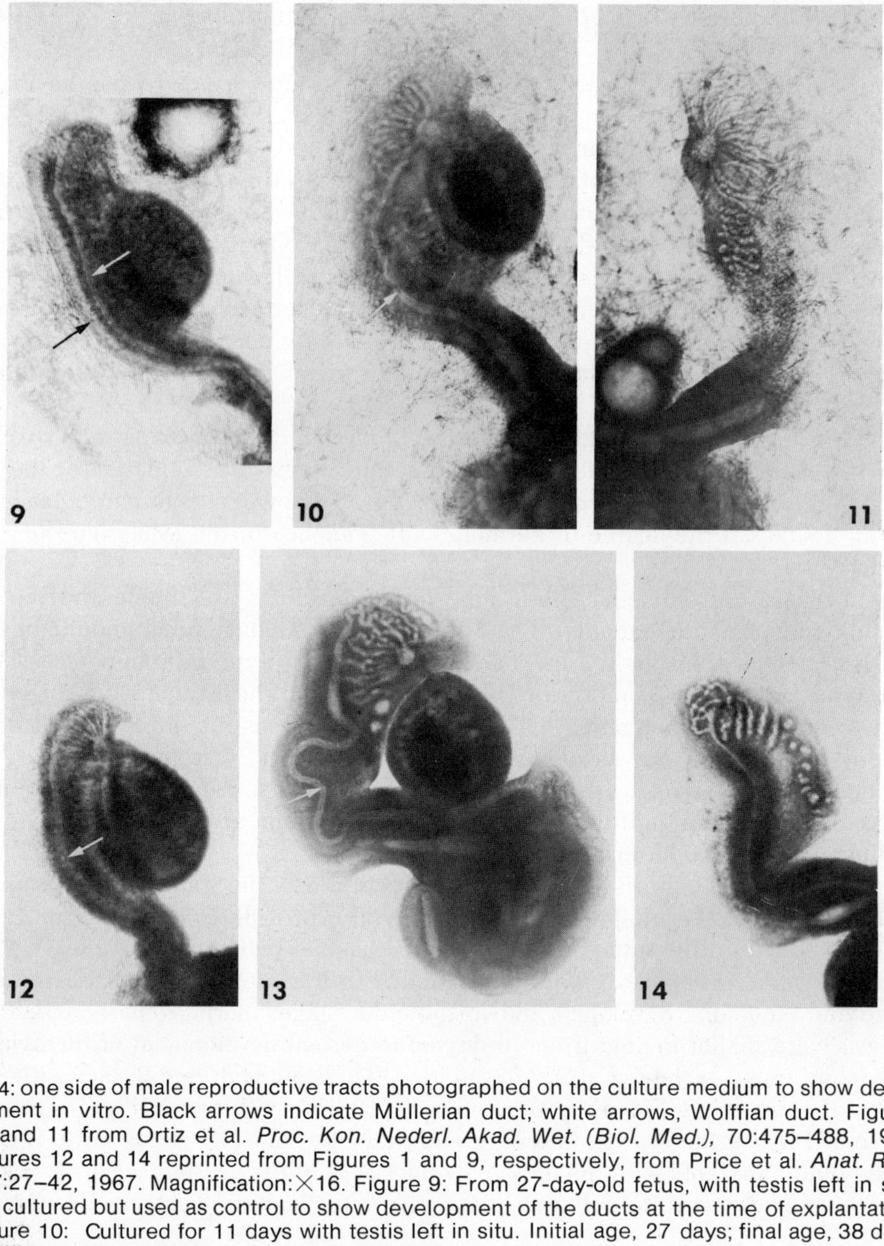

9-14: one side of male reproductive tracts photographed on the culture medium to show development in vitro. Black arrows indicate Müllerian duct; white arrows, Wolffian duct. Figures 10 and 11 from Ortiz et al. *Proc. Kon. Nederl. Akad. Wet. (Biol. Med.),* 70:475–488, 1967. Figures 12 and 14 reprinted from Figures 1 and 9, respectively, from Price et al. *Anat. Rec.,* 157:27–42, 1967. Magnification:×16. Figure 9: From 27-day-old fetus, with testis left in situ; not cultured but used as control to show development of the ducts at the time of explantation. Figure 10: Cultured for 11 days with testis left in situ. Initial age, 27 days; final age, 38 days. Wolffian duct completely maintained; Müllerian duct retrogressed to posterior remnant not visible in the photograph. Figure 11: Cultured for 11 days with testis removed. Initial age, 27 days; final age, 38 days. Note complete absence of Wolffian duct and retrogression of the Müllerian duct to a posterior remnant. Figure 12: From 29-day-old fetus with testis left in situ; not cultured but used as control to show development of the ducts at the time of explantation. The Müllerian duct is folded underneath and does not show in the photograph. Figure 13: Cultured for 9 days with testis cut away and replaced. Initial age, 29 days, final age, 38 days. Note well-developed Wolffian duct and the presence of a posterior remnant of the Müllerian duct. Compare with Figure 12. Figure 14: Cultured for 9 days with testis removed. Initial age, 29 days; final age, 38 days. Note maintenance of short cranial and caudal segments of the Wolffian duct, while the middle segment has degenerated. A short posterior remnant of the Müllerian duct is visible.

tion in vitro of fimbriae around the ostium was particularly striking (Fig. 8). This degree of differentiation was observed expecially in Müllerian ducts of tracts explanted at 29 to 30 days. Coiling failed to develop in a tract explanted at even a slightly younger stage (Fig. 6).

It was concluded that the Müllerian ducts in explanted female tracts did not retrogress in the presence of testicular androgen even when explanted at the ambisexual stage. Furthermore, hormone from adrenals or 60-day-old ovaries did not, apparently, inhibit or stimulate the Müllerian duct or its oviductal segment. The presence of androgen from these organs was effectively demonstrated by the stimulation of Wolffian ducts in the tracts (Figs. 6 and 8). There was no evidence of stimulation from estrogenic hormone which, if present, might have caused precocious coiling of the oviduct as was observed in oviducts in explanted tracts of fetal female rats when estrogen was added to the medium(8).

The findings from the culture of reproductive tracts of male and female guinea pig fetuses are in agreement with earlier in vitro findings in the rat and mouse as to the morphogenetic influence of testicular androgen in the male and the anhormonic or hormone-independent development of the Müllerian ducts in females. A review of the comparative evidence from studies on the rat and mouse in vitro and experiments on fetal gonadectomy in the rabbit points up some differences in results(2). In vitro, the Müllerian ducts of male rats and guinea pigs are not retained when testes are removed.* In the mouse, a few cases of retention or partial retention were observed but in the rabbit, Müllerian duct retention appeared to be a consistent finding when castration was performed in vivo at an early stage. In vitro, testes placed against the female tracts of the rat, mouse, and guinea pig did not cause retrogression of Müllerian ducts but in the rabbit a testis introduced into a female as an intraperitoneal graft caused retrogression of the Müllerian duct on the side close to the graft. These differences may possibly be resolved when tracts of younger guinea pigs are explanted.

The observation in the guinea pig that adrenocortical androgen stimulates the Wolffian ducts of males and females in vitro certainly does not suggest that the same is normally true in vivo. As was pointed out long ago(24), adrenal androgen does not maintain the Wolffian ducts in castrated rabbit fetuses. In addition, there is convincing evidence that adrenal androgen does not affect seminal vesicle growth in castrated male rat fetuses(25). It is obvious that the Wolffian ducts of female guinea pigs would not retrogress in vivo if they were subjected to the level, and possibly the type, of adrenocortical androgen that reaches them in vitro.

SUMMARY AND CONCLUSION

The main results of in vitro studies of the guinea pig fetus can be summarized as follows:

*Subsequent to the preparation of this manuscript and its submission at the time of the Conference in Nottingham in 1968, evidence was presented by Dr. R. Picon for the inhibitory influence of fetal testis hormone on cultured Müllerian ducts of male and female rats (Arch. Anat. Microsc. Morph. Expt. 58: 1-19, 1969; chapter 5 in this book). These results are not in agreement with the earlier findings of Price and Pannabecker in the rat and discrepancies may require further research for clarification.

Bioindicator Test for Androgen

1. The primordium of the testis and the adrenal cortex are already secreting androgens at 22 days of fetal age. This secretory activity precedes the beginning of sex differentiation of the reproductive ducts. The character of the androgens secreted at the earliest stage is unknown, but the possibility exists that they may be steroidal.
2. The ovary does not become metabolically active as judged by androgen secretion until the last third of fetal life. This is long after the critical stages in sex differentiation of the reproductive ducts have been completed.

Explanation of Male and Female Reproductive Tracts

1. Wolffian ducts of both sexes become androgen-dependent at the end of the ambisexual stage and retrogress unless androgen is available from testes or other organs—adrenals or 60-day-old ovary. The Wolffian ducts in the two sexes seem equally responsive to androgen.
2. Müllerian ducts of males retrogress even when the testes are removed at explantation at the ambisexual stage. Androgen from other fetal organs neither hastens nor retards retrogression.
3. Müllerian ducts of females are retained under all experimental conditions. They are definitely not inhibited by testicular androgen or androgen from other sources. The oviductal segment is capable of a high degree of hormone-independent morphogenesis and histogenesis.

These findings answer some of the questions that were posed. The fetal testis is indeed secreting an androgen at a very early stage when the Müllerian duct is just beginning to form (Fig. 1). Testicular androgen must be available to the Wolffian ducts of males in vivo by diffusion, via the circulation, or by both routes by the end of the ambisexual stage. Testicular androgen does not cause retrogression of Müllerian ducts of females under the conditions imposed in our experiments. The Müllerian ducts of females are hormone independent (as has been shown in some other species) and the oviduct of the guinea pig is strikingly so.

REFERENCES

1. Burns, R. K. Role of Hormones in the Differentiation of Sex. *In* Sex and Internal Secretions, 3rd ed. Young, W. C., ed. Baltimore, The Williams and Wilkins Co., pp. 76–158, 1961.
2. Price, D., and Ortiz, E. The Role of Fetal Androgen in Sex Differentiation in Mammals. *In* Organogenesis. Dehaan, R. L., and Ursprung, H., eds. New York, Holt, Rinehart and Winston, Inc., pp. 629–652, 1965.
3. Bloch, E. Hormone production by the foetal adrenocortical gland. Proc. 2nd Int. Congr. Endocr., Excerpta Medica Int. Congr. Ser., 83:785–789, 1965.
4. ——— *In vitro* steroid synthesis by gonads and adrenals during mammalian fetal development. Proc. 2nd Int. Congr. Hormonal Steroids, Excerpta Medica Int. Congr. Ser., 132:675–679, 1967.

5. Jost, A. Gonadal Hormones in the Sex Differentiation of the Mammalian Fetus. *In* Organogenesis. DeHaan, R. L., and Ursprung, H., eds. New York, Holt, Rinehart and Winston Inc., pp. 611–628, 1965.
6. ——— Steroids and sex differentiation in the mammalian foetus. Proc. 2nd Int. Congr. Hormonal Steroids, Excerpta Medica Int. Congr. Ser., 132:74–81, 1967.
7. Price, D., and Pannabecker, R. Organ culture studies of foetal rat reproductive tracts. Ciba Found. Colloquia on Ageing, 2:3–13, 1956.
8. ——— and Pannabecker, R. Comparative responsiveness of homologous sex ducts and accessory glands of fetal rats in culture. Arch. Anat. Micr. Morph. Exp., 48:223–243, 1959.
9. Brewer, N. L. Sex differentiation of the fetal mouse *in vitro*. Ph. D. Thesis, University of Chicago, 1962.
10. Zaaijer, J. J. P., Price, D., and Ortiz, E. Organ culture studies of hormone secretion in endocrine glands of fetal guinea pigs. I. Androgenic secretion as demonstrated by a bioindicator method. Proc. Kon. Nederl. Akad. Wet. (Biol. Med.), 69:389–399, 1966.
11. Ortiz, E., Price, D., and Zaaijer, J. J. P. Organ culture studies of hormone secretion in endocrine glands of fetal guinea pigs. II. Secretion of androgenic hormone in adrenals and testes during early stages of development. Proc. Kon. Nederl. Akad. Wet. (Biol. Med.), 69:400–408, 1966.
12. Price, D., Ortiz, E., and Zaaijer, J. J. P. Organ culture studies of hormone secretion in endocrine glands of fetal guinea pigs. III. The relation of testicular hormone to sex differentiation of the reproductive ducts. Anat. Rec., 157:27–42, 1967.
13. Ortiz, E., Zaaijer, J. J. P., and Price, D. Organ culture studies of hormone secretion in endocrine glands of fetal guinea pigs. IV. Androgens from fetal adrenals and ovaries and their influence on sex differentiation. Proc. Kon. Nederl. Akad. Wet. (Biol. Med.), 70:475–488, 1967.
14. Price, D., Zaaijer, J. J. P., and Ortiz, E. Prenatal Development of the Oviduct *in Vivo* and *in Vitro*. *In* The Mammalian Oviduct. Hafez, E. S. E., and Blandau, R. eds. Chicago, University of Chicago Press, 1969.
15. ——— and Williams-Ashman, H. G. The Accessory Reproductive Glands of Mammals. *In* Sex and Internal Secretions, 3rd ed. Young, W. C., ed. Baltimore, The Williams and Wilkins Co., pp. 366–448, 1961.
16. ——— Actions in vivo et in vitro des hormones sur les glandes accessoires de l'appareil genital male. J. Suisse Med., 83:835–891, 1953.
17. Lasnitzki, I. The effect of hydrocortisone on the ventral and anterior prostate gland of the rat grown in culture. J. Endocr., 30:225–233, 1964.
18. Price, D., Ortiz, E., and Deane, H. W. The presence of Δ^5-3β-hydroxysteroid dehydrogenase in fetal guinea pig testes and adrenal glands. Amer. Zool., 4:327, 1964.
19. Black, V. H. Differentiation of interstitial cells in fetal guinea pig testes. Anat. Rec., 157:214–215, 1967.
20. Christensen, A. K. The fine structure of testicular interstitial cells in guinea pigs. J. Cell Biol., 26:911–935, 1965.
21. Hoare, R. M. The normal maternal and fetal peripheral blood plasma and adrenal cortical levels of hydrocortisone throughout pregnancy in guinea pigs. Anat. Rec., 151:362, 1965.
22. MacArthur, E., Short, R. V., and O'Donnell, V. J. Formation of steroids by the equine foetal testis. J. Endocr., 38:331–336, 1967.
23. Weniger, J. P., and Ehrhardt, J. Sécrétion de testostérone et d'androstènedione par les testicules foetaux de veau cultivés in vitro. C.R. Acad. Sci. (Paris), 264:1911–1912, 1967.
24. Jost, A. Problems of fetal endocrinology: the gonadal and hypophyseal hormones. Recent Progr. Hormone Res., 8:379–418, 1953.
25. Wells, L. J. Fetal hormones and their role in organogenesis. *In* Organogenesis. DeHaan, R. L., and Ursprung, H., eds. New York, Holt, Rinehart and Winston Inc., pp. 673–680, 1965.

[illegible] Hormones in the sex differentiation of the [illegible] (ed.), [illegible] New York [illegible]

[illegible]

[illegible] Nature [illegible]

[illegible] Univ. of Chicago Press, Chicago [illegible]

[illegible] Proc. Natl. Acad. Sci. U.S.A. [illegible]

[illegible] Amer. J. Anat. [illegible]

[illegible]

[illegible] J. Embryol. Exp. Morphol. [illegible]

[illegible]

[illegible] J. Exp. Zool. [illegible]

50

Effect of the Rat Fetal Testis on the Regression of the Müllerian Ducts in Vitro

Régine Picon

Laboratory of Comparative Physiology,
Faculty of Sciences, Paris, France

INTRODUCTION

The problem of regression of male Müllerian ducts and persistence of female ones was clarified by Jost in the rabbit(1): early castration of male fetuses resulted in complete persistence of Müllerian ducts; testicular graft against a female tract caused local degeneration of the Müllerian duct. In the mouse, irradiation of fetal testes in vivo(2,3) or removal of testes from the tract before explanation in vitro(4,5), resulted in partial or total persistence of the male Müllerian ducts; female mouse Müllerian ducts partly degenerated in vitro, when cultured with testes(6). But in other species, the inhibitory action of fetal testes on Müllerian ducts could not be shown(7,8) and the hypothesis of a very brief period of responsiveness for Müllerian ducts was advanced(9).

The aim of the present work was to obtain a bioindicator method for the inhibitory action of testes of various ages and to study the responsiveness of Müllerian ducts according to their age and genetic sex constitution. Our method of approach was the culture of fetal rat reproductive tracts in an ahormonal medium.

MATERIALS AND METHODS

In these experiments, Wistar and Sherman female rats were used. Gestational age was based on the timing of coitus(10). Pregnant females were anaesthetized with ether and fetuses were dissected under sterile conditions. The whole fetal reproductive tract was excised and transferred in Hanks solution. Explants were cultured on grids, by the Trowell technique(11) in a synthetic medium, medium 199(12) and incubated at 37°C.

Histochemical Study. Δ5-3β-Hydroxysteroid dehydrogenase (HSD) activity was sought for in testes cultured from 14½, 15½ and 16½ days to 21 days post coitum (p.c.); The testes were frozen, sectioned in a cryostat at a thickness of 20μ and treated according to the Wattenberg technique(13); the substrate was dehydroepiandrosterone.

Histological Study. The reproductive tracts were explanted at 14½, 15½, 16½, and 17½ days and cultured for three days. There were three categories of cultures: tracts were explanted with (1) their own gonads (2) no gonads and (3) without their own gonads, but a testis was placed against one of the Müllerian ducts. The explants and control fetal tracts were fixed in Bouin's fluid, embedded in paraffin, sectioned at 7μ and stained with hemalum-eosin.

RESULTS

Explants of Male Reproductive Tracts

Histochemical Results. The first Δ5-3β-HSD active interstitial cells were found in control testes 15 days old. The intensity of the reaction increased during the whole gestational period.

In vitro, Δ 5-3β-HSD activity, absent or weak at the time of explantation, was very strong on day 21 and similar to that of control fetuses of the same age. Nevertheless, the cytoplasm of the interstitial cells was less abundant(10).

Histological Results. In the control fetuses 14½ and 15½ days old, the Müllerian ducts have not yet reached the urogenital sinus. When explanted, they grow posteriorly and generally fuse with the sinus wall. This was observed in all male and female reproductive tracts, explanted with or without gonads (Table 1).

When the whole male reproductive tracts were explanted, there was antero-posterior regression of the Müllerian ducts (Table 2) similar to that observed in controls (Table 3).

When the tracts were explanted without gonads, results varied according to the time of explanation (Table 2). Maintenance of the Müllerian ducts occurred

TABLE 1

Growth of Müllerian ducts in vivo and in vitro

	Age (days p.c.)	Length[a] mean	Length[a] extr. val.	No. of cases
Control tracts	14½	41	35-50	6
	15½	81	63-88	4
	16½	100	—	4
Cultured tracts	14½-17½	87	57-100	45
	15½-18½	99	95-100	33

[a]The length of the Müllerian ducts was evaluated as the percentage of the distance ostium-sinus.

TABLE 2

Regression of Müllerian ducts in vitro

Categories of explants	Culture period (days p.c.)	Males: Regressed part[a] mean	Males: Regressed part[a] extr. val.	Males: No. of cases	Females: Regressed part mean	Females: Regressed part extr. val.	Females: No. of cases
With their own gonads	14½-17½	69	54-80	4	0	—	9
	15½-18½	80	—	1	—	—	—
	16½-19½	89	87-91	3	0	—	3
	17½-20½	100	—	1	0	—	1
Without gonads	14½-17½	0	—	5	0	—	4
	15½-18½	55	15-81	4	0	—	4
	16½-19½	48	26-59	5	0	—	3
	17½-20½	96	91-100	3	0	—	2
Without their own gonads + testis[c]	14½-17½	70	53-100	10	63	42-85[b]	13
	15½-18½	75	56-93	10	21	0-78	14
	16½-19½	84	61-94	6	0	—	6
	17½-20½	93	88-100	6	0	—	7

[a]The regressed part includes the completely regressed segment and the blind cord which follows. The length of the regressed part is expressed as the percentage of the distance ostium-sinus.
[b]In one case, the length of the regressed part was only 18.
[c]The value noticed in each explant only relates to the Müllerian duct close to the testis.

TABLE 3

Regression of Müllerian ducts in control male fetuses

Age (days p.c.)	Regressed part[a] mean	Regressed part[a] extr. val.	No. of cases
14½	0	—	6
15½	0	—	2
16½	60	59-71	3
17½	66	62-75	3
18½	81	—	1
19½	86	82-91	2
20½	95	94-96	3

[a]Cf. Table 2.

when the tracts were explanted at 14 ½ days. When explanted later, Müllerian ducts underwent regression of the cranial part; when explantation was done at 15 ½ or 16 ½ days, the regressed segment was shorter than in controls; when explantation was done still later, regression was similar to that observed in controls.

When a testis was placed against one of the male Müllerian ducts, regression occurred on the side of the testis. On the other side, results were similar to those of male tracts explanted without gonads.

Testes exerted an inhibitory effect on male Müllerian ducts from 15 days onward; they must be present until about 17 ½ days to induce full regression.

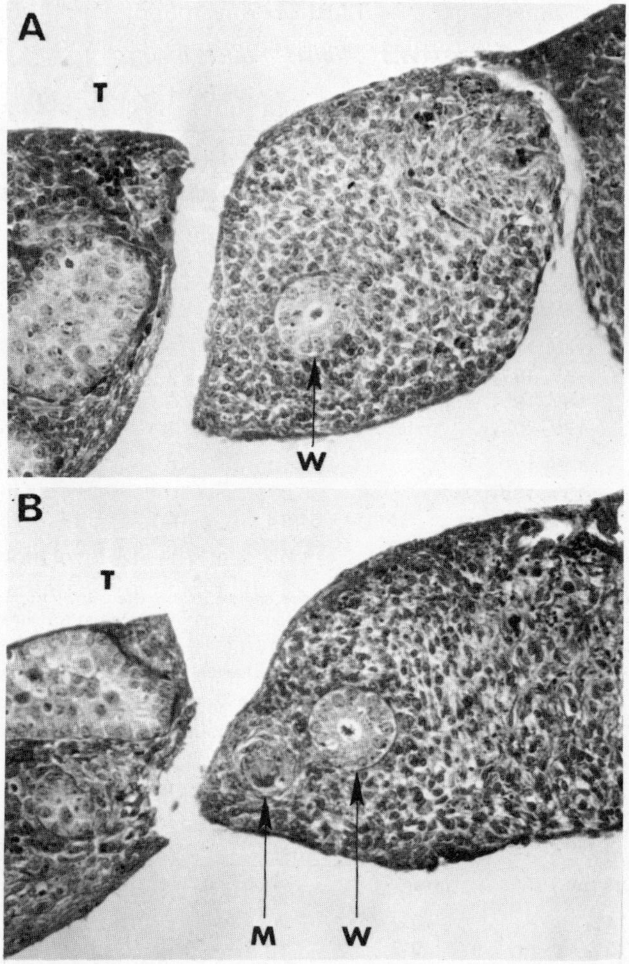

Fig. 1. Female reproductive tract cultured from 14½ to 17½ days p.c. with a testis. ×22. *W:* Wolffian duct; *M:* Müllerian duct; *T:* Testis. *A:* Cranial segment. The Müllerian duct is not visible. *B:* Middle segment. The Müllerian duct is a blind cord.

Explants of Female Reproductive Tracts

When female reproductive tracts were explanted with or without ovaries, maintenance of the Müllerian ducts was observed. Growth towards the sinus occurred in the younger explants.

When testes were placed against the female tracts from 14½ days onward, anteroposterior regression of the Müllerian ducts was always obtained. Histological aspects (Fig. 1) and length of the regressed part (Table 2) were similar to those of the male Müllerian ducts, both in vivo and in vitro. Moreover, the results were the same, whatever the age of the fetal testis: 14½ days (4 cases), 15½ days (7 cases), 18½ days (2 cases).

When testes were placed against the Müllerian ducts at a later stage, different results were observed. When association was made at 16½ or 17½ days, concentric accumulation of conjunctive fibers around the Müllerian duct was the only reaction noticed. When association was made at 15½ days, results were variable.

So, a critical stage for inhibiting Müllerian ducts can be shown in the rat fetus at 14½ days. Their genetic sex constitution is not a limiting factor for responsiveness. The action of the testis was always localized to the duct against which it was placed.

DISCUSSION

In the rat, the shortness and precocity of the period of responsiveness of the Müllerian ducts to the testicular inhibitory action account for the difficulties in obtaining this effect. The critical period lies at about 14½ days p.c., at the very moment when the ducts are growing towards the urogenital sinus.

To prevent regression of Müllerian ducts in male fetuses, testes must be taken off at that moment. In these conditions, Price and Pannabecker(7) noticed regression of male Müllerian ducts, but female ones regressed too, and instability of the youngest explants under culture conditions was suggested(9). The maintenance of male Müllerian duct fragments explanted without testes at 15 or 16 days(14,15) is apparently in conflict with our data. In those experiments the only part explanted was the caudal one; we also noticed persistence of the caudal half of the male Müllerian ducts when explanted without testes at 16½ days.

In the same way, to obtain regression of female Müllerian ducts, testes must be placed against female tracts from 14½ days onward. Early stabilization of the ducts might explain the failure of the fetal testis to inhibit female Müllerian ducts in some species(4,8).

The use of this bioindicator method has brought out some preliminary observations. The local inhibitory effect of testes on Müllerian ducts is in agreement with earlier observation(1). Elaboration of the inhibitory factor is not limited to the very short period of responsiveness of the Müllerian ducts; it is still obvious in the testes at 18½ days. Further, it can be obtained in a synthetic anhormonal medium. In the same culture conditions, steroidogenic activity also increases; hypophyseal stimulation is not absolutely necessary. However, the level of that activity probably stands below that of the controls(10,16).

Although recent observations show some steroidogenic activity in fetal mammal testes at early stages of sex differentiation(9,17,18,19,20), the question of the identity of the fetal testicular hormones, especially the inhibitory hormone, still remains unanswered.

REFERENCES

1. Jost, A. Recherches sur la differenciation sexuelle de l'embryon de lapin. III. Rôle des gonades foetales dans la différenciation sexuelle somatique. Arch. Anat. Micr. Morph. Exp., 36:271–315, 1947.

2. Raynaud, A., and Frilley, M. Irradiation, au moyen des rayons X, des ébauches des glandes génitales de l'embryon de souris, au 15ème jour de la vie intra-utérine. C.R. Acad. Sci. (Paris), 223:1187, 1946.
3. ——— and Frilley, M. Développement des canaux de Müller chez des foetus mâles de souris dont les testicules ont été irradiés, au moyen des rayons X, à l'age de 14 jours. Ann. Endocr. (Paris), 11:32–48, 1950.
4. Brewer, N. L. Sex differentiation of the fetal mouse in vitro. Ph. D. Thesis, University of Chicago, 1962.
5. Weniger, J. P. Régression du canal de Müller d'embryon de souris mâle soumis in vitro à l'action de l'hormone testiculaire. C.R. Acad. Sci. (Paris), 259:1899–1901, 1964.
6. ——— Etude comparée des actions homonales des testicules embryonnaires de poulet et de souris en culture in vitro. Arch. Anat. Micr. Morph. Exp., 54:909–919, 1965.
7. Price, D., and Pannabecker, R. Comparative responsiveness of homologous sex ducts and accessory glands of fetal rats in culture. Arch. Anat. Micr. Morph. Exp., 48:223–243, 1959.
8. ——— Ortiz, E., and Zaaijer, J. J. P. Organ culture studies of homone secretion in endocrine glands of fetal guinea pigs. III. The relation of testicular hormone to sex differentiation of the reproductive ducts. Anat. Rec., 157:27–41, 1967.
9. ——— and Ortiz, E. The role of fetal androgen in sex differentiation in mammals. *In* Organogenesis. Dettaan, R., and Ursprung, H., eds. New York, Holt, Rinehart and Winston, Inc., pp. 629–652, 1965.
10. Picon, R. Activité Δ5-3-β-hydroxystéroide déshydrogénasique du testicle foetalde rat in vitro. Arch. Anat. Micr. Morph. Exp., 56:281–290, 1967.
11. Trowell, O. A. A modified technique from organ culture in vitro. Exp. Cell Res., 6:246–248, 1954.
12. Morgan, J. F., Morton, H. J., and Parker, R. C. Nutrition of animal cells in tissue culture. I. Initial studies on a synthetic medium. Proc. Soc. Exp. Biol., 73:1–8, 1950.
13. Wattenberg, L. W. Microscopic histochemical demonstration of steroid-3β-ol-dehydrogenase in tissue sections. J. Histochem. Cytochem., 6:225–232, 1958.
14. Jost, A., and Bergerard, Y. Culture in vitro d'ébauches du tractus génital du foetus de rat. C.R. Soc. Biol. (Paris), 143:608–609, 1949.
15. ——— and Bozic, B. Données sur la différenciation des conduits génitaux du foetus de rat, étudiée in vitro. C.R. Soc. Biol. (Paris), 145:647–650, 1951.
16. Picon, R., and Ottowicz, J. Recherches histochimiques sur le testicule du foetus de rat, normal ou décapité. Arch. Anat. Micr. Morph. Exp., 56:1–10, 1967.
17. Niemi, M., and Ikonen, M. Steroid-3β-ol-dehydrogenase activity in foetal Leydig's cells. Nature, 189:592–593, 1961.
18. Lipsett, M. B., and Tullner, W. W. Testosterone synthesis by the fetal rabbit gonad. Endocrinology, 77:273–278, 1965.
19. Baillie, A. H., Ferguson, M. M., and Hart, D. McK. Histochemical evidence of steroid metabolism in the human genital ridge. J. Clin. Endocr., 26:738–741, 1966.
20. Hart, D. McK., Baillie, A. H., Calman, K. C., and Ferguson, M. M. Hydroxysteroid dehydrogenase development in the mouse adrenal and gonads. J. Anat., 100:801–812, 1966.

51

The Influence of Androgen Antagonists and Progestogens on the Sex Differentiation of Different Mammalian Species*

W. Elger, F. Neumann, and R. von Berswordt-Wallrabe

Main Laboratory of Schering AG,
Berlin, West Germany

INTRODUCTION

The use of well-defined steroids would seem to be a simple and effective method of obtaining information about the principles of sex differentiation. Since Neumann(1) discovered the antiandrogenic actions of cyproterone acetate (Cpa) from its feminizing effect on the external genitalia in male rat foetuses, many experiments have been carried out with antiandrogens in order to explain the part played by the androgens in the foetal or adult organism.

Apart from work on the development of the external genital organs of the rat (1,2), the development of the external and internal genital organs in the rabbit(3) was also investigated. An analogous study on the dog is still unpublished(4). I shall include the results from these last two works in this chapter. The development of the mammary gland in rats(5) and mice(6) under the influence of antiandrogens, has also been extensively studied. Similarly, Cpa was tested to determine its influence on the differentiation of hypothalamic centers for the control of gonadotropin secretion and sexual behavior in male rats in the early postnatal phase(7,8).

MATERIALS AND METHODS

In the present account I shall keep basically to the differentiation of the internal sex organs. Table 1 shows the steroids investigated in the experiment. Cy-

*The authors wish to thank Mrs. Stock, Miss Sperling, Miss Bonisch, and Miss Blank for skilled technical assistance.

TABLE 1

The steroids used in the experiments and their progestogenic activity

Drug	Progestational activity (threshold dose in the Clauberg-test)
Cyproteroneacetate (Cpa)	3.0 μg
Cyproterone	10,000.0 μg without activity
Progesterone	100.0 μg
19-Nor-17α-OH-progesterone-acetate	3.0 μg
19-Nor-17α-OH-progesterone-caproate	30.0 μg

TABLE 2

Plan of the experiments in the different species (Treatment = treatment of pregnant mother animals)

Species	Drug	Dose (mg/kg/d)	Period of treatment	Killed at day (p.c.)	Number of foetuses examined
Rabbit	Cpa	5.0[a]	day 13-24 p.c.	day 30 p.c.	28
	Cyproterone	10.0			7
	Progesterone	30.0			3
	19-nor-17α-OH-progesterone-acetate	30.0			2
	19-nor-17α-OH-progesterone-caproate	30.0			8
Dog	Cpa	10.0	day 23-43 p.c.	day 44 p.c.	8
Rat	Cpa	100.0 per animal per day	day 14-21 p.c.	day 22 p.c.	13

[a]Doses ranging from 0.1-100.0 mg/kg/day i.m. were tested in other experiments. Some results presented in this paper originate from these experiments.

proterone and cyproterone acetate[1] (Cpa) were used as antiandrogens. Cpa has a much more powerful antiandrogenic action than cyproterone. However, the threshold value in the Clauberg test indicates that Cpa also has extremely powerful progestogenic actions. (The data given in Table 1 were determined roughly by routine experiments.) On the other hand, even in high dosages, cyproterone has no progestational actions at all on the rabbit uterus.

[1]Cyproterone = 1,2α-methylene-6-chloro-$\Delta^{4,6}$-pregnadiene-17α-ol-3,20-dione. Cyproterone acetate = 1,2α-methylene-6-chloro-$\Delta^{4,6}$-pregnadiene-17α-ol-3,20-dione-17α-acetate. These steroids were synthesized by Dr. R. Wiechert of Schering AG, Berlin, West Germany.

Supplementary experiments with progesterone and two potent 19-nor-17α-hydroxy-progesterone esters which, basically, do not have any androgenic, antiandrogenic, or estrogenic properties, were carried out in order to examine the importance of progestogenic action in sex differentiation. Those control experiments were limited to rabbits; this also applies to the experiment with cyproterone. The action of Cpa on sex differentiation in male foetuses was investigated in three species, e.g., rabbits, dogs, and rats.

Table 2 contains information about the period and type of treatment carried out during pregnancy, in the various species.

Rabbits were treated from day 13 to day 24 post coitum (p.c.). The foetuses were removed on day 30. Pregnant dogs were treated from day 23 to 43 p.c. and the foetuses were removed on the 44th day p.c. by caesarian section. Pregnant rats were treated from day 14 to day 21 p.c. The mothers were killed and the foetuses removed on the 22nd day.

In all the experiments, the drugs were dissolved in oil and administered by intramuscular injection.

RESULTS

Gonadal Differentiation

There was no disturbance of testicular development in any of the species investigated. Figure 1 shows the gonad of a rabbit foetus; the mother had received

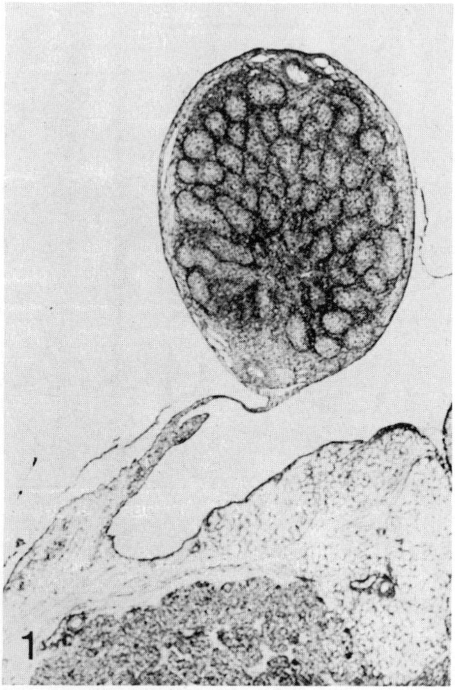

Fig. 1. Cpa and gonadal differentiation in a 30-day male rabbit foetus. Note the normally developed testis in spite of regressed Wolffian duct. 48×.

5.0 mg Cpa/kg/day. Although the Wolffian duct had undergone regression, the morphology of the testis is normal. The only effect that was constantly observed was a reduction in the size of the gonads. Figure 2 shows the testis of a dog foetus, the mother of which had received 10.0 mg Cpa/kg/day.

Here again, as in the rabbit, in spite of the antiandrogenic action (the Wolffian duct is absent) the histology of the testis is completely normal. These results suggest that under physiological conditions, androgens are not involved in the differentiation of the testes.

The Proportions of the Internal Genitalia (Descensus Testiculorum)

Figure 3 shows, on the 30th day of embryonic development the whole genital tract from the gonads to the sinus urogenitalis is higher in the female rabbit foetuses than in the male foetuses. Under the influence of Cpa, in the male foetuses the genital cord is raised considerably although the overall height of the genital tract (genital folds and genital cord) is practically unchanged. Basically, there is a change in the proportions of the internal genitalia. The ratio genital folds: genital cord in male and in female foetuses in about 1.67 and 1.4, respectively. However, under the influence of Cpa this becomes 0.9 in male foetuses. It may be concluded from this that it is only the height of the genital cord which is influenced in the female direction. However, descensus testiculorum remains undisturbed. This is sup-

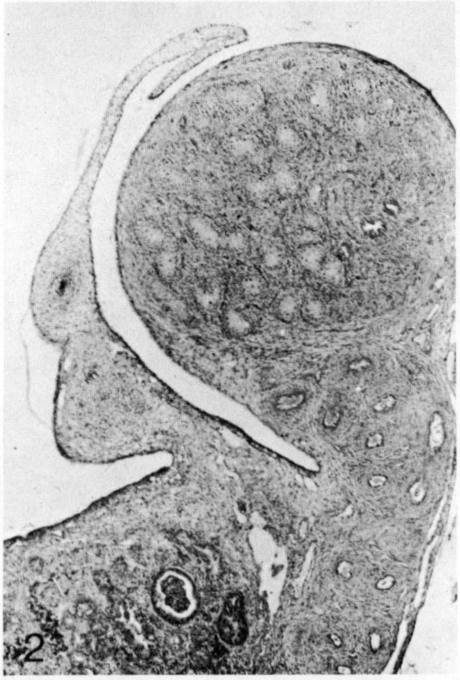

Fig. 2. Cpa and testicular morphology in a 44-day male dog foetus. Note the normal testicular histology and the complete regression of the Wolffian duct. 48×.

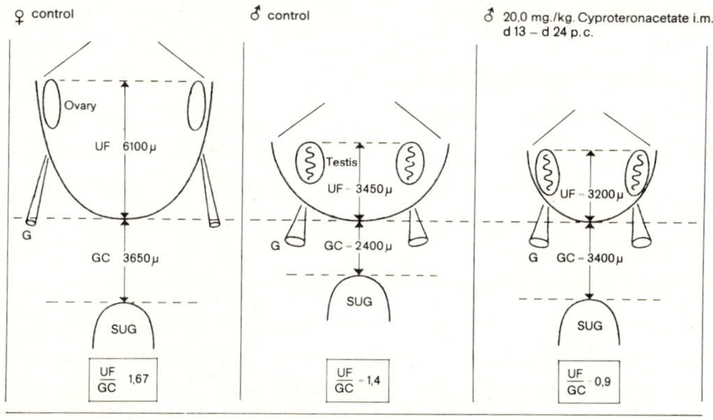

Fig. 3. Drawings showing the proportions of the genital tract in 30-day rabbit foetuses based on serial sections. The curved line represents the margin of the mesogenitalia which normally contain the gonoducts. Each group represents the combined date of 9 and 10 foetuses, respectively. *UF*=genital fold; *GC*=genital chord; *SUG*=sinus urogenitalis; *G*=gubernaculum.

ported by the fact that the gubernaculum testis was always normally developed (see Fig. 4).

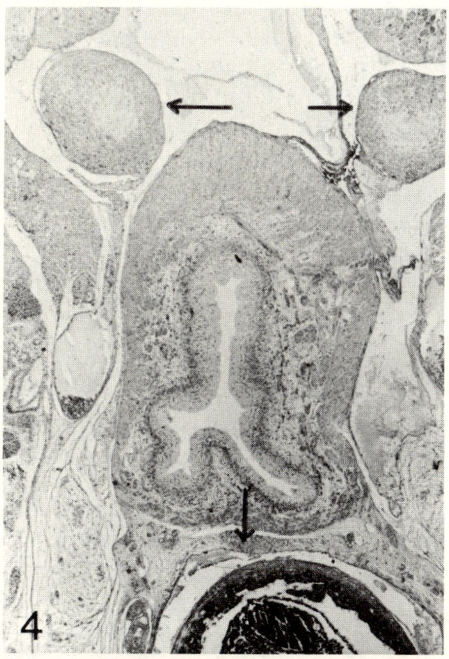

Fig. 4. Cpa and the gubernacula in 30-day rabbit foetuses. Note: Arrows at the top point to the well developed gubernacula; Arrow at the bottom points to the former position of the gonoducts between intestines and bladder. 30×.

From these results it is probable that androgens do not play an important part in descent of the testes in the rabbit.

The Differentiation of the Wolffian Ducts

Under Cpa the most important change in male rabbits(3) and dog foetuses(4) is the regression of the Wolffian ducts. However, even with excessively high doses of Cpa (100.0 mg Cpa/animal/day) it is not possible to achieve a comparable effect in male rat foetuses.

Rabbits. In male rabbits there is apparently a more or less complete regression of the Wolffian duct when a critical concentration of Cpa is reached. It is practically certain that there will be regression of both Wolffian ducts with a dose of 10.0 mg Cpa/kg/day or more (see Fig. 5). Up to now, in this dose range we have only found a single foetus still retaining a continuous Wolffian duct on one side. On the other hand, up to now only one foetus has responded with regression of the Wolffian duct when doses of 2.5 mg/Cpa/kg/day and less have been used. As Figure 5 shows, the critical dose for the Wolffian ducts is 5.0 mg Cpa/kg/day. Twenty-eight male foetuses were investigated in this group. Of a total of 56 Wolffian ducts, 28 were interrupted and had undergone more or less complete regression. The regression is usually bilateral. Unilateral regression was only found in four foetuses. As a rule, the male foetuses from one litter react in the same way.

Figures 6, 7, and 8 show unilateral and bilateral involution of the Wolffian ducts at various levels of the internal genitalia.

Dogs. Of eight dog foetuses investigated, only two had intact Wolffian ducts; the mothers had been treated with 10.0 mg Cpa/kg/day. In four foetuses only the right Wolffian duct was destroyed, although practically all the ducts were interrupted in the immediate vicinity of the sinus urogenitalis. In two foetuses, both

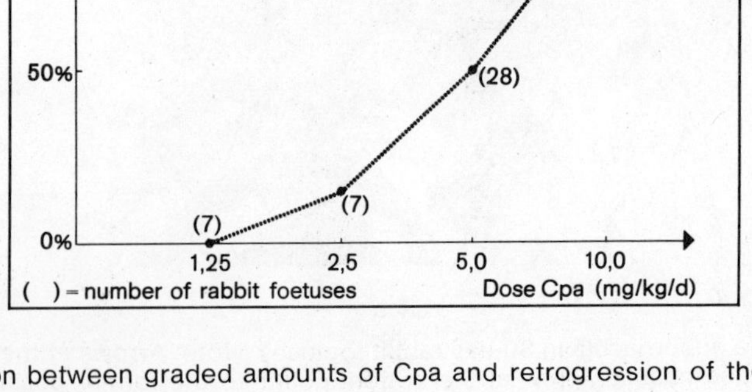

Fig. 5. Correlation between graded amounts of Cpa and retrogression of the Wolffian ducts in male rabbit foetuses.

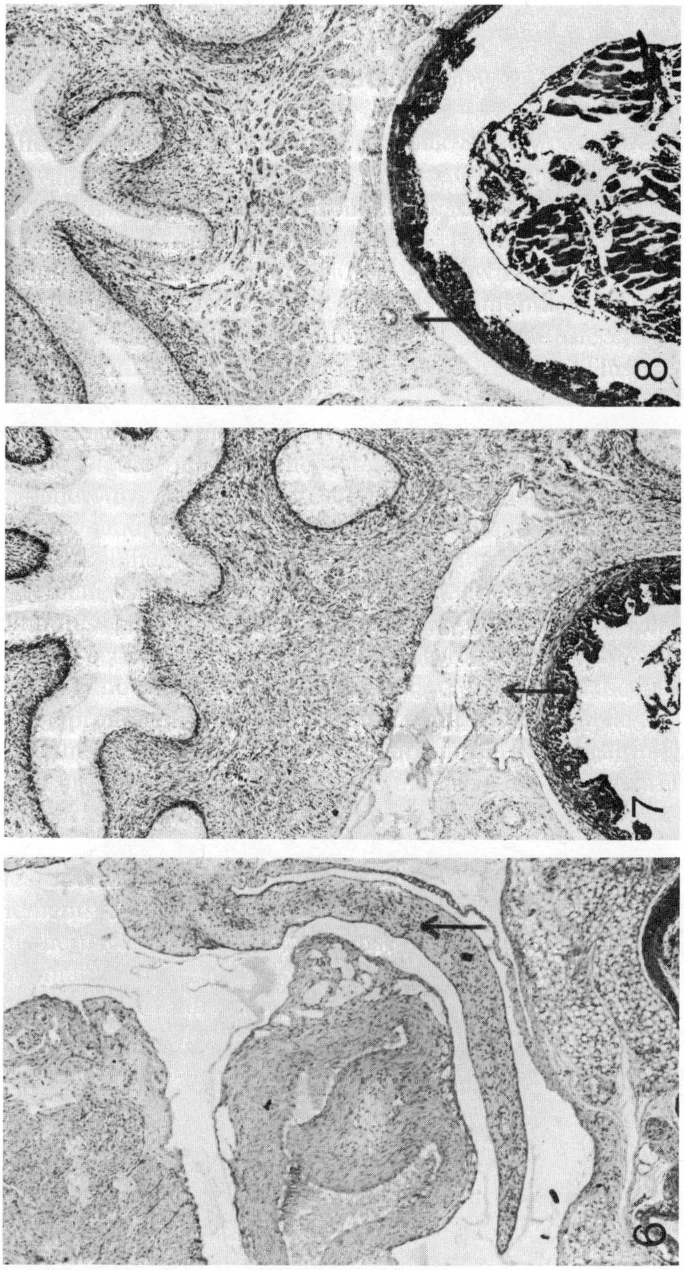

Figs. 6, 7, and 8. Various aspects of the retrogression of Wolffian ducts in Cpa-treated 30-day male rabbit foetuses. Figures 6 and 7: Note that there is no gonoduct within the mesogenitalia (arrows). 48×. Figure 8: Note the left Wolffian duct (arrow); the opposite one is regressed. 48×.

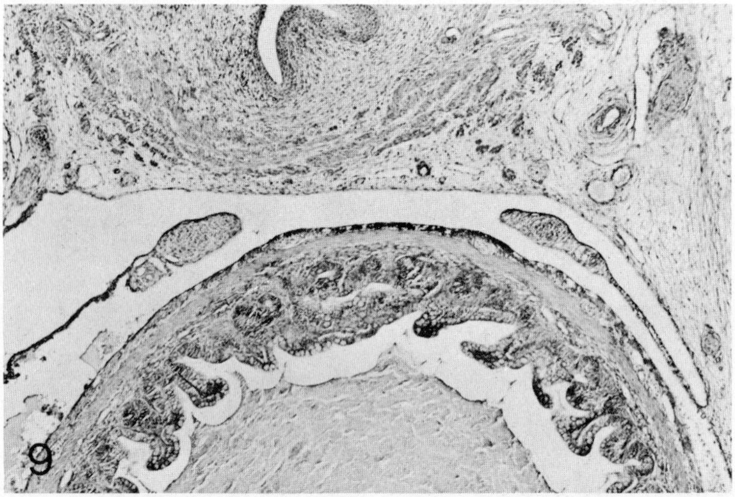

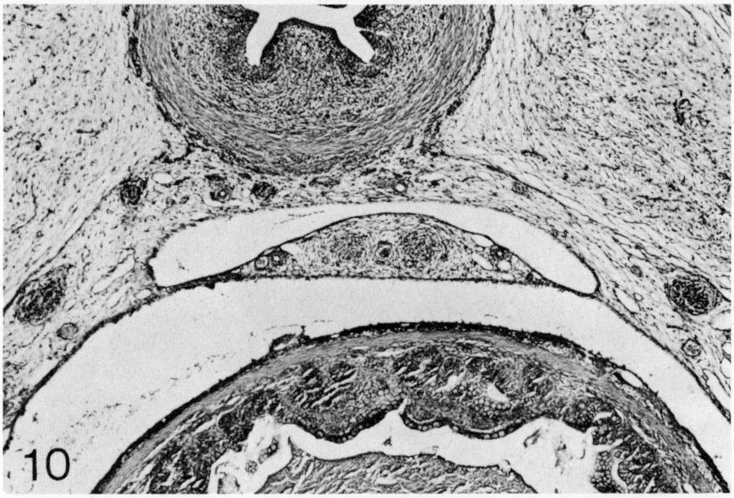

Figs. 9 and 10. Cpa-induced retrogression of Wolffian ducts in 44-day male dog foetuses. Figure 9: There are no gonoducts detectable within the mesogenitalia. 48×. Figure 10: Note the traces indicating the former positions of Wolffian and Müllerian ducts within the genital cord. This section is from a posterior level as compared with Figure 9. 48×.

ducts were extensively destroyed. Figures 9 and 10 show the empty mesogenitalia of a dog foetus at two different levels.

The original position of the gonoducts is indicated by a thickening of the mesenchymal tissue.

Rats. In the rat, in spite of the action of Cpa, the Wolffian ducts are differentiated normally. The ducts were continuous in 13 male foetuses which were examined more closely. Seminal vesicles and epididymis were established.

As Figures 11a and b show, the diameter of the ducts is often reduced. There are often degenerative changes in the epithelium.

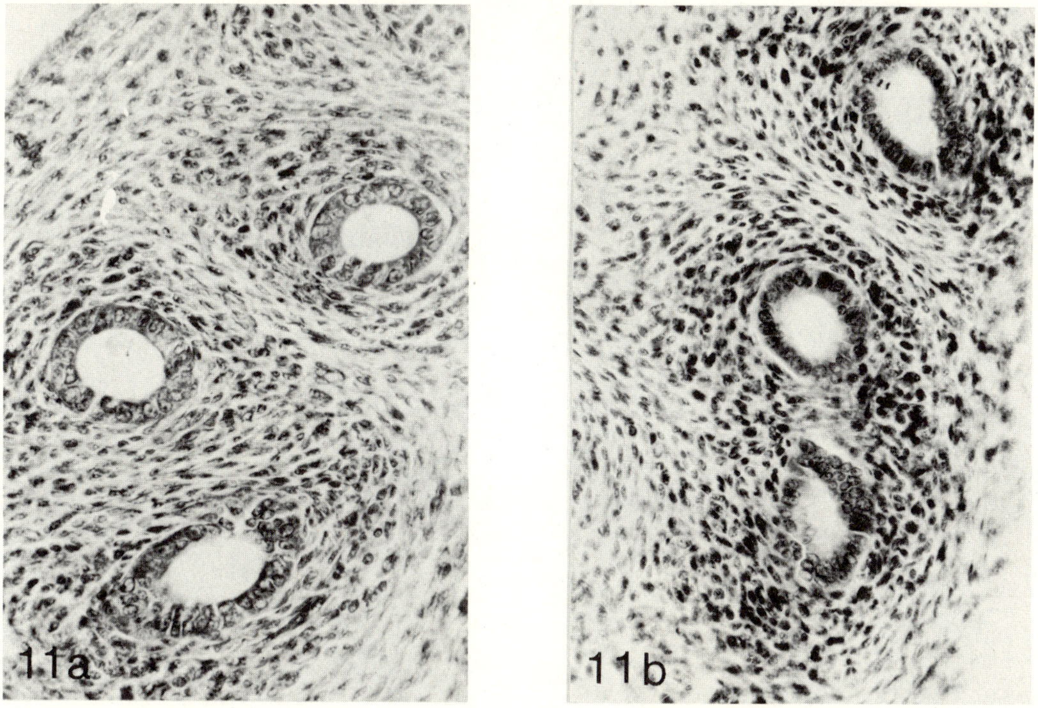

Fig. 11. The influence of Cpa on the development of the Wolffian ducts in 22-day male rat foetuses. *a.* Epididymal duct of a male control. *b.* Epididymal duct of a Cpa-treated male. Note: The diameter of this organ seems to be slightly reduced as compared with the control. 300×.

The Differentiation of the Müllerian Ducts

Although the Wolffian ducts are lost under the antiandrogenic influence of Cpa in dog and rabbit foetuses, there is no alternative development of the Müllerian ducts in the tubal and uterine portions. This applies to all the species investigated up to now. Thus, androgens appear to play no essential part in the regression of these ducts.

The situation is complicated in the vicinity of the sinus urogenitalis since here, in all the species investigated, rudiments of the Müllerian ducts were stimulated by Cpa. Figures 12 and 13 show the response in the rat and dog, respectively.

The development of Müllerian rudiments under increasing concentrations of Cpa in male rabbit foetuses is a process which deserves attention.

Under low doses of Cpa which do not interfere with the development of the Wolffian ducts, Müllerian duct rudiments proliferate forwards very irregularly from the posterior part of the seminal vesicle into the surrounding tissue. The situation seen in Figure 14 is typical.

Under higher concentrations of Cpa, the Müllerian rudiments lying outside the seminal vesicle are usually less developed. Instead, they spread into the seminal vesicle and finally often displace the epithelium of the Wolffian ducts alto-

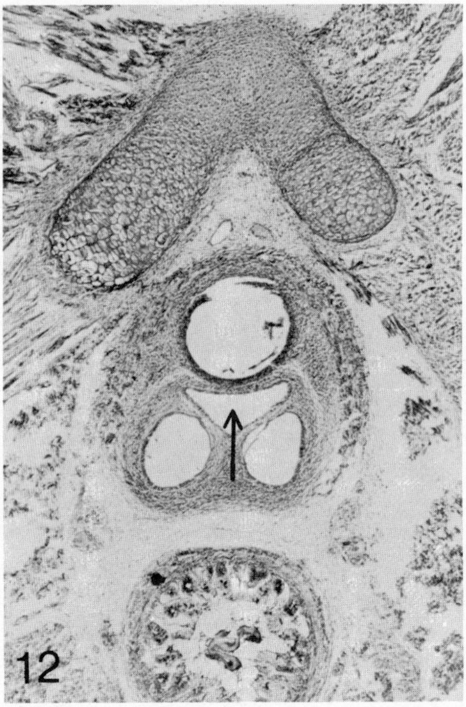

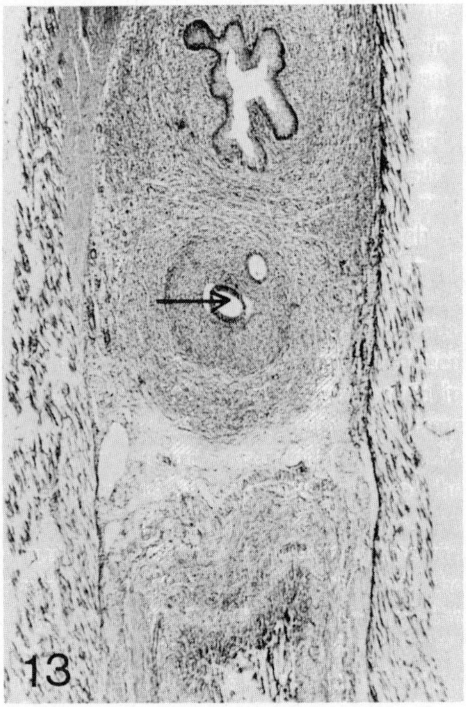

Fig. 12. Stimulated rudiment of the Müllerian ducts next to the sinus in a 22-day male rat foetus under Cpa influence. Note the typical triangular rudiment of the Müllerian ducts (arrow). 48×.

Fig. 13. A stimulated rudiment of the Müllerian ducts next to the sinus in a 44-day male dog foetus under Cpa influence. Note: A particularly well preserved segment of the Müllerian ducts (arrow). 48×.

gether (see Fig. 15). The ampullae which descend from the epithelium of the seminal vesicle(9) are often retained (see Fig. 15). Very frequently there is an organ formed which is comparable in every respect to the posterior portion of a vagina (see Fig. 16a and b).

Figures 17a and b serve to recapitulate a well-known finding in female rabbit foetuses which has already been described by Jost(10). Under the influence of

---

Figs. 14, 15, and 16. Stimulation of rudiments of the Müllerian ducts always restricted to the area next to the sinus in 30-day male rabbit foetuses under the influence of Cpa. Figure 14: Stimulated rudiments of the Müllerian ducts, under the influence of a rather small dose of Cpa (0.625 mg/kg/day; days 13 to 24 of pregnancy). Note: Arrows point to those rudiments which form irregular buds sprouting between the seminal vesicle *(VS)* and the sinus. 48×. Figure 15: Under an elevated Cpa dose as indicated in Table 2 the Wolffian epithelium of the seminal vesicle is replaced by the Müllerian epithelium which is chromophilic and undulated. *A*=ampulla. 48×. Figure 16: *a.* Cross section through a vagina-like and blind-ending organ, originating from the Müllerian ducts. The arrow points to a small Wolffian rudiment. 48×. *b.* Cross section through another male foetus having a Müllerian vagina-like structure. 48×. Note the hypertrophied Müllerian epithelium. The two individual ducts end in more posterior sections in one single lumen. No rudiments at all from the Wolffian ducts are detectable. The vesicular gland, normally androgen-dependent, is paradoxically well developed.

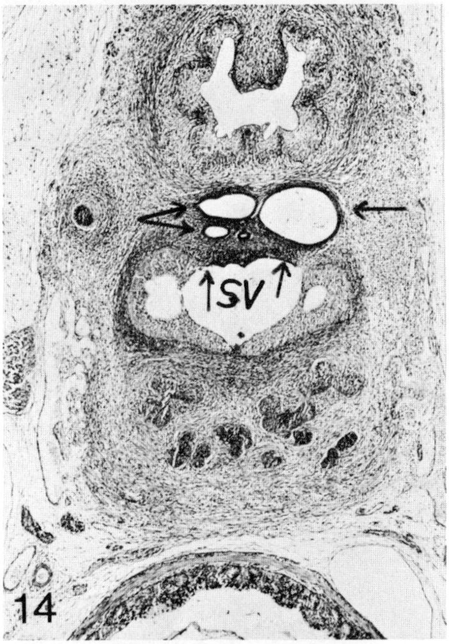

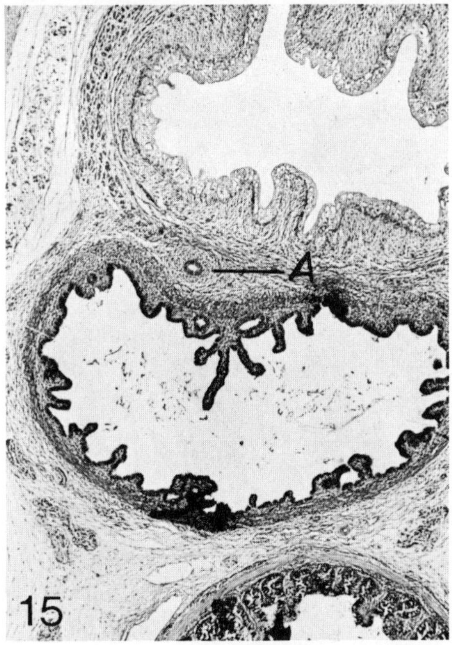

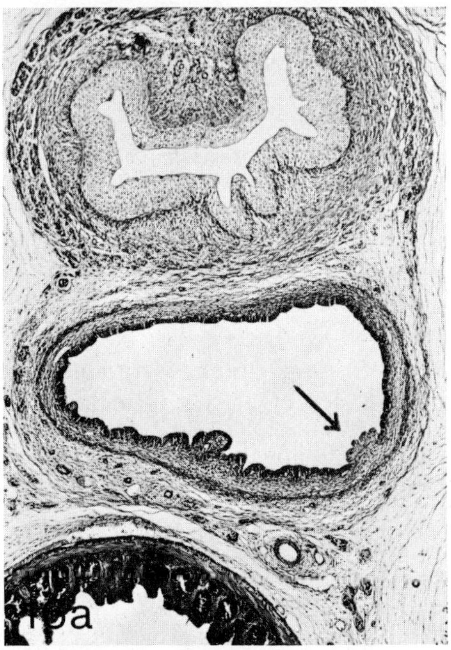

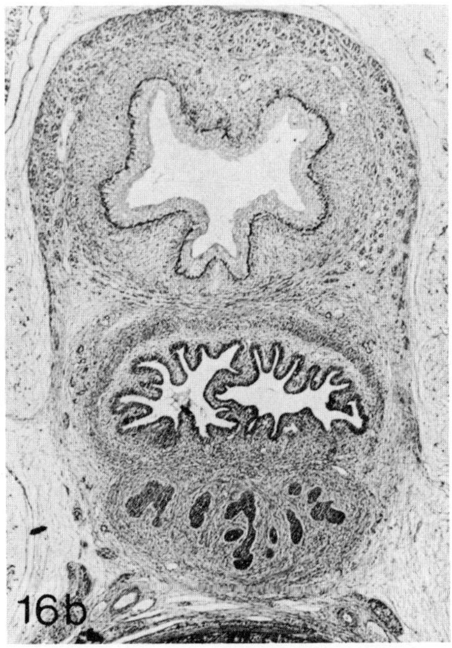

methyltestosterone the Müllerian ducts, here at the level of the anterior end of the seminal vesicle (see Fig. 17a), remain separated. At the same time the diameter is very much reduced. The same can be seen in another more posterior section in the vicinity of the sinus urogenitalis (see Fig. 17b).

From our own results with Cpa and from the known actions of androgens, we concluded that, in contrast to the more anterior part, the inhibition of these portions of the Müllerian ducts is normally induced by androgens. However, ac-

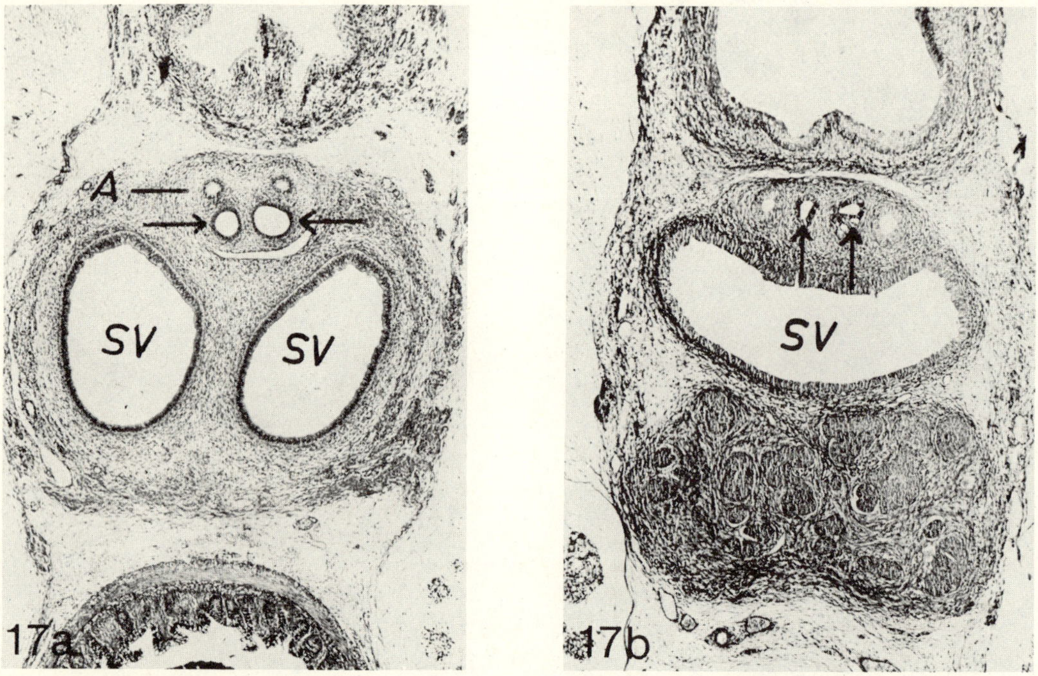

Fig. 17. Effects of methyltestosterone (MT) on Müllerian ducts in 30-day female rabbit foetuses. The mother had been injected with 1.5 mg MT/kg per day s.c.; days 13 to 24 of pregnancy. *a.* Arrows point at the Müllerian ducts with heavily reduced diameters. Note absence of a joining of both Müllerian ducts. *A*=ampulla; *SV*=horns of the seminal vesicle. *b.* Note in a more posterior cross section the rudimentary Müllerian ducts (arrows). *SV*=seminal vesicle.

cording to recent results, this conclusion appears to have been rather hasty. Thus, we found that progesterone in a dose of 30.0 mg/kg/day stimulated rudiments of the Müllerian ducts in the sinus region (see Fig. 18) of a (single) male rabbit foetus.

Two other synthetic potent progestogens, 19-nor-17α-hydroxy-progesterone caproate and 19-nor-17α-hydroxy-progesterone acetate also stimulate strong rudiments of the Müllerian ducts without feminization of other portions of the genital tract.

Of ten male rabbit foetuses investigated, where the mother animals had been treated with one of the two synthetic progestogens, nine had, in part, strongly stimulated rudiments of the Müllerian ducts.

Figure 19 illustrates a typical result. In contrast to the situation under the influence of Cpa, it was only exceptionally that epithelia of the Müllerian ducts appeared in the seminal vesicle wall, as is shown in Figure 20.

We never encountered a Müllerian vagina, such as is frequently found under Cpa treatment.

Surprisingly, in contrast to Cpa and the progestogens investigated, it was not possible to stimulate rudiments of the Müllerian ducts with cyproterone, which

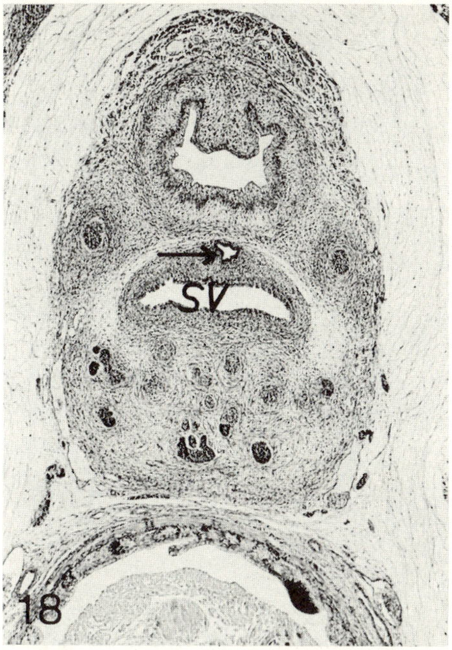

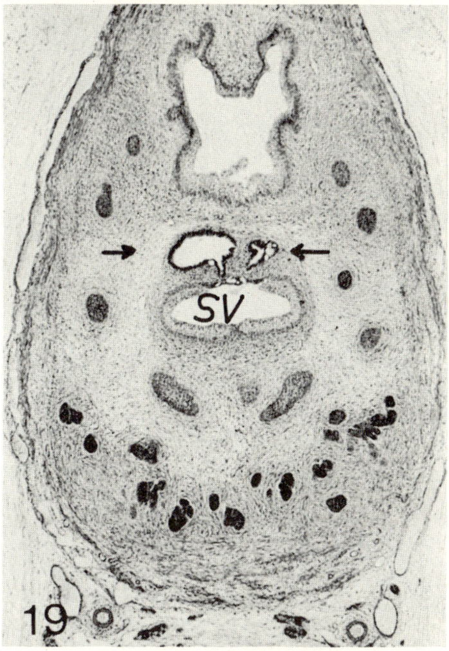

Fig. 18. Progesterone-stimulated Müllerian rudiments in a 30-day male rabbit foetus. Arrow points to this structure located between the sinus and the posterior portion of the seminal vesicle *(SV)*.

Fig. 19. A potent synthetic progestogen, 19-nor-17α-OH-progesterone caproate stimulates Müllerian rudiments (arrow) in a 30-day male rabbit foetus. Note: The origin of the buds spreading out from the ventral lining of the posterior portion of the seminal vesicle *SV*.

has—as already mentioned—no progestogenic effects, although there were often clear signs of an antiandrogenic effect on the external genitalia, prostate, vesicular gland, and seminal vesicle.

Figure 21 shows a section at the level of the sinus urogenitalis of a male foetus feminized with cyproterone. The prostate is completely suppressed and there are only rudiments of the vesicular gland.

In this foetus it was not possible to find any rudiment of the Müllerian ducts.

Experiments with higher dosages of cyproterone are necessary and definitive conclusions are still not possible. However, it appears that the regression of the posterior portions of the Müllerian ducts physiologically is also not caused by androgens. The strong development of the posterior rudiments under Cpa is probably related to the progestogenic action of this drug. In any case, there must be other important factors involved since progestogenic activity alone is not sufficient for the induction of a Müllerian vagina.

If androgens (methyltestosterone) do not have a direct influence on the Müllerian ducts in the vicinity of the sinus—as the experiments with cyproterone indicate—it must follow that an inhibitory effects is exerted indirectly, via stimulation of the Wolffian ducts. Conversely, the development of a Müllerian vagina

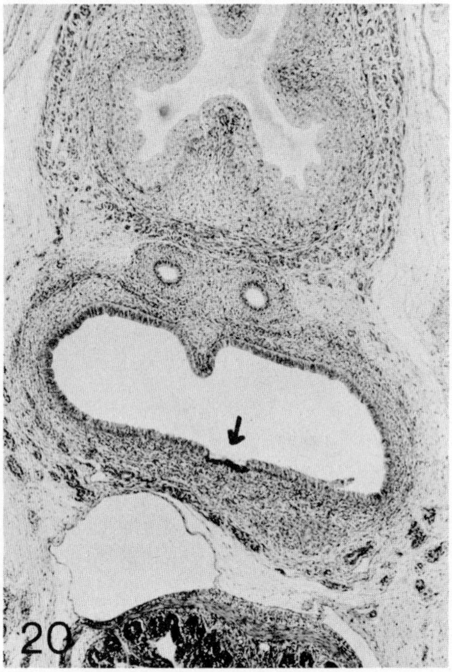

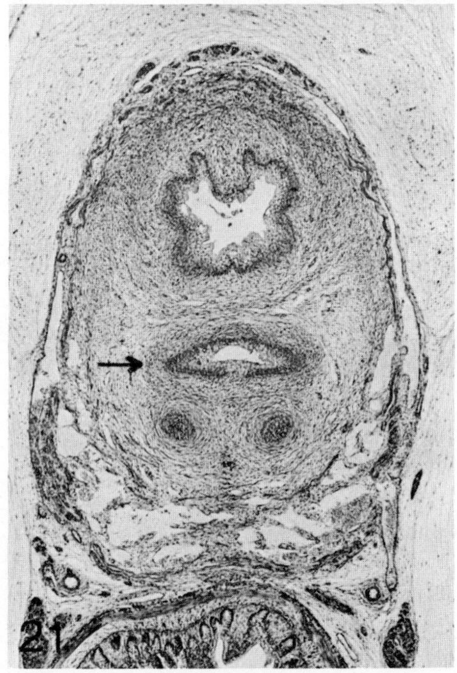

Fig. 20. Induction of infiltration of the seminal vesicle by Müllerian epithelium in a 30-day male rabbit foetus under the influence of a powerful synthetic progestogen, 19-nor-17α-OH-progesterone acetate. Note: Although the rudiment is rather small, it is nevertheless exceptional (it cannot be compared with the very pronounced infiltration of the seminal vesicle by Müllerian epithelium under Cpa).

Fig. 21. Noneffect of cyproterone, a specific antiandrogen without progestational side effects, on the Müllerian rudiments next to the stalk of the seminal vesicle (arrow) in 30-day male rabbit foetus. Note: No prostate buds as a consequence of the antiandrogenic influence.

under Cpa in male foetuses (after primary stimulation by the progestogenic activity of Cpa) seems to be possible only after regression of the Wolffian ducts, and particularly of the seminal vesicle.

Paradoxical Effects

These effects under the influence of Cpa have only been investigated in detail in the rabbit.

Male Foetuses. Paradoxical conditions are limited to the structures in the vicinity of the sinus. In particular, the development of the vesicular gland is nearly always in sharp contrast to that of the rudiments of the seminal vesicle which often are still present. In other words, complete regression of the seminal vesicle is nearly always combined with pronounced development of the vesicular gland. In contrast, the vesicular gland is nearly always suppressed where the seminal vesicle is comparatively strongly developed (see Figs. 22a and b).

Female Foetuses. It is interesting that in female foetuses relatively low doses of Cpa stimulate the vesicular gland and Wolffian rudiments. Often, there was also

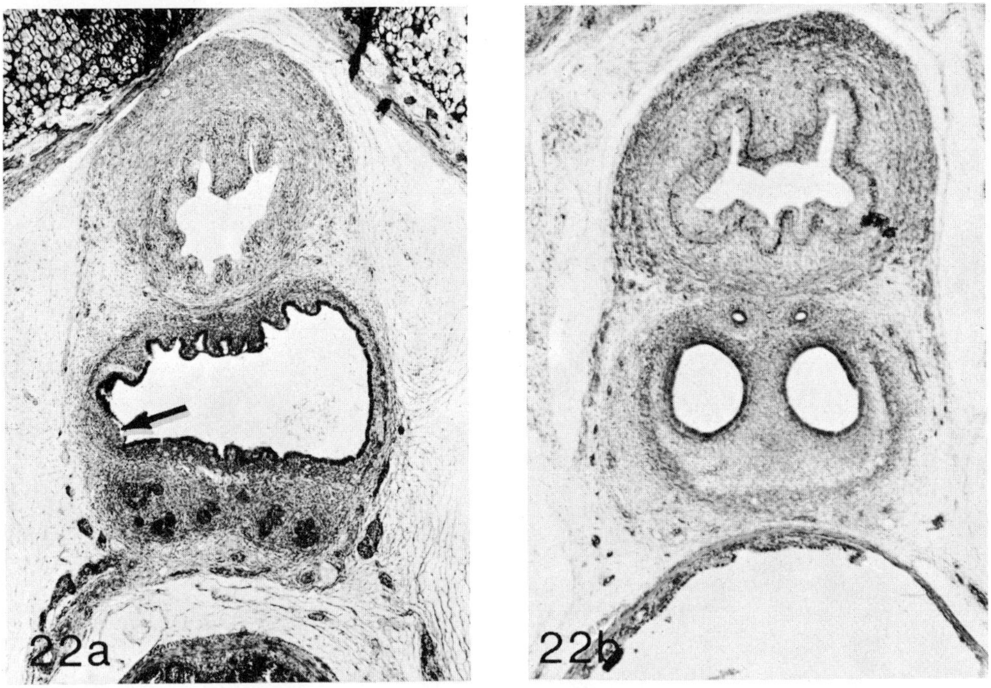

Fig. 22. Paradoxical effects of Cpa in 30-day male rabbit foetuses. *a.* A blind ending Müllerian vagina *(M).*, as found in a male rabbit foetus. The mother had been under the influence of 20 mg Cpa/kg/day i.m.; days 13 to 24 of pregnancy. The Wolffian ducts are rudimentary (arrow). The vesicular gland is well developed in the typical way. *b.* Exceptionally well developed seminal vesicle horns with two ampullar rudiments in a 30-day male rabbit foetus. The mother had been under the influence of 40 mg Cpa/kg/day i.m.; days 13 to 24 of pregnancy. Note: Under this high dose of Cpa and in spite of the well developed Wolffian rudiments, there was paradoxically no vesicular gland developed. From Elger. *Arch. Anat. Micr. Morph. Exp.,* 55: 657-743, 1966.

stimulation of the Wolffian ducts in the anterior portions of the genital cord (see Figs. 23a,b,and c). These effects of Cpa are not dose-related and only occur sporadically.

Because of these paradoxical actions under Cpa it is possible, for example, that in some cases the vesicular glands or Wolffian ducts of male foetuses are less developed than those of female foetuses from the same litter.

CONCLUSIONS

The experiments with Cpa permit certain statements to be made on the part played by androgens in sex differentiation of male foetuses, and they also permit a distinction to be drawn between this and other factors.

By means of supplementary experiments with cyproterone, which does not possess any progestogenic action, and various nonantiandrogenic progestogens, it has been possible to demonstrate which effects of Cpa are due to its antiandrogenic properties and which are progestogen-induced.

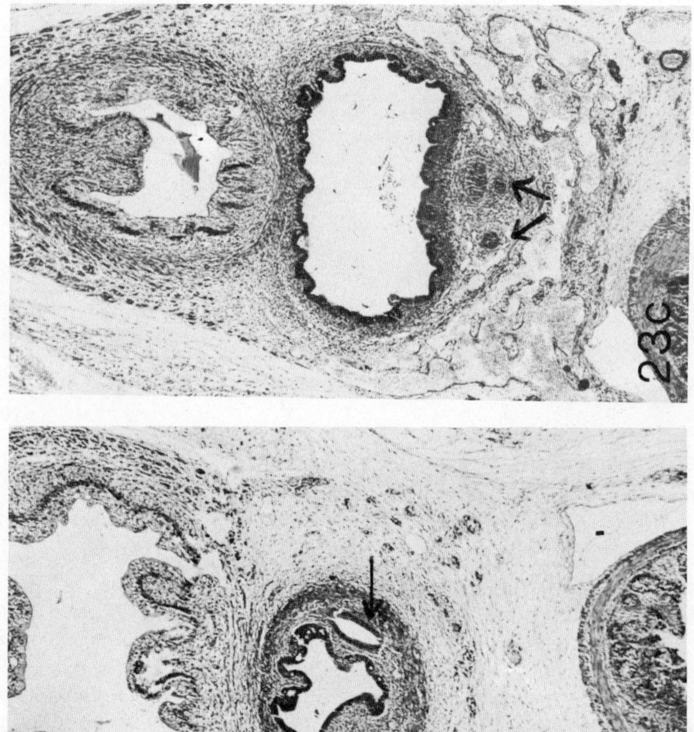

Fig. 23. Paradoxical effects of Cpa in 30-day female rabbit foetuses. *a.* Wolffian ducts (arrows) are maintained in the anterior part pf the genital cord. There are no other signs indicating androgenic actions in this foetus. *b.* Another foetus with identical treatment at a more posterior level: Again, the Wolffian ducts are stabilized (arrow). *c.* A stimulated vesicular gland spreading out in the vicinity of the posterior portion of the vagina.

1. The following proved to be androgen-dependent:
   a. Stabilization and differentiation of the Wolffian ducts (the results in the rat do not necessarily lead to this conclusion).
   b. The development of the vesicular gland and prostate.
   c. The height of the genital cord.
2. The following are not dependent on androgens:
   a. Gonadal differentiation.
   b. The regression of the Müllerian ducts, including the posterior portions.
   c. The descent of the testes.
3. Progestogenic activity is responsible for the stimulation of Müllerian rudiments in the vicinity of the sinus.
4. Where there is stimulation of the Müllerian ducts in the vicinity of the sinus, the presence or absence of a seminal vesicle is the decisive factor in determining further development of a Müllerian vagina in the rabbit.
5. Paradoxical effects with reference to the divergent responses of vesicular gland and seminal vesicle in male rabbit foetuses, and stimulation of the vesicular gland and posterior portions of the Wolffian ducts in female rabbit foetuses, cannot be explained at present.

REFERENCES

1. Hamada, H., Neumann, F., and Junkmann, K. Intrauterine antimaskuline Beienflussung von Rattenfeten durch ein stark gestagen wirksames Steroid. Acta Endocr. (Kobenhavn), 44:380–388, 1963.
2. Neumann, F., Elger, W., and Kramer, M. Development of a vagina in male rats by inhibiting androgen receptors with an anti-androgen during the critical phase of organogenesis. Endocrinology, 78:628–632, 1966.
3. Elger, W. Die Rolle der fetalen Androgene in der Sexualdifferenzierung des Kaninchens und ihre Abgrenzung gegen andere hormonale und somatische Faktoren durch Anwendung eines starken Antiandrogens. Arch. Anat. Micr. Morph. Exp., 55:657–743, 1966.
4. Steinbeck, H., Neumann, F., and Elger, W. Effect of an antiandrogen on the differentiation of the internal genital organs in dogs. J. Reprod. Fertil., in press, 1970.
5. Neumann, F., and Elger, W. The effect of the anti-androgen 1,2α-methylene-6-chloro-$\Delta^{4,6}$-pregnadiene-17α-ol-3,20-dione-17α-acetate (cyproterone acetate) on the development of the mammary glands of male foetal rats. J. Endocr., 36:347–352, 1966.
6. Elger, W., and Neumann, F. The role of androgens in the differentiation of the mammary gland in male mouse fetuses. Proc. Soc. Exp. Biol. Med., 123:637–640, 1966.
7. Neumann, F., and Elger, W. Permanent changes in gonadal function and sexual behavior as a result of early feminization of male rats by treatment with an antiandrogenic steroid. Endokrinologie, 50:209–225, 1966.
8. ——— Hahn, J. D., and Kramer, M. Hemmung von testosteronabhängigen Differenzierungsvorgängen der männlichen Ratte nach der Geburt. Acta Endocr. (Kobenhavn), 54:227–240, 1967.
9. Jost, A. Recherches sur la différenciation sexuelle de l'embryon de lapin: I. Introduction et embryologie génitale normale. Arch. Anat. Micr. Morph. Exp., 36:151–200, 1946/47.
10. ——— Recherches sur la différenciation sexuelle de l'embryon de lapin: II. Action des androgènes de synthèse sur l'histogenèse génitale. Arch. Anat. Micr. Morph. Exp., 36: 242–270, 1946/47.

52

The Incomplete or Delayed Early-Androgen Syndrome

J. J. van der Werff ten Bosch, W. E. Tuinebreijer, and J. Th. M. Vreeburg

Department of Endocrinology, Growth and Reproduction, Rotterdam Medical Faculty, Rotterdam, Netherlands

One of the sex differences that is not immediately apparent upon inspection of members of the two sexes, is the difference in gonadotropin secretion. In experimental animals this difference can be demonstrated after transplantation of gonads. If an ovary is implanted into an adult male rat, the normal cyclic changes within that ovary will stop; after some time such an ovary contains large follicles but no corpora lutea. Removal of the testes at the time of grafting will not change this course. The male pituitary gland probably secretes gonadotropins at a steady rate, while the ovary requires cyclic fluctuations in blood gonadotropin levels. The female pituitary provides such fluctuations.

Classic experiments by Pfeiffer(20) have shown that the sex difference in gonadotropin secretion in the rat is established shortly after birth under the influence of testicular secretions. Some of Pfeiffer's work is summarized in Table 1. Harris and Jacobsohn(11) have demonstrated that the neonatal influence of testes is not exerted at the pituitary but at a higher level, i.e., in the brain. Hypophysectomized adult female rats were found to resume cyclic ovarian activity when pituitary tissue was implanted under the median eminence of the hypothalamus, so that portal vessel connections could be established between the hypothalamus and the implant. The sex (and the age) of the donor of the implant were found to be immaterial.

Over the past decade, it has become clear that "masculinization" of the brain, with regard to the gonadotropin secretion pattern, can be brought about in the female and in the neonatally castrated male rat by a single injection of the androgen testosterone propionate, given between a few days before birth and about ten days after birth (see 10, 6, and 2). The doses used initially were of the order of one mil-

TABLE 1

Summary of some of the findings published by Pfeiffer (20). Note that corpora lutea are absent in animals which were exposed to testicular influences in the neonatal period

Sex of rat	Gonads present in neonatal period	Gonads present in adulthood	Corpora lutea in ovary of adult animal
Male	testis	testis + ovary	absent
	testis	ovary	absent
	ovary	ovary	present
	none	ovary	present
Female	ovary	ovary + testis	present
	ovary + testis	ovary + testis	absent
	testis	ovary + testis	absent
	none	ovary	present

ligram. The "masculinizing" effect of neonatal androgen is not limited to gonadotropin secretion. There occur also changes in the organization of future sexual behaviour patterns (see 10, 12) in running activity(10) and emotional behaviour pattern(9), in general body growth(22), and in the size of nerve cell nuclei and nucleoli in various parts of the brain(19).

The now classic results of such a single perinatal dose of testosterone propionate (TP) in the adult female rat are sterility and relatively small ovaries which contain large follicles but no corpora lutea; they are presumably anovulatory, probably through lack of a cyclic massive discharge of pituitary LH. A great deal of work has been performed to analyze the alterations in such animals on different levels of the hypothalamopituitary-ovarian system(see also 26).

In experiments designed to establish the threshold dose of TP when administered systemically (preliminary to experiments on intracerebral administration), the phenomenon of the incomplete or delayed early androgen syndrome was encountered. The work involved the testing of the effects of variations in the dose and the time of administration, of TP. The effects of these variables were studied in terms of the fertility and the histological aspect of the ovaries, of adult females treated perinatally.

In one experiment different doses were given on postnatal day 5 (day of birthday 1), and the right ovary was removed at the age of 10 weeks or at the age of 21 or 24 weeks (Table 2). It emerged that at 10 weeks there were many ovulatory rats, while the animals examined 11 or 14 weeks later were nearly all anovulatory. All of the animals which had received 5 μg and were examined at 10 weeks were ovulatory, but all but one of those examined at the later time were anovulatory. This dose was clearly a threshold dose when given on day 5. In another experiment rats were given 5 or 10 μg on either day 5 or day 2 (Table 3). One ovary was removed at 10 weeks and the animals were then placed with males to test their fertility. Twelve of 13 animals injected on day 5 had corpora lutea, and 8 of these 12 proved fertile and reared healthy litters. At 21 weeks, 10 of these 12 animals mated again, as shown by the presence of sperm in vaginal smears, but

TABLE 2

Condition of the right ovary of adult female rats which had received a single injection of testosterone propionate on day 5[a]

Dose of TP (μg)	No. of rats	No. of rats with an ovary containing corpora lutea at age indicated (total no. of rats) At 10 weeks	At 21-24 weeks
100	15	3 (6)	0 (9)
50	18	2 (8)	1 (10)
10	21	7 (10)	1 (11)
5	15	8 (8)	1 (7)
	All	20 (32)	3 (37)

[a]Data from Swanson and Van der Werff ten Bosch. Acta Endocr. (Kobenhavn), 45:1-12, 1964.

TABLE 3

Fertility and the condition of the two ovaries, examined at different ages, of adult female rats which had received a single injection of testosterone propionate on day 2 or day 5[a]

Day of injection	Dose of TP (μg)	No. of rats	No. of rats fertile at weeks		No. of rats with corpora lutea in the ovary indicated	
			10	21	Right ovary at 10 weeks	Left ovary at 27 weeks
2	10	8	0	0	0	0
2	5	5	1	0	2	0
5	10	7	4	1	6	2
5	5	6	4	0	6	3

[a]Data from Swanson and Van der Werff ten Bosch. Acta Endocr. (Kobenhavn), 47:37-50, 1964.

only one became pregnant and reared another litter. At 27 weeks the rats were autopsied and only 2 animals now had an ovary that looked normal, whereas the others had "anovulatory" ovaries. As shown in Table 3, the animals injected on day 2 became anovulatory sooner. These findings showed that the "masculinization" caused by a single dose of TP in the female was not an all-or-nothing effect. In current experiments in our laboratory, designed to study changes in ovarian histology and steroid secretion between the ages of 10 and 24 weeks, females have been given 5 μg TP on day 5. The incidence of persistent vaginalestrus has been found to increase steadily from 45 percent at 11 weeks to 88 percent at 19 to 23 weeks (Table 4). This finding demonstrates that many ovulatory animals subsequently become anovulatory.

The conclusion drawn from such results was that it is difficult to study the extent of the critical period for brain-masculinization, since it was clearly impossible to

TABLE 4

The incidence of persistent vaginal estrus in adult female rats after a single injection of 5 μg testosterone propionate on day 5. Smears were taken of all animals at the age of 11 weeks, and again prior to death at a later age[a]

Age at examination	No. of rats	Persistent estrus	
		No.	**Percent**
11 weeks	78	35	45
11-14 weeks	23	12	52
15-18 weeks	22	15	68
19-23 weeks	32	28	88

[a]Unpublished data of J.Th.J. Uilenbroek.

define satisfactorily the endpoint of the response. The age at which ovarian function is studied affects the results. The early-androgen syndrome is clearly a dynamic entity in that it may show up at one time but not at an earlier time. This is further illustrated by an experiment in which pregnant rats received a single injection of 10 or 25 mg TP. All the female young were found to be ovulatory at 8 weeks, but most had become anovulatory at 21 weeks(25).

The mechanisms underlying the initial alterations in the brain wrought by TP, and those involved in the inhibition (or lack of promotion) of ovulation when the animal is sexually mature, are unknown. Do ovulations fail to take place because of an abnormality in the ovaries, or because the ovaries do not receive appropriate stimuli? Ovulations may occur in previously anovulatory ovaries after their transplantation into cycling rats(4); such results have been taken to indicate that in the early-androgen rat the gonadotropic stimuli necessary to induce ovulation, are lacking, and that the brain is responsible for the pituitary failure. This interpretation has derived support from work in which it was shown that ovulations which can be induced by pregnant mare serum (PMS) in prepuberal rats, do not occur if such rats had been androgenized shortly after birth(5). The available evidence indicates that the normal response to PMS requires the release of an ovulation-inducing dose of pituitary LH(6).

If the failure to ovulate following a large dose of neonatal androgen can be attributed to an alteration in the central nervous system at the time of administration of the androgen, then it seems logical to assume that small doses which are followed by a delayed anovulatory state, likewise act through the brain. However, since in animals with the latter condition the central nervous system permitted ovulations prior to the onset of the anovulatory state, the central nervous system must have undergone an alteration at some time between roughly 10 and 20 weeks after the administration of 5 μg TP.

The activity of the central nervous system with regard to its control of gonadotropin secretion is subject to the feedback action of ovarian steroids. It is possible that a neonatal dose of TP alters the sensitivity of the brain to sex steroids.

It has been shown that the uterus and the pituitary gland of adult rats treated shortly after birth with TP take up less of a labeled estrogen than controls(8). The responsiveness to administered estrogen has been observed to be diminished in androgenized females, in comparison with controls, for the vagina(14), for the uterus(28,12,21,17), for pituitary weight(17) and pituitary LH content(21), and for the inhibition of the gonadotropic response to hemi-spaying(18). The widespread occurrence of the fall in estrogen sensitivity suggests that this is a general phenomenon which also embraces central nervous sites of estrogen action. Other data indicate that neonatal androgenization causes a rise in the sensitivity to androgen, at later ages, in the male accessories and in the brain-pituitary complex(16,3).

The onset of puberty of normal individuals is believed to be characterized by a change in the feedback system, the net result being a resetting of the "gonadostat" in such a manner that more sex steroid (in females presumably estrogen) is required for the inhibition of gonadotropin secretion(6). In other words, the central nervous system becomes less sensitive to the negative feedback action of estrogens. If the decrease in estrogen sensitivity in androgenized female rats is brought about by, and at the time of administration of the androgen, it is then to be expected that indications of such a change can be found at and before puberty. It is, in fact, common experience that vaginal opening is accelerated by neonatal androgen, but this might be due to a direct effect of the androgen itself(27). Other findings, however, suggest that ovarian steroid production may be increased. There is the observation that before puberty the uterus is larger in androgenized animals than in controls (5,2) and the finding that the onset of vaginal cornification is accelerated (13). These data seem compatible with the view that the maturation of the feedback system for gonadotropin secretion may be accelerated by early androgen administration. Since most of the evidence concerning the site at which maturation takes place points to the central nervous system(6), it is logical to assume that early androgen administration accelerates maturation of the central nervous system.

At this point it may be fruitful to draw some parallels between early androgen rats, and rats with the anovulatory state caused by different conditions.

Persistence of follicles, without formation of corpora lutea but with persistent vaginal estrus, occurs normally in old rats, i.e., in the course of their second year of life. In the DA strain of rats such a condition sets in much earlier, after a short reproductive span, and can be reversed by the administration of progesterone or by manipulating (reducing) the daily period of illumination(7). The finding that the anovulatory state is reversible, indicates that the abnormality is not primarily localized in the ovaries, but in the mechanisms which control ovarian activity. Aschheim(1) ovariectomized anovulatory "senile" rats and then implanted prepuberal or cyclic ovaries; these failed to ovulate, so that their new environment appears to have been too old. If old rats had been ovariectomized at an earlier age (when they were still cycling) and had lived without ovaries into old age, ovaries then implanted would undergo complete cycles; the donor animals of these grafts were prepuberal or persistently estrous old rats. The implication of these results is clear: the presence of ovaries during the reproductive period of life (possibly including the prepuberal period) seems to cause the brain-pituitary complex to age, in that the brain loses its capacity to support ovarian cyclicity. This view is supported by findings of Kikuyama and Kawashima(15) in rats which were given 10 μg TP per gram body weight (i.e., approximately 150 μg per animal) at the age of 10 days. When such rats were ovariectomized at 140 days, an implanted (ovulatory) ovary would fail to continue to cycle. If the ovariectomy was carried out at 40 days, and an implant made at 140 days, then new corpora lutea would appear.

It is tempting to speculate that an effect of TP given to a female rat at the perinatal time of life, is to accelerate aging of the central nervous system from the point of view of its gonadotropic control of ovulation. Instead of simply "masculinizing" the female brain, the androgen causes the brain to grow prematurely old. With a high dose of androgen, "senility" in terms of ovarian function is reached at or before puberty, while a low dose would stimulate the aging process to a lesser degree and therefore become manifestly effective some time later.

A demonstrable aspect of this "aging" may be the fall in the sensitivity of the brain to estrogens and the rise in its sensitivity to androgens. In fact, this aging process may be a feature common to all tissues which are responsive to sex steroids.

REFERENCES

1. Aschheim, P. Résultats fournis par la greffe hétérochrone des ovaires dans l'étude de la régulation hypothalamo-hypophyso-ovarienne de la ratte sénile. Gerontologia, 10: 65–75, 1964/65.
2. Barraclough, C. A. Modifications in the CNS regulation of reproduction after exposure of prepubertal rats to steroid hormones. Recent Progr. Hormone Res., 22:503–539, 1966.
3. Boyd, R., and Johnson, D. C. Gonadotrophin patterns in male and female rats: Inhibition of LH release by testosterone propionate in animals gonadectomized at puberty. Acta Endocr. (Kobenhavn), 58:600–612, 1968.
4. Bradbury, J. T. Permanent after-effects following masculinization of the infantile female rat. Endocrinology, 28:101–106, 1941.
5. Brown-Grant, K., Quinn, D. L., and Zarrow, M. X. Superovulation in the androgen-treated immature rat. Endocrinology, 74:811–813, 1964.
6. Donovan, B. T., and van der Werff ten Bosch, J. J. Physiology of Puberty. London, Arnold, 1965.
7. Everett, J. W. The mammalian female reproductive cycle and its controlling mechanisms. *In* Sex and Internal Secretions, 3rd ed. Young, W. C., ed. Baltimore, The Williams and Wilkins Co., pp. 497–555, 1961.
8. Flerkó, B., and Mess, B. Reduced oestradiol-binding capacity of androgen sterilized rats. Acta Physiol Acad. Sci. Hung., 33:111–113, 1968.
9. Gray, J. A., Levine, A., and Broadhurst, P. L. Gonadal hormone injections in infancy and adult emotional behaviour. Anim. Behav., 13:33–45, 1965.
10. Harris, G. W. Sex hormones, brain development and brain function. Endocrinology, 75: 627–648, 1964.
11. ——— and Jacobsohn, D. Functional grafts of the anterior pituitary gland. Proc. Roy. Soc. (B), 139:263–276, 1952.
12. ——— and Levine, S. Sexual differentiation of the brain and its experimental control. J. Physiol. (London), 181:379–400, 1965.
13. Jacobsohn, D. Development of female rats injected shortly after birth with testosterone of "anabolic steroids." Acta Endocr. (Kobenhavn), 45:402–414.

14. Kikuyama, S. Response of vaginal epithelium to estrogen in rats given injections of androgen in early postnatal life. Ann. Zool. Jap., 38:64–69, 1965.
15. ——— and Kawashima, S. Formation of corpora lutea in ovarian grafts in ovariectomized adult rats subjected to early postnatal treatment with androgen. Sci. Papers Coll. Gen. Educ. Univ. Tokyo, 16:69–74, 1966.
16. Morrison, R. L., and Johnson, D. C. The effects of androgenization in male rats castrated at birth. J. Endocr., 34:117–123, 1966.
17. Petrusz, P., and Flerkó, B. Effect of ovariectomy and oestrogen administration on the pituitary and uterine weight in androgen-sterilized rats. Acta Biol. Hung., 19:159–162, 1968.
18. ——— and Nagy, E. On the mechanism of sexual differentiation of the hypothalamus. Decreased hypothalamic oestrogen sensitivity in androgen-sterilized female rats. Acta Biol. Hung., 18:21–26, 1967.
19. Pfaff, D. W. Morphological changes in the brains of adult male rats after neonatal castration. J. Endocr., 36:415–416, 1966.
20. Pfeiffer, C. A. Sexual differences of the hypophyses and their determination by the gonads. Amer. J. Anat., 58:195–225, 1936.
21. Van Rees, G. P., and Gans, E. Effect of gonadectomy and oestrogen on pituitary LH-content and organ weights in androgen-sterilized female rats. Acta Endocr. (Kobenhavn), 52:471–477, 1966.
22. Swanson, H. E., and van der Werff ten Bosch, J. J. Sex differences in growth of rats, and their modification by a single injection of testosterone propionate shortly after birth. J. Endocr., 26:197–207, 1963.
23. ——— and van der Werff ten Bosch, J. J. The "early-androgen" syndrome: Its development and the response to hemi-spaying. Acta Endocr. (Kobenhavn), 45:1–12, 1964.
24. ——— and van der Werff ten Bosch, J. J. The "early-androgen" syndrome; Differences in response to pre-natal and post-natal administration of various doses of testosterone propionate in female and male rats. Acta Endocr. (Kobenhavn), 47:37–50, 1964.
25. ——— and van der Werff ten Bosch, J. J. The "early-androgen" syndrome: Effects of pre-natal testosterone propionate. Acta Endocr. (Kobenhavn), 50:379–390, 1965.
26. Szentágothai, J., Flerkó, B. Mess, B., and Halász, B. Hypothalamic Control of the Anterior Pituitary, 3rd ed. Budapest, Akadémiai Kiadó, 1968.
27. Tramezzani, J. H., Voloschin, L. M., and Nallar, R. Effect of a single dose of testosterone propionate on vaginal opening in the rat. Acta Anat., 52:244–251, 1963.
28. Wilson, J. G., Hamilton, J. B., and Young, W. C. Influence of age and presence of the ovaries on reproductive function in rats injected with androgens. Endocrinology, 29: 784–789, 1941.

53

Sexual Behaviour in the Golden Hamster and its Modification by Neonatal Administration of Testosterone Propionate

Heidi H. Swanson and D. A. Crossley

Department of Anatomy, University of Birmingham, Birmingham, England

INTRODUCTION

The sexual differentiation of the genital organs is hormonally controlled. Androgens secreted by the foetal testis act on the structural organization of the male reproductive tract(1-6), as well as that of neural tissues mediating sex-typical patterns of gonadotropin secretion and sexual behaviour in the adult(7-10). This conclusion derives from the observation that masculinization of females results from the injection of testosterone during critical periods of development, while castration of males results in the development of female structures and behaviour patterns(11-18). The period during which target structures are sensitive to endogenous and exogenous androgen lasts about 10 days from when the foetal testis starts to secrete androgens(4). Similar results are obtained with guinea pigs (68 days gestation) treated prenatally with testosterone propionate (TP) as with rats treated within a few days after birth (22 days gestation). Androgen-treated females when ovariectomized as adults and then given estrogen and progesterone show much less sexual response to a male than control females. Moreover they display a degree of male-type behaviour comparable to normal males(7,11,12,14,15,17-21).

The effect of neonatal sex steroid treatment has been studied in hamsters, a species with very short gestation (16 days). Administration of TP to females within a few days after birth prevents the appearance of estrus cycles and corpora lutea (17). In nonsexual situations, females treated neonatally with TP showed behaviour characteristic of the male, i.e., decreased ambulation in the open-field test and more rapid emergence from shelter(22). This work has now been extended to include a study of the sexual behaviour of the hamster, about which little is known apart from the descriptions given by Reed and Reed(23) and Beach and Rabedeau (24).

METHODS

The sequence of treatments and tests are summarized in Table 1. The neonatal treatment consisted of subcutaneous injections of TP or oil vehicle (0.05 ml). Whole litters were injected with the same solution to prevent contamination between pups. After weaning at 3 weeks, the animals were caged in groups of 2 to 5 with littermates of the same sex, and weighed weekly.

At 10 weeks of age all females were examined daily for at least 12 consecutive days for the presence of a postestrous discharge(25). They were then tested on 8 consecutive evenings for female mating behaviour. Males, whose spermatic cords had been ligated, were used to prevent undesired pregnancies. Five TP-treated females with regular vaginal cycles were also mated with intact males to test for fertility. The experimental males (control and TP-treated) were tested for male behaviour with receptive females. After completion of this preliminary series of tests, all subjects were castrated and 3 weeks later retested for sexual behaviour.

A test was then carried out to measure the degree of female behaviour which could be elicited in animals of either sex who had or had not received neonatal

TABLE 1

Treatment and Testing Sequence

Age	Females	Males
48-72 hours	Subcutaneous injection with hormone or oil (No. of animals)	
	Oil (11) 10 μg TP (13) 300 μg TP (12)	Oil (24) 10 μg TP (7) 300 μg TP (7)
Weeks 10	Vaginal examination for estrus cycles	
12	Test for female behaviour with male	Test for male behaviour with female
14	Gonadectomy	
17	Test for residual female behaviour	Test for residual male behaviour
19	Test for induced female behaviour	
	25 μg estradiol benzoate and 0.5 mg progesterone 24 hours later. 6-8 hours later, test with male.	
		Test for male behaviour with female
23	Test for induced male behaviour	
	Testosterone propionate (0.5 mg/animal/day) After 10, 17, and 27 days of treatment test with female	

androgen treatment. Their own gonads having been removed, females and males were treated with a regime of estrogen and progesterone which had been found satisfactory in inducing estrus in normal spayed females(26), and tested with intact males. The reaction of estrogen-progesterone primed males to receptive females was also observed.

Two weeks after the last priming with female hormones, both males and females received 0.5 mg TP daily for 4 weeks. Mating tests for male behaviour only, using receptive females as stimulus objects, were carried out 10, 17, and 27 days after commencement of treatment.

Mating Tests

The room in which the animals were housed was illuminated with white light from 06.00 to 20.00 hours, and with red light during the remaining time. Mating tests were carried out under red light at least one-half hour after the beginning of the dark cycle.

A transparent plastic cage (9× 15× 6 inches) with sawdust on the floor and a flat wire grid roof was used for observation. Three minutes after the male was placed in the mating cage, the female was introduced and timing was begun. The latency to first mount with pelvic thrusts was recorded, and the test continued for a further 10 minutes. The test was discontinued if the male failed to mount within 10 minutes after introduction of the female.

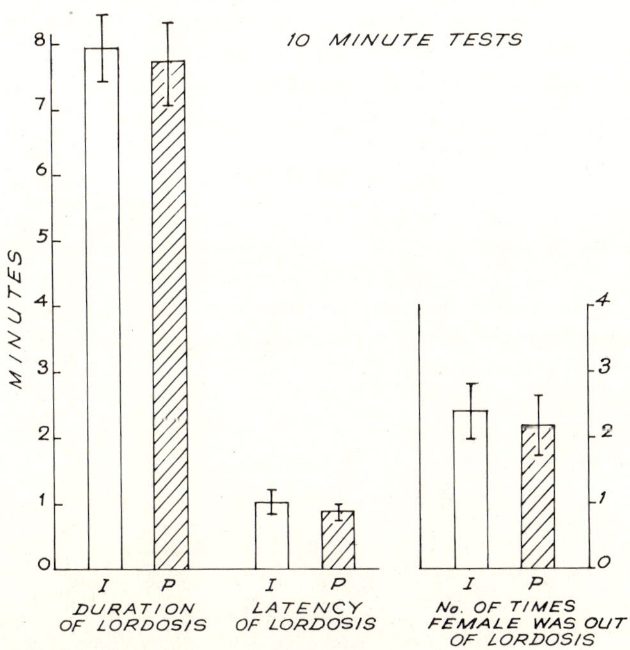

Fig. 1. Comparison of sexual behaviour of intact females in estrus and spayed females primed with 25 μg estradiol benzoate followed by 0.5 mg progesterone 24 hours later. *I*, intact, *P*, primed. The columns represent mean + S.E.M.

The following measures were recorded in each test: (1) Females: latency to lordosis, number of times lordic posture was lost; (2) Males: latency to first mount, intromission and ejaculation; number of mounts without intromission, with intromission and with ejaculation (in 10-minute period after first mount); duration of post ejaculatory intervals. Other behaviour shown by animals of either sex, e.g. sniffing and following, solicitation or rejection by female, aberrant mounts, grooming, displacement behaviour, or fighting was also noted.

Sexually experienced intact or ligated males were used in tests for female behaviour. Preliminary tests had shown that ligation of the spermatic cord did not significantly affect any of the measures of sexual performance (total mounts: df=15, t=1.3472, ns). Ovariectomized females were made receptive for use in tests for male behaviour by the administration of 25 μg estradiol benzoate and 0.5 mg progesterone 24 hours later. Testing began 4 to 8 hours after the progesterone injection. No significant difference was found between intact estrous and ovariectomized estrogen-progesterone primed females on any measure of sexual behaviour (Fig. 1) (latency to lordosis: t=0.5673, ns; duration of longest lordosis: t=0.2663, ns; number of times out of lordosis: t=0.3462, ns, df for all tests =38).

At least one week elapsed between repeated tests on the same experimental subjects as well as between reuse of the same stimulus animal.

Histology

The ovaries of experimental females were fixed in alcoholic Bouin, embedded in paraffin, sectioned at 10 μ and stained with hemotaxylin and eosin. The sections were examined for the presence of corpora lutea.

RESULTS

Normal Mating Behavior

The mating behaviour of hamsters has been described in detail(23,24,27); our data are consistent with previous findings. A typical mating sequence is shown

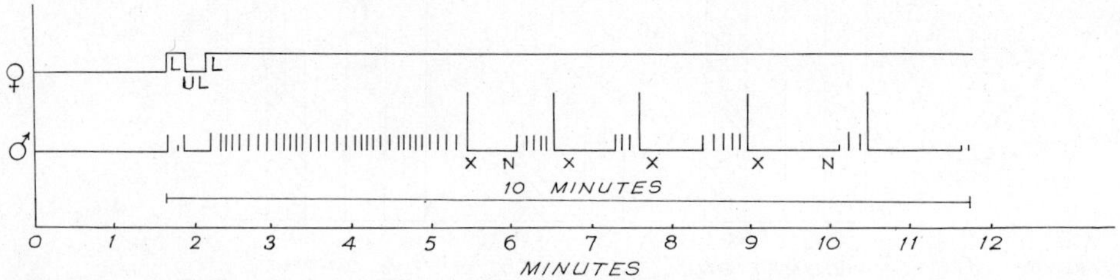

Fig. 2. Typical response of male and receptive female in a 12-minute test. *Female*: *L*, lordosis; *UL*, unlordotic. *Male*: Small bar, mount; intermediate bar, intromission; large bar, ejaculation; *X*, groom; *N*, nuzzle.

in Figure 2. Lordosis is initiated by sniffing and nuzzling by the male and is often held throughout the test period. The readiness of the receptive female to assume the lordotic posture and to hold it even when not being mounted is characteristic of the hamster and in contrast to the rat and guinea pig(8,10,14,18). The male hamster was found to perform 51.6 ± 2.89 total mounts, of which 40.6 ± 3.41 were intromissions, and 4.3 ± 0.45 were ejaculations, in a 10-minute test. A conspicuous decrease in the number of intromissions preceding ejaculation in the second and subsequent copulatory series is evident as compared with the first. There is a trend towards prolongation of the postejaculatory interval as the test proceeds. Analysis of variance showed that latency to first mount decreased (df=2, F=5.66, $P<0.01$) and the total number of mounts increased (df=2, F=7.34, $P<0.01$) on three successive tests, suggesting a beneficial effect of experience.

Effect of Neonatal TP Treatment on Sexual Behavior of Females

Control females had regular 4-day cycles with a copious postestrus discharge (Table 2). On the evening of estrus they were highly receptive, but on other evenings they were completely unreceptive and often aggressive towards their partner. Receptivity was completely abolished after ovariectomy but was fully restored by estrogenprogesterone priming (Fig. 3). Prolonged treatment with testosterone propionate failed to produce male-like behaviour in response to receptive stimulus females (Figs. 3 and 4). Only 2/11 animals mounted on any test; one of these showed the intromission pattern with pelvic thrusts on a single occasion.

The external genitalia were masculinized in females treated with TP as infants. The genitoanal distance was elongated and the clitoris enlarged, but the vaginal orifice did not remain closed (as in rats[28]). Masculinization was more marked in the subjects which had received the larger dose of testosterone. Three females which had received 10 μg and 2 treated with 300=g TP showed a vaginal discharge at 4-day intervals but the discharge was less copious than in controls. (Table 2). Cyclicity persisted until gonadectomy at 14 weeks but fewer corpora lutea were evident in their ovaries. The rest of the TP treated animals had no or a scanty discharge at irregular intervals or on several consecutive days; very few corpora lutea were found in their ovaries.

Sexual responsiveness was correlated with vaginal cyclicity (Fig. 3). TP-treated females with regular 4-day cycles were fully receptive to males when mated on the evening of estrus. Three became pregnant, delivered and reared litters and a

TABLE 2

Vaginal cycles of females treated with oil vehicle (control) or TP neonatally

Neonatal treatment		Total no. of animals	No. of animals with cycles: Regular	Irregular	Absent	Postestrous discharge
Control		11	11	0	0	Copious
TP	10 μg	13	3	5	5	Scanty or absent
	300 μg	12	2	5	5	Scanty or absent

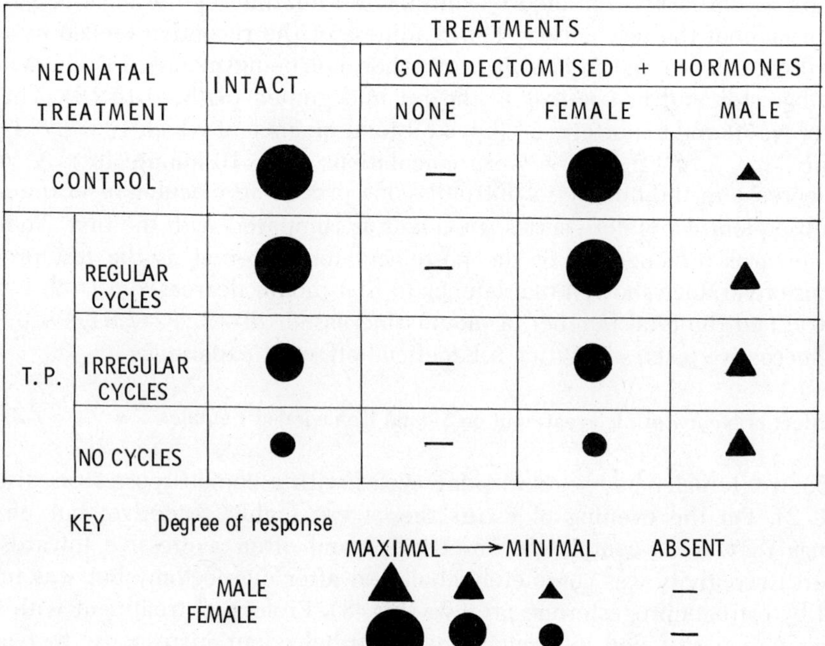

Fig. 3. Mating behaviour of females under various experimental conditions. *Female behaviour*: *Maximal*: Persistent lordosis, elicited promptly by nuzzling or the first mount. *Intermediate*: Persistent lordosis elicited only after vigorous mounting. *Minimal*: Transient lordosis, after vigorous mounting. *Absent*: No lordosis, often aggressiveness. *Male behaviour*: *Maximal*: Mounting, intromission and ejaculation. *Intermediate*: Mounting and intromission: *Minimal*: Precopulatory behaviour (sniffing and following), occasional mounting. *Absent*: no interest in female.

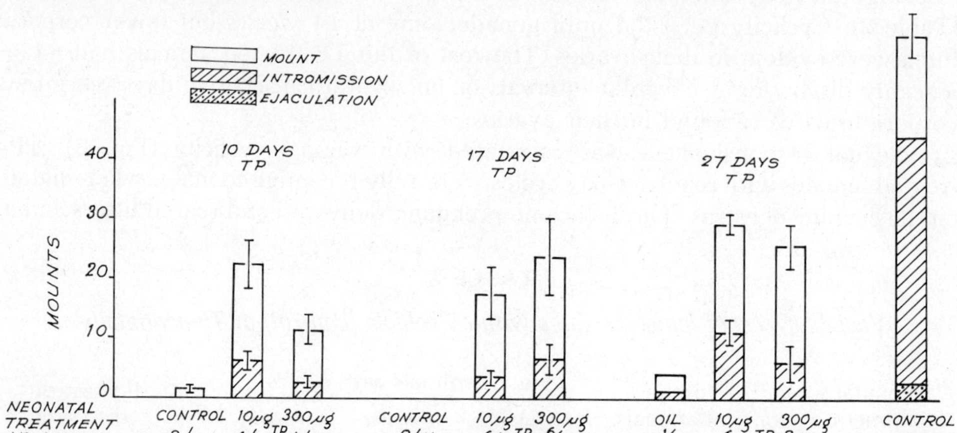

Fig. 4. Male behaviour of females treated with oil vehicle (control) or TP neonatally after 10, 17, and 27 days of TP treatment as adults (0.5 mg/animal/day). The scores are compared to control males. The columns represent mean ± S.E.M.

fourth became pseudopregnant. Subjects with irregular vaginal cycles showed receptivity of low order on several evenings (not 4-day periodicity). Lordosis was only elicited after vigorous mounting and was usually transient. Those with no vaginal discharge showed the least receptivity. None of the animals showing submaximal responsiveness became pregnant or pseudopregnant.

Spaying abolished sexual receptivity. The regime of estrogen and progesterone used to restore full receptivity in controls elicited the same degree of sexual responsiveness in the TP-treated females which they had shown before ovariectomy.

Daily TP treatment as adults induced male-like behaviour in the females which had received TP as infants (Figs. 3 and 4.). When confronted with stimulus females, they sniffed, followed and attempted mounting. Although the mounts were often aberrant and directed at the side or head, many ended in an intromission pattern; the ejaculation pattern was not seen. In contrast to female receptivity, masculine behaviour was not correlated with vaginal cyclicity, since some cyclic subjects showed male behaviour while some acyclic ones did not. The number of mounts and intromissions during adult TP treatment was significantly more than in spayed control females and significantly less than in castrated males on the same TP regime. There was no difference in the number of mounts (df=14, t=0.2021, ns) nor in the number of intromission patterns (df=14, t=1.4048, ns) between females which had received 10 or 300 μg TP neonatally.

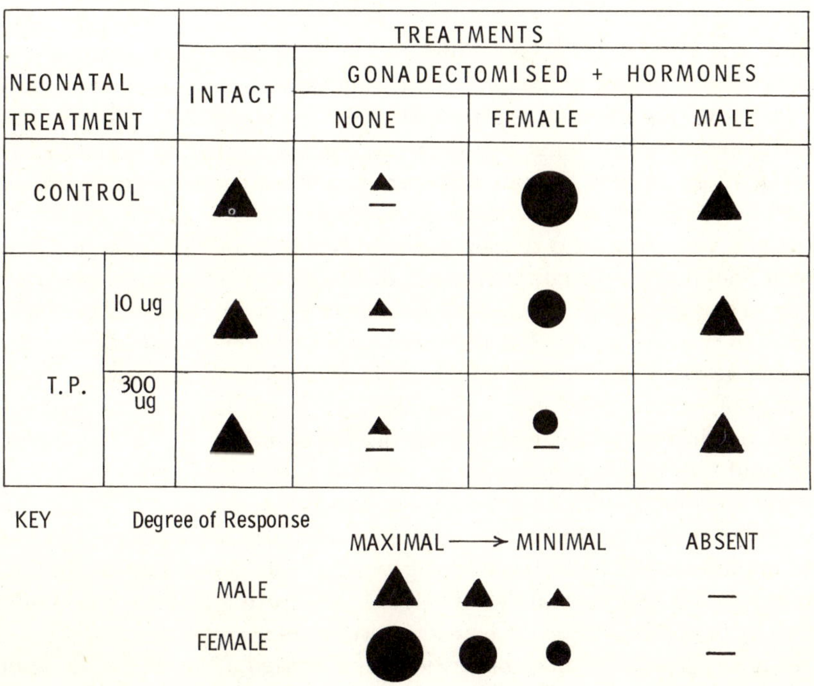

Fig. 5. Mating behaviour of males under various experimental conditions. Key same as for Figure 3.

Effect of Neonatal TP Treatment on Sexual Behavior of Males

All males showed normal mating activity with stimulus females which was virtually abolished by gonadectomy and restored by adult TP replacement therapy. Neonatal TP had no effect on adult sexual behaviour of males.

The most interesting findings were related to the reactions of male hamsters, castrated as adults, to female hormones. Control males treated with estrogen and progesterone showed prompt receptivity to the stimulus male. They adopted the lordotic posture and held it for a prolonged period. The males treated them in all respects as if they were females, although of course penile insertion was impossible. Males treated with TP in infancy showed less female behaviour than control males. The males which had received 10 μg TP showed lordosis on vigorous stimulation like TP-treated females with irregular cycles. Males which had received 300μg TP were nonreceptive and similar to acyclic TP-treated females.

After completion of the test with stimulus males, receptive females were presented to the estrogen-progesterone treated males. All showed some male behaviour—in some subjects this was limited to precopulatory sniffing while about one-third of each treatment group showed mounting and intromission.

DISCUSSION

Male hamsters castrated as adults and given female hormones showed a high degree of female sexual behaviour. Male rats show female behaviour only if they are castrated within a few days after birth(8,13,15,16,18,29). Also it is much more difficult to induce male behaviour in female hamsters than in female rats or guinea pigs(11,12,18,20,30). Spayed hamsters given sufficient testosterone to restore castrated males to full activity show virtually no interest in the stimulus females. In hamsters, therefore, the female pattern is much more easily and completely elicited in males than is male behaviour in females. The administration of as little as 10 μg of exogenous TP decreased responsiveness to female hormones in both sexes, which would suggest that the amount of endogenous androgens secreted by the infant's testis must be very small, and insufficient to abolish the capacity for the display of female behaviour (cf. rat,28). Although the level of hormone secretion may be different, the mechanism of "defeminization" of behaviour during development seems to be similar in the various species studied.

In previous work with hamsters it was found that 300 μg TP injected between 24 and 48 hours after birth was completely effective in abolishing estrus cycles and ovarian luteinization(17). In the present experiments some animals which had received 300 μg TP had normal cycles, mated, and bore a litter, while others which had received only 10 μg showed no estrus cycles and would not mate.

The age of injection may be critical: this varied from 48 to 72 hours after birth. Rats were found to be most susceptible to exogenous androgen between birth and the age of 2 days. Prenatal administration of TP had no effect on the function of the ovaries although the dose was sufficient to cause genital masculinization, and threshold doses of TP given postnatally were more effective the earlier they were given after birth(20,28,31). It seems that also in hamsters, the external genitalia and the neural centers controlling the rhythm of gonadotropin secretion are most susceptible to androgens at different ages. The scanty postestrous discharge seen in cyclic fertile TP-treated females suggests a local effect on the vagina (32); clitoral enlargement and anogenital elongation were not related to cyclicity, but to TP dosage. It has been observed that with relatively low doses of TP, some rats were cyclic and fertile at 10 weeks but by 21 weeks they were no longer fertile and their ovaries did not contain any fresh corpora lutea(28,33). The TP-treated hamsters which had a litter were still cyclic at 14 weeks but their ovaries (extirpated at this age) seemed to contain fewer corpora lutea than controls. This may indicate that cycles were about to stop.

Depression of female sexual behaviour was associated with a disturbance in the rhythm of estrus cycles. The low level of receptivity could not be attributed to a deficiency in ovarian secretion but rather to a decreased responsiveness to female hormones. After spaying and treatment with a regime of estrogen and progesterone sufficient to restore full estrous behaviour in controls, the level of receptivity induced in TP-treated females was similar to that which they displayed spontaneously before ovariectomy. It therefore seems that the neural centers controlling cyclicity and female sexual behaviour are simultaneously influenced by androgens during development. In view of the marked behavioural response to female hormones of male hamsters castrated as adults, experiments are in progress to see whether ovaries implanted into such animals would become luteinized and capable of inducing female behaviour (see note added in proof). In male rats luteinization only occurs if castration is carried out within 1 or 2 days after birth(8,34).

Irrespective of the level of female behaviour which they showed in response to estrogen and progesterone, female hamsters which had received TP in infancy all showed more male behaviour under the influence of TP as adults than control females. This suggests a separate mechanism for the inhibiting action of neonatal androgen on female behaviour and its potentiating action on male behaviour. Male sexual behaviour has an appetitive and a consummatory component. Both may be influenced by the presence of androgens during development. Most control females showed little interest in stimulus females, even after 4 weeks of daily testosterone treatment. Almost all the females treated with TP neonatally (and as adults) responded to the stimulus female by sniffing and following. Their consummatory behaviour, however, was deficient both in quantity and quality in comparison with males. There were fewer mounts and intromissions; many mounts were aberrant and the ejaculation pattern was not seen. This may be attributed to lack of penile development. It has been suggested that the arousal of sexual interest and preliminary mounting must be under control of the CNS but that copulation may be largely a reflex reaction to local sensory feedback(13,16,29,35,36). The centers controlling male and female behaviour may be susceptible to androgens at different ages. Endogenous secretion of androgens in the male hamster may effectively "masculinize" (without "defeminizing") him either because it is secreted for a limited time or because less is needed for a potentiating action on the male than for an inhibiting action on the female center.

This study suggests that endogenous androgen secreted during development may have less influence on the central organization of sexual behaviour in hamsters than in rats or guinea pigs. Although exogenous TP has similar effects in all species studied, it seems that endogenous secretion in hamsters is very low. The unusual responsiveness of adult castrates to female hormones indicates that hormones produced by the infant testis have little or no "defeminizing" effect.

NOTE ADDED IN PROOF

Males castrated at birth and later implanted with ovaries showed lordosis on consecutive days when tested with active males, while those castrated as adults showed lordosis only every fourth day. All ovarian implants were luteinized(37).

REFERENCES

1. Burns, R. K. Role of hormones in the differentiation of sex. *In* Sex and Internal Secretions, 3rd ed. Young, W. C., ed. Baltimore, Williams & Wilkins, pp. 76–158, 1961.
2. Deanesley, R. Foetal endocrinology. Brit. Med. Bull., 17:91–95, 1961.
3. Jost, A. Gonadal hormones in the sex differentiation of the mammalian fetus. *In* Organogenesis. DeHaan, R. L., and Ursprung, H., eds. New York, Holt, Rinehart and Winston, pp. 611–628, 1965.
4. Price, D., and Ortiz, E. The role of fetal androgen in sex differentiation in mammals. *In* Organogenesis. DeHaan, R. L., and Ursprung, H., eds. New York, Holt, Rinehart and Winston, pp. 629–652, 1965.
5. Wells, L. J. Experimental studies of the role of the developing gonads in mammalian sex differentiation. *In* The Ovary. Zuckerman, S., ed. London, Academic Press, Inc., Vol. 2, pp. 131–153, 1962.
6. ——— Fetal hormones and their role in organogenesis. *In* Organogenesis. DeHaan, R. L., and Ursprung, H., eds. New York, Holt, Rinehart and Winston, pp. 673–680, 1965.
7. Harris, G. W., and Levine, S. Sexual differentiation of the brain and its experimental control. J. Physiol., 181:379–400, 1965.
8. ——— Sex hormones, brain development and brain function. Endocrinology, 75:627–647, 1964.
9. Young, W. C. The organization of sexual behaviour by hormonal action during the prenatal and larval periods in vertebrates. *In* Sex and Behaviour. Beach, F. A., ed. New York, John Wiley and Sons, Inc., pp. 89–107, 1965.
10. ——— Goy, R. W., and Phoenix, C. H. Hormones and sexual behavior. Science.143: 212–217, 1964.
11. Phoenix, C. H., Goy, R. W., Gerall, A. A., and Young, W. C. Organizing action of prenatal administered testosterone propionate on the tissues mediating behaviour in the female guinea pig. Endocrinology, 65:369–381, 1959.
12. Gerall, A. A. Hormonal factors influencing masculine behaviour of female guinea pigs. J. Comp. Physiol. Psychol., 62:365–369, 1966.
13. ——— Hendricks, S. E., Johnson, L. L., and Bounds, T. W. Effects of early castration in male rats on adult sexual behaviour. J. Comp. Physiol. Psychol., 64:206–212, 1967.
14. ——— and Ward, I. L. Effects of prenatal exogenous androgen on the sexual behaviour of the female albino rat. J. Comp. Physiol. Psychol., 62:370–375, 1966.
15. Feder, H. H., Phoenix, C. H., and Young, W. C. Suppression of feminine behaviour by administration of testosterone propionate to neonatal rats. J. Endocr., 34:131–132, 1966.
16. Grady, K. L., Phoenix, C. H., and Young, W. C. Role of the developing rat testis in differentiation of the neural tissues mediating mating behaviour. J. Comp. Physiol Psychol., 59:176–182, 1965.

17. Swanson, H. H. Modification of the reproductive tract of hamsters of both sexes by neonatal administration of androgen or oestrogen. J. Endocr., 36:327–328, 1966.
18. Whalen, R. E., and Edwards, D. A. Hormonal determinants of the development of masculine and feminine behaviour in male and female rats. Anat. Rec., 157:173–180, 1967.
19. Barraclough, C. A., and Gorski, R. A. Studies on mating behaviour in the androgen-sterilized female rat in relation to the hypothalamic regulation of sexual behaviour. J. Endocr., 25:175–182, 1962.
20. Kennedy, G. C. Mating behaviour and spontaneous activity in androgen-sterilized female rats. J. Physiol., 172:393–399, 1964.
21. Feder, H. H. Specificity of testosterone and oestradiol in the differentiating neonatal rat. Anat. Rec., 157:79–86, 1967.
22. Swanson, H. H. Alteration of sex-typical behaviour of hamsters in open field and emergence tests by neonatal administration of androgen or oestrogen. Anim, Behav., 15: 209–216, 1967.
23. Reed, C. A., and Reed, R. The copulatory behaviour of the golden hamster. J. Comp. Psychol., 39:7–12, 1946.
24. Beach, F. A., and Rabedeau, R. G. Sexual exhaustion and recovery in the male hamster. J. Comp. Physiol. Psychol., 52:56–61, 1959.
25. Orsini, M. The external vaginal phenomena characterizing the stages of the estrous cycle, pregnancy, pseudopregnancy, lactation and the anestrous hamster, *mesocricetus auratus*. Proc. Anim. Care Panel, 11:193–206, 1961.
26. Frank, A. H., and Fraps, R. M. Induction of estrus in the ovariectomized golden hamster. Endocrinology, 37:357–361, 1945.
27. Beach, F. A., and Pauker, R. S. Effects of castration and subsequent androgen administration upon mating behaviour in the golden hamster (Cricetus Auratus). Endocrinology, 45:211--221, 1949.
28. Swanson, H. H., and van der Werff ten Bosch, J. J. The "early-androgen" syndrome: differences in response to pre-natal and post-natal administration of various doses of testosterone propionate in female and male rats. Acta Endocr. (Kobenhavn), 47:37–50, 1964.
29. Morrison, R. L., and Johnson, D. C. The effects of androgenization in male rats castrated at birth. J. Endocr., 34:117–123, 1966.
30. Beach, F. A. Male and female mating behavior in prepuberally castrated female rats treated with androgens. Endocrinology, 31:673–678.
31. Revesz, C., Keraghan, D., and Bindra, D. Sexual drive of female rats "masculinized" by testosterone during gestation. J. Endocr., 25:549–550, 1963.
32. Takewaki, K. Hormone-independent persistent changes in reproductive organs in female rats induced by early post-natal treatment with androgens. Proc. Japan. Acad. 41: 312–315, 1965.
33. Kikuyama, S., and Kawashima, S. Formation of corpora lutea in ovarian grafts in ovariectomized adult rats subjected to early postnatal treatment with androgens. Scientific Papers of the College of General Education, Univ. of Tokyo, 16:69–74, 1966.
34. Takewaki, K. Some aspects of the normal mechanism involved in persistent ostrus in the rat. Experientia, 18:1–6, 1962.
35. Beach, F. A., and Levinson, G. Effects of androgen on the glans penis and mating behavior of castrated male rats. J. Exp. Zool., 114:159–171, 1950.
36. Adler, N., and Bermant, G. Sexual behaviour in male rats: effects of reduced sensory feed-back. J. Comp. Physiol. Psychol., 61:240–243, 1966.
37. Swanson, H. H. Effects of castration at birth in hamsters of both sexes on luteinization of ovarian implants, oestrus cycles and sexual behaviour. J. Reprod. Fertil., 21:183–186, 1970.

[illegible] 1960.
[illegible]
[illegible]
[illegible] 1964.
[illegible]
[illegible]
[illegible]
[illegible]
[illegible]
[illegible]
[illegible]
[illegible]
[illegible]
[illegible]
[illegible]
[illegible]
[illegible]
[illegible]
[illegible]
[illegible]
[illegible]

54

Changing Pattern of the Uptake of Estradiol by the Anterior Hypothalamus, the Medial Eminence, and the Hypophysis in the Developing Female Rat*,†

Junzo Kato,[1] Noriko Sugimara,[1] and Takashi Kobayashi[2]

[1]Department of Obstetrics and Gynecology, Faculty of Medicine, University of Tokyo, Hongo, Bunkyo, Tokyo, Japan
[2]University of Tokyo and The Red Cross Lying-in Hospital, Tokyo, Japan

INTRODUCTION

There is a body of evidence that estrogen is an important factor in the maturation process of the brain-pituitary unit in developing female rats(1-5). The mechanisms by which estrogen leads to maturation of the brain-pituitary-gonad interrelationships responsible for puberty and the ensuing cyclicity of reproductive function are incompletely understood. A specific estradiol uptake mechanism with limited capacity has been demonstrated in the hypothalamus(6) and in the anterior hypothalamus of adult rats(7,8) with the use of physiological doses of tritiated estradiol. Estrogen concentration in the hypothalamus was also detected with radioautographic procedures(9-11). The binding mechanism may be involved in the steroid on the central nervous system in adult animals. The accumulation of ^{3}H-estradiol in the anterior hypophysis and central nervous system was shown to be similar in 100 g immature and mature female rats, leading to the conclusion that the physiological changes of puberty could not be controlled by changes in the estradiol binding sites(12). However, steroid sensitivity of rat brain is variable during brain maturation(13,14). The pattern of uptake of ^{14}C-estradiol by rat hy-

*Supported by Grant M67-121 from the Population Council, New York, N.Y.
†The authors wish to express their gratitude for the review and comments given by Dr. Dwain D. Hagerman and Dr. Claude A. Villee, and for the suggestions made by Dr. T. Nakayama, Dr. T. Jimbo, and Dr. T. Fujita.

pophysis was shown to be different before and after puberty(15). Woolley et al. (16) and Presl et al.(17) have reported changes in ^{3}H-estradiol distribution in brain, pituitary, and plasma with development in 4 to 15-day-old rats and in intact female rats.

In an attempt to elucidate the development of the estradiol uptake mechanism in the hypothalamus, the pattern of distribution of radioactive estradiol in various parts of the brain and hypophysis was studied with time in the young female rats from neonatal to postpuberal stages.

MATERIALS AND METHODS

Female rats of the Wistar strain (Animal Breeding Laboratories, Ohmiya, Saitama) were used in all the experiments. The rats at 3, 7, 12, 25, and 45 days of age weighed 5.5 to 6.7, 8.8 to 11.0, 21 to 24, 38 to 43, and 100 to 122 g, respectively. Rats at 45 days of age were examined for vaginal opening by smear each morning, and only animals in diestrus were used. Vaginal opening was 38.3 ± 0.7, mean ± standard error, (n=29), (33 to 42) days of age.

The animals were fasted for 12 hours, and then injected intraperitoneally with 0.013 μg of 17 β-estradiol-6, 7-^{3}H per 10 g of body weight. The steroid was dissolved in 0.1 ml of physiological saline containing 10 percent ethanol for the animals at 3 and 7 days of age, in 0.2 ml of the same solvent for the animals at 12 days of age, and in 0.3 ml of the same solvent for the animals at 25 and 45 days of age. After an appropriate interval, the animals were killed by decapitation.

Dissection of the Brain. The anterior, middle, and posterior hypothalamus, the cortex cerebri, the cerebellum, the median eminence, and the anterior hypophysis were dissected as described previously(7,8), and the tissues were weighed on an analytical balance. The whole cerebellum or the left half cerebellum was used at 3, 7, and 12 days of age and at 25 and 45 days of age. Muscle was taken from the thigh. Blood was taken in a heparinized tube and the plasma was separated by centrifuging.

Extraction of the Radioactivity. Each tissue was homogenized in 5 ml of ethano:acetone, 1:1, volume/volume for 2 minutes in a glass homogenizer and the residue then was extracted at least three more times with at least 5 ml of the same solvent by vigorous shaking in a stoppered centrifuge tube for at least 10 minutes. The extracts were combined and dried in a scintillation counting vial at a temperature less than 55°C.

Chemicals. Redistilled and purified solvents were used in all experiments. Estradiol-6, 7-^{3}H was obtained from New England Nuclear Corporation, Boston, Mass., and tested for impurities by thin-layer chromatography in two different systems. The steroids used as standards were supplied by Steraloids, Inc., Pawling, N.Y.; Sigma Chemical Co., St. Louis, Mo.; and Ikapharm, Israel.

Chromatography. Silica gel GF_{254} (Merck) was used. The thin-layer chromatography was carried out in benzene:ethanol, 9:1, volume/volume and 4 percent methanol in methylene chloride.

Measurement of Radioactivity. Tritium was measured in an Aloka model scintillation counter (Nihon Musen Co., Mitaka, Tokyo) with a scintillation fluid containing 4 g of 2, 5-diphenyloxazole and 0.4 g of 1, 4-bis-2-(4 methyl-5-phenyloxazole) benzene in 1 litter of toluene:ethanol, 98:2, volume/volume. Sufficient counts were accumulated to reduce the probable error of counting to less than 1 percent. Quenching, if present, was corrected for by the external standard ratio method. The tissue residue after extraction with organic solvents was dissolved in 0.5 ml of hydroxide of Hyamine 10-X ([p-diisobutyl-cresoxyethyl] dimethyl benzyl ammonium hydroxide) (Packard Instrument Co.) with constant shaking at a temperature of 37°C for 12 hours. Its radioactivity was measured using the same scintillation liquid. All counts were corrected for quenching by the external standard ratio method. The radioactivity remaining in the residue after extraction of the brain and hypophysis or plasma was less than 1 percent or 6 percent respectively of the initial radioactivity of the tissue. Results for all tissues for the time course (alphabetical superscript in tables), and the brain tissues, muscle, and plasma for each time interval (numerical superscript in tables) were subjected to analysis of variance, Duncan's multiple range test(18) as modified by Kramer(19), respectively, and for the anterior hypophysis and the median eminence at each time interval in the tables, the *t* test.

RESULTS

Changes in the Distribution of the Radioactivity in the Brain and Hypophysis at Various Times after Injection into the Rat at 3 Days of Age

The distribution of radioactivity in various parts of the brain and hypophysis in the 3-day-old rat was examined 1/2, 1, 2, and 4 hours after the injection of tritiated estradiol (Table 1). There were no differences in the concentration of radioactivity between any of the several pairs of combinations of the anterior, middle, and posterior hypothalamus, the cortex cerebri, and the cerebellum in the respective hourly groups. The median eminence also incorporated as much as radioactivity as the cortex cerebri. Thus, no differentiation was found in the brain tissue in terms of accumulation of radioactivity. The hypophysis contained three to four times as much radioactivity as the cortex cerebri ($P<0.005$).

The concentration of radioactivity in the brain and hypophysis was compared with respect to time (Table 1). Following the injection of labeled estradiol, the concentration of radioactivity in the anterior and posterior hypothalamus, the median eminence and the hypophysis tended to increase until 2 hours and then remained at almost the same level at 4 hours. The pattern of the time course in the middle hypothalamus and the cortex cerebri was similar to that in the anterior hypothalamus but an increase in the concentration of radioactivity found at 1 hour after the injection. In the cerebellum and the muscle the concentration of radioactivity increased gradually until 2 hours and was maintained for another 2 hours. The level of the concentration of radioactivity in blood remained unchanged up to 4 hours after the injection.

Distribution of Radioactivity in the Brain and Hypophysis at 7 Days of Age with Graded Dosages of Tritiated Estradiol

One hour after the injection of 0.13μg of tritiated estradiol per 10 g of body weight into the 7-day-old rat, the distribution of radioactivity was basically similar

TABLE 1

*Time course of the distribution of radioactivity in the brain and hypophysis in the 3-day-old rat following the injection of tritiated estradiol**

Tissue	½ hour		1 hour		2 hours		4 hours	
	dpm/mg wet wt	t/c†	dpm/mg wet wt	t/c	dpm/mg wet wt	t/c	dpm/mg wet wt	t/c
Anterior hypothalamus	74.1 ± 13.1‡ ^{1,2,b} (4)§	1.2	105.4 ± 11.5 ^{2,a,b} (8)	0.8	146.9 ± 16.9 ^{2,a} (7)	0.9	132.6 ± 15.1 ^{2,a} (8)	1.0
Middle hypothalamus	65.5 ± 7.4 ^{1,2,b} (4)	1.0	113.8 ± 11.5 ^{1,2,a} (8)	0.9	145.4 ± 17.0 ^{2,a} (7)	0.9	144.2 ± 11.1 ^{2,a} (6)	1.1
Posterior hypothalamus	91.4 ± 15.6 ^{1,2,b} (4)	1.4	126.3 ± 13.5 ^{1,2,a,b} (8)	1.0	160.6 ± 19.0 ^{2,a} (7)	1.0	139.3 ± 11.3 ^{2,a} (8)	1.0
Cerebral cortex	64.1 ± 8.0 ^{1,2,b} (4)	1.0	131.2 ± 11.4 ^{1,2,a} (8)	1.0	165.1 ± 20.6 ^{2,a} (7)	1.0	135.9 ± 10.8 ^{2,a} (8)	1.0
Cerebellum	74.1 ± 11.9 ^{1,2,b} (3)	1.2	124.5 ± 10.0 ^{1,2,b} (7)	1.0	176.6 ± 21.8 ^{2,a} (6)	1.1	145.6 ± 16.1 ^{2,a} (6)	1.1
Median eminence	88.4 ± 3.3 ^{b} (4)	1.4	133.3 ± 21.8 ^{a,b} (7)	1.0	216.8 ± 29.9 ^{a} (6)	1.3	215.0 ± 40.3 ^{a} (8)	1.6
Hypophysis	252.0 ± 39.0 ^{b} (4)	3.9	420.5 ± 57.8 ^{a,b} (8)	3.2	645.2 ± 133.0 ^{a} (6)	3.9	444.6 ± 54.6 ^{a} (8)	3.3
Muscle	100.8 ± 9.4 ^{1,b} (4)	1.6	160.4 ± 31.3 ^{1,b} (8)	1.2	317.7 ± 28.1 ^{1,a} (7)	1.9	247.9 ± 38.7 ^{1,a} (8)	1.8
Whole blood§§	835.8 ± 113.1 ^{a} (4)	13.0	1126.2 ± 137.4 ^{a} (6)	8.6	1373.2 ± 328.6 ^{a} (6)	8.3	1180.8 ± 171.1 ^{a} (5)	8.7

*0.013 μg of estradiol-6,7-^{3}H per 10 g of body weight (42 c per mmole) was injected intraperitoneally into 3-day-old rats. The animals were killed at ½, 1, 2, and 4 hours after the injection.

†Ratio of the concentration of the radioactivity between the tissue and the cortex cerebri.

‡Mean ± standard error.

§Number of animals.

§§Dpm/μliter. Ten μliter of whole blood was dissolved into 0.5 ml of hydroxide of Hyamine 10-X and its radioactivity was measured using the scintillation liquid.

^{1,2,a,b}The means having the same superscripts are not significantly different ($\underline{P} > 0.05$) from each other. The numerical and alphabetical figures are used for each hour group and tissue, respectively.

TABLE 2

*Distribution of radioactivity in the brain and hypophysis in the 7-day-old rat with graded dosages of tritiated estradiol**

Tissue	Tritiated estradiol administered (per 10 g of body weight)				
	0.013 μg	**0.026 μg**	**0.052 μg**	**0.104 μg**	**0.208 μg**
Anterior hypothalamus	76.7†	125.5	210.0	666.0	1055.9
Middle hypothalamus	62.7	129.2	213.4	676.6	1217.6
Posterior hypothalamus	74.3	171.2	268.3	782.0	1162.4
Cortex cerebri	65.3	168.8	277.1	947.2	1281.4
Cerebellum	85.6	183.8	288.2	1018.7	1475.5
Median eminence‡	107.5	192.6	343.7	974.4	—
Hypophysis	195.5	485.7	1086.3	2746.0	3183.8
Liver	701.7	1334.5	2823.0	5660.0	8257.6

*0.013 to 0.208 μg of estradiol-6,7-^{3}H per 10 g of body weight (42 c per mmole) was injected intraperitoneally into the 7-day-old female rat. The animal was killed by decapitation 1 hour after the injection.

†Mean dpm per mg of wet weight of the tissue from two samples, except that one sample was determined in a dose of 0.208 μg of tritiated estradiol per 10 g.

‡Pooled from two animals.

to that at 3 days of age, although lower concentrations were found (Table 2). These animals were also examined to determine whether or not the brain and hypophysis could be saturated with increasing amounts of tritiated estradiol. The pattern of distribution of radioactivity in the brain and hypophysis was virtually identical with five graded dosages of tritiated estradiol: 0.013, 0.026, 0.052, 0.104, and 0.208 μg 10 g of body weight. Moreover, the incorporation of radioactivity into the brain and hypophysis was found to be approximately proportional to the dose injected. The data thus suggest that no saturation of the uptake of radioactive material by the brain and hypophysis was possible in the range of doses used at this age.

Changes in the Distribution of the Radioactivity in the Brain and Hypophysis at Various Times after Injection into 12-Day-Old Rats

The distribution of radioactivity in various parts of the brain and hypophysis in 12-day-old rats was examined 1/2, 1, 2, and 4 hours after the injection of ^{3}H-estradiol (Table 3). There was no evidence that the hypothalamus could accumulate more radioactivity than the other brain tissue at this age. However, the anterior hypophysis incorporated seven to ten times as much radioactivity as the cortex cerebri ($P<0.005$), and seven to nine times as much radioactivity as plasma ($P<0.005$). Moreover, it was noted that the pattern of distribution of radioactivity in the brain and hypophysis was almost identical at all of the times studied. In the median eminence accumulation of radioactivity was found in comparison to the middle and posterior hypothalamus and the cerebellum at one hour ($P<0.025$, $P<0.05$, and $P<0.05$, respectively), the middle and posterior hypothalamus, the cortex cerebri, and the cerebellum at 2 hours ($P<0.025$, $P<0.025$, $P<0.025$, and $P<0.01$, respectively) and the middle hypothalamus and the cerebellum at 4 hours ($P<0.05$ and $P<0.025$).

TABLE 3

*Time course of the distribution of radioactivity in the brain and hypophysis in the 12-day-old female rat following the injection of tritiated estradiol**

Tissue	½ hour		1 hour		2 hours		4 hours	
	dpm/mg (5)[†] wet wt	t/c[‡]	dpm/mg (5) wet wt	t/c	dpm/mg (7) wet wt	t/c	dpm/mg (7) wet wt	t/c
Anterior hypothalamus	75.2 ± 8.0[§][1,a]	1.0	63.1 ± 8.4[1,a,b]	1.1	69.5 ± 6.5[1,a]	1.1	46.2 ± 7.4[1,b]	1.1
Middle hypothalamus	66.3 ± 7.3[1,a]	0.9	56.7 ± 2.8[1,a,b]	0.9	55.7 ± 9.1[1,a,b]	0.9	38.5 ± 7.3[1,b]	0.9
Posterior hypothalamus	71.6 ± 8.7[1,a]	1.0	55.8 ± 4.9[1,a,b]	0.9	58.4 ± 6.2[1,a,b]	1.0	41.8 ± 6.2[1,b]	1.0
Cortex cerebri	75.7 ± 8.6[1,a]	1.0	60.3 ± 4.2[1,a]	1.0	61.1 ± 6.3[1,a]	1.0	40.8 ± 6.8[1,b]	1.0
Cerebellum	71.8 ± 7.8[1,a]	1.0	57.3 ± 3.4[1,a]	1.0	58.4 ± 5.9[1,a]	1.0	36.9 ± 6.2[1,b]	0.9
Median eminence	88.1 ± 10.3[a]	1.2	72.0 ± 5.2[a,b]	1.2	84.2 ± 6.4[a]	1.4	62.0 ± 9.0[b]	1.5
Anterior hypophysis	668.7 ± 59.3[a]	8.8	405.1 ± 43.8[b]	6.7	468.5 ± 43.0[b]	7.7	417.1 ± 49.2[b]	10.2
Muscle	186.3 ± 12.1[a]	2.5	136.5 ± 9.5[b]	2.3	134.5 ± 20.7[b]	2.2	107.0 ± 12.8[b]	2.6
Plasma[§§]	84.1 ± 5.8[1,a]	1.1	59.4 ± 3.6[1,b]	1.0	52.7 ± 2.3[1,b]	0.9	48.2 ± 9.2[1,b]	1.2

*0.013 μg of estradiol-6,7-^{3}H per 10 g of body weight (42 c per mmole) was injected intraperitoneally into the 12-day-old female rat. The animals were killed by decapitation at ½, 1, 2, and 4 hours after the injection.

†Number of animals.

‡Ratio of the concentration of radioactivity between the tissue and the cortex cerebri.

§Mean ± standard error.

§§Dpm/μ liter.

1,a,bThe means having the same superscripts are not significantly ($\underline{P} > 0.05$) different from each other. The numerical and alphabetical figures are used for each hour group and tissue, respectively.

When the concentration of radioactivity in the brain and hypophysis in the rat at 12 days of age was compared with respect to time, the three parts of the hypothalamus showed no decrease up to 2 hours, declining thereafter. The time pattern of the median eminence, the cerebral cortex, and the cerebellum resembled that of the hypothalamus. In plasma, the concentration of radioactivity remained at the same level after an initial increase one-half hour after the injection. Maintenance of radioactivity in the brain might therefore be ascribed to radioactive material in plasma. In the anterior hypophysis the concentration of radioactivity was constant from 1 up to 4 hours after the injection. The ratio of the concentration of radioactivity in the hypophysis to the cortex cerebri rose gradually up to 4 hours after an initial drop in the ratio at 1 hour.

Changes in the Distribution of Radioactivity in the Brain and Hypophysis at Various Times after Injection into 25-Day-Old Rats

The radioactivity present in various parts of the brain and the hypophysis at 1/2, 1, 4, and 8 hours after the injection of tritiated estradiol into the 25-day-old rat is summarized in Table 4. At one-half hour the anterior hypophysis contained much more radioactivity than the tissues of the brain, muscle, and plasma. The radioactivity in the anterior hypothalamus was significantly higher than that in the other parts of the hypothalamus and in the cerebral cortex, cerebellum, and muscle. The concentration of radioactivity in the anterior hypothalamus was 1.9 times that in the cerebral cortex. There was no difference in the concentration of radioactivity present in the middle and posterior hypothalamus, the cerebral cortex, the cerebellum, and muscle. The median eminence collected 2.5 times as much radioactivity as the cerebral cortex ($P<0.005$) and contained a greater amounts of radioactivity than the anterior hypothalamus ($P<0.05$).

One hour after the injection, the concentration of radioactivity in the anterior hypothalamus was greater than the middle and posterior hypothalamus, the cerebral cortex, the cerebellum, and the muscle. The concentration in the anterior hypothalamus was four times that in the cerebral cortex. A preferential accumulation of radioactivity in the anterior hypothalamus was clearly observed. In addition, the middle hypothalamus had higher mean value of radioactivity than the cerebral cortex. The concentration of radioactivity in the anterior hypophysis was 45 and 10 times greater than in the cerebral cortex and plasma, respectively ($P<0.005$ and $P<0.005$). At this time the median eminence also incorporated greater amounts of radioactivity than the middle and posterior hypothalamus, the cortex cerebri, and the cerebellum ($P<0.025$, $P<0.005$, and $P<0.005$, respectively).

The pattern of distribution of radioactivity at 2 hours was similar to that in the 1-hour group but the differences became more pronounced. The concentration of radioactivity in the anterior hypothalamus was greater than in other brain tissue and was 5.4 times that in the cerebral cortex. The concentration of radioactivity in the anterior hypophysis at this time was 93 times and 14 times that in the cerebral cortex and plasma, respectively. A preferential accumulation in the median eminence was evident ($P<0.025$ for the middle and posterior hypothalamus, $P<0.01$ for the cortex cerebri and the cerebellum).

The concentration of radioactivity in all tissues tended to decrease at 4 hours, but the anterior hypothalamus again contained more radioactivity than the other brain tissues except for the median eminence. The ratio of the concentration in the anterior hypothalamus to the cerebral cortex was still 4.9 times that in the cerebral cortex. The ratio of the radioactivity in the median eminence or in the anterior hypophysis to that in the cortex cerebri was also great.

Eight hours after the injection, the concentration radioactivity in all tissues was greatly decreased, but a distribution of radioactivity similar to that seen earlier was found in the brain and the hypophysis. The anterior hypothalamus contained more radioactivity than the cortex cerebri and the cerebellum. The ratio of radioactivity in the anterior hypophysis to the cerebral cortex or plasma was still high even at this time.

The ability of the tissues to retain the radioactive material taken up at 25 days of age is shown in Table 4 and Figure 1. The concentration of radioactivity in the anterior hypophysis was retained for 2 hours, gradually decreasing thereafter. The median eminence showed maintenance of radioactivity up to 4 hours after an initial decrease at 1 hour. The ratios of the concentration of radioactivity in the anterior hypophysis or the median eminence to the cortex cerebri increased with time up to 2 hours after the injection. Thus the anterior hypophysis continued to incorporate radioactivity and retain it for a longer time than did the other brain tissues. The median eminence tended to show a similar ability, although in a lesser degree. In contrast, the mean value for the cortex cerebri and cerebellum decreased suc-

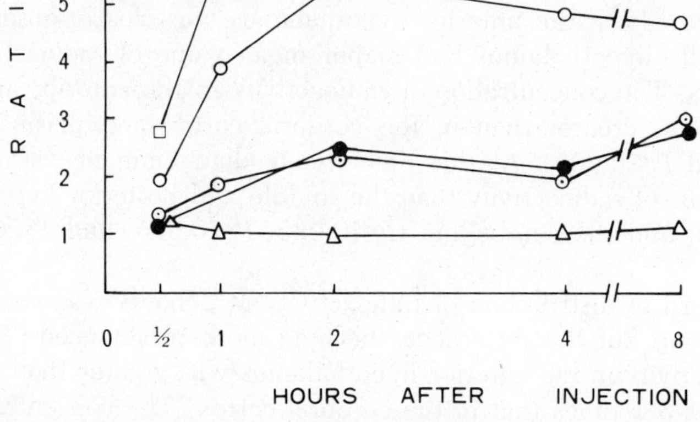

Fig. 1. Ratio of concentrations of radioactivity at various times after injecting tritiated estradiol in female rats at 25 days of age. The value for the cerebral cortex was taken as 1.0 and used for comparison with other parts of brain. ○ =Anterior hypothalamus; ⊙ =middle hypothalamus; ● =posterior hypothalamus; △=cerebellum; □=median eminence.

TABLE 4

*Time course of the distribution of radioactivity in the brain and hypophysis in the 25-day-old female rat following the injection of tritiated estradiol**

Tissue	½ hour		1 hour		2 hours		4 hours		8 hours	
	dpm/mg† wet wt	t/c§§	dpm/mg wet wt	t/c	dpm/mg wet wt	t/c	dpm/mg wet wt	t/c	dpm/mg wet wt	t/c
Anterior hypothalamus	59.2 ± $5.3^{2,a}$ (6)§	1.9	32.9 ± $3.2^{1,a}$ (9)	4.0	22.1 ± $2.7^{1,b}$ (7)	5.4	17.5 ± $1.8^{1,b,c}$ (7)	4.9	8.5 ± $3.4^{1,2,c}$ (3)	4.7
Middle hypothalamus	41.6 ± $3.0^{3,a}$ (6)	1.4	15.6 ± $1.6^{2,b}$ (9)	1.9	9.9 ± $1.2^{2,c}$ (7)	2.4	7.1 ± $0.9^{2,c}$ (7)	2.0	5.8 ± $2.2^{2,3,c}$ (3)	3.2
Posterior hypothalamus	36.6 ± $4.1^{3,a}$ (6)	1.2	15.1 ± $2.3^{2,3,b}$ (9)	1.8	10.1 ± $1.3^{2,b}$ (7)	2.5	7.8 ± $0.8^{2,b}$ (7)	2.2	5.6 ± $0.8^{2,3,b}$ (3)	3.1
Cortex cerebri	30.7 ± $2.7^{3,a}$ (6)	1.0	8.3 ± $0.7^{3,b}$ (9)	1.0	4.1 ± $0.5^{3,c}$ (7)	1.0	3.6 ± $0.4^{3,c}$ (7)	1.0	1.8 ± $0.1^{3,c}$ (3)	1.0
Cerebellum	35.7 ± $3.4^{3,a}$ (6)	1.2	9.0 ± $0.8^{2,3,b}$ (9)	1.1	3.9 ± $0.5^{3,c}$ (7)	1.0	3.8 ± $0.5^{3,c}$ (7)	1.1	2.1 ± $0.3^{3,c}$ (3)	1.2
Median eminence	77.9 ± 5.9^{a} (6)	2.5	48.4 ± 12.1^{b} (8)	5.8	32.7 ± $8.4^{b,c}$ (7)	8.0	22.9 ± $3.8^{b,c}$ (6)	6.4	12.1 ± 4.5^{c} (3)	6.7
Anterior hypophysis	469.3 ± 82.9^{a} (6)	15.3	372.5 ± 42.7^{a} (8)	44.9	379.6 ± 33.9^{a} (7)	92.6	223.5 ± 28.8^{b} (7)	62.1	89.6 ± 9.4^{b} (3)	49.8
Muscle	27.1 ± $1.7^{3,a}$ (6)	0.9	13.4 ± $2.5^{2,3,b}$ (9)	1.6	6.9 ± $0.5^{2,3,c}$ (7)	1.7	5.4 ± $0.7^{2,3,d}$ (7)	1.5	2.8 ± $0.3^{3,e}$ (3)	1.6
Plasma‡	107.9 ± $11.0^{1,a}$ (6)	3.5	38.2 ± $3.5^{1,b}$ (9)	4.6	26.6 ± $2.3^{1,b,c}$ (7)	6.5	16.1 ± $1.7^{1,c}$ (7)	4.5	11.8 ± $0.5^{1,c}$ (3)	6.6

*0.013 μg of estradiol-6,7-^{3}H per 10 g of body weight (42 c per mmole) was injected intraperitoneally into 25-day-old female rats. The animals were killed by decapitation at ½, 1, 2, 4, and 8 hours after the injection.

†Mean ± standard error.

‡Dpm/μ liter.

§Number of animals.

§§Ratio of the concentration of the radioactivity between the tissue and the cortex cerebri.

1,2,3,a,b,c,d,e The means having the same superscripts are not significantly ($P \geq 0.05$) different from each other. The numerical and alphabetical figures are used for each hour group and tissue, respectively.

cessively during the time course up to 2 hours. While the pattern in the middle hypothalamus resembled that in the cortex cerebri, the anterior hypothalamus retained radioactivity up to 1 hour, followed by no successive reduction after 2 hours. A great reduction in the concentration of radioactivity in plasma was found at 1 hour with a further decrease at 4 hours. It was noted that only muscle showed a gradual decrease during the entire course of time. The ratio of the anterior hypothalamus to the cortex cerebri increased with time in a way similar to the anterior hypophysis and the median eminence. The two parts of the hypothalamus showed a lesser rise and the cerebellum showed no rise in the ratio at all.

Changes in the Distribution of Radioactivity in the Brain and Hypophysis at Various Times after Injection into the Rat at 45 Days of Age

The distribution of radioactivity in the brain and the hypophysis was examined at 1/2, 1, 2, and 4 hours after the injection of tritiated estradiol into the 45-day-old female rat (Table 5). One-half hour after the injection the anterior and middle hypothalamus contained a greater concentration of radioactivity than the cortex cerebri, the cerebellum, and muscle. There was no difference in the concentration of radioactivity between the anterior and middle hypothalamus and the posterior hypothalamus. Much more radioactivity was accumulated in the median eminence than in the cortex cerebri, the cerebellum, and the middle and posterior hypothalamus ($P<0.005$, $P<0.005$, $P<0.005$, and $P<0.005$, respectively). The anterior hypophysis incorporated 12 and 2.4 times as much radioactivity as the cortex cerebri and plasma ($P<0.005$ and $P<0.005$) respectively.

At 1 hour after the injection, the anterior hypothalamus took up more radioactive material than the middle and posterior hypothalamus and the other brain tissue. Differentiation in the uptake of radioactivity by the hypothalamus became evident at this time. There was no difference in the concentration of radioactivity ($P<0.005$ and $P<0.005$), the cortex cerebri, the cerebellum ($P<0.005$, $P<0.005$), and plasma ($P<0.025$). The ratio of the concentration of radioactivity of the anterior hypophysis to the cortex cerebri or plasma was 71.4 or 16.4. Thus at 2 hours between any of the pairs of combinations of the middle and the posterior hypothalamus, the cortex cerebri, the cerebellum, and muscle. The median eminence, the value for which was almost equal to that for the anterior hypothalamus, incorporated much more radioactivity than the posterior hypothalamus ($P<0.025$), the cortex cerebri and the cerebellum ($P<0.005$ and $P<0.025$). Great accumulation of radioactivity was found in the anterior hypophysis. The ratio of the concentration of radioactivity in the anterior hypophysis to the cortex cerebri or plasma was 23.1 or 6.4.

At 2 hours after the injection, the pattern of distribution of radioactivity in the brain and the hypophysis was similar to that of the 1-hour group. The anterior hypothalamus contained more radioactivity than the two other portions of the hypothalamus, the cortex cerebri, and the cerebellum. At this time the concentration of radioactivity in the anterior hypothalamus was greater than that in plasma. In contrast to the 1-hour group, the middle and posterior hypothalamus incorporated more radioactivity than the other brain tissue. The median eminence also accumulated more radioactivity than the middle and posterior hypothalamus

TABLE 5

*Time course of the distribution of radioactivity in the brain and hypophysis in the 45-day-old female rat following the injection of tritiated estradiol**

Tissue	½ hour dpm/mg wet wt†	½ hour t/c§§	1 hour dpm/mg wet wt	1 hour t/c	2 hours dpm/mg wet wt	2 hours t/c	4 hours dpm/mg wet wt	4 hours t/c
Anterior hypothalamus	$29.1 \pm 4.0^{\S 2,a}$ (5)	2.4	$23.6 \pm 4.2^{1,a}$ (6)	3.8	$20.3 \pm 2.3^{1,a}$ (7)	5.8	$10.1 \pm 1.8^{1,b}$ (7)	7.2
Middle hypothalamus	$20.6 \pm 2.5^{2,a}$ (5)	1.7	$12.6 \pm 1.4^{2,b}$ (6)	2.0	$11.3 \pm 2.2^{2,3,b}$ (7)	3.2	$4.9 \pm 1.6^{2,c}$ (7)	3.5
Posterior hypothalamus	$19.1 \pm 2.4^{2,3,a}$ (5)	1.6	$11.0 \pm 0.9^{2,b}$ (6)	1.7	$7.5 \pm 1.7^{3,4,b,c}$ (7)	2.1	$3.8 \pm 0.6^{2,c}$ (7)	2.7
Cortex cerebri	$12.3 \pm 1.3^{3,a}$ (5)	1.0	$6.3 \pm 0.7^{2,b}$ (6)	1.0	$3.5 \pm 0.5^{4,c}$ (7)	1.0	$1.4 \pm 0.2^{2,d}$ (6)	1.0
Cerebellum	$14.5 \pm 1.6^{3,a}$ (5)	1.2	$7.5 \pm 0.9^{2,b}$ (6)	1.2	$3.8 \pm 0.5^{4,c}$ (7)	1.1	$1.5 \pm 0.3^{2,c}$ (6)	1.1
Median eminence	34.2 ± 2.4^{a} (5)	2.8	20.6 ± 3.5^{b} (6)	3.3	$29.1 \pm 3.8^{a,b}$ (7)	8.3	10.1 ± 1.8^{c} (6)	7.2
Anterior hypophysis	145.7 ± 5.2^{b} (5)	11.8	145.4 ± 18.0^{b} (6)	23.1	250.0 ± 24.3^{a} (7)	71.4	136.3 ± 11.7^{b} (6)	97.4
Muscle	$11.3 \pm 1.0^{3,a}$ (5)	0.9	$6.8 \pm 0.9^{2,b}$ (6)	1.1	$5.3 \pm 0.8^{4,b}$ (7)	1.5	$2.8 \pm 0.5^{2,c}$ (6)	2.0
Plasma‡	$61.9 \pm 10.5^{1,a}$ (5)	5.0	$22.9 \pm 2.5^{1,b}$ (6)	3.6	$15.2 \pm 2.6^{2,b}$ (7)	4.3	$11.0 \pm 1.2^{1,b}$ (6)	7.9

*0.013 μg of estradiol-6,7-^{3}H per 10 g of body weight (42 c per mmole) was injected intraperitoneally into the 45-day-old female rat. The animals were killed by decapitation at ½, 1, 2, and 4 hours after the injection.

†Mean ± standard error.

‡Dpm/μ liter.

§Number of animals.

§§Ratio of the concentration of radioactivity between the tissue and the cortex cerebri.

1,2,3,4,a,b,c,d The means having the same superscripts are not significantly ($P > 0.05$) different from each other. The numerical and alphabetical figures are used for each hour group and tissue, respectively.

the anterior hypothalamus showed an ability to take up radioactive material preferentially after injection of labeled estradiol. The anterior hypophysis and the median eminence showed characteristic abilities similar to that of the anterior hypothalamus at this time.

In comparison with the earlier times, although there was less concentration of radioactivity in all the tissues tested, the pattern of distribution of radioactivity was identical to that of the 1-hour group at 4 hours after the injection. The median eminence, with as much radioactivity as the anterior hypothalamus, contained much more radioactivity than the posterior hypothalamus ($P<0.001$), the cortex cerebri and the cerebellum ($P<0.005$ and $P<0.005$). A very high concentration of radioactivity was again found in the anterior hypophysis: 97 and 12 times that of the cortex cerebri and plasma, respectively.

The ability of the tissues to retain the radioactive material taken up was shown in the 45-day-old rat (Table 5 and Fig. 2). The concentration of radioactivity in the anterior hypothalamus was retained up to 2 hours and then decreased at 4 hours. In contrast, the mean value for the cortex cerebri decreased steadily. Gradual decrease in concentration was found in the cerebellum up to 2 hours. The cerebral cortex and the cerebellum gave no evidence of an ability to retain radioactive material after the injection of tritiated estradiol. As shown in Figure 2, the ratio of the concentration of radioactivity in the anterior hypothalamus and in the cortex cerebri increased greatly with time up to 4 hours. The mid-

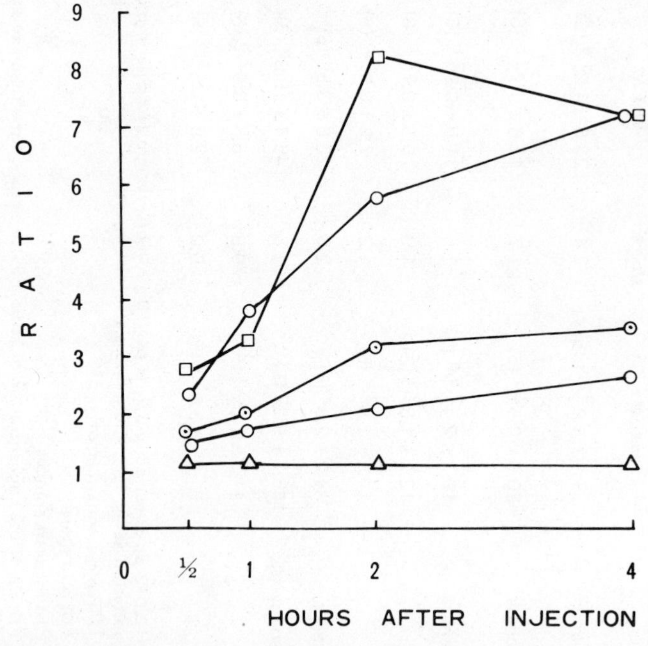

Fig. 2. Ratio of concentrations of radioactivity at various times after injecting tritiated estradiol in female rats at 45 days of age. The value for the cerebral cortex was taken as 1.0 and used for comparison with other parts of brain. O =Anterior hypothalamus; ⊙=middle hypothalamus; • =posterior hypothalamus; △=cerebellum: □=median eminence.

dle and posterior hypothalamus showed a lesser rise, and the cerebellum showed no increase in the ratio with time. In the anterior hypophysis, a peak was present at 2 hours, followed by a decrease to the initial level at 4 hours. The mean value for the median eminence decreased at 1 hour and again increased up to the initial level at 2 hours before decreasing at 4 hours. The ratio of the concentration of radioactivity of the median eminence to the cortex cerebri showed a rise similar to the anterior hypothalamus. The ratio for the anterior hypophysis increased to a much greater degree with time.

The anterior hypothalamus continued to incorporate radioactivity and retained it for a longer time than did other brain tissues. The median eminence and the anterior hypophysis took up and retained radioactivity.

DISCUSSION

Decreases with age in the amount of tritiated estradiol in various brain areas and high uptake by the tissues in the rats in the earliest ages of life were demonstrated in agreement with the findings of Woolley et al.(16) and of Presl et al.(17). Moreover, it was noticeable that no differential ability of the hypothalamus for uptake was recognizable in young rats up to 12 days after birth, being thereby different from young rats at later stages of development and adult rats(6-8). The high brain concentrations in the very young rats may be ascribed to several factors such as little or no blood-brain barrier(20,21), lower lipid content(16,22), or greater total water content in the brains of these animals(21). The much higher blood levels of radioactivity found in very young rats compared to adults following injection of tritiated estradiol is in agreement with the results obtained by Woolley et al.(16). Change in the concentration of radioactivity in the brain with time resembled that in blood. Estrogens might be excreted less efficiently by the kidney in these animals, which may be reflected in the prolonged presence of radioactivity in the brain and hypophysis in the rat at this age. The anterior hypothalamus of adult rats is saturated by a dose of estradiol six times the present smallest one of 0.0013 μg per 10 g of body weight, which suggests an estradiol receptor with limited capacity. In contrast, at 7 days of age, even with an ^{3}H-estradiol dosage 16 times greater than the same least dosage, no plateau in the concentration of radioactivity is seen (Table 2). These findings suggest that the estradiol uptake mechanism in the hypothalamus is not functional in early stages of the developing rat brain. It is interesting that the single injection of cumulative dosage of estrogen (1 μg/rat) at 5 days of age does not modify the normal onset of puberty, but at 26 days of age it does(4).

Marked decreases in the concentrations of tritiated estradiol in the brain tissues was recognized at 12 days of age in comparison to those in 3-or-7-day-old rats. It is claimed that after the 10th day the blood-brain barrier does not seem to be operative against estrogens(17).

In 25-day-old rats, the anterior hypothalamus showed preferential accumulation of radioactive estradiol with a tendency for prolonged retention after the injection of tritiated estradiol, neither of which were observed at earlier ages (Table 4). Since the major portion of the radioactivity was shown to be from estradiol in the anterior hypothalamus and the median eminence in adult rats(7) and developing rats(16), the anterior hypothalamus can concentrate estradiol in contrast to the other parts of the brain. Intracerebral differences in the distribution of radioactive estrogen were first evident at this age, as shown clearly by the concentration ratios (Fig. 1). In postpuberal rats at 45 days of age, the characteristic nature of the anterior hypothalamus on the preferential uptake of estradiol and its retention is almost identical to that in adults(7,8) (Table 5 and Fig. 2). Specific estradiol binding capacity or a possible estradiol receptor in the anterior hypothalamus of adult rats has been demonstrated by the gland's preferential accumulation and retention of estradiol(7,8). Thus, it seems possible that the estradiol binding capacity or possible estradiol receptor appears in the anterior hypothalamus in female rats at about 25 days of age, although still in incomplete form, but after vaginal opening at 45 days of age it becomes functionally "mature" enough to bind estradiol vigorously. In this context it is interesting to mention the findings of Ramirez and Sawyer(4) that puberty in the immature female rat was advanced by 7 to 10 days by treatment with physiological dosages of estrogen, regardless of the term of treatment. Our previous data(24,25) together with the findings of McGuire and Lisk(23) have shown changes in the amount of radioactive estrogen taken up by the anterior hypothalamus in phase with the sexual cycle in rats, suggesting a role for anterior hypothalamic estrogen binding in the feedback control by estrogen under physiological conditions. From these findings, along with the knowledge that the anterior hypothalamus is critical for the normal sexual cycle in rats(26-31), it is suggested that when the estrogen binding capacity or possible receptor in the anterior hypothalamus becomes functionally mature in synchronization with increasing estrogen levels in blood, its physiological function is initiated and becomes operational in the maintenance of cyclicity of reproductive function, serving as a site of action for feedback of estrogen.

A high accumulation of radioactive estrogen was observed in the anterior hypophysis even in 3-day-old rats with low tissue:blood concentration ratios, and in 12-day-old rats with much higher ratios. The median eminence tends to accumulate the radioactive material without retaining it for long at 12 days of age (Table 3), and it retained it more strongly at 25 days of age (Table 1). Fig. 2). The estradiol uptake mechanism developed more rapidly in the anterior hypophysis and the median eminence than in the anterior hypothalamus. Presl et al.(17) have suggested that the median eminence and the anterior hypothalamus show the first appearance of specific binding capacity or possible receptor at the age of 15 and 20 days, respectively. Since estrogen can affect the content of LH-RF in the median eminence and of pituitary gonadotropin in the pituitary of immature rats(4,32), it may be suggested that the estradiol binding capacity in the tissues can already respond to estrogen administration.

Immature female rats are reported to be more sensitive to the inhibitory effect of estrogen than adult ones(33,34). When the same amount of tritiated estradiol relative to body weight is injected in rats from neonatal to postpuberal stages, the brain and hypophysis in younger animals contained greater concentrations of radioactivity than did these tissues in older ones. Thus, in agreement with the statement of Woolley et al.(16), the differences in uptake in developing rats must be accounted for in any interpretation of the relative physiological effectiveness of estradiol at different stages.

REFERENCES

1. Everett, J. W. *In* Sex and Internal Secretion. Young, W. C., ed. Baltimore, Williams and Wilkins, Vol. 1, p. 497, 1961.
2. Greep, R. O. *In* Sex and Internal Secretions. Young, W. C., ed. Baltimore, Williams and Wilkins, Vol. 1, p. 240, 1961.
3. Donovan, B. T. Sci. Basis Med. Ann. Rev., p. 53, 1963.
4. Ramirez, V. D., and Sawyer, C. H. Endocrinology, 76:1158, 1965.
5. Smith, E. R. and Davidson, J. M. Endocrinology, 82:100, 1968.
6. Eisenfeld, A. J., and Axelrod, J. J. Pharmacol. Exp. Ther., 150:469, 1965.
7. Kato, J., and Villee, C. A. Endocrinology, 80:567, 1967.
8. ——— and Villee, C. A. Endocrinology, 80:1133, 1967.
9. Stumpf, W. E. Science, 162:1001, 1968.
10. Pfaff, D. W. Endocrinology, 82:1149, 1968.
11. Anderson, C. H., and Greenwald, G. S. Endocrinology, 85:1160, 1969.
12. Eisenfeld, A. J., and Axelrod, J. Endocrinology, 79:38, 1966.
13. Barraclough, C. A., and Gorski, R. A. Endocrinology, 68:68, 1961.
14. Gorski, R. A. Amer. J. Physiol., 205:842, 1963.
15. Nakayama, T. The Metabolism of Estrogen and Its Significance in Female Reproductive Function Report on the Theme Commissioned by the Japanese Obstetrical and Gynecological Society, Tokyo, pp. 14–18, 1965.
16. Woolley, D. E., Holinka, C. F., and Timiras, P. S. Endocrinology, 84:157, 1969.
17. Presl, J., Röhling, S., Horsky, J., and Herzmann, J. Endocrinology, 86:899, 1970.
18. Duncan, D. B. Biometrics, 11:1, 1955.
19. Kramer, C. Y. Biometrics, 12:307, 1956.
20. Donovan, B. T., and van der Werff ten Bosch, J. J. Physiology of Puberty. London, Edward Arnold, p. 147, 1965.
21. Vernadakis, A., and Woolley, D. E. Arch. Neurol., 12:284, 1965.
22. Sperry, W. M. *In* Biochemistry of the Developing System. Waelsch, H., ed. New York, Academic Press, p. 261, 1955.
23. McGuire, J. L., and Lisk, R. D. Proc. Nat. Acad. Sci., 61:497, 1968.
24. Kato, J., Inaba, M., and Kobayashi. T. Acta Endocr. (Kobenhavn), 61:585, 1969.
25. ——— Acta Endocr. (Kobenhavn), 63:577, 1970.
26. Dey, F. L. Endocrinology, 33:75, 1943.
27. Hillarp, N. A. Acta Endocr. (Kobenhavn), 2:11, 1949.
28. Greer, M. A. J. Clin. Endocr., 12:1259, 1952.
29. Critchlow, B. V., and Sawyer, C. H. Fed. Proc., 14:32, 1955.
30. Everett, J. V. *In* Control of Ovulation. Villee, C. A., ed. Oxford, Pergamon Press, p. 101, 1961.
31. Barraclough, C. A. *In* Control of Ovulation. Villee, C. A., ed. Oxford, Pergamon Press, p. 113, 1961.
32. Ramirez, V. D., and Sawyer, C. H. Endocrinology, 78:958, 1966.
33. Hohlweg, W., and Junkmann, K. Klin. Wschr., 11:321, 1932.
34. Ramirez, V. D., and McCann, S. M. Endocrinology, 72:452, 1963.

55

Biosynthesis of Steroid Hormones by the Chick Embryonic Gonads

Katy Haffen,[1] Lise Cèdard,[2] and Denise Scheib[1]

[1]Institute of Experimental Embryology and Teratology of the C.N.R.S., Nogent-Sur-Marne, France
[2]Laboratory of Maternal Hormonal Chemistry of Port-Royal, Paris, France

INTRODUCTION

Wolff and Haffen(27) have shown that genotypically female gonads explanted before the onset of gonadal differentiation (6 to 7 days of incubation) feminize cocultured male gonads of the same age. These results were corroborated in the 5-day-old chick embryo by Weniger(20). Furthermore, this author showed that an intimate contact between the two gonads is not a prerequisite for this transformation: the ovarian secretion diffuses into the medium and is thus capable of feminizing genetically male gonads. Estrogenic activity was demonstrated in extracts of these media(21,22) using the biological assay procedure of Allen and Doisy.

A study of the biosynthesis of steroids from labeled precursors by embryonic chick gonads growing in organ culture was undertaken by Cedard and Haffen(6,7, 11,12) and by Weniger et al.(1,23,24). In these studies, embryonic gonads were cultured on nutrient medium containing labeled precursors of steroids. After 24 to 72 hours of incubation, the culture media were collected and analysed for radioactive steroid hormone synthesized by the gonads. The results of these experiments will be presented in the first part of this paper.

In the second part of this paper, the main findings of Scheib and Haffin(18,19) concerning the histochemical localization of the Δ_5-3β-hydroxysteroid dehydrogenase enzyme system in the gonads will be reported.

MATERIALS AND METHODS

Organ Culture and Determination of Metabolic Transformation Products

The different techniques have been described in detail(12,7), and will only be summarized here.

Gonads were excised from untreated chick embryos (6 to 10½ or 18 days old) and from 10- and 12-day-old embryos which had been injected with 100 μg of estradiol dipropionate at 3 days of incubation. The gonads were explanted in vitro into culture medium consisting of agar, Tyrode solution, extract of 9-day-old chick embryos, and labeled precursors, according to the technique of Wolff and Haffen (26). The gonadal explantations were incubated for 24, 48, or 72 hours. The numbers of gonads explanted are listed in Table 1.

Before use, the radioactive hormones were purified by thin-layer chromatography on silica gel in solvent system A, according to the method of Lisboa and Diczfalusy(14).

For each experiment, radioactive precursors were added in the following concentrations:

Dehydroepiandrosterone-4-^{14}C (specific activity: 27.5 mc/mM)—0.6 to 2 μc per assay;

Dehydroepiandrosterone-6, 7-^{3}H (specific activity: 0.50c/mM)—5 μc per assay;

Testosterone-6, 7α-^{3}H (specific activity: 0.85 c/mM)—1.5 to 5 μc per assay;

Sodium acetate-1-^{14}C (specific activity: 41 mc/mM)—100 to 200μc per assay.

In some experiments, dehydroepiandrosterone-4-^{14}C and testosterone-6, 7α-^{3}H or sodium acetate-1-^{14}C and dehydroepiandosterone-6, 7-^{3}H were added together.

The effect of human chorionic gonadotropin was tested at a concentration of 15 IU per assay.

In the experiments with carbon-labeled precursors, estradiol-17β-6, 7-^{3}H (specific activity: 0.30c/mM), estrone-6, 7-^{3}H (specific activity: 0.51c/mM), and testosterone-6, 7α-^{3}H were added immediately after the extraction to permit correction for losses during isolation and purification. In all experiments, nonradioactive steroids were added as carriers.

The culture media from which the organs had been removed were extracted three times with 10 ml diethyl ether/chloroform. The extracts were separated into phenolic and neutral fractions by partition between ether and NaOH. The phenolic fraction, obtained after acidification of the NaOH phase and extraction with ether, was partitioned between benzene/petroleum ether and water containing 1 percent ethanol into two fractions. One fraction contained estrone and estradiol-17β, the other, the more polar estrogens. Estrone and estradiol were separated by celite column chromatography(5) and were further purified by methylation followed by column chromatography on alumina, according to the technique of Brown(4). The quantities of estrogens produced were estimated on the basis of recovery of tritiated or nonradioactive estrogens added at the time of extraction.

TABLE 1

Synthesis of estrone, estradiol-17β, and testosterone from dehydroepiandrosterone-4-^{14}C (A) and from testosterone-6,7α-^{3}H (B) by chick gonads during incubation in vitro and recovered in the culture media

Synthesis is expressed as radioactivity incorporated into each steroid (dpm) and as a percentage of radioactivity in the ether-soluble fraction.[a]

Exp. no.	Amount of precursor in μc	Age in days	Sex of the gonads	Number of gonads per series	Culture time in days	Estrone[b]		Estradiol-17β[b]		Estrone + Estradiol-17β		Testosterone[c]	
						dpm	%	dpm	%	dpm	%	dpm	%
(A) PRECURSOR DHEA-4-^{14}C													
H I	1.2	6	undifferentiated	297	3	417	0.03	440	0.10	857	0.13	null	null
J K	2	7	undifferentiated	332	3	408	0.5	1,156	0.14	1,564	0.19	3,080	0.40
L	2	7	undifferentiated	156	1	1,560	0.02	12,998	0.11	14,558	0.13	16,830	0.30
G	0.6	9	♀	145	3	5,200	0.34	16,500	1.12	21,700	1.46	null	null
16	1.2	10	♀	50	3	1,623	0.09	14,906	0.90	16,529	0.99	null	null
1	0.7	10	♀	63	3	1,840	0.19	10,406	1.12	12,246	1.31	null	null
9	0.6	10	♀	62	3	4,774	0.33	27,541	1.94	32,315	2.27	null	null
G	0.6	9	♂	145	3	419	0.023	530	0.028	949	0.05	24,850	1.3
17	1.2	10	♂	80	3	74	—	427	0.03	501	0.03	19,180	1.3
2	0.7	10	♂	98	3	95	0.01	349	0.03	444	0.04	55,000	6.2
8	0.6	10	♂	96	3	45	—	110	0.03	155	0.03	64,900	9.0
3	0.7	10	♀ and ⚥	44	3	1,214	0.15	5,660	0.71	6,874	0.86	null	null
19	1.2	10	♀ and ⚥	69	3	676	0.05	3,647	0.28	4,323	0.33	null	null
10a	0.8	10	⚥	35	3	778	0.07	4,329	0.37	5,707	0.44	null	null
10b	0.8	10	♀ > ⚥	71	3	3,329	0.45	10,591	1.43	13,920	1.88	null	null
(B) PRECURSOR TESTOSTERONE-6,7α-^{3}H													
1	1.5	10	♀	63	3	2,866	0.10	16,675	0.58	19,541	0.68		
2	1.5	10	♂	98	3	230	0.01	2,008	0.08	2,238	0.09		
3	1.5	10	♀ and ⚥	44	3	1,677	0.07	9,203	0.42	10,880	0.49		

[a]Corrections were made for losses during separation, purification, and derivative formation by estimation of the recovery of carrier or ^{3}H-labeled steroid added immediately after the extraction of the culture media.

[b]After formation of 3-methyl-estrone and 3-methyl-estradiol-17β.

[c]After oxidation to Δ_4-androstenedione.

The extract containing the neutral steroids was submitted to a series of paper chromatographies in different solvent systems. Based on the location of carrier and radioactive tracers, the metabolites on each paper chromatogram were eluted and identified by derivative formation (oxidation, acetylation).

Recrystallization of carbon-14 labeled estradiol-17β and testosterone was carried out by the technique of Axelrod et al.(2). These two steroids were obtained in a state of radiochemical purity.

Detection of radioactivity was made with a Packard model 7201 radiochromatogram scanner. The radioactivity was counted with a Packard Tri-Carb Model 3342 liquid scintillation spectrometer.

Histochemical Method for the Detection of the Δ_5-3β-hydroxysteroid dehydrogenase Enzyme System

Gonads and other tissues (adrenal, mesonephros, liver, spleen, and intestine) were removed from 6-to-18-day-old chick embryos. The method used for the detection of the Δ_5-3β-hydroxy steroid dehydrogenase enzyme system was that of Levy et al.(13) as modified(19). The tissues were briefly fixed with formol-calcium or buffered formol at 0° C. After freezing, they were cut at 20 μ with a Leitz-Kryotome, and the sections were incubated at 38.5: C in a medium containing nicotinamide-adenine dinucleotid (NAD), nitro-blue tetrazolium (NBT) and one of the following 3β-hydroxysteroids: 3β-hydroxy-pregn-5ene-20-one (pregnenolone), dehydroepiandrosterone, 3β-hydroxy-5α-androstan-17-one (epiandrosterone) or androst-5-ene-3β, 17β-diol. These were dissolved in dimethylformamide, a solvent which is suitable for giving negative controls in absence of steroids.

BIOCHEMICAL STUDY

Biosynthesis of Steroid Hormones from Dehydroepiandrosterone-4-^{14}C by Gonads from 6- to 10½-day-old Chick Embryos

C^{14}-Estrone and C^{14}-estradiol-17β were isolated from media on which ovaries from 7½- to 10½-day-old chick embryos and morphologically undifferentiated gonads from 6-and 7-day-old embryos had been cultured (Tables 1 and 2). The amount of estrogen formed correlated with the age of the donor embryo; i.e., gonads from older embryos formed more estrogen. All stages studied produced more estradiol-17β than estrone. Estradiol-17β was recrystallized to constant $^3H/^{14}C$ ratios (Table 2). Testes from embryos in the same age range did produce significant quantities of estrogens. In addition the formation of derivatives has confirmed the absence of estrogen production. Only traces of ^{14}C could be found in the 3-methyl-estradiol-17β.

The analysis of the neutral fraction revealed that only testes (7½ - to 10½ - day-old) are able to produce testosterone from the precursor under the influence of a Δ_{4-5}-isomerase and a 3β-hydroxysteroid-oxidoreductase. Male gonads formed this hormone in increasing amounts when taken from progressively older embryos (Table 1). The testosterone was crystallized to constant isotopic ratio (Table 2).

TABLE 2

Crystallization to constant $^3H/^{14}C$ ratio of testosterone and estradiol

Crystallization no.	Testosterone[a] formed by male gonads (after oxidation to Δ_4-androstenedione). $^3H/^{14}C$ before crystall.: 4.5		Estradiol-17β[b] formed by female gonads $^3H/^{14}C$ before crystall.: 25.1		Estradiol-17β[c] formed by ♀ and ⚥ gonads (after methylation). $^3H/^{14}C$ before crystall.: 12.8	
	Crystals	**Mother liquor**	**Crystals**	**Mother liquor**	**Crystals**	**Mother liquor**
1	4.5	4.6	25.3	23.2	10.8	11.6
2	3.7	4.4	24.6	22.8	12.7	10.7
3	4.5	4.2	26.2	20.7	11.2	10.7
4	4.2	4.3	26.7	—	—	—

[a]Crystallization of testosterone formed from DHEA-4-^{14}C after collection of the extracts of media on which male gonads of 7½ to 8 days (190 pairs), of 9 to 9½ days (118 pairs), and of 10 to 10½ days (123 pairs) have been cultivated for 3 days.

[b]Crystallization of estradiol-17β formed from DHEA-4-^{14}C after collection of the extracts of media on which female gonads of 7½ to 8 days (173 pairs) and of 10 to 10½ days (124 pairs) have been cultivated for 3 days.

[c]Crystallization of estradiol-17β formed from DHEA-4-^{14}C after collection of the extracts of media on which ♀ and ⚥ gonads have been cultivated for 3 days (Series no. 3, 10a, and 19, Table 1).

Other Metabolics Formed from Dehydroepiandrosterone-4-^{14}C

Among other products of dehydroepiandrosterone-4-^{14}C metabolism detected in the culture media of gonads of both sexes and of morphologically undifferentiated 7-day gonads, were androst-5-ene-3β,17β-diol, androst-5-ene-3β,17α-diol, and 3α-hydroxy-5β-androstan-17-one (ethiocholanolone). These products indicate that the 17β- and 17α-hydroxysteroid-oxidoreductases and 5β-reductase are present in these gonads. The relative amounts of these hormones detected in the neutral fraction are depicted in Figure A. The determinations were made on 145 pairs of female gonads and on the same number of male gonads taken from 9-day-old embryos and explanted for 3 days on 3 ml of medium containing 0.6 μc of dehydroepiandrosterone-4-^{14}C.

Control Experiments

Measurements of the radioactivity in the neutral and in the phenolic fractions extracted from media incubated without gonads showed that the major part of ^{14}C radioactivity was located in the neutral fraction (85.7 percent ± 14.1 percent). Chromatographic analysis revealed the presence of residual dehydroepiandrosterone-4-^{14}C and the formation of a small quantity of androst-5-ene-3β,17β-diol. It should be noted that the culture medium included chick embryo extract, which undoubtedly contained active enzymes. The presence of a 17β-hydroxysteroid-oxidoreductase would adequately explain the formation of androst-5-ene-3β-diol.

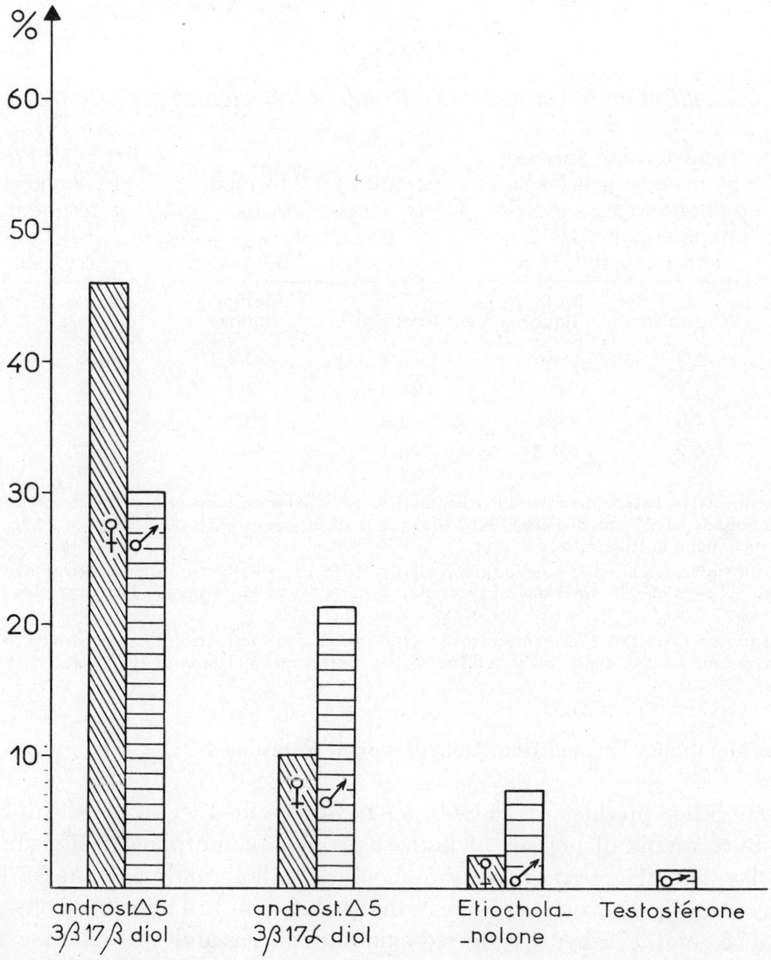

Fig. A. Radioactivity of the different neutral steroids formed from dehydroepiandrosterone-4-^{14}C by ovaries and testes from 9-day-old chick embryos. The radioactivity recovered after repeated chromatographic separations and oxidation of the metabolites is given as a percentage of the whole ether-soluble fraction. Correction for losses during separation, purification, and derivative formation was made by estimation of carrier or ^{3}H-labeled steroids added immediately after extraction.

Metabolism of Dehydroepiandrosterone-4-^{14}C and Testosterone-6,7α-^{3}H by Feminized Male Gonads

Wolf and Haffen(28), using the coelomic graft technique, have shown that testes feminized by the injection of estradiol into the embryo (after the technique of Wolff and Ginglinger[25]) are endowed with the same capacities as an embryonic ovary: an ovotestis grafted into the coelom of a male host induces the development of an ovarian cortex and the formation of a female medulla on part of the host testis. Thus, an estrogenic hormone caused the secretion of a feminizing substance by transforming an organ which would normally secrete a masculinizing hormone.

Is it possible to demonstrate an action of the female hormone injected into the embryo on the gonadal enzymes involved in the metabolism of dehydroepiandrosterone-4-^{14}C? To answer this question the following experiments were carried out: left gonads were taken from 10-day-old (Table 1) and 12-day-old (Fig. B) embryos injected 7 or 9 days earlier with 100 mg of estradiol dipropionate; these gonads were cultivated on media containing dehydroepiandrosterone-4-^{14}C. The feminized male gonads were chosen for the experiments only when it was possible to recognize them by morphological criteria. When the distinction between genetic females and totally transformed males was impossible, all the left gonads were cultured (theoretically half are genetic males if one accepts a 1:1 sex ratio for a large number of embryos).

Estrone and estradiol-17β (major product) were found in the media after culture of ovotestes or of female plus feminized male gonads. Estradiol-17β could be brought to radiochemical purity (Table 2). The results demonstrate that the feminized testes had acquired the capacity to convert dehydroepiandrosterone-4-^{14}C into estrogens, a capacity normally attributed to ovaries.

It should be emphasized that testosterone synthesized from dehydroepiandrosterone-4-^{14}C by testes in vitro usually accumulates in the culture media. Such

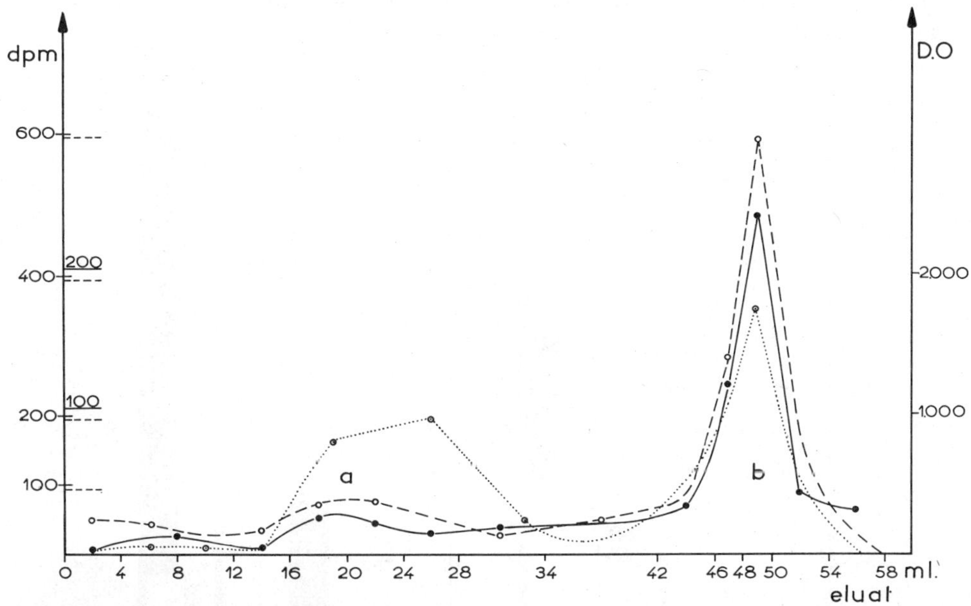

Fig. B. Chromatography (according to Brown's method[4]) of the methyl ethers of estrone (*a*) and estradiol-17β(*b*) fractions obtained from the media on which 62 ovotestes of 12-day-old chick embryos were cultivated in the presence of dehydroepiandrosterone-4-^{14}C and testosterone-6,7α-^{3}H for 3 days. Dashed line—dpm ^{3}H; solid line—dpm ^{14}C. Dotted line—optical density of the nonradioactive methyl ether of estrone and estradiol-17β added at the time of extraction.

an accumulation is not found with ovotestes. The following question arises: Do ovotestes cease to produce testosterone, or is testosterone converted into estrogens as soon as it is formed?

The capacity of ovotestes and of ovaries to aromatize testosterone into estrogens was tested. Explantation of both left ovaries and ovotestes of 10-day-old (Table 1) or 12-day-old (Fig. B) chick embryos on media containing both dehydroepiandrosterone-4-^{14}C and testosterone-6,7α-^{3}H resulted in synthesis of estrogens, especially estradiol-17β. These results suggest that estradiol injected into the embryo acts on the aromatizing enzyme system by eliciting or stimulating its production.

Action of Chorionic Gonadotropin on Steroid Biosynthesis from Sodium Acetate-1-^{14}C

Weniger et al.(1,23,24) have studied estrogen formation from sodium acetate-1-^{14}C by chick embryonic gonads cultivated in vitro for at least 24 hours. They have shown labeled estrone and estradiol synthesis in 7- to 9-day-old female gonads and in 18-day-old ovotestes. When progesterone-4-^{14}C was supplied to 5- and 6-day-old undifferentiated gonads, labeled estrone and estradiol were found in the culture media.

Connell et al.(9) have shown that the 2-day-old chicken testis synthesizes testosterone from sodium acetate-1-^{14}C. This synthesis is accomplished via dehydroepiandrosterone and is stimulated by gonadotropins.

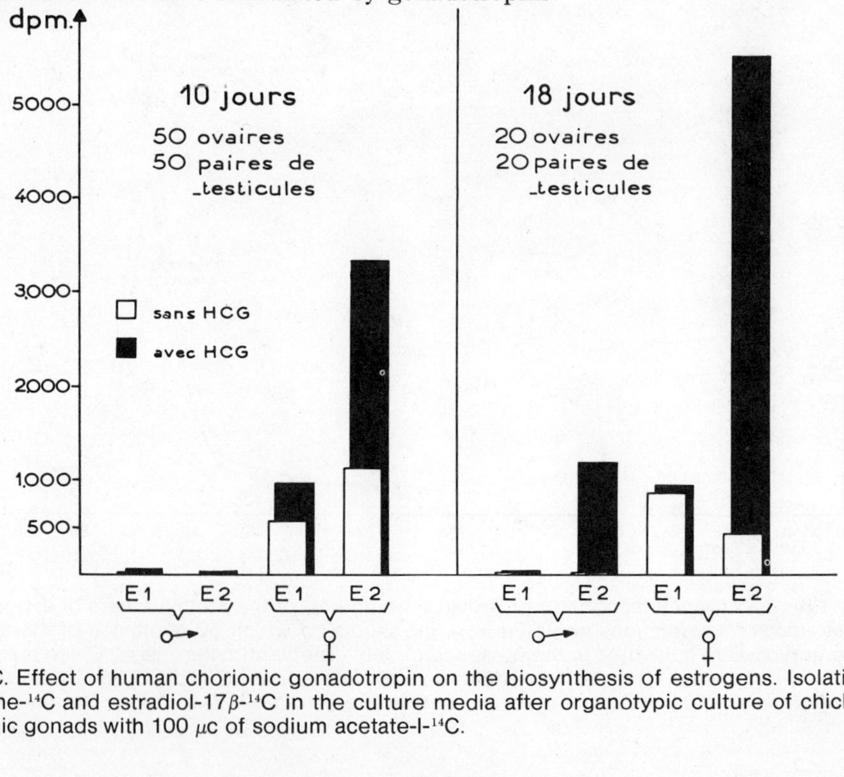

Fig. C. Effect of human chorionic gonadotropin on the biosynthesis of estrogens. Isolation of estrone-^{14}C and estradiol-17β-^{14}C in the culture media after organotypic culture of chick embryonic gonads with 100 μc of sodium acetate-I-^{14}C.

TABLE 3

Radioactivity in dpm of estrone and estradiol-17β synthesized from labeled dehydroepiandrosterone under the influence of human chorionic gonadotropin, related to 20 left ovaries and 20 pairs of testes (18 days) or 50 left ovaries and 50 pairs of testes (10 days)

Age of gonads and amount of precursor	Number of experiments	Sex	Estrone[a]	Estradiol-17β[a]
10 days	1	♂	76	Null
(1 μc)	1	♂ + HCG	465	Null
DHEA-4-^{14}C	2	♀	1,632	5,597
	2	♀ + HCG	3,391	9,548
18 days	2	♂	74,405	35,601
(5 μc)	2	♂ + HCG	161,718	49,571
DHEA-6,7-^{3}H	2	♀	98,288	40,191
	2	♀ + HCG	35,011	177,867

[a]After formation of 3-methyl-estrone and 3-methyl-estradiol-17β. From the specific activities of DHEA-6,7-^{3}H and DHEA-4-^{14}C, estrogen formation by ovaries was estimated as follows:
18 days—^{3}H: 0.9 X 10^{-4} mM (without HCG); 1.5 X 10^{-4} (with HCG);
10 days—^{14}C: 0.3 X 10^{-4} mM (without HCG); 1.2 X 10^{-4} (with HCG).

We tried to see if gonadotropin exerts the same stimulatory action on the biosynthesis of steroids from sodium acetate-1-^{14}C in 10- and 18-day-old chicken embryonic gonads. Gonadotropin, when added to the culture medium in which testes and ovaries were cultivated for 48 hours, enhances the incorporation of radioactivity into the ether soluble fraction. This increase partially represented cholesterol formation(12a). In spite of the presence of peaks of similar polarity, we could not identify (in the media) ^{14}C-labeled progesterone, pregnenolone, dehydroepiandrosterone, or testosterone. However, human chorionic gonadotropin stimulated the production of estrogens, especially of estradiol, by ovaries of 10- and 18-day-old chick embryos (Fig. C). It also provoked a small amount of estradiol synthesis by 18-day-old testes. A stimulatory effect of human chorionic gonadotropin on estradiol synthesis by ovaries also was seen with labeled dehydroepiandrosterone as the steroid precursor (Table 3). Further, it should be mentioned that 18-day-old testes are able to aromatize dehydroepiandrosterone-6,7α-^{3}H into estrogens (Table 3), whereas at earlier stages (Table 1) they are devoid of this capacity.

HISTOENZYMOLOGICAL LOCALIZATION OF THE Δ_5-3β-HYDROXYSTEROID DEHYDROGENASE ENZYME SYSTEM

The Δ_5-3β-hydroxysteroid dehydrogenase is an enzyme which, by dehydrogenation and isomerization, converts pregnenolone to progesterone and dehydroepiandrosterone to androstenedione; these steroids being precursors common to

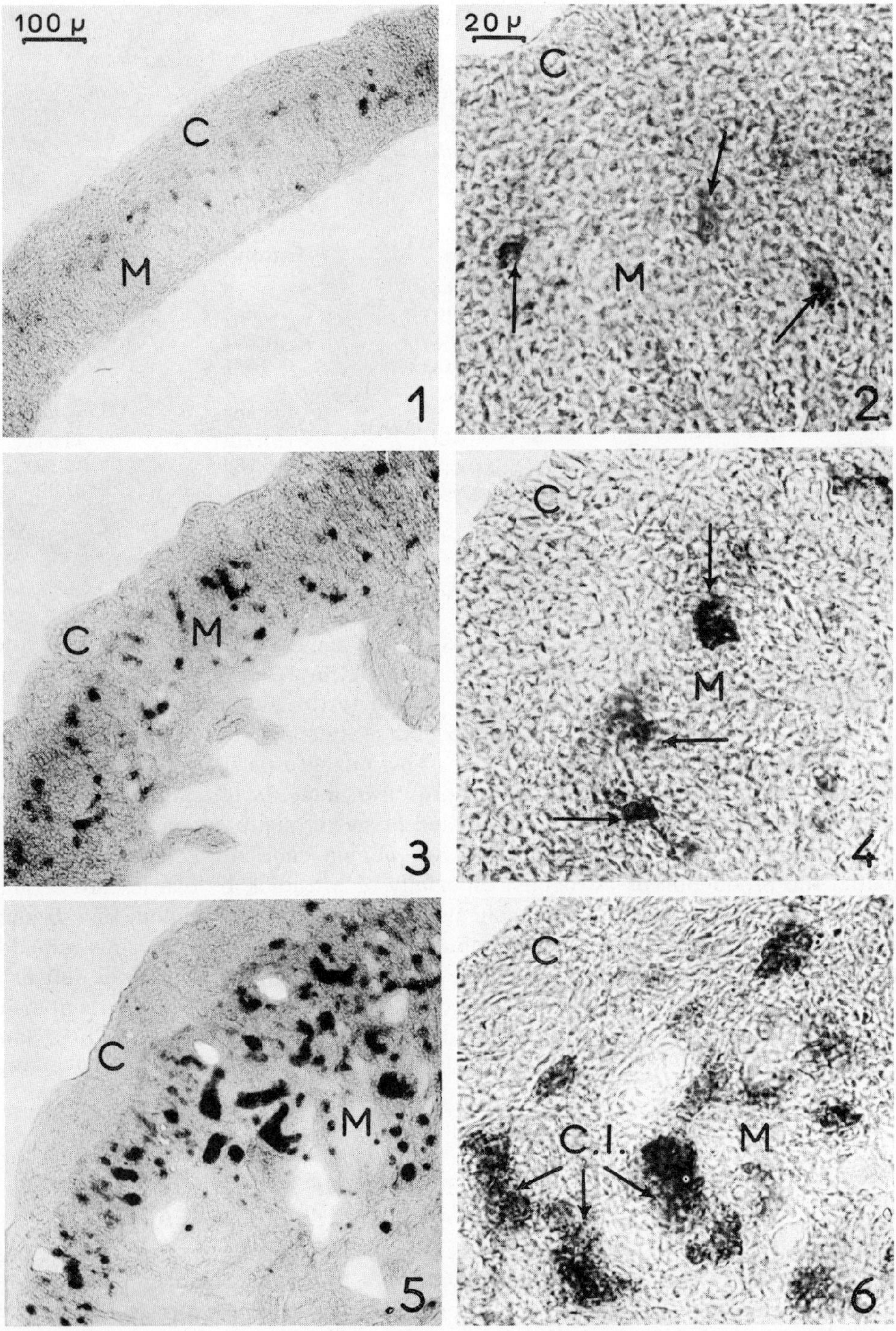
100 μ
C
M
1
20 μ
C
M
2
C
M
3
C
M
4
C
M
5
C
C.I.
M
6

both androgens and estrogens. This enzyme is active in presence of nicotinamide-adenine dinucleotide, a coenzyme which accepts the H from a Δ_5-3β-hydroxysteroid and transfers it to an acceptor, nitro-blue tetrazolium. The enzymatic reaction is visualized by the deposit of blue diformazan (reduced nitro-blue tetrazolium) and is specific for the tested steroid if the controls (incubated without steroid) are negative.

The observations described here are based on the enzymatic reaction produced with dehydroepiandrosterone as substrate, but they are also valid for that given with other 3β-hydroxysteroids(19).

The activity of the Δ_5-3β-hydroxysteroid dehydrogenase first appears in the primary cords at 6 ½ days of incubation, when the gonads are still undifferentiated (Figs. 1 and 7). However, in some cases, we could predict the sex of the gonads by the distribution of the reactive cells in the sections, the medulla having an aspect which foreshadows that of an ovary or a testis of 7 days.

In the ovary, the Δ_5-3β-hydroxysteroid dehydrogenase activity is confined to small groups of the medulla (Figs. 1 and 2). During development, this activity increases as does the number of reactive cells of the medulla; there is no activity in the cortex (Figs. 3 and 4). The initial reaction is probably present in the same cells which later differentiate into interstitial cells (Figs. 5 and 6). In the ovary of 10-day-old embryos, these cells can be identified histologically by their characteristic spongy structure and their organization in cords.

In the right female gonad, which in the chick is reduced to the medullary component, the enzymatic reaction is present at the same stages and is comparable to that of the ovarian medulla.

Enzymatic activity in the medulla has also been reported by others (3,8,16) and the data are in agreement with other experimental results demonstrating that the medulla is the site of the ovarian secretion(10,15,29).

In the testis, the reaction appears gradually in three different sites. The initial reaction which begins in the somatic cells of the primary cords (Fig. 8) persists throughout embryonic development, an observation also reported by Narbaitz and Kolodny(16) and by Chieffi et al.(8). The latter expressed some doubts about the specificity of the reaction.

From 8 days of incubation, some strongly reactive cells appear inside the cords, and clusters of such cells can be seen at every embryonic stage. The first signs of Δ_5-3β-hydroxysteroid dehydrogenase activity in the interstitial cells in the stroma between the cords can be detected at 8 ½ days (Figs. 9 and 10). The groups of interstitial cells increase in number and size as the gonads develop

---

Figs. 1-6. *Chick embryonic ovary:* Reaction of Δ_5-3β-hydroxysteroid dehydrogenase in presence of dehydroepiandrosterone.

Figs. 1 and 2. Undifferentiated gonad from 6½-day-old chick embryo (presumed to be ovary). General view (Fig. 1) and detail (Fig. 2). The formazan deposit is visible in some cells of the medulla.

Figs. 3 and 4. Ovary at 8½ days. General view (Fig. 3) and detail (Fig. 4). Note the strong reaction of medulla cell groups and the absence of reaction in the cortex.

Figs. 5 and 6. Ovary at 18 days. General view (Fig. 5). Detail of cortex and underlying medulla (Fig. 6). Intense reaction in the interstitial cells of the medulla. *M*—medulla; *C*—cortex; *CI*—interstitial cells.

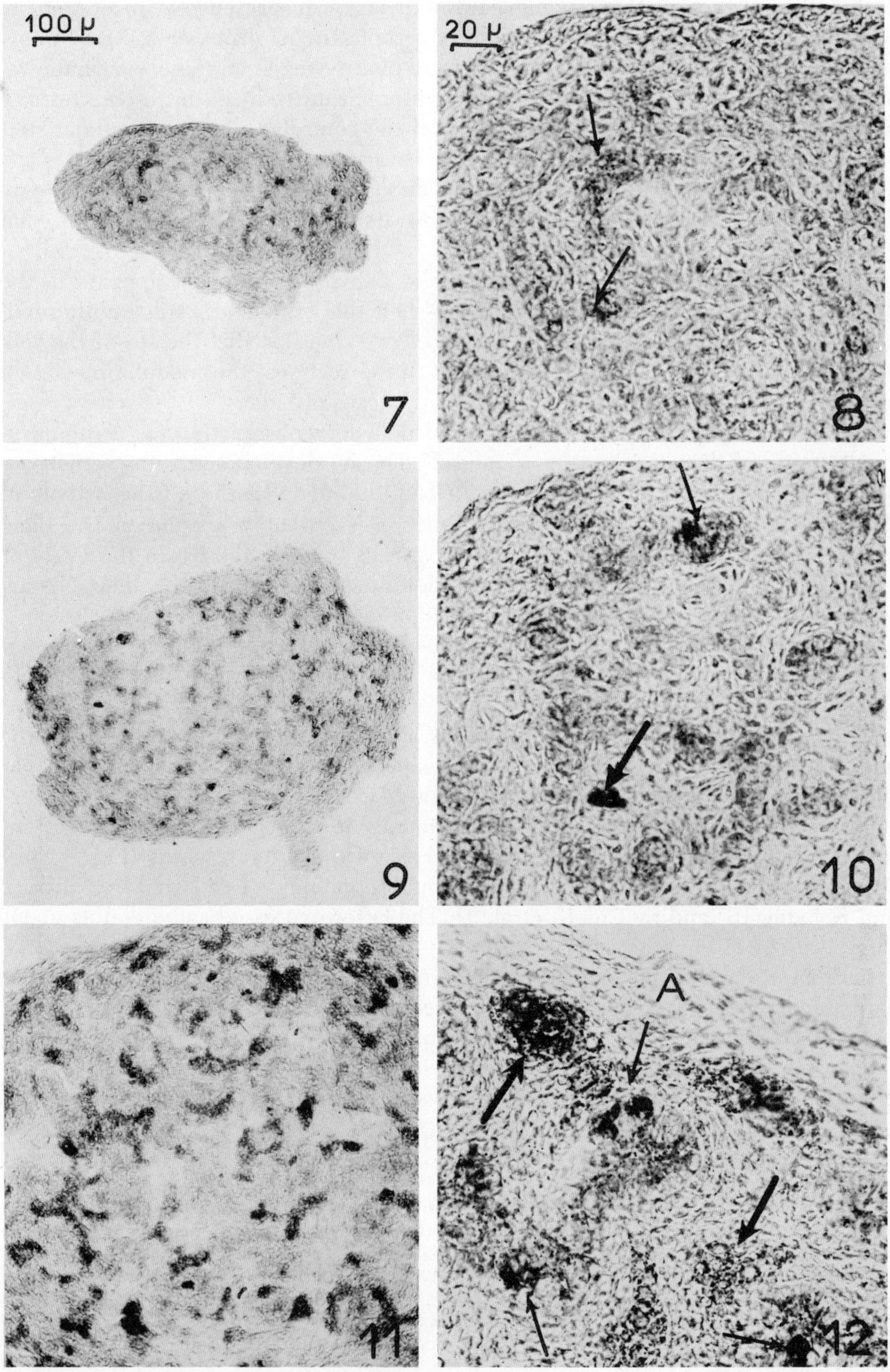
100 μ
20 μ
7
8
9
10
A
11
12

(Figs. 11 and 12). It must be emphasized that the localization of the Δ_5-3β-hydroxysteroid dehydrogenase activity in the stromal cells coincides with that of birefringence in the interstitial cells(17). The structural similarity between the interstitial cells and the strongly reactive cord cells is confirmed by some similarity in their enzymatic activities.

The appearance of Δ_5-3β-hydroxysteroid dehydrogenase activity in the adrenals precedes that in the gonads and has been observed at 4 days by Chieffi et al. (8). All the other organs examined do not show Δ_5-3β-hydroxysteroid dehydrogenase activity.

CONCLUSION

The biochemical data indicate that chick embryonic gonads are able to synthesize sex steroids in vitro and to produce estrogens (6 days) and testosterone (7 days) from a precursor, dehydroepiandrosterone-4-^{14}C. The enzymatic reaction detected in sections of 6½-day-old gonads represents one of the first indications of the localization of a functional specific activity.

REFERENCES

1. Akram, H., and Weniger, J. P. Sécrétion d'oestrone et d'oestradiol par le testicule féminisé de l'embryon de poulet. C.R. Acad. Sci., 264:1806–1807, 1967.
2. Axelrod, L. R., Matthijssen, C., Goldzieher, J. W., and Pulliam, J. E. Definitive identification of microquantities of radioactive steroids by recrystallization to constant specific activity. Acta Endocr., 49 (Suppl.):1–77, 1965.
3. Boucek, R. J., Gyori, E., and Alvarez, R. Steroid dehydrogenase reactions in developing chick adrenal and gonadal tissues. A histochemical study. Gen. Comp. Endocr., 7: 292–303, 1966.
4. Brown, J. B. A chemical method for the determination of estriol, estrone and estradiol in human urine. Biochem. J., 60:185–193, 1955.
5. Cèdard, L. Les oestrogénes naturels. Paris, Masson, 1964.
6. ——— and Haffen, K. Transformations de la déhydroépiandrostérone par les gonades embryonnaires de poulet, cultivées *in vitro.* C.R. Acad. Sci., 263:430–433, 1966.
7. ——— Haffen, K., and Guichard, A. Influence de l'hormone gonadotrope chorionique sur la production d'oestrogènes à partir d'acétate de Na et de déhydroépiandrostérone radioactifs par les gonades embryonnaires de poulet, cultivées *in vitro.* C.R. Acad. Sci., 267:118–120, 1968.
8. Chieffi, G., Manelli, H., Botte, V., and Mastrolia, L. Il differenziamento istochimico dell' interrenale e dei tessuti somatici della gonade embrionale di pollo: comportamento della 3β-olo-deiohogenasi. Acta Embryol. Morph. Exp., 7:89–91, 1964.

Figs. 7-12. *Chick embryonic testis:* Reaction of Δ_5-3β-hydroxysteroid dehydrogenase in presence of dehydroepiandrosterone.

Figs. 7 and 8. Undifferentiated gonad from 6½-day-old chick embryo (presumed to be testis). General view (Fig. 7) and detail (Fig. 8). A fine formazan deposit is visible in some of the cords.

Figs. 9 and 10. Testis at 8½ days. General view (Fig. 9) and detail (Fig. 10). Note the strong reaction in some small cell clusters in the cords (fine arrows) and in the stroma (heavy arrows).

Figs. 11 and 12. Testis at 18 days. General view (Fig. 11) and detail (Fig. 12). Intense reaction in the interstitial cells of the stroma (heavy arrows) and in some cells in the cords (fine arrows). *A*—albuginea.

9. Connell, G. M., Connell, C. Y., and Eik-Nes, K. B. Testosterone synthesis by the two day old chick testis *in vitro*. Gen. Comp. Endocr., 7:158–165, 1966.
10. Haffen, K. La culture *in vitro* de l'épithélium germinatif isolé des gonades mâles et femelles de l'embryon de canard. 2. Influence de la médullaire sur la différenciation de l'épithélium germinatif. J. Embryol. Exp. Morph., 8:414–429, 1960.
11. ——— and Cedard, L. Métabolisme de la déhydroépiandrostérone et de la testostérone radioactives par les gonades mâles intersexuées de l'embryon de poulet, cultivées *in vitro*. C.R. Acad. Sci., 264:1923–1926, 1967.
12. ——— and Cedard, L. Étude, en culture organotypique *in vitro*, du métabolisme de la déhydroépiandrostérone et de la testostérone radioactives par les gonades normales et intersexuées de l'embryon de poulet. Gen. Comp. Endocr., 11:220–234, 1968.
12a. ——— Cedard, L., and Guichard, A. Formation de cholésterol radio-actif à partir d'acétate de Na-1-^{14}C par les embryonnaires de poulet, cultivées *in vitro*. C.R. Acad. Sci., 268:2914–2917, 1969.
13. Levy, H., Deane, H. W., and Rubin, B. L. Visualization of steroid-3β-ol-dehydrogenase activity in tissues of intact and hypophysectomized rats. Endrocrinology, 65:932–943, 1959.
14. Lisboa, B. P., and Diczfalusy, E. Separation and characterization of steroid estrogens by means of thin layer chromatography. Acta Endocr., 40:60–83, 1962.
15. Mintz, B., and Wolff, E. The development of embryonic chick ovarian medulla and its feminizing action in intracoelomic grafts. J. Exp. Zool., 126:511–536, 1954.
16. Narbaitz, R., and Kolodny, L. Δ5-3β hydroxysteroid dehydrogenase in differentiating chick gonads. Z. Zellforsch., 63:612–617, 1964.
17. Scheib, D. Sur la détection des lipides figurés dans les gonades de l'embryon de poulet. Distribution, nature et signification de ces substances. Ann. Histochim., 4:33–50, 1959.
18. ——— and Haffen, K. Étude histochimique de la 3-β-hydroxystéroide-déshydrogénase dans les jeunes gonades embryonnaires de poulet. C.R. Acad. Sci., 264:161–164, 1967.
19. ——— and Haffen, K. Sur la localisation histoenzymologique de la 3β-hydroxystéroide-déshydrogénase dans les gonades de l'embryon de poulet; apparition et spécificité enzymatique. Ann. Embryol. Morph., 1:61–72, 1968.
20. Weniger, J. P. Diffusion des hormones gonadiques de l'embryon de poulet dans le milieu de culture. Arch. Anat. Micr. Morph. Exp., 51:325–336, 1962.
21. ——— Extraction d'une substance oestrogène (test d'Allen et Doisy positif) de milieux sur lesquels les gonades d'embryons de poulet femelles de 7 jours furent cultivées pendant 24h. C.R. Acad. Sci., 261:809–812, 1965.
22. ——— Activité oestrogène de la fraction "oestrone-oestradiol" de milieux incubés avec des gonades d'embryons de poulet femelles. C.R. Acad Sci., 262:578–580, 1966.
23. ——— Sur la précocité de la sécrétion d'oestrogènes par les gonades embryonnaires de poulet, cultivées *in vitro*. C.R. Acad. Sci., 266:2277–2279, 1968.
24. ——— Ehrhardt, J. D., and Fritig, B. Sur la formation d'oestrone et d'oestradiol par les gonades de l'embryon de poulet femelle cultivées *in vitro*. C.R. Acad. Sci., 264: 838–841, 1967.
25. Wolff, E., and Ginglinger, A. Sur la transformation des poulet mâles en intersexués par injection d'hormone femelle (folliculine) aux embryons. Arch. Anat. Histol. Embryol., 20:219–278, 1935.
26. ——— and Haffen, K. Sur le développement et la différenciation sexuelle des gonades embryonnaires d'oiseau en culture *in vitro*. J. Exp. Zool., 119:381–399, 1952.
27. ——— and Haffen, K. Sur l'intersexualité expérimentale des gonades embryonnaires de canard, cultivées *in vitro*. Arch. Anat. Micr. Morph. Exp., 41:184–207, 1952.
28. ——— and Haffen, K. Sur la féminisation induite par les gonades mâles intersexuées chez l'embryon de poulet. Arch. Anat. Histol. Embryol., 154 (Suppl.):274–302, 1961.
29. Wolff, E., and Wolff, Em. Mise en évidence d'une action féminisante de la gonade droite chez l'embryon femelle des oiseaux par les expériences d'hemicastration. C.R. Soc. Biol., 145:1218–1219, 1951.

56

Ultrastructure and Metabolic Studies on Sensory and Sympathetic Nerve Cells Treated with the Nerve Growth Factor and its Antiserum*

Rita Levi-Montalcini[1] and Pietro U. Angeletti[2]

[1]Department of Biology, Washington University,
St. Louis, Missouri
[2]Istituto Superiore di Sanita,
Rome, Italy

INTRODUCTION

The biochemical properties of a protein particle isolated in great quantities from snake venom and mouse salivary glands and present in trace amounts in other tissues and body fluids and its morphological and metabolic effects on sensory and sympathetic nerve cells have been the object of a large number of reports and review articles in this last decade years(1-5). The recent developments of these studies on the chemical properties of this factor, which became known as the nerve growth factor (NGF), are reported in another article in this volume(6).

By way of introduction to the present report, we shall briefly summarize the main features of the morphological effects elicited by the NGF and its antiserum.

Upon injection of the NGF into the yolk of developing chick embryos from the 7th to the 10th day of incubation, sensory and sympathetic ganglia of the host undergo a marked size increase. Both effects are due to an increase in the number and size of sensory and sympathetic nerve cells. Injections of the NGF in newborn mammals result in an even more marked volume increase of para- and prevertebral sympathetic chain ganglia, while no size change is apparent in the sensory ganglia or in any other sectors of the developing nervous system. In newborn mice, the animals which proved to be best suited for these experiments in view of their small size and high receptivity to the NGF, daily injections of the NGF for a 10-

*This work was supported by grants from The John A. Hartford Foundation, the National Institutes of Health (NB-03777), the National Science Foundation (GB-7304), and the Consiglio Nazionale delle Ricerche (Rome).

day period result in a volume increase of the sympathetic ganglia of 6 to 10 times the volume of control ganglia. In adolescent and adult specimens the NGF elicits hypertrophic, but no hyperplastic effects in the same neurons. The in vitro effects of NGF were studied on whole sensory and sympathetic ganglia dissected out from 8-day chick embryos and explanted in semisolid media and on nerve cells dissociated from these ganglia by trypsin and cultured in liquid media(7,8). Intact ganglia cultured in semisolid media produce in a 10- to 12-hour period a dense halo of fibers which spread radially around the explant. Attempts to correlate the size and density of the fibrillar halo with the amount of NGF present in the medium met with difficulties discussed in previous articles(4). An excess of NGF in the medium results in a circular rather than radial nerve fiber outgrowth(9). Dilutions by a factor of 10 of this "NGF excess" call forth the production of a dense and even fibrillar halo which at each subsequent higher dilution decreases in density until eventually only a few fibers grow out from the explant. It was agreed to consider a fibrillar halo of even density and length as the product of one biological unit of NGF, which in our experiments corresponds to 1 ng of the purified protein (10^{-9} g). In other laboratories the "halo effect" was also obtained with much higher dilutions of the NGF, as will be reported in another article in this volume(10).

Sensory and sympathetic nerve cells dissociated through trypsin and cultured in Eagle Basal Medium survive indefinitely and produce a dense fibrillar nerve net when the NGF is added to the culture medium in the amount of 10 or more units per ml. In the absence of the growth factor, nerve cells deteriorate and undergo complete disintegration in 2 to 3 days. These results indicate that the NGF not only stimulates nerve fiber outgrowth and produces an increase in the size of the target cells but also exerts a vital role in the life of these cells(8).

Another strong indication of the vital role played by the NGF came from studies with antibodies to NGF produced by injecting the purified NGF in rabbits and then assaying the effect of the antiserum in vitro and in vivo. In vivo the antiserum injected in newborn mammals results in the destruction of 95 to 98 percent of sympathetic immature nerve cells of para- and prevertebral ganglia. This effect is of a permanent nature. Large colonies of mice and rats deprived in this way of the adrenergic sympathetic nerve cells of these ganglia were thus produced. The "immunosympathectomized" animals are as vital as controls(11-13).

Histological studies of sympathetic ganglia in antiserum treated mice 12 hours after the first injection and in the subsequent days showed the precocity and extent of the adverse effects of the antiserum. Mitotic activity, which is still rather high in these ganglia at birth, sharply decreases in antiserum-treated ganglia. Neuroblasts at early differentiative stages show a complete range of degenerative marks. Disintegrating and dead cells increase progressively in number until toward the fourth day almost all nerve cells have disappeared and only glial and satellite cells are present in the ganglia. Also these cells undergo atrophy in the subsequent days and the ganglia persist throughout life as sclerotic microscopic nodules barely detectable at the dissecting microscope(12). Evidence that these dramatic effects are the result of a cytotoxic reaction elicited in sympathetic nerve cells by antibodies to the NGF was presented in previous articles(14-16). In the following pages we will consider at first the effects of the NGF and of its antiserum on the fine structure of its target cells and then we will report on the metabolic and morphological effects of RNA and protein inhibitors in these same cells.

RESULTS

NGF Effects on the Fine Structure of Embryonic Sensory Nerve Cells

Sensory ganglia dissected out from 8-day chick embryos were cultured in vitro in a plasma clot with or without the NGF. The NGF was added to the experimental cultures in the amount of 0.1 μg/ml, which corresponds to about 100 biological units of the growth factor. The experimental ganglia were examined after 2, 12, and 24 hours of incubation at 38° C while the control ganglia were examined at zero time and after 2 and 4 hours of culture. This shorter incubation time for the controls was suggested by the observation that in the absence of the growth factor, nerve cells show regressive marks already after 6 hours of culture and at the end of the first day the ganglia consist of a large number of degenerating cells and only a few well-preserved neurons(9). At the end of the incubation time, control and experimental ganglia were gently removed from the plasma clot and fixed for 1 hour at room temperature in glutaraldehyde and then for 2 hours in 2 percent osmium tetroxide; they were then dehydrated in ethanol solution for 5 minutes, placed in toluene for 1 hour, and embedded in Epon. The ganglia were then sectioned with an MT 1 microtome, stained with uranyl acetate and lead citrate, and examined in an RCA-3H electron microscope.

The most prominent NGF effect at the ultrastructural level, as already reported(17), is the massive production of neurofibrillar material in the cytoplasm of immature sensory nerve cells (Fig. 1). This effect is already apparent 2 hours after the beginning of the incubation period and is in fact so extensive as to overshadow other ultrastructural changes such as a marked increase in the Golgi complex, and in ribosomes associated with an increase and changes in configurations of the membranous elements of the endoplasmic reticulum. In the nuclear compartment one notices that the nucleoplasm is more dense than in control nerve cells and aggregates of chromatin are irregularly dispersed in the nuclear space. All the above effects, and in particular the production of neurofibrils, become more accentuated in the following hours. At the end of the first day of culture, masses of neurofibrils fill large parts of the cytoplasmic areas.

Antiserum Effects on the Fine Structure of Sympathetic Neurons

Previous work showed that the destructive effects of the antiserum to the NGF are restricted to sympathetic neuroblasts of newborn mammals(2,12). Studies at the ultrastructural level by Sabatini and co-workers(14) indicate the early appearance of degenerative changes in the nuclear compartment of sympathetic cells exposed in vivo or in vitro to the action of the antiserum. A reinvestigation of the same problem performed in our laboratory added new information on this effect. The results are reported in detail elsewhere(16). Here we will only consider the

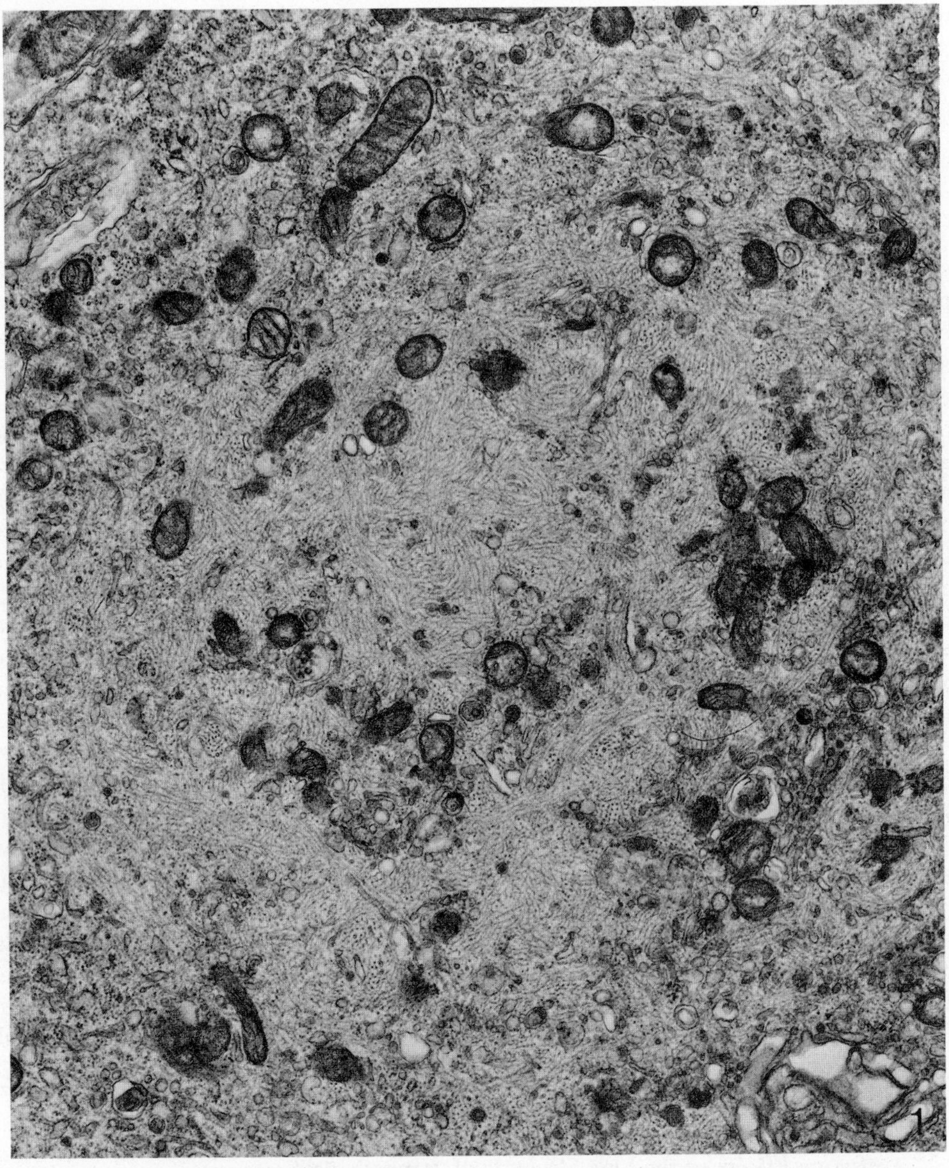

Fig. 1. Nerve cell from 8-day sensory ganglion cultured for 12 hours in the presence of NGF. The microphoto shows part of the cell cytoplasm filled with compact masses of neurofilaments. ×24,000.

main features of the cytotoxic effect elicited by the antiserum in the receptive nerve cells.

The object of this investigation was the superior cervical ganglion of newborn mice injected soon after birth with 0.05 ml of antiserum per g of body weight.

Twenty-four hours later the ganglia were removed from experimental and control mice of the same litter. They were fixed for 2 hours in 3 percent glutaraldehyde in phosphate buffer 0.1 M pH 7.4 with added CaCl . The specimens were washed for 2 hours in the same buffer (4 changes) and postfixed in osmic hydroxide 1.33 percent for 2 hours. After dehydration in ethanol, the tissue was immersed for 1 hour in toluene and then embedded in Epon 812. Sections were cut with a Potter-Blum-Servall MT microtome with a diamond knife and then mounted on copper grid 200 mesh without any supporting film. The specimens were stained with uranyl acetate in 50 percent ethanol or in lead hydroxide, according to Millonig (18). They were examined in an RCA 3 H electron microscope at 8,000 up to 50,000 magnification.

In control ganglia, large and small neurons are readily recognizable with the

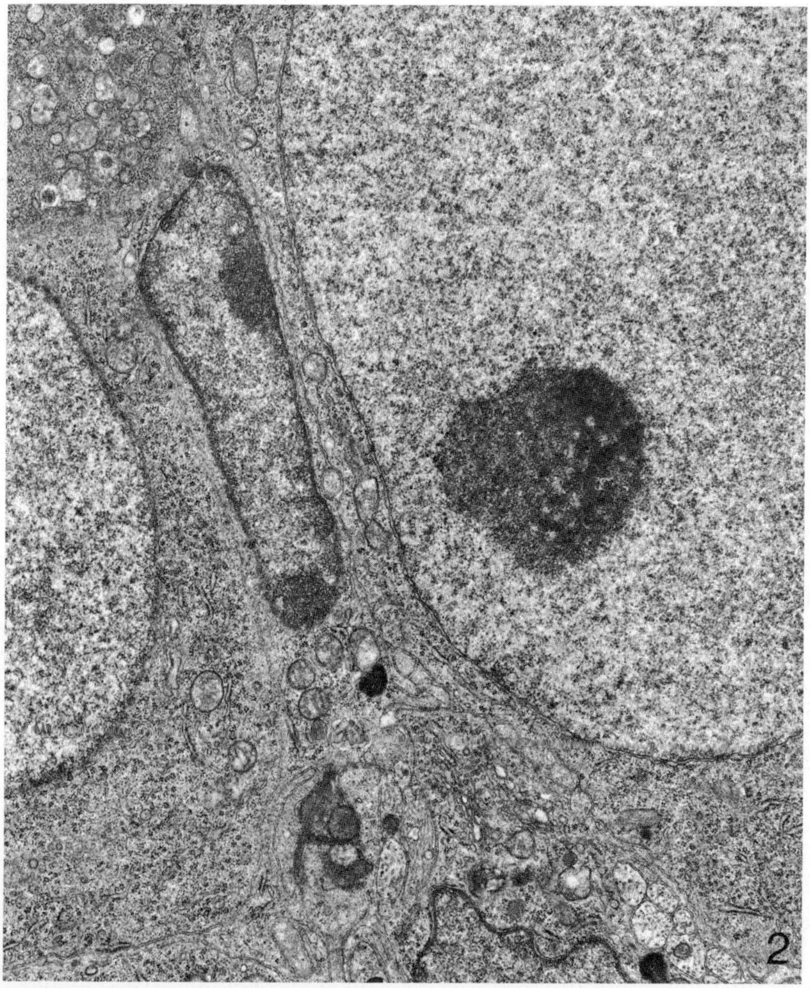

Fig. 2. Large neuron from control ganglion of the sympathetic superior cervical ganglion of 1-day-old mouse.

electron microscope. All neurons have a very pale nucleus, round or with irregular contour, limited by a distinct double membrane. The nucleolus is oval or round and presents a typical structure consisting of nucleolonema and a less dense pars amorpha. The cytoplasm is dense and very rich in organelles. Mitochondria are evident, as well as a Golgi apparatus in the form of a unique large complex or of multiple small areas composed of vacuoles and interconnected tubules. Droplets of lipids are found in proximity to the nucleus. The endoplasmic reticulum is represented by short and disorganized cisternae with a few ribosomes attached. Neurofilaments run in all directions, sometimes sparse and sometimes in small bundles (Fig. 2).

In experimental ganglia examined 12 hours after the injection of the antiserum, one sees cells with incipient regressive marks and others in an advanced state of deterioration. This differential response to the antiserum, already seen in our earlier studies with the light microscope(12), is due to the fact that in the same ganglion are found immature nerve cells and cells in a more advanced stage of differentiation; the former are far more vulnerable than the latter to the action of the antibodies to the NGF.

The first detrimental effects of the antiserum are observed in the fine structure of the nucleoli, which appear denser than controls and internally disorganized; the nucleolonema is no longer distinguishable from the pars amorpha in view of the presence of clumping irregular aggregations of coarse, granular material. More advanced nucleolar lesions are seen in other neurons also 12 hours after the antiserum injection; the typical organization of the nucleolar material is completely changed and abnormal structures are apparent within the nucleus. Some of the nucleolar material is condensed to form one or two very dense round bodies which are surrounded by coarse and dense granular material. In the vicinity of the altered nucleolus the chromatin is denser than normal and forms large clumps. In the cytoplasm the ribosomes are no longer grouped in rosettes but are sparse and uniformly distributed throughout the cytoplasm.

Twenty-four hours after the antiserum injection, all nucleoli of nerve cells present various degrees of disorganization. Three distinct formations are now observed in most of the nuclei: the two zones *a* and *b*, originated from separation of the nucleolar material, and a third large area of less dense granular material (zone *c*) (Fig. 3). At subsequent stages, zone *c* tends to expand progressively and finally occupies most of the nuclear space, while the nuclear envelope breaks in one or more points and the cytoplasmic and nuclear material become intermixed. At the same time, the cytoplasmic alterations become more pronounced: the free ribosomes are completely disorganized and the mitochondria show signs of degeneration. Vacuolization, pyknosis, rupture of the plasma membrane, and other signs of cytolysis are seen in many neurons (Fig. 4).

Metabolic Effects of NGF on Ganglia Cultured in Vitro

Since the time of the discovery of NGF in snake venom, it was found that the formation of the fibrillar halo by ganglia explanted in vitro in the presence of NGF is accompanied by a marked enhancement of synthetic and oxidative processes.

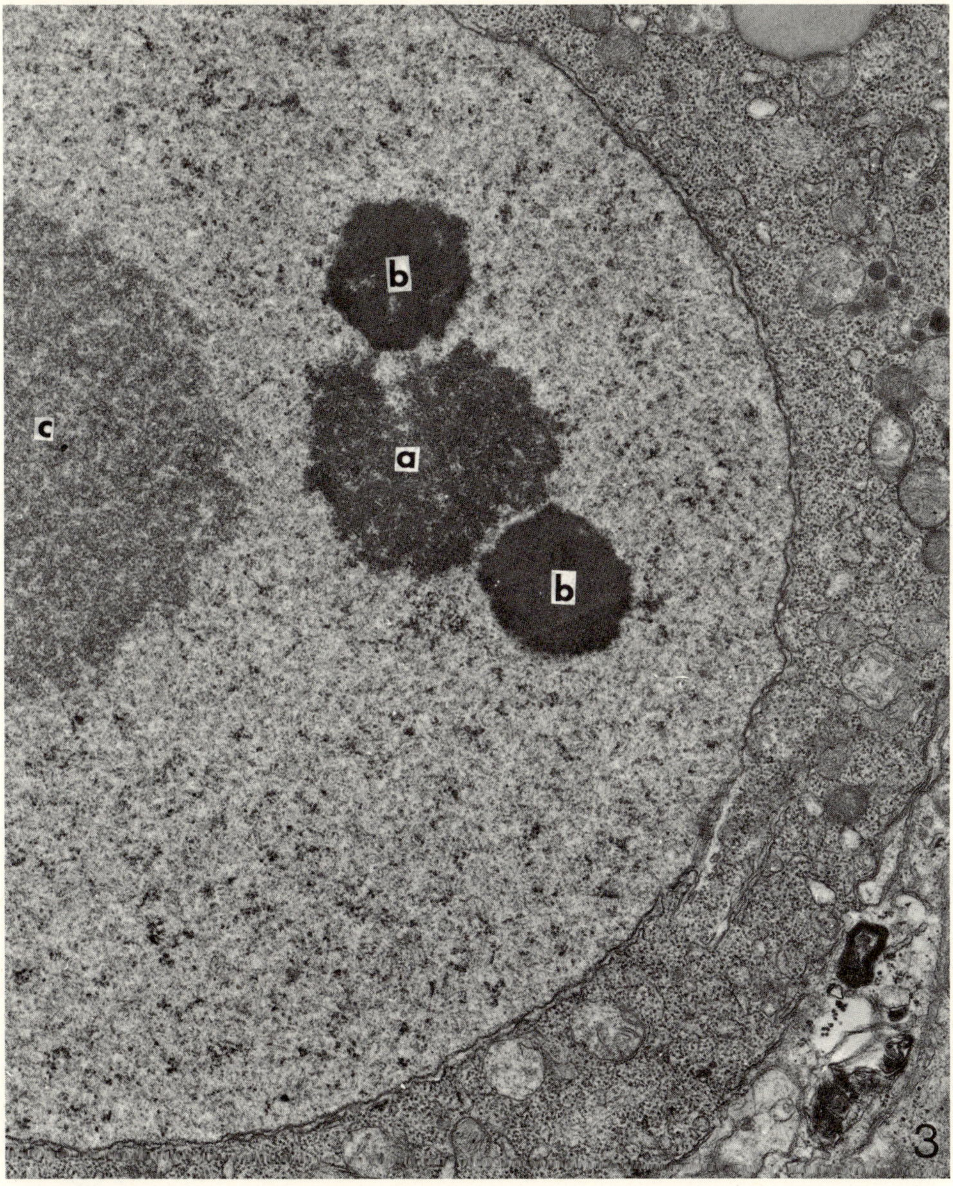

Fig. 3. Neuron from the superior cervical ganglion of a 1-day-old mouse, 12 hours after the injection of the antiserum. The nucleolus consists of one large granular zone (*a*) and two dense round bodies (*b*). Condensation of the nuclear material in a large zone somewhat less dense than zone *a* (*c*).

The presence of a proper energy source is required for the outgrowth of nerve fibers; in its absence the growth of fibers starts, but ceases almost immediately(19). Glucose utilization in sensory ganglia is stimulated by NGF; experiments with specifically labeled glucose indicate that this stimulation is mainly through a direct

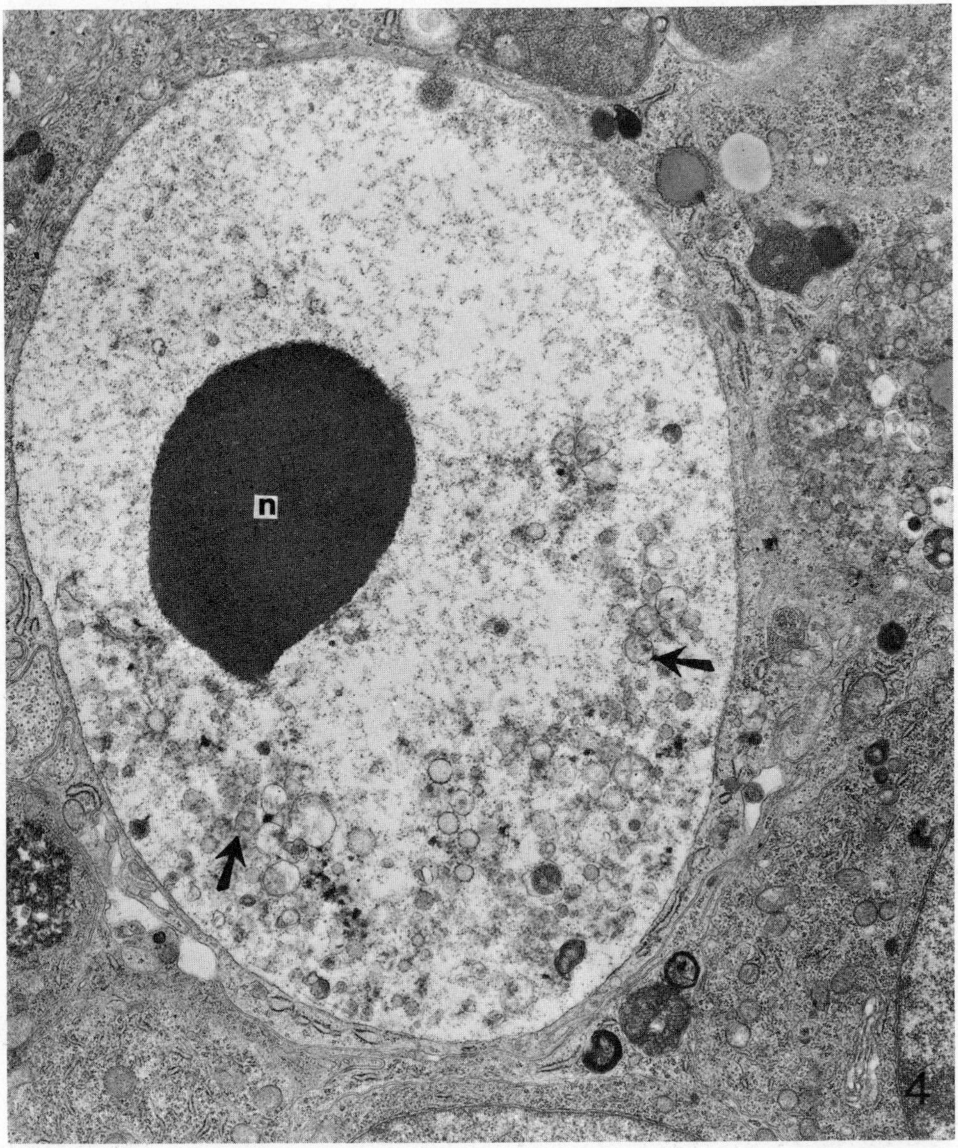

Fig. 4. Neuron from the superior cervical ganglion of 1-day-old mouse, 24 hours after the antiserum injection. Terminal stage of lesion by the antiserum. Pyknotic nucleus (*n*), disappearance of nuclear membrane. Arrows point to mitochondria in advanced stage of disintegration.

oxidative pathway. These results are in line with the finding that the presence of metabolic inhibitors such as fluoride or cyanide in the culture medium do not prevent the nerve fiber outgrowth nor the increase in glucose utilization(20).

The increased energy requirements of the cells stimulated by the NGF correlate with the enhanced synthetic activities of cells which undergo rapid growth.

Lipid synthesis is markedly stimulated(21,22). Studies still in progress show that a selective stimulation of some, but not other classes of lipids takes place in the target cells.

Likewise the protein synthesis is activated, as indicated by incorporation of labeled precursors. Quantitative measurements of the overall protein content in ganglia cultured in the presence or absence of NGF show net synthesis of protein. Preliminary experiments indicate that the synthesis of some classes of proteins is preferentially stimulated by the growth factor(23); conceivably these are the proteins required for the production of axonal material (neurofilaments, neurotubules). Inhibition of protein synthesis by using amino acid analogues in the medium (*p*-fluorophenylalanine) causes a strong reduction in the growth of the fibrillar halo; puromycin and cyclohexamide, when present in the medium at concentrations which practically block amino acid incorporation, inhibit the outgrowth of nerve fibers from the explanted ganglia(24). It appears, therefore, that the cell response to the NGF requires that the cell synthetic machinery be at its best efficiency. The experiments with puromycin, however, indicate that some of the metabolic effects elicited by NGF (RNA synthesis) are still present even when the cells have become unable to give the typical response which is the outgrowth of nerve fibers in vitro. RNA synthesis is rapidly and markedly stimulated by the NGF, as indicated by uridine incorporation studies and by comparison of the total RNA content in ganglia cultured with and without NGF(25).

Actinomycin D added to the incubation medium at a concentration which almost completely inhibits RNA synthesis, prevents most of the metabolic events elicited by NGF, thus suggesting that the growth effect in the target nerve cells is dependent upon the synthesis of some RNA species(25). The hypothesis that these species could be of a messenger RNA type was not supported by subsequent experiments(26). It is of interest to note that by using lower doses of actinomycin D (0.5 to 1 μg/ml) which are still effective in cutting down uridine incorporation (Table 2), the outgrowth of fibrillar halo induced by the NGF in cultured ganglia is not completely prevented (Table 1). Irregular sprouts of nerve fibers are still

TABLE 1

Effect of various metabolic inhibitors on the outgrowth of nerve fibers from embryonic sensory ganglia cultured for 18 hours with NGF (2 BU)[a]

Inhibitor		Nerve-fiber outgrowth	
None		regular halo	(3+)
Fluoride	0.01 M	regular halo	(3+)
Cyanide	0.01 M	regular halo	(3+)
Puromycin	20 μg/ml	no growth	—
Cyclohexamide	5 μg/ml	no growth	—
p-Fluorophenylalanine	0.01 M	markedly reduced	(1+)
Arouridine	0.01 M	reduced	(1-2+)
Actinomycin D	10 μg/ml	markedly reduced	(1+)
Actinomycin D	1 μg/ml	reduced	(1-2+)
Mitomycin	5 μg/ml	regular halo	(3+)

[a]Ganglia preincubated for 60 minutes in Eagle medium containing the inhibitor (concentration as indicated) and then transferred in semisolid medium with the inhibitor at the same concentration.

TABLE 2

Uridine H_3 incorporation into sensory ganglia cultured in semisolid medium for 18 hours, with and without NGF and in the presence of actinomycin D (1 μg/ml)[a]

	Uridine H_3 cpm/10 ganglia
Control	15.294
NGF	54.516
Actinomycin D (1 μg/ml) + NGF	132

[a]The nerve fiber growth response of these ganglia is shown in Figure 6.

clearly growing out from the explanted ganglia, although restricted to only some sectors of the circumference of the ganglia (Fig. 5). The possibility that the stimulation of RNA synthesis, like many other metabolic effects elicited by NGF, might be secondary to a triggering action of NGF exerted in different cell compartments, ought to be considered. Finally, the in vitro experiments seem to exclude a direct action of NGF on the DNA synthesis. Ganglia incubated with the NGF produce in fact a regular outgrowth of nerve fibers without any appreciable change in thymidine incorporation; the growth effect is also normally elicited even in the presence of a DNA inhibitor such as mitomycin (Table 1).

CONCLUSIONS

The studies reported in this article were aimed at gaining information on various aspects of the growth response to NGF by combining different techniques and exploring the target cells at the ultrastructural level and with the light microscope in the presence of different metabolic inhibitors. The experiments reported in the first part of this article give evidence for a remarkable enrichment of the endoplasmic reticulum and an abundance of ribosomes and massive neurofibrillar material in the cytoplasm of NGF-treated cells after only a few hours of incubation. The effect of the antiserum explored at the ultrastructural level showed the extensive nuclear lesions brought about by the antibodies to the NGF. These are of a different and much higher order of magnitude than other antiserum effects investigated at the ultrastructural level. The question remains whether this irreversible and extensive damage to the cell nucleolus results from a direct action of the antibodies to the NGF on the nucleolus or if the primary site of action of the antibodies is at other cell levels. Ultrastructural studies with ferritin-labeled antibodies are planned to investigate this point.

Experiments with metabolic inhibitors, as reported in the second part of this article, suggest the possibility that the enhancement of metabolic processes in the target cells by NGF might result from an indirect action on RNA synthesis, perhaps through a triggering action of the NGF in different cell compartments.

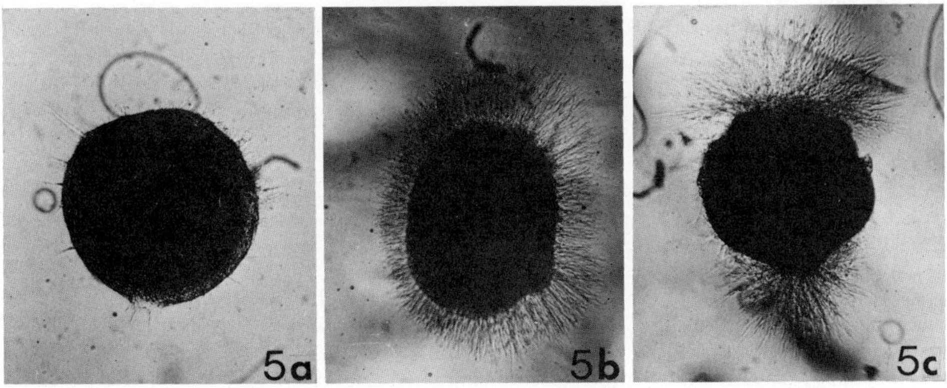

Fig. 5. All figures taken from living unstained sensory ganglia from 8-day chick embryo. *(a)* control; *(b)* ganglion cultured for 24 hours in a medium containing 1 BU of NGF; *(c)* ganglion preincubated for 60 minutes in Eagle medium containing 0.5 μg of actinomycin D and then cultured in the presence of 1 BU of NGF for 24 hours. Notice nerve fibers still growing out from the two opposite poles of the ganglion which correspond to the points of emergence of the afferent and efferent nerve from the ganglion.

REFERENCES

1. Levi-Montalcini, R. The nerve growth factor. Ann. N.Y. Acad. Sci., 118:149–168, 1964.
2. ——— The nerve growth factor: its mode of action on sensory and sympathetic nerve cells. Harvey Lect., 60:217–259, 1966.
3. ——— Differentiation and growth control mechanisms in the nervous systems. *In* Experimental Biology and Medicine. Morphological and Biochemical Aspects of Cytodifferentiation. Hagen, E., Wechsler, W., and Zilliken, P., eds. Basel, Karger, Vol. 1, pp. 170–182, 1967.
4. ——— and Angeletti, P. U. Nerve growth factor. Physiol. Rev., 48:534–569, 1968.
5. Angeletti, P. U., Calissano, P., Chen, J. S., and Levi-Montalcini, R. Multiple molecular forms of the nerve growth factor. Biochim. Biophys. Acta, 147:180–182, 1967.
6. ——— Levi-Montalcini, R., and Zanini, A. Immunochemical properties of the nerve growth factor. *In* Hormones in Development. Hamburgh, M., ed. New York, Appleton-Century-Crofts, 1971.
7. Levi-Montalcini, R., Meyer, H., and Hamburger, V. In-vitro experiments on the effects of mouse sarcoma 180 and 37 on the spinal and sympathetic ganglia of the chick embryo. Cancer Res., 14:49–57, 1954.
8. ——— and Angeletti, P. U. Essential role of the nerve growth factor in the survival and maintenance of dissociated sensory and sympathetic nerve cells in vitro. Develop. Biol., 7:653–659, 1963.
9. ——— and Angeletti, P. U. Biological aspects of the nerve growth factor. *In* Ciba Found. Symp. Growth of the Nervous System. Wostenholme, G. E. W., and O'Connor, M., eds. London, Churchill, pp. 126–147, 1968.
10. Schenkein, I. New aspects on the chemistry of the nerve growth promoting factor. *In* Hormones in Development. Hamburgh, M., ed. New York, Appleton-Century-Crofts, 1971.
11. Cohen, S. Purification of a nerve growth promoting protein from the mouse salivary gland and its neurocytotoxic antiserum. Proc. Nat. Acad. Sci. U.S.A., 46:302–311, 1960.
12. Levi-Montalcini, R., and Booker, B. Destruction of the sympathetic ganglia in mammals

by an antiserum to the nerve-growth promoting factor. Proc. Nat. Acad. Sci. U.S.A., 42:384–391, 1960.
13. ——— and Angeletti, P. U. Immunosympathectomy. Pharmacol. Rev., 18:619–628, 1966.
14. Sabatini, M. R., de Iraldi, A. P., and de Robertis, E. Early effects of the antiserum (AS) against the nerve growth factor (NGF) on the structure of sympathetic neurons. J. Exp. Neurol.,12:370–383, 1965.
15. ——— Levi-Montalcini, R., and Angeletti, P. U. Effetti precoci del siero anti-nerve growth factor (anti-NGF) sulla cellula nervosa. Ann. Ist. Sup. Sanita, 2:349–355, 1966.
16. Levi-Montalcini, R., Caramia, F., and Angeletti, P. U. Alterations in the fine structure of nucleoli in sympathetic neurons following NGF-antiserum treatment. Brain Res., in press.
17. ——— Caramia, F., Luse, S. A., and Angeletti, P. U. In vitro effects of the nerve growth factor on the fine structure of the sensory nerve cells. Brain Res., 8:347–362, 1968.
18. Millonig, G. A modified procedure for lead staining of thin sections. J. Biophys. Biochem. Cytol., 11:736–739, 1961.
19. Cohen, S. A nerve growth-promoting protein. *In* Chemical Basis of Development. McElroy, W. D., and Glass, S., eds. Baltimore, Johns Hopkins Press, pp. 665–667, 1958.
20. ——— Purification and metabolic effects of a nerve growth-promoting protein from snake venom. J. Biol. Chem., 234:1129–1137, 1959.
21. Liuzzi, A., Angeletti, P. U., and Levi-Montalcini, R. Metabolic effects of a specific nerve growth factor (NGF) on sensory and sympathetic ganglia enhancement of lipid biosynthesis. J. Neurochem., 12:705–708, 1965.
22. ——— and Foppen, F. H. Sterol-like compound from sensory ganglia: effect of nerve growth factor (NGF) and insulin on its biosynthesis. Biochem. J., 107:191–196, 1968.
23. Gandini-Attardi, D., Calissano, P., and Angeletti, P. U. Protein synthesis in embryonic sensory ganglia: effect of the nerve growth factor on soluble proteins. Brain Res., 6: 367–370, 1967.
24. Angeletti, P. U., Levi-Montalcini, R., and Calissano, P. The nerve growth factor: chemical properties and metabolic effects. Advances Enzym., 31:51–75, 1968.
25. ——— Gandini-Attardi, D., Toschi, G., Salvi, M. L., and Levi-Montalcini, R. Metabolic aspects of the effect of nerve growth factor on sympathetic and sensory ganglia: protein and ribonucleic acid synthesis. Biochim. Biophys. Acta, 95:111–120, 1965.
26. Toschi, G., Dore, E., and Sapera, O. Studies on RNA synthesized in vitro by sensory ganglia of chick embryo. *In* Ist Intern. Meeting Neurochemistry (Strasbourg). Abstr. of Communications, 1967.

57

Immunochemical Properties of the Nerve Growth Factor*

Pietro U. Angeletti,[1] Rita Levi-Montalcini,[2] and Antonia Zanini[1]

[1]Istituto Superiore di Sanita,
Rome, Italy
[2]Department of Biology, Washington
University, Saint Louis, Missouri

INTRODUCTION

The unique property of the nerve growth factor (NGF) is its ability to stimulate the growth of two types of nerve cells: the embryonic sensory cells and the sympathetic nerve cells throughout their life. The biological activity of NGF on its target cells was explored in vivo and in vitro. In vitro, the NGF produces the outgrowth of a dense halo of nerve fibers from explanted embryonic ganglia. In vivo, daily injections of the purified NGF into newborn mammals produce a marked hypertrophy of the sympathetic ganglia(1,2). The purified NGF is antigenic and a specific antiserum produced in rabbits by Cohen was found to inhibit the biological activity of NGF in vitro(3). When injected in vivo into newborn mammals, the NGF antiserum produces a dramatic neurocytotoxic effect which results in the almost complete and permanent destruction of the sympathetic nervous system(4,5).

Neither the mechanism of action of NGF nor the neurocytotoxic effect of the antiserum are yet clearly understood. Conflicting reports have also appeared in the literature regarding the chemical structure of NGF. In our laboratory the original findings by Cohen have been repeatedly confirmed: namely, the NGF activity is carried by a single protein moiety both in mouse salivary glands and in snake venom. This protein may appear in different molecular forms due, in part, to interactions with other proteins and, in part, to association-dissociation of a fundamental unit.

*This work was supported by grants from the National Institutes of Health (NB-03777) and from the Consiglio Nazionale delle Richerche (Rome).

In the present paper we shall present immunochemical data supporting this point. The serological relatedness of NGF from snake venom and from mouse submaxillary gland will be discussed and, finally, changes of the biological activity and of the immunological properties of NGF under various experimental conditions will be presented.

CHARACTERIZATION OF THE ANTIGEN: NGF FROM MOUSE SUBMAXILLARY GLAND

Purification of NGF from submaxillary gland extracts was achieved by following a modification of Cohen's procedure, briefly summarized in Figure 1. The purified CM fraction showed one single cathodic band on electrophoresis at pH 4.3. At pH 6.6, two or three closely-spaced bands were seen, probably due to polymerization of NGF. At the ultracentrifuge in acetate buffer pH 5, a single symmetrical

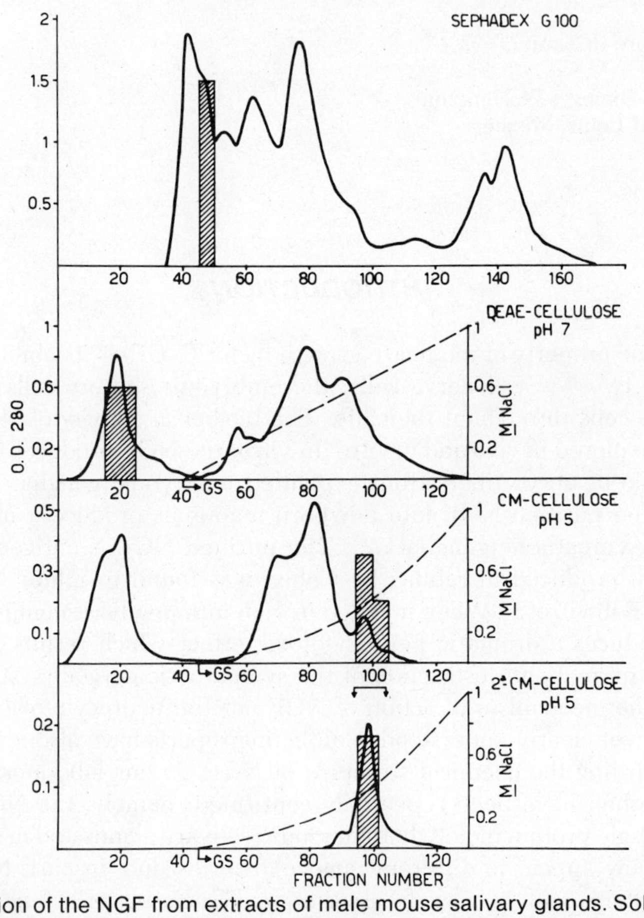

Fig. 1. Purification of the NGF from extracts of male mouse salivary glands. Solid line: protein; hatched areas: NGF activity. From Zanini, Angeletti, and Levi-Montalcini. *Proc. Nat. Acad. Sci., U.S.A.*, in press.

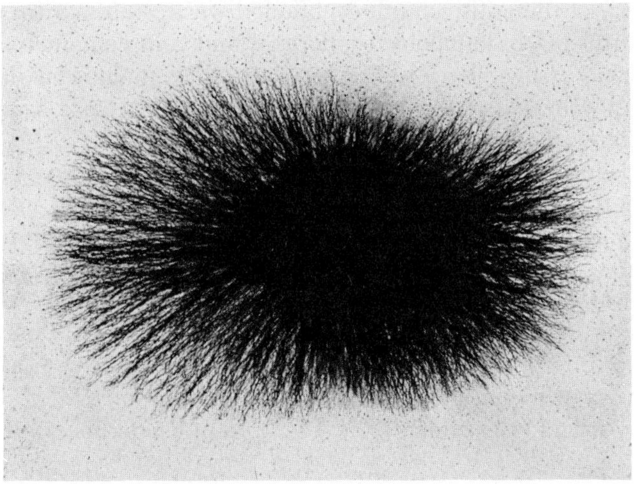

Fig. 2. Typical 3+response from an 8-day chick embryo, sensory ganglion explanted in semisolid medium and cultured 18 hours in the presence of NGF (0.5 ng=1 BU).

peak was seen, with an $S_{20,w}$ of 2.3. By gel filtration on Sephadex G-100 at pH 6, the NGF was eluted as a single peak with an elution volume corresponding to molecular weight of about 28,000. There is evidence, however, that this NGF molecule may dissociate into smaller molecular weight forms. Evidence that these multiple molecular forms may arise from association-dissociation equilibrium of a fundamental monomer will be discussed elsewhere. From the amino acid analysis of the NGF preparation after 24 hours of hydrolysis, a minimum molecular weight of about 15,000 was calculated.

The specific activity of the preparation when assayed in tissue culture, using saline plus albumen as diluent, was of the order of 0.001/μg/ml; thus one BU would correspond to about 0.05 ng, an amount sufficient to give a 3+ response (Fig. 2). It should be pointed out that because NGF is easily absorbed on glass walls, dilutions have to be made with different pipettes; by using the same pipette, even after repeated washings, a prolonged trailing of the activity or reappearance of activity at very high dilutions is obtained.

IMMUNOLOGICAL STUDIES

Antiserum to this apparently homogenous NGF preparation was obtained by injecting rabbits according to the following schedule:

1. 1 mg NGF plus complete Freund adjuvant.
2. After a week, 0.5 mg NGF plus incomplete Freund.
3. After 10 days, a cycle of 3 intravenous injections of 200 μg NGF each.
4. Four days after the last intravenous injection, the animals are bled, and once every second week thereafter.

The potency of the antiserum thus obtained was tested in vitro by: inhibitition test, immunodiffusion, immunoelectrophoresis, and complement fixation.

In vivo, it was assayed by injecting newborn mice daily for 7 days. At the end of the treatment, the experimental and control animals were sacrificed and the sympathetic chain ganglia dissected out, fixed, and studied with the light microscope. The antiserum used in the present experiments gave a 95 percent reduction of the sympathetic ganglia as judged by cell count. When tested in agar plate with various NGF preparations, one single precipitin band was obtained; likewise a single precipitin arc was observed by immunoelectrophoresis (Figs. 3,4).

The antiserum diluted 1:1000 gave maximum C′-fixation with about 50 to 100 ng of homologous antigen; as little as 10 to 20 ng of NGF could be measured with this antiserum dilution. The C′-fixation curve was consistent with a single antigen-antibody system over a wide range of antigen concentrations. Micro C′-fixation was then used to investigate the effects of several treatments on the molecular structure

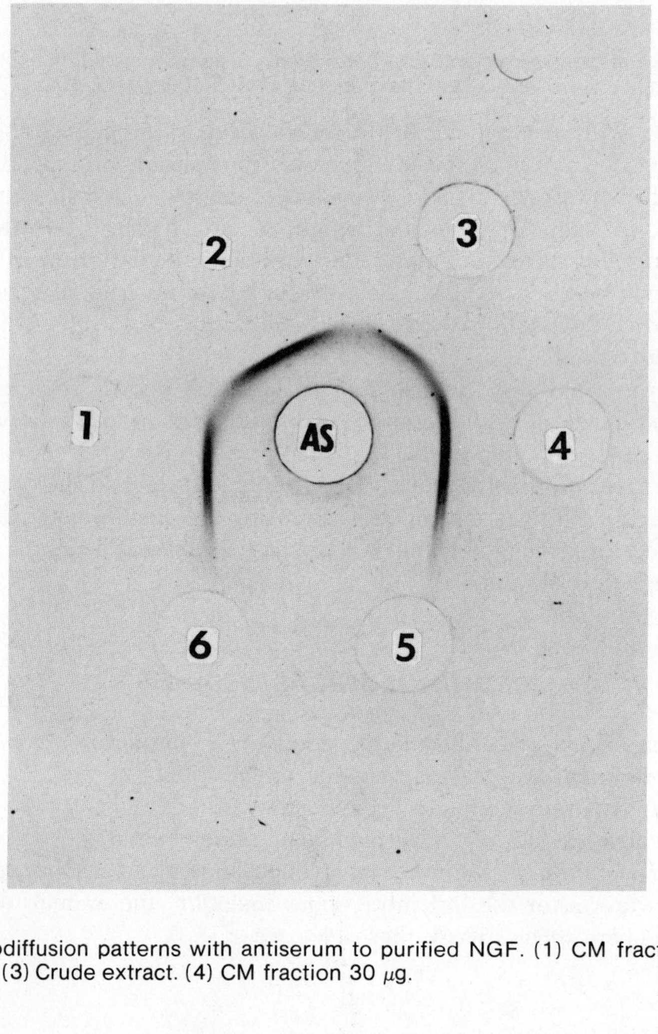

Fig. 3. Immunodiffusion patterns with antiserum to purified NGF. (1) CM fraction 10 μg. (2) DEAE fraction. (3) Crude extract. (4) CM fraction 30 μg.

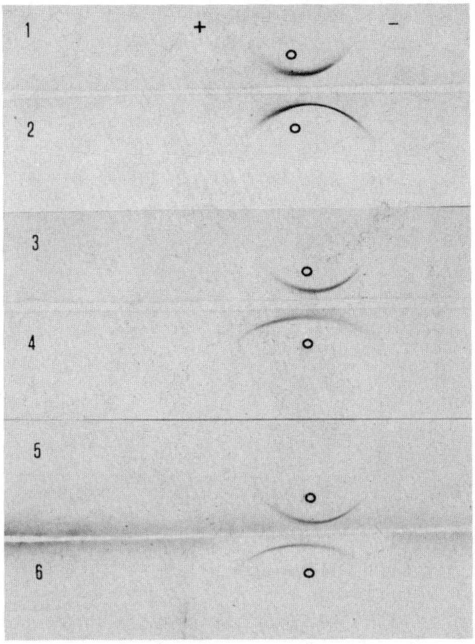

Fig. 4. Immunoelectrophoretic patterns with antiserum to purified NGF. (1) CM fraction 10 μg. (2) CM Fraction 30 μg. (3) CM fraction 10 μg. (4) G-100 fraction 50 μg. (5) CM fraction 10 μg. (6) DEAE fraction 50 μg.

of NGF, in order to correlate the antigenic and the biological properties of the NGF.

Figure 5 shows the C′-fixation curves obtained with NGF-treated with urea at various concentration. Under the experimental conditions (1 hour in urea at pH 7.4), the biological activity of NGF was not affected as judged by the tissue culture assay. Likewise, no changes were detected in the antigenic structure, the various C′-fixation curves being superimposable. If any change was observed, this was a shift of the peak to the left with urea 8 M, which could be interpreted as due to some exposure of antigenic sites.

Treatment of NGF for 2 hours in 0.1 M NaOH and 0.1 M formic acid did not result in appreciable changes in the biological activity; Figure 6 shows the respective C′-fixation curves, which again appear to be superimposable. After standing in NaOH 1 M for 10 hours, the biological activity of NGF is sharply reduced. The C′-fixation curves show a marked combined shift; after 20 hours both the serological and biological activity appear to be completely destroyed.

Preliminary experiments now in progress show that neither the antigenic nor the biological properties of NGF are significantly altered by standing in 0.1 and 1 N HCl for as long as 2 hours. To test the effect of thermal denaturation of the antigenic and biological activity of NGF, aliquots of a solution of NGF at a concentration of 1 mg/ml were treated for 10 minutes at different temperatures and then immediately cooled and diluted for micro C′-fixation. With increasing temperature, there is

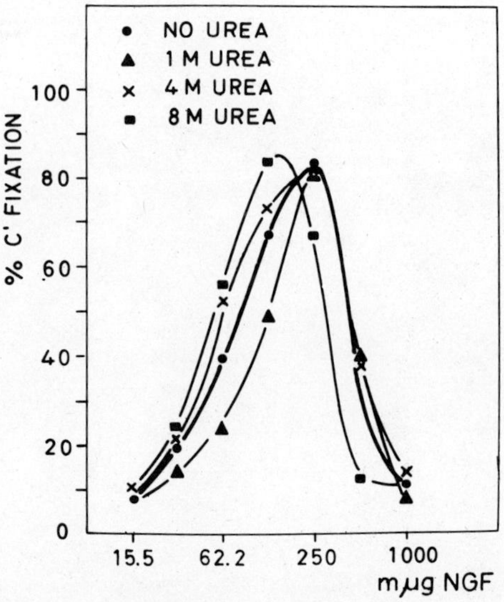

Fig. 5. Micro C'-fixation curves with NGF antiserum and NGF treated with urea at various concentrations.

a progressive shift of the curves toward the region of higher antigen concentration accompanied by a modest decrease in the peak height (Fig. 7). At the same time, a progressive decrease of the biological activity was observed. It is remarkable, however, that neither the antigenic nor the biological activity completely disappeared

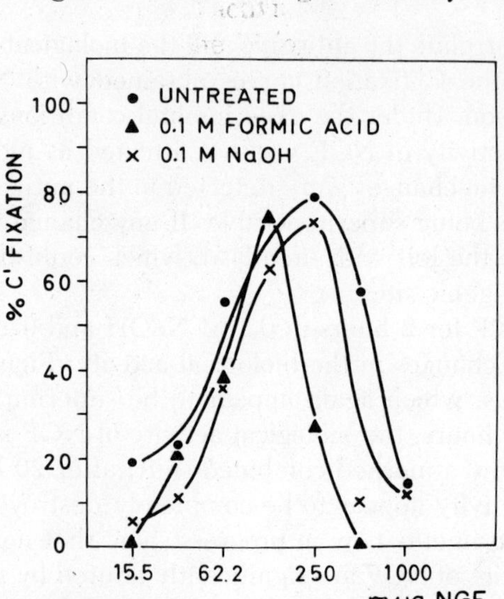

Fig. 6. Micro C'-fixation curves with MGF antiserum and MGF treated with formic acid and NaOH 0.1 M.

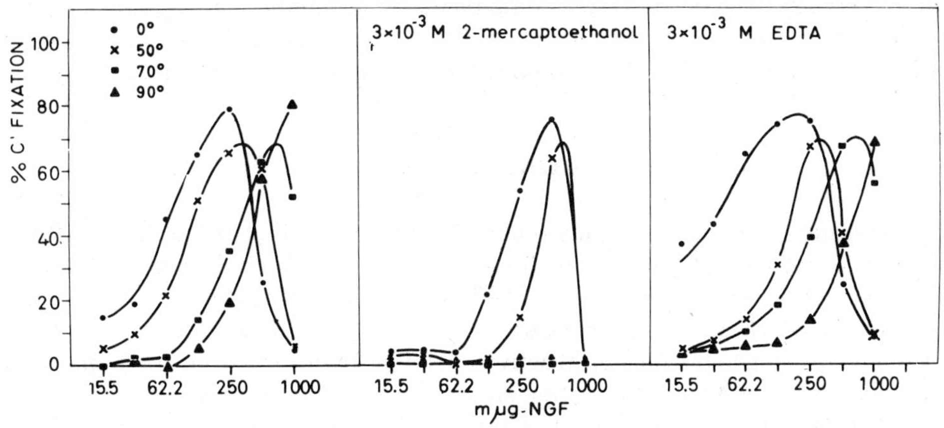

Fig. 7. Effect of thermal denaturation alone and in the presence of mercaptoethanol and EDTA, on micro C'-fixation with purified NGF and its antiserum.

even after prolonged heating (2 hours at 80°C). In the presence of 2-mercaptoethanol 3×10^{-3} M, however, the NGF was completely inactivated already after 10 minutes at 70°C (Fig. 7). At the same concentration, 2-mercaptoethanol alone (without heating) had no appreciable effect on the C′-fixation curve of NGF. This probably means that in the native protein molecule, the sites affected by mercaptoethanol are masked or protected. Experiments are now in progress to study the effect of partial or complete reduction of the disulfide bridges present in the NGF molecule, as reflected in its biological and serological properties. When EDTA at a concentration of 3×10^{-3} M was added to the buffer, the C′-fixation curves using antigen treated in this solution are very similar to those obtained in the absence of EDTA (Fig. 7).

CROSS REACTIVITY OF NGF FROM SNAKE VENOM AND SALIVARY GLAND

Cohen reported the first evidence for a cross reaction between the snake venom and salivary gland NGF's. By using the inhibition test he had in fact observed that the antiserum to salivary gland NGF was also able to inhibit, although to a much smaller extent, the NGF from snake venom.

These data were confirmed thereafter. With the antiserum and NGF preparation used in the present experiments, it was found that 1 ml of salivary gland NGF antiserum can inhibit about 150,000 BU of homologous antigen and only about 5,000 BU of snake venom NGF purified from *Bothrops jararaca.* The extent of the cross reaction was further investigated by using micro C′-fixation. At antiserum concentration of 1:1000, no C′-fixation was obtained with salivary gland NGF; however by increasing the antiserum concentration (10 to 50 times) the C′-fixation becomes positive. At an antiserum concentration about 50 times higher, a peak of

C′-fixation comparable to that obtained with the homologous antigen was obtained with salivary gland NGF. If one takes the factor by which the antiserum concentration had to be raised in order to obtain C′-fixation as a measure of the immunological distance for the two NGF's, this figure is about 50 (index of dissimilarity).

CONCLUSION

The results of the experiments reported above show that the NGF activity is carried by a protein component which can be isolated from crude salivary extracts to a degree of immunochemical purity. We have no evidence that other protein components with which NGF may be associated in crude or partially purified extracts contribute anything to the biological activity of NGF. In our experience, the purified NGF which we refer to as CM fraction shows the highest specific activity.

Antibodies to this protein moiety are those responsible for the inhibition of the biological activity and, more importantly, for the dramatic cytotoxic effect produced in vivo on the target nerve cells. A close relationship between the biological activity and the antigenic activity of the NGF molecule has been demonstrated. Any treatment which causes partial or complete destruction of antigenic sites in the NGF molecule as indicated by C′-fixation would also cause a partial or complete destruction of the biologically active sites. Conversely, treatment with those agents (such as urea, 0.1 NaOH–0.1 M formic acid) which do not alter the antigenic properties of NGF, does not result in any appreciable change of the biological activity. The immunochemical data suggest that the NGF from snake venom and from mouse salivary gland are similar but not identical. It is quite conceivable that, as is the case with many other proteins, marked changes in NGF took place in the evolution from reptiles to mammals. On the basis of their immunological distance, one should expect to find some difference in the primary structure of the NGF molecule from the two sources. This study is now under way in our laboratory.

Finally, another conclusion to be derived from the present experiments is that a combined approach to the study of NGF, utilizing physical methods, immunochemistry, and biological assay should prove particularly useful in obtaining additional information on the structured functional relationship of this unique protein.

REFERENCES

1. Levi-Montalcini, R., and Booker, B. Excessive growth of the sympathetic ganglia evoked by a protein isolated from mouse salivary glands. Proc. Nat. Acad. Sci. U.S.A., 42:373–384, 1960.
2. ——— and Cohen, S. Effects of the extract of the mouse salivary glands on the sympathetic system of mammals. Ann. N.Y. Acad. Sci., 85:324–341, 1960.
3. Cohen, S. Purification of a nerve growth promoting protein from the mouse salivary gland and its neurocytotoxic antiserum. Proc. Nat. Acad. Sci. U.S.A., 46:302–311, 1960.
4. Levi-Montalcini, R., and Booker, B. Destruction of the sympathetic ganglia in mammals by an antiserum to the nerve-growth promoting factor. Proc. Nat. Acad. Sci. U.S.A., 42: 384–391, 1960.
5. ——— and Angeletti, P. U. Immunosympathectomy. Pharmacol. Rev., 18:619–628, 1966.

58

Some Aspects of the Chemistry of Nerve Growth Factor*

Isaac Schenkein,[1,3] Edward Tokarsky,[1] Elmer D. Bueker,[2] and Milton Levy[1]

[1]Department of Biochemistry, New York University College of Dentistry, New York, New York
[2]Department of Anatomy, New York University College of Dentistry, New York, New York
[3]Present Address: Irvington House Institute, Department of Medicine, New York University School of Medicine, New York, New York

INTRODUCTION

It is now some twenty years since Bueker performed a series of experiments from which the "field" of nerve growth factor (NGF) has developed(1,2,3).

Bueker observed the extensive invasion of fragments of sarcoma 180, implanted into the body wall of the growing chick embryo, by nerve fibers from the spinal ganglia. He felt that the concommitant ipsilateral enlargement of the spinal ganglia was due to "Histochemical properties of sarcoma 180 which favor sensory innervation and hence enlargement of spinal ganglia"(1,2). Several years later Levi-Montalcini and Hamburger(4) confirmed Bueker's findings, and added the important observation that the sympathetic centers were also extensively involved. In general agreement with Bueker's hypothesis, these investigators wrote "All available data indicate that sarcoma 180 and 37 produce specific growth of some types of nerve fibers but not of others"(4). The effect was thought to be mediated by the nerve fibers to their respective centers. These observations were followed by an elegant series of experiments by Levi-Montalcini and Hamburger which showed that the observed hypertrophy and hyperplasia of the spinal and sympathetic centers was due to a "diffusible agent," since implantation of sarcoma 180 on the

*This work was supported by grants to E.D.B. and M.L. from the National Institutes of Health.

chorioallantoic membrane of the chick embryo evoked a bilateral growth response, although no nerve fibers could be traced into the tumor mass. No correlation was found between the ganglionic response and the small neoplastic nodules that in some experiments were scattered as metastases in mesentery and mesonephros(5, 6).

ISOLATION OF NGF

An attempt by Bueker and Hilderman in 1953 to obtain a cell-free preparation (Berkefeld filtrate) that had biological activity failed(7). Later studies by Cohen et al. with cell-free fractions from sarcoma 180 revealed that little or no active material could be obtained from tumors that had not been transplanted and allowed to grow in the chick embryo prior to the attempted isolation(8,9). This remarkable requirement in the isolation of the active material from sarcoma 180 has never been elucidated, but may well be related to the observation by Burdman and Goldstein that considerable NGF activity can be detected in the sera of children with neuroblastomas, though virtually no active material can be obtained from the tumors(10). In 1954 Levi-Montalcini et al.(11) demonstrated that explanted spinal or sympathetic ganglia from the 7 to 8-day-old chick embryo responded, when grown in the presence of sarcoma 180 or 37 tumor fragments, with what has aptly been referred to as a "halo of neurites"(8). This dramatic radial outgrowth indeed became the basis of an in vitro bioassay which throughout the research on NGF has provided a semiquantitative measurement for the evaluation of the activity of cell-free preparations. The same year Cohen et al.(8) published a method for the isolation of a biologically-active fraction from the microsomal particulates of sarcoma 180, which had had at least one passage as a graft on the chick embryo. The active material was a water-soluble, heat-labile, nondialyzable nucleoprotein. This, then, was the beginning of the study of the chemical nature of the material (NGF) responsible for the growth-promoting effect on spinal and sympathetic ganglia.

Subsequently, there have been isolated, often from apparently unrelated starting materials, some dozen proteins which each elicit an in vitro response not distinguishable from that provoked by a tumor fragment in the in vitro assay.

It can be seen from Table 1 that proteins with NGF activity have, in the main, been isolated from certain neoplasias, spinal and sympathetic ganglia, submaxillary glands of mice, as well as secretions (venom) from the poison glands of snakes.

The active material, where more closely studied, is chemically described as protein. Acid and heat lability as well as complete or partial loss of biological activity after exposure to proteolytic enzymes attest to this. The stability to alkali (1 hour at 26° C in 0.1 N NaOH) reported by Cohen for the highly purified materials from either snake venom or salivary glands(12,13,16) is remarkable.

The molecular weights (MW) reported vary widely for the purified proteins from venom or mouse gland. S_{20} values range from 1.5 to 7.1S, from which molecular weights have been estimated to be from 12,000 to 150,000.

Monomer-dimer and multimer relationships among identical as well as among nonidentical subunits have been inferred from data of analytical acrylamide gel

TABLE 1

Nerve growth factors reported

Source	Authors	Characteristics
Tumor	Cohen et al. (8)	1 BU = 250 μg $\epsilon_{260}/\epsilon_{280} = 1.78$
Venom	Cohen and Levi-Montalcini (12)	1 BU = 3×10^{-1} μg
	Angeletti et al. (15)	$S_{20,w} = 2.2$ MW = 20,000-80,000
MSG[a]	Cohen (16)	1 BU = 1.5×10^{-2} μg $S_{20,w} = 4.2$ MW = 44,000
MSG	Schenkein and Bueker (18)	1 BU = 3×10^{-2} μg $S_{20,w}$ = 2.41 and 4.24 MW = 20,000 and 40,000 Subunit structure (?) A + C = AC
MSG	Levi-Montalcini (25)	MW = 36,500 on basis of A.A. composition
MSG	Angeletti et al. (15)	1 BU = 10^{-2} μg Subunit, MW 20,000-40,000; Trimer, MW 120,000
MSG	Varon et al. (21)	1 BU = 1.3×10^{-2} μg Subunit structure: $S_{20,w} = 7.1$ Subunit MW = 30,000 Polymer = 140,000
MSG	Schenkein et al. (22)	1 BU = 1×10^{-2} μg 1 BU = 1×10^{-8} μg } Heterogenous fraction
Venom	Banks et al. (14)	1 BU = 10^{-8} μg $S_{20,w} = 2.85$ MW = 40,000; Other forms of MW: 24,000, 38,000
		Biological data only
Spinal ganglia Chick embryo	Bueker et al. (26)	No physiochemical data Activity demonstrated
Sympathetic ganglia Mouse, rat, Cat, Dog, Cow	Levi-Montalcini et al. (27)	Demonstrated by Antiserum to NGF from Mouse Submaxillary which cross reacts — Snake Venom; Granuloma Tissue; Sympathetic Ganglia
Sympathetic ganglia Human	Levi-Montalcini and Booker (24)	(same as above)
Granuloma Tissue	Levi-Montalcini and Booker (24)	(same as above)
Serum of Children with Neuroblastomas	Burdman and Goldstein (10)	Tumor itself has no NGF. Antiserum to mouse submaxillary gland NGF does not inhibit growth of tumor cells
Amphibia Bony Fish Frog Human Fetus	Winnick and Greenberg (28) Shulman-Weils (29)	All these NGFs are immunologically similar to NGF from mouse

[a]Mouse submaxillary gland.

electrophoresis and "molecular sieving" procedures(15,17-21). With the exception of the data from Banks et al.(14) for highly purified NGF from snake venom, no direct values have been reported for either the partial specific volume or the diffusion constant.

DIFFERENT SPECIFIC ACTIVITIES OF NGF ISOLATES

The differences in the reported specific activities of the various isolates are large, at least when taken on the basis of protein concentration.

Banks et al. have demonstrated that with a relatively simple procedure, NGF can be obtained from the venom of *Vipera russelli* with a specific activity of $1\times10^{-8}\mu g/BU$[1]. 2g of crude venom yielded 4 mg of final material or 2×10^{11} BU/g of crude venom(14). Earlier, Cohen and Cohen and Levi-Montalcini(12,13) had reported that 25 percent of the NGF activity of 1 g of crude venom from *Agkistrodon piscivorus* could be isolated in 5.2 mg of protein with a specific activity of $3\times10^{-1}\mu g/$BU. This gives 6.5×10^{5} BU as total yield per gram of this starting material. The venom of *V. russelli*, on the basis of a comparison done by Cohen is some four times as active as *A. piscivorus* on a weight to weight basis(12,13). The *V. russelli* venom should then yield 2.6×10^{6} BU/g by the procedure of Cohen. It is clear that specific activities and yields of these isolated and purified NGF's from the two snake venoms are of entirely different orders of magnitude.

Similar differences can be seen to exist (Table 1) in material isolated from mouse submaxillary gland. We have recently reported on the apparent existence of two response levels of NGF (*A* and *B*) in mouse glands. The in vitro responses from the explanted ganglia were identical when inspected with the light microscope at the two levels(22). NGF *A* shows full biological activity at 3×10^{-2} $\mu g/ml$ (i.e., 1 BU), a value similar to that reported by Cohen(16). NGF *B* became apparent upon serial tenfold dilutions of the solution that gave NGF *A*. In terms of protein concentration, the NGF *B* activity is observed at doses from 2×10^{-8} to $2\times10^{-10}\mu g/ml$.

In a recent review, the findings of biological activity at extremely low protein concentrations have been dismissed as being due to: "A few NGF molecules adhering to the glass pipettes" when the dilutions are carried out(23). It is of some interest to analyze this: (1) it is clear that the observation (i.e., biological activity in a series of dilutions) must be controlled by using different pipettes for each dilution. We have done this and the results are not affected by the maneuver(22); (2) Assume however that this control was not done. Then, the "few molecules" have precisely generated the situation of NGF activity at very low concentration. If, on the other hand, the contamination from one dilution to another is considered more serious, then the following must occur: If optimally 1 BU=0.04 μg of NGF protein (or some value with that order of magnitude), and we assume a molecular weight of 40,000, then that BU is $4\times10^{-12}M$. Allowing for a fivefold deviation on either side of this value this would still represent about 10^{11} molecules. Since the dilutions are made serially and by decades (i.e., 1/10; 1/100; 1/1000 and so on) it is difficult to imagine that some 10^{11} molecules have somehow transmitted themselves on a pipette through a series of 8 or 10 dilutions. The intensity of the growth process obtained in our laboratory at $3\times10^{-8}\mu g/ml$ (in terms of protein) from an explanted spinal ganglion of an 8-day-old chick embryo can be seen in Figure 1. It,

[1]The biological unit is defined as the amount of protein that could elicit a 4+ positive response with the in vitro assay(11).

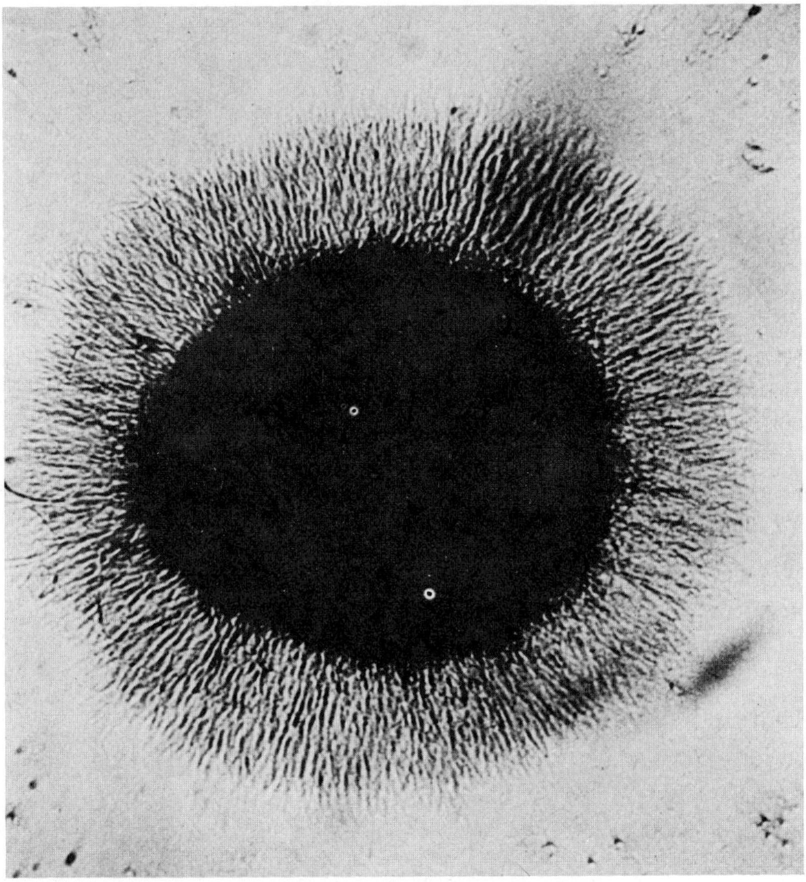

Fig. 1a. Nerve outgrowth obtained with spinal ganglia of 8-day-old chick embryos. (See text Table 3.) Photograph taken after 16 hours of incubation at 37°C in the presence of 3 x 10^{-3} μg protein of NGF.

in fact, is an example of the "stunting" phenomenon described by Levi-Montalcini et al.(11) and Cohen and Levi-Montalcini(12) at high doses.

SUBUNIT STRUCTURES

Recently, in studies performed by Varon et al. it was reported that the fully active NGF protein was a complex molecule of MW 150,000 consisting of subunits of MW 30,000 each(19,20,21). It was demonstrated that heterogeneity existed among these subunits. The specific biological activity of the fully assembled molecule (dissociation occurring below pH 3.8 and above pH 10.3) was of the same order of magnitude as the NGF isolated by Cohen from the same source, i.e., 1 to 3×10^{-2} μg(16).

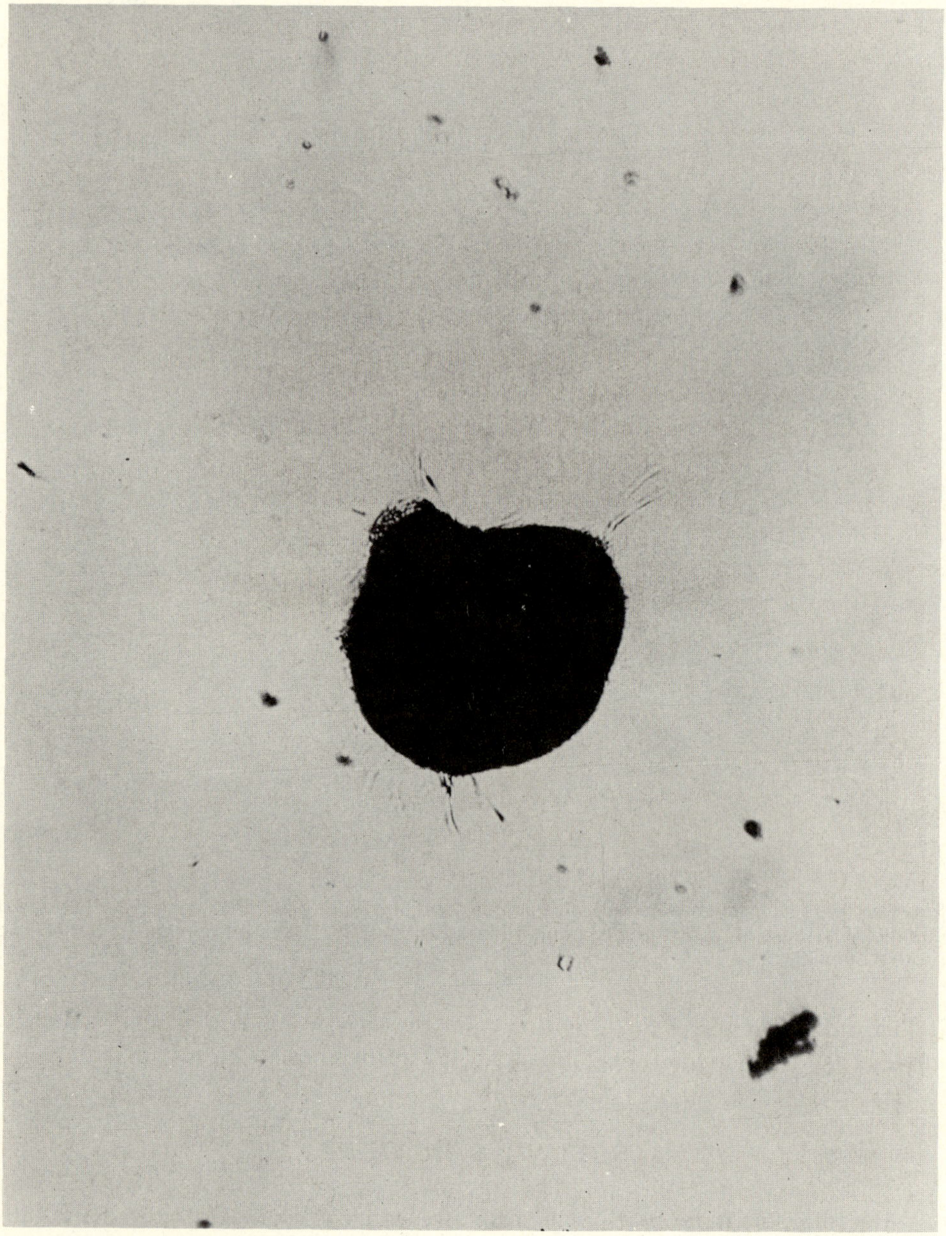

Fig. 1b. Control ganglia. Conditions identical to Figure 1a, but with MGF omitted.

The total yield of active material was however 14 times that of the Cohen procedure. These investigators found that the active material represented 2 percent of the total soluble protein of the gland. The purified material had 80 percent of the expressed activity of the (gland) homogenate. It was also noted that no NGF activity could be detected in proteins from the gland with molecular weights of 20,000 or 44,000(20,21).

As seen from Table 1, NGF from mouse salivary gland has also been found in a protein with MW 120,000, while a MW of 80,000 was reported for an NGF from snake venom(15).

The question arises as to how the large differences of specific activity and MW can be reconciled for the same (and in terms of target tissue) apparently highly specific biological activity, especially in view of the fact that for several of the isolated proteins powerful criteria of purity have been presented.

Some possible answers suggest themselves:

1. A given source has more than one growth factor but the biological responses are the same.
2. The different methodologies used for the isolation of NGF remove inhibitors with different degrees of effectiveness.
3. The proteins isolated are not the real "NGF," but merely serve as "carriers." NGF activity in a protein from mouse gland with an activity of 0.02 μg of protein for a BU could be due to contamination of that protein by the actual factor. It would, for example, take a chemically undetectable impurity (only "protein" determinations are ordinarily done) of NGF to account for the specific activity of NGF preparations where 1 BU = 0.01 to 0.1 μg/ml. The same would hold true for the NGF prepared from snake venom by the method of Banks et al.(1BU=$1 \times 10^{-8}\mu$) when compared with the factor from snake venom isolated by the method of Cohen and Montalcini(12)(1BU=$1 \times 10^{-1}\mu$) and is in fact suggested by these authors(14).

EXPERIMENTS ON NGF B

A series of experiments was performed recently in our laboratory in an attempt to obtain some information on the nature of the NGF *B*.

The method, which starting from adult male mouse submaxillary glands results in a preparation that has NGF *A* activity and contains or generates NGF *B* activity, has been described(22). The major steps involve precipitations by absolute alcohol, followed by salting out of the active material by ammonium sulfate. Column chromatography on carboxymethyl-cellulose and DEAE-sephadex are the last two steps[2](22).

The salient features of this NGF preparation are:

1. There appear (as described previously) upon serial tenfold dilutions of the material two "areas" of activity, referred to as NGF *A* and NGF *B*. (Table 2 shows yield and specific activity observed in the various steps; Table 3 shows the diphasic dose-response relationship.) The specific biological activity of NGF *B* is in the range of 3×10^{-8} to 3×10^{-10} μg/BU, when calculated on the basis of the protein concentration present in the solution from which the dilutions were made.

[2]Apart from NGF the method yields a group of esteroproteolytic enzymes, a detailed description of which is being prepared for publication(22).

TABLE 2

Fractionation of submaxillary glands of mature male mice (25 g and over)

Totals are calculated for 100 mice. The biological unit (BU) is the amount in 1 ml of solution, 0.04 ml of which added to 0.02 ml of rooster plasma and a spinal ganglion of a 7- to 9-day chicken embryo produces 3+ outgrowth of neurites in 18 hours. The esterase unit (tu) catalyzes hydrolysis of 1 μmole of tosylarginine methyl ester per minute for 4 to 8 μmole in 10 ml of solution containing 40 μmole of ester and maintained at 37.0 and pH 8.

Fraction	Total amount			Specific activities	
	Protein (mg)	NGF (BU)	Esterase (tu)	NGF (BU/μg)	Esterase (tu/μg)
Homogenate	2,200	2.6×10^6	3×10^5	1.1	0.14
35 to 75 percent $(NH_4)_2SO_4$	320	8×10^5	2.2×10^5	2.5	0.70
CM[b]	180	6×10^5	2×10^5	3.0	1.1
DS[c]I	64	—[a]	1.8×10^5	—[a]	3.1
DS II	19	2×10^{13}	6×10^2	1×10^9	0.0003
DS III	78	5×10^4	2.6×10^4	0.6	0.04
Cohen CM III (3)	3-4	2×10^5		67.0	

[a]Could not be determined because of clot lysis.
[b]CM = Carboxymethyl cellulose.
[c]DS = DEAE-sephadex.

TABLE 3

Dose of NGF and response of ganglia

Chick embryonic ganglia (7 to 8 days of incubation) with 0.04 ml of medium 199, containing the indicated quantity of protein per milliliter, and 0.02 ml of rooster plasma were incubated for 16 to 18 hours and scored. The arbitrary scale of response is based on references 11 and 13. The numbers are the average scores of two to four embryos at each dose.

Dose (μg)	Response
3000	No clot
300	No clot
30	1
3	1
0.3	Stunting
0.03	3.5
3×10^{-3}	2.5
3×10^{-4}	1.8
3×10^{-5}	0.9
3×10^{-6}	1.9
3×10^{-7}	2.0
3×10^{-8}	1.9
3×10^{-9}	2.3
3×10^{-10}	1.0
3×10^{-11}	0.5
3×10^{-12}	0.5
3×10^{-13}	0.5
3×10^{-14}	0
3×10^{-15}	0
3×10^{-16}	0

2. As shown in Figure 2, the material has not less than five components on analytical acrylamide gel electrophoresis when stained with amido black.

These observations raise a number of questions:

1. An earlier interpretation of similar data(22) led to the tentative conclusion that two levels of NGF existed in this fraction (*A* and *B*) and that both were materials (presumably protein) present in the original solution from which the serial tenfold dilutions were made. The high specific activity of NGF *B* (i.e., biological activity of dilutions where actually only extremely small numbers of protein molecules could be present) was thought to be due to an amplification mechanism at the level of the responding nerve cells(22). It appeared that (on the basis of the amount of protein measured in the original dilution) the observed number of neurites growing out from an explanted spinal ganglion at 1×10^{-11} or $^{-12}$ μg/ml exceeded the number of protein molecules that could possibly be present. Therefore one molecule of NGF *B* must somehow affect more than one ganglionic cell. It is of course impossible to maintain that the NGF *B* re-

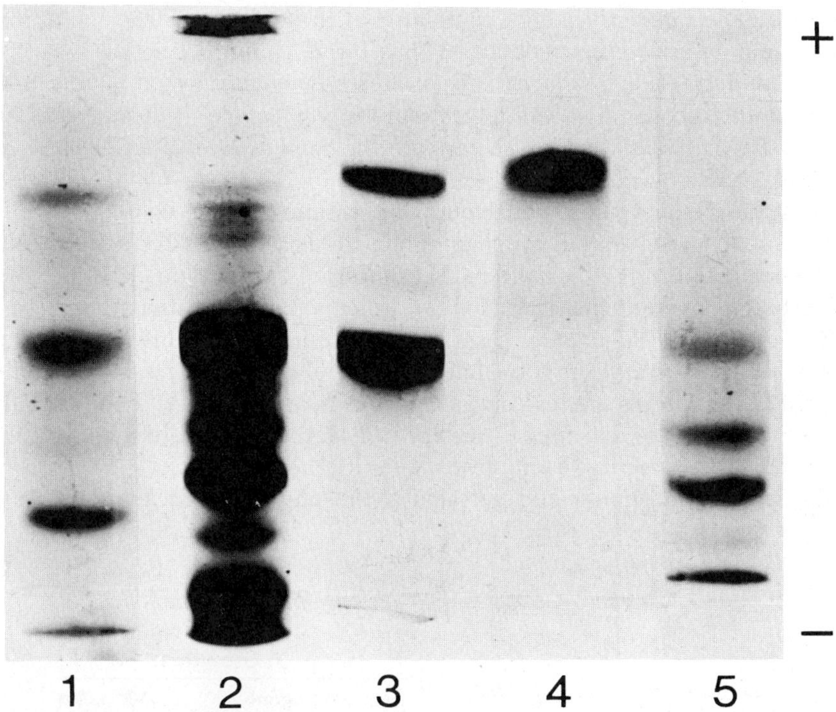

Fig. 2. Analytical acrylamide (7.5 percent) gel electrophoresis patterns, run with tris-glycine buffer at pH 8.5 and stained with amido black. Anode at the bottom. Fractions from left to right are respectively: (1) NGF prepared by the method of Cohen(16); (2) CM fraction of preparative procedure discussed; in text. (3) and (4) Esteroproteolytic enzymes discussed in text; and (5) NGF fraction (DS II) prepared by method given in text. It is this fraction that shows the biphasic dose-response relationship, i.e., NGF *A* and NGF *B*.

gion contains some 0.02 μg/ml, since this would imply (in a series of 8 or 10 tenfold dilutions) that the first solution used some *20g* of the protein per ml.

2. An alternate possibility to the "amplification" theory would be that NGF *B* is not present from the beginning but is the result of the transformation (presumed catalytic) of one (or more) of the components of medium 199, which is the "solvent" used for these dilutions. This process would then allow for significant numbers of molecules of the newly-formed substance to be present even after a series of 8 or 10 tenfold dilutions of the first solution. It of course requires that one (or more) of the materials present in the DS II fraction is the "transformer" or catalyst.

   The diphasic dose-response relationship could then be interpreted by saying that the two NGF's are active independently of each other and have widely different optimum concentrations.

A number of experiments suggested themselves in the probing of the last hypothesis. The components of medium 199 are chemically defined. It should therefore be possible, through systematic screening experiments, to identify the component(s) that are transformed by the DS II fraction into NGF *B*.

These experiments were not completed at the writing of this manuscript, but the following data have been obtained thus far: The DS II fraction was taken up in either medium 199, minimum essential medium (MEM) or Hanks basal salt solution (with glucose). Serial tenfold dilutions were made in these media and the results of the in vitro bioassay scored for the appearance of a diphasic activity curve (i.e., NGF *A* and NGF *B*). (See Table 4.)

These results and additional groupwise screening of the components of MEM led to the observation that in order to obtain the second peak of activity (NGF *B*), the "solvent" used for the serial tenfold dilution of DS II could be limited in composition to four amino acids (methionine, isoleucine, lysine, and tyrosine) plus the salts and glucose of the Hanks basal solution. Further simplification of this mixture has led to erratic results, so that it is not now possible to state whether the minimum requirements are less than this. As indicated in Table 4 the high dilution effect sometimes does not appear even with normally adequate media. It is, however, never present with the simplest medium (Hanks).

The second question asked pertains to the physicochemical characteristics of

TABLE 4

Diphasic responses with various dilution media

Medium of dilution[a]	Test number	Growth at 10^{-2} μg/ml number	Growth at 10^{8} μg/ml number
199	44	44	37
MEM	22	22	19
Hank's	8	8	0

[a]In all cases the medium of incubation contained 0.9 of the standard volume of medium 199, added at the last tenfold dilution.

NGF *B*. Heat stability was the only parameter studied in any detail, and this was done on the DS II fraction taken up in various media and heated immediately at 100°C for 5 minutes in a water bath. The results indicate that: (1) Total loss of biological activity occurs when heating is done with fresh DS II material dissolved in Hanks basal salt solution; (2) When heating is done in either medium 199 or MEM, dilutions can be made such that 3+ growth is obtained with 1.0 μg/ml of nominal protein in the in vitro assay.

As mentioned, (see Fig. 2) the DS II fraction shows some five bands when stained with amido black after analytical acrylmide gel electrophoresis.

Utilizing a preparative electrophoresis unit designed in our laboratory[3], we were able to separate the five visible components in reasonable purity. When tested in vitro, individually or in any combination over a considerable range of protein concentration (from 0.2 to 2×10^{-10} μg), no biological activity could be elicited from any of the materials. Medium 199 was used as diluent in these experiments.

DISCUSSION

The data available indicate that there is a possibility that the actual NGF is not to be identified with any of the proteins thus far implicated, but is an unidentified substance that can be "carried" by several of the proteins from the mouse glands. This would explain why it is possible to devise purification schemes (which are evaluated by the bioassay) that select for proteins of apparent widely divergent characteristics each showing NGF activity.

The concept of "carriage" of an NGF of small molecular weight by various proteins would have to be reconciled with both the observed complete or partial loss of biological activity after proteolysis, and with the following immunological data, it is possible to provoke a near total and highly selective destruction of the sympathetic neurons after injection of antiserum (rabbit) to NGF in the newborn mouse. The NGF that served as antigen was isolated as a protein by the method of Cohen(16) and had a MW of 44,000. It can also be demonstrated that the in vitro activity is abolished when this NGF fraction is incubated with the rabbit antiserum(16).

Neither the susceptibility to proteolytic enzymes nor the immunological properties is perfect proof of the protein nature of the growth factor. Proteolytic enzymes often show marked esterolytic activity with suitable substrates. Moreover, susceptible peptide linkages may occur in nonproteins. From the immunological data we may postulate that unrelated proteins are capable of binding the same hapten (in this case true NGF). The rabbit antiserum then contains antibodies to this haptenic group, recognizing it rather than the protein. It is perhaps a weakness in this hypothesis that haptenic behavior has been demonstrated with covalently bound haptenic groups mainly. We postulate a noncovalent bonding to allow for the isolation of a variety of protein-NGF complexes from the same material. The main immunological attraction of the haptenic hypothesis is that it removes the need to accept the presence of related proteins, derived from the submaxillary

[3]This unit is now commercially available under the name "Fractophorator" and is manufactured and distributed by Buchler, Inc., Fort Lee, New Jersey.

gland of the mouse (evoking antigen) on the one hand, and in the neuronal cells of the chick sympathetic and spinal ganglia (reacting antigen) on the other. In the haptenic hypothesis the effectiveness of the antibody relates to the presence of a small molecule (postulated NGF) tightly bound to some one or more submaxillary proteins and also tightly bound to the chick neuron in its functional action of promoting nerve growth.

In this connection it is pertinent that in our hands the antigen-antibody system of NGF with either the antiserum prepared in our own laboratory or that obtained from Abbot (using purified Cohen fractions) reveal considerable complexity. Both in Ouchterlony double-immunodiffusion and in immunoelectrophoresis we always observe several precipitin lines.

REFERENCES

1. Bueker, E. D. Implantation of tumors in the hind limb field of the embryonic chick and the developmental response of the lumbo-sacral nervous system. Anat. Rec., 102: 369–390, 1948.
2. ——— Hypertrophy of spinal ganglion cells in chick embryos after the substitution of mouse sarcoma 180 for the hind limb periphery. Anat. Rec., 100:735, 1948.
3. ——— Innervation of tumors transplanted in the hind limb of the embryonic chick. Yearbook Amer. Philosophical Soc., 185–186, 1950.
4. Levi-Montalcini, R., and Hamburger, V. Selective growth stimulatory effects of mouse sarcomas on the sensory and sympathetic nervous system of the chick embryo. J. Exp. Zool., 116:321–362, 1951.
5. ——— Effects of mouse tumor transplantation of the nervous system. Ann. N.Y. Acad. Sci., 55:2:330–343, 1952.
6. ——— and Hamburger, V. A diffusible agent of mouse sarcoma producing hyperplasia of sympathetic ganglia and hyperneurotization of the chick embryo. J. Exp. Zool., 123:233–258, 1953.
7. Bueker, E. D., and Hilderman, H. L. Gross stimulatory effects of mouse sarcoma 1, 37, 180 on spinal and sympathetic ganglia of chick embryos as contrasted with effects of other tumors. Cancer, 6:397–415, 1953.
8. Cohen, S., Levi-Montalcini, R., and Hamburger, V. A nerve growth stimulating factor isolated from sarcoma 37 and 180. Proc. Nat. Acad. Sci. U.S.A., 40:1014–1018, 1954.
9. ——— and Levi-Montalcini, R. Purification and properties of a nerve growth promoting factor isolated from mouse sarcoma 180. Cancer Res., 17:15–20, 1957.
10. Burdman, J. A., and Goldstein, M. N. Long term tissue culture studies of neuroblastomas. III. *In vitro* studies of a nerve growth-stimulating factor in the sera of children with neuroblastoma. J. Nat. Cancer Inst., 33:123–133, 1964.
11. Levi-Montalcini, R., Meyer, H., and Hamburger, V. *In vitro* experiments on the effects of mouse sarcomas 180 and 37 on the spinal and sympathetic ganglia of chick embryos. Cancer Res., 14:49–57, 1954.
12. Cohen, S., and Levi-Montalcini, R. A nerve growth stimulating factor isolated from snake venom. Proc. Nat. Acad. Sci. U.S.A., 42:571–574, 1956.
13. ——— Purification and metabolic effects of a nerve growth promoting protein from snake venom. J. Biol. Chem., 234:1129–1137, 1959.
14. Banks, B. E. C., Banthorpe, D. V., Berry, A. R., Davies, H. F. S., Doonan, S., Lamont, D. M., Shipolini, R., and Vernon, C. A. The preparation of nerve growth factors from snake venoms. Biochem. J., 108:157–158, 1968.
15. Angeletti, P., Callisano, P., Chen, J. S., and Levi-Montalcini, R. Multiple molecular forms of the nerve growth factor. Biochim. Biophys. Acta, 147:180–182, 1962.

16. Cohen, S. Purification of a nerve growth promoting protein from mouse salivary glands and its neurotoxic antiserum. Proc. Nat. Acad. Sci. U.S.A., 46:302–311, 1960.
17. Schenkein, I., and Bueker, E. D. The nerve growth factor as two essential components. Ann. N.Y. Acad. Sci., 118:3:171–182, 1964.
18. ——— and Bueker, E. D. Dialyzable cofactor in nerve growth promoting protein from mouse salivary gland. Science, 137:433–434, 1962.
19. Varon, S., Nomura, J., and Shooter, E. M. The isolation of the mouse nerve growth factor protein in a high molecular weight form. Biochemistry, 6:2202–2209, 1967.
20. ——— Nomura, J., and Shooter, E. M. Subunit structure of a high molecular weight form of the nerve growth factor from mouse submaxillary gland. Proc. Nat. Acad. Sci. U.S.A., 57:1782–1789, 1967.
21. ——— Nomura, J., and Shooter, E. M. Reversible dissociation of the mouse nerve growth factor protein into different subunits. Biochemistry, 7:1296–1303, 1968.
22. Schenkein, I., Levy, M., Bueker, E. D., and Tokarsky, E. Nerve growth factor of very high yield and specific activity. Science, 159:640–643, 1968.
23. Levi-Montalcini, R., and Angeletti, P. U. The nerve growth factor. Physiol. Rev., 48: 534–569, 1968.
24. ——— and Booker, B. Destruction of the sympathetic ganglia in mammals by an antiserum to the nerve growth promoting factor. Proc. Nat. Acad. Sci. U.S.A., 46:384–391, 1960.
25. ——— Tentori, L., Vivaldi, G., Angeletti, P. U., and Marin-Bettolo, G. B. Composizione in ammino acidi del fattore di crescita del sistema nervoso. Gazz. Chim. Ital., 95: 333–337, 1965.
26. Bueker, E. D., Schenkein, I., and Bane, J. The problem of distribution of a nerve growth factor specific for spinal and sympathetic ganglia. Cancer Res., 20:1220–1228, 1960.
27. Levi-Montalcini, R., and Angeletti, P. U. Growth control of the sympathetic system by a specific protein factor. Quart. Rev. Biol., 36:99–108, 1961.
28. Winnick, M., and Greenberg, R. Appearance and localization of a nerve growth promoting protein during development. Pediatrics, 35:221–229, 1965.
29. Shulman-Weis, J. The occurrence of nerve growth factor in teleost fishes. Experientia, 24:736–737, 1968.

16. Cohen, S. [illegible] growth promoting protein from [illegible] salivary gland [illegible] Proc. Nat. Acad. Sci. USA, 46: [illegible] (1960).

17. [illegible] and [illegible] The [illegible] nerve growth [illegible] Ann. N.Y. Acad. Sci. [illegible]

18. ——— and [illegible] R. [illegible] nerve growth-promoting protein from mouse salivary gland. [illegible] (1960).

19. [illegible] separation of the [illegible] nerve growth [illegible] factor from [illegible] (19[illegible]).

20. [illegible] and [illegible] submaxillary gland of male [illegible] molecular weight [illegible] nerve growth factor [illegible] gland. Proc. Nat. Acad. Sci. USA [illegible] (1967).

21. ——— [illegible] and Shooter, E.M. [illegible] of the nerve growth factor [illegible] Biochemistry [illegible] (1968).

22. [illegible] The [illegible] growth factor of [illegible] (1969).

23. [illegible] and [illegible] F. B. The [illegible] growth factor. [illegible] (1968).

24. ——— and [illegible] H. [illegible] of the [illegible] components of nerve growth factor [illegible]

25. ——— [illegible] and [illegible] Levi-Montalcini, R. [illegible] nerve growth factor [illegible] (1969).

26. [illegible] Morphological [illegible] of nerve growth factor [illegible] (1969).

27. [illegible] and [illegible] The [illegible] nerve growth factor [illegible] (1969).

28. [illegible] nerve growth factor [illegible]

59

Studies on the Mechanism of Action of Epidermal Growth Promoting Factor (EGF)*

Stanley Cohen

Department of Biochemistry, Vanderbilt
University School of Medicine,
Nashville, Tennessee

INTRODUCTION

The first indications of the existence of an epidermal (epithelial?) growth factor (EGF) were observed during our studies on the nerve growth factor. It was noted that the daily injection of partially purified extracts of the submaxillary gland of male mice into newborn mice resulted in some gross anatomical changes, in addition to effects on nerve cells. These were (a) precocious opening of the eyelids, (b) precocious eruption of the incisors, and (c) a marked stunting of the animals with an inhibition of hair growth(1). For the purpose of this symposium, I am outlining the progress we have made concerning the chemical nature and biological effects of EGF.

ISOLATION AND CHEMICAL PROPERTIES OF EGF

The biological assay devised for purification of the factor was based on its ability to induce precocious opening of the eyelids in newborn mice. The factor responsible for this effect was isolated(2), and was found to be a heat-stable, nondialyzable, antigenic polypeptide.

The isolation procedure involved standard methods of protein fractionation. From 22 g wet weight of submaxillary glands derived from 150 adult male mice, 5 to 7 mg of the polypeptide was obtained. This represented a yield of approximately 20 percent, with a purification of about 150-fold based on protein content.

*These investigations were supported by research grant HD 00700 from the National Institutes of Health. The author is a Research Career Development Awardee of the U.S. Public Health Service.

The daily subcutaneous injection of EGF (1 μg per g of body weight per day) into newborn mice resulted in the opening of the eyelids in 9 days instead of the normal 12 to 13 days. In addition, the purified factor still caused precocious incisor eruption in both newborn mice and rats.

The purity and properties of the isolated material were studied and the results obtained are indicated here. The details of the procedures have been published(2).

Only a single component having a sedimentation constant of 1.25 $S_{20,w}$ was detectable in the preparation upon ultracentrifugation. The ultraviolet absorption spectrum showed a maximum at 277 to 278 mμ; the ratio of A_{280} to A_{260} was 1.6. Only a single component was detected (protein stain) after either paper chromatography or paper electrophoresis. The isoelectric point was found to be at pH 4.2. The factor was antigenic and upon examination by immunoelectrophoresis with cellulose acetate strips, only one precipitating band could be demonstrated.

The biological activity of the factor was destroyed by incubation with crystalline chymotrypsin or bacterial proteinase and was partially inactivated with trypsin. The activity was stable to boiling in distilled water for 30 minutes, but was destroyed when heated in 0.1 N NaOH for 1 hour or in 0.2 N HCl for 2 hours in a boiling water bath.

It was originally reported that all of the common amino acids were present in the preparation except phenylalanine and lysine. The minimal molecular weight, as determined from the amino acid ratios with the assumption of one alanine residue per molecule, was 14,638. In collaboration with Dr. W. Mitchell and J. Taylor we have recently reexamined the question of the molecular weight of the factor, prepared with a slightly improved procedure. We now conclude that in addition to the absence of phenylalanine and lysine, alanine is also absent, and that the small amount of alanine in our previous preparations was due to contamination. The result of one analysis is shown in Table 1. The minimal molecular weight, as determined from these amino acid ratios and the assumption of one methionine and one histidine residue per molecule, was 6,210.

The molecular weight of the polypeptide has now been measured, both by gel filtration and by equilibrium sedimentation (unpublished experiments). From both procedures a molecular weight for EGF on the order of 6,000 to 7,000 has been estimated.

In preliminary gel filtration experiments (with J. Taylor) we have obtained evidence that in crude homogenates of the submaxillary gland EGF activity is associated not only with this polypeptide but also with a much higher molecular weight component. We are attempting to isolate and characterize this material.

IN VIVO EFFECTS OF EGF

The mechanism by which EGF induced the premature opening of the eyelids was clarified when it was noted in a histological study(3) that the observed eyelid separation was a consequence of a more generalized biological effect, i.e., an enhancement of epidermal keratinization and an increase in the overall thickness of the epidermis.

TABLE 1

Amino acid composition of EGF

Amino acid	Residues per molecule: Calculated	Assumed
Lysine	0.08	0
Histidine	1.07	1
Arginine	4.13	4
Aspartic acid	7.73	8
Threonine	2.34	2
Serine	7.00	7
Glutamic acid	3.13	3
Proline	2.18	2
Glycine	6.90	7
Alanine	0.10	0
Half cystine	5.46	6
Valine	2.06	2
Methionine	0.97	1
Isoleucine	2.00	2
Leucine	4.20	4
Tyrosine	5.25	5
Phenylalanine	0.00	0
Tryptophan	2.05	2
Total		56
Estimated molecular weight		6,210

Following the daily subcutaneous injection of EGF into newborn mice or rats, enhanced keratinization and an increase in the thickness of the epidermis both in the eyelid area (Fig. 1) and in back skin were observed. When the factor is in-

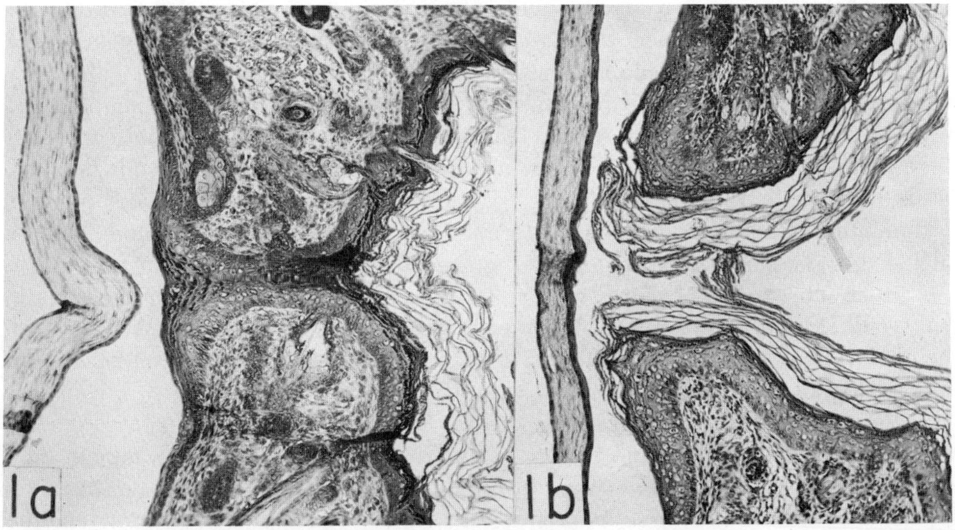

Fig. 1. Sections of the eyelid area from control (*a*), and experimental (*b*), 8-day-old rats. The experimental animal had received daily injections (1 μg/g body weight) of the active protein. ×100. From Cohen and Elliott. *J. Invest. Derm.*, 40:1-5, 1963.

jected into weanling mice for several weeks, gross observation of the mice revealed a clearly visible increase in the diameter of the tail in all of the experimental animals. Histologically there is a marked increase in the thickness of the epidermal layers of the tail and of the plantar surface of the feet. However, in contrast to the newborn animal, no differences were noted in sections of the skin of the back from these older control and experimental animals.

We have confirmed the histological picture by making a number of chemical measurements of pure epidermis obtained by trypsinization of standard areas of skin from 5-day-old control and experimental rats (injected with 2 μg/g/day of EGF). The ratios of the average values for the experimental animals to those for the control animals were: dry weight, 1.65; percent total N, 1.0; DNA content, 1.33; RNA, content, 1.30; histidase activity, 1.8; and acid phosphatase activity, 1.74.

A number of other in vivo effects of EGF have been observed. It will induce a fatty liver when injected into neonatal rats and mice(5). This fatty liver appears to to be, almost exclusively, the result of the net accumulation of triglycerides. Histological changes have also been noted in the lungs and thymus of these animals and the suggestion has been made(6) that the factor may have more widespread effects on epithelial cells than had been previously thought.

EFFECTS OF EGF ON CELLS IN CULTURE

Our next series of experiments were designed to see whether the factor acts directly on the skin or indirectly by way of some other systemic (hormonal?) process. The initial experiments in this direction were carried out in collaboration with Drs. Levi-Montalcini and Attardi.

Back skins derived from 7-day chick embryos were cultured for several days in the presence and absence of EGF (5 μg/ml). The epidermis of the control cultures remained almost unchanged in appearance during the period of cultivation whereas a marked increase in the number of layers of epidermal cells had occurred in the experimental culture (Fig. 2). It was found by autoradiographic techniques (with [^{3}H] thymidine) that in the presence of EGF many more basal cells had synthesized DNA than in control cultures. Upon culturing for longer periods of time, a stimulating effect of EGF upon epidermal keratinization was also noted. The details of these experiments have been reported(7).

In view of the extensive literature on mesenchymal-epithelial interactions in embryonic skin, it was of interest to determine whether the EGF acted directly on the epidermis or whether the epidermal effects were secondary consequences of an initial effect of EGF on the dermis.

We have therefore examined the effects of the factor on dermisfree epidermal sheets derived from chick embryo skin cultured on millipore filters. It appears that EGF does stimulate the growth of the epidermis directly, but that the exact response of the cells varies to some extent with the age of the embryo from which the epidermis is derived and the culture environment. In epidermis derived from 7- to 8-day embryos, EGF induces hypertrophy of the cells and subsequently stimulates

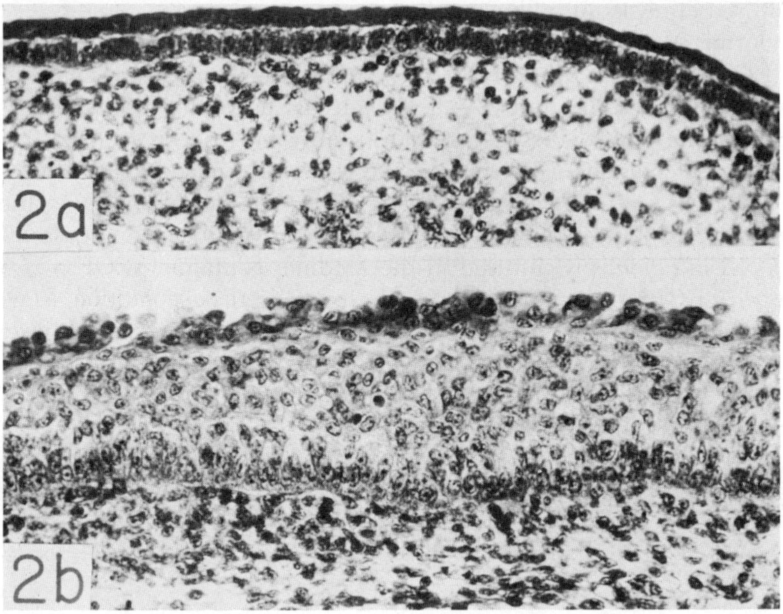

Fig. 2. Control (*a*), and EGF-treated (*b*), explants of back skin from 7-day chick embryos after 3 days of incubation. ×380. From Cohen. *Develop. Biol.,* 12:394-407, 1965.

DNA synthesis as measured by autoradiographic methods with [^{3}H] thymidine(7). In epidermis derived from the back skin of 9-day embryos, EGF stimulates the net accumulation of protein and RNA with no detectable net accumulation of DNA(8). We have preliminary evidence that when epidermis is recombined with the underlying mesenchyme, DNA synthesis is also stimulated in the older cells.

Jones(9) has examined the effects of extracts of the submaxillary gland of male mice on the histological appearance of rat organ cultures. The preparation stimulated the growth of epithelium in organ cultures of rat ureter, ductus deferens, trachea, uterus, vagina, prostate gland, salivary glands, kidney, and lung. Cultures of a number of other organs such as liver, thyroid, adrenal, and pituitary were unaffected. It was suggested that the "epidermal growth factor" be called "epithelial growth factor." I think this probably is valid, but a final conclusion must await the repetition of the experiments of Jones with the most purified preparations of EGF; the preparation used in the original experiments was relatively crude and may have contained similar, but not identical, factors.

STUDIES ON THE MECHANISM OF ACTION OF EGF

I would now like to turn to our studies on some of the biochemical events accompanying the morphological effects of EGF. In all of these experiments epidermal sheets were derived from the back skin of 8.5- to 9-day old chick embryos and were cultured on disks of millipore filter.

Epidermis was cultured in a medium containing chick embryo extract with or without EGF. Tissues cultured for 0, 24, 48, and 72 hours were analyzed for the contents of protein, RNA, and DNA(8). Figure 3 shows the results of a typical experiment. In the control tissues no significant changes occurred in the total amounts of these components during the 3-day period of incubation. However, total protein and RNA in the experimental cultures increased twofold over their initial values during the first 48 hours of incubation. No significant increase in the content of DNA in epidermis cultured in the medium containing EGF was observed. The failure of EGF to stimulate DNA synthesis under these condition, as well as the clear stimulation of protein and RNA synthesis, has been confirmed in experiments in which the appropriate radioactive precursors (thymidine, orotic acid, uridine, and lysine) were added to the medium in the presence and absence of EGF, and their incorporation into the tissues studied(8).

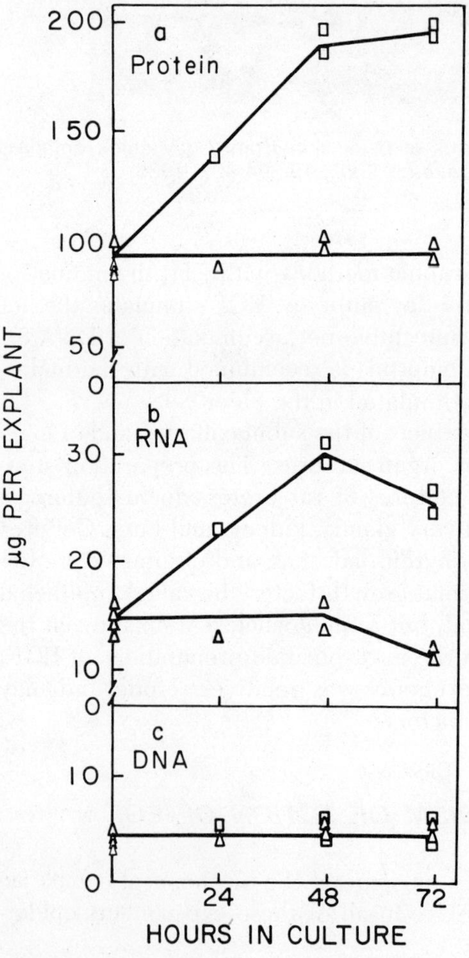

Fig. 3. Changes in (*a*) total protein, (*b*) RNA, and (*c*) DNA of 8.5-day chick embryo epidermis during culture with (□) and without (△) EGF. Tissues were cultured for the times indicated. Three explants were analyzed together, and each point is the average value per explant. From Hoober and Cohen.) *Biochim. Biphys. Acta.*, 138:347-356, 1967.

As one approach to the problem of the mechanism of action of EGF, we have examined the possibility that the growth factor may affect the protein-synthesizing capacity of the cell at the translational level(10). We have been able to prepare and partially characterize a cell-free protein-synthesizing system from chick embryo epidermis. The requirements for protein synthesis in these extracts were similar to those reported for a variety of animal, plant, and microbial systems.

Freshly isolated epidermis was cultured in the presence and absence of EGF. After 4.5 hours of incubation, the tissues from each group of explants were homogenized and tested for their ability to incorporate labeled amino acids into protein. Figure 4 shows that the extract prepared from tissues cultured with EGF was over two times as active as the extract of tissue cultured without EGF, with respect to both the rate of incorporation and the total amount of amino acids incorporated.

As a first approach toward understanding the reasons for this observed difference, the homogenates were fractionated into their ribosomal and soluble components. The activity of each fraction was then assayed independently. The specific activities of the *soluble fractions* from experimental and control tissues to support poly U-directed incorporation of [^{14}C] phenylalanine were compared using an excess of ribosomes obtained from freshly isolated epidermis. No differences could be detected in the soluble fractions derived from these cells.

Therefore, the specific activities of the *ribosomal fractions* from experimental

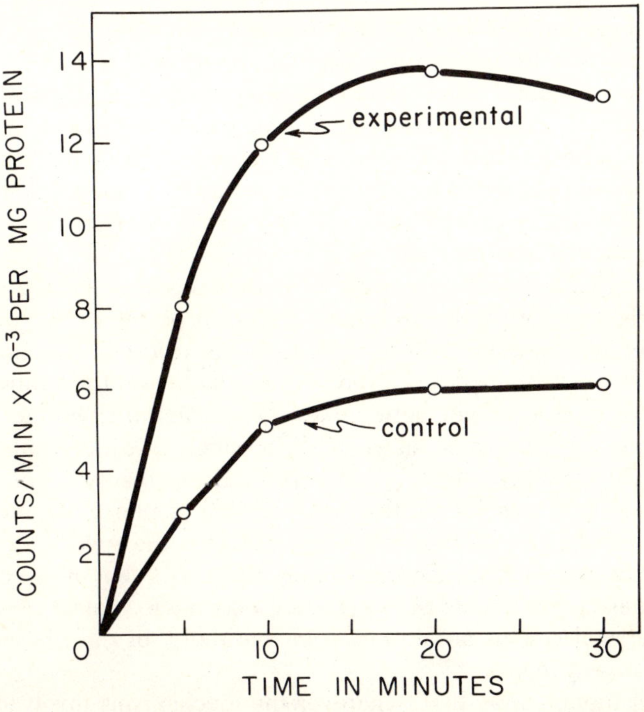

Fig. 4. Amino acid incorporation by 1,300 *g* supernatants of epidermis cultured 4.5 hours with (•) and without (O) EGF. A reconstituted ^{14}C-labeled protein hydrolysate was employed. From Hoober and Cohen. *Biochim. Biophys. Acta,* 138:357-368, 1967.

TABLE 2

Synthesis of protein by ribosomes from control and EGF-treated epidermis[a]

The complete reaction mixture (50 μl) contained 75 mM Tris-HCl (pH 7.5), 50 mM KCl, 8 or 10 mM $MgCl_2$, 2 mM ATP, 0.5 mM GTP, 10 mM phosphoenolpyruvate, 40 μg/ml pyruvate kinase, 300 μg/ml sRNA, 0.2 mM CTP, and 14 mM glutathione. [^{14}C] Lysine (0.15 μc), [^{14}C] phenylalanine (0.15 μc) and poly U (1 mg/ml) were added as noted. An excess of soluble fraction and ribosomes containing 4 to 8 μg of rRNA were employed. The reaction mixtures were incubated for 15 minutes.

Exp. No.	Amino acid	Source of ribosomes	Counts/minute per 10 μg rRNA – Poly U	Counts/minute per 10 μg rRNA + Poly U
1	[^{14}C] Phenylalanine	Control	75	260
		EGF-treated	147	486
2	[^{14}C] Phenylalanine	Control	61	349
		EGF-treated	133	698
3	[^{14}C] Lysine	Control	86	73
		EGF-treated	173	151

[a]Adapted from Hoober and Cohen (10).

and control tissues were assayed with an excess of soluble fraction prepared from freshly isolated epidermis. Table 2 shows that when endogenous mRNA was used to direct protein synthesis, the incorporation observed with ribosomes from experimental tissues was about 2 times that with ribosomes from control tissue. Thus the different activities initially observed in extracts of control and EGF-treated tissues was accounted for by a difference in the activities of the ribosomal fractions. One possible explanation for this result is that the ribosomal fraction from the experimental tissue contained more mRNA, and presumably more polysomes. It will be demonstrated in a later section that this condition does occur.

The following experiment suggested to us that this was not the complete explanation. The response of ribosomes to poly U in incorporating [^{14}C] phenylalanine into polypeptides was measured to determine if the ribosome fraction of experimental tissues simply contained more mRNA (in which case the control ribosomes should respond as well as, if not better than, experimental ribosomes) or if a difference existing in the ribosomes themselves (in which case the same twofold difference in specific activities might be observed). Experiments 1 and 2, Table 2, show that when poly U was added, the incorporation of phenylalanine was stimulated in both control and experimental ribosomes, but the specific activity of the experimental ribosomes was still approximately two times that of the control preparation. From these experiments the suggestion was made that at least one effect of EGF on epidermal cells is an alteration in the ability of the ribosomes to bind mRNA in a functional manner.

These experiments imply that, whatever the mechanisms involved, treatment of the epidermis with EGF should result in a polysomal fraction that was either greater in quantity or functioning with a greater efficiency (or both) than was contained in the control cells.

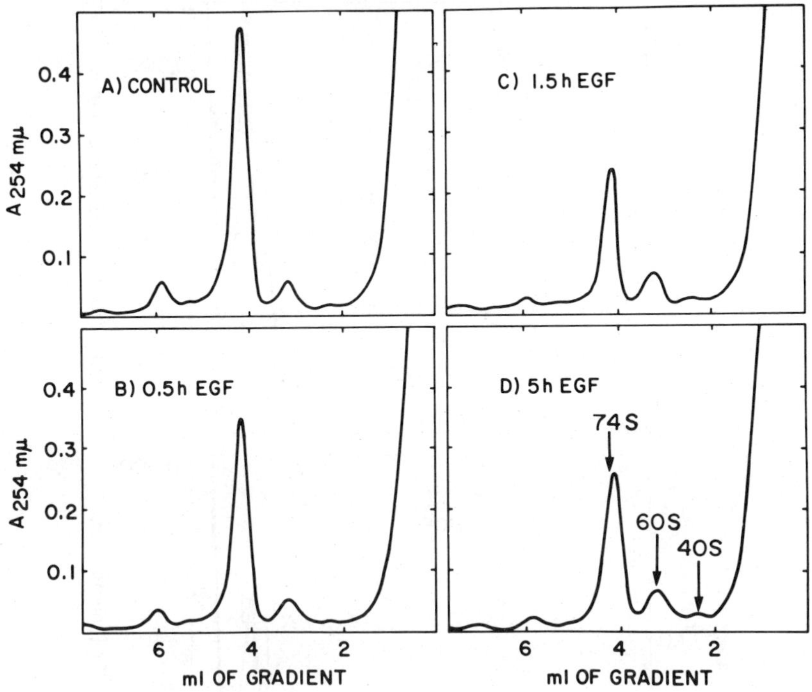

Fig. 5. Sucrose gradients of cytoplasmic extracts of epidermis. (*A*) control; (*B*) tissue exposed to EGF for 0.5 hours; (*C*) tissue exposed to EGF for 1.5 hours; and (*D*) tissue exposed to EGF for 5 hours. Centrifugation was carried out for 2.5 hours at 35,000 revolutions per minute. From Cohen and Stastny. *Biochim. Biophys. Acta,* 166:427-437, 1968.

We have recently been able to demonstrate (by sucrose density gradient centrifugation of postmitochondrial supernatants) that when epidermal cells are treated with EGF there is a conversion of preexisting ribosomal monomers into functional polysomal structures(11). Within 0.5 hours after exposure of the tissue to EGF, the ribosomal monomer concentration had fallen to approximately 70 percent of the control value. This fall in monomer concentration continued for about 1.5 hours, after which period it stabilized at a level of 30 to 50 percent of the control value. These results are illustrated in Figure 5. In the gradient pictured, the polysomes had been sedimented to the bottom of the tube. In a similar experiment after exposure of the experimental tissue to EGF for 1.5 hours, the tissue was pulsed for 5 minutes with a ^{14}C-labeled protein hydrolysate. After homogenization and preparation of a postmitochondrial supernatant, the centrifugation was carried out in a manner allowing for the visualization of the polysomal fraction. The radioactivity of each fraction was also determined. The results are shown in Figure 6. It may be seen that the total polysomal area was increased in the presence of EGF, and that considerably more radioactivity was associated with both the polysomal fraction and soluble protein fractions of the EGF-treated cells as compared to the control cells.

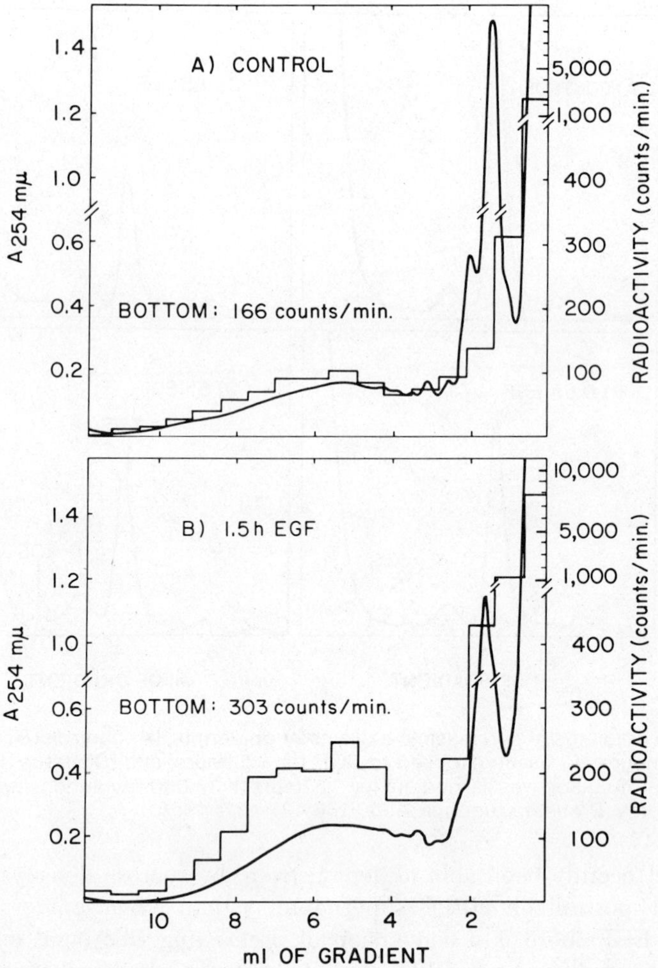

Fig. 6. Distribution of ^{14}C-labeled protein in extracts of control and EGF-treated tissues fractionated on 15 to 45 percent (w/w) sucrose gradients. (*A*) control; (*B*) tissue treated with EGF for 1.5 hours. The incubations were carried out in Earle's balanced salt solution. The tissues were exposed to a ^{14}C-labeled protein hydrolysate for a 5-minute period. The radioactivity of each fraction is indicated on the bar graph; ——— $A_{254m\mu}$. From Cohen and Stastny. *Biochim. Biophys. Acta,* 166:427-437, 1968.

We had demonstrated that EGF stimulated the uptake of uridine in epidermis, resulting finally in the net accumulation of RNA. In the following experiments the effect of EGF on the appearance of newly-synthesized RNA in the cytoplasm was studied in more detail.

Epidermis was cultured for a 4-hour period in the presence of [^{14}C] uridine, with and without the addition of EGF. Cytoplasmic extracts were then prepared and were analyzed on sucrose density gradients to delineate the ribosomal subunits. All fractions from the experimental tissue were considerably more radio-

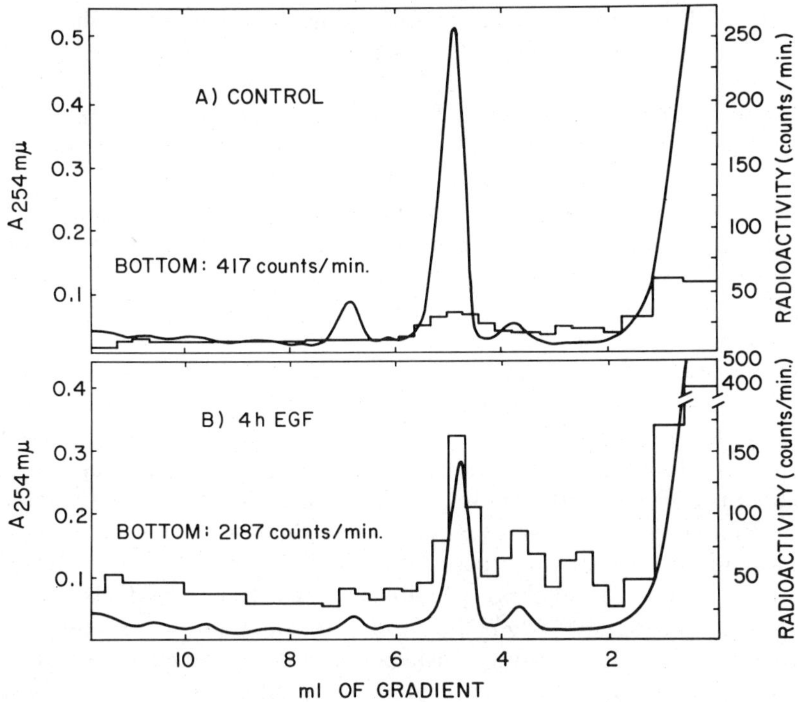

Fig. 7. Distribution of [^{14}C] uridine in extracts of control and EGF-treated tissue, fractionated on sucrose gradients. (*A*) Control tissue exposed to [^{14}C] uridine for 4 hours; (*B*) tissue exposed to [^{14}C] uridine and EGF for 4 hours. The radioactivity of each fraction (or pooled fraction) is indicated on the bar graph and is expressed as counts/min. per 0.3 ml fraction; ——— $A_{254m\mu}$. From Cohen and Stastny. *Biochim. Biphys. Acta*, 166:427-437, 1968.

active than the corresponding fractions of the control tissue (Fig. 7). The transfer RNA (tRNA) region of the gradient (top 1 ml) prepared from EGF-treated tissue was approximately five times as radioactive as that of control tissue; more radioactivity was also present in the 40S, 60S, monomer, and polysome (sedimented to the bottom of the tube) regions of the gradient prepared from the experimental tissue. A number of experiments have been performed which partially characterize the nature of the newly-formed RNA which appears in the cytoplasm after EGF addition. After 90 minutes' incubation in the presence of EGF the synthesis of all of the types of RNA discernible on sucrose gradients (4S, 16S, 18S, and heterogeneous RNA) had been stimulated in the presence of EGF.

These data show that the conversion of ribosomal monomers into functional polysomal structures is accompanied by an increase in protein synthesis and an increase in the synthesis of all of the classes of cytoplasmic RNA.

The central problem of the mechanism by which EGF induces these changes is not known. We have, as yet, not been able to demonstrate any effect of EGF on broken-cell preparations. No experimental evidence has been obtained against the possibility that EGF acts in some manner at a cell membrane and that the observed changes are all secondary responses.

Some possible mechanisms have, however, been examined(11). It does not appear that EGF induces the cell to make a new protein which then stimulates polysome formation since, even in the presence of cycloheximide or cycloheximide plus puromycin, EGF still induces polysome formation. The possibility that the residual protein synthetic activity (1 to 3 percent) which occurs in the presence of these inhibitors is responsible for the observed effects of EGF has not been categorically eliminated, but is considered unlikely.

Although EGF does appear to stimulate the transport of certain metabolites (demonstrable with aminoisobutyric acid, unpublished experiments, Stastny and Cohen) such effects do not appear to be essential for the monomer-polysome conversion since cells incubated for 90 minutes in the absence of both amino acids and glucose respond to EGF by polysome formation.

Another possibility is that EGF may initially stimulate the synthesis of new mRNA. The observation that the incorporation of uridine into RNA is increased by EGF even in the presence of inhibitors of protein synthesis (Table 3) lends some support to this. However, the quantitative interpretation of such data in terms of actual amounts of RNA synthesized is rendered difficult by observations indicating that EGF stimulated the uptake of radioactive uridine into the acid-soluble fraction of the cells and thus probably increased the specific activity of the nucleotide pool. This, however, cannot be solely responsible for the observed effects on RNA synthesis, since in experiments where the uridine nucleotide pool is prelabeled with radioactive uridine and the tissue transferred to a nonradioactive medium, the addition of EGF for 1 to 2 hours again results in a stimulation of RNA synthesis. EGF may cause an alteration in the ribosomes which enable them to bind mRNA more efficiently as suggested by our experiments using poly U in the cell-free protein-synthesizing system. Conceivably EGF may make mRNA more available to the ribosomes by, for example, increasing its rate of transport into the cytoplasm.

TABLE 3

Effect of EGF on the incorporation of [^{14}C]uridine into epidermis

Epidermis was cultured in the standard medium for 18 hours, washed, and then transferred to 1 ml of Earle's balanced salt solution. The cultures were preincubated for 4 minutes in the presence of the indicated inhibitors when those were tested. [^{14}C]Uridine (0.5 μc/ml) and unlabeled cytidine (10^{-3} M) were added to each culture. EGF was then added to the experimental dish and after 1.5 hours of incubation the tissues were removed and washed repeatedly with cold 5 percent trichloroacetic acid, dissolved in 0.05 M NaOH, and aliquots were counted. The concentrations of inhibitor used were: cycloheximide, 5×10^{-4} M and puromycin, 10^{-4} M.

Exp. No.	Inhibitor	EGF	Counts/min/mg protein
1	None	−	475
		+	1,510
	Cycloheximide	−	556
		+	1,680
2	None	−	702
		+	1,700
	Cycloheximide plus puromycin	−	780
		+	1,780

Finally, I would like to summarize briefly our information on another aspect of the problem: the role of the submaxillary gland. EGF is detectable only in extracts of the submaxillary gland of the adult male mouse. Extracts prepared from the submaxillary gland of female mice are inactive. However, within 2 days after the injection of testosterone into adult female mice, the factor may be detected in the gland, and after 1 to 2 weeks of such treatment, the level approximates that in the adult male. In the adult male mouse, the factor could not be detected in the submaxillary gland until about 30 days after birth. This distribution parallels that of the nerve growth factor. The question of whether the salivary gland is the site of actual synthesis of either of these two factors is still open. As far as the normal biological function of EGF is concerned, we have as yet no direct evidence that its presence is essential to the organism, but the experiments of Jones indicating that EGF may stimulate the growth of a variety of epithelial tissues appears to open new avenues for investigation. For example, it is conceivable that some of the diverse effects of the treatment of an organism with testosterone are not direct ones but are mediated through a stimulation of the production of EGF.

NOTE ADDED IN PROOF

We have recently found(12) that EGF induces a marked, but transient, increase of ornithine decarboxylase activity in cultures of chick embryo epidermis. The induction of the enzyme is prevented by inhibitors of protein synthesis. In the presence of EGF and labeled ornithine there is an accumulation of labeled putrescine and, to a lesser extent, of spermidine in the epidermal cells. The injection of EGF into mice causes an increase of ornithine decarboxylase activity in skin.

REFERENCES

1. Cohen, S. Purification of a nerve-growth promoting protein from the mouse salivary gland and its neuro-cytotoxic antiserum. Proc. Nat. Acad. Sci. U.S.A., 46:302–311, 1960.
2. ——— Isolation of a mouse submaxillary gland protein accelerating incisor eruption and eyelid opening in the newborn animal. J. Biol. Chem., 237:1555–1562, 1962.
3. ——— and Elliott, G. A. The stimulation of epidermal keratinization by a protein isolated from the submaxillary gland of the mouse. J. Invest. Derm., 40:1–5, 1963.
4. Angeletti, P. U., Salvi, M. L., Dhesanow, R. L., and Cohen, S. Azione dell' "epidermal growth factor" sulla sintesi di acici nucleici e protine dell'epitelio cutaneo. Experientia, 20:146–148, 1964.
5. Heimberg, M., Weinstein, I., LeQuire, V. S., and Cohen, S. The induction of fatty liver in neonatal animals by a purified protein (EGF) from mouse submaxillary gland. Life Sci., 4:1625–1633, 1965.
6. Cohen, S. Growth factors and morphogenic induction. *In* Developmental and Metabolic Control Mechanisms and Neoplasia. Baltimore, Maryland, The Williams and Wilkins Co., pp. 251–272, 1965.
7. ——— The stimulation of epidermal proliferation by a specific protein (EGF). Develop. Biol., 12:394–407, 1965.
8. Hoober, J. K., and Cohen, S. Epidermal growth factor. I. The stimulation of protein and

ribonucleic acid synthesis in chick embryo epidermis. Biochim. Biophys. Acta, 138:347–356, 1967.
9. Jones, R. O. The *in vitro* effect of epithelial growth factor on rat organ cultures. Exp. Cell Res., 43:645–656, 1966.
10. Hoober, J. K., and Cohen, S. Epidermal growth factor. II. Increased activity of ribosomes from chick embryo epidermis for cell-free protein synthesis. Biochim. Biophys. Acta, 138:357–368, 1967.
11. Cohen, S., and Stastny, M. Epidermal growth factor. III. The stimulation of polysome formation in chick embryo epidermis. Biochim. Biophys. Acta, 166:427–437, 1968.
12. Stastny, M., and Cohen, S. Epidermal growth factor. IV. The induction of ornithine denarboxylase. Biochim. Biophys. Acta, 204:578–589, 1970.

60

Development of the Human Fetal Thyroid*†

Thomas H. Shepard

Central Laboratory for Human Embryology, School of Medicine, University of Washington, Seattle, Washington

INTRODUCTION

This presentation describes the growth, organogenesis, and histogenesis of the human fetal thyroid gland together with a correlation between these morphological features and the biochemical maturation of the gland. Some effects of in vitro exposure of developing human thyroid to thyrotropin will be presented. For these studies length of pregnancy has been estimated from measure of the crown-rump length of the fetus. For the period under study the crown-rump length can be compared to the estimated gestational age as reported by Streeter(1) and by Shepard et al.(2). The estimated gestational age is obtained in most cases by subtracting 14 days from the period extending from the last normal menstrual period until the interruption of pregnancy.

Three general periods of development of the thyroid can be designated: precolloid, beginning colloid, and follicular growth. Table 1 illustrates the relationship of these periods to crown-rump length and age. An earlier stage of development during which the descent of the thyroid anlage occurs is omitted here but publications by Weller(3), Boyd(4), and Shepard(5) describe this event.

Weight Increase

The ratio (×100) of the weight of the thyroid gland to body weight gradually increases during the precolloid period and near the 80 mm crown-rump stage

*Supported by Grants HD-00449 and HD-00836 of the National Institute of Child Health and Human Development, National Institutes of Health, U.S. Public Health Service, and by the State of Washington Initiative 171 Funds for Research in Biology and Medicine.

†Prepared by invitation for presentation at the symposium on Hormones in Development held at Nottingham University, Nottingham, England on September 9 to 12, 1968.

TABLE 1

Stages of development of human thyroid

Stage	Crown-rump length (mm)	Estimated gestation age (days)	Obstetrical weeks
Precolloid	22-64	47-72	8½-12½
Beginning colloid	65-80	73-80	12½-13½
Follicular growth	> 80	> 80	> 13½

(80 days gestation) it attains a plateau at an average of 0.046, which is close to that found in the newborn child (0.049[6]) and in the adult (0.036[7]). This illustrated in Figure 1 which is based on 53 human fetal thyroids obtained mostly after therapeutic abortion(2).

Histological and Histochemical Studies

During the precolloid stage, the thyroid gland consists of strands of compact epithelial cells separated by loose connective tissue and relatively few blood vessels. The number of blood vessels between follicles increases during this period. During the last period of follicular growth the spaces between follicles are reduced progressively and almost completely filled with blood vessels. No change in average cell height has been found during the later two periods.

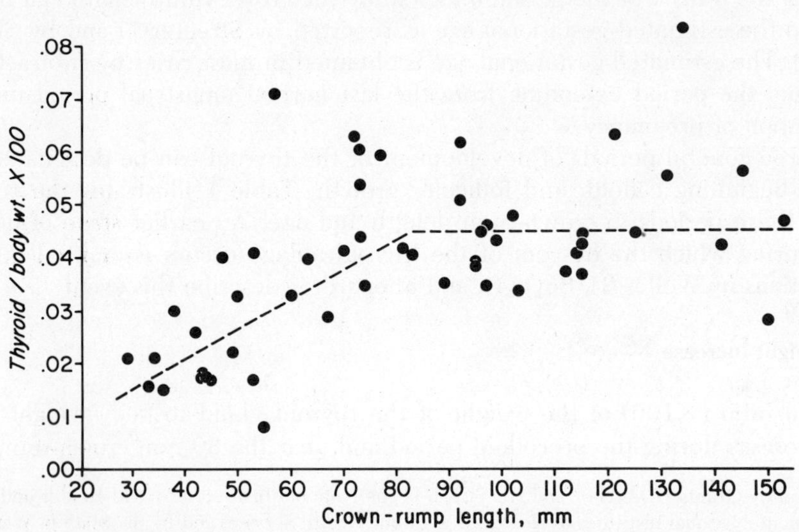

Fig. 1. The ratio of the thyroid weight to body weight (×100) is plotted against crown-rump length. From Shepard, *Gen. Comp. Endocr.*, 10:174-181, Courtesy of Academic Press, Inc.

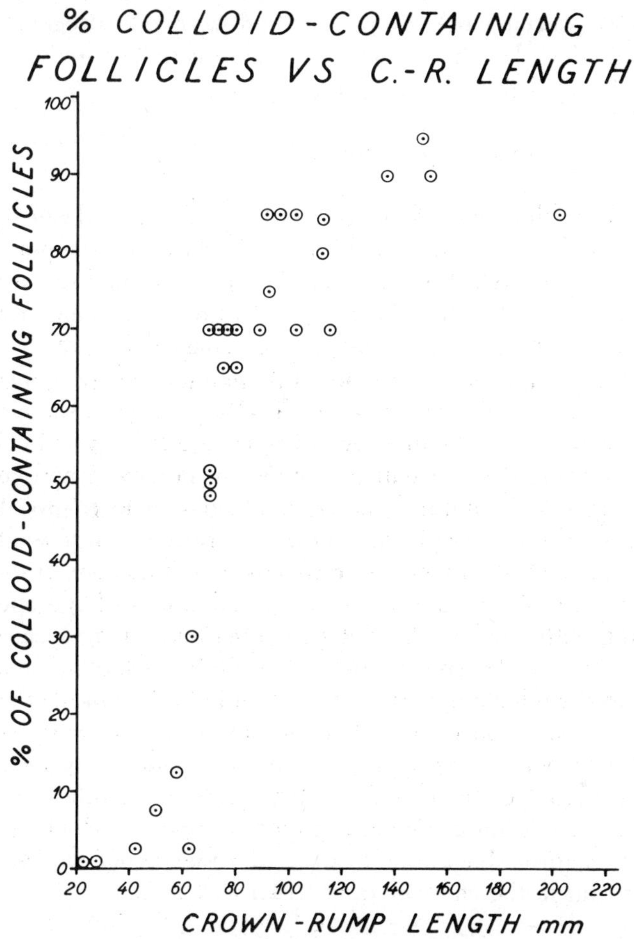

Fig. 2. Percentage of cells that were part of colloid-containing follicles from 29 thyroids is plotted against their respective crown-rump lengths in millimeters. From Shepard et al., *Anat. Rec.*, 149:363-380, 1964. Courtesy of the Wistar Institute of Anatomy and Biology.

The abrupt appearance of colloid in the follicles of fetuses between 60 and 80 mm in crown-rump length has been reported previously(8,9) and is illustrated in Figure 2. Small accumulations of periodic-acid-Schiff-positive material first appear in the centers of the follicles. This luminal material is resistant to digestion with amylase. The staining characteristics of the early colloid are similar to those seen at later fetal stages except that the larger colloid spaces are more deeply stained by toluidine.

With the onset of colloid production the nuclei of the epithelium becomes less compact appearing and the RNAase-removable cytoplasmic material increases(10). Also during this stage there is an abrupt decrease in the preexisting high content of glycogen(9,10,11). It is conceivable that this glycogen serves as a source of energy for the process of differentiation.

The cytoplasm of developing thyroid epithelium contains a moderate amount of acid phosphatase, succinic dehydrogenase, and lactic dehydrogenase activity but no evidence of alkaline phosphatase activity. These activities undergo little change during the periods of further differentiation(10).

Electron Microscopic Analysis of Thyroid Differentiation

During the precolloid period of development the cytoplasm of the thyroid epithelium is scanty. There is relatively little endoplasmic reticulum and the ribosomal particles are mostly unattached. During and following the early colloid period there is an increase in rough-surfaced endoplasmic reticulum. Although a Golgi apparatus is present in these cells it is at no time conspicuous(12).

Based on electron microscopic studies of human fetal thyroid, a mechanism for colloid cavity formation has been described by Bierring and Shepard(12) and by Shepard(5). As early as the 28-mm crown-rump stage the thyroid cells contain an intracellular canaliculus (Fig. 3) which is lined by microvilli and contains a thin granular substance. With maturation these cavities, which are thought to be formed from smooth-lined endoplasmic reticulum, become more dilated, contain a denser substance, and gradually move toward the eventual apex of the cell (Fig. 4). Subsequent stages are observed when these apical cavities have opened extracellularly, coalescing with similar adjacent structures and giving a cloverleaf outline on cross section (Fig. 5). These extruded intracellular spaces are maintained in a central location in the follicle by desmosomes that hold the apical edges of the cells together. From this mechanism of colloid formation, a very close correlation between colloid production by the cell and central colloid cavity formation can be made; in fact, it might be stated that the first secretory product of these cells is mechanically responsible for their organization around a colloid cavity. Figure 6 illustrates this mechanism diagrammatically. No evidence of a similar sequence has been observed so far in the chick or rat(13,14,15,16,17).

Biochemical and Radioautographic Studies from Organ Culture

Previous studies of the onset of function in the human fetus have required administration of relatively large amounts of ^{131}I to the mother and recovery of the fetus by subsequent therapeutic abortion(18,19,20,21). The earliest evidence of iodide concentration was furnished by Hodges et al.(20) who observed radioautographic concentration in the thyroid of a fetus of 73 mm crown-rump length. Yamazaki et al.(18), the only investigators to study iodinated end-products in the human fetus, found the first evidence for thyroid hormone production at the 142-mm stage.

The studies to be summarized in this paper were made with thyroid glands removed in a fresh state from fetuses obtained most usually following therapeutic abortions. This material has been published previously (22). The thyroid, and a piece of liver as control, were placed in a complete tissue culture medium to which 1 to 2 μc of ^{125}I was added and incubated for 18 hours at 38°C. A small sample of this tissue was fixed for radioautographic histology and the remainder of the tis-

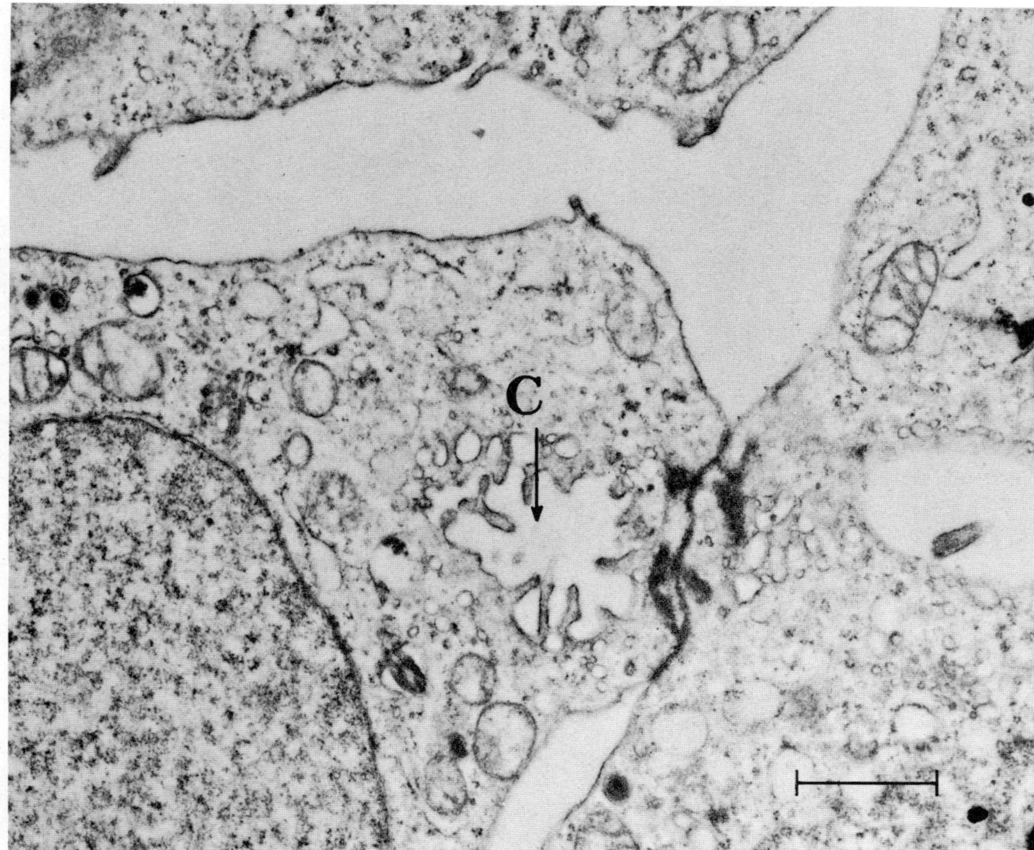

Fig. 3. Electronmicrograph from the thyroid of a 28-mm human fetus to illustrate an intracellular cavity *C* which is lined by microvilli. Note the free ribosomes. The marker in Figures 3 through 6 signifies one micron. From Shepard, Organogenesis, 1965. Courtesy of Holt, Rinehart and Winston, Inc.

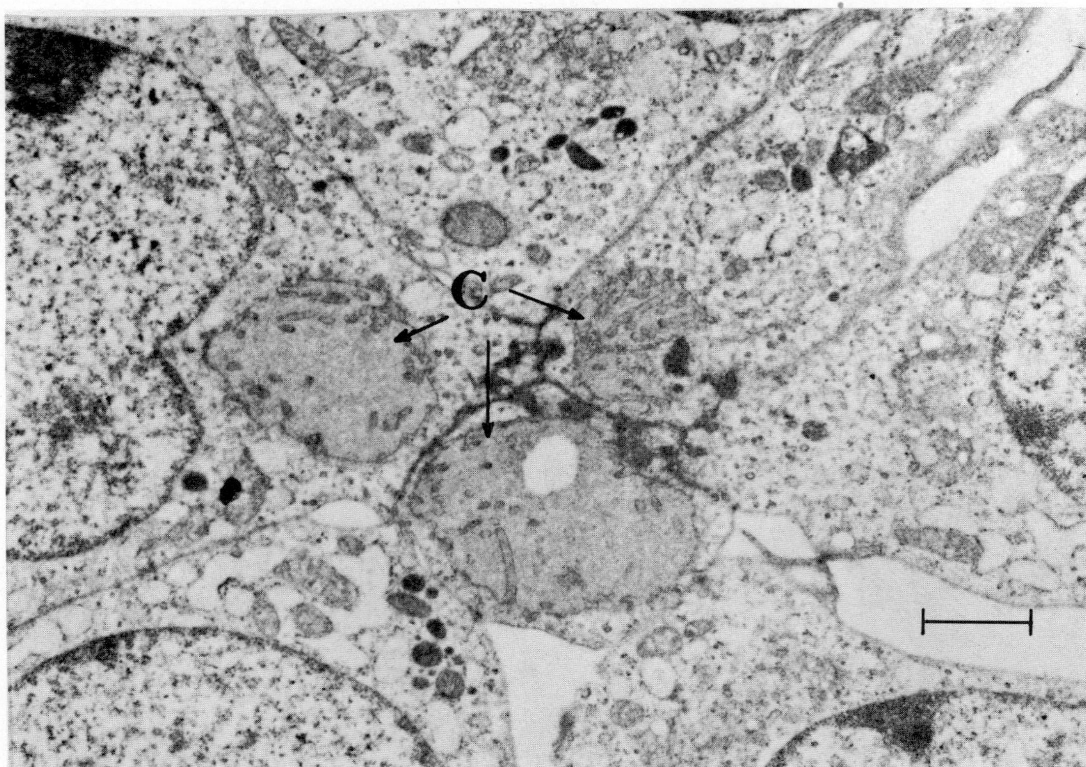

Fig. 4. Fetal thyroid from a 61-mm human fetus. The intracellular spaces *C* have enlarged and are now located close to the apex of the cell. Note the increased density of their contents. From Shepard, Organogenesis, 1965. Courtesy of Holt, Rinehart and Winston, Inc.

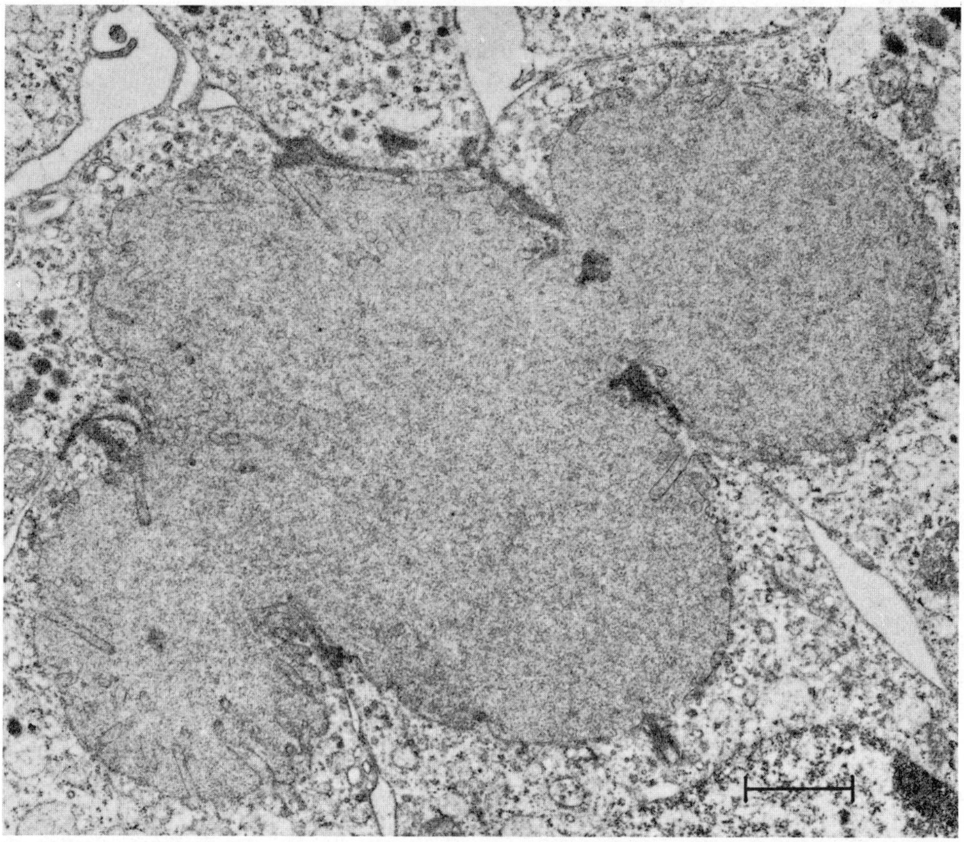

Fig. 5. Fetal thyroid from a 98-mm human fetus. The intracellular cavities are in the process of being disgorged into a central colloid cavity, which gives a clover-leaf pattern on cross section. Note the desmosomes which cause central retention of the colloid. The ribosomes are now mostly attached to endoplasmic reticulum. From Shepard, Organogenesis, 1965. Courtesy of Holt, Rinehart and Winston, Inc.

sue specimen was washed, homogenized, digested, and analyzed for formation of iodinated compounds using paper chromatography and liquid scintillation counting.

To date 20 specimens have been studied. Specimens from fetuses of 60 mm crown-rump length and less did not contain more radioactivity than the control liver and only iodide was identified. Starting at the 68-mm stage, which is equivalent to a gestational age of 74 days, iodinated end-products were identified and these end-products were the same as those found in the adult gland (Fig. 7 and 8). The relative proportions of iodinated amino acids in fetuses larger than 68 mm did not change and only small and variably present amounts of triiodothyronine were found. No evidence for sequential stages of biochemical maturation have been obtained from this relatively small group of glands studied.

The tissue which was removed for histological studies after organ culture was fixed in osmium tetroxide buffered with bicarbonate buffer and embedded in ep-

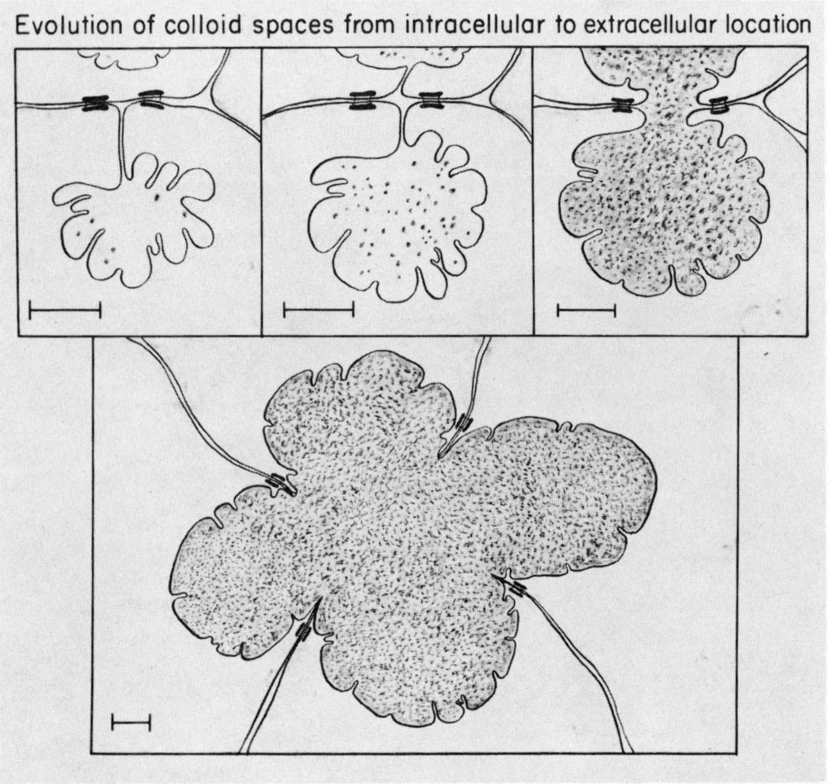

Fig. 6. Diagrammatic drawing of the histogenesis of the human thyroid central colloid cavity. In the bottom panel the intracellular canaliculi are shown in the final phase of union. Desmosomes indicated by the short parallel marks play an important part in the final organization of these cells by containing the extruded contents of the canaliculi. The marker signifies l μ. The final stage resembles in cross-section a clover leaf. From electron micrographs taken by Dr. Franz Bierring, Department of Anatomy, University of Odense, Denmark.

oxy resin. The blocks were then sectioned at 1 μ thickness using a glass knife. This technique was found to be much superior to previous standard preparations and, furthermore, allowed for the identification of the incipient stages of colloid cavity formation described from electron micrographs in the preceding part of this paper. The tissue sections were exposed to a liquid photographic emulsion, and after appropriate lengths of incubation time the emulsion was developed and the tissue stained with toluidine. In all specimens where organic iodinated products were identified, mature colloid was present and overlying the colloid in most but not all areas were numerous exposed silver grains (see Fig. 9). On some sections in which the colloid cavities were in the process of coalescing (cloverleaf stage) there was evidence of iodine concentration (Fig. 10).

Effect of Thyrotropin on Developing Thyroid-Antrum Formation

Explanted prefollicular embryonic chick thyroid, when exposed to thyroid-stimulating hormone (TSH) or pituitary explants, develops conspicuous intracellu-

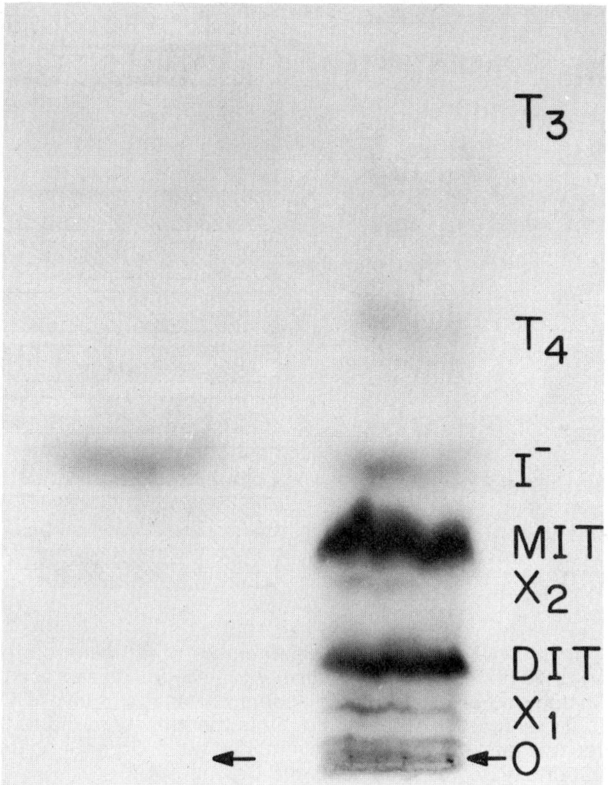

Fig. 7. Radioautograph of dried paper chromatogram (n-butanol-absolute ethanol-0.5 N NH_4OH, 5:1:2) from organ cultures of thyroid and liver of a human fetus of 68-mm crown-rump length. The arrow and 0 stand for origin; the solvent front was just above the top. X_1 and X_4 were unidentified. Diiodotyrosine (DIT), monoiodotyrosine (MIT), iodide (I^-), Thyroxine (T_4), and triiodothyronine (T_3) were identified. The Rf for thyroxine was 0.48. From Shepard, *J. Clin. Endocr.*, 27:945-958, 1967. Courtesy of J. B. Lippincott and Co.

lar dilations called antra shown in Figure 11(13,23,24,26). This phenomenon occurs as early as 8 minutes after TSH exposure and disappears after 12 to 24 hours. The antra are larger than the nuclei and their clear contents have not been identified. The chick thyroid is responsive to as little as 10^{-5} units per ml of culture media and after the 10th day of incubation there is no antrum response to the stimulation. The antra can be identified by phase-contrast microscopy and also in vivo after TSH is added to the chorioallantoic membrane. Electron micrographic studies by Petrovic and Porte(13) and Daems and Thesingh(26) have shown that the antra are formed from small microvillous-lined cavities and are located extracellularly. In unpublished work, Shepard has found antra in human thyroid explanted from fetuses of 30 to 65 mm crown-rump length (Fig. 11). A tempting explanation for the antrum formation is that small intracellular canaliculi, as described above in man (Figs. 3 to 6), are present also in the chick, and that they are stimulated to imbibe water at a rate which is greater than the rate of thyroglobin production. As

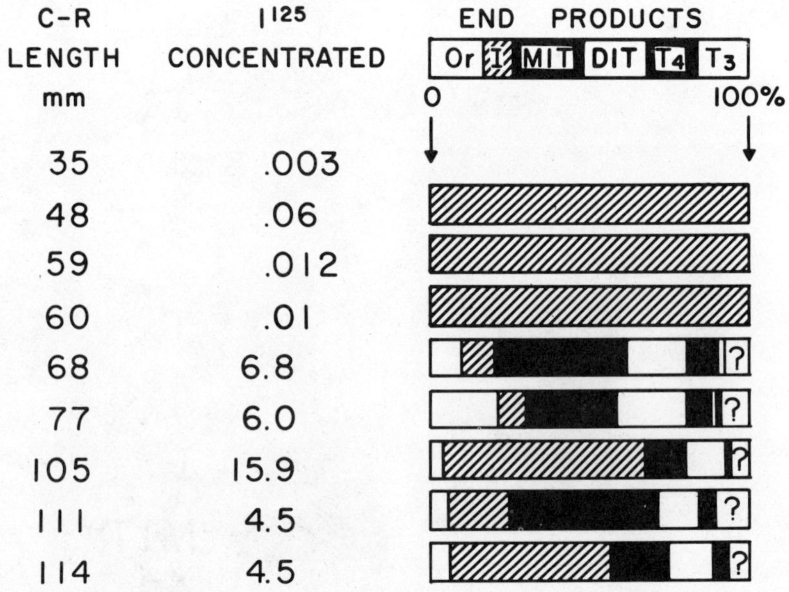

Fig. 8. Summary of the crown-rump length and percentage of I^{125} incorporated from the media into the thyroid during 18 hours of organ culture and a schematic representation of the type and amount of iodinated end product found by paper chromatography. In the last column on the right the entire bar represents 100 percent; *Or* is the activity found at the origin, and the question mark represents total unidentified amount of isotope. From Shepard, *J. Clin. Endocr.*, 27:945-958, 1967. Courtesy of J. B. Lippincott and Co.

a result, the small microvillous-lined cavities could, by the same mechanism as diagrammed in Figure 6, be disgorged into the intercellular spaces. Thesingh(25) has shown that sodium is pumped into explants of chick thyroid undergoing antrum formation, and that ouabain inhibits the production of antra; these findings support her theory that antrum formation is an expression of TSH stimulation of a sodium pump in the cell membranes bordering the antra.

The antrogenic effect of TSH is relatively resistant to heat. In addition, Thesingh(25) has studied a biologically inactive TSH preparation which retains its antrum-inducing activity.

DISCUSSION

The weight of the thyroid gland in relation to body weight gradually increases and reaches a relative plateau between the 65- to 80-mm crown-rump stages. At this same time the histogenesis of the follicular colloid cavities is in an active phase and produces by the end of this period a predominantly colloid-containing follicular thyroid gland. The emergence of the ability of the gland to bind iodide organically during organ culture first appears also during this 65- to 80-mm crown-rump

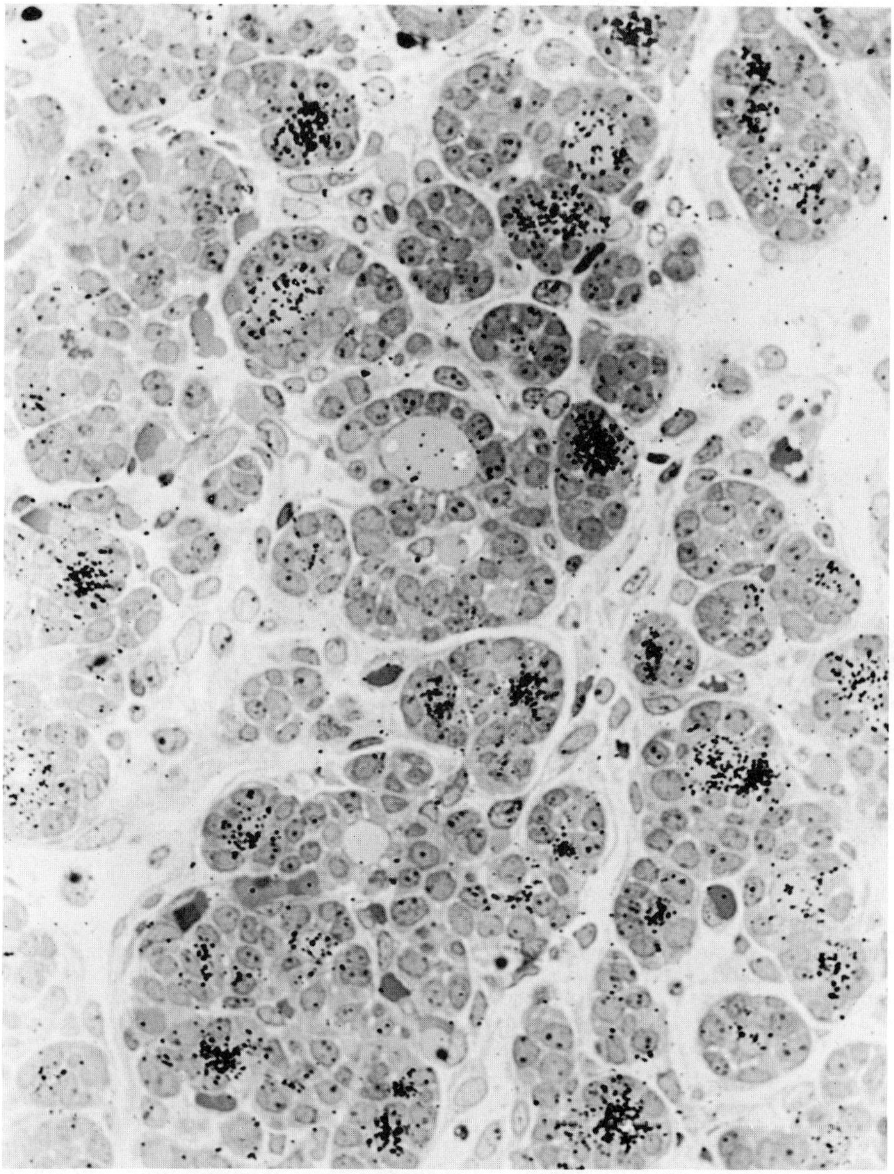

Fig. 9. Tissue radioautograph from the thyroid of a fetus of 77-mm crown-rump length. It was exposed to nuclear film photographic emulsion (NTB-3) for 51 days. The major number of grains are over the colloid, although some colloid spaces are inactive (×480). From Shepard, *J. Clin. Endocr.*, 27:945-958, 1967. Courtesy of J. B. Lippincott and Co.

stage. Therefore, the attainment of mature relative weight and the appearance of colloid in central cavities appears to be closely related to onset of specific functional capabilities.

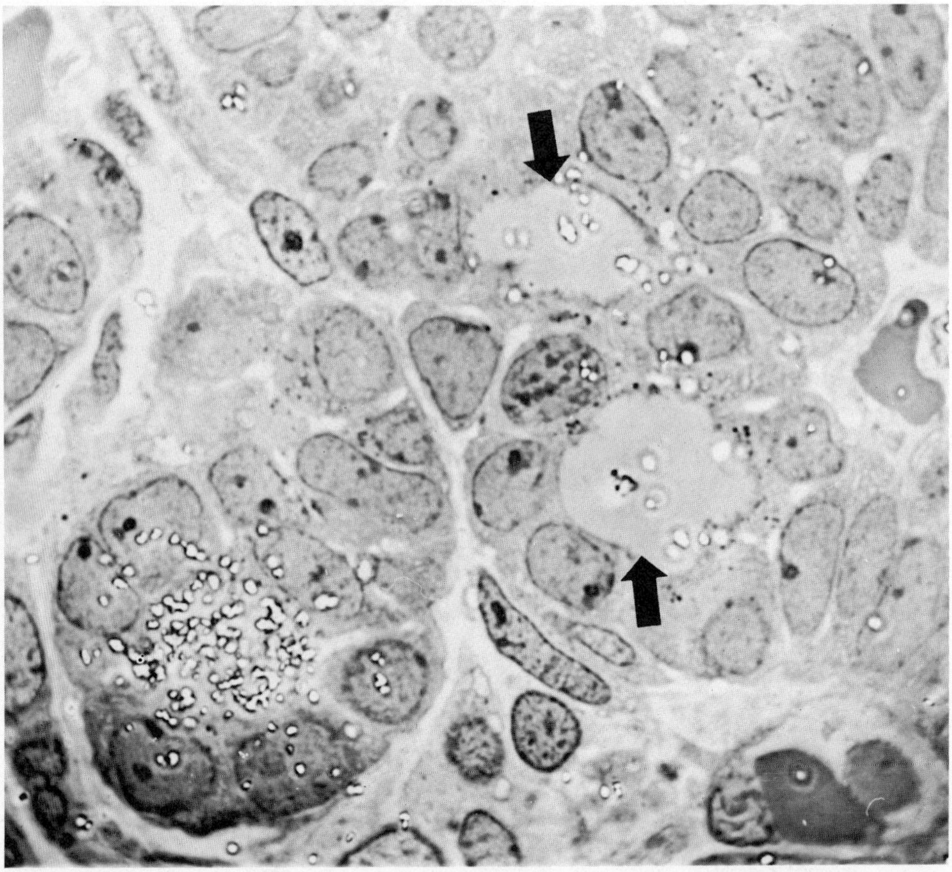

Fig. 10. Photomicrograph from the thyroid of a 77-mm fetus to show clover-leaf processes with overlying silver grains. Arrows indicate two follicles where the intracellular canaliculi are in the process of union. The light areas over these structures are silver nitrate grains as seen under phase microscopy. Epon embedded, one-micron thick sections stained with toluidine after 9 weeks exposure to nuclear emulsion (×1,300). From Shepard, *J. Clin. Endocr.,* 27: 945-958, 1967. Courtesy of J. B. Lippincott and Co.

Precision in establishment of the time of onset of function is of special practical importance in view of a number of recorded cases where therapeutic amounts of ^{131}I inadvertently administered to pregnant women resulted in radiochemical thyroidectomy of the fetus(27,28,29). The possibility of mass population exposure to radioiodine is further linked to the demonstration that the primate fetal thyroid attains higher radioiodine concentrations than the mother(30,31) and could lead to a major public health hazard. Although there can be no innocuous period during gestation for administration of ^{131}I, the earliest stage of development where iodide was found to be concentrated was at a crown-rump length of 68 mm which is equivalent to 74 gestational days(32).

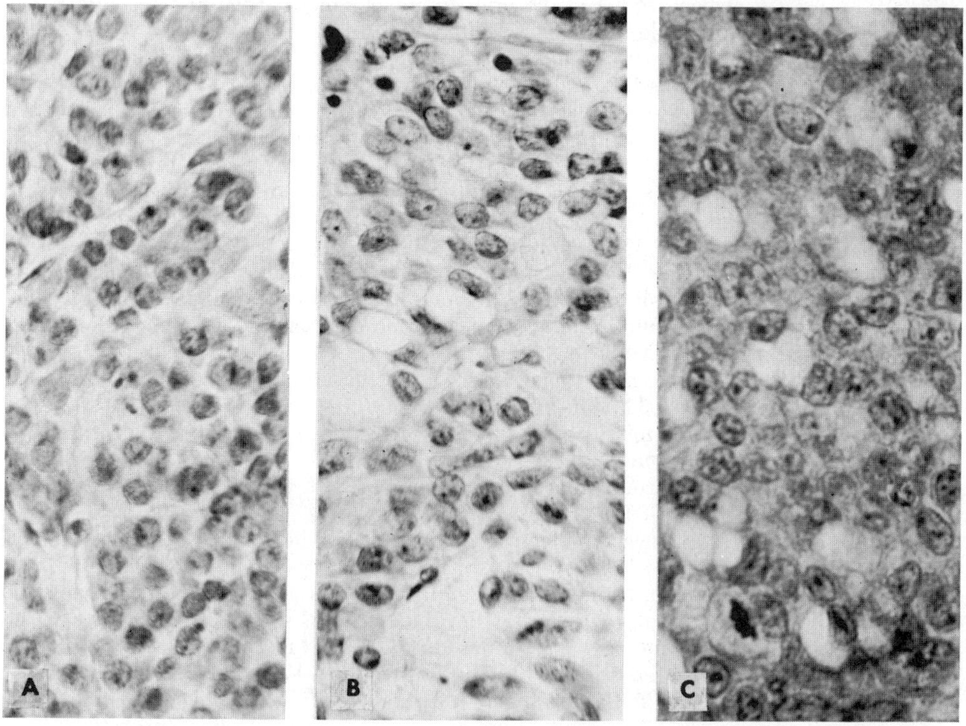

Fig. 11. Antrum formation in human and chick fetus. *A*. Control human thyroid from a 49-mm fetus maintained in vitro for 24 hours. *B*. Same thyroid exposed for 24 hours in vitro to 0.01 unit thyrotrophic hormone per ml of culture media. Note dilated spaces which displace the nucleus. *C*. Eight-day chick thyroid exposed to 0.1 unit thyrotrophic hormone per ml of culture media.

REFERENCES

1. Streeter, G. L. Weight, sitting height, head size, foot length and menstrual age of human embryos. *In* Contributions to Embryology. Carnegie Inst. Wash., Pub. 274, No. 55, 2: 143–170, 1920.
2. Shepard, T. H., Andersen, H. J., and Andersen, H. The human fetal thyroid. I. Its weight in relation to body weight, crown-rump length, foot length and estimated gestation age. Anat. Rec., 148:123–128, 1964.
3. Weller, L. G., Jr. Development of the thyroid, parathyroid and thymus glands in man. *In* Contributions to Embryology. Carnegie Inst. Wash., No. 141, 24:93–139, 1933.
4. Boyd, J. D. Development of the human thyroid gland. *In* The Thyroid. Pitt-Rivers, R., and Trotter, W. R., eds. Washington, Butterworths and Co., Ch. 2, pp. 9–31, 1964.
5. Shepard, T. H. The thyroid. *In* Organogenesis. DeHaan, R. L., and Ursprung, H., eds. New York, Holt, Rinehart and Winston, Inc., Ch. 19, pp. 493–512, 1965.
6. Potter, E. L. Pathology of the Fetus and Infant, 2nd ed. Chicago, Year Book Medical Publishers, Inc., p. 14, 1961.
7. Huber, G. C., ed. Piersol's Human Anatomy, 9th ed. Philadelphia, J. B. Lippincott, p. 1790, 1930.
8. Norris, E. H. The early morphogenesis of the human thyroid gland. Amer. J. Anat., 24: 433–465, 1918.

9. Taki, A. Histological studies of the prenatal development of the human thyroid gland. Okajimas Folia Anat. Jap., 32:65–85, 1958.
10. Shepard, T. H., Andersen, H. and Andersen, H. J. Histochemical studies of the human fetal thyroid during the first half of fetal life. Anat. Rec., 149:363–380, 1964.
11. McKay, D. G., Adams, E. C., Hertig, A. T., and Donziger, S. Histochemical horizons in human embryos. II. 6 and 7 millimeter embryos—Streeter horizon XIV. Anat. Rec., 126:433–463, 1956.
12. Bierring, F., and Shepard, T. H. Electron microscopic studies on early follicle and colloid formation in the human fetal thyroid. Acta Path. Microbiol. Scand., in press, 1970.
13. Petrovic, A., and Porte, A. Sur la formation en culture organotypique, de lacunes intercellulaires dans la thyroid d'embryon de poulet de six jours et demi. Étude sous l'influence de la thyréostimuline. Étude au microscope electronique. C.R. Soc. Biol. (Paris), 155:1848–1855, 1961.
14. Feldman, J. D., Vazquez, J. J., and Kurtz, S. M. Maturation of rat fetal thyroid. J. Biophys. Biochem. Cytol., 11:365–383, 1961.
15. Ishikawa, K. Electron microscopic studies of the thyroid gland of the rat in embryonic life. Okajimas Folia Anat. Jap., 41:295–311, 1965.
16. Hilfer, R. S. Follicle formation in the embryonic chick thyroid. I. Early morphogenesis. J. Morph., 115:135, 1964.
17. Fujita, H., and Tanizawa, Y. Electron microscopic studies on the development of the thyroid gland of the chick embryo. Z. Anat. Entwicklungsgesch., 125:132, 1966.
18. Yamazaki, E., Naguchi, A., and Slingerland, D. W. The development of hormonal biosynthesis in human fetal thyroids. J. Clin. Endocr., 19:1437–1439, 1959.
19. Chapman, E. M., Corner, G. W., Robinson, D., and Evans, R. D. Collection of radioiodine by the human fetal thyroid. J. Clin. Endocr., 8:717–720, 1948.
20. Hodges, R. E., Evans, T. C., Bradbury, J. T. and Keettel, W. C. The Accumulation of radioactive iodine by human fetal thyroids. J. Clin. Endocr., 15:661–667, 1955.
21. Andersen, H. J. Studies of hypothyroidism in children. Acta Pediat. Scand., Suppl. 125: 106–111, 1960.
22. Shepard, T. H. Onset of function in the human fetal thyroid: Biochemical and radioautographic studies from organ culture. J. Clin. Endocr., 27:945–958, 1967.
23. Bakker-Sauer, E. K. Some effects of thyrotropic hormone on thyroid tissue *in vitro*. Koninkl. Nederl. Acad. Wetenschap. Proc. Ser. C., 64:86–95, 1961.
24. Thesingh, C. W. The antrum phenomenon: A reaction of embryonic chick thyroid tissue to TSH stimulation. Gen. Comp. Endocr., 2:621–622, 1962.
25. ——— Antrumvorming in Embryonaal Kippeschildklierweefsel: een Reactie op Thyreotroop Hormoon. Thesis, Leiden, pp. 1–82, 1964.
26. Daems, W. T., and Thesingh, C. W. Electron microscopy of TSH-stimulated embryonic chick thyroids. *In* 5th Internat. Congr. Electron Microscopy. New York, Academic Press, Inc., pp. WW-2, 1962.
27. Russell, K. P., Rose, H., and Starr, P. The effects of radioactive iodine on maternal and fetal thyroid function during pregnancy. Surg. Gynec. Obstet., 104:560–564, 1957.
28. Hammill, G. C., Jarman, J. A., and Wynne, M. D. Fetal effects of radioactive iodine therapy in a pregnant woman with thyroid cancer. Amer. J. Obstet. Gynec., 81:1018–1023, 1961.
29. Fisher, W. K., Voorhess, M. L., and Gardner, L. I. Congenital hypothyroidism in infant following maternal I^{131}. J. Pediat., 62:132–135, 1963.
30. Pickering, D. E., and Kontaxis, N. E. Thyroid function in the fetus of the macaque monkey. II. Chemical and morphological characteristics of the foetal thyroid gland. J. Endocr., 23:267–275, 1961.
31. Beierwaltes, W. H., Hilger, M. T. J. and Wegst, A. Radioiodine concentration in fetal human thyroid from fallout. Health Phys., 9:1263–1266, 1963.
32. Shepard, T. H. Development of the human fetal thyroid. Gen. Comp. Endocr., 10:174–181, 1968.

61

Effect of Thyrotropic Hormone on Thyroid Glands of the Fetal and Newborn Rat*

B. M. Nataf,[1] J. Harel,[2] and J. Imbenotte[2]

[1]Isotope Laboratory, Institut Gustave Roussy, Villejuif, France
[2]Laboratory of Physiopathology, Institut Gustave Roussy, Villejuif, France

INTRODUCTION

It is well known that thyroid function in the rat does not begin until late in gestation(1-12) and that the thyroid of the rat at term can concentrate iodine(3-8, 12-17) and incorporate it into organic compounds(12,14,16,17).

This paper reports the effect of injected TSH on the in vitro uptake and metabolism of ^{131}I by thyroid glands of fetal and newborn rats, and the results obtained with organ cultures of fetal thyroid glands to which TSH was added to the culture medium. The comparative effects of TSH on ^{131}I metabolism on incorporation of ^{14}C aminoacids into protein, and on biosynthesis of RNA are presented.

EFFECT OF INJECTED TSH ON THE IN VITRO UPTAKE AND METABOLISM OF ^{131}I BY THYROID GLANDS OF FETAL AND NEWBORN RATS

Fetal, newborn, and adult rats were all injected with TSH in doses equivalent in terms of relative body weights. The intact thyroid lobes of fetal, newborn, and adult rats, as well as slices from adult glands were incubated for 2 hours in a

*The technical assistance of Miss M. Jennequin and Mr. G. Frezouls is gratefully acknowledged.

Krebs-Ringer bicarbonate buffer containing ^{131}I and the incorporation of the labeled iodine into iodoamino acids was compared(18).

The surviving glands of fetal and newborn rats are indeed extremely sensitive to injected TSH: effects on iodine metabolism were observed as early as a half hour after injection of the hormone.

The in vitro uptake of ^{131}I by glands of newborn and fetal rats was not appreciably affected when the lobes were excised either 30 minutes or 3 to 5 hours after injection of TSH (Figs. 1 and 2), but rose significantly in glands excised 20 to 24 hours after TSH injection. Despite the fact that TSH injection at early intervals failed to augment ^{131}I uptake, the disposition of ^{131}I within the lobes was altered. An effect was observed principally on the proportions of the chromatographically separated ^{131}I present as T_4 (Figs. 1 and 2). Because of the increased uptake of the thyroid gland 24 hours after TSH injection, the amount of ^{131}I incorporated into T_4 was actually more than four times that of controls.

Thus, under in vitro conditions, two phases of TSH action on the thyroid glands of young animals can be distinguished: (1) an initial one (30 minutes to 5 hours) in which only the disposition of the labeled iodine within the tissue was affected, consisting mainly of an increased formation of T_4; and (2) a later effect (20 to 24 hours) in which both the magnitude of the uptake of the ^{131}I from the medium as well as its internal disposition was involved.

In the experiments with adult rats, TSH injection did not significantly (P >

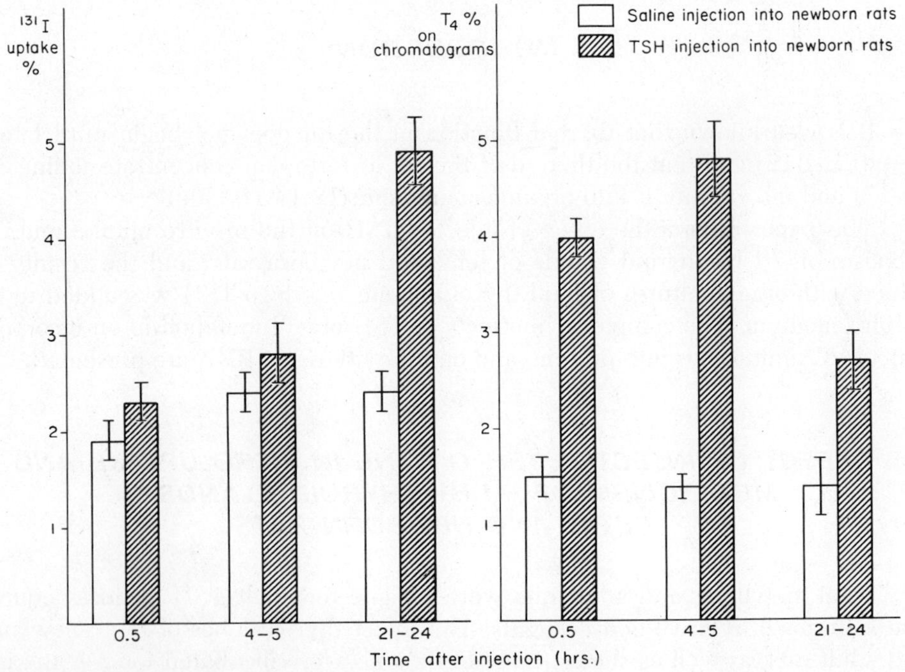

Fig. 1. Effect of TSH, injected into newborn rats, on ^{131}I uptake and its incorporation into thyroxine by their thyroid glands incubated in vitro.

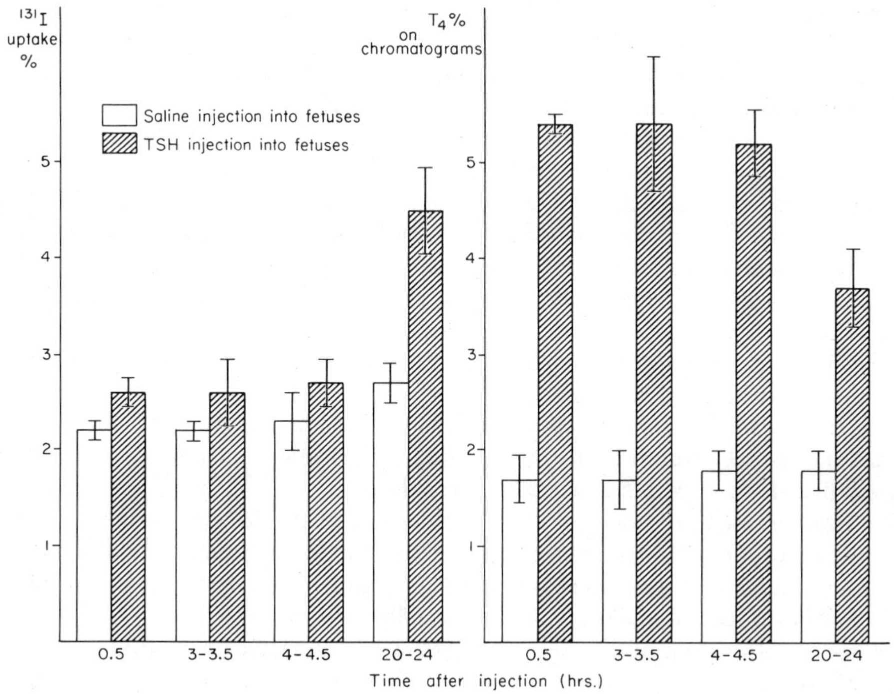

Fig. 2. Effect of TSH, injected into fetal rats, on ^{131}I uptake and its incorporation into thyroxine by their thyroid glands incubated in vitro.

0.05) alter ^{131}I incorporation into T_4, regardless of whether the glands were incubated in the form of intact lobes or slices.

One might seek to explain the differences between adult rats and fetal and newborn rats by the fact that endogenous TSH is low in young animals(19); however, injected TSH does not increase in vitro labeling of T_4 in hypophysectomized adult rats(20). The fact that thyroid glands of fetuses and newborn rats are still undergoing differentiation, is most likely responsible for the difference in response between immature and adult rats.

ACTION OF TSH ADDED TO THE CULTURE MEDIUM ON FETAL RAT THYROID GLANDS IN ORGAN CULTURE

The organ-culture method permits organ development and function to be studied independently of the effects exerted by the internal environment of the intact organism.

Using the organ-culture technique we were able to show that thyroid glands of 16- to 16.5-day-old fetal rats do not significantly utilize in vitro ^{131}I that is added to a synthetic medium(21,22). However, when TSH was added to the medium,

^{131}I utilization by thyroid explants was enhanced, and labeled T_4 formed only in its presence.

Previous in vivo observations(6,8-12) have shown that iodine metabolism is first detected at 17 to 17.5 days of fetal life in the rat. When thyroid glands were explanted into organ culture at this stage, the addition of TSH to the medium apparently accelerated not only the conversion of monoiodotyrosine (MIT) to diiodotyrosine (DIT), but the formation of thyroid hormones as well(21,22).

Similar effects were achieved in organ culture of thyroid glands of 21-day-old rat embryos. The ^{131}I uptake, as well as the incorporation of the isotope into protein, were higher in the presence of TSH. The formation of ^{131}I-labeled DIT or T_4 was principally affected, the most striking response being the increase in incorporation of ^{131}I into T_4(22-24).

TSH produces a slight increase in the recoveries of radioactive iodide in the glands that were treated with methimazole before addition of the isotope to the medium(22,24). Inasmuch as methimazole blocks the organification of iodide without affecting its accumulation in the gland(25), TSH apparently has an influence on the ability of the explants to concentrate iodide.

It seems reasonable, however, to view the effect of TSH on organ culture of embryonic thyroid glands as not wholly dependent on the stimulation of the iodide-concentrating ability of the explants since TSH significantly increases the incorporation of ^{131}I into iodoamino acids, and especially into thyroxine.

In spite of the profound effect of TSH on ^{131}I metabolism of the embryonic thyroid glands in organ culture, the effect of TSH on the incorporation of ^{14}C-labeled leucine into protein was negligible(22,24). It was found that ^{14}C-leucine added to the basal medium was readily incorporated into protein. The 6-hour incorporation of leucine into thyroid protein was about the same in the basal medium as in the TSH-containing medium(22,24). It was also observed that puromycin, a potent inhibitor of protein synthesis, almost completely blocked the 6-hour incorporation of leucine carbon into protein in glands cultured for 3 days in basal medium. The inhibition was the same when the medium containing TSH was used (22,24).

However, under similar conditions, this antibiotic did not affect either the 6 hour uptake of ^{131}I or its organification by the explants cultured in the basal medium. It seems therefore that iodination of tyrosine must have taken place within the matrix of preformed thyroglobulin molecules, as reported by several investigators using various experimental conditions different from our own(26-33).

Puromycin also fails to abolish the TSH-induced increase in incorporation of ^{131}I into iodotyrosines and thyroxine that usually takes place within 6 hours. It could be concluded, therefore, that thyroid hormone synthesis is not dependent on de novo formation of protein.

With Harel and co-workers(34) we have been able to isolate, from 21-day-old fetal thyroid glands cultured in a hormone-free chemically defined medium, a very heavy DNA-like RNA (dRNA), under conditions in which the synthesis of thyroxine was very low.

Accordingly, it was of interest to study the action of TSH both on the synthesis of various fractions of RNA as well as on the formation of thyroid hormones.

In the two foregoing studies, the ribosomal RNA and the soluble RNA were separated from nuclear messenger RNA and ribosomal RNA precursors by a method of extraction described by Harel et al.(35).

The RNAs referred to as phenol RNA were extracted by four successive phenol treatments. The RNAs referred to as interphase RNA were extracted with 1 percent SDS and phenol and precipitated with ethanol and sodium citrate or carrier RNA from mouse cells. Centrifugations were performed in glycerol gradients (36). The S values of the ^{32}P-labeled fractions were estimated(37). The fractions were precipitated with carrier RNA and their base composition was determined as described by Harel et al.(38).

When fetal glands were cultured for 48 hours in a chemically defined medium and ^{32}P added 24 hours before the end of the experiment, phenol RNA represented uniformly labeled rRNA and sRNA, as shown by gradient centrifugation (Fig. 3) and base composition; a very small proportion sedimented above 60S. In the interphase RNA a major proportion sedimented between 20 and 45S (Figs. 3 and 4) and had a G-C content lower than that of rRNA. Fifteen to 25 percent of total interphase RNA was represented in heavier fractions (>60S) and was distinctly more DNA-like. From such experiments (Fig. 4), high S values did not seem to correspond to RNA associated with DNA or proteins or RNA aggregates.

It appears, then, in agreement with Georgiev and Mantieva(39), the phenol RNA consists mostly of cytoplasmic RNA, and interphase RNA represents nuclear RNA. Base composition indicates that in the latter, 30 to 45S fractions may correspond to a mixture of dRNA and rRNA precursors(40) and that the heavier ones represent mostly dRNA. In our own experiments 50 to 60 percent of nuclear RNA and 8 to 12 percent of total RNA consisted of dRNA as judged by the relative proportion and G-C content of the various fractions. It is noteworthy that in conditions

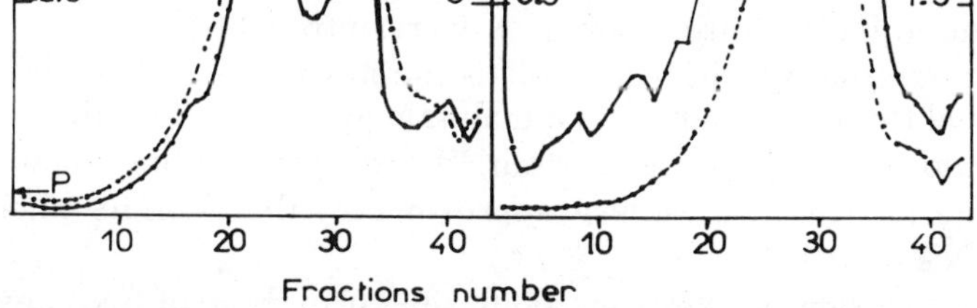

Fig. 3. *Left*: Labeled phenol RNA centrifuged in 5 to 20 percent glycerol gradient, PO^4 0.01 M at 39,000 rpm for 210 minutes in the SW 39 of Spinco ultracentrifuge. *Right:* Labeled interphase RNA centrifuged in 5 to 30 percent glycerol gradient, PO^4 0.01 M for 150 minutes. Solid lines: radioactivity; dotted lines: optical density of carrier RNA. *P*: cpm in pellets. From Harel et al. *Biochem. Biophys. Res. Commun.,* 25:573-578, 1966. Courtesy of Academic Press, Inc.

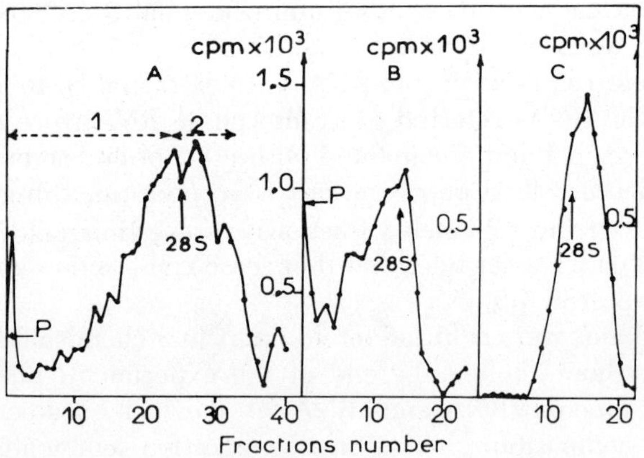

Fig. 4. *A*: Interphase RNA treated with DNAase (2 μg/ml, 30 minutes) and pronase (100 μg/ml, 60 minutes), reextracted with phenol, centrifuged as in Figure 3. Pooled fractions were precipitated with citrate and ethanol, dialyzed overnight against PO^4 0.01 M, EDTA 0.01 M. *B*: Fraction 1, *C*: fraction 2, centrifuged in 5 to 30 percent glycerol, EDTA 0.01 M for 225 minutes. From Harel et al. *Biochem. Biophys. Res. Commun.*, 25:573-578, 1966. Courtesy of Academic Press, Inc.

which allow the synthesis of only a small amount of thyroxine, heavy nuclear dRNA, presumably polycistronic, was found.

Furthermore, Cartouzou et al.(41-43) have extracted from sheep thyroid messenger nuclear RNAs(30 to 40S) which stimulated incorporation of amino acids in vitro; RNAs of the same types are also present in thyroid polysomes.

It has been reported that injections of TSH increase the RNA content of thyroid glands in chickens(44) and guinea pigs(45). More recently, Kerkof and Tata (46) have noted that administration of TSH to guinea pigs significantly increases the incorporation of ^{32}P into RNA and phospholipid as well as the incorporation of ^{125}I into iodoprotein.

TSH, when added in vitro, also stimulates the synthesis of RNA in thyroid slices of various animals(47-55) and in isolated nuclei of the thyroid gland(50-52) incubated with various labeled precursors of RNA.

Inasmuch as such incubation experiments cannot be conducted over a long period of time, organ culture techniques were introduced to study the effects of TSH on the biosynthesis of RNA as well as the metabolism of ^{131}I(56).

When fetal thyroid explants were maintained in organ culture in a medium to which TSH had been added, the 24-hour ^{32}P incorporation into total RNA was increased (Table 1). The increase was approximately the same for phenol RNA and interphase RNA.

The sedimentation profiles were quite similar to those of RNAs extracted from glands cultured in a hormone-free medium, when phenol RNA and interphase RNA were centrifuged in glycerol gradient as described previously. Indeed, the relative proportions of the various fractions of RNA do not seem to be altered by the presence of TSH in the culture medium; that is, TSH increases quite similary the 24-hour ^{32}P incorporation into ribosomal RNA, transfer RNA, and RNAs of the

TABLE 1

24-hour incorporation of ^{32}P in phenol RNA and in interphase RNA in thyroid lobes of embryonic rats cultured in the absence and presence of TSH[a]

RNA fractions	^{32}P incorporation (cpm) Culture in medium 109	Culture in medium 109 + TSH	% increase over control
Phenol RNA	43,000	100,000	233
Interphase RNA	25,000	55,000	220

[a]Results of a typical experiment (experiment 5).

messenger type. Concurrently, TSH has a stimulatory action on the metabolism of ^{131}I, which is reflected mainly in the incorporation of ^{131}I into thyroxine (Table 2).

Preliminary experiments on hybridization between thyroid RNAs labeled for 24 hours with ^{32}P and homologous DNA were performed, using the techniques described by Nygaard and Hall(57) and by Gillespie and Spiegelman(58). These experiments show that the degree of hybridization of nuclear messenger RNAs is higher than that of cytoplasmic messenger RNAs, which is not affected by the presence of TSH in the culture medium.

Although TSH has been shown to influence the transport of numerous anions into thyroid tissue(59-62), the effects of TSH cannot be explained wholly on the basis of increased transport or availability of labeled phosphate. Kerkof and Tata (46) have shown that total uptake (tissue content) of ^{32}P was not effectively influenced by TSH injections into guinea pigs. In our organ-culture studies, however, when fetal lobes were incubated for 48 hours in the presence of TSH and ^{32}P added 1 to 3 hours prior to the end of the experiment, total ^{32}P incorporation was found to be significantly increased (Fig. 5).

Yet, the 1-hour ^{32}P uptake was proportionately greater in the acid-precipitable

TABLE 2

Effect of TSH on 24-hour ^{131}I uptake and metabolism of fetal rat thyroid glands maintained in organ culture

^{131}I measurement	Culture in medium 109	Culture in medium 109 + TSH
Uptake %	0.73 ± 0.21[a]	4.08 ± 2.69
Distribution (%) on chromatograms		
MIT	32.65 ± 7.15	37.61 ± 4.78
DIT	15.25 ± 4.78	27.79 ± 4.42
$T_4 + T_3$	1.84 ± 0.36	12.36 ± 5.90
I^-	47.45 ± 11.37	17.63 ± 6.65

[a]Each value is the mean ± S.E. of the results of nine separate experiments.

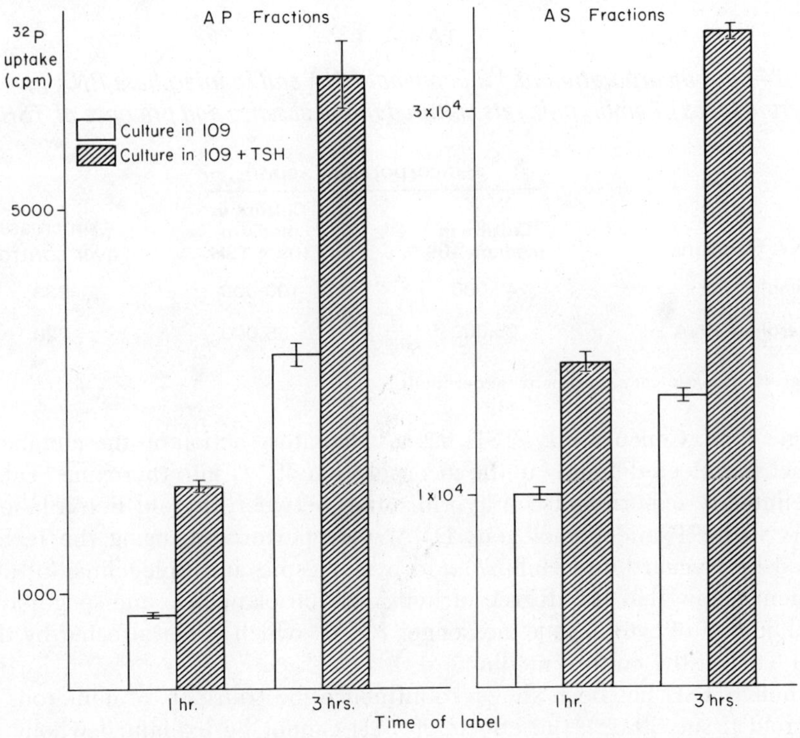

Fig. 5. ^{32}P-incorporation into trichloracetic acid precipitable fractions (*AP*) and acid-soluble fractions (*AS*) in thyroid lobes of fetal rats cultured in the absence and presence of TSH. Mean and S.E. of the results of 4 experiments.

fractions than in the acid-soluble fractions, which seems to confirm the view that the stimulatory effect of the hormone cannot be explained solely by its effect on active transport of precursors. Similarly, as far as ^{131}I metabolism is concerned, the effect of TSH on organ culture of embryonic thyroid glands does not seem to be dependent wholly on the stimulation of iodide-concentrating ability of the explants, since TSH accelerates the conversion of MIT to DIT and the formation of thyroid hormones(22,24).

The mode of action of TSH on the biosynthesis of RNA is not yet completely understood.

According to Hall and Tubmen(54) and Lindsay et al.(55), the effect of TSH on RNA synthesis in thyroid gland in vitro cannot be blocked by puromycin and therefore does not appear to be dependent on the formation of new protein. Their findings, along with the reports of others(47-49,63), indicate that one site at which TSH may affect nucleic acid synthesis is the formation of nucleotides, which is mediated by stimulation of glucose oxidation via hexose monophosphate pathway, thereby increasing the supply of available ribose.

However, the results of Begg and Munro(50,51), Shimada and Yasumasu(52), and Lecocq and Dumont(53) strongly indicate another site, the polymerization of nucleotides, at which TSH can effect the synthesis of nucleic acids. Furthermore, according to Begg and Munro(50,51), the stimulatory action of TSH on RNA synthesis was not observed in the presence of puromycin; it would thus appear that the action of TSH on RNA metabolism requires preliminary formation of protein. These investigators suggested that the first effect of TSH on nuclear RNA metabolism may involve synthesis of a messenger RNA for a specific nuclear protein.

From these contradictory results, it is not yet possible to affirm that TSH acts by a process of "derepression" of one or many messenger RNAs; however, this hypothesis cannot be excluded.

Our own results showed that TSH stimulated both the synthesis of ribosomal RNA and apparently that of the messenger RNAs as a whole. This increase in the replication of RNA is paralleled by an increase in the metabolism of ^{131}I, concerned mainly in the augmentation of the synthesis of thyroid hormones. We have shown (22,24) that the stimulatory action of TSH on 6-hour uptake and metabolism of ^{131}I is not inhibited by puromycin and, consequently, does not seem to be dependent on de novo synthesis of protein.

In conclusion, we cannot exclude the view that TSH might induce or stimulate the synthesis of specific messenger RNAs which could play a role in the formation of thyroid proteins. Additional experiments are needed to elucidate further the mechanism of action of TSH.

REFERENCES

1. Kull, H. A. The late embryonic development of the thyroid gland of the albino rat. Anat. Rec., 32:133–141, 1926.
2. Hall, A. R., and Kaan, H. W. Anatomical and physiological studies on the thyroid gland of the albino rat. Anat. Rec., 84:221–239, 1942.
3. Gorbman, A., and Evans, H. M. Beginning of function in the thyroid of fetal rat. Endocrinology, 32:113–115, 1943.
4. Vidovic, V. Beginning of the ability of the thyroid to take up iodine in the embryonic development of the rat, determined with iodine.^{131} Helv. Physiol. Pharmacol. Acta, 9: 438–441, 1951.
5. Mitskevich, M. S. Glands of internal secretion in the embryonic development of birds and mammals. Acad. Sci., Moscow, USSR, 1957.
6. Carpenter, E., and Rondon-Tarchetti, T. Differentiation of embryonic rat thyroid *in vivo* and *in vitro*. J. Exp. Zool., 136:393–405, 1957.
7. Noumura, T. Development of the hypophyseal-thyroid system in the rat embryo and its relation to the maternal system. Japan. J. Zool., 12:301–318, 1959.
8. Carpenter, E. Development of fetal rat thyroid with special reference to uptake of radioactive iodine. J. Exp. Zool., 142:246–257, 1959.
9. Nataf, B., and Sfez, M. Début du fonctionnement de la thyroide foetale du rat. C.R. Soc. Biol. (Paris), 155:1235–1238, 1961.
10. Geloso, J. P. Date de l'entrée en fonction de la thyroide chez le foetus de rat. C.R. Soc. Biol. (Paris), 155:1239–1244, 1961.
11. Feldman, J. D., Vazquez, J. J., and Kurtz, S. M. Maturation of the rat fetal thyroid. J. Biophys. Biochem. Cytol., 11:365–383, 1961.
12. Geloso, J. P. Fonctionnement de la thyroide et corrélations thyréo-hypophysaires chez le foetus de rat. Ann. Endocr. (Paris), 28 (Suppl. 1):1–79, 1967.
13. Nikitovitch, M., and Knobil, E. Placental transfer of thyrotropic hormone in the rat. J. Clin. Endocr., 15:837–838, 1955.

14. Geloso, J. P. Recherches sur le métabolisme de l'iode radioactif par la thyroide du foetus de rat, C. R. Soc. Biol. (Paris), 150:2140–2145, 1956.
15. Knobil, E., and Josimovich, J. B. Placental transfer of thyrotropic hormone, thyroxine, triiodothyronine and insulin in the rat. Ann. N.Y. Acad. Sci., 75:895–904, 1959.
16. Sfez, M., and Nataf, B. Étude chromatographique du fonctionnement de la thyroide foetale du rat. Bull. Soc. Chim. Biol., 42:419–425, 1960.
17. Nataf, B. M., and Chaikoff, I. L. Effect of hypophysectomy of the pregnant rat on ^{131}I metabolism of the fetal thyroid gland. Endocrinology, 73:518–521, 1963.
18. ——— and Chaikoff, I. L. Effect of injected thyrotropic hormone on the *in vitro* uptake and metabolism of ^{131}I by thyroid glands of fetal and newborn rats. Endocrinology, 75:547–553, 1964.
19. Levey, H. A. Progressive age-related increase in pituitary thyrotropin level and T/S iodine ratio in the rat. Endocrinology, 71:763–767, 1962.
20. Taurog, A., Tong, W., and Chaikoff, I. L. Thyroid ^{131}I metabolism in the absence of the pituitary: the hypophysectomized rat treated with thyrotropic hormone. Endocrinology, 62:664–676, 1958.
21. Nataf, B. M., Rivera, E. M., and Chaikoff, I. L. Role of thyrotropic hormone in iodine metabolism of embryonic rat thyroid glands in organ culture. Endocrinology, 76:35–42, 1965.
22. ——— Fetal rat thyroid gland in organ culture. Gen. Comp. Endocr., 10:159–173, 1968.
23. ——— and Chaikoff, I. L. The effect of insulin on iodine matabolism of fetal thyroid glands in organ culture. Biochim. Biophys. Acta, 111:422–428, 1965.
24. ——— Chaikoff, I. L., and Freeman, W. E. Mechanism of action of TSH studied by organ culture of fetal rat thyroid gland. Proc. Soc. Exp. Biol. Med., 124:7–12, 1967.
25. Astwood, E. B. Mechanism of action of antithyroid compounds. Brookhaven Symp. Biol., 7:61–73, 1954.
26. Taurog, A., and Howells, E. M. Contrasting effect of puromycin on ^{131}I metabolism and on ^{14}C amino acid metabolism of thyroid slices. Fed. Proc., 23:149, 1964.
27. Soodak, M., Maloof, F., and Sato, G. Effect of thiourea, puromycin and actinomycin D on iodination and amino acid incorporation in normal rat thyroid. Fed. Proc., 23:268, 1964.
28. Goldberg, I. H., Seed, R. W., Schneider, A. X., and Sellin, H. G. Thyroglobulin biosynthesis. Fed. Proc., 23:434, 1964.
29. Cartouzou, G., Aquaron, R., and Lissitzky, S. Purification of soluble RNA from sheep thyroid gland. Absence of acceptor activity for the iodotyrosine. Biochem. Biophys. Res. Commun., 15:82–86, 1964.
30. Lissitzky, S., Roques, M., Torresani, J., Simon, C., and Bouchilloux, S. Iodination and biosynthesis of rat thyroglobulin. Biochem. Biophys. Res. Commun., 16:249–253, 1964.
31. Alexander, N. M. Studies on amino acid activating enzymes in thyroid glands. Endocrinology, 74:273–278, 1964.
32. Tong, W. Thyrotropin stimulation of thyroxine synthesis in isolated thyroid cells treated with puromycin. Endocrinology, 76:163–165, 1965.
33. Tishler, P. V., and Ingbar, S. H. Correlative effects of puromycin on ^{131}I metabolism and amino acid incorporation by calf thyroid slices. Endocrinology, 76:295–300, 1965.
34. Harel, J., Nataf, B., Harel, L., and Imbenotte, J. High molecular weight DNA-like RNA in rat thyroid. Biochem. Biophys. Res. Commun., 25:573–578, 1966.
35. Harel, L., Harel, J., Boer, A., Imbenotte, J., and Carpeni, N. Persistance d'une synthèse de D-RNA dans le foie de rat traité par l'actinomycine. Biochim. Biophys. Acta, 87: 212–218, 1964.
36. Huppert, J., Lacour, F., Harel, J., and Harel, L. High molecular weight RNA from avian myeloblastosis virus. Cancer Res., 26:1561–1568, 1966.
37. Nomura, M., Hall, B. D., and Spiegelman, S. Characterization of RNA synthesized in Escherichia coli after bacteriophage T2 infection. J. Molec. Biol., 2:306–326, 1960.
38. Harel, J., Harel, L., Lacour, F., Boer, A., and Imbenotte, J. Fractions with differing base composition in RNA from malignant cells of mouse. J. Molec. Biol., 7:645–651, 1963.

39. Georgiev, G. P., and Mantieva, V. L. The isolation of DNA-like RNA and ribosomal RNA from the nucleolo-chromosomal apparatus of mammalian cells. Biochim. Biophys. Acta, 61:153–154, 1962.
40. Scherrer, K., Latham, H., and Darnell, J. E. Demonstration of an unstable RNA and of a precursor to ribosomal RNA in HeLa cells. Proc. Nat. Acad. Sci. U.S.A., 49:240–248, 1963.
41. Cartouzou, G., Mante, S., and Lissitzky, S. Purification and some properties of sheep thyroid and rat liver messenger RNAs. Biochem. Biophys. Res. Commun., 20:212–217, 1965.
42. ——— Attali, J. C., and Lissitzky, S. Acides ribonucleiques messagers de la glance thyroide l. RNA à marquage rapide des noyaux et des polysomes. Europ. J. Biochem., 4:41–54, 1968.
43. ——— Aquaron, R., Mante, S., and Lissitzky, S. Acides ribonucléiques messagers de la glande thyroide de mouton 2. RNA messagers. Europ. J. Biochem., 4:55–67, 1968.
44. Creek, R. O. Effect of thyrotropin on the weight, protein, ribonucleic acid, and the radioactive phosphorus uptake of chick thyroids. Endocrinology, 76:1124–1133, 1965.
45. Ekholm, R., and Pantic, V. Effect of thyrotropin on nucleic acids and protein contents of the thyroid. Nature, 199:1203–1204, 1963.
46. Kerkof, P. R., and Tata, J. R. Simultaneous acceleration *in vivo* of the formation of thyroid ribonucleic acid, phospholipid and iodoprotein by thyroid-stimulating hormone. Biochem. Biophys. Res. Commun., 28:111–116, 1967.
47. Hall, R. Stimulation of purine synthesis *in vitro* in the calf thyroid by thyroid-stimulating hormone. J. Biol. Chem., 238:306–310, 1963.
48. ——— and Tubmen, J. Further studies on effects of thyroid-stimulating hormone on thyroid nucleotide biosynthesis. J. Biol. Chem., 240:3132–3135, 1963.
49. ——— and Tubmen, J. The mechanism of action of thyroid-stimulating hormone on nucleotide biosynthesis in the thyroid. *In* Current Topics in Thyroid Research. Cassano, C., and Audreoli, N., eds. Proc. 5th Intern. Thyroid Conf. New York, Academic Press, Inc., pp. 564–571, 1965.
50. Begg, D. J., and Munro H. N. Site of action of thyroid-stimulating hormone on RNA metabolism in the thyroid gland. Biochem. J., 96:33–34 p, 1965.
51. ——— and Munro, H. N. Action of thyroid-stimulating hormone on ribonucleic acid synthesis in thyroid slices and in isolated thyroid nuclei. Nature, 207:483–485, 1965.
52. Shimada, H., and Yasumasu, I. Effect of thyroid stimulating hormone *in vitro* on the thyroid nuclear RNA synthesis. Gunma Symp. Endocr., 3:47–59, 1966.
53. Lecocq, R. E., and Dumont, J. E. Stimulation by thyrotrophin of [^{3}H] uridine incorporation in the polysomes of thyroid slices. Biochem. J., 104:13–14 C, 1967.
54. Hall, R., and Tubmen, J. Effect of thyroid-stimulating hormone on ribonucleic acid synthesis in the calf thyroid *in vitro* in the presence of puromycin. J. Biol. Chem., 243: 1598–1602, 1968.
55. Lindsay, R. H., Cash, A. G., and Hill, J. B. TSH stimulation of nucleotide and RNA synthesis in bovine thyroid slices. 3rd Intern. Congr. of Endocr., Mexico, June 30–July 5, 1968. Excerpta Medica Intern. Congr. Ser. No. 157, Abstr. 35, p. 14.
56. Imbenotte, J., Nataf, B., and Harel, J. Action de la TSH sur la biosynthèse du RNA dans des thyroïdes de foetus de rat en culture organotypique. Bull. Soc. Chim. Biol., 51:428–432, 1969.
57. Nygaard, A. P., and Hall, B. D. A method for the detection of RNA-DNA complexes. Biochem. Biophys. Res. Commun., 12:98–104, 1963.
58. Gillespie, D., and Spiegelman, S. A quantative assay for DNA-RNA hybrids with DNA immobilized on a membrane. J. Molec. Biol., 12:829–842, 1965.
59. Solomon, D. H. Enhancement of thyroidal water content and uptake of ^{24}Na after TSH administration. Biochim. Biophys. Acta, 43:346–347, 1960.
60. ——— Effects of thyrotropin on thyroidal water and electrolytes in the chick. Endocrinology, 69:939–954, 1961.
61. Turkington, R. W. Thyrotropin-stimulated adenosine triphosphatase in isolated thyroid cell membranes. Biochim. Biophys. Acta, 65:386–388, 1962.

62. Wolff, J. Transport of iodide and other anions in the thyroid gland. Physiol. Rev., 44: 45–90, 1964.
63. Lindsay, R. H., Cash, A. G., and Hill, J. B. Thyrotropin stimulation of pyrimidine nucleotide synthesis in bovine thyroid. Biochem. Biophys. Res. Commun., 29:850–855, 1967.

62

Role of the Hypophysis in the Initiation and Evolution of Thyroid Function in Mammals

J. P. Geloso

Laboratory of Comparative Physiology,
Faculty of Sciences, Paris, France

INTRODUCTION

It is now well-established that in mammalian fetuses at the end of gestation, the hypophysis controls the activity of thyroid gland: hypophysectomy results in a decrease of the thyroid activity which is restored to a normal level by administration of thyrotropic hormone.

For a more detailed study of the role of the pituitary gland in the functional differentiation of the thyroid, the fact that fetuses are changing day after day must be taken into account. In adult animals, the only variable in hypophysectomy is the time interval between operation and sacrifice. In fetuses, the rapid evolution of thyroid function in the course of gestation brings in another parameter, i.e., the day of gestation on which hypophysectomy is performed.

In the present paper, results of a systematic inquiry on the evolution of thyroid activity in fetuses hypophysectomized at various stages of their development are reported(1).

MATERIALS AND METHODS

Rat fetuses were deprived of their hypophysis by partial decapitation between 21.5 and 20.5 days of gestation and sacrificed between 18.5 and 21.5 days, 24 hours after a previous administration of radioiodide (^{131}I) to the mother. Adult female rats were fed with a commercial diet rich in iodine (U.A.R.).

In each of these operated fetuses the ^{131}I/mg thyroid to ^{131}I/mg plasma ratio (T/P), as a test of thyroid activity, was determined and compared with that of controls of the same age (Tables 1 and 2). In the case of fetuses decapitated on day

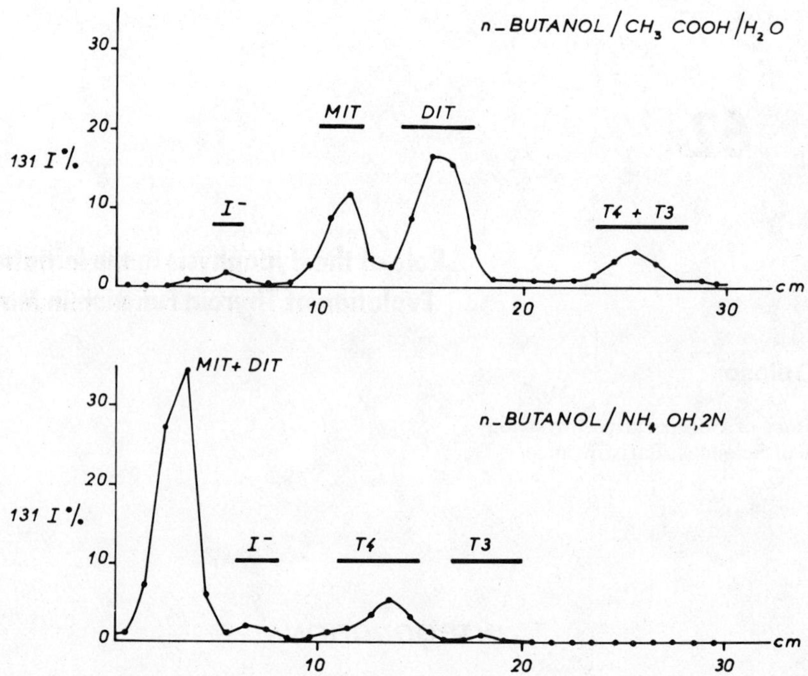

Fig. 1. Chromatograms of thyroid hydrolysate of 21,5-day-old fetuses decapitated on day 16.5 of gestation. ^{131}I has been injected 18 hours before sacrifice. From Geloso. *Ann. Endocr.* (Paris), 28 (Suppl. 1):1-79, 1967.

16.5 and sacrificed on day 21.5, a chromatographic analysis of thyroid extract was done (Fig. 1).

RESULTS

Results of these experiments (summarized in Fig. 2) are as follows:

1. On day 16.5 of gestation, the fetal thyroid does not concentrate iodide (T/P=0.7 and the ^{131}I/mg thyroid ^{131}I/mg muscle ratio is close to 1) and histologically is yet undifferentiated. Fetuses decapitated at this time exhibit, on day 18.5, a slight trapping of radioiodide which becomes greater at the end of gestation. Moreover, on day 21.5, thyroids of these operated fetuses disclosed radiothyroxine (Fig. 1).

2. In fetuses decapitated on day 17.5 (at this time, thyroid begins to concentrate iodide and to elaborate thyroxine [2,3,4]), there is, on the following day, a slight increase of the T/P ratio. The thyroid activity of these fetuses, compared with that of fetuses decapitated on day 16.5 is significantly higher on day 18.5 but quite similar on day 21.5

3. When decapitation is done on day 18.5, the T/P ratio does not change on the following days and remains at the normal level reached on day 18.5. Differ-

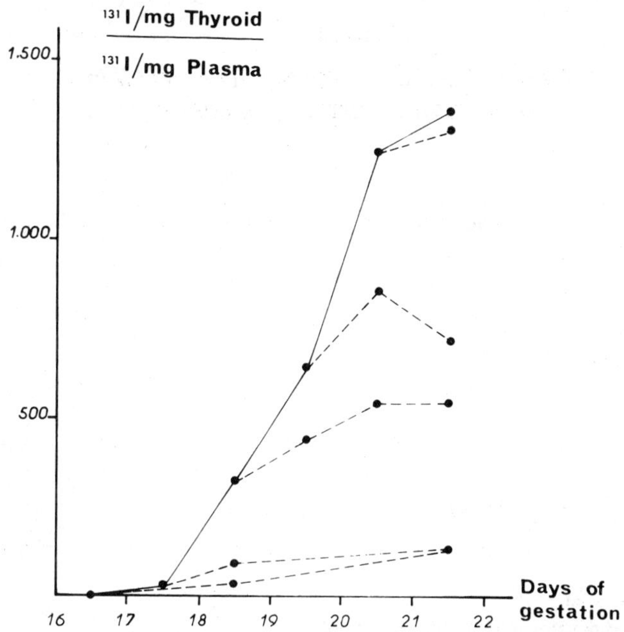

Fig. 2. Role of the hypophysis in the evolution of radioiodide trapping in the course of gestation. Full line: normal fetuses; broken line: decapitated fetuses. From Geloso. *Ann Endocr.* Paris),28 (Suppl. 1):1-79, 1967.

ences between values of the T/P ratio on days 19.5, 20.5, or 21.5 (operated fetuses) and on day 18.5 (normal fetuses) are not statistically significant.

The thyroid activity of these fetuses is clearly higher than that of fetuses hypophysectiomized on day 16.5 or 17.5 whatever their age may be.

4. Effects of decapitation on day 19.5 are similar to the preceding ones; T/P

TABLE 1

Ratio of thyroidal to plasma iodine (T/P) in rat fetuses

Date of gestation (days)	Number of fetuses	T/P[a,b]
16.5	9	0.7 ± 0.2
17.5	16	12 ± 2
18.5	16	317 ± 64
19.5	11	642 ± 111
20.5	12	1246 ± 320
21.5	35	1358 ± 232

[a] $T/P = \frac{\text{radioactivity/mg thyroid}}{\text{radioactivity/mg plasma}}$.

[b] Mean and confidence interval ($\underline{P}$ = 0.05).

TABLE 2

Ratio of thyroidal to plasma iodine (T/P) in rat fetuses decapitated at various ages

Date of decapitation[a]	Date of sacrifice[a]	Number of fetuses	T/P[b]
16.5	18.5	7	16 ± 5
	21.5	7	145 ± 56
17.5	18.5	7	96 ± 67
	21.5	7	118 ± 58
18.5	19.5	7	443 ± 106
	20.5	7	547 ± 100
	21.5	9	545 ± 122
19.5	20.5	7	863 ± 169
	21.5	11	720 ± 70
20.5	21.5	12	1303 ± 250

[a]Days of gestation.
[b]Mean and confidence interval (P = 0.05).

ratio on the following days does not significantly differ from the level reached at the time of operation.

5. In normal fetuses, the T/P ratio remains constant between 20.5 and 21.5 days of gestation; ablation of the hypophysis on day 20.5 does not change the T/P ratio on day 21.5.

DISCUSSION AND CONCLUSIONS

From these results one may conclude the following points:

1. Since fetuses hypophysectomized on day 16.5 exhibit at the end of gestation a slight trapping of iodide, the initiation of thyroid function does not appear to depend on the presence of fetal hypophysis. A slight development of the gland may occur without any fetal pituitary stimulus, but functional activity of the thyroid remains very low compared with that of normal fetuses. This conclusion is in agreement with histological data quoted in the literature. In mice(5) and in rats (6,7,8), the early hypophysectomy (destruction by x-rays or ablation by decapitation) does not prevent a further development of vesicular structures. With regard to the hormonal production, a synthesis of thyroxine by the gland of an anencephalic human fetus deprived of its hypophysis was observed(9).

However, these results do not imply that the initiation of thyroid function is quite autonomous. Factors other than fetal TSH may act on the undifferentiated thyroid; for example it is possible that the maternal TSH passes through the placenta in amounts too small to palliate the effects of fetal hypophysectomy in the late gestation, but in amounts large enough to elicit the onset of thyroid function.

2. When the hypophysis is removed just before day 18 (on day 17.5), the thyroid activity remains low and does not differ from the level reached without any hypophyseal stimulus (decapitation on day 16.5); when hypophysectomy is carried out after the day 18 (on day 18.5) the thyroid is much more active on following days than that of earlier operated fetuses. Thus, day 18 of gestation appears to be a crucial turning point in thyroid development. Such a notion of "*point de virage physiologique*" has been introduced by Jost(6) following a histological study of the thyroid development in early decapitated rabbit fetuses.

3. When the decapitation is performed between 18.5 and 20.5 days of gestation, the T/P ratio on the following days remains close to the level reached at the time of operation. In hypophysectomized fetuses, the activity of the thyroid gland depends mainly on the day of gestation on which decapitation is done and not on the time elapsed between operation and sacrifice. Jost and Cohen(10) studying the effects of hypophysectomy on adrenal organogenesis in rat fetuses have come to the same conclusion.

4. In normal fetuses, the T/P ratio remains constant between days 20.5 and 21.5, and the ablation of hypophysis on day 20.5 is ineffective. These facts may result from either a pause of the thyrotropic hypophyseal activity or from a lack of sensitiveness of thyroid gland to TSH.

A choice between these two alternatives is provided by data from other experiments(1), which are briefly analyzed below.

The percentage of protein-bound ^{131}I (PB^{131}I %=[radioactivity of PBI/radioactivity of plasma]$\times$100) has been determined 24 hours after radioiodide administration in the plasma of 20.5- and 21.5-day-old fetuses, the mothers of which having been fed either the preceding iodine-rich diet (U.A.R.) or an iodine-poor diet (wheat). Details of the experimental scheme and results are given in Table 3.

When the diet is rich in iodine, the PB^{131}I % in fetal blood drops between 20.5 and 21.5 days (20 percent and 9 percent, respectively), and decapitation on

TABLE 3

Changes in protein-bound iodine with diet, hypophysectomy, and TSH administration

Mother's diet	Treatment of fetuses	Date of sacrifice[a]	Number of fetuses	PB^{131}I%[b,c]
U.A.R. (see text)	Controls	20.5	10	20.9 ± 4.2
900 μg ^{127}I/100 g		21.5	10	8.9 ± 2.0
	Decapitated on day 20.5	21.5	8	8.7 ± 1.8
	Decapitated on day 20.5 + TSH (1 IU)	21.5	8	26.0 ± 7.5
Wheat	Controls	20.5	13	55.5 ± 3.6
1 μg ^{127}I/100 g		21.5	10	63.5 ± 4.6

[a]Days of gestation.
[b]PB^{131}I% = (radioactivity of PBI/total radioactivity of plasma)%.
[c]Mean and confidence interval (P = 0.05).

day 20.5 is without effect. These results are in agreement with those obtained by measuring the T/P ratio.

Administration of TSH to decapitated fetuses clearly increases the $PB^{131}I$ %, which means that the thyroid tissue remains sensitive to thyrotropic hormone. Thus, the absence of effects of hypophysectomy at the end of gestation must be attributed to a pause in thyrotropic hormone secretion. In rat fetuses, a decrease in adrenocorticotropic secretion at the same stage of gestation has been observed (11).

The question arises as to whether the decrease in thyrotropic hormone secretion is the result of an autonomous hypophyseal function or is due to some feedback mechanism. In mothers fed with an iodine-poor diet, the fetal PB ^{131}I % is clearly greater than in those with mothers fed an iodine-rich diet, and it does not decrease at the end of gestation, suggesting that the hypophyseal pause previously observed is mainly dependent on regulatory endocrine mechanisms. Such a feedback mechanism operates as well at earlier stages of gestation(1,12), and the degree of intervention of the hypophysis in fetal thyroid maturation seems to depend on the equilibrium between iodine supply, fetal plasma thyroxine degradation, and degree of maturity of thyroid and pituitary glands.

In conclusion, one may state that the initiation of thyroid function does not depend on the presence of the fetal hypophysis, but the hypophyseal stimulus is essential to the normal maturation of thyroid. Ablation of the pituitary gland during the course of thyroid development results in only a partial maturation of the latter, and the level of activity already reached under hypophyseal influence prior to the operation remains practically constant until birth. Such a picture of the role of hypophysis in thyroid differentiation is not peculiar to the hypophysis-thyroid system but is applicable to other systems (e.g., hypophysis-adrenal weight, hypophysis-hepatic glycogen, and so on). In a recent paper(13) and in the introductory lecture of this conference, Jost has given a tentative generalization of the role of pituitary gland in the maturation of endocrine structures.

REFERENCES

1. Geloso, J. P. Fonctionnement de la thyroïde et corrélations thyréohypophysaires chez le foetus de rat. Ann. Endocr. (Paris), 28 (Suppl. 1):1–79, 1967.
2. Feldman, J. D., Vazquez, J. J., and Kurtz, S. M. Maturation of the rat fetal thyroids. J. Biophys. Biochem. Cytol., 11:365–383, 1961.
3. Geloso, J. P. Date de l'entrée en fonction de la thyroïde chez le foetus de rat. C.R. Soc. Biol., 155:1239–1244, 1961.
4. Nataf, B., and Sfez, M. Début du fonctionnement de la thyroïde foetale du rat. C.R. Soc. Biol. (Paris), 155:1235–1238, 1961.
5. Raynaud, A., and Frilley, M. Procédé d'irradiation localisée au moyen des rayons X, des embryons de souris dans l'utérus maternel, et principaux résultats obtenus. Bull. Soc. Zool. Franc., 73:186–188, 1948.
6. Jost, A. Sur le développement de la thyroïde chez le foetus de lapin décapité. Arch. Anat. Micr. Morph. Exp., 42:168–183, 1953.
7. Noumura, T. Development of the hypophyseal-thyroidal system in the rat embryo and its relation to the maternal system. Jap. J. Zool., 12:301–318, 1959.

8. Slipka, J. Development of the thyroid gland in decapitated embryos of the rat. Csl. Morfol., 9:397–405, 1961.
9. Yamazaki, E., Noguchi, A., and Slingerland, J. W. The development of hormonal biosynthesis in human fetal thyroids. J. Clin. Endocr., 19:1437–1439, 1959.
10. Jost, A., and Cohen, A. Signification de l'"atrophie" des surrénales foetales du rat provoquée par l'hypophysectomie (décapitation). Develop. Biol., 14:154–168, 1966.
11. Cohen, A. Corrélation entre l'hypophyse et le cortex surrénal chez le foetus de rat. Le cortex surrénal du nouveau-né. Arch. Anat. Micr. Morph. Exp., 52:277–407, 1963.
12. Mistkevich, S. M. Glands of internal secretion in the embryonic development of birds and mammals. Published by the Academy of Science of USSR. Moscow, 1957. Translation published for the Nat. Sci. Found. and the Dept. of Health, Education and Welfare, U.S.A., by the Israel Program for Scientific Translation, 1959.
13. Jost, A. Full or partial maturation of fetal endocrine systems under pituitary control. Perspect. Biol. Med., 11:371–375, 1968.

63

The Effect of Premature Weaning on Thyroid and Adrenal Gland Functions in the Rat*

L. Macho,[1] V. Štrbák,[1] and A. Strážovcová[2]

[1]Department of Developmental Endocrinology,
Institute of Experimental Endocrinology,
Slovak Academy of Sciences, Bratislava,
Czechoslovakia
[2]Department of Histology, Institute
of Developmental Endocrinology, Slovak
Academy of Sciences, Bratislava,
Czechoslovakia

INTRODUCTION

The ontogenic development of mammals has some periods in which interference with the environment changes the characteristics of the individual in such a way that its further development is permanently affected(1). The development of the regulation of gonadal activity can be changed by the administration of sex hormones during the first five postnatal days(2). Further, it was observed that handling or shocking rats in infancy resulted in a reduction in emotional reactivity when animals are tested in adulthood(3). The first week of postnatal life is considered a critical period in the rat for the development and maturation of behavior patterns.

According to Křeček(5,6), the weaning period also consitutes a critical period. Premature weaning affects reproductive functions, impairs the fertility of rats(7), disturbs spermatogenesis and androgen production in adult male rats(8,9), and changes adrenal function(10). Further, the longevity of prematurely weaned female rats is shortened and higher nervous activity is impaired(19).

In the present work, the effect of premature weaning on the development of thyroid and adrenal gland functions was studied. Thyroid function was evaluated by estimating the radioiodine uptake in the thyroid gland and thyroxine plasma

*The authors wish to acknowledge the technical assistance of Mr. Š. Kaločai and Mrs. E. Šeligová.

concentration. The production of UV-absorbing corticosteroids by adrenal tissue in vitro and the plasma corticosterone concentration were used as indicators for studies of adrenal gland function.

MATERIALS AND METHODS

White Wistar rats from the Dobrá Voda breeding station were kept in a temperature-controlled (24°C) environment that was illuminated during normal daylight hours, and they were fed the standard pellet diet. All litters were reduced to 8 or 9 animals. Premature weaning (PW) was initiated by removing the mother on the 18th postnatal day, according to the method described by Hahn(11). Normally weaned (NW) rats remained with the mother for 30 days and on day 45 were redistributed according to sex. Animals in one experimental group were derived from at least 2 to 4 different litters to exclude any effects due to differences between litters. In all, 42 litters were used. For in vitro studies, animals were killed on days 19, 20, 22, 23, 30 to 31, 45, 90, and 120.

^{131}I Thyroid Uptake. The young animals were injected intraperitoneally with 0.5 μc of Na^{131}I in 0.2 ml saline solution, the NW groups being kept separately from their mothers. Four hours after injection, the rats were decapitated, their thyroid glands removed and weighed, and the radioactivity was measured. The ^{131}I uptake was expressed as a percentage of administered dose per 10 mg of thyroid tissue.

Plasma Thyroxine Determination. The modification of Murphy's method as described by Siersbeak-Nielsen(12) was used. In this method, a Dowex 1 x 8, 200/400 mesh is used instead of Sephadex for separation of free and bound thyroxine-^{131}I.

Determination of Corticosterone in Plasma. The plasma corticosterone concentration was determined according to the method of van der Vies et al.(13).

Corticosteroid Production in Vitro. The adrenals were rapidly removed from the decapitated animals, cleaned, weighed, and cut into halves or quarters. Tissue was preincubated for 30 minutes in Krebs-Ringer bicarbonate solution containing 200 mg percent glucose (KRBG solution) at 37°C in 95 percent O_2 and 5 percent CO_2. After this period, the medium was exchanged for a fresh KRBG solution and the tissue was incubated for 2 hours. The incubation medium was then extracted with dichloromethane, and UV-absorbing corticosteriods were determined according to the technique of Saffran and Schally(14).

To simplify the comparison of the results from these experiments and to avoid sex differences in the function of adrenals, the values from prematurely weaned rats were expressed as a percentage of those of normally weaned animals.

RESULTS

The Effect of Premature Weaning on Thyroid Gland Function

The body weights of animals and relative weights of thyroid glands are given in Table 1. The relative weights of thyroid glands gradually decreased in both

TABLE 1

Body weights and relative thyroid weights of normally and prematurely weaned rats

Age (days)	Sex	Body weight (g)		Thyroid weight (mg/100 g BW)	
		NW	PW	NW	PW
19	—	27.1 ± 0.5[a]	27.0 ± 0.5	15.6 ± 0.4	15.0 ± 0.6
20	—	31.1 ± 0.9	26.8 ± 0.8	15.1 ± 0.4	13.9 ± 0.7
22	—	29.3 ± 0.9	27.4 ± 0.6	13.8 ± 0.5	11.7 ± 0.6[c]
31	—	46.2 ± 2.6	48.8 ± 2.7	9.09 ± 0.5	9.4 ± 0.4
45	—	88.2 ± 4.5	88.7 ± 3.9	7.7 ± 0.3	7.7 ± 0.7
90	M	253 ± 14	251 ± 15	5.3 ± 0.3	6.5 ± 0.2[b]
	F	172 ± 4	193 ± 5		
120	M	312 ± 17	291 ± 11	5.8 ± 0.2	7.7 ± 0.5[c]
	F	219 ± 7	218 ± 3		

[a]Mean ± S.E.
[b]$P < 0.05$.
[c]$P < 0.01$.

groups during the period of observation. In 22- and 90-day-old prematurely weaned rats, the relative weight of the thyroid was lower whereas at 120 days it was higher than in normally weaned animals (Fig. 1).

Premature weaning affects radioiodine uptake in the thyroid gland. During the first days after the removal of the mother, ^{131}I thyroid uptake is higher than in NW rats (Fig. 2). No differences were found in 45- and 120-day-old rats, but in 90-day-old PW animals ^{131}I thyroid uptake was significantly lowered.

The determination of plasma thyroxine concentration (Fig. 3) showed an increase of thyroxine in 22- and 31-day-old PW rats. These results suggested an activation of the thyroid gland in PW rats with a subsequent release of thyroid hormones. This finding was also supported by histological examination of the thyroid gland in 22-day-old rats (Fig. 4). An increase in the percentage of epithelium and

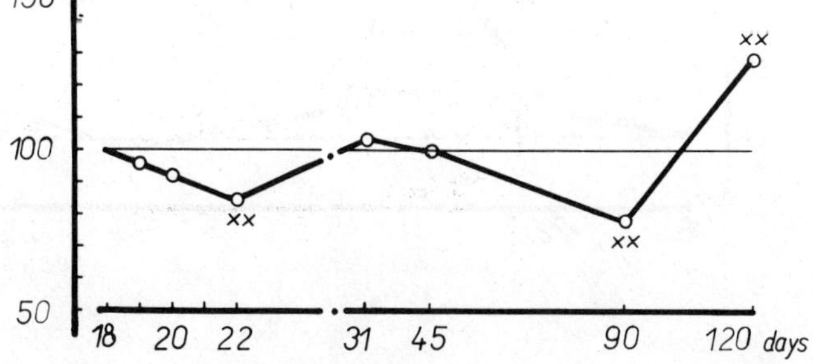

Fig. 1. Effect of premature weaning on the weight of thyroid gland expressed as a percentage of those in controls. Means ± S.E. 18 to 35 animals in each age group. XX=$P<0.01$.

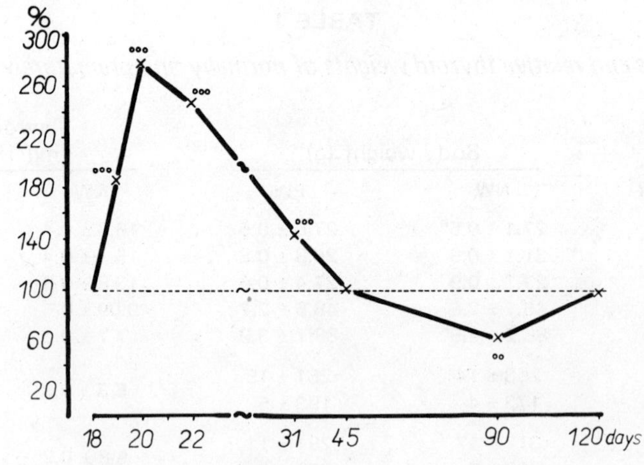

Fig. 2. ^{131}I thyroid uptake in premature weaned rats. ooo=$P<0.0001$, oo=$P<0.01$.

a decrease in colloid content was established using methods previously described (15).

Production of Corticosteroids in Normal and Prematurely Weaned Rats

There were no substantial differences in the relative weights of the adrenals in PW and NW animals. Determination of corticosteroid production by the adrenals in vitro showed a decrease on days 23, 30, and 90 in PW rats. In 45-day-old PW rats, a small increase in this production was noted in comparison with NW rats. No marked differences in plasma corticosterone concentration between NW and PW rats was observed except in 30-day-old animals, in which the PW rats had lower plasma corticosterone values (66 percent) than those normally weaned. The results of these

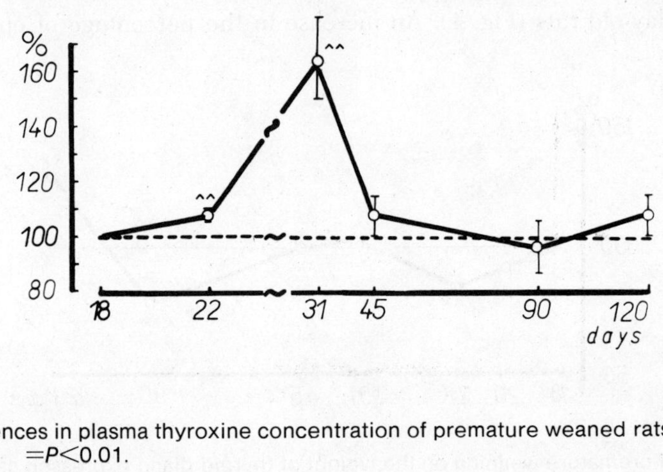

Fig. 3. Differences in plasma thyroxine concentration of premature weaned rats from controls. Means±S.E. =$P<0.01$.

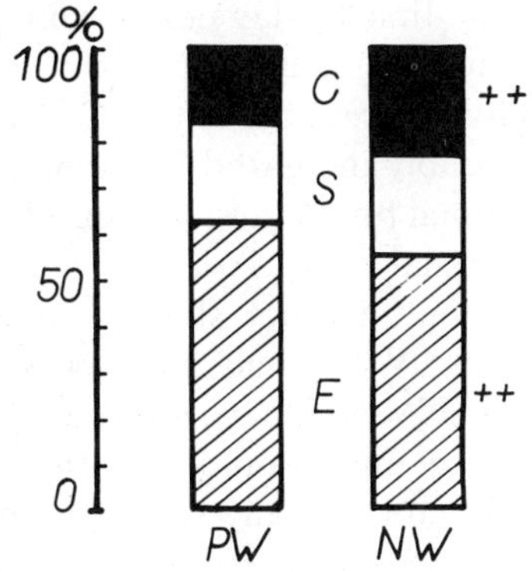

Fig. 4. The ratio of epithelium *(E)* to stroma *(S)* to colloid *(C)* in normally and prematurely weaned rats. $++=P<0.01$.

experiments suggest that during the first two weeks after premature weaning, the activity of the adrenals is decreased (Fig. 5).

DISCUSSION

During the weaning period, the method of feeding changes from suckling to the adult type. Premature weaning presents drastic interference with the environment of growing animals and causes several changes that express themselves even weaning as early as day 16 on the production of corticosterone and aldosterone, and they found a decrease in formation of these hormones in 23- and 30-day-old animals that had been prematurely weaned. In the older groups, the effect of premature weaning depended on the sex of the animals, female rats being more affected. In

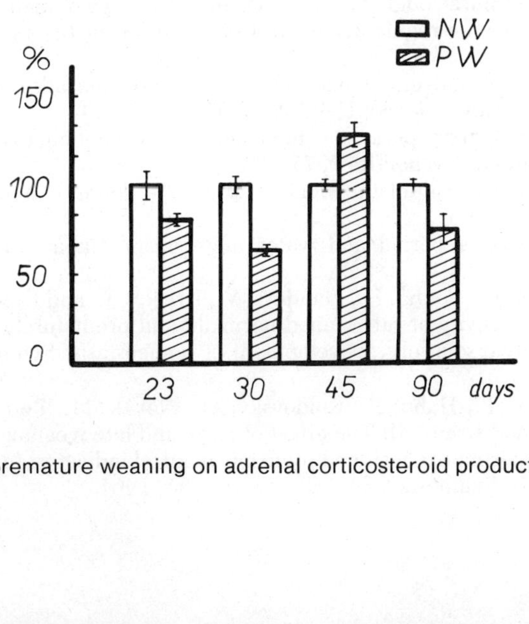

Fig. 5. The effect of premature weaning on adrenal corticosteroid production in rats of various ages.

our experiments we also found that premature weaning impaired corticosteroid production by rat adrenals in the first 14-day period and also in adult animals. This decrease of corticosteroid production cannot be due to impairment of steroid biosynthesis or decreased reactivity of target tissue to ACTH, for injecting corticotropin into 23-day-old PW rats considerably increased hormone production. It is very probable that the disturbance in adrenal function is a result of the changes in the activity of the system regulating the adrenal gland(10).

The results of our experiments further showed an increase of thyroid gland function during the first days in PW animals. One possible explanation of the increase in ^{131}I thyroid uptake in PW rats is suggested by the fact that premature weaning involves changes in the composition of the diet. Premature weaning stops the iodine supply from the milk and the only source of iodine is the solid diet. According to studies of Štolc et al.(18), during this period approximately one-half of the iodine intake is normally from milk. This transient iodine deficiency could explain the increase of radioiodine uptake on days 19 to 31. But, in addition, an increase of plasma thyroxine concentration was found in 31-day-old rats. This suggests that release of thyroid hormones from the gland is stimulated by an increased secretion of TSH from the pituitary.

Our results and the observations of other authors(9,10) indicate that in these young animals, the system regulating the function of the endocrine glands is fully operative and is sensitive to environmental conditions. Premature weaning affects the activity of several endocrine glands, increasing thyroid hormone secretion and decreasing adrenal and sexual hormone production. Further experiments are in progress to study the mechanism of these changes and their effects on the adult animal.

REFERENCES

1. Scott, P. Critical periods in behavioral development. Science, 138:949, 1962.
2. Gorski, R. A., and Baraclough, C. A. Effects of low dosage of androgen on the differentiation of hypothalamic regulatory control of ovulation in the rat. Endocrinology, 73: 210, 1963.
3. Levine, S. Maternal and environmental influences on the adrenocortical response to stress in weanling rats. Science, 156:258, 1967.
4. Denenberg, V. H. Critical periods, stimulus input and emotional reactivity. A theory of infantile stimulation. Psychol. Rev., 71:135, 1964.
5. Křeček, J. The physiology of the weaning period in young mammals. Advances Biol. Sci. (Praha) SZdN, 1962.
6. ——— The weaning period and water metabolism. Babáková Sbirka (Praha) 28, SZdN, 1962.
7. Kraus, M., Andel, J., Dlouhá, H., Flandera, V., Křeček, J., and Kreckova, J. Rkos J. Development and longevity of rat weaned normally and prematurely and living on a free choice of electrolyte solutions. Development of homeostasis. Symposia CSAV (Praha), p. 202, 1961.
8. Kubát, K., Fabry, P., Hahn, P., Koldovsky, O., Novák, M., Petrásek, R., Placer, Z., Slabochová, Z., and Kraus, M. The effect of early and late weaning to diets of different composition on spermatogenesis and some biochemical indicators in white rats. Development homeostasis. Symposia CSAV (Praha), p. 185, 1961.

9. Křeček, J., and Palatý, V. The effect of premature weaning on the development of androgenic activity in male rats. Physiol. Bohemoslov., 16:501, 1967.
10. Kraus, M., Křeček, J., and Popp, M. The development of corticosterone production by the adrenal gland in normally and prematurely weaned rats. Physiol. Bohemoslov., 16: 120, 1967.
11. Hahn, P. Vývoj metabolismu hlavních zivin u savcu. SZdN. Praha, 1966.
12. Siersbeak-Nielsen, K. Determination of serum thyroxine. Acta Med. Scand., 181:327, 1967.
13. Vies, J., Van der, Bakker, R. F. M., and Wied, D. De. Correlated studies on plasma free corticosterone and on adrenal steroid formation in vitro. Acta Endocr. (Kobenhavn), 34:513, 1960.
14. Saffran, M., and Schally, A. H. In vitro bioassay of corticotropine modification and statistical treatment. Endocrinology, 56:523, 1955.
15. Macho, L., Mitro, A., Drgonová, H., and Sokolová, V. Changes in the content of proteins, nucleic acids and the histological picture of the thyroid gland during the postnatal period in the rat. Physiol. Bohemoslov., 15:368, 1966.
16. Ader, R. Social factors affecting emotionality and resistence to disease in animals. III. Early weaning and susceptibility to gastric ulcers in the rat. A control for nutritional factors. J. Comp. Physiol., 55:600, 1962.
17. Kraus, M., Křeček, J., and Popp, M. The development of aldosterone production by the adrenal glands of normally and prematurely weaned rats. Biol. Neonat., 11:338, 1967.
18. Stolc, V., Knopp, J., and Stolcová, E. Iodine, solid diet, water and milk intake by lactating rats and their offsprings. Physiol. Bohemoslov., 15:219, 1966.
19. Nováková, V., Koldovský, O., Faltin, J., Hahn, P., and Flandera, V. Conditional reflex activity in male rats weaned normally and prematurely. Physiol. Bohemoslov., 12:325, 1963.

64

Analysis of the Action of Hormone on Growth and Function of the Adenohypophysis

V. Schreiber

Laboratory for Endocrinology and Metabolism,
Faculty of General Medicine, Charles University,
Prague, Czechoslovakia

INTRODUCTION

The adenohypophysis is a complex organ composed of at least four types of cells, equipped with many enzymes and producing at least six hormones (ACTH, TSH, LH, FSH, GH, and prolactin). Its function—and possibly its structure also—is probably regulated by hypothalamic releasing factors, each hormone having its own factor (CRF, TRF, LRF, FRF, GH-RF, and PIF=prolactin inhibiting factor). Considerable interest has always been devoted to the various functions of the adenohypophysis and to the structural parameters of these functions, but less attention has been paid to the adenohypophysis as a whole. This paper submits the results of a general study of the adenohypophysis under different experimental conditions characterized by overproduction of the adenohypophyseal hormones, both individually and in combination. Attention was concentrated on: (1) trophic (growth) reactions; (2) enzymatic (metabolic) reactions, and (3) basic hormonal reactions, in particular those occurring after unilateral removal of a gland in which compensatory hypertrophy of the remaining gland furnishes information on secretion or availability of the relevant adenohypophyseal tropin.

TROPIC ADENOHYPOPHYSEAL REACTIONS

We compared (1) the growth reaction of the adenohypophysis of male rats 3 weeks after adrenalectomy, castration, and thyroidectomy (supplemented by elimination of the thyroid residue with radioiodine) and after combined adrenalectomy and thyroidectomy and combined castration and thyroidectomy. The results (1) are shown in Table 1. Of the individual operations, maximum hypertrophy was produced by thyroidectomy, while of the combined operations, castration plus thyroidectomy produced a greater reaction than adrenalectomy plus thyroidectomy; adrenalectomy alone had very little effect on the weight of the hypophysis. Experiments in which methylthiouracil (MTU) feeding was substituted for surgical thyroidectomy (2) gave similar results (Table 2).

In further experiments(2), we tested the effect of excess amounts of the peripheral hormones on hypertrophy of the hypophysis after removal of the target gland.

TABLE 1

Weight of the adenohypophyses in various groups of male rats 3 weeks after operations (1)

Group	Number of rats	Final weight (g)	Weight of the adenohypophyses (mg)	(mg/100 g)
Controls	10	212 ± 4	5.9 ± 0.25	2.78 ± 0.10
Adrenalectomy	5	186 ± 9	5.8 ± 0.25	3.14 ± 0.15
Castration	9	196 ± 4	6.9 ± 0.23	3.53 ± 0.10
Thyroidectomy	10	200 ± 7	7.6 ± 0.16	3.85 ± 0.18
Adrenalectomy + thyroidectomy	10	185 ± 7	7.3 ± 0.36	4.00 ± 0.18
Castration + thyroidectomy	8	199 ± 9	8.4 ± 0.43	4.20 ± 0.23

TABLE 2

Weight of the adenohypophyses in control male rats, 3 weeks after castration and after 3 weeks MTU (0.2 percent in the diet) feeding (2)

Group	Number of rats	Final weight (g)	Weight of the adenohypophyses (mg)	(mg/100 g)
Controls	10	228 ± 6	4.6 ± 0.19	2.04 ± 0.09
Castrated	10	217 ± 4	6.1 ± 0.16	2.82 ± 0.09
MTU-fed	10	215 ± 7.5	7.4 ± 0.46	3.52 ± 0.16

TABLE 3

Weight of the adenohypophyses in control male rats, 3 weeks after castration and 3 weeks after castration with simultaneous dried thyroid (0.3 percent in the diet) feeding (2)

Group	Number of rats	Final weight (g)	Weight of the adenohypophyses (mg)	(mg/100 g)
1. Controls	10	228 ± 6.4	4.6 ± 0.19	2.04 ± 0.09
2. Castrated	10	217 ± 4.1	6.1 ± 0.16	2.82 ± 0.09
3. Castrated + dried thyroid	9	185 ± 6.8	4.8 ± 0.22[a]	2.60 ± 0.08[b]

[a]In comparison with group 2, $\underline{P} < 0.01$.
[b]In comparison with group 2, $\underline{P} < 0.05$.

TABLE 4

Weight of the hypophyses in control male rats, rats injected daily with 150 μg estradiol dipropionate for 3 weeks, rats fed 0.3 percent dried thyroid in the food for 3 weeks, and rats with both estradiol and dried thyroid (3)

Group	Number of rats	Final weight (g)	Weight of the hypophyses (mg)	(mg/100 g)
1. Controls	10	145 ± 3	5.5 ± 0.2	3.8 ± 0.1
2. Estradiol	10	96 ± 3	6.7 ± 0.4	7.0 ± 0.4
3. Thyroidine	5	114 ± 7	3.7 ± 0.4	3.2 ± 0.2
4. Estradiol + thyroidine	10	82 ± 3	4.1 ± 0.2[a]	5.0 ± 0.2[a]

[a]In comparison with group 2, $P < 0.01$.

We showed that (Table 3) dried thyroid feeding not only completely inhibited hypertrophy of the hypophysis induced by MTU, but that it also partly inhibited castration-induced hypertrophy. Testosterone completely inhibited hypertrophy after castration, but neither testosterone nor hydrocortisone modified MTU-induced hypertrophy.

Another type of hypophyseal hypertrophy studied was the one caused by the administration of excess estrogens. We investigated whether excess thyroid hormone would also inhibit this type of hypertrophy. It was found(3) that excess thyroid hormones in the diet largely inhibited the development of hypophyseal hypertrophy in male and female rats to which estrogens were administered (Table 4). In our opinion, therefore, hypertrophy of the hypophysis induced by excess estrogens is due, at least in part, to a relative deficiency of thyroxine. This may be a deficiency of free, biologically active thyroxine, since estrogens stimulate the binding of thyroxine to the plasma proteins (for recent review see 4). The deficiency of thyroxine is abolished by the administration of excess thyroid hormone and the degree of hypophyseal hypertrophy is therefore lower. This hypothesis is borne out by fur-

TABLE 5

Radioiodine uptake (4-hour test) in the thyroids of control male rats, rats injected daily with 150 μg estradiol dipropionate for 3 weeks, rats fed 0.3 percent dried thyroid in the food for 3 weeks, and rats with both (5)

Group	Number of rats	Percent dose per whole thyroid	Percent dose per mg of the thyroid
1. Controls	10	6.10 ± 0.82	0.77 ± 0.08
2. Estradiol	10	6.10 ± 0.58	0.80 ± 0.07
3. Dried thyroid	5	0.19 ± 0.02	0.03 ± 0.002
4. Estradiol + dried thyroid	10	0.45 ± 0.04[a]	0.08 ± 0.006[a]

[a]In comparison with group 3, $P < 0.01$.

ther evidence, in particular that the administration of estrogens(5) lowers the degree of inhibition of the thyroid (the decrease in radioiodine uptake) produced by the administration of excess exogenous thyroid hormone (Table 5). The increase in binding capacity (and hence the inactivation) of the thyroid hormones in the presence of excess estrogens thus seems to apply in the case of exogenously administered thyroid hormones also.

There may, however, be another explanation for the estrogen-induced deficiency of thyroid hormones: it could be caused by their *decreased* binding to plasma proteins and secondarily by their increased elimination. This hypothesis was recently reviewed and experimentally supported by Yamada and co-workers(4). We tried to clarify the controversy by combined estrogen and diphenylhydantoin (DPH) administration to rats, since DPH is said to release thyroid hormones from their bond with the plasma proteins(6). In our experiments, however, DPH—at least in the doses given—did not block adenohypophyseal hypertrophy after estrogen administration. This shows that alternative hypotheses must be taken into account.

METABOLIC REACTIONS OF THE ADENOHYPOPHYSIS

A study of the effects of hypothalamic extracts on the adenohypophysis in vitro, which came out in 1958 (reviewed by 7) revealed an increase in adenohypophyseal acid phosphatase activity. Since this might have been a manifestation of the effect of one of the releasing factors in the extracts, we wondered whether the activation of adenohypophyseal acid phosphatases might be associated with hypersecretion of one of the adenohypophyseal tropins. In the very first experiments, in which we compared the effect of adrenalectomy, castration, and thyroidectomy, we found that acid phosphatase activity of pituitary homogenates rose after thyroidectomy only(8).

Postthyroidectomy activation of adenohypophyseal acid phosphatases was also demonstrated histochemically(9). Activity was found to be localized mainly in the adenohypophyseal thyrotropic basophils. The histochemically demonstrated relationship between adenohypophyseal acid phosphatase and TSH secretion was confirmed by Sobel(10), who, however, found no significant changes on using a biochemical method of estimation in adenohypophyseal homogenate. The possibility of biochemical demonstration of this activation was confirmed in further experiments(11) and at the same time a correlation was found between the increase in the weight of the thyroid and the increase in adenohypophyseal acid phosphatase following a MTU block. We also demonstrated(12) a correlation between adenohypophyseal acid phosphatase activity and the interval after thyroidectomy and between the weight of the adenohypophysis and relative adenohypophyseal acid phosphatase activity (i.e., per mg of tissue) after thyroidectomy.

In histochemical studies we showed that nonspecific esterase, as well as adenohypophyseal acid phosphatase, was activated in the presence of TSH hypersecretion(13) and published a comprehensive study(14) on adenohypophyseal enzymatic activities after adrenalectomy, castration, and thyroidectomy in rats (Table 6). Of these three peripheral operations, thyroidectomy has the most pronounced

TABLE 6

Semi-quantitative histochemical examination of enzymatic activities in the adenohypophyses of control male rats and adrenalectomized, castrated, and thyroidectomized (or MTU-fed) rats (14) 3 weeks after operations

Enzyme	Controls	Adrenal-ectomized	Castrated	Thyroid-ectomized or fed MTU
Alkaline phosphatase	+	+	+	++
Acid phosphatase	+	+	++	+++
ATPase	+	+	+	+/++
5-nucleotidase	+	+	+	+/++
Thiamine pyrophosphatase	+	+	++	++/+++
Nucleoside diphosphatase	+	+	++	++/+++
Nonspecific esterase	+	+	+++	+++
E-600 resist. nonspecific esterase	+	+	++	+++
NAD-diaphorase	+	+	+/++	++
NADP-diaphorase	+	+	+/++	++
Succinic dehydrogenase	+	+	+/++	++
NAD-dependent dehydrogenases (lactic, malic, alpha-glycerophosphate)	+	+	+/++	++
NADP-dependent dehydrogenases (isocitric, glucose-6-phosphate)	+	+	+/++	++

effect on adenohypophyseal enzymatic activities and adrenalectomy has the least effect (or none at all), while castration occupies an intermediate position. This is in complete agreement with findings on adenohypophyseal tropic reactions.

HORMONAL REACTIONS OF THE ADENOHYPOPHYSIS

In order to study the interrelationships of some of the basic hormonal reactions of the adenohypophysis in general and thyrotropic function in particular, we chose compensatory hypertrophy of the peripheral endocrine glands following unilateral hemiglandectomy as a test situation. The question remains unsettled whether these forms of hypertrophy are caused by an increase in the amount of the relevant hypophyseal tropin available (per unit weight) to the remaining half of the gland, as assumed by Shellabarger(15) and—for compensatory hypertrophy of the adrenals—by Lipscomb and Critchlow(16), or whether they are due to the feedback stimulation in adenohypophyseal secretion of the relevant tropin responding to the decrease in the level of the peripheral hormone available after removal of the target glands as was demonstrated for the ovaries by Edgren and co-workers (17) and by Greenwald(18). Shellabarger(15), however, clearly demonstrated that the same compensatory hypertrophy of the peripheral glands develops both after single and combined hemiglandectomies, i.e., no preferential secretion of any of the adenohypophyseal tropins can be demonstrated under conditions of *partial* deficiency of the peripheral hormones.

In experiments combining *complete* deficiency of one peripheral hormone and *partial* deficiency of another, i.e., involving total elimination of one gland and hemielimination of another, the situation is quite different. Horn and LoMonaco(19) showed that compensatory hypertrophy of the adrenals was blocked in cretin rats (born to hypothyroid mothers and treated postnatally with thiouracil), and that compensatory hypertrophy of the gonads increase. As shown below, our own results are in agreement with these findings as far as the adrenals are concerned, but sharply conflict with them in the case of ovaries.

In this review I have intentionally confined myself to the manifestations of compensatory hypertrophy, passing over the extensive literature on the general effect of hormone excess and deficiency on secretion of the individual adenohypophyseal tropins. I should, however, like to point out that the decrease in ACTH secretion (20) and gonadotropin secretion(21) in hypothyroidism and the increase in ACTH secretion(22) and gonadotropin secretion(23) in the presence of excess exogenous thyroxine are observations that have been known for a long time.

The chief finding in our experiments(24) was that chemical thyroidectomy (MTU) completely inhibited compensatory hypertrophy of the adrenals (Table 7), while MUT-induced hypertrophy of the thyroid was not inhibited either by unilateral or total adrenalectomy; similarly compensatory hypertrophy of the thyroid after hemithyroidectomy was(25) likewise not inhibited by total adrenalectomy (Table 8). Chemical thyroidectomy(26) also completely inhibited compensatory hypertrophy of the ovaries after unilateral ovariectomy (Table 9) but total castration did not inhibit compensatory hypertrophy of the thyroid after unilateral thyroidectomy (Table 8). This conflicts with the findings of Horn and LoMonaco(19), who worked with cretin rats born to hypothyroid mothers. In our experiments, even long-term pretreatment with MTU did not enhance compensatory hypertrophy of the ovaries; on the contrary, it always led to a complete block.

The decrease in the secretion of endogenous gonadotropins in peripheral hypothyroidism mentioned above is manifested in raised sensitivity to exogenous gonadotropins(27,28). The sensitivity of adult hypothyroid animals to HCG does not attain the sensitivity of hypophysectomized or immature animals, however, so that hypothyroid animals cannot be substituted for the hypophysectomized or immature

TABLE 7

Weight of the right adrenals expressed as percentage of the weight of left adrenals in control male rats, MTU-fed rats, unilaterally (left) adrenalectomized rats, and rats with both MTU feeding and unilateral left adrenalectomy (24). (MTU was present in the diet (0.2 percent), animals were killed 2 weeks after operation and/or MTU feeding.)

1. Controls	96.6 ± 2.43 %
2. MTU-fed	91.5 ± 2.82 %
3. Unilat. adrenalectomy	137.3 ± 13.2 %
4. MTU + Unilat. adrenalectomy	95.0 ± 5.9 %[a]

[a]In comparison with group 3, $\underline{P} < 0.01$.

TABLE 8

Weight of the left and right thyroid lobe in control male rats, hemithyroidectomized rats, and after combination of hemithyroidectomy with total castration or adrenalectomy (25). (In all operated groups, left thyroid lobes were weighed at the time of operation, right lobes 2 weeks later.)

Group	Number of rats	Left lobe (mg/100 g)	Right lobe (mg/100 g)
Controls	10	1.96 ± 0.12	2.33 ± 0.08
Hemithyroidectomized	10	2.37 ± 0.14	3.32 ± 0.23
Hemithyroidectomized + castrated	10	2.20 ± 0.10	3.57 ± 0.21
Hemithyroidectomized + adrenalectomized	10	1.74 ± 0.14	3.50 ± 0.24

TABLE 9

Compensatory hypertrophy of rat ovaries after unilateral ovariectomy (26). Weights of left and right ovaries in control rats, rats fed 0.2 percent MTU in the diet for 5 weeks, rats after unilateral (left) ovariectomy (5 weeks after operation), and rats subjected to both unilateral ovariectomy (UOE) and MTU treatment

Group	Number of rats	Left ovary (mg/100 g)	Right ovary (mg/100 g)	Percentage of hypertrophy
Control	6	15.4 ± 1.7	16.4 ± 1.9	
MTU	6	9.2 ± 1.3	9.4 ± 1.0	
UOE	9	15.3 ± 0.49	25.5 ± 1.7	166 ± 8.8
UOE + MTU	9	17.0 ± 0.81	17.7 ± 1.7	104 ± 10.0[a]

[a]In comparison with group UOE, $P < 0.01$.

in bioassays of gonadotropins. On the other hand, hyperthyroidism induced by exogenous thyroid hormones is associated—as already mentioned—with hypersecretion of endogenous gonadotropins, with the result that ovaries hypertrophy(29,30).

DISCUSSION

In summing up the contents of the three parts of this paper, we are obliged to conclude that thyrotropic function occupies a special place—one might even say a "privileged" position—among the other adenohypophyseal functions. This conclusion is based on the following evidence:

1. Thyroidectomy produces the maximum tropic (growth) reaction of the adenohypophysis, castration a moderate reaction, and adrenalectomy the smallest or no reaction.

2. Metabolic (enzymatic) activation of the adenohypophysis is also greatest after thyroidectomy, moderate after castration, and absent (as far as we know today) after adrenalectomy. The hypophysis thus requires a pronounced tropic and metabolic reaction to effect maximum TSH secretion, while none is required for ACTH hypersecretion; the postcastration reaction occupies an intermediate position.
3. Thyroid hormone deficiency leads to involution of the adrenals and gonads, whereas corticoid and sex hormone deficiency does not result in involution of the thyroid. Involution of the thyroid is altogether difficult to demonstrate, even after the administration of excess thyroxine. This is evidently due to the relative independence of resting thyroid weight on TSH.
4. Compensatory hypertrophy of the remaining adrenal (ovary) after unilateral adrenalectomy (ovariectomy) can be blocked by total thyroid hormone deficiency, but compensatory hypertrophy of the thyroid after hemithyroidectomy is not blocked by total adrenalectomy (ovariectomy).

Under conditions of combined deficiency of the hormones of several peripheral glands, one of which is the thyroid, the adenohypophysis evidently preferentially secretes TSH in an attempt to restore the thyroid hormone level, even at the expense of its regulatory influence on the adrenals and the gonads. Nothing is known of the cause of this phenomenon. All we can do is to speculate on the general metabolic role of the thyroid hormones on one hand, or on the survival of ancient phylogenetic preferences on the other, as the cause.

In conclusion, it should be added that the theory of the special position of the thyrotropic function of the adenohypophysis does not preclude the possibility of simultaneous hypersecretion of several adenohypophyseal tropins, even when one of them is TSH. This was demonstrated for partial deficiency of peripheral hormones by Shellabarger(15) and strong evidence for this possibility is provided by experiments with amphenone B. Chronic administration of this substance, which simultaneously blocks biosynthesis of the thyroid hormones and of corticoids, leads to hypertrophy of the thyroid (i.e., TSH hypersecretion) and simultaneously to hypertrophy of the adrenals (i.e., ACTH hypersecretion), as found by a number of authors(31-34). The amphenone B block of biosynthesis of the thyroid hormones need not be absolutely complete, however, and it is not at present known whether amphenone B also produces adrenal hypertrophy after total thyroidectomy or in the presence of a simultaneous block of the thyroid with MTU.

Since amphenone B was not available in our laboratory, we have made several experiments with aminoglutethimide (α-ethyl-α-p-aminophenylglutarimide, Elipten [CIBA]). This substance simultaneously inhibits thyroid, adrenal, and ovarian hormone synthesis and thus stimulates (by means of adenohypophyseal feedback mechanism) growth of the thyroid, adrenals, and ovaries(35-37). Because aminoglutethimide alone did not completely block thyroid function (presumably because it produces simultaneously hypersecretion of TSH, ACTH, and gonadotropins) we have compared the effect of this drug on adrenal and ovarian weights in intact and thyroidectomized rats(38). Thyroidectomy (Table 10) blocked growth of the adrenals after aminoglutethimide completely and ovarian growth partially. This is in

TABLE 10

Weights of adrenals and ovaries in control female rats, rats fed 0.3 to .5 percent aminoglutethimide in the diet, rats completely thyroidectomized, and rats completely thyroidectomized and fed aminoglutethimide (38)

Group	Final body weight (g)	Adrenal weight (mg/100 g)	Ovarian weight (mg/100 g)
1. Control	154 ± 2.2	26.8 ± 1.7	26.3 ± 2.2
2. Aminoglutethimide	153 ± 3.8	55.0 ± 5.3	43.3 ± 3.1
3. Thyroidectomy	160 ± 2.4	20.2 ± 1.0	24.6 ± 1.7
4. Thyroidectomy + aminoglutethimide	152 ± 3.4	31.5 ± 1.6[a]	36.2 ± 2.5

[a]In comparison with group 2, $\underline{P} < 0.01$.

agreement with our hypothesis, that in a complete deficiency of thyroid hormones, the adenohypophysis produces only or preferentially TSH.

REFERENCES

1. Schreiber, V., and Kmentová, V. L'hypertrophie de l'hypophyse du rat après surrénalectomie, castration, thyroidectomie et leur combinaisons. C.R. Acad. Sci. (Paris), 258: 4151, 1964.
2. ——— and Kmentová, V. Partial block of castration hypertrophy of the adenohypophysis in male rats by thyreoidin. Physiol. Bohemoslov., 14:332, 1965.
3. Zbuzková-Kmentová, V., and Schreiber, V. Empéchement par les hormones thyroidiennes de l'hypertrophie hypophysaire causée par l'oestrogéne. C.R. Acad. Sci. (Paris), 262:1565, 1966.
4. Yamada, T., Takemura, Y., Kobayashi, I., and Shichijo, K. Re-evaluation of the effect of estrogen on thyroid activity in the rat and its mechanism. Endocrinology, 79:849, 1966.
5. Zbuzková-Kmentová, V., and Schreiber, V. Gonado-thyroidal relationships. Interaction of oestrogen and thyroid hormones in infantile rats. Endocr. Exp. (Bratislava), 1:29, 1967.
6. Oppenheimer, J. H., and Tavernetti, R. R. Studies on the thyroxine-diphenyl-hydantoin interaction: Effect of 5,5′-diphenylhydantoin on the displacement of L-thyroxine from thyroxine binding globulin. Endocrinology, 71:496, 1962.
7. Schreiber, V. The Hypothalamo-Hypophysial System. Prague, Publ. House of the Czech. Acad. Sci., 1963.
8. ——— and Kmentová, V. Bedeutung der Phosphatasenaktivität in der Hypophyse. Acta Biol. Acad. Sci. Hung., 9:285, 1959.
9. Lojda, Z., Schreiber, V., and Kmentová, V. Histochemie fosfatas v hypofyse krysy za experimentálni hypothyresoy. Cesk. Fysiol., 9:173, 1960.
10. Sobel, H. J. The localization of acid phosphatase activity in the rat pituitary and thyroid glands, and its relation to secretory activity. Endocrinology, 68:801, 1961.
11. Schreiber, V. Biochemical estimation of elevated acid phosphatase activity in anterior pituitaries of rats and golden hamsters fed methylthiouracil. Endocrinology, 70:924, 1962.
12. ——— and Kmentová, V. Biochemical determination of elevated acid phosphatase activity in the adenohypophyses of hypothyroid rats. Comparison of the King-Armstrong method and the method of Bessey, Lowry and Brock. Folia Biol. (Praha), 8:48, 1962.

13. Lojda, Z., and Schreiber, V. Enzymatické vybaveni hypothalamo-hypofysárniho systému. Cesk. Fysiol., 11:314, 1962.
14. ——— and Schreiber, V. Activités enzymatiques de l'adénohypophyse du rat dans diverses conditions expérimentales. J. Physiol. (Paris), 56:559, 1964.
15. Shellabarger, C. J. Compensatory hypertrophy of the thyroid gland, adrenal gland and gonad studied singly or in combination. Endocrinology, 73:124, 1963.
16. Lipscomb, H. and Critchlow, V. "Compensating" adrenal hypertrophy in the absence of reflex increase in ACTH release. Proc. 2nd Intern. Congr. Endocr., Excerpta Medica London, p. 1278, 1965.
17. Edgren, R. A., Parlow, A. F., Peterson, D. L., and Jones, R. C. On the mechanism of ovarian hypertrophy following hemicastration in rats. Endocrinology, 76:97, 1965.
18. Greenwald, G. S. Influence of one or two ovaries on ovulation and ovarian weight in the hypophysectomized rats. Endocrinology, 82:591, 1968.
19. Horn, E. L., and LoMonaco, M. B. Unilateral adrenalectomy in the cretin rat. Proc. Soc. Exp. Biol. Med., 98:817, 1958.
20. Zarrow, M. X., and Money, W. L. Involution of the adrenal cortex of rats treated with thiouracil. Endocrinology, 44:345, 1949.
21. Benoit, J. Relation between thyroid and growth of testes and penis when stimulated by electric light. Proc. Soc. Exp. Biol. Med., 36:782, 1937.
22. Wallach, D. P., and Reineke, E. P. The effect of varying levels of thyroidal stimulation on the ascorbic acid content of the adrenal cortex. Endocrinology, 45:75, 1949.
23. D'Angelo, S. A. Thyroid hormone administration and ovarian and adrenal activity in rats bearing hypothalamic lesions. Endocrinology, 64:685, 1959.
24. Schreiber, V., and Kmentová, V. Vztahy mezi TSH a ACTH. Změny při experimentálni hypothyreose a hypokortikalismu u krys. Sborn. Lék., 61:253, 1959.
25. ——— and Kmentová, V. L'hypertrophie compensatrice de la thyroide n'est bloquée ni par la castration ni par l'adrénalectomie complète. C.R. Acad. Sci. (Paris), 261:3672, 1965.
26. ——— Kmentová, V., and Zavadil, M. Relationships between gonadotrophin and thyrotrophin secretion: inhibition of compensatory hypertrophy of rat ovaries by methylthiouracil. Physiol. Bohemoslov., 13:554, 1964.
27. Kmentová, V. and Schreiber, V. Comparison of the sensitivity of the weight reaction of the ovaries to chorionic gonadotrophin in the adult intact, hypophysectomized, hypothyroid and immature rats. Folia Biol. (Praha), 11:305, 1965.
28. Brown, P. S., and Wells, M. Factors which influence assays of gonadotrophin based on the induction of ovulation in mice. J. Endocr., 33:507, 1965.
29. D'Angelo, S. A. Thyroid hormone administration and ovarian and adrenal activity in rats bearing hypothalamic lesions. Endocrinology, 64:685, 1959.
30. Kmentová, V., Schreiber, V., and Zavadil, M. Thyroidin induced hypertrophy of the ovaries in rats. Physiol. Bohemoslov., 12:506, 1963.
31. Heming, A. E., Holtkamp, D. E., Kerwin, J. F., Mansor, L. F., and Dacanay, J. G. Effect of Amphenone B on adrenal and thyroid function of adult ovariectomized rats. Proc. Soc. Exp. Biol. Med., 80:154, 1952.
32. Coste, F., Delbarre, F., and Lacrobique, F. Action de la bis-para-aminophényl-méthyl propane sur les endocrines. I. Action sur le poids et la structure des glandes endocrines du rat. C.R. Soc. Biol. (Paris), 150:959, 1956.
33. Barry, J., Racadot, J., and Lenys, D. Recherches concernant l'action de l'ampnénone B sur le corps thyroide chez le cobaye. Ann. Endocr. (Paris), 20:423, 1959.
34. Greer, M. A. Discussion, p. 45. *In* Thyrotropin. Werner, S. C., ed. Springfield, Charles C Thomas Pub., 1963.
35. Rallison, M. L., Kumagai, L. F., and Tyler, F. H. Goitrous hypothyroidism induced by amino-glutethimide, anticonvulsant drug. J. Clin. Endocr., 27:265, 1967.
36. Dexter, R. N., Fishman, L. M., Ney, R. L., and Liddle, G. W. Inhibition of adrenal corticosteroid synthesis by aminoglutethimide: studies of the mechanism of action. J. Clin. Endocr., 27:473, 1967.

37. Pittman, J. A., and Brown, R. W. Antithyroid and antiadrenocortical activity of aminoglutethimide. J. Clin. Endocr., 26:1014, 1966.
38. Schreiber, V., Zbuzková-Kmentová, V., and Zbuzek, V. The blocking effect of thyroidectomy on adrenal and overian hypertrophy produced by aminoglutethimide feeding in rats. J. Endocr., 42:349, 1968.

65

The Evolution of Vertebrate Endocrine Systems

D. Bellamy

Department of Zoology, University
College, Cardiff, Wales

INTRODUCTION

The evolution of endocrine systems is only one aspect of the evolution of organization at the physiological level. In order to consider this topic in a Darwinian fashion it is essential to know the detailed form and function of all past and present physiological systems. There is no fossil record and the only possible approach is through comparisons between various present-day organisms. As yet, the comparative method has not been applied rigorously to the field of physiological research and the number of available facts is negligible in relation to the data of comparative anatomy. Also, any conclusions must of necessity be set against the background of evolutionary trends that have been established on anatomical grounds.

Comparative physiology has made only a minor contribution to the field of animal taxonomy although physiological adaptations have been of overwhelming importance in the understanding of the phenomenon of adaptive radiation. Thus, it appears that speciation has occurred through adaptations of physiological systems as much as through large scale morphological changes. Structure, in terms of physiological adaptation, is often of secondary importance.

Facts that have a bearing upon physiological variation are scarce, but in comparison, the facts of comparative endocrinology are negligible. For this reason it is usual to present the data of comparative endocrinology as a catalogue of similarities, stressing trends of evolution wherever a sufficient number and range of animals have been investigated. This approach will be adopted in the first part of the article. Endocrine glands of vertebrates will be compared broadly in terms of their variations in structure and secretions. The second part will deal with the question of functional evolution with particular emphasis upon the origins of endocrine control. There is not the time or space to include all known variations in vertebrate endocrinology but an effort has been made to indicate the kinds of variations that may have been utilized in natural selection.

COMPARISONS OF STRUCTURE

Most endocrine tissues in vertebrates are composed of cells with epithelial or glandular characteristics. Five tissues of this kind have long been established as endocrine organs of widespread occurrence: the thyroid, gonadal interstitial tissue, adrenocortical tissue, islet cells, and the pars distalis of the pituitary. Lately there has been a renewed interest in the endocrine status of two other tissues of this type which are also ubiquitous throughout the vertebrates, the thymus(1,2) and the pineal gland(3,4), which have been subject to much speculation on the part of endocrinologists through the years.

All of the tissues mentioned above run throughout the vertebrate series and this fact alone has often prompted statements to the effect that there has been little evolution of vertebrate endocrine systems at a morphological level. This conclusion is not in accord with broad comparisons between the endocrine glands of the various vertebrate classes. Nor is the idea of a limited evolution borne out in detailed analysis within a class. Even the thyroid follicle, which is an easily recognizable morphological feature of the thyroid gland in all vertebrates, shows considerable structural variation within a narrow taxonomic range(5).

Variability in gland structure is most evident in those tissues that show an evolutionary tendency to gather together in a compact gland showing a high degree of structural complexity. This tendency is a consistent characteristic of adrenocortical tissue, where the ultimate mammalian condition is a division of labour expressed by the production of different hormones by several cell types(16). This differentiation of function is also a widespread feature of other glands such as the pars distalis, islet tissue, and the thyroid where it appears as an early characteristic. The advantages which underlie the tendency for endocrine cells to form glands is obscure. In this regard, one would anticipate that a system of widely scattered cells would be more efficient than a gland in ensuring the rapid and uniform dispersion of hormones. On the other hand, organ systems have extra possibilities for discrete control of function through their respective nervous and blood systems. Although it is established that direct nervous control does not play a predominant role in the function of many endocrine glands, there is evidence which indicates the importance of a special blood supply. This is especially noteworthy in the thyroid gland where an exceptionally high rate of blood flow is essential for normal function. The importance of control of secretion through localized changes in the vascular bed of a gland is apparent in the adrenal cortex under stimulation by adrenocorticotropin.

The possibilities of evolution through association of different cell types may be explored with reference to the comparative histology of islet tissue. This tissue is a well-defined histological unit in all vertebrates, being composed of a more or less diffuse collection of different kinds of cells. Representative members of all vertebrate groups have been found to possess three cell types: *A* cells, producing glucagon; *B* cells, forming insulin; and *D* cells, to which no endocrine function has yet been assigned but which may well produce a third pancreatic hormone. It is said that the various cells function independently but there is no firm physiological evidence to support this hypothesis. Some workers have described a fourth cell type, which has some similarities to the *D* cell, and this shows some degree of phylogenetic variation(7). In this connection, the attention of histologists has been directed towards another type of islet cell termed an amphiphil. This cell type appears to have been lost from pancreatic tissue at the level of the birds(8). Other claims that islet cells have undergone sporadic, but drastic, changes during the evolution of the vertebrates concern the absence of *A* cells from neotenous urodeles(9,10). Rerently, however, counterclaims have been made that the failure to detect *A* cells in these animals was due to inadequate cytological techniques(11).

On morphological evidence alone, there is much to be said with regard to major evolutionary changes in the organization of the vertebrate endocrine system. For example, parathyroid glands are found only in tetrapods. Suggestions have been made that the ultimobranchial bodies which, like the parathyroids, are of branchial origin, serve the same role in fishes as do the parathyroids in tetrapods. Such a simple displacement of function appears unlikely as ultimobranchial tissue persists in terrestrial vertebrates. In the mammals it may become closely associated with the thyroid. In this repsect, certain cells (*C* cells) in the mammalian thyroid have been identified as derivatives of ultimobranchial tissue. Current ideas on the functional evolution of the ultimobranchial bodies stress the widesp| pread role that these organs play in the maintenance of calcium and phosphate balance. Indeed, thyroid *C* cells and ultimobranchial tissue may have a common function: the secretion of calcitonin(12). Thus, it appears that the various branchial glands have undergone considerable evolutionary development associated with terrestrial life.

Colonization of the land has resulted in the loss of other glandular tissues. One of these, termed the corpuscles of Stannius, has some histological and physiological characteristics of an endocrine organ(13,14). This tissue does not seem to occur in either elasmobranchs or tetrapods and appears to be confined to teleosts and lungfish, indicating that it has been of importance in the evolution of the bony fishes.

The comparative anatomy of the pineal gland also points to marked evolution occurring at the morphological level. All vertebrates have an equivalent of the pineal gland. In fishes and amphibia, the pineal is eclipsed in histological terms by the medial parietal eye. With the trend towards the elimination of the latter structure in the higher vertebrates, the mass of the pineal assumes greater proportions in relation to the other endocrine organs and there is a consolidation of a highly complex cell pattern. Indeed, it would appear that it is within the lesser known and little investigated organs exemplified by the ultimobranchial bodies, the corpuscles of Stannius, and the pineal gland that the morphological evidence points to large evolutionary developments in the vertebrate systems for chemical coordination.

The preservation of the pituitary gland as an easily recognizable unit alongside considerable phylogenetic variations in gross external anatomy of the various vertebrate types is regarded by many workers as an indication of the early dominance of the pituitary in the vertebrate endocrine system. However, careful investigations indicate that there has been considerable anatomical evolution, so much so that comparisons of the structure of the pituitary within a single class is often of taxonomic value at the level of the genus.

The hypothalamic neurosecretory system which connects the par distalis with the brain also appears to be a common feature of all vertebrates. There are considerable variations in structure throughout the evolutionary series but these differences may owe more to the structural evolution of the brain than to changes in the pituitary. Of more phylogenetic interest is the caudal neurosecretory system found in elasmobranchs and teleosts. This feature is probably common to most vertebrate types below the amphibia. Recent work has pointed to an endocrine role for this tissue(15,14) but little is known about its evolutionary significance. Histological evidence points to evolution within elasmobranchs and teleosts, involving cell aggregation and increasing independence of the central nervous system. This trend probably featured in several lines of evolving teleosts(16).

Chrommaffin tissue is found in representatives of all vertebrate classes, arising from neural tissue which is common to the sympathetic system. In the lower vertebrates the tissue is distributed in isolated groups of cells, but there is a consistent evolutionary tendency for these cells to become associated with interrenal tissue (6). This phenomenon culminates in the mammals, where the adrenocortical tissue surrounds a central mass of chromaffin tissue to form the adrenal gland.

These evolutionary trends towards the aggregation of mixtures of endocrine cells are usually dismissed as being of minor importance in terms of vertebrate evolution. However, they occur frequently, and since we know little about the functional significance of cell aggregation in the stimulation and maintenance of hormone secretion it is better to reserve judgement on the selective advantages of such an arrangement. This holds particularly for the apparent haphazard association of different cell types in the pancreas as well as for the orderly arrangement of the various kinds of steroid-secreting cells in the mammalian adrenal cortex. Gross anatomy points to an independence of function but there are signs that cells may interact to produce subtle physiological and biochemical advantages that have so far escaped detailed study. Removal of the medulla from the mammalian adrenal gland is known to produce large, short-term abnormalities in adrenocortical function(17), and it may be that there are other less marked long-term changes reflecting a degree of interdependence. Bearing on this, it has been stated that interactions and transformations among the various islet cells and exocrine cells is a normal feature of the pancreas(8,11).

VARIATIONS IN HORMONES

There is a chemical identity of all hormones throughout the vertebrates that points to their early establishment at this level of organization. Thus, all vertebrate hormones are derivatives of either cholesterol or amino acids. The various steroid hormones derived from cholesterol differ in only a few atoms in the complex ring structure despite large differences in physiological function. Target cells therefore have the means of exploiting very small differences in molecular configuration for chemical control. Other small molecules with hormonal activity have a chemical unity in that they may be regarded as derivatives of the aromatic amino acid tyrosin. The various substituted thyronines belong to this class (two tyrosine units) together with adrenaline and noradrenaline (one tyrosine unit). These two groups of hormones, the steroid and tyrosine derivatives, are characterized by a dominant carbon ring structure and it is this which confers some similar physicochemical properties. The similarity is most evident in the readiness with which they form adsorption complexes with proteins.

Polypeptides make up the other major class of hormones. In these compounds, the amino acids are joined by peptide links but are otherwise unchanged chemically. The number of amino acid residues varies considerably according to the type of hormone. Little can be said about the amino acid composition in comparative terms because sequential analysis has only been carried out on hormones from a narrow range of animals. In view of the importance of the aromatic ring in the thyronines and catecholamines, the question may be raised as to whether the aromatic amino acids have a dominant position in the polypeptide series of hormones. This question cannot yet be answered from studies on a sufficiently comprehensive range of hormones, but already there is evidence that points to the frequent occurrence of tyrosine or phenylalanine in the terminal portions of polypeptide chains that have hormonal activity. When the last six residues of the N-terminus of seven different hormones are examined, phenylalanine and tyrosine appear together in four of them (Table 1). Of the others, one contains tyrosine and two, phenylalanine. In a total of 48 terminal residues in the seven hormones, about a quarter of the amino acids are accounted for by phenylalanine and tyrosine (Table 1). The importance of phenylalanine as the terminal residue in growth hormone is well established; all hormones from fish to man having a phenylalanine residue at both ends of the molecule(18). Recently, it has been reported that ovine and bovine luteinizing hormone have penultimate tyrosine residues in the *B* chain(19). Tyrosine is also a consistent feature of the cyclic peptides of the posterior pituitary which belong to the oxytocin and vasopressin series. The high frequency of occurrence of the aromatic amino acids in the terminal portion of the few polypeptide chains that have been analyzed may be purely fortuitous but, as the evidence stands, it is highly suggestive of a general rule in which attachment of polypeptide hormones to receptors in gland or target cells involves at least one aromatic residue.

Although vertebrate hormones have evolved little with regard to general chemical type, it is possible to trace variations in structure over a wide range of animals. The small molecules appear to have altered least. This applies particularly to the catecholamines and the thyronines. These are thought to occur in all vertebrates on the basis of either chemical analysis or the capacity of the relevant tissue to synthesize the hormone(20,21,22). There is considerable variation in the steroids both in the products of synthesis and the number of different circulating hormones (Tables 2 and 3). So far, between 15 to 20 variations in the basic steroid hormone ring structure have been identified in the plasma of vertebrates. This potential for making a wide range of steroids appears to have been established at the beginning of the vertebrate line. Subsequent evolution appears to have been concerned with restrictions in the number and quantity that are secreted from the various glands.

In the field of steroid analysis, the techniques for the isolation and identification of hormones do not always satisfy strict chemical criteria of specificity. This has sometimes resulted in particular claims to identification of a steroid being subse-

TABLE 1

N-terminal residues in polypeptide hormones

	Tyr	Phen	Ileu	Gly	Met	Asp	Thre	Pro	Lys	Ala	Leu	Glu	Cy	Arg	Va	His	Ser
Gastrin[a]	+	+	–	+	+	+	–	–	–	–	–	–	–	–	–	–	–
Insulin[b]																	
A chain	+	+	–	–	–	–	+	+	+	+	–	–	–	–	–	–	–
B chain	+	–	–	–	–	++	–	–	–	–	+	+	+	–	–	–	–
αMSH[c]																	
common effective core	+	+	–	+	+	–	–	–	–	–	–	+	–	+	–	+	–
ACTH[d]	+	–	–	–	+	–	–	–	–	–	–	+	–	–	–	+	++
Human growth[e] hormone	–	+	–	–	–	+	+	+	–	–	++	–	–	–	–	–	–
Parathyroid[f] hormone	–	+	–	–	–	–	–	–	–	+	–	++	–	–	+	–	+
Angiotensin[g] minimum core	+	+	+	–	–	–	–	+	–	–	–	–	–	–	+	+	–
Totals	6	6	1	2	3	4	2	3	1	2	3	5	1	1	2	3	3

[a]Bentley et al. (47).
[b]Harris and Lerner (46).
[c]Knobil and Hotchkiss (18).
[d]Li (45).
[e]Potts et al. (92).
[f]Sanger (93).
[g]Smeby (74).

TABLE 2

Vertebrate corticosteroids

	Corticos-terone	18(OH) Corticos-terone	Cortisol	Cortisone	Aldosterone	DOC
Agnatha						
Cyclostomata	+[a]		+[a]			
Gnathostomata						
Elasmobranchii	+[a]		+[a]	+[b]	+[b,c]	+[b]
Actinopterygii	+[a]	*[d]	+[a]	+[a]	*[d]	
Choanichthyes	*[e]		+[a]			
Amphibia	+[a]	*[g]	+[a]		+[f]	*[g]
Reptilia	+[a]	*[g]	+[a]		*[h]	*[i]
Aves	+[a]	*[j]			+[a]	*[j]
Mammalia	+[a]	+[k]	+[a]	+[n]	+[l]	+[m]

Plasma +
Synthesis or release *

[a]Phillips and Bellamy (23).
[b]Lupo di Prisco et al. (25).
[c]Bern et al. (94).
[d]Truscott and Idler (95).
[e]Janssens et al. (96).
[f]Crabbé (97).
[g]Chan et al. (98).
[h]Phillips et al. (99).
[i]Sandor et al. (100).
[j]Sandor et al. (101).
[k]Sandor and Lanthier (102).
[l]Coghlan and Scoggins (55).
[m]Vinson and Rankin (103).
[n]Gore and Baron (36).

TABLE 3

Vertebrate sex steroids

	Testos-terone	11-keto-testosterone	Proges-terone	Estriol	Estrone	Estradiol-17β
Agnatha						
Cyclostomata						
Gnathostomata						
Elasmobranchii	+[a,b]		+[a]	+[a]	+[a]	+[n]
Actinopterygii	+[b]	+[c]			*[m]	*[m]
Choanichthyes			*[l]	*[l]	*[l]	
Amphibia					+[h]	+[h]
Reptilia	*[i]		*[j]		*[j]	*[j]
Aves					+[d]	+[d]
Mammalia	+[e]		+[g,k]	+[f]	+[f]	+[f]

+Plasma
*Gonads or synthesized in vitro

[a]Lupo di Prisco et al. (25).
[b]Idler and Truscott (32).
[c]Schmidt and Idler (104).
[d]O'Grady (33).
[e]Forchielli et al. (105).
[f]Roy and Brown (106).
[g]Feder et al. (107).
[h]Gallien and Chalumeau Le Foulgoc (108).
[i]Lupo di Prisco et al. (109).
[j]Lupo di Prisco et al. (110).
[k]Short and Moore (111).
[l]Dean and Chester Jones (112).
[m]Gottfried et al. (113).
[n]Simpson et al. (114).

quently disputed on the grounds of poor technique. However, taking all analytical claims which have not been questioned by later work, it may be said that representatives of all vertebrate groups have been shown to contain or to make most use of the steroids that are found in mammals.

Because of the early availability of advanced techniques for dealing with isolation and identification, most of the comparative work on hormone variation refers to the distribution of the steroids which originate in adrenocortical tissue. In order to indicate the nature of hormone variation in the vertebrates the adrenocorticoids will be considered in some detail. Three steroids predominate throughout the vertebrates in terms of their concentration and frequency of occurrence. These are the closely related compounds cortisol, cortisone, and corticosterone. It is usual for either cortisol or corticosterone to predominate in plasma but there are considerable variations in the ratio of one to the other. These three corticosteroids have similar properties although, molecule for molecule, they have different potencies in various pharmacological tests. They are found in quantities ranging from around 2 to 50 μg/100 ml in peripheral plasma(23) and in this regard may be distinguished from two other major corticosteroids, aldosterone and deoxycorticosterone (DOC), which are found at a concentration of 1 μg/100 ml or less. Because of the low concentration of the latter hormones and the concomitant difficulties of measurement there is considerable controversy over their taxonomic distribution. Deoxycorticosterone has been found as a normal constituent of plasma from elasmobranchs and mammals (Table 2), and can be synthesized by adrenocortical tissue from birds, reptiles, and amphibia. Aldosterone has been identified in plasma from mammals, birds, and amphibia and is an adrenocortical product of reptiles. Some teleosts have the capacity to synthesize aldosterone but this is not general, and despite concerted efforts this hormone has not been shown to be an important plasma constituent in this group. With improvements in technique it is likely that 'new' corticosteroids will be detected at the lower end of the concentration range. In this regard, blood samples from two species of the elasmobranch, *Raja*, in which none of the commonly found corticosteroids could be detected, contained small quantities of 1-α-hydroxycorticosterone(24).

With respect to the distribution of plasma corticosteroids, it is customary to refer to an animal as a cortisone, cortisol, or corticosterone secretor depending on the major plasma steroid. On this basis the birds are the only group in which a predominant plasma corticosteroid has been found consistently. The main, and often the only detectable corticosteroid is corticosterone. Representatives of other vertebrate groups have been found to possess two or more steroids by analysis. Teleosts often contain the two closely related steroids, cortisol and cortisone. Within the elasmobranchs it is usual for cortisol to be found together with corticosterone. The ratio of these steroids one to another range from 3:1 through 1:1 to 1:3 depending on the species(23). Some of these variations may be due to sex differences but there are also large seasonal variations to be considered(25). In spite of these difficulties of interpretation it is correct to say that, at the moment, there is no consistent pattern of adrenocortical evolution at the level of the secretory products.

Within any group of animals the three major corticosteroids show modes of variation in both pattern and concentration. Changes in pattern sometimes occur over a small phylogenetic range and may be conveniently examined in mammals. For example, from the early work it appeared that within the order Rodentia members of the three suborders show a consistency of corticosteroid pattern. The suborder Hystricomorpha contains the corticol-secretors guinea pig, chinchilla, and coypu. On the other hand the members of the suborder Myomorpha (rats and mice) secrete mainly corticosterone(26). However, this is not a taxonomic character at the level of the suborder(27,28,29).

Species variation in concentration is probably common to all the vertebrate hormones. With the corticosteroids, and possibly other hormones, difficulties are encountered in the interpretation of concentration differences within different species because of the large and rapid increase of corticosteroids in stressed animals together with the diurnal rhythm in plasma corticosteroids, both of which appear to be widespread vertebrate features(30,32). Taking the minimum values recorded for unstressed animals, plasma corticosteroids mostly cover the range from about 1 to 10 μg/100 ml. However, it may be concluded from their low secretion rates that some surgically stressed species have considerably less than a tenth of the amount at the lower end of this scale(31). A similar degree of variability in concentration holds for the sex steroids. For example, the concentration of testosterone in elasmobranchs and teleosts is about 10 times or more higher than that in mammalian blood(32), and above the range in avian blood(33). Circulating growth hormone may also show phylogenetic variations in concentration(34). However, because of concomitant variations in hormone binding to plasma proteins the actual amounts of hormones that are biologically active may not be so different.

It is not possible to conclude that the kind of differences discussed above form the basis of important phylogenetic differences that have played a consistent role in vertebrate evolution. Quite often the species range is not very comprehensive and the possible effects of age, sex, and diurnal shifts are not usually taken into account. Also, evidence is accumulating that points to the prevalence of polymorphism in corticosteroid secretion and metabolism(35,36,37,38,39).

Endocrine polymorphism may underlie certain aspects of physiological adaptability, and studies of species variability in endocrine function are now being placed on an experimental footing. In this respect, several studies are in progress aimed towards an understanding of the genetics of comparative physiology with particular reference to the variability in the endocrine system(40).

With regard to the polypeptides, an awareness of the importance of the evolution of effector function has tended to obscure the great extent of structural variation within these hormones. For example, now that comparative data are available on a range of vertebrates, it is possible to suggest a sequence of changes in amino acid composition which lead to the evolution of the mammalian neurohypophyseal hormones. These changes involve substitutions of amino acids at three positions in the molecule(41,42).

Many of the larger polypeptides have been found to have a universal distribution throughout the vertebrates. However, this is not obvious from their biological properties. Based on the mammalian tibia test, pituitaries of a range of teleosts and elasmobranchs do not appear to contain a growth hormone. It seems that a mammalian-like growth hormone first made its appearance in the lungfish (43). This evolutionary pattern is due to changes in the structure of both the growth-controlling hormone and its targets, since fish pituitaries do contain a polypeptide hormone that has the growth-promoting properties of somatotropin when tested in fish. Similar changes appear to have taken place with respect to thyroid-stimulating hormone and prolactin. There is often a blurring of differences between most of the polypeptide hormones when they are tested in lower vertebrates, suggesting a phylogenetic origin in a common protein.

As the work slowly progresses on the secondary structure of the larger polypeptide hormones it is becoming increasingly clear that it is exceptional for vertebrates to have evolved without some change in the amino acid sequence of their hormones. Most of the evidence comes from comparison within the mammals, where evolutionary changes have been detected in insulin(44), growth hormone (18), adrenocorticotropin(45), melanophore-stimulating hormone(46), gastrin(47), luteinizing hormone(48), and angiotensin(49). Changes have occurred in the various polypeptide hormones by substitution of one amino acid residue for another. So far the evidence suggests that only about a dozen amino acids have been interchanged in this way. These amino acids differ widely in structure and it is perhaps more illuminating to examine the amino acids that do not appear to have been frequently replaced. Amongst these are those with nucleophilic side chains such as proline, hydroxyproline, tryptophan, tyrosine, and phenylalanine. It is also probably significant that cysteine with its reactive sulphydryl group has not been frequently lost. Thus it may be concluded tentatively that these amino acids confer an element of stability on the evolving polypeptide hormone, possibly as a site of attachment to the receptor.

It is not possible to see a consistent pattern emerging from sequence analysis and a good many deletions and substitutions appear to have occurred at random (44). However, immunochemical methods have indicated consistent trends in the evolution in the tertiary structure of some of the larger polypeptides. One of the most careful studies in this field has shown that within the primates, the cross reactivity of human growth hormone slowly decreases down the taxonomic series (50). Polypeptide hormones are not unique in showing this tendency towards diversity, for this type of gradual change in teriary structure has been a general feature of the evolution of nonhormonal protein(51) and will probably be found to be a general feature in the evolution of a good many other structural and enzymic proteins.

With regard to the evolution of hormones it is necessary to deal at this point with evolutionary changes in plasma proteins as they effect the physico-chemical state of circulating hormones. It is well established that the steroid hormones and the thyronines are reversibly adsorbed to certain proteins in the plasma. As the unbound hormone is responsible for the biological activity, phylogenetic variation in hormone concentration and binding capacity of the plasma proteins have an important bearing on variations in the hormone-target cell interaction.

A wide range of vertebrates have been examined in order to make phylogenetic comparison of plasma corticosteroid binding(52) and the binding values obtained vary over a one hundred-fold range. There is no obvious taxonomic pattern. For example, in man and rat more than 90 percent of the circulating corticosteroid is bound whereas less than l0 percent of circulating cortisol is bound in the salmon and sheep(53). However, since the sheep is characterized by having a concentration of circulating cortisol that is only about 10 percent of that in man, the concentration of free and biologically active hormone is probably little different in all four animals. There may also be a correlation between low plasma corticosteroids and low plasma binding in other groups(30,31,53,54,55). Thus it is possible to envisage coupled changes in the evolution of an endocrine system in which a low concentration of plasma hormone is selected along with a low capacity for plasma binding so that overall the two effects cancel each other out.

Apart from variation in hormones there may also be variations in their metabolic products, some of which may have endocrine activity(56). Little work has been done with regard to comparisons of metabolic pathways of hormones. However, from studies into the fate of various steroid hormones it appears that glucuronide formation developed late in the vertebrate seriies. Amphibia can only form steroid glucuronides to a limited extent whereas fish cannot achieve this type of conjugation(57).

FUNCTION

Studies on functional variation within the vertebrates have made a considerable impact upon comparative endocrinology, linked as they are with the two major vertebrate adaptations: to terrestrial life and placental reproduction. This point was taken up in a previous discussion on the evolution of adrenocortical hormones (58). It was also pointed out that we are largely ignorant of the biological role of a number of endocrine organs even in the higher vertebrates. The major evolutionary trends arising from most comparative studies appear in the form of pharmacological differences in potency(59,60) and the time and extent of response(61,18,62 63,64). Salient differences in function coincide with the gain or loss of physiological systems. For example, prolactin is a factor which controls sodium balance in euryhaline teleosts and regulates milk production in mammals. These differences are obviously connected with the possession of gills and an unspecialized kidney on the one hand and the development of mammary glands on the other. However, even here, the different physiological actions may be based on similrar cellular effects(65).

The current preoccupation with the details of physiological and biochemical effects of hormones has tended to obscure the basic needs of animals for endocrine systems. It is helpful in this respect to generalize these needs in relation to the control of digestion, polymer synthesis, polymer breakdown, maintenance of salt balance, sexual development, and maturation. Only polymer breakdown, as manifest in the starved animal, appears to be largely independent of the endocrine system (66). A major difficulty in understanding the process of evolution of the endocrine system with respect to these basic processes is that there is a considerable degree of overlap and mutual interaction between the various endocrine glands(67). Indeed, it is an important feature of the evolution of the endocrine system that one hormone often limits the action of another(68). From this point of view the original role of a hormone may be more obvious in a less complex vertebrate than in a higher form, where there has been an "overlayering phenomenon" by changes in other hormones or the parallel development of new endocrine organs. There are examples of this process in the comparative endocrinology of reproduction. A uniform requirement throughout the vertebrates appears to be a system which promotes the growth of the embryo to a greater or lesser degree at the expense of the utilization of stored lipids and dispensable cellular protein in the maternal tissues. This requires the participation of at least two hormones, one to stimulate the growth of eggs or embryos and the other to accelerate the breakdown of maternal polymers. The simplest vertebrate system may be seen in those animals in which yolky eggs develop in the absence of an external food source. In this respect the migratory euryhaline teleosts have been examined in greatest detail and a part of the endocrine requirement appears to be met by sex steroids and corticosteroids as in the reproducing mammal. Mammals, given an adequate amount of carbohydrate but no protein, do not produce sufficient ovarian hormones to maintain pregnancy(69). When this hormone deficit is corrected gestation proceeds, emphasizing the survival from fish to mammals of a primitive system for the provision of amino acids. The basis for the "new" endocrine pattern in mammals is the system of uterus and placenta. New requirements emerge, namely for the inhibition of yolk formation and stimulation of uterine growth. Also the embryo-placenta functions as a self-contained endocrine system, giving a large measure of metabolic independence to the embryo. How many of these evolutionary changes involve new hormones is not known. In this connection, available evidence suggests that the new systems developed utilizing the same steroid hormones and similar kinds of polypeptides. The main changes involve the timing of hormone release and the formation of new targets.

When considering the integration of endocrine control it is customary to think of vertebrate endocrine effectors as being of two types. There are those glands which secrete in response to a direct nonhormonal chemical stimulus and others where the gland reacts to a tropic hormone. The latter system of interlinked glands may have a complicated command-structure, as exemplified by the hypothalamic control of the release of tropic hormones. How did the multilayered endocrine system develop? It is possible that, originally, the secondary gland may have functioned independently. That is to say, this gland (*B*) may have responded in some way to the chemical stimulus that now evokes the entire response. Evolutionary development of the tropic gland required that a second type of cell (*A*) acquired the capacity of producing a substance which was a limiting factor (not necessarily a precursor) in the synthesis of hormone by *B* cells. If the product of *A* directed the metabolism of *B* in a way that was advantageous to the organism the way would be open to the ultimate capture of *B* cells by gland *A*. The alternative, that the two or three cell system appeared suddenly as a functional unit with the simultaneous coupling of stimulation and secretion, is more difficult to visualize. Evidence from the organization of the endocrine system of present-day vertebrates is in favour of the first alternative. Secondary glands do not cease to produce hormone or to lose complete control of secretion in the absence of the primary glands (70,71). Also there is evidence that in the lower vertebrates the components in a multilayered system may be capable of normal independent function(72) and that "capture" by a primary gland may occur during ontogeny(73). In an effort to understand the development of complex command-structure in the endocrine system, it is usual to consider variations in the pituitary neurosecretory system. However, it may be that in the mammalian angiotensin-aldosterone system(74), where tye master cells are not organized on a glandular basis and produce a hormone precursor, we have a multilayered endocrine system in its early stages of development.

Although even the most primitive living vertebrate presents a system of interacting endocrine glands, the pattern of evolution is not through large-scale changes in the function of each component hormone. Polypeptides from higher vertebrates have the same or similar qualitative effects in lower vertebrates which possess an appropriate target system. As yet, there is little comparative work on the action of this type of hormone isolated from lower vertebrates, but the few studies that have been made indicate that the polypeptides of the fishes already have signs of the increased versatility demanded by more advanced organisms(41,42,61,65,75). However, it would not be true to say that these hormones are fully preadapted in molecular terms to the subsequent evolution of function.

When various physiological and pharmacological effects of the corticosteroids are examined in different mammals it is seen that they always fall within the same activity spectrum but aldosterone and deoxycorticosterone tend to have a high potency with regard to effects on the kidney that are not so marked in the other three steroids. Thus one might argue that there are clear-cut roles for the two types of hormone. This is largely borne out by the physiological evidence for the relative activities of cortisol and cortisone compared with DOC(76,77); but the separation of corticosterone and deoxycorticosterone is not so clear cut. However, even within the three substances corticosterone, cortisol, and cortisone there are quite large differences in biological potency, particularly when different species are used for pharmacological tests(60) and this poses the question as to the selective advantage of having a predominance of one particular steroid over others. The great potential variability in the biological activity connected with minor variations in structure has been shown by artificial modifications of several types of hormones(78,79 80).

CONCLUSION

The five basic processes underlying the cycle of growth, maturation, and reproduction are common to all animals, both vertebrates and invertebrates. Therefore in a given metabolic situation linked with a particular stage of the life cycle one would anticipate that all animals would require the same basic endocrine system. This presupposes a consistent general metabolic pattern throughout the animal kingdom. Although this is taken by most biochemists as a likely possibility, the number of animals upon which it is based would not satisfy a zoologist.

It has been stressed that the neurosecretory system is of widespread occurrence in the animal kingdom(81) and this may be taken as evidence of a common ancestral endocrine system. However, as yet there is no evidence for an identity of vertebrate and invertebrate hormones. The work of isolation and identification of the invertebrate hormones has only just begun but already a close chemical similarity has been established between the hormones that control moulting in arthropods and the steroid hormones of vertebrates(82). Catecholamines have been identified in tissues from a wide range of invertebrates, and have been shown to have pharmacological effects akin to those in vertebrates(83), although as yet no definite endocrine role has been assigned to these compounds. Recently, work has begun on the study of the action of some vertebrate hormones in invertebrates(84). It is a striking fact that the administration of corticosteroids to insects, certain of which are known to synthesize corticosteroids, has the same specific effects as in the vertebrates(85). This suggests the presence of similar biochemical systems in both groups relating to the synthesis of hormones and target-cell response but the work, being of a pharmacological nature, does not define the invertebrate endocrine system.

Where did hormones originate? The answer concerns the potential for metabolic control in acellular organisms through changes in the concentration of small quantities of steroid-like molecules and polypeptides. At the moment there is some uncertainty about the details of biochemical control systems in microorganisms. Much is known in general about the regulation of enzyme synthesis, and the ability to vary enzyme patterns appears to be of great importance in the day-to-day life of bacteria. Enzyme synthessis is controlled in the first instance by changes in the concentration of metabolites which are often effective in very low concentrations. The little that is known about the mechanism of protein synthesis in Metazoa suggests that it is basically the same as in microorganisms. Thus, there is a likelihood that enzyme levels in Metazoa are also regulated in the same way. We know that the formation of key enzymes underlies the response of higher vertebrates to a variety of hormones, and it is possible that endocrine function developed in Metazoa making use of a preexisting system for the control of enzyme synthesis. That is to say, a primitive system in which synthesis of proteins was controlled by specific metabolites, was in turn controlled by hormones which regulated metabolite availability to selected intracellular pools. There is circumstantial evidence that the latter system functions with respect to the steroid hormones in present-day vertebrates but current research is more concerned with the possibility of a direct interaction between genes and hormones. If the latter system is found to exist, it would indicate that the appearance of the endocrine system of Metazoa required drastic evolutionary changes in the gene complex.

At the molecular level we are handicapped by the lack of knowledge as to how present-day systems function. Protein synthesis in microorganisms requires a very complex system of integrated structures and it is likely that many factors, as yet unidentified, are required to make codons available(86). There are consequently numerous theoretical points for the insertion of endocrine control both inside and outside the nucleus. In this connection it is known that nonenzymic cytoplasmic proteins are capable of influencing protein synthesis at the level of the ribosome(87). It only requires that the capacity for synthesizing protein molecules of this type be restricted to certain cells, and an endocrine system is established. On this basis we may reconsider the possibility that endocrine control represents a system that evolved after the establishment of the nervous system in early Metazoa. There is evidence that the development of Protozoa and rotifers is regulated by small quantities of proteins and lipids in the environment(88,89). Thus, molecules akin to present-day hormones may have been present in the acellular ancestors of the Metazoa which fulfilled the role of the intracellular activators of Huxley(90). Bearing on this, a vertebrate endocrine complex, the catecholamine glycogen system, has been tentatively identified in a microorganism(91) suggesting that at least one vertebrate endocrine system evolved together with, if not before, the establishment of nervous coordination.

REFERENCES

1. Clark, S. L. *In* The Thymus: Experimental and Clinical Studies. Wolstenholme, G. E. W., and Porter, R. eds., Ciba Found. Symp. London, J. & A. Churchill, 1966.
2. Jaanus, S. D., and Martine, C. R. An extra-adrenal response to sodium loading revealed by thymectomy. Gen. Comp. Endocr., 10:147–154, 1968.
3. Wurtman, R. J., Altschule, M. D., Greep, R. O., Falk, J. L., and Grave, G. The pineal gland and aldosterone. Amer. J. Physiol., 199:109–111, 1960.
4. Kinson, G., Wahid, A. K., and Singer, B. Effect of chronic pinealectomy on adrenocortical hormone secretion rates in normal and hypertensive rats. Gen. Comp. Endocr., 8:445–454, 1967.
5. Dawson, W. D. Comparative thyroid activity in two species of Peromyscus. Gen. Comp. Endocr., 8:267–272, 1967.
6. Chester Jones, I. The Adrenal Cortex. London, Methuen, 1957.
7. Ostberg, H., Hellerström, C., and Kern H. Studies on the A_1 cells in the endocrine pancreas of some cartilaginous fishes. Gen. Comp. Endocr., 7:475–481, 1966.
8. Epple, A. Cytology of pancreatic islet tissues in the toad *(Bufo bufo)*. Gen. Comp. Endocr., 7:191–196, 1966.
9. Wurster, D. H., and Miller, M. R. Studies on blood glucose and pancreatic islets of the salamander, *Taricha torusa*. Comp. Biochem. Physiol., 1:101–109, 1960.
10. Kern, H. Die zytologie des insel organs im pankreas einiger neotener urodeln *Megalobatrachus cryptobranchus* Amphiuma. Z. Zellforsch., 70:499–514, 1966.
11. Epple, A. Islet cytology in urodele Amphibians. Gen. Comp. Endocr., 7:207–214, 1966.
12. Chan, D. K. O., and Chester Jones, I. Regulation and distribution of plasma calcium and inorganic phosphate in the European ell *(Anguilla anguilla* L.). J. Endocr., 42: 109–117, 1968.
13. Hanke, W., and Chester Jones, I. Histological and histochemical studies on the adrenal cortex and the corpuscles of Stannius of the European eel *(Anguilla anguilla* L.). Gen. Comp. Endocr., 7:166–178, 1966.
14. Chester Jones, I., Chan, D. K. O., and Rankin, J. C. Renal function in the European eel *(Anguilla anguilla* L.). Effects of the caudal neurosecretory system, corpuscles of Stannius, neurohypophysial peptides and vasoactive substances. J. Endocr., 42:21–31, 1969.
15. Fridberg, G., Bern, H. A., and Nishioka, R. S. The caudal neurosecretory system of the isospondylous teleost *Albula vulpes*, from different habitats. Gen. Comp. Endocr., 6:195–212, 1966.
16. ——— and Bern, H. A. The urophysis and the caudal neurosecretory system of fishes. Biol. Rev., 43:175–199, 1968.
17. Gaunt, R., Renzi, A. A., Gisoldi, E., and Howie, N. C. A sodium retaining influence of enucleate rat adrenal glands. Endocrinology, 81:1331–1337, 1967.
18. Knobil, E., and Hotchkiss, J. Growth hormone. Ann. Rev. Physiol., 26:47–74, 1964.
19. Ward, D. N., Fujino, M., Showalter, M., and Ray, N. Comparative studies of luteinizing hormone from beef, pork and sheep pituitaries. II. The C-terminal amino acids. Gen. Comp. Endocr., 8:289–296, 1967.
20. Gorbman, A. Thyroid hormones. *In* Comparative Endocrinology. Heller, H., and von Euler, U. S., eds. London, Academic Press, Inc., pp. 291–324, 1963.
21. Donso, A. O., and Segurae, E. T. Seasonal variations of plasma adrenaline in toads. Gen. Comp. Endocr., 5:440–443, 1965.

22. Boehlke, K. W., Church, R. L., Tiemeier, O. W., and Eleftheriou, B. E., Diurnal rhythm in plasma glucocorticoid levels in channel fish (*Ictalurus punctatus*). Gen. Comp. Endocr., 7:18–21, 1966.
23. Phillips, J. G., and Bellamy, D. Adrenocortical hormones. *In* Comparative Endocrinology. von Euler, U. S., and Heller, H., eds. London, Academic Press, Inc., Vol. 1, pp. 208–257, 1963.
24. Idler, D. R., and Truscott, B. 1α-hydroxycorticosterone: synthesis *in vitro* and properties of an interrenal steroid in the blood of cartilaginous fish (Genus Raja). Steroids, 9:457–477, 1967.
25. Lupo di Prisco, C., Vellano, C., and Chieffi, G. Steroid hormones in the plasma of the elasmobranch, *Torpedo marmorata* at various stages of the sexual cycle. Gen. Comp. Endocr., 8:325–331, 1967.
26. Wilson, E. D., Zarrow, M. X., and Lipscomb, H. S. Bilateral dimorphism of the adrenal glands in the coypu (*Myocastor coypu* Molinn.). Endocrinology, 74:515–517, 1964.
27. Oliver, J. T., and Péron, F. G. 19-hydroxy-11-deoxycorticosterone, a major steroid secreted by the adrenal gland of the mongolian gerbil. Steroids, 4:351–363, 1964.
28. Vinson, G. P. Steroid production *in vitro* by the ovaries and adrenal glands of the thirteen-lined ground squirrel *(Citellus tridecemlineatus)*. J. Endocr., 32:401–402, 1965.
29. Whitehouse, B. J., Vinson, G. P., and Janssens, P. A. The effect of blood-borne factors on adrenal steroid synthesis in the golden hamster (*Mesocricetus auratus* Nehring). J. Endocr., 37:139–146, 1967.
30. Leloup-Hatey, J. Influence de l'agitation motrice sur la teneur du plasma en 17-hydroxycortocosteroids d'un Teleostean: la carpe *(Cyprinus carpio* L). C. R. Soc. Bid. (Paris), 246:1088–1091, 1958.
31. Weiss, M. and McDonald, I. R. Corticosteroid secretion in the monotreme *Tachyglossus aculeatus*. J. Endocr., 33:203–210, 1965.
32. Idler, D. R., and Truscott, B. Identification and quantification of testosterone in peripheral plasma of skate. Gen. Comp. Endocr., 7:375–383, 1966.
33. O'Grady, J. E. The determination of oestradiol and oestrone in the plasma of the domestic fowl by a method involving the use of labelled derivatives. Biochem. J., 106:77–86, 1968.
34. Schalch, D. S., and Reichlin, S. Plasma growth hormone concentration in the rat determined by radio immune assay: influence of sex, pregnancy, lactation, anesthesia, hypophysectomy and extracellular pituitary transplants. Endocrinology, 79:275–280, 1966.
35. Lombardo, M. E., McMorris, C., and Hudson, P. B. The isolation of steroidal substances from human adrenal vein blood. Endocrinology, 65:426–432, 1959.
36. Gore, M. B. R., and Baron, D. N. Estimation and identification of free and conjugated cortocosteroids in human plasma. J. Endocr., 21:303–314, 1960.
37. Cortes, J. M., Péron, F. G., and Dorfman, R. I. Secretion of 18-hydroxydeoxycorticosterone by the rat adrenal gland. Endocrinology, 73:713–720, 1963.
38. Eleftheriou, B. E. Bound and free corticosteroid in the plasma of two subspecies of deer mice (*Peromyscus maniculatus*) after exposure to a low ambient temperature. J. Endocr., 31:75–80, 1964.
39. Burstein, S., Bhavnani, B. R., and Kimball, H. L. Genetic aspects of cortisol hydroxylation in guinea pigs: urinary excretion and production rates of cortisol and 2-α-hydroxycortisol. Endocrinology, 76:753–761, 1965.
40. Spickett, S. G., Shire, J. G. M., and Stewart, J. Genetic variation in adrenal and renal structure and function. Mem. Soc. Endocr., 15:271–291, 1967.
41. Acher, R., Chauvet, J., and Chauvet, M. T. Phylogeny of the neurohypophysial hormones. Nature, 216:1037–1038, 1967.
42. ——— Chauvet, J., Chauvet, M. T., and Crepy, D. Les hormones neurohypophysaires des amphibiens. Isolement et caracterisation de la mesotocine et de la vasotocine chez le crapaud (*Bufo bufo*). Gen. Comp. Endocr., 8:337–343, 1967.
43. Geschwind, I. T. Growth hormone activity in the lungfish pituitary. Gen. Comp. Endocr., 8:82–83, 1967.

44. Hama, H., Titani, K., Sakaki, S., and Narita, K. The amino acid sequence in fin-whale insulin. J. Biochem., 56:285–293, 1964.
45. Li, C. H. Synthesis and biological properties of ACTH peptides. Recent Progr. Hormone Res., 18:1–40, 1962.
46. Harris, I., and Lerner, A. B. Amino acid sequence of the melanocyte-stimulating hormone. Nature, 179:1346–1347, 1957.
47. Bentley, P. H., Kenner, G. W., and Sheppard, R. C. Structures of human gastrins I and II. Nature, 209:583–585, 1966.
48. Ward, D. N., Adams-Mayne, M., Ray, N., Balke, D. E., Coffey, J., and Showalter, M. Comparative studies of luteinizing hormone from beef, pork and sheep pituitaries. I. Purification and physical properties. Gen. Comp. Endocr., 8:44–53, 1967.
49. Valloton, M. B., Page, L. B., and Haber, E. Radioimmunoassay of angiotensin in human plasma. Nature, 215:714–715, 1967.
50. Tashijian, A. H., Levine, L., and Wilhemi, A. E. Immunochemical relatedness of porcine, bovine, ovine and primate pituitary growth hormones. Endocrinology, 77:563–573, 1965.
51. Sarich, V. M., and Wilson, A. C. Rates of albumin evolution in primates. Proc. Soc. Exp. Biol. Med., 58:142–148, 1967.
52. Seal, U. S., and Doe, R. P. *In* Steroid Dynamics. Pincus, G., Nakau, T., and Tait, J. F., eds. London, Academic Press, Inc., pp. 63–68, 1957.
53. Freeman, H. C., and Idler, D. R. Transcortin binding of cortisol in atlantic salmon *(Salmo salar)*. Gen. Comp. Endocr., 7:37–43, 1966.
54. Weiss, M., and McDonald, I. R. Corticosteroid secretion in the Australian phalanger *(Trichosurus vulpecula)*. Gen. Comp. Endocr., 7:345–351, 1966.
55. Coghlan, J. P., and Scoggins, B. A. The measurement of aldosterone, cortisol and corticosterone in the blood of the wombat (*Vombatus hirsutus* Pemy) and the kangaroo (*Macropus giganteus*). J. Endocr., 39:445–448, 1967.
56. Granick, S., and Kappas, A. Steroid control of porphyrin and heme biosynthesis. A new biological function of steroid hormone metabolites. Proc. Nat. Acad. Sci. U.S.A., 57:1463–1467, 1967.
57. Ozon, R., and Breuer, H. Enzymic glucuronidation of estrogens by *Pleurodeles watlii Michah* (Amphibia, Urodela) *in vivo* and *in vitro*. Gen. Comp. Endocr., 6:295–302, 1966.
58. Bellamy, D., and Chester Jones, I. The evolution of adrenocortical hormones. Proc. 2nd Intern. Congr. Endocrinology, p. 153. London, Excerpta Medica, 1964.
59. Cartland, G. F., and Kuizenga, M. H. The bioassay of adrenocortical extracts. A direct comparison of rat and dog units. Amer. J. Physiol., 117:678–685, 1936.
60. Venning, E. H., Kazmin, V. E., and Bell, J. C. Biological assay of adrenal corticoids. Endocrinology, 38:79–89, 1946.
61. Pickford, G. E., and Wilhelmi, A. E. Comparative studies on the response of hypophysectomized killifish *Fundulus heteroclitus* to growth hormone preparations. Anat. Rec., 134:624–625, 1959.
62. Penhos, J. C., Houssay, B. A., and Lujan, M. A. Total pancreatectomy in lizards: effects of several hormones. Endocrinology, 76:989–993, 1965.
63. Wu, C. H., Reitman, M., Sodero, E., White, R., and Levine, R. Effects of several hormones after total pancreatectomy in alligators. Gen. Comp. Endocr., 8: 32–43, 1967.
64. Sundararaj, B. I., and Nayyar, S. K. Effects of exogenous gonadotrophins and gonadal hormones on the testes and seminal vesicles of hypophysectomised catfish, *Heteropneustes fossilis* (Bloch). Gen. Comp. Endrocr., 8:403–416, 1967.
65. Nicoll, C. S., and Bern. H. A. 'Prolactin' and the pituitary glands of fishes. Gen. Comp. Endocr., 4:457–471, 1964.
66. Cahill, G. F. The endocrine control of metabolism during fasting. Proc. 3rd Intern. Cong. Endocrinology. London, Excerpta Medica, in press, 1968.
67. de Bodo, R. C., and Sinkoff, M. W. Anterior pituitary and adrenal hormones in the regulation of carbohydrate metabolism. Recent Progr. Hormone Res., 8:511–570, 1953.

68. Chester Jones, I., and Bellamy, D. Hormonal mechanisms in the homeostatic regulation of the vertebrate body with special reference to the adrenal cortex. Symp. Soc. Exp. Biol., 18:195–236, 1964.
69. Hays, R. L., Hahn, E. W., and Kendall, K. A. Evidence of decreased steroidogenesis in pregnant rats fed a sucrose diet. Endocrinology, 76:771–776, 1965.
70. Civen, M., Ulrich, R., Trimmer, B. M., and Brown, C. B. Circadian rhythms of liver enzymes and their relationship to enzyme induction. Science, 157:1563–1564, 1967.
71. Frankel, A. I., Cook, B., and Graber, J. W. Determination of corticosterone in plasma by fluorimetric techniques. Endocrinology, 80:181–194, 1967.
72. Ball, J. N., Olivereau, M., Slicher, A. M., and Kallman, K. D. Functional capacity of ectopic pituitary transplants in the teleost *Poecilia formosa*, with a comparative discussion on the transplanted pituitary. Proc. Roy. Soc. B., 249:69–99, 1965.
73. Barrington, F. J. W., and Sage, M. On the response of the endostyle of the hypophysectomized larval lamprey to thiourea. Gen. Comp. Endocr. 7:463–474, 1966.
74. Smeby, R. R. *In* Angiotensin Systems and Experimental Renal Diseases. Metcoff, J, ed. London, J. & A. Churchill, pp. 29–40, 1963.
75. Saxena, B. B., and Heeneman, P. H. Preparation and properties of growth hormone from equine pituitary glands. Endocrinology, 78:561–567, 1966.
76. Ingle, D. J., and Kuizenga, M. H. The relative potency of some adrenal cortical steroids in the muscle work test. Endocrinology, 36:218–226, 1945.
77. Simpson, S. A., and Tait, J. F. A quantitative method for the bioassay of the effect of adrenal cortical steroid on mineral metabolism. Endocrinology, 50:150–161, 1952.
78. Dixon, H. B. F. Treatment of corticotropin with periodate and borohydride. Biochem. J., 83:91–94, 1962.
79. Heller, H. Neurohypophyseal hormones. *In* Comparative Endocrinology. Heller, H., and von Euler, U. S., eds. London, Academic Press, Inc., pp. 25–80, 1963.
80. Tolksdorf, S., Battin, M. L., Cassidy, J. W., MacLeod, R. M., Warren, R. M., and Perlman, P. L. Adrenocortical properties of Δ^{1-4}-pregnadiene-17α-21-diol-3, 11,20-trione (Meticorten) and Δ^{1-4}-pregnadiene 11β,17α-21-triol-3,20-dione (Meticortelone). Proc. Soc. Exp. Biol. Med., 92:207–215, 1956.
81. Novák, V. J. A. The phylogenetic origin of neurosecretion. Gen. Comp. Endocr., 4: 696–703, 1964.
82. Karlson, P., and Sekeris, C. E. Ecdysone, an insect steroid hormone and its mode of action. Recent Progr. Hormone Res., 22:473–502, 1966.
83. Florey, E. Comparative physiology: transmitter substances. Ann. Rev. Physiol., 23: 501–528, 1961.
84. Goodman, R. M., Goidl, J., and Richart, R. M. Larval development of *Sciara coprophila* without the formation of chromosome puffs. Proc. Nat. Acad. Sci. U.S.A., 58:553–559, 1967.
85. Mordue, W. Cortisol and growth in insects. Comp. Biochem. Physiol., 23:721–727, 1967.
86. Erbe, R. W., and Leder, P. Initiation and protein synthesis: translation of di- and tricodon messengers. Biochem. Biophys. Res. Commun., 31:798–803, 1968.
87. Brown, J. C., and Doty, P. Protein factor requirement for binding of messenger RNA to ribosomes. Biochem. Biophys. Res. Commun., 30:284–291, 1968.
88. Gilbert, J. C. Control of sexuality in the Rotifer, *Asplancha brightwelli* by dietary lipids of plant origin. Proc. Nat. Acad. Sci. U.S.A., 57:1218–1225, 1967.
89. Starr, R. C. Cellular differentiation in Volvox. Proc. Nat. Acad. Sci. U.S.A., 59:1082–1088, 1968.
90. Huxley, J. S. Chemical regulation and the hormone concept. Biol. Rev., 10:427–441, 1935.
91. Blum, J. J. An adrenergic control system in Tetrahymena. Proc. Nat. Acad. Sci. U.S.A., 58:81–88, 1967.
92. Potts, J. T., Aurbach, E. D., and Sherwood, L. M. Parathyroid hormone: chemical properties and structural requirements for biological and immunological activity. Recent Progr. Hormone Res., 22:101–151, 1966.

93. Sanger, F. Chemistry of insulin. Brit. Med. Bull., 16:183–188, 1960.
94. Bern, H. A., de Roos, C. C., and Biglieri, E. E. Aldosterone and other corticosteroids from chondrichthean interrenal glands. Gen. Comp. Endocr., 2:490–494, 1962.
95. Truscott, B., and Idler, D. R. Biosynthesis of aldosterone and 18-hydroxycorticosterone from corticosterone by interrenal tissue of a teleost (*Clupea harengus harengus*). J. Fish. Res. Board. Canad., 25:431–435, 1968.
96. Janssens, P. A., Vinson, G. P., Chester Jones, I., and Mosley, W. Amphibian characteristics of the adrenal cortex of the African lungfish (*Protopterus* sp.). J. Endocr., 32: 373–382, 1965.
97. Crabbé, J. Stimulation of active sodium transport across the isolated toad bladder after injection of aldosterone to the animal. Endocrinology, 69:673–682, 1961.
98. Chan, S. W. C., Huang, D. P., Vinson, G. P., and Phillips, J. G. Pathways of corticosteroid biosynthesis in snake (*Naja naja*) and frog (*Rana rugulosa*) adrenal glands. Acta Endocr. (Kobenhavn), Suppl., 119:70, 1967.
99. Phillips, J. G., Chester Jones, I., and Bellamy, D. Biosynthesis of adrenocortical hormones by adrenal glands of lizards and snakes. J. Endocr., 25:233–237, 1962.
100. Sandor, T., Lamoureux, J., and Lanthier, A. Adrenal cortical function in reptiles. The *in vitro* biosynthesis of adrenal cortical steroids by adrenal slices of two common North American turtles, the slider turtle (*Pseudemys scripta elegans*) and the painted turtle (*Chrysemys picta picta*). Steroids, 4:213–227, 1964.
101. ——— Lamoureux, J., and Lanthier, A. Adrenocortical function in birds. *In vitro* biosynthesis of radioactive corticosteroids from pregnenolone-7-H^3 and progesterone-4-C^{14} by adrenal glands of the domestic duck (*Anas platyrhynchos*) and the chicken (*Gallus domesticus*). Endocrinology, 73:629–636, 1963.
102. ——— and Lanthier, A. The *in vitro* biosynthesis of 18-hydroxycorticosterone-4-^{14}C by slices of zona glomerulosa of beef adrenals and by human adrenals. Acta Endocr. (Kobenhavn), 42:355–363, 1963.
103. Vinson, G. P., and Rankin, J. C. A comparison of steroid production *in vitro* and *in vivo* by the adrenal gland of the rat. J. Endocr., 33:195–201, 1965.
104. Schmidt, P. J., and Idler, D. R. Steroid hormones in the plasma of salmon at various stages of maturation. Gen. Comp. Endocr., 2:204–214, 1962.
105. Forchielli, E., Sorcini, G., Nightingale, M. S., Brust, N., Dorfman, R. I., Perloff, W. H., and Jacobsen, G. Testosterone in human plasma. Anal. Biochem., 5:416–421, 1963.
106. Roy, E. J., and Brown, J. B. A method for the estimation of oestriol, oestrone and oestradiol-17-β in the blood of the pregnant woman and of the foetus. J. Endocr., 21: 9–23, 1960.
107. Feder, H. H., Resko, J. A., and Goy, R. W. Progesterone concentrations in the arterial plasma of guinea-pigs during the oestrous cycle. J. Endocr., 40:505–513, 1968.
108. Gallien, L., and Chalumeau Le Foulgoc, M. Mis en évidence de steroides oestrogènes dans l'ovaire juvénile de *Xenopus laevis* Daudin, et cycles des oestrogènes au cours de la pont. C.R. Acad. Sci. (Paris), 251:460–462, 1960.
109. Lupo di Prisco, C., Chieffi, G., and Delrio, G. Identification of steroid hormone from *Lacerta sicula* testes. Experientia, 23:73–74, 1967.
110. ——— Delrio, G., and Chieffi, G. Sex hormones in the ovaries of the lizard, *Lacerta sicula*. Gen. Comp. Endocr., 10:292–295, 1968.
111. Short, R. V., and Moore, N. W. Progesterone in blood *V* Progesterone and 20a-hydroxypregn-4-en-3-one in the placenta and blood of ewes. J. Endocr., 19:288–293, 1959.
112. Dean, F. D., and Chester Jones, I. Sex steroids in the lungfish (*Protopterus annectens* Owen). J. Endocr., 24:425–430, 1959.
113. Gottfried, H., Hunt, V. H., Simpson, T. H., and Wright, R. S. Sex hormones in fish. The oestrogens of cod (*Gadus callarias*). J. Endocr., 24:425–430, 1962.
114. Simpson, T. H., Wright, R. S., and Hunt, S. V. Sex hormones in fish Part II. The oestrogens of *Scyliorhinus caniculua*. J. Endocr., 26:499–507, 1963.

95. [illegible] Ann. [illegible] 16: 185–186, 1960.
96. [illegible] M., de [illegible] C., and [illegible] L. R. [illegible] carotenoids [illegible] glands [illegible] 1361–1364, 1962.
97. [illegible] E., and [illegible] D. R. [illegible] and [illegible] A [illegible] Vitamin A [illegible] 35: 1489–1498, 1968.
98. [illegible] R. A. [illegible] and [illegible] V. [illegible] of the [illegible] 27: 322, 1962.
99. [illegible] Anal. Biochem. [illegible] 1963.
100. [illegible] L. [illegible] and [illegible] J. C. [illegible] 119–120, 1967.
101. [illegible] and [illegible] 1967.
102. [illegible] and [illegible] Am. J. Clin. Nutrition [illegible] 1960.
103. [illegible] J. [illegible] 1960.
104. [illegible] 1963.
105. [illegible] and [illegible] J. [illegible] plasma [illegible]
106. [illegible] Tokyo [illegible]
107. [illegible] 1966.
108. [illegible] 1967.
109. [illegible] Am. J. Clin. [illegible] 1966.
110. [illegible] 1966.
111. [illegible] Ind. [illegible] 1962.
112. [illegible] 1959.
113. [illegible] J. [illegible] 1963.
114. [illegible] 1960.

Index

4.95